Allyn and Bacon

General Science

Making science relevant to your students through readable content and unparalleled teaching support

from Prentice Hall

Comprehensive coverage with the balance of physics, chemistry, earth and space, and life science you want...

Comprehensive and Balanced Coverage of physics, chemistry, earth and space, and life science gives students a practical understanding of the world around them

Readable and Structured Content is organized into short, numbered sections with section goals, highlighted vocabulary, and checkpoint review questions

Comprehensive Chapter Reviews offer three pages of recall, comprehension, and critical thinking questions; includes "Challenge Your Understanding" for developing critical thinking skills and "Projects" for additional library, laboratory, and field research

11-3

Fossils

Goals

1. To define fossils.
2. To list conditions for the preservation of fossils and describe likely fossil locations.
3. To list some uses for fossils.

fossils

If you are interested in rock and mineral hunting, knowledge of fossils will add a new excitement to your hobby. **Fossils** are evidence of formerly-living things. Fossils can be the actual bones, teeth, or shells of animals. Or they can be indirect evidence of former life, such as footprints, trails, or imprints of shells in rocks.

Fossils are almost always found in sedimentary rocks. (See Figure 11-10.) Igneous or metamorphic rocks do not contain fossils because the heat that forms these rocks would destroy the fossils. In some rare cases fossils are found in layers of volcanic dust that have hardened to rock.

2-7

DNA

Goals

1. To describe the structure of DNA.
2. To identify the role of DNA in inheritance.

DNA

Chromosomes contain a substance called **DNA**, which stands for *deoxyribonucleic acid*. DNA contains the information that controls all cell activity. Scientists came up with a model of the structure of DNA that helps explain how DNA functions. According to the model, a DNA molecule is made up of small repeating units joined together in a long, twisted, ladderlike structure. (See Figure 2-17.)

The two sides of the "ladder" making up a DNA molecule contain alternating sugar (deoxyribose) and phosphate units. Pairs of chemicals called *bases* are connected to the sides of the ladder, forming "rungs." There are four types of bases: adenine (AD n een), thymine (THY meen), guanine (GWAH neen), and cytosine (SY toh seen). Adenine pairs with thymine and guanine with cytosine. (See Figure 2-18.)

Figure 2-17 *This model shows the twisted, ladderlike structure of a DNA molecule.*

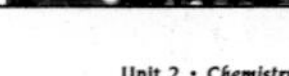

Figure 2-18 *Notice the repeating sugar-phosphate units and the paired bases.*

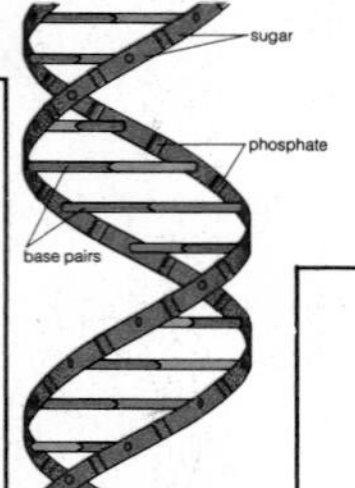

6-5

Elements and Compounds

Goals

1. To explain how chemical and physical changes differ.
2. To define elements and compounds.
3. To state the law of definite proportions.

physical changes

The methods used to separate mixtures involve **physical changes**, in which no new substances are formed. When the water evaporates from salt water, leaving solid salt behind, there is no change in the salt or water. The salt left behind has the same properties as the salt that was dissolved in water to form the solution. The water that evaporated is simply water in a different state. If the water vapor condenses, it will have the same properties as the water from which the solution was made. Physical changes include changes of state, shape, and size.

chemical changes

In **chemical changes**, new substances, with different properties, are formed. If electricity is passed through water, the water is changed into two new substances. (See Figure 6-13.) The new substances, hydrogen and oxygen, are gases. They have properties that are quite different from those of water.

If hydrogen and oxygen gas are mixed together, they do not produce water. However, if a spark is added to the mixture, an explosion takes place as hydrogen and oxygen

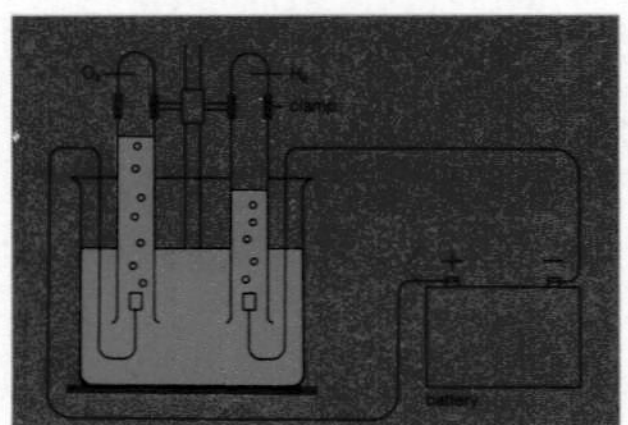

Figure 6-13 *When electricity goes through water, the water breaks into two new substances — hydrogen and oxygen.*

17-3

Simple Machines

Goals

1. To explain the difference between simple and complex machines
2. To give examples of simple machines.

simple machines

complex machine

The **simple machines** are devices from which all of the other machines are made. The lever is one of six simple machines. The others are the pulley, the wheel and axle, the inclined plane, the wedge, and the screw. A **complex machine** is made up of two or more simple machines.

You can change the direction of a force with a *fixed pulley*. You can lift an object by pulling downward as shown in Figure 17-7a. This is easier than bending to pick up the object in your arms. With a *movable pulley* the effort can be increased. A force less than the weight of the load can be used to lift it. (See Figure 17-7b.) Fixed and movable pulleys can be used together in a *block and tackle*, as shown in Figure 17-7c. Block and tackle machines are very important on a sailboat. Using them the crew can quickly change sails to take advantage of a shift in the wind. (See Figure 17-8.)

Figure 17-8 *Sails can be lifted by a block and tackle. The fixed pulley is on the left and the moving one on the right.*

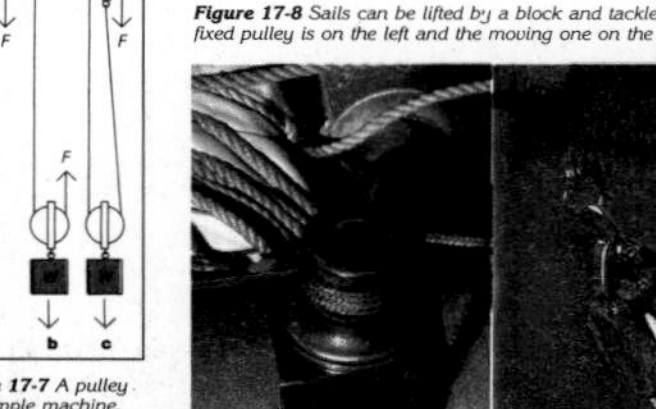

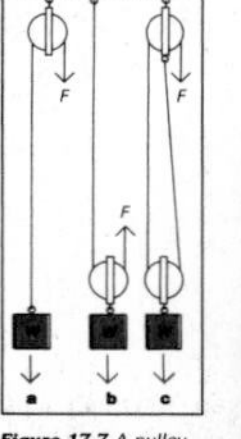

Figure 17-7 *A pulley is a simple machine. (a) Fixed pulley. (b) Movable pulley. (c) Block and tackle.*

...plus a wealth of special features

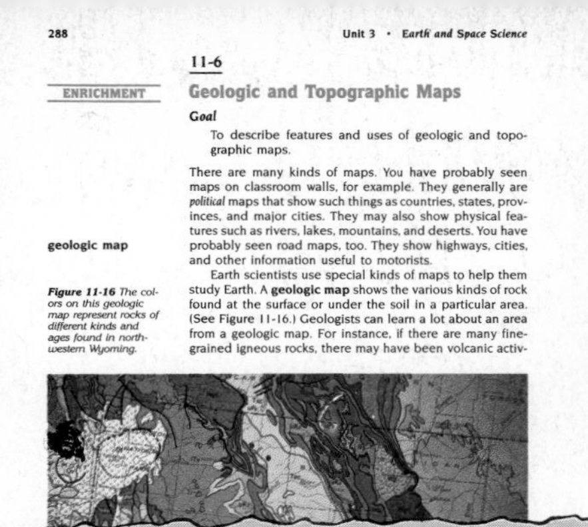
288 Unit 3 • Earth and Space Science

ENRICHMENT

11-6

Geologic and Topographic Maps

Goal

To describe features and uses of geologic and topographic maps.

There are many kinds of maps. You have probably seen maps on classroom walls, for example. They generally are *political* maps that show such things as countries, states, provinces, and major cities. They may also show physical features such as rivers, lakes, mountains, and deserts. You have probably seen road maps, too. They show highways, cities, and other information useful to motorists.

geologic map

Earth scientists use special kinds of maps to help them study Earth. A **geologic map** shows the various kinds of rock found at the surface or under the soil in a particular area. (See Figure 11-16.) Geologists can learn a lot about an area from a geologic map. For instance, if there are many fine-grained igneous rocks, there may have been volcanic activ-

Figure 11-16 The colors on this geologic map represent rocks of different kinds and ages found in northwestern Wyoming.

Enrichment lessons at the end of every chapter provide additional topics for more advanced students

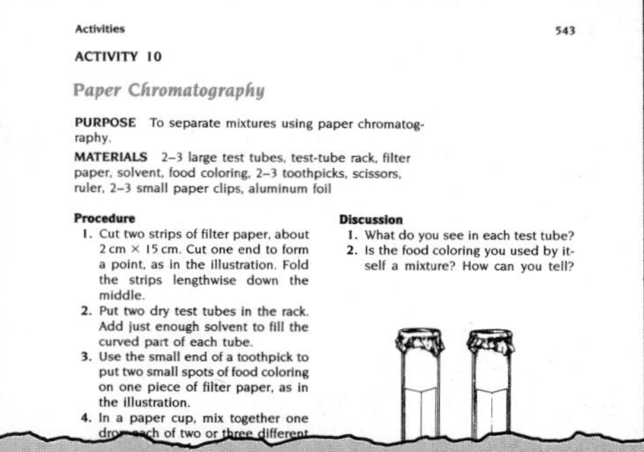
Activities 543

ACTIVITY 10

Paper Chromatography

PURPOSE To separate mixtures using paper chromatography.

MATERIALS 2–3 large test tubes, test-tube rack, filter paper, solvent, food coloring, 2–3 toothpicks, scissors, ruler, 2–3 small paper clips, aluminum foil

Procedure

1. Cut two strips of filter paper, about 2 cm × 15 cm. Cut one end to form a point, as in the illustration. Fold the strips lengthwise down the middle.
2. Put two dry test tubes in the rack. Add just enough solvent to fill the curved part of each tube.
3. Use the small end of a toothpick to put two small spots of food coloring on one piece of filter paper, as in the illustration.
4. In a paper cup, mix together one drop each of two or three different

Discussion

1. What do you see in each test tube?
2. Is the food coloring you used by itself a mixture? How can you tell?

Activities are easy-to-follow experiments organized in a laboratory manual format in the back of the student text

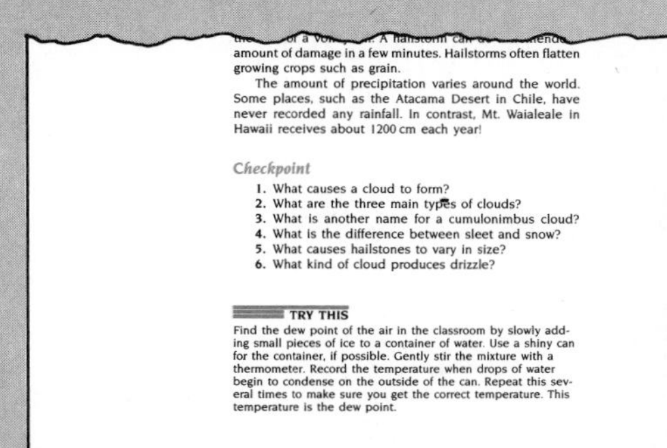
amount of damage in a few minutes. Hailstorms often flatten growing crops such as grain.

The amount of precipitation varies around the world. Some places, such as the Atacama Desert in Chile, have never recorded any rainfall. In contrast, Mt. Waialeale in Hawaii receives about 1200 cm each year!

Checkpoint

1. What causes a cloud to form?
2. What are the three main types of clouds?
3. What is another name for a cumulonimbus cloud?
4. What is the difference between sleet and snow?
5. What causes hailstones to vary in size?
6. What kind of cloud produces drizzle?

TRY THIS

Find the dew point of the air in the classroom by slowly adding small pieces of ice to a container of water. Use a shiny can for the container, if possible. Gently stir the mixture with a thermometer. Record the temperature when drops of water begin to condense on the outside of the can. Repeat this several times to make sure you get the correct temperature. This temperature is the dew point.

Try This are quick and easy hands-on activities in every chapter that reinforce concepts and develop science skills

388 FEATURE ARTICLE

Telescopes in Space

The study of astronomy will take a giant step forward when a telescope is launched into space. It is possible that a space telescope will help us to learn more about the universe in a few years than has been learned in all of past history! The greatest advantage of a space telescope is that it will be outside Earth's atmosphere. Earth's thick atmosphere limits the distance telescopes on Earth can "see" into space. The atmosphere also absorbs and refracts much of the light from the stars.

The space telescope that will be put into orbit by the Space Shuttle will be 5 m wide and 15 m long, with a mass greater than 10,000 kg. Its mirror will be 30 cm thick and 240 cm in diameter. Along with the telescope there will be two cameras, two spectrographs, and two instruments to measure distances between stars and changes in their brightness. All of this equipment will be in orbit 500 km above Earth. This is far above the thickest part of Earth's atmosphere. The largest telescopes on Earth can detect objects two billion light-years away. The space telescope will be able to detect light from galaxies 14 billion light-years away. This will extend our view of the universe by 350 times.

Feature Articles describe new scientific breakthroughs and explain how they relate to everyday life

394

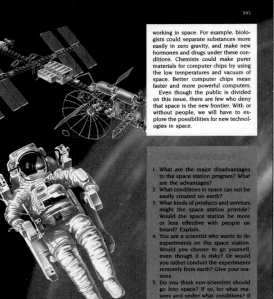

ISSUES IN SCIENCE

PEOPLE IN SPACE:

IS IT WORTH THE RISK?

Imagine living in a space station. You can't go out in the sunlight without protection because the strong radiation would damage your skin and eyes. You would suffocate in the vacuum of space if you left your life support systems. And long periods of weightlessness would cause your muscles to weaken.

Does it sound dangerous? It is. The 1986 accident of the Space Shuttle *Challenger* made it very clear that first getting people into space is dangerous. Yet, despite the dangers, the United States is continuing towards its goal of a permanent Space Station orbiting the earth, where people will live and work for months at a time.

The Space Station will have cost at least $14 billion by the time it is completed in the 1990s. In addition, more money must be spent to perfect the Space Shuttle necessary for carrying supplies to build and operate the Space Station.

Many scientists question the costs and risks involved in sending people into space. Some think that most of the operations in space could be performed more safely and at a lower cost by space stations remotely controlled from earth. They point out the success of the Voyager and Pioneer missions which explored the outer planets. But others argue that putting people into space is far more effective. They are concerned that we are falling behind the Soviet Union, which already has people working in a space station called Mir.

Supporters of the Space Station point out that the costs will be more than repaid by large profits. The Center for Space Studies predicts that by the year 2000 there will be 40 billion dollars in profits from renting the Space Station to industry.

Scientists and industries are interested in developing new technologies that may only be possible with people

395

working in space. For example, biologists could separate substances more easily in zero gravity, and make new hormones and drugs under these conditions. Chemists could make purer materials for computer chips by using the low temperatures and vacuum of space. Better computer chips mean faster and more powerful computers.

Even though the public is divided on this issue, there are few who deny that space is the new frontier. With or without people, we will have to explore the possibilities for new technologies in space.

1. What are the major disadvantages to the space station program? What are the advantages?
2. What conditions in space can not be easily created on earth?
3. What kinds of products and services might the space station provide? Would the space station be more or less effective with people on board? Explain.
4. You are a scientist who wants to do experiments on the space station. Would you choose to go yourself, even though it is risky? Or would you rather conduct the experiments remotely from earth? Give your reasons.
5. Do you think non-scientists should go into space? If so, for what reasons and under what conditions? If

Issues in Science challenges students to explore both sides of exciting science issues through thought-provoking articles followed by critical-thinking questions

Unique wraparound Teacher's Edition provides unmatched teacher support...

- **Overview** presents a brief summary of the basic concepts and topics that are presented in the chapter
- **Goals** provides a complete listing of all the section goals so teachers can easily see what will be expected of the students
- **Vocabulary Preview** lists all scientific terms students will encounter in the chapter

CHAPTER 19 Overview

The purpose of this chapter is to introduce basic concepts of electricity and magnetism. The chapter opens with a discussion of electric charge and a description of the electrical nature of matter, which is then related to static electricity and electrical conductivity. Charging by induction is described, followed by a definition of grounding and a discussion of related safety considerations. The fundamentals of current electricity are then introduced, and series and parallel circuits are described. A discussion of magnetism, magnetic fields, and the relationship between a magnetic field and an electric current follows. The differences between a generator and an electric motor are explained. Finally, the many uses of electricity in the home are described, followed by a discussion of conservation measures.

Goals

At the end of this chapter, students should be able to:

1. explain how electric charges are produced.
2. describe the behavior of charged objects.
3. explain how the atomic model of matter is related to electric charges.
4. use the atomic model to explain static electricity.
5. define and illustrate induction and grounding.
6. describe some practical uses of static electricity.
7. describe the parts of a simple electric circuit.
8. explain Ohm's law.
9. compare series and parallel circuits.
10. explain how fuses work.
11. describe magnetic fields, including Earth's magnetic field.
12. show that there is a magnetic field around a current-carrying wire.
13. explain how electricity can be produced by a changing magnetic field.
14. explain how an electric motor works.
15. explain how the amount of electrical energy used is measured.
16. list ways people can reduce their use of electricity.

Chapter 19

Electricity and Magnetism

This photo of the effects of electric charge was taken at the Ontario Science Center in Canada.

Vocabulary Preview

negative charge	amperes
positive charge	volts
conductors	resistance
insulators	ohms
static electricity	series circuit
induction	battery
grounding	parallel circuit
electric current	magnets
electric circuit	magnetic field
	electromagnet

Mainstreaming

A three-dimensional model of the lungs and chest cavity will help all handicapped students grasp their anatomy and function. In addition, listening to breathing action with a stethoscope will be instructive to all handicapped students except the hearing impaired.

Cross Reference

Refer to Chapter 13, section 13-1, for a discussion of atmospheric pressure.

3 THE HUMAN BODY

3-3 Breathing

Take a Breath!

Respiratory Tracking

trachea
alveoli
capillaries
inhalation
resuscitation
exhalation

Checkpoint Answers

1. The function of breathing is to put oxygen into the blood and remove carbon dioxide from it.
2. Answers will vary. See student text pages 74–75.
3. During inhalation the diaphragm contracts and moves downward while the chest muscles pull the ribs up. These movements increase the volume of the chest cavity and reduce air pressure in the lungs, causing air to rush in and fill the lungs. Exhalation occurs when the chest muscles and diaphragm relax, making the chest cavity smaller and forcing air out of the lungs.
4. Oxygen is used for respiration.

a b c

ribs

lungs

diaphragms

inhalation exhalation inhalation

Figure 3-16 (a) The ribs move up and the diaphragm moves down. (b) The diaphragm and chest muscles relax. Air is forced out of the lungs. (c) Air rushes in, filling the lungs.

During inhalation, the contractions of the chest muscles pull the ribs up. At the same time, the diaphragm contracts and moves downward. (See Figure 3-16a.) These two movements increase the volume of the chest cavity and reduce air pressure in the lungs. Air pressure inside the lungs becomes lower than air pressure outside the body. Air from outside rushes in, causing the lungs to expand.

Exhalation occurs when the chest muscles relax and the ribs move down. The diaphragm relaxes at the same time and moves upward. These movements make the chest cavity smaller. Air is forced out and the lungs return to their original size. (See Figure 3-16b.)

Exhaled air is not pure carbon dioxide. It contains a mixture of gases, including oxygen, and can be used by the body if inhaled a second time. This is why mouth-to-mouth resuscitation can save the life of someone who has stopped breathing. During resuscitation, the rescuer exhales air into the mouth of the victim.

Checkpoint

1. What is the function of breathing?
2. Trace the pathway of oxygen as it enters your body from the air until it reaches your cells.
3. Describe the role of the diaphragm and chest muscles during inhalation and exhalation.
4. What is oxygen used for in the cells?

- **Mainstreaming** enables the teacher to make the topic more meaningful for exceptional students
- **Cross Reference** is a handy cross-reference of the topic to other relevant sections of the text
- **Checkpoint Answers** for checkpoint questions at the end of every lesson helps you evaluate student mastery of the material

...right at your fingertips

19-1

Electric Charge

Goals

1. To explain how electric charges are produced.
2. To describe the behavior of charged objects.

You reach for a door knob after walking across a rug and a spark "jumps" between your hand and the knob. When you take off your sweater in a dark room, you see small flashes of light and hear crackling sounds. During the winter, you notice that your hair tends to "fly away." If you brush it, you may see sparks snapping between your hair and the brush. All of these events result from the buildup of *electric charge*. For example, your shoes moving across a rug cause electric charge to build up on your body. If the charge is large enough, a spark will jump between your hand and a metal object, such as a door knob.

People were aware of the effects of electric charge long before they understood it. The ancient Greeks knew that amber (shown in Figure 19-1) could be rubbed vigorously to produce a charge. In fact, our word *electricity* is derived from the Greek word *elektron*, which means amber.

Suppose two objects are charged in the same way, say by rubbing them with a woolen cloth. They will then move away from, or *repel*, each other. (See Figure 19-2a.) Figure 19-2b shows two objects that were charged in different ways. These two objects move toward, or *attract*, each other.

Charged objects always either attract or repel each other. This is because there are only two kinds of charge.

Figure 19-1 *A piece of amber. Rubbed amber was known to attract light objects (because it had a charge) as long ago as about 300 B.C. Amber is the hardened sap of extinct pine trees.*

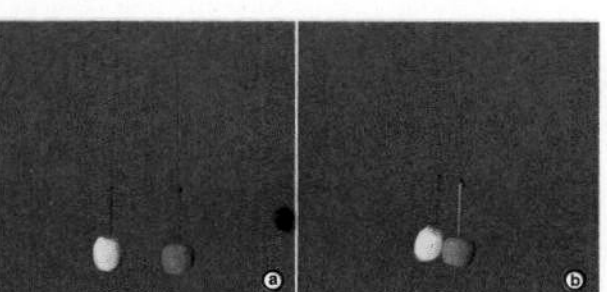

Figure 19-2 *(a) When two pith balls have the same charge, they repel each other. (b) Oppositely charged pith balls attract each other.*

Background

Benjamin Franklin assigned a "negative" charge to a rubber rod rubbed with fur, since two such rods repelled each other. Because a glass rod rubbed with silk attracted a rubber rod rubbed with fur, he knew the charges on the two rods were different. He called the charge on the glass rod "positive." These definitions of charge are still in use today.

Static electricity is discussed in greater detail in section 19-3.

Demonstration

Show how charged objects behave. Charge a rubber rod by rubbing it with a piece of fur. Hang the rod in a stirrup made of plastic-covered wire, or some other insulating material. Then bring an identically charged rod near one end of the suspended rod. The two rods will repel each other. Then approach the suspended rod with a glass rod that has been rubbed with silk; the two rods will attract each other. Test other charged objects, such as plastic rulers or pens rubbed with paper. If you cannot obtain rubber or glass rods, substitute plastic strips, etc. Note that this demonstration will work only if the air is dry; don't try it on a humid day.

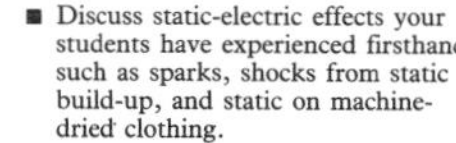

Teaching Tips

- Discuss static-electric effects your students have experienced firsthand, such as sparks, shocks from static build-up, and static on machine-dried clothing.
- Point out the similarities between electrical and gravitational forces. Both decrease markedly with physical separation and both are proportional to the product of two factors—the two masses in the case of gravitational force, the two charges in the case of electrical force.

eaching Suggestions 19-1

sk students to relate their own experi-nces of "electric shocks." Explain that ıch "shocks" are caused by the build-ıp of electric charge. Then explain and escribe positive and negative electric ıarge.

Be sure students understand that ɔpposite charges attract and like charges repel. Stress that the force of attraction or repulsion depends on two factors: the amount of charge and the distance between the charges. A demonstration of the behavior of charged objects can be very helpful in teaching this section. (See Demonstration.)

- **Background** supplies additional background information for you to make the topic more meaningful to students
- **Demonstration** for introducing concepts makes the topic come alive in the classroom
- **Teaching Suggestions** gives section-by-section ideas to reinforce and enrich the topics in the student text
- **Teaching Tips** includes further suggestions for developing teaching strategies

- **Reduced Teacher's Resource Book Page** makes your TRB fully integrated with the student text, providing hundreds of blackline masters that reduce teacher planning time
- **Using Try This** provides teaching tips for the Try This activities in each chapter
- **Facts At Your Fingertips** gives brief but fascinating facts about the topic, providing teachers with an exciting option to perk up the lesson

mountain
desert
tundra
prairie and grassland
forest
ice cap

Figure 12-4 *North American soils. Mountain soils are thin and rocky. Forest soils are thin with little humus. Grassland and prairie soils have a thick layer of rich, black topsoil with more humus than forest soils. Desert soils are high in minerals because there is little rain to wash the minerals downward. The lower layers of tundra soil remain frozen, reducing drainage in the top layers.*

type of bedrock in an area affects the soil that develops from it. The kinds of organisms that live in an area affect soil development, also. Since soil needs time to develop, time is another factor. The most important factor in the development of soil, however, is climate. In warm, moist climates rocks weather rapidly, and soil develops quickly. Desert areas, on the other hand, have little rainfall. Rock weathers much more slowly in deserts.

The North American continent has many different types of soil. Soils differ because of where they occur, and what the local climates are. (See Figure 12-4.)

TRY THIS

Find a spot in the schoolyard or at home where you can dig a hole deep enough to expose the depth of the topsoil. How does topsoil differ from the layer beneath it? How thick is the topsoil? Replace the soil when you are finished.

Checkpoint

1. What is soil?
2. How is topsoil formed?
3. How is subsoil formed?
4. What is the major influence on the type of soil that develops in an area?

12 EARTH'S CHANGING CRUST
12-2 Soils

Another Way to Classify Soils

Soil Type	Composition	Environment Needed	Location
Pedalfer			
Pedocal			
Laterite			

Using Try This

This activity is best done as a class project in the schoolyard, if appropriately sited, so that all can share in the observation. If students choose to do the activity at home, have them photograph or draw a diagram of their findings. Students in the inner city may find soil displays in museums, or can research and photocopy illustrations from texts in the library.

The type of soil in your area can be determined roughly from soil maps. If your soil seems different from what you expected, discuss factors that might account for the difference.

Checkpoint Answers

1. Soil is a mixture of weathered rock and decayed organic material.
2. Topsoil is formed when the decayed remains of organisms, called humus, mix with weathered rock material.
3. Subsoil is formed when clay and dissolved minerals are carried with water downward from the topsoil, forming a middle layer of soil.
4. Climate.

Teaching Tips

- Have a committee of students locate soil maps in the library and describe them to the class.
- Interested students could research soil conservation methods, such as contour plowing, minimum tillage, and windbreaks.

Facts At Your Fingertips

- It takes approximately 100 years to develop two centimeters of topsoil.
- Glacial deposits have contributed to the rich farmland in the northern part of the U.S. and Canada.

Total teaching support in the Teacher's Resource Book...

The Teacher's Resource Book offers a wide variety of options for reinforcing and enriching science concepts. Includes:

- Introductory Science Skills
- Review Worksheets
- Enrichment Worksheets
- Laboratory Safety Section
- Laboratory Activities
- Challenge Activities
- Computer Programs
- Projects
- Chapter Tests
- Unit Tests
- Graphing Skills
- Science Fair Manual

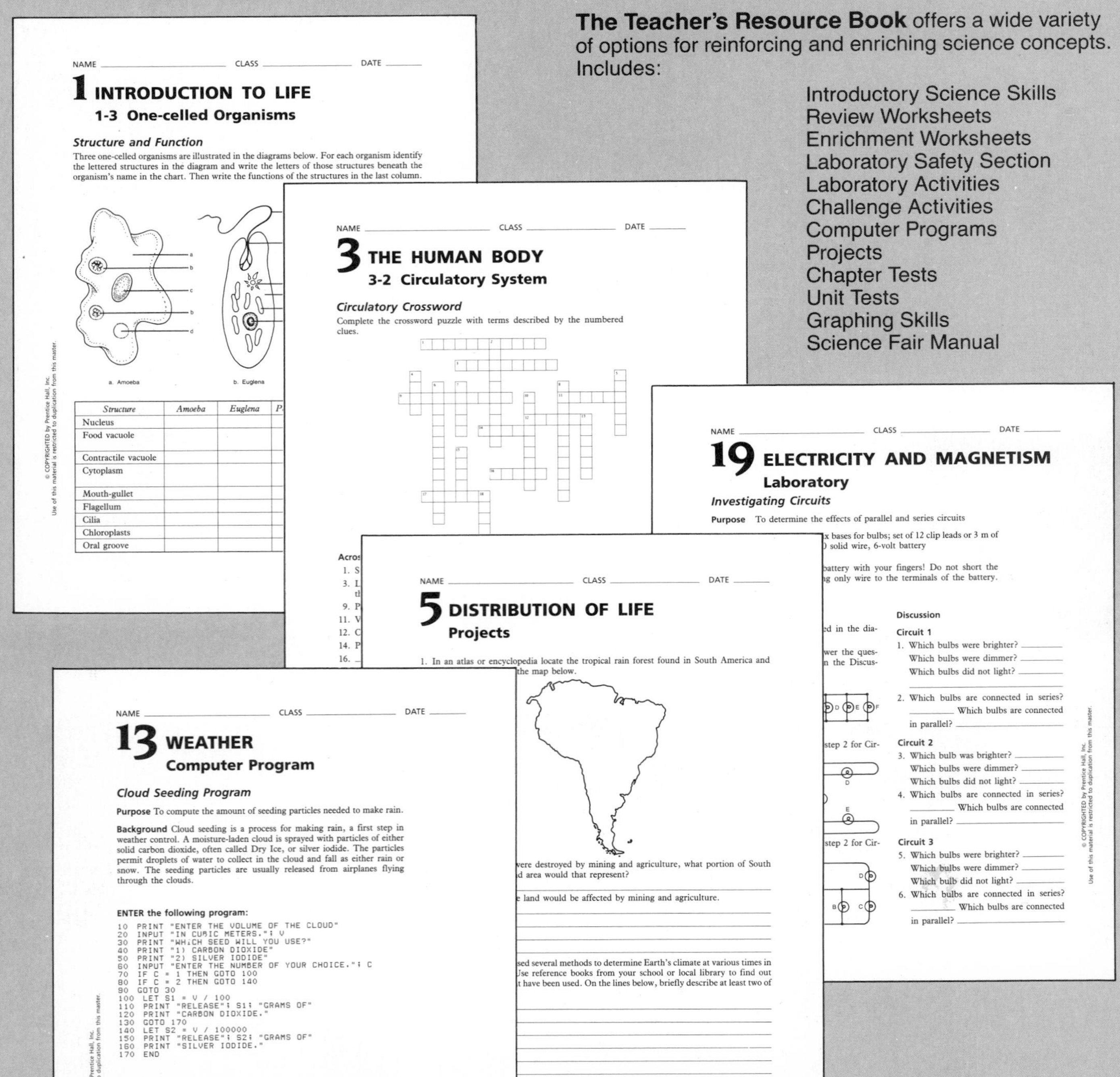

NAME ______ CLASS ______ DATE ______

1 INTRODUCTION TO LIFE
1-3 One-celled Organisms

Structure and Function

Three one-celled organisms are illustrated in the diagrams below. For each organism identify the lettered structures in the diagram and write the letters of those structures beneath the organism's name in the chart. Then write the functions of the structures in the last column.

a. Amoeba b. Euglena

Structure	Amoeba	Euglena	P
Nucleus			
Food vacuole			
Contractile vacuole			
Cytoplasm			
Mouth-gullet			
Flagellum			
Cilia			
Chloroplasts			
Oral groove			

NAME ______ CLASS ______ DATE ______

3 THE HUMAN BODY
3-2 Circulatory System

Circulatory Crossword

Complete the crossword puzzle with terms described by the numbered clues.

NAME ______ CLASS ______ DATE ______

19 ELECTRICITY AND MAGNETISM
Laboratory

Investigating Circuits

Purpose To determine the effects of parallel and series circuits

Discussion

Circuit 1

1. Which bulbs were brighter? ______ Which bulbs were dimmer? ______ Which bulbs did not light? ______
2. Which bulbs are connected in series? ______ Which bulbs are connected in parallel? ______

Circuit 2

3. Which bulb was brighter? ______ Which bulbs were dimmer? ______ Which bulbs did not light? ______
4. Which bulbs are connected in series? ______ Which bulbs are connected in parallel? ______

Circuit 3

5. Which bulbs were brighter? ______ Which bulbs were dimmer? ______ Which bulb did not light? ______
6. Which bulbs are connected in series? ______ Which bulbs are connected in parallel? ______

NAME ______ CLASS ______ DATE ______

5 DISTRIBUTION OF LIFE
Projects

1. In an atlas or encyclopedia locate the tropical rain forest found in South America and ... the map below.

NAME ______ CLASS ______ DATE ______

13 WEATHER
Computer Program

Cloud Seeding Program

Purpose To compute the amount of seeding particles needed to make rain.

Background Cloud seeding is a process for making rain, a first step in weather control. A moisture-laden cloud is sprayed with particles of either solid carbon dioxide, often called Dry Ice, or silver iodide. The particles permit droplets of water to collect in the cloud and fall as either rain or snow. The seeding particles are usually released from airplanes flying through the clouds.

ENTER the following program:

```
10 PRINT "ENTER THE VOLUME OF THE CLOUD"
20 INPUT "IN CUBIC METERS."; V
30 PRINT "WHICH SEED WILL YOU USE?"
40 PRINT "1) CARBON DIOXIDE"
50 PRINT "2) SILVER IODIDE"
60 INPUT "ENTER THE NUMBER OF YOUR CHOICE."; C
70 IF C = 1 THEN GOTO 100
80 IF C = 2 THEN GOTO 140
90 GOTO 30
100 LET S1 = V / 100
110 PRINT "RELEASE"; S1; "GRAMS OF"
120 PRINT "CARBON DIOXIDE."
130 GOTO 170
140 LET S2 = V / 100000
150 PRINT "RELEASE"; S2; "GRAMS OF"
160 PRINT "SILVER IODIDE."
170 END
```

Discussion

1. What is the weight of carbon dioxide required to seed cloud A of 132,000 cubic meters?

...and more!

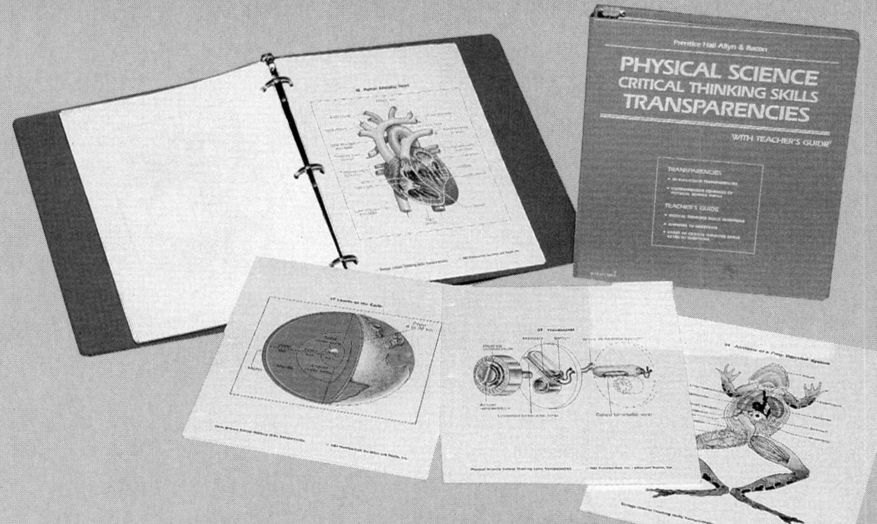

Life Science, Earth Science, and Physical Science Critical Thinking Skills Transparencies supply teachers with vivid full-color transparencies that bring new excitement into the classroom. With 50 transparencies in each set, teaching is easier because you no longer lose valuable time having to draw diagrams and pictures on the blackboard

Activity Book provides a convenient separately bound supplement for students that contains the same supplementary exercises referenced throughout the Teacher's Edition

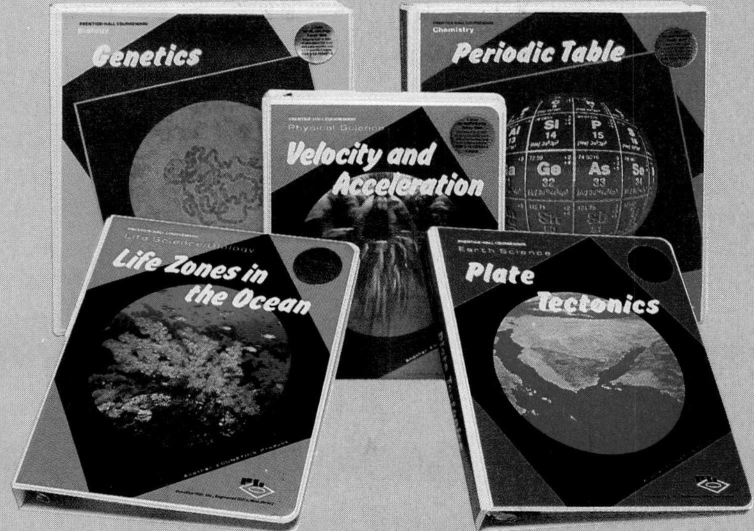

Prentice Hall Science Courseware presents highly interactive tutorials with extensive simulations that reinforce, enrich, and extend important science concepts. Over 40 titles available in physics, chemistry, earth and space, and life science

The program you'll be proud to teach

Allyn and Bacon **General Science**

Table of Contents

- **Fully Integrated Teaching Package**
  All components—Student Edition, Teacher's Edition, Teacher's Resource Book, and Activity Book—are totally integrated and cross-referenced in the Teacher's Edition for ease of planning and teaching
- **Wraparound Teacher's Edition**
  A format so helpful it provides all the information you ever wanted in a general science program right at your fingertips
- **Balanced and Comprehensive Coverage**
  For students—a perfect balance of the four major areas of science
  For teachers—a format where teachers can select units in any order they prefer
- **Solid Skills Development**
  Laboratory, writing, and thinking skills are developed in various types of activities throughout the program

For more information,
please call TOLL FREE
1-800-848-9500

Or write:

130-23862-7

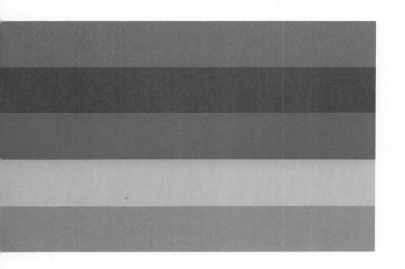

Teacher's Edition

Allyn and Bacon GENERAL SCIENCE

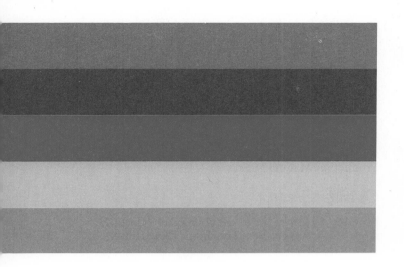

Teacher's Edition

Allyn and Bacon GENERAL SCIENCE

SECOND EDITION

Prentice Hall

Needham, Massachusetts
Englewood Cliffs, New Jersey

Carolyn Sheets Brockway
Science Education Center
University of Iowa
Iowa City, Iowa

Robert Gardner
Chair, Science Department
Salisbury School
Salisbury, Connecticut

Samuel F. Howe
Educational Consultant and Writer
formerly Science Teacher
Friends' Central School
Philadelphia, Pennsylvania

Staff Credits:

Lois B. Arnold, Joel R. Gendler
Editorial Development

L. Christopher Valente
Design Direction

Stuart Wallace
Design Coordination

Bill Wood
Production/Manufacturing

Outside Credits:

Prentice Crosier, Helen Reebenacker, Carol H. Rose, Susan Gerould, The Book Department
Production/Design Services

Laurel Anderson, Susan Van Etten
Photograph Research

Hannus Design Associates and L. Christopher Valente
Cover Design

 **A Simon & Schuster Company**

ISBN 0-13-350810-2

Printed in the United States of America

2 3 4 5 6 7 8 9 96 95 94 93 92 91 90 89 88

Acknowledgements

Consultants

Science Content

Richard L. Rotundo
Department of Anatomy and Cell Biology
University of Miami
Miami, Florida

W. T. Lippincott
Department of Chemistry
The University of Arizona
Tucson, Arizona

Lou Williams Page
Geologist and Author
Houston, Texas

Sharon C. Thomas
John G. Diefenbaker High School
Calgary, Alberta

Reading

Patricia Terrell
School of Education
East Carolina University
Greenville, North Carolina

Mainstreaming/Handicapped Adaptations

Dennis W. Sunal
Division of Education
West Virginia University
Morgantown, West Virginia

Activities Safety

Ronald J. Kendall
Institute of Wildlife Toxicology
Western Washington University
Bellingham, Washington

Teacher Reviewers

John J. Sheridan
Biology Teacher
Boca Ciega Senior High School
Gulfport, Florida

Ann S. Burden
Chair, Science Department
North Pitt High School
Bethel, North Carolina

Linda B. Knight
General Science/Earth Science Teacher
Bloomington High School North
Bloomington, Indiana

Larry A. Syron
Chair, Science Department
Gresham Union High School
Gresham, Oregon

Contents

TEACHER'S GUIDE

An Overview of the Program

GENERAL SCIENCE is designed to provide maximum flexibility with respect to the nature of the learning experiences included, time available, and teaching sequence selected. All of the authors are experienced and effective science teachers. Thus, there is a high level of credibility in the presentation of content and choice of learning activities. All activities have been proven effective in the classroom.

The text is divided into four units: Life Science, Chemistry, Earth and Space Science, and Physics. Each unit contains five chapters, and each chapter is subdivided into six to nine sections or lessons, each of which deals with one or more closely related concepts. One section in each chapter is designated as Enrichment, and is optional. Enrichment sections can be included for the entire class, assigned to challenge better students, or omitted.

Flexibility

Although the order of the units is the suggested teaching order, any order can be adopted. Each unit is self-contained. No unit depends on another being taught first. Some teachers may prefer to teach earth and space science after life science, followed by chemistry and physics. Others will see reasons for following chemistry and physics with life, earth, and space science, for the latter three make extensive use of principles of physical science. You may also choose to leave out some material. For example, the chemistry unit can be taught very effectively to less able students by going more slowly over chapters 6–9 and leaving out chapter 10 on Nuclear Reactions.

Several features in each chapter aid student learning and provide for individual differences. Each section begins with learning goals. These provide students with a precise statement of what they are expected to achieve. A set of Checkpoint questions at the end of each section then provides for immediate review of the content covered. Checkpoint questions test student recall.

Many of the sections contain "Try This" activities. These activities, for the most part, can be done in a short period of time with simple and readily obtainable materials. They are designed to clarify concepts presented in the section. The Try This activities are designed to be done without teacher supervision, at home or in school, and are optional.

Each chapter contains a short Feature Article.

These newsy, up-to-date pieces are optional. They are designed to present current topics related to the content of the chapter. They may be omitted, assigned as homework reading, or serve as the basis for classroom discussion.

The Main Ideas listed at the end of each chapter give students a way, during review, to make the necessary connection between learning goals and the major concepts taught. The section called Vocabulary Review allows students to test their understanding of new terms introduced. The terms included are those that appear in bold face in the text and also in the margins. Throughout the text new technical terms introduced are immediately defined and elaborated on to make their meanings clear.

The Chapter Review Questions at the end of each chapter are divided into three levels of difficulty. "Know The Facts" questions are the simplest to answer. They require only recognition of the correct short answer. "Understand The Concepts" questions are at a higher level of difficulty. These questions require recall of facts and the ability to define, describe, explain, and relate concepts. Students must write out answers to "Understand The Concepts" questions. "Challenge Your Understanding" questions are the most difficult to answer. They call for the ability to relate, compare, and contrast, and sometimes for analysis and/or synthesis of facts, concepts, or ideas. The answers to these questions too must be written out.

Chapter Review questions are best assigned in terms of student ability level. Clearly, only the most able students will be capable of answering "Challenge Your Understanding" Questions. Appropriate questions can be assigned as homework, for small group work in class, and as a means of preparing for chapter tests.

The Projects suggested at the end of each chapter are optional. They provide a variety of ways to focus and expand on concepts taught in the chapter. Many Projects require the use of community resources, the library, or experimentation, and some call for special reports, either oral or written. Projects can be assigned in terms of student interest and ability level, and time available.

Information on careers related to the content presented is provided at the end of each unit. Calling attention to this information and/or devising learning experiences using it will stimulate student interest, and perhaps lead students to further explore career possibilities with your Guidance Department.

A set of laboratory Activities closely related to the content of the text is provided at the back of the book. While these Activities are optional, it should be noted that the most effective science instruction includes an ongoing program of hands-on laboratory work. Such work not only reinforces and expands on concepts covered in the text, it also helps to develop student eye-hand coordination, manipulative skills, and problem solving skills.

The student ACTIVITY BOOK provides additional laboratory and written activities, and computer-oriented activities, for review and enrichment. These materials help reinforce and extend learning. Activities for all student ability levels are provided.

Research in science education shows that students learn best when they become involved physically, as well as mentally, in the learning process. GENERAL SCIENCE provides materials for a full-scale hands-on component to accompany traditional text-oriented learning. The teacher who schedules at least one laboratory Activity per chapter, and has students do many of the Try This activities, in addition to the reading, discussion, and homework of a typical text-oriented program, will be offering an excellent course.

The authors recognize, on the other hand, that it is not always possible to provide the best possible array of learning experiences. For this reason, GENERAL SCIENCE is structured to permit an almost limitless number of combinations of text-oriented and hands-on learning activities. A school with limited laboratory equipment and/or space can successfully incorporate meaningful hands-on activities by assigning most of the Try This activities and Projects. For those that are better equipped, including some or all of the laboratory Activities adds a more formal hands-on component to the program.

As demonstrated in the Prologue, a goal of GENERAL SCIENCE is giving students a basic understanding of the scientific methodology used by scientists in addition to providing comprehensive coverage of content. Teachers desiring a strong process orientation will find the needed basis in laboratory Activities, the Try This activities, and Projects. They should encourage students to view the activities as problems requiring possible solutions.

In addition, teachers can encourage students to explore any unanswered questions that come up in class discussions, as well as assign questions to which the answers are not known. For example, after describing the nature of acid rain, the teacher might ask, "What effect, if any, does acid rain have in our area?" Students tackling this problem will first have to determine experimentally if rainfall is acid, and what the degree of acidity is. Then, assuming a significant level of acidity has been found, it will be necessary for students to approach such community resource persons as the public health officer and agricultural agent to determine what, if any, are the effects of the acid rainwater experimentally confirmed.

Many of the Projects listed at the end of each chapter offer opportunities for students to use inquiry approaches on their own. The Projects are phrased as problems requiring solutions. Students will need to analyze the nature of a problem, do background research, and then formulate possible solutions for testing. For more on problem solving skills, see pages T28–T29.

Teaching Schedules

The suggested Pacing and Tracking Chart that follows is based on a school year of 36 weeks (180 days), with 50–55 minute periods every day. On this basis, if the entire text is to be covered during the school year, a maximum of nine class periods can be allocated for each chapter. And each unit will require about nine weeks. With planning, it should be relatively easy to compensate for time lost to assemblies, reduced teaching days, and other unavoidable interruptions.

The chart below provides an overview of the program and the accompanying laboratory Activities, along with recommended teaching times. Adjust the Chart entries to fit your requirements. Factors as time limitations, the availability of laboratory space and materials, and the needs and interests of you and your students should be taken into account. You may or may not want to include the Enrichment Sections with your students. In addition, the time allocated for each chapter will vary some in terms of other activities planned, such as the use of A-V materials, Activities, field trips, tests, and the amount of discussion time needed. If the student ACTIVITY BOOK is used, additional classroom time will be needed to discuss student results and responses.

Pacing and Tracking Chart

UNIT Chapter	Class Periods Allowed	Enrichment Section	Activity	Class Periods
PROLOGUE: Introduction to Science	3			
LIFE SCIENCE 1. Introduction to Life	9	Viruses	1. Using a Microscope 2. Observing Cells 3. Observing One-Celled Organisms	1 1 1–1½
LIFE SCIENCE 2. Heredity	9	Plant and Animal Breeding	4. Observing Onion Root-Tip Cells	1

LIFE SCIENCE 3. Human Body	9	Alcohol, Tobacco, and other Drugs	5. Measuring Lung Volume 6. Mapping Taste Areas of the Tongue	1 1
LIFE SCIENCE 4. Ecology	8	Endangered Species	7. Life in a Square Meter of Ground	1–2
LIFE SCIENCE 5. Distribution of Life	7	Mountain Biomes	8. Bird Beak Adaptations	1
CHEMISTRY 6. Properties of Matter	9	Combustion	9. Measuring Density 10. Paper Chromatography	1 1
CHEMISTRY 7. Atoms and Molecules	9	The Current Model of the Atom	11. Conservation of Mass 12. Some Model Reactions	1 1–1½
CHEMISTRY 8. Chemical Elements	9	Carbon: Element Six	13. Metals and Nonmetals	1–1½
CHEMISTRY 9. Chemical Reactions	9	Rates of Reaction	14. Unsuspected Electrochemical Cells 15. Acids and Bases	1–1½ 2
CHEMISTRY 10. Nuclear Reactions	9	Uses of Radioisotopes	16. Half-Life	1
EARTH and SPACE SCIENCE 11. Earth Materials	9	Geologic and Topographic Maps	17. Testing Mineral Hardness	1–1½

EARTH and SPACE SCIENCE 12. Earth's Changing Crust	9	Soils	18. Moving Land Masses	1
EARTH and SPACE SCIENCE 13. Weather	9	Climate	19. Making a Simple Hygrometer	1
EARTH and SPACE SCIENCE 14. Oceans	9	Resources of the Ocean	20. Investigating Density Currents	1
EARTH and SPACE SCIENCE 15. Astronomy	9	Tools of Astronomy	21. Seasons of the Year	1
PHYSICS 16. Motion and Force	9	Circular Motion	22. Measuring Walking Speed 23. Measuring Forces of Friction	1 1–1½
PHYSICS 17. Work and Energy	9	Perpetual Motion Machines	24. Measuring Power	1
PHYSICS 18. Heat	9	Specific Heat	25. Testing Insulation	1–2
PHYSICS 19. Electricity and Magnetism	9	Saving Electricity	26. Conductors of Electricity	1
PHYSICS 20. Light and Sound	9	Color	27. A Look into a Plane Mirror	1

Developing Science Skills

Two fundamental purposes of any science course are (a) the acquisition of sufficient knowledge to react intelligently to issues that relate to the content covered, and (b) the development of the type of inquiry skills used by scientists. The extent to which these goals are achieved depends in large measure on how well certain other learning skills are used by students. It seems clear that the greatest amount of learning will occur when students use time-tested and effective reading, writing, mathematics, and problem solving techniques.

Reading Skills

Careful attention was given to readability in the development of GENERAL SCIENCE. Both the Dale-Chall and Fry Readability formulas were applied to ensure that students of average ability will be able to handle the text. In addition, other factors favoring easy reading such as format, patterns of organization, style of writing, and content to be included were addressed. The result is a highly readable and manageable program.

The ability to read at different rates is essential to reading proficiency. Many students, however, read all material the same way. In fact, students should read in different ways and at different rates depending on the material.

This text is organized for flexibility in reading. Each chapter is broken up into sections that deal with just one or two important concepts. Learning goals are provided at the beginning of each section. These, plus the Main Ideas at the end of the chapter, constitute an overview of the chapter content. Together they provide a general idea of what the student is to achieve in the chapter.

The sections themselves contain new material that must be read slowly, carefully, and systematically (see discussion of SQ3R method below). How fast a student reads this material depends on his or her background, knowledge, and familiarity with the material, and on the material itself. The average reading rate at the ninth grade is 250 words per minute.

In addition to developing the ability to read at a rate most conducive to comprehension, students should become familiar with certain "glancing" techniques. *Skimming* is used to get a general impression of content by hitting high points at a high viewing speed. *Scanning* is a method of finding a single piece of information—a specific term, number, detail, or answer—by quickly looking through the material.

A reading and study system known as SQ3R (*S*urvey, *Q*uestion, *R*ead, *R*ecite, and *R*eview) has proven to be a most effective method for learning science concepts. Explain the system to your students and encourage them to use it.

The first step is to *survey* the chapter for an overall impression. Read the section titles and Goals and the Main Ideas. Briefly skim the new vocabulary introduced and the Try This activities.

Next, students should formulate *questions* to be answered during the reading. These can be derived from the chapter and section titles, the Checkpoint questions, end-of-chapter questions, the Main Ideas section, and the illustrations. A written list of the questions serves well as a study quiz after reading the chapter. The questions derived provide an overall purpose for the reading assigned.

With the purpose for reading established, the next step is to *read* to find answers to the questions formulated in the previous step. Reading with a purpose is a powerful aid to comprehension. Making a set of notes of the major concepts encountered is another useful tool for comprehension and retention of content.

After the reading step, students should *recite* answers to their questions from memory as a self-check. It should be made clear to students that they should not look back at the text for answers to these questions. Answers should be in their own words.

Finally, students should reinforce what they have learned by both an immediate and a delayed *review*. First, reread the text to verify the answers given in the previous step. Then, write out answers to any Checkpoint or end-of-chapter questions assigned.

Writing Skills

Students should be led to understand that effective writing is an important part of the scientific endeavor. Written reports are the major means of communication between scientists throughout the world. It should be emphasized that all written work in the course—from answers to Checkpoint questions to any reports assigned—should exhibit the qualities essential to good writing. Students should always write complete and grammatically correct sentences, and they should pay close attention to spelling, punctuation, paragraphing, organization, and other requirements of good written communication.

The Projects sections at the ends of chapters often call for written reports. Reports may often be quite short. Whether long or short, however, the best reports result from careful planning. Some time spent explaining to students how to go about planning and writing a report, and some written feedback on reports, will go a long way toward making students better communicators.

Every report begins with the development of an *outline*. The outline lays out the basic points in the report and presents them in the desired order. After an outline has been completed, it should be checked to make sure it contains all the necessary information and that it presents the information in the best possible order.

The second step in writing a report is producing a *rough draft* from the outline. The rough draft is a translation of the outline into text.

Rough drafts frequently contain errors and may not have the best organization and wording. Explain to students that almost no one can write a report directly from an outline, and that the rough draft is a stepping stone to the final draft.

After the rough draft has been completed, it must be carefully *edited.* Students should read through their rough drafts several times. Spelling errors should be corrected and awkward phrases should be rewritten clearly and directly. It helps to read the draft aloud, making note of any places where the train of thought is awkward or the meaning is unclear.

The edited rough draft should be copied over to obtain the *final draft.* If your school has word processing software for its computers, encourage students to compose rough drafts on the word processor. This will greatly simplify editing and preparation of the final draft. In addition, a word processor permits easy alterations after the graded paper has been returned.

Many scientific reports focus on experiments that test hypotheses (see Prologue). In such a report, the hypothesis should be stated at the beginning, and it should be followed by a description of the experiment whose results either support or do not support the hypothesis. Developing this type of report builds an understanding of scientific problem solving, which will be discussed in more detail later in this Guide.

You may wish to assign brief report writing exercises to strengthen writing skills. Offer a list of hypotheses for which ample experimental support exists. Have students choose a hypothesis and then prepare an outline. Collect the outlines, read them, and offer constructive criticism. Then have students write a rough draft. The rough draft too should be collected, read, and returned with appropriate feedback. Finally, a final draft should be prepared and handed in for grading.

Math Skills

Although this text was designed to require a minimum amount of mathematics, computational and other math skills are essential to scientific work. Measurement is one of the best places to begin building math skills. Appendix B contains valuable information on measurement.

In making any measurement there is always a certain amount of inaccuracy. The *greatest possible error* is equal to half the smallest measured increment, or unit of measure used. For example, suppose the mass of an object that has a mass between 24 grams and 25 grams is measured on a scale calibrated in increments of 1 gram. We may estimate the object's mass to be 24.5 g. However, because the smallest increment is 1 gram, the greatest possible error for the measurement is 0.5 g. The measurement is correctly expressed as 24.5 g ± 0.5 g, which is read "24.5 grams plus or minus 0.5 grams."

Students may not at first appreciate the importance of this discussion of errors in measurement. Scientists, however, must know the level of inaccuracy of their instruments. For example, even very low levels of error in position measurements made during a space flight can result in course errors of hundreds of kilometers. As an exercise, supply the class with metric rulers and a collection of objects, and have students measure each object and indicate the measurement errors.

Graphing too is an essential skill for scientific work. Graphs are often the best way to present data that are related in some way. Appendix D contains a thorough explanation of how to construct and use a graph. A series of brief exercises that requires both graphing and interpretation of graphs will help students master this skill.

When calculators are used to compute measured quantities, the concept of *significant digits* must be considered. All non-zero digits are significant, and all zeros are significant, *except* as follows: at the end of whole numbers and on either side of the decimal point in decimals between 0 and 1. Thus, all the zeros in both 64,500 and 0.0008 are *not* significant. Suppose we divide 4.030 cm by 0.027 cm. On a calculator with an 8-digit display, the answer will appear as 149.25925. However, any sum, difference, product, or quotient of numbers that comes from measurements must be rounded to the number of significant digits in the *least precise* measurement—that is, the one with the fewest significant digits. Because 4.030 has four significant digits and 0.027 has two significant digits, the answer must be rounded to two significant digits, or 150.

Problem Solving Skills

A goal of this program is to have students acquire an understanding of the hypothesis-testing process used by scientists to discover new knowledge. This process is clearly described in the Prologue and referred to throughout the text. Because of time and other limitations, it is not feasible to provide students with extensive experience developing and testing hypotheses. In many ways, it is sufficient for them to grasp how hypotheses arise, how experiments are designed to test hypotheses, and that positive experimental results merely support or fail to support a hypothesis. It should be made clear that there is no such thing as experimental proof of a hypothesis; students should be encouraged to think in terms of support rather than proof.

Problem solving in the broadest sense of the term, however, is a significant part of any science program. Students should acquire the skill to analyze a problem and develop possible solutions. Practice using basic problem solving skills occurs in many guises throughout this text—in the Activities at the back of the text, in Try This activities, in many of the end-of-chapter questions, and in some of the Projects listed at the end of each chapter. Because the skill of problem solving has application in virtually all walks of life, it should be introduced to students in a formal manner.

The first step in solving any problem is to clearly define the problem—that is, to produce an unambiguous statement of what the problem amounts to. This often requires breaking the problem down into individual components for examination and studying the problem's relationship to other issues.

The second step is to do background research on the problem. This usually occurs at a library and can include, in addition, interviews with experts. From the definition of the problem and the background research, possible solutions are then developed. These are carefully analyzed in terms of the problem, and the best solution is finally selected and implemented.

Other skills may be used in the analysis of problems and the development of solutions. Classification is often used to separate objects or data with similar properties. Observation is a skill important in the collection of data. For example, students can learn to observe and record animal behavior. Skill in making accurate observations can be strengthened by video taping animals, especially primates, at a zoo and then using the tape for exercises.

All problem solving skills are strengthened by regular use. Weekly or biweekly problem solving sessions are excellent for this purpose. When choosing a problem for a session, make sure background information is readily available and that there are several possible solutions. Many of the problems discussed in the section called Science, Technology and Society (page T48) are well suited to the type of session suggested. In addition, these problems will help provide a focus on the inevitable interaction between the results of scientific research and technological achievement and ordinary citizens.

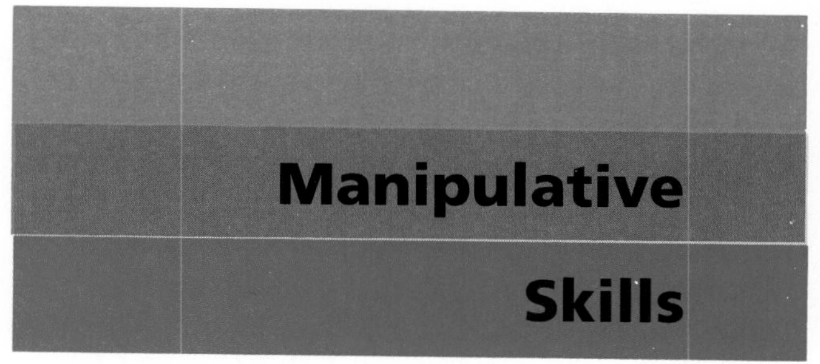

Manipulative Skills

Throughout this program, in laboratory Activities, Try This activities, and in some of the Projects students will be handling laboratory equipment, chemicals, and biological specimens. It is important that they learn and use proper methods for manipulating apparatus and materials, for both safety and cost reasons. It goes without saying that improper and/or careless use of laboratory equipment will almost certainly yield useless results, and thus have no learning value. In addition, the careful attention to detail required for the proper and effective use of apparatus will help students further develop their eye-hand coordination.

The most important single consideration in the development of manipulative skills is that all laboratory operations be done slowly and carefully, not with speed or haste. Each technique should be practiced before it is used in an Activity, and its introduction should be preceded by a description of safety considerations. Some procedures requiring special attention follow.

- All students should learn the proper operation of a microscope, following the directions given in Appendix A.
- Students should learn how to operate a balance, and how to adjust the zero level of the balance. In addition, they should learn how to compensate for the mass of a container. Have

them measure the masses of containers and objects and liquids in containers to acquire this skill.

- Be sure to demonstrate the safe operation of an alcohol and Bunsen burner. Students should learn the proper use of a burner, and how to extinguish one safely, and they should practice using a burner in an actual experiment. The gelatin exercise discussed under Lab Note and Report Skills below offers a good opportunity to work with a burner. In addition, the locations of all fire extinguishers in the classroom or laboratory should be pointed out to students.
- Students should be shown the proper way to obtain samples from stock solutions, so that the solutions are neither contaminated nor spilled. This is especially important when handling acids or strongly alkaline solutions.
- When carrying out physics experiments, students should learn to measure all experimental data at least twice, to minimize error.
- Take the time to demonstrate safe and correct wiring of meters and batteries to avoid short circuits that might damage meters. Students should assemble numerous simple circuits to become familiar with electrical circuitry.

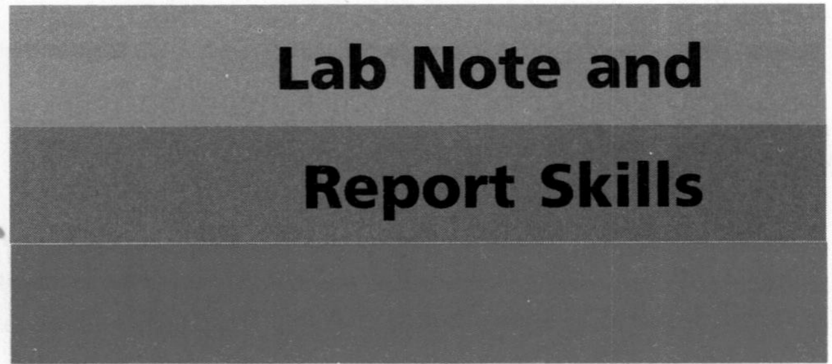

Lab Note and Report Skills

Careful note taking during experimental procedures is essential, for once an experiment has moved forward it is very difficult to recall exactly what occurred in earlier steps. Thus, the details of every step and the data resulting from each step must be recorded at once. Students should be cautioned not to wait until an experiment has been completed to make notes. Explain too that carefully made, thorough laboratory records have yielded many surprising discoveries when studied critically. And, of course, a complete record of procedures and data makes the discovery of errors much easier.

Here is a simple exercise that provides practice and also reinforces the importance of careful note taking: Pass out small samples of powdered gelatin and provide general instructions for making colloidal gelatin. Do not write any instructions on the chalkboard. Tell students their task is to successfully prepare a gelatin sample, and also to provide detailed instructions for a person who has no knowledge whatever of the properties of gelatin. Notes taken during each step of the procedure should permit development of the detailed instructions.

Written laboratory reports should begin with an introductory paragraph that sets out the purpose of the experiment. This should then be followed by a description of the experimental procedure—not the one given in the Activity, but the one the student recorded in his or her laboratory notebook during the experiment. The data resulting from the experiment should be listed in table or graph form, followed by a discussion of the results. Tentative conclusions may be added, if appropriate, but students should be cautioned to avoid statements of an absolute nature.

Skill in writing laboratory reports will develop quickly if this format is followed, starting with the first Activity.

Safety In Classroom and Laboratory

GENERAL SCIENCE is, in part, a course of study providing hands-on activities in both classroom and laboratory. Students will from time to time be handling such diverse objects and materials as preserved specimens, microscope slides, glassware, burners, other pieces of apparatus, and several different types of chemicals.

Every attempt has been made in designing the laboratory activities of the program to reduce to a minimum the use of hazardous substances. Nevertheless, there is always the potential, even though it may be quite small, for an accident in a student laboratory. For this reason, teachers and students both should be acutely conscious of safety considerations at all times.

A list of safety precautions for the student called Working Safely appears just before the laboratory Activities in the text (page 531). The teacher should take the time to review these guidelines carefully with the class before any laboratory work is undertaken by students. The conscientious teacher will then brief students at the beginning of every laboratory session on any potential hazards to be concerned with. There is good evidence that the safest school laboratories are those in which there is an ongoing concern for safety, on the part of both teachers and students.

The discussion that follows attempts to establish for the teacher just what a "safe" school laboratory environment is. All teachers are encouraged to read this material and to adopt an acute sensitivity to safety. Accidents, often silly and occasionally quite serious, do happen in student laboratories when teachers, although meaning well, relax their guard.

A school laboratory consists of three elements—the instructor, the students, and the physical facilities. The instructor and students have very definite safety responsibilities. The physical facilities must be intelligently planned and used with maximum safety.

The responsibility of the instructor can be summed up in one brief statement: The supervisor of laboratory activities must at all times prevent accidents in such activities. This, of course, requires the constant teaching of safety practices, as well as close supervision of student performed experiments.

To meet this responsibility, the teacher must: *Teach safety at all times,* not just during the first laboratory session of the year. *Circulate among the students* during laboratory periods. *Stay in the lab-*

oratory. Students often misbehave in bizarre ways when they are unsupervised. The teacher should reject the temptation to sit alone at one end of the laboratory for the entire period, or leave the laboratory unsupervised.

Students must accept the following obligations. They must become familiar with the health and safety hazards of all equipment and chemicals used. They must plan their laboratory exercises and experiments to minimize the possibility of accidents. They must record and investigate with the teacher any accidents or unplanned events that occur.

The safe school laboratory has the following characteristics:

1. Adequate first aid equipment, backed up by knowledge of what to do in the event of an injury.
2. Appropriate fire extinguishers. Different types of fires require different fire extinguishing substances.
3. Fire protection for flammable chemicals.
4. Two exits from the laboratory, if feasible, as far apart as possible.
5. A properly designed ventilation system and deluge showers, if deemed necessary.
6. Adequate space for both storage and laboratory work.
7. Safe methods for the disposal of waste chemicals and other materials.
8. Proper and sufficient safety equipment such as goggles.

Typical Laboratory Mishaps

Studies of laboratory injuries give an indication of the types of accidents most likely to occur. By far the most common injury is cut hands, caused by broken glass tubing, other broken glassware, and sharp hand tools. Not far behind are injuries due to chemical spills and uncontrolled reactions. A surprising number of internal injuries result from careless pipetting. Others result from electrical shocks, falls, foreign bodies in the eye, splashing hot liquids, careless handling of mechanical equipment, and exposure to vapors or gases.

Many accidents happen when least expected. For this reason, teachers and students alike must always be alert to the possibility of an accident. In particular, students and teachers must protect their eyes. Approved safety goggles must be worn whenever caustic chemicals or hot liquids or solids are in use.

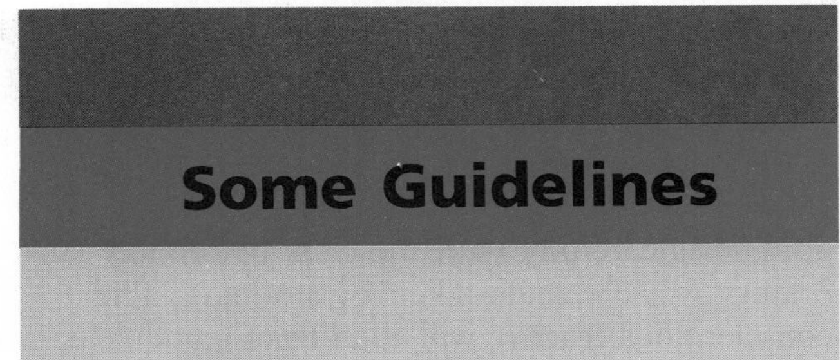

Some Guidelines

As pointed out earlier, some responsibilities fall on the teacher and others fall on the student.

The lists that follow identify some of the major responsibilities of each. Any overlap or repetition of guidelines given is intentional, for safety precautions cannot be overemphasized.

For The Student

The teacher should stress the following as well as the guidelines in the student text appearing just before the laboratory Activities (page 531).

- Never pour water into acid. Always add acid to water.
- When heating substances in glassware with a flame, protect the glassware with a wire gauze.
- Never taste a chemical to identify it unless specifically instructed to do so.
- Never use broken or chipped glassware.
- Never point the open end of a heated test tube toward another person.
- Always maintain a positive attitude about safety. It is not necessary to fear laboratory Activities, but it is necessary to respect potential hazards in performing them.
- Never force glass tubing into stoppers.
- Always use a suction bulb to draw liquid reagents into a pipette.
- Do not inhale large amounts of fumes or odors. If a smell must be detected, waft the vapor or gas toward the nose with a hand.
- Never eat or drink in a laboratory or from laboratory glassware.
- Never play practical jokes or indulge in horseplay in a laboratory.
- Always wear protective goggles when directed to do so.
- Always wear a laboratory apron if chemical reagents are in use.
- Always confine loose clothing and long hair.

For The Teacher

In addition to maintaining safe standards of behavior in the laboratory, the teacher should teach safe practices at all opportunities. The guidelines that follow will help with this as well as provide a checklist for a safe laboratory environment.

- The following items should be present in the laboratory, and both teachers and students should know where they are and how they are used: first aid kit, fire extinguishers, sand buckets, fire blankets, eye rinsing materials, emergency showers, safety goggles, aprons, and heat-resistant gloves.
- Maintain a clean and neat laboratory.
- Never allow students to carry out experiments unsupervised.
- Never leave the laboratory without first checking all gas jets and electrical equipment.
- Conduct periodic safety checks of all laboratories.
- Post emergency procedures for fires, chemical spills, and first aid in a conspicuous place, and make sure students understand them.
- Always carry out demonstrations and student activities ahead of time and alert students to any potential hazards discovered.
- Always set a good safety example when doing a demonstration or experiment. If you do not wear safety goggles, students too will ignore them.
- Try to become familiar with the medical problems of students, and excuse from laboratory activities those whose conditions might be aggravated by chemicals or some other aspect of an activity.

Materials List

The list below indicates the amount of materials needed to teach the Activities at the back of the student text to a class of thirty students. It is assumed that students will share major pieces of equipment such as microscopes and laboratory balances. Lesser amounts of other equipment and materials will be needed if students work on some activities in pairs or small groups. Materials needed for the Try This activities are not included here. Those activities utilize easily obtainable materials, do not require teacher supervision, and may even be done at home.

Activity Number	Materials	Amount Needed for 30 Students	Chapter Sections in Which Used
13	Alcohol burners	15	8-1
26	Alligator clips (optional)	60	19-3
10, 13, 23	Aluminum foil, pieces of	15	6-4, 8-1, 16-2
3	Amoeba culture	one vial	1-3
9, 11, 12	Balances, laboratory	15	6-1, 7-3
26	Batteries, 6-V	30	19-3
23	Blocks, wooden	30	16-2
23	Boards, wooden (30–40 cm. long)	30	16-2
5	Buckets	30	3-3
13, 26	Bulbs	30	8-1, 19-3
13, 26	Bulb Holders	30	8-1, 19-3
13	Bunsen burners	15	8-1
15	Calcium hydroxide solution	1 liter	9-5
13	Candles	30	8-1
5, 19	Cardboard, pieces (about 8 cm × 8 cm)	30	3-3, 13-6
10	Chromatography solvent	2 liters	6-4
27	Clay	1 pkg.	20-2
20	Clock	1	14-4

Activity Number	Materials	Amount Needed for 30 Students	Chapter Sections in Which Used
20	Columns (1 meter long), with plastic caps	30	14-4
16	Containers (for cubes)	15	10-3
25	Container, large (plastic, glass, or paper)	30	18-Feature
25	Container, small (plastic, glass, or paper)	30	18-Feature
6	Cotton-tip swabs	2 boxes	3-4
1, 2, 3	Cover slips	30	P-2, P-3, 1-2, 1-3
13	Cubes, sugar	30	8-1
16	Cubes, wooden (small)	1500	10-3
10	Cups, paper	30	6-4
15	Dishes, evaporating	30	9-5
1, 2, 3, 15	Droppers	30	P-2, P-3, 1-2, 1-3, 9-5
13	Dry cells, C or D	30	8-1
12	Fasteners, paper	180	7-3
10	Filter paper, 2×15 cm strips	60	6-4
21	Flexible rulers	30	15-1
10, 20	Food coloring, boxes of	30	6-4, 14-4
14	Fruits, assorted	30	9-4
14	Galvanometers	15	9-4
7	Garden tools, small	15	4-2
5, 9	Graduated cylinders	30	3-3, 6-1
19	Hair, pieces of human	30	13-6
15	Hydrochloric acid, dilute	1 liter	9-5
11	Ice, bucket of crushed	1	7-3
2	Iodine stain	1 liter	1-2
5	Jars, large	30	3-3
21	Lamps (with 25-watt bulbs)	30	15-1
15	Lamps, heat	30	9-5
11	Lead nitrate solution	1 liter	7-3
6	Lemon juice	1 bottle	3-4
18	Maps, world	30	12-5

Activity Number	Materials	Amount Needed for 30 Students	Chapter Sections in Which Used
9, 11, 12	Masses, set of standard	15	6-1, 7-3
13	Matches	1 box	8-1
7, 22, 24	Metersticks	30	4-2, 16-1, 17-7
1, 2, 3, 4	Microscopes	15	P-2, P-3, 1-2, 1-3, 2-2
7	Microscopes, dissecting (optional)	15	4-2
17	Mineral sample kit	1	11-1
25	Mineral wool, pieces of	30	18-Feature
27	Mirrors, small plain	30	20-2
14	Nails, aluminum	1 box	9-4
14	Nails, copper	1 box	9-4
14, 17	Nails, iron	1 box	9-4, 11-1
21	Needles, teasing	30	15-1
23	Newspapers	1 whole	16-2
14	Newspapers	3 whole	9-4
1	Newspapers	3 small pieces	P-2, P-3
2	Onion pieces	30	1-2
15	Paper, blue litmus	1 box	9-5
25	Paper, large pieces	30	18-Feature
15	Paper, *p*H (optional)	1 box	9-5
15	Paper, red litmus	1 box	9-5
25	Paper, small pieces	30	18-Feature
27	Paper, white (8½″ × 11″)	30	20-2
10, 23	Paper clips	30	6-4, 16-2
3	Paramecium culture	1 vial	1-3
5	Pencil, wax marking	30	3-3
17	Pennies, copper	30	11-1
15	Phenophthalein solution	1 liter	9-5
27	Pins, large straight	30	20-2
27	Pins, small straight	30	20-2
15	Potassium hydroxide solution	1 liter	9-5

Activity Number	Materials	Amount Needed for 30 Students	Chapter Sections in Which Used
11	Potassium iodide solution	1 liter	7-3
21	Protractors	30	15-1
6	Quinine solution	1 liter	3-4
20	Ringstands with clamps	15	14-4
15	Rods, stirring	30	9-5
13	Rubber bands, regular	1 box	8-1
23	Rubber bands, wide	1 box	16-2
5, 19, 27	Rulers, centimeter	30	3-3, 13-6, 20-2
6, 20	Salt (NaCl)	1 box	3-5, 14-4
9	Samples, solids and liquids	30	6-1
23	Sandpaper pieces (8 cm × 12 cm)	30	16-2
17, 18, 19	Scissors	30	11-1, 12-5, 13-6
27	Sheets, cardboard (21 cm × 27 cm)	30	20-2
19	Shoeboxes	30	13-6
1, 2, 3, 17	Slides, glass	30	P-2, P-3, 1-2, 1-3, 11-1
4	Slides, prepared onion root tip	15	2-2
15	Sodium hydroxide solution	1 liter	9-5
26	Solid objects	30	19-3
23	Spring scales	15	16-2
7	Stakes, wooden	60	4-2
22, 24	Stop watches	15	16-1, 17-7
7, 23	String	1 ball	4-2, 16-2
21	Styrofoam balls	30	15-1
25	Styrofoam pieces	30	18-Feature
13	Sulfur, roll pieces	30	8-1
18, 19	Tape, transparent	2 rolls	12-5, 13-6
6, 10, 15, 20	Test tubes (150 mm × 25 mm)	120	3-4, 6-4, 9-5, 14-4
6, 10	Test-tube racks	30	3-4, 6-4
25	Thermometers (0–100 C)	30	18-Feature
9	Thread	1 spool	6-1

Activity Number	Materials	Amount Needed for 30 Students	Chapter Sections in Which Used
13	Tin, small pieces	30	8-1
13	Tongs	30	8-1
2, 10	Toothpicks	1 lg. box	1-2, 6-4
14	Towels, paper	1 pkg.	9-4
18	Tracing paper	30	12-5
2	Tweezers	30	1-2
25	Vermiculite, pieces	30	18-Feature
11	Vials with covers	30	7-3
12	Washers	180	7-3
1, 2, 5, 6	Water		P-2, P-3, 1-2, 3-3, 3-5
13	Wax, pieces of candle	30	8-1
23	Waxed paper	30	16-2
13	Wire, copper, iron, and platinum	30	8-1
26	Wire, insulated electrical	30	19-3
21	Yarn, pieces	30	15-1
14	Zinc strips (6 cm long)	30	9-4

Materials Suppliers

EQUIPMENT

Carolina Biological Supply Co.
2700 York Road
Burlington, NC 27215

Central Scientific Co. (CENCO)
11222 Melrose Avenue
Franklin Park, IL 60131

Connecticut Valley Biological Supply Co., Inc.
82 Valley Road
Southampton, MA 01073

Fisher Scientific Co.
4901 W. LeMoyne Street
Chicago, IL 60651

Frey Scientific Co.
905 Hickory Lane
Mansfield, OH 44905

McKilligan Supply Corp.
435 Main Street
Johnson City, NY 13790

Nasco West Inc.
P.O. Box 3837
Modesto, CA 95352

Wards Natural Science Establishment, Inc.
5100 West Henrietta Road
P.O. Box 92912
Rochester, NY 14692

GEOLOGIC AND TOPOGRAPHIC MAPS

USA
Box 25286
Federal Center
Denver, CO 80225

Canada
Surveys and Mapping Branch
Department of Energy, Mines and Resources
615 Booth Street
Ottawa, ON K1A0E9

Audio-Visual Suppliers

A list of appropriate audio-visual materials, chapter by chapter, is provided at the beginning of each unit in the Teacher's Edition. (See pages 12–13, 144–145, 268–269, and 396–397.) These materials can be ordered using the addresses of the suppliers below.

Agency for Instructional Television
Box A
Bloomington, IN 47402

Carousel Film and Video
241 East 34th Street
New York, NY 10016

Coronet Films
108 Wilmot Road
Deerfield, IL 60015

CRM/McGraw-Hill
674 Via De La Valle
Solana Beach, CA 92075

Educational Activities, Inc.
P.O. Box 392
Freeport, NY 11520

Encyclopedia Brittanica Educational Corp.
425 North Michigan Avenue
Chicago, IL 60611

Films, Inc.
5547 North Ravenswood Avenue
Chicago, IL 60640

Guidance Associates
Communications Park
Box 3000
Mount Kisco, NY 10549

National Film Board of Canada
16th Floor
1251 Avenue of the Americas
New York, NY 10020

National Geographic Society
Educational Services Department
P.O. Box 1640
Washington, DC 20013

Phoenix Films, Inc.
468 Park Avenue South
New York, NY 10016

Pyramid Film and Video
Box 1048
Santa Monica, CA 90406

Random House School Division
400 Hahn Road
Westminster, MD 21157

Society for Visual Education, Inc.
1345 Diversey Parkway
Chicago, IL 60614

Sterling Educational Films
711 Fifth Avenue
New York, NY 10016

Time Life Video
Time and Life Building
New York, NY 10020

Computer Applications

The computer is an outstanding tool for the science classroom. A computer can be used to run software and perform calculations required in laboratory Activities. The most productive use of the computer, however, occurs when students develop their own programs. Encourage students to develop short programs that will perform lab calculations and display results on graphs. Many such programs can be written by the student before Activities, stored on a floppy diskette, and then used during lab periods.

Computer programs, designed to reinforce or extend information presented in the student text, have been included in each chapter of the GENERAL SCIENCE ACTIVITY BOOK. Students should read the introduction for each program, work through the program itself, and complete the discussion questions that follow. Encourage students who have successfully run the program to try the section "Upgrading the Program," which appears below the discussion questions.

The computer is a challenging medium, for it permits students to create programs as sophisticated as they can make them. The simulation is a type of program well suited to science courses. A computer simulation presents a user with a problem and provides a range of possible solutions. The user must analyze the data presented in the problem and then select and enter a solution. The computer then tells the user the impact of the solution. For example, a simulation of an epidemic might permit the user four or five options for eradicating the disease-causing organism. Each option results in the computer calculating a certain percentage reduction of the effects of the epidemic.

Creating simulations has two advantages for science instruction. First, the process of creating a simulation develops scientific problem solving skills. And second, a simulation requires background research, thus further strengthening a student's background in science.

A computer may also be used to develop Computer Assisted Instruction lessons. The simplest CAI lessons are drill and practice. Drill and practice exercises are well suited for reviews before tests and general skill building. They are not well suited for introducing new terms or other new material. CAI programs can be written in the language BASIC, or for the inexperienced programmer, in any one of the PILOT languages.

CAI games and simulations require much more time to develop and are best written in BASIC or PASCAL. Few teachers have the time to develop this type of software themselves. Games and simulations are most frequently purchased from outside software suppliers.

Here are some general guidelines for selecting educational software.

1. See and use a software package before you buy it.
2. Be certain the software will run on your computer. Be sure to check memory capacity and other requirements.
3. Check out the instructional objectives of the software and how they relate to the objectives of your course.

4. Make sure the software includes instructions that both you and your students can understand and follow. Look over the manual, often called documentation. Does it present all the information you need to run the software successfully?
5. Run the software and determine whether it will be interesting to students, whether it meets the goals it claims to meet, or if it is just boring. Is the software really designed to meet its educational objectives? Enter an incorrect response to see how the software handles errors.
6. Look up a review of the software package in an educational magazine or other publication such as *Classroom Computer Learning*.

Some Software to Consider

Life Science

Circulation, by Micro Power and Light, for Apple II+.
Ecology, Simulations I and II, by Creative Computing Software, for Apple II+.
Genetics, by Microcomputer Workshops, for Commodore 64.

Chemistry

Chem Lab Simulations 1 and 2, by High Technology, Inc., for Apple II+.
Heat of Fusion and Formulas of a Compound, by Microcomputer Workshops, for Commodore 64.
Temperature Grapher, by NSTA, for Apple II, II+, and IIe.

Earth and Space Science

Volcanoes, by Earthware Computer Service, for Apple II+.
The Star Gazer's Guide, by Synergistics, for Apple II+.
Diffusion, by Microcomputer Workshops, for Commodore 64.

Physics

Conservation Laws, by Cross Educational Software, for Apple II+.
Simple Machines I, by MicroEd, for Commodore 64.
Power Grid, by National Science Teachers Association (NSTA), for Apple II, II+, IIe, and TRS-80.
Energy Conversion, by NSTA, for Apple II, II+, IIe, and TRS-80.
Home Energy Savings, by NSTA, for Apple II, II+, IIe, and TRS-80.
Electric Bill, by NSTA, for Apple II, II+, IIe, and TRS-890.

Accommodating the Exceptional Student

In the past children with severe physical, mental, or emotional disabilities were segregated from their nondisabled schoolmates. Some attended special schools. Far too many grew up with less than adequate schooling.

Federal Law PL 94-142 completely changes this situation. Today every student with a disability or disabilities must be integrated into the school community. The disabling condition or conditions must be identified, analyzed, and evaluated, and a suitable educational plan must be devised for the student. Such plans may call for partial integration into the school community, with some instruction taking place in special education classes, or total integration, with selected special services, such as speech therapy.

In all cases, a fundamental goal is the integration of the child who is disabled into the mainstream of the school community, and ultimately integration into the mainstream of society. Disabled students within an appropriate educational plan are said to have been mainstreamed. The plans should reflect agreement among student, parents, and school authorities with respect to what are the most suitable learning activities.

It is the responsibility of the science teacher, indeed, of all teachers, to become familiar with the educational plan of each disabled student in a class. Then, whatever is needed to facilitate learning by the disabled student must be developed and put in place.

Nature of Disabilities

It may be useful to review the different types of disabilities that may turn up in the classroom. Poor sight, blindness, mechanical impairments involving the arms, legs, and spine, and other health problems are known as *physical impairments*.

Emotional disturbance is any condition involving the emotions, such as a tendency to unprovoked outbursts of anger, that interferes with normal functioning in social and educational settings.

Conditions that interfere with a student's ability to use language are called *communicative disabilities*. These include deafness, hearing loss, speech problems, and certain specific learning disabilities—dysfunction in one or more of the psychological processes involved in either using or understanding language.

Disorders such as dyslexia, developmental aphasia, perceptual problems, and minimal brain damage are called *specific learning disabilities*.

Finally, intellectual functioning slower than the range of function considered normal is referred to as *mild mental retardation*.

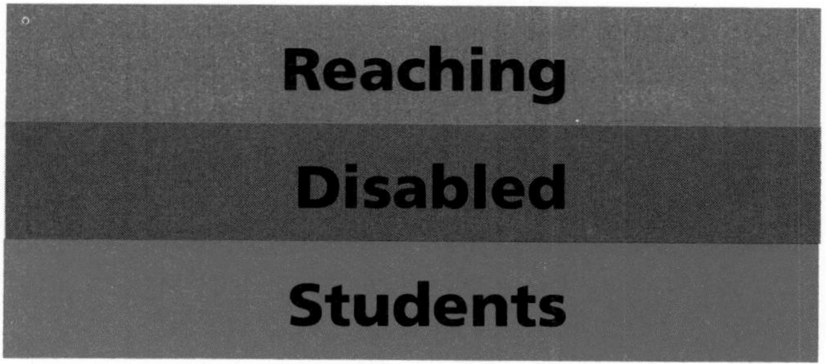

Reaching Disabled Students

Students with disabilities, like other classroom children, continuously interact with their environment while learning about it and developing skills. However, a considerable body of research has established that certain disabled children can be hindered in their growth when classroom instruction is based primarily on reading, listening and writing. Meaningful instruction of these children must meet their special needs. These needs depend upon the nature of their disability and their experiential background. Generally, what is required are appropriate experiences focused on student participation before, during, and after presentation of concepts.

Participatory experiences provide the background necessary for disabled students to structure, interpret and assimilate the concepts. The experiences affect the time for processing ideas, the form of sensory input, the quality of interpretive discussion, and other elements involved in a successful science lesson. For example, visually impaired students benefit greatly by being allowed to manipulate lesson objects and materials before and during the time of instruction. This provides a background of physical characteristics which the teacher can use to great advantage. Students with no visual disability can readily obtain the same background by viewing the objects as the lesson begins. It follows that carefully constructed experiential activities, allowing participation to the extent permitted by a student's disability, are essential to meaningful learning.

On the other hand, it is often possible to motivate students with disabilities to memorize the terms and definitions needed to answer textbook review questions. For the most part, unfortunately, this is nothing more than empty verbalism, that does not require an understanding of concepts. Such a teaching strategy should be avoided with students who are disabled.

A teaching strategy consistent with the definition of the disabled student given above is as follows. First, students should be permitted to solve simple open-ended problems by means of manipulative activities. Working out "Try This" activities, handling materials and equipment used for classroom demonstrations, and carrying out any other activities specially designed, with teacher assistance as needed, will meet this need. For some students, laboratory activities will be suitable. Second, students should read, or have the text paraphrased for them, as needed, and then interact directly with the teacher. As the concepts being taught are discussed, reference should be made to the concrete activities previously carried out to help students assimilate and integrate the desired content. Third, students should be given simple convergent problems that call for the use of concepts covered. A useful technique, when feasible, is to modify "Checkpoint" questions so they can be performed, rather than responded to in writing. For example, suppose a question calls for the definitions of the terms transparent, translucent, and opaque. Allowing the disabled student to select an object from among several and show the meaning of each concept both demonstrates achievement and reinforces learning. Finally, if needed, remedial learning activities can be devised and administered.

To supplement these general suggestions, numerous more specific ideas for helping disabled students are presented in the chapter annotations throughout this volume. These suggestions are included under the heading "Mainstreaming." In addition, teachers desiring more background information and ideas will find the following title useful: *Sourcebook: Science Education and the Physically Handicapped,* edited by H. Hofman and K. Ricker, National Science Teachers Association, 1979.

Reaching the Gifted Student

Far too often the gifted student is neglected on the grounds that bright children need little attention or guidance to achieve at a satisfactory level. Then, when such children fail to achieve up to their potential, teachers complain that their lack of performance is incomprehensible. In fact, bright children need as much attention as disabled students if they are to avoid becoming casualties of the system.

Left to his or her own resources and required to master only the content and skills presented to the average learner, the gifted child often tunes out and experiences anger and frustration. Such students very quickly master whatever is taught, and thus are bored by the lack of mental stimulation inherent in plodding through learning activities that fail to challenge them. Some gifted students rebel and become behavior problems.

GENERAL SCIENCE provides a sufficient number and variety of learning activities to meet the needs of most, if not all, gifted students. Each chapter contains an "Enrichment" section and a "Feature Article". These can be used to broaden and deepen a student's exposure to relevant content. Many "Try This" activities, in addition, lend themselves to expansion into more sophisticated problems. The same is true of the laboratory activities, which, with little effort, can be amplified and made more challenging.

As pointed out earlier, the "Chapter Review" questions are divided into three levels of difficulty. Bright students will find the "Challenge Your Understanding" questions interesting and stimulating. Finally, the "Projects" listed at the end of each chapter provide ample opportunity for the gifted as well as committed student to explore concepts and problems far beyond the scope of the text proper.

Learning Outside the Classroom

Many valuable science lessons can be taught outside of the classroom or laboratory. The school yard, for example, is an excellent location for a discussion on cloud formations or for examining local rocks or minerals. The best way to teach a lesson outside of the classroom is to organize the lesson into a series of steps.

The first step is to prepare the class for the activity. This may be accomplished in the classroom, or it can be done on a bus trip to a remote location. This preparation should include a description of both the objectives of the activity and the process of the activity. The most successful preparation will include a handout listing objectives and processes.

The second step is the actual activity. If a class is going to a lecture or other nonparticipatory activity, make sure students have a sheet of questions to respond to afterwards. The question sheet should provide specific guidelines for effective listening to ensure maximum absorption of ideas and information. Encourage note taking. The questions can help indicate the kinds of notes needed.

The third step is to hold a discussion on the activity. This is best done back in the classroom. If you have handed out questions, include a discussion of student responses.

Here are some ideas for activities outside the classroom. Many have worked well for other teachers. You'll find additional suggestions in this Teacher's Edition at Unit opening pages, in annotations, and in the "Projects" sections of the student text.

- Use the rock materials in buildings and bridges as the basis for a geology lesson. Learn to identify the types of rocks used, and, if possible, some of the minerals the rocks contain. Are any buildings in your town constructed from locally quarried rock? If you find some structures built from sedimentary rock, can you find evidence of fossils? If so, what types of fossils?
- Visit a nature center or arboretum and conduct a survey of the types of trees present. The best way to conduct such a survey is to run out 100 meters of cord, tie it between two stakes, then count the number and variety of trees that overhang and touch the cord. This method can be used to identify the number and variety of plants in a field also.

A nature center can also be used to keep track of seasonal changes. For example, what types of birds are found at the center at different times of the year? Which birds pass through in the spring and fall, and which ones are year-round residents?

- Visit a local zoo to examine variety in animal species, animal adaptations, and animal behavior. Have students work in pairs to make a record of animal behavior. One student should read off time intervals and describe behavior, while the other records the observations. Have the students switch duties from time to time. Primates offer the best behavior for such an observation activity.

Have students go through a zoo and record the biome for a set of animals, together with special adaptations for survival in that biome. Have a class examine each species of cat or primate in a collection, and then develop a classification system that can be used to identify each species. If you have a video camera, animal behavior can be recorded and then observed during post-activity discussions.

● Use community resources such as local science/technology businesses and industry. For example, visit a petroleum refinery as a way to learn about organic chemistry. Study the processes used to create gasoline and other fuels, and the petrochemical basis for medicines and plastics. Find out what the refinery is doing to control air and water pollution.

● If you live in a farm community, visit an agricultural extension agent to discuss pesticide control. What types of pesticides are being used in your area? Are they toxic to humans or animals? Have there been problems with pests becoming immune to the chemicals used? Are alternatives to chemical spraying, such as biological controls, being used or explored? What other alternatives are in use or under consideration?

Discuss erosion control with the agent. Is erosion a serious problem in your area? How much top soil can be lost in a year? What can farmers do to avoid loss of topsoil by erosion?

● Visit a chicken, pig, cattle, or dairy farm to discuss selective breeding. How has selective breeding improved meat or milk production in the past 50 years? Why is artificial insemination important to selective breeding? How has selective breeding been used to improve crops?

● Visit a local weather station and examine the instruments, weather satellite pictures, and forecasting process. Have a meteorologist describe a weather satellite and the transmission process. Ask for an explanation of each instrument and how readings are interpreted. Have a meteorologist explain the symbols on a weather map and how they are read.

● Visit a nearby electric power generating facility to discuss the physics of power generation and conservation of electric power. How much power is produced by the facility? How many homes does it serve? What methods are used to reduce air and water pollution? What can people do to conserve energy? Is local energy usage increasing or decreasing, and by how much? Ask for explanations of peak power and electric power grids.

● Visit a local planetarium to study star patterns and learn about the planets. Examine the changes of the stars through an individual night and through the seasons. Locate the north star. What other constellations can you find? Are there any planets visible to the naked eye at the time of your visit?

● If a telephone amplifier is available, arrange to conduct a telephone interview of a scientist, and have the class listen in. Such interviews function most effectively with prepared questions. Many local and state politicians can be interviewed in a similar manner. Try to keep the interview time to a maximum of 30 minutes, so students can record their impressions and discuss the interview during the same class period.

● Have students use the library. As mentioned above, the "Projects" sections at the ends of chapters provide ideas for a variety of outside-the-classroom activities, including library projects. Although encyclopedias and other standard reference texts may be adequate sources of information for some of these projects, encourage students to find other sources by using the card catalog and the *Readers' Guide to Periodical Literature*. Certain projects require students to research and report on various individuals who have made important contributions in areas of science. Several sources that will be useful in these instances include the following: *American Black Scientists and Inventors*, Edward S. Jenkins, ed., 1975. National Science Teachers Association;

Notable American Women, 1607–1950: A Biographical Dictionary, 3 vols., Edward T. James, ed., 1971. *Supplement,* 1980. Belknap Press of Harvard University Press, Cambridge, Mass.; *Pioneers of Science: Nobel Prize Winners in Physics,* Robert L. Weber. J.H. Lenihan, ed., 1980. Adam Hilger, Ltd., England; *The Women's Book of World Records and Achievements,* Lois D. O'Neill, ed., 1979. Doubleday, Garden City, New York; *Women Pioneers of Science,* Louis Haber, 1979, Harcourt Brace Jovanovich, New York; *Women Scientists in America: Struggles and Strategies to 1940,* Margaret W. Rossiter, 1984. The John Hopkins University Press, Baltimore.

Science, Technology and Society

An important component of any science course is the relationship between science and society. The National Science Teachers Association has recommended that 15 to 20 percent of classroom time be devoted to science-related social issues. Some ideas for topics are listed below; newspapers and current events magazines frequently contain articles dealing with issues of a scientific nature. The class discussions should emphasize objective evaluation of information, and viewing controversial subjects from many points of view—at least as many as are represented in the class.

Several approaches can be used to fuel class discussions. Conducting a survey of students and interviewing people in the community work well. Another effective technique is the debate. Divide the class into two groups with opposing viewpoints on an issue, and have students gather information to support their positions. Such debates will function best when students try to argue on the basis of scientific evidence.

Class discussions can be used as the starting point for action. For example, if your class demonstrates a concern for energy conservation, students might develop posters outlining energy conservation measures in the home and arrange to display them in local stores.

The list of potential topics for discussion is almost limitless. Those that follow should prove stimulating, for they are issues of great concern, and they are related to content presented in the text.

- *Genetic engineering* See Feature Article in Chapter 2. Discuss the processes employed in genetic engineering. Should there be restrictions on genetic research or specific safety measures for laboratories conducting this kind of research? Should genetic engineering be used to create lower organisms for use in experiments? Why or why not? Should humans be genetically adapted to prevent disease or for other purposes?
- *Acid rain* results primarily from burning sulfur-rich fuels. Discuss the steps involved in the formation of acid rain. Why is acid rain a problem in the Northeast and in eastern Canada? What are the effects of acid rain on lakes, plants and animals, buildings, and statues? How can acid rain be reduced? What alternatives are there to high-sulfur fuels? What should be done to improve matters in areas currently experiencing severe acid rain?
- *Nuclear power* is discussed in Chapter 10. What dangers, if any, accompany the use of nuclear power? Conduct a survey to see how students rank dangers in their lives, and include nuclear power and weaponry. Why do power companies build and operate nuclear power plants? How does a nuclear powered plant compare with those that use fossil fuels? What is done with a nuclear power plant after its life is ended? Discuss problems of nuclear waste disposal and the production of nuclear isotopes. What air and water pollution problems, if any, are related to the use of nuclear power?
- *Indoor air pollution* can result from the use of superinsulation in homes and other buildings. What gases or other substances constitute indoor air pollution? What can be done to control or reduce indoor air pollution?

- *Weather control* may harm as well as help. Who should have the right to control the weather? When should cloud seeding and other weather control measures be used? When should they not be used?
- Current study of the causes of earthquakes may lead to the ability *to predict earthquakes* within the foreseeable future. How should a prediction of an 80 percent chance of a quake in the next week be handled? How should a prediction be handled if there is a 90 percent chance of a quake within the next 12 hours? Who should be notified and by what means? How might mass panic be avoided? What if the probability is 50 percent? How accurate are any scientific predictions, and how should the predictions be handled by society?
- *Marine mineral resources* The oceans contain vast amounts of valuable minerals. Who should be permitted to mine these minerals? Should the minerals be taken on a first come, first served basis? Or should the minerals be divided up among all nations by treaty? Should there be environmental restrictions on the mining of oceanic mineral resources? What group should have jurisdiction over such mining activities?
- How should *toxic wastes* be handled? What problems are associated with burying them? Should a remote island be made a dump for toxic wastes? Should we continue to permit ocean dumping of toxic wastes? of raw sewage? What detoxification alternatives exist? Is anyone to blame for toxic wastes? See Feature, p. 266.
- *Endangered species* Should a nuclear or other power plant be denied construction rights because of the presence of an endangered plant or animal? What should constitute an endangered species, and how far should protection go? Should we permit mining and forestry in our national parks and recreation areas?
- *Fuels from grains*. Should grain be used for alcohol fuel production when millions of people are starving around the Earth? How much grain is used for alcohol production? How many people could be fed by that grain?
- *Grain products or meat?* Is it more efficient to feed grain to livestock or directly to people? What are the difficulties in providing grain to starving people?
- Examine and discuss the field of *bioengineering*. Who should pay for the implantation of an artificial device in a human? If an artificial kidney or other organ were produced, who should be permitted to have it implanted, or should it be made available to all who need it?

Evaluating Student Achievement

Evaluation of student achievement can serve several functions important to the student, the teacher, and the school administration. Evaluation can provide the student with a measure of his or her understanding of the content and skills presented. Often knowledge that evaluation will occur provides an incentive to complete assignments. Evaluation can help the teacher determine a student's progress, provide a measure of the teacher's success promoting learning, and, inevitably, assist in grading a student's work. Test scores also provide evidence to the school administration that the course curriculum is being carried out successfully and that students are indeed learning. Finally, a well designed evaluation program satisfies the public's requirement for educational accountability.

The guidelines for evaluation presented here are designed (a) to help the teacher understand the essentials of successful evaluation, and (b) to show how to use the tests developed for this course to best advantage.

It is important to understand that a complete program of student evaluation involves more than simply using the tests provided or writing a set of tests. Such tests are designed to assess a student's retention of information and, in some cases, his or her understanding of concepts presented. These aspects will be discussed below. Tests cannot assess important factors that should be included in any complete evaluation of student achievement. These factors include such matters as successful completion of homework, laboratory and classroom assignments, student participation in classroom discussion and activities, independent work on science projects such as those listed at the end of each chapter, any other pursuit of knowledge beyond that required for the course, and transfer of knowledge and skills to other areas of learning. Each of these factors, and possibly others, adds to the total level of achievement that must be evaluated by the teacher. All should be considered when grading or reporting student achievement.

No guide to evaluation can give the teacher a universal formula that will incorporate all desired aspects of student performance. It is incumbent upon individual teachers and/or schools to determine their own goals and objectives, and then to develop an evaluation model that assesses those goals. Only then can the weight that should be given to written tests be determined.

Grading

Final grading should be done using a procedure that properly weights each factor that was judged to be important in student performance. The easiest way to accomplish such weighting is to place all relevant factors on the same scale—for instance, a point scale from 0 to 10. Then, the numerical value of each factor on this scale is multiplied by the percentage weighting given to that factor, and the resulting products are added together to give a final numerical grade that can be translated into a letter grade, if desired.

For example, suppose a teacher felt that 25 percent of a grade should represent classroom participation, 25 percent should represent successful completion of laboratory and homework assignments, and 50 percent should represent test performance. Suppose also that a particular student earned a 5 out of 10 possible points for classroom participation, completed eight out of ten laboratory and homework assignments, and received 7 out of 10 possible points on test performance. The weighted grade for this student, on the basis of the 0 to 10 grade scale, would be given by the following equation:

$$0.25(5) + 0.25(8) + 0.5(7) = 6.75$$

If the letter grading scale is defined in terms of the numerical scale as

A = 9.0 to 10
B = 7.0 to 8.9
C = 5.0 to 6.9
D = 3.0 to 4.9, and
F = 0 to 2.9,

then this student would receive a grade of C+.

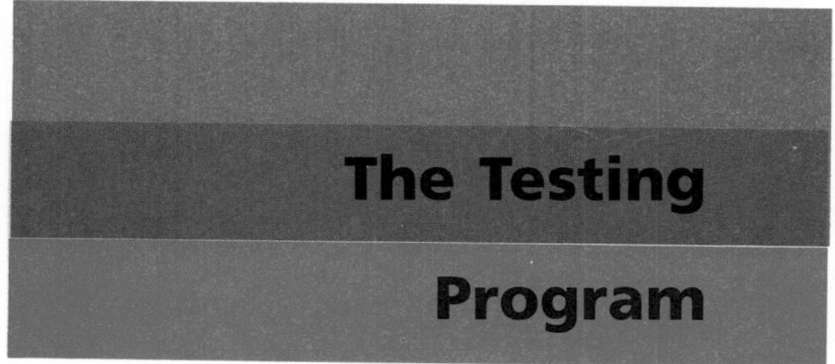

The Testing Program

The remainder of this discussion of evaluating student achievement assumes that the reader has given due thought to all aspects of evaluation and has reached a conclusion about the weight to be given to written tests in his or her evaluation plan. Accordingly, the discussion will focus on the tests provided for GENERAL SCIENCE. These are available as blackline masters for copying. The masters and answers to the tests are included in the GENERAL SCIENCE TEACHER'S RESOURCE BOOK, along with specific information on how to administer and score the tests. Topics covered here include test design, test format, the relationship of tests to the course text, and suggestions for use, administration, and scoring.

The tests for the course correspond directly to the major subdivisions of the text. The set includes 25 four-part sets of tests, one test devoted to each of the 20 chapters, one covering each of the four units, and one final examination encompassing topics from the entire course. There is also a two-part test devoted to the Prologue.

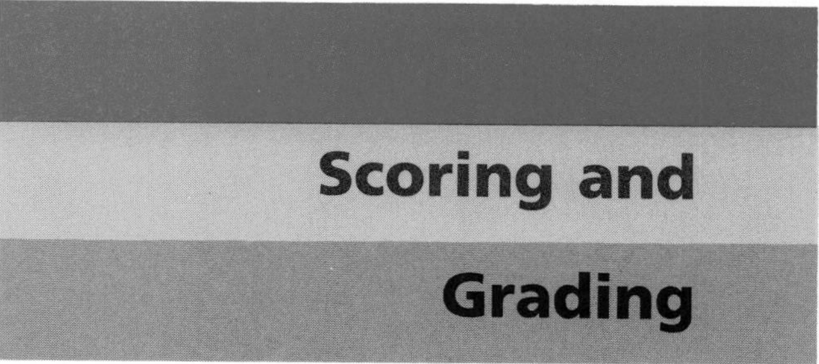

Scoring and Grading

The tests should be scored using the directions and answers provided in the TEACHER'S RESOURCE BOOK. Allow one point for each correct answer.

Grading based on test performance can be handled in one of two ways. The teacher may choose to review each test or test part and compare the emphasis given to various topics in the test to the emphasis given to the same topics in the course. Based on this approach, a decision can be made with respect to the percentage of questions a student must answer correctly to show the minimal understanding desired. This type of standard setting is appropriate if all that is desired is a pass/

fail grade. Of course, the standard can be expanded on to establish a set of grades.

An alternative approach to setting standards for grades involves the use of statistical criteria to determine grades based on the performance of the entire class or a group of similar classes. In this case the performance of an individual student is compared to the performances of a large number of students, and a grade is assigned in terms of relative position. A student achieving as well as 80 percent of the large group, for example, would be given a grade that reflects that level of achievement.

In either case, the setting of cutoff points for grades is a subjective task for the test administrator. Such cutoff points must be established by the teacher.

Teachers interested in a more comprehensive discussion of evaluation and grading should consult *Basic Measurement and Evaluation of Scientific Instruction*, Rodney L. Doran, National Science Teachers Association, 1980.

Allyn and Bacon

General Science

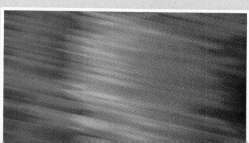

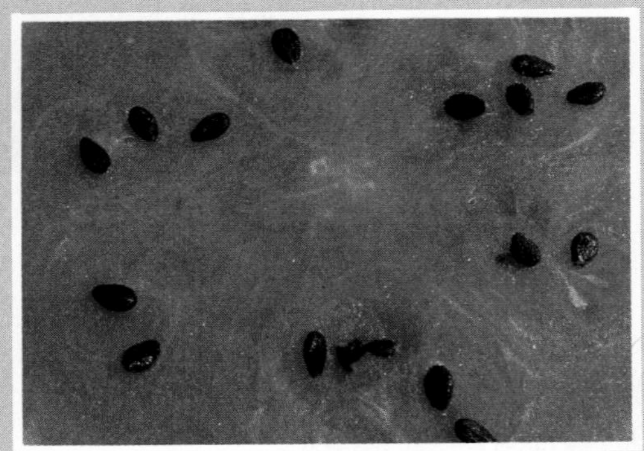

APPROXIMATE VOLUME
400
300

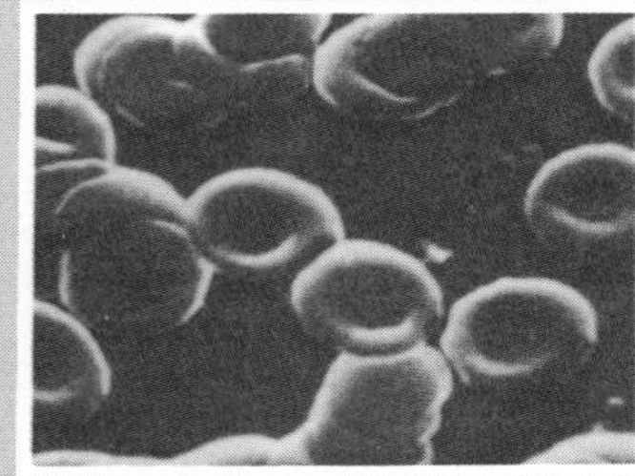

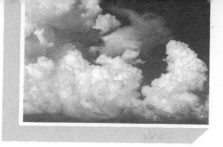

Allyn and Bacon

General Science

SECOND EDITION

CAROLYN SHEETS BROCKWAY
Science Education Center
University of Iowa
Iowa City, Iowa

ROBERT GARDNER
Chair, Science Department
Salisbury School
Salisbury, Connecticut

SAMUEL F. HOWE
Educational Consultant and Writer
formerly Science Teacher
Friends' Central School
Philadelphia, Pennsylvania

Prentice Hall
Needham, Massachusetts Englewood Cliffs, New Jersey

Staff Credits:

LOIS B. ARNOLD, JOEL R. GENDLER
Editorial Development

L. CHRISTOPHER VALENTE
Design Direction

STUART WALLACE
Design Coordination

BILL WOOD
Production/Manufacturing

***Outside Credits*:**

PRENTICE CROSIER, HELEN REEBENACKER, CAROL H. ROSE, SUSAN GEROULD
Production/Design Services

LAUREL ANDERSON, SUSAN VAN ETTEN
Photograph Research

***Cover Design*:**

HANNUS DESIGN ASSOCIATES AND L. CHRISTOPHER VALENTE

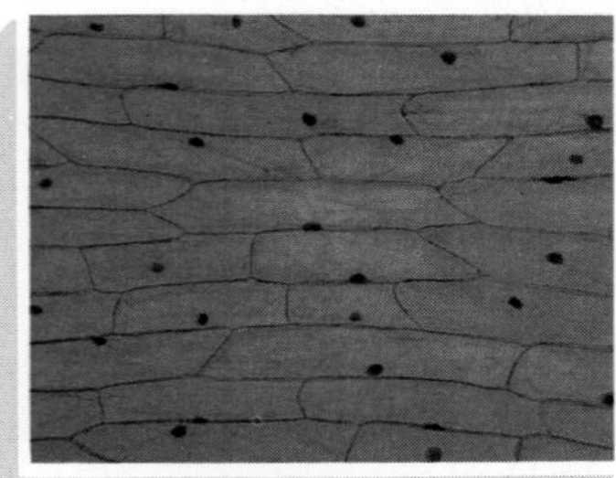

ISBN 0-13-350802-1

Printed in the United States of America

2 3 4 5 6 7 8 9   96 95 94 93 92 91 90 89 88

A Simon & Schuster Company

Acknowledgements

Teacher Reviewers

JOHN J. SHERIDAN
Biology Teacher
Boca Ciega Senior High School
Gulfport, Florida

ANN S. BURDEN
Chair, Science Department
North Pitt High School
Bethel, North Carolina

LINDA B. KNIGHT
General Science/Earth Science Teacher
Bloomington High School North
Bloomington, Indiana

LARRY A. SYRON
Chair, Science Department
Gresham Union High School
Gresham, Oregon

Consultants

SCIENCE CONTENT

RICHARD L. ROTUNDO
Department of Anatomy and Cell Biology
University of Miami
Miami, Florida

W.T. LIPPINCOTT
Department of Chemistry
The University of Arizona
Tucson, Arizona

LOU WILLIAMS PAGE
Geologist and Author
Houston, Texas

SHARON C. THOMAS
John G. Diefenbaker High School
Calgary, Alberta

READING

PATRICIA TERRELL
School of Education
East Carolina University
Greenville, North Carolina

MAINSTREAMING/ HANDICAPPED ADAPTATIONS

DENNIS W. SUNAL
Division of Education
West Virginia University
Morgantown, West Virginia

ACTIVITIES SAFETY

RONALD J. KENDALL
Institute of Wildlife Toxicology
Western Washington University
Bellingham, Washington

Unit One • Life Science 12–143

Unit Two • Chemistry 144–267

Contents

Unit Three • Earth and Space Science 268–395

Unit Four • Physics 396–529

PROLOGUE Overview

The purpose of the Prologue is to introduce the process used by scientists to discover new information about the natural world. The problem-solving method described is important to all people, for it has application in all disciplines, not just in the sciences.

The Prologue opens with a discussion of how scientists approach problems. The point is made that scientists are curious about natural phenomena, and that they are always seeking answers to questions about the world around them. While some answers are obtained by luck or accident, most of the knowledge gained is acquired by means of a systematic attack on problems. This method is often called scientific method.

To illustrate the scientific method, students will read about the peppered moths and how camouflage played a role in their evolution. The several steps, from initial observations through hypothesis formation to hypothesis testing, are described in some detail. The differences between hypothesis, theory, and scientific law are carefully explained. The importance of a controlled experiment is covered, and it is pointed out that positive results of hypothesis testing support, but do not prove, a hypothesis.

Prologue

Introduction to Science

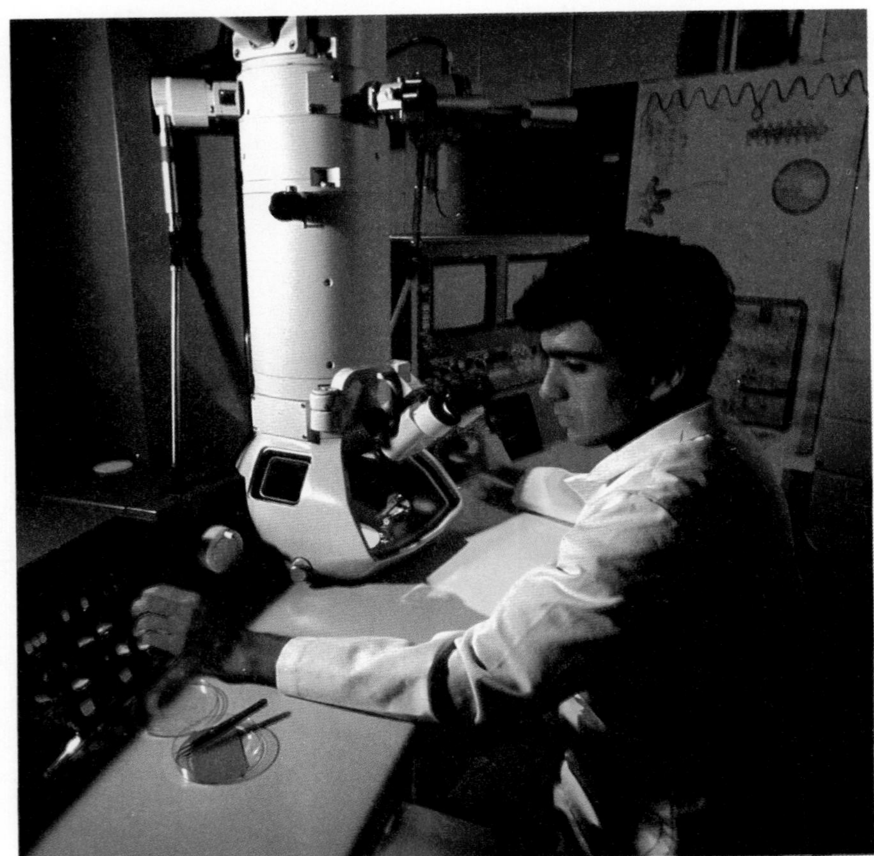

This life scientist is using a scanning electron microscope.

Vocabulary Preview

scientific method
evolution
evolve
hypothesis
experiment
observations
larvae
data
controlled experiment
experimental factor
experimental group
control group
theory
scientific law

P-1

Approaches to Problems

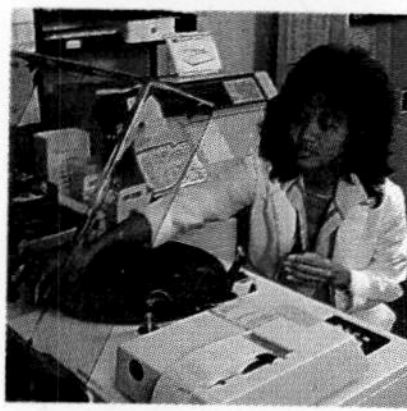

Figure P-1 *Scientists work in many different settings.*

Have you ever wondered how scientists are able to launch the space shuttle at just the right speed so it will orbit the Earth and return safely? Whether there is life on other planets? Why human organ transplants have become more successful than ever before? Or what causes cancer? If you have, then you have begun to think as a scientist does.

Scientists are curious. They are always asking questions and seeking answers. Scientists try to solve problems and learn new things about the world and universe. Often they seek solutions to problems that affect the lives of us all.

Scientists work in four major areas of study. Life scientists, or biologists, study living things. Chemists investigate the make-up, structure, properties, and reactions of matter. Earth and space scientists study the Earth and the objects in space. Physicists explore the interactions between matter and energy. Each area of study includes several specialties. For example, life science includes ecology, which is the study of the interactions between living things and their environment.

The process of science is complex, but it involves thinking, observing, experimenting, discovering and demonstrating. Often a scientist acquires a new insight into the workings of nature. He or she then finds a way to test the validity of that insight. Scientific discoveries have been made in many ways. Isaac Newton is said to have discovered the theory of gravity by thinking about why an apple fell from a tree and hit him. Alexander Fleming discovered penicillin by accident. But then, of course, he tested his discovery experimentally.

However, science usually involves a systematic approach to problem solving. The approach is sometimes called the **scientific method.** Actually, it is a planned, logical way to develop and test ideas. The ideas to be tested are tentative explanations for what scientists have observed. Carefully controlled testing is what makes knowledge gained in science different from knowledge acquired in other fields. But even after testing, results must be verified by other scientists.

scientific method

Teaching Suggestions P-1

Open discussion of the content of the section by asking students to name some of the problems they face in their daily lives. List a few on the chalkboard. Then have students describe how they go about solving the problems listed. Ask for detailed explanations of procedures followed, and stress that there is a reason for looking closely at the problem-solving methods being described. If there is a consensus about a method among the student approaches to problem solving, list the steps on the chalkboard.

Now discuss how scientists approach problems they identify in the natural world. Emphasize that scientists seeking facts must be objective, and that this requires a problem-solving approach that eliminates the possibility of bias. Point out that the scientist's "planned, logical way to develop and test . . . tentative explanations" for what has been observed does just this. In addition, make sure students understand that other scientists must be able to repeat experimental results if they are to stand up. Finally, point out that if a tested explanation is superseded by a better tested explanation, that the earlier one must be discarded.

Suggest to students that they will find it instructive to compare their approaches to problem solving with the scientific process they are about to study. Point out too that they will find this method can be adapted to other types of problems, and that it has the potential to serve them better than the approaches they are currently using.

Teaching Suggestions P-2

Because the major purpose of the Prologue is to understand the hypothesis-testing method used by scientists, it is suggested that only enough information about the peppered moth experiment be provided to achieve this goal. Avoid becoming too involved in a discussion of evolution and genetics at this point. However, be sure that students understand that moths are nocturnal insects. Start by asking them when they have seen moths.

The important concepts in this section are (a) the development of a testable hypothesis and (b) the method used to test that hypothesis. Stress that the key aspect of the experimental test is a prediction based on the hypothesis.

Have the students define *hypothesis*. Ask them for ideas about how hypotheses are formed. Answers will vary, but they should focus on the idea that a hypothesis is a possible explanation for a phenomenon consistent with observations previously made, and also that hypotheses should be testable.

Go over the details of Kettlewell's first experiment. Ask the students to come up with other tests of this hypothesis. They may suggest approaches similar to Tinbergen's work, or Kettlewell's second experiment.

Be sure the students understand the concept of a *controlled experiment*, and why the experimenter sets up both a control group and an experimental group. Explain why the only difference between groups is the experimental factor, that is, the variable under study. With everything else identical, the experimental outcome can only be the result of action (or lack of action) by the experimental factor.

In 1926, Heslop Harrison hypothesized that moth *larvae*, feeding on the pollution-covered foliage, became darker and then passed this trait on to their offspring. There are actually two hypotheses in Harrison's suggestion: (a) larvae that eat darkened foliage become darker; and (b) this trait then is passed on to their offspring. Point out that if the first hypothesis were not validated, the second could not be tested.

Ask the students to think up ways to test Harrison's hypotheses in the laboratory. A control group of moths could be fed normal foliage, while an experimental group could be given polluted foliage. By comparing the control and experimental groups, the first hypothesis could be tested. The offspring of the two groups could be compared to test the second hypothesis.

P-2

Dark-Colored Peppered Moths: Solving a Problem

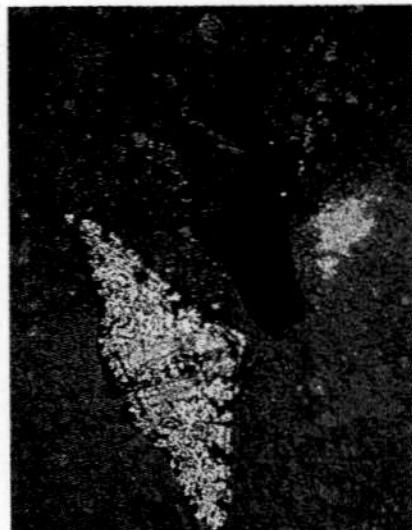

Figure P-2 *Peppered moths in England have a light form and a dark form.*

One way to understand the scientific method is to look at how scientists tried to solve an actual problem. One such problem concerned the peppered moth, a common moth found throughout the forests of Great Britain. There are over 760 kinds of large moths in the British Isles; many are known to have a light form and a dark form which are otherwise very similar. (See Figure P-2.) Before 1850, most peppered moths in England were light-colored. By the early 1900s, however, the dark-colored peppered moth had become much more common in the forests near factories and industrial cities.

In the 1950s, H.B.D. Kettlewell, a biologist at the University of Oxford, began investigating why this happened. The first step in scientific investigation is to identify, or clearly state, the problem. Kettlewell's problem was easy to define: why had the dark-colored moths become more common than the light-colored moths near cities?

The next step in scientific investigation is to collect information related to the problem. One way to do that is to read about the work done by other scientists. A second way is to communicate with others who have investigated the problem.

In his research, Kettlewell read the work of an earlier British biologist, Heslop Harrison. Harrison suggested that the increase in dark moths was caused by harmful substances they ate that came from polluted air. Polluted air was a fairly new problem in England, resulting from a great change that had taken place in the century prior to 1900. During that period of time, known as the *Industrial Revolution*, many new industries sprang up. Many of these industries burned coal, which produced black smoke and soot. (See Figure P-3.) Particles of soot settled on tree trunks and darkened them. Plants that grew on tree bark began to die off from chemicals in the soot. Harrison and other scientists thought the pollution might have caused the changes in the color of the moths.

Figure P-3 *During the Industrial Revolution in Great Britain, many new industries produced smoke and soot that darkened the skies over growing cities. Scientists wondered if the increase in dark peppered moths was caused by the increase in air pollution.*

How exactly did the pollution affect the moths? To find out, Harrison tried feeding polluted leaves to moth *larvae*, the young wormlike grubs which later turn into moths. He examined the resulting adults, and concluded that there were more dark moths than usual. Harrison also studied the next generation of moths. He claimed that the dark coloring the moths had acquired from eating polluted leaves was inherited by their offspring.

Kettlewell, however, was not convinced of this, and he continued to examine the problem. He studied the behavior of moths. He read reports about moths in parts of the world where there was no industry. In addition to reading about Harrison's work, Kettlewell read other scientists' research about the increase in numbers of dark-colored peppered moths. He also studied Charles Darwin's ideas about *evolution*.

Teaching Tip

- Have students observe Figure P-3 and ask them how air pollution has affected their lives. For example, they may have a friend with asthma who listens to the radio to find out whether the air quality will be dangerous. Students who live in a city have probably seen the reduction in visibility that air pollution can cause. Many students will also be familiar with the effects of acid rain, even in areas far from sources of air pollution.

Background

Forests were lighter in color prior to the Industrial Revolution because tree bark and rocks were often covered with light-colored *lichens*, plants that are a combination of a fungus and an alga. When peppered moths rested by day on the bark and lichen-covered rocks, they were camouflaged because of their protective coloring.

After the Industrial Revolution, much of the lichen growth died because these plants were sensitive to industrial pollutants in the atmosphere. While the pollutants themselves directly darkened the forests, the effect on lichens was also important.

Facts At Your Fingertips

- Moths differ from butterflies, which are active in daylight, more colorful, and rest with their wings folded. Moths are mostly nocturnal and rest with their wings in any position. Their antennae also differ. Butterflies have antennae with knobs on the ends, while moths have feathery antennae.
- The antennae of moths are used primarily for smelling. The male moths can smell the scent given off by females from a distance as great as 5 miles away. This scent is used to attract the males for mating.
- Sometimes it is beneficial to an animal or plant to be easily visible rather than camouflaged. This is called warning coloration. The monarch butterfly is an example of this evolutionary development. Birds that eat butterflies learn that monarchs are bad-tasting, and avoid them by recognizing their bright coloring.

Figure P-4 *Have you ever seen a walking stick? This insect is hard to spot in the wild.*

Darwin thought that plants and animals *evolve*, or change over long periods of time, in response to changes in the world around them. The plants and animals that are best suited to their environment will survive to produce a new generation with similar characteristics. The plants and animals that are not as well suited are not as likely to survive. Figures P-4 and P-5 show striking examples of animals which have evolved to blend into their environment.

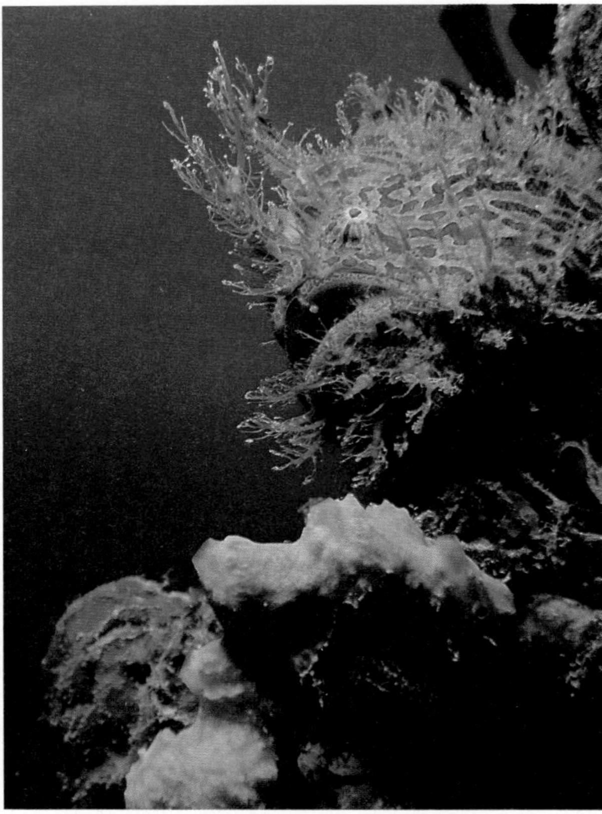

Figure P-5 *The spitlure frogfish is found in coral reefs near St. Croix. Can you find this exotic fish in the photo?*

Teaching Tip

- Ask students if they can think of other examples of camouflaged animals, such as tigers, leopards, or chameleons. Have the students guess which environmental features serve as a background for camouflage in each of their examples. The leopard's spots help it to hide among the shadows, and the tiger's stripes blend into a background of vertical foliage. The chameleon is able to change color to match its background.

Kettlewell believed that Darwin's concept of evolution explained the changes that had taken place in the peppered moths. He had enough information to take an important step in the problem-solving process: He was ready to make an educated guess about a possible solution. Such a guess, or tentative explanation, is called a **hypothesis.** A hypothesis is based on facts related to the problem. Kettlewell's hypothesis was that dark moths survived better than light moths in industrial areas because their camouflage hid them from birds that eat moths. (See Figure P-6.) In other words, the dark-colored peppered moths were better suited for survival.

hypothesis
(hy PAHTH uh sis)

If a hypothesis is to be useful, it must be testable. A hypothesis is tested by means of experiments. An **experiment** is a way of testing that makes use of exact procedures. A scientist carefully observes and measures any changes that occur during an experiment. Experiments may be done in a few simple steps or they may take years of complicated work.

experiment

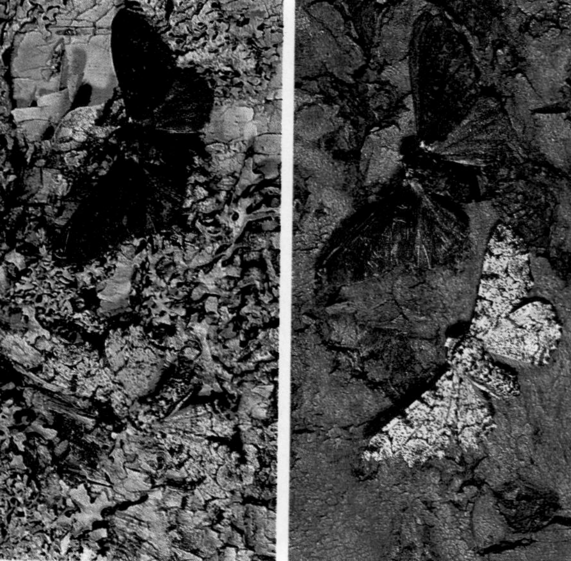

Figure P-6 *The light-colored moths are better camouflaged on light-colored trees (left). The dark-colored moths are better camouflaged on dark-colored trees (right).*

Cross References

Chapters 1, 2, and 5 of the *Life Science Unit* contain material that may help the students understand the *Prologue*. Section 5-1 gives more information on the theory of evolution and natural selection. Section 5-2 deals with adaptations, including camouflage. Also see Figure 5-8 for examples of camouflage. Section 1-7 introduces the plant and animal kingdoms, including the arthropods, to which moths belong. Chapter 2 gives information about how traits are passed from generation to generation.

Demonstration

This demonstration will give a concrete example of the process of hypothesis-testing. Start by asking the students what a candle needs to burn. They will probably answer fire, a wick or wax; some students might suggest air or oxygen. Tell the class that you are going to test the hypothesis that a candle will burn longer in a larger volume of air.

First describe the demonstration to the students:

1. Set up two jars, one smaller and one larger (about twice the size).
2. Two identical candles will be lit and then covered simultaneously, one with the larger jar turned upside down, and the other with the smaller jar.

Before conducting this experiment, ask the students to predict the results, based on the hypothesis. Since a larger jar holds more air than a smaller jar, the larger jar should allow the candle to burn longer.

Having predicted the results, proceed with the demonstration. Light the candles, and cover them simultaneously. Have the class write down their observations. The students should observe that the candle flame goes out first in the smaller jar. This trial confirms the prediction, and supports the hypothesis.

Repeat the experiment to see if the same results occur. This will give the students the opportunity to make further observations. Make sure that each time the experiment is run the jars are "refilled" with room air by waving them around gently.

After a discussion of the results, ask the students how they could further test the hypothesis. For example, they could record the length of time a candle burns in several sizes of jars to see if there is a correlation with the volume of the jar. The students could then make a graph of the burning time versus the jar volume to make any patterns in their data more obvious.

Background

Kettlewell released moths on different days. Before releasing them, the moths were marked and color-coded according to the day of the week. When they were recaptured, Kettlewell then knew how long the moths had survived in the wild.

Mercury vapor lights were used to attract moths for recapture. Traps also used female peppered moths to lure the male moths with their scent.

The larvae of dark-colored moths are hardier than those of light-colored moths. Kettlewell studied this and other differences between the two forms to avoid bias in his results.

First a scientist must predict the results of an experiment. If the results occur, the hypothesis is supported. If the results do not occur, then the hypothesis is rejected. In the case of the peppered moths, Kettlewell predicted that birds would eat fewer dark-colored moths in industrial environments.

Then Kettlewell conducted an experiment to test his hypothesis. He raised thousands of peppered moths in his laboratory and released an equal number of light and dark moths into a forest in an industrial area of England. Before releasing the moths, he marked them carefully. The moths were then recaptured in the evenings using a light to attract them. By counting the survivors, Kettlewell hoped to determine whether fewer of the dark moths had been eaten by birds.

controlled experiment

This is an example of a **controlled experiment.** In a controlled experiment, there are two test groups. Everything in the two groups is the same, except for the one factor that the scientists are testing. The factor being tested is called the **experimental factor.** In this case, the experimental factor was the dark color of the moths.

experimental factor

experimental group

control group

The group that has the experimental factor is called the **experimental group** (the dark-colored moths). The group that does not have the experimental factor is called the **control group** (the light-colored moths). The use of a control group helps to prove that the changes observed during the experiment are actually caused by the experimental factor. The control group provides a standard way of comparing the results.

observations

data

When conducting an experiment, a scientist must keep accurate records of **observations:** measurements, changes observed, and other information. These observations are called **data.** The purpose of keeping records is to help organize data so that patterns, if any exist, can be seen. Patterns in data support a hypothesis when they are consistent with it. Or, patterns can prove that a hypothesis is false when the results are different from those predicted by the scientist. If Kettlewell had found that an equal number of light and dark-colored moths survived, his hypothesis would have been proven incorrect. An incorrect hypothesis must be rejected or changed. In such a case, it may be necessary to make a new hypothesis to test.

Teaching Tips

The map of England in Figure P-6 will help students visualize the location of peppered moths relative to industrial areas in the latter half of the nineteenth century. More than 20,000 observations of peppered moths by 170 volunteers provided the data for this map.

Although Eastern England was not heavily industrialized, a high percentage of dark-colored moths were found there. Ask the students to hypothesize about this. They may have difficulty without the additional information that the weather patterns tend to move eastward in this region.

Kettlewell hypothesized that the high proportion of dark moths in the east was due to the blowing of smoke and soot eastward by the prevailing southwesterly winds. He demonstrated

P-3

Peppered Moth Study: Experimental Results

Kettlewell's results were encouraging. He recovered twice as many dark-colored moths as light-colored moths. At this point, the hypothesis became a **theory.** A theory is a hypothesis that has been supported by the results of experiments. It is an explanation of observations that leads to an even wider range of experimental testing. Kettlewell's theory was that the dark moths became more numerous than light moths in industrial areas because their camouflage protected them from being eaten by birds.

theory

Kettlewell tested his own theory by performing the same experiment in a wooded area far from industry, where

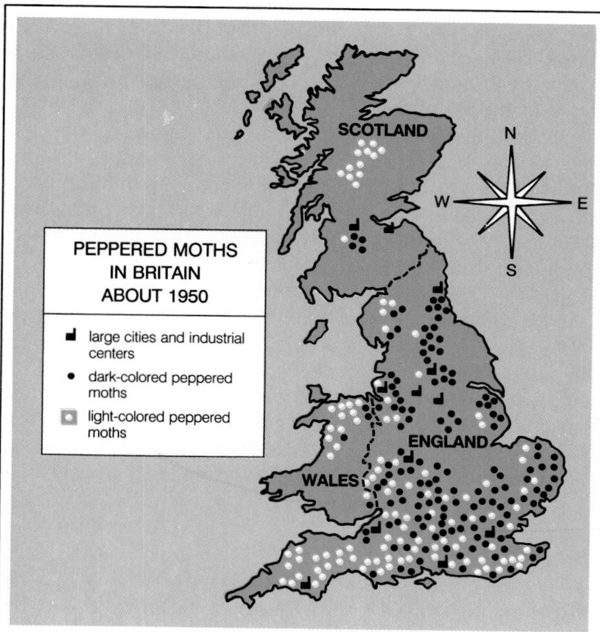

Figure P-7 *Kettlewell's theory explained the observation that dark moths had become more common near industrialized cities.*

Teaching Suggestions P-3

Review the elements that go into testing a hypothesis. The students should understand the process of conducting a controlled experiment.

Discuss the role of data and observations in the scientific method. Kettlewell's positive results, the recapture of twice as many dark-colored moths in industrial areas, led to further tests. Kettlewell tried the experiment in non-industrial areas and also directly observed birds eating the moths. The results from these additional experiments further supported his theory.

Kettlewell asked Niko Tinbergen to photograph birds catching peppered moths that were resting on tree trunks. He did this because the editor of one journal questioned whether birds captured resting moths at all.

Ask the students whether they think a dark-colored or a light-colored moth would be more visible while flying. Kettlewell studied this and found that the light-colored moths were better camouflaged in daytime flight, while dark-colored moths were better camouflaged during night flight.

Ask the students if they think this additional information concerning camouflage in flight would affect Kettlewell's data. Point out that it does not affect the data, which comes directly from observations, but it may affect the interpretation of the data because there is another variable in addition to the experimental factor. When uncontrolled variables are present, there is no longer any assurance that the experimental factor produced the changes observed in the data.

this by washing down leaves of trees in the eastern regions. Upon examination, he found pollutants in the wash.

Be sure students are aware of the two project suggestions on page 11. Both relate to the experimental work described in the Prologue and offer opportunities to extend the material covered.

Background

The phenomenon Kettlewell studied is known as industrial melanism, the darkening of organisms in adaptation to their darker environment. Approximately 70 out of the more than 750 species of moths in England have exhibited industrial melanism, usually in species which fly at night and spend the day resting on tree trunks and rocks.

Melanism is not solely a modern phenomenon; it has been documented in areas unaffected by pollution, such as rain forests, high mountains, and arctic regions. In each case mutations have been the source for change, and natural selection then determines whether the mutation survives.

In order to film birds capturing peppered moths without interfering in the process, Niko Tinbergen would spend the greater part of the day in a "hide," or hiding place. His films helped to show not only the capture of the moths, but also the order in which they were captured. Most of the time the birds selected the more visible moths first. The films also showed that the capture of one moth put the other moths in the area in danger. Presumably this happened because a bird would search more actively for moths after catching one.

A detailed discussion of the experiment with peppered moths can be found in "Darwin's Missing Evidence," H.B.D. Kettlewell, *Scientific American*, March, 1959.

Figure P-8 *This is one of Kettlewell's original photographs, showing a dark-colored peppered moth being eaten by a bird.*

the light-colored moths were better camouflaged. Again he released an equal number of light and dark-colored moths. In this experiment, he recovered many more light-colored moths. This supported his theory that the color of moths protected them from being seen and eaten by birds.

After completing his experiments, Kettlewell reported his findings in scientific journals and at scientific meetings. In this way scientists make their results available so that other scientists can repeat their experiments. If the same results are obtained, the theory being tested receives additional support. If the results differ, the theory must usually be changed, or even discarded. This is one way the body of scientific knowledge changes and grows. Table P-1 summarizes the steps that can lead to the formulation and testing of a theory, using the example of the dark-colored peppered moths.

Kettlewell's theory did receive additional support from the work of other scientists. Niko Tinbergen, a noted behavioral biologist, accompanied Kettlewell on field trips and filmed birds eating the moths. The films showed that birds had a much harder time catching dark moths in the forests affected by pollution.

Reporting experimental findings also allows other scientists to use them. Recently, biologists have found that the light-colored peppered moths are increasing in numbers in industrial areas of England. Scientists think this may indicate that pollution control measures are taking effect, changing the environment back to the way it once was.

The process of scientific investigation is never finished. Every ending has a new beginning. Each time a scientist answers a question about nature, many new questions arise. These questions lead to new hypotheses that are then tested experimentally. Kettlewell's approach to the problem of peppered moths raises many questions. Are there other plants and animals that can "change color" because of changes in the environment? Can camouflage be used reliably to judge the long-term effects of pollution? If so, this information could be useful in efforts to study and preserve the environment.

Kettlewell's work did not establish any new laws of nature; it mainly supported the theory of evolution. Some theories, however, become so well established by experimen-

Table P-1: Scientific Method

Steps	*Peppered Moth Example*
Carefully identify the problem under study.	The problem was to find out why dark-colored moths became more common than light-colored moths.
Collect all information that may help in solving the problem.	Kettlewell researched the peppered moth's habits and communicated with other scientists.
Form a hypothesis or possible solution to the problem.	The dark moths survived better due to camouflage.
Plan and conduct a controlled experiment or series of observations.	Kettlewell released equal numbers of light and dark moths into darkened woods and counted how many of each were recovered.
Record the data.	Twice as many dark-colored as light-colored moths were recovered.
Formulate a theory based on the facts discovered.	The dark-colored moths increased in numbers because of their camouflage.
Other, independent scientists test the theory again.	Tinbergen filmed birds eating peppered moths; the evidence showed camouflage helped moths to survive.

tal evidence in many different situations over many years that they become scientific laws. A **scientific law** is an exact statement of a relationship in nature—so exact that it can be expressed mathematically. For example, in climates where the weather varies greatly between summer and winter, variation in growth causes visible rings to form within the trunks of trees. It is a scientific law that each ring in a tree's trunk represents one year of its growth. Using this law, scientists reported one bristle-cone pine tree as being 4900 years old.

scientific law

Teaching Tips

- Have students name other scientific laws they may have heard about, e.g. the law of gravity, the law of conservation of energy, and Einstein's law of relativity. They will be learning about the first two of these laws in later chapters.
- To help students appreciate the relevance and power of scientific laws, use the familiar example of the law of gravity. Gravity affects us directly in many ways. It causes the earth to orbit near the sun, our primary source of energy. Gravity also keeps everything we need, including the atmosphere we breathe, near the surface of the earth. Because the law of gravity can be expressed mathematically, scientists and engineers have been able to make the calculations necessary to lift an airplane off the ground, and send astronauts to the moon.

Cross References

Newton's law of gravity is discussed in Chapter 16, section 16-5. The law of conservation of energy is covered in Section 18-5, *Transformation of Energy*. The Feature Article on *Carbon-14 Dating* on page 260 discusses how laws are independently confirmed, using the example of estimating the age of a tree by counting its rings.

Understanding The Prologue

Main Ideas

1. Scientists are always asking questions and trying to solve problems. They develop and test ideas in a planned, logical way called the scientific method.
2. Scientists often start to investigate a problem by identifying or stating the problem clearly, collecting information relating to the problem, and forming a hypothesis.
3. A hypothesis is a tentative explanation, or solution to a problem, that can be tested by means of an experiment. A hypothesis shown by an experiment to be incorrect must be discarded or revised.
4. An experiment is a way of testing that makes measurements or observations in the natural world.
5. A controlled experiment uses two test groups to test the effects of one experimental factor. The experimental group is subjected to the experimental factor; the control group is not.
6. During an experiment, scientists keep accurate records of data in order to see whether the data support the hypothesis.
7. A theory is a hypothesis backed by experimental results.
8. Scientists report their experimental results so they can be checked by other scientists.
9. A scientific law is a precise statement of a relationship in nature that can be expressed in mathematical terms.

Vocabulary Review

From the following list, choose the term that best completes each of the statements. Write your answers on a separate sheet of paper.

evolution	experimental group
biologists	larvae
control group	hypothesis
controlled experiment	observations
	organisms
data	scientific law
experiment	scientific method
experimental factor	theory

1. People who study living things are called __?__.
2. The observations recorded in an experiment are called __?__.
3. A(n) __?__ is a tentative explanation for a problem found in nature.
4. The part of an experiment that provides a standard against which to compare results is the __?__.
5. The process by which living things change over time in response to their environment is called __?__.
6. A precise statement of a relationship in nature that can be expressed mathematically is a(n) __?__.
7. A test of a hypothesis in which observations and measurements are made is a(n) __?__.
8. A(n) __?__ is a hypothesis supported by experimental results.
9. The part of a controlled experiment that varies is the __?__.
10. A planned, logical way to develop and test ideas is called __?__.

Vocabulary Review Answers

1. biologists
2. data
3. hypothesis
4. control group
5. evolution
6. scientific law
7. experiment or controlled experiment
8. theory
9. experimental factor
10. scientific method

Prologue Review

Write your answers on a separate sheet of paper.

Know The Facts

1. The peppered moths became darker over time in response to darker __?__. (birds, trees, larvae)
2. The Industrial Revolution brought a gradual change in the peppered moth's environment because of __?__. (mining, population increases, air pollution)
3. The moths that were most similar to their environment __?__. (survived, were eaten, left the region)
4. A hypothesis that fails to get experimental support must be __?__. (reported to other scientists, rejected or changed, solved)
5. Kettlewell's __?__ was supported by the experimental work of Niko Tinbergen. (theory, scientific law, hypothesis)
6. In Kettlewell's experiments in industrial areas, the control for the experiment was the __?__. (light moths, dark moths, birds)
7. In order for experimental results to be accepted, other scientists must be able to __?__. (solve them, repeat them successfully, control them)
8. Tinbergen supported Kettlewell's research by filming __?__. (moths being trapped, birds eating moths, moths breeding)

Understand The Concepts

9. What process did Kettlewell use to test his hypothesis that dark-colored moths increased due to industrialization?
10. Describe the experiment that Kettlewell performed to further support his theory.
11. What was the experimental factor in Kettlewell's experiments? In what way was the experimental factor related to the control group?
12. Why is controlled experimentation important in science?
13. In what ways might a test of a hypothesis lead to additional experimental work?
14. What is the relationship between hypothesis and theory?
15. Explain the meaning of the term scientific law.

Challenge Your Understanding

16. Describe the evidence that convinced Kettlewell that industrial pollution had resulted in the color change of the peppered moths.
17. Often there is more than one hypothesis to a problem. What other hypothesis to the problem of the color change in moths might exist and how might it be tested?

Projects

1. Investigate other plants or animals that respond to the environment with changes in color. Write a report about several of these and how their color changes correspond to environmental changes.
2. Scientific research is an ongoing process. Report to your class on a current research problem concerning pollution. What theories have been proposed? Have they been rejected or changed as research has progressed?

Prologue Review Answers

Know the Facts

1. trees
2. air pollution
3. survived
4. rejected or changed
5. theory
6. dark moths
7. repeat them successfully
8. birds eating moths

Understand the Concepts

9. He studied the peppered moths and their behavior. Then he conducted a controlled experiment by releasing dark-colored and light-colored moths in woods near industrialization.
10. He tried the same experiment in non-industrialized areas to see if in those areas the light-colored moths benefited from camouflage.
11. The experimental factor was the dark color of the moths. The control group was lacking the experimental factor.
12. The controlled experiment is a rigorous procedure that allows other scientists to verify experimental results. This type of objectivity is needed in scientific work.
13. Testing hypotheses often stimulates additional questions. These, in turn, may lead to additional hypotheses to test.
14. A theory is a hypothesis that has been supported by experimental testing.
15. A scientific law is a precise statement of a relationship in nature. Many scientific laws can be expressed mathematically.

Challenge Your Understanding

16. Kettlewell recovered many more dark-colored moths in darkened environments, and many more light-colored moths in lighter environments. He also observed the birds selectively eating the moths that were easier to see.
17. Answers may vary. An example of another hypothesis is found in Harrison's work: the moths became darker because of something they ate in the polluted environment. This could be tested by controlling the diet of laboratory moths.

Unit 1 · Life Science

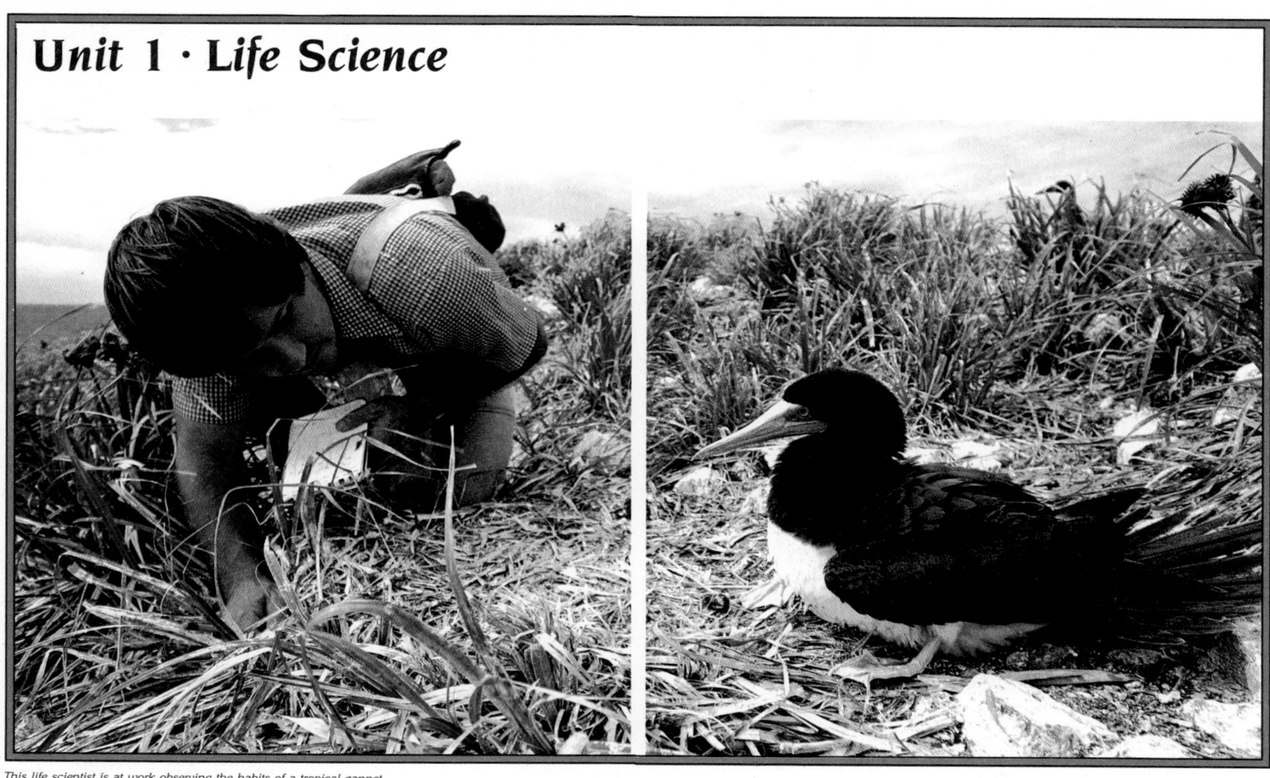

This life scientist is at work observing the habits of a tropical gannet.

13

Unit 1 · Life Science

The Unit In Perspective

The goal of this unit is to introduce basic concepts of biological science to students and to provide an understanding of the scope of the world of living matter.

Chapter 1, *Introduction To Life,* discusses the characteristics of living matter, the cell, levels of organization in multi-cellular organisms, photosynthesis, classification, and the wide variety of living forms.

Chapter 2, *Heredity,* describes cell division, asexual and sexual reproduction, the fundamentals of inheritance, DNA, mutations, and plant and animal breeding.

Chapter 3, *The Human Body,* explores the major organ systems of the body, and provides relevant material on the effects of alcohol, tobacco, and other drugs.

Chapter 4, *Ecology,* introduces ecosystems, food chains, food webs, food pyramids, populations succession, and habitat destruction.

Chapter 5, *Distribution of Life,* discusses plant and animal adaptations, the characteristics of biomes, and life zones in the seas.

Students will be interested in the life science careers featured on pages 140–141. The occupations selected illustrate a wide range of career possibilities.

Issues in Science, on pages 142–143 at the end of Unit 1, introduces the issue of animal experimentation. This feature focuses on critical thinking questions and includes a study skills exercise in the teacher pages.

Preparing To Teach The Unit

The advance preparation needed to teach this unit includes the acquisition and organization of equipment, materials, and living specimens for laboratory activities, planning field trips, ordering audio-visual aids, and collecting and/or placing on reserve in the library outside readings for students. A list of suggested readings will be found adjacent to the Main Ideas section at the close of each chapter.

Equipment and Materials

The listing below identifies the equipment, materials, and living specimens needed for the student Activities of this unit. The Activities are located at the back of the student book. Quantities required will depend on the nature of the Activity, the availability of materials, numbers of students, budgetary considerations, and so on. Some Activities proceed better if students work in pairs or threes. Check each Activity to make this decision.

ACTIVITY 2 (Ch. 1): *Observing Cells* (page 534). MATERIALS: microscope, slide, cover slip, iodine solution, water, dropper, toothpick, onion section, tweezers.

ACTIVITY 3 (Ch. 1): *Observing One-Celled Organisms* (page 535). MATERIALS: microscope, slide, cover slip, dropper, paramecium and amoeba cultures. *NOTE: Be sure to order paramecium and amoeba cultures from a biological supply house well in advance.* See page T-39 for a list of suppliers.

ACTIVITY 4 (Ch. 2): *Observing Onion Root-Tip Cells* (page 536). MATERIALS: microscope, prepared slides of onion root-tip cells.

ACTIVITY 5 (Ch. 3): *Measuring Lung Volume* (page 537). MATERIALS: large jar, water, bucket, cardboard, rubber tube, graduated cylinder, wax marking pencil, ruler.

ACTIVITY 6 (Ch. 3): *Mapping Taste Areas of the Tongue* (page 538). MATERIALS: 8 cotton-tipped swabs, dropper, one test tube each of sugar, salt, lemon juice, and quinine solution, test-tube rack.

ACTIVITY 7 (Ch. 4): *Life in a Square Meter of Ground* (page 539). MATERIALS: string, stakes, small garden tools, meterstick, dissecting microscope (optional).

ACTIVITY 8 (Ch. 5): *Bird Beak Adaptations* (page 540). MATERIALS: pencil, paper.

Field Trips

A well planned field trip is a powerful teaching tool. Plan as many as time and your local circumstances allow.

Use local habitats to collect specimens. Gather microorganisms from the muddy bottoms of ponds and streams, fungi from nearby woods, and shells and organisms from seashore habitats. Plan a trip to a nearby university or medical research laboratory so students can observe cell cultures.

Many science museums have special exhibits on DNA. Schedule a trip so students can study models of DNA and its functions.

Visit a local Red Cross office to have students learn CPR and the Heimlich Maneuver and to learn about the Red Cross's role in collecting blood.

Plan a trip to an aquarium or zoo to study efforts to make the exhibits complete and balanced ecosystems. If you live near the sea, and such trips are available, schedule a whale-watching expedition.

Audio-Visual Aids

The list of A-V materials below provides a wide selection of topics on sound filmstrips (SFS), videocassettes (VC), and 16 mm film. Choose titles in terms of availability, student interest, coverage of content vis-à-vis field trips and other learning activities, availability of projection equipment, and time limitations. A list of A-V suppliers is given on page T-39. Consult each supplier catalog for ordering information.

Chapter 1: Introduction To Life

Kingdom of Plants, SFS, National Geographic.
Taxonomy: How Living Organisms Differ, SFS, Guidance Associates.
Microscopic Life and Microscopes, SFS, National Geographic.
Bioscope: Cells, 15 min, VC, Agency for Instructional Television.
Cell Biology, 17 min, 16 mm, VC, Coronet.
Bacteria, 19 min, 16 mm, Britannica.
Single Celled Animals, 17 min, 16 mm, Britannica.

Chapter 2: Heredity

Cell Division: Mitosis and Meiosis, 20 min, 16 mm, VC, CRM/McGraw Hill.
The Riddle of Heredity, 30 min, 16 mm, CRM/McGraw-Hill.
Reproduction: The Continuity of Life, SFS, Guidance Associates.
Genetic Biology, 16 min, 16 mm, VC, Coronet.
Genetics: Mendel's Laws, 14 min, 16 mm, VC, Coronet.
Genetics: Improving Plants and Animals, 13 min, 16 mm, VC, Coronet.

Chapter 3: The Human Body

Muscle, 26 min, 16 mm, VC, CRM/McGraw-Hill.
Incredible Voyage, 26 min, 16 mm, CRM/McGraw-Hill.
Human Body: Circulatory System, 16 min, 16 mm, Coronet.
Human Body: Endocrine System, 15 min, 16 mm, Coronet.
Human Body: Skeletal System, 12 min, 16 mm, VC, Coronet.
The Human Body: Digestive System, 15 min, 16 mm, VC, Coronet.
The Human Body: Respiratory System, 12 min, 16 mm, VC, Coronet.

Chapter 4: Ecology

Ecology: Populations, Communities, and Biomes, SFS, Guidance Associates.
Pollution: Problems and Prospects, SFS, National Geographic.
Bioscope: The Natural Balance, 15 min, VC, Agency for Instructional Television.
Our Endangered Wildlife, 54 min, 16 mm, VC, CRM/McGraw-Hill.
Ecological Biology, 15 min, 16 mm, VC, Coronet.
Still Waters, 57 min, 16 mm, VC, Time.

Chapter 5: Distribution of Life

Bioscope: Where Plants and Animals Live, 15 min, VC, Agency for Instructional Television.
Small Animals of the Plains, 15 min, 16 mm, VC, Coronet.
Animals of the South American Jungle, 26 min, 16 mm, VC, Coronet.
Animals at Home in the Desert, 23 min, 16 mm, VC, Coronet.
The Temperate Deciduous Forest, 17 min, 16 mm, Britannica.

CHAPTER 1 Overview

This chapter introduces students to the characteristics of living matter. The cell and its parts are described, and plant and animal cells are compared. Some one-celled organisms are introduced, and the organization of cells into tissues, organs, and systems in multi-celled organisms is described. Photosynthesis and respiration are then compared. Next, the concept of classification is introduced, and the five kingdoms, with examples, are described. Finally, viruses and the concept of immunization are discussed.

Goals

At the end of this chapter, students should be able to:

1. identify the characteristics that define life.
2. describe a cell and its parts.
3. compare plant and animal cells.
4. describe the characteristics of some one-celled organisms.
5. describe the levels of organization in a multi-celled organism.
6. explain the process of photosynthesis.
7. compare respiration and photosynthesis.
8. describe the features of a scientific classification system.
9. explain why a classification system is used.
10. identify the general characteristics of the five kingdoms.
11. give examples of organisms in each kingdom.
12. describe viruses.
13. explain immunization.

Chapter 1

Introduction to Life

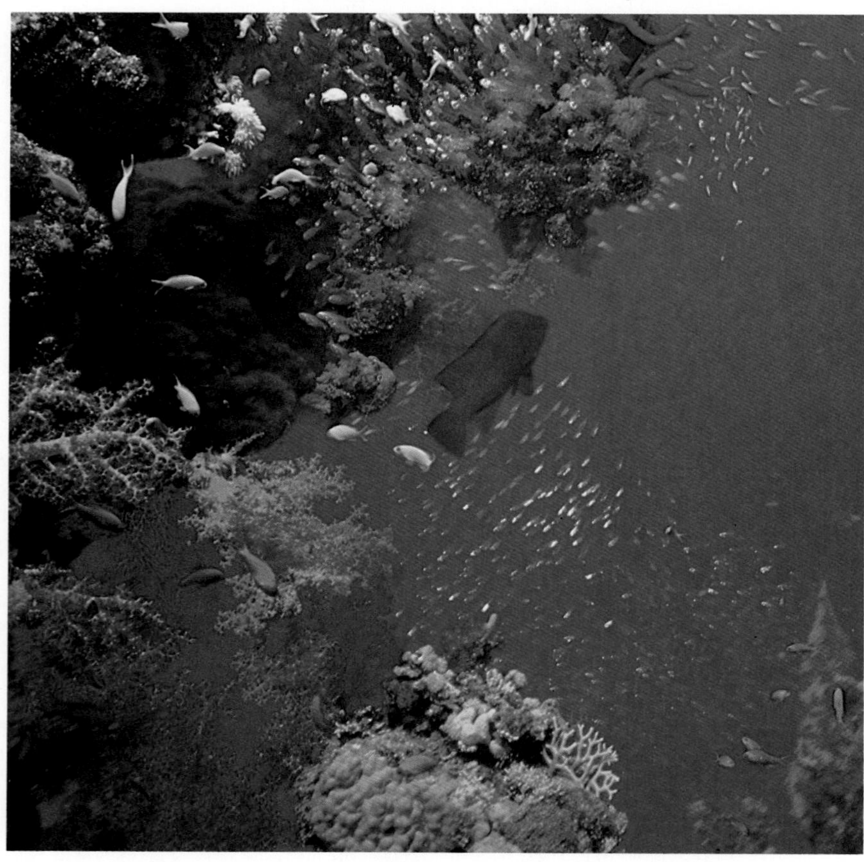

An underwater scene in the Red Sea.

Vocabulary Preview

organism
cell
cell membrane
diffusion
nucleus
chromosomes
cytoplasm
respiration
chlorophyll
chloroplasts
cell wall
cilia
tissue
organ
system
photosynthesis
classification
Moneran Kingdom
Protist Kingdom
Fungi Kingdom
Plant Kingdom
vascular plants
Animal Kingdom

1-1

Characteristics of Life

Goal

To identify the characteristics that define life.

Most people would say that the fish shown at the left are living and the rock is not. But what about the coral? Like a rock, the coral stays in one place most of its life and its hard, outer covering makes it appear lifeless. But if you watched the coral for a while you would notice long, slender *tentacles* reaching out of the rocklike covering and moving about in search of food. You would also discover that the coral reacted in different ways to the surroundings. For example, the coral would probably pull back its tentacles when touched by an unfamiliar object. You would also notice that the coral grew and produced new corals over time. Corals display all the characteristics that define life. They are one example of a living thing, or **organism.**

organism

Scientists agree that there is a set of characteristics that distinguishes living things from nonliving things. These are:

1. Movement (of the organism, its parts, or materials inside of it)
2. Getting and using food
3. Growth and repair of body parts
4. Reproduction
5. Response to changes in the surroundings

Nonliving things can have one or two of the characteristics of life. A car moves, a computer responds, and an icicle grows. But only organisms carry out *all* life activities.

Checkpoint

1. What is an organism?
2. Name the life activities carried out by organisms.
3. How do scientists know that corals are alive?

TRY THIS Set up a freshwater aquarium at home or in your classroom. Learn how to feed and care for the organisms. Observe how the various organisms carry out their life activities. Keep a record of your observations.

Teaching Suggestions 1-1

Have students study the photograph and then name the living and nonliving objects pictured. Discuss the characteristics used to distinguish living from nonliving. Make a list of the characteristics on the chalkboard. Explain that movement occurs in plant tissues although it is sometimes difficult to see.

Indicate the characteristics that are *life activities.* Point out that excretion of wastes, inheritance, and use of certain chemicals, such as oxygen, are life activities. Explain that green plants do not "get" food. They make it during the process of photosynthesis.

Background

Corals are animals that live in colonies. Each coral produces a hard outer covering, or exoskeleton, of calcium carbonate. When a coral dies, its exoskeleton remains. Be sure students do not confuse coral skeletons with living colonies of coral. Living corals generally attach themselves to the skeletons of dead corals. The build-up of thousands of coral skeletons is known as a coral reef.

Demonstration

Show students that movement occurs within plants. Add several drops of red food coloring to a half-full glass of water. Cut off the wide end of a celery stalk and place the stalk in the glass, cut end down. Have students check the celery the next day. They should see red color in the celery stalk above the water line. Students should conclude that the colored water moved through the tissues of the celery stalk.

1 INTRODUCTION TO LIFE

1-1 Characteristics of Life

Living Behaviors

Each sentence below is followed by a set of four choices for completing it. Decide which of the choices is best. Then place the letter of the correct choice on the numbered space.

___ 1. When something moves, you can assume a) it's alive b) it may be alive c) it's not alive d) it's reproducing.

___ 2. If a sugar crystal gets larger when it is suspended in a sugar solution, this can be interpreted to mean a) it's growing b) it's alive c) it's dead d) it will live.

___ 3. Growth and repair are made by a) animals b) plants c) bacteria d) all organisms.

___ 4. An amoeba dividing in two is an example of a) growing b) living c) reproducing d) responding.

___ 5. A thermostat responds to a drop in temperature and moves a switch. This is a) a living condition b) a mechanical action c) impossible d) sensitivity.

___ 6. When a paramecium moves out of a bright light, it is a) reproducing b) responding c) growing d) respiring.

___ 7. When onions take root, cells divide and the new cells push the root further into the ground. This activity is an example of a) responding b) reproducing c) growing

Using Try This

If possible, set up an aquarium before school starts and let students observe it the first day. In addition, have students with a pet at home observe the life activities of the pet.

Checkpoint Answers

1. An organism is a living thing.
2. Movement, getting and using food, growth and repair, reproduction, and response to surroundings.
3. They show all the characteristics of living things.

Teaching Suggestions 1-2

Ask students to compare and contrast the different kinds of cells shown in Figure 1-1. Have them attempt verbal descriptions of some of the structures shown. Point out that the cells are stained to make the structures more visible, and that the photomicrographs are of thin slices of cells. Thus, what is seen is a two-dimensional cross section of a three-dimensional cell.

Now introduce the major parts of cells. Refer students to Figure 1-2, and explain the structure and function of each part covered in the text.

Point out that diffusion occurs in two directions—substances diffuse into and out of cells. CO_2 and other wastes diffuse out of cells.

Refer students to the list of life characteristics on page 15 and to Figure 1-3. Point out that cellular respiration is the process by which energy is released from food matter. Add that "consisting of one or more cells" is a characteristic of living things.

Point out that animal cells have only a cell membrane, while plant cells have both a cell membrane and a cell wall. It is the rigid cell walls of plant tissues that give them their stiffness.

Laboratory Activity 2, Observing Cells, student text page 534, should be performed in conjunction with this section. Chapter-end projects 7 and 8 also apply to this section.

Background

Most organisms consist of one or more cells. This has been a very useful generalization for explaining the structure and function of organisms. There are exceptions, however. Viruses, for example, reproduce only inside host cells and otherwise exist as very tiny particles that do not resemble cells at all. In another instance, a stage in the life cycle of some slime molds consists of a slimy mass of cytoplasm containing thousands of nuclei. There are no cell walls. Such a multinucleated mass is called a plasmodium.

1-2

The Cell

Goals

1. To describe a cell and its parts.
2. To compare plant and animal cells.

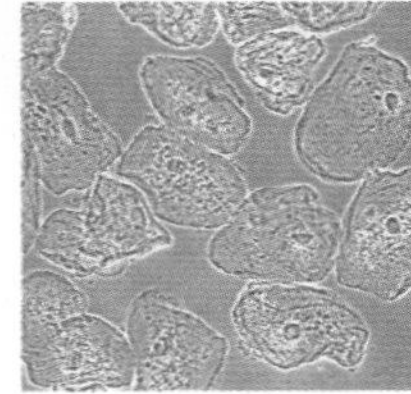

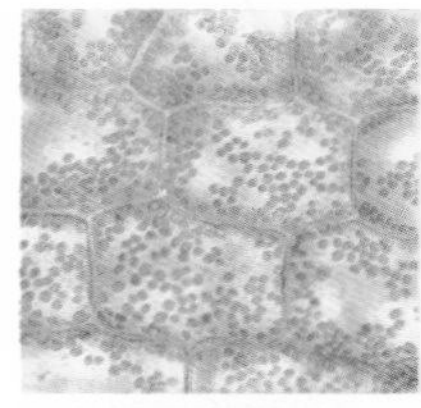

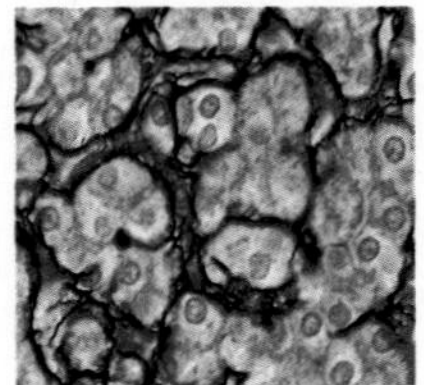

Figure 1-1 *Three kinds of cells. Human cheek cells, stained (top).* Elodea *leaf cells (middle). Monkey liver cells (bottom).*

The invention of the microscope in the 1600's by Anton van Leeuwenhoek (LAY vun hook) allowed people to observe what was once invisible to the unaided eye. An English scientist, Robert Hooke, examined cork bark through a microscope. He noticed that the bark was divided into tiny, boxlike compartments, which he called *cells.*

Many years after Hooke's observations, scientists concluded that all living things are made up of cells and that cells are produced only by other cells. A **cell** is the smallest unit able to carry out all the life activities of an organism. Having cells is a characteristic of life. In fact, it is the characteristic that makes all the other life activities possible. You yourself are made up of many different cells.

cell

Some cells, such as the yolk of an ostrich's egg, are quite large. Most cells, however, can be seen only with the aid of a microscope. This is because they are smaller than the point of a pin. Cells may differ in size, shape, and function, but all cells have certain features in common. Figure 1-1 shows cells from different organisms.

There are three main parts, or *structures*, in every cell: (1) the *cell membrane*, (2) the *nucleus*, and (3) the *cytoplasm*. Each structure carries out certain *functions*, or performs certain "jobs." Study the cells shown in Figure 1-2, as you read about the different cell structures.

cell membrane

The **cell membrane** is a thin layer that surrounds the parts of a cell and keeps them together. The cell membrane protects the contents of a cell and gives the cell its shape. Certain materials move into and out of the cell through spaces in the cell membrane. Food, water, gases, and other materials that are small enough, move through the cell membrane by diffusion. **Diffusion** is the movement of a substance from an area where there is more of it to an area where there is less. The cell membrane also acts like a "guard," preventing some substances from entering and others from leaving the cell.

diffusion

Demonstrations

1. Show students the phenomenon of diffusion. Spray perfume in one corner of the classroom, and have students raise their hands as the odor reaches them. Or, very carefully and gently add one or two drops of ink or food coloring to a large beaker of water. Students can see the color spread throughout the beaker. Have students describe what they have seen in terms of the definition of diffusion.
2. Demonstrate that the human body produces carbon dioxide during cellular respiration. Have a student blow through a straw into a saturated solution of limewater ($Ca(OH)_2$). CAUTION: The student should avoid taking into the mouth any limewater solution. The solution

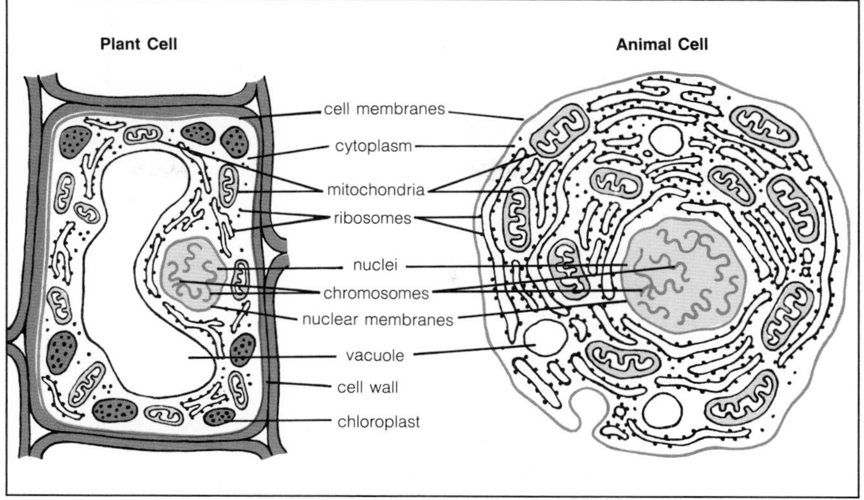

Figure 1-2 *A generalized plant cell and a generalized animal cell contain many of the same structures. Notice the large, liquid-filled vacuole in the plant cell.*

nucleus
(NOO klee us)

The **nucleus** is the "control center" of the cell. It is surrounded by a thin layer called the *nuclear membrane*. The nuclear membrane separates the nucleus from the other parts of the cell. Only certain substances pass through the nuclear membrane. Inside the nucleus are chromosomes. **Chromosomes** are long, thin fibers containing instructions that control the activities of the cell.

chromosomes
(KROH muh sohmz)

cytoplasm

Cytoplasm makes up most of the "body" of the cell. It contains many small structures, which carry out cell activities.

Ribosomes (RY buh sohmz) are structures in the cytoplasm that manufacture complex substances called *proteins* (PROH teenz). The "building materials" of cells and the substances that control many cell activities are made of protein.

Storage areas in the cytoplasm are called *vacuoles*. They may contain water, food, or waste materials. Animal cells usually have a few small vacuoles. Plant cells may contain many large, liquid-filled vacuoles or a single, large vacuole that fills most of the cell. Water-filled vacuoles help support a plant cell. When the vacuoles lose water, the plant wilts.

should turn cloudy, showing the presence of CO_2 in the exhaled breath. When fresh air is bubbled through a limewater solution, little or no cloudiness appears.

3. Use a knife, a fork, and a soup spoon to help students understand the difference between structure and function. Hold up the spoon and ask what kind of food students could eat with the spoon (soups, liquids). Then ask how the shape, or structure, of the spoon makes it possible to eat soup (large, bowl-shaped end holds liquids). Ask why a fork is not useful for eating soup (most of the soup passes through the tines). Finally, ask how the structure of the spoon is related to its function. Repeat, using the fork, then the knife, as needed, until the concepts are clear. Finish by asking students to give examples of how structure and function are related in objects, such as a hammer, scissors, and a tape measure.

Teaching Tip

- When teaching the concept of respiration, make sure students do not confuse the act of breathing with respiration, which takes place at the cellular level.

Facts At Your Fingertips

- Leeuwenhoek (1632–1723) was a Dutch merchant.
- Hooke used the term *cell* because the boxlike structures he saw reminded him of the small rooms, called cells, in a monastery.
- Cytoplasm was once thought to consist of a jellylike substance called protoplasm. Current thought is that cytoplasm contains a skeletonlike matrix called cytoskeleton that holds the cell structures and limits their movements.
- Cell structures in the cytoplasm are called organelles, meaning "little organs."
- Vacuoles are spaces surrounded by membranes and filled mainly with fluid. Vacuoles usually are more prominent in plant cells than in animal cells. In some one-called organisms, vacuoles are used to expel excess water and wastes.
- Cells, such as muscle cells, which require much energy, may contain thousands of mitochondria.

Mainstreaming

Obtain models of plant and animal cells from a biological supply house. Let visually-impaired students feel the models in order to gain a better idea of the structure of cells.

Cross Reference

Photosynthesis is discussed in section 1-5, Chapter 1.

1 INTRODUCTION TO LIFE
1-2 The Cell

Unit of Life

Decide what term is described by each of the following clues and write that term on the numbered space beside the clue.

_______ 1. Inventor of the microscope.
_______ 2. Person who looked at cork and saw "cells".
_______ 3. Smallest unit that carries on all life activities.
_______ 4. Thin layer that holds cell together.
_______ 5. The spreading out of a substance.
_______ 6. Control center of the cell.
_______ 7. Thin layer surrounding the nucleus.
_______ 8. Fibers inside nucleus that contain instructions for controlling cell activities.
_______ 9. Substance that makes up the body of a cell.
_______ 10. Structures that make proteins

Using Try This

Students may ask you to suggest materials that can be used to make a three-dimensional model of a cell and its parts. Cell models can be made from fruits, where different sized pits represent various cell structures. Cell models can also be created from fabrics. For example, an old nylon stocking can be used to represent the cell membrane. Inside it, students may place different kinds of fabrics to represent organelles. Finally, an "edible cell" model can be made with gelatin-dessert mix and various fruits and vegetables.

Students may want to describe their models and how they built them. If space is available, display the models.

Checkpoint Answers

1. A cell is the smallest unit able to carry out all the life activities of an organism.
2. (a) The cell membrane surrounds the parts of a cell and holds them together. It protects the contents of a cell and gives it its shape. It controls the movement of materials into and out of a cell. (b) Cytoplasm is the substance that makes up most of the body of a cell. (c) The nucleus is the "control center" of the cell. (d) Respiration takes place in the mitochondria. (e) Vacuoles are storage spaces in the cytoplasm.
3. Plant cells have cell walls and most contain chlorophyll.
4. Respiration is the process by which energy is made available from food. It takes place in the mitochondria.

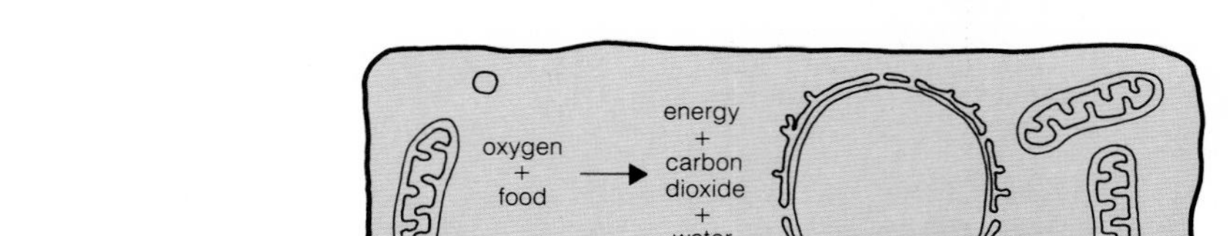

Figure 1-3 *Respiration takes place in the mitochondria. This process provides the energy for an organism's life activities.*

respiration

The energy for movement, growth, and other life activities comes from food. Energy is made available from food by a process known as **respiration.** Respiration takes place in cell structures called *mitochondria* (my tuh KAHN dree uh). The shape and number of mitochondria in different cells varies.

Oxygen and carbon dioxide are gases found in air and water. During respiration, oxygen combines with food in the mitochondria, releasing energy, carbon dioxide, and water. (See Figure 1-3.) Respiration goes on continually in all cells.

chlorophyll (KLOR uh fil)

chloroplasts (KLOR uh plastz)

Most plant cells contain a green substance called **chlorophyll** that enables them to produce food. Chlorophyll is found in structures called **chloroplasts.** Animal cells do not contain chlorophyll and, therefore, cannot produce their own food.

cell wall

Another structure found only in plant cells is the cell wall. The **cell wall** provides protection and support for the plant cell. Cell walls contain a tough, fibrous material called *cellulose* (SEL yuh lohs), which remains long after the cells have died. The cork "cells" that Robert Hooke studied were really the empty cell walls left behind after the cells of a tree died.

TRY THIS

Make a three-dimensional model of a cell and its parts using materials you can find at home.

Checkpoint

1. What is a cell?
2. Describe the functions of the following parts of a cell: (a) the cell membrane, (b) the cytoplasm, (c) the nucleus, (d) mitochondria, (e) vacuoles.
3. How are plant cells different from animal cells?
4. What is respiration? Where does it occur?

1-3

One-Celled Organisms

Goal

To describe the characteristics of some one-celled organisms.

Some organisms consist of only one cell. Special structures enable one-celled organisms to carry out all the life activities performed by multi-celled organisms.

The *paramecium* (pa ruh MEE see um) is a slipper-shaped, one-celled organism that lives in ponds and streams. (See Figure 1-4.) A stiff layer around the cell membrane gives the paramecium a definite shape. Paramecia (plural of paramecium) are propelled by rows of tiny, beating hairlike structures, called **cilia.** The movement of a paramecium is like a boat being propelled by many oars.

Paramecia respond to conditions in their surroundings. When a paramecium bumps into an object in its path, it turns and swims off in a different direction. Paramecia tend to move towards areas that contain food and away from areas of extreme heat or extreme cold.

Figure 1-5 *Imagine a paramecium's cilia propelling it like the oars of these Roman galleys.*

cilia
(SIL ee uh)

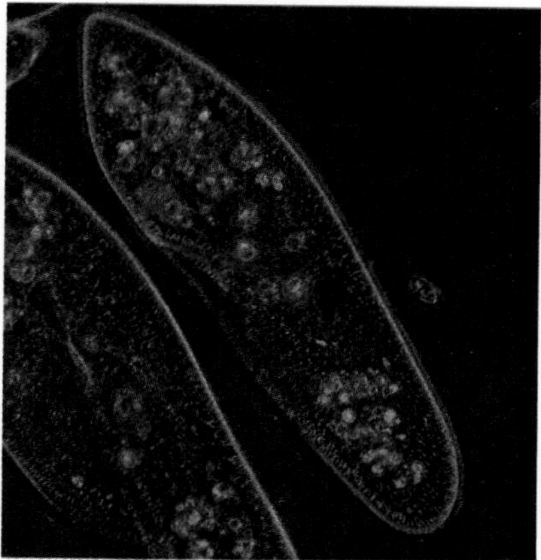

Figure 1-4a *Use Figure 1-4b to help you identify the structures of the paramecia shown here.*

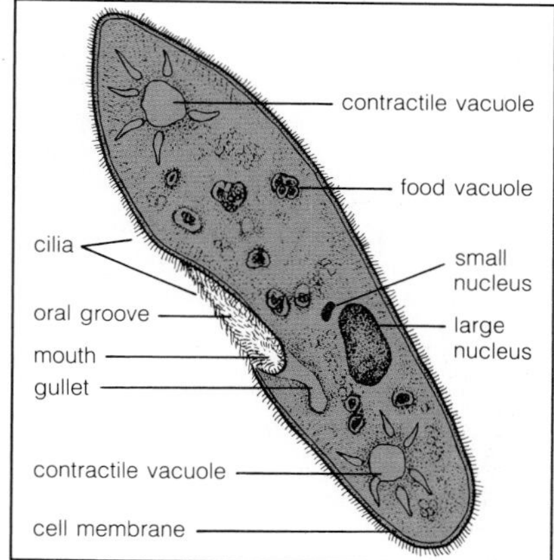

Figure 1-4b *Notice the contractile vacuoles, the oral groove, the gullet, and the many cilia on the paramecium.*

Teaching Suggestions 1-3

Have students review the list of life activities described in section 1-1. Point out that one-celled organisms too carry out all the life activities. Explain in some detail how the paramecium and the amoeba each carry out these activities. Be sure students understand that the methods used by one-celled organisms to perform these functions cannot be compared directly with the methods used by higher, multi-celled organisms to do the same things.

Explain the three methods of movement covered in this section: beating cilia—paramecium; flowing cytoplasm—amoeba; whipping flagellum—euglena. Point out that the movement of cilia is similar to that of a bank of oars working in unison. Some students may wonder about the differences between cilia and flagella. A flagellum is usually long in relation to the size of its cell, and there are usually only one or two per cell. A cilium, however, is very short compared to the size of its cell, and there are many per cell.

Explain that the fingerlike projections of an amoeba are called pseudopods. This means "false feet."

Be sure to make clear that paramecia and euglenas can propel themselves through water, but that amoebas need a solid surface to move about on. Amoebas tend to be bottom dwellers.

Laboratory Activity 3, Observing One-Celled Organisms, student text page 535, should be carried out by students in conjunction with this section.

Background

There are two types of one-celled organisms, those that belong to the Moneran Kingdom and those that belong to the Protist Kingdom. Moneran cells have an outer membrane, but no membrane around a nucleus. Thus, the nuclear material is not separated from the cytoplasm. Such cells are called prokaryotic cells. Bacteria and blue-green algae are monerans. Some biologists consider monerans a primitive cell type. The thought is that they existed before distinct nuclei evolved.

Protist cells have a cell membrane, cytoplasm, and a nucleus with a nuclear membrane. Cells with distinct nuclei are called eukaryotic cells. All organisms other than monerans are made up of one or more eukaryotic cells. Eukaryotic cells are much more complex than prokaryotic cells. Paramecia, amoebas, and euglenas each consist of one eukaryotic cell.

Demonstration

If projection equipment is available, obtain prepared slides or transparencies of a variety of one-celled organisms and show them to your class. A film or film loop will be even more effective in conveying the number and diversity of one-celled organisms.

Teaching Tip

- The material in this section is best mastered by students if they can observe live one-celled organisms. Thus, it is important that they perform laboratory Activity 3, Observing One-Celled Organisms (page 535), and/or the Try This on page 21.

Facts At Your Fingertips

- The stiff layer around the cell membrane of a paramecium is called the pellicle.
- A paramecium has two nuclei: a macronucleus and a micronucleus. The large nucleus controls cell metabolism; the small nucleus controls reproduction.
- Paramecia sometimes reproduce by conjugation—a process in which the material in the micronuclei is exchanged between two individuals.
- The process by which amoebas engulf food is called phagocytosis. White blood cells in human blood destroy foreign substances by a similar process.
- At one time, euglenas were classified as plants because they contain chloroplasts and thus are able to carry out photosynthesis. In most modern systems of classification, they are labeled protists.
- Euglenas have a light-sensitive eyespot near the base of the flagellum.

Cross Reference

The Moneran Kingdom and the Protist Kingdom are discussed in section 1-7.

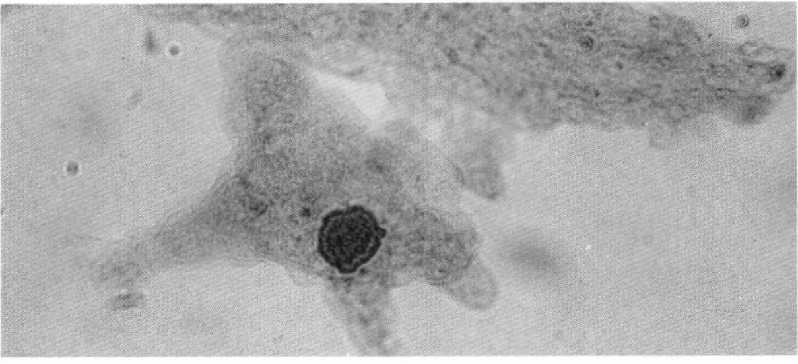

Figure 1-6 *An amoeba's shape is constantly changing as the flowing cytoplasm forms fingerlike projections called* pseudopods.

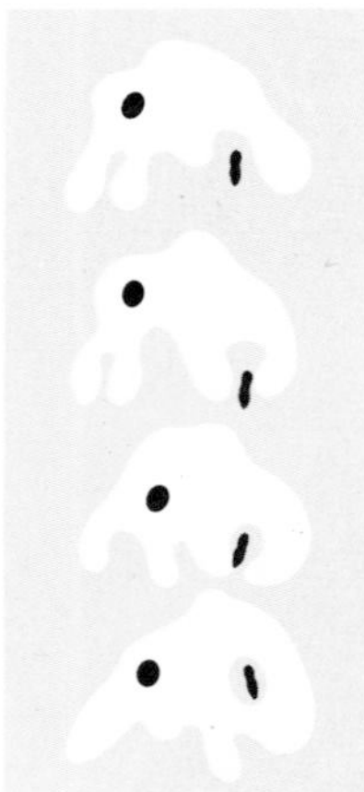

Figure 1-7 *An amoeba surrounds its food, trapping it in a food vacuole. Digestive juices change the food into a form that can be used by the amoeba for energy.*

Smaller one-celled organisms and other tiny bits of food are taken in by a paramecium through an *oral groove*. Beating cilia lining the groove pull in food and sweep it toward the *mouth* and *gullet*. At the end of the gullet, the food is enclosed in a food vacuole, where it is digested. The food vacuole floats through the cytoplasm, releasing food in different areas of the cell. Undigested food particles diffuse out through the cell membrane.

Water and waste products collect in two *contractile* (kun TRAK tul) *vacuoles*, located at opposite ends of the cell. These vacuoles stretch and swell to hold water and wastes. When full, the vacuoles contract, "pumping" their contents into the water outside the cell.

A paramecium reproduces by dividing into two identical cells that are copies of the parent.

Another one-celled organism that lives in pond water is the *amoeba* (uh MEE buh). (See Figure 1-6.) Unlike a paramecium, an amoeba does not have a definite shape. The cytoplasm of an amoeba is constantly moving. The moving cytoplasm pushes against the cell membrane, causing it to bulge out in fingerlike projections. The rest of the cell flows along behind the projections. An amoeba is constantly changing shape as it moves.

An amoeba takes in food by surrounding it. (See Figure 1-7.) Food is trapped in a food vacuole, which breaks away from the cell membrane and floats through the cytoplasm. The digested food is absorbed by the cytoplasm and is used for energy, growth, and repair. Wastes pass out through the cell membrane.

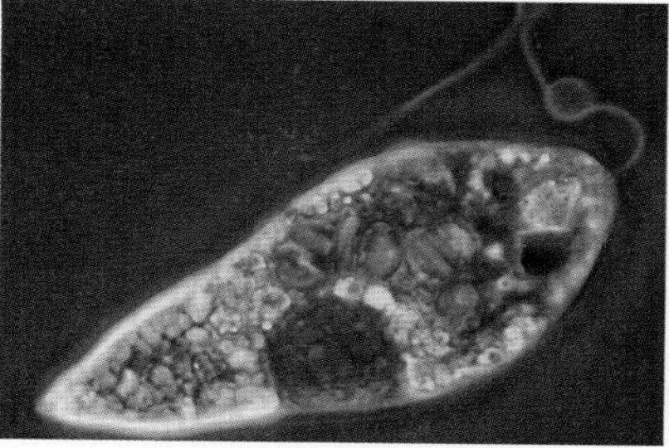

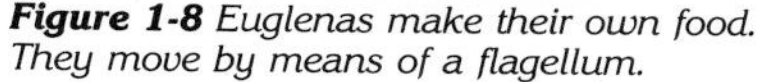

Figure 1-8 *Euglenas make their own food. They move by means of a flagellum.*

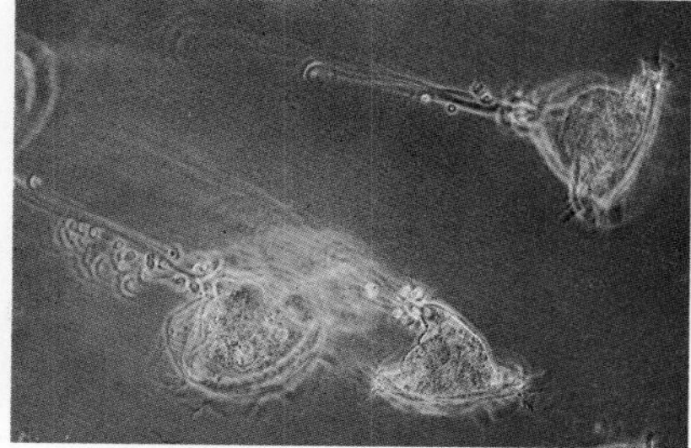

Figure 1-9 *A ring of beating cilia creates a whirlpool that pulls in food from the water.*

Amoebas respond to their surroundings. When conditions are unfavorable for life, such as when a pond dries up, an amoeba is able to form a protective covering called a *cyst* (sist). The amoeba becomes inactive inside the cyst and may remain this way for years. When conditions are favorable, the cyst dissolves and the amoeba carries on.

Like a paramecium, an amoeba reproduces by dividing. One *parent* cell divides, forming two identical cells in less than an hour.

There are thousands of different kinds of one-celled organisms. Each kind of organism has unique ways of carrying out many of its life activities. For example, *euglenas* (yoo GLEE nuhz), the one-celled organisms shown in Figure 1-8, have chlorophyll and can make their own food in sunlight. But during periods of darkness, euglenas take in food from their surroundings. They move by means of a whiplike tail, called a *flagellum* (fluh JEL um). Other one-celled organisms attach themselves to objects in the water. *Vorticella* (vor tuh SEL uh) "spring" up and down to capture food, as shown in Figure 1-9.

TRY THIS

Collect water from a nearby pond or stream. If you can, carefully take some water from the bottom. Make slides of the pond water and look at them under a microscope. Use books to identify the variety of one-celled organisms that you find.

Checkpoint

1. How does a paramecium move? How does it get food?
2. Describe how an amoeba takes in food.
3. How does an amoeba reproduce?
4. How are euglenas different from paramecia?

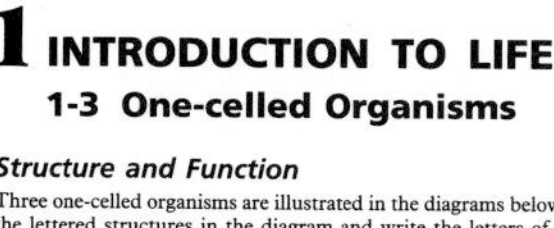

1 INTRODUCTION TO LIFE

1-3 One-celled Organisms

Structure and Function

Three one-celled organisms are illustrated in the diagrams below. For each organism identify the lettered structures in the diagram and write the letters of those structures beneath the organism's name in the chart. Then write the functions of the structures in the last column.

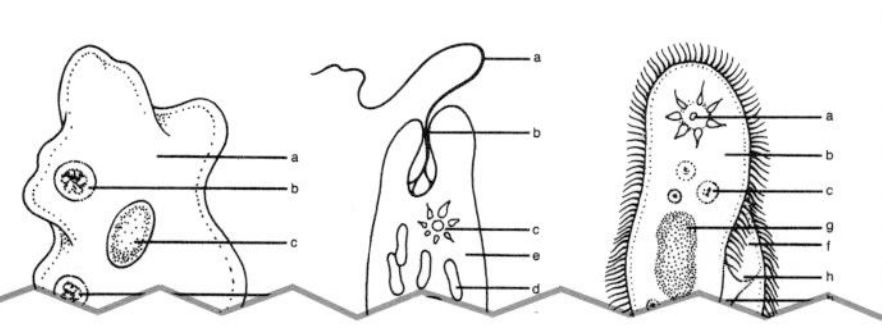

Using Try This

Because the number and variety of organisms in pond or stream water varies widely, it is best to have as many students as possible bring in samples. Instruct students to gather and label samples from the bottom as well as the top and middle layers of water.

Check the samples ahead of time to find those that contain the greatest number and variety of organisms. Different organisms are likely to appear in different samples.

Have students make sketches of the organisms they find. Discovering a variety of types is more important than identifying those found. Be sure to help students distinguish between one-celled and multi-celled organisms.

Keep the pond water samples for a period of time. Place the containers in an indirect source of light. Keep them loosely covered to avoid evaporation or spillage. Add fresh pond water as needed. Have students observe the samples periodically. Ask them how the makeup of the samples appears to change over time.

Checkpoint Answers

1. Beating cilia move paramecia through the water. The cilia sweep food into the oral groove.
2. Moving cytoplasm pushes an amoeba through the water along a surface. An amoeba takes in food by engulfing, or surrounding, it.
3. An amoeba reproduces by dividing into two identical cells.
4. Euglenas move by means of a flagellum. In addition, they have chlorophyll and can thus make their own food in sunlight.

Teaching Suggestions 1-4

Begin by asking students:

What steps are involved in building a house? (drawing a plan, estimating costs, selecting materials, hiring workers to build the house)

What kind of worker does each job? (architect, builder, supplier, masons, carpenters, plumbers, and so on)

By means of discussion, develop the concept that the finished product is built more efficiently through specialization and cooperation. Discuss the specialization of labor in society also—some people grow food, others are health care providers, some teach, and so on. Then have students discuss how specialization and "cooperation" of cells, tissues, and organs enable an organism to survive.

Ask students to describe similarities and differences among the cells shown in Figure 1-10. Discuss how the shape and overall structure of each kind of cell are related to its function. Ask a student to define tissue. Then explain how each type of tissue in multi-celled organisms has a specific function.

Describe how tissues make up organs. Discuss the tissues and their functions in an organ, such as the heart. Point out that the roots of a tree also constitute an organ. This organ anchors and supports the tree and absorbs water and minerals. Tissues in roots carry water and other chemicals from the roots to the stem or trunk.

1-4

Levels of Organization

Goal

To describe the levels of organization in a multi-celled organism.

Some organisms, such as blue whales or giant redwood trees, contain billions of cells. Multi-celled organisms are made up of many different kinds of cells. Each cell carries out the same basic life activities. But each *kind* of cell is *specialized,* or performs a special function, as well. Different kinds of cells have different structures, which allow the cells to specialize. Observe the similarities and differences among the cells shown in Figure 1-10.

tissue

In multi-celled organisms, cells function in groups. A **tissue** is a group of cells that are alike in appearance and function. Multi-celled organisms contain many kinds of tis-

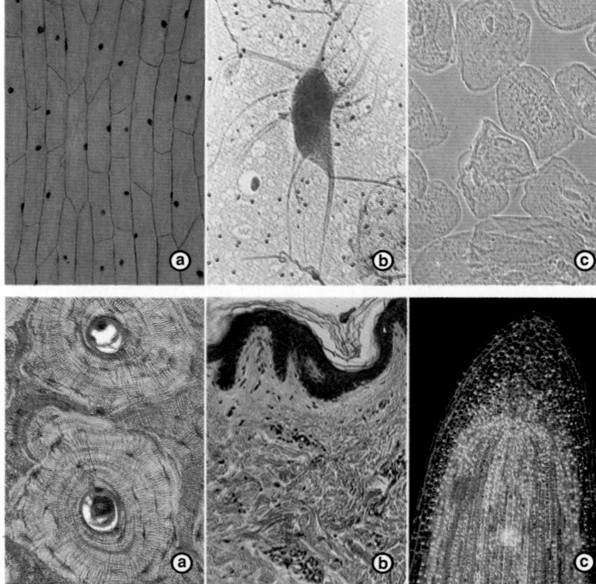

Figure 1-10 *Observe the similarities and differences among: (a) onion skin cells, (b) ox nerve cell, (c) human cheek cells.*

Figure 1-11 *Different kinds of tissue: (a) bone tissue, (b) skin tissue, (c) onion seedling root-tip.*

Ask students to explain how organs working together make up a system. Discuss the human circulatory system in some detail. Then have students name systems of the human body other than those mentioned in the text. These are the muscular, skeletal, excretory, nervous, reproductive, respiratory, and endocrine systems.

Figure 1-12 shows the levels of organization of a human. Be sure students understand the concept of levels of organization. Some may be surprised to learn that skin is an organ. It contains several different types of tissue, including epithelial, connective, muscle, nerve, and blood tissues.

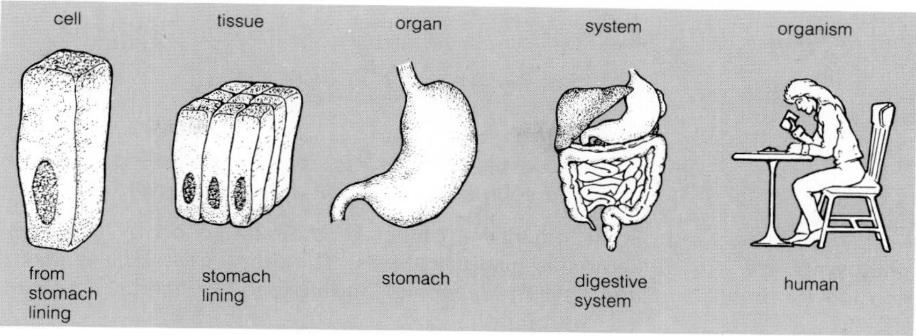

Figure 1-12 Levels of organization of a human.

sue. Blood, muscle, skin, and bone are examples of human tissue. Each kind of tissue has a specific function. For example, muscle tissue moves body parts and skin tissue protects the body from harmful substances. You can think of these specializations as a "division of labor."

Plants have different kinds of tissue, too. For example, leaves contain a skinlike protective tissue and tissues that transport water and minerals. Figure 1-11 shows some different kinds of tissue.

organ

Two or more different types of tissue functioning together make up an **organ.** Each organ has a specific function. The heart is an organ that pumps blood through the body. It contains muscle tissue, skin tissue, nerve tissue, connective tissue, and blood tissue. Plants have organs, too. Roots, leaves, flowers, and stems are examples of organs in plants.

system

Organs function together, making a **system.** The *circulatory system* is made up of the heart, blood vessels, and blood. It functions to move materials through the body. The *digestive system* contains organs that change food into a form in which it can be used by the body. In an organism, all the systems together enable the organism to survive. Figure 1-12 shows the levels of organization of a human.

Checkpoint

1. What is a tissue? Name two kinds of tissue.
2. What is meant by the term *specialized cell*?
3. Name the levels of organization in a multi-celled organism, beginning with a cell.

1 INTRODUCTION TO LIFE

1-4 Levels of Organization

Organizing Cells

The chart below contains a list of the levels of cellular organization. Describe each level of organization beneath its name in the left-hand column. In the right-hand column, write the names of three examples that fit the description.

Level Description	*Examples*
1. The Cell	a. b. c.
2. The Tissue	a. b. c.
3. The Organ	a. b. c.
4. The System	a. b. c.
5. Multi-Cell Organism	a. b. c.

Roots and Stems

Just as animals have organs to help them function, plants have parts that make life and growth possible. The questions below are about these plant parts. Answer each question in the space provided. You may need to use a biology or other reference book.

1. How does a tree carry water and food to its parts? ______________________
2. What are the tree's water tubes called? ______________________
3. How does a plant push its roots into the ground? ______________________
4. How is this like the growth of a leg in an animal? ______________________

Checkpoint Answers

1. A tissue is a group of cells alike in appearance and function. Answers to second part will vary.
2. A specialized cell performs a special function.
3. Cell, tissue, organ, system.

Demonstrations

1. If projection equipment is available, obtain prepared slides, transparencies, or motion-picture films of different types of tissues and show them to your class.
2. Bring an uncooked chicken leg into class and dissect it. Point out the different tissues and their functions.
3. Bring in a plant and demonstrate the shoot and root systems and the organs that comprise them.

Cross Reference

The major systems of the human body are the focus of Chapter 3.

Teaching Suggestions 1-5

Help students grasp the process of photosynthesis by leading them through the steps shown in Figure 1-13. Remind students that certain one-celled organisms too, such as euglenas, are capable of photosynthesis. Refer students to Figure 1-2, the plant cell, to discuss the locations of chloroplasts within green plant cells.

Make sure students do not confuse glucose, the sugar formed during photosynthesis, with table sugar, which is sucrose.

Point out that photosynthesis is a process vital to living organisms. All living things depend on green plants for the food they produce during photosynthesis. All animals feed either on plants or on other animals that eat plants. Moreover, the oxygen released during photosynthesis is required by almost all organisms for respiration.

Emphasize that all photosynthetic organisms carry on *both* photosynthesis and respiration. Copy the diagrams of respiration (Figure 1-3) and photosynthesis (Figure 1-13) on the chalkboard. Explain that the end products of respiration are the starting materials for photosynthesis, and vice versa.

Discuss the carbon cycle (Figure 1-14a) and the oxygen cycle (Figure 1-14b) and relate them to photosynthesis and respiration.

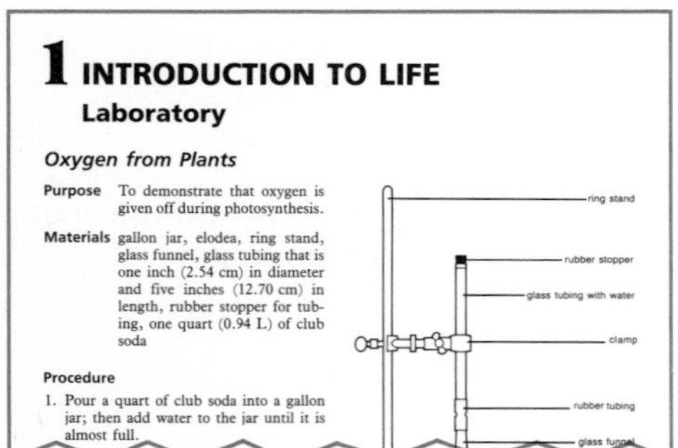
1 INTRODUCTION TO LIFE

Laboratory

Oxygen from Plants

Purpose To demonstrate that oxygen is given off during photosynthesis.

Materials gallon jar, elodea, ring stand, glass funnel, glass tubing that is one inch (2.54 cm) in diameter and five inches (12.70 cm) in length, rubber stopper for tubing, one quart (0.94 L) of club soda

Procedure

1. Pour a quart of club soda into a gallon jar; then add water to the jar until it is almost full.

1-5

Photosynthesis

Goals

1. To explain the process of photosynthesis.
2. To compare respiration and photosynthesis.

photosynthesis (foh toh SIN thuh sis)

Some organisms can make their own food by a process known as **photosynthesis.** The word *photosynthesis* comes from *synthesis*, meaning "putting together," and *photo*, meaning "with light."

Green plants and some one-celled organisms contain structures that trap energy from sunlight. The "solar collectors" are *chloroplasts*, which contain the substance *chlorophyll*. The trapped energy is used in converting carbon dioxide and water into a type of sugar called *glucose* (GLOO kohs). Plants use glucose as food. Figure 1-13 shows the steps involved in photosynthesis. Notice that oxygen is given off in the process.

When plants make more food during photosynthesis than they use, the extra food is stored in the form of *starch*. Plants use the stored starch when they run out of sugar or when photosynthesis cannot take place. Plants store energy in molecules of protein, fat, and oil also. Plant-eating animals use the substances stored in plants as food.

Compare the diagram of photosynthesis with the diagram of respiration, Figure 1-3. Note that the two processes seem to be "opposites." The materials produced by one process are the starting materials for the other.

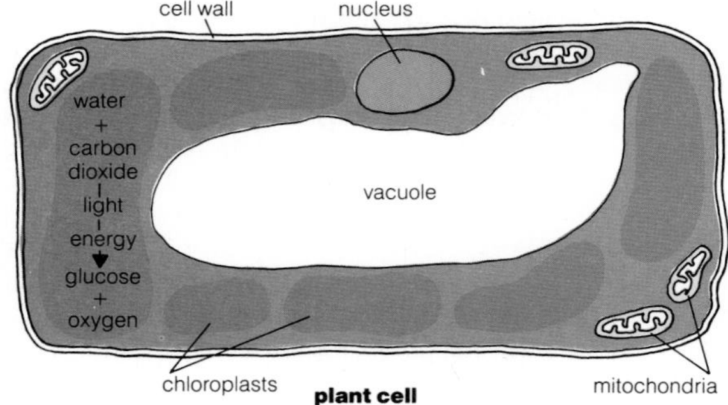

Figure 1-13 *Photosynthesis takes place in the leaves and stems of most plants.*

Background

A useful equation that summarizes photosynthesis is as follows:

$$6CO_2 + 6H_2O + \text{light energy} \xrightarrow{\text{chlorophyll}} C_6H_{12}O_6 + 6O_2$$

Respiration can be thought of as the reverse of photosynthesis:

$$C_6H_{12}O_6 + 6O_2 \longrightarrow 6CO_2 + 6H_2O + \text{energy}$$

The equations are simplifications of what occurs during photosynthesis and respiration. They are useful, however, for establishing the concept that carbon and oxygen are continuously cycled among living organisms.

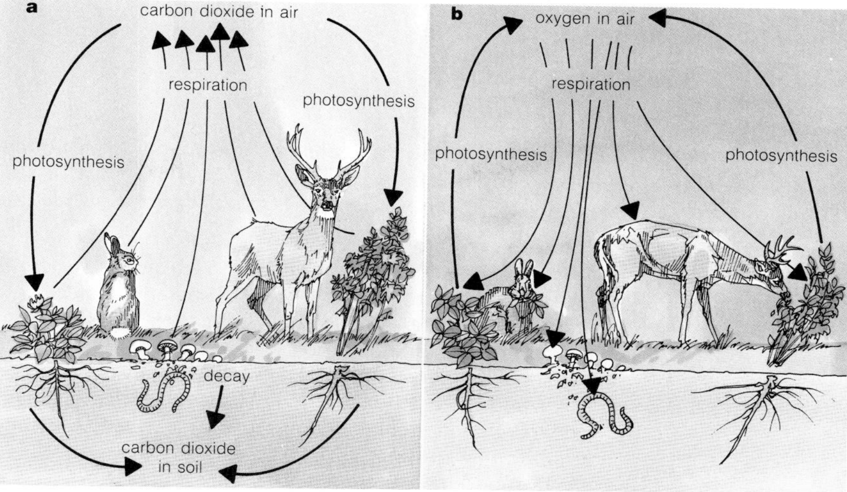

Figure 1-14a *The carbon cycle.*

Figure 1-14b *The oxygen cycle.*

There is a limited amount of oxygen and carbon on Earth. These materials are constantly being reused, or *recycled*. Figure 1-14 shows how carbon and oxygen are recycled by organisms. Notice that all organisms use oxygen given off in photosynthesis during respiration. Plants use the carbon dioxide produced during respiration for photosynthesis.

The exchange of oxygen and carbon dioxide among organisms is of great interest to space scientists, also. They are considering using plants in space colonies. A large number of plants might be able to provide astronauts with enough oxygen and fresh food to survive in space for a long time.

Checkpoint

1. What is photosynthesis? Which organisms are capable of photosynthesis?
2. Which substances are necessary for photosynthesis to take place? What are the products of photosynthesis?
3. Compare the processes of respiration and photosynthesis.

Demonstration

Show that plants store food as starch. Iodine solution stains starch a dark blue or black. Place several drops of iodine solution on a slice of raw potato. The results are quite dramatic. Explain that potatoes are underground stems, called tubers, that store starch.

Facts At Your Fingertips

- Most of the world's photosynthesis takes place in marine algae.
- The energy stored in fossil fuels, such as coal and oil, was captured by plants millions of years ago.
- Some parts of plants, such as roots and flowers, lack chlorophyll and do not carry out photosynthesis.

1 INTRODUCTION TO LIFE

1-5 Photosynthesis

Living Solar Collectors

Some sentences in the paragraph below are incomplete. Think of the term that best completes each numbered space. Then write the term on the corresponding numbered space at the left.

1. __________
2. __________
3. __________
4. __________
5. __________
6. __________
7. __________
8. __________
9. __________
10. __________

Solar energy is trapped by the 1. of green plants and some one-celled organisms. Since leaves contain a green substance called 2., the "solar collectors", known as 3., are able to use the trapped energy to convert 4. and water into the simple sugar 5. and to give off oxygen. When plants make extra food through the process of 6., they store the food in the form of 7.. Plants may use this stored food when they run out of 8. or when photosynthesis cannot take place. They may also store energy as 9. or as fat; these materials provide food for 10..

Comparing Photosynthesis and Respiration

Each of the phrases below describes either photosynthesis or respiration. To compare the two processes, write each phrase in the appropriate column.

- Occurs in the presence of chlorophyll
- Takes place day and night
- Results in the release of oxygen
- Results in the storage of energy as glucose
- Requires the intake of water and carbon dioxide
- Produces carbon dioxide and water
- Occurs in all living cells
- Results in the breakdown of food
- Produces glucose
- Occurs only in light
- Requires the energy of glucose

Photosynthesis	Respiration

Checkpoint Answers

1. Photosynthesis is the process by which organisms with chlorophyll make their own food.
2. Carbon dioxide, water, and chlorophyll are necessary. Glucose and oxygen are produced.
3. Respiration and photosynthesis may be thought of as opposite processes. Carbon dioxide and water, the end products of respiration, are the starting materials for photosynthesis. Glucose and oxygen, the end products of photosynthesis, are the starting materials for respiration.

Teaching Suggestions 1-6

To introduce the concept of classification, ask students to describe how goods are organized, or categorized, in a supermarket. Produce, canned goods, dairy products, and others may be mentioned. Write these categories on the chalkboard. Ask students what would happen if items were not categorized, but placed randomly throughout the store. Ask what happens when the storekeeper and customer have a different idea about how an item should be classified (the item will probably be difficult to find). Stress the need for a universally recognized organizational system. Discuss other examples of classification systems, such as those used for books in libraries.

Ask students to look at the list of supermarket categories again. Choose one category, such as canned goods. Ask how canned goods are further subdivided (canned vegetables, canned meat, canned fruit, and so on). Pick one category, such as canned vegetables, and ask how it is subdivided (canned vegetables: canned string beans, canned peas, canned carrots, and so on). Point out that each variety of canned vegetable can be further classified by brand or label. Draw a branching-tree diagram on the chalkboard, as shown below, to illustrate. Explain that, as in scientific classification, this system moves from a broad, general category to more specific categories. There are far more items in the first category than in those lower in the branching-tree diagram.

Tell the class that not all scientists agree on where some organisms "fit" in the classification scheme. Different authorities use either a three-, four-, or five-kingdom system of classification. All do agree, however, on the seven levels of classification shown in Table 1-1 in the student text. Review this plan of organization with the class. Many of the levels have sub-categories as well. Point out that classification is not a fact, but that it is a method of

1-6

Classification

Goals

1. To describe the features of a scientific classification system.
2. To explain why a classification system is used.

classification

When you go into a supermarket or variety store, you usually know where to look for a product because the goods are organized into groups or categories. Separating objects into groups by similarities is called **classification.** Classifying is a way of organizing information.

There are over a million known kinds of living things on Earth, and more are still being discovered. Scientists have *classified* all living things into groups to make the study and identification of them easier.

The modern system for classifying organisms is based on a system developed by an 18th-century Swedish scientist, Carolus Linnaeus (lih NEE us). Linnaeus studied the characteristics of thousands of organisms. He set up a classification system based on the similarities and differences he observed.

Linnaeus divided all the organisms into two large groups, which he called *kingdoms.* According to Linnaeus' system, all organisms belonged to the *Plant Kingdom* or to the *Animal Kingdom.* As more precise tools and methods for studying organisms became available, scientists realized that all organisms did not fit neatly into the two-kingdom system. Scientists agreed that additional categories had to be added. Most modern classification systems have more than two kingdoms. The system presented in the next section of this chapter has five kingdoms.

All scientific classification systems for organisms are divided into a series of levels. The first level is called a *kingdom.* A kingdom is a large group containing many organisms. The members of a kingdom have some broad, general characteristics in common. Each kingdom is divided into smaller groups called *phyla* (FY luh) (singular, *phylum,* FY lum). These are divided into classes; classes are divided into orders. Family, genus, and species follow. Table 1-1 shows the seven levels of classification.

Table 1-1: Levels of Classification

kingdom
phylum
class
order
family
genus
species

Members of a *species* (SPEE sheez) share many similarities. They are so much alike that they are able to mate and produce young, which in turn can produce young. Moving from the kingdom level to the species level, the groups become smaller and the organisms in each group more alike.

Scientists classify organisms based on similarities in structure. Note the similarities between the domestic housecat and the mountain lion, shown in Figure 1-15. They are both classified as members of the *Animal Kingdom* because they are multi-celled and get their food by eating other organisms. The cat and mountain lion are in the phylum *Chordata* (kor DAH tuh) and the subphylum *Vertebrata* (vur tuh BRAH tuh) because they have a series of bones, or *vertebrae* (VUR tuh bree), along their backs.

Mammalia (ma MAY lee uh) is the class of animals that includes the cat and mountain lion. Mammals have a body covered with hair or fur, they breathe with lungs, and give birth to live young who are fed with milk. Cats and mountain lions are in the order *Carnivora*. Carnivores have strong jaws and sharp teeth, which help them to catch and eat other animals.

Notice that the mountain lion and cat have low-slung bodies, small round heads, and claws. Scientists classify them in the family *Felidae* (FEE lih day), with other animals that have catlike features. But mountain lions and housecats do not belong to the same genus.

Figure 1-15a *A mountain lion.*

Figure 1-15b *A housecat.*

organization that results from the interpretation of facts.

Some students may notice that some of the scientific names contain familiar words. Explain that these names are based on Latin and Greek roots, as are many words in English.

An amusing way to teach the method of scientific naming is to assign an organism to each student, and have that student come up with an appropriate name. Tell students they must give their organisms two-part names, and that a name should describe one or more characteristics of the organism. Humorous names are acceptable. The new name for an elephant, for example, might be *Hosenose gigantea*.

Some students may wish to do projects 1 and 6 at the end of the chapter.

Teaching Tips

- The following activity should help students understand broad categories containing many organisms as opposed to specific categories containing far fewer individuals: Beginning with Earth, have students classify people based on where they live. For example, Earth, hemisphere, continent, country, state or province, county, city or town, street, house or apartment. Students can even specify the floor and room in a house or an apartment. Point out that the number of individuals per category decreases as the category becomes more specific.
- Give students practice with the classification scheme. Assign a different organism to each student. By means of reference materials, each student should classify his or her organism, beginning with kingdom and ending with the organism's genus and species names. With student help, prepare a bulletin-board display of pictures of organisms and their scientific classifications.

1 INTRODUCTION TO LIFE
Computer Program

A Classification System

Purpose To use a classification system as a means of identifying objects.

Background When scientists want to identify an object, they use a classification system. The system consists of a series of questions about the object. Each question narrows the choices for proper identification.

ENTER the following program:

```
10  INPUT "IS THE BALL ROUND (Y OR N)?"; C$
20  IF C$ = "Y" THEN GOTO 80
30  PRINT "IS THE BALL POINTED ON BOTH ENDS"
40  INPUT "AND MADE OF PIGSKIN (Y OR N)?"; C$
50  IF C$ = "N" THEN GOTO 180
60  PRINT "THE BALL IS A FOOTBALL."
70  GOTO 190
80  PRINT "IS THE BALL SMALL, HARD AND"
90  INPUT "PLASTIC (Y OR N)?"; C$
100  IF C$ = "N" THEN GOTO 130
110  PRINT "THE BALL IS A GOLF BALL."
120  GOTO 190
130  PRINT "IS THE BALL COVERED WITH WHITE"
140  INPUT "LEATHER (Y OR N)?"; C$
150  IF C$ = "N" THEN GOTO 180
160  PRINT "THE BALL IS A BASEBALL."
170  GOTO 190
180  PRINT "I DO NOT KNOW THAT BALL."
190  END
```

Discussion

1. Run through the system and answer the questions about a football. Round? ____ Pointed on both ends and made of pigskin? ____
2. Run through the system again and answer the questions about a baseball. Round? ____ Small, hard, and plastic? ____ Covered with white leather? ____

Upgrade the Program

Before you begin, note that all the classification questions are presented in INPUT, or combinations of PRINT and INPUT, statements. User answers are stored in the variable C$. Each time a new entry is made, the new value replaces the previous value of C$. Once a ball has been identified, a GOTO statement routes the program to the END statement in line 190.

1. Adapt the program to include questions about a basketball.
2. Rewrite the existing program for another group of common objects.

Facts At Your Fingertips

- Carolus Linnaeus was born Karl von Linné. He Latinized his own name.
- Modern taxonomists use biochemical and developmental data to aid in classification.
- A species can be further subdivided into varieties. The many breeds of dog, *Canis familiaris,* for example, are varieties.
- Some people call the housecat *Felis catus*.
- The scientific names of organisms should be printed in italics or underlined when handwritten.

1 INTRODUCTION TO LIFE

1-6 Classification

Describing Living Things

Column A contains terms related to the classification of living things. Column B contains phrases that describe the terms. Match each phrase from Column B to the term it describes in Column A by writing the letter of the appropriate phrase next to each term.

A	B
____ class	a. "Two names".
____ classification	b. Level of classification of which Vertebrata is an example.
____ kingdom	c. Smaller groups into which each kingdom is divided.
____ binomial	d. Level of classification whose members are able to mate.
____ subphylum	e. Scientist who created modern system for classifying organisms.
____ species	f. Level of classification whose names always end in "-ae".
____ genus	g. First level of classification for organisms.
____ phyla	h. Separation of objects into groups by similarities.
____ Linnaeus	i. Level of classification in which Carnivora is an example.
____ order	j. Group to which all mammals belong.
____ family	k. Group that can be classified into species.

What's in a Name?

The scientific names of several plants and animals appear below. Research each name in a key in your library and find out what well-known organism it is. Write the common name of the organism on the space next to its Latin name.

1. Tree: *Acer saccharum* ____________
2. Flower: *Daucus carota* ____________
3. Bird: *Geococcyx californianus* ____________
4. Mammal: *Canis latrans* ____________
5. Fish: *Sphyraena barracuda* ____________
6. Reptile: *Alligator mississippiensis* ____________
7. Insect: *Musca domestica* ____________

How many common names did you guess correctly without looking them up?

____

Using Try This

Many states and provinces have an honorary flower, tree, and animal. Students may be interested in the historic, geographic, economic, or other reasons behind the choice of an organism to represent a state or province.

The cat and mountain lion are similar enough to be classified in the same genus, *Felis* (FEE lis). Animals in this genus are night-hunters. They have claws that can be extended and pulled back inside a cover, large feet, and well-developed eyes and ears. The cat and mountain lion are different enough from each other, however, to be categorized in separate species. The cat is in the species *domesticus* (duh MES ti kus) and the mountain lion in the species *concolor*.

In some parts of North America, a mountain lion is called a puma (PYOO muh) or a cougar. A crayfish (which is not even a fish) is called a crawfish or crawdad, depending on where you live. To avoid confusion, and to make world-wide communication possible, scientists have given a scientific name to every known organism.

The scientific name of an organism is its genus name (beginning with a capital letter) followed by its species name (beginning with a small letter). When many of the scientific names were invented, Latin was understood and used by scientists all over the world. Latin names were given to many plants and animals at that time and are still being used today.

Scientific names sometimes include words that describe the organisms, such as *Sequoia gigantea* (si KWOY uh jy gan tee uh) for *giant* redwood, *Alligator mississippiensis* for American alligator, and *Megabombus pennsylvanicus* for bumblebee. Occasionally, an organism is named after the person who studied or discovered it. Milkweed plants are in the genus *Fernaldia* (fur nal DEE uh), after the botanist Merritt Lyndon Fernald. The system of naming organisms is called the *binomial* (by NOH mee ul) *system*. Binomial means "two names."

TRY THIS

Find out the name of the flower, tree, or animal representing your state or province. Use reference books to find out its scientific name.

Checkpoint

1. Define classification. What is the function of the scientific classification system for organisms?
2. Why do most modern classification systems have more than two kingdoms?
3. List the seven levels of the scientific classification system for organisms.
4. *Canis lupus* (KAY nis LOO pus) is the scientific name for a wolf. Which is the genus name? Which is the species name?

Checkpoint Answers

1. Classification is the separating of objects into groups by similar properties. Scientists have classified all known living things to make their identification and study easier.
2. All organisms do not fit neatly into a two-kingdom classification system.
3. Kingdom, phylum, class, order, family, genus, and species.
4. *Canis* is the genus name. The species name is *lupus*.

1-7

Variety of Life

Goals

1. To identify the general characteristics of the five kingdoms.
2. To give examples of organisms in each kingdom.

The **Moneran Kingdom** is made up of the smallest one-celled organisms, the blue-green algae (AL jee) and the bacteria. (See Figure 1-16, left.) Monerans have no nuclear membranes, mitochondria, or chloroplasts. Monerans do have nuclear material.

Moneran Kingdom (moh NAIR an)

The largest group of monerans is the *bacteria*. Bacteria can live almost anywhere. They have been found in deep soil, in hot springs, and in the bodies of animals. For example, *E. coli* (KOH ly) bacteria live in human intestines. They appear to be necessary for proper digestion and for elimination of waste products. Some bacteria are harmful. Different varieties of bacteria cause diseases such as pneumonia, strep throat, and food poisoning.

Protist Kingdom (PROH tist)

The **Protist Kingdom** contains one-celled organisms that have nuclear membranes, such as the paramecium and amoeba. (See Figure 1-16, right.) Most protists live in water

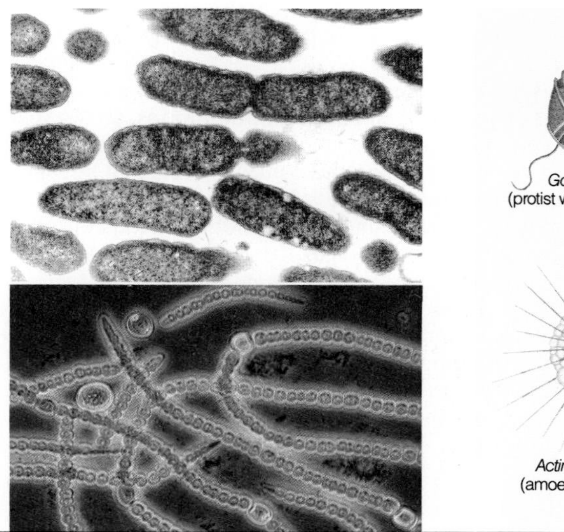

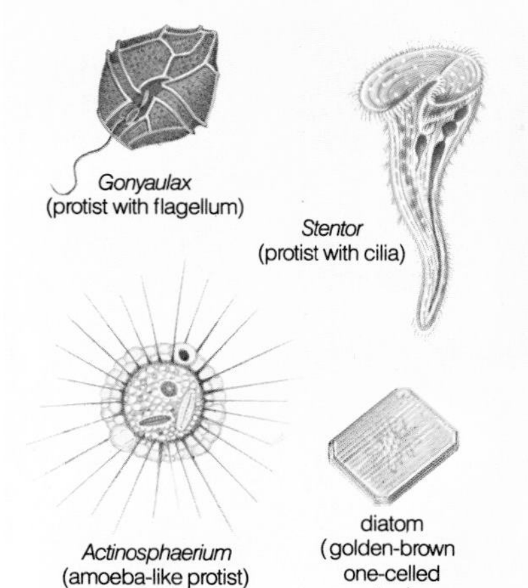

***Figure 1-16** Left, photographs of monerans:* E. coli *(top), a bacterium;* Anabaena *(bottom), a blue-green alga. Right, drawings of protists.*

Teaching Suggestions 1-7

Introduce the section by listing the five kingdoms on the chalkboard. Then describe the characteristics of each. Ask students for examples of organisms in each kingdom, but be prepared to supply them yourself.

Tell students they will not be expected to memorize the scientific names of a large number of organisms. They should, however, be able to name the five kingdoms, describe the general properties of organisms in each kingdom, and give two examples of organisms in each kingdom.

The protists shown in Figure 1-16 include the reddish *Gonyaulax*, the spiny-looking *Actinosphaerium*, the trumpet-shaped ciliate, *Stentor*, and a diatom. Point out *Gonyaulax* and explain that it periodically becomes abundant, turning the ocean water red. This is the source of the common name "red tide." *Gonyaulax* contains a toxin that does not harm molluscs that ingest it, but that can cause illness, or even death, in fish and in people that eat the molluscs. Explain to students that the spiny projections on *Actinosphaerium* are covered with flowing cytoplasm that traps food particles. The food is carried to the cell body by wavelike movements of the cytoplasm.

Figure TE 1-1 shows the locations of the xylem and phloem—the two types of transport tubes—found in a typical vascular plant stem. Copy the diagram on the chalkboard and explain the function of each network. Stress that many familiar plants, including grasses, trees, and some bushes, are classified as flowering plants. Point out that any higher plant with flowers and seeds is a flowering plant.

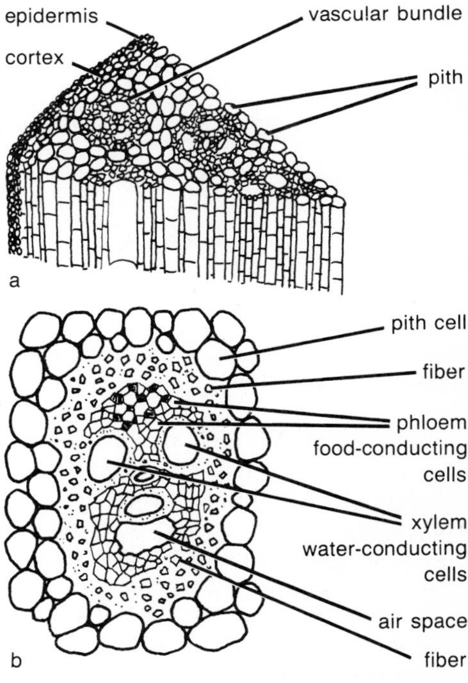

Figure TE 1-1 The locations of xylem and phloem in a corn stem. (a) Microscopic view of cells and vascular bundles in a wedge-shaped section of stem. (b) Enlarged cross section of one vascular bundle.

Many students are probably not aware that sponges are animals. Make sure there is no confusion between synthetic sponges, such as those used for household chores, and natural sponges. Most natural sponges that people use for bathing are the skeletonlike framework left behind when the sponge organisms die. Point out that loofah "sponges" are an exception. Loofahs are the fibrous substance of the pods of the *Luffa* plant.

The coelenterate body plan has radial symmetry—that is, similar body parts radiate from the center. This is sometimes described as a wheel-spoke shape. The following view of a jellyfish (Figure TE 1-2), seen from above, can be drawn on the chalkboard to help explain the radial symmetry of the coelenterate body plan.

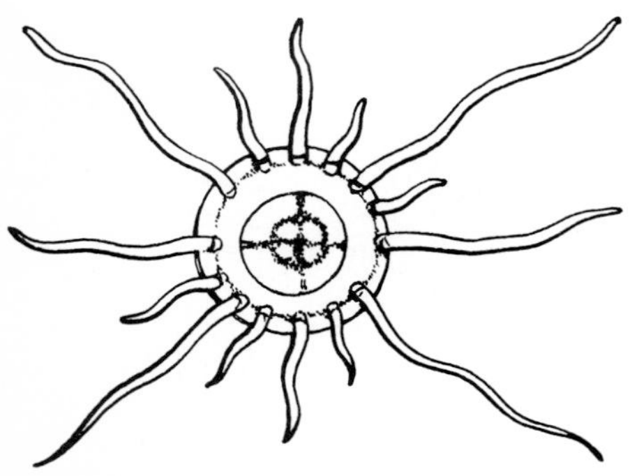

Figure TE 1-2 View of a jellyfish from above showing the radial symmetry of the coelenterate body plan.

Explain that bilateral symmetry is seen in animals that have a right and a left side that are mirror images. Have students study the pictures of animals shown in this section and name those that have bilateral symmetry.

Some of your more able students may be interested in the scientific names of the three phyla of worms: Platyhelminthes (flatworms, such as flukes and tapeworms); Nematoda (roundworms, such as the trichina worm and pinworms); Annelida (segmented worms, such as the common earthworm).

or in moist places. One-celled algae, a kind of protist, are an important food source for many water organisms. They give off large amounts of oxygen, through photosynthesis.

Fungi Kingdom (FUHN jy)

Mushrooms and *molds* are probably the best-known members of the **Fungi Kingdom.** Fungi (plural of fungus) were once classified as plants because they have cells with cell walls and they do not move. But fungi do not have chlorophyll and must obtain their food from other organisms. Figure 1-17 shows some examples of different fungi. Yeasts are fungi used by people in making bread and wine.

Figure 1-17 *Fungi. (top) Mushrooms growing on a rotting log. (bottom) Bracket, or shelf, fungi often grow and feed on old tree trunks.*

Plant Kingdom

The **Plant Kingdom** is made up of multi-celled organisms with chloroplasts. Examples are shown in Figure 1-18. Plants make their own food by photosynthesis and have cells with cell walls. The Plant Kingdom can be divided into two large groups—plants with transport tubes and plants without them.

Many-celled algae are plants without transport tubes. They are found in ponds, oceans, streams, and in damp places. Some algae may be made up of only a few cells, while others may grow to be 10 m in length. Mosses and liverworts are small plants that also do not have transport tubes. They grow in moist, shady places.

vascular plants (VAS kyuh lur)

Plants that do have a system of transport tubes are called **vascular plants.** Vascular plants have two types of transport tubes. Water and minerals move through one set of tubes from the roots to all parts of the plant. Food moves from the leaves to the rest of the plant through another set of tubes.

Ferns, gymnosperms (JIM nuh spurmz), and angiosperms (AN jee uh spurmz) are three groups of vascular plants. Ferns grow in moist, shaded areas and are commonly found in woods and along the banks of rivers and streams.

Most gymnosperms have needlelike leaves, which stay green all year long. Gymnosperms produce seeds that develop in cones. Pines, spruces, larches, and firs are some common gymnosperms.

Angiosperms are flowering plants. Grasses, fruit trees, garden flowers, and vegetable plants are just a few examples of angiosperms. Angiosperms make up a large part of the world's food supply. They provide food and shelter for many organisms and are used for products such as paper, cotton, and linen.

Ask students to list the many ways people use molluscs (food, jewelry, in the home aquarium, shell collections, research, and so on).

Draw a starfish on the chalkboard and point out the five-fold radial symmetry. Mention that, in addition to starfish and sea urchins, the echinoderms include brittle stars, sand dollars, and sea cucumbers. Point out that all echinoderms are marine animals.

Ask students to look at the five classes of arthropods illustrated in Figure 1-20. The five classes are arachnids (spiders, ticks, mites, daddy longlegs), crustaceans (crabs, lobsters, shrimp, crayfish, barnacles), centipedes, millipedes, and insects.

Tell students that animals without an internal skeleton are called invertebrates, and give several examples.

Have students compare the limbs of

Figure 1-18 *The Plant Kingdom. (a)* Spirogyra, *a green algae commonly found in ponds. (b)* Fucus, *a brown algae often found growing on northern rocky coasts. (c) Moss. (d) Liverworts. (e) Ferns. (f) A sunflower. (g) A lodgepole pine tree with cones. (h) A crabapple tree in bloom.*

seven classes of vertebrates. Discuss similarities and differences among the seven classes. Explain that some fish, such as sharks and rays, have a skeleton made of cartilage—a softer, more flexible material than bone. The human body has cartilage at the tip of the nose, in the ears, and between bones.

Chapter-end projects 2, 3, 4, and 5 apply to this section.

Demonstrations

1. Grow colonies of bacteria. Obtain several sterile Petri dishes containing solidified nutrient agar. Divide the dishes into four sections by marking the covers with grease pencil. Give students small strips of cellophane tape and have each student touch the sticky side to a different exposed surface in the room. Then have each student gently press the sticky side to a section of the agar surface in a Petri dish. Cover each dish and secure the cover with tape. Place tape labels on the cover, each with the name of the student and the location of the "smear." Be sure each label is above the section of agar touched by the student. The labels should be small, and the agar surface visible through the cover. Let the Petri dishes sit for several days in a warm place. A considerable variety of bacterial colonies should appear. CAUTION: Do not untape or open the Petri dishes. Explain that bacterial (and fungal) spores are everywhere, and that they simply need the right conditions to grow. Some of the organisms in the Petri dishes may be injurious to humans.
2. If possible, obtain one or more specimen organisms from each kingdom. Prepared slides and dried and preserved specimens of monerans, protists, fungi, plants, and animals are available from biological supply houses. Allow students to examine the specimens. In addition, have students bring in pieces of coral, dried sponges, shells, or any other specimens they have collected. Arrange a display of the specimens.
3. Set up an ant farm for students to observe. Have them develop the concept of social behavior among insects from their observations.
4. Observe the life cycle of an amphibian. Collect or order frog's eggs or young tadpoles and establish them in an aquarium. Have students observe the growth and development that ensues.
5. Set up a large woodland terrarium with mosses, ferns, salamanders, and other small organisms. Have students observe the interrelationships that develop.

6. Demonstrate the stages in the life cycle of *Tenebrio* beetles. Obtain the larvae, called mealworms, from an aquarium shop. Culture the mealworms in jars half filled with moist oatmeal or bran. Cover the jars with fine mesh to prevent adult beetles from escaping. Feed adult beetles bits of raw carrot or potato. Have students observe metamorphosis and the three stages of insect development: larva, pupa, and adult.

Teaching Tips

- To help develop the material in this section, you might want to divide the class into groups and assign each group the task of researching and reporting on one of the five kingdoms. Encourage students to prepare bulletin-board displays or to create a mural depicting the variety of life on Earth.
- Some students may be interested in maintaining plants and animals in the classroom. Ask for contributions of plant cuttings, plastic containers, old aquariums, cages, and so on. Make sure all used containers and other pieces of hardware are thoroughly cleaned before use. Provide reference materials on the care and maintenance of the organisms students will be working with. This is especially important for any animals brought in.

Facts At Your Fingertips

- Some bacteria can produce a tough shell, or spore, that enables them to survive unfavorable conditions in a state of dormancy for many years.
- Bacteria are often classified by shape: *coccus* (round), *bacillus* (rod-shaped), and *spirilla* or *spirochaetes* (spiral-shaped).
- The full name of the bacterium called *E. coli* (see Figure 1-16) is *Escherichia coli*.

Animal Kingdom

Members of the **Animal Kingdom** are multi-celled organisms that get food by eating other organisms. Animal cells do not contain chlorophyll or cell walls. The Animal Kingdom is divided into many phyla.

Sponges belong to the phylum of animals known as *Porifera* (puh RIF ur uh). Their bodies are made up of groups of cells attached to a skeletonlike framework. Sponges take in bits of food from the water flowing through their *pores*, or openings. (See Figure 1-19.) The phylum name *Porifera* means "having pores."

Animals with a saclike body and tentacles surrounding their mouth opening are in the phylum *Coelenterata* (sih LEN tuh rah tuh). Hydras, jellyfish, coral, and sea anemones (uh NEM uh neez) have the stinging cells on their tentacles that are characteristic of organisms in this phylum.

Figure 1-19 *Sponges: (a) yellow tube sponges. Coelenterates: (b) sea anemones and (c) a jellyfish. Worms: (d) a planarian flatworm. Molluscs: (e) a land snail, (f) a sea scallop, and (g) a squid.*

- The following diseases or conditions are caused by bacteria: diphtheria, tetanus, tuberculosis, leprosy, typhoid fever, whooping cough, botulism, staph infections, scarlet fever, some dysentery, gonorrhea, some meningitis, and some pneumonias.
- Bacteria are used to manufacture buttermilk, yogurt, wine vinegar, sauerkraut, pickles, sourdough bread, and other products.
- Bacteria are being used in recombinant DNA research to manufacture important substances, such as interferon and insulin.
- The White Cliffs of Dover, England, and the large chalk beds in Mississippi and Georgia are composed of the shells of billions of *Foraminifera*, an amoeba-like protist.
- Athlete's foot and ringworm are skin diseases caused by fungi.

Figure 1-20 Echinoderms: (a) starfish and (b) sea urchins. Arthropods: (c) a cucumber beetle. (d) a centipede— often found around rotting logs and under rocks. (e) a giant red millipede. (f) an orchard spider. (g) a green crab.

There are three phyla of worms: *flatworms, roundworms,* and *segmented worms.* Some worms, such as tapeworms or leeches, live in or on another organism, obtaining energy from it. They are called *parasites.*

Most animals classified in the phylum *Mollusca* (mah LUHS kah) have a hard shell surrounding their soft body parts. A few, like the squid, octopus, and sea slug do not have a shell. Molluscs live in water or in damp places. Figure 1-19 shows a few kinds of molluscs.

Animals with a star-shaped body and spine-covered skin are called *echinoderms* (ih KY nuh durmz). *Echinoderm* means "spiny-skin." (See Figure 1-20.)

About 75% of all known animals belong to the phylum *Arthropoda.* Arthropods have a hard, external skeleton, a segmented body, and jointed legs. Members of this phylum are very important to people. For example, people raise bees for the large quantities of honey they produce. Crabs, lobsters, and shrimp are food for many people. Figure 1-20 shows representatives of the five classes of arthropods.

- Vascular plants make up the phylum Tracheophyta, meaning "tube" (*tracheo*) "plants" (*phyta*).
- The term *gymnosperm* comes from Greek words meaning "naked seed." Gymnosperms have seeds that are not enclosed in fruits.
- *Ginkgo* is a gymnosperm with broad, fan-shaped leaves. It does not produce cones. Ginkgo trees are highly resistant to smog, fungi, and insects, and for this reason, ginkgoes are common in many urban areas.
- On Earth, the oldest known living things—bristle-cone pines—and the tallest living things—giant sequoias—are gymnosperms.
- The tough, horny outer skeleton, or exoskeleton, of an arthropod is composed of chitin, a complex carbohydrate material.
- Arthropods occupy a greater number and variety of niches than any other group in the Animal Kingdom.
- The duck-billed platypus and the spiny anteater are the only known egg-laying mammals.
- Pouched mammals, the marsupials, include kangaroos, koalas, opossums, and Tasmanian devils.

- The 19th-century Irish potato famine was caused by a parasitic fungus of potatoes that destroyed thousands of crops.
- Common bread mold is *Rhizopus stolonifer.*
- A lichen is an organism made up of an alga and a fungus living together in a mutually beneficial association. Lichens are the crusty-looking growths commonly seen on rocks and tree trunks. Lichens are sensitive to pollutants in the air, and for this reason are sometimes used as indicators of air-pollution levels.
- Seaweeds are kinds of algae.
- Some kinds of algae are used as thickeners in processed foods, such as soups and ice cream. Agar, a substance obtained from red algae, is used as a growth medium for bacteria and other organisms.

1 INTRODUCTION TO LIFE

1-7 Variety of Life

Kingdoms and Their Members

Supply the missing information to complete the chart below.

Name of Kingdom	*Characteristics of Kingdom, Phylum*	*Examples*
1. Moneran	made up of little known, small, simple organisms; widespread	a. ______ b. ______
2. Protist	made up of one-celled organisms with nuclear membranes; water dwelling; give off oxygen gas	a. ______ b. ______
3. Fungi	______ ______	a. ______ b. ______
4. ______	made up of multi-celled organisms with chloroplasts	
a. ______	no water tube	a. ______ b. ______
b. vascular	______	______
1. ______	______	
2. ______	______	a. ______ b. ______
3. ______	______	a. ______ b. ______
5. ______		
a. Porifera	______	______
b. ______	have tentacles and saclike bodies	a. ______ b. ______
c. ______	______	a. tapeworms b. earthworms
d. ______	______	a. ______ b. oysters
e. ______	have a star shape and spiny skin	______
f. ______	______	a. ______ b. ______
g. ______	have an internal skeleton, backbone	humans

Checkpoint Answers

1. The Moneran Kingdom.
2. Protists are one-celled organisms that have nuclear membranes.
3. Fungi have cells with cell walls, but no chlorophyll. They do not move, and must obtain food from other organisms. Mushrooms, molds, and yeasts are fungi.
4. Plants are multi-celled organisms whose cells contain chloroplasts. Plants make food by the process of photosynthesis and have cells with cell walls. The three groups of vascular plants are the ferns, the gymnosperms, and the angiosperms.
5. Animals are multi-celled organisms that get food by eating other organisms. Animal cells lack both chlorophyll and cell walls.

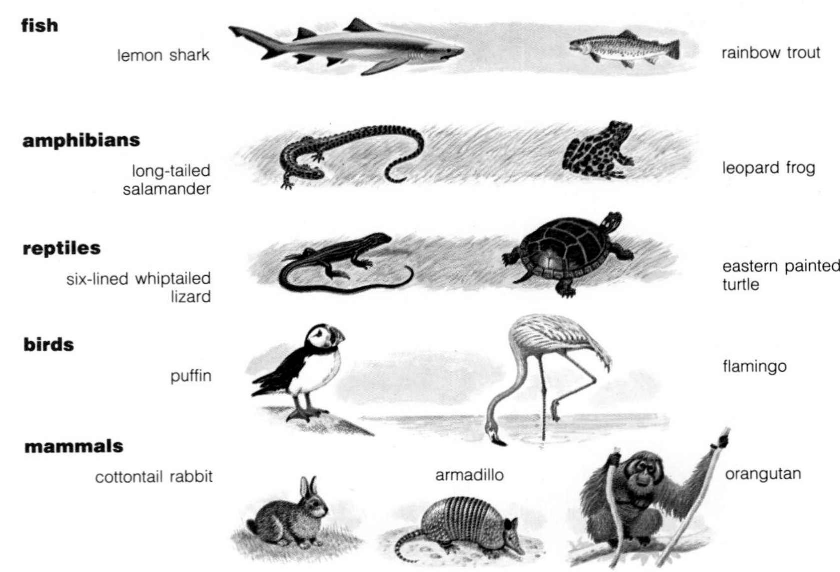

Figure 1-21 *Vertebrates include fish, amphibians, reptiles, birds, and mammals.*

Chordata is the phylum humans belong to. *Vertebrata*, a subphylum of the chordates, is divided into the seven classes shown in Figure 1-21. All vertebrates have an internal skeleton that includes a vertebral column (backbone) and limbs.

Checkpoint

1. Which kingdom contains one-celled organisms without a nuclear membrane?
2. Describe the general characteristics of protists.
3. What characteristics do members of the Fungi Kingdom have in common? Name three fungi.
4. Describe the general characteristics of the Plant Kingdom. What are the three groups of vascular plants?
5. What features do all animals have in common?

1-8

Viruses

ENRICHMENT

Goals

1. To describe viruses.
2. To explain immunization.

There is one group of tiny life forms that has puzzled scientists for many years. These life forms seem to have some of the properties of living things, but they do not have cells and they cannot carry out life activities on their own. These tiny "things," called *viruses*, are classified in a group by themselves.

Viruses are so small that they cannot be seen with a light microscope. For many years scientists could only guess about the existence of viruses, based on their observations of the damage the viruses caused. With the invention of powerful electron microscopes, scientists could finally study the structure of these mysterious life forms. Figure 1-22 shows several kinds of viruses.

Some scientists consider viruses to be alive, while others do not. Viruses seem to use neither food nor oxygen. They lack most of the parts of a cell and cannot reproduce on their own. Viruses need a host cell in order to become active. They do not grow, reproduce, or carry on respiration outside of the host cell.

Figure 1-22 *Three kinds of viruses: (a) bacteriophage attacks bacterium, (b) tobacco mosaic viruses infect tobacco leaves, (c) the viruses that cause measles.*

Teaching Suggestions 1-8

Ask students if any have ever had a viral infection. Ask those who responded "yes" what the infection was, and how they knew it was caused by a virus. Tell the class that measles, chicken pox, mumps, the flu, and the common cold are caused by viruses.

Discuss whether viruses should be considered living or nonliving. Compare the characteristics of viruses with those of cellular organisms. Explain that some scientists consider viruses to be nonliving.

Explain that viral diseases cannot be treated with antibiotics, such as penicillin, as bacterial diseases are. Point out that some viral diseases can be prevented by the type of immunization called vaccination. Explain that there also are vaccines against bacterial diseases, such as tetanus, diphtheria, and whooping cough.

Ask students to name the types of vaccinations they have received, and the diseases they were vaccinated against.

Tell the class that the first vaccination against a virus was given in 1796, before scientists knew about the existence of viruses. At that time, Dr. Edward Jenner demonstrated that a person vaccinated with the (virus-containing) fluid from a cowpox sore on a cow developed immunity to smallpox, a similar disease. The word vaccination comes from *vacca*, which means "cow" in Latin.

Explain that immunity is the body's natural defense against disease-causing organisms. Immunities may be present at birth or acquired. Vaccination is one way to acquire immunity. People also develop immunities to some diseases once they have had them.

Facts At Your Fingertips

- For many years, scientists assumed that viruses were very small bacteria. It was not until 1935 that it was conclusively demonstrated that viruses and bacteria are different.
- The tobacco mosaic virus was the first virus to be isolated and crystallized. Dr. Wendell Stanley won a Nobel prize for this discovery.
- There are three categories of viruses: bacterial viruses (also known as bacteriophages), animal viruses, and plant viruses (such as tobacco mosaic virus). Each type of virus affects a specific kind of cell.
- Interferon is a substance produced by cells in response to viral infection. It seems to protect uninfected cells in a host organism. Interferon is now being produced by recombinant DNA procedures.
- There is evidence that some types of cancer are caused by viruses.
- New outbreaks of polio have occurred in regions where immunization was stopped.

1 INTRODUCTION TO LIFE

1-8 Viruses

Fighting Viruses

Each paragraph below contains information about a substance that appears to combat viruses. Read each paragraph. Then write the answers to the questions that follow.

During the 1960's Linus Pauling recommended large doses of vitamin C to prevent and cure colds. Since then people have disagreed about its usefulness. During 1971–72 the University of Toronto School of Hygiene conducted a two-month study of cver 800 people. The people were divided into two groups and were matched according to age, sex, smoking habits, time spent in crowds, contact with children, and past problems with colds. Members of one group took two grams of vitamin C each day. Members of the other group took an equal amount of a placebo (inactive substance). People in both groups doubled the number of pills taken at the first sign of a cold. The first look at results seemed to indicate that vitamin C was of little value. A closer check of the data showed that people who took vitamin C were confined to their homes 30 percent fewer days than those who did not take the vitamin. Vitamin C seemed to reduce disagreeable symptoms such as muscular aches, fever, and chills. Later studies confirmed these findings.

1. What was the purpose of the University of Toronto's research? ________________
2. For what traits were the two groups paired? ________________
3. How much vitamin C was consumed by each person? ________________
4. According to the test data, what benefits are attributed to vitamin C? ________________

Interferon, a substance that fights viruses, is produced by white blood cells in response to tumor cells and viral infections. It appears to make some white cells called lymphocytes more effective in destroying tumor cells and invading viruses. Interferon seems to activate "killer" cells believed to be part of the body's immune system. But more research is needed. At present, the cost and scarcity of interferon are delaying studies.

1. How and why is interferon produced? ________________
2. What does interferon do? ________________
3. What problems are there with interferon? ________________

Checkpoint Answers

1. Viruses can invade other cells and reproduce inside them. They seem to use neither food nor oxygen; they lack most of the parts of a typical cell; they do not grow, reproduce, or carry on respiration outside a host cell.
2. The virus overpowers the host cell's control system and uses materials from the host cell to make new viruses. Polio virus attacks one kind of cell in the brain and spinal cord. Each kind of virus generally attacks a specific kind of cell.
3. The common cold, measles, polio, chicken pox, mumps, influenza, smallpox, and so on.
4. A vaccine is a small amount of dead or weakened virus. It is given to people to stimulate the production of antibodies.
5. Within a few days of receiving a vaccine, the body produces antibodies against the virus in the vaccine. Polio, mumps, measles, and smallpox have been prevented through immunization.

When a virus enters a cell, it seems to "overpower" the cell's control system. The infected cell begins making copies of the virus, instead of following the normal instructions on the cell chromosomes. The cell becomes a "virus factory," producing hundreds of new viruses. Eventually the host cell bursts, releasing the viruses, which then attack other cells.

Viruses generally invade specific kinds of cells. For example, the polio virus attacks one kind of cell in the brain and spinal cord. Viruses cause diseases in a number of animals and plants. They are responsible for the common cold, measles, polio, chicken pox, mumps, influenza, and smallpox. Scientists now believe that some viruses may remain inactive, or *dormant*, in a cell for many years, until a change in the cell causes the virus to become active. It is thought that some types of cancer may be caused by viruses such as these.

Doctors use a process, called *immunization*, to prevent certain diseases caused by viruses. When you are immunized, you swallow or receive an injection of a small amount of weakened or dead viruses, called a *vaccine* (vak SEEN). The vaccine causes the body to prepare its own defense system for fighting that particular type of virus. Within a few days, the body's natural defense system has produced substances called *antibodies* that can recognize and destroy a specific type of virus if it enters the blood. There are vaccines for polio, mumps, measles, and smallpox. Through the systematic use of vaccines in North America, smallpox, polio, and measles have been nearly wiped out.

Checkpoint

1. What characteristics make viruses seem alive? How are viruses different from living things? From nonliving things?
2. Describe what happens when viruses enter living cells. What kinds of cells do different viruses invade?
3. Name two diseases caused by viruses.
4. What is a vaccine?
5. How does immunization prevent disease? What diseases have been prevented through immunization?

Zeroing In On Life's Secrets

The first microscope used to magnify objects was a *simple* microscope. It consisted of a single lens, or magnifying glass, which magnified an object ten to twenty times its normal size. In the 1600's the *compound* microscope was invented, using two lenses mounted at each end of a tube. It magnified objects hundreds of times more than the simple microscope.

Most modern microscopes are compound microscopes. They use a source of light to illuminate the objects being viewed. Because of the properties of light, a light microscope cannot go beyond a magnification of 400–1000 times normal size without losing the clear details of the image.

The invention of the *electron* microscope made it possible to observe a specimen at magnifications 200,000 times its normal size. With the electron microscope, scientists could see details never before visible. Compare the pictures taken through a light microscope (Figure 1) and an electron microscope (Figure 2).

There are two kinds of electron microscope. Both use a beam of negatively charged particles, called *electrons*, instead of light.

The transmission electron microscope aims a beam of electrons at thin slices of the specimen being studied. The specimen is coated with chemicals and is inserted into a part of the microscope that has had all the air removed from it. A limitation of the transmission electron microscope is that things that are alive cannot be observed.

The scanning electron microscope makes possible three-dimensional images, such as the one shown in Figure 2. Its use is not limited to the viewing of thin slices of material. An electron beam is passed over the surface of a specimen, producing images on a screen. Using the scanning electron microscope, scientists have been able to work with some live organisms.

Figure 1

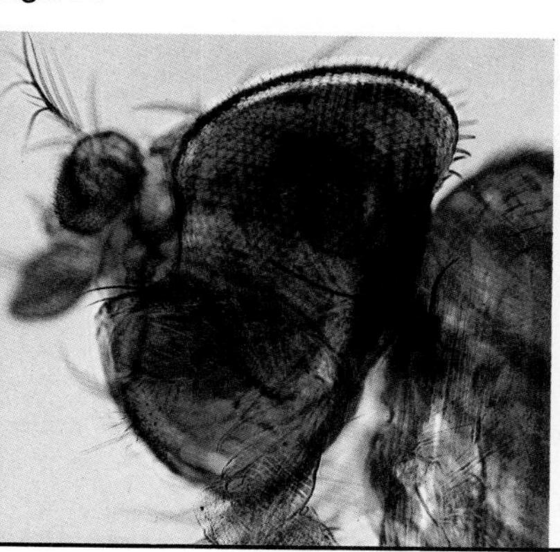

Figure 2

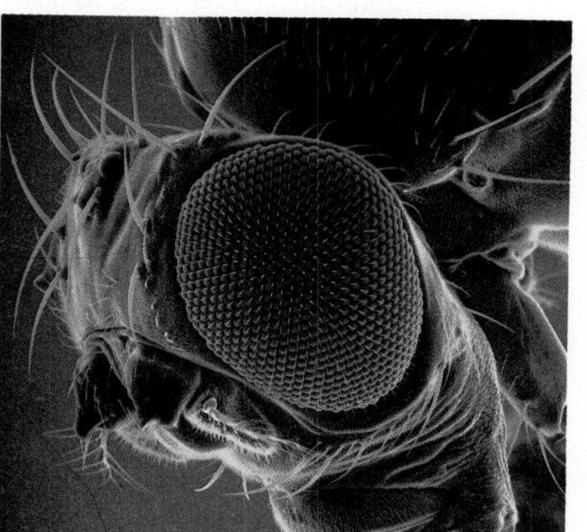

Using The Feature Article

Hold up a magnifying glass and ask the class if it is a microscope. Most students will probably answer "no." Explain that a single lens, or magnifying glass, is called a simple microscope. Tell students what the magnification of the lens is. If possible, have students compare the way common objects appear when viewed through a single lens and when seen through a compound microscope.

Figures 1 and 2 were taken through a light microscope and a scanning electron microscope, respectively. Remind students that viruses and many cell structures can be seen only with the aid of an electron microscope.

Many books containing collections of scanning electron microscope photographs are available (see page 38). If possible, bring in one for your students to examine. Try covering up the captions for the electron micrographs of common objects. See how many objects students recognize.

Reading Suggestions

For The Teacher

Arthur, Wallace. THEORIES OF LIFE. New York: Penguin, 1987. Major theories of what life is and how it evolves.

Borek, Ernest. THE ATOMS WITHIN US. Rev. ed. New York: Columbia University Press, 1980. Introduces subjects, such as cell structure, enzyme function, amino acids, and proteins.

deDuve, Christian. A GUIDED TOUR OF THE LIVING CELL. New York: W. H. Freeman, 1984. Two volumes. Covers every aspect of cell biology. Well-illustrated.

Goodfield, June. QUEST FOR KILLERS. New York: Hill and Wang, 1985. Fascinating book on microbial and virus diseases.

Margulis, Lynn and Karlene V. Schwartz. FIVE KINGDOMS: AN ILLUSTRATED GUIDE TO THE PHYLA OF LIFE ON EARTH. San Francisco: W. H. Freeman, 1982. A concise and accurate catalogue of the diversity of life.

Scheeler, Philip and Donald Bianchi. CELL BIOLOGY: STRUCTURE, BIOCHEMISTRY, AND FUNCTION. New York: Wiley, 1980. An exhaustive, illustrated survey of cell biology.

For The Student

Barr, Bonnie B. and Michael B. Leyden. LIFE SCIENCE. Reading, MA: Addison-Wesley, 1980. An ecological view of living systems.

Hoagland, Mahlon B. THE ROOTS OF LIFE. Boston: Houghton Mifflin Co., 1978. Some of the mysteries of life are explained based on Dr. Hoagland's broad knowledge and personal laboratory experiments.

Postgate, John. MICROBES AND MAN. New York: Penguin, 1986. Updated classic. Describes how microbes affect our lives.

Wolberg, Barbara J. ZOOMING IN: PHOTOGRAPHIC DISCOVERIES UNDER THE MICROSCOPE. New York: Harcourt Brace Jovanovich, 1974. A microscopic view of nature, with some information about microscopy.

Understanding The Chapter 1

Main Ideas

1. The characteristics that define life include: movement, growth and repair, getting and using food, reproduction, responding to changes in the surroundings, and having cells.
2. A cell is the smallest unit able to carry out all the life activities of an organism.
3. Cells contain a cell membrane, cytoplasm, and a nucleus. In addition, plant cells have cell walls and chloroplasts.
4. Respiration is the process of obtaining energy from food.
5. In multi-celled organisms, cells make up tissues, tissues form organs, and organs function together in systems.
6. Photosynthesis is the process by which some organisms make food, using water, carbon dioxide, chlorophyll, and energy absorbed from the sun.
7. Scientists have organized all known organisms into a seven-level classification system.
8. The scientific name of an organism is its genus name followed by its species name.
9. Organisms are classified in one of the five kingdoms: Moneran, Protist, Fungi, Plant, and Animal.
10. Viruses show some characteristics of life. They need a host in order to become active.

Vocabulary Review

From the following list, choose the term that best completes each of the statements. Write your answers on a separate piece of paper.

Animal Kingdom	Moneran Kingdom
cell	nucleus
cell membrane	organ
cell wall	organism
chlorophyll	photosynthesis
chloroplast	Plant Kingdom
chromosomes	Protist Kingdom
cilia	respiration
classification	ribosomes
cytoplasm	system
diffusion	tissue
Fungi Kingdom	vacuoles
mitochondria	vascular

1. __?__ is the process by which energy is released from food.
2. __?__ is the substance in a plant cell that absorbs energy from the sun during photosynthesis.
3. __?__ makes up the body of a cell.
4. Paramecia move as a result of the beating action of __?__.
5. Another name for a living thing is __?__.
6. A __?__ is the smallest unit able to carry on the life activities of an organism.
7. A __?__ is a group of specialized cells, which together carry out a specific function.
8. __?__ is the separation of objects into groups by similarities.
9. __?__ are storage areas in cells.

Vocabulary Review Answers

1. respiration
2. chlorophyll
3. cytoplasm
4. cilia
5. organism
6. cell
7. tissue
8. classification
9. vacuoles
10. mitochondria
11. cell membrane
12. cell wall
13. chromosomes
14. diffusion
15. Moneran or Protist Kingdom

10. Respiration takes place in structures called __?__ in cells.
11. The __?__ surrounds an animal cell and protects it.
12. The __?__ is the tough, outer-covering of a plant cell.
13. The __?__ are fibers containing instructions that control cell activities.
14. Some substances move through a cell membrane by __?__.
15. The __?__ is made up of one-celled organisms with cell membranes.

Chapter Review

Write your answers on a separate piece of paper.

Know The Facts

1. Respiration is a process that results in the production of __?__. (food, oxygen, carbon dioxide)
2. During photosynthesis, energy is absorbed by the substance called __?__. (chromosomes, chlorophyll, carbon dioxide)
3. __?__ is given off during photosynthesis. (carbon dioxide, oxygen, water)
4. Mushrooms are part of the __?__ Kingdom. (Moneran, Protist, Fungi)
5. __?__ are found in plant cells, but not animal cells. (vacuoles, mitochondria, chloroplasts)
6. Flowering plants are called __?__. (vascular plants, gymnosperms, angiosperms)
7. Amoebas get food by __?__. (surrounding it, beating their cilia, springing up and down)
8. A group of bone cells functioning together forms bone __?__. (tissue, organs, systems)
9. A __?__ is the largest division in a classification system. (species, kingdom, genus)
10. The scientific name of an organism is its __?__ and species name. (genus, order, kingdom)
11. __?__ do not have a membrane around their nucleus. (amoebas, bacteria, green algae)
12. A paramecium is an example of a __?__. (protist, fungus, moneran)
13. When a virus enters a cell, it causes the cell to __?__. (grow, produce more viruses, become inactive)
14. __?__ are given to people to protect them from disease-causing organisms. (antibodies, vaccines, vacuoles)
15. The __?__ phylum is made up of animals with hard shells and soft bodies. (moneran, protist, mollusc)
16. Animals have many cells and __?__. (eat other organisms for food, produce food by photosynthesis, produce oxygen by respiration)

Understand The Concepts

17. What are the characteristics a scientist uses to define life?
18. Draw a diagram of an animal cell. Label the parts. Explain the function of each part.
19. Draw a plant cell. Label the parts. Explain the function of each part.
20. Describe three structures and specialized areas found in the cytoplasm of a cell. What are their functions?

Chapter Review Answers

Know The Facts

1. carbon dioxide
2. chlorophyll
3. oxygen
4. Fungi
5. chloroplasts
6. angiosperms
7. surrounding it
8. tissue
9. kingdom
10. genus
11. bacteria
12. protist
13. produce more viruses
14. vaccines
15. mollusc
16. eat other organisms for food

Understand The Concepts

17. Movement, getting and using food, growth and repair of body parts, reproduction, response, having cells.
18. See diagram, page 17.
19. See diagram, page 17.
20. Ribosomes make proteins; vacuoles store materials; mitochondria release energy.

21. A paramecium moves by means of beating cilia. The cilia sweep tiny bits of food into the oral groove. An amoeba moves by streaming cytoplasm in fingerlike projections. Amoebas surround, or engulf, their food. A euglena moves by means of a whiplike flagellum. Euglenas contain chloroplasts; they can make their own food.
22. Cell, tissue, organ, system, and organism.
23. Respiration makes energy available from food. Photosynthesis stores energy in food. During respiration, carbon dioxide and water are given off; in photosynthesis, carbon dioxide and water are the starting materials. Oxygen is used in respiration and given off during photosynthesis.
24. Classification means separating objects into groups by similarities. Living things, books in a library, items in a supermarket.
25. Moneran, Protist, Fungi, Plant, and Animal. Examples will vary.
26. Viruses are tiny life forms that cannot grow, reproduce, or carry on respiration outside of a host cell. Vaccines prevent disease by stimulating the body to produce antibodies that destroy viruses.

Challenge Your Understanding

27. A life form would have cells and would carry out life activities. It would be classified based on its physical characteristics.
28. They have many similar characteristics and are able to mate and produce young, which can in turn produce young.
29. A five-kingdom system has more precise, clearly defined categories. Plants, animals, and organisms that are not clearly either plant or animal.
30. Root cells and cells in the flower do not carry on photosynthesis. Food is transported from other parts of the plant where food is either made or stored.

21. Compare the ways a paramecium, an amoeba, and a euglena move and get food.
22. List the levels of organization of a multi-celled organism, in order, beginning with a cell.
23. Describe the differences between photosynthesis and respiration.
24. Define classification. Give three examples of things that are classified.
25. Name the five kingdoms of organisms. Give an example of one in each kingdom.
26. Describe the characteristics of viruses. How do vaccines prevent viral diseases?

Challenge Your Understanding

27. If an unknown "thing" were discovered on a space expedition, how would scientists decide whether or not it was a form of life? If it was alive how would they go about classifying it?
28. All housecats do not look alike. Why are they in the same species?
29. What are the advantages of using a five-kingdom classification system instead of a two-kingdom system? Some biologists have used a three-kingdom system. What organisms do you think are in each of those kingdoms?
30. Which cells in a plant do not carry on photosynthesis? How do they get food?

Projects

1. Find out the scientific names of some organisms, such as a dog, mosquito, corn plant, and rhinoceros beetle. If possible, find out what the names mean and why the names were chosen.
2. Collect hydras, planarias, and other small organisms from a pond or stream. Find out how to feed and care for them. Observe the organisms through a microscope. Keep a record of your observations.
3. Do a report on fungi that are helpful and fungi that are harmful to people. Make a bulletin-board display.
4. Invite a representative from the local public health service to speak to your class about bacterial diseases, their cure, and how to prevent them.
5. Find out how penicillin was discovered and by whom.
6. Collect a variety of seashells. Sort them into groups. Use reference books or field guides to identify them. Label each with its common name and genus and species name. Display the shells in class.
7. Collect electron microscope photographs from magazines and newspapers. Prepare a bulletin-board display.
8. Some research scientists study life processes by growing and observing cells in a laboratory. Find out how cell cultures are maintained and what scientists can learn by studying them.
9. Research and report on the contributions to biology of any one of the following scientists: Libbie H. Hyman; Katherine Esau; Miriam Louis A. Rothschild.

1 INTRODUCTION TO LIFE
Projects

1. Research the mitochondria.
   a. How is this "power house" of the cell constructed? ______
   b. What is its work in the cell? ______
   c. How does it release energy? ______
2. How would you classify a human being? Use a biology book or your library. Capitalize all but the species name.
   kingdom ______
   phylum ______
   class ______
   order ______

Chapter 2

Heredity

Offspring resemble their parents because many of their traits are inherited.

Vocabulary Preview

traits
mitosis
replicate
reproduction
asexual reproduction
spores
sexual reproduction
gametes
meiosis
fertilization
zygote
pollination
gene
dominant gene
recessive gene
pure trait
hybrid trait
sex chromosomes
sex-linked trait
DNA
mutation

CHAPTER 2 Overview

The purpose of this chapter is to introduce students to basic concepts of heredity. The passage of inherited traits from generation to generation is described. Then mitosis and several forms of asexual reproduction are introduced. The processes of asexual and sexual reproduction are then compared, and meiosis is described and related to sexual reproduction. Next come the inheritance of dominant and recessive traits and their relationship to genes and chromosomes. The sex chromosomes and sex-linked traits are then described. The structure and role of DNA in inheritance follow, and the nature and effects of mutations are introduced. Finally, the improvement of plant and animal characteristics by selective breeding is described.

Goals

At the end of this chapter, students should be able to:

1. identify traits that are passed from generation to generation.
2. describe the process of mitosis.
3. describe several methods of asexual reproduction.
4. describe how sexual reproduction differs from asexual reproduction.
5. explain the function of meiosis.
6. describe reproduction in flowering plants.
7. describe the inheritance of dominant and recessive traits.
8. describe the function of genes.
9. explain the difference between a pure trait and a hybrid trait.
10. identify the sex chromosomes and explain the inheritance of sex-linked traits.
11. describe the structure of DNA.
12. identify the role of DNA in inheritance.
13. identify what mutations are and how they occur.
14. describe how people have developed breeds of plants and animals with desirable traits.

Teaching Suggestions 2-1

Introduce the concept of inherited traits by asking students to study the photo of plants in Figure 2-1. Discuss the similarities and differences among the plants. Point out that individual leaves may show slight differences, but the overall shape and structure of the leaves are similar.

Ask students to think of several common breeds of dogs. Have them suggest some traits that dogs inherit from their parents. Tell students that the passage of inherited traits from parents to offspring is known as heredity.

Point out that inherited traits are not limited to eye color, body shape, and other obvious characteristics. Inherited traits are responsible for an organism's overall structure and physiology. Moreover, a particular structure and physiology are unique to a particular species. This is so because each species has a unique set of chromosomes that differs from the sets of chromosomes found in the body cells of all other species of organisms.

Tell students that almost all cells in the mature human body contain 46 chromosomes. An exception are fully-developed sex cells—egg cells in females and sperm cells in males. Some students may notice that in males, the chromosomes in pair 23 (the XY, or sex chromosomes) are not identical. Explain that this pair of chromosomes will be discussed later.

Background

The two types of cells found in multicelled organisms that reproduce sexually are somatic, or body, cells and germ, or reproductive, cells. Egg and sperm cells are the germ cells. Heredity is solely the function of the germ cells, but the results of inheritance are usually apparent only in somatic cells.

Somatic cells contain the diploid ($2n$), or full, number of chromosomes that characterizes a particular species. The chromosome numbers in Table 2-1 are diploid numbers. Sperm and egg cells, which contain the haploid (n) number of chromosomes, form when special somatic cells undergo a reduction division.

Teaching Tips

- Some students may raise the question of inheritance of acquired characteristics. Discuss the difference between inherited and acquired characteristics, such as hair texture and color versus docked tails and/or

2-1

From Generation to Generation

Goal

To identify traits that are passed from generation to generation.

Look at the photograph of the plants in Figure 2-1. Pay particular attention to the shapes of the leaves, and the colors and shapes of the flowers. What similar features do you see? What differences do you see?

traits

Specific features of an organism, such as the shape and texture of a leaf or the structure of a flower, are called **traits** or characteristics. Traits that are passed from parents to their offspring are called *inherited traits*. Offspring resemble their parents because many of their traits are inherited.

Agricultural scientists often try to improve the traits of a species of farm animals or crops. They try to produce crops that have a better taste or that are more convenient. For example, the navel orange was bred to be seedless and the tangelo was developed for its sweet flavor and thin, easy-to-peel skin.

Information about every inherited trait is contained in the chromosomes. Remember that chromosomes are structures in the nucleus of each cell. The chromosomes function

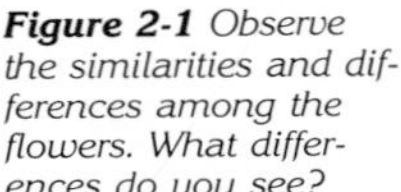

Figure 2-1 *Observe the similarities and differences among the flowers. What differences do you see?*

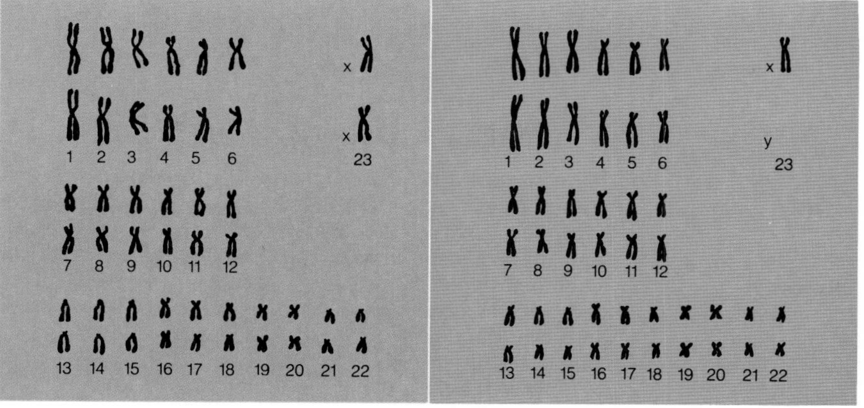

Figure 2-2 *The chromosomes of a human female (left) and a human male (right). The 46 chromosomes have been arranged in 23 pairs.*

like an architect's plan of a building. The plan tells where floors, walls, plumbing, and electric wires are to be located. It also specifies the building's shape and appearance. Like a building's plan, the chromosomes contain information that specifies the shape and structure of an organism. Your chromosomes contain information that controls inherited traits, such as eye color, height, shape of the mouth, and hair color.

Each species of organism has a certain number of chromosomes. Table 2-1 lists the number of chromosomes for some different kinds of organisms. Figure 2-2 shows the 46 human chromosomes arranged in 23 pairs.

Table 2-1: Chromosome Numbers

Organism	*Number*
mosquito	6
housefly	12
corn	20
hydra	32
human	46
potato	48
dog	78

Checkpoint

1. Where in each cell is information about inherited traits contained?
2. Name three inherited traits in people.
3. How many chromosomes are in human cells?

TRY THIS

One inherited trait is tongue-rolling, or the ability to roll the tongue into a U-shape. Find out how many members of your class have the tongue-rolling trait. Are there more rollers or nonrollers? Which appears to be the more common trait?

2 HEREDITY

2-1 From Generation to Generation

Identifying Characteristics

Complete each sentence below by unscrambling the letters that appear below the answer line.

1. Specific features of an organism are known as ______________. (saritt)
2. When traits are passed from parent to offspring, they are said to be ______________. (deethinir)
3. ______________ (smochsmoore) are structures in the nucleus of each cell that contain information about each inherited trait.
4. The ______________ (brunem) of chromosomes varies for each organism.
5. There are 46 human chromosomes arranged in 23 similar ______________. (rapis)

Of Peas and Grasshoppers

The work of Gregor Mendel is described in the paragraphs below. Read the paragraphs. Then write the answers to the questions that follow.

Gregor Mendel was an Austrian monk trained in mathematics and biology. As Mendel cared for the gardens of the monastery, he wondered how certain characteristics of sweet peas were passed on from generation to generation. Mendel studied pea traits such as seed shape, seed color, and pod shape. He bred and crossbred hundreds of pea plants to get round and wrinkled seeds as well as tall and short plants. He pollinated the plants by hand and kept careful records of his results.

Mendel believed that there were factors that controlled the traits in the peas he studied. But he did not know where the traits were located or how they were passed on. In 1865, Mendel wrote a research paper that described what happened in peas when he crossed them. At the time, no one seemed to notice his results. Some thirty years later, Walter S. Sutton, a graduate student who was working on grasshopper chromosomes, realized that Mendel's theories fit with his own. Sutton realized that Mendel's factors must be small particles, or genes, located on the chromosomes.

1. What did Gregor Mendel study? ______________
2. What were Mendel's factors? ______________
3. What are genes? ______________

Using Try This

The ability to roll the tongue into a U-shape appears to be a dominant trait in most populations. Assure nonrollers that there is no known correlation between the ability to roll the tongue and any other skill or capability.

Checkpoint Answers

1. This information is contained in the chromosomes within the nucleus of each cell.
2. Eye color, height, hair color, shape of mouth, and so on. Answers will vary.
3. Human body cells normally contain 46 chromosomes.

cropped ears in certain breeds of dogs. Point out that acquired characteristics are not inherited.

- Be sure during discussion of inherited traits that you are sensitive to students who are adopted or are from single-parent families.
- The PTC-tasting activity often found in chapters on genetics is not included in this text. Recent evidence implicates phenylthiourea as a carcinogen in rats. Although this substance does not appear to be a human carcinogen, there are no conclusive studies to document this.

Teaching Suggestions 2-2

Open by asking students if they know how growth in their bodies occurs. Explain that they began life as a single cell, and that as a result of individual cells dividing into two over and over again, their bodies now consist of millions of cells. Point out that the process of cell division includes mitosis, and that during mitosis the nuclear content of a cell is distributed equally to each new cell formed. Make sure students grasp the difference between cell division and mitosis.

Have the class follow in Figure 2-3 as you discuss each stage of mitosis. Be sure students understand clearly what occurs at each stage. Emphasize that mitosis is a continuous process. Each stage merges into the next. Students may find it helpful to think of the process as being similar to going around a circular track.

Laboratory Activity 4, Observing Onion Root-Tip Cells, student text page 536, should be conducted in conjunction with this section.

Background

Though mitosis is a continuous process, it is studied as if it were divided into four stages: prophase, metaphase, anaphase, and telophase. When cells are not undergoing mitosis, they are said to be in interphase.

During interphase, the chromosomes cannot be seen individually, although when stained the genetic material appears as tiny grains. It is during interphase that DNA is replicated.

The chromosomes gradually contract and coil up during prophase, until they appear relatively thick and short. At this time, each chromosome is composed of two chromatids attached at the centromere. Each chromatid is a complete and exact copy of the original chromosome. The chromosomes then begin to align themselves along the equator, or middle, of the cell.

The spindle microtubules begin to form during prophase. Soon spindle microtubules stretch across the cell, connecting the two poles. Other spindle microtubules attach themselves to the centromeres. The nuclear membrane disappears at the end of the prophase stage of mitosis.

By metaphase, the chromosomes are lined up at the equator, the spindle has formed, and the nuclear membrane and nucleoli have disappeared.

During anaphase, the centromeres separate and move with the chromosomes along the spindle microtubules toward their respective poles. These two groups of chromosomes will become the daughter cells' nuclei.

By telophase, the chromosomes have reached the poles and have begun to move back into their interphase state. The spindle microtubules have largely

2-2

Mitosis

Goal

To describe the process of mitosis.

All new cells arise from old ones. When you cut yourself, new skin cells grow, replacing the damaged ones. Oak trees grow new leaves each spring. A fish begins life as one cell and grows to many. Each new cell must have a complete set of chromosomes. **Mitosis** is the process by which the chromosomes in a cell duplicate themselves and then separate into two identical sets. The cell then divides, forming two identical cells. Mitosis occurs in recognizable stages. Figure 2-3 shows the stages of mitosis.

mitosis (my TOH sis)

replicate

The chromosomes **replicate**, or copy themselves in the period just *before* mitosis begins. Each chromosome attaches to its copy, forming a two-part chromosome. (See Figure 2-3a and Figure 2-3b.)

In the first stage of mitosis the nuclear membrane surrounding the chromosomes disappears. (See Figure 2-3c.) A series of thin strands, called *spindle fibers*, begins to form. The chromosomes take on a thicker appearance.

Figure 2-3 *Mitosis is a continuous process. It does not stop and start between stages. Mitosis can occur in minutes or several hours.*

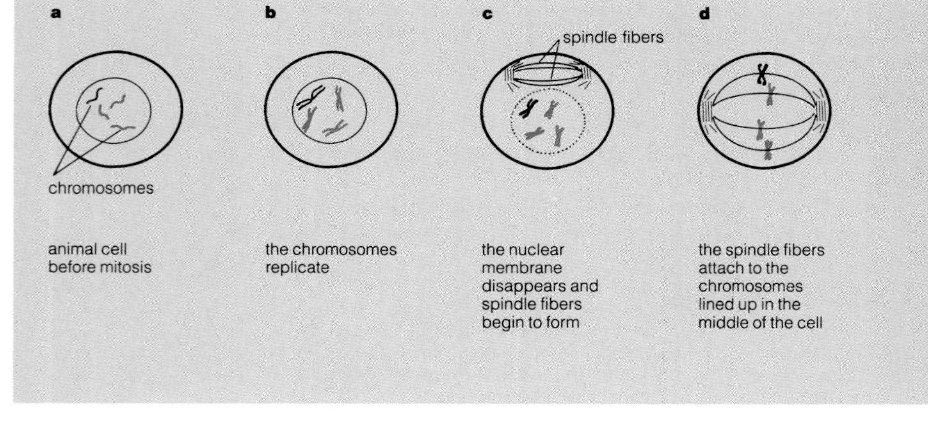
a
b
c
d
spindle fibers
chromosomes
animal cell before mitosis
the chromosomes replicate
the nuclear membrane disappears and spindle fibers begin to form
the spindle fibers attach to the chromosomes lined up in the middle of the cell

The second stage of mitosis begins with the chromosomes lining up in the middle, or *equator*, of the cell. The spindle fibers attach to the chromosomes, as shown in Figure 2-3d.

During the third stage, the two-part chromosomes separate and are pulled by the spindle fibers to opposite ends of the cell. At the end of this stage, there is an identical set of chromosomes at opposite ends of the cell. This stage is shown in Figure 2-3e.

The spindle fibers disappear in the final stage of mitosis. A new nuclear membrane forms around each group of chromosomes. Mitosis is usually followed by *cell division*. The cell splits in half along the equator, producing two identical cells. (See Figure 2-3f and Figure 2-3g.)

Mitosis and cell division are continuous processes. There are no stops and starts between different stages. The processes ensure that cell characteristics will be maintained from one cell generation to the next.

Checkpoint

1. What is mitosis?
2. When do the chromosomes in a cell replicate?
3. Draw and label the stages of mitosis, in order.
4. What happens during the last stage of mitosis?

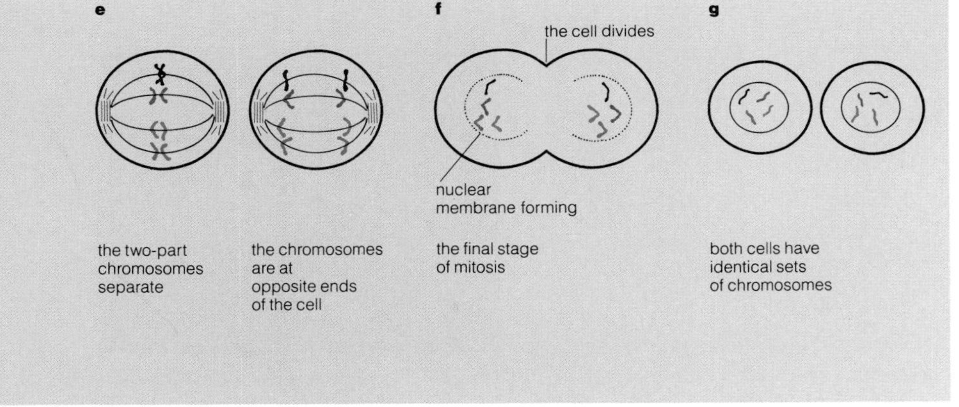
e
f
g
the cell divides
nuclear
membrane forming
the two-part chromosomes separate
the chromosomes are at opposite ends of the cell
the final stage of mitosis
both cells have identical sets of chromosomes

disappeared, and nuclear membranes have formed around the chromosomes at each pole. Finally, division of the cell into two daughter cells begins.

Plant and animal cells divide differently. In an animal cell, after mitosis, the furrow at the center of the dividing cell deepens until the cell splits in two. In a plant cell, a new cell wall appears along the metaphase plate.

Mitosis is often incorrectly used as a synonym for cell division. Mitosis refers only to the duplication and separation of nuclear material. Cell division is the physical separation of this newly separated material into two identical daughter cells.

Facts At Your Fingertips

- Most cellular biochemical research is done on cells during interphase because the chromosomes are uncoiled and spread out. This allows easier access to the genetic material. Also, this is the time when the genetic material is replicating, which allows study of that process.
- Metaphase is the best time to count chromosomes because they are lined up at the equator of the cell.
- The spindle "fibers" are now known to be microtubules.

Mainstreaming

The use of a three-dimensional model of mitosis will be helpful to visually-impaired students. Reinforce verbal descriptions of the stages of mitosis by encouraging visually-impaired students to feel the model of mitosis.

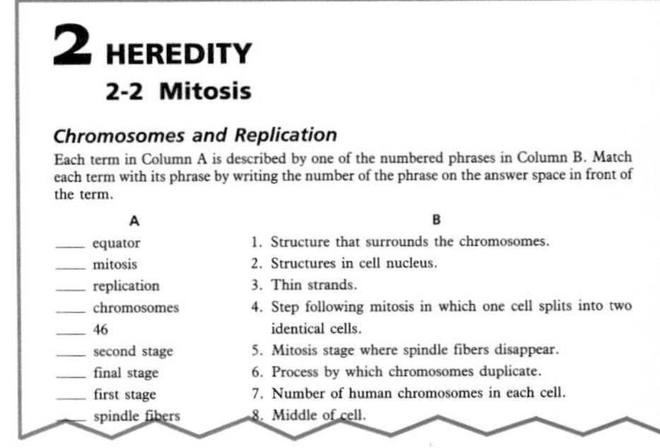

2 HEREDITY

2-2 Mitosis

Chromosomes and Replication

Each term in Column A is described by one of the numbered phrases in Column B. Match each term with its phrase by writing the number of the phrase on the answer space in front of the term.

A	B
____ equator	1. Structure that surrounds the chromosomes.
____ mitosis	2. Structures in cell nucleus.
____ replication	3. Thin strands.
____ chromosomes	4. Step following mitosis in which one cell splits into two identical cells.
____ 46	5. Mitosis stage where spindle fibers disappear.
____ second stage	6. Process by which chromosomes duplicate.
____ final stage	7. Number of human chromosomes in each cell.
____ first stage	8. Middle of cell.
____ spindle fibers	

Checkpoint Answers

1. Mitosis is the process by which the nuclear material duplicates and then separates into two identical sets.
2. The chromosomes in a cell replicate just before mitosis begins.
3. Drawings will vary. The order of the drawings should correspond with the order shown in Figure 2-3.
4. The spindle fibers disappear and new nuclear membranes form. Cell division usually follows.

Teaching Suggestions 2-3

Introduce the section by asking students for a definition of reproduction. Discuss all answers. Help students arrive at a definition that will apply to both sexual and asexual reproduction. Stress that in asexual reproduction only one parent is needed. Use the text illustrations to define and give examples of fission, budding, regeneration, and spore formation. Give examples of common plants that reproduce by regeneration. (See Background.)

Explain to students that spore formation is common in plants and protists, but not in animals. Point out that spores are not the same as seeds. Spores are far simpler than seeds. A seed contains an embryo and food to nourish the embryo, and is covered by a seed coat.

Chapter-end projects 4 and 6 apply to this section.

Background

Asexual reproduction is usually faster and less costly in terms of energy than sexual reproduction. Asexual reproduction occurs in organisms that produce large numbers of offspring to take advantage of rare and widely dispersed habitats. It also occurs in organisms that reproduce quickly to exploit a temporarily favorable environment.

Regeneration occurs in many plants. For example, many ferns produce new plants from underground stems. Spider plants, strawberry plants, and aquarium eel grass reproduce from special stems called runners. Onion, narcissus, and hyacinth plants reproduce from bulbs. Jack-in-the-pulpit, crocus, and gladiolus reproduce by means of corms. White potatoes and dahlias can be grown from buds on tubers, and carrots can be grown from fleshy roots.

A drawback to asexual reproduction is that it does not promote genetic diversity. Some diversity is needed to adapt to changing environments. Mutation is the only means of variability in asexual organisms, such as amoebas.

2-3

Asexual Reproduction

Goal

To describe several methods of asexual reproduction.

reproduction

asexual reproduction

Inherited traits are passed from parents to offspring as a result of reproduction. **Reproduction** is the process by which organisms produce new organisms. One type of reproduction is asexual reproduction. Through **asexual reproduction**, new organisms are produced by a single parent. The offspring is an exact copy of the parent.

Most one-celled organisms reproduce by a kind of asexual reproduction called fission. During *fission*, the parent cell undergoes mitosis and cell division, resulting in two identical *daughter cells*. (See Figure 2-4.)

Hydras, and other relatives of the jellyfish, reproduce by budding. During *budding*, cells in the side of an adult hydra form a young hydra, called a *bud*. (See Figure 2-5.) The bud is a copy of the parent. When it is large enough, the bud separates from the parent. Buds may go on to become parent cells themselves.

Some organisms can reproduce by a process known as regeneration. *Regeneration* is the ability to replace missing parts, or a whole new organism from a small body part. Planaria, starfish, and many plants can reproduce by regeneration. An example is shown in Figure 2-6.

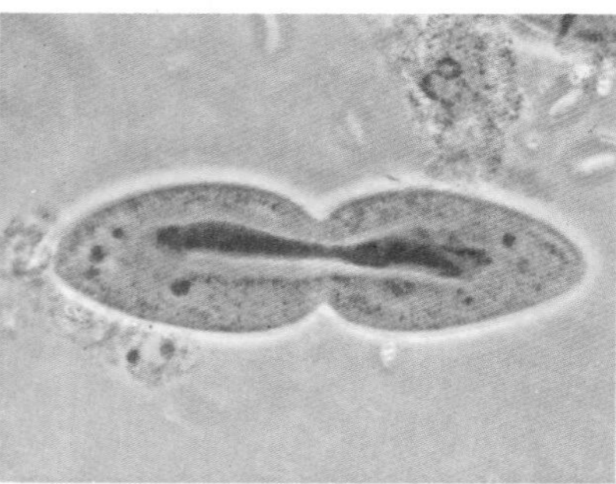

Figure 2-4 *A paramecium undergoing fission. Fission results in two identical cells.*

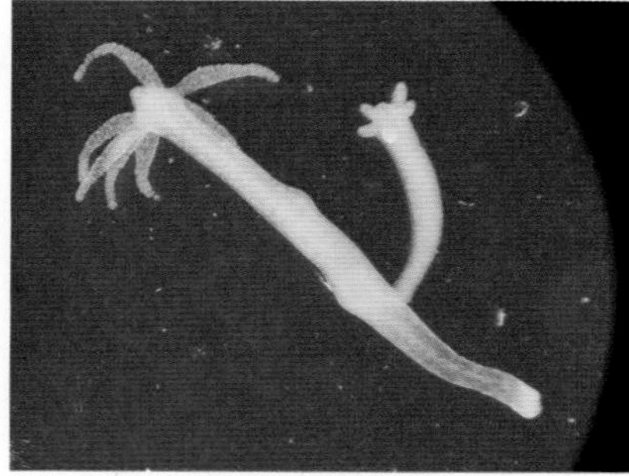

Figure 2-5 *Hydras often reproduce by budding. The young hydra is a copy of the parent.*

Demonstration

Show students some examples of asexual regeneration in plants. Place the new plantlets at the tip of a spider plant *(Chlorophytum)* runner on moist soil in another pot. Each plantlet will take root and grow into a mature plant. Establish a snake plant *(Sansevieria)* in a large pot. The plant will send out rhizomes, which are underground stems, and eventually new plants will develop from the rhizomes. Suspend the root end of an onion just below the surface of water in a jar. Support the onion on the rim of the jar. A good root system will develop in about a week. Using toothpicks for support, suspend the top portion of a carrot or white turnip in water. New shoots will appear quickly.

Figure 2-6 *This starfish is regenerating an arm. Some starfish can regenerate a whole new body from a small piece of one arm.*

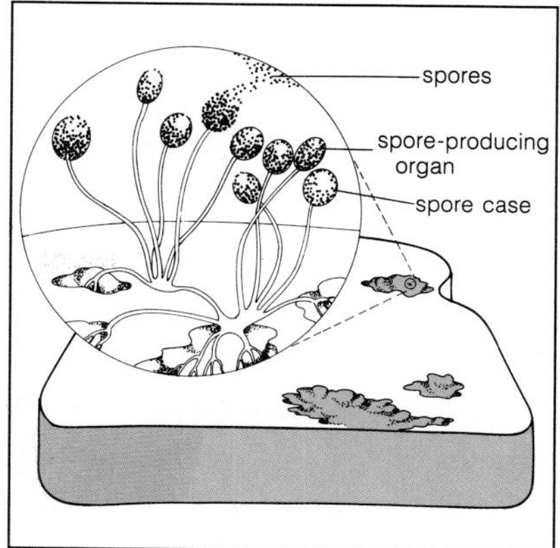

Figure 2-7 *When a spore case breaks open, millions of spores are released. Spores grow only when they are in the right surroundings.*

Bread molds and mushrooms form reproductive cells called **spores**. Millions of spores are produced and stored in *spore cases*. (See Figure 2-7.) When the spore cases break open, the spores are released. Each spore is protected by a tough covering. If a spore lands in an area with favorable conditions, the covering dissolves and the spore develops into a new organism. Spores can remain *dormant*, or inactive, for a long time until "activated" by the right conditions.

spores

Checkpoint

1. Define asexual reproduction.
2. Describe the differences between fission and budding. Give examples of each.
3. Describe how a bread mold reproduces.

TRY THIS

Carefully rub a piece of fresh bread on the floor, window, or desktop. Moisten the bread and place it inside a plastic bag. Seal the bag and put it in a dark place for several days. Moisten it from time to time if it dries out. Watch for the appearance of bread mold. CAUTION: Be careful not to touch the molds.

Facts At Your Fingertips

- Planaria and starfish can regenerate whole sections of their bodies.
- If a leg is removed from a salamander, a new leg will form. The new leg is a perfect duplicate of the lost one, in both structure and function.
- The human body has some regenerative powers. Skin wounds heal by regeneration of epidermis and dermis tissues. Some other organs too can partially regenerate. The tongue has good regenerative capacity, and the liver will regenerate to normal size when portions have been removed surgically.

Cross Reference

Refer to Chapter 1, section 1-7, for further discussion of the kinds of organisms mentioned in this section.

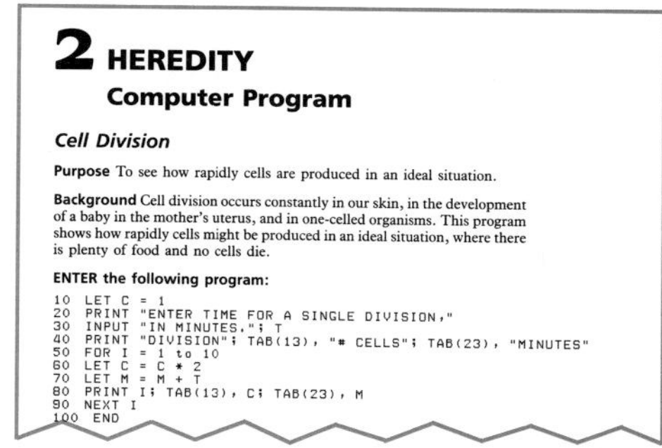

2 HEREDITY

Computer Program

Cell Division

Purpose To see how rapidly cells are produced in an ideal situation.

Background Cell division occurs constantly in our skin, in the development of a baby in the mother's uterus, and in one-celled organisms. This program shows how rapidly cells might be produced in an ideal situation, where there is plenty of food and no cells die.

ENTER the following program:

```
10 LET C = 1
20 PRINT "ENTER TIME FOR A SINGLE DIVISION,"
30 INPUT "IN MINUTES."; T
40 PRINT "DIVISION"; TAB(13), "# CELLS"; TAB(23), "MINUTES"
50 FOR I = 1 to 10
60 LET C = C * 2
70 LET M = M + T
80 PRINT I; TAB(13), C; TAB(23), M
90 NEXT I
100 END
```

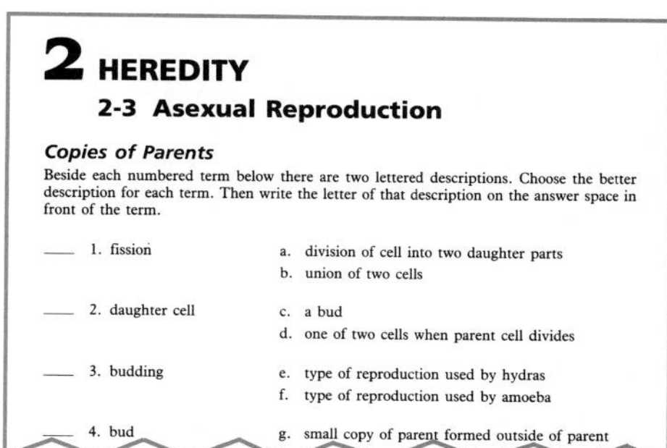

2 HEREDITY

2-3 Asexual Reproduction

Copies of Parents

Beside each numbered term below there are two lettered descriptions. Choose the better description for each term. Then write the letter of that description on the answer space in front of the term.

___ 1. fission
a. division of cell into two daughter parts
b. union of two cells

___ 2. daughter cell
c. a bud
d. one of two cells when parent cell divides

___ 3. budding
e. type of reproduction used by hydras
f. type of reproduction used by amoeba

___ 4. bud
g. small copy of parent formed outside of parent

Using Try This

CAUTION: Avoid opening the plastic bag containing the bread mold. Some students may be allergic to the spores and should not be exposed to them.

Checkpoint Answers

1. Asexual reproduction is the production of new organisms by a single parent.
2. In fission, a cell undergoes mitosis and cell division. In budding, a new organism grows as a bud off the parent organism. Examples will vary.
3. Bread mold forms spores. The spores are released. If one lands in a place suitable for growth, a new mold organism will develop.

Teaching Suggestions 2-4

Start by reviewing the definition of asexual reproduction. Point out that sexual reproduction differs in that the new organism is the result of the union of two cells, the male and female gametes. Where two parents are involved, the offspring show characteristics of both parents. Ask students to describe the ways they resemble their mothers and/or fathers. Take care to consider that some students may be adopted or from single-parent families and that these students may have no knowledge of one or both of their biological parents.

Refer students to Figure 2-8 and describe the process of meiosis stage by stage. Point out that both sperm and egg parent cells contain the diploid, or $2n$, number of chromosomes, and that sperm and egg cells, which arise from two reduction divisions after the chromosome replication shown in Figure 2-8, contain the haploid, or n, number of chromosomes. Explain again that in both sexual and asexual reproduction the body cells of adult organisms always contain the diploid number of chromosomes.

Using a typical flowering plant as an example (see Demonstrations), describe the structures and processes that occur in sexual reproduction. Emphasize the basic event that occurs—the union of an egg with a sperm cell. Point out that this is what occurs in all plants and animals that reproduce sexually.

Background

Meiosis is similar to mitosis except that it has two successive cell divisions and only one replication of the genetic material. Spermatogenesis produces four sperm cells, oogenesis one ovum, or egg cell. The other three cells, called polar bodies, produced during oogenesis are discarded. All their cellular material goes to the one ovum. This material supports the embryo during the first stages of development after fertilization. An animal sperm cell consists of three regions—the head, the mitochondrial sheath, and the tail.

Fertilization of an egg cell by a sperm cell takes very little time. After the sperm penetrates the egg, the egg releases proteins that change the egg's coating into the fertilization membrane. This prevents the entry of additional sperm cells. These events take place within 60 seconds after the sperm cell contacts the egg cell.

Plant fertilization, following pollination, is quite different from animal fertilization. Egg cells are stored in ovules, which are found in an ovary at the base of the pistil. A pollen grain containing two sperm nuclei must attach to the stigma and then grow a pollen tube down through the pistil to the ovule. The sperm nuclei are released into the ovule when the pollen tube breaks open. One sperm nucleus fuses

2-4

Sexual Reproduction

Goals

1. To describe how sexual reproduction differs from asexual reproduction.
2. To explain the function of meiosis.
3. To describe reproduction in flowering plants.

sexual reproduction

gametes (GAM eetz)

meiosis (my OH sis)

Sexual reproduction is the creation of a new organism by joining one male sex cell with one female sex cell. Male sex cells, or **gametes**, are called *sperm* and female gametes are called *eggs*. Sexual reproduction occurs in many plants and animals, including birds, mammals, fish, and flowering plants. Through sexual reproduction parents contribute chromosomes (in the sex cells) to offspring. Thus, offspring show characteristics of their parents.

Meiosis is a kind of cell division that results in cells with half the usual number of chromosomes. Gametes are produced through meiosis. All gametes have half the number of chromosomes as other cells in an organism. You may remember that humans have 46 chromosomes in all their

Figure 2-8 *Meiosis occurs only in egg and sperm parent cells. Each sperm or egg cell has half the number of chromosomes as the parent. This sequence shows meiosis occurring in a sperm parent cell.*

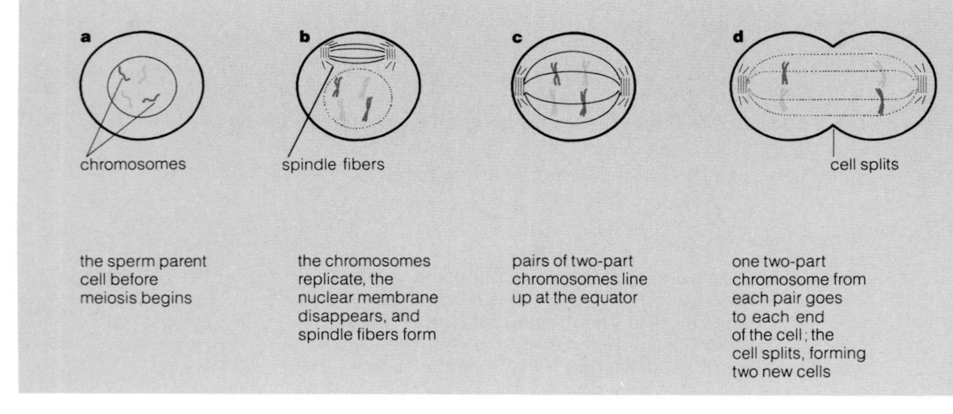
a
b
c
d
chromosomes
spindle fibers
cell splits
the sperm parent cell before meiosis begins
the chromosomes replicate, the nuclear membrane disappears, and spindle fibers form
pairs of two-part chromosomes line up at the equator
one two-part chromosome from each pair goes to each end of the cell; the cell splits, forming two new cells

body cells. The sex cells have only 23 chromosomes. Because of meiosis, when two sex cells join, the offspring have the same number of chromosomes as their parents.

Meiosis occurs only in cells that produce the sex cells. In males, sperm parent cells undergo meiosis and form sperm cells. (See Figure 2-8.) During meiosis, each sperm parent cell divides twice, producing four sperm cells. Each sperm cell has half the number of chromosomes as the parent cell.

In females, the egg parent cell undergoes meiosis. The parent cell divides twice, creating four cells. Only one cell becomes an egg. The other cells are discarded by the body. The egg cell contains half the number of chromosomes as the parent cell.

During sexual reproduction, a sperm cell joins an egg cell by a process known as **fertilization**. Fertilization is complete when the egg nucleus and sperm nucleus fuse. The fertilized egg is called a **zygote**. Following the "blueprint" in its chromosomes, the zygote undergoes mitosis many times, forming a multi-celled organism. After numerous cell divisions, the zygote is called an *embryo* (EM bree oh). Cells in the embryo grow and develop into specialized organs.

fertilization

zygote (ZY goht)

You can study an example of sexual reproduction by observing the process in flowering plants. Flowers contain the reproductive organs of flowering plants. Locate each part on the flower shown in Figure 2-9.

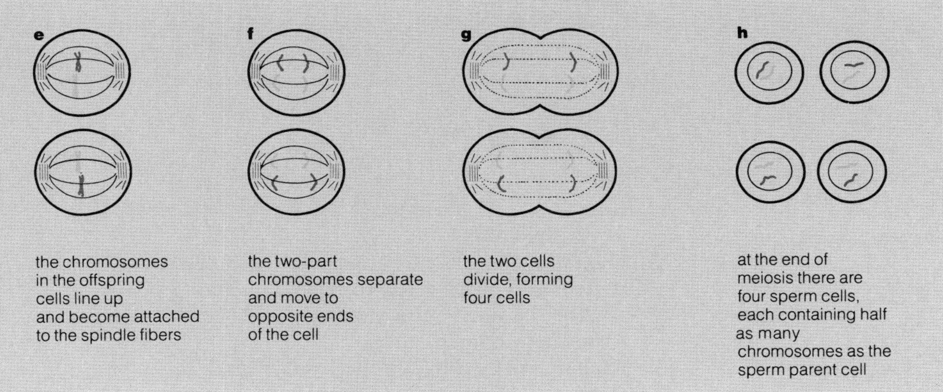

with the egg nucleus. The other sperm nucleus fuses with the polar nuclei, thus forming the primary endosperm nucleus. This double fertilization occurs only in flowering plants.

e
f
g
h
the chromosomes in the offspring cells line up and become attached to the spindle fibers
the two-part chromosomes separate and move to opposite ends of the cell
the two cells divide, forming four cells
at the end of meiosis there are four sperm cells, each containing half as many chromosomes as the sperm parent cell

The ovule with the fertilized egg, or zygote, then develops into a dormant seed. The developing plant embryo is nourished by surrounding active cells, rather than by yolk that nourishes many animal embryos.

Demonstrations

1. Show students the structural features of a typical flower. Use flowers that have both stamens and pistils. Distribute them. Have the students follow as you go through the anatomy of the flower. See Figure TE 2-1. You might want to do this demonstration in conjunction with the initial discussion of the flower as the reproductive structure of flowering plants.

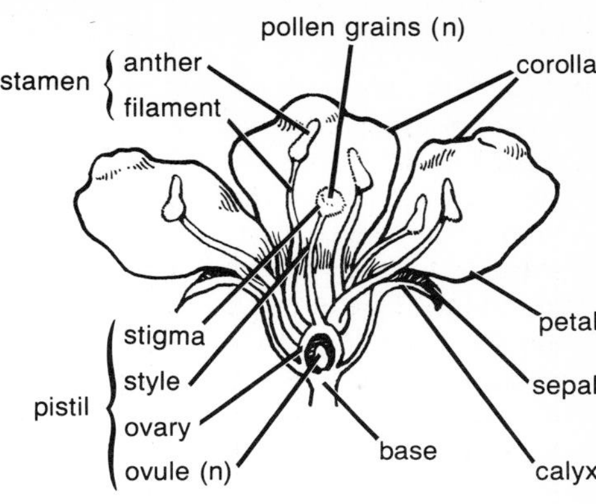

Figure TE 2-1 The anatomy of a flower.

2. Show students the primary structures in a typical seed. A seed is essentially an embryo plant surrounded by other tissues. The embryo develops from the zygote. Obtain enough large dried beans for the entire class. Soak the beans for several hours, then distribute them. Have the students carefully separate the two halves of the seed. Point out the embryo, the two cotyledons, which contain food for the young plant until it is established, and the protective seed coat. Sketch Figure TE 2-2 on the chalkboard.

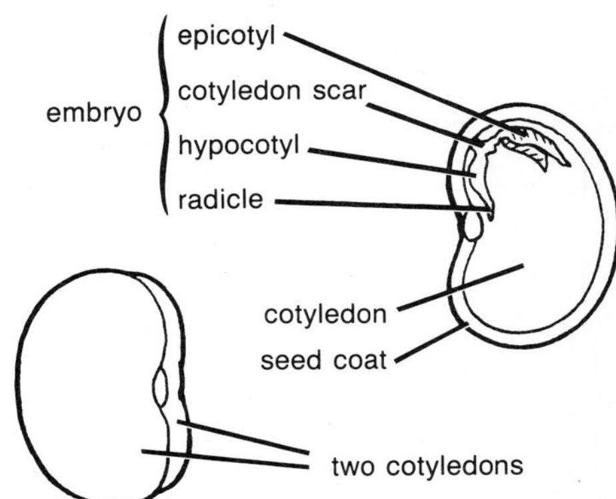

Figure TE 2-2 The anatomy of a seed.

3. Show students the relationship between the seeds and the other tissues in several fruits. Apples, tomatoes, and oranges are usually available all year long.

Teaching Tips

- Students may be interested in how seeds are dispersed. Several mechanisms exist. Maple seeds have wings and are carried by the wind. Dandelion seeds have a sort of feathery parachute; they too are carried by the wind. Cocklebur seeds have hooks that cling to the bodies of animals or to clothing. Birds disperse the seeds of grapes, apples, and cherries as they carry the fruits off for food. Squirrels carry nuts and acorns quite far before they bury them.
- The functions of the fruit in flowering plants are in general not well understood. These are (1) providing moisture and mineral matter for the germinating seed, (2) protecting against fungus and insect enemies, (3) reducing water loss in seeds, and (4) aiding seed dispersal.
- Some students will be more knowledgeable than others about sexual reproduction in animals, including humans. A sensitivity to student readiness and knowledge of community mores and standards will determine to what extent sexual reproduction can be discussed.

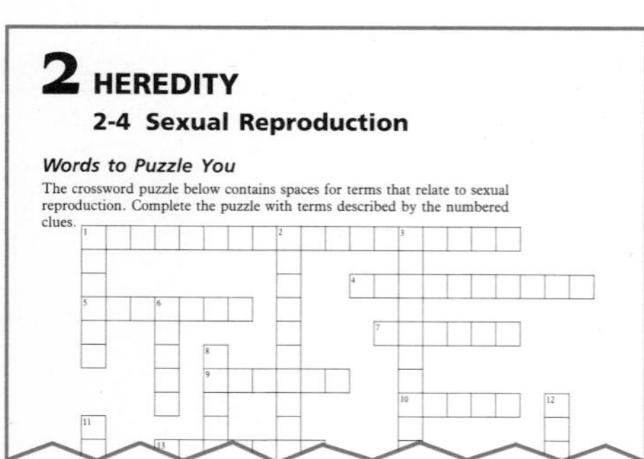

2 HEREDITY

2-4 Sexual Reproduction

Words to Puzzle You

The crossword puzzle below contains spaces for terms that relate to sexual reproduction. Complete the puzzle with terms described by the numbered clues.

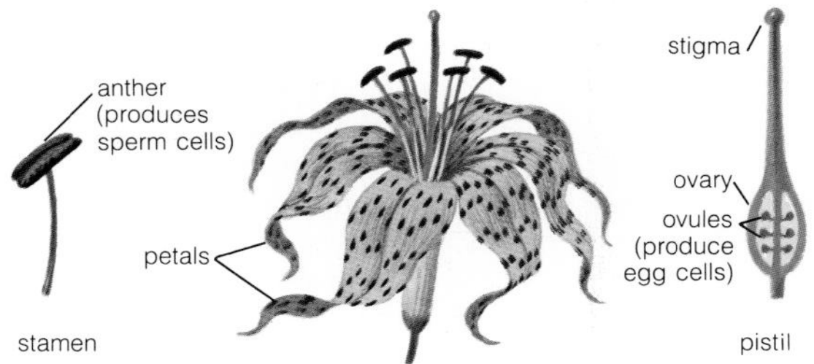

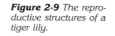

Figure 2-9 *The reproductive structures of a tiger lily.*

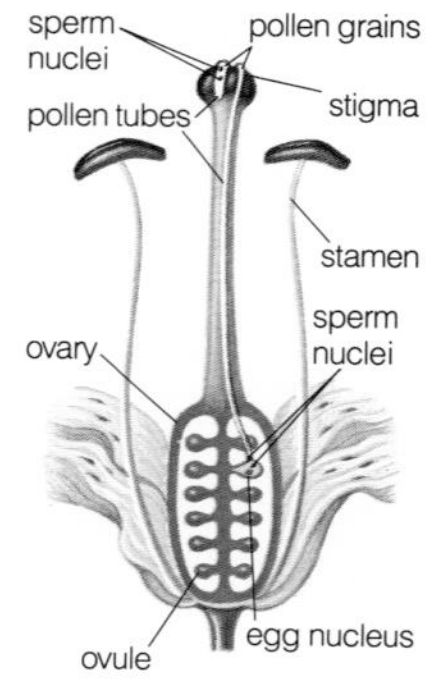

Figure 2-10 *Pollination makes fertilization of the egg possible.*

pollination

Figure 2-11 *A fruit is the ripened ovary of a plant. It may contain one or more seeds.*

The male reproductive structure is called the *stamen* (STAY mun). *Pollen grains* are produced by meiosis in the top of the stamen, called the *anther*. The sperm cells are contained in pollen grains. The female reproductive structure is called the *pistil*. Egg cells are formed by meiosis in the parts of the pistil called the *ovules* (OH vyoolz).

In order for fertilization to occur, a pollen grain (containing sperm) must be transported from the stamen to the pistil. This process is known as **pollination**. Wind or insects carry pollen to the sticky top of the pistil, called the *stigma*. (See Figure 2-10.) Each grain grows a *pollen tube* from the top of the stigma to the *ovary*. The sperm nuclei in the pollen grain travel down the tube and into an ovule. Fertilization occurs when a sperm nucleus and egg nucleus fuse.

Tissues in the ovule form a food-storage layer and a protective covering, or *seed*, around the zygote. The ovary develops into a fruit around the seed. (See Figure 2-11.) A seed remains dormant until it is transported by wind, water, or animals to a place where conditions are right for growth. When a seed *germinates*, the life cycle begins again.

Checkpoint

1. How is sexual reproduction different from asexual reproduction?
2. What are gametes? What are male gametes called? What is the name given to female gametes?
3. What is the function of meiosis? How is meiosis different from mitosis?
4. Explain the process of pollination and the steps leading to seed-formation in a flowering plant.

Checkpoint Answers

1. Sexual reproduction is reproduction by the union of two sex cells, or gametes. Asexual reproduction does not involve the union of a male and a female sex cell.
2. Gametes, or sex cells, are the reproductive cells of organisms that reproduce sexually. Male gametes are called sperm cells. Female gametes are called egg cells.
3. Meiosis is division of the nuclear material of a cell so as to produce gametes with half the number of chromosomes of body cells. Mitosis is division of the nuclear material so as to produce the same number of chromosomes in daughter cells as in parent cells.
4. Answers will vary somewhat. The basic events are: pollination is transfer of a pollen grain to the top surface of a pistil; a pollen tube then grows to the ovary at the base of the pistil; a sperm nucleus enters the ovule and fuses with an egg nucleus. The zygote thus formed develops into the seed.

2-5

Dominant and Recessive Traits

Goal

To describe the inheritance of dominant and recessive traits.

In the mid-1800's an Austrian monk named Gregor Mendel examined inherited traits in pea plants. Mendel discovered that certain traits occurred in two forms. He studied the way certain trait pairs, such as yellow versus green pods and round versus wrinkled peas, were inherited. Figure 2-12 shows some of the traits Mendel observed. Mendel was able to understand and predict the pattern of these inherited traits.

In his experiments, Mendel *crossbred* pea plants by placing pollen from plants with green pods on the stigmas of plants with yellow pods. He crossbred plants with other traits as well. Mendel grew the seeds that resulted from each cross and observed the pattern of inherited traits. He discovered that when a pea plant with green pods was

Figure 2-12 *Gregor Mendel studied these inherited traits in pea plants. Shown here are five of the seven pairs of traits that Mendel studied.*

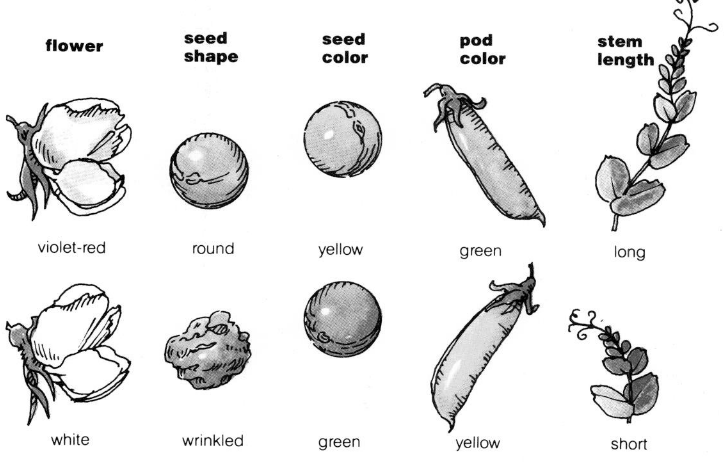

Teaching Suggestions 2-5

Remind students that in the previous section they discussed how they resembled their parents. At this point, list on the chalkboard some of the characteristics that seem to be passed from parents to offspring in humans. Such traits as eye color, height, weight, body shape, hair color and texture, and skin color probably will be volunteered by students. Tell the class that we now know how these traits are inherited, and that it helps in understanding the process to study the work of Mendel.

Explain that Mendel's most important work was with pea plants. Call the traits he worked with (Figure 2-12) to the students' attention. Then describe in detail his cross of pea plants that have green pods with pea plants that have yellow pods. Emphasize the meanings of dominant and recessive. List on the chalkboard some of the traits in humans that are inherited in simple Mendelian fashion (see Background).

Point out that Mendel referred to whatever controlled inheritance as "factors." Today, we know that these "factors" are genes. They will be studied in greater detail in the next two sections. Some students may wish to do chapter-end project 5 at this point.

Background

After Mendel's time, several useful terms relating to inheritance were introduced. *Phenotype* refers to the appearance of the trait being studied, *genotype* to the genetic basic of that trait. For example, in the cross discussed in section 2-5 the phenotype is green or yellow pods, and the genotype is the pair of "factors" that determine each type of trait.

Two other important terms are *homozygous* and *heterozygous*. A pea plant with the genotype *GG* for pod color (green) is a homozygote, as is a pea plant with the genotype *gg* for pod color (yellow). A pea plant with the genotype *Gg*, on the other hand, is a heterozygote. An additional important term is *allele*. An allele is simply one of the members of a gene pair. For example, in the gene pair *Gg*, *G* is an allele of *g*, and vice versa.

A number of traits in humans are inherited in simple Mendelian fashion. Some of these are:

Dominant	*Recessive*
nonred hair	red hair
white forelock	normal
premature gray	normal
very myopic	normal
normal	albinism
normal	sickle cell
free ear lobes	attached ear lobes
normal	diabetes mellitus

Teaching Tip

- You may want to reinforce the meanings of dominance and recessiveness by explaining *incomplete dominance* and *codominance*. In the former, two genes (alleles) are both expressed in the offspring by blending. For example, when red- and white-flowered four-o'clocks are crossed, all of the offspring flowers are pink. (Skin color and other human traits, incidentally, are not the result of incomplete dominance.) In codominance, two alleles are both expressed at the same time without blending. When red and white shorthorn cattle, for example, are mated, the offspring are roan colored. This coloration is the result of both red and white hairs in the animals' coats.

Facts At Your Fingertips

- Mendel's crossbreeding of plants and waiting for results to grow the following year was very time-consuming. It took years for him to fully understand the results of his experiments. Today scientists often do genetic experiments with fruit flies, because the time from one generation to the next in fruit flies is only a couple of weeks.
- Mendel had good luck in choosing the pea plant. Its traits are well defined and they do not blend. The traits are inherited independently of one another. And, because peas are self-fertilizing, pure lines could be developed relatively easily.
- Precise understanding of Mendel's "factors" had to wait until the 1950's, when Watson and Crick developed a model for the structure of the DNA molecule.

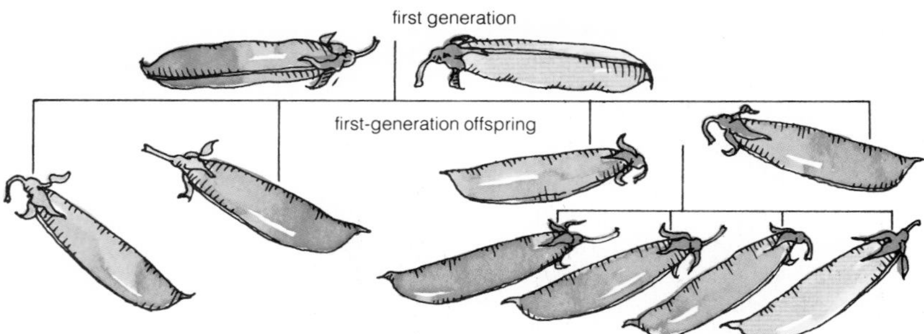

Figure 2-13 *A summary of the results of Mendel's studies. The "lost" trait reappeared in second-generation offspring.*

crossbred with a pea plant that had yellow pods, the first-generation offspring all had green pods. The yellow-pod trait seemed to "disappear." Mendel called the green-pod trait *dominant* because the trait "dominated" or overpowered the yellow-pod trait. He called the yellow-pod trait *recessive* because the trait receded or disappeared.

When first-generation offspring (all green-pod plants) were crossed with each other, the second-generation offspring included green-pod plants and yellow-pod plants. The "lost" yellow-pod trait had reappeared. Figure 2-13 summarizes the results of Mendel's experiments.

From his observations, Mendel concluded that something in pea plants controlled each trait. He called the controllers "factors." Mendel also concluded that some traits were controlled by a *pair* of factors.

Mendel was one of the first scientists to suggest the existence of factors that are responsible for the inheritance of traits. Though he described certain actions of the factors accurately, he never discovered that the factors were contained on the chromosomes.

Checkpoint

1. Give examples of dominant and recessive traits.
2. What happened to the recessive traits in the first-generation offspring Mendel experimented with?
3. Where in a cell are Mendel's "factors" found?

2 HEREDITY

2-5 Dominant and Recessive Traits

Investigating Fruit Flies

Read the following paragraphs describing the work of Thomas Hunt Morgan. Then write the answers to the questions that follow.

In the early 1900's Thomas Hunt Morgan decided to investigate inherited characteristics in fruit flies. Fruit flies are tiny insects often found on decaying fruit. Morgan chose the fruit fly because it has only four pairs of chromosomes. Since the four pairs are very large in the fly's salivary glands, they can be easily seen under a microscope. Males and females are easy to tell apart. Fruit flies reproduce quickly so that many generations can be studied in a short time.

Under his microscope, Morgan saw eight chromosomes in the fruit fly cell. Six of the chromosomes were matched pairs known as autosomes. The other two chromosomes did not match. This unmatched pair is known as sex chromosomes. Genes on the sex chromosomes determine the sex of the fly.

Generally, fruit flies have red eyes. In one large group, however, Morgan found a male fruit fly with white eyes. Morgan decided to mate the white-eyed male with a red-eyed female. Their offspring all had red eyes. Accord-

Checkpoint Answers

1. Answers will vary. Most will be taken from student text Figure 2-12.
2. Their effects were overpowered, or dominated, by the other trait, the dominant one.
3. Mendel's factors are located on the chromosomes.

2-6

Genes and Chromosomes

Goals

1. To describe the function of genes.
2. To explain the difference between a pure trait and a hybrid trait.
3. To identify the sex chromosomes and explain the inheritance of sex-linked traits.

Scientists now know that the "factors" Mendel described are genes. A **gene** is a segment of a chromosome. There are hundreds of genes on each chromosome. All inherited traits are controlled by one or more genes. These traits include those that you yourself inherited from your parents and grandparents. A human gamete carries thousands of different genes. In sexually reproducing organisms, such as humans, half of the genes in a zygote come from each parent through the gametes.

gene (jeen)

In body cells, almost all genes occur in pairs. Letters are used to represent the different forms of a gene in a gene pair. A **dominant gene** is a gene that is always expressed when it is present in an organism. A **recessive gene** is expressed only when the dominant form of the gene is absent. A dominant gene is represented by a capital letter. A recessive gene is represented by the small letter of the dominant gene. The ability to roll your tongue in a U-shape is a dominant trait, symbolized by the letter *T*. Small *t* is used to represent the non-tongue-rolling trait.

dominant gene
recessive gene

If both genes of a pair are identical, the trait that is expressed is called a **pure trait**. People with two genes for tongue-rolling, *TT*, are tongue-rollers. People with the two recessive genes, *tt*, are non-tongue-rollers. If the two genes are different, the trait is called a **hybrid trait**. People with the hybrid gene pair, *Tt*, are tongue-rollers because tongue-rolling is a dominant trait.

pure trait

hybrid trait

Mendel realized that some pea plants exhibited pure traits. The traits shown in Figure 2-12 were pure traits in the first generation Mendel studied. The plants with green pods had the gene pair *GG*. Yellow-pod plants had the gene pair *gg*. When Mendel crossed a pure green-pod plant with a pure yellow-pod plant, all the offspring had green pods.

Teaching Suggestions 2-6

Ask students to volunteer definitions of the term gene. After several ideas have been considered, point out that the molecular structure of the gene will be discussed in the next section, and that for now a gene should be considered to be (a) a segment of a chromosome, and (b) the carrier of the code for inherited traits, such as eye and hair color.

Explain how the Punnett square permits working out possible gene combinations. Draw on the chalkboard the Punnett squares for Mendel's F_1- and F_2-generation crosses of pea plants with green and yellow pods. Mendel's results for the *Gg* × *Gg* cross were 428 green to 152 yellow pods, or a ratio of 2.82 to 1.

Make sure students study the X and Y chromosomes in Figure 2-2 before you explain sex determination in humans. Stress that the sex of human offspring is determined by the father. Referring to Figure 2-15, point out that the sex of a child depends on whether a sperm cell with an X chromosome or a sperm cell with a Y chromosome fertilizes the egg cell.

Draw Figure 2-16 on the chalkboard and point out how the recessive gene for colorblindness is expressed in males because the Y chromosome does not carry any gene for color vision. There are far fewer colorblind females than males because the gene combination X_nX_n is much more unlikely than X_nY.

Chapter-end project 3 applies to this section.

Demonstrations

1. Show students an actual example of the ratios that result from a hybrid-hybrid cross. The purple and yellow aleurine kernel colors in ears of corn work well. Kits containing dried ears are available from biological supply houses.

   When corn that has purple (*RR*) kernels, the dominant color, is crossed with corn that has yellow (*rr*) kernels, ears that have all-purple hybrid (*Rr*) kernels result:

   Parents: *RR* × *rr*
   Gametes: *R* and *r*
   Offspring: all *Rr*

	r	*r*
R	*Rr*	*Rr*
R	*Rr*	*Rr*

When plants grown from hybrid (*Rr*) kernels are crossed, a ratio of about 3 purple kernels to 1 yellow kernel appears in each resulting ear:

Parents: *Rr* × *Rr*
Gametes: *R* *r* *R* *r*
Offspring: *RR*, *Rr*, *Rr*, *rr*

	R	*r*
R	*RR*	*Rr*
r	*Rr*	*rr*

Have students count the purple and yellow kernels per ear of corn in the F_2 generation. They should get approximately three purple kernels to each yellow kernel. Point out that students cannot see the difference between the heterozygotes (Rr) and the dominant homozygotes (RR) because the gene for purple (R) is dominant over the gene for yellow (r). However, one third of the purple kernels are pure for color (RR) and two thirds are hybrid for color (Rr). Point out also that the actual ratios seldom match the theoretical ratios perfectly, but that the match usually improves as the size of the sample being counted increases.

2. Show students that the 3 : 1 ratio among the offspring of a hybrid-hybrid cross is the result of random combinations of egg cells and sperm cells. Mix 100 green and 100 yellow dried peas in a large bowl. Blindfold a student, and have him/her take two peas at a time out of the bowl until the bowl is empty. Keep a tally of the combinations drawn: two green; one green, one yellow; two yellow. Let each pea represent an egg or sperm cell, carrying one gene of a pair. The combinations drawn thus represent the result of fertilization. The combinations drawn will approach the ratio of 1 green-green, 2 green-yellow, and 1 yellow-yellow. Of course, the larger the number of selections, the closer the ratios will approach the theoretical 1 : 2 : 1. If green is regarded to be dominant over yellow, the result is about 75% green and 25% yellow.

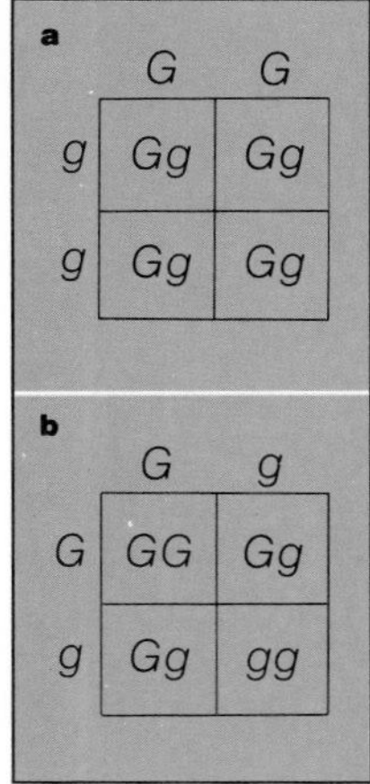

Figure 2-14 *(a) All first-generation offspring are hybrid. (b) Yellow-pod color appears in plants with two recessive genes.*

Figure 2-14a shows the results of Mendel's first-generation crosses in a table called a *Punnett square*. The symbols for the genes of one parent are listed along the side of the square. The symbols for the genes from the other parent are written across the top of the square. Each possible combination of genes that could be inherited by the offspring is listed inside the Punnett square. You can count the number of times a combination appears in the Punnett square to predict the likelihood of that trait occurring in the offspring.

All the offspring of Mendel's first-generation crosses were *hybrid* for the pod-color trait. Since the green-color gene (G) is dominant, all the first-generation offspring had green pods.

Figure 2-14b summarizes the results of the crosses between first-generation offspring. Plants with yellow pods "reappeared" in second-generation offspring. Recessive traits, such as yellow-pod color, are expressed only when *both* genes in a pair are recessive.

sex chromosomes

In many plants and animals one pair of chromosomes contains genes that determine the sex of the organism. These chromosomes are called the **sex chromosomes**. Look back at the human chromosomes pictured in Figure 2-2. Pair 23 is the sex-chromosome pair.

In females, both sex chromosomes appear identical. Females have an XX combination. In males, the sex-chromosome pair contains two chromosomes that appear different. The larger one is an X chromosome and the smaller, hook-shaped one is a Y chromosome.

The sex of a new organism depends on the sex chromosome in the sperm cell at fertilization. Since the sex chromosomes of a female are identical, all eggs produced by a female contain an X chromosome. In males, meiosis results in the production of sperm cells containing either a single X chromosome or a single Y chromosome. The sex of the offspring is determined by whether an X- or Y-carrying sperm fertilizes the egg. (See Figure 2-15.) A sperm carrying an X chromosome will result in a female (XX) offspring when joined with an egg cell. A sperm carrying a Y chromosome will result in a male (XY) offspring when joined with an egg cell.

Some traits are controlled by genes that are found on the sex chromosomes. In humans, color vision is one exam-

2 HEREDITY
Laboratory

Examining Beans for Inherited Traits

Purpose To demonstrate how traits are passed on.
Materials Dark and light dried beans, plastic containers with tops.

Procedure.

1. Open the plastic container labeled *Parents* and examine the color of each bean. Record your observation. Each parent bean represents a pure strain. Return the beans to their container. Then imagine that after you plant these beans the pollen from the dark bean plant is brushed on the stigma of the light bean plant's flower. After the flowers dry up, bean pods appear. The pods can be picked up and opened, and you have the first generation from the light bean. The light

3. Open the container labeled *Second Generation* and examine the color of each bean. Notice what color reappeared from the parents and what color shows most often.
4. Take out five scoopfuls of beans and separate them into color piles. Then count the beans in each pile and record the number of dark beans and the number of light beans that you have.

Discussion

1. Complete the chart below. Then divide

Teaching Tips

- Remind students that the dominant trait appears in both pure strains and hybrid strains because of the presence of the dominant gene.
- It helps to have students make a diagram of meiosis that shows the sex chromosomes. The diagram can be extended to show the possible combinations of sperm cell and egg cell

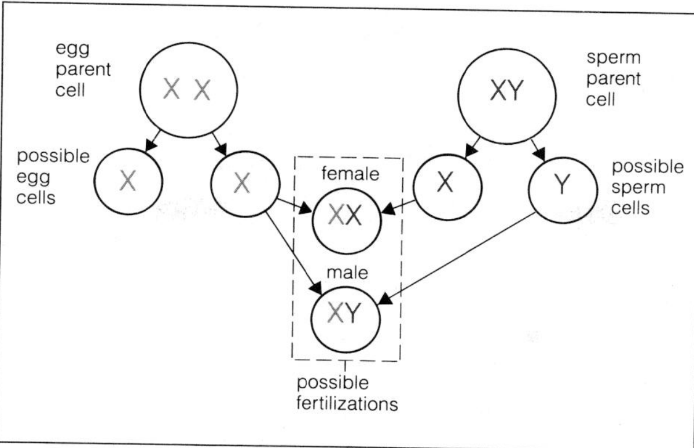

Figure 2-15 *The sex of a new organism depends on whether an X- or Y-carrying sperm fertilizes the egg.*

ple of a **sex-linked trait**. The gene for color vision is carried on the X chromosome only. Y chromosomes have no gene for color vision. Color vision in males is inherited from the mother.

sex-linked trait

Since the gene for colorblindness is recessive, a female with only one normal gene for color vision would still have normal color vision. The woman would be a *carrier* of colorblindness — she would not show the trait, but she could pass it on to her offspring. (See Figure 2-16a.) A male with the gene for colorblindness would be colorblind. (See Figure 2-16b.) Colorblind females are rare. About 20 times as many males as females are colorblind.

Hemophilia is a blood disease that occurs mostly in males, as a result of a recessive, sex-linked gene. People with hemophilia lack a necessary substance that helps blood to clot. A person with hemophilia bleeds heavily, even from a small cut or scratch.

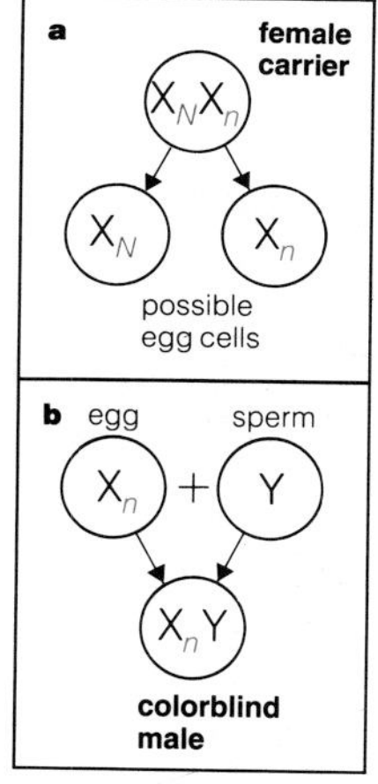

Figure 2-16 *(a) A carrier has normal color vision. (b) A colorblind male inherits the trait from his mother.*

Checkpoint

1. What is a gene?
2. What is the difference between a pure and a hybrid trait?
3. Which chromosome pair is different in males and females? How are the pairs different?
4. Give one example of a sex-linked trait.

at fertilization. This shows that, all other factors disregarded, there is an equal chance of creating a male or a female.

Facts At Your Fingertips

- There are many more genes than chromosomes.
- There are several types of colorblindness. Some people can see some colors, but not others. The most common form of colorblindness is to the color red. This is one reason why the color of fire engines is gradually being changed from red to a bright lime green.
- Hemophilia is a group of diseases called the "bleeders' disease." People who suffer from hemophilia have slow clotting response time. They will usually not bleed to death from a small cut. The most common problem is damage to bone joints from bleeding into the joints. Hemophilia appeared in many of the royal houses of Europe as a result of intermarriage among royalty who carried a recessive gene for the trait.

Mainstreaming

Instead of using green and yellow dried peas in demonstration 2, you can toss two pennies and keep a tally of "Heads," "Head-Tail," and "Tails." Then the visually-impaired can tell "Head" from "Tail" by feeling the surface of each penny.

Cross Reference

Refer to Table 3-1 in Chapter 3 for mention of hormones that control the development of female and male sex characteristics.

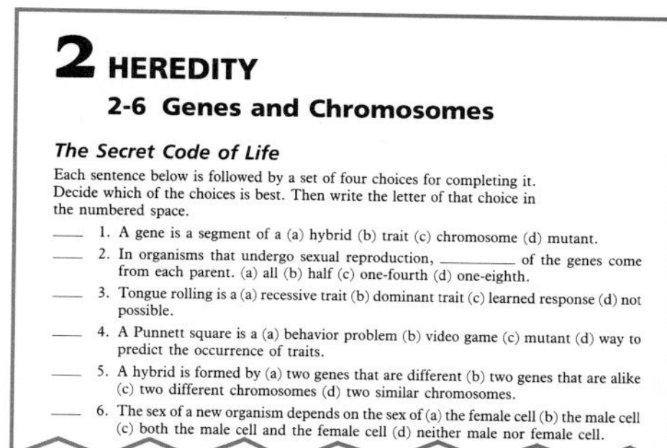

2 HEREDITY

2-6 Genes and Chromosomes

The Secret Code of Life

Each sentence below is followed by a set of four choices for completing it. Decide which of the choices is best. Then write the letter of that choice in the numbered space.

___ 1. A gene is a segment of a (a) hybrid (b) trait (c) chromosome (d) mutant.

___ 2. In organisms that undergo sexual reproduction, ______ of the genes come from each parent. (a) all (b) half (c) one-fourth (d) one-eighth.

___ 3. Tongue rolling is a (a) recessive trait (b) dominant trait (c) learned response (d) not possible.

___ 4. A Punnett square is a (a) behavior problem (b) video game (c) mutant (d) way to predict the occurrence of traits.

___ 5. A hybrid is formed by (a) two genes that are different (b) two genes that are alike (c) two different chromosomes (d) two similar chromosomes.

___ 6. The sex of a new organism depends on the sex of (a) the female cell (b) the male cell (c) both the male cell and the female cell (d) neither male nor female cell.

Checkpoint Answers

1. A gene is a segment of a chromosome that controls an inherited trait.
2. If both genes of a pair are identical, the trait is said to be pure. If the gene pair consists of one dominant and one recessive gene, the trait is hybrid.
3. The 23rd chromosome pair, called the sex chromosomes, is different in males and females. In females, the sex chromosomes are XX; in males, XY.
4. Colorblindness and hemophilia are two examples of sex-linked traits.

Teaching Suggestions 2-7

Proceed slowly and carefully when presenting this section. The structure of DNA and its role in inheritance are abstract concepts difficult for some students to grasp. Stress repeatedly when you discuss the structure of DNA that you are describing something at the molecular level that is far too small to see, even under a microscope. To put DNA in perspective, take students mentally from organism to system to organ to tissue to cell to nucleus to chromosome. Then point out that some chromosomes contain thousands of genes, and that genes consist of unique arrangements of base pairs within the DNA molecule.

Point out again that it is the order of a certain number of base pairs in DNA that makes up a gene. Emphasize that DNA indirectly directs all growth and function in an organism. DNA controls the synthesis of proteins in cells. But enzymes, the substances that catalyze *all* living reactions, are proteins. Thus, DNA is the source for all the chemicals and processes found in living matter.

Take students slowly and carefully through the steps in the replication of DNA, and explain that replication occurs before mitosis and cell division take place.

You may want to describe protein synthesis in the cell. A summary of this process is given in the Background section. Diagrams accompanied by more complete explanations appear in almost all first-year biology texts. A well-made film may be the best way to introduce this process to your students.

2-7

DNA

Goals

1. To describe the structure of DNA.
2. To identify the role of DNA in inheritance.

DNA

Chromosomes contain a substance called **DNA**, which stands for *deoxyribonucleic acid*. DNA contains the information that controls all cell activity. Scientists came up with a model of the structure of DNA that helps explain how DNA functions. According to the model, a DNA molecule is made up of small repeating units joined together in a long, twisted, ladderlike structure. (See Figure 2-17.)

The two sides of the "ladder" making up a DNA molecule contain alternating sugar (deoxyribose) and phosphate units. Pairs of chemicals called *bases* are connected to the sides of the ladder, forming "rungs." There are four types of bases: adenine (AD n een), thymine (THY meen), guanine (GWAH neen), and cytosine (SY toh seen). Adenine pairs with thymine and guanine with cytosine. (See Figure 2-18.)

Figure 2-17 *This model shows the twisted, ladderlike structure of a DNA molecule.*

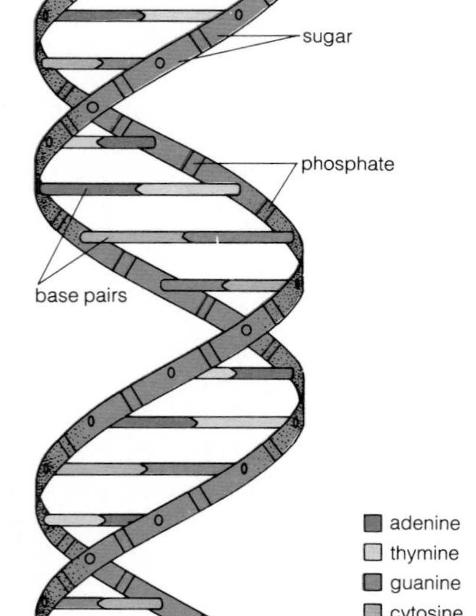

Figure 2-18 *Notice the repeating sugar-phosphate units and the paired bases.*

Background

The structure of DNA is related to its function. DNA directs all life by giving specific instructions for protein formation. Proteins are long chains of amino acids bonded together. The components of DNA significant for information storage and direction are the bases adenine, guanine, thymine, and cytosine. Each DNA unit of base, sugar, and phosphate is called a nucleotide.

The DNA molecule is made up of two strands of nucleotides. Each nucleotide is met by its complementary nucleotide on the opposite strand. The two strands are held together by relatively weak hydrogen bonds; thus, they separate easily during replication.

A string of nucleotides forms a gene, and a string of genes forms a chromosome. Each three nucleotides forms a code for a single amino acid. Though there are only 20 amino acids, there are 64 possible code words. Thus, one amino acid may have several different codes. Some nucleotide triplets don't actually code for an amino acid, but rather mark the end of one gene and the beginning of another.

Proteins are manufactured on ribosomes in the cytoplasm of the cell. A

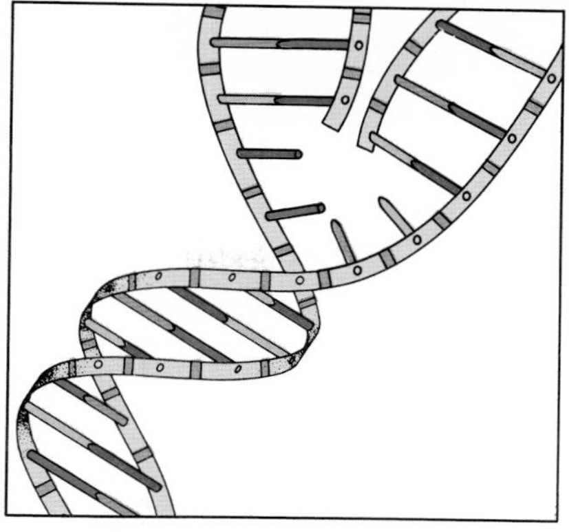

Figure 2-19 *A DNA molecule unwinds and a copy of each "missing" half is made.*

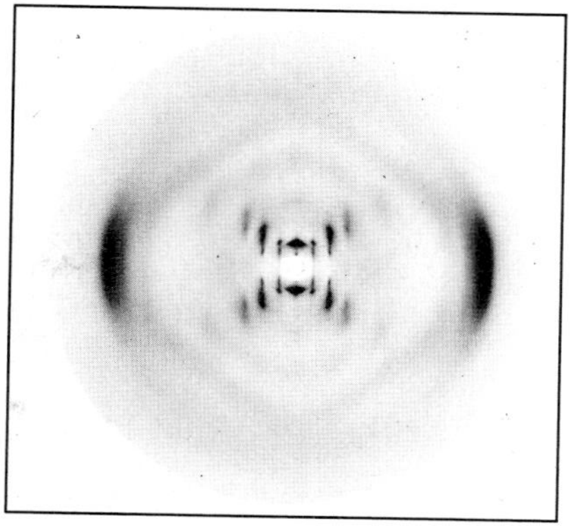

Figure 2-20 *X-ray studies of DNA helped scientists figure out its structure.*

A molecule of DNA contains millions of base pairs. The order in which the base pairs occur is responsible for the traits of an organism. Each gene is a portion of the long series of base-pair rungs in the DNA ladder. Genes control the production of proteins, which are responsible for the growth, development, and functioning of an organism. Every organism has a different arrangement of genes (base pairs). This makes each organism different from any other.

During mitosis, the strands of DNA reproduce, or replicate. In this way, each new cell contains all the necessary genetic information. In the replication process, the DNA molecule unwinds and separates into two separate parts, as shown in Figure 2-19. Each half of the DNA molecule acts as a *template*, or pattern, for the missing half. (See Figure 2-19.) Two copies of the original molecule are formed.

James Watson and Francis Crick first proposed the model of DNA in 1953. They described the *double helix* (HEE liks) after examining x-ray pictures of DNA taken by the chemist, Rosalind Franklin. Figure 2-20 is an x-ray picture of DNA.

TRY THIS

Build a model of a DNA molecule out of objects you can find at home, such as colored toothpicks and styrofoam balls. Try to follow the illustration in Figure 2-18.

Checkpoint

1. Draw a diagram of a model of a DNA molecule.
2. How do DNA molecules determine inherited traits?
3. What happens to DNA during mitosis?

substance called messenger RNA carries the code for the amino acid sequence from DNA in the nucleus to the ribosome. In the cytoplasm, transfer RNA picks up amino acids and matches them to the code on messenger RNA. The amino acids are then joined by enzymes on the ribosomes to form the protein molecule.

Demonstration

Show students a model of the three-dimensional structure of DNA's double helix. Many different types of models are available. Some are designed to show how DNA replicates itself.

Facts At Your Fingertips

- From x-ray crystallography pictures taken by Rosalind Franklin, James Watson, Francis Crick, and their colleagues figured out a three-dimensional model of the DNA molecule.
- Identical twins have identical arrangements of DNA.

Mainstreaming

Visually-impaired students will have difficulty with the Try This. They should be given a simple model of DNA that can be handled and explored tactually and that can be used as a guide for constructing their own model of the DNA molecule.

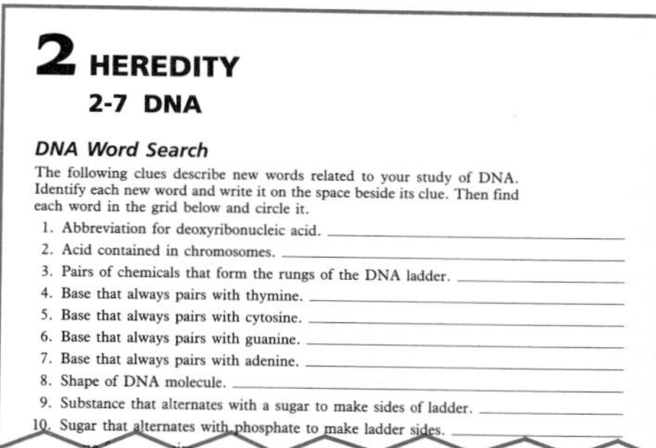

2 HEREDITY
2-7 DNA

DNA Word Search

The following clues describe new words related to your study of DNA. Identify each new word and write it on the space beside its clue. Then find each word in the grid below and circle it.

1. Abbreviation for deoxyribonucleic acid. ______
2. Acid contained in chromosomes. ______
3. Pairs of chemicals that form the rungs of the DNA ladder. ______
4. Base that always pairs with thymine. ______
5. Base that always pairs with cytosine. ______
6. Base that always pairs with guanine. ______
7. Base that always pairs with adenine. ______
8. Shape of DNA molecule. ______
9. Substance that alternates with a sugar to make sides of ladder. ______
10. Sugar that alternates with phosphate to make ladder sides. ______

Using Try This

Have students do flat drawings of DNA's structure before they attempt a three-dimensional model. Before they begin, it's important that they grasp the notion that the base pairs combine in a particular way.

Checkpoint Answers

1. Student drawings will vary.
2. Small portions of DNA molecules are the genes. Genes control all growth, development, and function in an organism.
3. Before mitosis actually occurs, DNA replicates itself. As a result, each daughter cell has a complete array of genetic information.

Using The Feature Article

This section is important because it allows you to discuss some potential applications of DNA research. When you describe how a piece of "foreign" DNA can be inserted into a bacterium's DNA, point out that genetic engineers *hope* someday to carry out this procedure with many kinds of organisms, including humans. For example, perhaps in the distant future, it may be possible to transfer specific genes to the cells of persons whose genes are defective. The new genes would then produce the chemicals the patient lacks. This process, although obviously on a microscopic scale, is similar to such accepted procedures as inserting intraocular lenses into defective eyes, pacemakers to correct heart action, and artificial hip and knee joints.

Genetically correcting body cells does not affect the patient's reproductive cells. However, some genetic engineers look ahead to the day when copies of normal genes can be transferred into sperm and egg cells. Some people find this possibility, as well as the research leading to it, disturbing. They feel it is wrong to interfere with natural processes, such as reproduction.

Hardly a month passes without some new development in genetic engineering being announced. Have your students watch the print media for such articles and bring them in. Discussing them as they come out and then posting them on the bulletin board is a good way to begin developing responsible attitudes toward genetic engineering in your students. Some students also may wish to do project 1 in conjunction with the Feature Article.

Genetic Engineering

Scientists have learned how to slice, splice, and rearrange DNA from different organisms. The DNA produced is called *recombinant* DNA because different parts have been "recombined," or rejoined, to form a new kind of DNA molecule. Recombinant DNA technology has made it possible to isolate specific human genes and attach them to bacterial DNA. Researchers have been able to get the bacterial cells to produce important substances.

One success that researchers have had is with the production of *insulin*. Insulin is normally produced in humans by the *pancreas*. People who cannot produce enough insulin on their own must receive injections of insulin every day. Using recombinant DNA techniques, researchers have been able to remove the insulin-producing gene from a sample of human DNA and join it to the DNA inside E. *coli* bacteria. During cell division, the insulin-producing gene is replicated, along with the bacterial DNA, and is passed on to future generations of bacteria. "Fooled" by the new DNA, the E. *coli* cells follow the "instructions" in the genes and produce human insulin. Until now, the only source of insulin was from an animal pancreas.

The photographs show a piece of "foreign" DNA being inserted into the DNA of a common type of bacteria. Bacteria are being used for most recombinant DNA research because they are easy to grow and they reproduce rapidly. Colonies of bacteria can be made to function like tiny "chemical factories," programmed to produce any one of a variety of important substances.

Researchers envision many ways in which recombinant DNA techniques can be used to solve medical problems. Bacteria have been developed that can produce human *interferon* (in tur FEER ahn), a substance that protects cells from attack by viruses. Another strain of bacteria has been "programmed" to break up oil and other petroleum products. These bacteria may someday be used to clean up large oil spills.

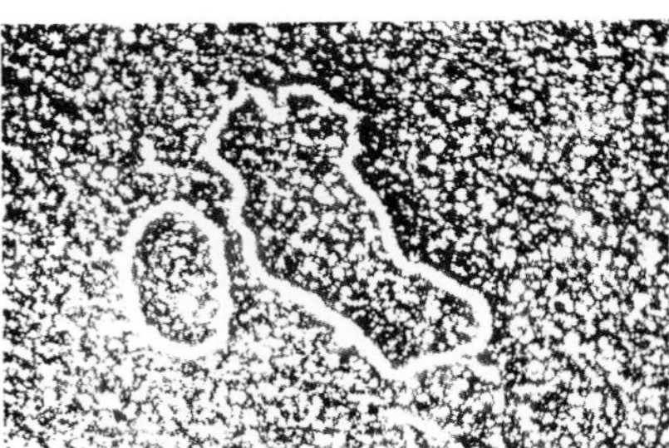

Figure 1a

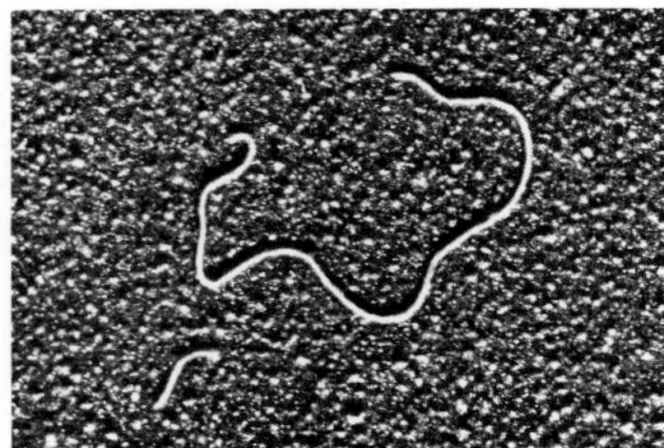

Figure 1b

2-8

Mutations

Goal

To identify what mutations are and how they occur.

A **mutation** is a change in a gene or chromosome. Mutations usually happen during the replication of DNA. A *gene mutation* usually involves a change in one base in a DNA molecule. *Albinism* is caused by a gene mutation. Albinos are missing the substance that gives color to the hair, skin, and eyes. (See Figure 2-21.)

mutation
(myoo TAY shuhn)

The expression of a gene with a mutation may result in a trait that affects the survival of an organism. For example, a plant with a mutation may be more susceptible to diseases.

Occasionally, the expression of a mutation results in a trait that increases the survival ability of an organism. For example, mutations in some insects have made them less susceptible to chemical pesticides. Other mutations have resulted in bacteria that are resistant to certain *antibiotics*, or medicines. If an organism with a gene mutation survives and reproduces, it will pass the mutation on to its offspring.

Figure 2-21 *This albino deer is missing the substance that gives color to the hair, skin, and eyes.*

Chromosome mutations usually involve many genes. During mitosis or meiosis, parts of chromosomes may be broken off and "lost" or they may get reattached in the wrong place. Chromosome mutations are often fatal.

Some mutations occur spontaneously. In most organisms the mutation rate is very low. Certain substances have been found to increase the rate of mutation in cells. Radioactive materials, x rays, ultraviolet light, tars (in cigarette smoke), and some chemicals are known causes of mutations in laboratory animals.

Checkpoint

1. What is a mutation? What causes mutations to occur in organisms?
2. What is the difference between a gene mutation and a chromosome mutation?
3. How can a mutation affect the survival ability of an organism?

Teaching Suggestions 2-8

Briefly review the structure of DNA, and remind students that genes consist of small molecular segments of DNA. Point out that there are often thousands of genes in a sperm or egg cell. A human gamete, for example, contains more than 20,000 genes.

Explain that a gene mutation is essentially a change in the chemical structure of the gene's DNA. Contrast this with chromosome mutations, which are the rearrangement or loss or addition of chromosome segments. Be sure students grasp the difference. In either case, the mutation is usually harmful, and sometimes fatal.

It is useful to summarize our knowledge of mutations. This will help students maintain a sense of perspective. (a) *Mutations are rare.* Mutation rates vary in different organisms. In humans, for example, albinism occurs in about one in 10,000 individuals. (b) *Mutations are stable.* Mutant genes, like all genes, are passed from generation to generation. Of course, if an organism carrying a mutated gene dies without reproducing, the gene dies with it. (c) *Most mutations are harmful.* Apparently because they are random events and usually have no relation to an organism's adaptive needs, the vast majority of mutations are harmful. Very, very few increase an organism's ability to adapt to its environment.

Chapter-end project 7 applies to this section.

Demonstration

Show your students some examples of mutant forms. Slides of mounted specimens of mutant forms of fruit flies are available from biological supply houses.

2 HEREDITY

2-8 Mutations

Genetic Changes

Each sentence below is followed by a set of four choices for completing it. Decide which of the choices is best. Then write the letter of the correct choice on the space at the left.

___ 1. A gene mutation involves the change in one ___ in the DNA. (a) acid (b) base (c) chromosome (d) gene

___ 2. A ___ is a change in a gene or a chromosome. (a) mutation (b) transmigration (c) mitosis (d) meiosis

___ 3. ___ is caused by a gene mutation. (a) Multiple sclerosis (b) Heart attack (c) Mononucleosis (d) Albinism

___ 4. A mutation may affect an organism's ___. (a) survival (b) health (c) length of life (d) all of the above

___ 5. An organism that has a ___ mutation will pass the mutation on to its offspring. (a) gene (b) jean (c) brain (d) nucleus

Checkpoint Answers

1. A mutation is a change in a gene or chromosome. Some mutations occur spontaneously. Others appear to be caused by chemicals, x rays, ultraviolet light, and other agents.
2. A gene mutation is a change in one gene's DNA. A chromosome mutation involves many genes.
3. A mutation usually results in a trait that decreases an organism's ability to adapt to its environment.

Teaching Suggestions 2-9

Ask students if they are familiar with any plant or animal products that are the result of special breeding to emphasize certain traits. Answers will vary, but some students may suggest navel oranges, high-milk-yielding cattle, thick-wool-producing sheep, new hybrids of day lilies, roses, or other flowers, and so on. Point out that we use many plant and animal products developed by careful breeding. Nectarines, tangelos, rust-resistant wheat plants, and eggs from hens bred to lay large numbers are good examples.

Be sure students understand that plant and animal breeders use the laws of Mendelian heredity to produce organisms or plant and animal products with desirable characteristics. Explain selective breeding and contrast it with hybridization, or crossbreeding. Point out that mutations too can produce desirable traits. For example, the navel orange first appeared as a mutant. The Ancon (shortlegged) sheep, now extinct, also was a mutant. This animal was once widely used in New England because farmers wished to avoid building high stone walls. Stress that a desirable animal or plant product appears by mutation very rarely.

You may wish to stress some of the traits breeders work toward. Often increased yield is sought. Resistance to disease is another desirable trait in valuable plants and animals. Appearance, as in show animals, is another. Speed in race horses, obedience in farm and sheep dogs, and gentle dispositions in pets are other examples. Breeders attempt to emphasize a particular trait without detracting from the animal's general condition.

Chapter-end project 2 applies to this section.

2-9

ENRICHMENT

Plant and Animal Breeding

Goal

To describe how people have developed breeds of plants and animals with desirable traits.

Agricultural scientists and farmers often try to improve the traits of crops or farm animals by a process called *breeding*. In the breeding process, a scientist crosses, or mates, two organisms in order to produce offspring with the desirable traits of both parents.

Selective breeding involves choosing, or selecting, individuals with the most desirable traits in a species and breeding them over several generations. By selectively breeding organisms over a period of years, the desirable traits of many organisms are combined in a few individuals. The white Leghorn chicken, a variety that lays many eggs, was produced by selective breeding. This method of breeding has a drawback. The continued breeding of offspring for many generations may result in the passing along (and maintenance) of some undesirable traits along with the favored ones.

Figure 2-22 *The beefalo is a cross between a buffalo and a cow. Someday people may be eating beefalo burgers.*

Hybridization, or *crossbreeding*, is the crossing of two different, but related, kinds of organisms to produce a new kind (hybrid) of organism. The desired traits of both parents are combined in the hybrid. For example, scientists have crossed cattle and buffalo to produce steer with the desirable traits of both. The resulting *beefalo*, seen in Figure 2-22, can live on ordinary grass, like the buffalo. Beefalo fatten faster and yield more meat than standard breeds of cattle.

Agricultural breeding is not a new technique. Figure 2-23 shows the improvements that have occurred in corn plants as a result of many years of breeding. Scientists believe that the modern corn plant was once similar to the teosinte (tee uh SIN tee) plant that grows wild in Southern Mexico. It is believed that Native Americans bred teosinte in fields with another type of corn, one with softer kernels and more kernels per ear. The two types of corn probably crossbred naturally, producing a better-tasting hybrid. As a result of continued breeding, the sweet corn eaten today contains approximately 1000 kernels per ear.

Facts At Your Fingertips

- Farmers commonly mate their best livestock to try to improve their herds or flocks. Selective breeding is common in pure-bred show dogs. Scientists use selective breeding to create pure strains of mice that are genetically susceptible to cancer or other diseases. Such mice are often used in experiments aimed at saving human life.
- Santa Gertrudis cattle are a cross between Texas Shorthorn and Indian Brahmin cattle. These hybrids produce excellent beef, are resistant to Texas cattle fever, and thrive in hot climates on sparse foliage.
- Some valuable fruits are hybrids. The tangelo, for example, is a cross between the tangerine and grapefruit. The Spygold apple is a cross between Northern Spy and Golden Delicious apples.

Figure 2-23 *The teosinte plant (left) is believed to be an ancestor of modern corn (right).*

Recent advances in plant breeding have resulted in the production of crops with greater nutritional value. Fruits and vegetables have been developed that can withstand mechanical harvesting with little damage.

The breeding process can often be quite slow. For example, farmers attempting to grow plants that produce more soybeans must wait until the crop is fully grown to see if the desired improvement has occurred.

Researchers in Canada are looking for ways to speed up the breeding process. Left on its own, a cone-bearing tree may take 20 to 30 years to produce cones. Using a method called *cone induction*, the scientists "fool" pine, fir, spruce, and hemlock trees into sprouting cones in three to five years. This method may someday be used to repopulate forests that have been cut down or lost to forest fires.

Checkpoint

1. What is meant by selective breeding?
2. Give an example of a plant or animal variety that has been improved as a result of breeding.
3. How was the beefalo developed?

2 HEREDITY

2-9 Plant and Animal Breeding

Crossing Organisms

In each sentence below a term is missing. Find the missing term; then write it on the space at the left.

______ 1. Farmers use a process called __1.__ to improve the traits of crops or farm animals.

______ 2. The crossing of two different, but related, kinds of organisms results in a __2.__

______ 3. The process in which individuals with the most desirable traits are chosen for breeding over several generations is called __3.__

______ 4. The __4.__, a plant that grows wild in Southern Mexico, is believed to have been bred with another type of corn by Native Americans.

______ 5. Scientists have crossed cattle and buffalo to produce __5.__, steer with the desirable traits of both kinds of animals.

______ 6. Selective breeding may result in the passing along of undesirable __6.__ along with the desired ones.

______ 7. __7.__ is a method used by scientists to repopulate forests in a relatively short period of time.

______ 8. The process in which two different, but related, kinds of organisms are crossed to produce a new kind of organism is called __8.__

______ 9. The __9.__, a variety that lays many eggs, was produced by selective breeding.

______ 10. Advances in plant breeding have resulted in crops with greater __10.__ value.

Hybrids

Using library reference books, research one hybrid not mentioned in your textbook. Then write a brief summary that includes the name of the hybrid, the names of the organisms that were crossed, and the desirable traits that result from the crossbreeding.

Checkpoint Answers

1. Selective breeding is the breeding of individuals with certain desirable traits for many generations to combine the desirable traits in a single organism.
2. Answers will vary. Most will be selected from the student text.
3. Beefalo was developed by crossing cattle and buffalo to produce fertile steers and cows with the most desirable characteristics of each species.

- Most hybrid organisms are infertile. The mule, for example, is a cross between the horse and the donkey. Mules are incapable of producing offspring.
- In many agricultural areas, insects have become resistant or immune to the effects of pesticides as a result of mutations. Increased crop damage is the result. Bacteria resistant to antibiotics cause great concern among physicians who must treat infectious diseases. Recently, some scientists have wondered if the regular addition of antibiotics to cattle feed is contributing to the occurrence of resistant bacteria.

Reading Suggestions

For the Teacher

Hartl, Daniel L. A PRIMER OF POPULATION GENETICS. Sunderland, MA: Sinauer, 1981. A readable book that covers genetic variation, mutation, migration, selection, genetic drift, and quantitative genetics.

Scientific American. GENETICS. San Francisco: W. H. Freeman, 1981. A collection of articles on all aspects of heredity.

Stebbins, G. Ledyard. DARWIN TO DNA, MOLECULES TO HUMANITY. San Francisco: W. H. Freeman, 1982. A clear, thorough discussion of biological and cultural evolution.

Zimmerman, Burke K. BIOFUTURE. New York: Plenum Press, 1984. Present and future aspects of genetic engineering.

For The Student

Arnold, Caroline. GENETICS: FROM MENDEL TO GENE SPLICING. New York: Watts, 1986. Easy-to-read history of genetics research.

Berger, Gilda. ALL IN THE FAMILY: ANIMAL SPECIES AROUND THE WORLD. New York: Coward McCann and Geoghegan, 1981. Presents the basic concepts underlying the theory of evolution.

Marshall, Kim. THE STORY OF LIFE FROM THE BIG BANG TO YOU. New York: Holt, Rinehart and Winston, 1980. An outstanding introduction to the theory of evolution.

Morrison, Velma Ford. THERE'S ONLY ONE YOU: THE STORY OF HEREDITY. New York: Messner, 1978. An easy-to-read introduction to human heredity.

Silverstein, Alvin and Virginia Silverstein. THE GENETIC EXPLOSION. New York: Four Winds, 1980. An introduction to genetics that includes an examination of the potential for good and bad in genetic engineering.

Sylvester, Edward J. and Lynn C. Klotz. THE GENE AGE. New York: Scribner's, 1983. Discussion of the genetics engineering industry, present and future.

Understanding The Chapter

2

Main Ideas

1. Inherited traits are passed on to offspring through the genes.
2. Mitosis results in the production of two identical cells that are copies of the parent cell.
3. Reproduction is the process by which organisms produce offspring.
4. Asexual reproduction, through fission, budding, regeneration, or spore formation, is carried out by a single parent.
5. Sexual reproduction is the creation of a new organism by joining a male sex cell with a female sex cell.
6. Meiosis is the cell division process that produces gametes with half the number of chromosomes as body cells.
7. A dominant gene is always expressed when it is present in an organism. A recessive gene is expressed only when the dominant gene is absent.
8. Sex chromosomes determine the sex of an offspring.
9. Chromosomes contain DNA.
10. DNA is a long molecule with a twisted, ladderlike structure. The order of base pairs in DNA determines an organism's traits.
11. A mutation is a change in a gene or chromosome.
12. The traits of crops or farm animals can be improved by selective breeding and hybridization.

Vocabulary Review

From the following list, choose the term that best completes each of the statements. Write your answers on a separate piece of paper.

asexual reproduction	pollination
budding	pure trait
DNA	recessive gene
dominant gene	regeneration
fertilization	replicate
fission	reproduction
gametes	sex chromosomes
gene	sex-linked trait
hybrid trait	sexual reproduction
meiosis	spores
mitosis	traits
mutation	zygote

1. A _?_ is created when a sperm nucleus and an egg nucleus unite.
2. A change in a gene or chromosome is called a _?_.
3. A segment of a chromosome that determines a specific trait is called a _?_.
4. _?_ is the process in which chromosomes are duplicated and then divided between two cells.
5. A pair of identical genes is expressed as a _?_.
6. _?_ is the general process by which a single parent produces offspring.
7. Another name for reproductive cells is _?_.

Vocabulary Review Answers

1. zygote
2. mutation
3. gene
4. mitosis
5. pure trait
6. asexual reproduction
7. gametes
8. pollination
9. meiosis
10. fission
11. dominant gene
12. sexual reproduction
13. hybrid trait

8. __?__ is the transfer of pollen from a stamen to a pistil.
9. __?__ results in sex cells with half the number of chromosomes as in body cells.
10. An amoeba reproduces by a type of asexual reproduction known as __?__.
11. A __?__ is always expressed when it is present in an organism.
12. The type of reproduction in which organisms produce offspring by joining two gametes is called __?__.
13. A trait that is expressed when the two genes in a pair are different is known as a __?__.

Chapter Review

Write your answers on a separate piece of paper.

Know The Facts

1. In a hybrid gene pair, such as W*w*, the W stands for the __?__ trait. (recessive, pure, dominant)
2. Budding and fission are examples of __?__ reproduction. (asexual, sexual, both sexual and asexual)
3. __?__ produces cells with half the number of chromosomes found in most body cells. (mitosis, fission, meiosis)
4. Information about all inherited traits is contained on __?__. (spindle fibers, genes, mutations)
5. The joining of an egg cell nucleus and a sperm cell nucleus is known as __?__. (fertilization, regeneration, mitosis)
6. The beefalo is an example of an animal developed by __?__. (selective breeding, crossbreeding, cross-pollination)
7. In flowering plants, sperm cells are contained in the __?__. (stigma, pollen, ovary)
8. The symbol *ss* stands for a __?__ trait. (pure dominant, hybrid recessive, pure recessive)
9. In humans the sex chromosomes are X and Y. The __?__ combination produces a female. (XX, YY, XY)
10. Sex-linked genes are responsible for __?__. (colorblindness, hemophilia, both)
11. The arrangement of __?__ on DNA molecules determines an organism's traits. (bases, sugars, phosphates)
12. After meiosis, human gametes normally contain __?__ chromosomes. (43, 23, 46)

Understand The Concepts

13. Draw and describe the stages of mitosis.
14. Identify each of the following: genes, chromosomes, DNA.
15. Draw a diagram of the DNA model.
16. Explain the differences between sexual and asexual reproduction. Give an example of an organism that reproduces each way.
17. How is the process of meiosis different from mitosis?
18. In the gene pair Ss, which letter represents the dominant gene? Which letter stands for the recessive gene? Is this a hybrid or a pure-gene pair?

Chapter Review Answers

Know The Facts

1. dominant
2. asexual
3. meiosis
4. genes
5. fertilization
6. crossbreeding
7. pollen
8. pure recessive
9. XX
10. both
11. bases
12. 23

Understand The Concepts

13. Drawings and answers will vary. See Figure 2-3 on page 44 of the student text.
14. Genes are segments of chromosomes that carry traits from parent to offspring. Chromosomes are long strings of genes found in the nuclei of cells. DNA is the chemical substance that makes up genes.
15. Drawings will vary. See page 56 of the student text.
16. Asexual reproduction is the production of offspring by one parent only. Sexual reproduction involves the joining of two sex cells, a male sex cell and a female sex cell, to produce offspring. Examples of organisms that reproduce asexually include bacteria, amoebas, hydras (budding), yeasts, and fungi (spores). Examples of organisms that reproduce sexually include flowering plants, conifers, amphibians, fish, and mammals.
17. In meiosis, the daughter cells have the haploid (n) number of chromosomes. In mitosis, each daughter cell has the diploid ($2n$) number of chromosomes.
18. *S* represents the dominant gene. The recessive gene is *s*. *Ss* is a hybrid gene pair.

19. A recessive trait is expressed only when both genes of the gene pair are recessive, or in the case of a sex-linked gene, when one recessive occurs on either the X or the Y chromosome in males.
20. Selective breeding is the breeding of individuals with certain desirable traits for many generations to combine those traits in a single organism. Crossbreeding is the crossing of two different, but related, types of organisms to produce a new hybrid organism.
21. Drawings and explanations will vary. See student text page 50.
22. Pollination is the transfer of pollen grains from the stamen to the pistil. Fertilization is the union of the sperm nucleus from a pollen grain with the egg nucleus in the ovule of the flower.

Challenge Your Understanding

23.

	V	*v*
V	*VV*	*Vv*
v	*Vv*	*vv*

VV and *vv* are the two possible pure gene combinations. The two *Vv*'s are the possible hybrid pairs. *VV* and *Vv* will result in violet-red flowers. Only *vv* results in white flowers.

24. Knowledge of reproduction and gene-inheritance allows scientists to use the processes of selective breeding and hybridization to improve crops and farm animals.
25. Each parent contributes half of the genetic makeup of the offspring. Different combinations of genes thus give the offspring some similar traits and some different traits.
26. The recessive gene in question is on an X chromosome: X_n. Suppose the mother is hybrid for the clotting factor (X_NX_n) and the father is normal (X_NY). (No allele appears on the Y chromosome.) If an egg cell with an X chromosome carrying the recessive gene (X_n) is fertilized by a Y-carrying sperm cell, the resulting boy (X_nY) will have hemophilia. The single recessive gene is expressed in the absence of any allele.
27. Genetic counselors have either a medical degree or a master's degree in genetic counseling.

19. Under what conditions will a recessive trait be expressed?
20. Describe two methods agricultural scientists use for developing new varieties of plants and animals.
21. Draw and label the reproductive structures in a flower. Explain the function of each part.
22. Describe the process of pollination and fertilization.

Challenge Your Understanding

23. Use a Punnett square to show the possible outcomes of a cross between two flowers, each with the same gene pair for color, V*v*. How many pure combinations result? How many hybrid pairs are there? Suppose V represents the gene for the violet-red form of the flower and *v* represents the gene for the white flower form. How many combinations result in plants with violet-red flowers? Which are they? How many combinations result in white flowers?
24. How can a knowledge of reproduction and gene-inheritance help scientists to improve the traits of crops and farm animals?
25. Explain why individuals with the same parents have some traits that are alike and some traits that are different.
26. A sex-linked gene controls the clotting ability of blood. Explain how a couple with normal clotting factors can have a son with hemophilia.
27. Genetic counselors do research and counsel couples about their chances of having a child with a genetic disorder. What kind of background and training would genetic counselors need to have?

Projects

1. Do a report on the ways genetic engineering is being used to produce substances that are important to people.
2. Prepare a report on the new varieties of crops and livestock produced by selective breeding and crossbreeding. Discuss the advantages of some of the new varieties.
3. Invite a local public health official to speak to your class about inherited diseases. Find out about tests to detect carriers of genetic diseases.
4. Collect water from a pond or stream. Make slides of the water and observe them under a microscope. Look for examples of organisms undergoing fission, budding, or regeneration.
5. Obtain red- and white-eyed varieties of fruit flies (*Drosophila melanogaster*) from a biological supply house. Mate them over several generations. Keep a record of the number of offspring with each trait. Which trait do you think is dominant? Which is recessive? Explain, based on your observations.
6. Hold a plant exchange day in class. Learn how to produce new plants from parts of old plants.
7. Research and report on the contributions to cell biology and genetics of any two of the following scientists: Barbara McClintock; Ernest E. Just; Francisco Ayala; F. Agnes N. Stroud Lee; Elizabeth Neufield.

2 HEREDITY
Projects

1. Use reference books in your library to find out about three careers in the field of genetics. You may also want to write to the American Genetic Association, 1028 Connecticut Ave. NW, Washington, DC 20036.
   a. What are the three careers that you researched?
   b. What kind of training is needed for each career?
   c. What kinds of personal traits would each career require?

Chapter 3

The Human Body

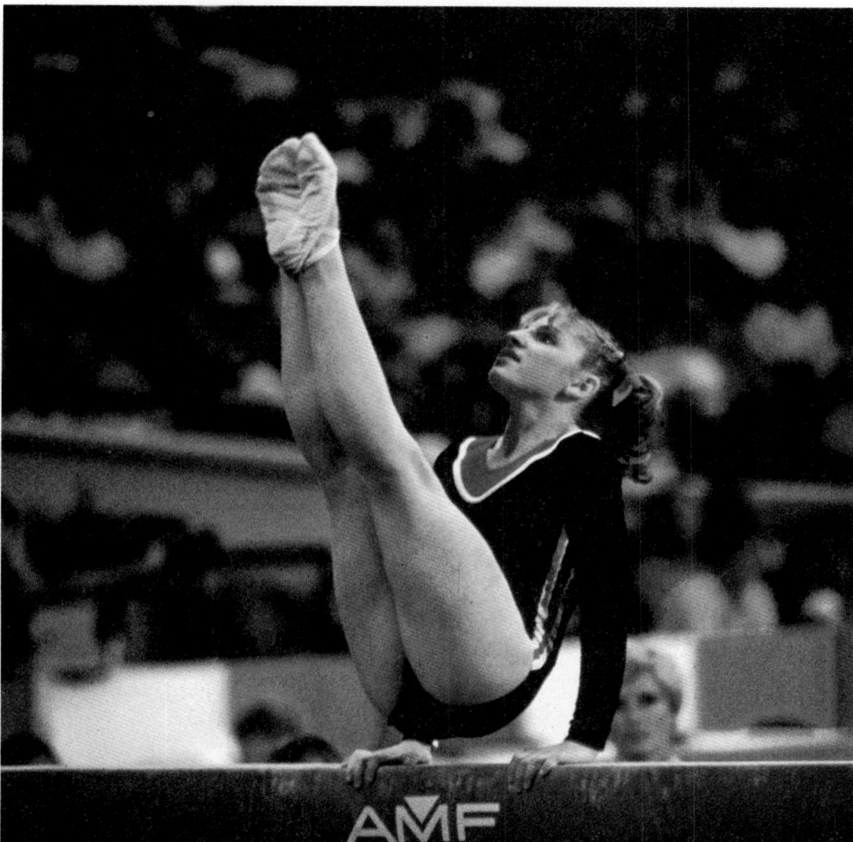

This gymnast must exercise control over the muscles in her body.

CHAPTER 3 Overview

The purpose of this chapter is to give students a basic understanding of the major systems of the human body. The systems covered are treated separately, but stress is placed on the fact that in a healthy body they function together in an efficient way. The skeletal and muscular systems are described first, followed by the circulatory system. Next, respiration, digestion, and excretion are discussed. Then, the nervous and endocrine systems are covered. Finally, the effects of alcohol, tobacco, and other drugs on the body are treated in the enrichment section.

Goals

At the end of this chapter, students should be able to:

1. describe what bones and muscles are like and how they work in the body.
2. identify parts of the circulatory system and their functions.
3. describe the actions involved in breathing.
4. explain the function of breathing.
5. explain how food is digested and absorbed by the body.
6. describe the structure and function of the excretory system.
7. identify the parts of the nervous system.
8. explain the function of the nervous system.
9. describe the function of different hormones.
10. identify the effects of alcohol, tobacco, and other drugs on the body.

Vocabulary Preview

cartilage
joint
ligaments
voluntary muscles
involuntary muscles
tendons
blood
plasma
arteries
capillaries
veins
trachea
alveoli
diaphragm
digestion
esophagus
villi
kidneys
urine
stimulus
neurons
impulse
synapse
cerebrum
cerebellum
medulla
reflexes
glands
hormones

Teaching Suggestions 3-1

Begin the lesson by having students point to and name as many bones as they can in their skeletons. Have them look at Figure 3-1 to see if they missed any. Ask students to imagine what a person would look like if he or she didn't have a skeleton. (A shapeless bag of organs would be the result.)

Discuss the functions of a skeletal structure. Point out examples of structural elements in buildings going up in your area. Encourage students to make comparisons between a building's framework and the human skeleton.

Discuss some injuries that are likely to occur to the skeletal and muscular systems. Tell students that a fracture is any break in a bone. Explain that a fracture must be set—the bone must be held in place so it will heal properly. When a bone break is healing, bone cells produce material that repairs the break.

Emphasize that bone is living tissue that stores and uses up materials throughout most of a person's life. Blood vessels carry nutrients to, and wastes away from, living bone cells.

Explain that pads, or discs, of cartilage between the vertebrae act as cushions, allowing flexible movement of the spinal column. Have students find these discs on the drawing of vertebrae in Figure 3-3. You might want to discuss the common vertebral ailment called "slipped disc."

Ask students to point to and name as many muscles in their bodies as they can. They'll be more successful pointing to muscle groups than in naming any muscles. Stress that all movement of the body is produced by muscular action.

Point out that muscles move by contracting, and that in contracting they become shorter and fatter. Describe the three types of muscle tissue, and emphasize that normally only striated, or skeletal, muscles are under the direct control of one's will. It's important to make clear the antagonistic nature of most skeletal muscles. Use motion at the elbow or any other joint to point out that each member of a pair of muscles acts in opposition to each other. When one contracts, the other relaxes. For motion in the opposite direction, the reverse occurs.

Be sure to point out that cardiac muscle, although striated, is involuntary muscle. You may wish to mention that biofeedback techniques allow a certain amount of control over some involuntary muscles.

Some students may wish to do project 4 at the end of this chapter.

Demonstrations

1. Show students examples of muscular contraction. Have them place their hands on the tops of their thighs near the knees, while sitting. Tell them to raise their lower legs while feeling the muscular contractions

3-1

Skeleton and Muscles

Goal

To describe what bones and muscles are like and how they work in the body.

As you turn a page in this book, the bones and muscles in your hand and arm work together. Your hand and arm bones are part of your body's framework, or *skeleton*. (See Figure 3-1.) A tall building contains a similar framework that supports the floors and walls. Your skeleton supports the muscles and organs of your body. Without a skeleton, humans would have a very different form.

The skeleton protects certain organs of the body. The skeleton serves, also, as a point of attachment for the muscles, making movement possible.

Figure 3-1 *Some bones protect soft body parts. For example, the lungs are protected by the ribs and the brain is protected by the skull. Other bones, such as the thigh bones, support the body and aid movement.*

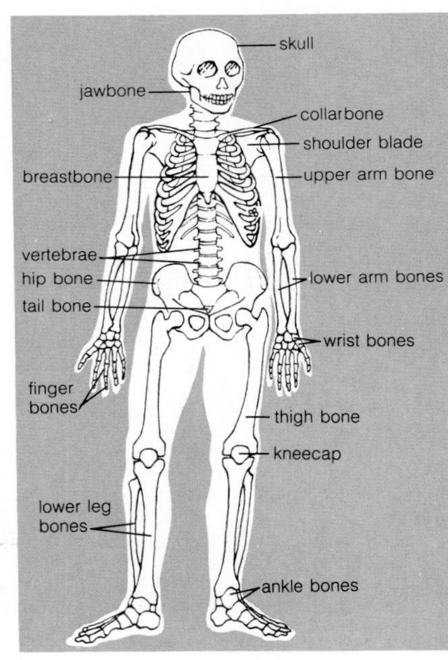

Figure 3-2 *Bones are light in weight in comparison to their strength because they are porous, not solid.*

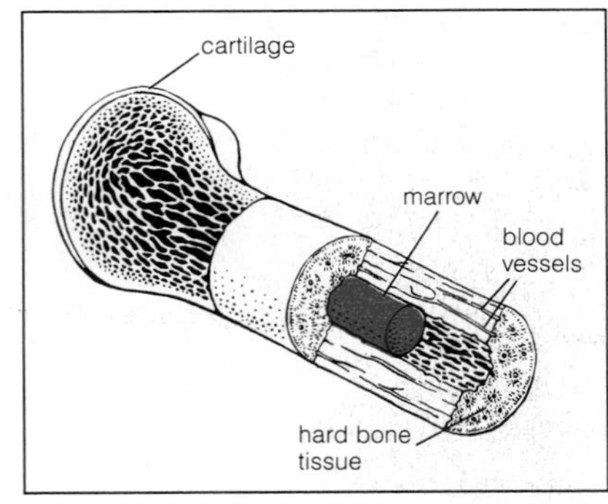

Your skeleton contains 206 bones. Bone tissue is made up of living cells surrounded by a hard, calcium-rich covering. Blood vessels provide nourishment to the bone cells and remove wastes from them. (See Figure 3-2.) **Cartilage** makes up the stiff, yet flexible, part of the skeleton. It is found in the tip of the nose, in the outer ear, and at the ends of many bones. Cartilage cushions the ends of bones, and allows bones to move smoothly against one another.

cartilage

The point at which two bones meet is called a **joint**. There are many types of joints in the skeleton. Some are shown in Figure 3-3. Bones that have grown together form *immovable joints*. Figure 3-3 shows the fused bones that make up the skull. *Hinge joints*, such as the elbow and knee, operate like the hinge on a door. They permit back-and-forth movement only. *Ball-and-socket joints*, located in the shoulder and hip, permit movement in many directions. *Gliding joints* are found between the individual bones of the back, allowing the back to bend and twist.

joint

Most bones are held together at movable joints by bands of tough, stretchy tissue called **ligaments**. (See Figure 3-4.) Ligaments enable bones to move, yet keep bones from separating from one another at the joint. The bone-ends in a movable joint are lubricated by a fluid that keeps them from grinding against each other.

ligaments

A sprain occurs when ligaments are stretched or torn. A muscle strain usually involves soreness, but no damage to the tissue.

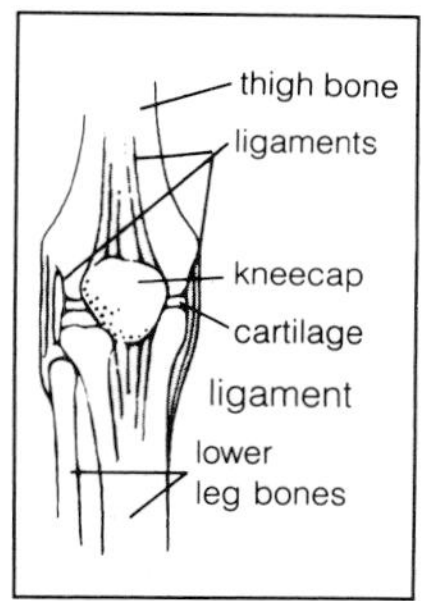

Figure 3-4 *Ligaments enable bones to move.*

Figure 3-3 *The different joints make a variety of movements possible. Joints in the skull begin to seal at age twenty.*

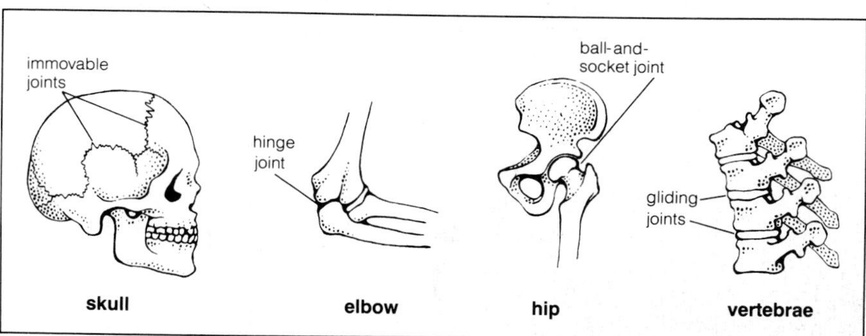

under their fingers. The muscle that they feel contracting is the quadriceps femoris.

Contractions of two of the muscles of mastication are easy to detect. Have students place the fingertips over the temporal regions at the sides of the head, and then clamp the jaws. The muscle contracting is the temporalis. Contraction of the masseter can be felt by placing the fingers on the sides of the jaw and clamping the jaws. These two muscles elevate the lower jaw to close the mouth and bring the lower teeth up against the upper teeth.

2. Demonstrate how muscles function in pairs. Have students place their left hands on their right biceps muscles while curling the left thumbs around to feel the triceps. Tell students to squeeze the right arm gently with the left hand. Now have them flex and extend their lower right arms several times slowly, noting which muscles are contracting, or thickening, with each movement. Ask students to locate and identify other areas on their bodies where muscles operate in pairs.
3. Boil a chicken for several hours or until the meat falls off the bones easily. Have students identify the bones and compare them to bones with similar functions in humans. Point out the cartilage at the ends of bones. If possible, try to obtain a thigh bone and leg bone that are still attached. Show students the elastic ligaments. Demonstrate how the bones move at joints.

Teaching Tips

- Ask students to look carefully at Figure 3-3. Point out that the lower jaw is connected to the skull by hinge joints.
- Borrow some x rays from a local health center. Tape them to a window and have students identify the bones and the joints. Point out any fractures or dislocations.
- Cardboard skeletons are often available in stores at Halloween time, but note that they are often very inaccurate. One can be hung in your classroom as students study the skeletal system. Some students may be interested in learning the scientific names of bones and finding the errors in the cardboard skeletons.
- Invite a physical education teacher or coach to speak to your class about the benefits of an exercise program. Ask the speaker to give the class general guidelines for a fitness program that works, and to discuss precautions to take when beginning a new exercise regimen.
- Assign students the following research topics: sports injuries and their treatment; new treatments for arthritis; and new developments in artificial limbs and joints.

Facts At Your Fingertips

- Bones act as mineral "banks," storing extra calcium and phosphorus taken in through the diet. When there is a need for calcium and phosphorus, old bone tissue breaks down, making the minerals available to body cells. People must take in a sufficient amount of calcium, phosphorus, and vitamin D in their diets in order to replenish bone tissue. These substances are present in milk products, eggs, fish, and whole-grain cereals.
- Muscles account for about 40% of body weight.

Mainstreaming

If possible, obtain a model of a skeleton. Allow visually-impaired students to feel the different bones and joints as they learn their names.

Physically handicapped students may not be able to walk well enough to do the Try This. Such students should analyze whatever movements are most convenient for them.

Cross Reference

Bone marrow has been included in the diagram of bone in Figure 3-2. The function of marrow will be discussed in section 3-2.

Your body contains over 600 muscles. All your body movements are made possible by the action of muscles. Some muscles move the skeleton. Others help to move blood and food through the body. All muscles function by *contracting*, becoming shorter and thicker.

voluntary muscles

involuntary muscles

Muscles are classified as voluntary or involuntary. **Voluntary muscles** function when you want them to. The muscles of the arm, leg, and tongue are voluntary muscles. **Involuntary muscles** are not under your conscious control. Involuntary muscles are found in the stomach, blood vessels, and pupil of the eye. Some muscle fibers in the body are always contracted, giving the body muscle *tone*. One group of fibers contracts, while another relaxes.

There are three types of muscle cells in the body: (1) striated (STRY ay tid), (2) smooth, and (3) cardiac. (See Figure 3-5.)

Voluntary muscles are made up of cells that are striped, or striated, in appearance. *Striated muscle* moves the skeleton and body parts, such as the lips, tongue, and face. Striated muscles are capable of two kinds of actions. They can move in relatively short and rapid contractions or they contract more slowly, producing a strong force.

Muscles that move bones work by pulling. When a muscle contracts it pulls a bone in one direction. Pairs of muscles make movement of bones in two directions possible. For example, the *biceps* and *triceps* are a pair of muscles located in the upper arm. (See Figure 3-6.) Each muscle is

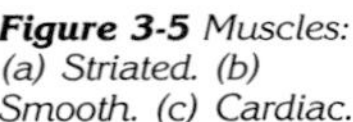

Figure 3-5 *Muscles: (a) Striated. (b) Smooth. (c) Cardiac.*

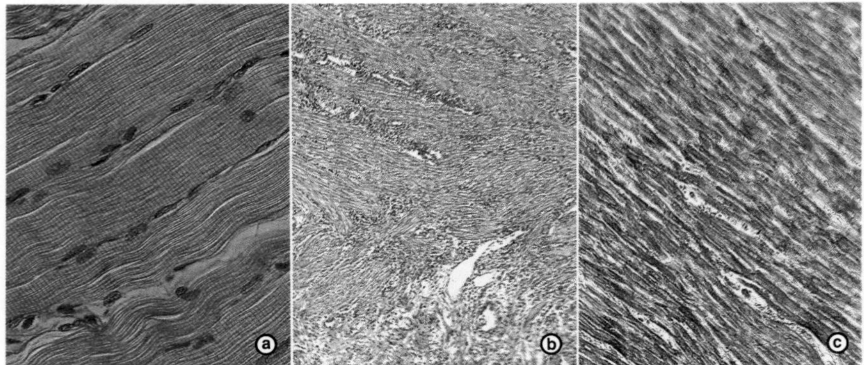

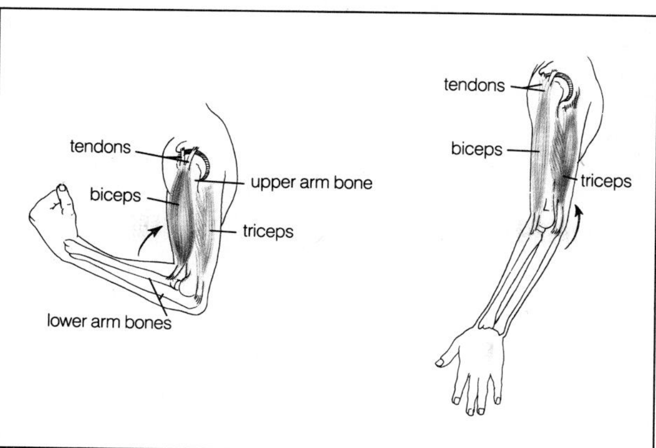

Figure 3-6 *Muscles that move bones work in pairs. When one muscle contracts, the other stretches. The muscles move the bone in opposite directions.*

attached to a bone by bands of tissue called **tendons**. When the biceps contracts, the forearm is pulled up and the arm bends. When the triceps contracts, the arm straightens.

tendons

Most involuntary muscles are made up of smooth muscle cells. *Smooth muscle* does not have bands or stripes. Smooth muscle is capable of slow contractions over a long period of time. For example, smooth muscle in the digestive system may contract continually for hours, churning and grinding the food inside.

A third type of muscle cell occurs in the heart only and is called *cardiac muscle*. Cardiac muscle has the strength and appearance of striated muscle and the endurance of smooth muscle. It contracts about 70 times a minute and 100,000 times each day. It slows down when you sleep but it does not stop as long as you are alive.

TRY THIS

Analyze your body as you walk slowly across the room. Which muscles are contracting and which joints are moving as you walk?

Checkpoint

1. Describe three functions of the human skeleton.
2. Name two bones that protect soft body parts.
3. What two types of tissue is the skeleton made of?
4. Name two kinds of joints. Tell where they occur in the skeleton.
5. Explain the differences between voluntary and involuntary muscles.
6. Describe how voluntary muscles work in pairs.

Using Try This

Students should mention that muscles in the legs and feet are contracting and relaxing as they walk. If they are moving their arms, muscles in their shoulders may be contracting. Contraction of other muscles contributes to overall muscle tone. Joints at the hips, knees, shoulders, and neck are probably moving during walking.

Checkpoint Answers

1. Support of muscles and organs; protection of soft body parts; and points of attachment for muscles, making movement possible.
2. The ribs, skull, breastbone, and vertebrae protect soft body parts.
3. The skeleton is made of bone tissue and cartilage.
4. Answers will vary. Among them are immovable joints in the skull, hinge joints at the elbow and knee, ball-and-socket joints at the hips and shoulders, and gliding joints between the bones of the back.
5. Voluntary muscles function when you want them to. They are made up of striated muscle tissue. Many voluntary muscles work in pairs. Involuntary muscles are not normally under your conscious control. They are made up of smooth muscle tissue. Smooth muscle may contract continually for hours. Cardiac muscle, although striated, is classified as involuntary muscle.
6. When one of the muscles contracts and its partner relaxes, it moves a bone at a joint. When the other muscle contracts and the first one relaxes, the bone is moved in the opposite direction.

3 THE HUMAN BODY
3-1 Skeleton and Muscles

Secrets in Bones and Muscles

Did you ever use a secret code to write to a friend? Each term below is written in code and accompanied by a phrase that describes it. Read each numbered phrase and think of the term that is being described. Before you write that term on the answer space, check yourself by using the key below to decode the term.

1. Material that cushions ends of bones.
2. Place where two bones meet.
3. Structure that contains 206 bones.
4. Joints found in back.
5. Skull joint.
6. Description of the way muscles function.
7. Muscles found in stomach.
8. Heart muscle cells.
9. ...

3 THE HUMAN BODY
3-1 Skeleton and Muscles

Bone Identification

Certain bones on the skeleton below have been numbered for you. Identify each numbered bone. Then write the name of that bone on the answer space with the corresponding number. Note: Bones in the fingers and toes have the same name.

1. ______
2. ______
3. ______
4. ______
5. ______
6. ______
7. ______
8. ______
9. ______

Teaching Suggestions 3-2

Open the section by asking students if they know how nutrients and oxygen get to the body's cells, and how carbon dioxide and chemical wastes are taken away from the body's cells. Explain that the first part of this question will be answered in this section, and the second part a bit later. Now give students an overview of the circulatory system and its functions.

Ask students if they have ever had any blood tests, and for what reasons. Explain that by counting the number of red and white blood cells in a sample of blood, a physician can learn a lot about a person's health. For example, the presence of a large number of white blood cells (a "high WBC count") usually indicates that the body is fighting an infection or infectious disease. A low number of red blood cells (RBC) means that the patient may have a condition known as anemia. Since the red blood cells carry oxygen, a typical patient with anemia may lack sufficient oxygen for the body's cells. There are many kinds of anemia and causes vary.

Be sure students understand that the heart diagram in Figure 3-11 is labeled according to the heart's position in the body, not its position on paper. The right side of the heart will be on the reader's left. Students can hold the heart diagram up to their chests to see the proper position and to better understand why it is labeled as it is.

Some students will have difficulty learning the pathway of blood through the heart and other parts of the circulatory system. Draw a large outline of the heart, showing the four chambers, and the rest of the circulatory system on the chalkboard. Have students take turns tracing through the system with colored chalk, naming each part as the chalk line passes through it.

Point out that a thick, muscular wall separates the right and left sides of the heart, preventing oxygen-poor blood from mixing with oxygen-rich blood. This insures that blood leaving the heart from the left ventricle carries the maximum amount of oxygen to the body's cells.

Stress that the heart is referred to as a double pump because it pumps blood through two major circulatory pathways. One is from the right side of the heart to the lungs and back to the left side of the heart. The second is from the left side of the heart to the body cells and back to the right side of the heart. The coronary arteries, capillaries, and veins may be considered a third circulatory loop. Blood is pumped through all pathways simultaneously.

To avoid confusion, emphasize that the pulmonary arteries are the only arteries in the body that carry oxygen-poor blood and that the pulmonary veins are the only veins that carry oxygen-rich blood. Trace the pulmonary circulation loop so students understand why this is so. Explain that these ves-

3-2

Circulatory System

Goal

To identify parts of the circulatory system and their functions.

Blood is a liquid that carries various substances to and from body cells. For example, it supplies cells with food and oxygen and carries away wastes. The tubes through which blood flows are the *blood vessels*. The muscular pump that pushes, or circulates, blood throughout the body is the *heart*. The heart, blood, and blood vessels are the major parts of the *circulatory system*.

The circulatory system of an average adult contains five liters of blood. About half of the blood is a clear liquid called **plasma**, made up mostly of water. (See Figure 3-7.) Plasma carries food, wastes, and other materials.

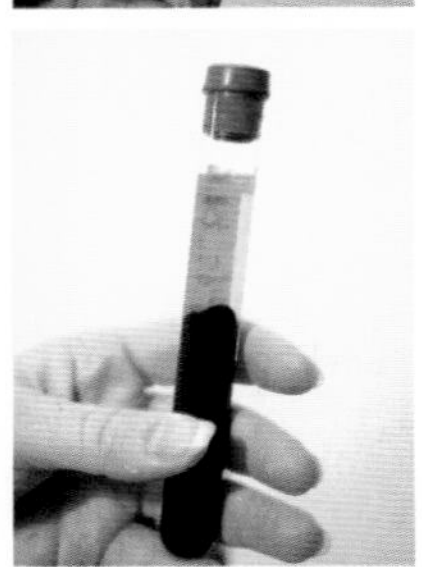

Figure 3-7 *(a) Fresh, whole blood (top). (b) Separated blood (bottom). The pale-yellow liquid is plasma.*

blood

plasma (PLAZ muh)

In addition to plasma, blood contains three kinds of cells: (1) red blood cells, (2) white blood cells, and (3) platelets. The 25 trillion *red blood cells* give blood its characteristic red color. These cells carry oxygen from the lungs to all cells in the body. Red blood cells also carry some carbon dioxide from the cells to the lungs. The rest of the carbon dioxide is dissolved in the plasma. Red blood cells are made in *bone marrow*—the soft tissue in the center of many bones. A red blood cell lives for about 120 days. New red blood cells are being made all the time, replacing the ones that are "worn out." A highly magnified view of red blood cells can be seen in Figure 3-8.

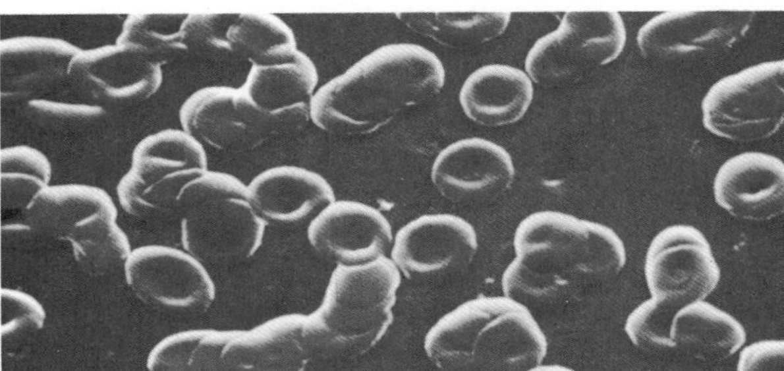

Figure 3-8 *Red blood cells are doughnut-shaped cells without nuclei.*

Your body contains approximately 35 billion *white blood cells*, one for every 600 to 800 red blood cells. There are several kinds of white blood cells. One kind uses pseudopods to surround and destroy foreign bacteria in the body. (See Figure 3-9.) Another kind produces *antibodies*, which combine with foreign bacteria, viruses, and harmful substances. White blood cells and antibodies enable the body to "fight" infection and disease.

Platelets are tiny pieces of cells produced in the bone marrow. They play an important role in the clotting of blood. When tissues are damaged, platelets release a substance that starts a chain of chemical reactions in the damaged area. A network of fibers forms that traps platelets and other blood cells, creating a clot. The clot stops the loss of blood from the damaged area. (See Figure 3-10.)

Your heart is a muscular pump about the size of your fist. It is located in the center of your chest with the bottom end tilted toward the left side of your body. Your heart pumps blood through a network of more than 100,000 km of blood vessels. This is more than twice the distance around Earth at the equator.

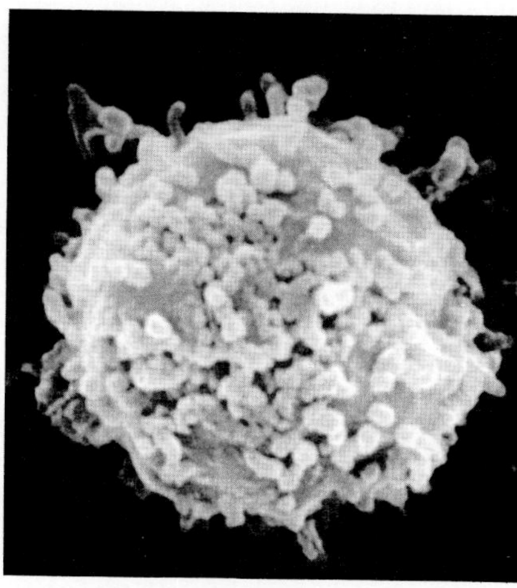

Figure 3-9 *Like the amoeba, a white blood cell has pseudopods to surround objects, such as foreign bacteria.*

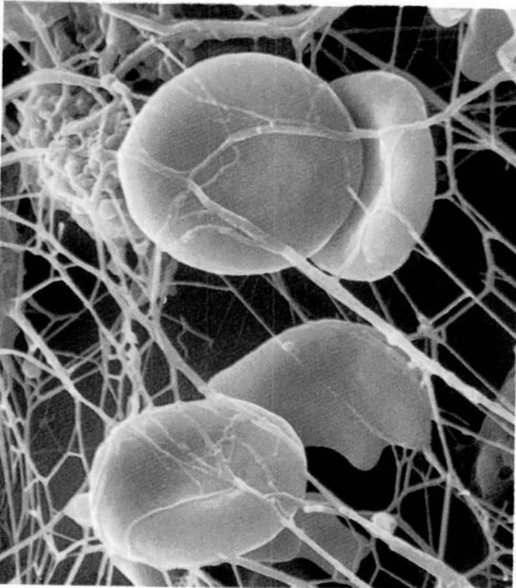

Figure 3-10 *Red blood cells are trapped in a fiber network. The clot that is created closes off the cut and stops it from bleeding.*

sels are the only exception to the general rule about the blood-carrying roles of arteries and veins.

Chapter-end projects 3 and 5 apply to this section.

Demonstrations

1. Show students capillary circulation in the tail of a small goldfish. This is observed easily under a microscope. Figure TE 3-1 shows how to set up the demonstration. Be sure the cotton is wrapped loosely around the head and gills of the goldfish, and add a small amount of water to the Petri dish. Add water to the cotton if it starts to dry out. The slide is placed over the fish's tail to hold it flat and spread it out. Place the goldfish back in its container within 15 minutes. It may be more convenient to set up two or three goldfish tails, so the fish are returned to water quickly.

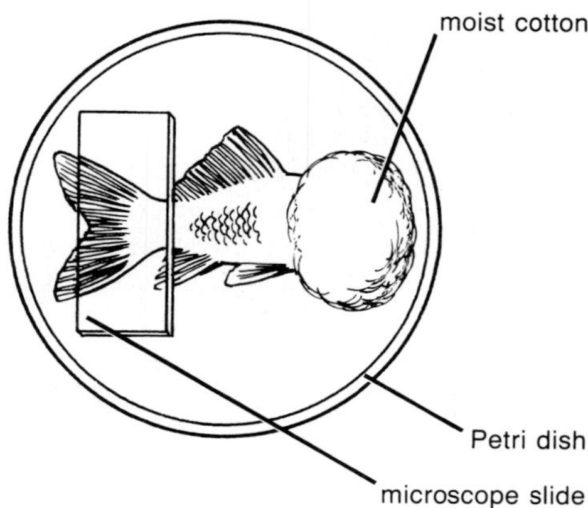

Figure TE 3-1 Demonstration showing capillary circulation in the tail of a goldfish.

2. Show students a model of the capillary bed between an arteriole and a venule. Separate and spread out the central strands of a length of coarse hemp cord. See Figure TE 3-2. Dip one unfrayed end in red ink and the other in blue ink. Display the cord segment against a sheet of white paper for the best effect.

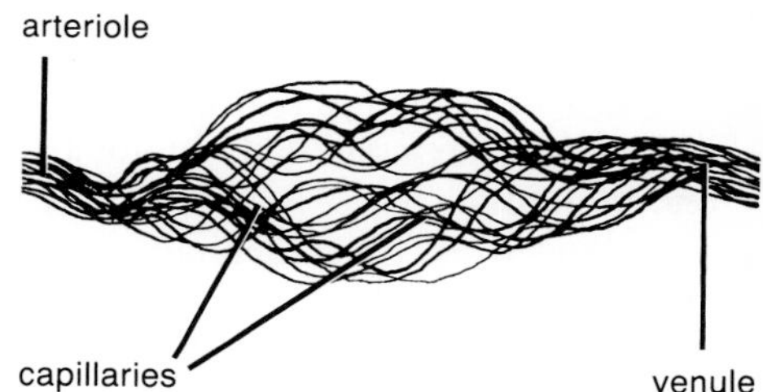

Figure TE 3-2 Model of a capillary bed between an arteriole and a venule.

Teaching Tips

- Figure 3-8 shows red blood cells magnified 6,000 × . Have students compare and contrast the red blood cells with the white blood cell, shown in Figure 3-9.
- When discussing the origin of red blood cells and platelets, have students look back at Figure 3-2 to remind themselves of the location of bone marrow.

- Arrange to borrow some stethoscopes from your school nurse, local public health department, or local physicians. Show students where to place the stethoscope tip on the chest or back to hear the heart beat. CAUTION: Students should not tap on or blow into the stethoscope tip when the earpieces are in a student's ears.
- Invite the school nurse to your class to demonstrate how blood pressure is measured. Include a discussion of the causes, treatment, and prevention of high blood pressure.
- Arrange for a Red Cross representative to speak to your class about blood typing, transfusions, and the Red Cross Blood Donor Program.
- Have interested students research diseases of the circulatory system, and report their findings to the class.
- Have interested students research careers related to the circulatory system. They can interview a cardiologist, an electrocardiogram technician a laboratory technologist, a blood-bank technician, a hematologist, and others, at a local medical center.

Facts At Your Fingertips

- The heart pumps the entire volume of blood through the body in about one minute.
- Arteriosclerosis, or hardening of the arteries, results in loss of arterial elasticity, impeding blood flow through the body. The loss of "push" in the arteries places an extra burden on the heart.
- When blood is drawn for a blood test or transfusion, it is taken from the veins. A hypodermic needle can penetrate the thin walls of a vein easily; the blood flows out gently, not in spurts as it would if drawn from an artery. A "fingerstick" test is when a small amount of blood is drawn from the fingertip.
- Some capillaries are so narrow that the red blood cells travel through them single file.

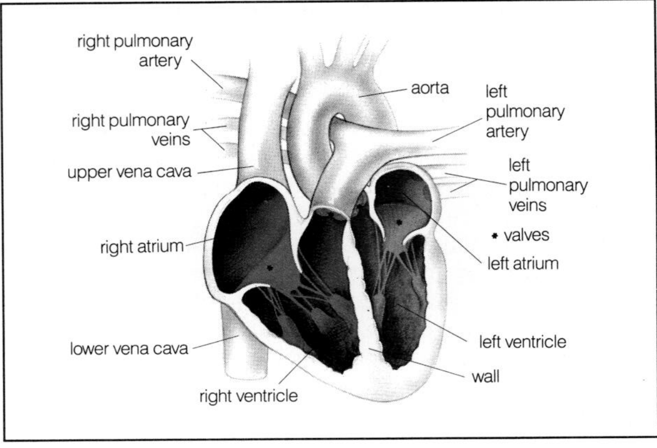

Figure 3-11 *Vena cavas bring oxygen-poor blood to the heart. Pulmonary arteries carry blood to the lungs. Pulmonary veins bring oxygen-rich blood from the lungs to the heart.*

Look at the diagram of a human heart in Figure 3-11. As you read about each part, try to locate it in the diagram. The heart has a thick wall separating the right side from the left. Each side of the heart is divided into two areas or chambers. Each upper chamber is called an *atrium* (A tree um, plural, *atria*). The two lower chambers are called *ventricles*. Each chamber acts as a pump, pushing the blood through the body. Valves between each atrium and ventricle allow blood to flow through the heart in one direction and keep blood from flowing backward in the heart. There are also valves between the ventricles and the blood vessels leading away from the heart. The closing of these two sets of valves creates a lub-dub sound, which can be heard with a stethoscope.

During a normal heartbeat, the right and left atria contract at the same time, pumping blood into the relaxed right and left ventricles. (See Figure 3-12a.) While the atria relax, the right and left ventricles contract. The right ventricle forces blood to the lungs. The left ventricle pumps blood to the rest of the body. (See Figure 3-12b.)

You can trace a drop of blood as it flows through the body. Refer to Figure 3-11 as you follow the path of blood. Begin your journey at the right atrium. Here, carbon dioxide-rich blood from the body enters the heart. A contraction pumps it into the right ventricle. The atrium relaxes as the ventricle contracts, pumping the blood to the lungs. Oxygen

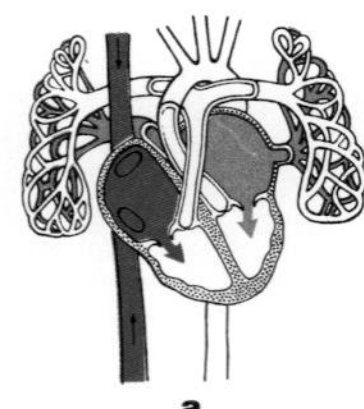

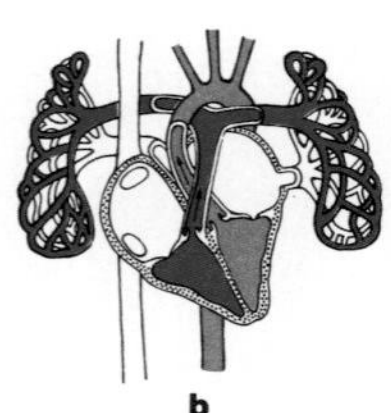

Figure 3-12 *The heart is called a double pump because it sends blood through two circulatory "loops."*

- All veins, except the pulmonary veins, contain deoxygenated, dark red blood. A pigment in the skin makes venous blood appear blue when seen through the skin.

Mainstreaming

A three-dimensional model of a mammalian heart will greatly help visually-impaired students grasp its anatomy.

Cross References

The blood disease, hemophilia, is discussed in section 2-6 of Chapter 2. The effects of drugs on the heart are mentioned in section 3-7, Chapter 3.

and carbon dioxide are exchanged between the blood and the air in the lungs. The oxygen-rich blood leaves the lungs and enters the left atrium. From here, a contraction pumps the blood into the left ventricle. Finally, a strong contraction pumps the blood out through the *aorta* (a OR tuh), a large, branching blood vessel. Blood vessels branching from the aorta carry blood to all parts of the body.

Blood flows throughout the body in a series of vessels that are like one-way streets in a city. There are three types of blood vessels in the body: (1) arteries, (2) capillaries, and (3) veins (vaynz). Except for the pulmonary arteries, **arteries** carry oxygen-rich blood away from the heart toward the cells. Arteries have thick, muscular, elastic walls, which must withstand the pressure, or force, of the blood being pumped from the heart. The thick walls of an artery stretch and contract as the blood is pushed by the rhythmic contractions of the heart. The contraction of the elastic artery walls pushes the blood along in spurts.

Millions of narrow, thin-walled **capillaries** connect the arteries and veins. This network reaches every cell. Food, oxygen, and wastes are exchanged across the thin walls of the capillaries and the body cells.

Except for the pulmonary veins, **veins** carry blood containing little oxygen and a lot of carbon dioxide back toward the heart. The walls of veins are thinner and less muscular than the walls of the arteries. By the time blood reaches the veins, the pressure is very low. Movement of blood in the veins is helped along by the contractions of nearby muscles. (See Figure 3-13.) Veins contain tiny valves that prevent blood from flowing backwards. These valves allow blood to move only toward the heart.

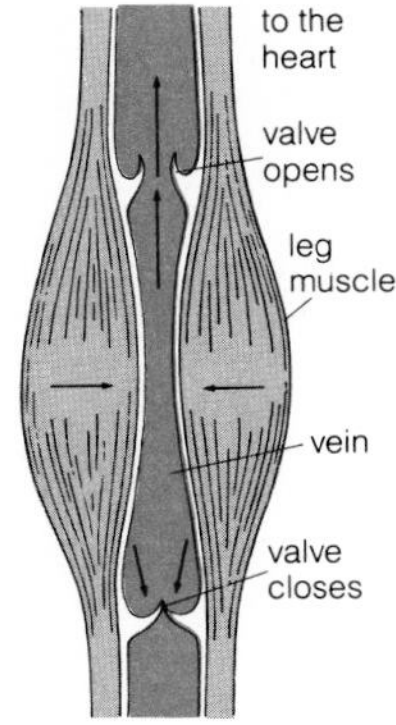

Figure 3-13 *When leg muscles contract, the vein is squeezed, pushing the blood along. The valves slam shut if the blood begins to flow backwards.*

arteries

capillaries

veins

Checkpoint

1. Name the major parts of the circulatory system.
2. Name the components of human blood. Describe the function of each.
3. Trace a drop of blood through the body beginning at the right side of the heart.
4. What is the function of the capillaries?
5. How do arteries and veins differ?

TRY THIS

Find your pulse at your wrist using your first two fingers but not your thumb. Count the number of times the artery wall stretches in one minute to find your heart rate.

Using Try This

Some students may need aid finding their pulses. The thumb should not be used because it has a pulse that can interfere with the counting of the wrist pulse. Be sure students understand that pulse rates vary a great deal. A normal resting pulse can vary from 40 to 90 beats per minute. CAUTION: Do not allow students to take their pulses at the neck or temple. Too much pressure on arteries in these regions can cause a drop in blood pressure.

Try having an athletic student jog in place for several minutes and then take his or her pulse. Compare the pulse rate before and after the exercise. CAUTION: Check on medical restrictions before having a student jog.

Checkpoint Answers

1. The heart, blood vessels, and the blood.
2. Red blood cells carry oxygen from the lungs to the body cells. They also carry some carbon dioxide from the cells to the lungs. White blood cells surround and destroy bacteria and produce antibodies. Platelets aid in blood clotting.
3. Answers will vary. See student text, bottom of page 72.
4. Food, oxygen, wastes, and carbon dioxide are exchanged between the capillaries and the body cells.
5. All arteries, except the pulmonary arteries, carry oxygen-rich blood away from the heart toward the cells. Arteries have thick, muscular, elastic walls that stretch and contract, pushing the blood along in spurts. All veins, except the pulmonary veins, carry blood containing little oxygen and much carbon dioxide toward the heart. The walls of veins are thinner and less muscular than the walls of arteries. Veins contain tiny valves that permit blood to move only toward the heart.

3 THE HUMAN BODY
Computer Program

Aerobic Exercise

Purpose To determine your ideal pulse rate during aerobic exercise.

Background In aerobic exercise, you exercise and then measure your pulse rate to determine how hard your heart is working.

ENTER the following program:

```
10 DIM P(5)
20 INPUT "ENTER YOUR AGE."; A
30 INPUT "ENTER RESTING PULSE RATE."; RP
40 FOR I = 1 TO 5
50 INPUT "ENTER PULSE READING"; P(I)
60 IF P(I) < RP THEN GOTO 100
70 NEXT I
80 LET PP = .60
90 GOTO 120
100 IF I < 4 THEN LET PP = .80
110 IF I = 4 THEN LET PP = .70
120 LET PE = (220 - A) * PP
```

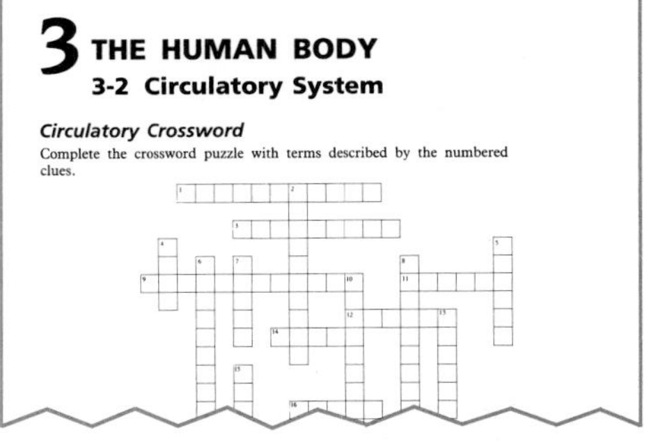

3 THE HUMAN BODY
3-2 Circulatory System

Circulatory Crossword

Complete the crossword puzzle with terms described by the numbered clues.

Teaching Suggestions 3-3

Open the section by reminding students that respiration and breathing are not the same thing. Breathing is carried out by the respiratory system and involves the exchange of oxygen and carbon dioxide between an organism and its environment. Respiration is a cellular process. It is a series of reactions between nutrients and oxygen in the mitochondria, resulting in the release of energy and the production of carbon dioxide and water. The act of breathing and the circulation of blood bring a fresh supply of oxygen to the cells for respiration. Respiration provides the energy for all life activities.

Emphasize that breathing is made possible by the action of the chest muscles and diaphragm. The combined movements of these change the volume of the chest cavity, causing air to flow into the lungs or be expelled. See Demonstrations for a model that shows the mechanics of breathing.

Point out that the nicotine in tobacco smoke has an adverse effect on the respiratory system. Nicotine stops the movement of cilia in the trachea, inhibiting their filtering action. Ask students what effect this would have on the health of the respiratory system. Explain that tobacco smoke dries out the moist mucous membranes of the respiratory system also.

Laboratory Activity 5, Measuring Lung Volume, student text page 537, should be done in conjunction with this section. Chapter-end project 2 also applies to this section.

3-3

Breathing

Goals

1. To describe the actions involved in breathing.
2. To explain the function of breathing.

You breathe twelve to sixteen times each minute, yet most of the time you are unaware of it. Though you can command yourself to breathe or to hold your breath, most of the time your brain controls your breathing automatically.

The function of breathing is to put oxygen into the blood and remove carbon dioxide from it. Oxygen is used by the cells to obtain energy from food through cellular respiration. Carbon dioxide and water are given off as a result.

During breathing, air is warmed as it enters the nose or mouth. (See Figure 3-14.) Dust and other particles are trapped by tiny hairs and moist surfaces as the air passes through the nose and throat and enters the **trachea**. The trachea, or windpipe, is an air tube that is held open by rings of stiff cartilage. It divides into two *bronchi* (BRAHNG ky, singular, *bronchus*), each of which enters one of the two lungs. The bronchi branch into smaller and smaller tubes called *bronchioles*.

trachea (TRAY kee uh)

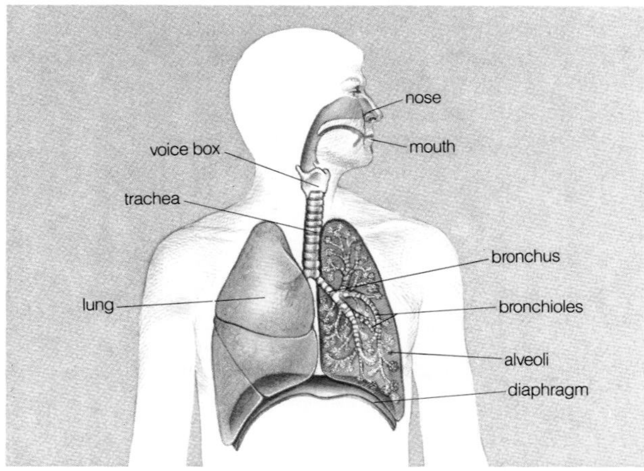

Figure 3-14 *This view shows a cut-away picture of one of the lungs. The lungs are part of the respiratory system.*

Demonstrations

1. Show students a way to feel the movements of breathing. Have them place the hands on the sides of the chest while taking two or three deep breaths with the stomach muscles held in. Ask them to describe the movements they feel.
2. Demonstrate that water vapor is one component of exhaled air. Have a student breathe out on a small pocket mirror held close to his or her mouth. Water will condense on the glass.
3. Show students a model of the capillary circulation around an alveolus. Try strands of fine cord wrapped around a Florence flask, or draw "capillaries" on the flask with a glass-marking pencil.

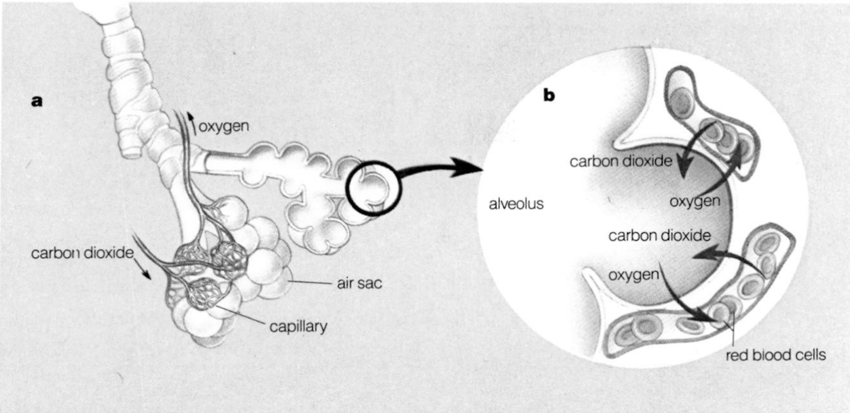

Figure 3-15 *There are over 300 million alveoli in each lung. Alveoli provide a large surface area for the exchange of gases. Oxygen and carbon dioxide are exchanged between the air in an alveolus and the blood in the surrounding capillaries.*

A cluster of moist, thin-walled air sacs called **alveoli** (singular, *alveolus*) lies at the end of each bronchiole. There are hundreds of millions of alveoli in each lung. The thin-walled alveoli are surrounded by tiny capillaries. (See Figure 3-15a.) The exchange of carbon dioxide and oxygen occurs between the thin walls of the alveoli and the capillaries. (See Figure 3-15b.)

alveoli
(al VEE uh ly)

In the exchange process, carbon dioxide brought from all parts of the body diffuses from the blood in the capillaries into the air in the alveoli. At the same time, oxygen diffuses from the air in the alveoli to the blood. The oxygen attaches to red blood cells in the capillaries. The red blood cells transport the oxygen to all the body cells.

Fresh air must be brought in to the alveoli constantly. Breathing makes this possible. The process of breathing consists of two actions, inhalation and exhalation. *Inhalation* is movement of air into the lungs. *Exhalation* is the movement of air out of the lungs. The lungs have no muscles and are unable to move air in and out on their own. The **diaphragm**, a strong flat muscle located below the lungs at the base of the ribs, and the chest muscles control inhalation and exhalation.

diaphragm
(DI uh fram)

4. Demonstrate the mechanics of breathing by means of a model. See Figure TE 3-3. The handle at the center of the rubber sheeting "diaphragm" is made by inserting a small cork and tying it off. When the "diaphragm" is pulled down, the volume of trapped air inside the bell jar increases, and its pressure decreases below atmospheric pressure. The two "lungs" thus expand. The reverse occurs when the "diaphragm" is pushed up into the interior space. Point out that movement of the chest walls is not shown by this model.

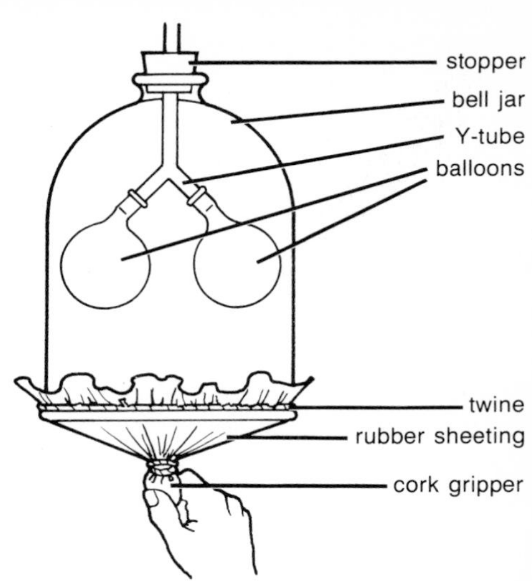

Figure TE 3-3 Model of the mechanics of breathing.

Teaching Tips

- Invite an emergency medical technician, nurse, or Red Cross instructor to your class to demonstrate artificial and mouth-to-mouth resuscitation.
- Have interested students research respiratory diseases, their causes, cures, and prevention, and report to the class.
- Have students investigate careers related to the respiratory system, and report their findings to the class.

Facts At Your Fingertips

- The scientific name for the voice box is larynx. Laryngitis is an inflammation of the larynx; it causes hoarseness and sometimes the temporary loss of speech.
- The two lungs vary in shape and size. The right lung is usually larger than the left lung.

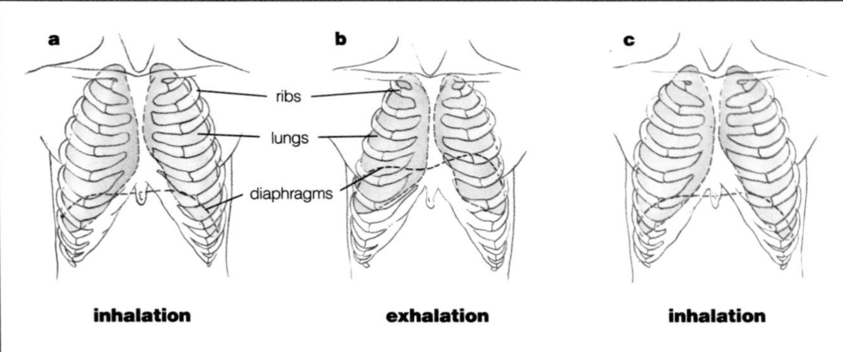

Figure 3-16 *(a) The ribs move up and the diaphragm moves down. (b) The diaphragm and chest muscles relax. Air is forced out of the lungs. (c) Air rushes in, filling the lungs.*

During inhalation, the contractions of the chest muscles pull the ribs up. At the same time, the diaphragm contracts and moves downward. (See Figure 3-16a.) These two movements increase the volume of the chest cavity and reduce air pressure in the lungs. Air pressure inside the lungs becomes lower than air pressure outside the body. Air from outside rushes in, causing the lungs to expand.

Exhalation occurs when the chest muscles relax and the ribs move down. The diaphragm relaxes at the same time and moves upward. These movements make the chest cavity smaller. Air is forced out and the lungs return to their original size. (See Figure 3-16b.)

Exhaled air is not pure carbon dioxide. It contains a mixture of gases, including oxygen, and can be used by the body if inhaled a second time. This is why mouth-to-mouth resuscitation can save the life of someone who has stopped breathing. During resuscitation, the rescuer exhales air into the mouth of the victim.

Checkpoint

1. What is the function of breathing?
2. Trace the pathway of oxygen as it enters your body from the air until it reaches your cells.
3. Describe the role of the diaphragm and chest muscles during inhalation and exhalation.
4. What is oxygen used for in the cells?

Mainstreaming

A three-dimensional model of the lungs and chest cavity will help all handicapped students grasp their anatomy and function. In addition, listening to breathing action with a stethoscope will be instructive to all handicapped students except the hearing impaired.

Cross Reference

Refer to Chapter 13, section 13-1, for a discussion of atmospheric pressure.

3 THE HUMAN BODY
3-3 Breathing

Take a Breath!

The sentences in the paragraphs below are incomplete. Complete each sentence by unscrambling the letters that appear in the numbered parentheses. Write the resulting term on the corresponding numbered line at the left.

1. ______
2. ______
3. ______
4. ______
5. ______
6. ______
7. ______
8. ______
9. ______
10. ______
11. ______
12. ______
13. ______
14. ______
15. ______
16. ______
17. ______

The function of breathing is to put [1](nyogex) into the blood and remove [2](brocan iddoiex). Air passes through the nose and throat into the [3](charate). Air then flows into two [4](chirbon), each of which enters one of the lungs. In the lungs the bronchi branch into smaller tubes called [5](slochinbore). Each is then attached to a cluster of moist, thin-walled sacs called [6](lleavio). The exchange of gases takes place through the thin walls of the [7](lleavio) and the [8](sicrealpail). [9](ithonalain) is a movement of air into the lungs; [10](theaxinoal) is a movement out. These motions are controlled by muscles of the chest and a strong flat muscle beneath the lungs called the [11](midhaparg). When the chest muscles relax and the [12](midhaparg) moves up, air is [13](dexheal). When the lung muscles tighten, they raise the rib cage and the [14](midhaparg) moves down; air is [15](leadinh).

Air that leaves the lungs still contains [16](nyogex). This fact makes it possible to save someone's life by mouth-to-mouth [17](suttainsoirec).

Respiratory Tracking

The terms in the left-hand column are related to breathing. Think of a word or phrase that has the same meaning as the term in the left-hand column. Write this word or phrase in the middle column. In the right-hand column write the name of the place or places where the part is located or the activity occurs.

trachea	______	______
alveoli	______	______
capillaries	______	______
inhalation	______	______
resuscitation	______	______
exhalation	______	______

Checkpoint Answers

1. The function of breathing is to put oxygen into the blood and remove carbon dioxide from it.
2. Answers will vary. See student text pages 74–75.
3. During inhalation the diaphragm contracts and moves downward while the chest muscles pull the ribs up. These movements increase the volume of the chest cavity and reduce air pressure in the lungs, causing air to rush in and fill the lungs. Exhalation occurs when the chest muscles and diaphragm relax, making the chest cavity smaller and forcing air out of the lungs.
4. Oxygen is used for respiration.

3-4

Digestion and Waste Removal

Goals

1. To explain how food is digested and absorbed by the body.
2. To describe the structure and function of the excretory system.

digestion

All living things need food. Food supplies materials for cell growth and repair, and energy for all life activities. Food cannot be used by cells in the form in which it is eaten. It must be broken down into simpler substances first.

The breaking down of food into smaller and simpler substances is called **digestion**. There are two types of digestion, mechanical and chemical. *Mechanical digestion* breaks food into smaller pieces. The teeth digest food mechanically, by tearing and grinding it up. *Chemical digestion* changes the small pieces of food into simpler substances. Chemical digestion occurs in the stomach (and other organs) when digestive juices break food down from large molecules to smaller ones. The digestive juices are aided by chemicals called *enzymes* that speed the breakdown of food.

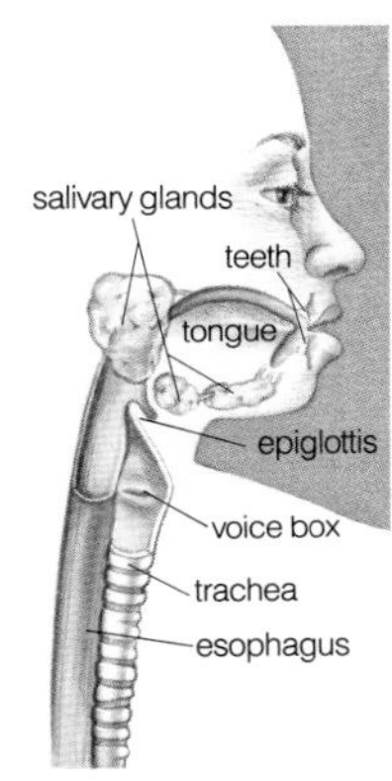

Figure 3-17 *Digestion begins in the mouth.*

Digestion and absorption of food take place in the *digestive system*. Food enters the digestive system through the mouth. (See Figure 3-17.) There, teeth start the digestive process by grinding food into small pieces. The tongue mixes the food with *saliva* (suh LY vuh), a watery liquid released into the mouth by the *salivary* (SAL uh vehr ee) *glands*. Saliva moistens and lubricates food, making it easier to swallow. An enzyme in saliva begins the digestion of starches.

esophagus
(ih SAHF uh gus)

When food is swallowed, it goes down the throat and into the **esophagus**, a long tube that connects the throat and the stomach. A flap called the *epiglottis* (ep ih GLAHT is) drops down like a trap door, covering the trachea, to keep food from getting lodged there. (Within the trachea is the voice box.) The esophagus is about 25 cm long. Its walls are made of smooth muscle and can stretch to accommodate large pieces of food. The walls of the esophagus contract rhythmically, moving food to the stomach. (See Figure 3-18.)

Teaching Suggestions 3-4

Begin with a discussion of food and the need for food. Ask students why eating is necessary. Emphasize that food supplies the energy an organism needs to carry out its life activities. It provides the raw materials for operating, building, and repairing cells also.

Explain that saliva is secreted in response to the sight, smell, taste, or thought of food. Saliva softens food and acts as a lubricant, making swallowing easier. Point out that the tongue contains specialized sensory neurons (the taste buds) that enable people to differentiate between foodstuffs. When one swallows, the tongue acts to push food into the esophagus.

Be sure students understand that the organs of the digestive system actually lie quite close together within the peritoneal cavity. They are shown slightly separated in Figure 3-18 so that the relative sizes, shapes, and positions of the organs are clear.

Point out that no digestion takes place in the esophagus or large intestine. Emphasize that no food travels to the pancreas, liver, or gall bladder.

Students may have difficulty grasping how the arrangement of the villi increases the ability of the small intestine to absorb digested nutrients. Point out that with its inner surface folded into villi, the small intestine has a much greater surface area than if it were a smooth tube. This is like a bath towel with thick pile as compared with a smooth sheet. In addition, the villi are in constant motion, thus mixing and bringing digested nutrients into contact with the absorbing surface cells.

Some students may be curious about the appendix. Explain that this is a small sac near the juncture of the small and large intestines. This organ appears to have no function in humans.

Be sure students understand that urine formation is not the only method of excretion in the human body. The skin (sweat), the lungs (gaseous wastes), and the intestinal tract (solid wastes) also have excretory functions.

Plan to do laboratory Activity 6, Mapping Taste Areas of the Tongue, student text page 538, in conjunction with this section.

Demonstrations

1. For a dramatic demonstration of the effects of hydrochloric acid on a substance, place a few drops of concentrated HCl on a piece of cloth. Show students how the acid corrodes the cloth fibers. Relate this to the situation within the stomach, where dilute HCl breaks down food. Point out that without its mucous lining the stomach would be eaten away by digestive juices.

2. Show students a model of a villus that permits explanation of its internal circulation. See Figure TE 3-4 for the anatomy of the villus. Obtain two test tubes, one of which just slips inside the other. Using a glass-marking pencil, draw a capillary net on the surface of the smaller tube. Insert this into the larger tube. The two together represent a villus with an outer layer of epithelial cells. The cavity inside the smaller tube represents a lacteal. See Facts At Your Fingertips.

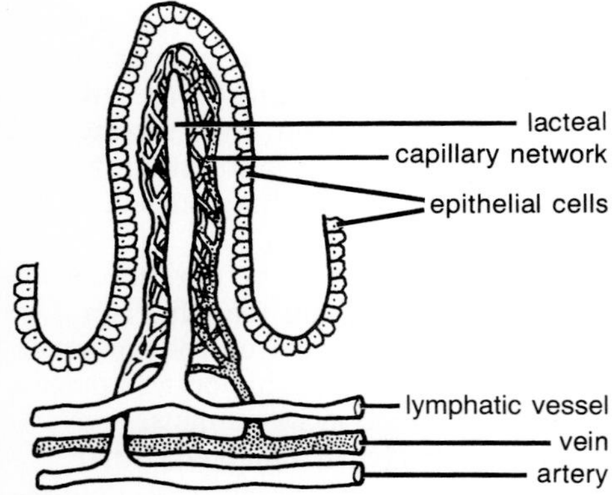

Figure TE 3-4 Model of a villus.

Teaching Tips

- Introduce the term nutrients—substances in foods needed for the organism to stay alive—when discussing digestion. List the nutrients on the chalkboard: proteins, carbohydrates, fats and oils, vitamins, minerals, and water. (Water is considered a nutrient because it is needed by all cells to survive.) Discuss examples of foods that are rich in each nutrient.
- Have interested students research and report on the recommended amounts of the nutrients in a balanced diet. Some may be interested in researching the eating disorders anorexia and bulimia.
- Explain that the term "diet" means the usual food and drink consumed by a person each day. Caution students not to follow any special diets without first consulting a physician. Bring in newspaper and magazine ads that promote so-called sure-fire weight-loss diets. Help students analyze the unhealthy aspects of these diets. Discuss some of the hazards of fad diets. Invite a dietician to talk about developing good eating habits.
- Discuss the causes and prevention of digestive disorders, such as indigestion, gas, and ulcers.
- Have an interested student research dialysis and kidney transplants and report to the class.

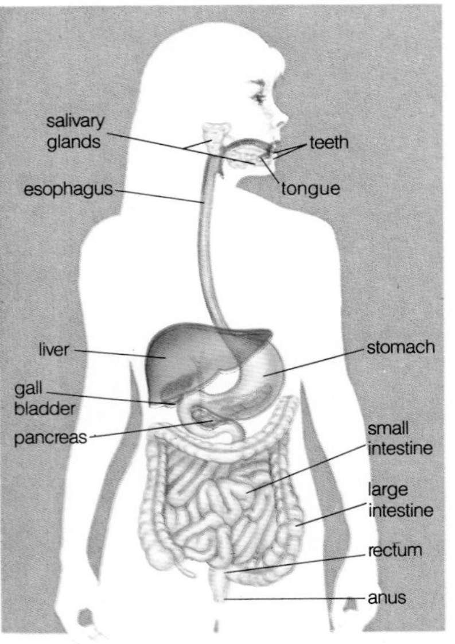

Figure 3-18 The digestive system is made up of organs that form a long tube. The food tube would be about 9 m long if it were stretched out.

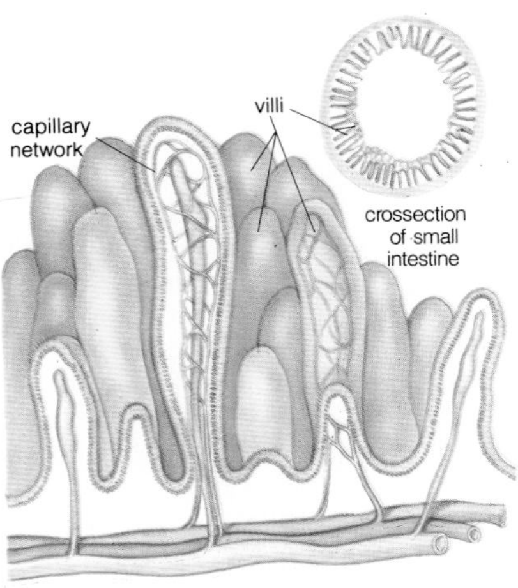

Figure 3-19 The inner surface of the small intestine is folded into villi. Many more cells are exposed to digested food than if it were a smooth tube.

TRY THIS

Enzymes in your saliva digest large starch molecules into smaller sugar molecules. Place a piece of unsalted cracker in your mouth and leave it there for three minutes. How does the taste change?

The walls of the stomach contain smooth muscle. Muscle contractions churn and grind up the food, mixing it with *gastric juice*, given off by cells in the stomach lining. Gastric juice contains hydrochloric acid and an enzyme that digests protein. As the digestive juices mix with the food, it begins to look like a thick milkshake. This partially digested food is squeezed out of the stomach in spurts by muscular contractions and moved into the small intestine.

A thick mucous lining in the stomach protects it from being digested. If the stomach releases too much hydrochloric acid, ulcers can result. Ulcers are areas where the stomach lining is "eaten away" and sore.

Most digestion takes place in the small intestine. Juices from the pancreas (PANG kree us) and liver flow into the small intestine through tiny tubes. One of the juices from the liver is stored in a baglike organ called the *gall bladder*.

Facts At Your Fingertips

- Food and stomach juices are sometimes regurgitated into the esophagus, causing "heartburn."
- Food usually remains in the human stomach for two to three hours.

The gall bladder releases *bile* which helps break up fats in the small intestine. All the juices combine with enzymes given off in the small intestine and complete the digestion of food.

The walls of the small intestine are lined with fingerlike projections called **villi**. (See Figure 3-19.) The villi contain networks of capillaries. Digested food molecules diffuse into the capillaries at the villi. The food is carried by the blood to all body cells.

villi (VIL eye)

Not all the food you eat can be digested. Undigested food enters the two-meter long large intestine as a watery mass. Most of the water is absorbed by the large intestine and reused by the body cells. The undigested food collects at the end of the large intestine before being passed out through the rectum and the anus (AY nus).

Respiration of food produces waste products. If these wastes build up, they could poison the body. Cellular waste products are removed from the body by the *excretory system*.

Urea (yu REE uh) is a waste product formed in the liver. Blood carries urea from the liver to the **kidneys**, two bean-shaped organs located on each side of the spine at the lower back. (See Figure 3-20.)

kidneys

The kidneys contain millions of tiny tubes surrounded by capillaries. Urea and water pass out of the blood and into these tubes, forming **urine**. Urine passes through the *ureters* (yu REE turz) and into the *urinary bladder*. The bladder, which has walls of smooth muscle, stores urine. When it is full, it contracts and urine is excreted through the *urethra*.

urine (YUR in)

Checkpoint

1. Define digestion. Explain the differences between mechanical and chemical digestion.
2. What are enzymes? Name two places in the body where they are released.
3. Trace a piece of food through the digestive system, beginning at the mouth.
4. Where does diffusion of food molecules into the blood take place?
5. What is the function of the excretory system?
6. List the parts of the excretory system and their function.

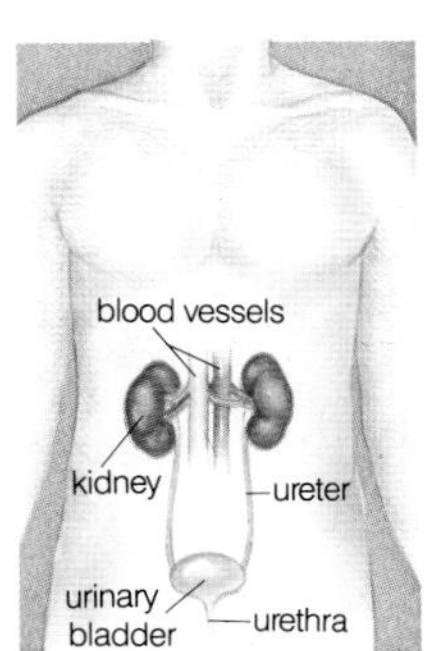

Figure 3-20 *The excretory system.*

- Food does not usually enter the appendix. Occasionally, however, food particles may become trapped there, and cause infection. Inflammation of the appendix (appendicitis) may result. An appendectomy, or the surgical removal of the appendix, eliminates the site of infection and the danger of rupture.
- The central portion of each villus contains a lacteal (a lymph vessel), which absorbs digested fat molecules. The lacteals are shown in gold on the diagram of the villi, Figure 3-19. See demonstration 2 and Figure TE 3-4 on page 78 for a simple model of a villus.
- People can survive with one healthy kidney. If both are diseased or damaged, however, the individual must use a so-called artificial kidney or receive a kidney transplant.

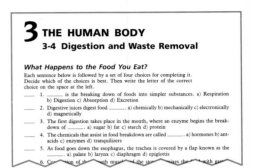

Mainstreaming

A three-dimensional model of the digestive system will be helpful to all handicapped students. The sense of taste is usually undamaged in the handicapped. Have these students use other starchy foods in the Try This activity.

3 THE HUMAN BODY
3-4 Digestion and Waste Removal

What Happens to the Food You Eat?

Each sentence below is followed by a set of four choices for completing it. Decide which of the choices is best. Then write the letter of the correct choice on the space at the left.

___ 1. _____ is the breaking down of foods into simpler substances. a) Respiration b) Digestion c) Absorption d) Excretion

___ 2. Digestive juices digest food _____. a) chemically b) mechanically c) electronically d) magnetically

___ 3. The first digestion takes place in the mouth, where an enzyme begins the breakdown of _____. a) sugar b) fat c) starch d) protein

___ 4. The chemicals that assist in food breakdown are called _____. a) hormones b) antacids c) enzymes d) tranquilizers

___ 5. As food goes down the esophagus, the trachea is covered by a flap known as the _____. a) palate b) larynx c) diaphragm d) epiglottis

Using Try This

Be sure to use unsalted crackers. Students should find that the matter in the mouth begins to taste sweet as a result of salivary amylase digesting starch (tasteless) to sugar (sweet).

Checkpoint Answers

1. Digestion is the breaking down of food into simpler substances. Mechanical digestion breaks food particles into smaller pieces. Chemical digestion changes small particles of food into simpler substances.
2. Enzymes are substances that bring about and speed the breakdown of food. Enzymes are released by glands in the mouth, by the stomach lining, by the pancreas, and by glands in the small intestine.
3. Answers will vary. See student text, pages 77–79.
4. Diffusion takes place at the villi.
5. The excretory system removes cellular waste products from the body.
6. The kidneys remove urea and water from the blood and form urine. Urine passes through the ureters and into the urinary bladder. The bladder stores urine, which is then excreted through the urethra.

Teaching Suggestions 3-5

An interesting way to open this lesson is to ring a bell, bang on a desk, or perform some activity that will produce an obvious response from your students. Introduce the terms "stimulus" and "response" in connection with what you did and then have students give other examples of stimulus-response situations.

Be sure students understand clearly that neurons are not attached to each other, and that the transmission of a nerve impulse from neuron to neuron is a chemical event.

Emphasize the three types of neurons and the roles they play in the actions of the nervous system. It's important, in addition, that students grasp the difference between a reflex action and a brain-controlled response to a stimulus. Selecting a ripe piece of fruit in terms of its appearance (sight) or texture (touch), for example, is a case of the latter. Have students suggest other examples.

Point out that many reflex responses, such as eye-blinking, coughing, and sudden withdrawal from heat, protect a person from harm. Because the reaction is automatic rather than a thinking event, much less time is required for the response.

Explain that during a reflex response, such as the one shown in Figure 3-25, impulses are sent to the brain while the response is taking place. The pain is sensed when the brain receives the impulses.

Demonstration

Show students a reflex response. Have a student stand and hold a sheet of clear, stiff plastic in front of his or her face. Have another student throw crumpled wads of paper at the plastic. No matter how hard the first student tries, he or she will not be able to avoid blinking when the paper wads strike the sheet of plastic. Discuss the importance of this reflex response in terms of the safety of the eye.

3-5

Nervous System

Goals

1. To identify the parts of the nervous system.
2. To explain the function of the nervous system.

The *nervous system* controls and coordinates all the activities of the body. The brain and spinal cord make up the *central* nervous system. Nerves that branch from the central nervous system to the rest of the body make up the *peripheral* (puh RIF ur ul) nervous system. (See Figure 3-21.) The nervous system is similar to a telephone system. Messages are received and then responded to. Any "message" that produces a response is called a **stimulus** (plural, *stimuli*).

stimulus (STIM yuh lus)

neurons (NUR ahnz)

impulse

Neurons, or nerve cells, respond to stimuli from inside or outside the body. There are different kinds of neurons in the body, but all have some features in common. (See Figure 3-22.) A message, or **impulse**, enters a neuron through its branchlike parts called *dendrites* (DEN drytz). From the dendrites, the impulse passes through the *cell body* and along the *axon* (AK sahn). Impulses move in only one direc-

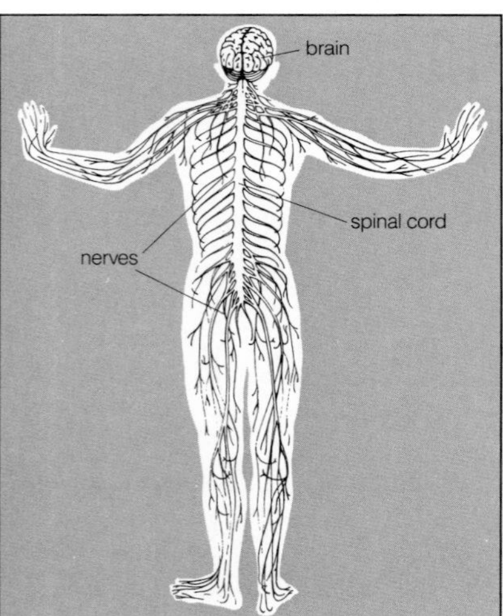

Figure 3-21 *Nerves go from the spinal cord to almost every part of the body and back again.*

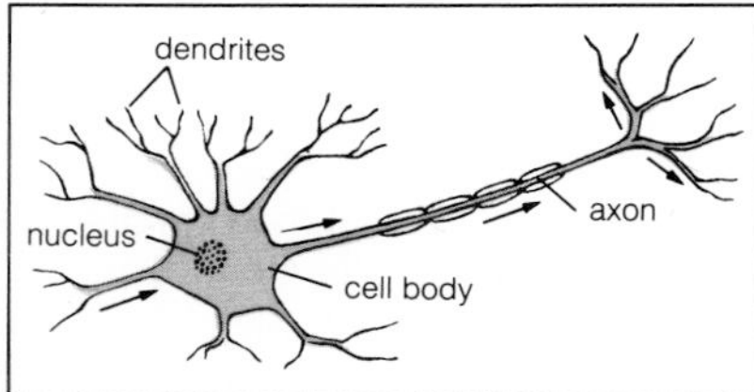

Figure 3-22 *Impulses move along a neuron from the dendrites, through the cell body, and out along the axon.*

Teaching Tips

- Be sure students do not confuse a conditioned response with a reflex response. A conditioned response, such as answering to one's name, appears to be automatic, but unlike a reflex response, it is not innate. Conditioned responses must be learned. Discuss other examples of conditioned responses.
- Have interested students look up recent developments in brain function research. A bulletin-board display of pictures and articles should interest students.
- Arrange for your class to visit a CAT-scan laboratory at a local hospital to view the equipment and perhaps to see a CAT-scan examination in progress. Have students read the Feature Article before the visit.

tion through a neuron. Some impulses move at speeds of 130 meters per second. That is like traveling across the length of a football field in one second!

Neurons are not attached to one another. There is a gap where the ends of the axon of one neuron meet the ends of the dendrites of another neuron. This gap is called a **synapse** (See Figure 3-23.) Impulses are transferred from one neuron to the next across the synapse by chemicals. When the chemicals produced in the axon ends of one neuron reach the dendrites of the next neuron, the transfer is complete. The chemicals carrying the impulse across the synapse break apart quickly, making way for chemicals carrying next impulse. Impulses move in a fraction of a second.

synapse (SIN aps)

The nervous system is made up of three types of neurons: (1) sensory, (2) association, and (3) motor. *Sensory neurons* carry impulses from *sense organs* to the spinal cord and brain. Sense organs include the eyes, ears, nose, tastebuds on the tongue, and the skin. Sensory neurons carry impulses from internal organs, too. *Association neurons* are "connectors." They receive impulses from sensory neurons or other association neurons and send the impulses to motor neurons. Most *motor neurons* transmit impulses to muscles from the brain and spinal cord.

Most responses are controlled by the brain. Three regions of the brain are shown in Figure 3-24. The **cerebrum** is

cerebrum (suh REE brum)

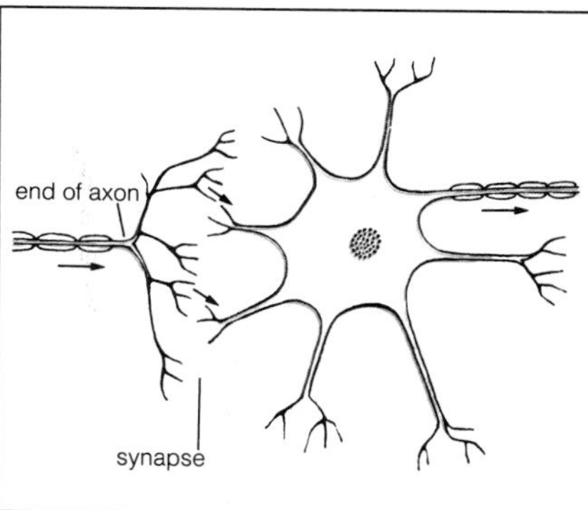

Figure 3-23 *A small electric charge can be detected during impulse transmission.*

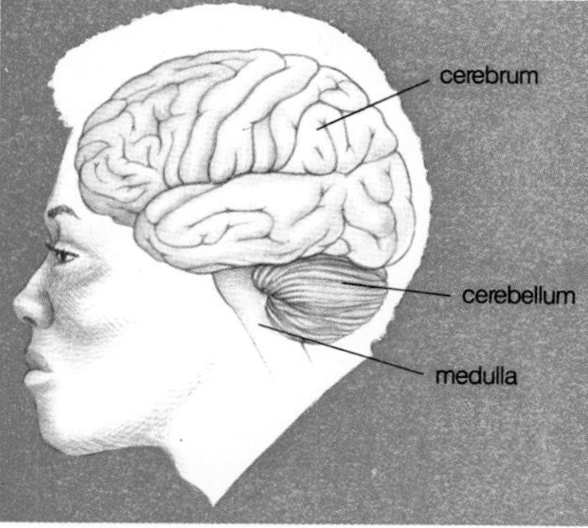

Figure 3-24 *Three main parts of the brain. Each part controls certain functions.*

Facts At Your Fingertips

- The brain is a 1.4-kilogram organ consisting of soft tissues surrounded by tough, protective membranes called the meninges. The brain is protected by the skull and is cushioned by fluid.
- A severe blow to the head can bruise brain tissue. This is a concussion.
- The brain has a rich supply of blood vessels carrying food and oxygen to its cells at all times. Severe brain tissue damage occurs if the brain is deprived of oxygen for longer than three minutes.
- The cerebrum is divided longitudinally into a right and left half. Each half, or hemisphere, is responsible for certain functions. Researchers have mapped the areas of brain function. It is thought that the right side of the cerebrum is responsible for musical, artistic, and perceptual skills, while the left side is responsible for mathematical ability, logic, and language.
- Research suggests that most people have a dominant side of the brain—they are either right-brained or left-brained. Some experts feel that a person learns best when information or skills are taught in the mode most compatible with the dominant side of the individual's brain. For example, it is suggested that people who are right-brained learn most effectively when taught visually rather than aurally.

Mainstreaming

Use a three-dimensional model of the brain to help handicapped students grasp its structure. A raised model of a neuron also will be helpful to visually-impaired students.

- Invite a speech therapist or physical therapist to discuss rehabilitation work with patients who have neurological problems. Have the speaker demonstrate techniques and discuss career opportunities in the field.
- Have an interested student contact the local Association for the Blind to find out about aids available for the visually impaired. Obtain samples of Braille and explain how it is read.

3 THE HUMAN BODY
Laboratory

What You Don't See

Purpose To locate your blind spot.

Materials ruler

Background

In the retina, or innermost layer of the eye, there are light-sensitive receptors called rods and cones. The rods and cones are connected to neurons, which become the optic nerve leading to the brain. The optic nerve comes out behind the eye at the optic disc. Here there are no rods or cones. Since no sensations of light or color can be received without rods and cones, the optic disc is also called the blind spot. The blind spot does not affect normal vision when both eyes are

the page toward you until the filled-in circle disappears. You have found your blind spot.

3. Now close your right eye and focus your left eye on the filled-in circle. Then follow the procedure in Step 2. What happens to the letter *X*? ______

Discussion

1. Does this experiment work if you focus the right eye on the *X* and the left eye on the filled-in circle? ______
2. Have a friend measure the distance

3 THE HUMAN BODY
3-5 Nervous System

Structures of the Nervous System

Column A contains terms that are related to the nervous system. Column B contains phrases that describe the terms. Match each phrase in Column B with its term in Column A by writing the letter of the phrase next to the correct term.

A	B
____ motor	a. Largest part of brain.
____ stimulus	b. Part of nervous system that consists of brain and spinal cord.
____ dendrite	c. Nerve cells.
____ axon	d. Substance that allows transfer of impulses.
____ nervous system	e. Branch-like part of nerve cell.
____ synapse	f. Kind of neuron that carries messages from sense organs.
____ association	g. Part of brain that controls involuntary activities.
____ chemical	h. System that controls and coordinates body activities.
____ central nervous system	i. Place where impulse leaves the neuron.
____ neurons	j. Message that produces body response.
____ cerebellum	k. Gap between axon and dendrite.
____ sensory	l. "Connector" neurons.
____ cerebrum	m. Type of neurons that transmit impulses to muscles.
____ reflexes	n. Part of brain that coordinates nerve impulse.
____ medulla	o. Responses controlled by spinal cord.

Communication

The nervous system is similar to a telephone network with messages flying from place to place. Imagine that your finger accidentally touches a hot pan. How would you describe the path of the message that must be sent to relieve the situation? ____________

Checkpoint Answers

1. The nervous system controls and coordinates all the activities of the body.
2. Sensory, motor, and association neurons are the three types of neurons that make up the nervous system. Sensory neurons carry impulses from sense organs to the spinal cord and brain. Association neurons receive impulses from sensory or association neurons and send the impulses to motor neurons. Motor neurons transmit impulses to muscles from the brain and spinal cord.
3. Impulses are transferred from one neuron to the next across a synapse by chemicals. When the chemicals produced in the axon ends of one neuron reach the dendrites of the next neuron, the transfer of an impulse is complete.
4. A reflex is an automatic response to a stimulus. During a reflex response, a sensory neuron receives an impulse and transmits it to an association neuron in the spinal cord. The impulse is then transferred rapidly to motor neurons and a response, such as blinking, occurs.

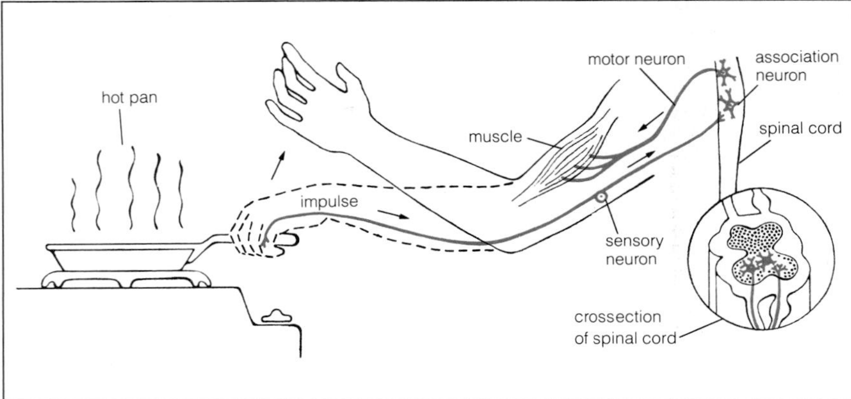

Figure 3-25 *Many reflexes, such as this one, protect people from harm. Time is saved by reacting automatically rather than thinking about what to do.*

the largest part of the brain. It controls the senses, memory, learning, thought, speech, and voluntary movement. The **cerebellum** is just below the cerebrum. It coordinates nerve impulses to and from the cerebrum. Muscle coordination and balance are controlled by the cerebellum, too. The **medulla**, located at the base of the brain, controls involuntary activities such as heartbeat, breathing, and the contraction of muscles in the digestive system.

cerebellum (sehr uh BEL um)

medulla (muh DUHL uh)

reflexes

All responses do not originate in the brain. Some responses, called **reflexes**, are controlled by the spinal cord. A reflex is an automatic response to a stimulus. Figure 3-25 shows what happens during a reflex response. The person does not have to stop and think about what to do. A sensory neuron receives the impulse. It transmits the impulse to an association neuron in the spinal cord. The impulse is transferred rapidly to the motor neuron controlling the arm muscles and a response occurs.

Checkpoint

1. What is the function of the nervous system?
2. What are the 3 types of neurons that make up the nervous system? Describe the function of each.
3. Describe how an impulse is transmitted from one neuron to another.
4. What is a reflex?

Windows on the Brain

The human brain has always been a difficult part of the body to study. It is an organ that cannot be touched or looked at without dangerous surgery. But in recent years, scientists have developed several safer, less painful ways to "look inside" the skull.

One new method is a form of x-ray photography called a *computerized axial tomography scan*, or CAT scan. During a CAT-scan examination, a fan-shaped beam of x rays is passed through the organ being studied. The x rays are detected by a scanner as they emerge at different angles from the body. A computer analyzes the x-ray pattern and "translates" it to produce a detailed cross-section map of the organ. Figure 1 shows a CAT-scan picture of a human brain.

Another tool that doctors use, the *electroencephalogram*, or EEG, takes advantage of the fact that neurons in the brain give off electrical energy in certain patterns. Doctors paste electrodes on the patient's scalp that detect the tiny amount of electrical activity given off. The measurements are recorded on strips of paper like the ones shown in Figure 2. A trained physician can recognize patterns such as those caused by epilepsy.

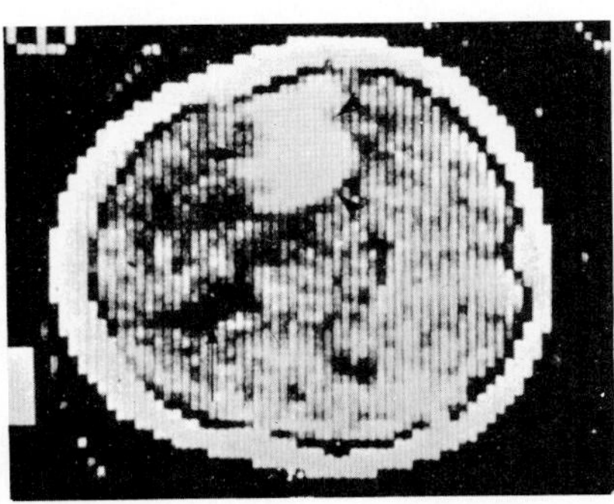

Figure 1

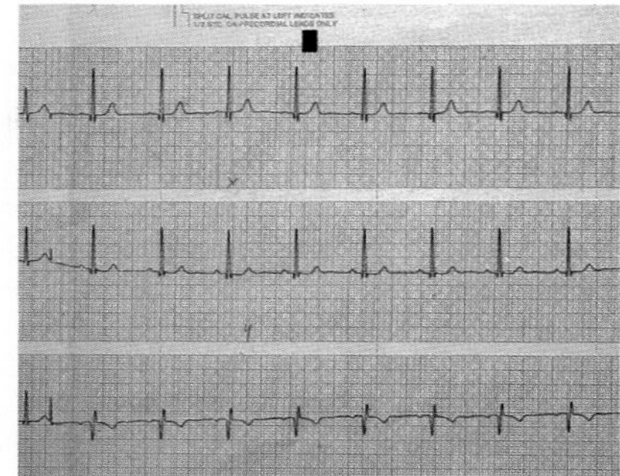

Figure 2

The newest test for problems inside the brain is called the *positron-emission tomography scan*, or PET scan. Like the CAT scan, it maps certain areas of the brain. But like the EEG, it can measure the level of activity occurring within the brain.

For the PET-scan test, a patient either inhales or is injected with a radioactive substance used by the brain. As the brain cells use the substances, high-energy gamma rays are given off, which are detected by scanners. A computer "translates" the information from the scanners into a picture of brain-cell activity.

Using The Feature Article

The title of the Feature Article refers to new ways to examine the brain. Explain that before the invention of CAT-scan and PET-scan technologies, most methods of examining the brain and other organs were invasive and potentially dangerous. Even with these new diagnostic tools, however, the brain remains one of our least understood organs.

Have students compare x-ray prints with CAT-scan pictures. Point out that x rays show differences in density—light and dark areas—and do not give a three-dimensional picture as a CAT-scan does. CAT-scan pictures are actually made in black-and-white. Computer-enhanced color is added to improve the clarity of the pictures.

Teaching Suggestions 3-6

Open the section by asking students how they feel when frightened. Try to elicit some responses that refer to pulse rate, rate of breathing, and sense of anxiety. Point out that these responses are the result of a hormone, adrenalin, being released into the blood stream.

Next, ask students if they are familiar with the condition called diabetes. Not all will be aware that diabetes is the result of too little of the hormone insulin. Explain that the hormones insulin and glucagon, both secreted by the pancreas, work together to maintain the proper level of sugar in the blood. Insulin converts excess sugar into starch and fat, which are stored in the body. Glucagon changes stored starch to sugar and releases it into the blood steam. When there is too little insulin, there is too much sugar in the blood. This is diabetes.

Summarize the endocrine glands, their hormone secretions, and their functions. See Table 3-1. Stress that both the nervous system and the endocrine system have control and coordination functions, and that they work together, not independently. Use the "fight or flight" response as an example of how the nervous and endocrine systems work together.

Be sure students understand clearly that the circulatory system allows hormones to travel from the sites of secretion to the places of activity.

Teaching Tips

- Have an interested student research the effects on the body of too little or too much of the hormones listed in Table 3-1, and report to the class.
- Draw an outline of the human body on the chalkboard and sketch in the locations of the endocrine glands. See Figure TE 3-5.
- Have a student obtain information about the insulin pump, a new device being used to treat diabetics, and report to the class.

3-6

Endocrine System

Goal

To describe the function of different hormones.

glands

hormones

Your body has two control and coordination systems. One is the nervous system. The other is the *endocrine* (EN duh krin) *system*. The endocrine system is made up of **glands**, structures that produce chemicals called **hormones**. (See Figure 3-26.) Hormones are released directly into the blood by the glands and travel to different body parts where they have specific effects. Hormones act as chemical "messengers." Each hormone causes changes in specific areas of the body. Table 3-1 describes some of the body's hormones, where they are made, and their function. For example, *thyroxin* is made by the thyroid gland. It controls metabolism. *Metabolism* is the process by which energy is produced and used in the body.

Figure 3-26 *The hypothalamus is part of the brain. It releases substances that turn endocrine glands "on" and "off."*

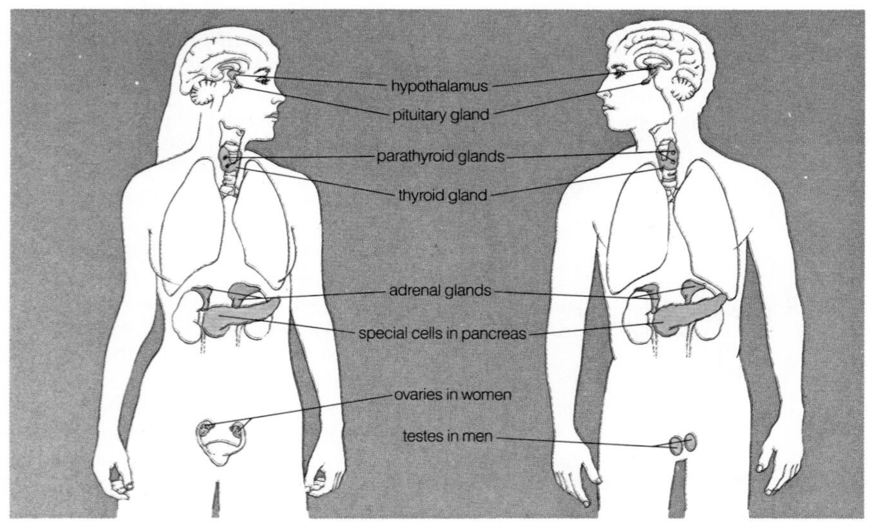

- Assign a student the task of researching pheromones (hormonelike substances secreted by many animals), and reporting to the class.

Facts At Your Fingertips

- Each hormone produces a response that is different from the response of every other hormone.
- The hypothalamus, located at the base of the brain, is an important link between the nervous and endocrine systems. Neurons of the hypothalamus secrete chemicals called releasing factors. These substances travel to the nearby pituitary gland via the blood stream, and stimulate the pituitary to secrete its hormones. Some pituitary hormones act on

Table 3-1: Some Glands and Their Hormones

Glands	Hormones	Functions
pituitary	growth hormone	controls growth of bones too little: dwarfism too much: giantism
	other pituitary hormones	control release of hormones from other glands; regulate water balance in the body; help regulate blood pressure
thyroid	thyroxin	controls release of energy from food too little: sluggishness, overweight too much: nervousness, overactivity
parathyroid	parathyroid hormone	controls the balance of calcium in the body
pancreas	insulin, glucagon	regulates blood-sugar level and use of sugar by cells too little insulin: diabetes (sugar cannot be used properly by cells)
adrenals	adrenalin	maintains "fight or flight" response by increasing blood pressure and heartbeat rate and making more food available to cells for energy
	corticoids	regulate minerals and water level in body
ovaries (OH vuh reez)	estrogen (ES truh jun)	controls development of female sex characteristics
testes (TES teez)	testosterone (tes TAHS tuh rohn)	controls development of male sex characteristics

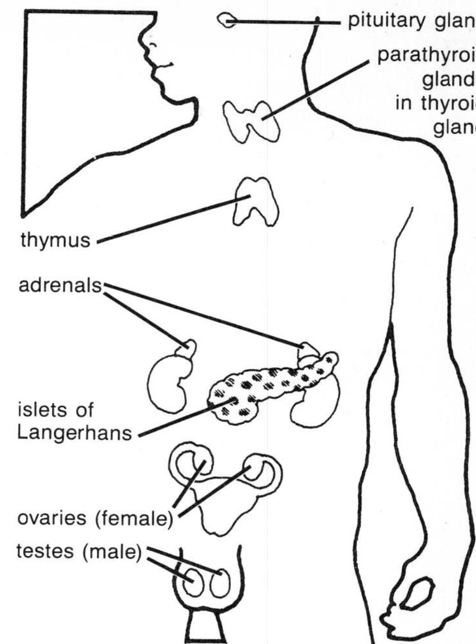

Figure TE 3-5 Locations of the endocrine glands in the human body.

other endocrine glands. For example, the pituitary secretes a thyroid-stimulating hormone.

- When the pituitary secretes thyroid-stimulating hormone, the thyroid secretes its hormone. This hormone then circulates via the blood stream through the pituitary, and suppresses secretion of thyroid-stimulating hormone. This is a feedback loop.

3 THE HUMAN BODY

3-6 Endocrine System

The Role of Glands

Read the following statements. If the statement is true, circle the *T* and leave the space blank. If the statement is false, circle the *F* and replace the underlined word with the term that makes the statement true.

1. ________________ T F The endocrine system is made up of nerves, structures that produce chemicals called hormones.
2. ________________ T F Adrenalin speeds up the heartbeat and increases blood flow to the brain and muscles.
3. ________________ T F The "fly a kite" response is an example of how the endocrine and nervous systems work together.
4. ________________ T F The pituitary gland is often called the "master gland."
5. ________________ T F Testosterone controls the development of male sex characteristics.
6. ________________ T F Too little estrogen may result in diabetes.

Glands and Hormones

Some information is missing from the chart below. Complete the chart by supplying the missing terms and functions.

Gland	*Hormone Produced by Gland*	*Function of Hormone*
1.	growth hormone	
2.		controls energy release from food
3. Parathyroid		
4. Pancreas		
5.		maintains "fight or flight" response

Checkpoint Answers

1. The pituitary is called the "master gland" because it controls the release of hormones from other endocrine glands.
2. A hormone is a chemical released directly into the blood by endocrine glands. Hormones act as chemical messengers, causing changes in specific parts of the body.
3. Growth hormone, produced by the pituitary gland, controls growth.
4. The endocrine system and nervous system together produce the "fight or flight" response. The rapidly acting nervous system responds first. The slower-acting endocrine system strengthens and maintains the response. The pituitary is an endocrine gland, affected by the release of substances from the hypothalamus, a part of the brain.

The endocrine system does not act as swiftly as the nervous system. However, hormones can be released continuously, in tiny amounts, causing long-term effects on the body. The production of too little or too much of a hormone can cause serious problems. For example, the production of too little growth hormone in children results in a condition called *dwarfism*. The production of too much growth hormone results in *giantism*.

The endocrine and nervous systems work together in coordinating body activities. The pituitary (pih TOO uh tehr ee) gland, located in the brain, is one link between the nervous and endocrine systems. It is affected by the release of substances from the part of the brain called the *hypothalamus* (hy poh THAL uh mus). The pituitary gland controls the release of hormones from other endocrine glands and is often called the "master gland."

If an angry dog chased after you, your nervous system would respond by sending impulses to your heart and blood vessels. Your heartbeat would speed up and blood flow to the striated muscles would increase. At the same time, nerves would send impulses to the adrenal glands to release *adrenalin*. Adrenalin speeds up the heartbeat and increases the flow of blood to the brain and muscles also. The changes caused by the nerves and the adrenalin would enable you to run faster. This reaction by the body is called the "fight or flight" response. It is another example of how the endocrine and nervous systems work together. The rapidly acting nervous system responds first. The slower-acting endocrine system strengthens the response of the nervous system and maintains it for as long as necessary for you to escape the dog.

Checkpoint

1. Why is the pituitary gland called the "master gland"?
2. What is a hormone?
3. Which hormone regulates growth? Where is it produced?
4. Give one example of how the nervous system and endocrine system work together to coordinate body activities.

3-7

Alcohol, Tobacco, and Other Drugs

ENRICHMENT

Goal

To identify the effects of alcohol, tobacco, and other drugs on the body.

Sometimes people use substances to alter their mood in some way. These substances are called *drugs*. Some people use drugs to relax, or to become more alert. Others use drugs to increase their self-confidence or block out unpleasant feelings. Substances that can be useful as medicines are often called drugs, also. They are known as *medicinal* drugs. Aspirin, penicillin, and antihistamines are some common medicinal drugs.

Drugs may be legal or illegal. Legal drugs have been researched by scientists to determine safe dosages and side effects. Legal drugs are obtained by a physician's prescription or by purchasing them "over the counter." No drug should be used without careful thought. People should find out what the side effects are and whether the drug might affect their ability to drive, operate machinery, or concentrate.

Drugs can become habit-forming. If a person becomes physically dependent on a drug, that is called an *addiction*. He or she will experience discomfort if the drug is removed. Sometimes a person with an addiction develops a tolerance for a drug. The person has to increase the amount of the drug each time to get the same effect.

One of the most commonly used drugs is *alcohol*. It is found in beer, wine, and "hard liquor." Even in small quantities alcohol depresses, or slows down, the central nervous system, decreasing the number and intensity of nerve impulses. Alcohol slows reaction time and decreases muscular coordination.

Drinking too much alcohol can damage the liver and may lead to a disease called *alcoholism*. Alcoholism is an addiction to alcoholic beverages.

Nicotine is another widely used drug. It is most commonly found in tobacco. Nicotine is habit-forming and has harmful effects on the cells in the trachea. It increases the blood pressure as well. Tars and certain gases in tobacco smoke

Teaching Suggestions 3-7

Most students respond well to a discussion of the use of tobacco, alcohol, and drugs. They are usually eager for information, but often skeptical of adult motivation, for they are well aware that many adults, often including their own loved ones, use and sometimes abuse the use of one or more of these substances. Thus, any discussion held in class should be completely objective and devoid of emotional content.

Some questions worth exploring are: Is a "high" necessary? Can people feel good about themselves without using drugs? Are there personality types associated with drug use or abuse? What alternatives to the use of drugs are there for coping with stress?

Students should be helped, as much as possible, to sort out for themselves why people use drugs, alcohol, and/or tobacco. In addition, you should explore how young people can learn to resist peer pressure to use any potentially dangerous substance. Point out the legal age limits for the use of alcohol in the different states, and why such limitations exist. Stress also that possession of many drugs is illegal.

In recent years, the number of highway deaths caused by drunk drivers has risen dramatically. Stress the effects of alcohol on mental alertness, muscular coordination, and reaction time during your discussion of this section. Ask students to suggest alternatives to driving under the influence, such as using a taxi or public transportation, staying at a friend's home overnight, or having a sober friend drive. Remind students that a person does not have to be intoxicated to have his or her judgment impaired by alcohol.

Discuss the effects of cigarette smoke on nonsmokers, and stress the fact that recent research shows that so-called second-hand smoke is more harmful than previously suspected.

Chapter-end projects 6 and 9 apply to this section.

Teaching Tips

- Discuss the differences between drug use and abuse. Even legal over-the-counter drugs can be abused, if they are taken in large quantities or used inappropriately.
- Invite a physician, nurse, or substance-abuse counselor to speak to your class about drugs.
- Have an interested student research the hazards of smoking, why people begin smoking, and how they can avoid succumbing to peer pressure. Have the student include successful methods for giving up smoking. Any student report on this topic should stimulate a lively discussion.
- Discuss the methods required by the U.S. Food and Drug Administration for testing the effectiveness and safety of new drugs before they are released for use.

Mainstreaming

Many handicapped students must take medicinal drugs to control or alleviate problems. Point out that the side effects of some drugs may include altering the user's mood, and that this effect must be tolerated in order to obtain the beneficial effects.

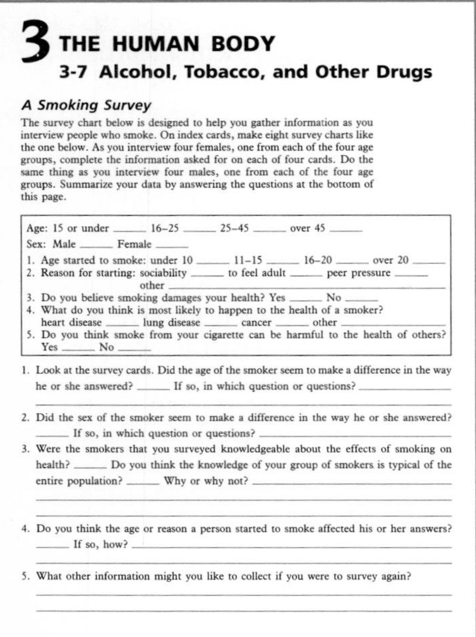

3 THE HUMAN BODY

3-7 Alcohol, Tobacco, and Other Drugs

A Smoking Survey

The survey chart below is designed to help you gather information as you interview people who smoke. On index cards, make eight survey charts like the one below. As you interview four females, one from each of the four age groups, complete the information asked for on each of four cards. Do the same thing as you interview four males, one from each of the four age groups. Summarize your data by answering the questions at the bottom of this page.

Age: 15 or under ______ 16–25 ______ 25–45 ______ over 45 ______
Sex: Male ______ Female ______
1. Age started to smoke: under 10 ______ 11–15 ______ 16–20 ______ over 20 ______
2. Reason for starting: sociability ______ to feel adult ______ peer pressure ______ other ______
3. Do you believe smoking damages your health? Yes ______ No ______
4. What do you think is most likely to happen to the health of a smoker? heart disease ______ lung disease ______ cancer ______ other ______
5. Do you think smoke from your cigarette can be harmful to the health of others? Yes ______ No ______

1. Look at the survey cards. Did the age of the smoker seem to make a difference in the way he or she answered? ______ If so, in which question or questions? ______
2. Did the sex of the smoker seem to make a difference in the way he or she answered? ______ If so, in which question or questions? ______
3. Were the smokers that you surveyed knowledgeable about the effects of smoking on health? ______ Do you think the knowledge of your group of smokers is typical of the entire population? ______ Why or why not? ______
4. Do you think the age or reason a person started to smoke affected his or her answers? ______ If so, how? ______
5. What other information might you like to collect if you were to survey again?

Table 3-2: Some Commonly Used Drugs

Drugs	*Effects on the Body*	*Harmful Side Effects*
nicotine	increases heartbeat rate, blood pressure	increases risk of heart disease, mouth and lung cancer, and emphysema
alcohol	relaxant; slows reaction time; may cause drowsiness, loss of coordination, and slurred speech	excessive use can cause alcoholism, liver and kidney damage, nutritional deficiencies, and emotional problems
caffeine	increases heartbeat rate and blood pressure; reduces feeling of fatigue	loss of sleep; upset stomach; increases blood pressure
amphetamines ("uppers")	reduce appetite, stimulate nervous system	nervousness, lack of sleep, severe weight loss
barbiturates ("downers")	relieve tension; cause sleep	cause drowsiness, poor muscular coordination, and decreased reaction time; excessive use causes liver or kidney damage; mixing with certain other drugs or accidental overdose may cause death
morphine codeine opium heroin	cause drowsiness; relief from pain; cause sleep	addictive; produce loss of consciousness; overdose can cause death
marijuana	relaxes; causes drowsiness; decreases muscular coordination	interferes with concentration and muscular coordination; smoke may cause lung cancer
cocaine	increases heartbeat rate; raises blood pressure	causes feelings of depression and fatigue and desire for more shortly after use; sleeplessness; loss of appetite; behavior changes

damage the sensitive alveoli of the lungs. (See Figure 3-27.) The tars also contain a variety of harmful chemicals. Some of them have been linked to cancer, a disease that involves the abnormal reproduction of certain body cells. Cancer cells grow quickly, often destroying and replacing healthy tissue. Smokers have a higher risk of getting lung cancer than nonsmokers.

Marijuana is a commonly used illegal drug. Since it is usually smoked, marijuana can cause damage to the lungs similar to that caused by tobacco. Marijuana can interfere with short-term memory and learning. Substances in marijuana affect coordination and concentration. It can be unsafe to drive a car or operate machinery after smoking marijuana. Marijuana contains over 300 different chemicals. Some of them have been linked to cancer. Scientists have only begun to understand the effect of marijuana on the body.

Certain risks are involved in using drugs illegally. First, there is the risk of overdose. When physicians prescribe medications they take into account the patient's age, weight, and general health. The dosage is adjusted according to these factors. Drug abusers often do not know safe dosages, nor can they be certain of the purity of a drug bought on the street. Street drugs might contain other harmful substances. Certain combinations of drugs, such as barbiturates and alcohol, or an overdose could be deadly. Table 3-2 summarizes the effects of commonly used drugs.

Research continues on the short- and long-term effects of many drugs. Because of the dangerous nature of many drugs, much of the research is done with mice and rats. People who abuse drugs are making experimental subjects of themselves.

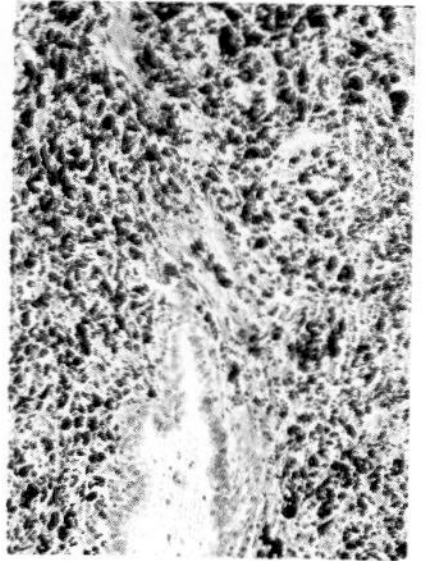

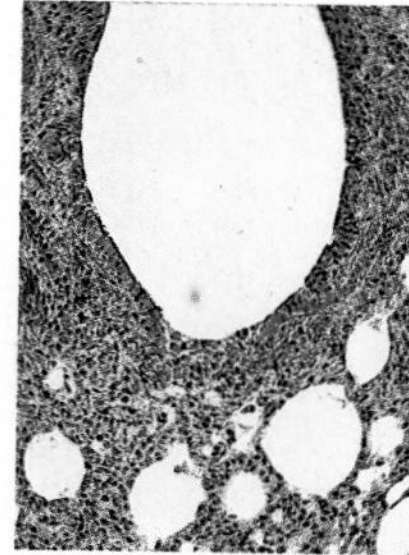

Figure 3-27 *Compare lung tissue from a nonsmoker (top) with lung tissue from a smoker (bottom).*

Checkpoint

1. What is an addiction? Name one drug that can be addictive.
2. Which substances in tobacco smoke are harmful? What effects do they have on the body?
3. What factors does a physician consider when prescribing a medication? Why can drugs taken without a doctor's advice be dangerous?

Checkpoint Answers

1. An addiction is a need to have a certain substance in order to feel good. Alcohol is addicting.
2. Nicotine, tars, and certain gases in tobacco smoke are harmful. Nicotine increases heartbeat rate, blood pressure, and the risks of heart disease, emphysema, and cancer. It also has harmful effects on the cells in the trachea. The other substances in tobacco smoke damage the alveoli and have been linked to cancer.
3. When a physician prescribes medication, he or she takes into account the patient's age, weight, and general health. The dosage is adjusted according to these factors. Drugs taken without a doctor's advice can be dangerous because the user does not know the purity and may not know the safe dosage of a drug.

Reading Suggestions

For The Teacher

Bodonis, David. THE BODY BOOK. Boston: Little, Brown, 1984. Human physiology, with high-tech illustrations.

Cohen, Sidney. DRUG ABUSE AND ALCOHOLISM: CURRENT CRITICAL ISSUES. New York: Haworth, 1981. A clearly written, comprehensive discussion of legal and illegal drugs.

Crapo, Lawrence. HORMONES: THE MESSENGERS OF LIFE. New York: W. H. Freeman, 1985. Endocrine Systems. Comprehensive.

Hooper, Judith and Dick Teresi. THE 3-POUND UNIVERSE. New York: Dell, 1986. Current brain research. Good for gifted students.

Lumberg, Peter. THE STORY OF YOUR HEART. New York: Coward McCann and Geoghegan, 1979. An informative book on how the heart works and how it is repaired when it malfunctions.

Smith, Anthony. THE BODY. New York: Penguin, 1985. Detailed reference book on human physiology.

For The Student

Claypool, Jane. ALCOHOL AND YOU. New York: Watts, 1981. Concise, readable book about alcohol use and abuse.

Geller, Judith. INNER SPACE: THE WONDER OF YOU. New York: Richards Rosen, 1979. Covers metabolism, nutrition, digestion, respiration, circulation, and excretion; includes experiments.

Hyde, Margaret O. and Bruce G. Hyde. KNOW ABOUT DRUGS. New York: McGraw-Hill, 1979. A factual, easy-to-read explanation of common drugs, their effects, and the reasons for their use or abuse.

Karlen, Arno. NAPOLEON'S GLANDS. New York: Warner Books, 1984. True medical detective stories.

Nilsson, Lennart. THE BODY VICTORIOUS. New York: Delacorte Press, 1985. Fine photo essay on the immune system.

Understanding the Chapter 3

Main Ideas

1. The skeleton is a framework of bones and cartilage that provides protection and support, and enables us to move.
2. Striated, smooth, and cardiac are the three types of muscle cells.
3. Muscles move bones and other body parts by contracting.
4. Blood moves food, oxygen, and waste products through the body.
5. The heart pumps blood through arteries, capillaries, and veins.
6. Blood consists of plasma, red blood cells, white blood cells, and platelets.
7. Oxygen and carbon dioxide are exchanged between the capillaries and the alveoli in the lungs.
8. Digestion is the breakdown of food into simpler substances that can be used by cells.
9. The kidneys remove waste and water from the blood and form urine. Urine is excreted through the urethra.
10. The brain and spinal cord control all the activities of the body.
11. Impulses are transmitted by the nervous system along neurons.
12. Hormones are released into the blood stream by the endocrine glands. Hormones control growth, development, and other activities.
13. There are serious health risks involved in using alcohol, tobacco, and other drugs.

Vocabulary Review

From the following list, choose the term that best completes each of the statements. Write your answers on a separate piece of paper.

alveoli	kidneys
arteries	ligaments
blood	medulla
capillaries	motor neurons
cartilage	neurons
cerebellum	plasma
cerebrum	reflexes
diaphragm	sensory neurons
digestion	stimulus
esophagus	synapse
glands	tendons
hormones	trachea
impulse	urine
involuntary muscles	veins
	villi
joint	voluntary muscles

1. Your __?__ filter wastes and water from the blood.
2. __?__ are air sacs in the lungs.
3. __?__ attach muscles to bones.
4. The gap between two neurons is called the __?__.
5. Tough, stretchy, bands of tissue called __?__ connect bone to bone.
6. __?__ are single nerve cells.
7. The __?__ is the part of the brain that controls the senses.
8. __?__ is the clear liquid in blood.
9. The __?__ is the muscle at the base of the ribs that aids in breathing.
10. __?__ usually carry oxygen-rich blood away from the heart.

Vocabulary Review Answers

1. kidneys
2. alveoli
3. tendons
4. synapse
5. ligaments
6. neuron
7. cerebrum
8. plasma
9. diaphragm
10. arteries
11. hormones
12. cartilage
13. glands
14. esophagus

11. ? are chemicals released directly into the blood, which cause changes in specific areas of the body.
12. ? is the stiff, flexible material found in the tip of the nose, outer ears, and ends of bones.
13. ? produce hormones.
14. The ? is a tube that carries food to the stomach.

Chapter Review

Write your answers on a separate piece of paper.

Know The Facts

1. Food is moved to the ? from the stomach. (esophagus, small intestine, liver)
2. ? muscle moves bones in the skeleton. (striated, smooth, cardiac)
3. ? are the blood vessels where the exchange of food, oxygen, and carbon dioxide takes place at the cells. (arteries, veins, capillaries)
4. ? are the part of blood that carries oxygen. (platelets, white blood cells, red blood cells)
5. Information carried by all nerve cells is called ?. (reflexes, synapses, impulses)
6. The ? system removes waste products from the body. (excretory, endocrine, nervous)
7. Muscles move body parts by ?. (ligaments, contracting, stretching)
8. Inhalation and exhalation are controlled by the contraction and relaxation of the ?. (lungs, alveoli, diaphragm)
9. Digested food diffuses into the blood at the ?. (pancreas, villi, large intestine)
10. A ? is anything that causes a reaction by the body. (stimulus, neuron, reflex)
11. The ? gland controls the release of hormones from other endocrine glands. (salivary, pituitary, adrenal)
12. ? slows reaction time and affects muscular coordination. (alcohol, nicotine, adrenalin)
13. ? are waste products that are excreted by the body. (oxygen and carbon dioxide, urea and platelets, carbon dioxide and urea)
14. ? is a commonly used drug that reduces fatigue and increases heartbeat rate and blood pressure. (alcohol, caffeine, bile)

Understand The Concepts

15. Explain how striated muscles work in pairs.
16. Describe the differences between voluntary and involuntary muscles.
17. Compare the functions of red blood cells, white blood cells, and platelets.
18. Describe the actions involved in breathing.
19. Name three places in the digestive system where food is acted upon by enzymes.
20. Draw and label the parts of the digestive system.
21. Beginning at the right atrium, trace a drop of blood on its journey through the heart, to a cell in the big toe, and back to the heart.
22. Explain the functions of the bladder and urethra.
23. Draw a diagram of two neurons and a synapse. Label the axons, dendrites, cell bodies, and synapse.

Chapter Review Answers

Know The Facts

1. small intestine
2. striated
3. capillaries
4. red blood cells
5. impulses
6. excretory
7. contracting
8. diaphragm
9. villi
10. stimulus
11. pituitary
12. alcohol
13. carbon dioxide and urea
14. caffeine

Understand The Concepts

15. Most striated muscles work in opposing pairs. Each muscle in the pair is attached to a bone. When one muscle contracts, a bone is moved one way. When the second muscle contracts, the bone is moved the opposite way.
16. Voluntary muscles function when a person wants them to. They are made up of striated cells. Voluntary muscles move the skeleton and some body parts. Involuntary muscles normally are not under conscious control. With the exception of cardiac muscle, involuntary muscles are made up of smooth muscle cells.
17. Red blood cells carry oxygen from the lungs to body cells. They carry carbon dioxide from the cells to the lungs. White blood cells surround and destroy foreign bacteria and produce antibodies. Platelets aid in blood clotting.
18. During inhalation, the contractions of the chest muscles pull the ribs up and the diaphragm contracts and moves downward. This results in an increase in the volume of the chest cavity and air from outside rushes in. The opposite actions occur during exhalation.
19. Food is acted upon by enzymes in the mouth, in the stomach, and in the small intestine.
20. See Figure 3-18 on page 78 of the student text.
21. Answers will vary. See student text, pages 72–73.
22. The bladder stores urine. When it is full, it contracts and urine is excreted through the urethra.
23. See Figure 3-23 on page 81 of the student text.

24. The cerebrum controls the senses, memory, learning, thought, speech, and voluntary movement. The cerebellum coordinates nerve impulses to and from the cerebrum and controls muscle coordination and balance. The medulla controls involuntary activities, such as heartbeat and breathing.
25. These systems work together to produce the "fight or flight" response. See page 86.
26. Answers will vary. See student text, page 88, Table 3-2.

Challenge Your Understanding

27. Sensory neurons in the ear would detect the stimulus and send a series of impulses to the brain. The brain would process the information and send impulses to motor neurons in different parts of your body. You might respond by turning your head in the direction of the sound or by speaking.
28. A lack of red blood cells would mean that the blood could not carry as much oxygen as it should. As a result, the cells of the body would not have as much oxygen as they need for respiration. Decreased cellular respiration would result in less energy being made available to the body, causing a feeling of tiredness and weakness.
29. Lungs rid the body of carbon dioxide. They are part of the breathing system. The kidneys rid the body of urea and water and are part of the excretory system. The large intestine eliminates solid wastes from the body and is part of the digestive system.
30. Mechanical digestion breaks food into small pieces, thus increasing the surface area acted upon by enzymes and digestive juices.
31. The villi provide a large surface area for absorption. If the small intestine were smooth, absorption would be decreased.
32. You could test the blood or the urine for the presence of normal levels of certain hormones.

Use arrows to show the direction in which an impulse would move along both neurons.

24. Name three regions of the brain. Describe some of the functions carried out by each region.

25. Give an example of how the nervous system and endocrine system work together.

26. What are some of the dangers of using drugs illegally?

Challenge Your Understanding

27. Pretend that you heard someone call your name. How would you respond to that stimulus? Trace the path of the impulses that lead to your response, beginning with sensory neurons in one or more sense organs.

28. Anemia is a condition in which the blood does not contain enough red blood cells. What effects would anemia have on the body?

29. The lungs, kidneys, and large intestine rid the body of certain waste products. Tell (a) what waste products each gets rid of and (b) what system each of these organs is part of.

30. Food goes through a mechanical and chemical digestive process. How does mechanical digestion aid chemical digestion?

31. Imagine that your small intestine was smooth instead of lined with villi: How would absorption of food be different?

32. Some diseases are caused by too little of a certain hormone. How would you test for such a condition?

Projects

1. Artificial parts are being used to replace many worn-out or damaged organs. Find out about replacement parts that are being used safely now in humans.

2. Arrange for a representative of your local Red Cross to teach your class how to perform CPR and the Heimlich Maneuver.

3. Find out about the effects of diet and exercise in preventing heart disease.

4. Using materials you can find at home, construct models of hinge and ball-and-socket joints. Include ligaments in your models.

5. Research and write a report about blood diseases. Find out how they are contracted, how each affects the body, the treatment or cure, and how each can be prevented.

6. Driving a car under the influence of alcohol is very dangerous. Find out how your state or province treats drunk drivers.

7. Write to the American Hearing Association or Canadian Council on Deafness for information on sign language. Demonstrate signing to your class.

8. Sleep is an important aspect of health. Look for recent articles in magazines and newspapers about sleep. Prepare a report on sleep for your class.

9. Research and report on the contributions to human biology of any two of the following scientists: Rita Levi-Montalcini; Florence Sabin; Charles Drew; Rebecca Lancefield; Gerty Cori; Samuel Kountz.

3 THE HUMAN BODY
Projects

1. Imagine that you broke your radius or your ulna and had to wear a cast for some time. Interview the school physician or nurse to get the answers to the following questions.
   a. What would happen to the bone you broke? ____________
   b. What would happen to the surrounding muscles? ____________
2. Take the pulse of several athletes and several nonathletes.
   a. What difference did you find? ____________
   b. Explain the difference. ____________
3. Call the Red Cross in your area to find out about requirements and restrictions for giving blood.
   a. What requirements are there for giving blood? ____________
   b. How is blood stored? ____________

3 THE HUMAN BODY
Challenge

Careers

The names of ten careers related to the human body are hidden among the letters below. The names may be written up, down, backward, forward, or diagonally. Circle each career name as you find it. Pick three of the names and find out what each does in a typical day. Your library and your school counselor will have information to help you.

```
C D E N A I C I N H C E T G K E O A I R
A P P Q R S T E E D H E A L K N P T T E
R C A T S T C H C A I E N L L H G H I P
A A C H C F A N T H R I W I Y N O L T I
P P Q N X U H I X U O G D S C A T E I O
H S R O M N L K J V P E I H O I O T N R
Y R A Z F E J I X U R C R L O C T I O T
S A R Q Z R H A Z Y A C D K L I V C L N
I P B G B A G F B L C B E X W T U T N O
```

Chapter 4

Ecology

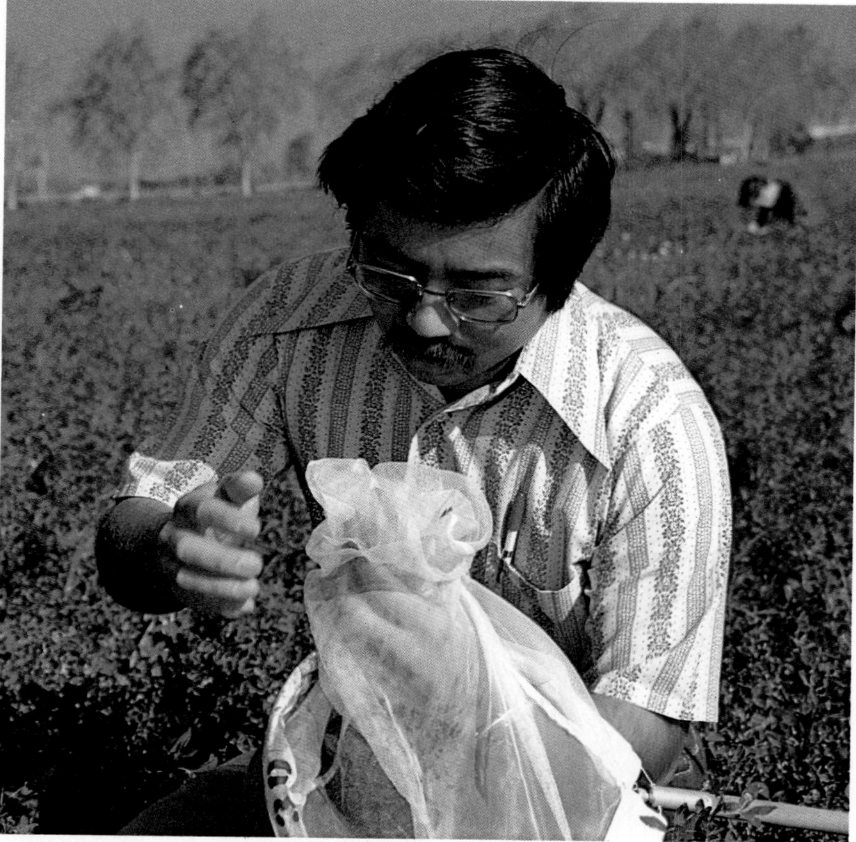

A biologist examines insects found in an alfalfa-field ecosystem.

Vocabulary Preview

environment
community
ecosystem
habitat
niche
food chain
producers
consumers
food web
decomposers
food pyramid
population
limiting factor
predators
prey
succession
pioneer community
climax community
pollution
extinct
endangered species

CHAPTER 4 Overview

The purpose of this chapter is to introduce the basic principles of ecology to students. The chapter opens by introducing the concept of the ecosystem. The parts of an ecosystem are described, and habitat and niche are defined. Food chains and food webs are then described, and the roles of producer, consumer, and decomposer organisms are discussed. The transfer of energy in a food pyramid is then covered, and the notion that energy is "lost" as one moves up to the top of a food pyramid is introduced.

The concept of population is introduced next. Following this, the processes that cause population change are described. Then some factors that limit population size are discussed. The next topic is succession of populations in an ecosystem.

The chapter then identifies the factors, both natural and human, that can threaten and harm habitats. Types of pollution and their effects are considered. Finally, extinction of species, endangered species, and efforts to protect endangered species are described in the enrichment section.

Goals

At the end of this chapter, students should be able to:

1. identify the parts of an ecosystem.
2. define and give examples of food chains and food webs.
3. identify the roles of producers, consumers, and decomposers in food chains.
4. trace the transfer of energy in a food pyramid.
5. describe four processes that cause a population to change in size.
6. identify limiting factors in an ecosystem.
7. describe the process of succession in an ecosystem.
8. identify the factors that can harm a habitat.
9. identify factors that lead a species to become endangered or extinct.

Teaching Suggestions 4-1

Open the section by asking students to define the term environment. Accept and discuss all answers. Then, explain the term from the ecological point of view. Use the pond, shown in Figure 4-1, and the rotting log, shown in Figure 4-2, as examples of ecosystems.

Be sure students understand the meaning of community, and describe how a community and the nonliving parts of an environment constitute an ecosystem. Stress that an ecosystem is a mental model that makes it easier for people to study and discuss the interrelationships between organisms and their environment.

Be sure students are clear in their minds about the difference between habitat and niche. Ask them to name the habitats of a polar bear (the Arctic ice), a tuna fish (ocean water), a frog (a lake or pond). Then point out that an organisms's niche is its way of living, or its role, in its habitat. One can think of habitat as an organism's "address." Its niche is then its "occupation" or that part of the habitat where it pursues its "occupation."

Chapter-end projects 1 and 4 apply to this section.

Teaching Tips

- Develop the concept of community further by having students identify the interacting organisms in the following: back-yard community, inner-city park community, and school-yard community.
- Arrange a field trip to a nearby pond, and have students collect samples of organisms in the pond community or record signs—animal tracks, burrows, nests, and so on—of members of the community.
- Have students list all the ecosystems they can think of and find a drawing or photo of each ecosystem listed. Let students arrange this material in a bulletin-board display.

4-1

Ecosystems

Goal

To identify the parts of an ecosystem.

environment

A family of beavers depends upon a pond and its surroundings for survival. (See Figure 4-1.) An organism's surroundings, consisting of both living and nonliving things, are called an **environment**. All organisms interact with their environment.

community

Air, water, and soil are three nonliving parts of the beavers' environment. They provide the things beavers need to stay alive. For example, the beavers take in oxygen from the air, drink water from the pond, and feed on plants that grow in the soil. Living parts of the environment, such as the small trees and bushes, provide food and building material for the beavers.

ecosystem (EK oh sis tum)

Many of the organisms in an area, such as a beaver pond, interact with one another. These interacting organisms make up a **community**. Trees, reeds, fish, beavers, and one-celled organisms are part of the pond community.

The pond community and the nonliving parts of the pond environment can be thought of as one interacting unit. The unit is called an *ecosystem*. An **ecosystem** is made up of a community and the nonliving parts of the environment.

An ecosystem is a model invented by people as a way of talking about a particular area. Materials such as food, water, and oxygen are exchanged, and energy is transferred throughout an ecosystem. A tiny ecosystem can exist in a

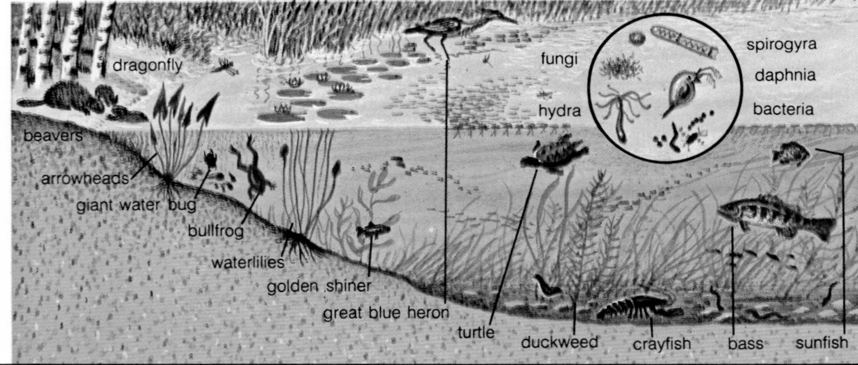

Figure 4-1 *Many different organisms live in and around a pond. Some live at the pond's edge, others live on the surface of the water, and some live in the muddy pond bottom.*

- Stress that ecosystems change constantly, and ask students to suggest ways in which familiar ecosystems undergo change.

Facts At Your Fingertips

- Communities can sometimes be very large and complex. In a study of the animals of the forest floor in Panama, for example, it was estimated that at least 40 million organisms (representing over 400 species) lived on, above, or under the ground per 0.4 hectare of land. Protozoa, bacteria, fungi, and higher plants were not counted.
- Ecosystems consists of two parts—an autotrophic part and a heterotrophic part. Energy from the sun is used to convert simple chemicals to the complex materials needed by the living

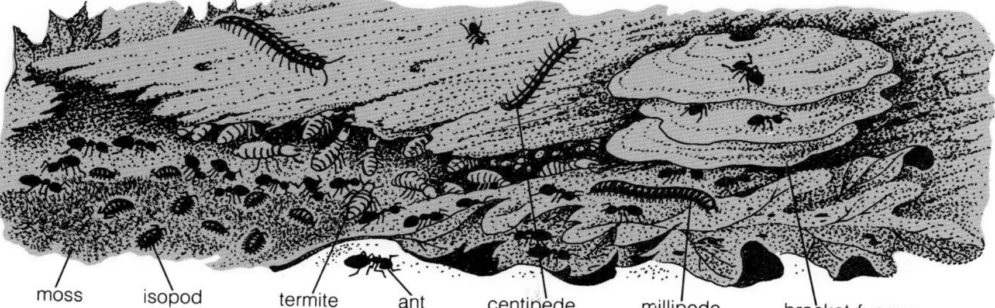

puddle in your backyard. Or an ecosystem can be very large, such as a forest. Figure 4-2 shows an ecosystem you are likely to find in and around a rotting log.

Figure 4-2 *Many different organisms live in and around a rotting log. The ecosystem contains several plants and animals.*

The place where an organism lives in an ecosystem is called a **habitat**. Each ecosystem contains many different habitats. The habitat of a crayfish in a pond ecosystem is along the bottom of the pond. The habitat of a pine tree may be among other trees in a forest ecosystem.

habitat

Different kinds of organisms can live in the same habitat. For example, tadpoles, minnows, and duckweed may share the same habitat—the edge of a pond.

Within a habitat, each species has its own way of living. Each different way of living is called a **niche**. A niche includes the kind of food an organism uses, the places it goes to find food, and the way it protects itself from enemies. No two species have the same niche, but niches can overlap. For example, in a pond ecosystem, a fish may eat insects that move across the surface of the water and insects that move below the surface. A frog may eat insects that land on small plants along the pond's edge and insects that move on the surface of the water. The feeding locations of the frog and the fish overlap, but each has other sources of food as well. Each has a different niche.

niche
(nich)

Checkpoint

1. What is an ecosystem? Give two examples of ecosystems.
2. Name three nonliving parts of an ecosystem.
3. Define community.
4. Explain the difference between a habitat and a niche.

TRY THIS

Describe the ecosystem in which you live. Name as many living and nonliving things as you can think of that are part of your ecosystem.

world in the autotrophic part. The heterotrophic part is comprised of organisms that use the materials produced by organisms in the autotrophic part. The most important autotrophs are green plants.

- In any community, living things are both interrelated and interdependent. There are countless interactions—plants with plants, plants with animals, and animals with animals.

4 ECOLOGY
4-1 Ecosystems

Organisms and Their Surroundings

A series of numbered squares forms the spiral below. Read the clues to the left of the spiral and decide what term is being described. Write the letters of that term in the corresponding numbered squares of the spiral.

1. An organism's surroundings.
12. Organisms that interact with each other.
21. Each species' different way of living.
25. A unit made up of a community plus the nonliving parts of its environment.
34. Place where an organism lives in an ecosystem.
41. A nonliving part of the environment.
44. Liquid part of the environment.
49. Solid nonliving part of the environment.
53. The environment is made up of living and ______ parts.

One Animal's Habitat

Research the habitat of the Monarch butterfly. In the space below, sketch and label this animal's habitat. Then write a description of a favorable climate and appropriate food for the animal.

Using Try This

In the broadest sense, the ecosystem of humans consists of all the parts of Earth that humans inhabit. Included in the ecosystem of humans would be the community of organisms with which humans interact and all the physical factors of the environment. Students' lists should be quite long. Stop when a representative sample of living and nonliving things has been compiled.

Checkpoint Answers

1. An ecosystem is a mental model of a community of organisms and the nonliving parts of the environment that affect and interact with those organisms. A lake, a pond, a rotting log, a strip of beach, and a forest are examples of ecosystems.
2. Air, water, and soil are three nonliving parts of an ecosystem.
3. A community is a group of interacting organisms in a region.
4. An organism's habitat is the place where an organism lives. The organism's niche is its way of living within the habitat.

Teaching Suggestions 4-2

Introduce the concept of food chains by asking students to relate their fishing experiences. Then, have the class respond to the following questions:

Are the kinds of fish you eat also food for other organisms? (yes)

Do such fish feed on smaller organisms? (yes)

Stress that fish, such as perch, bluegill, or minnow, both feed on smaller organisms and are, in turn, themselves food for other organisms. They are thus a part of a food chain. Direct students to study the examples of food chains in Figure 4-3. Emphasize that every food chain begins with the sun.

Stress that in aquatic ecosystems algae are the main producers. The algae trap the energy of the sun and use it to produce complex materials. The other organisms in food chains are either consumers, scavengers, or decomposers. Referring to Figure 4-3, explain first-order consumer, second-order consumer, third-order consumer, scavenger, and decomposer. Point out that crayfish are scavengers, which eat *dead* fish. Scavengers occur within food chains. Decomposers, such as bacteria or fungi, end food chains.

To establish the relationships in a food web, sketch the organisms in a simple food web, like the one shown in Figure TE 4-1, on the chalkboard. Leave out the arrows. Ask students to identify the producers (grass). Then ask them to name the first-order consumers (mouse, rabbit, sheep). Draw in the arrows from grass to these organisms. Next, have students identify the second-order consumers (mountain lion, owl, hawk, fox), and insert the appropriate arrows from the organisms they feed on. Ask if any third-order consumers are shown (there are none). Finally, have students name the decomposers (bacteria) and tell you from which of the organisms arrows should be drawn (all of them). Stress that the arrows within the food web show the direction of energy movement from organism to organism. Discuss the food web shown in Figure 4-4.

4-2

Food Chains and Food Webs

Goals

1. To define and give examples of food chains and food webs.
2. To identify the roles of producers, consumers, and decomposers in food chains.

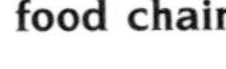

food chain

producers

consumers

Energy is trapped in ecosystems by plants and some protists during photosynthesis. The energy is transferred when one organism eats another. In a pond ecosystem algae capture and store energy from the sun. The stored energy is passed on to minnows that eat the algae. A heron feeding on the minnows obtains the energy stored in the small fish. A kind of chain is set up.

A set of eating interactions within an ecosystem is called a **food chain**. A food chain shows how the energy stored in food is passed from one organism to another. Figure 4-3 shows some examples of pond food chains.

Every food chain begins with the sun. Energy from the sun is converted to food during photosynthesis. Organisms that manufacture food are called **producers**. Plants and some protists are producers. Organisms that do not produce their own food are called **consumers**. Consumers eat other organisms.

Figure 4-3 All food chains begin with the sun. Energy from sunlight is trapped during photosynthesis. The arrows show how energy is transferred when one organism eats another. Food chains usually end with bacteria or fungi.

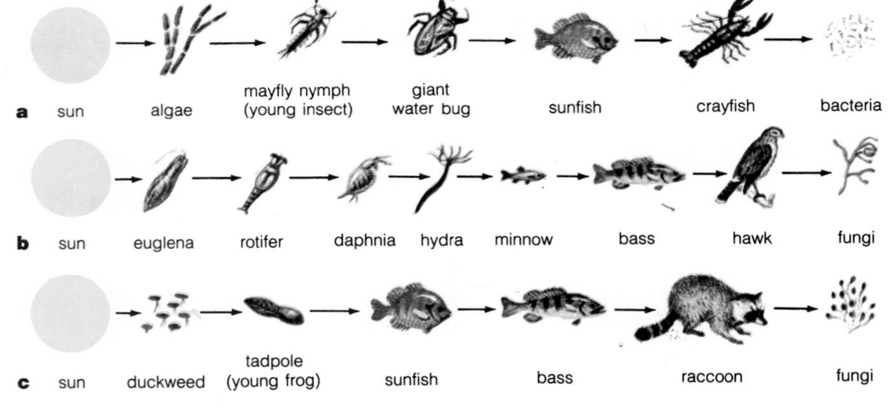

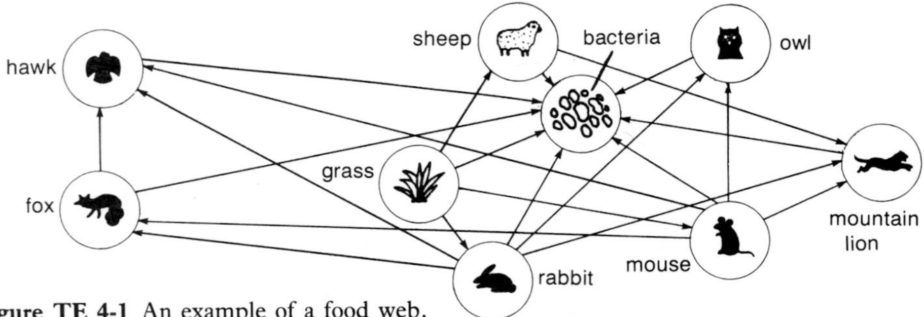

Figure TE 4-1 An example of a food web.

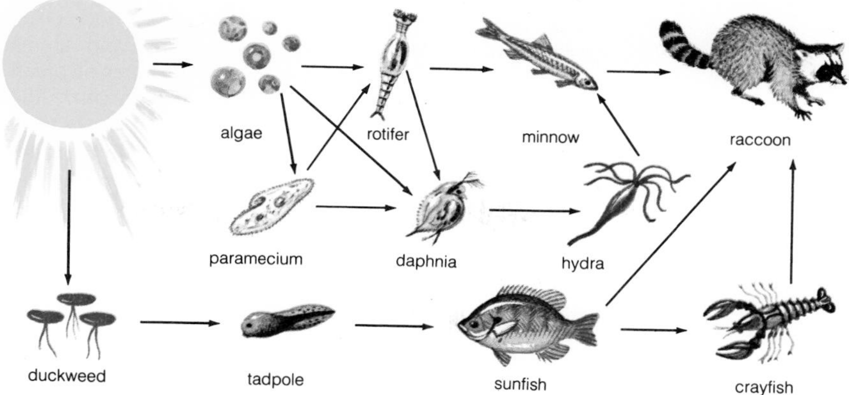

Figure 4-4 *Most organisms eat more than one type of food, as shown in this simplified food web. The complete food web would contain many more organisms.*

First-order consumers eat producers. The young insect, rotifer, and tadpole, shown in Figure 4-3, are first-order consumers. *Second-order consumers* feed on first-order consumers. The giant water bug, daphnia, and sunfish are second-order consumers. *Third-order consumers*, such as the sunfish, hydra, and bass, eat second-order consumers. Most organisms eat more than one kind of food. An organism, such as a sunfish, may be a third-order consumer in one food chain and a second-order consumer in another.

Look at the food chains shown in Figure 4-4. Note that some species of organisms are in more than one food chain. Food chains that overlap make up a **food web**. A food web shows the complex interactions among producers and consumers.

food web

Some organisms feed on plants only. Plant-eaters are known as *herbivores* (HUR buh vorz). Caterpillars, rabbits, sheep, and mice are common herbivores. All herbivores are first-order consumers.

Carnivores are organisms that eat protists or animals. Wolves, frogs, spiders, and snakes are carnivores. All carnivores are second- or higher-order consumers. Organisms that feed on both plants and animals are known as *omnivores* (AHM nuh vorz). Bears, chimpanzees, and raccoons are omnivores. Human beings eat both plants and animals too. They are omnivores also. An omnivore can be a first-order consumer in one food chain and a second- or third-order consumer in another.

Discuss the nitrogen cycle, shown in Figure 4-6. Point out that the nitrogen in dead plants as well as in dead animals is returned to the soil by decomposers. Then tell your class that other substances vital to life are cycled. These include water, oxygen, carbon, calcium, and phosphorus.

Chapter-end projects 1 and 4 apply to this section.

Background

There are three basic types of food chains. The first is the predator-prey chain. In such a chain, each successive organism feeds on a living organism. A predator feeds on a living animal. The animal fed on is called the prey. An example of such a food chain is producer ⟶ insect larva ⟶ frog ⟶ bird ⟶ mountain lion.

In the second type of food chain, the parasite chain, a parasitic organism obtains energy from a larger life form, the host. An example is producer ⟶ pig ⟶ trichina. The pig is the host. The trichina is a parasitic worm. If the parasite consumes too much of the host's energy, the host may die.

The third type of food chain is the saprophyte chain. Saprophytes, such as bacteria and fungi, are decomposers. They obtain their energy from living forms that have died, and recycle minerals and other materials important to life. Some simple saprophyte food chains are as follows: producer ⟶ fungus; producer ⟶ mouse ⟶ bacteria; and producer ⟶ rabbit ⟶ hawk ⟶ bacteria. These chains are oversimplified, for a saprophyte will feed on the dead remains of all organisms in each chain.

The Earth is a closed ecosystem. No materials essential for life either enter or leave it. Thus, without recycling, essential substances would eventually accumulate in forms not usable by living organisms, and life would cease.

4 ECOLOGY
Laboratory

Constructing Food Webs

Purpose To show relationships among members of a food web.

Materials pencil, paper

Procedure

In the space below, draw a diagram with arrows to indicate the relationships among members of each of the following food webs:

a. fruits and nuts; bear; deer; squirrels
b. insects; raccoons; snakes; toads; leaves, twigs, and bark
c. mice; skunks; rabbits; salamanders; insects; red foxes; leaves and twigs

Discussion

1. Compare your diagrams with those of your classmates. If everyone cannot agree on details, your teacher may have books available for you to consult.
2. What members are common to more than one of the given food webs?

____

3. Do you think that a complete food web for a given ecosystem might be more complex than the food webs given in this laboratory? ____ Explain.

____

Teaching Tip

- Ask interested students to research and report on one of the following cycles: water, oxygen, carbon, calcium, and phosphorus. Suggest that students use Figure 4-6 as a guide for developing their reports. These cycles are thoroughly explained in most high school biology texts.

Facts At Your Fingertips

- A food chain is a model of a series of eating interactions. Food chains end when the final living organisms are not themselves fed on. Interference by humans sometimes shortens a food chain and upsets the balance of the prey-predator relationship. For example, in the forest food chain, starting with green plants and going through deer to carnivores such as the mountain lion, many of the deer's natural predators have been eliminated by human intervention. Now humans must periodically step in and reduce deer populations to prevent mass starvation.
- All herbivores are first-order consumers, but not all first-order consumers are herbivores. Bears and pigs, for example, are first-order consumers, but they are omnivores, not herbivores.
- In a food web, each organism pictured represents a species, not an individual organism, for an individual can be eaten only once.
- Individual decomposers shown in food webs do not end them, because food webs represent entire populations. When a decomposer dies, its matter too is recycled by the actions of other decomposers.
- The hyena is a scavenger that is also an efficient predator. Many carnivores, such as lions and other big cats, will scavenge an easy meal. The bald eagle, the national symbol of the United States, is primarily a scavenger, although occasionally it does hunt some of its food.

decomposers

Decomposers are the final link in a food chain. **Decomposers** break down dead organisms and animal wastes, using them as food. A food chain ends when an organism dies and is decomposed. *Scavengers* are animals that feed on dead organisms also. Some common scavengers are vultures, coyotes, crayfish, and crows. The action of scavengers is fast, while the action of decomposers is slow.

Decomposers do not use all materials in dead organisms or wastes as food. Through the action of decomposers, important materials are returned to the soil, water, and air. Nitrogen is a substance needed by all organisms for making proteins. It is released into the air and soil through the process called *decomposition*. Most organisms cannot use nitrogen gas from air directly. *Nitrogen-fixing* bacteria in the soil and in the roots of some plants change nitrogen gas into a substance that can be used by plants.

Plants take up the nitrogen substance that has been added to the soil by bacteria. The nitrogen is used in building proteins. Animals obtain the plant proteins when they eat the plants. Some of this nitrogen is returned to the air and soil through animal wastes. The remainder is stored in the animal and will be returned to the soil by decomposers when the animal dies. The reuse, or *recycling*, of nitrogen is called the *nitrogen cycle*. (See Figure 4-6.)

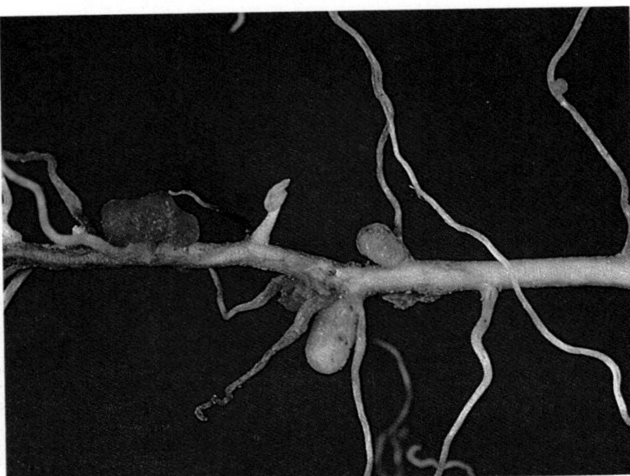

Figure 4-5 *Nitrogen-fixing bacteria live in the nodules, or lumps, on the roots of this clover plant. Beans, peas, peanuts, and soybeans also contain nodules. Farmers use these plants to replace nitrogen removed from soil by other plants.*

Cross Reference

Refer to Figures 1-14a and 1-14b in section 1-5 of Chapter 1 for diagrams of the carbon cycle and the oxygen cycle.

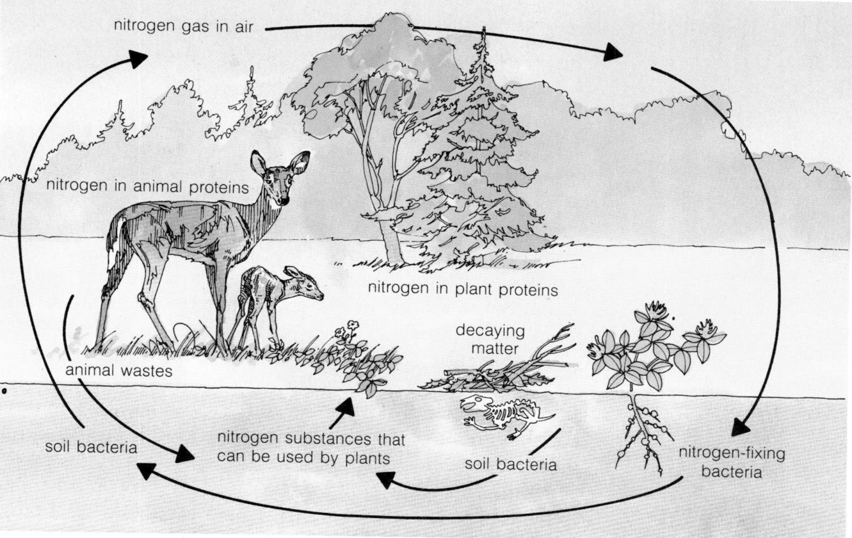

Figure 4-6 *The nitrogen cycle on land. Plants take in nitrogen substances from the soil. Animals get nitrogen from the food they eat.*

Checkpoint

1. What is a food chain?
2. Draw a complete food chain. Label the producer, first-order consumer, second-order consumer, third-order consumer, and decomposer.
3. Explain the difference between a food chain and a food web.
4. What is the role of a producer in a food chain? a consumer? a decomposer?
5. Describe how nitrogen is recycled in an ecosystem.

TRY THIS

Discover which organisms live in your backyard, schoolyard, or local park. Draw a quick sketch or write a description of each one. Use reference books to identify the organisms. Then draw a food web that is likely to occur in the area you studied.

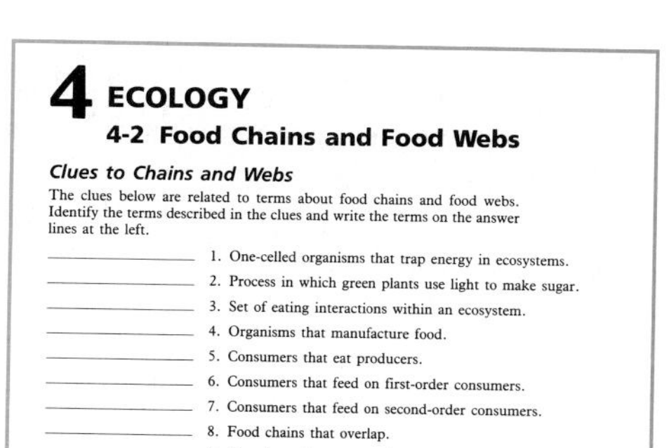

4 ECOLOGY
4-2 Food Chains and Food Webs

Clues to Chains and Webs

The clues below are related to terms about food chains and food webs. Identify the terms described in the clues and write the terms on the answer lines at the left.

________ 1. One-celled organisms that trap energy in ecosystems.
________ 2. Process in which green plants use light to make sugar.
________ 3. Set of eating interactions within an ecosystem.
________ 4. Organisms that manufacture food.
________ 5. Consumers that eat producers.
________ 6. Consumers that feed on first-order consumers.
________ 7. Consumers that feed on second-order consumers.
________ 8. Food chains that overlap.
________ 9. Organisms that scavenge in streams.

4 ECOLOGY
4-2 Food Chains and Food Webs

How Are Organisms Related?

The organisms illustrated below are all part of one food web. Identify the organisms on the lines provided. Draw arrows to tie the organisms together in a food web. Then complete the chart by describing the food that each organism eats and placing check marks in the appropriate columns.

Name of Organism	*Food for Organism*	*Producer*	*First-Order Consumer*	*Second-Order Consumer*	*Third-Order Consumer*
1. plants	—				
2. grasshopper					
3. millipede					
4. dung beetle					
5. hawk					
6. lizard					
7. snake					
8. toadstool					
9. mouse					
10. bacteria					

Using Try This

Encourage students to sketch the organisms they observe. Many who believe they have no artistic skill will discover they can do satisfactory drawings. Make sure students understand that the food webs they come up with are probably incomplete. You may want to show them how to add some of the organisms they did not observe.

Checkpoint Answers

1. A set of eating interactions within an ecosystem is a food chain.
2. Answers will vary.
3. A food chain is a linear set of eating interactions. A food web is produced when two or more food chains overlap.
4. A producer captures energy from the sun and uses it to produce materials essential to life. Consumers feed on producers and/or other consumers. Decomposers feed on the dead remains of producers and consumers.
5. Answers will vary. See student text, Figure 4-6.

Teaching Suggestions 4-3

Some students may find the ten percent law difficult to comprehend. Stress that the 10-to-1 ratio from layer to layer in a food pyramid has been found to be a useful average. Actual figures vary. In the wild, for example, the ratio is often greater. Whereas for domestic animals, it is often less because of highly enriched feeds.

It may help to relate humans, big cats, and other higher-order consumers to the producers at the bottom of Earth's food pyramid. Humans, for example, are atop the pyramid. The life we feed on is greater in mass and numbers than we are. The life that supports the life we consume is even greater in numbers and mass. This pattern continues back to the first producer plants consumed, where the total mass and number are tremendous.

Teaching Tips

- Discuss the following in relation to the ten percent law and food pyramids. Much grain grown in industrialized nations is fed to steers to produce beef. But only 1/10 of the energy of the grain is available to the steer, and only 1/10 of the energy of the animal's meat is available to humans. More energy is obviously available to people in grain than in beef. Should there be a reduction in the amount of grain fed to steers and other livestock to help alleviate world hunger problems? Point out that in many countries the traditional diet is grains and legumes with a "garnish" of meat, not the opposite, as in the U.S. and other western nations.
- Have a student weigh the food fed to a young mouse, kitten, or puppy over a period of time. At the same time, the student should weigh the animal and keep track of its weight changes. The total weight of food consumed compared to the animal's weight gain will be indicative of the amount of energy that was used for the animal's life activities.

4-3

Food Pyramids

Goal

To trace the transfer of energy in a food pyramid.

food pyramid

A **food pyramid** is another way of showing how energy is passed along in an ecosystem. (See Figure 4-7a.) The transfer of energy is not 100% efficient. On the average, nine tenths of the energy trapped by plants during photosynthesis is used for their life activities. The remaining one tenth of the energy is stored in materials that become part of the plants' cells.

When a rabbit eats the plants, it takes in the materials and stored energy. The rabbit uses up most of the energy gained from the food it eats in carrying out its life activities. A small amount of the energy from the food is stored in the rabbit's body.

About one tenth of the energy acquired by an organism is available to an organism at the next level of a food pyramid. Scientists refer to this "loss" of energy as you move up the pyramid as the *ten percent law*.

A food pyramid can also be used to represent the *number* of organisms it takes to meet the energy needs at each level of a particular food chain. (See Figure 4-7b.) For example, tens of thousands of plants are required to keep alive several hundred rabbits during a single year. Several hundred rabbits keep alive a single fox. Notice that the organisms become fewer in number and larger in size at each higher level in a food pyramid. Since each organism in any food chain takes in only a small amount of energy from every organism it eats, a consumer must eat a large number of producers or lower-order consumers to survive.

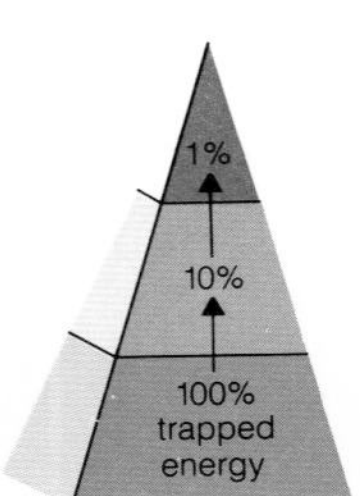

Figure 4-7a *Each level of a food pyramid contains one tenth of the energy of the level below it.*

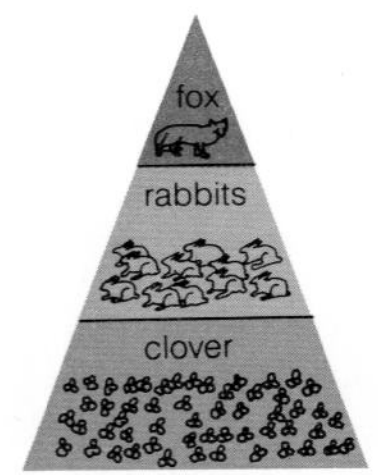

Figure 4-7b *The number of consumers a food pyramid can support depends on the number of producers at the base.*

Checkpoint

1. What is a food pyramid?
2. The transfer of energy in a food chain is not 100% efficient. Explain.
3. How much energy is available to each higher level of a food pyramid? Explain.

4 ECOLOGY

4-3 Food Pyramids

Passing Energy Along

Each sentence below is followed by a set of four choices for completing it. Decide which of the choices is best. Then write the letter of the correct choice on the space at the left.

____ 1. Plants use about ______% of the energy they get from the sun. a) 10 b) 50 c) 90 d) 100

____ 2. As you move up the food pyramid ______ energy becomes available. a) no b) less c) slightly more d) much more

____ 3. The "loss" of energy as you move up the pyramid is known as the ______ Law. a) Diminishing b) Ten Percent c) No Gain d) Ninety Percent

____ 4. The "lost" energy as you go up the food pyramid is used for ______. a) reproduction b) heat c) respiration d) carrying on life activities

____ 5. The ______ organisms are on the bottom of the food pyramid. a) largest b) most active c) smallest d) most unusual

____ 6. There must be ______ producers to supply the consumers in the world. a) fewer b) the same number c) a few more d) many more

Checkpoint Answers

1. A food pyramid shows how energy is transferred in an ecosystem.
2. About 9/10 of the energy of food consumed at each level of a food chain is used up by life activities.
3. Only 1/10 of the energy of food is converted to living substance; this is all that is available to the next level in the pyramid.

4-4

Populations

Goals

1. To describe four processes that cause a population to change in size.
2. To identify limiting factors in an ecosystem.

A meadow community may contain hundreds of field mice. These mice make up a population. A **population** is the number of organisms of the same kind, or species, living within a certain area at a particular time. Some other examples of populations in a meadow are the population of meadow grass, the population of grasshoppers, and the population of garter snakes.

population

Most populations change in size over time. There are four processes that can cause a population to change: (1) immigration, (2) emigration, (3) birth, and (4) death. *Immigration* means movement of organisms into a particular area. *Emigration* means movement out of an area. Immigration increases the size of a population. Emigration decreases it.

The population of fur seals on the Pribilof (PRIB uh lawf) Islands off the North American coast increases each spring due to immigration. (See Figure 4-8.) Adult females usually give birth shortly after arriving on the islands, increasing the population even more. The seals leave the island (emigrate) in late September, moving south for the winter. By December the population of seals on the islands is nearly zero. The seals return to the breeding grounds again the following spring.

Figure 4-8 *Adult male seals, called bulls, arrive on the island first and establish their territories. Female seals, called cows, arrive later. The young seals, called pups, are born on the island.*

Teaching Suggestions 4-4

Introduce the concept of population by asking students to volunteer their ideas of what a population is. Answers will vary, and will probably lack the precision needed by biologists. Help students arrive at a scientific definition that includes the following parameters: (a) specific type of individual, (b) time of count, and (c) space limitations.

Thus, a biologist will speak of a population of grasshoppers in a particular field on a particular date, or the population of paramecia in a particular culture on a specific day.

Stress the four processes that act on population numbers by thoroughly discussing the fur seal, yeast, and rabbit and lynx populations on pages 101–102. Be sure students understand both the individual and combined roles of immigration, emigration, birth, and death in population change.

It will be necessary to carefully develop the concept of limiting factors on population growth and size. Make sure students understand the examples given in the text. A logical extension of the concept is a discussion of the predictions made by Thomas Malthus in 1798 and their influence on current thinking about human population growth. Malthus hypothesized that food limitations would eventually halt the growth of the human population because population growth is geometric whereas it is only possible to increase food production arithmetically (see Figure TE 4-2).

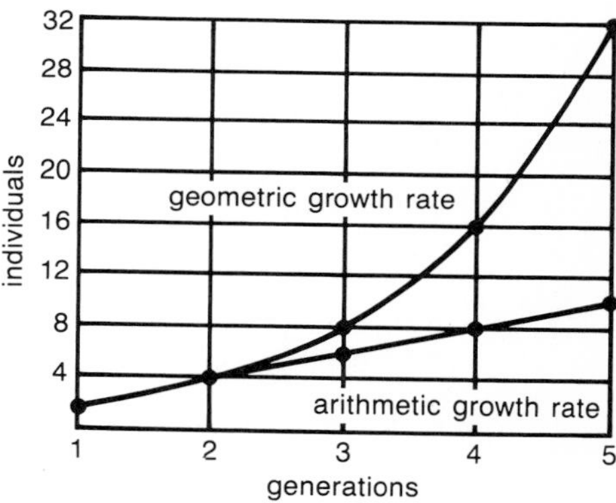

Figure TE 4-2 The difference between geometric and arithmetic growth rates.

One aspect of Malthus's hypothesis alarmed people when his work was published and is still debated today. This is the idea that when the human population outruns Earth's food-producing capacity there will be widespread misery, sickness, and starvation. Was Malthus correct? Despite many investigations and studies in the past, and some still under way, the evidence is inconclusive. Many factors other than food supply seem to be instrumental in controlling population size.

Recall that Earth is a closed ecosystem. It is clear also that there is a finite limit to Earth's food-producing capacity. Moreover, although population growth has been brought under control in some countries, it continues at an alarming rate for Earth as a whole. Figure TE 4-3 shows past world population growth and expected world population growth. As the curve indicates, if total population growth continues unchecked, at some point in time in the relatively near future, humankind will face an unwanted and potentially disastrous test of Malthus's hypothesis.

Be sure students understand that biological factors are not the only ones affecting human population growth. Powerful social, economic, and political forces too are involved. Nevertheless, well-informed citizens should be conversant with the issues facing humankind as rapid population growth continues in parts of the world.

The field Activity 7, Life in a Square Meter of Ground, on student text page 539, should be performed in conjunction with this section. Also, some students may wish to do project 5 at the end of the chapter.

4 ECOLOGY
Computer Program

Pronghorn Trap

Purpose To demonstrate one method of preserving a herd of wild animals.

Background The pronghorn antelope is a herbivore living in certain areas of western North America. The antelope's only natural predators are the wolf and cougar, both of which are endangered species. With so few natural predators, pronghorn populations may grow rapidly, exceeding the amount of available food. To prevent large-scale deaths due to starvation, herds must be maintained at a constant size.

ENTER the following program:

```
10  INPUT "WHAT IS THE CURRENT HERD SIZE?"; HS
20  INPUT "WHAT IS MAXIMUM ALLOWABLE SIZE?"; MS
30  INPUT "WHAT NUMBER WILL YOU REMOVE BY TRAPPING?"; TN
40  FOR I = 1 TO 10
50  LET HS = (HS * .4) + HS - TN
60  IF HS > MS THEN GOTO 110
70  IF HS < 0 THEN GOTO 140
80  PRINT "YEAR"; I; "HERD SIZE"; HS; "."
90  NEXT I
100  GOTO 150
110  PRINT "HERD SIZE EXCEEDED MAXIMUM."
120  PRINT "ALL PRONGHORN DIED."
130  GOTO 150
140  PRINT "NO MORE PRONGHORN LEFT."
150  END
```

Discussion

1. Wildlife Refuge A contains 200 pronghorn. The range can support a maximum herd of 750. What is the lowest transport number that will keep herd size below maximum?

____

2. Wildlife Refuge B has a herd size of 20 pronghorn and a maximum of 320. If no animals are transported away from the refuge, when will the herd exceed the maximum?

____

Upgrade the Program

Note that lines 10 through 30 obtain all the necessary entries. The FOR . . . NEXT loop in lines 40 through 90 calculates the herd size for each of ten years. The calculation in line 50 assumes that 80% of the herd are adults and that half of these are females, producing one offspring each year. Line 60 checks to see if herd size has exceeded the maximum.

1. Expand the listing of herd size to 15 years.
2. Rewrite the calculation to include a 40% offspring mortality rate.

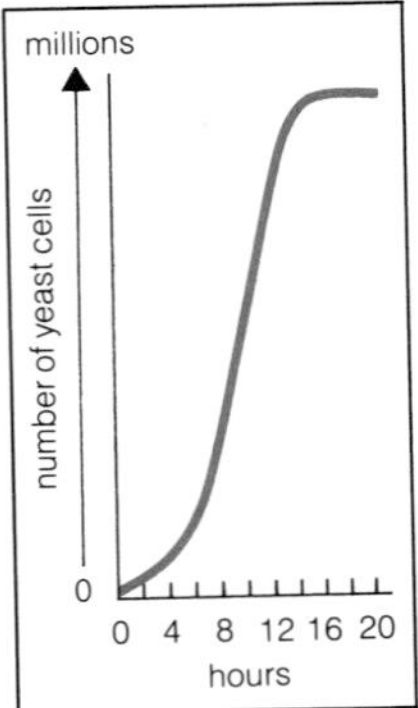

***Figure 4-9** A graph of a yeast-cell population. The curve rises when the population is increasing and flattens out when the population levels off.*

The amount of food available places a limit on the number of organisms an ecosystem can support. If food is plentiful, many individuals are likely to live long enough to produce offspring. The offspring are likely to survive to an age at which they, too, can produce offspring. New individuals might move into the area to take advantage of the food supply. So the population will increase.

If the population increases faster than the food supply, some organisms will not have enough to eat. They may weaken, become susceptible to disease, and die. Some individuals may move out of the area in search of food. Others may die of starvation.

limiting factor

Any factor that keeps a population from growing indefinitely is called a **limiting factor**. Food, space, climate, the presence of enemies, and disease are some limiting factors in an ecosystem. In most ecosystems, the supply of resources needed for survival is limited. Organisms may compete for a share of soil, sun, water, food, oxygen, or shelter.

Most populations do not grow indefinitely. Eventually one or more limiting factors slow the growth rate. Figure 4-9 shows a graph of a yeast-cell population in a closed container. As wastes built up and food ran out, the population size leveled off. The number of yeast cells being produced became equal to the number of yeast cells that were dying.

***Figure 4-10** Changes in population size usually occur around a set point (dotted line).*

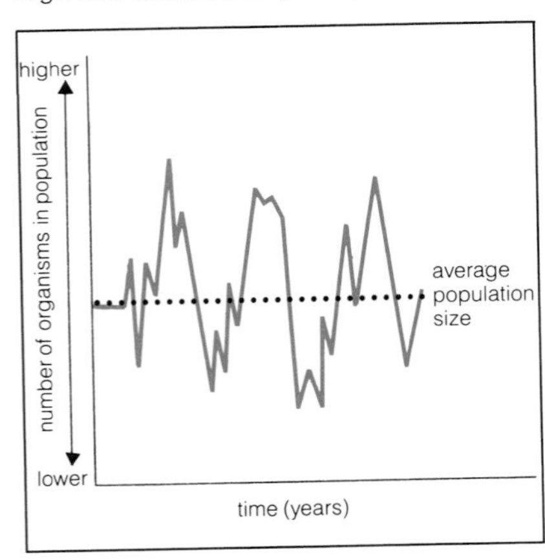

***Figure 4-11** Changes in rabbit and lynx populations usually occur in a ten-year cycle.*

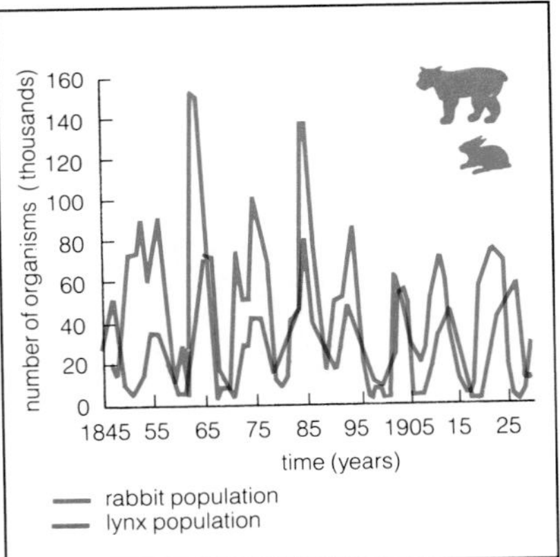

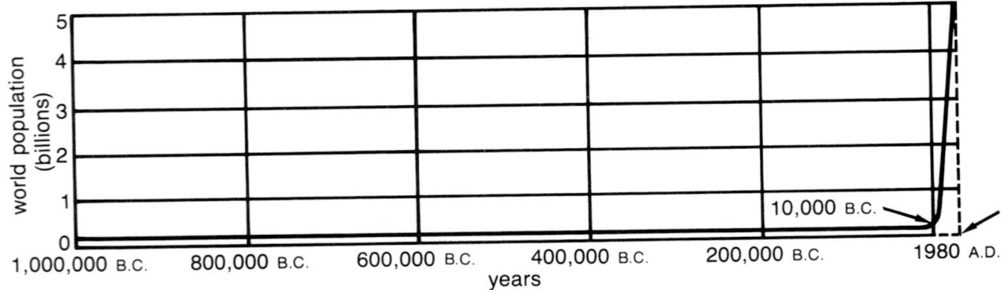

Figure TE 4-3 Past world population growth and expected world population growth.

Most populations rise and fall over time but the *average* population size stays the same from year to year. Figure 4-10 shows how a graph of most populations would look if they were measured over time.

Populations in a community affect each other. If one population changes greatly in size, other populations in the community may change also. For example, if the number of field mice in a meadow increased because of a plentiful food supply, more snakes or owls might come to the area to feed on the mice. The snake or owl population in the meadow would increase. Organisms, such as snakes and owls, that catch and eat other organisms are called **predators**. The organism that is caught and eaten is called the **prey**. As predators increase, prey decreases.

predators
(PRED uh turz)

prey
(pray)

In communities that have only a few populations, a change in the size of one population will have a dramatic effect on the size of the other populations. Figure 4-11 shows what happens in a community where there are just a few populations. The lynx is a predator of rabbits. As the rabbit population grows, the lynxes have more to eat. More lynxes survive and produce young, increasing the lynx population. As more and more rabbits are eaten, the rabbit population drops off. Fewer rabbits live long enough to produce offspring. With a smaller rabbit population, the lynx population has less food. Some lynxes die of starvation. As the lynx population decreases, the rabbit population starts to increase, beginning the cycle again.

Checkpoint

1. What are the four processes that cause a population to change in size?
2. Name two limiting factors that might keep a population from growing.
3. What is a predator? What is a prey?

TRY THIS

Obtain two identical pots with soil. In the first pot, plant 2 radish seeds. In the second pot, plant 12 radish seeds. Place the pots in a sunny window. Water the plants when the soil feels dry. Compare the plants in both pots after several weeks. Explain the results in terms of limiting factors.

Background

In some cases predators control prey populations. In other cases the opposite is true. The first situation often develops when the predator-prey interaction is of recent origin. For example, a biologist introduced a population of small rodents to a small island in the middle of a lake. The population grew. By tagging, the biologist kept track of births and deaths. Clearly, there was no immigration or emigration. One day, however, the investigator was unable to find any live rodents on the island.

A thorough search turned up a new mink den containing the bodies of many rodents. Because the rodents could not escape, one mink was able to track and kill them all. Here a predator was able to wipe out a prey population. This often occurs when humans disturb an ecosystem.

The population of the red fox in the wild, on the other hand, is controlled by the number of small rodents and rabbits that the fox preys on. Good rodent and rabbit years mean plenty of fox pups. But the numbers of rodents and rabbits seem to depend on certain chemicals in the seeds and grasses they eat. These chemicals affect the secretion of reproductive hormones. The chemicals, in turn, appear to be dependent on sunspot activity. In this case, the red fox population ultimately depends on sunspot activity, but in the intermediate sense, on rodent and rabbit populations.

4 ECOLOGY

4-4 Populations

Population Limits

Some terms are missing in the paragraphs below. Complete the sentences by writing the correct term on the matching numbered space at the left.

1. ____________
2. ____________
3. ____________
4. ____________
5. ____________
6. ____________
7. ____________
8. ____________
9. ____________
10. ____________

A limiting factor is any factor that can keep a population from 1. indefinitely. In yeast-cell populations, for example, the limiting factors are the 2. that build up and the 3. that runs out. Eventually, the number of yeast cells being produced 4. the number of yeast cells that die, and the population 5..

In a community, populations affect each other. If, for example, the population of field mice increases, the population of owls may 6.. Organisms, such as owls, that catch and eat the field mice are called 7.. The field mice that are caught by the owls are called the 8.. Generally, the population of prey 9. as the predator population increases. As the predator population decreases, the population of prey increases, starting the 10. again.

Tracking Populations

Using Try This

Be sure that each pot in all pairs contains identical amounts of soil and that each experimental set of plants receives the same amount of water at each watering. Students should observe that growth in the pot containing 12 plants is less luxuriant than growth in the pot with just two plants. The most important limiting factor is probably space for adequate root development.

Checkpoint Answers

1. Immigration, emigration, births, and deaths.
2. Food, space, climatic conditions, disease, and the presence of enemies are all limiting factors.
3. A predator is an organism that captures and eats other organisms. The organism captured and eaten is the prey organism.

Teaching Suggestions 4-5

Tell students about the island called Surtsey that lies off the coast of Iceland. Surtsey was formed in 1963 by the eruption of an undersea volcano. At first, it was totally barren of life. Two years later, the island had small plants growing on it. Point out to students that this is similar to the starting point for any succession.

To illustrate the stages of succession, carefully lead students through the events shown in Figure 4-12. Make sure they understand what a pioneer and a climax community are. Stress that a climax community can be regarded as a balanced ecosystem.

You may wish to trace the succession steps that occur to the meadow, or grassland, that is the final stage in the pond example. Grasslands, of course, will develop first wherever there is barren land with adequate nutrients, water, and light. Sketch Figure TE 4-4 on the chalkboard and describe each stage, culminating in the mature hardwood forest.

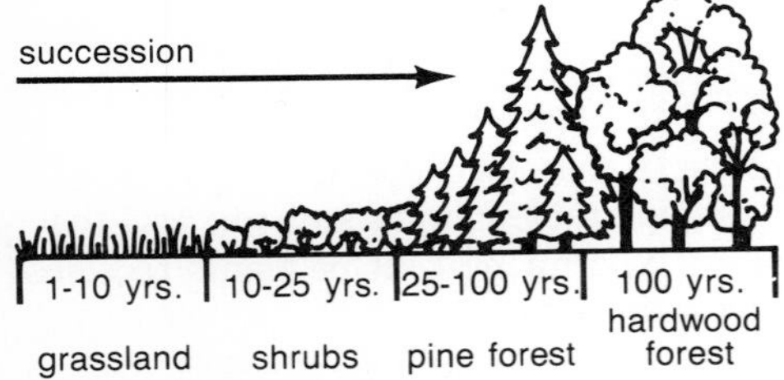

Figure TE 4-4 Land succession from grassland to mature hardwood forest.

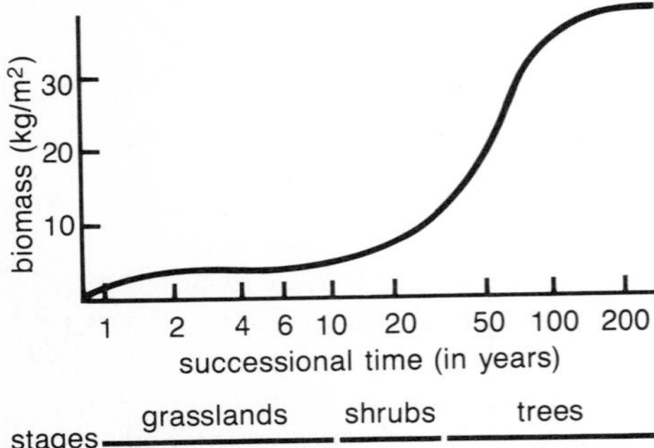

Figure TE 4-5 The increase in biomass as a land succession proceeds.

4-5

Succession

Goal

To describe the process of succession in an ecosystem.

An ecosystem may change over time. New populations can move into an area, replacing old ones. The new community that is formed has an effect on the nonliving parts of the ecosystem. For example, if thick grasses began to grow on a dry, abandoned field, they would form a protective cover over the surface of the soil. The grasses would keep the soil from being blown away by wind and would increase the soil's ability to hold water. As some grass plants died and decomposed, they would enrich the soil with valuable materials. Eventually the soil would become a suitable habitat for a variety of plant species. The dry field might become a moist, rich meadow or forest over time. The grass plants cause changes in the ecosystem that eventually lead to their own replacement. The process by which an ecosystem

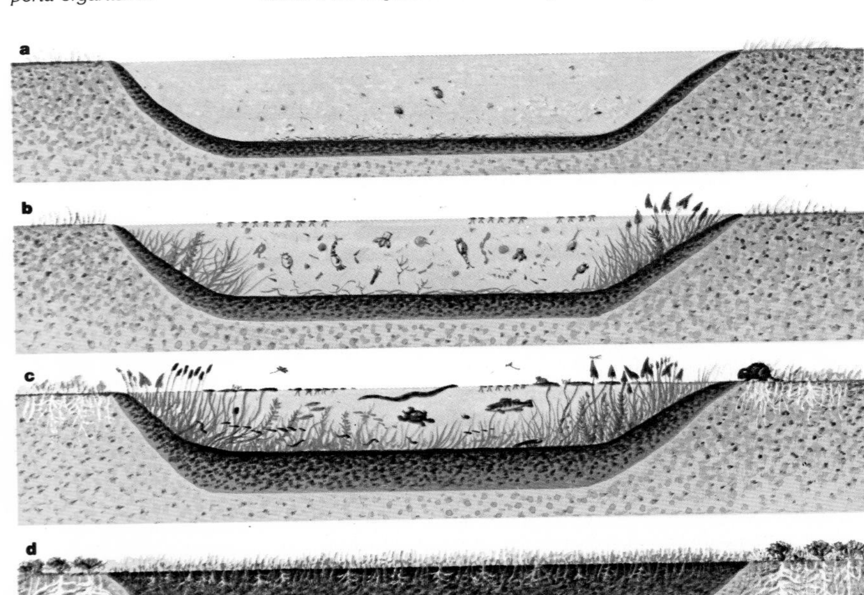

Figure 4-12 *(a) Pioneer organisms establish the first community. (b) Floating pondweeds and rooted plants appear. (c) As the pond bottom builds up, thick growths of plants take hold. A wide variety of organisms live in and around the pond. (d) After many years, land organisms replace the pond organisms.*

Students may be interested in the total plant biomass that accumulates during the many years of succession. See Figure TE 4-5. Plant biomass remains low during the early years, when grasses are the dominant plants, then increases significantly when shrubs and trees take over as the dominant species.

Teaching Tips

- Locate small ponds in your area and try to determine where they are in the succession process. Lakes too undergo succession, but more slowly.
- Discuss succession in bodies of water created by dams.

changes over time is called **succession**. Succession takes place gradually over many years. Figure 4-12 shows succession in a pond ecosystem.

succession

The first community to develop in a bare area is called the **pioneer community**. Spores of fungi and bacteria, one-celled organisms, and plant seeds carried by wind probably established the pioneer community in the pond.

pioneer community

As the community multiplied and grew, it had an effect on the pond. Organisms died and were decomposed, adding mineral-rich matter to the bottom of the pond. The accumulated decayed matter acted as fertilizer, permitting the growth of algae and other vegetation. Animals immigrated to the pond to feed on the plants. Animal wastes and decaying material began to fill in the pond. It grew more shallow. Reeds, cattails, and lilies grew at the edge of the shallow water. Spores and eggs of other organisms were transported on the fur and feathers of animals. The community eventually contained a variety of plants, one-celled organisms, insects, fish, amphibians, reptiles, birds, and mammals.

After many years, the pond became filled in with sand, mud, and materials that had decayed into soil. Grasses and small shrubs began to grow where the pond once was. As land organisms gradually replaced pond organisms, the pond changed into a meadow. The meadow may change into a forest through continued succession.

Succession goes on in stages. Each stage may last for a short time or for many years. Eventually a *climax community* is formed. A **climax community** is a community that does not change any more. Individuals are replaced, but the same kinds of organisms remain in a climax community for many, many years. Only natural disasters, sudden large climate changes, or human intervention change a climax community. A climax community usually contains many habitats and many different species making up large, complex food webs. Oak-hickory and beech-maple forests are two examples of climax communities.

climax community

Checkpoint

1. What is a pioneer community?
2. Describe the succession of a pond to meadow.
3. What is the final stage in succession?

- Have an interested student prepare a report on the role of glaciers or the Ice Age in succession.
- Investigate succession in empty lots in developed areas. Have students look for examples of bare, leveled earth, and examples of succeeding stages. Often several can be found in a relatively restricted area. Photographs taken with an instant camera will show succession graphically.
- Have an interested student research the current state of plant and animal habitation of Surtsey Island, and report to the class.

Facts At Your Fingertips

- Succession may be as brief as 25 years for a small bulldozed pond.
- Succession of birds accompanies the succession stages of plants in a forest. It is not unusual for the number of species of birds to go from one or two to about 20 during the 100 years or so of development from grassland to mature hardwood forest.

4 ECOLOGY

4-5 Succession

New Replaces Old

Each step below describes one event in the succession of a pond to a meadow community. Number each step to show the order in which it occurs in the succession.

____ Reeds and cattails grow at the edge of the water.
____ Algae grow in the pond.
____ The area becomes a meadow.
____ The pond becomes shallow.
____ Seeds and spores are brought to the pond by the wind.
____ Animal wastes and decaying matter partly fill in pond.
____ Land organisms replace pond organisms.
____ Dead organisms add mineral-rich material to the bottom of the pond.
____ Animals come to the pond to eat plants.
____ Pond becomes completely filled with sand, mud, and decayed material.
____ Grasses and shrubs grow where the pond was.

Changing Ecosystems

The diagrams below illustrate stages in the succession of a pond. Number each diagram in order from pond to forest. Then answer the questions.

1. What organisms are found in a pioneer community? ____
   Why, then, is it impossible to show a pioneer community? ____
2. Why did you choose the diagram you did as first in the succession? ____
3. Which diagram represents the climax community? (Note: Trees are cut off in the picture.) ____
   Why? ____

Checkpoint Answers

1. A pioneer community is the first community of organisms to develop in a barren, or bare, region.
2. Answers will vary. See student text, page 105.
3. A climax community is the final stage of succession. Unless some drastic change occurs, the climax community remains balanced.

Teaching Suggestions 4-6

Open the section by asking students if any of the wastes they dispose of affect habitats in any way. Help them understand that many substances discarded by humans can have adverse effects on habitats and the organisms they contain. For example, you may want to cite raw sewage dumped into rivers, lakes, and harbors; smoke particles from burned trash; roadside litter of all kinds; toxic household chemicals dumped in yards and fields; and so on. It's important to establish that the threat to habitats from human activities is far greater than the threat from natural disasters.

Review the many human activities given in the text that somehow hold the potential to harm habitats, and try to give an example or two for each one. Students will benefit most from examples they can relate to personally.

Emphasize that the disposal of radioactive and toxic wastes, the effect of acid rain, the control of automobile exhaust pollutants, and the "greenhouse effect" are environmental problems that students will no doubt have to cope with when they are adults (see Teaching Tips). Have students suggest the kinds of problems created by these factors that they might have to contend with as adults. Stress that knowledge of pollutants and their effects is necessary if citizens are to vote in a responsible way.

Chapter-end project 3 applies to this section.

4-6

Habitat Destruction

Goal

To identify the factors that can harm a habitat.

When one habitat in an ecosystem is damaged or destroyed, it can affect all the organisms in the ecosystem. The "balance" of an ecosystem may be threatened by two things: natural disasters and the actions of people.

A volcano can damage or destroy an ecosystem. Ash, lava, and poisonous gases from a volcano may kill plant and animal populations. It may take months for organisms to slowly reappear in an area affected by a volcano. (See Figure 4-13.)

In 1883 a volcanic eruption left the island of Krakatoa, in the Pacific Ocean, covered with lava and ash. No life remained on the island. Twenty-five years later, 263 species of animals were part of the island ecosystem. Forest fires, earthquakes, tidal waves, and floods are other naturally occurring events that can damage or destroy an ecosystem. Some ecosystems, such as forests, may take hundreds of years to become re-established.

Figure 4-13a *This is the area around a volcano in Washington state, Mount St. Helens, immediately after it erupted in 1980.*

Figure 4-13b *Once a pioneer community is established, succession will lead to repopulation of the area.*

Teaching Tips

- Have students investigate the dumping of toxic and radioactive wastes. Their reports should include information about measures being taken to control these waste problems.
- Have an interested student research the threat of industrial and residential development to coastal wetlands and report to the class. Be sure the student includes information about the role of estuaries and salt marshes in maintaining life in the seas.
- Have a team of students investigate the current status of state and federal legislation to control acid rain. Have the students look into any agreements on acid-rain control that may exist between the U.S. and Canada.
- Have a student research the "greenhouse effect" and report to the class on its projected influence on global climate and crop production.
- As a class activity, investigate recycling programs that are being implemented in your community.

Human activities, such as the building and operating of factories and machinery, mining, agriculture, and the search for fuel can result in the production of harmful substances in the environment. The accumulation of these harmful substances in the environment is known as **pollution**. Pollution is generally categorized by the type of resource that is affected. Air pollution, water pollution, and land pollution are the three major kinds of pollution.

pollution

Air pollution may occur as gases, solid particles, or liquids. A major cause of air pollution is the burning of *fossil fuels*, such as coal, gas, and oil, for energy. Air pollution damages the tissues of organisms that are exposed to it. (See Figure 4-14.)

Figure 4-14 *Air pollution can have a harmful effect on plants and trees.*

Sulfur dioxide is a *pollutant* produced during the burning of coal and high-sulfur oil. When sulfur dioxide reacts with water in the atmosphere, it forms dilute sulfuric acid which falls to Earth as *acid rain*. In many parts of the world acid rain has built up in lakes and streams, changing the chemical balance of the water, and killing fish and other organisms that live there. Salamanders are disappearing in some areas because their eggs and skin are harmed by acid rain. The pollutants that cause acid rain can be carried by strong winds far away from where they were formed.

Automobile exhaust produces many harmful gases. Some of these gases react in the presence of sunlight and form *smog*. Smog contains a dangerous mixture of gases. Carbon monoxide, one gas in smog, prevents oxygen from getting into the blood stream. Other substances in smog are believed to cause cancer. In some cities, people with breathing problems cannot go outside when the smog is particularly bad.

Water pollution is caused by the dumping of industrial wastes, human wastes, garbage, and other substances into ponds, streams, rivers, and oceans. Some of the wastes are *biodegradable*. That means that they can be decomposed by bacteria and fungi and become part of the environment. Wastes that are *nonbiodegradable* do not decay. Often they harm the habitats of the ecosystem where they are dumped.

TRY THIS Cover several slides with petroleum jelly and leave them outside in different areas for 24 hours. Examine the slides and account for the differences.

Facts At Your Fingertips

- In addition to sulfuric acid, dilute nitric acid occurs in acid rain. The nitric acid comes from oxides of nitrogen that are released when gasoline is burned in automobiles.
- Although major forest fires are usually disastrous to ecosystems, small forest fires may be beneficial. They remove dead wood and kill some insect pests without seriously damaging trees. Also, some species of pine require fire to force open their cones and permit seeds to germinate.

Using Try This

Have students use both hand lenses and microscopes to examine the particles trapped on the petroleum jelly-covered slides. Students can speculate on the identity and source of individual particles, but precise identification is very difficult, if not impossible. Stress that the density of particles in particles/cm^2 is a good measure of the severity of pollution, all other factors being equal. This is easily calculated by dividing the total number of particles on a slide by the area of the slide in cm^2.

Mainstreaming

When setting up teams for the two Try This activities, have nonhandicapped students work with visually-impaired or special-needs students.

Cross References

See section 9-5, Chapter 9, for more information about acids. Refer to section 10-7, Chapter 10, for discussion of nuclear wastes. Volcanoes are the subject of section 12-7, Chapter 12.

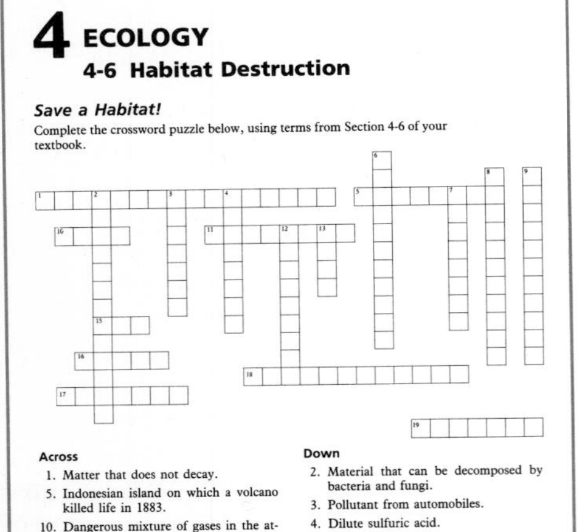

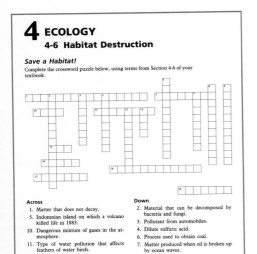

4 ECOLOGY
4-6 Habitat Destruction

Save a Habitat!
Complete the crossword puzzle below, using terms from Section 4-6 of your textbook.

Across
1. Matter that does not decay.
5. Indonesian island on which a volcano killed life in 1883.
10. Dangerous mixture of gases in the atmosphere.
11. Type of water pollution that affects feathers of water birds.
15. Pollution caused by burning fossil fuels.
16. Pollution caused by dumping wastes.
17. Natural phenomenon that emits poisonous gases.
18. Measure that will help resources last longer.
19. Waste metal that poisons organisms.

Down
2. Material that can be decomposed by bacteria and fungi.
3. Pollutant from automobiles.
4. Dilute sulfuric acid.
6. Process used to obtain coal.
7. Matter produced when oil is broken up by ocean waves.
8. Organisms that are disappearing because of acid rains.
9. Coal, gas, oil.
12. Accumulation of harmful substances in the environment.
13. Strip mining has polluted the ______.

Some of the harmful pollutants in water are taken in by the plants and other producers that live there. When the plants are eaten by consumers, the harmful substances are passed along to them. Lead and mercury are two waste materials that have been found in the tissues of organisms living in polluted waters. People have become seriously ill and some have died as a result of mercury poisoning from eating contaminated fish. Figure 4-15 shows how a pollutant can get passed along in a food chain.

Cooperation among government agencies, industry, and communities has resulted in the cleanup of several rivers that were once thought to be hopelessly polluted. For example, today people swim and fish in the Willamette River in Oregon—a river once filled with sludge and slime.

The transportation of oil and the drilling for oil at sea have caused a serious type of water pollution known as an *oil spill*. Since oil and water do not mix, the oil forms a slick on the surface of the water. The feathers of birds that land on the water often get covered with the oil, preventing the birds from flying and causing many to drown. Fish die when their gills become covered with the thick, black oil. When the oil is finally broken up by the action of ocean waves, it sinks to the bottom as black tar balls. The tar balls may cover and kill clams, shrimps, lobsters, and the eggs of many

Figure 4-15 *Notice how mercury is passed along a food chain. The mercury collects in the tissues of organisms and is transferred when one organism eats another.*

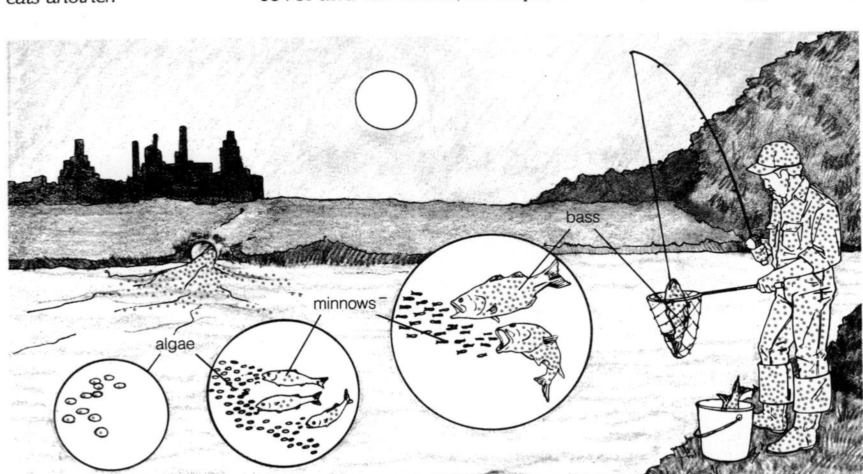

sea animals. Oil spills are very expensive and difficult to clean up.

People have changed ecosystems in other ways. *Strip mining* is a process used to obtain coal that is buried in the ground. In this process, topsoil and plants are removed from the surface of the Earth. Very few organisms can survive in an area that has been stripped. Some old strip mines have been restored with improved soil, grass, and other plants. (See Figure 4-16.)

Figure 4-16 *A reclaimed coal strip mine in West Virginia.*

Laws have been passed in an effort to control the kinds and amount of pollutants that are put into an ecosystem. Factories and power plants must now have antipollution devices to reduce the amount of pollutants that go into the air. Automobile engines have been redesigned to reduce pollutants in exhausts and to burn fuel more efficiently. Chemical dumping is prohibited in many areas and some companies are now required to pay the cost of cleaning up waterways they have polluted.

Efforts are being made to use less fossil fuel and other limited resources. *Conservation* means that resources will last longer and less damage will be done to ecosystems. For example, people are using nonpolluting sources of energy, such as the sun and wind, instead of relying on gas and oil only. Many companies are recycling glass, paper, and metals in an effort to conserve limited resources and to reduce the amount of waste that gets put into the environment.

Checkpoint

1. Describe one natural event that can destroy an ecosystem.
2. Define pollution. Name one source of air pollution. Describe what steps are being taken to prevent this type of pollution.
3. What is acid rain? How does it affect ecosystems?
4. What steps have been taken to protect ecosystems?

TRY THIS Place a clean bucket outside during a rainstorm. Measure the acidity of the rainfall with *p*H paper. If it measures below 7.0, the rain is acidic. If it is below 5.6, it may be having a harmful effect on the environment. Try to find out the source of the acid rain.

Students may ask you what "*p*H" means. The chemist's symbol *p*H comes from the French *pouvoir hydrogène*, meaning "hydrogen power". Pure water ionizes slightly,

$$H_2O \rightleftharpoons H^+ + OH^-,$$

and contains 1×10^{-7} moles per liter of hydrogen ion (H^+). Instead of using this cumbersome value, however, the chemist says the *p*H of pure water is 7. The *p*H is thus defined as the negative logarithm of the hydrogen ion concentration. When some acidic substances dissolve in water the concentration of H^+ is greater than 1×10^{-7} moles/L, and the *p*H is lower. The H^+ concentration in vinegar, for example, is about 1×10^{-3} moles/L, and the *p*H is 3. When some alkaline substances dissolve in water, the H^+ concentration is less than 1×10^{-7} moles/L, and the *p*H is higher. The *p*H of ammonia solution, for example, is about 12.

Checkpoint Answers

1. A volcanic eruption, a large-scale earthquake, and an extensive forest fire are examples.
2. Pollution is the accumulation of harmful substances in the environment. Answers will vary. See student text, pages 107–109.
3. Acid rain is rainwater that contains dilute sulfuric acid. The acid is produced when sulfur dioxide in the atmosphere reacts with moisture. Answers will vary. See student text, page 107.
4. Answers will vary. See student text, pages 107–109.

Using Try This

Figure 9-12 on page 231 shows the *p*H scale. The *p*H scale is a measure of acidity. Rainwater is always slightly acidic. When it has a *p*H of less than 5.6, however, it may be harmful to the environment. Use Figure 9-12 and Table 9-6 on page 232 to relate the students' *p*H findings to the *p*H of other common substances.

A natural extension of this activity is measurement of the *p*H values of streams, lakes, and ponds in your community. If acid precipitation is a serious problem in your area, consider having a local expert come in and discuss it with your class.

Teaching Suggestions 4-7

To dramatize the severity of the problem of endangered species, ask students to volunteer their ideas about the rate of extinction of mammals worldwide. Answers will vary. Few, if any, will be aware that on the average one mammalian species has become extinct each year since 1900, or that today about 600 species of mammals are in danger of extinction. Nor will they be aware that plant extinction is an even greater problem. (See Facts At Your Fingertips.)

You may want to explore why it is important to protect endangered species and to try to prevent extinction. To begin with, the natural world is a precious heritage that should be preserved for future generations. In addition, there are good biological arguments against allowing extinctions to occur. For example, as long as the diversity of species of organisms exists, there is an enormous resource of genetic characteristics. When a species becomes extinct, however, its genotype vanishes and is not recoverable.

Another argument is ecological. A basic principle of ecology is that simplified ecosystems are unstable. The formation of new species is far, far slower than the rate at which humans are capable of destroying species. Thus, the tendency worldwide is for ecosystems to become simpler. As ecosystems become simpler, the task of maintaining a stable biosphere becomes greater.

A classic example is what happened to portions of the great grasslands of central North America when European humans invaded. Prior to that time the grassland was an ecosystem in balance. The new human inhabitants, however, rapidly converted parts of the region into plowed and cultivated fields. Gone were the grasses that were capable of surviving drought and thus preventing erosion. Then, in the 1930's, when one of the severe droughts that is characteristic of grasslands occurred, winds swept away the dry topsoil, producing desert conditions. This was the United States' depression-era dust bowl. Because the ecosystem of the grasslands was greatly simplified, it was unstable, and could not survive drought conditions.

Finally, point out to students that there is general agreement that animals bred in zoos may look like their wild counterparts, but that they are vastly different in terms of physical condition, and also that they have none of the behavioral attributes of animals bred in the wild. Without the influence of its natural habitat, a zoo animal is a mere shadow of what it might be in the wild.

Chapter-end project 2 applies to this section.

Teaching Tips

- As a class investigate the whaling industry and worldwide efforts to halt whale hunting. How are current

ENRICHMENT

4-7 Endangered Species

Goal

To identify some of the factors that lead a species to become endangered or extinct.

extinct

endangered species

The growing need for land and resources has placed people in competition with many other species on Earth. In some cases, this has resulted in the *extinction* of a species. When a species no longer exists it is said to be **extinct**. Those species that are close to extinction are called **endangered species**. Figure 4-17 shows some endangered species.

A species may become extinct because of natural changes in the environment. But in the last few hundred years human activities have brought about the extinction of many species. The clearing of forests, and the filling in of swamps and marshes, have resulted in the loss of breeding sites, hiding places, and food supplies for many animals.

Figure 4-17 *Human activities, such as the clearing of land and unrestricted hunting, have led to the near extinction of thousands of species such as these.*

The destruction of habitats has contributed to the extinction of the eastern cougar and the passenger pigeon.

Many plant species are threatened by extinction also. The Nile lotus and the Egyptian papyrus are two plants that became endangered as a result of changes in the irrigation system of the Nile River.

Pollution can also threaten the survival of a species. DDT is a chemical that was widely used to kill insects. People did not realize that DDT would also harm other organisms. DDT stays in the tissues of organisms that come in contact with or ingest it. Figure 4-18 shows how DDT was passed along in one food pyramid.

Consumers that ate large numbers of DDT-contaminated organisms accumulated dangerous amounts of DDT in their tissues. For example, scientists discovered that DDT in the tissues of peregrine falcons and bald eagles caused the females to produce eggs with weak, thin shells. Their eggs broke easily, killing the chicks inside. Through the efforts of conservationists, the birds were saved from extinction. DDT is now outlawed in many countries.

Many steps have been taken to protect our environment and endangered species. Books about ecology, like *Silent Spring* by Rachel Carson, have made the public aware of the threat posed by the release of toxic chemicals into the environment. Laws have been enacted making it illegal to kill, trap, or collect most endangered species. The breeding grounds of many species have been protected, also. Other laws restrict the import and production of products made from parts of endangered species, such as their fur, hide, and tusks. Many wildlife reserves and zoo-breeding programs have been established to protect and increase the number of individuals in some endangered species.

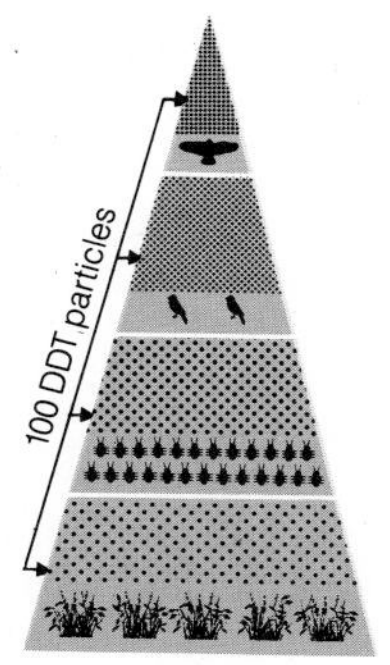

Figure 4-18 *DDT becomes more concentrated moving up the food pyramid. At each higher level, the same amount of DDT particles is distributed among fewer organisms. So each organism receives a larger amount.*

Checkpoint

1. What factors can lead to the extinction or endangering of a species?
2. Name one species that has become extinct. Name an endangered species.
3. What steps have been taken to protect endangered species?

TRY THIS

Find out if there is an endangered plant or animal species in your state or province. What is being done to help the species survive?

---

hunting practices affecting endangered species of whales? What measures are being taken to protect endangered whales?

- Have an interested student research recent theories about the extinction of the dinosaurs and report the findings to the class.
- Have a team of students research and report on current efforts to bring back the osprey and the bald eagle to areas where they once were abundant. The populations of both birds declined sharply when DDT was in wide use.

Facts At Your Fingertips

- Plant extinction is a greater problem than higher-animal extinction, for all life ultimately depends on photosynthesis. Today some 20,000 species of vascular plants are endangered.
- Although elephants and rhinoceros are protected, there is significant illegal traffic in ivory from elephant tusks and rhinoceros horn. In addition, there is a large traffic in objects made from endangered reptiles and in endangered species, such as parrots, which are desirable as pets.
- In a few years, some endangered species will no longer be seen in their natural habitats, prompting the fear that such species will survive only in zoos or on game farms.
- Volcanoes, earthquakes, and tidal waves are natural phenomena that can influence extinction.

Cross Reference

Plant and animal breeding is discussed in section 2-9 of Chapter 2.

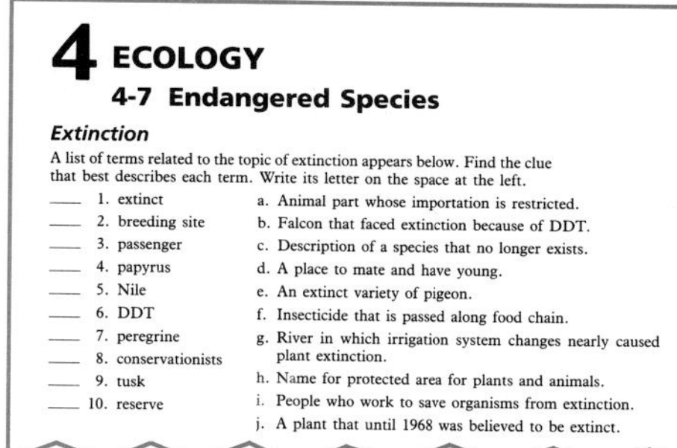

4 ECOLOGY

4-7 Endangered Species

Extinction

A list of terms related to the topic of extinction appears below. Find the clue that best describes each term. Write its letter on the space at the left.

___ 1. extinct	a. Animal part whose importation is restricted.
___ 2. breeding site	b. Falcon that faced extinction because of DDT.
___ 3. passenger	c. Description of a species that no longer exists.
___ 4. papyrus	d. A place to mate and have young.
___ 5. Nile	e. An extinct variety of pigeon.
___ 6. DDT	f. Insecticide that is passed along food chain.
___ 7. peregrine	g. River in which irrigation system changes nearly caused plant extinction.
___ 8. conservationists	
___ 9. tusk	h. Name for protected area for plants and animals.
___ 10. reserve	i. People who work to save organisms from extinction.
	j. A plant that until 1968 was believed to be extinct.

Using Try This

Students can contact their state Audubon Society, federal and state agencies (such as conservation, environmental protection, and fish and wildlife), and the National Wildlife Federation for information about endangered species.

Checkpoint Answers

1. Natural changes in the environment and many human activities, such as clearing land, filling in swamps and marshes, and polluting, can endanger species.
2. Answers will vary. See student text, pages 110–111.
3. Answers will vary. See student text, page 111 for possible answers.

Using The Feature Article

Efforts to save the whooping crane from extinction and bring its population back up to normal levels is a good case study in the area of endangered species. After you cover the situation that prompted action to save the cranes, you may wish to explore with your class why this unusual bird is so valuable, and also why the expenditure of large sums of money on it is justified.

Explore the value of the animal as an integral part of its ecosystem, what would be lost in terms of genetic material by its extinction, and what might happen to its ecosystem once the bird is gone. Refer back to Teaching Suggestions for section 4-7.

An interesting project for a student is researching the captive breeding program for cranes at the U.S. Fish and Wildlife Service's Patuxent Wildlife Research Center in Maryland. The student also should report on plans to restock the bird's habitat. Similar efforts on behalf of the bald eagle are underway at Patuxent. This too would make a fine student research project.

Whooping Cranes

The secretive whooping crane is an endangered species that nearly became extinct because of unrestricted hunting and the loss of feeding and breeding sites.

A bird refuge was created in 1937 in Texas to provide a place where the cranes could live unharmed. The cranes' winter migration route was protected by legislation making it illegal to hunt the cranes.

Despite the protective measures, the population of cranes did not increase. Scientists began to study the cranes' life cycle, hoping to find out how the cranes could be saved from extinction. They discovered that the cranes formed tight family structures known as *pair bonds.* Most males and females mated for life. Females laid two eggs each season, but usually only one survived.

Each winter the crane pairs and their offspring migrated to a warm climate and established winter feeding territories. However, many young birds did not survive the long, slow migration. Competition for nesting sites and the difficult migration kept the crane population from growing.

Recently, scientists have been experimenting with ways to increase the population of cranes. (See Figure 1.) United States and Canadian wildlife officials have gathered extra eggs from whooping-crane nests in Canada and have placed them in incubators at a wildlife center in Maryland. The chicks that have hatched and survived make up a small flock of captive cranes. Scientists hope to raise, and eventually breed, the captive cranes.

Researchers have also used sandhill cranes, a close relative of the whooper, as "foster parents" for whooping cranes. Whooping-crane eggs, placed in the nests of sandhill cranes, have been successfully hatched. Several whoopers have been raised by the sandhills. (See Figure 2.)

Figure 1

Figure 2

Understanding The Chapter

4

Main Ideas

1. An ecosystem is made up of a community and the nonliving parts of the environment.
2. The place where an organism lives is its habitat. Its way of living there is its niche.
3. A food chain shows how the energy stored in food is passed from one organism to another. Overlapping food chains form food webs.
4. Decomposers break down dead organisms and animal wastes. Through the action of decomposers, materials are returned to an environment.
5. Only one tenth of the energy acquired by an organism is available to an organism at the next level of a food pyramid.
6. A population is the number of organisms of the same species living in a certain area at a particular time.
7. The four processes that cause a population to change in size are immigration, emigration, birth, and death.
8. The process by which an ecosystem changes over time is called succession. The climax community is the final stage in succession.
9. Pollution is the accumulation of harmful substances in the environment.
10. The changes people have made in ecosystems have led to the extinction and endangering of some plant and animal species.

Vocabulary Review

From the following list, choose the term that best completes each of the statements. Write your answers on a separate piece of paper.

climax community	habitat
community	limiting factor
consumer	niche
decomposer	pioneer community
ecosystem	pollution
endangered	population
environment	predator
extinct	prey
food chain	producers
food web	succession
food pyramid	

1. A(n) __?__ breaks down dead organisms and animal wastes.
2. __?__ make food by photosynthesis.
3. In an ecosystem, each organism's way of living is called its __?__.
4. A(n) __?__ is made up of all the living things in an area.
5. A(n) __?__ is a group of overlapping food chains in an ecosystem.
6. The place where an organism lives in an ecosystem is called a(n) __?__.
7. __?__ is the presence of harmful substances in the environment.
8. A species that no longer exists is said to be __?__.
9. The process by which an ecosystem changes over time is called __?__.
10. The organism that a predator eats is called its __?__.

Vocabulary Review Answers

1. decomposer
2. producers
3. niche
4. community
5. food web
6. habitat
7. pollution
8. extinct
9. succession
10. prey
11. consumer
12. ecosystem

Reading Suggestions

For The Teacher

Berger, Joel. WILD HORSES OF THE GREAT BASIN. University of Chicago Press, 1986. Ecology and behavior of wild horse populations in Nevada.

Gondie, Andrew. THE HUMAN IMPACT: MAN'S ROLE IN ENVIRONMENTAL CHANGE. Cambridge, MA: MIT Press, 1982. The impact that human societies have had on the environment.

Janovy, John, Jr. ON BECOMING A BIOLOGIST. New York: Harper and Row, 1985. Chapters on Practice of Biology and Teaching and Learning are outstanding.

Owen, D. F. WHAT IS ECOLOGY? 2nd ed. New York: Oxford University Press, 1980. A brief and very readable survey of ecology.

Smith, Robert Leo. ECOLOGY AND FIELD BIOLOGY. 3rd ed. New York: Harper & Row, 1980. Covers basic ecological topics with examples from all parts of the world.

For The Student

Day, David. THE DOOMSDAY BOOK OF ANIMALS: A NATURAL HISTORY OF VANISHED SPECIES. New York: Viking, 1981. Magnificently illustrated and clearly written account of animals that have become extinct over the last 300 years. Includes species now endangered.

Miles, Hugh and Mike Salisbury. KINGDOM OF THE ICE BEAR. University of Texas (Austin), 1985. Ecological research on polar bears. Good on field research.

Ricciuti, Edward R. PLANTS IN DANGER. New York: Harper & Row, 1979. Covers the history of plants, including endangered plants and conservation efforts.

Stuart, Gene S. WILDLIFE ALERT! THE STRUGGLE TO SURVIVE. Washington, DC: National Geographic, 1980. A beautifully illustrated presentation of the impact of human activities on wildlife.

Van Lawick, Hugo. AMONG PREDATORS AND PREY. San Francisco: Sierra Club Books, 1986. Excellent pictorial essay.

Chapter Review Answers

Know The Facts

1. community
2. niche
3. first-order consumer
4. the sun
5. omnivores
6. one tenth
7. pyramid
8. immigration and birth
9. a limiting factor
10. pioneer
11. succession
12. biodegradable
13. extinct
14. an ecosystem
15. bacteria

Understand The Concepts

16. A population is the number of organisms of the same kind living within a certain area at a given time. Answers will vary. See student text, page 94.
17. Answers will vary. See student text, pages 96–97.
18. A producer uses the energy of sunlight to convert certain raw materials into substances usable as food.
19. A decomposer breaks down the dead remains of organisms into substances that can be reused by other organisms.
20. A food chain is a set of eating interactions within an ecosystem.
21. A food pyramid shows how much energy is available to each succeeding level in a food chain.
22. Immigration, emigration, birth, and death.
23. An organism's habitat is the place in an ecosystem where the organism lives, whereas its niche is its manner of living.
24. Answers will vary. See student text, pages 102–103.
25. Succession is the process by which an ecosystem changes over time. A pioneer community is the first group of organisms to become established in a barren region. Fungi, bacteria, one-celled organisms, and plants might all be members of a pioneer community.
26. A climax community is one that no longer changes. Natural disasters, major climate changes, and human intervention can change climax communities.
27. Acid rain is rainwater made acidic by the presence of industrial pollutants. Acid rain can interfere chemically with stages of the life cycles of numerous organisms.
28. Pollution is the accumulation of harmful substances in the environment. Automobile exhaust is a form of air pollution. Answers will vary.
29. Competition from humans, natural changes in the environment, pollution, and overhunting are all factors that have endangered or led to the extinction of organisms.

Challenge Your Understanding

30. All plant eaters are first-order consumers because plants are producers. However, some first-order consumers are omnivores—they feed on both plants and animals.

11. An organism that eats producers is called a(n) __?__.
12. A(n) __?__ is made up of a community and the nonliving parts of the environment.

Chapter Review

Write your answers on a separate piece of paper.

Know The Facts

1. All the living things in an area make up a __?__. (population, niche, community)
2. In an ecosystem, different kinds of organisms cannot have the same __?__. (habitat, niche, community)
3. Another name for a plant-eater is __?__. (producer, second-order consumer, first-order consumer)
4. All food chains begin with __?__. (a plant, the sun, a producer)
5. Animals that are both plant-eaters and animal-eaters are called __?__. (omnivores, carnivores, prey)
6. About __?__ of the energy acquired by an organism is available to one at the next level of a food pyramid. (one fourth, one half, one tenth)
7. A food __?__ shows how energy is "lost" when it is passed from organism to organism. (chain, pyramid, web)
8. Two processes that can *increase* the size of a population in an area are __?__. (immigration and birth, immigration and emigration, increased food supply and more predators)
9. Anything that keeps a population from growing indefinitely is called __?__. (a limiting factor, succession, a niche)
10. After a fire turns a forest into a bare area, the first community to develop is known as the __?__ community. (predator, pioneer, climax)
11. A pond may change into a meadow through __?__. (extinction, a climax community, succession)
12. Wastes that can be decomposed are said to be __?__. (nonbiodegradable, biodegradable, pollutants)
13. A species that no longer exists is __?__. (endangered, a climax community, extinct)
14. All the interacting populations in an area and their environment form __?__. (a community, an ecosystem, a habitat)
15. Nitrogen in soil is changed into a substance that can be used by plants through the action of __?__. (fungi, scavengers, bacteria)

Understand The Concepts

16. Define population. Name two populations that may be found in a pond ecosystem.
17. Draw a pond food chain. Label the producers, the consumers, and the decomposers.
18. What is the role of a producer in a food chain?
19. What is the role of a decomposer in a food chain?
20. What is a food chain?
21. What is a food pyramid designed to show?
22. What are the four processes that cause a population to change in size?

23. Explain the difference between an organism's habitat and its niche.
24. Name three limiting factors in a forest ecosystem. Explain how each factor can keep a population from growing.
25. Define succession. Explain what is meant by the term *pioneer community*. Name two organisms that might belong to a pioneer community.
26. What is a climax community? What would cause a climax community to change?
27. What is acid rain? What causes it? How does it affect wildlife?
28. Define pollution. Name one kind of air pollution. Tell how it is caused and describe its effect on the environment.
29. Name three factors that have led to the extinction of a species.

Challenge Your Understanding

30. All herbivores are first-order consumers, but not all first-order consumers are herbivores. Explain this statement.
31. The number of species living in an ecosystem is limited by the number of available niches. Explain.
32. Which group probably has a greater variety of niches—insects or whales? Explain.
33. Which food chain can feed the most people? Explain. A. sun⟶soybeans⟶people. B. sun⟶soybeans⟶cattle⟶people. (Hint: Draw two food pyramids. Put the same amount of soybeans in each food pyramid.)
34. Explain why the changes brought about by organisms in an ecosystem during succession often lead to their own replacement.

Projects

1. Obtain an old aquarium or a large, wide-mouthed jar. Collect the organisms needed to set up a small ecosystem. Observe the interactions among the organisms. Keep a record of your observations. Note the changes that take place over time. Return the organisms to where you found them when you have completed the project.
2. Investigate what is being done to protect endangered species of whales. Prepare a report for your class. Find out what individuals can do to help save whales.
3. Find out if there is a recycling program in your town or city. Find out how your school can participate in it. Arrange to have bottles and cans from the school cafeteria brought to the recycling center each week.
4. Do a report on the following kinds of interactions that take place in ecosystems: mutualism, commensalism, and parasitism. Look for examples of these interactions in your community.
5. Find out how the human population has changed in your community, state or province, and country in the last 50 years. What projections are being made about the size of each population in the future?
6. Research and report on the contributions to ecology of any one of the following scientists: Ruth Patrick; Anna Comstock; Rosa Smith Eigenmann.

4 ECOLOGY
Projects

1. Research the status of international agreements on whaling.
   a. Which nations are the leading whalers? ______
   b. Why is whaling so important to them? ______
   c. Why is it so difficult for nations to agree on whaling regulations? ______
2. Investigate the recycling of aluminum cans.
   a. Which states have laws requiring the recycling of aluminum cans? ______
   b. What good results do these states report? ______
   c. Are there problems with recycling? ______
   d. Do you think that your state should have such a law? ______ Why or why not? ______
3. Find out about the work of a game warden, a fisheries biologist, or a national park ranger. Which kind of work did you choose to research? ______
   a. What training is needed? ______
   b. What are his or her duties? ______
   c. Which duties would you enjoy? ______ Why? ______
4. Why does acid rain kill lakes in the Northeast and not in the Midwest? Look for answers in your school library. The following questions should help you arrive at your answer.
   a. What kind of bedrock is found in the Northeast? ______ In the Midwest? ______
   b. How could the difference in bedrock affect the water? ______

31. A niche includes the kind of food an organism uses, and behaviors that allow an organism to survive. The number of available niches in an ecosystem is limited, and no two species have the same niche. Thus, the number of niches limits the number of species.
32. Insects. Because there are far more species of insects, there must be many more niches.
33. A. Because the energy from soybeans that would be lost feeding the cattle is retained in the first food chain, more energy is available for people.
34. The changes brought about by organisms in an ecosystem often change the physical features of the ecosystem, making it attractive to other organisms. These then move in and replace the first organisms.

CHAPTER 5 Overview

The purpose of this chapter is to understand the diversity of living things in the context of the theories of evolution and of natural selection, and in relation to the ways organisms are adapted to live in certain habitats. Examples are given of adaptations in plants and animals. The concept of biomes, and the relationship of biomes to climate, are explained. Tundra and tropical rain forest biomes are described, followed by discussions of forest biomes, the desert biome, the grassland biome, and life zones in the oceans.

Goals

Upon completing this chapter, students should be able to:

1. define the theory of evolution.
2. explain Darwin's theory of natural selection.
3. identify some plant and animal adaptations.
4. explain how climate affects biomes.
5. describe the features of tundra and tropical rain forest biomes.
6. describe the characteristics of deciduous and evergreen forest biomes.
7. identify some adaptations of forest organisms.
8. describe the features of the desert biome.
9. identify some adaptations of desert plants and animals.
10. identify the features of the grassland biome.
11. describe life zones in the ocean.
12. identify adaptations that help organisms survive in the ocean.

Chapter 5

Distribution of Life

Life is distributed worldwide. This is an African savanna.

Vocabulary Preview

theory of evolution
natural selection
adaptation
instincts
climate
biomes
tundra
permafrost
deciduous
hibernation
evergreen forest
savannas

5-1

Development of Life

Goals

1. To explain the theory of evolution.
2. To describe Lamarck's explanation of evolution.
3. To compare Lamarck's explanation of evolution with Darwin's theory of natural selection.

In Chapter 1, you were introduced to the variety of life on Earth. Why are there so many different forms of life? Where did they all come from? These are questions that have puzzled scientists for centuries.

A theory that explains the great variety of living things and the basic similarities among them is the **theory of evolution**. This states that, over time, new species develop, or evolve, from earlier species. It is believed this process has been occurring since organisms first appeared on Earth.

theory of evolution

In the early 1800's, the French scientist Jean Lamarck proposed an explanation of how evolution occurs. He believed that organisms evolve, or change, according to how they respond to their environment. He thought that species evolve when organisms develop new traits through using or not using certain organs or structures. The more an organ or structure is used, the more it develops. The less it is used, the less it develops. Moreover, Lamarck believed that such *acquired traits* are inherited. For example, Lamarck would

Figure 5-1 *(a) Early giraffes had short necks. (b) When low-growing plants became scarce, giraffes stretched their necks to reach food. (c) The giraffes with stretched necks passed on their long-neck trait to their offspring.*

Teaching Suggestions 5-1

Begin by pointing out that the theory of evolution can account for the diversity as well as the similarities among all living things. Although Darwin is the person most closely identified with this theory, he was not the first to propose it. Other individuals, including Lamarck, had earlier written about organisms changing over the course of many generations. However, Darwin was the first person to propose an explanation of how evolution occurs that was based on vast amounts of data and sound scientific principles. His theory of natural selection forms the basis for much of modern science's view of evolution.

Using the giraffe example shown in Figures 5-1 and 5-3, review with students the differences between Lamarck's and Darwin's views of how evolution occurs. The example of the body builder can be used to show why Lamarck was wrong. The large muscles of a body builder are not inherited, but develop in response to a special diet and exercise. This trait, like all acquired traits, does not affect the person's genetic makeup. It can not be passed on to offspring and therefore can not be a mechanism for evolution. Ask students to suggest additional examples of acquired traits.

The peppered moth of England (see Prologue) illustrates the process of natural selection. Prior to the Industrial Revolution, light-colored moths, because they were camouflaged while on tree trunks, had a much better chance of surviving than dark moths. After the Industrial Revolution, the reverse became true in urban areas with high levels of air pollution. In each case, natural selection, acting through the activity of moth predators, favored the survival of individuals with one trait over those with the other trait. Emphasize that natural selection for moth color would not have occurred unless there had been genetic variation, due to previous mutations, preexisting in the moth populations. This kind of selection, acting on this or other traits over a long enough period of time, could lead to the development (evolution) of a new species of moth.

To demonstrate how extensive Darwin's travels were, point out on a map of the world, or on a globe, the areas that Darwin visited during the 5 year voyage of the *Beagle*. These included the Cape Verde Islands off the west coast of Africa; Brazil; Argentina; Tierra del Fuego, a group of islands off the extreme southern tip of South America; Chile; the Andes Mountains; the Galapagos Islands off the coast of Ecuador; Tahiti, in the South Pacific; New Zealand; Australia; the Cocos Islands southwest of Singapore in the Indian Ocean; Mauritius, in the Indian Ocean east of Madagascar; South Africa; St. Helena, off the southwest coast of Africa; and Ascension Island in the South Atlantic. It was on the Galapagos Islands, each with its own species of finch, that Darwin saw most clearly evidence to support the idea of natural selection.

Background

Darwin did not reach his conclusions about evolution and natural selection in isolation from the ideas of other scientists. In fact Darwin's thinking was influenced by other scientists, among them the two Englishmen Charles Lyell and Thomas Malthus. In Lyell's *Principles of Geology*, Darwin read that the earth was much older than previously thought, and that it had evolved, i.e., undergone dramatic changes through the slow and steady forces of nature. Darwin realized that the long and dynamic history of the earth provided the conditions in which evolution in living things could have occurred. In *An Essay on the Principles of Population*, Malthus explained that human populations tended to grow faster than the supply of available resources. Darwin applied this principle to all living things, since it correlated with his own observations that organisms compete with each other. Darwin's knowledge of artificial selection used in the breeding of domesticated plants and animals no doubt also contributed to his idea of a selection process operating in nature. In the tradition of good science, Darwin formulated his theory of natural selection by integrating the ideas and knowledge of other individuals with those of his own.

The five basic principles on which Darwin's theory of natural selection is based are:

1. Variation. The individuals of a species vary with regard to many traits.
2. Inheritance of variations. The individuals that survive and reproduce pass their traits on to their offspring.
3. Excess reproduction. All species are capable of producing more offspring than can survive.
4. Struggle for survival. Members of the same and of different species compete with one another for limited resources, *e.g.*, food, shelter.
5. Survival of the fittest. The individuals that survive and reproduce are those with the traits best suited to the environment.

Fossils have been very important in establishing the theory of evolution. For Darwin, fossils proved the existence, and showed the relative ages, of previous life forms, and, in some cases, revealed similarities with certain modern day organisms. Scientists now can analyze fossils to determine how long ago organisms lived, and what conditions were like at the time.

Facts at Your Fingertips

- When Darwin proposed his theory of natural selection, Alfred Russell

Figure 5-2 *The English naturalist Charles Darwin proposed the theory of natural selection to explain the great variety of life.*

argue that giraffes evolved from short-necked ancestors as shown in Figure 5-1.

Today, we know that Lamarck's theory is incorrect. Acquired traits, such as well-developed muscles, or an amputated leg, cannot be passed on to offspring. As you learned in Chapter 2, only traits that are encoded in the DNA of chromosomes can be inherited.

The year that Lamarck published his theory to explain the variety of life, another scientist, Charles Darwin, was born in England. When Darwin was 22, he took a job as a naturalist aboard a ship, the H.M.S. *Beagle*. During the five-year voyage of the *Beagle*, Darwin was struck by the variety of species distributed in different parts of the world. Throughout his travels, Darwin kept detailed records about what he observed. He noticed that organisms were constantly struggling to get enough food, space, and shelter. He also noted the likenesses and differences between the *fossils*, or preserved remains, of extinct organisms and the bodies of living organisms.

For more than twenty years, Darwin gathered, organized, and analyzed data. Then, in 1859, he published his findings in a book, *On the Origin of Species*. In this book, Darwin proposed his own theory of evolution.

Darwin claimed that individuals of a species are born with differences among themselves. These different traits, or *variations*, are inherited. In addition, some of these traits affect the ability of each individual to survive. Those individuals who survive tend to be the ones best suited to the

Figure 5-3 *Darwin's theory: (a) Adult giraffes' necks varied in length. (b) When the environment changed, only the long-necked giraffes could reach food. (c) The short-necked giraffes died, leaving only the long-necked giraffes to reproduce.*

environment. Because only they live to reproduce, the traits important for survival are passed on to offspring. Over the course of many generations, these traits become characteristic of a new species. Darwin called this process whereby the fittest survive **natural selection**. Figure 5-3 illustrates how Darwin would have explained the evolution of modern day giraffes.

natural selection

Darwin's theory of natural selection was debated from the time it was published. Only after evidence was provided from the developing science of *genetics* did Darwin's theory gain wide acceptance. This supporting evidence came from Hugo DeVries, a Dutch botanist. DeVries explained how mutations produce new characteristics that can be passed on to succeeding generations. Mutations cause the inherited variations within species that permit natural selection to take place.

Islands provide good examples of how natural selection works. New species tend to thrive on an island because there are many new niches to occupy, and few predators to limit the population. The Hawaiian Islands provided such a habitat for the honeycreepers, birds whose ancestors arrived there 5 million years ago. Over many years, variations in their beaks evolved for eating different foods. Figure 5-4 shows honeycreepers that have evolved long, curved beaks to drink nectar from flowers; other honeycreepers have short, strong beaks to crack and eat hard nuts. Because birds with different beaks were not competing in the same niches, most survived to pass these traits on to their offspring. As a result, there are now 22 species of honeycreepers in Hawaii, all evolved from a common ancestor.

Figure 5-4 *(top) Some honeycreepers have evolved long beaks for drinking nectar from flowers. (bottom) Other honeycreepers have short beaks for eating nuts.*

Checkpoint

1. How is the evolution of long-necked giraffes explained using Lamarck's ideas about evolution?
2. Use Darwin's theory of natural selection to explain the evolution of long-necked giraffes.
3. What is the importance of variation in Darwin's theory of natural selection?
4. How did DeVries' discovery about genetic mutations support Darwin's theory?

Wallace, a British naturalist, independently proposed a nearly identical theory. Because Darwin had written about his ideas earlier in private correspondence, and subsequently wrote so extensively on the subject of natural selection, he is credited with first developing the theory.

- *Adaptive radiation* is the evolution of a single, unspecialized species into many species with specialized adaptations. Adaptive radiation occurs when animals or plants enter previously unpopulated environments where they find few predators and no competitors for food. Minor mutations in the gene pool can prevail under these circumstances, leading to a proliferation of new species. This is why honeycreepers evolved so extensively in the isolation of the Hawaiian Islands.

- The controversy that arose when Darwin proposed his theory of evolution had to do with the challenge it presented to prevailing beliefs. A similar controversy took place in the late sixteenth century. At that time Nicolaus Copernicus proposed the heliocentric theory that the sun was at the center of the universe. It had been believed that the earth was at its center.

Cross References

Chapter 2 covers the topics of genetics, DNA, chromosomes, and mutations. A discussion of how fossils are formed is found in section 11-3, pp. 279–281. The carbon-14 method for dating fossils is explained on page 260.

Checkpoint Answers

1. Early giraffes had short necks that grew longer as they were stretched to reach leaves on trees. According to Lamarck's theory, the acquired trait of a long neck was passed on to the offspring.
2. Variation in neck length occurred in the ancestors of giraffes. When food near the ground became scarce, only the individuals with the longest necks could get food by browsing on trees. Since only those individuals lived and reproduced, all the offspring had long necks.
3. Variation accounts for some individuals being better suited than others.
4. The discovery of genetic mutations by De Vries provided a basis for the inherited variations in species that Darwin had proposed.

Teaching Suggestions 5-2

Open the section by asking students to volunteer definitions of adaptation. Answers will vary, and will probably not stress that adaptive traits are inherited. Make sure students understand that organisms do not "become adapted" to their environments. Instead, an organism that is adapted to its environment already has the inherited trait that enables it to live and reproduce in that environment. Stress that adaptation does not mean adjusting to the environment. Instead, it is the advantage conferred on an organism by the existence of one or more inherited traits.

Discuss each of the examples of adaptation given in the section. Stress that individuals having adaptive traits have a better chance of surviving and reproducing than individuals lacking adaptive traits.

You may want to describe how adaptations, whether structural or otherwise, are the result of natural selection acting on the variability present in a population of organisms. Many instances of natural selection have been observed (Figure 5-4), and the process has been demonstrated in the laboratory. In one experiment, about 100 million bacterial cells were exposed to a weak dose of penicillin. Fewer than 10 cells survived and reproduced. When the colony produced by the survivors was treated with a dose of penicillin twice as strong, again most of the cells were killed, but a very few survived to reproduce. Finally, a strain of bacteria was obtained that could withstand a dose of penicillin 2500 times as strong as the original dose. This strain was selected by the change in environment produced by gradually increasing the concentration of the antibiotic penicillin. Point out that the common expression, the bacterium "developed a resistance" to penicillin, is not the case at all. Instead, the resistant bacteria were the offspring of those few that had inherited a favorable advantage.

Another example of selection occurred when the insecticide DDT was introduced. When it was first used on houseflies, it killed most of the flies treated. A few, however, lived because of some inherited variation. These resistant flies reproduced and soon outnumbered the less-resistant type of fly in the area. As a result, the next time DDT was applied, it was less effective. More flies survived. In areas where DDT was used for many years, a population of resistant flies emerged and became the usual form. DDT in the environment was the factor that selected the surviving population of flies. Stress that DDT did not *give* resistance to the flies, as is commonly believed.

These are examples of adaptation through selection that occurred in a relatively short period of time. For the most part, the examples given in the student text occurred over very long periods of time. Make sure students

5-2

Adaptations

Goal

To identify some plant and animal adaptations.

adaptation

Any trait that enables an organism to live in a particular environment is called an **adaptation.** Certain body parts, or structures, are adaptations that help organisms get food. Teeth and mouth parts are examples of *food-getting adaptations* in animals. Herbivores, such as horses and cows, have large, flat, grinding teeth. Carnivores, such as dogs and cats, have sharp, pointed teeth that help them tear meat into small pieces. (See Figure 5-5.) Omnivores, such as humans, have sharp, cutting teeth in the front of the mouth and flat, grinding teeth in the back. Look at the pictures of birds' beaks in Figure 5-6. The shape of each bird's beak determines the type of food it eats and how it gets the food.

Figure 5-5 *Teeth are an adaptation that help an animal get and use a particular food.*

Figure 5-6 *Three different kinds of feeders: a water-feeder, a seed-eater, and a meat-eater.*

Plants also have adaptations that enable them to get food. Chloroplasts, in plant cells, make photosynthesis possible. Some plants, such as the Venus' flytrap, grow in areas where there is little nitrogen in the soil. When an insect lands on its leaves, the leaves close up, trapping the insect. (See Figure 5-7.) The insect is then digested. This provides the plant with nitrogen and other nutrients not otherwise available.

Figure 5-7 *The Venus' flytrap captures insects with its leaves. Digestion and absorption take place slowly.*

Some adaptations protect organisms. For example, quills are an adaptation that protect a porcupine from enemies. Contrary to popular belief, a porcupine does not shoot its quills, but it can imbed them deeply in an enemy with the slap of its tail. Claws, shells, and spines are other *protective adaptations* in animals. So is the foul-smelling spray of a skunk.

Some animals have colors, shapes, or textures that make it possible for them to blend in with their surroundings. Examples of *camouflage* are shown in Figure 5-8.

Behavior is the way an organism responds to its internal and external environments. Behaviors can be adaptations. **Instincts** are behaviors an organism is born with. Instinctive behavior does not have to be learned. For example, a baby albatross pecks at its parent's bill when it feels hungry. The parent responds by spitting up food. (See Figure 5-9 on page 122.) Both the pecking of the baby and the providing of food by the parent are examples of instinctive behavior. The instincts help baby albatrosses survive.

instincts

Figure 5-8 *The animals in these photographs are hard to distinguish from their surroundings because of camouflage.*

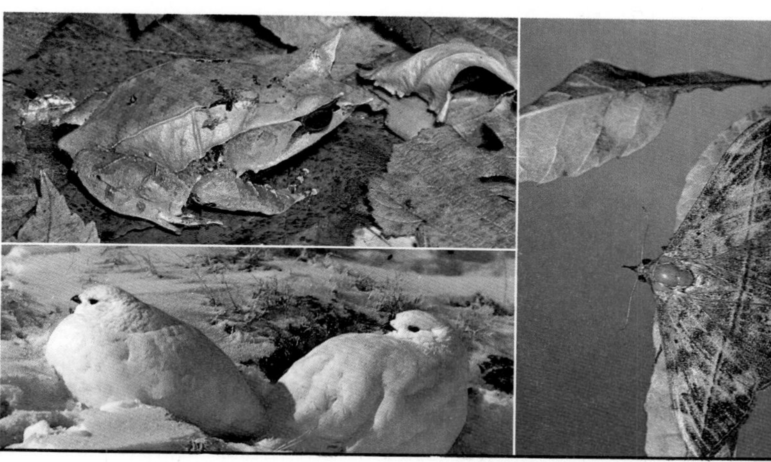

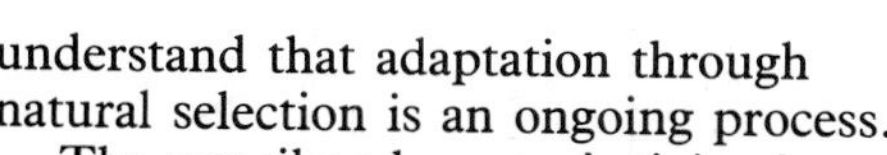

understand that adaptation through natural selection is an ongoing process.

The pencil-and-paper Activity 8, Bird Beak Adaptations, student text page 540, should be done in conjunction with this section. Chapter-end projects 2, 3, and 5 also apply.

Background

Camouflage can help prey avoid the notice of predators; or it can help predators ambush prey. The examples of camouflage in Figure 5-8 are (top left) a horned, or leaf, frog in a Malaysian rain forest; (bottom left) two white-tailed ptarmigans in winter; and (right) a leaf-mimicking moth in Georgia. See also Figure 5-4.

The bright patterning of the monarch butterfly (Figure 5-10) serves as warning coloration. Predators learn to associate the coloration with bad taste and avoid it. Because of the similar coloration of the viceroy butterfly, predators also avoid it. Thus, the bright coloration of the viceroy helps it survive. This is the opposite of camouflage.

Demonstration

Show students skulls of several different animals and have them determine if each animal was a carnivore, an herbivore, or an omnivore. Skulls or skull models are available from biological supply houses.

Teaching Tips

- Have a team of students research and report on adaptive radiation and adaptive convergence.
- Ask students if they know of two species of birds living in the same area. Robins and woodpeckers are two good examples. Point out that if the two species thrive in the same setting, they must have different adaptations, or ecological niches. Review the definition of niche, and discuss how the food-getting adaptations of the two birds enable them to survive in the same setting.

Facts At Your Fingertips

- Wings in birds and large leg muscles in frogs are structural adaptations for movement.
- Some additional adaptations in plants are growth toward sunlight, air pockets in certain seaweeds, and the thick, waxy covering on certain cacti.

- The viceroy butterfly's similarity in appearance to the monarch butterfly is an example of mimicry.
- The sucking response in human infants is instinctive.

Mainstreaming

Provide visually-impaired students with examples of an herbivore's and a carnivore's teeth. Let them tactually explore the differences in the teeth. The same kind of activity can de done by letting the students feel different bird's beaks if they can be made available.

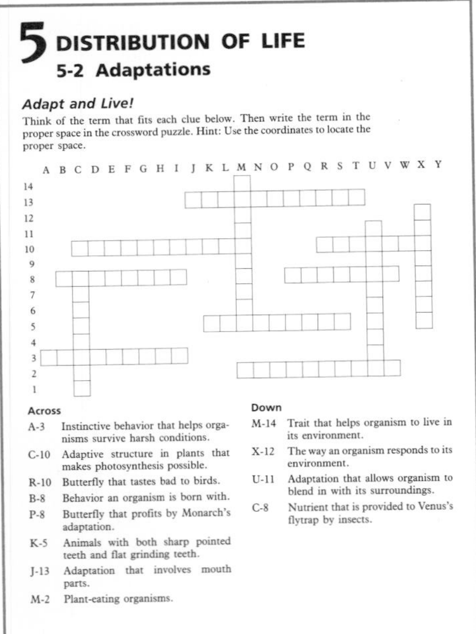

5 DISTRIBUTION OF LIFE
5-2 Adaptations

Adapt and Live!

Think of the term that fits each clue below. Then write the term in the proper space in the crossword puzzle. Hint: Use the coordinates to locate the proper space.

Across

- A-3 Instinctive behavior that helps organisms survive harsh conditions.
- C-10 Adaptive structure in plants that makes photosynthesis possible.
- R-10 Butterfly that tastes bad to birds.
- B-8 Behavior an organism is born with.
- P-8 Butterfly that profits by Monarch's adaptation.
- K-5 Animals with both sharp pointed teeth and flat grinding teeth.
- J-13 Adaptation that involves mouth parts.
- M-2 Plant-eating organisms.

Down

- M-14 Trait that helps organism to live in its environment.
- X-12 The way an organism responds to its environment.
- U-11 Adaptation that allows organism to blend in with its surroundings.
- C-8 Nutrient that is provided to Venus's flytrap by insects.

Using Try This

It will help students if you brief them on the sorts of adaptations you are looking for, and if you limit the number sought to those that are obvious. Make sure students understand that any behavioral adaptations observed in a zoo setting are probably not behaviors typical of the animal in the wild.

Figure 5-9 *A baby albatross pecks at its parent's bill when it feels hungry.*

Figure 5-10 *The viceroy buttlerfly (on the left) looks like the monarch butterfly (on the right).*

In areas that have cold, snowy winters, many animals travel, or *migrate,* to warmer places. Migration is an instinctive behavior that helps organisms survive harsh conditions.

The ability to learn is an adaptation. Sometimes learned behavior becomes automatic by being repeated a number of times. For example, you learned to write your name by practicing. Now you do it without effort. Learning how to do something saves time. Learning can also protect an organism from harm. For example, a bird gets sick and vomits when it eats a monarch butterfly. It learns to recognize and avoid the monarch. Birds also avoid the nonpoisonous viceroy butterfly because it looks like the monarch. (See Figure 5-10.)

Checkpoint

1. What is an adaptation?
2. Describe two food-getting adaptations in animals.
3. Give one example of an adaptation in plants.
4. Describe one behavioral adaptation in animals. Explain how it helps the animal survive.

TRY THIS

Visit a zoo or wildlife preserve. Select one animal for observation. Make a list of as many adaptations as you can. Pay special attention to structural adaptations for movement, protection, and feeding. What behavioral adaptations do you observe?

Checkpoint Answers

1. An adaptation is any trait that enables an organism to live in a particular environment.
2. Answers will vary. See student text, page 120.
3. Answers will vary. See student text, page 121.
4. Answers will vary. See student text, pages 121–122.

5-3

Biomes

Goals

1. To explain how climate affects biomes.
2. To describe the features of tundra and tropical rain forest biomes.

Picture a "zone of life" that covers the entire Earth. It extends approximately 100 m above the land and several meters underground. In areas that are covered by water, the zone of life begins at the surface of the water and goes to a depth of 200 m. Conditions in the zone of life vary.

The **climate** of an area is the average weather over a long period of time. Climate is one factor that determines the kinds of organisms that can live in a given area. Scientists have divided Earth into three major climate zones: *polar, temperate,* and *tropic* zones.

climate

Regions in the polar zones have long, cold winters and short, cool summers. The coldest winter temperature on Earth, −88°C, was reported in the polar zone. The tropic zone is warm or hot most of the time. Seasons in the tropic zone are marked by changes in the amount of rainfall more than by changes in temperature. The average temperature may change by only a degree or two between summer and winter. The temperate zone has four seasons during which the weather gradually shifts from colder to warmer and then back again.

Within each climate zone there are large areas of land scattered around the globe that contain similar communities. These areas are called **biomes.** Each biome contains plants and animals that have adaptations that help them survive in that particular area.

biomes

Climate zones and biomes are related. The climate determines the types of plants growing in an area. The plants provide food for certain animals. The climate, then, determines the characteristics of a biome. Wherever the climates are the same, the biomes are similar. For example, the tropics are located within about 25° north and south of the equator. Here, the average temperature is about the same year round. Here, too, are found similar biomes. Each biome may contain one or more ecosystems.

TRY THIS

Look at the map in Figure 5-11 on page 124. Locate your state or province. In which biome do you live? Find out the average yearly temperature and amount of rainfall for your area.

Teaching Suggestions 5-3

Open the section by describing the so-called "zone of life" that covers the entire Earth. Point out that the dimensions given in the student text are averages, and that variations occur throughout the world. For example, only the top few centimeters of soil support life in the tundra, but the average depth of life-supporting soil is about one meter.

Ask students if they are aware of the technical term for the zone of life. Answers will vary, and "biosphere" may be volunteered. You may want to give your class the definition of biosphere: all living things on Earth and the water, air, soil, and other matter that surrounds them. Using this definition as a springboard, stress that humans are a part of the biosphere, and explain how human activities affect both its living and nonliving parts.

Describe the weather characteristics of Earth's three major climate zones—polar, temperate, and tropic—and stress the relationship between climate and the communities of living things within climate zones. Make sure students understand that similar biomes occur within each climate zone, and that the key to the success of organisms within a biome is adaptations to the climate and other conditions of the region. It may help to explain that biomes can be thought of as large patches within the biosphere that consist of collections of communities. The patches are not necessarily connected, but a patch in one part of the world can be very similar to one in another part of the world. For example, deserts throughout the world are remarkably alike. Point out that the tundra biome shown on the map in Figure 5-11 is arctic tundra, not alpine tundra.

Stress that climate is one of the most important factors that determines the characteristics of a biome. The diagram in Figure TE 5-1 shows the mean annual temperature and the mean annual precipitation in different biomes of North America.

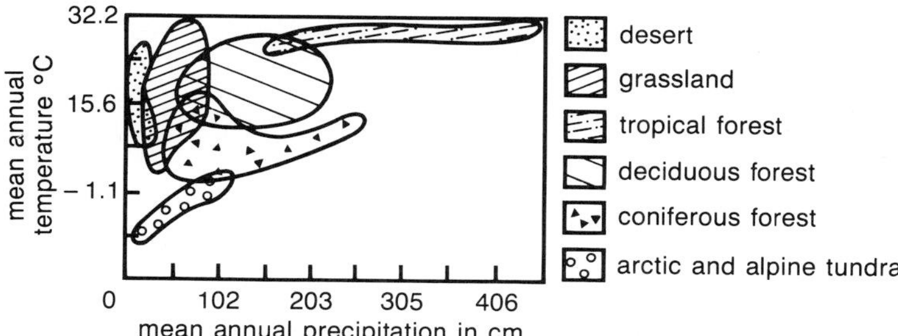

Figure TE 5-1 The mean annual temperature and precipitation of six biomes.

Point out that whenever climate is the same, the organisms tend to have similar adaptations. Such organisms may not be related genetically, but they look and act alike. This is an example of adaptive convergence. Make sure students grasp how climate influences the existence of producers, and therefore the consumers of the region. This is a good time to review the definition of ecosystem, and point out that a biome will often contain numerous ecosystems.

Describe the physical characteristics of the tundra and tropical rain forest biomes. Compare these properties, and discuss some of the more evident adaptations to them. The organisms shown in Figures 5-12 and 5-13 should provide enough information for a lively discussion. The organisms in Figure 5-13 include caribou, an arctic hare, lemmings, musk oxen, a snowy owl, a weasel, an arctic fox, ptarmigans, reindeer moss, grasses, dwarf willows, and sedges. Typical of the organisms in a tropical rain forest, Figure 5-12, are boa constrictors, spider monkeys, marmosets, toucans, parrots, numerous kinds of insects, tree frogs, jaguars, bromeliads, orchids, ferns, lianas, and tropical trees.

Review the definition of niche. Discuss why some of the niches in the tropical rain forest could not exist in the tundra. Give examples of the niches occupied by some of the organisms shown in Figures 5-12 and 5-13.

Chapter-end projects 4, 5, 7, and 8 apply to this section.

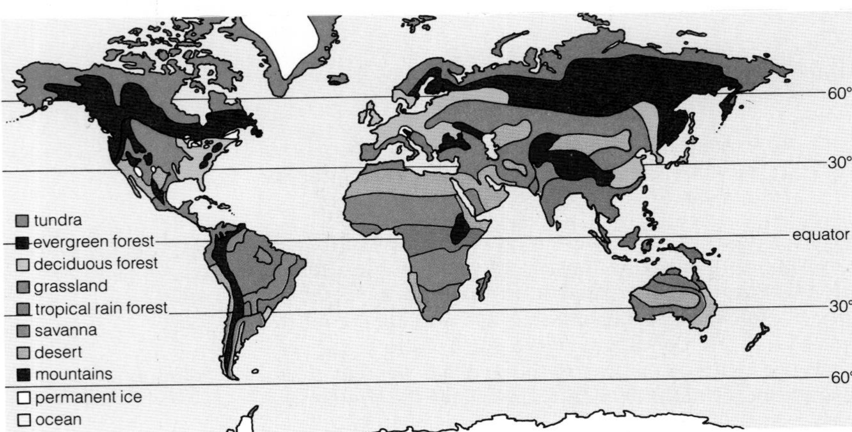

Figure 5-11 The world's biomes are shown with distinct boundaries. In nature, biomes overlap.

tundra

permafrost

Figure 5-12 The tropical rain forest biome is characterized by tall trees, climbing vines, and epiphytes.

Find the location of the tundra biome on the map in Figure 5-11. The **tundra** is an area in the polar climate zone that receives little energy from the sun. The tundra has low temperatures and little rainfall. During the winter months, much of the tundra receives *no* sunlight. Tundra soil is frozen solid in winter. During the summer, only the top few centimeters of soil thaw, permitting plants with shallow root systems to take hold. The frozen soil below the surface is called **permafrost.** The tundra has a short growing season for plants. Most tundra plants grow close to the ground—they never become very tall. Large plants would not survive the cold, strong winds of the tundra. Figure 5-13 shows some of the organisms that live in the tundra biome.

The tropical rain forest biome is located in the tropic zone. Direct sunlight, heavy rainfall, and constant high temperatures provide a climate that is well suited for a great variety of plant species. Tall trees form a thick canopy, shading the ground below. Shade-tolerant trees and long, climbing vines make up a layer of life beneath this canopy. Rootless plants, which absorb water from the moisture-rich air, live on the bark of trees. Such plants are called *epiphytes* (EP uh fyts). See Figure 5-12.

There is a tremendous variety of niches in the tropical rain forest. Many animals, including insects and tree-dwelling varieties of reptiles, amphibians, and mammals, live in

Teaching Tips

- Select a biome near your school and do a class study of the interrelationships within the biome. What organisms are present? What adaptations to climate exist? How does the climate affect the producers? the consumers? the decomposers?
- Have an interested student research and report on plant, bird, and mammal adaptations to tundra and tropical rain forest conditions.
- If there is a modern zoo in your area, make a class visit and study the zoo's attempts to duplicate the climate and physical properties of the animals' habitats. Begin planning now and make the trip when you complete work on the chapter, so students can make reference to all of the biomes studied.

Using Try This

Have students refer to an almanac or contact a nearby weather service for temperature and rainfall data.

Figure 5-13 *The plants and animals shown in this portion of the tundra biome are typical of those organisms found in the tundra in the summer.*

"layers" among the thick vegetation. Vast amounts of dead plant and animal matter constantly fall to the forest floor. Huge numbers of insects, fungi, and bacteria live on this material. The fungi and bacteria cause decomposition. This releases minerals back into the soil for reuse by plants.

Checkpoint

1. Name the three climate zones on Earth and describe each zone.
2. Define biome.
3. Explain how the climate of an area affects a biome.
4. Describe some of the principal characteristics of the tundra biome.
5. Describe the climate of the tropical rain forest biome.
6. Describe one animal adaptation and one plant adaptation to life in the tropical rain forest biome.

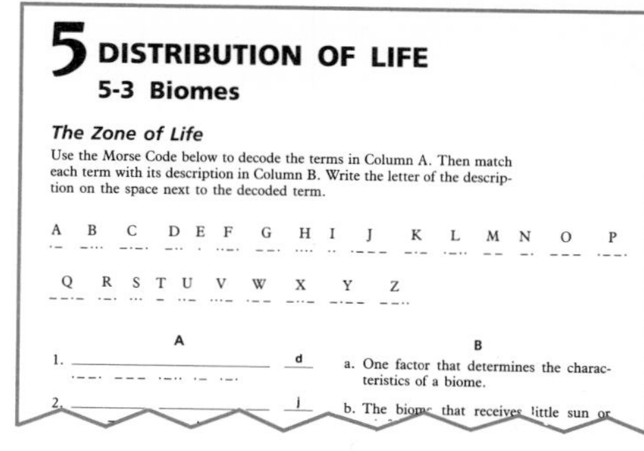

5 DISTRIBUTION OF LIFE
5-3 Biomes

The Zone of Life
Use the Morse Code below to decode the terms in Column A. Then match each term with its description in Column B. Write the letter of the description on the space next to the decoded term.

A B C D E F G H I J K L M N O P

Q R S T U V W X Y Z

A		B
1. ________	d	a. One factor that determines the characteristics of a biome.
2. ________	i	b. The biome that receives little sun or

Facts At Your Fingertips

- Biome is an abstract concept. When we speak of the tundra (or any other) biome, we are not referring to a particular geographic region, but rather to all tundra on Earth.
- Climate includes such factors as amount and intensity of sunlight, wind patterns, rainfall, and temperature variations.
- Competition for sunlight in the tropical rain forest biome leads to a stratification of the plant community. This accounts for the many vines climbing trees toward the light and epiphytes, plants that perch in trees and have no connection with the soil below.
- Biomes differ from ecosystems. Biome refers to large global areas of land that contain similar communities. An ecosystem is a community of organisms and the nonliving parts of the environment that affect the community.

Mainstreaming

Maps and map activities will be difficult for visually-impaired and special-needs students. Relief maps, if available, will help. In addition, pair handicapped students with nonhandicapped students who can help them.

Cross References

Section 13-7, Chapter 13, gives additional information about climate.

Checkpoint Answers

1. The three climate zones on Earth are the polar, temperate, and tropic zones.
2. A large area of land containing communities similar to those of other areas with the same type of climate.
3. Answers will vary. See student text, page 123.
4. Answers will vary. See student text, pages 124–125.
5. Answers will vary. See student text, page 124.
6. Answers will vary. See student text, pages 124–125.

Teaching Suggestions 5-4

Open the section by reviewing the major characteristics of the tundra and tropical rain forest biomes. Then tell students they are going to study the deciduous and evergreen forest biomes, and compare these to the tropical rain forest. Be sure students are clear in their minds about the major differences between deciduous and evergreen trees. After explaining the rainfall, temperature, and light parameters that apply in the temperate regions, contrast these with those conditions that apply in the tropics.

Many students think that all forests are alike. Thus, you should carefully specify the differences between tropical and temperate forests. Another common misconception is that all tropical rain forests are so dense it is necessary to hack one's way through them. In fact, the forest floor in the tropics is often just trunks, stems, and litter because so little light penetrates. It is usually necessary for scientists working in tropical rain forests to use ladders to reach the forest canopy. Many deciduous forests in the temperate climate zone, on the other hand, are extremely brushy and dense.

You may want to have students start a large chart to organize the major facts of this chapter. Set up three columns. In the far left column name and describe the major climate zones. In the middle column name and describe the biomes in each climate zone. In the right-hand column list the countries or continents where the biomes are found. The chart can be filled in as you progress through the chapter.

Using Figures 5-14 and 5-15, discuss the plants and animals typically found in deciduous and evergreen forest biomes and also some of the more evident adaptations present in each of these biomes. Stress animal adaptations that aid in survival, and deciduous and evergreen adaptations to winter conditions. Point out that the brown coat of the snowshoe hare in summer changes to white in the winter.

5-4

Forest Biomes

Goals

1. To describe the characteristics of deciduous and evergreen forest biomes.
2. To identify some adaptations of forest organisms.

Forest biomes in the temperate zone are characterized by ample rainfall, seasonal temperature changes, and day length that varies with the season. There are two types of forest biomes in North America: *deciduous forest* and *evergreen forest* biomes.

deciduous (dih SIJ oo us)

Trees that lose their leaves in response to shortening periods of daylight are called **deciduous** trees. The deciduous forest biome contains many trees, such as maple, oak, and hickory, that lose their leaves each autumn. The fallen leaves form a thick layer of "forest litter" on the ground, which is slowly broken down by decomposers.

Trees use large amounts of water during photosynthesis. Some water escapes through openings in the leaves. During the winter, when the ground is frozen and cannot absorb water, the leafless trees use and lose very little moisture. Losing leaves is an adaptation that helps deciduous trees stay alive through the winter.

A variety of wildflowers and shrubs grow in the deciduous forest. These plants grow and bloom early each spring, before the tree leaves have grown back. The canopy of trees shades much of the sunlight from the forest floor in late spring and summer.

In the deciduous forest, each "layer" of plant life has different adaptations. The adaptations enable plants to survive the given amounts of sunlight and moisture in each layer of the forest. For example, mosses and ferns have structures that allow them to grow successfully on the damp, shady, forest floor. (See Figure 5-14.)

The large number of producers in the deciduous forest provide food for a large number of consumers. Deer, mice, pheasants, and quail feed on the leaves, berries, and seeds of plants on the forest floor. Second-order consumers, such as woodpeckers, feed on insects living in the bark of trees.

Demonstration

Set up one or more miniature biomes for students to observe and study on an ongoing basis. An aquarium or a terrarium, or both, will work well. It is only necessary to include a few of the most commonly used organisms for the demonstration to have learning value. Thus, almost any aquarium or terrarium setup you are already familiar with will be satisfactory. Once the biomes are set up, students should study them for a period of time and then prepare reports. Have them list the features of the biome, if you set up only one, or compare the characteristics of the biomes, if you set up both. In addition, the students should list the plants and/or animals included, describe at least one interrelationship between organisms, and discuss two interrelation-

Figure 5-14 *Plants and animals typically found in a deciduous forest biome are shown in this view of a deciduous forest in the early fall. Try to identify some of the plants and animals.*

ships between living organisms and nonliving factors in the environment.

Teaching Tips

- Have a team of students research and report on reproduction in coniferous and deciduous trees.
- Gather leaves and needles from a variety of deciduous trees and evergreens in your area and have students decide which biome the trees live in.
- Make a class visit to both a deciduous forest and an evergreen forest if they are within reasonable distance. In some areas, students will be confused by forestlands that contain both deciduous and evergreen trees. In most such instances, the deciduous trees have grown in areas that were once covered with pines. The deciduous trees probably became established after many pines were cut down for lumber. Such mixed forests are common in eastern North America, Europe, and parts of China, Japan, and Australia.

Facts At Your Fingertips

- In many areas of North America, the soil determines whether a forest is deciduous or evergreen. If the soil is sandy and acid, evergreens will predominate. If the soil is more loamy and alkaline, deciduous trees will be more common.
- Some deciduous forests have layers of plants at different heights. Each layer is adapted to conditions at its height.
- The deciduous forest biome is not continuous around the world. It occurs in eastern North America, Western Europe, eastern Asia, and a small area in Chile.
- The thin needles of evergreens are an adaptation that prevents significant water loss.
- The northern coniferous forest is called the taiga. The taiga is a belt of evergreen forestlands just south of the arctic tundra biome.
- The openings in leaves through which water vapor escapes are called stomates.
- Researchers have discovered that a very high level of thyroid hormone in Richardson ground squirrels is responsible for their very low body temperature and heart and breathing rates during hibernation. This knowledge may be of use in the space program. Hibernation by astronauts would slow their metabolic rates and reduce food requirements. All body processes would be slowed down, including the process of aging. This conclusion is supported by a Harvard study that found that hibernating animals live longer than other animals of the same species that do not hibernate.

Mainstreaming

Provide visually-impaired students with specimens of leaves from deciduous and coniferous trees. Let them tactually explore the differences.

Cross Reference

Succession is described in section 4-5, Chapter 4.

5 DISTRIBUTION OF LIFE

5-4 Forest Biomes

What's Your Biome?

Use the information in Section 5-4 of your textbook to complete the chart.

Organism	Biome	Adaptation
mosses, ferns		
		hunt at night
		have layer of fat under skin
		migrate to warmer area
		hibernate
pine tree		
snowshoe hare		

Layered Living

The deciduous forest can be divided into three horizontal layers called life zones. Find out about the three zones. Then complete the chart below.

Life Zone	Description of Zone	Organisms in Zone
1. Canopy	• • •	
2. Middle Layer	• • •	
3. Ground	• • •	

Checkpoint Answers

1. Answers will vary. See student text, pages 126–128.
2. Answers will vary. See student text, page 128.
3. Answers will vary. See student text, pages 126–128.

Hawks, owls, and foxes hunt at night for small animals such as mice, shrews, and rabbits.

hibernation

In the fall, mammals, such as mice and squirrels, gather and store food. Woodchucks and skunks develop thick layers of fat. These adaptations, and others, help many animals survive the cold winter months when food is scarce. Some birds and insects migrate from the forest to warmer climates where food is more plentiful. Small animals, such as snakes and chipmunks, spend the winter in burrows in a sleeplike state called **hibernation.** During hibernation an animal's body temperature is lower and its heartbeat and breathing rates decrease. Hibernation allows an animal to survive the winter on very little energy. In the spring, the animal "wakes up."

evergreen forest

The evergreen forest biome occurs in the northern parts of the North American continent, Europe, and Asia. The **evergreen forest** is made up mostly of trees that stay green all year long, such as pine, fir, and spruce. Evergreen trees have thin needles, instead of broad leaves. This adaptation prevents the loss of water during cold winter months. In many parts of the forest, evergreens grow close together, blocking out sunlight from the forest floor year round. Few ground plants grow in the very dense shade of an evergreen forest. (See Figure 5-15.)

Figure 5-15 *Trees growing in an evergreen forest create a shaded environment where few ground plants can grow.*

Elk, moose, squirrels, and finches are common herbivores in the evergreen forest. Caribou, migrating from the northern tundra, make their home in the evergreen forest in winter. The snowshoe hare, another evergreen forest dweller, has a brown coat in spring. The coat gradually turns white in winter, helping the hare blend in with its snowy surroundings. The hare's main predator is the lynx. Hawks, wolves, and owls are some other predators in the evergreen forest.

Checkpoint

1. Compare and contrast the deciduous forest biome and the evergreen forest biome.
2. Describe two adaptations that help animals survive the winter in the deciduous forest biome.
3. Describe some of the adaptations that help deciduous and evergreen trees survive in winter.

5-5

The Desert Biome

Goals

1. To describe the features of the desert biome.
2. To identify some adaptations of desert plants and animals.

The most obvious feature of the desert biome is its lack of moisture. In some desert areas of North America, rainfall may be only 12 cm a year. (Compare this to the average 100 cm of rainfall in a forest biome.) When it does rain, the water evaporates very quickly. The desert biome generally contains sandy soil that does not absorb much water. The soil is often blown about by strong winds because it has little, or no, vegetation holding it down.

Temperatures in the desert tend to vary dramatically from very hot in the afternoon to very cold at night. There might be as much as a 35°C temperature difference in a twelve-hour period.

To avoid the heat of the blazing sun, many desert animals sleep in the shadows of cacti and rocks or in cool, underground burrows. They become active at night when the desert cools off.

Figure 5-17 on page 130 shows some organisms of the desert biome. All desert organisms have adaptations for living in an area with very little moisture. For example, some desert plants have a wide network of shallow roots that quickly absorbs large quantities of water when it rains. Many desert plants have a thick, waxy, outer layer that reduces water loss.

Leaves are absent in many desert plants, reducing the possibility of water loss to the atmosphere. Photosynthesis occurs in the rounded and flattened stems of many desert plants. Plants such as cacti are covered with sharp spines that prevent animals from robbing moisture that has been stored inside. (See Figure 5-16.) The harsh desert conditions cause plants to grow slowly.

Desert animals also have adaptations that help them obtain and conserve water. Some animals, such as mice, have kidneys that produce highly concentrated urine, containing little water. Beetles and mice store food under-

Figure 5-16 *The prickly pear cactus is wide-spread in the western United States. Some kinds have dangerous spines.*

Teaching Suggestions 5-5

Open discussion of the desert biome by describing its features and comparing them to the features of biomes previously discussed. If students are developing the chart of climate zones and biomes, have them enter the appropriate information for the desert biome. Make clear that lack of moisture is the principal reason for the physical conditions that prevail in deserts.

Some students may have the erroneous idea that deserts are simply vast regions of dry sand, devoid of life. Point out that, with just a few exceptions, this is not the case at all. Most deserts receive up to 25 cm of rain per year, and support many organisms, including insects, spiders, reptiles, birds, rodents and other mammals, and plants. Have students study the organisms shown in Figure 5-17. Included in this illustration are cactus beetles, mice, a kit fox, a sidewinder, a spadefoot toad, a tortoise, a scorpion, a Gila monster, a roadrunner, an owl, a desert cottontail, saguaro cacti, Joshua trees, and prickly pears.

With reference to Figure 5-17, discuss adaptations of plants and animals to desert conditions. For example, burrowing is an animal adaptation. You also may want to describe the three ways plants survive desert conditions: (1) storing water (cacti and others); (2) using very little water (creosote and sagebrush); (3) growing only when moisture is present (annuals that bloom and set seed very quickly after a rainstorm).

Point out that desert conditions are the result of wind patterns. The prevailing winds determine how much rain a region will receive. Below a certain level of annual rainfall, a region will not be able to support as much plant and animal life, and desert conditions will develop.

Some students may wish to do chapter-end project 1 in conjunction with this section.

Teaching Tips

- Have an interested student research and report on how the jerboa of African and Asian deserts conserves moisture, and how this compares to the processes used by the kangaroo rat of North America's southwestern deserts. These two animals show adaptive convergence in their ways of dealing with low moisture levels.
- Acquire some small cactus plants from a local nursery and describe for your students the plants' adaptations to desert conditions.
- Find out the average annual rainfall in your area and compare it to that of desert areas. Use 12 to 20 cm a year as the desert figure.
- *Arizona Highways* magazine is a fine source of color photos of desert con-

ditions and desert life. Obtain several copies and make them available to your students.

Facts At Your Fingertips

- The scarcity of plants in the desert biome means there is very little decaying organic matter in the soil.
- Saguaro cacti *(Cereus giganteus)* may live to be 200 years old.
- Many desert animals are active at night, and thus avoid the heat of the day. Animals usually reduce body heat by evaporative cooling. Some desert animals, however, such as the kangaroo rat, must conserve as much moisture as they can. Kangaroo rats reabsorb water from respiration, produce very concentrated urine, and seal their burrows to trap water vapor lost during breathing.

Cross References

See section 13-4, Chapter 13, for the lowest amount of precipitation received by a desert. Refer to section 13-7, Chapter 13, for an explanation of why deserts are usually found on the leeward side of mountains.

Checkpoint Answers

1. Lack of moisture, sandy soil, and large temperature differences between night and day are features of the desert biome.
2. The waxy outer layer of cactus plants reduces water loss. Answers will vary. See student text, page 129 for possible answers.

Figure 5-17 *Plants and animals that are well adapted to living in a desert biome must be able to survive on a limited supply of water. Some plants and animals found in a desert biome are shown here.*

ground, where it often absorbs small amounts of moisture from the soil. Some desert animals survive without drinking water. Their cells use the water produced during cellular respiration.

Checkpoint

1. Name two features of the desert biome.
2. What is the function of the waxy, outer layer of a cactus? Name another adaptation in desert plants.

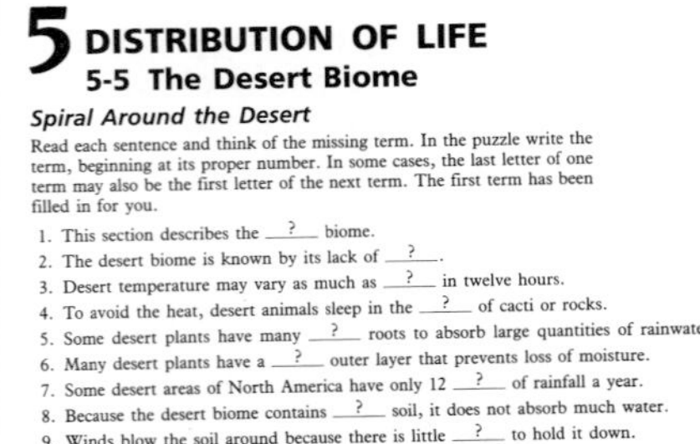

5 DISTRIBUTION OF LIFE
5-5 The Desert Biome

Spiral Around the Desert

Read each sentence and think of the missing term. In the puzzle write the term, beginning at its proper number. In some cases, the last letter of one term may also be the first letter of the next term. The first term has been filled in for you.

1. This section describes the __?__ biome.
2. The desert biome is known by its lack of __?__.
3. Desert temperature may vary as much as __?__ in twelve hours.
4. To avoid the heat, desert animals sleep in the __?__ of cacti or rocks.
5. Some desert plants have many __?__ roots to absorb large quantities of rainwater.
6. Many desert plants have a __?__ outer layer that prevents loss of moisture.
7. Some desert areas of North America have only 12 __?__ of rainfall a year.
8. Because the desert biome contains __?__ soil, it does not absorb much water.
9. Winds blow the soil around because there is little __?__ to hold it down.
10. Some animals stay in underground __?__ during the day and feed at night.

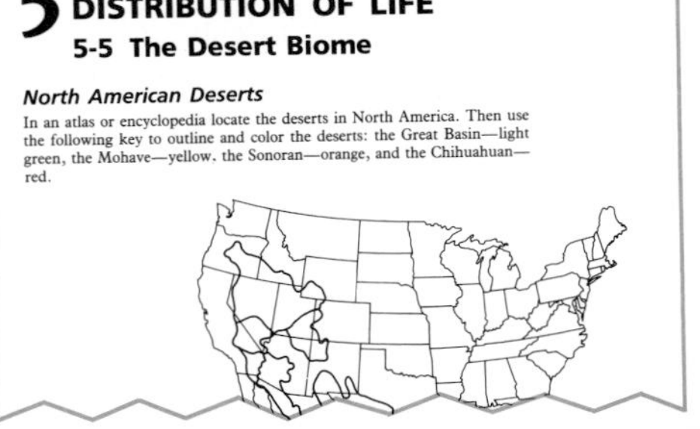

5 DISTRIBUTION OF LIFE
5-5 The Desert Biome

North American Deserts

In an atlas or encyclopedia locate the deserts in North America. Then use the following key to outline and color the deserts: the Great Basin—light green, the Mohave—yellow, the Sonoran—orange, and the Chihuahuan—red.

5-6

The Grassland Biome

Goal

To identify the features of the grassland biome.

The grassland biome is found throughout the temperate zone. Grasslands receive seasonal rainfall averaging 25–75 cm each year—less than a forest, but more than a desert.

Few trees grow in the grassland biome. Grasses, such as wheat, corn, clover, and barley, are the dominant vegetation. Most of the world's grain crops are grown on what used to be wild grasslands. Parts of the grassland biome are known as the "breadbaskets" of the world.

Large numbers of herbivores make their home in the grassland biome. Harvest mice, prairie dogs, and ground squirrels eat the seeds and leaves of grassland plants. They live in underground burrows, protected from the coyotes, badgers, snakes, and owls that hunt them. (See Figure 5-18.)

Antelopes and wild buffalo were once the dominant grassland mammals. They have been replaced by domesticated sheep and cattle.

Figure 5-18 *Some of the plants and animals that are typically found in a grassland biome are shown here. Grasses provide food for a number of animals. Other animals feed on the herbivores.*

Teaching Suggestions 5-6

Open discussion of the content of this section by describing the principal features of the temperate grassland biome, and contrast these with features of biomes already covered. Add information about this biome to the chart, if students are compiling one. Refer students to Figure 5-18, and discuss the types of organisms found in grasslands. Included in the illustration are prairie dogs, ground squirrels, harvest mice, coyotes, owls, pronghorn antelopes, a badger, a snake, a jackrabbit, a prairie falcon, and grasses. Talk about the implications of replacing wild organisms with domesticated organisms, such as sheep, cattle, corn, and wheat.

Conduct a similar discussion of savannas, and stress how climate determines the types of organisms that will be present. Encourage students to study Figure 5-19. Included in the illustration are an elephant, a giraffe, a lion, a leopard, a kori bustard, zebras, gazelles, antelopes, hyenas, ostriches, vultures, grasses, and umbrella acacia trees. Students should now be acquiring a stronger sense of the relationship between climate, physical conditions of a biome, and the types of animals and plants found in a biome.

As in previous sections in this chapter, stress the adaptations that permit plants and animals to flourish in grasslands and savannas. Add data to the climate and biome chart.

Teaching Tips

- Have a team of students research and report on the principal food chains and webs found in temperate grasslands and African savannas.
- Have an interested student research and report on how the introduction of rabbits into Australia devastated the grasslands and severely disrupted the balance of nature.

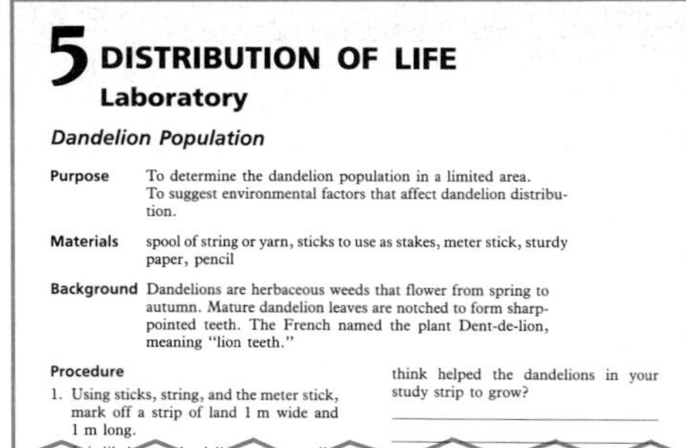

5 DISTRIBUTION OF LIFE
Laboratory

Dandelion Population

Purpose To determine the dandelion population in a limited area. To suggest environmental factors that affect dandelion distribution.

Materials spool of string or yarn, sticks to use as stakes, meter stick, sturdy paper, pencil

Background Dandelions are herbaceous weeds that flower from spring to autumn. Mature dandelion leaves are notched to form sharp-pointed teeth. The French named the plant Dent-de-lion, meaning "lion teeth."

Procedure

1. Using sticks, string, and the meter stick, mark off a strip of land 1 m wide and 1 m long.

think helped the dandelions in your study strip to grow?

Facts At Your Fingertips

- All of Earth's continents have vast areas of grasslands.
- Trees are found in those parts of grasslands located close to streams and rivers.
- The change from deciduous forest to grassland in the temperate zone is gradual, not abrupt.

Mainstreaming

Let visually-impaired students feel some specimens of grassland plants. If possible, also provide students with taxidermic models of animals that typify the grassland biome.

5 DISTRIBUTION OF LIFE
5-6 The Grassland Biome

Breadbaskets of the World

In each statement below a term is missing. Choose the best answer to complete each statement. Then write the letter of that answer on the numbered space.

____ 1. The grassland biome is found in the ______ zone. a) temperate b) frigid c) tropical d) neutral

____ 2. Grasslands receive rainfall averaging ______ cm each year. a) 0–25 b) 25–75 c) 50–100 d) 75–150

____ 3. In the grassland biome there are ______ trees. a) no b) few c) many d) pine

____ 4. Examples of grasses grown in the grassland are ______. a) clover, bean, flax b) sunflower, flax, cotton c) tobacco, cotton, pea d) wheat, corn, barley

____ 5. Parts of the ______ biome are known as the "Breadbaskets of the World." a) deciduous forest b) coniferous forest c) rainforest d) grassland

____ 6. Herbivores that live in a grassland biome might be ______. a) coyotes, badgers, owls b) bison, antelope, buffalo c) mice, prairie dogs, ground squirrels d) zebras, gazelles, antelopes

____ 7. Predators in the grassland biome are likely to be ______. a) bison, antelope, buffalo b) mice, prairie dogs, ground squirrels, c) zebra, gazelles, antelopes d) badgers, snakes, owls

____ 8. Animals of the savannas are ______. a) bison, antelope, buffalo b) mice, prairie dogs, ground squirrels, c) coyotes, snakes, owls, d) zebras, gazelles, antelopes

____ 9. Savannas are grasslands that contain ______ trees. a) no b) many c) scattered d) deciduous

____ 10. Plants in the savannas have to survive ______ winters. a) hot b) cold c) dry d) wet

____ 11. Herbivores of the savannas survive because they can ______. a) run fast b) fight c) burrow d) hide

____ 12. Hyenas and vultures are ______ in the savannas. a) herbivores b) scavengers c) producers d) predators

savannas (suh VAN uhz)

Where tropical dry seasons are long and severe, trees grow far apart. **Savannas** are grasslands that contain scattered trees. They are found in South America and Africa. Plants in the savanna have adaptations that help them survive the long, dry winter. Zebras, gazelles, and antelopes are large hoofed mammals that feed on grasses in the savanna. Figure 5-19 shows some other large herbivores of the savanna. Many of them are able to run at high speeds, helping them escape the leopards, cheetahs, and other predators that hunt there. Hyenas and vultures survive as scavengers in the savanna.

Checkpoint

1. Where is the grassland biome found?
2. What is the dominant vegetation type in the grassland biome?
3. Name two herbivores that live in the grassland biome. Name two predators in the grassland biome.
4. Why are parts of the grassland biome known as the "breadbaskets" of the world?
5. Name two animals that live in the savanna. Describe some of the adaptations that help these animals survive there.

Figure 5-19 *Some of the plants and animals that are found in a savanna are shown in this illustration. A savanna is a grassland that contains scattered trees.*

Checkpoint Answers

1. The grassland biome is found throughout the temperate zone.
2. Grasslike plants, such as wheat, corn, clover, and barley, are the dominant vegetation in grasslands.
3. Certain mice, prairie dogs, and ground squirrels are herbivores found in grasslands. Coyotes, badgers, snakes, and owls are grassland predators.
4. Because most of the world's grain crops are now grown in what was once wild grasslands.
5. Zebras, gazelles, leopards, and hyenas live in the African savanna. Answers will vary. See student text, page 132.

5-7

Life Zones in the Ocean

ENRICHMENT

Goals

1. To describe life zones in the ocean.
2. To identify adaptations that help organisms survive in the ocean.

Biome is a term used to describe *land* areas with a particular climate and communities. Since oceans are water environments with no definable climate, they are not usually called biomes. Oceans do, however, contain a large variety of life. Each species has adaptations that enable it to live in one of the life zones in the ocean.

The *coastal* zone is an area of shallow water containing an abundance of different organisms. (See Figure 5-20.) Some varieties of seaweed in the coastal zone contain air-filled bladders that allow them to float near the surface of the water where light is plentiful. Burrowing animals, such as crabs and clams, and bottom-dwelling fish are other examples of coastal-zone organisms.

Every day the area around the shore is covered and uncovered by tides. At low tide, the ocean floor near the shore is exposed. At high tide, the area is covered with water. The tides rise and fall twice a day. The zone between high and low tides is called the *intertidal* zone.

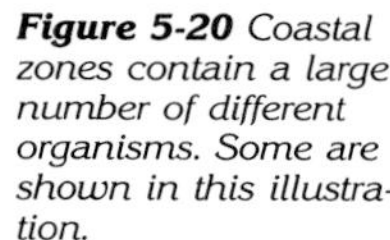

Figure 5-20 *Coastal zones contain a large number of different organisms. Some are shown in this illustration.*

Teaching Suggestions 5-7

To establish the differences between biomes and the ocean's life zones, start by redefining biome for the class, and then explain how the several life zones of the ocean do not meet the criteria set for a biome. Be sure students understand what a climate is and why the concept of climate does not apply to the oceans. Point out to students that they will discover that there are adaptations unique to each life zone in the world's ocean waters.

When discussing the coastal zone, stress that ample sunlight penetrates for numerous producer organisms to flourish. Point out that producer organisms do not exist in the ocean's deepest waters, but that millions of tiny, floating plants carry out photosynthesis in the upper levels of the open-ocean zone. Such organisms are called phytoplankton. It is estimated that 70% of the oxygen produced worldwide by photosynthesis comes from marine plants and protists, such as algae.

Refer students to Figure 5-20. The illustration represents an Atlantic coastal zone. See how many organisms students can identify in the illustration. Some of the organisms shown include a horseshoe crab, a herring gull, a sanderling, crabs, sand dollars, clams, sea cucumbers, a flounder, silversides, a lugworm, a sea urchin, shrimp, a starfish, a sea snail, kelp, and beach grass.

Be sure students clearly understand the mechanical, temperature, and light conditions that prevail in each of the ocean's life zones, and how these parameters affect the type of life that can exist in each zone.

Figure 5-21 represents a Pacific coast tidepool. Students may enjoy contrasting the conditions and communities in a tidepool with the conditions and communities that characterize a sandy beach. Some of the organisms in Figure 5-21 are a hermit crab, sea lettuce, starfish, sea urchins, limpets, sea snails, mussels, *Fucus,* and gooseneck barnacles. Encourage students to talk about the adaptations of these organisms to their environment.

Stress the conditions that prevail in estuaries, harbors, and intertidal zones. Explain why it is important to preserve, not destroy, these vital marine life zones. Most marine aquariums have interesting and easy-to-understand exhibits on the value of coastal waters. A field trip to an aquarium with a set of preplanned objectives is an excellent way to reinforce the facts and concepts developed in this section.

Teaching Tips

- Have an interested student research and report on how the bony fishes are thought to have originated in small freshwater streams and then spread to other aquatic habitats, including salt water. This is an example of adaptive radiation.
- Hold a class debate on the pros and cons of offshore oil drilling, which is done in the shallow waters of the coastal zone.
- Have a team of students research and report on such endangered marine mammals as the sea otter, dugong, and manatee.
- Have students prepare reports on mariculture, the commercial growth in the sea itself of marine organisms used as food for humans.
- Have an interested student research and report on the effects of commercial fishing on the coastal zone.
- Set up a class project to investigate and report on the effects of industrial and other pollutants on commercially-valuable marine species.
- Discuss the short- and long-range effects of dumping toxic and radioactive wastes into the oceans.
- Have an interested student research and report on the organisms that live in the ocean's great depths, or benthos. Some of these benthic organisms—the angler fish, the viper fish, the gulper, and the hatchet fish—are shown in Figure 5-22.

Facts At Your Fingertips

- The waters over the continental shelf are known as the littoral zone.
- The barnacle (see Figure 5-21) has feather-like projections that it uses to pull in food such as algae.
- Many organisms that live in deep water cannot be raised to the surface without dying. The change in pressure produces fatal changes in the organisms' internal tissues.
- About 60% of commercially-valuable invertebrates and fishes require the wetlands of estuaries at some point in their life cycles. Without the estuary's unique blend of fresh and saline water and rich supply of nutrients, many, if not all, of these organisms would face extinction.
- The skeletons of whales and sea lions show vestiges of limbs that indicate an earlier adaptation to land.

Organisms that live in the intertidal zone have adaptations that help them survive the daily changes in their environment. Part of the day, the organisms are exposed to the hot sun, air, and wind. Intertidal-zone organisms, such as barnacles and mussels, have tough, outer shells that remain closed when exposed, keeping in moisture. Many organisms live in *tide pools*—small areas near the shore that are left filled with water when the tide uncovers the rocks. (See Figure 5-21.) At times, intertidal-zone organisms must withstand the pounding surf. Some organisms, such as seaweed, barnacles, and snails, have structures that help them cling tightly to rocks. When covered with water, barnacles and mussels open their shells in order to feed.

Millions of tiny producers carry on photosynthesis in the upper layer of the *open-ocean* zone. Algae form the beginning of long ocean food chains. The abundance of producers in the open ocean supports large, complex food webs, which include animals as small as the microscopic copepod and as large as the gray whale.

Figure 5-21 *Tide pools are small water filled areas that are often present in the intertidal zones. A number of different organisms live in and around tide pools. Some of these organisms are shown in the illustration.*

Figure 5-22 A number of unusual adaptations are found among the organisms that live in the deep-water zone of the ocean.

The *deep-water* zone of the ocean contains organisms with adaptations for surviving in the cold and dark. Because sunlight does not reach the deep water, producers cannot survive there. Many organisms in the deep water live on the remains of organisms that drift down from the upper ocean.

Some deep-water organisms produce and give off a type of light through chemical reactions in their bodies. This process of giving off light is known as *bioluminescence*. Some bioluminescent fish use light to attract food or to frighten off predators. Figure 5-22 shows some deep-water organisms.

Checkpoint

1. Why is the ocean not called a biome?
2. Name four life zones found in an ocean. Describe each one.
3. Pick one ocean animal and one ocean plant. Describe the adaptations that help these organisms to survive in their environment.

TRY THIS

Visit a beach or an aquarium or pet store that has saltwater fish tanks. Observe the different kinds of organisms. Describe the adaptations different organisms have for food-getting, movement, and protection.

Cross References

Refer to Chapter 12, section 12-5, for information about the formation and topography of the ocean floor. All of Chapter 14 is devoted to oceanography.

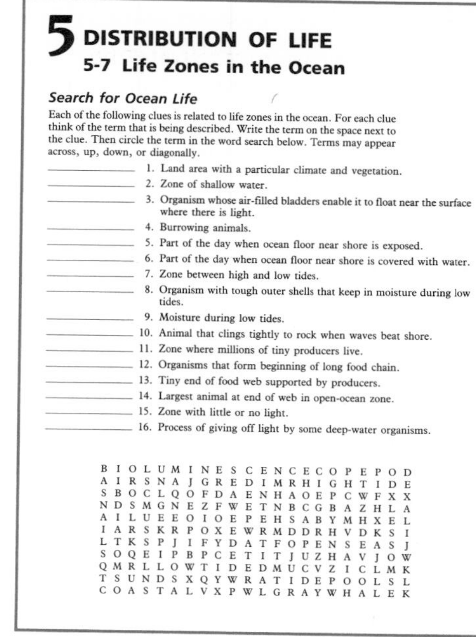

5 DISTRIBUTION OF LIFE

5-7 Life Zones in the Ocean

Search for Ocean Life

Each of the following clues is related to life zones in the ocean. For each clue think of the term that is being described. Write the term on the space next to the clue. Then circle the term in the word search below. Terms may appear across, up, down, or diagonally.

______ 1. Land area with a particular climate and vegetation.
______ 2. Zone of shallow water.
______ 3. Organism whose air-filled bladders enable it to float near the surface where there is light.
______ 4. Burrowing animals.
______ 5. Part of the day when ocean floor near shore is exposed.
______ 6. Part of the day when ocean floor near shore is covered with water.
______ 7. Zone between high and low tides.
______ 8. Organism with tough outer shells that keep in moisture during low tides.
______ 9. Moisture during low tides.
______ 10. Animal that clings tightly to rock when waves beat shore.
______ 11. Zone where millions of tiny producers live.
______ 12. Organisms that form beginning of long food chain.
______ 13. Tiny end of food web supported by producers.
______ 14. Largest animal at end of web in open-ocean zone.
______ 15. Zone with little or no light.
______ 16. Process of giving off light by some deep-water organisms.

B I O L U M I N E S C E N C E C O P E P O D
A I R S N A J G R E D I M R H I G H T I D E
S B O C L Q O F D A E N H A O E P C W F X X
N D S M G N E Z F W E T N B C G B A Z H L A
A I L U E E O I O E P E H S A B Y M H X E L
I A R S K R P O X E W R M D D R H V D K S I
L T K S P J I F Y D A T F O P E N S E A S J
S O Q E I P B P C E T I T J U Z H A V J O W
Q M R L L O W T I D E D M U C V Z I C L M K
T S U N D S X Q Y W R A T I D E P O O L S L
C O A S T A L V X P W L G R A Y W H A L E K

Using Try This

It will help to brief students on the types of organisms that will be seen, especially if the trip is to a marine aquarium, and the sorts of adaptations for food-getting, movement, and protection they should look for.

Checkpoint Answers

1. The ocean is not called a biome because it has no definite climate and it has numerous life zones.
2. The coastal zone, intertidal zone, open-ocean zone, and deep-water zone are the life zones of the ocean. Descriptions will vary. See student text, pages 133–135.
3. Answers will vary. See student text, pages 133–135.

Using the Feature Article

This article serves to illustrate the dynamic nature of science. Scientists are continuously gathering new information, reevaluating old theories, and proposing new theories, always for the purpose of more accurately describing nature. You may want to review this process, which was described in the Prologue of this book. Although the theory of natural selection remains today much as it was conceived by Darwin back in the 19th century, certain modifications to some of his ideas have been made as new information has become available. For example, the discovery of mutations and of the basic principles of genetics (see Chapter 2) served to expand and modify science's view of natural selection.

Another significant modification of Darwin's theory was proposed in 1972. Called the theory of punctuated equilibria, this theory attempts to explain why, in the fossil record, some species appear abruptly whereas other fossil species remain relatively unchanged throughout their history. According to Darwin, evolution occurs slowly and at a steady rate over the course of millions of years. In contrast, this new theory states that the rate of evolutionary change is not constant and that major evolutionary changes can occur rapidly during periods lasting only thousands of years. According to this theory, the long history of life on earth is one mostly of evolutionary "quiet" interrupted occasionally by short periods of dramatic change. Evolutionary change may occur in small, isolated populations, where the effects of selection pressure would be most dramatic.

The Theory of Punctuated Equilibria

Charles Darwin set forth his theory of natural selection over 125 years ago. He believed that new species evolved very slowly and gradually over the course of millions of years. Since Darwin's time, however, much more has been learned about past life from the study of fossils. This new information has caused some scientists to question the slow and gradual pace of natural selection.

The study of fossils has shown two main patterns that are contrary to some of Darwin's ideas. First, fossil species often show little or no change from the time when they first appear in the fossil record to the time when they disappear. Second, new species sometimes appear abruptly in the fossil record.

In 1972, two American scientists, Niles Eldredge and Stephen Jay Gould, proposed a theory to explain these patterns. Their theory is called the *theory of punctuated equilibria*. According to this theory, large, central populations of a species tend to stay the same, in equilibrium (plural, *equilibria*). This would explain why many fossil species show little change over millions of years.

The theory of punctuated equilibria also states that new species can arise very rapidly when small populations become isolated from large "parent" populations. This would explain the sudden appearance of new species in the fossil record. These new species "punctuate," or interrupt, the general equilibria of large, stable populations. According to this theory, evolution occurs rapidly in isolated populations before they are large enough to reach equilibrium.

Evidence of punctuated equilibria has been found near Lake Turkana in Africa. In the fossil record there, several new species of snails appear within a period that lasted only 50 thousand years. This period was at a time when the lake waters dropped. It is supposed that some of the snails became isolated from the main population at that time. The isolated populations varied and, through natural selection, several new species rapidly arose. The illustration below shows the change (left to right) from one species of fossil snail to another.

Understanding The Chapter

5

Main Ideas

1. The theory of evolution is a way of explaining the variety of life.
2. Adaptations are traits that enable an organism to live in a particular environment.
3. Biomes are large land areas in different parts of the world that have similar climates and communities.
4. The climate of an area determines the features of a biome.
5. The tundra biome has long cold winters, short cool summers, and a very short growing season.
6. Heavy rainfall, direct sunlight, and constant high temperatures are characteristics of the tropical rain forest biome.
7. The deciduous forest biome contains trees that lose their leaves each autumn in response to shortening periods of daylight.
8. The evergreen forest biome contains trees with needles that stay green all year long.
9. Desert-biome organisms have adaptations for living in an area with little moisture and daily large temperature changes.
10. Grassland biomes support large numbers of herbivores. People use grasslands for growing crops and raising cattle.
11. The coastal zone, intertidal zone, open ocean, and deep ocean are four life zones in the ocean.

Vocabulary Review

From the following list, choose the term that best completes each of the statements. Write your answer on a separate piece of paper.

adaptation	natural selection
biomes	permafrost
climate	savannas
deciduous	theory of evolution
evergreen forest	tide pools
hibernation	tundra
instincts	

1. The process by which populations of organisms change over many generations is called _?_
2. Trees that lose their leaves each autumn are called _?_ trees.
3. Behaviors that an animal is born with are called _?_.
4. _?_ is a sleeplike state that enables animals to conserve energy in the winter.
5. _?_ is soil that is frozen all year long.
6. The flat, grinding teeth of an herbivore are an example of a structural _?_.
7. Large land areas with similar climates and communities are called _?_.
8. The average weather of an area over a long period of time is _?_.
9. The biome that contains trees with needles that stay green all year long is the _?_.

Vocabulary Review Answers

1. natural selection
2. deciduous
3. instincts
4. hibernation
5. permafrost
6. adaptation
7. biomes
8. climate
9. evergreen forest
10. tundra

5 DISTRIBUTION OF LIFE

Computer Program

Biome Quiz

Purpose To strengthen your understanding of biomes.

Background You will be given three levels of hints about the biome and one guess of the correct biome after each hint. You score the most points if you guess correctly after one hint. You get fewer points for later guesses.

ENTER the following program:

```
10  PRINT "HERE IS THE FIRST HINT:"
20  PRINT "PLANTS WITH THICK OR WAXY OUTER LAYERS."
30  LET S = 100
40  GOSUB 150
50  PRINT "HERE IS THE SECOND HINT:"
60  PRINT "ANIMALS ACTIVE LARGELY AT NIGHT."
70  LET S = 50
80  GOSUB 150
90  PRINT "HERE IS THE THIRD HINT:"
100  PRINT "CLIMATE IS HOT AND DRY."
110  LET S = 25
120  GOSUB 150
```

Reading Suggestions

For The Teacher

Beazley, Mitchell, ed. THE INTERNATIONAL BOOK OF THE FOREST. New York: Simon & Schuster, 1981. Discusses the ecological importance of forests.

Eldredge, N. and Gould, S.J. *Punctuated Equilibria: an alternative to phyletic gradualism.* In: Schopf, Thomas J., ed. MODELS IN PALEOBIOLOGY. San Francisco: Freeman, Cooper and Co., 1972, pp. 82–115.

Gould, Stephen J. THE FLAMINGO'S SMILE. New York: Norton, 1985. Essays on evolution and natural history.

Moorehead, Alan. DARWIN AND THE BEAGLE. New York: Harper and Row, 1969. A fascinating account of Darwin's voyage on the *Beagle* with beautiful illustrations.

Scientific American. LIFE IN THE SEA. San Francisco: W. H. Freeman, 1982. A compilation of articles on topics such as oceanic life, marine food chains, and behavior of sea life.

For The Student

Ayensu, Edward S., ed. JUNGLES. New York: Crown, 1980. Highlights the variety of relationships occurring within Earth's tropical forests.

Durrell, Gerald. AMATEUR NATURALIST. New York: Knopf, 1982. Excellent field encyclopedia, with projects.

Jaspersohn, William. A DAY IN THE LIFE OF A MARINE BIOLOGIST. Boston: Little, Brown, 1982. Fascinating account of the workings of the Woods Hole Marine Biological Laboratory.

Lerner, Carol. SEASONS OF THE TALLGRASS PRAIRIE. New York: Morrow, 1980. An outstanding summary of the grasslands biome.

Roessler, Carl. THE UNDERWATER WILDERNESS. New York: McGraw-Hill, 1986. Introduces coral reef ecosystems.

Watson, James Werner. DESERTS OF THE WORLD: FUTURE THREAT OR PROMISE? New York: Philomel, 1981. A concise, readable, comprehensive account of desert climate, evolution, plants, animals, and human economics.

Chapter Review Answers

Know The Facts

1. deep ocean
2. carnivores and omnivores
3. desert
4. deciduous forest
5. tropical rain forest
6. temperate
7. waxy layers and needles
8. grassland
9. natural selection
10. permafrost and a short growing season
11. is born with
12. camouflage
13. desert

Understand The Concepts

14. Variations in teeth and mouth parts in mammals and beaks in birds are food-getting adaptations in animals.
15. Climate is the average weather of an area over a long period of time. The polar, temperate, and tropic are three climate zones.
16. polar—tundra biome; temperate—deciduous forest biome; tropic—tropical rain forest biome.
17. Answers will vary. See student text, pages 128–129.
18. Instinct is a behavior that an organism is born with. Migration and hibernation are instinctive behaviors that help organisms survive.
19. Camouflage, claws, shells, teeth, spines, and heavy fur are protective adaptations in animals. Spines, bark, and waxy outer coatings are protective adaptations in plants.
20. Climate is average weather for an area over a long period of time. Biomes are areas with similar communities of organisms within climate zones. Climate determines the characteristics of a biome.
21. Hibernation is a sleeplike state. It allows animals to survive winter conditions on very little energy.
22. Hibernation and migration are behaviors that help organisms survive. Thus, they are adaptations.
23. Lamarck: evolution occurs when organisms acquire traits and pass them on to their offspring. Darwin: evolution results from natural selection. Only individuals best suited to the environment survive and reproduce. The offspring inherit the traits of their parents.
24. Only traits encoded in the DNA of chromosomes are inherited.
25. Answers will vary. See student text, pages 133–134.

Challenge Your Understanding

26. A tropical rain forest is likely to have more niches than a desert because the climate of the tropical rain forest is warmer and moister and because there are more variations in physical conditions.
27. Biomes are large areas of land containing similar communities of organisms. An ecosystem, however,

10. The biome that has a short growing season and soil that is frozen year round is the ___?___.

Chapter Review

Write your answers on a separate piece of paper.

Know The Facts

1. Organisms with bioluminescence are found in the ___?___. (open ocean, deep ocean, intertidal zone)
2. Sharp, pointed teeth is an adaptation found in ___?___. (carnivores and herbivores, herbivores and omnivores, carnivores and omnivores)
3. A leafless plant covered with spines would most likely be found in the ___?___ biome. (desert, grassland, tropical rain forest)
4. Migration and hibernation are adaptations most likely found in animals of the ___?___ biome. (desert, deciduous forest, tropical rain forest)
5. Animal niches are most likely to occur in "layers" in the ___?___ biome. (tundra, grassland, tropical rain forest)
6. The climate zone with four distinct seasons is the ___?___ zone. (temperate, tropic, polar)
7. Two adaptations that prevent water loss in plants are ___?___. (spines and no roots, waxy layers and needles, lack of leaves and tallness)
8. Wheat, corn, barley, and clover are the dominant vegetation in the ___?___ biome. (grassland, tundra, deciduous forest)
9. According to Darwin, species evolve by means of ___?___. (acquired traits, adaptation, natural selection)
10. The tundra biome has ___?___. (much rainfall and a long growing season, little rainfall and a short growing season, permafrost and a long growing season)
11. Instincts are behaviors an animal ___?___. (learns, is born with, learns by practicing)
12. ___?___ enables animals to blend in with their environment. (camouflage, migration, bioluminescence)
13. Organisms in the ___?___ must withstand large daily changes in the environment. (tundra, tropical rain forest, desert)

Understand The Concepts

14. Describe two food-getting adaptations in animals.
15. Define climate. Name three climate zones.
16. Name a biome in each of the three climate zones.
17. Describe some of the adaptations that help desert plants and animals survive.
18. What is instinct? Name one instinctive behavior that helps an organism survive.
19. Describe three protective adaptations in plants or animals.
20. Explain how climate and biome are related.
21. What is hibernation?
22. Explain how hibernation and migration are adaptations.
23. Compare and contrast Lamarck's theory of evolution with Darwin's theory of evolution.

24. Explain how DeVries' discovery of mutations helped support Darwin's theory of evolution.

25. What adaptations help organisms survive the changing conditions in the intertidal zone?

Challenge Your Understanding

26. Which biome is likely to have a greater variety of niches, a desert or a tropical rain forest? Explain.

27. A biome and an ecosystem both contain communities of living things. How does a biome differ from an ecosystem?

28. Compare the tundra biome to the desert biome. How are they alike? How do they differ?

29. What might happen to the peppered moths in Manchester, England if the air were cleaned up?

30. What are some adaptations that deep-sea organisms need in order to survive the special conditions there? Explain.

Projects

1. Use an old aquarium or a wide-mouthed jar to set up a model of the desert biome. Obtain several cacti and other desert organisms for your terrarium. Find out how to care for them.

2. Insect mouth parts are adaptations that help insects get specific kinds of food. Find out what kinds of food different insects eat. Make a poster showing a close-up of each insect and its mouth parts. Explain how the mouth parts help each insect get food.

3. Prepare a report on insect-eating plants and their food-getting adaptations. Put together a bulletin-board display.

4. Read magazines and newspapers to find out about the effects of mining and agriculture on the Amazon tropical rain forest biome in South America. Report back to your class.

5. Monarch butterflies undergo a long-distance migration each year. Using library reference materials, trace the annual migration route of the monarch. Draw a map of the migration route. Label it to show where the monarch is during each season. Find out where the monarch mates and lays eggs. How does the monarch sustain itself during the migration?

6. Find out what precautions were taken to prevent the destruction of the North American tundra biome during the construction of the Alaskan oil pipeline.

7. Read the journals of explorers and naturalists who first crossed the North American continent, such as Powell, Peary, Muir, Sacagawea and Lewis and Clark, and Henson. What were North American biomes like at the time of their explorations? How has the continent changed?

8. Prepare a report on the kinds of scientific evidence that support the theory of evolution.

---

5 DISTRIBUTION OF LIFE
Projects

1. In an atlas or encyclopedia locate the tropical rain forest found in South America and shade in this area on the map below.

   a. If the rain forest were destroyed by mining and agriculture, what portion of South America's total land area would that represent?

   ______

   b. List three ways the land would be affected by mining and agriculture.

   ______

2. Paleontologists have used several methods to determine Earth's climate at various times in its geologic history. Use reference books from your school or local library to find out about the methods that have been used. On the lines below, briefly describe at least two of these methods.

   ______

---

5 DISTRIBUTION OF LIFE
Challenge

Climb Mt. Biome

If you have ever climbed a mountain, you may have noticed that the temperature fell as you neared the top. In general, the temperature falls as altitude increases. The graph at the bottom of the page shows temperature on the X axis and altitude on the Y axis. If the height of Mt. Biome is 3,500 m and the temperature is 23° C at the foot, calculate and graph the temperature of Mt. Biome at each 500-m distance to the peak. Hint: Temperature falls 2.7° for every 500 m.

Altitude vs Temperature

0 m ______
500 m ______
1,000 m ______
1,500 m ______
2,000 m ______
2,500 m ______
3,000 m ______
3,500 m ______

1. What is the temperature at 1000 m? ______
2. What is the temperature at 2,250 m? ______
3. What factors in addition to temperature would affect the nature of the biomes along the height of a mountain? ______

Altitude in Meters: 4,000, 3,500, 3,000, 2,500, 2,000, 1,500, 1,000, 500, 0

Temperature in °C: 25 24 23 22 21 20 19 18 17 16 15 14 13 12 11 10 9 8 7 6 5 4 3 2 1 0

---

consists of nonliving factors and the community of organisms those factors affect.

28. Answers will vary. See student text, pages 121–122 and 128–129.
29. Light individuals would become abundant, and dark ones would become scarce.
30. Answers will vary. See student text, pages 135–136.

Unit 1 Careers

Teaching Suggestions

Some of the parents of your students may be employed in occupations that are related to life science. Obtain the names of these parents and invite them to speak to your students about their careers.

Have students research life-science careers in the library. Some careers that do not require a four-year college degree include horticultural inspector, gamekeeper, farmer, electrocardiograph technician, tree surgeon, lab assistant, and fish-and-game warden. Careers that require advanced degrees include museum curator, geneticist, life-science teacher, microbiologist, plant physiologist, dentist, veterinarian, biologist, and hospital administrator.

Encourage students to display the results of their research on the bulletin board. Discuss the investment in training and education for each position and compare it with the expected average salary for the position. Invite some persons who hold some of the positions to speak to your class.

Have students arrange to conduct taped interviews with employees of local hospitals or health-care centers. Questions should center around employees' feelings about their jobs, the drawbacks and rewards of their positions, and the effects of their career choices on their lifestyles. The interviews can be played back to the class and discussed. If possible, keep the tapes for future reference and use.

Have students bring in classified advertisements for jobs. Show them where to look for jobs that relate to life science. Encourage students to cut out advertisements that relate to life science and use them to prepare a bulletin-board display. Discuss the personal requirements, educational background, and salaries for each kind of position. Which jobs are most in demand?

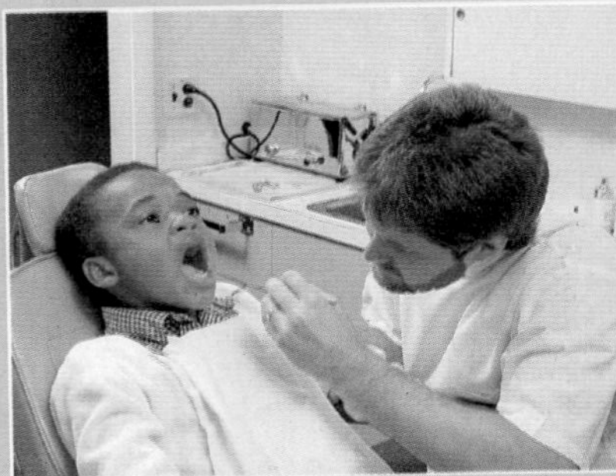

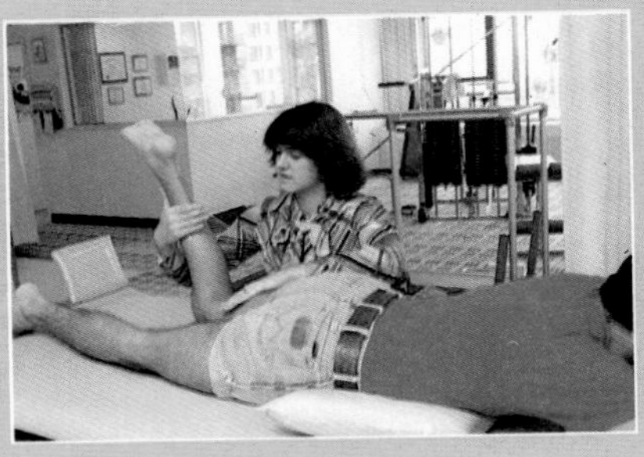

Dental Assistant

Community clinic seeks dental assistants to help examine and treat patients. Assistants put patients at ease and assist dentists by preparing equipment, suctioning patients' mouths, processing x rays, and instructing patients in oral hygiene. In addition, some assistants may be asked to help manage the office by arranging appointments, processing bills, and ordering supplies.

A high school diploma is required. A two-year college degree is preferred. Courses in biology, chemistry, health, typing, and office practice would be helpful. Applicants should be able to handle expensive equipment and work in a standing position. They also should be friendly and supportive with patients. As needed, some on-the-job training will be offered.

Physical Therapist

Rehabilitation center seeks physical therapists to treat disabled persons. Therapists work with the patient's physician to establish a program of therapy that will restore as much as possible the normal body functions of the patient. Therapists also help patients and their families adjust to limitations and continue treatment at home.

This job requires patience, tact, and understanding. Therapists must be able to care for the same patients for long periods of time. The work can be tiring, since therapists may have to lift or support patients.

Applicants must have a college degree in physical therapy with course work in health, biology, chemistry, physics, and mathematics. Prior experience with the disabled would be helpful.

Help Wanted

Additional Careers

A brief description and job outlook for three occupations related to life science follow:

1. *Food technologists* work in universities, government, and industry studying the nature of food so as to improve the quality of food and the methods of preservation. Many work in research and development to assure that foods meet government standards and taste good when they reach the consumer's table. In the future, food technologists will be needed to develop more nutritious, reasonably-priced foods.
2. *Biomedical engineers* carry out research to solve medical problems. They design and develop artificial organs and parts for humans. They also adapt computer hardware and

Forestry Technician

Federal government needs forestry technicians to help foresters protect national forest and park resources. Technicians plant trees, thin overgrown areas, supervise road-building, and oversee timber-cutting. Technicians also check areas for disease, fire, and erosion. Most of the work is done outdoors and is physically demanding.

Completion of a one-year or a two-year forestry technician's program is preferred. Knowledge of mathematics and biology plus experience in fighting forest fires or working in a tree nursery would be helpful. Successful applicants will begin as supervised trainees. Forestry technicians should expect to work in all kinds of weather and for long hours during emergencies, such as fires and floods.

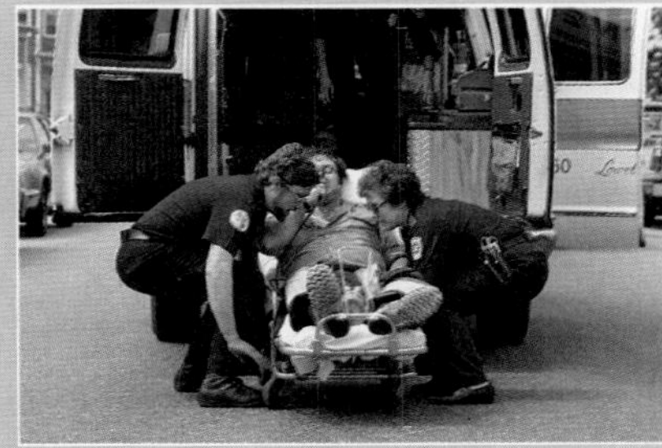

Emergency Medical Technician

Hospital seeks emergency medical technicians (EMT's). EMT's drive ambulances to the site of an emergency and give immediate treatment to victims. EMT's must be able to administer first aid, restore breathing, and assist in emergency births.

Applicants must have completed a training course for EMT's. Applicants also must be able to think quickly under pressure and to withstand stress. Courses in driver education, health, and the sciences plus knowledge of local roads would be helpful.

EMT's must be ready to treat victims indoors and outdoors in all types of weather. EMT's also should be able to lift heavy loads. All EMT's must work some night and weekend shifts.

software to monitor and assist patients. However, even though spending for medical research is likely to increase, the demand for biomedical engineers is not expected to grow appreciably because the field is new and relatively small.

3. *Range managers* help to preserve and protect rangelands. With a deep concern for the environment, they manage soil and plants for optimum use for wildlife, recreation, and timber. A lot of the work is outdoors, but administrative tasks and written reports mean indoor work as well. The employment outlook for range managers in federal and state governments and private industry is good. Realizing that land is a limited resource, people are demanding better care of wildlife habitats, mining lands, and water supplies.

For More Information

You or your students can write to the following organizations or agencies for more information about careers related to life science:

1. American Institute of Biological Sciences, 1401 Wilson Blvd., Arlington, VA 22209.
2. U.S. Dept. of Agriculture, Forest Service, P.O. Box 2417, Washington, DC 20013.
3. Institute of Food Technologists, Suite 2120, 221 North LaSalle St., Chicago, IL 60601.
4. Occupational Outlook Service, U.S. Dept. of Labor, Science Careers, 441 6th St., N.W., Rm. 2734, Washington, DC 20212.
5. American Dental Association, Div. of Career Guidance, Council on Dental Education, 211 East Chicago Ave., Chicago, IL 60611.

Using Issues In Science

This Feature helps students develop critical thinking skills by having them consider both sides of the issue of animal experimentation. The students can make a decision in favor of either side based on facts and personal values.

Have the students take sides in a classroom debate. Possible topics include:

- Animal experimentation should be restricted to experiments that directly save human lives.
- Many important scientific discoveries will not be made if animal experimentation is limited.

Have the students write down their positions before the debate, and then again following the debate. They will see how a discussion of the issues affected their opinions.

Have students analyze the arguments they used in the debate. Make a list of pros and cons on the blackboard, listing each argument in the suitable column. The students can see how many arguments fall into each category. Ask them if they would weight certain arguments more heavily than others. They should see that their personal values affect how they emphasize arguments.

Background

Many scientists think that some tests cannot be done without live animals. For example, the first antibiotic, sulphanilamide, was discovered when a dye, Prontosil Red, was found to control *Streptococcus* infections in mice. The antibiotic effect would never have been discovered by testing the dye on laboratory cultures because the active agent, sulphanilamide, was released only in the live animal.

Animal welfare advocates have made some legislative gains. New regulations passed by the Public Health Service in 1985 call for stricter monitoring of experimental animal use, including accreditation of laboratories by the American Association for Accreditation of Laboratory Animal Care.

ISSUES IN SCIENCE

ANIMAL EXPERIMENTATION: RIGHT OR WRONG?

Do you know anyone with diabetes? Chances are that you do. This disease, which affects one out of every twenty Americans, was once fatal. Then in 1921, experimenting with dogs, two scientists discovered a treatment. Dr. Charles Best and Dr. Fred Banting found a hormone called insulin that is produced in the pancreas of dogs. They knew that dogs with their pancreas surgically removed showed all of the signs of diabetes. Usually these dogs lived only a short time, but daily injections of insulin kept them alive indefinitely. The doctors soon found that this treatment also worked for humans. Today millions of diabetics live full, productive lives because of this research.

Scientists have made other major medical advances using many different kinds of animals in research. Primates are used for many experiments because of their similarity to humans. For example, a polio vaccine was first developed by growing the polio virus in monkeys. Antibiotics and techniques of heart and eye surgery were also developed using experimental animals. In fact, the Food and Drug Administration now requires that all new medicines be tested on several species of animals before being tried on humans.

However, some people think that it is wrong to use animals for research. People concerned about animal welfare feel that too many animals are used in research, and further, that it is unfair to make any animal suffer to benefit humans. They have cited several cases in which laboratories have been proven neglectful and cruel in their treatment of animals.

Animal rights supporters also point out that many animal experiments are done for purely commercial reasons. For example, some cosmetics companies use the Draize test on animals. Solutions containing chemicals to be used in eye make-up and shampoos are dropped into the eyes of rabbits.

Teaching Study Skills

Listening To Others

Listening is an important skill in the classroom and in life. Ask students to define *listening* and *hearing*. Hearing is a physical process, while listening requires the mind to interpret the meaning of the sounds that reach the ears. Students can listen *passively*, by just absorbing the meaning of what they hear, or they can listen *actively*, which requires them to respond to what they have heard.

To practice active listening, have students bring in news articles about science issues. Have one student read a section of an article out loud. The other students should listen for the main ideas, and write them down. This is equivalent to skimming while read-

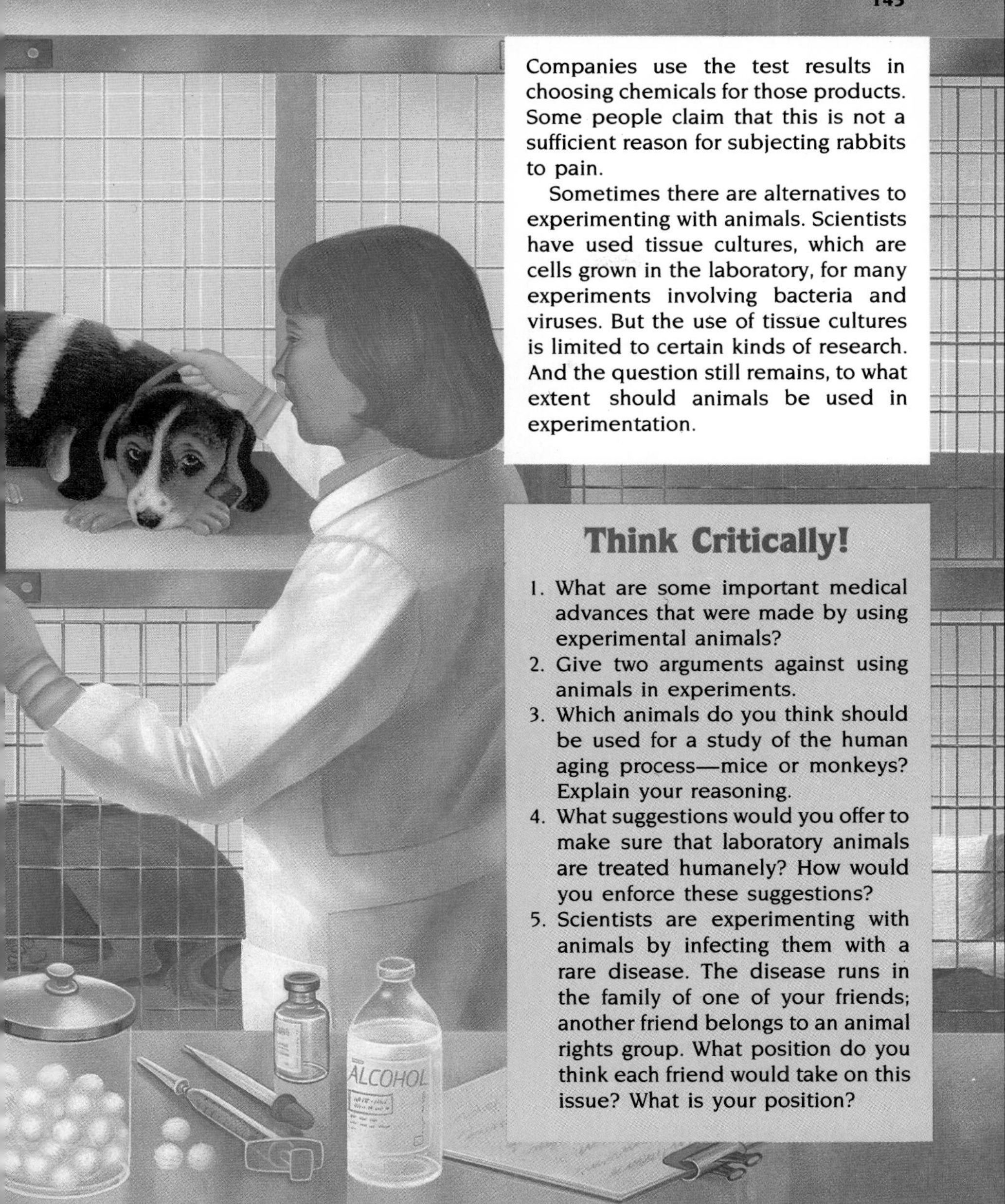

Companies use the test results in choosing chemicals for those products. Some people claim that this is not a sufficient reason for subjecting rabbits to pain.

Sometimes there are alternatives to experimenting with animals. Scientists have used tissue cultures, which are cells grown in the laboratory, for many experiments involving bacteria and viruses. But the use of tissue cultures is limited to certain kinds of research. And the question still remains, to what extent should animals be used in experimentation.

Think Critically!

1. What are some important medical advances that were made by using experimental animals?
2. Give two arguments against using animals in experiments.
3. Which animals do you think should be used for a study of the human aging process—mice or monkeys? Explain your reasoning.
4. What suggestions would you offer to make sure that laboratory animals are treated humanely? How would you enforce these suggestions?
5. Scientists are experimenting with animals by infecting them with a rare disease. The disease runs in the family of one of your friends; another friend belongs to an animal rights group. What position do you think each friend would take on this issue? What is your position?

Think Critically Answers

1. Medical advances include vaccines, antibiotics, and eye and heart surgical techniques.
2. Animals can suffer in experiments; some laboratories have been proven neglectful and cruel. Also, some experiments have no medical benefits.
3. Answers will vary. Some students will say mice, because they have a short life span and reproduce rapidly. This means the experiment will take less time to produce results. Other students may argue that monkeys must be used because they are more like humans.
4. Some rules that ensure humane treatment include: adequate food and drink; adequate space and hygiene; painless procedures or full anesthesia otherwise; well-trained animal caretakers. Some ways to enforce regulations are: require regular inspections; license animal research laboratories; license animal caretakers; fine or withhold funds from facilities found in violation.
5. Answers will vary. Students should "try on" different points of view to answer this question. Answers should note that when experimentation is necessary, discomfort and pain should be minimized.

ing. Then call on students to paraphrase and summarize what they have heard. This will improve verbal as well as listening skills.

Some other techniques that will help students to listen actively include: (1) asking themselves questions while they listen; (2) relating what they hear to things they already know; (3) trying to picture what the other person is saying.

Chemists make measurements with many different tools such as those shown here.

145

Unit 2 · Chemistry

The Unit In Perspective

The purpose of this unit is to present basic concepts of chemistry and nuclear reactions to students, and also to provide an understanding of the role of chemistry and nuclear science in modern civilization.

Chapter 6, *Properties of Matter,* introduces physical and chemical properties, discusses the three states of matter, and explores solutions, mixtures, separating mixtures, and elements and compounds.

Chapter 7, *Atoms and Molecules,* describes the development of the modern atomic model, and introduces chemical symbols, formulas, and equations.

Chapter 8, *Chemical Elements,* discusses metals, nonmetals, the noble gases, the halogens, the alkali metals, and relates these and other elements to the Periodic Table. The principles of chemical bonding are then introduced, and the chemistry of carbon compounds is described.

Chapter 9, *Chemical Reactions,* relates energy to changes of state and chemical reactions, discusses oxidation and reduction, electrochemical cells, acids and bases, and rates of reaction.

Chapter 10, *Nuclear Reactions,* introduces radioactivity, radioisotopes, radioactive decay, and the uses of radioisotopes. Also discussed are nuclear fission and fusion and atomic power.

Be sure to call student attention to the chemistry-oriented careers featured on pages 264–265. The careers chosen illustrate a wide range of occupations in the field of chemical science.

Issues in Science, on pages 266–267, introduces the issue of toxic waste disposal. Questions at the end of the article focus on critical thinking skills. The teacher pages include a study skills exercise.

Preparing To Teach The Unit

The advanced preparation needed to teach this unit includes the acquisition and organization of equipment, materials, and chemicals for laboratory activities, planning field trips, ordering audio-visual aids, and collecting and/or placing on reserve in the library outside readings for students. A list of suggested readings will be found adjacent to the Main Ideas section at the close of each chapter.

Equipment and Materials

The listing below identifies the equipment, materials, and chemicals you will need for the student Activities of this unit. The Activities are located at the back of the student book. Quantities required will depend on the nature of the Activity, the availability of materials, numbers of students, budgetary considerations, and so on. Some Activities proceed better if students work in pairs or threes. Check each Activity to make this decision.

ACTIVITY 9 (Ch. 6): *Measuring Density* (page 541). MATERIALS: laboratory balance with standard masses, thread, centimeter ruler, graduated cylinder, solid and liquid samples.

ACTIVITY 10 (Ch. 6): *Paper Chromatography* (page 543). MATERIALS: 2–3 large test tubes, test-tube rack, filter paper, solvent, food coloring, 2–3 toothpicks, scissors, ruler, 2–3 small paper clips, aluminum foil.

ACTIVITY 11 (Ch. 7): *Conservation of Mass* (page 544). MATERIALS: laboratory balance with standard masses, vial about one-third full of lead nitrate solution, vial about one-third full of potassium iodide solution or sodium iodide solution, vial full of crushed ice, three covers for the vials.

ACTIVITY 12 (Ch. 7): *Some Model Reactions* (page 545). MATERIALS: paper fasteners and washers (at least six of each), laboratory balance with standard masses.

ACTIVITY 13 (Ch. 8): *Metals and Nonmetals* (page 546). MATERIALS: candle, Bunsen burner, alcohol burner, small pieces of candle wax, sugar cubes, roll sulfur (pea-sized pieces), tin (foil or mossy), copper wire, iron wire, aluminum foil, platinum wire, pair of tongs, C or D cell, connecting wires, rubber band, small bulb, bulb holder, matches.

ACTIVITY 14 (Ch. 9): *Unsuspected Electrochemical Cells* (page 548). MATERIALS: lemons and other available fruits, such as apples, pears, olives, pickles, and oranges; galvanometer or microammeter; two insulated wires, preferably with alligator clips; copper, aluminum, and iron nails; zinc strips; centimeter ruler; paper towels and newspapers.

ACTIVITY 15 (Ch. 9): *Acids and Bases* (page 550). MATERIALS: stirring rod, 4 test tubes, medicine dropper, evaporating dish, graduated cylinder, phenolphthalein solution, red and blue litmus paper, *p*H paper (optional), safety glasses, dilute hydrochloric acid, dilute solutions of sodium, potassium, and calcium hydroxide, heat lamp.

ACTIVITY 16 (Ch. 10): *Half-Life* (page 552). MATERIALS: 100 cubes (sugar or wooden), each with a dot on one face, or 100 dice; container to hold cubes or dice.

Field Trips

Look into field trip possibilities in your area. Plan a trip to a water purification plant or a desalination plant to observe the processes used. Visit a state or federal food and drug laboratory to learn about food purification.

If one is nearby, visit a plant that manufactures photovoltaic equipment. Have students write reports on how the photovoltaic effect generates an electric current. Plan a trip to a steel mill or some other metal fabricating plant to observe how metals are worked.

Have teams of students set up rain traps in different parts of your community to measure the acidity of rainfall. Visit a swimming pool to learn how the water is kept free of polluting chemicals and how the proper *p*H is maintained.

Audio-Visual Aids

The list of A-V materials below provides a wide selection of topics on sound filmstrips (SFS), videocassettes (VC), and 16 mm film. Choose titles in terms of availability, student interest, coverage of content vis-à-vis field trips and other learning activities, availability of projectors, and time limitations. A list of A-V suppliers is given on page T-39. Consult each supplier catalog for ordering information.

Chapter 6: Properties of Matter

Particles in Motion: States of Matter, SFS, National Geographic.
Solutions, 16 min, 16 mm, VC, Coronet.
How Things Dissolve, 16 min, 16 mm, CRM/McGraw-Hill.

Chapter 7: Atoms and Molecules

The Nature of Matter, 24 min, 16 mm, VC, CRM/McGraw-Hill.
Radiant Energy and the Electromagnetic Spectrum, 10 min, 16 mm, VC, Coronet.
Atoms and Molecules: Building Blocks of Matter, SFS, Guidance Associates.
An Introduction to the Atom, SFS, National Geographic.

Chapter 8: Chemical Elements

Chemistry: A Basic Approach, SFS, Educational Activities.
Introducing Chemistry: How Atoms Combine, 10 min, 16 mm, VC, Coronet.
The Halogens, 16 min, 16 mm, VC, Coronet.
Metals and Non-Metals, 13 min, 16 mm, VC, Coronet.
Hydrocarbons and Their Structures, 13 min, 16 mm, VC, Coronet.

Chapter 9: Chemical Reactions

Introducing Chemistry: Types of Chemical Change, 13 min, 16 mm, VC, Coronet.
Acids, Bases, and Salts, 21 min, 16 mm, VC, Coronet.
Combustion: An Introduction to Chemical Change, 16 min, 16 mm, VC, Phoenix.
Chemical Change and Temperature, 14 min, 16 mm, VC, Phoenix.

Chapter 10: Nuclear Reactions

The Atom: A Closer Look, 30 min, 16 mm, VC, Coronet.
Radioactive Dating, 12 min, 16 mm, VC, Coronet.
Fusion: The Energy Promise, 56 min, 16 mm, VC, Time.
Nuclear Energy: The Question Before Us, 25 min, 16 mm, VC, National Geographic.

CHAPTER 6 Overview

The purpose of this chapter is to introduce students to matter and its properties. Matter is defined and the concept of density is explained. Some basic properties of matter are then introduced, and their use in identifying substances is described. Elements and compounds are then defined, and the differences between physical and chemical changes are discussed. The law of definite proportions is then introduced and its importance in chemistry is stressed. Finally, combustion as an example of chemical change is discussed, and some basic fire safety rules are given.

Goals

At the end of this chapter, students should be able to:

1. explain that mass and volume are properties of matter and to describe ways of measuring them.
2. define density.
3. describe the three states of matter.
4. define melting point and boiling point.
5. predict the state of a substance at a certain temperature, given its melting and boiling points.
6. discuss the properties of mixtures and solutions.
7. define solubility and describe how it can be used to identify a substance.
8. describe the effect of temperature on solubility.
9. distinguish between mixtures and pure substances.
10. give examples of the ways different types of mixtures can be separated.
11. explain how chemical and physical changes differ.
12. define elements and compounds.
13. state the law of definite proportions.
14. define and give examples of combustion.
15. discuss the dangers of fire and some fire safety rules.

Chapter 6

Properties of Matter

This photograph, taken in Yosemite National Park, shows several forms of matter.

Vocabulary Preview

property	density	boiling point	physical changes
matter	states of matter	evaporation	chemical changes
mass	solid	condensation	elements
kilogram	liquid	mixture	compounds
volume	gas	solution	law of definite proportions
liter	melting point	solubility	

6-1

Mass, Volume, and Density

Goals

1. To explain that mass and volume are properties of matter and to describe ways of measuring them.
2. To define density.

You can tell just by the color of a glass of milk that it is not water or orange juice. Even if you could not see it, you could probably tell it was milk by its taste. Color and taste are two properties that help you tell things apart. Generally, a **property** is any quality that can be useful in identifying something.

property

Milk, water, and orange juice are just three of the many kinds of matter. All matter has two properties in common. **Matter** is anything that takes up space and has mass. **Mass** is a measure of the amount of material in an object. Mass is normally measured using an equal-arm balance. An object of unknown mass is placed on the left side of the balance. Objects of known mass, or *standard masses*, are placed on the right side until both sides just balance. (See Figure 6-1.)

matter

mass

Standard masses are based on the standard **kilogram** (kg), a block of platinum metal that is kept in France. A number of exact copies have been made for use all over the world. One kilogram is equal to 1000 *grams* (g). The mass of

kilogram
(KIL uh gram)

Figure 6-1 *An equal-arm balance is used for finding the mass of an object.*

Background

Early in the history of science, it was discovered that the volume occupied by a particular substance is proportional to the mass of that substance; that is, 2 cm^3 of gold has twice the mass of 1 cm^3 of gold, and 10 cm^3 of gold has 10 times as much mass as 1 cm^3 of gold. Later, the ratio of mass to volume, which is a constant for any given substance, was defined as the density of that substance.

Because most pure substances have different densities, density is a characteristic property that can be used to help identify matter. On the other hand, because the densities of some substances are very similar and because density often varies with temperature, we cannot always use density to identify a substance. Before a substance can be identified, it must first be isolated and then tested for a number of characteristic properties, such as density, color, melting point, boiling point, and reactions with other substances.

Teaching Suggestions 6-1

You may wish to open this section by discussing the properties of one or more familiar objects (see Demonstrations). Then discuss the relative masses of various objects. When you are sure that students have a good grasp of the concept of mass, introduce the concept of volume and let students practice computing the volume of a cube or a rectangular solid. Next introduce the concept of density and go over the examples on student page 149. You also may wish to demonstrate how density is computed (see Demonstrations).

Once students have both a firm understanding of the meaning of density and a sense of how to determine it, have them do laboratory Activity 9, Measuring Density, student text pages 541–542.

Demonstrations

1. Use a metal cylinder to illustrate the concept of density as a property of matter. An aluminum cylinder large enough to be seen throughout the room works well. Don't tell students the metal is aluminum. Ask what might be done to identify the material the cylinder is made of. Answers will vary, and someone may suggest finding the substance's density. Before finding the density, point out that other, visible properties of the material rule out many possibilities. The material is not a gas or a liquid at room temperature. Its color tells us that it is not gold or copper, and so on. Ask students to suggest other possibilities that can be ruled out. Then point out that its color, hardness, and sheen indicate that the material is a metal.

   To see if the metal can be identified by its density, first determine its mass on a balance. Then find its volume, either by displacement of water using a graduated cylinder or by measuring its dimensions—radius of base (r) and height (h)—and using the formula for the volume (V) of a cylinder: $V = \pi r^2 h$. The density (d) of the metal in the cylinder then can be derived by dividing its mass (m) by its volume; that is, $d = m/V$. The result will show that the metal is probably aluminum. Ask students to check the result with the data in Table 6-1. Some variation is possible, but the result should be close enough to 2.7 g/cm^3 to identify the substance as aluminum.
2. Use water to show that the density of a substance is independent of the mass and the volume of the sample. Using a balance, measure the masses of various volumes of water in graduated cylinders. (Be sure to subtract the mass of the cylinder in each case.) Then, show that the ratio of mass to volume is constant for all sets of measured data. A good way to show this graphically is to plot, or have students plot, mass as a function of volume (mass on the vertical axis, volume on the horizontal axis). After four or five points have been plotted, students will see that the points lie on a straight line. The slope of the line is the density of the substance. If you repeat the experiment using different liquids, such as alcohol and glycol, the slope plotted for each kind of liquid will be different.

this book is about 1 kg, or 1000 g. Two raisins have a mass of about 1 g.

volume

The space taken up by an object is its **volume**. The volume of a cube or a rectangular solid can be found by multiplying length × width × height. (See Figure 6-2.) Most volumes cannot be measured using a ruler. You would need a measuring cup to find the volume of water or milk, for example. Suppose you filled a cube that measured 10 cm × 10 cm × 10 cm with water. The volume of the water would be the same as that of the container, 1000 *cubic centimeters* (cm^3). A volume of 1000 cm^3 is called a **liter** (L). One liter is equal to 1000 *milliliters* (mL). (See Figure 6-3.) A can of motor oil has a volume of about 1 L. The volume of a raindrop is about 0.1 mL.

liter (LEE tur)

You can measure out any volume of water that you might want. Changing the volume (or mass) does not change the basic nature of a substance. Some properties of a substance are not so easily changed, however. Suppose you measured out 10 mL of water, for example, and found its mass. Then suppose you measured out 50 mL of water and found its mass. You would find the mass of the second sample to be *five* times the mass of the first. The ratio of the mass of a sample of matter to its volume is the **density** of the sample.

density

$$\text{density} = \frac{\text{mass}}{\text{volume}} \quad \text{or} \quad d = \frac{m}{v}$$

Figure 6-2 *The volume of a rectangular solid can be found by multiplying length times width times height.*

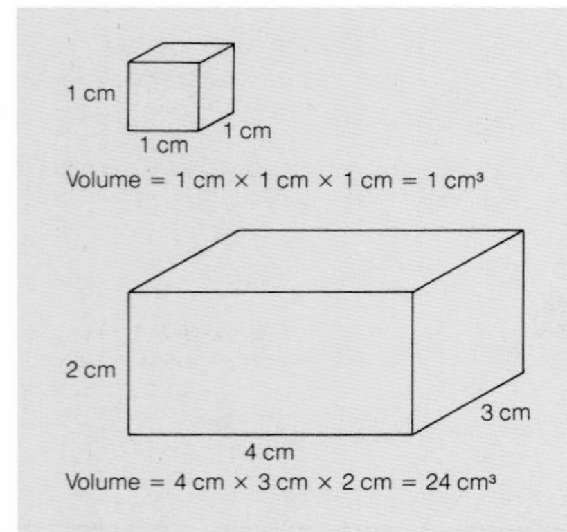

Figure 6-3 *A liter is a little more than a quart. Note that a milliliter and a cubic centimeter are the same volume.*

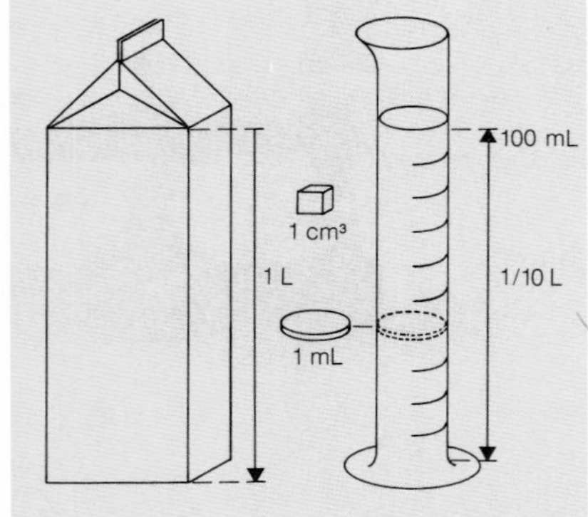

EXAMPLE 1

A 10-mL sample of water has a mass of 10 g. What is its density?

$$d = \frac{m}{v}$$

$$d = \frac{10\ \text{g}}{10\ \text{mL}} = 1.0\ \text{g/mL}$$

EXAMPLE 2

A sample of gold has a volume of $10.0\ \text{cm}^3$ and a mass of 193 g. What is its density?

$$d = \frac{m}{v}$$

$$d = \frac{193\ \text{g}}{10.0\ \text{cm}^3} = 19.3\ \text{g/cm}^3$$

The density of a substance does not depend on how much of it there is. A small piece of gold has the same density as a large piece. A small sample of water has the same density as a large amount. The density of a substance does not change. Therefore, density can be used to identify a substance. Suppose you find a shiny, yellow object and wonder if it is gold. You measure its mass and volume and find that its density is $5.0\ \text{g/cm}^3$. Since gold has a density of $19.3\ \text{g/cm}^3$, the object cannot be gold. Table 6-1 lists the densities of several substances.

Table 6-1: Densities of Some Substances (at 20°C)

Substance	*Density* (g/cm^3)	*Substance*	*Density* (g/mL)
ice	0.91	helium	0.00018
magnesium	1.74	air	0.00123
glass	2.4–2.8	carbon dioxide	0.00196
aluminum	2.7	alcohol (ethanol)	0.79
iron	7.9	gasoline	0.7
copper	8.9	water	1.0
silver	10.5	mercury	13.6
gold	19.3		
platinum	21.5		

Teaching Tips

- Be sure students know how to measure both mass and volume.
- Pass around three or four solids of the same size but different composition and have students heft them. Point out that students' perceptions of mass are indicative of the densities of the substances because the volumes of the samples are the same. This is a good way to acquire a sense of the meaning of density.
- Emphasize that chemists do not usually identify a substance on the basis of just one property. Identification is usually based on a variety of characteristic properties. By establishing a consistent set of characteristic properties, the chemist can be reasonably sure that she or he has identified a substance correctly.

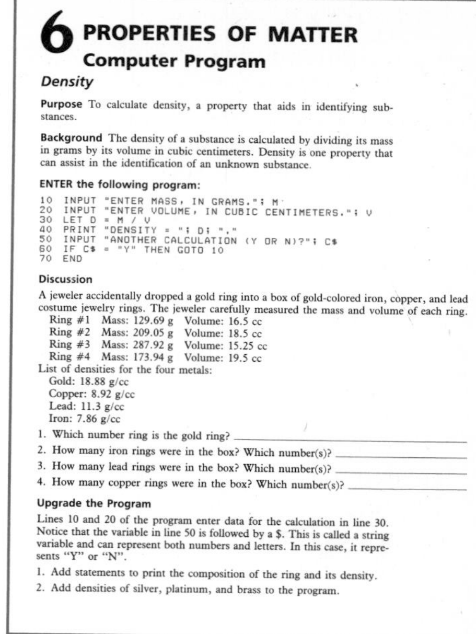

6 PROPERTIES OF MATTER
Computer Program
Density

Purpose To calculate density, a property that aids in identifying substances.

Background The density of a substance is calculated by dividing its mass in grams by its volume in cubic centimeters. Density is one property that can assist in the identification of an unknown substance.

ENTER the following program:

```
10 INPUT "ENTER MASS, IN GRAMS."; M
20 INPUT "ENTER VOLUME, IN CUBIC CENTIMETERS."; V
30 LET D = M / V
40 PRINT "DENSITY = "; D; "."
50 INPUT "ANOTHER CALCULATION (Y OR N)?"; C$
60 IF C$ = "Y" THEN GOTO 10
70 END
```

Discussion

A jeweler accidentally dropped a gold ring into a box of gold-colored iron, copper, and lead costume jewelry rings. The jeweler carefully measured the mass and volume of each ring.

Ring #1 Mass: 129.69 g Volume: 16.5 cc
Ring #2 Mass: 209.05 g Volume: 18.5 cc
Ring #3 Mass: 287.92 g Volume: 15.25 cc
Ring #4 Mass: 173.94 g Volume: 19.5 cc

List of densities for the four metals:

Gold: 18.88 g/cc
Copper: 8.92 g/cc
Lead: 11.3 g/cc
Iron: 7.86 g/cc

1. Which number ring is the gold ring? ______
2. How many iron rings were in the box? Which number(s)? ______
3. How many lead rings were in the box? Which number(s)? ______
4. How many copper rings were in the box? Which number(s)? ______

Upgrade the Program

Lines 10 and 20 of the program enter data for the calculation in line 30. Notice that the variable in line 50 is followed by a $. This is called a string variable and can represent both numbers and letters. In this case, it represents "Y" or "N".

1. Add statements to print the composition of the ring and its density.
2. Add densities of silver, platinum, and brass to the program.

Facts At Your Fingertips

- The mass of one liter of water is one kilogram. This fact makes it easy to manufacture reasonably good standard masses with simple materials.
- The mass of a body is a measure of its inertia, or its resistance to being moved. Weight is a measure of the pull of gravitational force on a body. The weight of a body, which is proportional to its mass, varies according to the strength of the gravitational field present. Thus, an object will have the same mass on the moon as on Earth but will weigh less on the moon, due to the weaker gravitational field of the moon.
- The symbol cc is commonly used to represent cubic centimeters, but cm^3 is the preferred symbol in scientific writing.

Mainstreaming

The Try This activity is particularly useful for handicapped students, especially those who are visually-impaired.

Cross References

For a discussion of how specific gravity is used to identify minerals, refer to Chapter 11, section 11-1. For another definition of mass, see Chapter 16, section 16-3. The difference between mass and weight is explained in Chapter 16, section 16-6. Measuring volume and mass is discussed in Appendix B.

6 PROPERTIES OF MATTER

6-1 Mass, Volume, and Density

Name That Property!

A term is missing in each of the sentences below. Read each sentence and decide what term should go in the numbered blank. Then write the term on the answer space at the left.

______ 1. Water, trees, rocks, and air are all examples of __1.__

______ 2. The __2.__ of an object is a measure of the amount of material in the object.

______ 3. __3.__ masses are objects of known mass.

______ 4. The standard unit of mass is the __4.__

______ 5. One kilogram is equal to __5.__ grams.

______ 6. The __6.__ of an object is the space that the object occupies.

______ 7. The standard unit of volume is the __7.__

______ 8. The __8.__ of a substance is the ratio of the mass of a sample of substance to its volume.

______ 9. From Table 6-1 in your textbook one substance that is more dense than gold is __9.__

______ 10. From Table 6-1 in your textbook one substance that is less dense than air is __10.__

Let's Find Solutions

Solve each of the density problems below. Express your answers in the proper units.

______ 1. If the mass of a substance is 20 g and its volume is 5 cm^3, what is the density of the substance?

______ 2. Given the density of gasoline in Table 6-1 of your textbook, what would be the mass of a 6-mL sample of gasoline?

______ 3. The mass of a substance is 54 g and its volume is 20 cm^3. What is the density of the substance?

______ 4. The substance in problem 3 is one of the substances listed in Table 6-1. What is the symbol for this substance?

Using Try This

Students will probably be able to distinguish between the spoons because of the difference in mass. If the spoons have about the same volume, it should be clear that the plastic spoon is less dense, because it has less mass than the metal spoon.

Figure 6-4 *Icebergs in the ocean float because liquid water is more dense than ice. These icebergs were photographed near Carey Island, Greenland.*

TRY THIS

Hold your hands in front of you, with the palms up. Place a piece of paper on each hand. Then close your eyes and have someone place a metal teaspoon on one hand and a plastic teaspoon on the other. Can you tell which spoon is which? Which spoon has more mass? Which is denser?

Notice in Table 6-1 that ice is less dense than water. This is why ice floats in water. (See Figure 6-4.) Copper is denser than water, so a piece of copper wire sinks in water. But copper floats in mercury, a very dense liquid. Helium, a gas less dense than air, is used in balloons. Metals with low densities, such as aluminum, are used in airplanes. A lump of iron sinks in water, but an iron ship floats. This is because the ship is a container. Most of its volume is occupied by air. The density of the ship is less than the density of water.

Checkpoint

1. What is matter?
2. Name two units of mass. How are they related?
3. Describe a method for measuring mass.
4. Which is more, a mL or a cm^3?
5. Water has a density of 1.0 g/cm^3. Solid X has a density of 2.7 g/cm^3. Solid Y has a density of 0.8 g/cm^3. Which of these solids will float in water?

Checkpoint Answers

1. Anything that has mass and takes up space is matter.
2. The gram and the kilogram are units of mass. 1000 g = 1 kg.
3. Place an object of unknown mass on the left side of an equal-arm balance. Place objects of known masses on the right side until the two sides balance. The two sides now hold equal masses.
4. A mL and a cm^3 have the same volume.
5. *Y* will float in water because its density is less than water's.

6-2

States of Matter

Goals

1. To describe the three states of matter.
2. To define melting point and boiling point.
3. To predict the state of a substance at a certain temperature, given its melting and boiling points.

Water can be a solid (ice), a liquid, or a gas. Solid, liquid, and gas are the three **states of matter**. A **solid** has a volume and shape that do not change. The shape of a **liquid** changes, but the volume does not. You can change the shape of a liquid simply by pouring it into a different container. A **gas** does not have a definite shape or volume. It fills whatever container it may be in. The densities of gases are much less than those of solids or liquids. (Look again at Table 6-1.)

states of matter

solid

liquid

gas

melting point

boiling point

evaporation

condensation

The state of a substance can be changed. An ice cube in a glass of lemonade melts as it absorbs heat from the lemonade. The change from solid to liquid happens at a temperature called the **melting point**. The melting point of water (ice) is 0°C. At 0°C, liquid water freezes to solid water (ice). The *freezing point*, the temperature at which a liquid changes to a solid, is the same temperature as the melting point of the solid.

At sea level liquid water boils and changes to a gas at 100°C. The temperature at which a liquid changes to a gas, when bubbles form in the liquid, is its **boiling point**.

At lower temperatures, water can change from a liquid to a gas without boiling. This change is called **evaporation**. Water evaporates from puddles on sunny days, and from clothes hung up to dry. The liquid water changes into a gas that mixes with air. Water in the gas state is called *steam* if it is hotter than 100°C. At lower temperatures it is called *water vapor*.

The change from a gas to a liquid is called **condensation**. Water vapor condenses when it cools. The drops of water that appear on a bathroom mirror when you take a shower have condensed from the air. So have the water drops that form on the outside of a cold drinking glass. The cloud above a pan of boiling water is sometimes called

Figure 6-5 *A dripping icicle is a good example of a change of state.*

Teaching Suggestions 6-2

Open this section by discussing familiar examples of changes of state. Ask students to describe situations where they have observed changes in the state of matter. To show the changes in state of water, do the suggested demonstration. (See Demonstration.) Chapter-end projects 1 and 3 apply to this section.

Background

Change of state depends on the direction of heat flow. If heat is added to a mixture of ice and water, the ice in the mix will melt. If heat is flowing from the mixture to cooler surroundings, the water in the mix will freeze. The same phenomenon occurs in a mixture of water and steam at water's boiling, or condensation, point (100°C). Adding heat makes the remaining water boil; removing heat makes the remaining steam condense. Of course, liquids evaporate even when the temperature is below their boiling points. For example, a puddle of water can disappear on a cool day. The average kinetic energy, or energy of motion, of the water molecules in the puddle is less than that of water molecules at the boiling point. However, there will be some molecules that have enough kinetic energy to escape from the surface and become water vapor.

Demonstration

Show changes of state by heating some crushed ice in the apparatus shown in Figure TE 6-1. Be sure the thermometer is not in contact with the test tube. After the solid melts, the temperature will climb to about 100°C. (There may be some variation because of thermometer characteristics, air pressure, or altitude.) Then, the water will boil and the water vapor will condense in the glass tubing and collect in the test tube surrounded by ice water. Stress the cha.ges of state observed: solid to vapor to liquid. Be sure students understand that the state of any substance depends on its temperature.

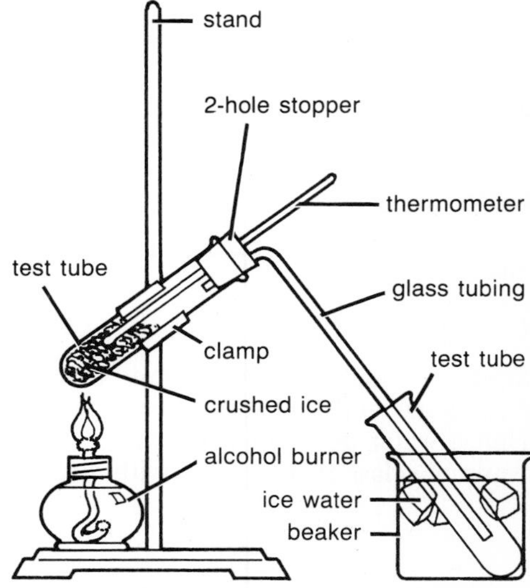

Figure TE 6-1 A demonstration of changes in states of matter.

Mainstreaming

Physically handicapped students and those with some form of visual-impairment will need help with the Try This activity. Pair visually-impaired students with students who can describe the activity and the results.

Cross References

For a discussion of the water cycle, see Chapter 11, section 11-4. Clouds and precipitation are discussed in Chapter 13, section 13-4. Refer to Chapter 17, section 17-5, for an explanation of kinetic energy. See Chapter 18, section 18-1, for further discussion of heat and temperature.

6 PROPERTIES OF MATTER
6-2 States of Matter

Unknown States

Some of the following statements are false. If a statement is true, leave the answer space blank. If the statement is false, rewrite the italicized portion of the statement to make it true. Hint: Table 6-2 of your textbook may be helpful.

_______________ 1. A solid has a volume and shape that *do not change*.
_______________ 2. A liquid has a *shape* that does not change.
_______________ 3. The melting point of a substance is the temperature at which a solid changes into a *gas*.
_______________ 4. Evaporation is the process in which a liquid changes into a *gas*.
_______________ 5. Condensation is the change from *liquid to gas*.
_______________ 6. Helium is a *liquid* at room temperature.
_______________ 7. Mercury is a *liquid* at room temperature.
_______________ 8. Gold is a *liquid* at room temperature.

The Fourth State of Matter

Read the paragraphs below. Then write the answers to the questions.

Plasma, the fourth state of matter, is made of electrically charged particles called ions. These particles are produced when the atoms of a substance break up. Although plasma cannot be seen under ordinary conditions, scientists know that lightning consists of plasma. They also know that plasma is found inside stars.

In the laboratory, lasers are sometimes used to make plasma. Scientists study plasma in the laboratory in hopes of learning more about nuclear fusion, the process that produces energy inside stars.

1. What are the four states of matter? _______________
2. What is plasma? _______________
3. List two places where plasma may be found. _______________
4. What instrument is sometimes used to produce plasma in the laboratory? _______________
5. Why do scientists study plasma? _______________

Using Try This

Students should be able to state that air is a gas and that it occupies space. Hence water does not enter the "empty" glass that is filled with air.

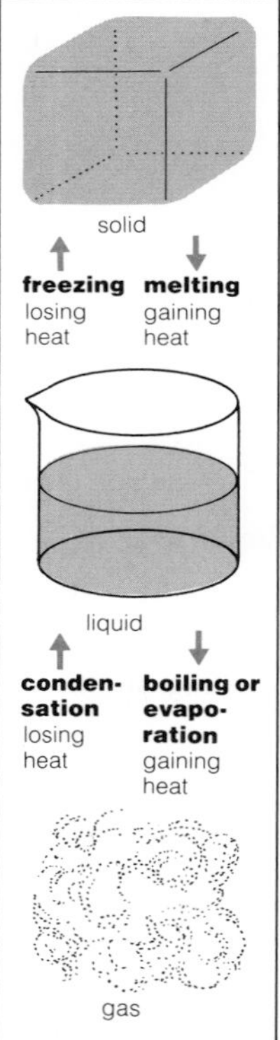

Figure 6-6 *Water gains or loses heat as it changes state.*

"steam," but the cloud is made up of tiny drops of water that have condensed from the invisible steam. Figure 6-6 summarizes water's changes of state.

The melting and boiling points of a substance can be used to identify it. Table 6-2 lists the melting and boiling points of some substances.

Table 6-2: Melting and Boiling Points

Substance	*Melting point* (°C)	*Boiling point* (°C)
helium	− 271	− 268
oxygen	− 219	− 184
nitrogen	− 210	− 195
mercury	− 39	357
water	0	100
moth flakes	53	174
lead	327	1525
gold	1063	2500
iron	1530	2450

On the Celsius scale, room temperature is about 20°C. At room temperature all substances with a boiling point less than 20°C are gases. Oxygen and nitrogen, which make up more than 99% of the air, are gases. Helium, which is less dense than air, is also a gas. Substances with boiling points above 20°C and melting points below 20°C are liquids at room temperature. Of the substances in Table 6-2, only water and mercury are liquids at 20°C. At 60°C, moth flakes would be a liquid, too.

TRY THIS

Fill a sink or large bowl with water. Turn a drinking glass upside down and lower it into the water. Why doesn't water enter the glass?

Checkpoint

1. Describe the properties of each of the three states of matter.
2. What is a melting point? a boiling point?
3. What is evaporation? condensation? Give an example to illustrate each one.
4. In what state (solid, liquid, gas) are the following at 300°C: (a) water, (b) nitrogen, (c) lead, (d) iron, (e) mercury?

Checkpoint Answers

1. A solid has a fixed shape and a fixed volume. A liquid has a fixed volume and takes the shape of its container. A gas has neither definite shape nor definite volume.
2. A melting point is the temperature at which a particular solid substance changes from the solid state to the liquid state. A boiling point is the temperature at which a liquid changes to a gas, forming bubbles within the liquid.
3. Evaporation is a change from liquid to gas without boiling. Condensation is a change from gas to liquid. Examples will vary. See student text, page 151.
4. (a) gas, (b) gas, (c) solid, (d) solid, (e) liquid.

Storing Heat

The sun is being used as a source for heating an increasing number of homes. Such homes must have some way of storing heat for use at night and on cloudy days. Heat can be stored in heavy walls or slabs made of stone or concrete, or in large containers of water. (In the photograph, the sun shines on solar collectors on a monastary in New Mexico.)

Storage space can be a problem, however. One 1940's solar house had a collector that covered half the roof and a hot-water storage tank that filled the whole basement!

Maria Telkes, a chemical engineer, decided that using large tanks of water or tons of rock to store heat was not practical. She found what she thought would be a perfect material for storing the sun's heat. It is Glauber's salt, which melts above room temperature, at about 32°C. When the salt melts, it absorbs heat. When it freezes, it gives off heat. It can store seven times as much heat as the same volume of water.

Telkes found a woman from Boston, Amelia Peabody, who agreed to furnish the money to build a new kind of solar house. Then Telkes hired an architect, Eleanor Raymond, to design the house, which was built in 1948. The walls between the rooms contained cans of Glauber's salt. When the sun shone, air was blown across the solar plates, then past the cans of salt. The warm air melted the salt.

When the temperature of the air around the walls decreased to 32°C, the liquid salt began to freeze and released the heat it had absorbed when it was melting. As long as there was any liquid salt, heat was available to warm the house. Even after all the Glauber's salt had frozen, it was still warmer than room temperature. Cool air blown over the warm salt was heated. The warmed air could be circulated throughout the house.

Unfortunately, the Glauber's salt worked well for only two winters. During the third New England winter, the salt separated into solid and liquid layers. The salt would no longer release the heat it absorbed. However, later experimenters solved this problem. Maria Telkes continued studying methods for using solar energy for many years.

Using The Feature Article

Glauber's salt is able to store heat because it takes a certain amount of heat to melt each gram of salt. Then, when the melted salt freezes as the house cools, the heat that was absorbed in the melting process is released. As long as any melted salt remains, heat can be generated to warm the house. After all the salt has solidified and its temperature has dropped to room temperature (the point at which heat will no longer flow between it and the air), some other means of heating will be required until sunlight can again warm and melt some salt. Today, coal- and wood-burning stoves and conventional gas, oil, and electric heating systems are used as backup.

Teaching Suggestions 6-3

Introduce the concept of solution by having students work with solutions. Give each pair of students 200 mL of water in a clean glass or beaker. Also provide sugar, plastic spoons, and stirring rods. Ask them to predict how many spoonfuls of sugar will dissolve in the 200 mL of water. Record estimates on the chalkboard. Then have students prepare the solution, adding the sugar a bit at a time. Caution them about adding too much. The idea is to find a solubility figure in terms of spoonfuls per 200 mL of water. The students should all get similar results.

Now pass out 400-mL samples of water and ask students to predict how much sugar will dissolve in this amount of solvent. They should be able to deduce that twice as much water will dissolve twice as much sugar. Have them test this prediction.

Students might also try to dissolve an insoluble substance, such as sand, in water. Or, they can try dissolving salt in water to show that different solutes have different solubilities. Some students may wish to do chapter-end project 2 in conjunction with this section.

Background

A solution is a homogeneous mixture. It is homogeneous because it consists of just one phase; that is, it is the same throughout. There are no clear separation boundaries, as in oil-and-water, or sand-and-water, or salt-and-pepper mixtures. Moreover, because a solution is a mixture, we can combine the materials in any proportion. Unlike a compound, in which elements combine in a fixed ratio, a solution can contain any amount of solute per unit of solvent up to the point of saturation.

Both solubility and solubility curves (solubility as a function of temperature) are characteristic properties for many substances. If a saturated solution is cooled, the excess solute generally will come out of solution (precipitate). Sometimes, if the solution is cooled carefully, the excess solute will remain dissolved. This results in a supersaturated solution—one that holds more solute than can normally be dissolved at that temperature.

A gas is less soluble in a given liquid when the liquid is warm than when it is cool. A gas is, of course, a more chaotic (disordered) state than a liquid. Consequently, as temperature, which promotes disorder, rises, the tendency is to the gaseous (more disorderly) state, and the gas solute bubbles out of solution.

The solid state, however, is less disordered than the liquid state. Therefore, temperature increases tend to make solids dissolve.

6-3

Solutions

Goals

1. To discuss the properties of mixtures and solutions.
2. To define solubility and describe how it can be used to identify a substance.
3. To describe the effect of temperature on solubility.

mixture

A **mixture** is a physical combination of two or more substances. Often you can tell that something is a mixture just by looking at it. A salad is clearly a mixture. Other mixtures are harder to recognize. Salt water is a mixture even though it looks like pure water. Salt water does not look different from pure water because the tiniest particles of salt are distributed evenly throughout the water.

solution

A mixture in which the smallest particles of two or more substances are evenly distributed is a **solution**. The most familiar solutions are those made by dissolving a solid in a liquid. However, solutions may include any of the three states of matter. Carbonated drinks, such as soda water, are solutions of a gas (carbon dioxide) in water. One popular drink is a solution containing water, cranberry juice, corn syrup, and vitamin C (a solid). Common solid solutions, or *alloys*, include steel, brass, and bronze. (See Figure 6-7.)

A substance in which another substance dissolves is called a *solvent*. The substance that dissolves in a solvent is called a *solute*. Water is a good solvent for many substances. It is a poor solvent for others. Water does not dissolve tar or plastics, for example. This is why solvents other than water are used to dry-clean clothes.

The largest solution of solid in liquid is the ocean. Its volume is 1.3 billion cubic kilometers. It contains enough salt to cover all the land on Earth with a layer of salt 150 meters deep.

There is a limit to how much salt can be dissolved in a given amount of water at a given temperature. If you continue to add salt to water after that limit has been reached, the salt does not dissolve. It simply sinks to the bottom of the container. A solution that contains the maximum amount of a substance that will dissolve at a given temperature is a *saturated solution*.

Figure 6-7 *Uses of alloys: (a) stainless steel surgical instruments, (b) a bronze statue of Mary McLeod Bethune, (c) a brass tuba.*

The **solubility** of a substance is the maximum amount of that substance that will dissolve in a given amount of solvent at a given temperature. The solubility of common salt (sodium chloride) in water at 20°C is 37 g per 100 g of water. Solubility is one property that can be used to identify a substance. For example, sodium nitrate looks like sodium chloride. But the solubility of sodium nitrate is 85 g per 100 g of water at 20°C.

A solution that is saturated at 20°C may not be saturated at higher temperatures. At 100°C a saturated solution of sodium chloride has 40 g of sodium chloride in 100 g of water. The solubility of sodium chloride increases by only 3 g per 100 g of water when the temperature rises from 20°C to 100°C. But the solubility of sodium nitrate more than doubles with the same rise in temperature. It goes from 85 g per 100 g of water at 20°C to 180 g per 100 g of water at 100°C.

solubility
(sahl yuh BIL uh tee)

6 PROPERTIES OF MATTER
Laboratory

Mixtures and Solubility Properties

Purpose To find the solubility properties of three substances.

Materials Water, acetone, three test tubes that are labeled each with a different number, stirring rod, spatula, and samples of sand, salt, and iodine

Procedure

1. Make a chart like the one below.
2. Examine each of the three substances and record your observations in the last three columns of your chart. CAUTION: Do not handle or heat the iodine.
3. Put enough water in each of three test tubes to half-fill the tube. Then place a small sample of sand in test tube 1, salt in test tube 2, and iodine in test tube 3...

Discussion

1. From your results in the Procedure, describe how you would recognize a mixture of each of the following pairs of substances:
   a. sand and salt: ____________
   b. sand and iodine: ____________

Demonstrations

1. Show students two beakers containing clear liquids. (One should be a salt solution, and the other, water.) Ask students if the two liquids are the same. Many will think so, but the answer cannot be determined by observation alone. Ask a student to taste each liquid. CAUTION: Remind students that as a rule they should not taste any unknown solution. He or she will now tell you they are different; one liquid has a salty taste. This should demonstrate that a solution looks the same throughout and is clear. Point out that solutions may also be colored, and, perhaps, show some that are. Solutions of copper sulfate, nickel chloride, cobalt chloride, potassium permanganate, and potassium or sodium chromate and/or dichromate are quite colorful.
2. Show that different solutes behave independently in the same solvent. Saturate some water with salt. Pour off some of the solution into a clean, empty glass or beaker. Ask students if more salt can be dissolved. Then show that any additional salt added will simply sit on the bottom of the beaker, despite prolonged stirring. Now pour the saturated salt solution into still another beaker and ask if any sugar can be dissolved in the solution. Students may be surprised to see that some sugar will dissolve.
3. Open one cold and one warm bottle of carbonated drink. Have students pay close attention to any gases that bubble out. Make sure they grasp that gases (carbon dioxide, in this case) are more soluble in cold than in warm solvents. This will help to underscore that the carbon dioxide in carbonated beverages is under pressure in the sealed bottle.

Teaching Tips

- In preparing salt solutions, use laboratory or Kosher salt. Ordinary table salt often contains impurities that make solutions cloudy.
- Point out that in Figure 6-9 on student text page 156, the minerals are precipitating, or coming out of solution, but may not be "falling" immediately. Precipitating does not necessarily involve "falling."

Facts At Your Fingertips

- Seawater is about 3.5% salt by mass. Most of the salt solute is sodium chloride.
- Automobile bodies are made of steel, an alloy (mixture) of iron and other elements.

Mainstreaming

Visually-impaired and special-needs students will need many concrete examples of solutions and solubility. In addition, they might "dry out" a salt water solution to feel the salt that remains after the water has evaporated.

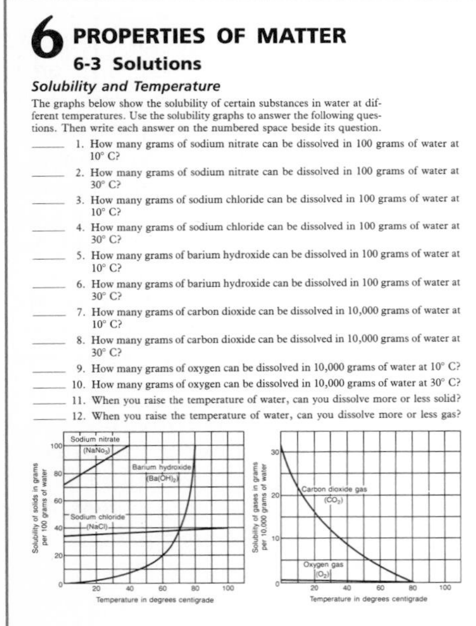

6 PROPERTIES OF MATTER

6-3 Solutions

Solubility and Temperature

The graphs below show the solubility of certain substances in water at different temperatures. Use the solubility graphs to answer the following questions. Then write each answer on the numbered space beside its question.

_____ 1. How many grams of sodium nitrate can be dissolved in 100 grams of water at 10° C?

_____ 2. How many grams of sodium nitrate can be dissolved in 100 grams of water at 30° C?

_____ 3. How many grams of sodium chloride can be dissolved in 100 grams of water at 10° C?

_____ 4. How many grams of sodium chloride can be dissolved in 100 grams of water at 30° C?

_____ 5. How many grams of barium hydroxide can be dissolved in 100 grams of water at 10° C?

_____ 6. How many grams of barium hydroxide can be dissolved in 100 grams of water at 30° C?

_____ 7. How many grams of carbon dioxide can be dissolved in 10,000 grams of water at 10° C?

_____ 8. How many grams of carbon dioxide can be dissolved in 10,000 grams of water at 30° C?

_____ 9. How many grams of oxygen can be dissolved in 10,000 grams of water at 10° C?

_____ 10. How many grams of oxygen can be dissolved in 10,000 grams of water at 30° C?

_____ 11. When you raise the temperature of water, can you dissolve more or less solid?

_____ 12. When you raise the temperature of water, can you dissolve more or less gas?

Using Try This

Students will find that sugar is more soluble in hot water than in cold water. You might want to have them repeat the experiment using salt as the solute. In this case, temperature has little effect on solubility.

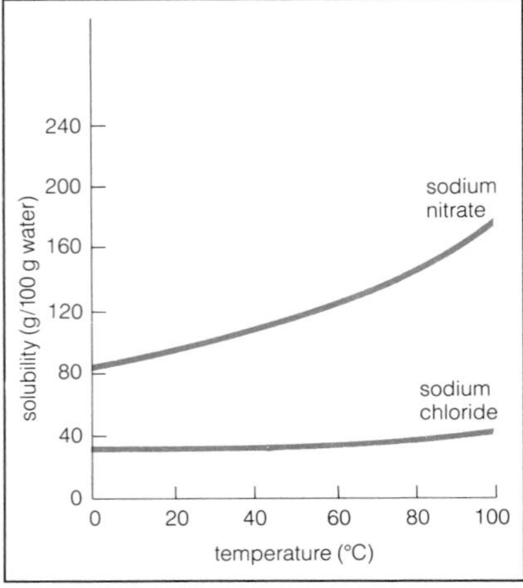

Figure 6-8 *As the temperature increases, the solubility of sodium nitrate changes much more than that of sodium chloride.*

Figure 6-9 *Hot, mineral-rich solutions erupt from cracks in the ocean floor. The minerals precipitate in the cold water.*

(See Figure 6-8.) The change in solubility as the temperature changes can be used to identify a substance.

Hot water generally dissolves more of a solid than cold water. As a hot solution cools, the excess solid *precipitates*, or falls out of the solution. (See Figure 6-9.)

Cold water generally holds more gas in solution than hot water. If a glass of carbonated drink is warm, the dissolved carbon dioxide gas escapes faster.

Fish take in oxygen dissolved in water through their gills. Sometimes the water in a lake or stream becomes very warm because of hot weather or because hot water is dumped into it from a factory or power plant. Fish may die because less oxygen can dissolve in the warm water.

TRY THIS

Add sugar, a teaspoon at a time, to a cup of cold water. How many teaspoons of sugar dissolve? Repeat the same procedure using hot tap water. What effect does the temperature of the water have on the solubility of the sugar?

Checkpoint

1. How do solutions differ from other mixtures?
2. Define solubility. Describe how this property can be used to identify a substance.
3. Describe the effect of temperature on the solubility in water of a typical (a) solid, (b) gas.

Checkpoint Answers

1. A solution is a mixture in which the smallest particles of each substance are evenly distributed. A solution appears to be the same throughout. This is not true of other mixtures.
2. The solubility of a substance is the maximum amount of that substance that will dissolve in a given amount of solvent at a given temperature. Because different substances have different solubilities, the solubility of an unknown can be a clue in identifying it. In addition, the change in solubility of a substance as the temperature changes is a clue to its identity.
3. (a) Solids are usually more soluble at higher temperatures. (b) Gases are less soluble at higher temperatures.

6-4

Separating Mixtures

Goals

1. To distinguish between mixtures and pure substances.
2. To give examples of the ways different types of mixtures can be separated.

Mixtures can be separated into the substances of which they are made. When a substance is no longer part of any mixture, it is known as a *pure substance*. A pure substance has a set of properties that identify it. Water is a pure substance. It freezes at 0°C. It boils at 100°C. Its density is 1.0 g/mL. It is colorless, odorless, and tasteless. One hundred grams of water will dissolve 37 g of sodium chloride at 20°C. These properties of water will not change no matter how many times it is frozen, boiled, or separated from mixtures.

If water is mixed with other substances, the properties of the mixture are different from those of water alone. For example, the freezing point of a solution of sodium chloride in water is less than 0°C and its boiling point is more than 100°C. Antifreeze is mixed with the water in a car's cooling system to keep the water from freezing. The solution that is made when antifreeze is added to the water does not freeze until the temperature is well below 0°C.

Some mixtures are easier to separate than others. The easiest ones to separate contain substances that are clearly different. The monkeys in Figure 6-10 separate a mixture of grain and sand by putting it in water. The sand sinks to the bottom, and the grain floats on top.

Figure 6-10 *Japanese macaque monkeys have been seen separating a mixture of grain and sand by putting the mixture in water.*

Teaching Suggestions 6-4

Begin this lesson by showing students a set of four unidentified test tubes, as follows: tube 1—cooking oil and water; tube 2—sand and water; tube 3—salt water; tube 4—water. Ask students which of the tubes contain heterogeneous mixtures. They should be able to see that the oil and water and the sand and water are heterogeneous.

Then ask which tube or tubes might contain a pure substance. Of course, observation alone cannot distinguish a homogeneous mixture (the salt water) from a pure substance (the water), but students should recognize that one of the two clear liquids might be a pure substance. Have a student taste the two liquids. It should now be clear that the salty liquid contains dissolved salt and is therefore not a pure substance.

Next, ask how the salt mixture might be separated. Someone will probably suggest boiling off the liquid. Do this with a small sample. Place a drop or two of the solution on a glass slide and warm it cautiously over a flame. When the solvent has evaporated, the solid that remains will look and taste like salt.

But what about the clear liquid? Is it a pure substance? Tell students that the liquid boils at 100°C, freezes at 0°C, and has a density of 1.0 g/cm^3, and that 100 g of the liquid will dissolve 37 g of sodium chloride at 20°C. Point out also that all attempts to separate it by boiling and freezing have resulted in the formation of a liquid that has all of the same properties. This evidence suggests that the liquid is a pure substance. Its properties suggest that it is water.

Have students perform laboratory Activity 10, Paper Chromatography, student text page 543, in conjunction with this section. Chromatography is an interesting and colorful method for separating substances. Try separating the colored pigments in autumn leaves, if some are available. These colors can be extracted from the leaves using alcohol warmed in a water bath. **CAUTION:** Never expose alcohol to a flame.

End-of-chapter projects 4, 5, and 8 apply to this section.

Background

The purity of substances is generally determined experimentally by trying to separate the substances as if they were a mixture. Once components have been separated, the chemist attempts to further separate these components. She or he may try to melt, distill, dissolve in different solvents, and so on, to see if each substance can be separated into components with different properties. If the characteristics of a component remain unaltered and no further separation is found, the component can be regarded as a pure substance.

Many properties of a mixture are intermediate between the properties of

In a mixture of oil and vinegar you can see the droplets of oil. If you let the mixture stand for a while it separates itself. The less dense oil forms a layer on top of the denser vinegar. A mixture that clearly does not have the same properties throughout is called a *heterogeneous* (het ur uh JEE nee us) mixture.

Most heterogeneous mixtures are easy to separate. To separate an oil-and-vinegar mixture, you can simply let it stand for a while and then pour off the layer of oil. Another way to separate a heterogeneous mixture is to dissolve one or more of the substances in water. A mixture of salt and sand can be separated in this way. The salt will dissolve in water and the sand will settle out. Then the solution of salt and water can be poured off.

A mixture that has the same properties throughout is called a *homogeneous* (hoh muh JEE nee us) mixture. Solutions are homogeneous mixtures. In general, homogeneous mixtures are more difficult to separate than heterogeneous ones. But it is not very difficult to separate the substances in a mixture of salt and water. If there is no need to recover the water, simply let it evaporate away, leaving the salt behind.

Sometimes the water is the substance that people want to recover from salt water. In places where drinking water is scarce and the ocean is nearby, *solar stills* may be used for this purpose. (See Figure 6-11.) As the sun heats it, the water evaporates. Pure water then condenses on the cooler plastic cover.

Another way to separate water from salt water is by freezing. Normally when salt water freezes, the ice that forms contains only water. Tests are being made to see whether towing icebergs from cold regions to places that are

Figure 6-11 *A solar still is a large container partly filled with seawater. The sun evaporates the water, which collects on the plastic cover. These stills are at the University of California.*

the pure substances that make up the mixture. For example, the density of air lies between the density of oxygen and the density of nitrogen, the two major components of air. A mixture of water and alcohol will smell like alcohol, have a density between 0.8 g/cm^3 and 1.0 g/cm^3, and may or may not burn depending on the ratio of alcohol to water in the mixture.

If the components of a simple mixture are recombined, the original mixture is restored. But if one tries to reconstitute wood by recombining the products formed when wood is distilled, wood is not obtained. Wood, therefore, is not a mixture of substances that can be separated by simple distillation.

Heating mercuric oxide, which is a pure substance, results in the formation of a gas and a silvery, dense liquid. The gas is oxygen and the liquid mercury. Mixing the two does not re-form mercuric oxide. Hence, mercuric oxide is not a mixture of oxygen and mercury.

Ideally, a pure substance contains only one kind of matter. In practice, this is not possible. Generally a substance is considered pure if the amounts and kinds of impurities do not interfere with the purpose for which the substance was prepared. For example, "pure" drinking water may contain dissolved minerals and air, as well as traces of bacteria and other substances.

Demonstration

Give students a practical separation problem. Mix together some sand, salt, and sawdust from a pencil sharpener and ask how this mixture might be separated. Have students volunteer methods, then select one that is workable. For example, you might add the mixture to water and stir. The sand will sink, the sawdust will float, and the salt will dissolve. The sawdust can be skimmed from the surface and dried. The solution can be poured off, leaving the sand, which will dry. The solution can then be allowed to evaporate or the water boiled off to recover the salt. The original mixture can be obtained by recombining these components.

Teaching Tips

- Spend some time having students distinguish between homogeneous mixtures, heterogeneous mixtures, and pure substances. Students should realize that the properties of a pure substance remain unchanged, while those of a mixture change depending on its percent composition.
- Solar stills (see Figure 6-11) may be of interest to some students. Such stills are of growing importance in arid countries that are close to the sea. By using solar energy to distill seawater, these countries can produce drinking water at low cost.

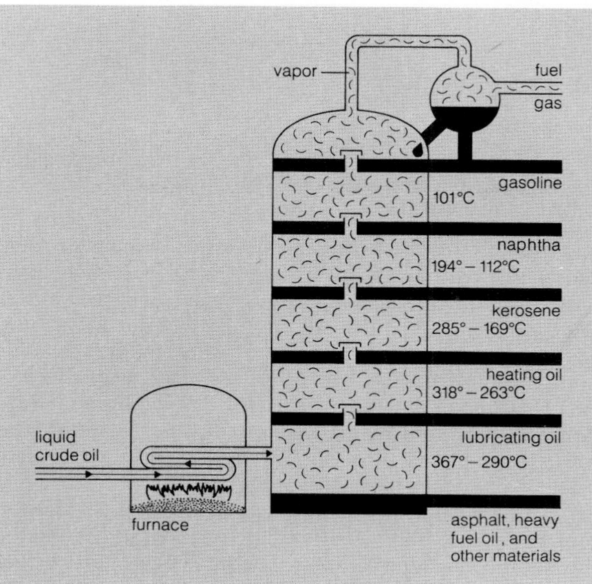

Figure 6-12 *A diagram of a fractionating tower at an oil refinery. As the hot vapor from the furnace rises and cools, different substances condense at different temperatures.*

short of drinking water is economical.

Petroleum, the thick liquid pumped from oil wells, is a mixture of many different substances. To be useful, petroleum must first be separated into simpler mixtures such as gasoline, kerosene, and asphalt. The petroleum is heated in a giant *fractionating column*. (See Figure 6-12.) Different heights in the column are kept at different temperatures. Each substance in petroleum condenses at a different temperature. So groups of substances, or fractions, condense at different heights in the column and flow into separate pipes.

Checkpoint

1. How does a pure substance differ from a mixture?
2. What is a heterogeneous mixture? a homogeneous mixture? Give an example of each.
3. Explain how you could separate (a) a mixture of salt and water, (b) a mixture of salt and sand.

Mainstreaming

Pair visually-impaired students and special-needs students with other students to perform laboratory Activity 10, Paper Chromatography, and any other hands-on activities.

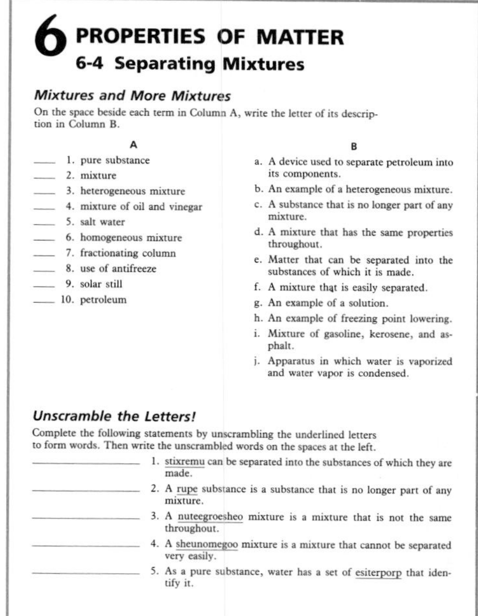

6 PROPERTIES OF MATTER
6-4 Separating Mixtures

Mixtures and More Mixtures

On the space beside each term in Column A, write the letter of its description in Column B.

A	B
____ 1. pure substance	a. A device used to separate petroleum into its components.
____ 2. mixture	b. An example of a heterogeneous mixture.
____ 3. heterogeneous mixture	c. A substance that is no longer part of any mixture.
____ 4. mixture of oil and vinegar	d. A mixture that has the same properties throughout.
____ 5. salt water	e. Matter that can be separated into the substances of which it is made.
____ 6. homogeneous mixture	f. A mixture that is easily separated.
____ 7. fractionating column	g. An example of a solution.
____ 8. use of antifreeze	h. An example of freezing point lowering.
____ 9. solar still	i. Mixture of gasoline, kerosene, and asphalt.
____ 10. petroleum	j. Apparatus in which water is vaporized and water vapor is condensed.

Unscramble the Letters!

Complete the following statements by unscrambling the underlined letters to form words. Then write the unscrambled words on the spaces at the left.

________ 1. stixremu can be separated into the substances of which they are made.

________ 2. A rupe substance is a substance that is no longer part of any mixture.

________ 3. A nuteegroesheo mixture is a mixture that is not the same throughout.

________ 4. A sheunomegoo mixture is a mixture that cannot be separated very easily.

________ 5. As a pure substance, water has a set of esiterporp that identify it.

Checkpoint Answers

1. A pure substance is only one substance; a mixture is made of more than one substance. The properties of a pure substance cannot be changed. The properties of a mixture are different from those of the individual substances of which it is made.
2. A heterogeneous mixture clearly does not have the same properties throughout. Examples are oil and water, and salt and sand. A homogeneous mixture has the same properties throughout. Solutions are homogeneous mixtures.
3. (a) Allow the water to evaporate. (b) First add water to dissolve the salt. Then, pour off the solution, leaving the sand. Finally, evaporate the water from the solution to obtain the salt.

- Discuss the fractionating tower in Figure 6-12, pointing out that gasoline, heating oil, kerosene, and so on, are obtained by separating the mixture of compounds in petroleum into less complex mixtures on the basis of their boiling points.

Fact At Your Fingertips

- Crystals are not always pure substances. In fact, it is the impurities in transistor crystals that make them work.

Teaching Suggestions 6-5

A good way to introduce the concepts covered in this section is to let zinc react with hydrochloric acid to form the compound zinc chloride. If you have balances that will allow mass determinations to ±0.01 gram, students can find that the ratio of zinc reacted to zinc chloride formed is a constant.

CAUTION: Students should wear safety glasses throughout this experiment. Also, they should never add water to acid. (Always add acid to *cold* water.) If they do spill acid on themselves, have them rinse the area immediately with water.

Give each group of students about 0.50 g–1.00 g of zinc. Use zinc chips cut from sheet zinc, not mossy zinc. Have students find the mass of the zinc chip and then add it to 10 mL of 6-molar hydrochloric acid. (Prepare the acid before class by adding a volume of concentrated hydrochloric acid to an equal volume of cold water.)

Measure the prepared acid in 10-mL graduated cylinders and pour the samples into test tubes that rest in beakers of cold water. The zinc then should be added carefully to the acid. It will react vigorously. A burning splint brought to the mouth of a test tube will produce a small explosion, showing that hydrogen (a new substance) has been formed.

Allow the reaction to continue overnight to be sure all the zinc has reacted. During the next class, have each group of students pour the resulting liquid into a clean, dry evaporating dish of known mass. Then have them rinse each test tube with 5 mL of water and add it to the evaporating dish.

The evaporating dish should then be placed on an asbestos pad on a ring stand and heated gently with an alcohol or Bunsen burner. Avoid spattering. Heat the liquid until a dry, whitish salt appears. Continue heating until the solid melts and forms a pool in the bottom of the dish. As soon as the dish cools, measure the combined mass of it and its contents.

By subtracting the mass of the evaporating dish from the combined mass of the dish and the melted product, students can determine how much zinc chloride was produced in the reaction. Finally, have each group calculate the ratio of zinc reacted to zinc chloride formed (mass of zinc/mass of zinc chloride). All teams should get a ratio of about 0.48, indicating that zinc and chlorine combine in a fixed ratio by mass.

This experiment gives students firsthand experience with a chemical reaction and experimental evidence that elements combine in fixed ratios. If you do not have enough materials to have the students do this experiment, do two runs as a demonstration using different masses of zinc, say 0.5 g and 1.0 g.

6-5

Elements and Compounds

Goals

1. To explain how chemical and physical changes differ.
2. To define elements and compounds.
3. To state the law of definite proportions.

physical changes

The methods used to separate mixtures involve **physical changes**, in which no new substances are formed. When the water evaporates from salt water, leaving solid salt behind, there is no change in the salt or water. The salt left behind has the same properties as the salt that was dissolved in water to form the solution. The water that evaporated is simply water in a different state. If the water vapor condenses, it will have the same properties as the water from which the solution was made. Physical changes include changes of state, shape, and size.

chemical changes

In **chemical changes**, new substances, with different properties, are formed. If electricity is passed through water, the water is changed into two new substances. (See Figure 6-13.) The new substances, hydrogen and oxygen, are gases. They have properties that are quite different from those of water.

If hydrogen and oxygen gas are mixed together, they do not produce water. However, if a spark is added to the mixture, an explosion takes place as hydrogen and oxygen

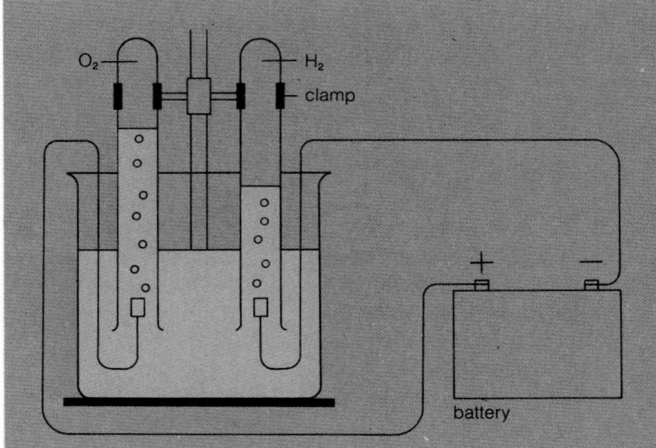

Figure 6-13 *When electricity goes through water, the water breaks into two new substances — hydrogen and oxygen.*

react, forming water. This is a chemical change, also. Chemical changes often produce noticeable heat or light, or both. Whenever something burns, chemical changes are taking place.

Chemical changes can be slow and unspectacular, too. The rusting of iron is a chemical change. As iron combines with oxygen to form rust, there is a color change. Other chemical changes involve color changes. The ripening of fruit and the fading of clothing are two examples.

Sometimes bubbles are a sign of a chemical change. When vinegar is added to baking soda, the carbon dioxide gas that is formed bubbles out of the mixture. When two liquid mixtures are poured together, a solid sometimes precipitates out. This happens when vinegar is added to milk. (The milk is said to "curdle.") Physical changes can produce bubbles and precipitates, too. But with physical changes the bubbles and precipitates usually appear during heating or cooling, not during mixing. Figure 6-14 summarizes some common signs of chemical changes.

Figure 6-14 *Some chemical changes: (a) a seltzer tablet in water, (b) curds forming in milk during the making of cheese, (c) a fire, (d) ripening blueberries.*

Background

The physical changes used to separate the components in mixtures involve changes in size, shape, or state. There is no change in the molecular nature of the substances involved, though.

Chemical changes involve the formation of new substances—substances whose molecular makeup is different. Since we can't see molecules, chemists have to look at the properties of substances in order to tell whether new substances have been formed.

A pure substance that can be broken down into two or more pure substances is called a compound. The properties of the new pure substances are quite different from those of the original pure substance. In addition, the new substances cannot be mixed together to form the original pure substance. Mercuric oxide is a compound. It can be decomposed into mercury and oxygen by heating. Simply mixing mercury and oxygen, however, does not produce mercuric oxide.

Water is a compound because it can be decomposed into two pure substances, hydrogen and oxygen. Mixing these two gases together does not produce water. However, if the mixture is sparked, an explosion occurs. Heat and light are emitted, indicating a chemical reaction. The product of this reaction is water. The properties of water are very different from the properties of the hydrogen and oxygen gases that reacted.

Pure substances that cannot be broken down by heat, electricity, reaction with acids or bases, or other methods are called elements. Any reactions that elements undergo result in compounds, which can be broken down by such methods.

Demonstrations

1. Show students experimental evidence that water is not a mixture. Collect in test tubes the gases that result when water is subjected to an electric current, or undergoes electrolysis. That is, place hydrogen in one test tube and oxygen in another. (See Figure 6-13.) Put the mouths of the test tubes together so the gases will mix. No reaction occurs. Clearly, this mixture of gases does not form water. We have not separated water into the types of components that make up a mixture because we can't get back water by simply mixing the components. This demonstration shows that water is not a mixture.
2. Show students that the formation of bubbles of gas is a sign of chemical change: Add a couple of milliliters of vinegar to some baking soda in a test tube. Bubbles of carbon dioxide gas will form immediately.
3. Show students that the formation of a precipitate is a sign of chemical change: Add a few drops of a dilute solution of silver nitrate to a solution of sodium chloride (salt). A white precipitate of silver chloride will appear. For a more dramatic and colorful reaction, add a few drops of a sodium iodide solution to a solution of lead nitrate. (You will need soft or distilled water to prepare the lead nitrate solution.) A bright orange precipitate will appear.
4. Show students that heat change is a sign of chemical change: Add a few drops of concentrated sulfuric acid to some cold water in a test tube. The heat produced is readily felt by simply holding the tube.

elements

Neither hydrogen nor oxygen can be broken down into simpler substances. They are two of more than 100 such pure substances, which are called **elements**. Elements do not break down into simpler substances in physical or chemical changes. Figure 6-15 shows some elements.

compounds

Pure substances that are a combination of two or more elements are called **compounds**. Most pure substances are compounds. The properties of a compound are very different from the properties of the elements from which it was made. Sodium is a metal that reacts violently with water. Chlorine is a poisonous green gas. These two elements combine vigorously, forming sodium chloride, or common salt. (See Figure 6-16.)

Figure 6-15 *Some elements: (a) copper, (b) mercury, (c) carbon, (d) iodine, (e) gold, (f) oxygen.*

Teaching Tips

- Explain that both elements and compounds are pure substances because their properties are constant. We can't make compounds from elements by simply mixing them together. Nor can we combine elements to form compounds in any ratio, as we can to make mixtures. Elements combine to form compounds only in fixed percentages by mass, as indicated by the law of definite proportions.
- Explain also that a scientific law is based on repeated experimental support. We accept the law of definite proportions because thousands of experiments have supported it. Such definite proportions are, of course, a regularity in nature.

Figure 6-16 *(a) Sodium is stored in kerosene to keep it from reacting with oxygen and water in the air. (b) A tank of chlorine gas. (c) The combination — table salt, sodium chloride.*

When elements combine to form a compound, they always combine in the same percentages by mass. Water is always found to be 89% oxygen and 11% hydrogen. For example, the amount of oxygen and hydrogen in a 10-g sample of water are 8.9 g and 1.1 g, respectively.

$$\text{grams of oxygen} = (0.89)(10\ \text{g}) = 8.9\ \text{g}$$
$$\text{grams of hydrogen} = (0.11)(10\ \text{g}) = 1.1\ \text{g}$$

Sodium chloride is 39% sodium and 61% chlorine by mass. Table sugar (sucrose) is 42.1% carbon, 51.5% oxygen, and 6.4% hydrogen.

Thousands of experiments have shown that each compound contains a definite percentage by mass of the elements that make it up. The results of these experiments are stated as the **law of definite proportions**: A compound consists of two or more elements combined in definite percentages by mass.

law of definite proportions

Checkpoint

1. How does a chemical change differ from a physical change?
2. What are two common signs of chemical change?
3. What is an element? a compound?
4. How do you know that water is a compound?
5. State the law of definite proportions.

6 PROPERTIES OF MATTER
6-5 Elements and Compounds

Changes, Changes Everywhere!

Each of the following questions may be answered with one or two words. Read each question. Then write the letters of your answer on the spaces below the question. Notice that some of the letters have numbers beneath them. You will be using these letters to find the answer at the end.

1. What type of change occurs when the size, shape, or state of matter is altered?

   11 7 1 2

2. What type of change occurs when new substances are formed?

   13 3 5 16

3. When sodium burns in chlorine, what substance is formed?

   14

4. When iron combines slowly with oxygen, what substance is produced?

   10

5. What type of substance cannot be separated into simpler substances?

   17 8

6. The ripening of fruit is an example of one change that often occurs during a chemical reaction. What is this change?

   9 6

7. What type of substance can be separated into simpler substances?

   12 4 15

8. Two students mix two liquids in a test tube. A small genie appears, jumps out of the test tube, and quickly disappears in a puff of smoke. What type of change is described here?

   1 2 3 4 5 6 7 8 9 10 11 12 13 14 15 16 17

Checkpoint Answers

1. In a chemical change, new substances are formed. This does not happen in physical changes, which involve changes only in size, shape, or state.
2. Release of heat or light, formation of gas bubbles, precipitation, and color changes are signs of chemical change.
3. An element is a pure substance that cannot be broken down into simpler substances during a chemical or physical change. A compound is a pure substance that is a combination of two or more elements in fixed percentages by mass; the properties of a compound are different from the properties of the elements from which it was made.
4. Water is a compound because it is a pure substance that can be broken down into two simpler substances.
5. A compound consists of two or more elements combined in definite percentages by mass.

Teaching Suggestions 6-6

Approach the concepts covered in this section by having students study Table 6-3. Then have them explain why the do's and don't's listed in the table are valid. Do they see that rolling or wrapping a burning body in a blanket will smother the flames, and that running will only add more oxygen to the combustion reaction? Do they realize that adding water to an oil flame will spread the fire because oil floats on water? Covering an oil fire is necessary to exclude oxygen from the combustible material.

Do students recognize that the smoke from a wood fire can fill their lungs with liquids and particles that replace the air normally there, and that water will lower the temperature of the burning wood and therefore the rate of burning, while excluding oxygen as well?

Having covered the fire safety rules given, define and give several examples of combustion. The demonstrations and the Try This activity will reinforce the concepts introduced in the section. Chapter-end projects 6 and 7 apply to this section.

Demonstrations

1. Show how to extinguish a fire by blocking the oxygen supply. Use a match to ignite some small pieces of paper on a can lid. Put out the fire by covering the flames with a small can. Explain to students that you prevented oxygen from reaching the paper.
2. Show an example of a flammable vapor. Light a normal-sized household candle. After it has burned for several minutes, light a match, blow out the candle, and quickly bring the match flame to a point 5–10 cm above the wick so that it meets the trail of whitish wax vapor that is rising from the wick. The flame will suddenly "jump" onto the wick; the candle is burning again. Repeat several times so students can see the effect clearly. The wax vapor is flammable; once ignited, it burns right back to the wick.

ENRICHMENT

6-6

Combustion

Goals

1. To define and give examples of combustion.
2. To discuss the dangers of fire and some fire safety rules.

Many substances react with the oxygen in air so rapidly that noticeable amounts of heat and light are produced. This happens when wood or kerosene burns. When a substance reacts with oxygen, releasing heat and light, the process is called *combustion*. Matter that will burn is said to be *combustible* or *flammable*. Combustion is a chemical change. Oxygen and the combustible substance form other substances in the process.

Because wood burns, it is used as a fuel. But wood is also used in construction. Any wooden building is combustible. Curtains, furniture, and other household items will burn, too. Some materials, including many plastic products, release poisonous gases when they burn.

Scientists are studying how fires start and spread. (See Figure 6-17.) These studies will help engineers design safer materials and find ways to prevent and control fires.

The production of heat and smoke and the consumption of oxygen are three reasons why fires are dangerous. Thousands of people die in fires every year. Most of them die from smoke inhalation. The tiny drops of liquid and solid particles that make up smoke fill their lungs, taking the place of oxygen.

Deaths from fires could be reduced if smoke detectors were installed in every home. These sensitive detectors signal the presence of smoke, allowing people to get out of burning buildings quickly and safely.

A fire needs fuel, oxygen, and heat. It can be put out by removing any one of these. Water will cool a fire and also keep oxygen from the fuel. Fires can be smothered by adding a dense gas, such as carbon dioxide, that will not support combustion. Small fires can be smothered by simply covering the flames to keep oxygen from the fuel. Table 6-3 provides some do's and don't's for fire safety.

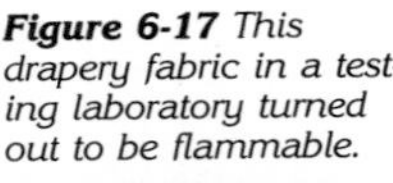

Figure 6-17 *This drapery fabric in a testing laboratory turned out to be flammable.*

Table 6-3: Some Fire-safety Tips

If	*Do*	*Don't*
Clothing catches fire.	Lie down and roll; wrap body in blanket.	Run.
Oil is the fuel.	Smother by covering; use chemical extinguisher.	Put water on the fire.
Wood is the fuel.	Add water.	Breathe the smoke.
Flammable matter is in your home.	Read labels and warnings.	Allow flames or sparks near the substance.

Checkpoint

1. What is combustion? Give two examples of it.
2. What are three reasons why fires are dangerous?
3. What can you do if someone's clothing catches fire?
4. What can you do if an oil fire starts on a stove?

TRY THIS

Light a birthday candle and place it on a can lid in the sink. Mix some vinegar with baking soda in a glass. "Pour" the gas produced by the mixture onto the flame as shown in Figure 6-18. The gas is carbon dioxide. Why is carbon dioxide used in fire extinguishers?

Figure 6-18 *Carbon dioxide gas will put out the flame.*

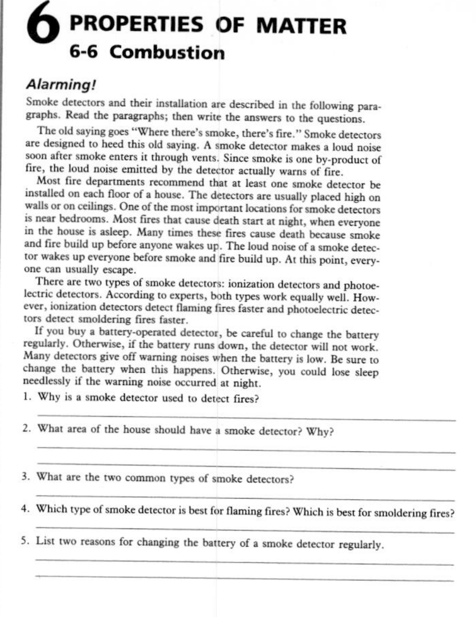

6 PROPERTIES OF MATTER
6-6 Combustion

Alarming!

Smoke detectors and their installation are described in the following paragraphs. Read the paragraphs; then write the answers to the questions.

The old saying goes "Where there's smoke, there's fire." Smoke detectors are designed to heed this old saying. A smoke detector makes a loud noise soon after smoke enters it through vents. Since smoke is one by-product of fire, the loud noise emitted by the detector actually warns of fire.

Most fire departments recommend that at least one smoke detector be installed on each floor of a house. The detectors are usually placed high on walls or on ceilings. One of the most important locations for smoke detectors is near bedrooms. Most fires that cause death start at night, when everyone in the house is asleep. Many times these fires cause death because smoke and fire build up before anyone wakes up. The loud noise of a smoke detector wakes up everyone before smoke and fire build up. At this point, everyone can usually escape.

There are two types of smoke detectors: ionization detectors and photoelectric detectors. According to experts, both types work equally well. However, ionization detectors detect flaming fires faster and photoelectric detectors detect smoldering fires faster.

If you buy a battery-operated detector, be careful to change the battery regularly. Otherwise, if the battery runs down, the detector will not work. Many detectors give off warning noises when the battery is low. Be sure to change the battery when this happens. Otherwise, you could lose sleep needlessly if the warning noise occurred at night.

1. Why is a smoke detector used to detect fires?
2. What area of the house should have a smoke detector? Why?
3. What are the two common types of smoke detectors?
4. Which type of smoke detector is best for flaming fires? Which is best for smoldering fires?
5. List two reasons for changing the battery of a smoke detector regularly.

Using Try This

Students should conclude that carbon dioxide is used in fire extinguishers because it is a dense gas that does not support combustion. It settles over the combustible material, thus preventing oxygen from reaching the burning material.

Checkpoint Answers

1. Combustion is a reaction with oxygen that releases heat and light. Burning wood and a lit candle are examples of combustion.
2. The production of heat and smoke and the consumption of oxygen make fires dangerous.
3. Wrap the person in a heavy blanket or coat to smother the flames. Or place the person on the ground and have him or her roll about to put out the flames.
4. Smother the fire with a blanket, pan, or whatever else is available. Or spray the fire with carbon dioxide from an extinguisher.

Facts At Your Fingertips

- The terms flammable and inflammable are synonomous, although flammable is preferred in technical material. Nonflammable is the antonym.
- While most ordinary combustion reactions involve burning in oxygen, many substances will burn in other gases, such as chlorine.

Cross Reference

For a more detailed discussion of oxidation, see Chapter 9, section 9-3.

Reading Suggestions

For The Teacher

Asimov, Isaac. ASIMOV ON CHEMISTRY. New York: Doubleday, 1974. Essays on inorganic chemistry, nuclear chemistry, organic chemistry, biochemistry, and geochemistry.

Haber-Schaim, Uri and others. INTRODUCTORY PHYSICAL SCIENCE. Englewood Cliffs, NJ: Prentice-Hall, 1987. A textbook that develops the atomic model through experiments involving the basic properties of matter.

Mulvey, J. M., ed. THE NATURE OF MATTER: WOLFSON COLLEGE LECTURES, 1980. New York: Oxford University Press, 1981. These lectures, aimed at a general audience, explain the progress made in recent years toward understanding the basic components of matter and the forces that determine their behavior.

For The Student

Allman, William F. "Are No Two Snowflakes Alike?" *Science 83*, December 1983, p. 24. An amusing discussion of the factors involved in the formation of the symmetrical ice crystals we know as snowflakes.

Gardner, Robert. SCIENCE AROUND THE HOUSE. New York: Messner, 1985. A variety of activities that can be done with materials found in the home.

Lapp, Ralph E. MATTER. New York: Time-Life, 1974. A classic, illustrated introduction to the properties of matter, including the history of physical investigation, solid, liquid and gaseous states, atomic structure, and nuclear reactions. Part of the *Life Science Library*.

Trefil, James S. "Supersteel of the Ancients." *Science Digest*, February 1983, p. 38. Stanford research scientists experiment to develop a "superplastic" metal similar to those used in medieval weaponry.

Understanding The Chapter

6

Main Ideas

1. Matter is anything that has mass and takes up space.
2. Density is the ratio of mass to volume.
3. Matter exists in three states: solid, liquid, and gas.
4. Substances melt and boil at characteristic temperatures called their melting and boiling points.
5. Properties such as density, melting point, boiling point, color, and solubility can be useful in identifying substances.
6. Matter exists in mixtures or as pure substances.
7. Mixtures may be heterogeneous or homogeneous. Solutions are homogeneous mixtures.
8. Pure substances are either elements or compounds.
9. Changes in matter may be either physical or chemical. There are some key signs that indicate chemical change: noticeable heat and light; a color change; the formation of bubbles and precipitates during mixing.
10. A compound consists of two or more elements combined in definite percentages by mass. This statement is known as the law of definite proportions.
11. Combustion is a chemical change in which a substance reacts with oxygen, producing heat and light.

Vocabulary Review

From the following list, choose the term that best completes each of the statements. Write your answers on a separate piece of paper.

boiling point	liter
chemical change	mass
compound	matter
condensation	melting point
density	mixture
element	physical change
evaporation	property
gas	solid
kilogram	solubility
law of definite proportions	solution
	states of matter
liquid	volume

1. The state of matter that has no definite shape or volume is a(n) __?__.
2. The __?__ is a unit of volume.
3. A(n) __?__ is a pure substance that cannot be changed into a simpler pure substance.
4. The temperature at which a substance freezes is also its __?__.
5. The __?__ of water is 1 g/mL.
6. The __?__ of a certain substance is 32 g per 100 g of water at 20°C.
7. When steam comes into contact with a cold surface __?__ occurs.
8. A(n) __?__ is a pure substance that can be broken down into two or more simpler pure substances.
9. The rusting of iron is an example of a(n) __?__.
10. The freezing of water is an example of a(n) __?__.

Vocabulary Review Answers

1. gas
2. liter
3. element
4. melting point
5. density
6. solubility
7. condensation
8. compound
9. chemical change
10. physical change
11. evaporation
12. mixture
13. matter
14. solution

11. The change from a liquid to a gas without boiling is __?__.
12. A salad is obviously a(n) __?__.
13. __?__ is defined as anything that takes up space and has mass.
14. A mixture in which the smallest particles of two or more substances are evenly distributed is a(n) __?__.

Chapter Review

Write your answers on a separate piece of paper.

Know The Facts

1. The temperature at which water changes to steam is its __?__. (melting point, freezing point, boiling point)
2. The standard unit of mass is the __?__. (liter, kilogram, cubic centimeter)
3. A __?__ has a definite volume, but no definite shape. (solid, liquid, gas)
4. The solubility of oxygen in water __?__ when the water temperature increases. (increases, decreases, stays the same)
5. An equal-arm balance is normally used to measure __?__. (mass, volume, density)
6. The solubility of a solid in water usually __?__ as the temperature of the water increases. (increases, decreases, stays the same)
7. Ice floats on water because it has a lower __?__ than water. (mass, volume, density)
8. A solid may precipitate from a solution when the solution is __?__. (warmed, cooled, stirred)
9. A mixture that has the same properties throughout is __?__. (homogeneous, heterogeneous, impossible)
10. A __?__ is made up of two or more elements combined in definite percentages by mass. (mixture, solution, compound)
11. A fire can be put out by adding __?__. (oxygen, carbon dioxide, heat)

Understand The Concepts

12. A substance has a density of 0.00134 g/mL. Is the substance a solid, a liquid, or a gas?
13. Some sugar is completely dissolved in water. Is the mixture homogeneous or heterogeneous? Explain.
14. List four key signs that a chemical change may be taking place.
15. Is water an element or a compound? How do you know?
16. Give two examples of a physical change.
17. List four properties that can be used in identifying a substance.
18. Salt (sodium chloride) is added to water at 80°C until the solution is saturated. What will happen as the solution cools down to room temperature?
19. What is the volume of a plastic box that is 8 cm × 4 cm × 5 cm? How many mL of water will just fill the box?
20. Give two examples of a chemical change.
21. People cannot drink salt water, yet icebergs that form in salt water could be used for drinking water. How do you explain this?

Chapter Review Answers

Know The Facts

1. boiling point
2. kilogram
3. liquid
4. decreases
5. mass
6. increases
7. density
8. cooled
9. homogeneous
10. compound
11. carbon dioxide

Understand The Concepts

12. Gas
13. Homogeneous. It has the same properties throughout.
14. Release of light and/or heat, color change, production of a gas, and formation of a precipitate are all evidence of a chemical change.
15. Water is a compound. It is a pure substance that can be decomposed into two other pure substances—hydrogen and oxygen.
16. Answers will vary. Examples include melting ice and boiling water.
17. Density, solubility, boiling point, and melting point.
18. A little of the salt will precipitate.
19. 160 cm^3. It will hold 160 mL.
20. Burning wood, rusting of iron, and so on. See student text, page 161.
21. When salt water freezes, very little salt freezes with the water.

22. (a) 510 g, (b) 0.510 kg.
23. Oxygen, fuel, and heat. Oxygen can be removed by smothering the fire. If oil is *not* the fuel, a fire can be cooled by adding cold water.

Challenge Your Understanding

24. Add sugar until no more will dissolve. Pour off the solution and find its mass. Evaporate the water, dry the sugar, and measure its mass. By subtraction, determine the mass of the water that was present in the saturated solution. Divide the mass of sugar by the mass of water to find the solubility.
25. Shape the clay into the form of a boat or make it into a hollow ball so that the combined density of clay and air is less than the density of water.
26. 8.9 g/cm^3. It could be copper. (See Table 6-1 on student text page 149.)
27. Air is a homogeneous mixture. It is the same throughout but it contains several gases.
28. 180 g for every 100 g of water, or 1.8 g per 1 g of water.
29. Oil floats on water. The water will spread the burning oil.

6 PROPERTIES OF MATTER
Projects

1. Read the labels on various materials found around your home. Find out if any of these materials are dangerous, especially if they give off toxic gases when burned. To answer the questions, you may want to contact the fire department or write the manufacturer.
   a. What materials did you find to be dangerous? ______
   b. How should these materials be stored or safely used? ______
2. If you are an outer space fan, imagine a world that has an ocean made of mercury.
   a. What density must such a world have in order to allow mercury to float on its surface in oceans? ______
   b. From Table 6-1 in your textbook, what are two elements that could be found in this imaginary world?

22. An object of unknown mass is placed on the left pan of an equal-arm balance. It is just balanced when a 500-g mass and a 10-g mass are placed on the right pan. What is the mass of the object in (a) grams, (b) kilograms?
23. What three things are needed to support combustion? Describe two ways in which these things can be removed in order to put out a fire.

Challenge Your Understanding

24. Describe how you could prepare a saturated solution of sugar in water. Then describe how you would determine the solubility of sugar in water at room temperature.
25. Explain what you could do to a lump of clay in order to make it float on water.
26. A block of metal has the dimensions 2 cm × 5 cm × 10 cm. The mass of the block is 890 g. What is the density of the block? Of what metal may the block be made?
27. Is air a pure substance, a heterogeneous mixture, or a homogeneous mixture? Explain your answer.
28. The solubility of potassium nitrate in water at 90°C is 200 g/100 g water. Its solubility at 20°C is 20 g/100 g water. If a potassium nitrate solution that is saturated at 90°C is cooled to 20°C, what mass of potassium nitrate will precipitate out of the solution?
29. Why do you think it is dangerous to add water to an oil fire?

Projects

1. Prepare a mixture of crushed ice and water or snow and water. What is the temperature of the melting ice or snow? Add salt to the mixture and stir. What happens to the temperature?
2. Determine the solubility of sugar in water at several different temperatures. Organize your results in a data table. Then plot a graph showing the number of grams that dissolve at different temperatures.
3. Prepare some ice "cubes" that have the same volume but different surface areas. How does the surface area affect the time it takes the ice to melt? (To speed up the melting you could put the ice in a pan of water.)
4. Find out what methods are being used to make salt water drinkable. Make a scale model of one of the plants currently operating.
5. Investigate the uses of *chromatography* to separate mixtures.
6. Visit a firehouse. Describe the function of each major piece of equipment that is used in fighting fires.
7. The purity of food and drugs is a concern to all consumers. Use the library or interviews to prepare a report on some aspect of food and drug inspection.
8. Research and report on the contributions to chemistry of any one of the following scientists: Percy L. Julian; Ellen H. Richards; Lloyd N. Ferguson.

6 PROPERTIES OF MATTER
Challenge

Questions to Ponder

1. Substance A has a mass that is greater than substance B, but a volume equal to substance B. Substance B, on the other hand, has a mass that is equal to substance C, but a volume that is less than substance C.
   a. Arrange the three substances from greatest density to least density. ______
   b. If the substances were sold at the same price per kilogram, with which substance would you get the most volume? ______
   c. If the substances were sold at the same price per liter, with which substance would you get the most mass? ______
2. You have a heterogeneous mixture of sand, salt, wood shavings, and iron filings. Describe how you would separate each material.
   a. iron filings ______
   b. wood shavings ______

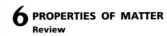

6 PROPERTIES OF MATTER
Review

Puzzling Substances

Use the clues below to complete the crossword puzzle.

Across

3. The process that occurs when a substance reacts with oxygen, releasing heat and light.
4. A measure of the amount of material in an object.
5. Anything that takes up space and has mass.
6. A change in which no new substances are formed.
9. The space that an object occupies.
11. A pure substance that is a combination of two or more elements
8. The change from liquid to gas without boiling.
10. A substance that will not separate into simpler substances during physical or chemical change.
14. A mixture in which the smallest particles of two or more substances are evenly distributed.
16. The state of matter whose volume and shape do not change.

Chapter 7

Atoms and Molecules

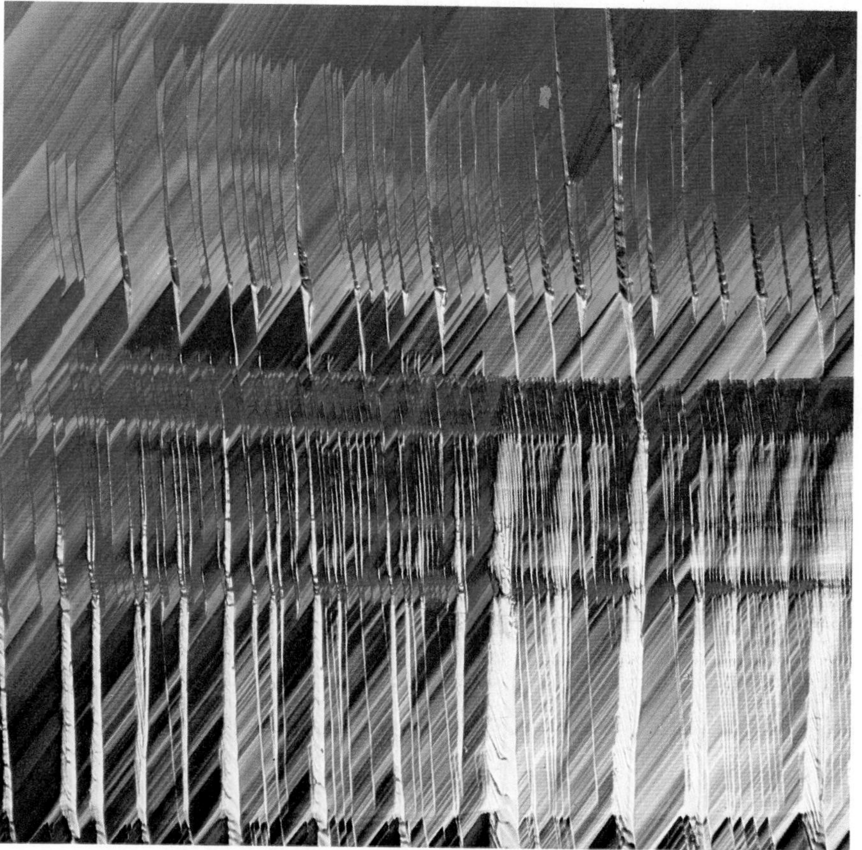

Imagine the surface of this silicon crystal made up of regularly arranged particles called atoms.

Vocabulary Preview

scientific model
atoms
molecule
chemical symbols
chemical formula
chemical reaction
chemical equation
law of conservation of mass
balanced equation
electrons
nucleus
protons
neutrons
atomic mass
atomic number
energy levels

CHAPTER 7 Overview

This chapter introduces the concepts of atoms, molecules, and chemical reactions. The modern view of atomic structure and chemical reactions is explained. Chemical symbols, formulas, and equations are introduced, and both the significance and use of the law of conservation of mass are stressed. The structure of the nucleus of the atom is then described, and the relationship between electron energy levels and line spectra is discussed. The chapter ends with a brief enrichment section on the modern quantum model of the atom.

Goals

At the end of this chapter, students should be able to:

1. explain what is meant by a scientific model.
2. describe Dalton's atomic model.
3. define atoms and molecules.
4. describe how chemical symbols are used.
5. define a chemical formula.
6. determine the number of atoms of each element in a chemical formula.
7. explain what happens during a chemical reaction.
8. identify the parts of a chemical equation.
9. state the law of conservation of mass.
10. tell whether a chemical equation is balanced.
11. describe Thomson's model of the atom.
12. describe Rutherford's model of the atom.
13. define electrons, neutrons, and protons.
14. explain how elements can be identified by the light they give off.
15. describe the Bohr model of the atom.
16. compare the current model of the atom with the Bohr model.

Teaching Suggestions 7-1

Introduce the concepts covered in this section by providing each pair or small group of students with a sealed container housing two or three objects. (See the Try This activity on student text page 172.) You might use cylindrical oatmeal boxes with a marble and a washer inside each, or small gift boxes containing similar objects. Other possible contents are a short pencil and a block, a nail and a domino, and so on.

Ask the student groups to create models of the contents of their boxes, based on the evidence they collect when examining and manipulating the boxes. Some will notice a sliding or rolling sound when they tip a box. Some may ask for a magnet to see if they can make the objects inside move by passing the magnet over the outer surface.

Some students may want to look inside after they have completed their model development. Discourage this. Point out that a scientist never knows if a model is true. She or he can only test the consequences of the model. This is a good time to mention that scientists often modify their models on the basis of new experiments and observations. Ask students to discuss how often they did this when examining the sealed boxes.

Use the Try This activity as a jumping-off point for introducing the basic postulates of Dalton's atomic theory: (1) Matter consists of indivisible atoms. (2) Each element is made up of identical atoms, and there are as many different kinds of atoms as there are elements. (3) Atoms cannot be changed. (4) When different elements combine to form a compound, the smallest portion, or molecule, of the compound is made up of a definite number of atoms of each element. (5) In chemical reactions, atoms are neither created nor destroyed; they are only rearranged. Then discuss how Dalton's atomic model explains the law of definite proportions. Stress that atoms act as if they are indivisible during ordinary chemical reactions; that is, they retain their integrity. End-of-chapter projects 1 and 4 apply to this section.

7-1

Dalton's Atomic Model

Figure 7-1 Nineteenth-century scientist John Dalton proposed an atomic model of matter.

Goals

1. To explain what is meant by a scientific model.
2. To describe Dalton's atomic model.
3. To define atoms and molecules.

A scientific hypothesis is an educated guess. If you noticed that something had been eating the lettuce in your garden, you might guess that a rabbit had done it. Other indirect evidence, such as rabbitlike footprints, would support your guess. But you would not know for certain unless you saw a rabbit eating the lettuce.

Sometimes a scientific hypothesis is a guess about something that has never been seen or that cannot be seen. Such a hypothesis must be based on indirect evidence only. Suppose that the footprints near the lettuce were not rabbitlike, nor like those of any known animal. You might try to imagine some kind of creature that would eat lettuce and leave the kind of footprints that you observed. Your imaginary creature would be a **scientific model**. A scientific model explains something that is not familiar or that cannot be seen in terms of something that is familiar or visible.

scientific model

Over 2000 years ago some Greek thinkers, or philosophers, devised a model of matter. They said that all matter is made up of invisible particles so small that they cannot be divided any further. They called these particles **atoms**, which means "indivisible." In 1808, John Dalton, an English schoolteacher, described matter in a strikingly similar way. Dalton's model, unlike that of the Greek philosophers, was based on the results of scientific experiments.

atoms

Dalton said that all elements are made up of small, invisible, dense particles that cannot be divided into anything smaller. The particles were like very tiny billiard balls or marbles. Like the Greek philosophers, Dalton called the particles atoms. According to Dalton's model, atoms of a particular element are alike in mass and every other property.

Since Dalton's time, the atomic model has undergone many changes. However, many of its features are still valid. In today's model, for example, atoms are the smallest particles of individual elements. Also, different atoms have dif-

Background

Scientific models are developed to explain the observations and experimental results obtained in the real world. To be of any use to the scientific community, a model must also lead to predictions that can be tested by experiment. Dalton's atomic model explained why compounds consist of elements united in fixed percentages by mass (the law of definite proportions), but it also predicted that elements might combine in more than one way to form compounds. For example, hydrogen might combine with oxygen in a ratio of one

ferent properties. For example, the mass of an oxygen atom is about 16 times as great as that of a hydrogen atom.

Dalton also assumed that the atoms of different elements have different properties. He thought that in a chemical change, atoms are neither created nor destroyed too. Instead, they are only rearranged. When oxygen and hydrogen combine, two hydrogen atoms can combine with one oxygen atom, forming a water **molecule**. (See Figure 7-2a.) In general, a molecule is a particle of matter made up of chemically joined or bonded atoms. Molecules are extremely small. There are many trillions of water molecules in the tiniest drop of water, for example.

molecule
(MAHL uh kyool)

Oxygen and hydrogen can form another compound, called hydrogen peroxide. When hydrogen peroxide is dissolved in water, the resulting solution can be used as a bleach or an antiseptic. Hydrogen peroxide molecules each contain two hydrogen atoms and two oxygen atoms, as shown in Figure 7-2b.

Dalton's atomic model explains what is called the *law of definite proportions*. Suppose one atom of element A combines

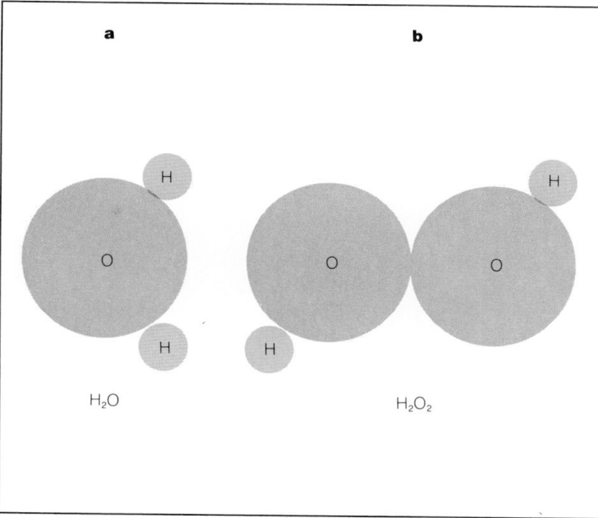

Figure 7-2 *Hydrogen and oxygen can combine to form two different kinds of molecules: (a) water, (b) hydrogen peroxide.*

atom to one atom, as in hydrogen peroxide (H_2O_2). Or, hydrogen might combine with oxygen in a 2 : 1 atomic ratio, as in water (H_2O).

Stress that models, such as the atomic model of matter, can never be tested directly. We cannot see atoms, so we cannot see in what ratio the atoms combine. All we can do is determine if the model satisfactorily explains the evidence derived through experiment and observation, and if it can make predictions that have consequences that can be tested.

A hypothesis is similar to a model. It, too, is an attempt to explain some phenomenon, but it is generally limited to a much smaller domain than a model or theory.

Demonstration

Show the formation of compounds and how this union of elements, according to Dalton's atomic model, gives rise to the law of definite proportions. Use colored spheres to represent different atoms. Stress that you are using the spheres to illustrate the model, and that real atoms cannot be seen. You might let blue spheres represent the atoms of one element (*B*), while red spheres represent the atoms of another element (*R*). Assign a mass to each "atom." For example, blue "atoms" might have 1 unit of mass and red "atoms" 3 units of mass.

Join a blue "atom" and a red "atom" to form a "molecule" of the "compound *BR*." Then, ask what the ratio of the mass of "element *R*" to "element *B*" is in the "compound." Make sure students see that it is 3 : 1. Continue joining blue and red "atoms" to form "molecules" of "*BR*." Find out if students grasp that the mass ratio will remain 3 : 1 no matter how many "molecules" are formed. If they do not, have them add up the total mass of red and blue "atoms" as the "molecules" are put together to see that the ratio does indeed remain 3 : 1. Then have students predict the mass ratios if the "atoms" combine in 2 : 1 ratios: 2 blue to 1 red and 2 red to 1 blue. (2 : 3 and 6 : 1.) Again, if students have difficulty, build the "molecules" one at a time and keep a running record of the masses of each "element."

Teaching Tip

- Be sure students understand that scientific models are not models in the sense of a model airplane, which is really a miniature copy of something whose structure and function are thoroughly understood.

Facts At Your Fingertips

- The diameter of an atom is about 10^{-8} cm.
- If atoms were enlarged to the size of marbles (1 cm across), the atoms in one gram of hydrogen, lined up side by side, would encircle Earth about 15 trillion times.

Mainstreaming

Pair visually-impaired and special-needs students with other students to carry out the Try This activity.

7 ATOMS AND MOLECULES

7-1 Dalton's Atomic Model

The Atom: So Small Yet So Important

A list of scrambled words from Section 7-1 of your textbook appears below. Unscramble the words. Then use each word to complete one of the sentences that follow. Write the unscrambled word on the space next to the sentence.

sledom	comelleu
spenremitex	slidenibivi
smato	ticniscefi

_______ 1. In order to explain what cannot be seen directly, people use __1.__ to describe how something looks or behaves.

_______ 2. A __2.__ model is used by scientists to explain scientific laws and observations.

_______ 3. According to the Greeks, matter is made up of particles called __3.__ that are so small they cannot be divided any further.

_______ 4. The main difference between Dalton's model and the Greek model is that Dalton's model was based on the results of scientific __4.__.

_______ 5. According to Dalton, the particles of matter are small, dense, and __5.__.

_______ 6. A __6.__ is a particle of matter made of two or more chemically joined atoms.

Models Galore!

The rectangles below are "frames" for models that you will create. Read the description below each rectangle. Based on the description, draw a model inside the frame. Don't be afraid to use your imagination!

A. Model of the ultra-modern car of the year 2000.

B. Model of an extraterrestrial creature.

Using Try This

This activity is best used to introduce the concepts covered in the section. See Teaching Suggestions.

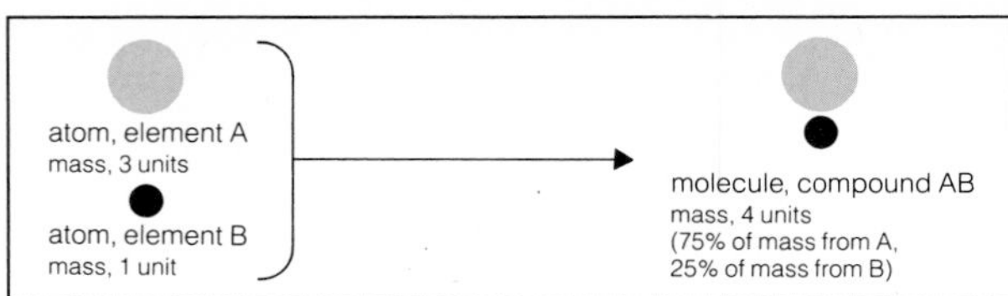

***Figure 7-3** According to the law of definite proportions, compound AB will always have the same percentages by mass of A and B.*

with one atom of element B, forming one molecule of compound AB. (See Figure 7-3.) Suppose, further, that one atom of A has a mass three times as great as that of an atom of B. If we call the mass of a B atom 1 unit, the mass of an A atom must be 3 units. Then ¾ of the mass of every molecule of AB must come from A and ¼ must come from B. AB is 75% A and 25% B, by mass.

$$\frac{\text{mass of A}}{\text{mass of AB}} = \frac{3}{4} = 75\% \qquad \frac{\text{mass of B}}{\text{mass of AB}} = \frac{1}{4} = 25\%$$

All molecules of AB contain these percentages by mass because all of them contain exactly one A atom and one B atom, and cannot contain fractions of these atoms. The law of definite proportions is a consequence of the fact that only whole atoms combine. Atoms are indivisible in chemical changes.

Checkpoint

1. What is a scientific model?
2. Decribe Dalton's atomic model.
3. Define (a) an atom, (b) a molecule.
4. How does Dalton's atomic model explain the law of definite proportions?

TRY THIS

Scientific models can be developed by imagining and experimenting. Get a sealed container that has been prepared by your teacher or another student. Try to develop a model of the inside of the box and its contents. Since you cannot actually see the inside, develop your model by gathering indirect evidence. Then sketch the inside of the box and its contents, based on the evidence you have collected.

Checkpoint Answers

1. A scientific model is an explanation of something unfamiliar or invisible in terms of something familiar or visible.
2. Dalton's atomic model described an element as consisting of identical, indivisible atoms. The atoms of different elements are different. In a chemical change, atoms are neither created nor destroyed; they are only rearranged.
3. (a) An atom is the smallest particle of an element. (b) A molecule is a particle of matter made up of chemically joined atoms.
4. See student text, pages 171–172. See also answer to Chapter Review question 15 (b) on page 190.

7-2

Symbols and Formulas

Goals

1. To describe how chemical symbols are used.
2. To define a chemical formula.
3. To determine the number of atoms of each element in a chemical formula.

A *symbol* is a picture or sign that stands for something else. Highway signs contain symbols, as shown in Figure 7-4. If you play a musical instrument, you read symbols, called notes, that stand for certain musical tones. If you take shorthand, you write symbols that stand for words or phrases. Each time you solve a math problem, you use symbols—numerals and signs—that stand for certain quantities and mathematical operations.

chemical symbols

As a convenience, scientists use **chemical symbols** to stand for the names of the elements. Each chemical symbol is made up of one or two letters, the first of which is capitalized. Many of the chemical symbols are abbreviations for the English names of the elements. For example, S stands for sulfur and He stands for helium. Other symbols come from Latin names. For example, the symbol for lead (Pb) comes from the Latin word *plumbum*. Lead is a soft metal that was used by the Romans to make pipes. The English word *plumber* ("one who makes pipes") comes from the Latin word for lead. The symbol for gold is Au, from the Latin word for

Figure 7-4 *Highway signs, such as these, contain symbols that stand for certain objects.*

Teaching Suggestions 7-2

Begin this lesson by writing several symbols on the chalkboard, such as H, P, O, and your first and last initials (in capital letters without periods). Ask students what these symbols represent. If your students are well prepared, they should know that each symbol represents one atom of an element, as well as the element itself. In addition, they should be able to state that your initials are not a valid symbol because both letters are capitalized. When changed so the first letter is capitalized and the second one is lower-case, your initials may or may not serve as the symbol for an element.

Now write some formulas, such as O_2, $2I_2$, CH_4, SO_2, $3NH_3$, $5NO$, and N_2O_5, on the chalkboard, and ask:

How many atoms are represented in each formula? (2;4;5;3;12;10;7)

How many molecules are represented in each? (1;2;1;1;3;5;1)

Incorrect answers may indicate a need for more drill. Continue until all students show mastery of the concepts covered. Chapter-end project 2 applies to this section.

Background

The symbol for an element is more than just an abbreviation of its name; it is also used to represent one atom of the element. By using symbols, we can represent the molecules that result when elements combine to form compounds. Subscripts in the formula for a molecule of a compound tell us the number of atoms of each type in the molecule. Coefficients, which are numbers written in front of the formulas, indicate the number of molecules. Thus, $4CH_2O_2$ contains four molecules, each consisting of one carbon, two hydrogen, and two oxygen atoms bonded together. Lack of a subscript, as after C in CH_2O_2, means there is just one atom of that element present.

Note that the term *molecule* may be used to describe atoms of the same element, as well as of different elements, chemically joined together. Thus, O_2 is a diatomic molecule.

7 ATOMS AND MOLECULES
Computer Program

Chemical Symbols

Purpose To improve your recognition of chemical names and symbols.

Background Chemistry is sometimes a difficult subject to learn because the symbols for the chemicals are unfamiliar.

ENTER the following program:

```
10  DIM N$(20)
20  DIM S$(20)
30  FOR I = 1 TO 10
40  READ S$(I)
50  NEXT I
60  DATA O, N, H, S, P, I, F, C, B, K
70  FOR I = 1 TO 10
80  READ N$(I)
90  NEXT I
100  DATA OXYGEN, NITROGEN, HYDROGEN, SULPHUR, PHOSPHOROUS
110  DATA IODINE, FLUORINE, CARBON, BORON, POTASSIUM
120  FOR I = 1 TO 10
130  PRINT "NAME THE CHEMICAL THAT MATCHES THE"
140  PRINT "FOLLOWING SYMBOL: "; S$(I); "."
150  INPUT C$
160  IF C$ = N$(I) THEN GOTO 190
170  PRINT "WRONG, THE CORRECT ANSWER WAS "; N$(I); "."
180  GOTO 200
190  PRINT "CORRECT!!!"
200  NEXT I
210  END
```

Discussion

Record your scores for the first and third run through the program. Is there a change in your score? What kind of change?

Upgrade the Program

Notice that the DIM statements in lines 10 and 20 tell the computer to save room for lists of up to twenty items. The FOR...NEXT loops in lines 30 to 50 and lines 70 to 90 READ the words and symbols from the DATA statements into the lists. The main loop of the program is in lines 120 through 200. The program displays a symbol from the S$ list and checks the entry with the corresponding name from the N$ list.

1. Enter a new series of symbols and chemical names in the DATA statements. Follow the same format.
2. Add a random number selector of symbols on the list.

gold, *aurum*. Mercury's symbol, Hg, comes from the Latin word *hydrargyrum*, which means "silver water." Table 7-1 presents the symbols for some of the most common and useful elements.

Chemical symbols serve as more than abbreviations for the elements. Each symbol can stand for one atom of a particular element. Thus, Au stands for one atom of gold, and Hg stands for one atom of mercury. Si stands for an atom of silicon and W for an atom of tungsten.

Chemical symbols can also be used to show the number of atoms in a molecule. A molecule of water is made up of two atoms of hydrogen (H) and one atom of oxygen (O). Its symbol is H_2O. The subscript, or small number written to the right of H, shows that there are two H atoms. The symbol for oxygen has no subscript. Since it represents a single atom, the subscript 1 is understood. The number and types of atoms that are combined in a molecule of a substance are given by its **chemical formula**. The chemical formula for a molecule of ordinary table sugar is $C_{12}H_{22}O_{11}$. The formula shows that there are 12 atoms of carbon (C), 22 atoms of hydrogen, and 11 atoms of oxygen in each molecule of ordinary table sugar.

chemical formula

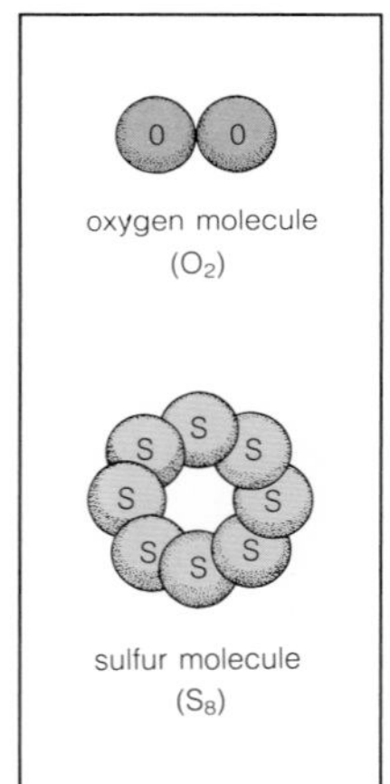

Figure 7-5 *Different elements may be made up of molecules that have different numbers of atoms.*

So far, we have looked only at single molecules of a substance. Suppose you want to represent three molecules of water. You know that the chemical formula for one molecule of water is H_2O. To represent three molecules, the coefficient 3 would be written in front of the chemical formula for water. Three molecules of water are then represented as $3H_2O$. To find the total number of atoms of each kind in the three molecules, multiply the coefficient times the number of each kind of atom shown in the formula. In three molecules of water, there are 3×2, or 6, atoms of hydrogen and 3×1, or 3, atoms of oxygen.

Scientists have found that in nature, atoms of hydrogen, oxygen, and several other elements usually exist in pairs. For this reason, molecules of these elements are called *diatomic*, or two-atom, molecules. Thus, the formula for the naturally occurring element hydrogen is H_2, and the formula for oxygen is O_2. Bromine (Br_2), iodine (I_2), nitrogen (N_2), chlorine (Cl_2), and fluorine (F_2) are also diatomic molecules. The formula for sulfur (S), a yellow solid, is S_8. The formula shows that each sulfur molecule contains eight sulfur atoms. (See Figure 7-5.)

Teaching Tip

- Refer students to the periodic table on page 204 (Chapter 8, section 8-5) so they can see the symbols for all the elements. It is useful to have a periodic table in the classroom.

Cross Reference

The properties of different groups of elements are explained in Chapter 8.

Table 7-1: Some Elements and Their Symbols

Name	Symbol
aluminum	Al
calcium	Ca
carbon	C
chlorine	Cl
copper (Latin name: *cuprum*)	Cu
fluorine	F
gold (*aurum*)	Au
helium	He
hydrogen	H
iodine	I
iron (*ferrum*)	Fe
lead (*plumbum*)	Pb
mercury (*hydrargyrum*)	Hg
neon	Ne
nickel	Ni
nitrogen	N
oxygen	O
phosphorus	P
platinum	Pt
potassium (*kalium*)	K
silicon	Si
silver (*argentum*)	Ag
sodium (*natrium*)	Na
sulfur	S
tin (*stannum*)	Sn
uranium	U
zinc	Zn

Checkpoint

1. In what two ways are chemical symbols used?
2. What information does a chemical formula provide?
3. List the number of atoms of each kind in the following: (a) H_2SO_4, (b) $2CO_2$, (c) $3CH_4$.
4. Which of the following are diatomic molecules: $C_{12}H_{22}O_{11}$, Br_2, CO_2, S_8, H_2O, Cl_2?

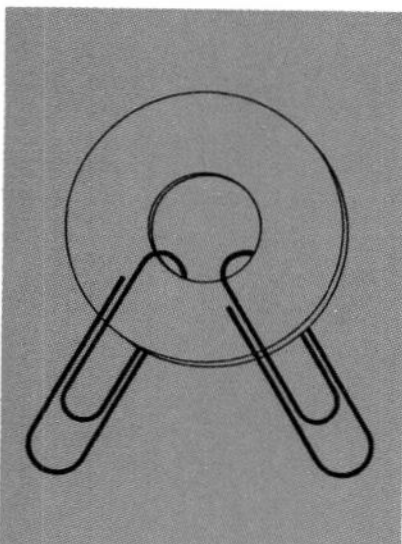

Figure 7-6 A washer-paper clips model of a molecule of water (H_2O).

TRY THIS Let paper clips represent hydrogen atoms; washers, oxygen atoms; and bolts, carbon atoms. Use these "atoms" to represent "molecules" of carbon monoxide (CO), carbon dioxide (CO_2), and hydrogen peroxide (H_2O_2).

7 ATOMS AND MOLECULES

7-2 Symbols and Formulas

Chemistry Is Elementary!

Identify each symbol in the following series as an element or a compound. If the symbol represents an element, place it under the heading *Element*. If the symbol represents a compound, place it under the heading *Compound*. Then answer the summary question.

$MgCl_2$, Ag, Br_2, Co, CO, HI, Ne, NO_2, NaF, P_4

Element	Compound
1.	1.
2.	2.
3.	3.
4.	4.
5.	5.

What is the key to whether a symbol represents an element or a compound?

____

____

____

Interpret the Formulas

Fill in the chart below by placing the name of the underlined element in the second column and the number of atoms of that element in the third column. The first compound is done for you as an example.

Chemical Formula	*Name of Underlined Element*	*Number of Atoms of Underlined Element*
$H_2\underline{S}O_4$	sulfur	1
$C_{12}H_{22}\underline{O}_{11}$		
$3\ \underline{Pb}O_2$		
$2\ Sn\underline{Cl}_4$		
$\underline{C}_9H_8O_4$ (aspirin)		
$4\ \underline{Al}_2S_3$		
$5\ \underline{Ag}_2CrO_4$		

Using Try This

Students should be able to create molecular models similar to those shown in Figure TE 7-1.

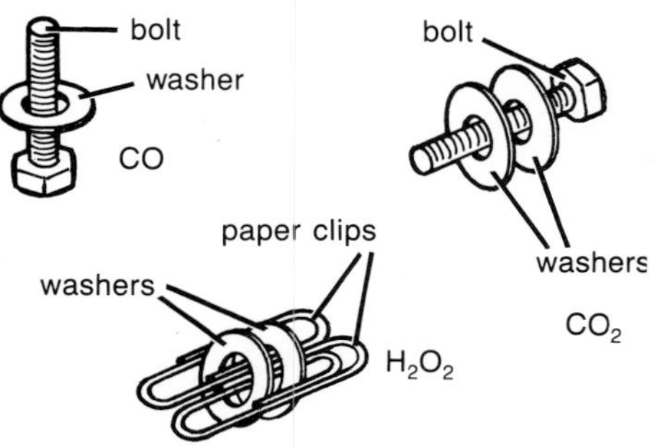

Figure TE 7-1 Models of molecules.

Checkpoint Answers

1. Symbols are used to represent the names of elements and the atoms of the elements.
2. A formula reveals both the number and kinds of atoms in a molecule.
3. (a) H_2SO_4: 2 atoms of hydrogen, 1 atom of sulfur, and 4 atoms of oxygen. (b) $2CO_2$: 2 carbon atoms and 4 oxygen atoms. (c) $3CH_4$: 3 carbon atoms and 12 hydrogen atoms.
4. Br_2 and Cl_2 are diatomic molecules.

Teaching Suggestions 7-3

A good way to introduce the concepts covered in this section is to demonstrate the reaction between copper and sulfur illustrated in Figure 7-7 on student text page 177. (See Demonstration.) When discussing the reaction, write the formulas for the reactants and products on the chalkboard, as described in the text. Point out that we write copper as Cu and sulfur as S_8 because other experiments indicate that copper atoms occur singly, whereas sulfur molecules are clusters of 8 sulfur atoms. Then proceed to balance the equation, as described in the student text on page 178.

Next, choose some other equations to balance. Select a few not found in the text. The burning of alcohol ($C_2H_5OH + O_2$) and the burning of glucose ($C_6H_{12}O_6 + O_2$) are two good examples.

$$C_2H_5OH + O_2 \longrightarrow CO_2 + H_2O$$
$$(C_2H_5OH + 3O_2 \longrightarrow 2CO_2 + 3H_2O)$$

or

$$C_6H_{12}O_6 + O_2 \longrightarrow CO_2 + H_2O$$
$$(C_6H_{12}O_6 + 6O_2 \longrightarrow 6CO_2 + 6H_2O)$$

Laboratory Activity 11, Conservation of Mass, student text page 544, and laboratory Activity 12, Some Model Reactions, student text page 545, should be done in conjunction with this section. End-of-chapter project 3 also applies to this section.

7-3

Chemical Equations

Goals

1. To explain what happens during a chemical reaction.
2. To identify the parts of a chemical equation.
3. To state the law of conservation of mass.
4. To tell whether a chemical equation is balanced.

chemical reaction

chemical equation

Have you ever watched a burning match or baked a loaf of bread? If so, you have witnessed a **chemical reaction**, a process in which chemical change takes place. The changes that take place in a chemical reaction can be represented by a set of symbols called a **chemical equation**. An arrow is placed between the substances that change chemically, the *reactants* (ree AK tunts), and the substances formed by the chemical change, the *products*. The arrow is a symbol for the word *forms*, or yields. Reactants undergo chemical change to form products. The reaction in which elements A and B combine, forming compound AB, can be written as follows:

$$\underset{\text{reactants}}{A + B} \longrightarrow \underset{\text{product}}{AB}$$

A chemical reaction occurs when the elements copper (Cu) and sulfur (S) are heated together. The two elements combine, forming the compound copper sulfide (CuS). Suppose that this reaction is carried out as shown in Figure 7-7. The photo at the left shows a crucible, a small container used to heat chemicals. The crucible, which contains copper and sulfur, is placed on a device used to measure mass. The mass of the crucible and its contents is recorded. To start the reaction, the mixture of copper and sulfur is then heated, as shown in the middle photo. After the reaction has stopped, the crucible and its contents are allowed to cool. Their mass is measured again, as shown at the right. The mass of the crucible and its contents is the same as it was before the reaction.

law of conservation of mass

Experiments such as this had led to the discovery of a scientific law before Dalton published his atomic model of matter. The law, called the **law of conservation of mass**,

Background

Before he developed his atomic model, Dalton knew that mass is conserved. Indeed, it was this knowledge that led him to conclude that atoms cannot be changed and are indivisible. They can be rearranged to form various molecules, but not created or destroyed, because matter, which is made up of atoms, cannot be created or destroyed in ordinary reactions. A properly balanced chemical equation must also show conservation of mass. There must be as many atoms of each element on

states that mass is neither created nor destroyed during a chemical reaction. Dalton's model explains why this law holds. He assumed that atoms are neither created nor destroyed in a chemical reaction, but only rearranged.

A chemical equation, then, must reflect the law of conservation of mass. It must show that there is no change in the number of atoms involved. A chemical equation in which the reactants and products contain the same number of each kind of atom is called a **balanced equation**. The following examples illustrate a step-by-step procedure for writing and balancing equations.

balanced equation

EXAMPLE 1

Write a balanced equation for the reaction between copper and sulfur that produces copper sulfide, CuS.

Step 1 Write the correct formulas for the reactants and product.

$$Cu + S_8 \longrightarrow CuS$$

Note that the equation is *not* balanced. The number of sulfur atoms on each side are unequal.

Figure 7-7 *When a mixture of copper and sulfur is heated, copper sulfide is formed. As this sequence shows, the mass remains the same throughout the process. This illustrates the law of conservation of mass.*

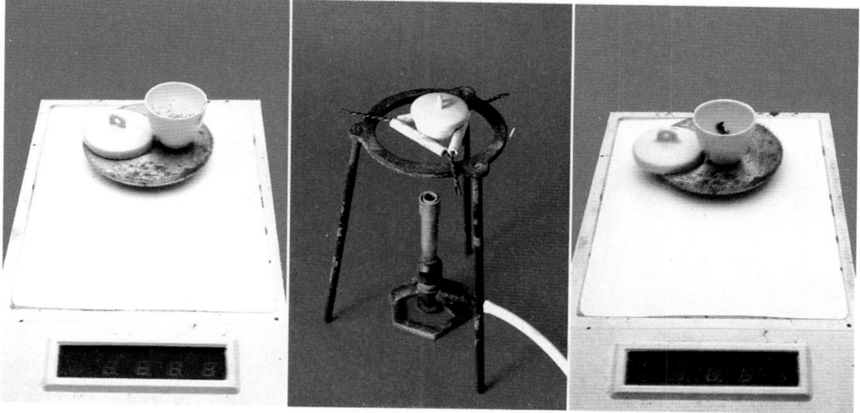

Demonstration

You can use a test tube instead of a crucible to show the reaction between copper and sulfur illustrated in Figure 7-7 on student text page 177. Thoroughly mix about 10 g of powdered copper and 5 g of flowers of sulfur, and pour the mixture into a test tube. (If you cover the opening of the test tube and measure its mass before you heat it, you can demonstrate that mass does not change during the reaction.) Heat it with a Bunsen burner in a darkened room. Once the mixture begins to glow, remove the flame so students can see that the reaction releases light and heat as it proceeds. (Afterward, measure its mass to show that it has not changed.)

Cross Reference

Chemical reactions are discussed in detail in Chapter 9.

the left side of the arrow as on the right side.

The molecular formulas in a reaction are written first. They cannot be changed, because experiment has shown that they are correct. Also, a deficiency of one atom of oxygen in a product in a correctly written set of formulas cannot be balanced by changing H_2O to H_2O_2, because that would mean that hydrogen peroxide and not water was formed. Nor can diatomic molecules, such as O_2, be changed to O, because there is evidence that oxygen normally occurs as diatomic molecules.

7 ATOMS AND MOLECULES

7-3 Chemical Equations

The Juggling Act

For each reaction below, underline the reactant(s) and circle the product(s). Then check the equation to see if it is balanced. If the equation is balanced, write *OK* on the numbered space to the left of the equation. If the equation is not balanced, write the symbol of the element that is not balanced.

___ 1. $CO + O_2 \longrightarrow CO_2$

___ 2. $H_2O + SO_3 \longrightarrow H_2SO_4$

___ 3. $K + F_2 \longrightarrow KF$

___ 4. $C_2H_2 + O_2 \longrightarrow 2\,CO_2 + H_2O$

___ 5. $Fe_2O_3 + 3H_2 \longrightarrow Fe + 3\,H_2O$

___ 6. $6\,Li + N_2 \longrightarrow 2\,Li_3N$

___ 7. $CH_4 + 2\,Br_2 \longrightarrow CHBr_3 + HBr$

___ 8. $4\,NaOH + 2\,F_2 \longrightarrow 2\,O_2 + 4\,NaF + 2\,H_2O$

___ 9. $Al + 2\,HCl \longrightarrow AlCl_3 + H_2$

___ 10. $K_2O + CaCl_2 \longrightarrow CaO + 2\,KCl$

Find the Coefficient

In the following equations the missing coefficient for each formula is indicated with the letter *a*, *b*, *c* or *d*. Balance the equations by finding the correct coefficient for each formula. Then write the coefficient on the space next to the appropriate letter.

a ___ *b* ___ *c* ___ 1. $a\text{Na} + b\text{Cl}_2 \longrightarrow c\text{NaCl}$

a ___ *b* ___ *c* ___ 2. $a\text{Fe} + b\text{I}_2 \longrightarrow c\text{FeI}_3$

a ___ *b* ___ *c* ___ 3. $a\text{H}_2\text{CO}_3 \longrightarrow b\text{H}_2\text{O} + c\text{CO}_2$

a ___ *b* ___ *c* ___ *d* ___ 4. $a\text{Li} + b\text{BaF}_2 \longrightarrow c\text{LiF} + d\text{Ba}$

a ___ *b* ___ *c* ___ 5. $a\text{O}_2 + b\text{Cl}_2 \longrightarrow c\text{Cl}_2\text{O}$

a ___ *b* ___ *c* ___ 6. $a\text{K} + b\text{O}_2 \longrightarrow c\text{K}_2\text{O}_2$

a ___ *b* ___ *c* ___ 7. $a\text{Na} + b\text{O}_2 \longrightarrow c\text{Na}_2\text{O}$

a ___ *b* ___ *c* ___ *d* ___ 8. $a\text{HgCl}_2 + b\text{H}_2\text{S} \longrightarrow c\text{HgS} + d\text{HCl}$

a ___ *b* ___ *c* ___ 9. $a\text{BaO}_2 \longrightarrow b\text{BaO} + c\text{O}_2$

a ___ *b* ___ *c* ___ *d* ___ 10. $a\text{SO}_2 + b\text{C} \longrightarrow c\text{CS}_2 + d\text{CO}$

Checkpoint Answers

1. A chemical change takes place.
2. The reactants are CH_4 and O_2; the products are CO_2 and H_2O.
3. $CH_4 + 2O_2 \longrightarrow CO_2 + 2H_2O$
4. Mass is neither created nor destroyed during a chemical reaction.

Step 2 Since eight atoms of sulfur react, there must be eight atoms of sulfur in the product. We cannot change the chemical formulas, but we can add coefficients.

$$Cu + S_8 \longrightarrow 8CuS$$

Step 3 Since the product side now contains a total of eight copper atoms, eight copper atoms must have reacted.

$$8Cu + S_8 \longrightarrow 8CuS$$

The equation is now balanced. Note that a coefficient of 1 is understood in the case of S_8.

EXAMPLE 2

When heated, a certain oxide of mercury (HgO) decomposes, forming mercury and oxygen. Write the balanced equation for the reaction.

Step 1 Write the correct formulas for the reactant and products.

$$HgO \longrightarrow Hg + O_2$$

Step 2 Since two atoms of oxygen are produced, two atoms of oxygen must be present in the reactant.

$$2HgO \longrightarrow Hg + O_2$$

Step 3 Since two atoms of mercury react, two must be produced.

$$2HgO \longrightarrow 2Hg + O_2$$

The equation is now balanced.

Checkpoint

1. What takes place during a chemical reaction?
2. Identify the reactants and the products in the following chemical equation for the burning of methane, CH_4, found in "natural gas."

$$CH_4 + O_2 \longrightarrow CO_2 + H_2O$$

3. Balance the equation in Question 2.
4. State the law of conservation of mass.

7-4

The Atomic Model Is Modified

Goals

1. To describe Thomson's model of the atom.
2. To describe Rutherford's model of the atom.
3. To define electrons, neutrons, and protons.

Scientific models are useful—for example, in predicting and explaining scientific laws. But all models have some limitations. As scientists do more experiments, models must often be altered to account for new observations. Dalton's billiard-ball model of an atom, for example, helped explain why elements combine in definite proportions. One of its limitations was that it did not explain *how* atoms combine as they do. That is, it did not explain what caused the "billiard-ball atoms" to unite chemically.

In 1897, almost 100 years after Dalton proposed his atomic model, J. J. Thomson, an English physicist, suggested a new model of the atom. A new model was needed to explain the results of experiments that Thomson had done. These experiments had shown interesting facts about what is called the *electric charge* of matter.

Scientists had known for some time that there are two kinds of electric charge—positive charge and negative charge. Two objects with the same charge repel each other. Two objects with opposite charges attract each other. In a series of experiments, Thomson observed that when atoms are strongly heated, they give off particles that are even smaller than atoms. This was a surprising fact, considering that atoms had been thought to be indivisible. The particles were attracted to objects having a positive electric charge. Since oppositely charged objects attract, Thomson reasoned that the particles being given off must be negatively charged. He called the tiny negatively charged particles **electrons**.

electrons
(ih LEK trahnz)

Ordinary matter is *neutral*. That is, it has no electric charge. If matter contains electrons (negatively charged particles) and is neutral, there must be an equal amount of positive charge to *balance* the negative charge of the electrons. Thomson modified the Dalton atom to include positive and negative charge. He proposed that the electrons

Teaching Suggestions 7-4

Emphasize that both Thomson and Rutherford changed the model of the atom, as proposed by earlier workers. Make sure students understand that the modification of models in science is very common. The new models, based on new knowledge, may be more complete, but are still subject to change on the basis of additional new evidence. Often the new evidence, as in the case of Rutherford's experiment, is totally unexpected.

The demonstration, a simulation of Rutherford's experiment, is highly recommended and a cornerstone of modern science. (See Demonstration.) Also, end-of-chapter projects 1 and 4 apply to this section.

Background

J. J. Thomson found that if he removed the air from a tube that had electrodes at both ends and placed a strong (high-voltage) electrical potential across the electrodes, rays (called cathode rays, because they came from the negative electrode, or cathode) would appear in the tube. The rays, like light, could cast shadows, but, unlike light, they were deflected by a magnetic field in the same way that a current of negative charge is deflected in an electric circuit.

This led Thomson to the idea that the particles were negative. Since they came from the cathode, they must come from atoms in the cathode. Thus, atoms must contain negative charges. Because pure substances, which consist of atoms, normally have no charge, atoms must contain equal amounts of positive and negative charge (electrons). From his observations of the way cathode rays were deflected in a magnetic field, Thomson found that the mass of the electrons must be very small relative to the mass of even the lightest atom (hydrogen). Thomson concluded that most of the atom's mass must reside in the positive part of the atom, which he imagined to be a puddinglike material that filled the entire atom. The electrons were like very tiny raisins spread throughout the positive pudding.

In his experiments, Ernest Rutherford used alpha particles, which are positively charged helium ions (helium atoms that have lost electrons), and which were released by radioactive plutonium. Rutherford knew that alpha particles are positive because he had "seen" these particles, which leave tracks in a detection device called a cloud chamber, bent in a direction opposite to that of cathode rays in a magnetic field. The curvature of their paths indicated that each alpha particle has a mass equal to that of a helium atom.

Thomson's model led Rutherford to predict that most of the alpha particles would pass straight through a thin sheet of gold foil, being slowed down only slightly by the puddinglike atoms. The deflection of a few particles through large angles could be explained only by assuming that there were tiny centers of positive charge within the atoms of gold. Since Thomson had already shown that electrons have very little mass, the positive charge must carry the bulk of the atom's mass and be concentrated in a very small portion of the atom. This region Rutherford called the nucleus.

Further experiments indicated that the charge on the nucleus varies from one element to another. Atoms with greater atomic masses seem to have nuclei with more positive charge. Other experiments indicated that atomic mass and charge are not related in a 1 : 1 fashion, but that additional mass relative to charge is present. This led to the notion of neutrons. The total mass of an atom is the sum of the masses of its protons and neutrons, which have virtually the same mass. The mass of the electrons can be ignored because it is so small.

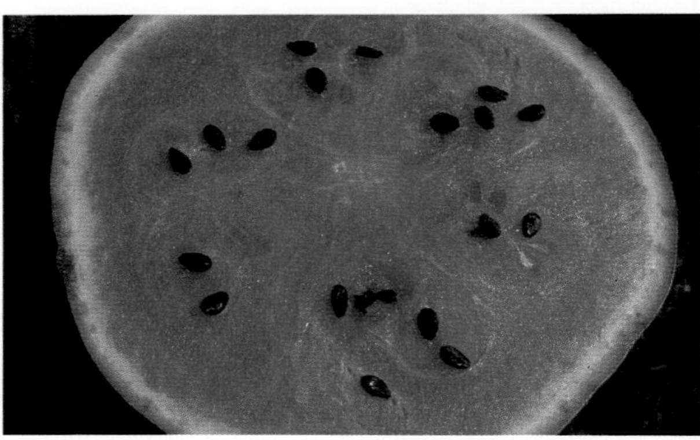

Figure 7-8 *The inner structure of a watermelon, with its seeds embedded throughout its red flesh, is similar to that of the atom proposed by J. J. Thomson. The "seeds" in Thomson's model are the negatively charged electrons, and the "flesh" is the positive part of the atom.*

were embedded within the atoms. The positive charge was spread uniformly throughout the atoms. He thought this explained why electrons are readily removed from atoms, but the positive portion is not. You might picture Thomson's atom as being somewhat like a watermelon in structure. (See Figure 7-8.) The seeds stand for the electrons. The red fleshy part of the melon stands for the positive part of the atom.

In 1911, Ernest Rutherford, a physicist from New Zealand who worked with Thomson, tested Thomson's model. He tested it by aiming a beam of fast-moving, positively charged particles at a very thin sheet of gold foil. (Even a thin sheet of gold foil is hundreds of atoms thick, and the atoms are packed closely together). Rutherford expected that most of the positive particles would pass through the gold foil with a very slight change in direction. This was because the positive charge in the atoms was thought to be thinly spread out, as Thomson had suggested. It would therefore have little effect on the bulletlike particles being fired through it.

The results of the experiment were very different from what Rutherford had expected. In the first place, most of the particles passed through the foil with no change of direction, just as though they were not passing through anything at all. However, a few of the particles swerved greatly from their straight-line path. Some even came straight back toward the source of the beam. Rutherford said that this was as surprising as if he had fired a cannon ball at a piece of tissue paper and it had come back and hit him!

To explain these strange results, Rutherford suggested yet another model of the atom. In Rutherford's model, an atom is mostly empty space. At the atom's center is a tiny, positively charged **nucleus**. (The plural of nucleus is *nuclei.*) All the atom's positive charge and almost all its mass are located in the nucleus, which is much more massive than the very light electrons found outside the nucleus. Rutherford knew that the atom's nucleus must be very small because most of the positively charged particles fired at the atoms passed right through the gold foil. To get an idea of the size of the nucleus compared to that of the entire atom, picture a tiny bug in the center of a large athletic stadium. The bug represents the nucleus. The stadium represents the entire atom. (See Figure 7-9.)

nucleus
(NOO klee us)

Rutherford knew that the nucleus was positively charged because the few positive particles that swerved or came straight back must have been repelled, rather than attracted, by the nucleus. (Remember, like charges repel

Figure 7-9 *Compared to the atom, the nucleus is about as small as an insect would be when compared to a sports stadium.*

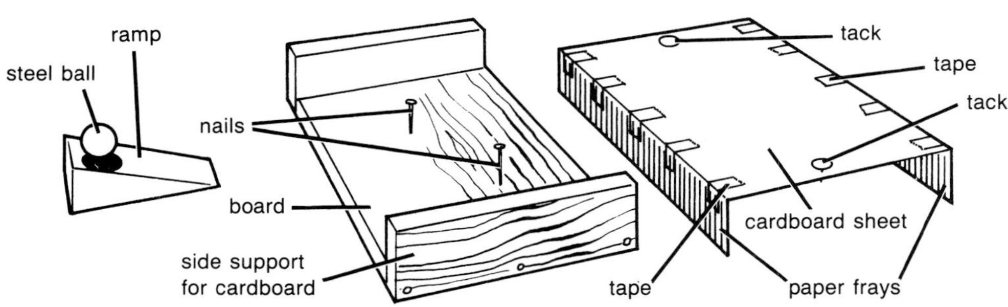

Figure TE 7-2 Setup for a simulation of the Rutherford experiment.

Demonstration

Show students a simulation of the Rutherford experiment. Place the apparatus shown in Figure TE 7-2 so that all students can observe. The surface with the nails should be covered with the sheet of cardboard that has a frayed paper curtain on each open side. The paper should be cut so that it will hide the nails after it has been tacked to the side supports.

Let a steel ball roll down the ramp and through the covered space a number of times. Start at one edge of the board and gradually move the ramp after each run so you cover the entire width. On a few runs students will hear a clicking sound and see that the path of the ball is significantly deflected.

Challenge students to create a model of the unseen space beneath the cardboard top. Make sure they understand that this setup is similar to Rutherford's experiment. The unseen area is mostly empty space, through which the ball passes undisturbed. There are, however, sites that give rise to significant changes in the ball's path. Ask for comments on how this experiment differs from Rutherford's.

Facts At Your Fingertips

- A sheet of gold only 1/10,000 cm thick is still about 10,000 atoms thick.
- An electron's mass is about 1/1840 the mass of a proton or neutron.

Cross Reference

Electric charges and the electrical nature of matter are discussed in Chapter 19, sections 19-1 and 19-2.

7 ATOMS AND MOLECULES

7-4 The Atomic Model Is Modified

Unlocking the Atom

Read the questions below. In the word search grid, find and circle the words that answer the questions. Then write each word beside its number.

```
N D O P R O N R
X E A F I L E M
O Y U C A L K N
P N O T O R P U
N E U T R O N C
N H O B V A P L
N O R T C E L E
E U S W E L S U
N T G O W U J S
```

__________ 1. Ordinarily, what is the charge on matter?

__________ 2. What particle in an atom has a charge of −1 and a very small mass compared to the mass of the proton?

__________ 3. What particle in an atom is uncharged and has a mass that is close to the mass of a proton?

__________ 4. What particle in an atom has a charge of +1?

__________ 5. In what part of the atom are the protons located?

Historic Models

The pictures below represent different models of the atom. Write the name of each model on the space provided. Then label the parts requested.

1. The __________ model

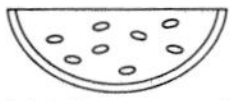

Label the electrons and the positive area.

2. The __________ model

How are atoms like tiny billiard balls?

____

____

3. The __________ model

Label protons and electrons.

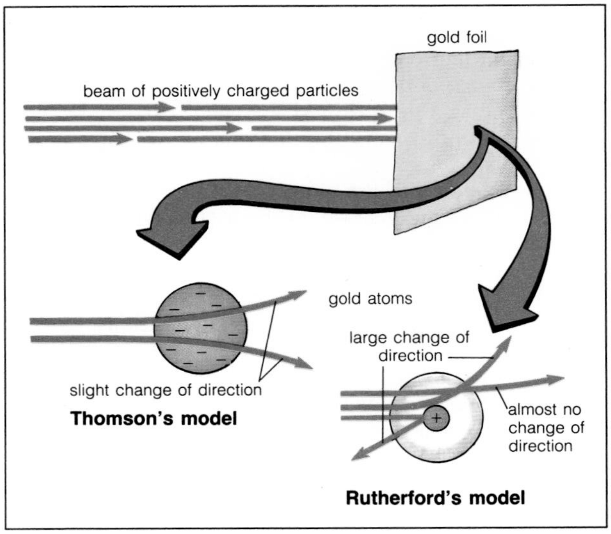

Figure 7-10 *The atoms proposed by Thomson and by Rutherford would behave differently when beams of positively charged particles are passed through them. Rutherford's idea that the atom has a nucleus turned out to be correct.*

each other.) Figure 7-10 shows how Rutherford's model of the atom accounts for the observed results.

Experiments with metals other than gold showed that the positive charge of the nucleus varies from element to element. Nuclei of heavier atoms carry more charge than do nuclei of lighter atoms. This led to a model of the atom in which the nucleus includes small positively charged particles called **protons**. The more protons an atom contains, the greater the charge on its nucleus, and, in general, the heavier the atom.

protons

However, careful measurements showed that the number of protons in a nucleus alone did not determine the mass of the nucleus. Rutherford's model therefore had to be modified. In addition to protons, the nucleus also contains particles called **neutrons**. Neutrons have roughly the same mass as protons but they carry no charge. They are neutral. The number of protons plus the number of neutrons is the **atomic mass**.

neutrons

atomic mass

Table 7-2: The Charges and Masses of Protons, Neutrons, and Electrons

Particle	*Charge* (proton or electron = 1 unit)	*Approximate atomic mass (amu)* (proton or neutron = 1 unit)
proton	+1	1
neutron	0	1
electron	−1	1/1840

Protons and neutrons each have a mass that is more than 1840 times that of an electron. A proton is assigned 1 unit of atomic mass and 1 unit of positive charge. A neutron is assigned 1 *atomic mass unit* (or 1 amu) and 0 units of charge. Thus, a helium nucleus with two protons and two neutrons has a mass of 4 and a charge of +2. A gold nucleus contains 79 protons and 118 neutrons, giving it a mass of 197 units and a charge of +79. The number of protons in a nucleus is called the **atomic number**. The atomic number of helium is 2. That of gold is 79. Forces act within nuclei to stabilize their structure. These forces have been investigated by Nobel laureate Maria Goeppert Mayer.

atomic number

We may sum up by saying that atoms were once thought to be the tiniest particles of matter. Later, evidence from experiments led scientists to believe that atoms were made up of even smaller particles. Electrons were eventually pictured by Rutherford as tiny, negatively charged particles that existed in the space surrounding the tiny, heavy, positively charged nucleus.

Checkpoint

1. What are the two kinds of electric charge and how do the kinds of charge affect one another?
2. Describe Thomson's model of the atom and tell why Thomson's model is somewhat like a watermelon in structure.
3. Describe Rutherford's model of the atom. What evidence led to Rutherford's model?
4. Compare the masses and charges of protons, neutrons, and electrons.

Checkpoint Answers

1. Positive and negative. Like charges repel each other. Unlike charges attract each other.
2. Basically, the Thomson model views atoms as tiny electrons embedded in a positive sphere, much like seeds in a watermelon.
3. Rutherford's model of the atom consists of a tiny, massive, positively charged nucleus surrounded by electrons. Rutherford was led to this model by bombarding thin sheets of gold foil with positively charged particles. Most of the particles passed through the gold foil with little change in direction, but a few swerved greatly.
4. See Table 7-2 on student text page 183.

Teaching Suggestions 7-5

Introduce the concepts of this section by showing students the light emitted by flame tests or electrically excited gases. (See Demonstrations.) For example, if you do demonstration 1, students will see that different salts provide different colors. Lithium salts produce a crimson color; strontium salts, red; calcium, orange; potassium, lavender; and copper, green. It should be pointed out that the chlorine, nitrogen, and oxygen in chlorides and nitrates seem to have no effect; the colors arise from the metals. Both sodium chloride and sodium bicarbonate give off a yellow light in the flame test and all lithium salts give off a crimson-colored light.

End-of-chapter projects 1, 4, and 6 apply to this section.

7 ATOMS AND MOLECULES
Laboratory
Flame Tests

Purpose To group compounds according to flame test colors.
Materials Bunsen burner; stainless steel spatula; distilled water; 12 small beakers; samples of sodium nitrate, lithium nitrate, potassium nitrate, barium nitrate, copper nitrate, calcium nitrate, sodium chloride, lithium chloride, potassium chloride, barium chloride, calcium chloride, copper chloride

Procedure

1. Make a chart like the one shown below:

Color of Compound in Flame Test	*Compounds Showing This Color*

Background

Scientists were able to identify elements by looking at their line spectra long before they could explain the cause of the spectra. Bohr's explanation argues that there are only certain allowed orbits for electrons. The closer the electron is to the nucleus, the less energy it has. This arises from the fact that the negatively charged electron is attracted to the positive nucleus. Moving an electron farther from the nucleus increases its electrical potential energy, just as lifting a body from Earth increases the gravitational potential energy of the body. Therefore, to move an electron farther from the nucleus (from one allowed orbit to another), the electron must receive a fixed amount of energy. When the electron falls back to the lower allowed orbit, it gives up this energy in the form of light. With many atoms present, all of the various changes in electron orbits are represented, and we see all the spectral lines characteristic of the element. If not enough energy is fed into the element to excite electrons to their most energetic levels, then we will see only part of the spectrum; the most energetic light will be missing.

The energy of light may be thought of as photons, or tiny bundles of light energy. In fact, the energy of a photon is given by the following equation:

$$E = hf \quad \text{or} \quad E = h/\lambda$$

E is the energy, f is the frequency, which is the reciprocal of λ (the Greek letter lambda), the symbol for the

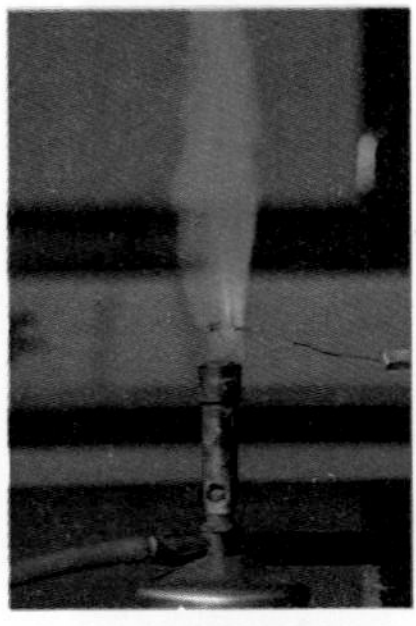

Figure 7-11 *From top to bottom: Lithium compounds give off red light; copper compounds, green light; and sodium compounds, yellow light in a Bunsen burner flame.*

7-5

Line Spectra

Goals

1. To explain how elements can be identified by the light they give off.
2. To describe the Bohr model of the atom.

When compounds of certain metals are heated in the flame of a Bunsen burner, they give off light of a characteristic color. The resulting flame color helps chemists identify the atoms present in the compound. Chemists know, for example, that lithium compounds give off red light in the flame. Copper compounds give a green light, and sodium compounds give a yellow light. (See Figure 7-11.)

By looking at a flame through a *spectroscope*, an instrument that spreads light out into a spectrum, or band of colors, a chemist can tell which element is present. The light from a hot, glowing solid, such as tungsten, can produce a *continuous spectrum*, (plural, *spectra*), in which colors blend into each other. (See Figure 7-12a.) The light given off by a glowing gaseous element, such as hydrogen, does not give off a continuous spectrum. Its spectrum is a series of distinct, colored lines. (See Figure 7-12b.) No two elements have the same *line spectrum*. An element's line spectrum can therefore be used to identify the element. You might think of an element's line spectrum as its "fingerprint."

Niels Bohr, a Danish physicist working in Rutherford's laboratory in 1913, wanted to account for the line spectrum produced by hydrogen. Hydrogen is the simplest and lightest element, having only one proton in the nucleus and one electron. Its line spectrum should therefore be the easiest to understand. Bohr reasoned that the closer the negative electron is to the positive nucleus, the less energy it has. It must gain energy to move away from the positively charged nucleus that attracts it. The farther away it moves, the higher its energy must become. (If it gains enough energy, an electron can completely "escape" from an atom, as Thomson's experiments showed.)

In order to explain the line spectrum, Bohr assumed that the electron in a hydrogen atom could have only certain "allowed" energies. Bohr thought that each energy corre-

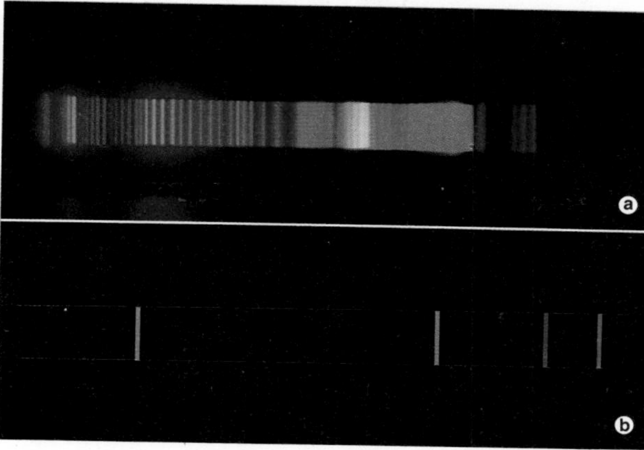

Figure 7-12 *Elements such as tungsten produce a continuous, rainbowlike spectrum, as shown in (a). Elements such as hydrogen produce a spectrum made up of separate, colored lines, as shown in (b).*

sponded to a certain average distance from the nucleus. You might think of the electrons as airplanes in assigned holding patterns above an airport. A plane may circle at the lowest "allowed" level, or altitude, or it may circle at one of the higher "allowed" levels. A plane gains energy in order to move from a lower level to a higher level. It loses energy in moving from a higher to a lower level.

energy levels

The energies that an electron can have in an atom are called **energy levels**. Energy levels were pictured as orbits, or (in three dimensions) as shells, by Bohr. (See Figure 7-13.) Normally an electron occupies the lowest available energy level. When it gains energy—by being heated, for example—it moves to a higher energy level.

A ball at the top of a hill tends to roll back down if there is nothing to prevent it from doing so. Similarly, an electron at a higher-than-normal energy level tends to move back to a lower level. When an electron moves from a higher to a lower energy level, it gives off energy that it had originally absorbed. The energy is given off in the form of light. The color of the light depends on the amount of energy lost by the electron. For example, blue light has more energy than does red light. Electrons giving off blue light are therefore losing more energy than those giving off red light.

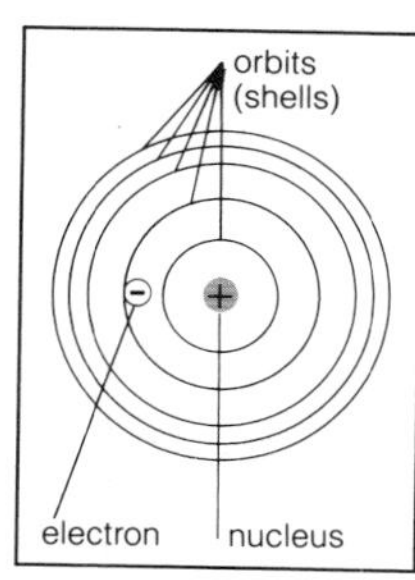

Figure 7-13 *In Bohr's model of the atom, an electron could only be found in certain definite energy levels, which Bohr pictured as orbits or shells.*

wavelength, and h is a constant (Planck's constant), which is equal to $6.624 \times 10^{-34} J \cdot s$ (joule seconds). Because the energy is proportional to the frequency (and inversely proportional to the wavelength), blue light, which has a higher frequency and a shorter wavelength than red light, has more-energetic photons than red light.

Some spectral lines are in the infrared and ultraviolet regions of the spectrum and can be seen only by photographic means.

Demonstrations

1. Darken your classroom. Place small amounts of several different salts in clean vials or medicine cups. Possible salts include sodium chloride (NaCl), potassium nitrate (KNO_3), copper nitrate ($Cu(NO_3)_2$), strontium nitrate ($Sr(NO_3)_2$), calcium chloride ($CaCl_2$), lithium chloride (LiCl), lithium nitrate ($LiNO_3$), and sodium bicarbonate ($NaHCO_3$). Beside each salt sample place a piece of clean Nichrome wire, the end of which has been inserted into a rectangular-shaped piece of balsa wood. Heat each piece of Nichrome wire until it produces no color and dip it into a salt. When the salted wire is placed in a flame, a characteristic color will be produced. For example, sodium salts give yellow light. This test is so sensitive that simply touching the wire will transfer sufficient salt from the human body, which is rich in sodium chloride, to produce yellow light when placed in the flame. For this reason, students should not touch any of the Nichrome wires.
2. Show students the glowing vertical line filament of a tube-shaped showcase bulb through a diffraction grating. They will see that the bright light produces the entire visible spectrum, which should appear to either side of the grating when the light is viewed through it. If students have difficulty seeing the spectrum, it is probably because the lines of the grating are not parallel to the filament. By rotating the grating through 90 degrees, the spectrum should become clearly visible.

   Repeat the demonstration using a fluorescent bulb. Superimposed on the spectrum will be the bright lines of the mercury spectrum because mercury vapor is present in fluorescent tubes.
3. If your school has spectral tubes containing other gases and vapors, as well as power sources to excite the gases, and simple, inexpensive spectroscopes, students will be able to see the line spectra produced by various glowing tubes in a darkened room.

Teaching Tips

- The concept of line spectra is potentially difficult for students. You may need to proceed slowly. However, the variety of viewable spectra should provide good motivation.
- Many physics and chemistry texts contain color photos of line spectra.

Cross References

The use of line spectra in astronomy is discussed in Chapter 15, section 15-6. The nature of light is the subject of section 20-4, Chapter 20.

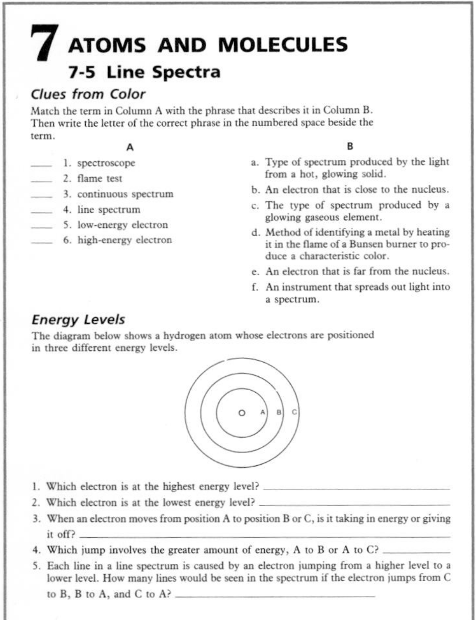

7 ATOMS AND MOLECULES

7-5 Line Spectra

Clues from Color

Match the term in Column A with the phrase that describes it in Column B. Then write the letter of the correct phrase in the numbered space beside the term.

A	B
____ 1. spectroscope	a. Type of spectrum produced by the light from a hot, glowing solid.
____ 2. flame test	b. An electron that is close to the nucleus.
____ 3. continuous spectrum	c. The type of spectrum produced by a glowing gaseous element.
____ 4. line spectrum	d. Method of identifying a metal by heating it in the flame of a Bunsen burner to produce a characteristic color.
____ 5. low-energy electron	e. An electron that is far from the nucleus.
____ 6. high-energy electron	f. An instrument that spreads out light into a spectrum.

Energy Levels

The diagram below shows a hydrogen atom whose electrons are positioned in three different energy levels.

1. Which electron is at the highest energy level? ____________
2. Which electron is at the lowest energy level? ____________
3. When an electron moves from position A to position B or C, is it taking in energy or giving it off? ____________
4. Which jump involves the greater amount of energy, A to B or A to C? ____________
5. Each line in a line spectrum is caused by an electron jumping from a higher level to a lower level. How many lines would be seen in the spectrum if the electron jumps from C to B, B to A, and C to A? ____________

Using Try This

For this activity to work, the air must be dry enough for static charge to build up on the fluorescent tube. Attraction of the positively charged glass for electrons within the tube will cause the electrons to lose energy (as they strike the glass) in the form of ultraviolet light. This, in turn, causes the fluorescent coating on the glass to release light, just as it does under normal operation when bombarded by electrons.

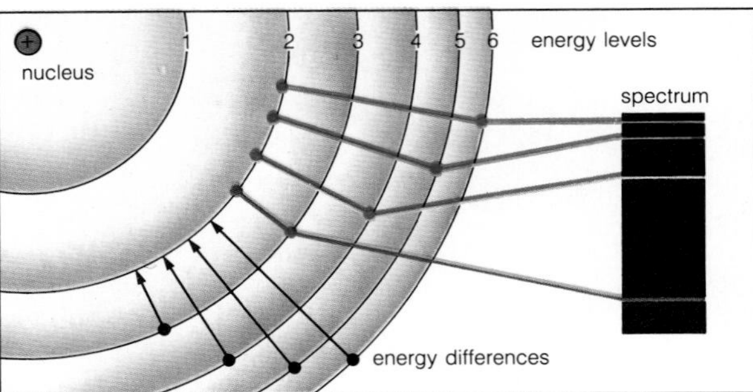

Figure 7-14 *The colored lines of the hydrogen spectrum are produced when electrons move down from various higher energy levels.*

Each line in the hydrogen spectrum corresponds to the energy difference between two energy levels. Figure 7-14 shows the energy differences for the visible lines. Note that these result from the electron's movement from higher levels to lower levels.

Bohr's explanation worked well in accounting for the hydrogen spectrum. It formed the starting point for the understanding of other atoms. Eventually, scientists were able to describe the arrangements of electrons in more complicated atoms. They found that each energy level could be occupied by only a certain number of electrons. The arrangement of electrons in the atom is what determines the chemical properties of a given element.

Checkpoint

1. How can the light that elements give off be used to identify the elements?
2. Describe the Bohr model of the atom.
3. How did Bohr explain the hydrogen line spectrum?

TRY THIS

Light a fluorescent lamp tube in the following way, without plugging it in. On a dry day, darken a room. Gently rub the tube up and down several times on your sleeve. (CAUTION: Be careful not to break the tube.) This rubbing removes electrons from the glass, and the resulting charge on the glass produces electricity inside the tube.

Checkpoint Answers

1. Each element emits light that is characteristic of that element. By looking at the known spectra of elements and comparing them with the spectral lines produced by an unknown sample, the elements in the unknown can be determined.
2. See student text, pages 184–186.
3. See student text, pages 184–186.

7-6

The Current Model of the Atom

ENRICHMENT

Goal

To compare the current model of the atom with the Bohr model.

The atomic model used by scientists today includes features of the Bohr model. Electrons in an atom are still thought to exist outside a tiny central nucleus. But they are not thought to be moving in definite "orbits." Werner Heisenberg, an atomic physicist, first concluded that it is impossible to know both the exact location and direction of movement of an electron in an atom, and that definite paths could not be assigned to electrons. His conclusion is often referred to as Heisenberg's uncertainty principle. Scientists can calculate only the *probability*, or likelihood, of finding an electron at a certain distance from the nucleus. Figure 7-15 shows that the electron in a hydrogen atom is most likely to be found fairly close to the nucleus.

Scientists can calculate the energies that electrons have in atoms. As Bohr assumed, the electrons can have only certain "allowed" energies. In moving from one energy level to another, an electron must gain or release an exact amount, or "packet," of energy. Each packet of energy is called a *quantum* (plural, *quanta*). The current atomic model is generally referred to as the *quantum model*. It is a mathematical model that gives no very clear "picture" of electron motion. The predictions that scientists have made based on the quantum model have been borne out by experiments, however. Therefore, that model is the accepted one, even though it is very difficult to visualize.

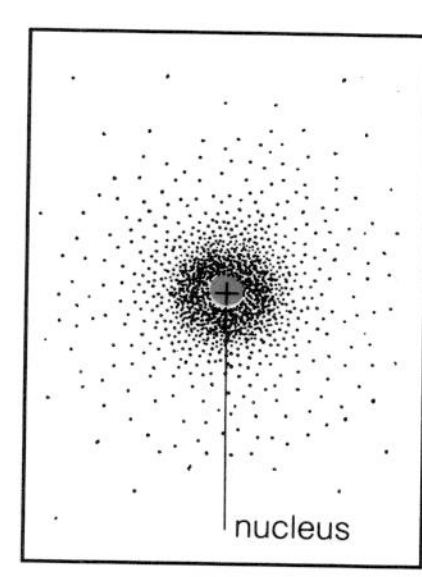

Figure 7-15 *According to the current model of the atom, the electron in a hydrogen atom is most likely to be found in the more dense regions of dots shown here.*

Checkpoint

1. In what ways is the quantum model of the atom like the Bohr model? In what ways is it different?
2. What information does the quantum model of the atom give concerning the location of an electron in an atom?
3. Why is the quantum model accepted by scientists?

Teaching Suggestions 7-6

This is difficult material. Students should, however, be able to grasp the basic ideas presented. You may wish to treat it as a reading assignment, or discuss it with the Checkpoint questions as a frame of reference, or even omit it entirely.

Background

The quantum model of the atom is the result of attempting to ext·nd the Bohr model of the hydrogen atom to other, more-complex atoms. The result, as described briefly in section 7-6, is a mathematical, statistical model that provides predictability but is very difficult to visualize. The "packets" of energy, or quanta, are the energy differences between allowed energy levels in the atom. These allowed levels differ from atom to atom, and therefore cause the atoms of each element to emit spectral lines unique to that element.

7 ATOMS AND MOLECULES

7-6 The Current Model of the Atom

An Electron Fantasy

Read the following piece of fiction. Then write answers to the questions.

Let's take a trip. You will not need a coat or swimsuit, but be sure to bring a camera with a flash! First you need to shrink from five or six feet to five or six angstroms. (An angstrom is 0.0000000001, or 10^{-10} meters.)

Now let's visit a hydrogen atom. The hydrogen atom is a sphere, about one angstrom in diameter. It is very dark, so you'll need to use the flash when you take pictures.

Now snap a picture. You have a record of where the electron was in the atom at the instant the picture was taken. Whoops, the camera is not working properly; the film will not advance. It doesn't matter. Just keep taking pictures on the same exposure. Quickly now, take as many pictures of the hydrogen atom as you can. Each time, you are making a record of where the electron was at the instant the picture was taken. Before leaving, you will have taken one hundred pictures, all on that one single part of exposed film!

Like most vacations, this one is over before you know it. You have now returned to normal size, and the hydrogen atom has moved away. When you have the film developed, this is what you see. Each dot shows where the electron was at the time of one of the snaps of the camera.

Notice that the electron in the picture occupies a spherical space, or orbital, around the nucleus. An orbital is the volume in which the electron is most likely to be found. Scientists do not know what path the electron travels, but they do know that it is usually somewhere within the orbital. Notice the closeness of the dots. The dots are closer near the nucleus. This tells you that the electron must be near the nucleus more often than it is far away from the nucleus. Farther away from the nucleus, the dots are farther apart. Finally, at a certain distance the dots stop. They stop at about the same distance in any direction around the nucleus. This end of dots defines the edge of the orbital.

1. What is the shape of the orbital of the electron in the hydrogen atom? __________
2. In the orbital of the hydrogen atom, where does the electron spend most of its time—far from the nucleus or near it? __________
3. How do we know where the orbital ends? __________

Checkpoint Answers

1. Both models agree that electrons exist outside a tiny central nucleus only at certain allowed energy levels, but the quantum model argues that the exact position and direction of movement of an electron cannot be determined. As a result, the location of an electron is known only statistically, that is, as a probability.
2. The quantum model cannot give an exact location for an electron in an atom. It can only show the most-likely positions for electrons at allowed energy levels.
3. The quantum model is accepted because it provides predictions that have been borne out by many experiments.

Using The Feature Article

If you can find and purchase some cold light devices, students will enjoy seeing them in operation. They are available under various trade names.

As the article describes, emission of cold light is an example of the release of light by electrons in excited energy states as they fall back to lower, more-stable energy levels. In the case of these chemicals, the electrons may stay in the elevated energy levels for relatively long periods of time. Thus, they can provide light for several hours or longer. If spectroscopes are available, students will enjoy examining the spectra produced by the cold light sources. For further enrichment, some students may wish to do project 5 at the end of the chapter.

Cold Light

Nearly all natural light comes from the sun. You can feel the heat it produces. Light from artificial sources is produced in several ways. Materials may be heated until they glow, or become *incandescent*, as in an ordinary electric light bulb. If you put your hand next to a lighted bulb, you can feel its warmth. Fluorescent lights also produce heat, but not as much. Light without heat is unusual.

Imagine that you are standing in a dark room, holding a slim plastic tube. When you bend the tube, an eerie green light suddenly fills the room. You might think the tube is some kind of electric light bulb. However, there are no wires leading to the tube. The tube also remains cool to your touch, even as it releases light.

Before it was bent, the tube contained a liquid and a thin-walled capsule filled with a second liquid. When the tube was bent, the capsule was broken, releasing the second liquid. The two liquids reacted chemically, forming a substance that gives off light as a result of the reaction.

Some of the atoms of which the light-giving substance is made contain electrons that have moved to higher-than-normal energy levels. The electrons absorbed energy during the chemical reaction and moved up to higher levels. After the reaction, the electrons give off the absorbed energy as they fall back to their normal energy levels. This loss of energy appears as light. The light from this particular substance is green, but other substances release different colors of light.

Cold light sources are used as emergency lights in boats and around explosive materials, where flames or electric sparks would be dangerous. Cold light sources are now expensive compared to conventional light sources. However, cold light sources may eventually be even less expensive and more useful than fluorescent lights.

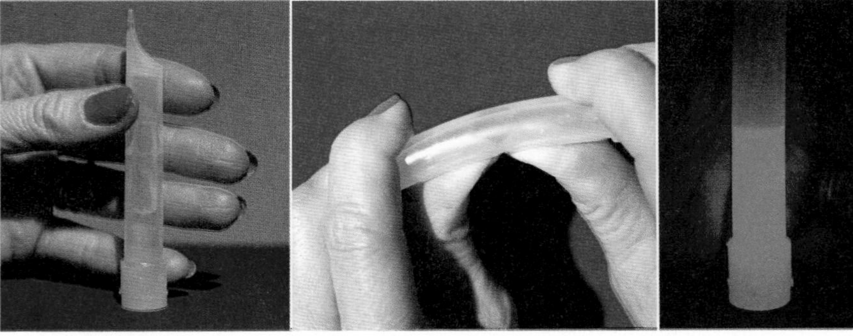

Understanding The Chapter

7

Main Ideas

1. A scientific model explains something unfamiliar or invisible in terms of something that is familiar or visible.
2. An atom is the smallest particle of an element. A molecule is a particle of matter made up of chemically joined atoms.
3. Chemical symbols stand for the names of the elements and also stand for one atom of an element.
4. A chemical formula gives the number and types of atoms that are combined in a molecule.
5. The changes that take place in a chemical reaction can be represented by a chemical equation.
6. During a chemical reaction, mass is neither created nor destroyed.
7. The nucleus of an atom contains positively charged protons and uncharged neutrons, and its mass accounts for most of the mass of the atom.
8. Each element has a unique line spectrum that can be used to identify it.
9. Electrons exist outside the nucleus in certain energy levels.
10. It is impossible to determine both the exact location and direction of movement of an electron in an atom.
11. Electrons in an atom can gain or release energy only in certain exact amounts called quanta.

Vocabulary Review

From the following list, choose the term that best completes each of the statements. Write your answers on a separate piece of paper.

atom	law of conservation of mass
atomic mass	molecule
atomic number	neutron
balanced equation	nucleus
chemical equation	products
chemical formula	proton
chemical reaction	reactants
chemical symbol	scientific model
electron	spectroscope
energy level	

1. The __?__ for carbon dioxide is CO_2.
2. A(n) __?__ is found in the nucleus of an atom, and has a charge of +1.
3. A(n) __?__ has very little mass and a charge of −1.
4. A(n) __?__ of water contains two atoms of hydrogen combined chemically with one atom of oxygen.
5. In chemical formulas, each kind of atom is represented by a(n) __?__ that is different from those of all other elements.
6. When hydrogen and oxygen react, forming water, hydrogen is one of the __?__.
7. The reactants in a chemical reaction change into the __?__.
8. A(n) __?__ has a mass equal to that of a proton, and carries no charge.

Reading Suggestions

For The Teacher

Feinberg, Gerald. WHAT IS THE WORLD MADE OF? Garden City, NY: Doubleday, 1977. A historical survey of atomic structure, up to and including discovery of the newest particles.

Haber-Schaim, Uri and others. INTRODUCTORY PHYSICAL SCIENCE. Englewood Cliffs, NJ: Prentice Hall, 1987. A textbook that develops the atomic model through experiments involving the basic properties of matter.

Scientific American. PARTICLES AND FIELDS. San Francisco: W. H. Freeman, 1980. Outstanding papers that explain the search for the ultimate building blocks of matter.

For The Student

Asimov, Isaac. HOW DID WE FIND OUT ABOUT ATOMS? New York: Walker, 1976. An easy-to-read historical account of atomic exploration.

Berger, Melvin. ATOMS, MOLECULES, AND QUARKS. New York: Putnam, 1986. Some basic chemistry with step-by-step instructions for simple experiments that can be done at home.

Chester, Michael. PARTICLES: AN INTRODUCTION TO PARTICLE PHYSICS. New York: Macmillan, 1978. A look at the atom and the gaps in our knowledge about it.

Fisher, Arthur. "First Movies of Atoms in Color." *Popular Science*, February, 1979, p. 75. A photographic essay about atoms.

Vocabulary Review Answers

1. chemical formula
2. proton
3. electron
4. molecule
5. chemical symbol
6. reactants
7. products
8. neutron
9. chemical equation
10. balanced equation
11. nucleus
12. energy level

Chapter Review Answers

Know The Facts

1. equation
2. O_2
3. molecules
4. 1
5. gold
6. nucleus
7. positive
8. model
9. 6
10. 3
11. 12
12. quanta

Understand The Concepts

13. They are identical.
14. The nucleus is very small and carries a positive charge.
15. (a) Atoms are neither created nor destroyed. Hence the mass of matter, which consists of atoms, cannot be changed in chemical reactions. (b) Because all atoms of the same element have the same mass and combine wholly with atoms of other elements in fixed ratios, the percentages of the masses of the various elements in compounds must be fixed.
16. The lines in the hydrogen spectrum show that energy of only certain discrete amounts is released by the atoms. This led Bohr to suggest that electrons occupy only certain allowed orbits or energy levels. The energy differences between these levels are just the energies of the light emitted by hydrogen atoms when electrons in excited atoms fall back to lower energy levels.
17. When heated, atoms release electrons. Atoms in thin gold foil repel positively charged particles. (Students may offer other evidence, such as conductivity of metals and solutions, and magnetic deflection of electrons and alpha particles.)
18. On each side of the equation there are 2 K atoms, 2 Cl atoms, and 6 O atoms.

9. The changes that take place in a chemical reaction can be represented by a(n) ?.
10. A chemical equation in which the reactants and products contain the same number of each type of atom is a(n) ?.
11. The ? of a gold atom has a greater charge than that of a hydrogen atom because it contains more protons.
12. Normally, an electron in an atom occupies the lowest available ?.

Chapter Review

Write your answers on a separate piece of paper.

Know The Facts

1. A chemical ? is used to represent a chemical reaction. (equation, formula, product)
2. ? is an example of a diatomic molecule. (CO_2, H_2O, O_2)
3. Atoms of oxygen and carbon combine chemically, forming ? of carbon dioxide. (nuclei, atoms, molecules)
4. ? atom(s) of sulfur is (are) represented in the formula H_2SO_4. (3, 2, 1)
5. Au is the symbol for ?. (gold, mercury, lead)
6. Most of an atom's mass is in its ?. (electrons, protons, nucleus)
7. In Thomson's model of the atom, electrons are embedded throughout a ? sphere that makes up the rest of the atom. (positive, negative, neutral)
8. A scientific ? can be used to explain and predict scientific laws and observations. (method, model, investigation)
9. The total number of atoms represented by "$2CO_2$" is ?. (3, 6, 8)
10. $2Fe$ + ? $Cl_2 \rightarrow 2FeCl_3$. (2, 3, 6)
11. The number of carbon atoms in a molecule of sugar ($C_{12}H_{22}O_{11}$) is ?. (11, 12, 22)
12. ? correspond to the energy differences between energy levels in atoms. (electrons, nuclei, quanta)

Understand The Concepts

13. According to Dalton's atomic theory, how do atoms of the same element compare with one another?
14. What did Rutherford's experiments show about the size and charge of the nucleus?
15. How can Dalton's atomic model be used to explain (a) the law of conservation of mass, (b) the law of definite proportions?
16. What evidence led Bohr to propose his atomic model? How did his model explain this evidence?
17. What evidence is there for the presence of electric charges in matter?
18. Using the balanced equation $2KClO_3 \rightarrow 2KCl + 3O_2$, show that mass is conserved in this reaction.
19. Where in an atom are located protons, neutrons, electrons?
20. What is the explanation for the different colors in an element's line spectrum?
21. A neutral gold atom has 79 protons and 118 neutrons in its nucleus. How many electrons does the atom have? How do you know?

19. Protons and neutrons in nucleus; electrons outside of nucleus.
20. Each color represents the different energy levels from which electrons lose energy.
21. The atom has 79 electrons. In a neutral atom, the number of protons equals the number of electrons.
22. The quantum model of the atom gives only the probability of an electron's location within the atom because it is impossible to know both the exact location and direction of movement of the electron.

22. Why does the quantum model of the atom give only the probability of an electron's location in the atom?

Challenge Your Understanding

23. Balance the following equations:
    (a) $H_2 + O_2 \rightarrow H_2O$
    (b) $H_2 + O_2 \rightarrow H_2O_2$
    (c) $C + O_2 \rightarrow CO$
    (d) $N_2 + H_2 \rightarrow NH_3$
    (e) $H_2 + Cl_2 \rightarrow HCl$
    (f) $Na + Cl_2 \rightarrow NaCl$
24. A compound is found to consist of 1 part by mass hydrogen (atomic mass = 1 unit) and 16 parts by mass sulfur (atomic mass = 32 units). What is the simplest formula for the compound?
25. How does the quantum model differ from the Bohr model of the atom? What features do the two models have in common?

Projects

1. Write a brief biography of one of the following scientists, emphasizing his or her contribution to science: John Dalton, Ernest Rutherford, Niels Bohr, Maria Goeppert Mayer.
2. At a library, find out about the history of the symbols used to represent elements.
3. Use styrofoam balls and spray paint to prepare some atomic and molecular models. Use your models to represent the chemical reactants and products of certain chemical reactions. Write the reactions on a card to be displayed beneath the models.
4. Build a series of models to represent the various atomic models that were developed before the 1930's.
5. Obtain three identical tubes that will produce cold light. Bend all three tubes in a warm, dark room to produce light. Then leave one tube in the room, place one tube in a refrigerator, and place the third tube in a freezer. What differences do you observe among the tubes during the next few hours?
6. Obtain a diffraction grating, and use it to view light from various sources. The grating spreads light out into a spectrum. View (a) the straight vertical filament of a clear (unfrosted) light bulb (continuous spectrum), (b) an upright fluorescent bulb (mercury line spectrum superimposed on a continuous spectrum), (c) other light sources, such as neon signs, sodium lamps, and mercury vapor lamps (line spectra). You might also try looking at a white light source through a grating, with a colored filter placed between the grating and the source.
7. Research and report on the contributions to chemistry of any one of the following scientists: Florence Seibert; Mary L. Caldwell; Mary L. Petermann; Helen Ranney.

7 ATOMS AND MOLECULES
Projects

1. Many elements, such as those listed in the charts below, are beneficial to people. Use a chemistry textbook or references in the library to find out about these elements.
   a. Complete the following chart of benefits.

Element	*Benefit*
oxygen	
fluorine	
iodine	
calcium	
zinc	
magnesium	
sodium	
potassium	

   b. Complete the following chart to show where the elements are commonly found in nature.

Element	*Where Found*
oxygen	
fluorine	
iodine	
calcium	
zinc	
magnesium	
sodium	
potassium	

2. One of the responsibilities of a dietitian is to include the elements listed in the previous charts in people's diets. At the library, or from the American Dietetic Association, 620 N. Michigan Ave., Chicago, IL 60611, find out about the occupation of dietitian.
   a. What are the responsibilities of a dietitian?

Challenge Your Understanding

23. (a) $2H_2 + O_2 \longrightarrow 2H_2O$
    (b) $H_2 + O_2 \longrightarrow H_2O_2$
    (c) $2C + O_2 \longrightarrow 2CO$
    (d) $N_2 + 3H_2 \longrightarrow 2NH_3$
    (e) $H_2 + Cl_2 \longrightarrow 2HCl$
    (f) $2Na + Cl_2 \longrightarrow 2NaCl$
24. Because H atoms have an atomic mass of 1 and S atoms an atomic mass of 32, the simplest formula that gives a mass ratio of 1 : 16 in the compound is H_2S (2 : 32 is the same as 1 : 16).
25. See student text, page 187. See also answer to Checkpoint question 1 on page 187.

CHAPTER 8 Overview

This chapter begins by classifying the elements as metals, nonmetals, and metalloids. Three major families of elements—the noble gases, the halogens, and the alkali metals—also are discussed. The periodic table is then introduced. Following this, the mechanisms of bonding are discussed, and the nature of polar molecules is described. Finally, the enrichment section focuses on carbon compounds.

Goals

At the end of this chapter, students should be able to:

1. describe typical physical properties of metals and nonmetals.
2. give examples of metals and nonmetals and state some of their uses.
3. define and give examples of metalloids.
4. identify similarities in properties among the noble-gas elements.
5. describe some uses of the noble gases.
6. explain why the noble gases are unreactive.
7. identify similarities in properties among the family of elements called the halogens.
8. describe some uses for halogens and halides.
9. explain why the atoms of halogens tend to gain or share electrons.
10. identify similarities in properties among the family of elements called the alkali metals.
11. describe some uses for the alkali metals.
12. explain why the atoms of alkali metals tend to lose one electron.
13. explain the use of the periodic table.
14. describe how ionization energy varies.
15. explain how ionic bonding occurs.
16. state how covalent bonding occurs.
17. describe polar molecules.
18. explain why there are so many carbon compounds.
19. tell what a structural formula represents.

Chapter 8

Chemical Elements

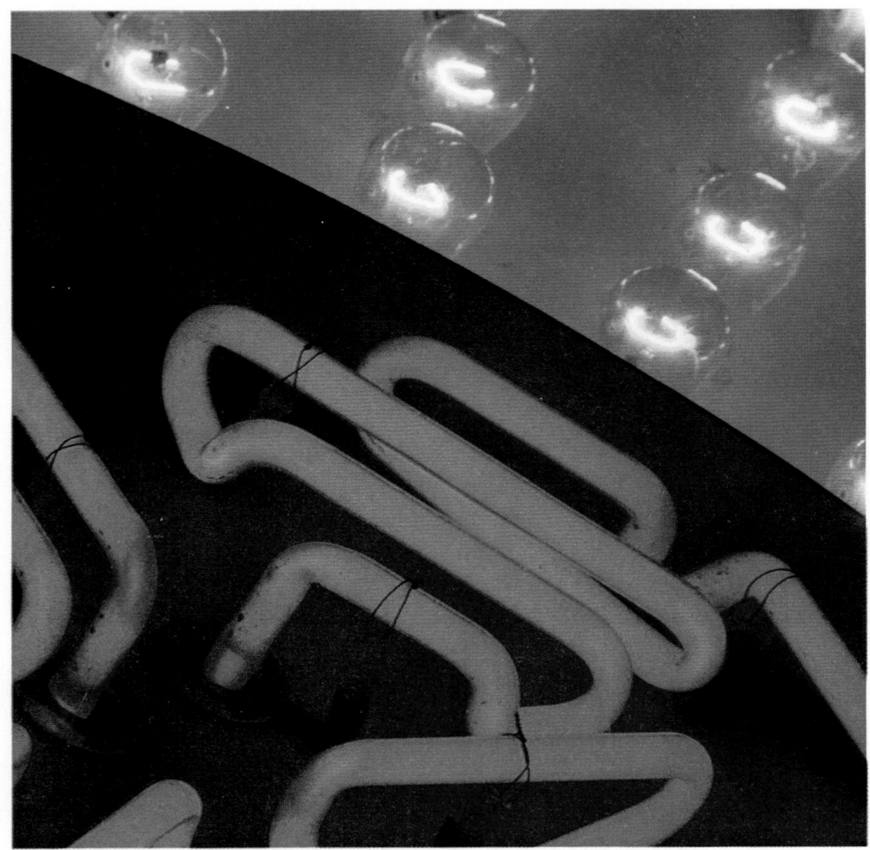

The chemical element neon is used for signs.

Vocabulary Preview

metals
nonmetals
noble gases
halogens
ion
alkali metal
periodic table
groups
period
ionization energy
ionic bonds
covalent bonds
polar molecules
hydrocarbons
structural formulas

8-1

Metals and Nonmetals

Goals

1. To describe typical physical properties of metals and nonmetals.
2. To give examples of metals and nonmetals and state some of their uses.
3. To define and give examples of metalloids.

There are more than 100 chemical elements. Each element is unique, with its own set of properties. For example, hydrogen, a *nonmetal*, is a very light gas that will explode if mixed with oxygen and ignited with a spark. Gold, on the other hand, is a very dense, soft *metal* that can be hammered into sheets so thin that light will pass through them. Sheets like these can be used for lettering on glass, buildings, and leather book bindings.

Metals have many different uses. Copper and aluminum can be stretched into wires for electrical appliances, industrial machines, telephones, and cables. Iron can be melted and mixed with small amounts of carbon and other elements to make the steel used in automobiles, bridges, skyscrapers, and cooking utensils. Lead is a soft, dense metal that melts at 328°C.

Figure 8-1 *The metallic properties of gold make it possible to hammer the element into thin sheets, as this goldsmith is doing.*

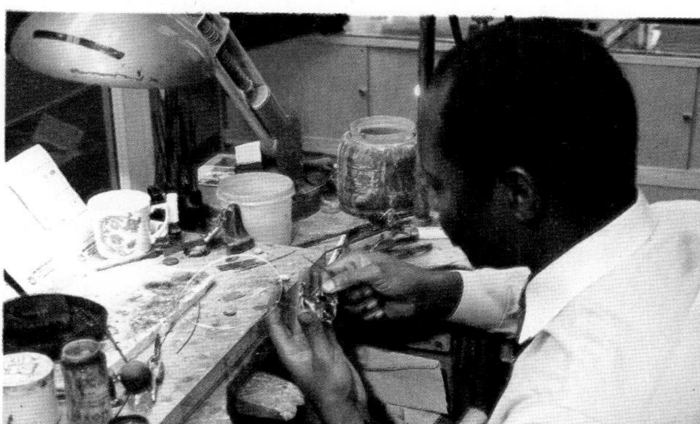

compare it with the density of a nonmetal, such as sulfur. You can demonstrate differences in the malleability, or pliability, and conductivity of metals and nonmetals by doing demonstrations 2 and 3. (See Demonstrations.)

Laboratory Activity 13, Metals and Nonmetals, student text pages 546–547, should be carried out in conjunction with this section. Some students also may wish to do projects 1 and 2 at the end of the chapter.

Background

The properties of metals and nonmetals can be explained by differences in their structure. For example, the conductivity of metals is related to their crystalline structure. The electrons of the atoms in these crystals are loosely held by the nuclei. Thus, the electrons can move freely through the metal. Because of this, electricity, which is the flow of electrons, is conducted readily in metals. The jostling electrons of metals also readily conduct heat. In nonmetals, where electrons are not as free to move, heat and electricity are not conducted very well.

Because electrons in metals can move quite freely, metallic bonds may be thought of as "pliable." This makes metals malleable. The more rigid, fixed bonds in nonmetals cause solid nonmetals to shatter when hammered.

Teaching Suggestions 8-1

Start this section by showing students samples of metals, such as aluminum, copper, lead, iron, and zinc; of nonmetals, such as lump sulfur and phosphorus; and of metalloids, such as silicon and germanium. Point out that the gases that make up air are also nonmetals.

Discuss the physical properties of metals and nonmetals. From the samples of metals and nonmetals, students should see that metals are shiny and nonmetals are dull. (If you have any dull samples of metals, rub them with steel wool to bring out the shine.) Let students hold a sample of a metal and compare its heft with a similar-sized sample of a nonmetal. You may also wish to have students measure the density of a metal, such as copper, and

Demonstrations

1. Use zinc and dilute hydrochloric acid to generate hydrogen. The hydrogen can be ignited with a flaming splint, as shown in Figure TE 8-1. However, a bubble-making solution of the kind that children use will be needed to "capture" the hydrogen within a bubble.

   Notice that, after removing the bulb, the narrow end of an eyedropper is inserted into the end of the tubing. The wide end of the eyedropper is then dipped into the bubble-making solution in order to pick up a bubble-making film. The hydrogen-filled bubbles that form on the wide end of the eyedropper will rise to be ignited with the splint.
2. The conductivity of metals, nonmetals, and metalloids can be tested with the apparatus shown in Figure TE 8-2. Metals, when placed between the clip leads, allow the bulb to glow. Nonmetals, such as sulfur and air, do not conduct electricity well, so the bulb does not light. Metalloids, such as silicon, also do not allow the bulb to light because they are very weak conductors.
3. Show that different metals melt at different temperatures. Place small samples of lead, tin, aluminum, copper, and iron on a can lid. Hold the lid over an alcohol flame. The lead and the tin should melt. Then, hold the lid over the flame of a Bunsen burner. The aluminum, and perhaps the copper, will melt, but the iron will not.

Fact At Your Fingertips

- The greater density of metals occurs not because their atoms are more massive, but because the atoms are packed more closely together than in nonmetals.

Figure 8-2 *Sulfur is a nonmetallic yellow element that is mined for a variety of uses.*

Oxygen, the world's most abundant element (by mass), is a nonmetal. Combined oxygen accounts for nearly nine-tenths of the mass of Earth's water and nearly half the mass of Earth's minerals. Elemental oxygen accounts for more than one-fifth of the mass of Earth's air. Most living organisms need oxygen. The element sulfur (S) is a yellow nonmetallic solid that is used in producing sulfuric acid (H_2SO_4), which is widely used in industry. The element phosphorus (P) is yet another nonmetal. Ordinary phosphorus is a white, waxy solid that bursts into flame when exposed to air. (See Figure 8-3.) For this reason, it is stored under water.

Some elements share enough similar properties that they can be grouped together. The oldest way of classifying elements is to divide them into the two groups described here: **metals** and **nonmetals**. Such a division is based on differences in physical properties that are easy to observe. For example, samples of metals tend to be heavy for their size. That is, they tend to have a higher density than do nonmetals. The physical properties of metals and nonmetals are listed in Table 8-1. Notice that metals are good conductors of heat and electricity. Nonmetals are poor conductors. Metals and nonmetals also differ in their chemical properties, or the way they react.

metals

nonmetals

A few elements, the *metalloids*, have properties that are intermediate between those of metals and nonmetals. The metalloids include silicon, germanium, arsenic, and antimony. Though not good conductors of electricity, their crystals are better conductors than nonmetals. Because of their

Figure 8-3 *White phosphorus is a nonmetal that bursts into flame when it comes into contact with air.*

Table 8-1: Physical Properties of Metals and Nonmetals

Metals	*Nonmetals*
1. Good conductors of heat and electricity	1. Mostly poor conductors
2. Shiny surface when polished	2. Dull
3. Almost all solid at 20°C	3. Many are liquids or gases at 20°C
4. Can be hammered, stretched, bent, or molded into sheets, wires, or other shapes	4. Solids tend to be brittle
5. Tend to be relatively dense	5. Tend to have low densities

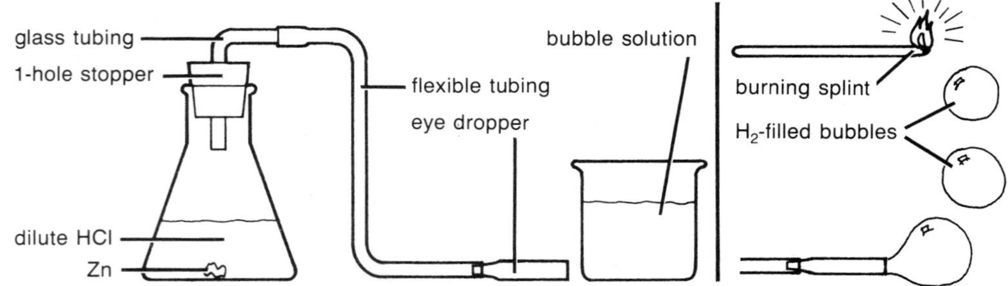

Figure TE 8-1 Demonstration showing the flammability of hydrogen.

Figure 8-4 *Crystals of the metalloid element silicon are often used in the electronics industry.*

moderate ability to conduct electricity, metalloids are called *semiconductors*, and are often used in the electronics industry. Such semiconductor crystals are found in transistors and in solar cells that convert sunlight to electricity. A *crystal* is a regular, repeated arrangement of the particles (such as atoms) of which it is made. (See Figure 8-4.)

Checkpoint

1. How do metals and nonmetals differ?
2. Give two examples each of metals and nonmetals. What are some of their uses?
3. What are metalloids? Give two examples.

TRY THIS

Using a hand lens, look closely at the backs of some pieces of real gold and silver jewelry. What words or numbers do you see? Do research in the library to find out what sterling silver is. What is the difference between 14-carat gold and 18-carat gold?

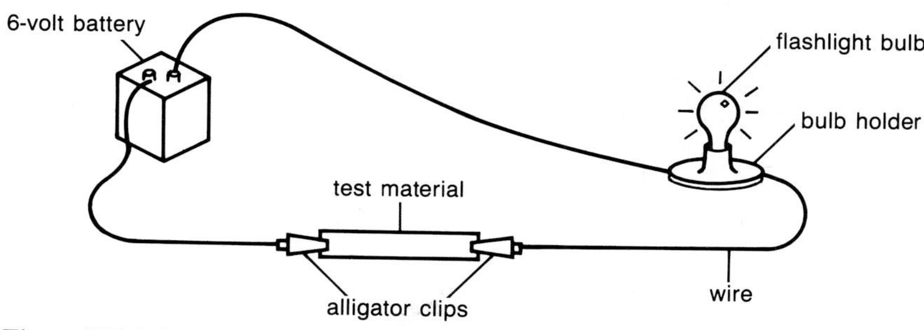

Figure TE 8-2 Apparatus for testing conductivity of a solid.

Mainstreaming

Pair handicapped students and special-needs students with other students to conduct Activity 13 and the Try This activity.

Cross References

The definition of density is given in Chapter 6, section 6-1. More is said about the properties of metals and nonmetals in Chapter 11, section 11-1. Conductor is defined in Chapter 18, section 18-4, and current electricity is discussed in Chapter 19, section 19-4.

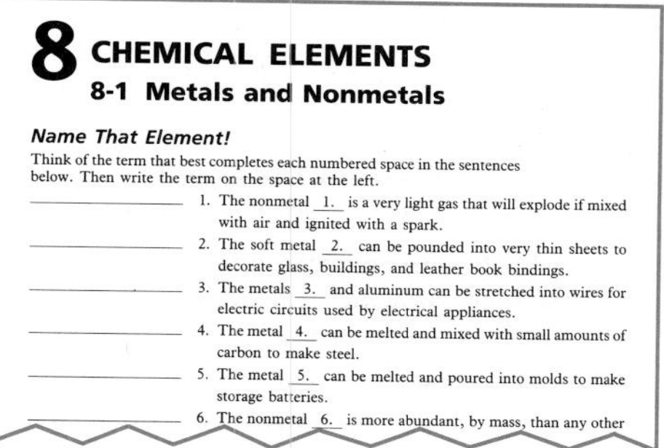

8 CHEMICAL ELEMENTS

8-1 Metals and Nonmetals

Name That Element!

Think of the term that best completes each numbered space in the sentences below. Then write the term on the space at the left.

______ 1. The nonmetal __1.__ is a very light gas that will explode if mixed with air and ignited with a spark.

______ 2. The soft metal __2.__ can be pounded into very thin sheets to decorate glass, buildings, and leather book bindings.

______ 3. The metals __3.__ and aluminum can be stretched into wires for electric circuits used by electrical appliances.

______ 4. The metal __4.__ can be melted and mixed with small amounts of carbon to make steel.

______ 5. The metal __5.__ can be melted and poured into molds to make storage batteries.

______ 6. The nonmetal __6.__ is more abundant, by mass, than any other

Using Try This

Sterling silver is 92.5% silver and 7.5% copper. Pure gold is designated "24-carat." 14-carat gold is 14 parts gold and 10 parts another metal or metals (often silver or copper). 18-carat gold is 18 parts gold and 6 parts another metal.

Checkpoint Answers

1. Answers will vary. See Table 8-1, student text page 194.
2. Answers will vary. Two metals are copper and aluminum, both of which are used to make wires. Two nonmetals are sulfur, used to make sulfuric acid, and oxygen, a gas required by most living things.
3. Metalloids are elements with properties that are intermediate between those of metals and nonmetals. Examples include silicon and germanium.

Teaching Suggestions 8-2

A good way to begin the study of this section is to have students examine Table 8-2 on student text page 197. Then, ask the following questions:

Which noble gases are in a gaseous state at room temperature? (All)

How cold would it have to be for xenon to freeze? (−112°C)

Which of the gases would be the most dense, assuming that all noble gases contain equal numbers of molecules in equal volumes? (Radon)

After discussing the properties of the noble gases, refer students to Table 8-3 on student text page 198. Ask:

What information in Table 8-3 best explains why all the noble gases are unreactive? (The fact that the outermost electron shells in the atoms are all filled.)

Figure 8-5 *The welding shown in this photograph is called heliarc welding. The noble gas helium is used in the process.*

8-2

The Noble Gases

Goals

1. To identify similarities in properties among the noble-gas elements.
2. To describe some uses of the noble gases.
3. To explain why the noble gases are unreactive.

The metals and nonmetals make up two large groups of elements. The elements can also be classified in other ways. They can be placed in small groups, called *families*. Elements in a family have similar properties.

A member of one important family of nonmetals was discovered in the nineteenth century. During an eclipse of the sun in 1868, a French astronomer took photographs through an instrument called a spectroscope. The photographs showed the spectral lines given off by glowing gases in the sun. A few of the lines had never been seen before. In 1871, scientists proposed that the lines belonged to a new element, which they named helium. (*Helios* in Greek means "sun.")

Helium does not react with other substances. At ordinary temperatures, it is a gas that is less dense than air. That is why helium balloons rise. Helium is also used in blimps and weather balloons. The most important industrial use of helium is in heliarc welding, a type of electric-arc welding shown in Figure 8-5. Liquid helium is used as a cooling agent. It has the lowest boiling point (−269°C) of any substance known. When substances are placed in liquid helium, they become very cold. At the very low temperatures produced in this way, many metals become *superconductors*. These are substances that offer practically no resistance to the flow of electricity. Such a property is very useful in large computers, where currents must flow rapidly through many circuits.

A gas similar to helium was detected in 1892 when the British physicist Lord Rayleigh removed all the oxygen, carbon dioxide, nitrogen, and water from an air sample. The small amount of gas that remained would not react with anything. The gas was thought to be a new element and was named argon from the Greek word meaning "lazy." Like he-

Background

It is interesting that helium was first found in the sun by means of spectroscopic analysis, because the element is found beneath Earth's surface too. Helium arises from the decay of radioactive elements. At the time that helium was discovered in the gases of the sun, radioactive elements were unknown; helium had gone undetected for so long because it is unreactive.

When Lord Rayleigh, John William Strutt, discovered argon in 1892, he first removed all the oxygen, carbon dioxide, and water from a sample of air. He assumed the remaining gas was nitrogen, but, to his surprise, the gas had a density slightly greater than that

Figure 8-6 *The noble gas krypton is used in airport runway lights.*

lium, argon does not react with other elements. Because of this low reactivity, argon is used as the gas in light bulbs.

During the next few years, other stable, unreactive elements of the same family as helium and argon were found in the atmosphere. These elements are krypton, xenon, and radon. They are all gases under normal conditions.

Chemists group all these unreactive elements in a family called the **noble gases**. They are listed in Table 8-2, along with some of their properties.

noble gases

Neon signs contain gaseous neon, which emits reddish-orange light when electricity is passed through it. The light from krypton penetrates fog so well that krypton lamps are used to illuminate airport runways, as shown in Figure 8-6.

Table 8-2: Some Properties of the Noble Gases

Noble gas	*Atomic number (number of protons)*	*Atomic mass (amu)*	*Melting point* (°C)	*Boiling point* (°C)
helium (He)	2	4.0	−272	−269
neon (Ne)	10	20.2	−249	−246
argon (Ar)	18	39.9	−189	−186
krypton (Kr)	36	83.8	−157	−152
xenon (Xe)	54	131.3	−112	−107
radon (Rn)	86	222.0	−71	−62

Facts At Your Fingertips

- The noble gases are sometimes called the inert gases because they resist reacting with other substances. In fact, however, these gases have been made to react with the very reactive gas fluorine and with other elements. The unstable compounds formed, such as xenon fluoride and krypton fluoride, can be isolated.
- Even when helium boils at −269°C, it is colder than any other substance. Because it is so cold, other substances lose heat to boiling helium, just as warm water loses heat to ice.

of pure nitrogen. He proceeded to remove nitrogen from the remaining gas by reacting it with hot magnesium to form magnesium nitride:

$$3Mg + N_2 \longrightarrow Mg_3N_2$$

Argon, another noble gas, later found wide use as a nonreactive gas in light bulbs. It serves to reduce the evaporation and reaction of metallic filaments.

It is well known that an element is stable when its atoms have eight electrons in the outer energy level. However, it is not fully understood why such a configuration is so stable.

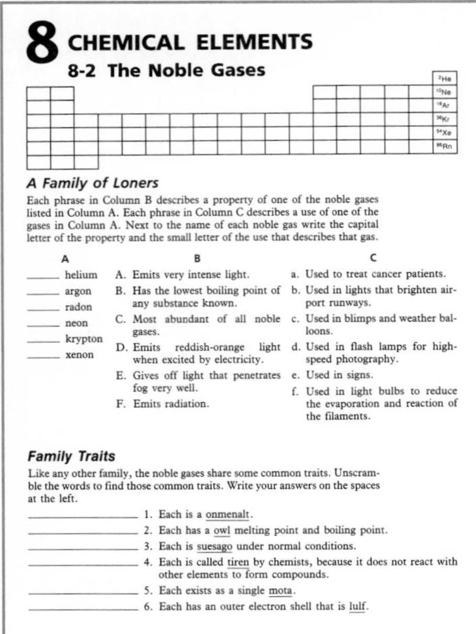

8 CHEMICAL ELEMENTS

8-2 The Noble Gases

A Family of Loners

Each phrase in Column B describes a property of one of the noble gases listed in Column A. Each phrase in Column C describes a use of one of the gases in Column A. Next to the name of each noble gas write the capital letter of the property and the small letter of the use that describes that gas.

A	B	C
______ helium	A. Emits very intense light.	a. Used to treat cancer patients.
______ argon	B. Has the lowest boiling point of any substance known.	b. Used in lights that brighten airport runways.
______ radon	C. Most abundant of all noble gases.	c. Used in blimps and weather balloons.
______ neon	D. Emits reddish-orange light when excited by electricity.	d. Used in flash lamps for high-speed photography.
______ krypton	E. Gives off light that penetrates fog very well.	e. Used in signs.
______ xenon	F. Emits radiation.	f. Used in light bulbs to reduce the evaporation and reaction of the filaments.

Family Traits

Like any other family, the noble gases share some common traits. Unscramble the words to find those common traits. Write your answers on the spaces at the left.

________________ 1. Each is a onmenalt.
________________ 2. Each has a owl melting point and boiling point.
________________ 3. Each is suesago under normal conditions.
________________ 4. Each is called tiren by chemists, because it does not react with other elements to form compounds.
________________ 5. Each exists as a single mota.
________________ 6. Each has an outer electron shell that is lulf.

Checkpoint Answers

1. The noble gases are He, Ne, Ar, Kr, Xe, and Rn. All are gases under normal conditions, have low boiling and melting points, and are extremely unreactive.
2. Helium was found in spectroscopic photographs of the sun. Argon was found after all the oxygen, carbon dioxide, nitrogen, and water vapor were removed from a sample of air.
3. Helium is used in lighter-than-air balloons; neon, in signs that glow; argon, in light bulbs; krypton, in fog-penetrating lights; xenon, in flash lamps; and radon, in treating cancer.
4. Valence electrons are the electrons in the outermost energy level, or shell, of an atom.
5. Their outermost shells are filled with electrons.

Table 8-3: Electron Arrangement in the Noble Gases

Noble gas	*Total number of electrons*	*Electrons in each shell*	*Electrons in outermost shell*
helium	2	2	2 (filled)
neon	10	2-8	8 (filled)
argon	18	2-8-8	8 (filled)
krypton	36	2-8-18-8	8 (filled)
xenon	54	2-8-18-18-8	8 (filled)
radon	86	2-8-18-32-18-8	8 (filled)

Because xenon emits a very intense light, it is used in flash lamps made for high-speed photography. Radon, like radium, emits radiation and is used to treat cancer patients.

The arrangement of the electrons in the outermost energy level, or shell, determines the chemical properties of an element. Those electrons are sometimes called *valence* (VAY luns) electrons. The outermost electron shell of any atom can hold a maximum of 8 electrons (except for the first shell which can only hold 2). If this shell is full, the atom is unreactive. It does not tend to gain or lose electrons or combine with other atoms, even atoms of its own kind.

Table 8-3 indicates the total number of electrons in each noble-gas atom, the number in each shell of the atom, and the number in the outermost shell. Note that helium's outermost shell (the first) is filled. It contains two electrons. Each of the other noble-gas atoms has eight electrons in its outermost shell. These shells are also filled. It is the filled outermost shell that gives each noble gas its special stability, or unreactivity.

Checkpoint

1. What are the noble gases? What properties do they have in common?
2. How were helium and argon discovered?
3. List a use for each of the noble gases.
4. Define valence electrons.
5. What property of their atoms makes the noble gases unreactive?

8-3

The Halogens

Goals

1. To identify similarities in properties among the family of elements called the halogens.
2. To describe some uses for halogens and halides.
3. To explain why the atoms of halogens tend to gain or share electrons.

Fluorine, chlorine, bromine, and iodine make up a family of elements called the **halogens**. (See Figure 8-7.) Only traces of a fifth family member, astatine (At), occur naturally. Some physical properties of the halogens are listed in Table 8-4 on page 200.

The halogens are highly reactive. This is because each halogen atom has seven electrons in its outermost shell. That is one less electron than the noble-gas atoms have. By gaining one electron, each halogen atom can fill its outermost shell, becoming more stable. An atom (or group of atoms) that has gained or lost one or more electrons is called an **ion**. Since the number of electrons is no longer equal to the number of protons, an ion has an electric *charge*. When a fluorine atom gains an electron it becomes a fluoride ion, F^-, with 8 electrons in its outermost shell and a charge of −1. (See Figure 8-8.) Generally, when a halogen atom gains an electron, it becomes a hal*ide* ion.

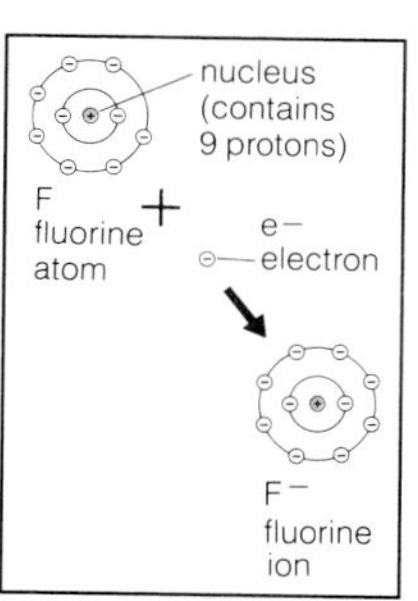

Figure 8-8 *By gaining one electron, a neutral fluorine atom becomes a stable fluoride ion, with a filled outermost shell.*

halogens
(HAL uh junz)

ion
(EYE un)

Figure 8-7 *The halogen elements include yellow-green gaseous chlorine, reddish-brown liquid bromine, and blue-black solid iodine.*

Teaching Suggestions 8-3

Open discussion of the section by having students examine Table 8-4 on student text page 200. Discuss how the boiling and melting points can be used to determine the physical state of a substance at a particular temperature. Ask what the physical state of each halogen is at room temperature (F and Cl—gas; Br—liquid; I—solid). If possible, show samples of the halogens, such as iodine and bromine.

CAUTION: All halogens are toxic, reactive substances and should be handled with care. Bromine fumes are noxious; keep the bottle closed so students do not breathe the vapor.

Point out that because the halogens are very reactive, they are found primarily in compounds. You may wish to perform Demonstration 1 to underscore this fact. (See Demonstrations.) Point out that silver bromide is the salt used in the coating of photographic film and paper.

Chapter-end projects 3 and 5 apply to this section.

Background

The least reactive halogen is iodine, which exists as a solid at 20°C. The halogen elements are normally found in compounds as halide ions, all of which carry a charge of −1.

Astatine, the fifth halogen, was a blank space on the periodic table until 1947. Then it was found that this element is one in a chain of the radioactive series containing neptunium. Astatine is most commonly produced by bombarding bismuth with alpha particles to form the 211-isotope ($^{211}_{85}At$). Astatine can also be produced by bombarding gold with carbon nuclei to form the 205-isotope ($^{205}_{85}At$). Since its half-life is only 7.5 hours, this element is short-lived.

Demonstrations

1. Show students the white crystals of sodium or potassium chloride, bromide, and iodide. Dissolve the crystals in water to form dilute solutions of the salts. Place a sample in each of three test tubes and to each add a drop or two of dilute silver nitrate solution. All will form silver halide precipitates that settle out slowly.
2. If possible, have students visit a darkroom where photographic prints are being made so they can see how silver is deposited and how it darkens paper upon exposure to light during the development process.
3. Show the test for detecting bromine or iodine in solutions. Place about 10 mL of chlorine water, bromine water, and iodine water in separate test tubes. (If you have trouble dissolving enough iodine to produce an effect, add a little KI to increase the solubility of iodine.) Add about 2 mL of carbon tetrachloride to each

tube. (CAUTION: Avoid inhaling carbon tetrachloride fumes; they are toxic.) Stopper the tubes and shake. The carbon-tet layer, which will lie below the solutions in the tubes, will remain clear in the case of chlorine. It will turn an orangish color as bromine leaves the water and enters the carbon tet, in which it is more soluble. In the iodine test, the carbon-tet layer becomes violet upon shaking.

Repeat the tests using potassium bromide (KBr) and potassium iodide (KI), or use the sodium salts of these halide ions. The carbon-tet layer will remain clear in both cases, because there is no elemental bromine or iodine present; there are only the ions of these elements in compounds. If you now add chlorine water to each tube and shake, the bromine and iodine ions will give their extra electrons to chlorine, forming chloride ions and releasing elemental bromine and iodine:

$$Cl_2 + 2Br^- \longrightarrow 2Cl^- + Br_2$$
$$Cl_2 + 2I^- \longrightarrow 2Cl^- + I_2$$

4. If you have demonstration tubes of bromine, you can freeze the bromine by immersing the lower end of a tube into a styrofoam cup containing alcohol and dry ice. The temperature of this mixture will be about −40°C, which is well below bromine's freezing point.

Teaching Tips

- The demonstrations suggested for this section should help make students aware of the differences between the halogens as elements (Cl_2, Br_2, I_2) and as halide ions (Cl^-, Br^-, I^-).
- Have students find out if fluoride ions are present in their drinking water. Ask, too, if they use a fluoride toothpaste or rinse.
- Be sure students can locate the halogens on the periodic table on student text page 205.

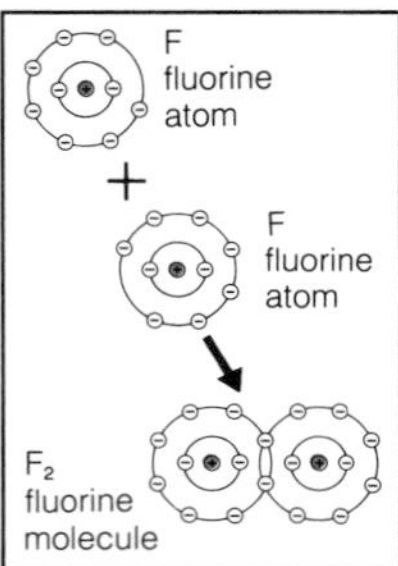

Figure 8-9 By combining to form a diatomic molecule, two halogen atoms, such as fluorine (F), share electrons and achieve filled outermost shells.

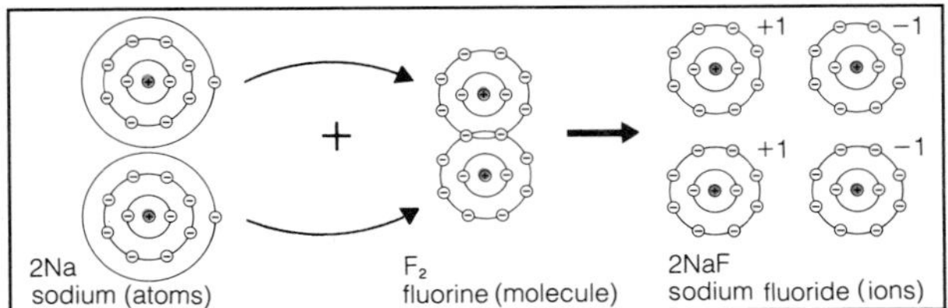

Figure 8-10 Sodium atoms can combine with a halogen molecule, such as fluorine, to produce stable sodium and halide ions.

Another way for halogen atoms to become more stable is to *share* electrons. Under ordinary conditions, the halogens exist as diatomic, or two-atom, molecules. They are more stable in this form than as individual atoms. By sharing electrons, the halogen atoms achieve filled outermost shells, as shown in Figure 8-9. You will learn more about such sharing and transfer of electrons later in this chapter.

Any compound of a halogen and a metallic element is called a hal*ide*. Thus, NaCl is sodium chlor*ide* and NaI is sodium iod*ide*. The sodium halides are solid *salts*. The sodium salts are made up of ions. In a reaction between sodium and a halogen, each halogen atom gains an electron from a sodium atom. (See Figure 8-10.) A typical equation is:

$$2Na + F_2 \longrightarrow 2NaF$$

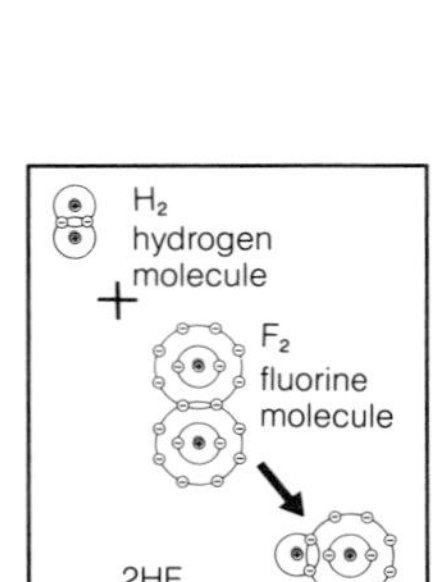

Figure 8-11 A hydrogen molecule can combine with a halogen molecule, such as fluorine, to produce stable hydrogen halide molecules.

The halogens react with hydrogen to form hydrogen halides. (See Figure 8-11.) A typical equation is:

$$H_2 + F_2 \longrightarrow 2HF$$

Table 8-4: Some Properties of the Halogens

Halogen	*Atomic number*	*Atomic mass (amu)*	*Melting point (°C)*	*Boiling point (°C)*
fluorine (F)	9	19.0	−220	−188
chlorine (Cl)	17	35.5	−101	−35
bromine (Br)	35	80.0	−7	59
iodine (I)	53	127.0	114	184

Facts At Your Fingertips

- Despite strong evidence that fluoride ions effectively reduce tooth decay, some people oppose the addition of fluoride to drinking water. In high concentrations fluorides can be poisonous. The safety of the concentrations used in drinking water, which are only 0.6–1.7 ppm (parts per million), is questioned by such people. Studies show that $1.00 spent on fluoridation reduces dental bills by $36.00.
- Because of the high cost of silver, some photographic development firms use chemical techniques to recover the silver lost in film processing.

Figure 8-12 *A photographic negative contains dark silver deposits that result from the breakdown of silver bromide by light. The resulting positive is light in those areas that correspond to the dark areas of the negative.*

The hydrogen halides are gases that form substances called *acids* when they dissolve in water.

All the halogen elements are poisonous. However, many halogen compounds are not harmful, and some are necessary for good health. Table salt (NaCl) occurs naturally in many foods. We use additional salt on food to improve flavor. Small amounts of fluoride salt in toothpastes and drinking water help strengthen tooth enamel and reduce tooth decay. Iodide salts, often added to table salt, help the thyroid gland function.

Photographic film (and paper) are coated with a thin layer of gelatin containing tiny crystals of light-sensitive silver bromide (AgBr). In those places where light strikes the gelatin layer, the compound changes to silver and bromine. The silver deposits form the picture on the negative. To make prints from the film, light is shone through the negative onto paper that is coated with a light-sensitive layer. The places that were darkest on the negative are lightest on the print, and vice versa. (See Figure 8-12.)

Checkpoint

1. Why are the halogens grouped together in a family?
2. Why are halide ions stable?
3. Using a diagram, explain how hydrogen and fluorine bond to form HF.
4. List five uses for halogens and halides.

8 CHEMICAL ELEMENTS
8-3 The Halogens

Solve a Mystery!

For each mystery below there is a set of four clues. Each clue describes one of four halogens—fluorine, chlorine, bromine, and iodine—which are represented by the letters *A*, *B*, *C*, and *D*. As you decide which halogen is being described by a particular clue, keep in mind that no halogen should be used more than once in a set. If, for example, A has an atomic number of 9, it must be fluorine. If B is a halogen, it cannot be fluorine since A is already fluorine. Element B must be chlorine, bromine, or iodine. After finishing each group of four clues, start the next group with the same four halogens. Use your textbook and Table 8-4 as well as reference books. Place your answers on the spaces at the left.

Set 1:

____ Element A is a liquid at 20° C, or about room temperature.
____ Element B is added to water to kill bacteria that cause disease and infection.
____ Element C has an atomic number of 53.
____ Element D is a halogen.

Set 2:

____ Element A is used in toothpaste to strengthen tooth enamel.
____ Element B is a solid at 20° C.
____ Element C has an atomic mass of 35.5 amu.
____ Element D is a diatomic molecule.

Set 3:

____ Element A is added to the diet to help the thyroid gland function properly.
____ Element B is produced by dissolving its sodium salt in water and then bubbling chlorine gas through the solution.
____ Element C is a gas at 20° C.
____ Element D has a melting point of −101° C.

Set 4:

____ Element A is produced by electrolysis of one of its compounds.
____ Element B has a boiling point of 184° C.
____ The silver compound of element C is used as a coating for photographic film because the compound is light-sensitive.
____ Element D is a poisonous yellow-green gas.

Checkpoint Answers

1. Because they have similar chemical properties.
2. Halide ions have filled outermost electron shells.
3. See Figure TE 8-3.
4. Answers will vary. Here are five possibilities: (1) Silver bromide is used in photographic film and paper. (2) Fluoride salts are added to toothpastes to prevent cavities. (3) Chlorine is added to water to kill bacteria. (4) Sodium chloride is table salt. (5) Iodide salts are added to table salt to help the thyroid gland function properly.

- Although halogens are poisonous, halides such as NaCl and NaI are commonly used in foods and are necessary to life.

Cross Reference

For more information on radioactivity, see Chapter 10.

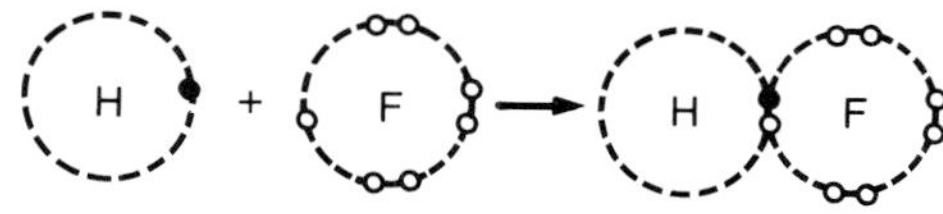

Figure TE 8-3 Diagram of hydrogen and fluorine atoms bonding to form HF.

Teaching Suggestions 8-4

Begin the study of this section by dropping a very small piece of potassium metal into a beaker of water containing a few drops of phenolphthalein. Suddenly, as the potassium darts about on the water's surface, it will burst into flame and the water will turn magenta. This is the first of a series of demonstrations that students will enjoy and that will show the major properties of the alkali metals. For details, see Demonstrations section.

Background

The atoms of the alkali metals have large radii when compared to other elements of similar mass. This large atomic radius accounts for the very low densities of the alkali metals, some of which (Li, Na, K) float in water. Alkali metals also are very reactive, because the energy required to remove an electron is low, and they are never found free in nature. The heaviest of the six, francium, occurs only in very small quantities as a short-lived radioactive isotope.

Alkali-metal compounds have many uses. For example, sodium-potassium alloys have low densities, low viscosities, high boiling points, and low melting points. For these reasons, they are liquid over a wide temperature range and have a high heat capacity, which make sodium-potassium alloys useful as heat-transfer media. Sodium hydroxide, often called caustic soda, is widely used in industry for producing other chemicals, rayon, cleansers, textiles, soap, paper, and pulp. Sodium carbonate, or soda ash, is used in making glass, caustic soda, and water softeners, and in the refining of petroleum. Sodium bicarbonate, also called bicarbonate of soda, is the familiar baking soda used as a source of carbon dioxide in the baking process. Table salt—sodium chloride—is vital to all living things. Also, in industry, it is the usual source of all other sodium and chlorine compounds, particularly elemental chlorine,

8-4

The Alkali Metals

Goals

1. To identify similarities in properties among the family of elements called the alkali metals.
2. To describe some uses for the alkali metals.
3. To explain why the atoms of alkali metals tend to lose one electron.

alkali metal (AL kuh ly)

Each atom of the **alkali metal** family has only one electron in its outermost shell. The alkali metals include lithium, sodium, potassium, rubidium, and cesium. Only traces of a sixth alkali metal, francium (Fr), occur naturally. Some properties of the alkali-metal elements are given in Table 8-5. The alkali metals are very soft and can be cut with a knife.

An alkali-metal atom attains the stability of the noble-gas electron arrangement by losing its single outermost electron. It is thus left with a filled shell and becomes a positive ion. (See Figure 8-13.) Very little energy is required

Table 8-5: Some Properties of the Alkali Metals

Alkali Metal	*Atomic number*	*Atomic mass (amu)*	*Melting point (°C)*	*Boiling point (°C)*
lithium (Li)	3	6.9	181	1342
sodium (Na)	11	23.0	98	892
potassium (K)	19	39.1	64	774
rubidium (Rb)	37	85.5	39	686
cesium (Cs)	55	133.0	28.7	669

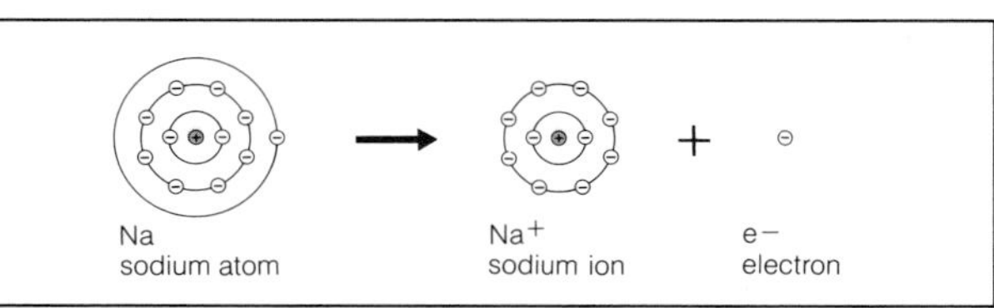

Figure 8-13 *By losing one electron, a neutral sodium atom becomes a stable sodium ion, with a filled outermost shell.*

hydrochloric acid, sodium hydroxide, and sodium carbonate.

Chapter-end projects 3 and 4 apply to this section.

Demonstrations

1. Show the reaction between an alkali metal and water. Carefully cut a *very small sliver* of sodium or potassium from a chunk of the metal that has been stored under kerosene. Add the metal to a beaker containing water with a few drops of phenolphthalein. The metal will react violently as it skims along the surface of the water. If potassium is used, the heat generated will be sufficient to ignite the hydrogen gas produced. The liquid will turn magenta as the reaction proceeds. This color change indicates the presence

Figure 8-14 *When exposed to the air, alkali metals react vigorously with the oxygen and water vapor in the air.*

to remove the outermost electron from an alkali-metal atom. This is another way of saying that alkali metals are very reactive. They can react, sometimes explosively, with halogens to form halides, and with oxygen to form oxides. Alkali metals react with water, releasing hydrogen gas. The energy given off in the reaction can ignite the hydrogen and produce a dangerous explosion. To prevent alkali metals from reacting with air or moisture in the air, they are usually stored submerged in kerosene.

Many alkali-metal compounds, such as sodium chloride, sodium hydroxide ($NaOH$), sodium carbonate (Na_2CO_3), sodium bicarbonate ($NaHCO_3$), and potassium nitrate (KNO_3), are abundant and widely used in homes and industry. For example, sodium bicarbonate—baking soda—is used in cooking. The uncombined elements themselves have limited use. Sodium's excellent conductivity has led to its use in underground cables. Sodium metal is also used as a cooling agent in nuclear reactors. It has a low melting point, a high boiling point, and a good capacity for absorbing the large amounts of heat released in such reactors.

Checkpoint

1. What elements are in the alkali-metal family? What properties do they have in common?
2. Why are alkali metals found in compounds instead of by themselves?
3. What gives alkali-metal ions their stability?

TRY THIS

Look at the compounds in your kitchen and bathroom cabinets. Which compounds contain alkali metals? What are the compounds used for?

of hydroxide ions. You may show that phenolphthalein turns this color when added to solutions of NaOH, KOH, LiOH, and so on.

Lithium also reacts with water, but with considerably less violence than potassium or sodium does. You might ask students what the reaction would be like if you added rubidium or cesium to water. Make sure they grasp that the reaction becomes more violent, or rapid, as the atomic mass of the alkali metal increases. All these reactions can be represented by the following general equation, in which A represents any alkali metal:

$$2A + 2H_2O \longrightarrow 2AOH + H_2$$

2. Use a Nichrome wire and a handle to show students the flame tests for the alkali metals in compounds. The bright yellow light in the flame of sodium is easily seen. Clean the wire after each test by dipping it in HCl and reheating. You might test NaCl, $NaHCO_3$, KCl, and LiCl.

Cross Reference

Isotopes are discussed in greater detail in Chapter 10, section 10-2.

8 CHEMICAL ELEMENTS

8-4 The Alkali Metals

Let's Summarize

Complete the chart below. You may want to use Section 8-5 of your textbook.

Name of Compound or Element	*Formula*	*Use or Property*
		Alkali metal with the greatest density.
sodium		
		Alkali metal with the highest melting point.
sodium carbonate		
potassium nitrate (salt peter)		
	$NaHCO_3$	

The Truth Serum

Read the paragraphs below. Then write the answers to the questions.

Sodium pentothal is a compound of the alkali metal sodium. Its use depends on the size of the dose. If a small dose of this drug is given to someone, the person tends to talk more freely. A psychiatrist may give the drug to a patient to help the patient discuss emotions more easily.

In the past, a physician might have given the drug to a suspect in a police investigation. The suspect might then talk about the crime more freely.

If a large dose of sodium pentothal is given to someone, it produces a deep sleep. Physicians use the drug in this way to produce the deep sleep required for surgery.

1. What happens when a small dose of sodium pentothal is given to a person?
2. What are two ways small doses of sodium pentothal have been used?
3. What is sodium pentothal used for when given in large doses?

Using Try This

Responses will vary, but all students will probably find sodium bicarbonate (baking soda) in pure form and in some antacids. Sodium chloride and sodium or potassium citrate are other common compounds.

Checkpoint Answers

1. Li, Na, K, Rb, Cs, and Fr are the alkali metals. They are all very reactive and are soft solids (at room temperature).
2. They are very reactive because of their single outermost electron.
3. The loss of an electron leaves a filled outermost shell.

Teaching Suggestions 8-5

If possible, display a large periodic table that is visible to all students. Otherwise, have students refer to the one on student text page 205. Make sure students understand that the elements are arranged in the order of their atomic numbers. Ask students to name the elements in the first three rows of the table (hydrogen through argon). Make sure students see that the elements become more nonmetallic from left to right along a row. Ask a student to point out the noble-gas family (column VIII), the alkali metals (column I), and the halogens (column VII).

As a class activity, construct a table similar to Table TE 8-1. Point out that the atomic number is also the number of electrons in the atom.

Table TE 8-1: Electrons in Elements

Element	Atomic Number	Electrons in Each Level		
		1	2	3
H	1	1	0	0
He	2	2	0	0
Li	3	2	1	0
Be	4	2	2	0
B	5	2	3	0
C	6	2	4	0
N	7	2	5	0
O	8	2	6	0
F	9	2	7	0
Ne	10	2	8	0
Na	11	2	8	1
Mg	12	2	8	2
Al	13	2	8	3
Si	14	2	8	4
P	15	2	8	5
S	16	2	8	6
Cl	17	2	8	7
Ar	18	2	8	8

Once the table has been constructed, have students relate the electron arrangements of these atoms to their chemical properties and to their positions in the periodic table.

8-5

The Periodic Table

Goals

1. To explain the use of the periodic table.
2. To describe how ionization energy varies.

periodic table

For many years chemists tried different ways of classifying elements on the basis of their properties. In 1871, the Russian chemist Dmitri Mendeleev arranged the elements in order of increasing atomic mass. When he did, he found that similar properties occurred over and over again at fairly regular intervals. The table he set up was the first **periodic table**, an arrangement in which elements that have similar properties are grouped together. At that time, some of the elements known today had not been discovered. For example, there was no known element with an atomic mass of 72. So Mendeleev left spaces for that and other missing elements in his table. He even correctly predicted what the properties of the missing elements would be.

groups

In the modern periodic table, the elements are ordered by increasing atomic number, rather than atomic mass. The vertical columns in the table are called **groups**, or *families*. The alkali metals, with one outermost electron, are in group I. The halogens, with seven outermost electrons, are in group VII and the noble gases, with eight outermost electrons, form group VIII. In 1984, Arabic numbers for labelling the 18 groups in the periodic table were approved.

period

Each horizontal row in the periodic table is a **period**. From left to right along a period, the properties of the elements change dramatically. The first period contains only hydrogen and helium. The second period goes from lithium to neon. Neon atoms, with 10 electrons, have filled outermost energy levels. Sodium, the next element, has 11 electrons and lies below lithium, which it resembles in many ways. Magnesium, with properties similar to beryllium, is the eighth element after beryllium, just as sodium is the eighth element after lithium. This progression continues across the third period up to the noble gas argon, the eighth element after neon. It is the repetition of similar properties that gives the periodic table its name. The lengths of the periods depend on the number of electrons in each shell.

Background

Although each element is unique, similarities in properties make it possible to group elements into families. These are the vertical columns in the periodic table.

Dobereiner, a German chemist, first suggested a relationship between properties and atomic masses. He noted that the atomic mass of strontium lies about midway between those of calcium and barium, and that strontium's properties are intermediate between those of calcium and barium. He later identified two other *triads* of elements: chlorine, bromine, and iodine; and lithium, sodium, and potassium.

By 1854, other chemists had found more extensive families of elements, such as the oxygen family, consisting of oxygen, sulfur, selenium, and tellu-

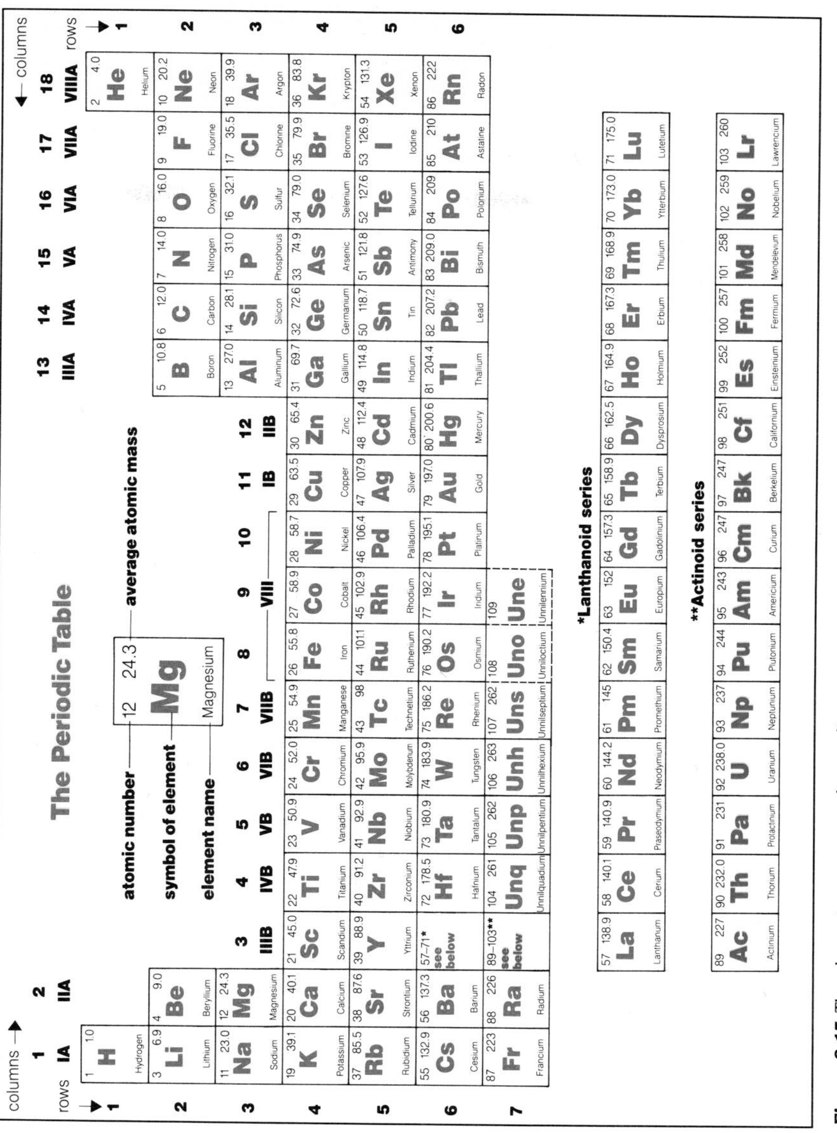

Figure 8-15 The elements are arranged according to atomic number in the modern periodic table.

rium; and the nitrogen family, consisting of nitrogen, phosphorus, arsenic, antimony, and bismuth. By 1870, Mendeleev in Russia and Meyer in Germany had independently developed the idea that the properties of the elements are a periodic function of their atomic masses.

Mendeleev bravely left spaces in his table for elements yet to be discovered. He even predicted the properties of these elements. When the elements scandium, gallium, and germanium were discovered a few years later, their properties closely matched Mendeleev's predictions.

After Rutherford's discovery that most of an atom's mass is concentrated in its positively charged nucleus, a better way of ordering the elements was developed. Henry Moseley arranged them in order of increasing atomic number, instead of atomic mass. As you can see, this arrangement places tellurium (atomic number 52) before iodine (atomic number 53), which has greater atomic mass. Also, it places argon, which was unknown to Mendeleev, ahead of the lighter element potassium, and cobalt before the slightly less massive nickel.

It is the similarities in electron configuration that give rise to similar chemical properties. As we move to the right across the periodic table from the alkali metals in column I to the noble gases in column VIII, an additional electron is added to the valence energy level in each column.

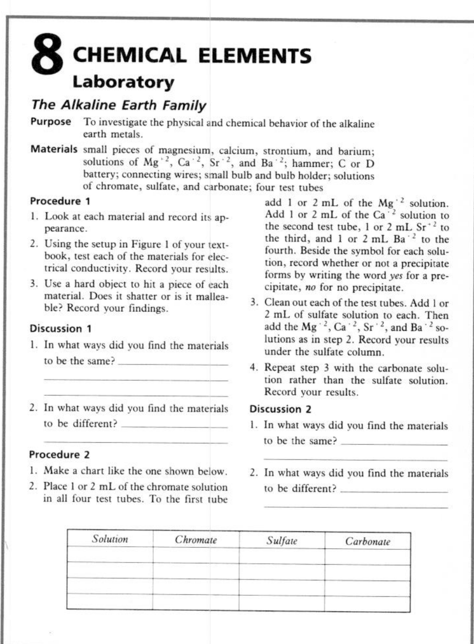

8 CHEMICAL ELEMENTS
Laboratory

The Alkaline Earth Family

Purpose To investigate the physical and chemical behavior of the alkaline earth metals.

Materials small pieces of magnesium, calcium, strontium, and barium; solutions of Mg^{+2}, Ca^{+2}, Sr^{+2}, and Ba^{+2}; hammer; C or D battery; connecting wires; small bulb and bulb holder; solutions of chromate, sulfate, and carbonate; four test tubes

Procedure 1

1. Look at each material and record its appearance.
2. Using the setup in Figure 1 of your textbook, test each of the materials for electrical conductivity. Record your results.
3. Use a hard object to hit a piece of each material. Does it shatter or is it malleable? Record your findings.

Discussion 1

1. In what ways did you find the materials to be the same? ______
2. In what ways did you find the materials to be different? ______

Procedure 2

1. Make a chart like the one shown below.
2. Place 1 or 2 mL of the chromate solution in all four test tubes. To the first tube add 1 or 2 mL of the Mg^{+2} solution. Add 1 or 2 mL of the Ca^{+2} solution to the second test tube, 1 or 2 mL Sr^{+2} to the third, and 1 or 2 mL Ba^{+2} to the fourth. Beside the symbol for each solution, record whether or not a precipitate forms by writing the word *yes* for a precipitate, *no* for no precipitate.
3. Clean out each of the test tubes. Add 1 or 2 mL of sulfate solution to each. Then add the Mg^{+2}, Ca^{+2}, Sr^{+2}, and Ba^{+2} solutions as in step 2. Record your results under the sulfate column.
4. Repeat step 3 with the carbonate solution rather than the sulfate solution. Record your results.

Discussion 2

1. In what ways did you find the materials to be the same? ______
2. In what ways did you find the materials to be different? ______

Solution	*Chromate*	*Sulfate*	*Carbonate*

Teaching Tips

- A revolving block periodic table is a useful teaching aid. Construction of one is a good student project.
- Relate the graph of ionization energies in Figure 8-16, on student text page 206, to the periodic table. Show students that elements on the left side of the table are more likely to form positive ions because relatively little energy is required to remove electrons from their atoms. Elements on the right side of the table are much less likely to form positive ions because of their high ionization energies. (They are more likely to *gain* electrons, thus forming negative ions.)

Fact At Your Fingertips

- The repetition of chemical properties with every eighth element led some scientists to relate the periodicity seen in atomic structure to octaves in music.

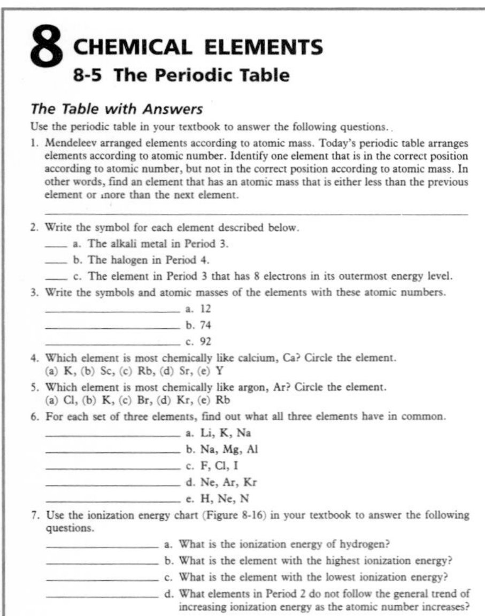

8 CHEMICAL ELEMENTS
8-5 The Periodic Table

The Table with Answers

Use the periodic table in your textbook to answer the following questions.

1. Mendeleev arranged elements according to atomic mass. Today's periodic table arranges elements according to atomic number. Identify one element that is in the correct position according to atomic number, but not in the correct position according to atomic mass. In other words, find an element that has an atomic mass that is either less than the previous element or more than the next element.

   ____

2. Write the symbol for each element described below.
   ____ a. The alkali metal in Period 3.
   ____ b. The halogen in Period 4.
   ____ c. The element in Period 3 that has 8 electrons in its outermost energy level.
3. Write the symbols and atomic masses of the elements with these atomic numbers.
   ____________ a. 12
   ____________ b. 74
   ____________ c. 92
4. Which element is most chemically like calcium, Ca? Circle the element.
   (a) K, (b) Sc, (c) Rb, (d) Sr, (e) Y
5. Which element is most chemically like argon, Ar? Circle the element.
   (a) Cl, (b) K, (c) Br, (d) Kr, (e) Rb
6. For each set of three elements, find out what all three elements have in common.
   ____________ a. Li, K, Na
   ____________ b. Na, Mg, Al
   ____________ c. F, Cl, I
   ____________ d. Ne, Ar, Kr
   ____________ e. H, Ne, N
7. Use the ionization energy chart (Figure 8-16) in your textbook to answer the following questions.
   ____________ a. What is the ionization energy of hydrogen?
   ____________ b. What is the element with the highest ionization energy?
   ____________ c. What is the element with the lowest ionization energy?
   ____________ d. What elements in Period 2 do not follow the general trend of increasing ionization energy as the atomic number increases?

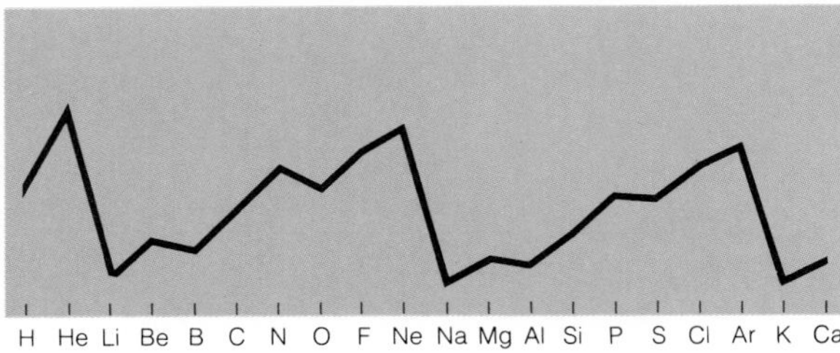

Figure 8-16 *The ionization energies for the first 20 elements of the periodic table are plotted in this graph.*

Although Mendeleev did not know it, the periodic arrangement of the elements corresponds to their electron arrangements. Elements in a group, such as the alkali metals or halogens, have similar chemical properties because they have similar electron arrangements. The changes in chemical properties across a period correspond to the changes in the elements' electron arrangements.

ionization energy (eye uh nuh ZAY shuhn)

The energy required to remove an electron from an atom is called the **ionization energy**. A graph of the ionization energies for the first 20 elements is shown in Figure 8-16. Notice that the energy required to remove an electron tends to increase as you go from lithium to neon or from sodium to argon. All the alkali metals have very low ionization energies. This explains why they are so reactive. It takes very little energy for these metals to lose their one outermost electron, forming an ion with a charge of +1. Elements in group II tend to lose two electrons, and so form ions with a charge of +2. Those in group VI have six electrons in the outermost shell of their atoms. Rather than losing six electrons (which would require a great deal of energy), they tend to gain two, forming ions with a charge of −2.

Checkpoint

1. How are elements ordered in the periodic table?
2. What is (a) a period, (b) a group?
3. Identify three chemical families in the periodic table.
4. What is ionization energy? How does it vary across a period in the periodic table?

Checkpoint Answers

1. According to atomic number. (Mendeleev arranged them according to atomic mass.)
2. (a) A period is a horizontal row of the periodic table. (b) A group is a vertical column of the periodic table.
3. Answers will vary. Three possibilities are alkali metals, halogens, and noble gases.
4. Ionization energy is the energy required to remove an electron from an atom. From left to right, the ionization energy increases across a period in the periodic table.

8-6

Bonding of Elements

Goals

1. To explain how ionic bonding occurs.
2. To state how covalent bonding occurs.
3. To describe polar molecules.

An understanding of the electron arrangements of the elements can lead to an understanding of how the atoms react to form compounds. Recall the reaction that produces sodium chloride, discussed earlier. In reaching the stable noble-gas arrangement, sodium gives up an electron and chlorine receives one. After the electron transfer takes place, sodium has a positive charge and chlorine a negative charge. Oppositely charged particles attract each other. This attraction holds the ions together in **ionic bonds**. Many such ions may come together in a regular repeating arrangement to form a crystal. Figure 8-17 shows that in sodium chloride crystals, each ion is surrounded by ions of opposite charge. Ionic bonds usually form between atoms that have very different ionization energies. One kind of atom tends to lose electrons, and the other kind tends to gain electrons.

ionic bonds

Atoms sometimes form chemical bonds by sharing electrons rather than by gaining or losing them. Such bonds formed by sharing valence electrons are called **covalent bonds**. You have already seen an example of covalent bonding in the case of the fluorine (F_2) molecule. By sharing electrons, atoms may acquire enough to fill their outermost shells. Covalent bonds usually form between atoms that have similar ionization energies. Neither atom in the molecule tends to give up electrons to the other atom.

covalent bonds (koh VAY lunt)

When atoms in a compound have ionization energies that are neither very similar nor very different, an unequal sharing of electrons may result. Such a bond is called a *polar covalent bond*. The electrons in such bonds are held more closely to one atom than to another. One side of the bond is slightly positive, while the other is slightly negative. The bond between hydrogen and chlorine in the hydrogen chloride (HCl) molecule, for example, is a polar covalent bond. The chlorine has a greater attraction for electrons than hydrogen does. Thus the chlorine end of the HCl bond is

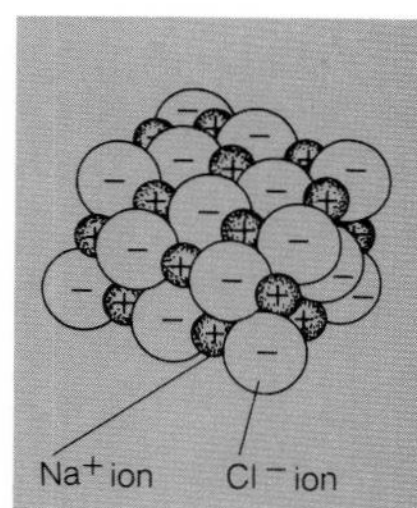

Figure 8-17 *In a sodium chloride crystal, ionic bonds hold together particles of opposite charge.*

Teaching Suggestions 8-6

A good way to open this section is to show students some table salt (NaCl). Ask them to locate sodium and chlorine in the periodic table on student text page 205. Then have them examine the ionization energies of these elements in Figure 8-16 on student text page 206. Ask what type of bonding they might expect in NaCl (ionic).

Repeat for water, H_2O (covalent); sugar, $C_{12}H_{22}O_{11}$ (covalent); and alcohol, C_2H_6O (covalent). You may wish to test these substances for the presence of ions by performing Demonstration 1.

Demonstrations

1. Use the setup in Figure TE 8-4 to test NaCl, water, sugar, and alcohol for the presence of ions. Place the electrodes in a small beaker of distilled water. The light will not go on because there are too few ions. Dry the electrodes and repeat the demonstration, using solid salt. Students may be surprised to see no current flow, but they should recognize that in the solid state the ions cannot move. Some may suggest dissolving the salt in water. Dissolved table salt will conduct electricity, as will other salt solutions. Then, try solid sugar, a sugar solution, and, finally, alcohol. (None of these will conduct electricity.)

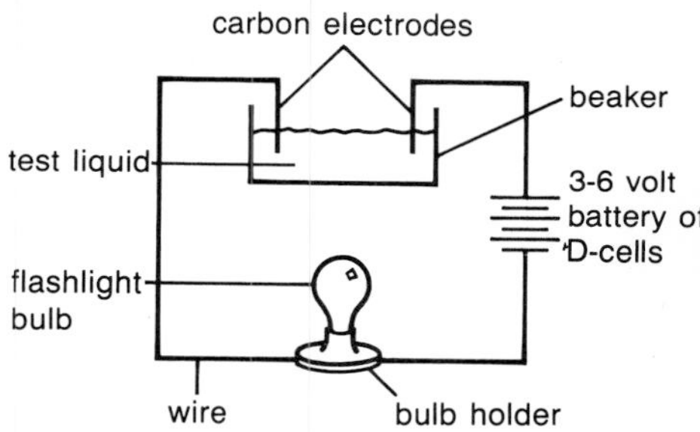

Figure TE 8-4 Setup to test conductivity of a liquid.

2. Show the polarity of water molecules by means of the demonstration described in the last paragraph on page 208 of the student text. Then repeat the experiment using a nonpolar liquid such as benzene. (**CAUTION:** Avoid inhaling benzene fumes. Use in a well-ventilated area.) Let the liquid flow from a burette or pipette into a beaker. The charged rod will have no effect when held near this stream, but it will still attract a column of water.

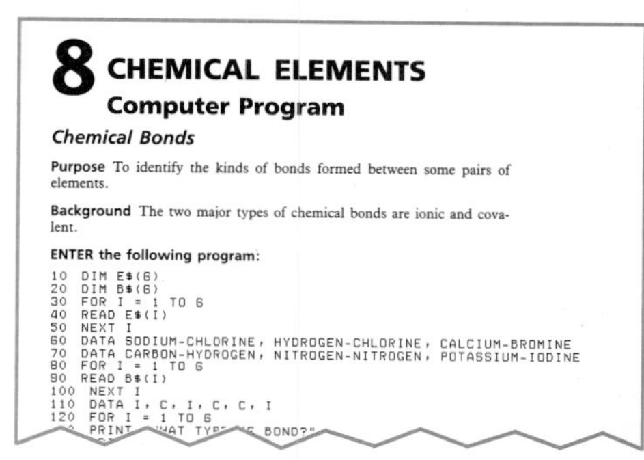

8 CHEMICAL ELEMENTS
Computer Program

Chemical Bonds

Purpose To identify the kinds of bonds formed between some pairs of elements.

Background The two major types of chemical bonds are ionic and covalent.

ENTER the following program:

```
10  DIM E$(6)
20  DIM B$(6)
30  FOR I = 1 TO 6
40  READ E$(I)
50  NEXT I
60  DATA SODIUM-CHLORINE, HYDROGEN-CHLORINE, CALCIUM-BROMINE
70  DATA CARBON-HYDROGEN, NITROGEN-NITROGEN, POTASSIUM-IODINE
80  FOR I = 1 TO 6
90  READ B$(I)
100  NEXT I
110  DATA I, C, I, C, C, I
120  FOR I = 1 TO 6
[illegible]  PRINT "[illegible]AT TYP[illegible] BOND?"
```

Teaching Tip

- With Demonstration 1 students observe that water does not conduct electricity, presumably because it has no ions to carry charge. A salt solution, however, does conduct, suggesting that salt consists of ions. Of course, it might be that the ions form as a result of interaction between the salt and the water. Testing a melted salt, such as silver nitrate, is a good way to show that salt ions exist independently of the water in which they are dissolved.

Cross Reference

Energy and chemical reactions are discussed further in Chapter 9, section 9-2.

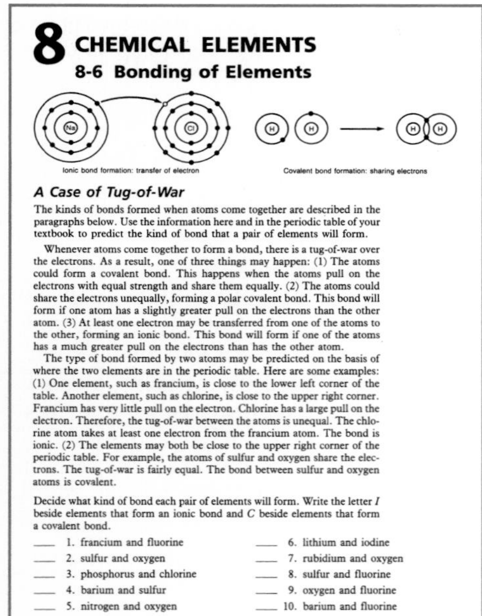

8 CHEMICAL ELEMENTS
8-6 Bonding of Elements

A Case of Tug-of-War

The kinds of bonds formed when atoms come together are described in the paragraphs below. Use the information here and in the periodic table of your textbook to predict the kind of bond that a pair of elements will form.

Whenever atoms come together to form a bond, there is a tug-of-war over the electrons. As a result, one of three things may happen: (1) The atoms could form a covalent bond. This happens when the atoms pull on the electrons with equal strength and share them equally. (2) The atoms could share the electrons unequally, forming a polar covalent bond. This bond will form if one atom has a slightly greater pull on the electrons than the other atom. (3) At least one electron may be transferred from one of the atoms to the other, forming an ionic bond. This bond will form if one of the atoms has a much greater pull on the electrons than has the other atom.

The type of bond formed by two atoms may be predicted on the basis of where the two elements are in the periodic table. Here are some examples: (1) One element, such as francium, is close to the lower left corner of the table. Another element, such as chlorine, is close to the upper right corner. Francium has very little pull on the electron. Chlorine has a large pull on the electron. Therefore, the tug-of-war between the atoms is unequal. The chlorine atom takes at least one electron from the francium atom. The bond is ionic. (2) The elements may both be close to the upper right corner of the periodic table. For example, the atoms of sulfur and oxygen share the electrons. The tug-of-war is fairly equal. The bond between sulfur and oxygen atoms is covalent.

Decide what kind of bond each pair of elements will form. Write the letter *I* beside elements that form an ionic bond and *C* beside elements that form a covalent bond.

____ 1. francium and fluorine	____ 6. lithium and iodine
____ 2. sulfur and oxygen	____ 7. rubidium and oxygen
____ 3. phosphorus and chlorine	____ 8. sulfur and fluorine
____ 4. barium and sulfur	____ 9. oxygen and fluorine
____ 5. nitrogen and oxygen	____ 10. barium and fluorine

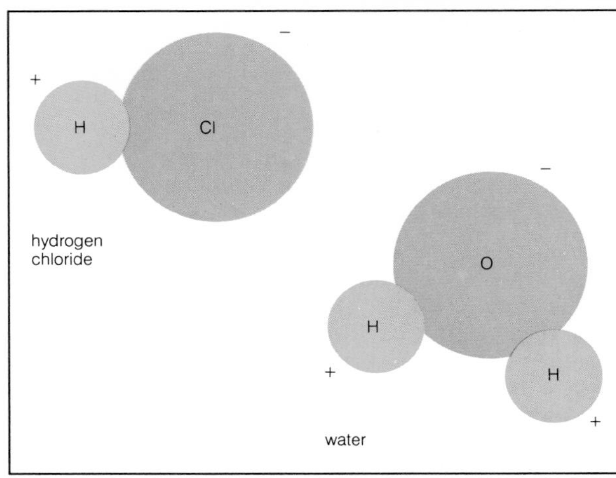

Figure 8-18 *Hydrogen chloride and water are examples of polar molecules.*

slightly negative and the hydrogen end slightly positive.

polar molecules

Molecules such as hydrogen chloride are called **polar molecules**. Water molecules are also polar. The two hydrogen atoms in water are slightly positive, and the oxygen is slightly negative. Note that the atoms in water are not in a line, as is shown in Figure 8-18. If all three atoms in H_2O were in a line, the two polar bonds would "cancel out" and the molecule would not be polar. Thus, the polarity of a molecule depends upon its shape as well as its bonds.

An effect of the polarity of water molecules can easily be observed. A charged rubber rod held near a thin stream of water causes the stream to bend. The negatively charged rod attracts the positive ends of the water molecules, and the molecules are pulled sideways.

Checkpoint

1. What kind of chemical bond holds sodium chloride crystals together?
2. What are covalent bonds? Give some examples.
3. What are polar covalent bonds? Give at least two examples.

Checkpoint Answers

1. Ionic bonds.
2. Covalent bonds are bonds in which electrons are shared. The bonds between Cl and Cl in Cl_2 or between F and F in F_2 are examples.
3. Polar covalent bonds are covalent bonds in which the electrons are closer to one atom or group of atoms, so that one side of the bond is slightly positive and the other is slightly negative. Examples are found in the molecules HCl and H_2O.

8-7

Carbon: Element Six

ENRICHMENT

Goals

1. To explain why there are so many carbon compounds.
2. To tell what a structural formula represents.

The element carbon can occur in the form of graphite or diamonds. Carbon forms many more different compounds than any other element. Carbon compounds make up fossil fuels—coal, petroleum, and natural gas. The energy in these fuels is used to heat homes, to drive turbines, and to turn the wheels of automobiles. Coal and petroleum also serve as the basic materials for making many plastics, fertilizers, perfumes, synthetic fabrics, medicines, and other substances. (See Figure 8-19.)

The large number of carbon compounds is primarily due to the ability of carbon atoms to form covalent bonds with other carbon atoms. The linking of carbon atoms can produce chains of carbon atoms of almost any length.

Figure 8-19 *Fossil fuels serve as the basic materials for making compounds found in a wide variety of products.*

Teaching Suggestions 8-7

Begin by showing the class some sugar in a large beaker. Have a student or two taste the white crystals to verify that they are sugar. (CAUTION: Students should never taste an unidentified substance on their own.) Then ask what elements are present in sugar (C, H, O). Some students may know the formula for sucrose, $C_{12}H_{22}O_{11}$. Next, ask how we know there is carbon in sugar. Some students may know that heating sugar strongly drives off water, forming caramel first and, eventually, carbon. There is another way to remove water from sugar. Concentrated sulfuric acid's dehydrating ability will remove the $H_{22}O_{11}$ portion of the sugar molecule. (See Demonstration section.)

Use the spheres and sticks of a molecular model kit to show the structures of some of the molecules discussed in this section. Depending on what is available in the kit, you may be able to include the double and triple bonds in such compounds as C_2H_4 and C_2H_2, respectively.

Background

The study of carbon and its compounds constitutes organic chemistry. This is a discipline of its own because of the importance and vast number of such compounds.

Millions of years ago, Earth's climate was very warm and humid. Plants and many kinds of animals, including the now-extinct dinosaurs, flourished. As these plants and animals died, their remains formed thick layers of decaying matter. The build-up of these layers produced high pressures and temperatures that squeezed out moisture and generated chemical changes leading to the formation of coal, oil, and natural gas. Because these fuels are the remains of ancient plants and animals, they are called fossil fuels.

Demonstration

To remove water from sugar, add some concentrated sulfuric acid to some sugar in the bottom of a large beaker. (CAUTION: Concentrated sulfuric acid is a powerful oxidizing and dehydrating agent. Avoid contact with skin; wear safety goggles.) Black carbon, which takes on a light, air-filled structure, will seem to grow upward from the bottom of the beaker.

Teaching Tip

- Students may be interested in discussing the worldwide consumption of fossil fuels, the finite nature of such fuels, and alternative sources of energy, such as nuclear, solar, and wind energy.

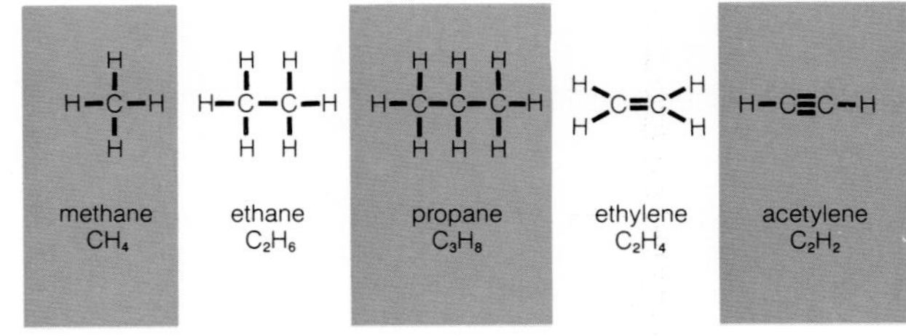

Figure 8-20 *Single lines represent single bonds, double lines double bonds, and triple lines represent triple bonds in the structural formulas of the hydrocarbons shown here.*

hydrocarbons

structural formulas

Compounds that contain only hydrogen and carbon are called **hydrocarbons**. Methane, ethane, and propane are three simple hydrocarbons. They are gases that are commonly used as fuels. Their molecular formulas are CH_4, C_2H_6, and C_3H_8, respectively. *Molecular formulas* show only the type and number of atoms in a molecule. **Structural formulas** are used to show the arrangement of atoms in a molecule. In a structural formula, each pair of shared electrons is shown as a single line. Such a pair of electrons is called a *single bond*. (See Figure 8-20.) Each carbon atom can form four single bonds. Therefore, each carbon atom in a structural formula must be surrounded by four lines.

Atoms may sometimes share more than a single pair of electrons. If atoms share two pairs of electrons, they form a *double bond*. If they share three pairs of electrons, they form a *triple bond*. In structural formulas, double bonds are represented by two lines, and triple bonds by three.

The simplest hydrocarbon having a double bond is ethene (C_2H_4), or ethylene. Ethylene molecules can react with each other, forming very long chains, or *polymers*, such as polyethylene, a common plastic. The simplest hydrocarbon having a triple bond is ethyne (C_2H_2), or acetylene. This compound is commonly burned in oxygen, producing a very hot flame for welding.

Checkpoint

1. Why are there so many carbon compounds?
2. What information does a structural formula give that a molecular formula does not give?

Facts At Your Fingertips

- Until the nineteenth century it was believed that organic compounds could be produced only by living organisms. Then, in 1828, Frederick Wöhler synthesized urea, a component of urine, in his laboratory. Since that time chemists have synthesized a vast number of organic compounds, including molecules as complex as insulin.
- More than 90% of the energy used in the United States comes from fossil fuels. In 1976, some 47.3% of the energy came from oil. In the same year, coal supplied 18.5% of the energy, and 27.4% came from natural gas.
- Nearly 60% of the oil consumed in the U.S. is used to transport people and materials in cars, trucks, buses, trains, and planes.
- About 35% of the natural gas consumed in the U.S. is used to heat homes and businesses; 45% is consumed by industry. About 20% of the coal dug from U.S. soil is used in industry. A large fraction of this is required to manufacture steel.

8 CHEMICAL ELEMENTS

8-7 Carbon: Element Six

Regular octane rating of 88
Unleaded 87
Super Unleaded 92 (alcohol added)

Help Stamp Out Knocking!

The paragraphs below describe how gasoline is rated. Read the paragraphs. Then write the answers to the questions that follow.

The sign reads, "Regular, octane rating of 88; unleaded, 87; super unleaded, 92, alcohol added." The type of fuel is fairly well understood. But what about the octane rating?

The main ingredients of gasoline are hydrocarbons with seven, eight, or nine carbons. Some of these hydrocarbons burn quite well in automobile engines. They burn in a way that causes the pistons of the engine to move smoothly up and down in the cylinder. Other hydrocarbons burn too quickly, upsetting the motion of the pistons. This rapid burning causes engine knock—a kind of pinging noise. The engine loses power.

The octane number is based on an arbitrary scale used to show you how well the fuel will perform in the engine. N-heptane is an example of a hydrocarbon that burns too rapidly. It has been assigned an octane number of zero. The other extreme is isooctane. Isooctane burns smoothly in the engine. It has been assigned an octane number of 100. The octane number posted on the pump indicates how well the fuel will perform compared to these two standards. A gasoline rated at 88 will burn in a manner similar to a mixture made up of 88% isooctane and 12% n-heptane. But the fuel is not just a mixture of these two hydrocarbons; it simply burns in a manner similar to such a mixture.

Most fuels are rated with an octane number between 90 and 100. (Some fuels used in special situations like racing are rated even greater than 100.) Fuels with higher octane numbers will burn more smoothly, but they will also cost more. The owner's manual for a car may discuss the octane number range that is best for the car. The car owner may want to use this information to experiment with different fuels to find the one that performs well, but does not cost too much.

1. What are the main ingredients of gasoline? ______
2. What causes engine knock? ______
3. What is the octane number? ______
4. What is the octane number of n-heptane? ______ Of isooctane? ______
5. What is meant when we say that a fuel has an octane number of 90? ______

Checkpoint Answers

1. The ability of carbon atoms to form covalent bonds with other carbon atoms gives rise to carbon chains of almost any length.
2. A structural formula shows the arrangement of atoms in a molecule, as well as the types and relative numbers of atoms in the molecule.

Photoelectric Cells

Have you ever reached out to push open a supermarket door when, suddenly, as if by magic, the door springs open? The device that makes this happen is often a photoelectric cell, sometimes called an electric "eye." It is the same device used to open garage doors, to turn on lights automatically after sunset, and to set off burglar alarms.

Electrons can be made to flow from a negatively charged *cathode* to a positively charged *anode*, within a tube from which most of the air has been removed. Such a tube is called a *vacuum tube*. Modern photoelectric cells may be vacuum tubes. When light shines on the cathode, electrons are released. (See Figure 1.) The cathode is often coated with a thin layer of an alkali metal, such as cesium or potassium. These metals release electrons easily.

Ultraviolet light, an invisible high-energy light, shining on the cathode in a burglar alarm system causes a constant flow of electrons from the cathode to the anode. Ultraviolet light, sometimes called black light, is used because it cannot be seen by a burglar. If the burglar crosses the ultraviolet light beam, electrons stop flowing because there is no light to drive the electrons out of the alkali-metal atoms on the cathode. This causes the alarm to go off. The diagram in Figure 2 shows a burglar-alarm system. Some smoke alarms work in much the same way. A tiny ray of light inside is interrupted by smoke particles and the alarm sounds.

The solar, or photovoltaic, cells that convert sunlight to electricity are also photoelectric cells. Solar cells are often seen in a group on the roofs of houses. Solar cells may also power your watch or calculator. The electrons usually come from a metalloid semiconductor rather than from an alkali-metal coating on a vacuum-tube cathode.

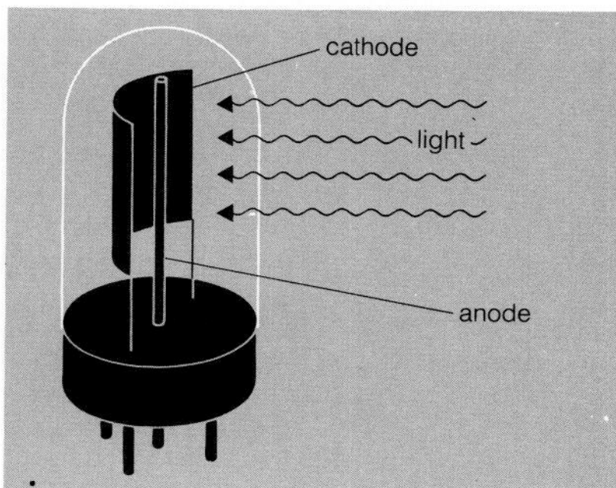

Figure 1

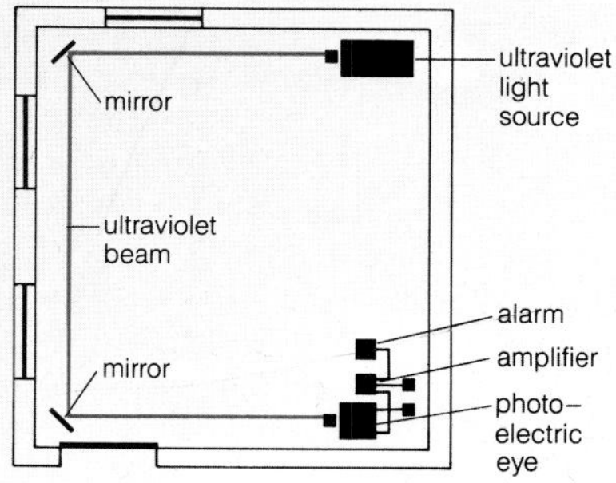

Figure 2

Using The Feature Article

If you have a photoelectric cell or a set of solar cells, show students how they operate. Because the currents involved are quite small, you will need a sensitive ammeter to detect them. In addition, try to bring in a camera with a built-in light meter. Your class will enjoy learning how the meter functions to control exposure.

Reading Suggestions

For The Teacher

Asimov, Isaac. THE NOBLE GASES. New York: Basic Books, 1976. Discusses the composition, sources, and uses of the noble gases.

Bickel, Lennard. THE DEADLY ELEMENT: THE STORY OF URANIUM. New York: Stein and Day, 1979. The history of the development of uranium and the implications for the future.

Sagan, Dorion. "Sulfur: Toward a Global Metabolism." *The Science Teacher*, January 1986, pp. 15–20. The element sulfur on Earth, including its role in certain microorganisms.

Weeks, Mary E. DISCOVERY OF THE ELEMENTS. 7th ed. Easton, PA: Journal of Chemical Education, 1968. A classic history that discusses the discovery of the chemical elements beginning with the ancient Greeks.

For The Student

______ "New Alloys to Replace Rare Metals." *Science Digest*, May 1983, p. 28. Efforts of scientists to find more-abundant alloys for use as special metals.

Asimov, Isaac. BUILDING BLOCKS OF THE UNIVERSE. Rev. ed. New York: Harper & Row, 1974. Describes the elements, including information about their discovery, naming, structure, unique properties, uses, and, in some cases, their dangers.

Ridgeway, James. "Stalking Strategic Metals." *Science Digest*, February 1983, pp. 42–43. Dramatic new discoveries of scarce minerals.

Taylor, Ron. THE INVISIBLE WORLD. New York: Facts on File, 1985. Deals with matter and forces that must be sensed and interpreted, not things that can be seen or touched.

Understanding The Chapter 8

Main Ideas

1. Elements differ greatly in physical and chemical properties.
2. Elements can be classified as metals, nonmetals, and metalloids on the basis of their properties.
3. The noble gases are a group of very unreactive elements.
4. The unreactive, stable nature of the noble gases is due to filled outermost electron shells.
5. The halogens are a very reactive family of elements that have seven electrons in their outermost shell. Halogen atoms can become halide ions by gaining one electron, giving them the stable electron arrangement of the noble-gas atoms.
6. Alkali metals are a very reactive family of elements that have one electron in their outermost electron shell. They attain the noble-gas electron arrangement by losing an electron.
7. Elements are arranged in the periodic table according to their atomic number. The vertical columns of the table contain families of elements that have similar properties.
8. Some compounds form when electrons are transferred from the atoms of one element to the atoms of another. Ionic bonds result.
9. Some compounds form when atoms of different elements share electrons, to form covalent bonds.
10. If one atom in a covalent bond has a greater attraction for electrons than the other, a polar covalent bond results.
11. Carbon atoms form four covalent bonds with one another as well as with other elements.

Vocabulary Review

From the following list, choose the term that best completes each of the statements. Write your answers on a separate piece of paper.

alkali metal	ionization energy
covalent bond	metal
crystal	metalloid
family	noble gas
group	nonmetal
halide	period
halogen	periodic table
hydrocarbon	polar covalent
ion	polar molecule
ionic bond	structural formula

1. Elements arranged vertically in the periodic table constitute a(n) __?__, or __?__, of elements.
2. The bond that forms when an alkali metal combines with a halogen to form a compound is a(n) __?__.
3. A(n) __?__ is a compound made from a halogen and another element.
4. Bromine is a(n) __?__.
5. Most __?__ elements have atoms with eight electrons in the outermost shell.

Vocabulary Review Answers

1. group, family
2. ionic bond
3. halide
4. halogen
5. noble gas
6. nonmetal
7. alkali metal
8. polar molecule
9. covalent bond
10. period
11. metal
12. polar covalent
13. crystal
14. ion

6. ___?___ elements are poor conductors of heat and electricity.
7. ___?___ elements have atoms with one electron in the outermost shell.
8. The water molecule is an example of a(n) ___?___.
9. A(n) ___?___ forms when atoms share electrons.
10. A horizontal row in the periodic table is called a(n) ___?___.
11. ___?___ elements are good conductors of heat and electricity.
12. If shared electrons are attracted more strongly by one atom than another, the bond is said to be ___?___.
13. A(n) ___?___ is a regular, repeated arrangement of the particles of which it is made.
14. An atom, or group of atoms, that has gained or lost one or more electrons is a(n) ___?___.

Chapter Review

Write your answers on a separate piece of paper.

Know The Facts

1. ___?___ is a nonmetal. (lead, sodium, fluorine)
2. The elements lithium, carbon, and fluorine are all found in the same ___?___ of the periodic table. (period, column, group)
3. ___?___ is a metalloid. (sodium, fluorine, silicon)
4. Metals tend to release electrons ___?___ nonmetals. (more readily than, less readily than, about as readily as)
5. ___?___ atoms have eight electrons in their outermost shell. (helium, chlorine, argon)
6. Atoms of the ___?___ elements have seven electrons in their outermost shell. (noble-gas, halogen, second-period)
7. Halogens are not found in the ___?___ period of the periodic table. (first, second, fourth)
8. When a halogen atom gains an electron it becomes a(n) ___?___. (noble gas, alkali metal, ion)
9. The noble gases have ___?___ ionization energies. (high, low, intermediate)
10. ___?___ formulas show the arrangement of atoms in a molecule. (chemical, molecular, structural)

Understand The Concepts

11. Explain why argon is commonly used in light bulbs.
12. All the halogen elements are poisonous. Describe three halogen compounds that are not.
13. List the alkali metals in terms of their (a) increasing atomic mass, (b) increasing melting point, (c) increasing boiling point.
14. Explain how the reaction of an alkali metal with a halogen produces ions with the same outermost electron arrangement as in a noble-gas atom.
15. Why is liquid helium used in preparing very low-temperature environments?
16. List five ways in which the properties of metals and nonmetals are different.

Chapter Review Answers

Know The Facts

1. fluorine
2. period
3. silicon
4. more readily than
5. argon
6. halogen
7. first
8. ion
9. high
10. structural

Understand The Concepts

11. Argon is used in light bulbs because of its low reactivity.
12. Answers will vary. Three possibilities are sodium chloride, or table salt; fluoride salts in toothpastes; iodide salts, which are added to table salt to aid in the functioning of the thyroid gland.
13. (a) Li, Na, K, Rb, Cs; (b) and (c) Cs, Rb, K, Na, Li.
14. Alkali-metal atoms have one electron in the outermost, or valence, shell of electrons. The next inner shell has eight electrons. Halogen atoms all have seven electrons in the outermost shell. By transferring an electron from an alkali-metal atom to a halogen atom, the alkali-metal atom becomes a positively charged ion, the resulting halide ion has a negative charge, and both ions end up with eight electrons in their outermost shells. This is the arrangement of outermost electrons found in a noble-gas atom.
15. Because liquid helium has the lowest boiling point of any substance known.
16. Answers will vary. See Table 8-1, student text page 194.

17. Answers will vary. See student text, page 201. See also answer to Checkpoint question 4 on page 201.
18. A covalent bond is formed by the sharing of valence electrons by the bonding atoms, whereas in an ionic bond, one or more electrons have been transferred from one atom to another, forming oppositely charged ions that are held together by their attraction to each other.
19. Answers will vary. Some possibilities are: (a) NaCl, NaI, and NaF; (b) Br_2, F_2, and I_2.
20. Answers will vary. See student text, page 209.

Challenge Your Understanding

21. Covalent. Chlorine atoms have seven electrons in the outermost shell. Thus, by sharing one pair, each atom in Cl_2 acquires the stable 8-electron arrangement.
22. $2K + Cl_2 \longrightarrow 2KCl$
    $H_2 + Br_2 \longrightarrow 2HBr$
    $2Na + 2H_2O \longrightarrow 2NaOH + H_2$
23. The charge on the nucleus increases across a period, while each electron added is in the same shell. Thus, the attractive force pulling each electron toward the nucleus is greater. As a result, it takes more energy to remove an outer electron from an element at the right side of the periodic table.
24.

```
        H  H  H
        |  |  |
(a) H—C—C—C—H
        |  |  |
        H  H  H

        H  H
        |  |     H
(b) H—C—C=C/
        |      \H
        H

        H
        |
(c) H—C—C≡C—H
        |
        H
```


17. List three practical uses for the halogens and halides.
18. Explain the difference between a covalent bond and an ionic bond.
19. Give two examples of compounds that have (a) ionic bonds, (b) covalent bonds.
20. List four products made from carbon compounds.

Challenge Your Understanding

21. Would you expect the bonds between halogen atoms in the molecule Cl_2 to be ionic, covalent, or polar covalent? Explain your answer.
22. Write the chemical equations that describe the reactions between (a) potassium and chlorine, (b) hydrogen and bromine, (c) sodium and water.
23. Why do you think the ionization energy shown in Figure 8-16 increases across a period from lithium to neon or sodium to argon?
24. Draw the structural formula for a three-carbon hydrocarbon having (a) all single bonds, (b) a double bond, (c) a triple bond.

Projects

1. Research and report on the contributions to chemistry of any one of the following scientists: Marguerite Perey; Dorothy Hodgkin; Isabella Karle.
2. Visit a steel plant or do some research in your library to find out how steel is made.
3. Go to the supermarket and look at the lists of ingredients on various packages of salt. How does kosher salt differ from other table salt? What is "lite salt"? What is "no salt"?
4. The alkali metals in group I of the periodic table will react with water to form hydrogen. Will metals of group II react with it to form hydrogen? To find out, add small chips of magnesium to a test tube of cold water, with your teacher's supervision. Try the same thing using hot water. Do you see any evidence of a reaction? Repeat the experiment, using calcium. What do you find? How do you explain your results?
5. Investigate the controversy regarding the addition of fluoride compounds to public water supplies. What are the positions involved? What is your position after your investigation?
6. Compounds that have the same molecular formulas but different structural formulas and different properties are known as *isomers*. Do some library research to find examples of isomers. Make a poster showing the names, molecular formulas, and structural formulas of these isomers.

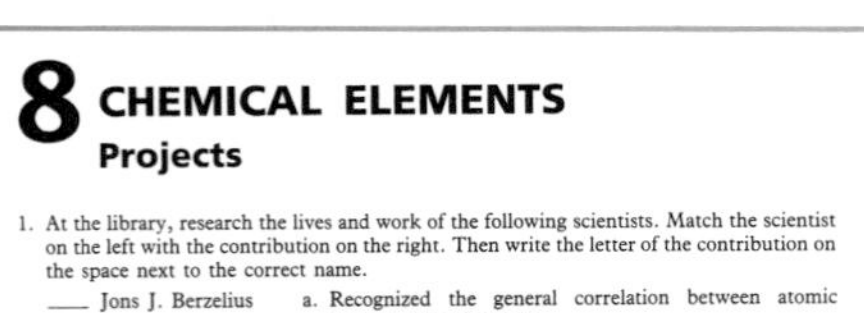

8 CHEMICAL ELEMENTS
Projects

1. At the library, research the lives and work of the following scientists. Match the scientist on the left with the contribution on the right. Then write the letter of the contribution on the space next to the correct name.

___ Jons J. Berzelius
___ Dimitri Mendeleev
___ J. Lothar Meya
___ H. G. J. Moseley
___ John Newlands

a. Recognized the general correlation between atomic weights and properties.
b. Developed the specific relationship between atomic weight and properties of elements.
c. Devised the periodic table, which organized the elements and allowed the prediction of properties of new elements.
d. Discovered the relationship between an x-ray spectrum and the atomic number of an element emitting a spectrum.
e. Originated the system of writing chemical symbols and formulas.

2. In a reference book, find the information called for below.
a. Identify the element described by the following properties.

8 CHEMICAL ELEMENTS
Review

Find That Term!

Column B contains clues that describe terms introduced in Chapter 8. Read each clue and decide what term is being described. Then write the term on the corresponding numbered space in Column A.

A	B
1. ________	1. Substances that are good conductors of heat and electricity.
2. ________	2. Substances that are dull, brittle, and light for their size.
3. ________	3. Kind of conductor that a metal becomes at very low temperatures.
4. ________	4. Type of number that describes the number of protons in the atom.
5. ________	5. Word associated with electrons in the outermost energy level.
6. ________	6. Name for a compound made up of a halogen and another element

Chapter

9

Chemical Reactions

A chemist working at American University in Washington, D.C.

Vocabulary Preview

exothermic
endothermic
oxidation
reduction
acids
indicator
bases
salt
neutralization
*p*H scale
catalyst

CHAPTER 9 Overview

This chapter introduces the nature of chemical reactions and the factors that influence chemical reactions. Energy considerations are discussed first. The energy changes during physical changes of state are contrasted with those that occur during chemical reactions. Exothermic and endothermic reactions are discussed, and oxidation and reduction are explained. Electrochemical cells are then described, and corrosion and the relative chemical activities of metals are explained. Acids and bases are then defined, and the *p*H scale and neutralization reactions are explained. Finally, the factors that control the rates of chemical reactions are discussed.

Goals

At the end of this chapter, students should be able to:

1. contrast the particle motions in the three states of matter.
2. describe the energy changes that occur when matter changes state.
3. explain what happens to its particles when matter changes state.
4. explain why the energy associated with chemical changes is, in general, greater than that associated with physical changes.
5. explain the difference between exothermic and endothermic reactions.
6. define oxidation and reduction.
7. give examples of oxidation-reduction reactions.
8. describe how electrochemical cells work.
9. explain how unwanted electrochemical reactions can cause corrosion.
10. describe some ways of preventing corrosion.
11. state the properties of acids and bases.
12. describe what happens when an acid and a base react.
13. explain what is meant by the *p*H of a solution.
14. state the factors that affect the rate of a chemical reaction.
15. explain why these factors have their effects.

Teaching Suggestions 9-1

One way to open the chapter is with the demonstration (see Demonstration), which shows quantitatively the energy changes involved in heating and boiling water. Be sure students understand that the heat added to water after it has begun to boil is used to separate the molecules of water, which is changing from the liquid state to the vapor state. At the boiling point, the temperature of the water remains the same because the average energy of motion of the molecules in the liquid and vapor states remains unchanged.

Make certain students understand that temperature is a quantitative measure of the average energy of motion of the molecules in a substance. A drop in temperature indicates a slowing of molecular motion; a rise means a speedup.

Background

To convert a liquid to a gas at the liquid's boiling point requires energy, because molecules attract one another. To separate molecules, work must be done. A force equal to, and opposite in direction from, the attractive forces between the molecules must be exerted through the distance the molecules move as they separate. This is similar to lifting an object. The work required to lift the object is equal to the gravitational pull on the object times the distance the object was lifted (weight × height). The lifted object has potential energy, just as the molecules of a vaporized liquid have additional energy.

If the object is released, it will fall, gaining an amount of kinetic energy (energy of motion) equal to the potential energy lost. When the object strikes the ground, thermal energy is released in a quantity equal to the potential energy lost (weight × height). Similarly, when a gas condenses, the potential energy stored in the separated molecules of gas (the heat of vaporization) is released as thermal energy (the heat of condensation) when the gas condenses to a liquid.

9-1

Energy and Changes of State

Goals

1. To contrast the particle motions in the three states of matter.
2. To describe the energy changes that occur when matter changes state.
3. To explain what happens to its particles when matter changes state.

The smallest particles of elements and compounds are constantly in motion. The molecules of water in a glass of water, the sodium and chloride ions in a shaker of salt, and the neon atoms in a neon sign are all moving. In the solid state, particles only vibrate back and forth about fixed positions. In a liquid, the particles are so close together that they do not move very far before colliding with other particles. In a gas, the particles are so far apart that they move nearly independently of each other. In Figure 9-1, you can compare the motions of particles in the three states of matter.

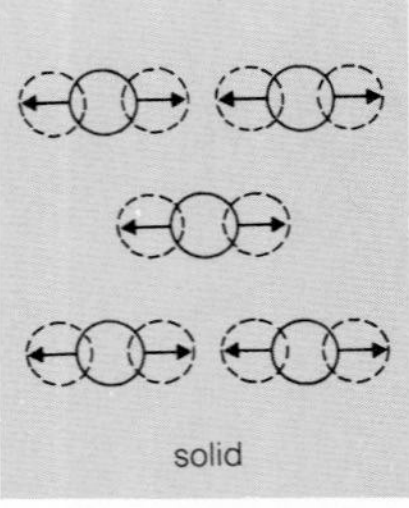

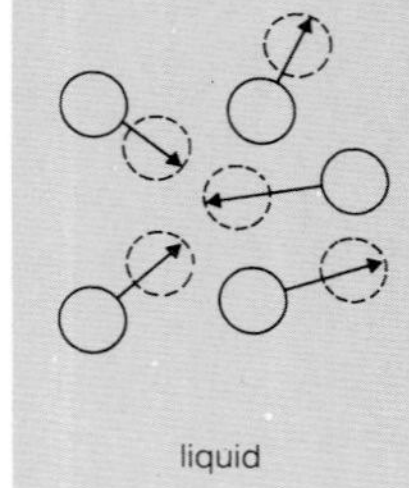

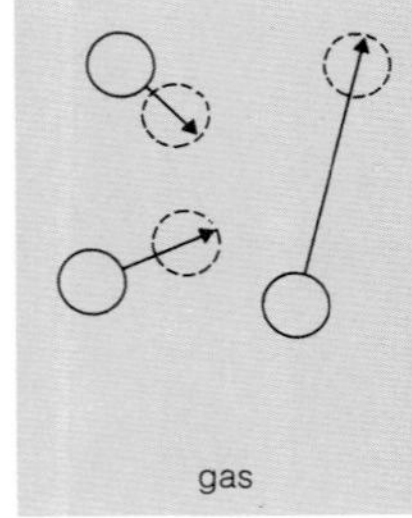

Figure 9-1 *The motions of particles in solids, liquids, and gases are represented above, using arrows to suggest movement.*

The *temperature* of a substance is a measure of the average energy of motion of its particles. The higher the temperature of a substance, the greater the average energy and average speed. When heat is applied to a substance, however, change in temperature does not always result. For example, as a container full of water is heated, a thermometer suspended in the water shows an increase in temperature only until the water starts to boil. At that point, the temperature stops rising. It remains at 100°C as long as the water continues to boil. If the temperature of the steam given off by the boiling water is measured, it is found to be the same as that of the boiling water, 100°C. At the boiling point, all of the energy being supplied to the water is used to change the water from liquid to gas.

Water molecules attract each other strongly because they contain polar bonds. Although a water molecule is neutral, the charge within the molecule is not distributed evenly. The hydrogen atoms have a slightly positive charge and the oxygen atom has a slightly negative charge. As a result, the hydrogen atoms in a water molecule are attracted to the oxygen atoms in nearby water molecules, and vice

Demonstration

Show quantitatively the energy changes involved in heating and boiling water.

First, calibrate a 200- to 300-watt immersion heater. To do this, place the heater in 100 mL (100 grams) of cold water in an insulated coffee cup. Then, measure the temperature of the water (t_1). Plug in the heater for 30 seconds. After pulling the plug, use the heater coil to stir the water until it reaches its final (maximum) temperature. Record this temperature (t_2).

Starting at t_2, bring the water to a boil. Continue heating the water for 6 minutes. Use a thermometer to show students that the boiling water's temperature remains constant at about 100°C. Disconnect the heater, stir, and then remove the heater from the cup. Cover the cup with a glass plate.

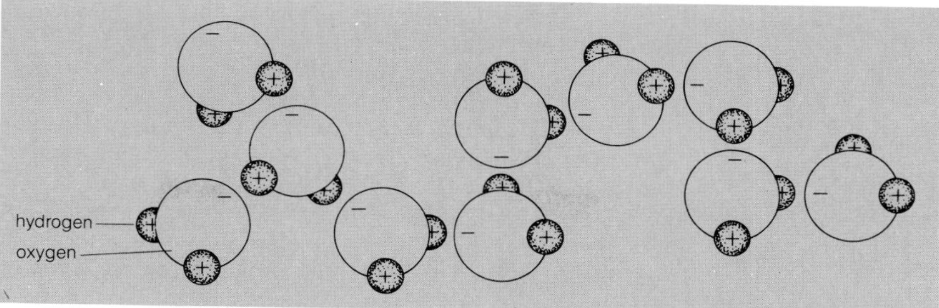

Figure 9-2 The hydrogen atoms in a water molecule have a slightly positive charge. They attract the slightly negative oxygen atoms in other water molecules.

versa. (See Figure 9-2.) The closer two oppositely charged objects are, the stronger the attraction between them.

Water molecules are much closer together in liquid water than in steam. When water changes to steam, the volume becomes more than 1500 times as great. Since water molecules attract each other strongly in liquid water, energy is needed to move them apart.

More energy is needed to change water into steam, or *vaporize* it, than is needed to vaporize many other substances. The energy needed to change one gram of a liquid to a gas at the boiling point is the *heat of vaporization* (vay pur uh ZAY shuhn). Of the substances listed in Table 9-1, water has the greatest heat of vaporization. Each gram of water that changes to steam absorbs 2300 joules of energy. The *joule* (J) is a unit of energy, or heat, used by scientists. (The energy needed to lift an apple a distance of one meter is

Table 9-1: Some Heats of Vaporization

Substance	*Heat of vaporization (J)*	*Boiling point (°C)*
water	2300	100
air	214	−191
carbon dioxide	365	−56
grain alcohol	860	78
helium	25	−269
oxygen	214	−184
phosphorus	546	200
sulfur dioxide	400	−10
wood alcohol	1100	65

The quantity of heat needed to raise the temperature of one gram of water one degree Celsius is 4.2 J. It follows that the heat transferred to the water in 30 seconds (H) is the product of the mass (m) in grams, the constant (4.2 J), and the temperature change in degrees Celsius ($t_2 - t_1$).

$$H = 4.2m(t_2 - t_1)$$

Compute the heat that was transferred in 30 seconds. Multiply by 12 to find the heat transferred in the 6-minute period. Then compute the heat transferred in raising the temperature of the water from t_2 to 100°C. Find the difference between this figure and the heat transferred in the 6-minute period. This "missing heat" is the heat used to vaporize the volume of water that has boiled away. Pour the remaining water into a graduated cylinder to measure its volume. Subtract this figure from 100 mL to find the volume of water that was vaporized.

The heat needed to vaporize this volume of water can be assumed to be the "missing heat" calculated above. It follows that the heat needed to boil away 1 g of water (the heat of vaporization) is the ratio of the missing heat to the mass of water vaporized. Since the density of water is approximately 1 g per mL, the mass in grams is numerically equal to the volume in milliliters.

Point out to students that the value calculated for the heat of vaporization is only a rough approximation. Some heat is lost to the surrounding air and some droplets of water may have spattered out of the cup.

Teaching Tip

- When students study Table 9-1 on student text page 217, be sure they realize that heat of vaporization is higher for water than for most substances. This is an indication of the strength of the attractive forces between water molecules.

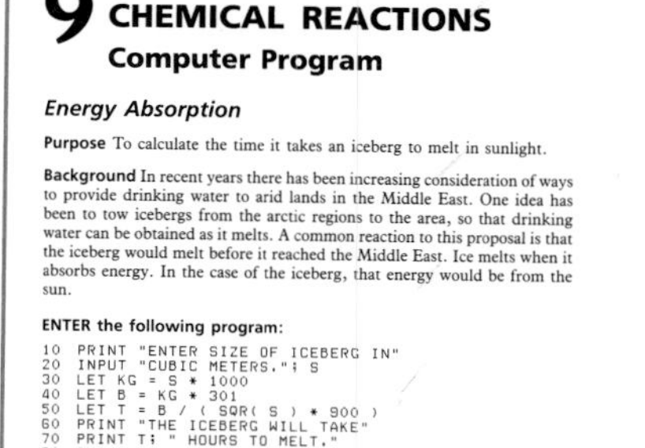

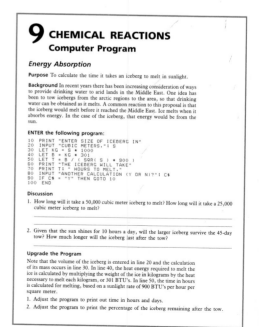

9 CHEMICAL REACTIONS
Computer Program

Energy Absorption

Purpose To calculate the time it takes an iceberg to melt in sunlight.

Background In recent years there has been increasing consideration of ways to provide drinking water to arid lands in the Middle East. One idea has been to tow icebergs from the arctic regions to the area, so that drinking water can be obtained as it melts. A common reaction to this proposal is that the iceberg would melt before it reached the Middle East. Ice melts when it absorbs energy. In the case of the iceberg, that energy would be from the sun.

ENTER the following program:

```
10  PRINT "ENTER SIZE OF ICEBERG IN"
20  INPUT "CUBIC METERS."; S
30  LET KG = S * 1000
40  LET B = KG * 301
50  LET T = B / ( SQR( S ) * 900 )
60  PRINT "THE ICEBERG WILL TAKE"
70  PRINT T; " HOURS TO MELT."
80  INPUT "ANOTHER CALCULATION (Y OR N)?"; C$
90  IF C$ = "Y" THEN GOTO 10
100  END
```

Discussion

1. How long will it take a 50,000 cubic meter iceberg to melt? How long will it take a 25,000 cubic meter iceberg to melt?

2. Given that the sun shines for 10 hours a day, will the larger iceberg survive the 45-day tow? How much longer will the iceberg last after the tow?

Upgrade the Program

Note that the volume of the iceberg is entered in line 20 and the calculation of its mass occurs in line 30. In line 40, the heat energy required to melt the ice is calculated by multiplying the weight of the ice in kilograms by the heat necessary to melt each kilogram, or 301 BTU's. In line 50, the time in hours is calculated for melting, based on a sunlight rate of 900 BTU's per hour per square meter.

1. Adjust the program to print out time in hours and days.
2. Adjust the program to print the percentage of the iceberg remaining after the tow.

Cross References

You may wish to review states of matter with your students. If so, refer to Chapter 6, section 6-2. This section provides the conceptual groundwork for the discussion of the water cycle in Chapter 11, section 11-4. Refer to Chapter 18, section 18-1, for more information about heat and temperature.

9 CHEMICAL REACTIONS

9-1 Energy and Changes of State

A Heated Matter

Read the following statements. If the statement is true, circle the *T* and leave the numbered space blank. If the statement is false, circle the *F* and replace the italicized term with a term to make the statement true.

1. ________ T F In a solid, the particles are very *far apart*.
2. ________ T F In a gas, the particles move nearly *independently* of each other.
3. ________ T F In a gas, the particles are moving *faster* than the particles are moving in a solid or liquid.
4. ________ T F The temperature of a substance is a measure of the *greatest* speed of the particles of the substance.
5. ________ T F The higher the temperature, the *greater* is the speed of the particles of a substance.
6. ________ T F When heat is applied to a substance, the particles in the substance *lose* energy.
7. ________ T F Steam burns do *the same amount of damage as* hot water burns.
8. ________ T F Evaporation of perspiration has a cooling effect because the perspiration *releases energy to* the skin as it evaporates.

Can You Solve These?

An energy change occurs in each of the following problems. On a separate sheet of paper, solve each problem. Then write your answer on the space at the left of the problem. Hint: Use Table 9-1 in your textbook.

________ 1. How many joules of energy are needed to change 1 g of phosphorus liquid to phosphorus gas?

________ 2. How many joules of energy are given off when 1 g of oxygen gas is changed to oxygen liquid?

________ 3. How many joules of energy are given off when 3 g of steam are changed to 3 g of water?

________ 4. How many joules of energy are given off when 2.5 g of helium condense?

________ 5. If 4,600 J of energy are released when steam changes to water, how many grams of steam are involved?

Using Try This

Students should recognize that the heat required to evaporate the water comes from the body. The flow of heat from the hand to the water makes the hand feel cool.

about one joule.) The energy absorbed during the change is stored in the steam molecules.

Steam burns can do much more damage than hot-water burns because steam has much more energy. Every gram of steam at 100°C that condenses on the skin releases nearly ten times the amount of energy released when one gram of water at 100°C cools to normal body temperature (37°C). When water on the skin evaporates, it absorbs energy from the skin. That is why the evaporation of perspiration has a cooling effect. For the same reason, a person feels cooler after swimming or taking a shower.

When one gram of steam changes back to liquid water, or condenses, it releases 2300 J. The energy gained by the water molecules when they moved apart to form the steam is given up when they move together again, forming liquid water. The *heat of condensation* (kahn dun SAY shuhn), the energy given up when a gas condenses to a liquid at its boiling point, is equal to the liquid's heat of vaporization.

When a solid melts, energy is absorbed. When a gram of ice melts at 0°C, 335 J of energy are absorbed. The energy needed to melt ice is much less than that needed to boil the same amount of water. When a liquid freezes, the particles lose the same amount of energy that they would gain on melting. Each gram of water that freezes at 0°C releases 335 J of energy.

Checkpoint

1. Contrast the particle motions in the three states of matter.
2. What energy change occurs when water (a) boils, (b) condenses?
3. What energy change occurs when (a) ice melts, (b) water freezes?
4. What happens to water molecules when water (a) boils, (b) condenses? How does your answer explain the observed energy change in each case?

TRY THIS

Sprinkle some warm water on your palm. Then wave your hand through the air. Why does this cause your hand to feel cool?

Checkpoint Answers

1. In solids, particles vibrate about fixed positions; in liquids, they move short distances before colliding with one another; in gases, they move nearly independently.
2. (a) When water boils, 2300 J of energy are absorbed to convert 1 g of liquid to gas. (b) When 1 g of steam condenses, 2300 J of energy are released.
3. (a) Each gram of ice that melts at 0°C absorbs 335 J of energy. (b) Each gram of water that freezes at 0°C releases 335 J of energy.
4. (a) The water molecules move apart when water boils. Energy is required to overcome the forces of attraction that hold the molecules together in the liquid state. (b) When water condenses, the energy gained by the separated molecules is released when the molecules come close enough together to be in the liquid state.

9-2

Energy and Chemical Reactions

Goals

1. To explain why the energy associated with chemical changes is, in general, greater than that associated with physical changes.
2. To explain the difference between exothermic and endothermic reactions.

The energy required to *decompose*, or to break down, one gram of water into the elements hydrogen and oxygen is 15,900 J. Since energy is absorbed in this process, you might guess that it is "stored" in the molecules of hydrogen and oxygen that are formed. To test this idea, elemental hydrogen and oxygen can be ignited to form water. When this is done, energy is released. For every gram of water that is formed, 15,900 J of energy are released. The energy that was stored in the hydrogen and oxygen molecules is released when they combine, forming water.

The energy involved in the decomposition or the formation of water is nearly seven times as large as the heat of vaporization of water. Table 9-2 shows that there is a large difference between these values for other substances, as well. In general, more energy is released or absorbed when a compound is formed or decomposed than when it changes state. In other words, chemical changes tend to involve more energy than do physical changes.

The energy associated with a physical change involves the physical forces of attraction between particles that are not chemically bonded. When water boils, the molecules

Table 9-2: Energies of Some Physical and Chemical Changes

Substance	*Heat of vaporization (J/g)*	*Heat released in formation of compound (J/g)*	*Heat needed to decompose compound (J/g)*
water	2300	15,900	15,900
carbon dioxide	365	8900	8900
sulfur dioxide	400	4600	4600
hydrogen chloride	416	4500	4500

Background

When 1/9 g of hydrogen and 8/9 g of oxygen react to form 1 g of water, 15,900 J of thermal energy are released. This indicates that the energy in water molecules is significantly less than the energy in the molecules of hydrogen and oxygen that combine to form water. This is confirmed when we convert water to hydrogen and oxygen by electrolysis. Some of the electrical energy used appears as thermal energy in the solution, but 15,900 J per gram of water decomposed does not appear as heat. Because this amount of energy is equal to the energy released when water is formed from its elements, we assume the "missing" energy is stored in the molecules of the products—hydrogen and oxygen.

In the case of unstable compounds, energy is absorbed in their formation. Their high energy content makes them subject to relatively easy decomposition; stable compounds, such as water, have a low energy content and thus require significant energy input before they will decompose.

In the breaking and forming of chemical bonds, the evidence clearly indicates that the energy changes involved are generally much larger than those associated with physical changes.

Teaching Suggestions 9-2

You can start a discussion of energy and chemical reactions by igniting some hydrogen. See Demonstration 1.

Point out that the product of this combustion reaction is water and that energy was released during the reaction. If the hydrogen in the tube is reasonably pure, the gas will burn only where it is in contact with the oxygen of the air. If hydrogen is mixed with air, it explodes, because all the hydrogen can react at once.

A good way to convey the concepts of endothermic and exothermic reactions is to demonstrate them. See Demonstration 2. Have students touch the test tube after a reaction to determine if it has heated or cooled. This helps clarify the concepts of energy release and energy absorption.

Demonstrations

1. Show the explosive nature of hydrogen. Collect a tubeful of hydrogen using zinc and dilute hydrochloric acid. (See Demonstrations for section 8-1.) Simply run the rubber tubing from the flask into a water-filled test tube inverted in a container of water. The gas will displace the water. When the test tube is filled with gas, turn it upright for a moment to allow air to enter and then bring a burning splint to the mouth of the tube. A pop will be heard as the hydrogen reacts with the oxygen in the air-hydrogen mixture.
2. Show temperature changes when certain substances dissolve. Dissolve some ammonium chloride in a test tube full of water. Have students feel the tube before and after dissolving the salt. Ask them if the dissolving of ammonium chloride is endothermic or exothermic. Repeat the experiment, using calcium chloride.

gain enough energy to overcome the forces of attraction between them. As a result, they move farther apart. When steam condenses, the energy stored in the molecules is released, and the molecules move closer together.

The energy associated with a chemical change involves the breaking of chemical bonds. When water is decomposed, the bonds between the hydrogen and oxygen atoms in the water molecules are broken. The energy associated with the breaking or forming of chemical bonds within molecules is generally much greater than that needed to overcome physical forces of attraction between molecules.

exothermic (ek soh THUR mik)

endothermic (en doh THUR mik)

In a chemical reaction, the reactants may either gain or lose energy. When hydrogen and oxygen molecules react, forming water, energy is released. The product molecules therefore have less energy stored in them than did the reactant molecules. Similarly, the burning of sodium in chlorine releases energy. The reaction produces 18,000 J per gram of sodium. The combustion of fuel in a furnace releases about 46,000 J per gram of fuel. (See Figure 9-3.) All such chemical reactions, which release energy, are called **exothermic** reactions.

Chemical reactions that require energy in order to take place are called **endothermic** reactions. The decomposition of water is an example of an endothermic reaction. In such reactions, the product molecules have more energy stored in them than did the reactant molecules. Thus, in the decomposition of water, molecules of hydrogen and oxygen have more energy than the reactant molecules of water.

Chemical equations often include an energy term that makes it clear whether the reaction is endothermic or exothermic. For example, the first equation below shows that the decomposition of water is endothermic. In endothermic reactions, the energy term appears with the reactants on the lefthand side of the equation. The second equation shows that the burning of hydrogen in oxygen is exothermic. In exothermic reactions, the energy term appears next to the product molecules on the righthand side of the equation.

$$\text{energy} + 2H_2O \longrightarrow 2H_2 + O_2$$

$$2H_2 + O_2 \longrightarrow 2HO + \text{energy}$$

Teaching Tips

- Call attention to the relative magnitudes of the heats of vaporization and the heats of formation and decomposition in Table 9-2 on student text page 219. Be sure students are aware that vaporization is a physical process and that formation and decomposition are chemical changes.
- Use the equations on student text page 220 to stress that energy is either released or absorbed during a chemical reaction.

Figure 9-3 *When fuels burn, they release energy. Energy-releasing reactions are called exothermic reactions.*

Checkpoint

1. Why is the energy associated with chemical changes generally larger than that associated with physical changes?
2. Which has the greater amount of stored energy: (a) two water molecules or (b) two hydrogen molecules plus one oxygen molecule? Explain your answer.
3. Define exothermic and endothermic reactions.

TRY THIS

Add some vinegar to a small glass. Then use a thermometer to find the temperature of the liquid. Add a teaspoon of baking soda. The chemical reaction that occurs releases carbon dioxide gas. Is the reaction exothermic or endothermic? How can you tell?

9 CHEMICAL REACTIONS

9-2 Energy and Chemical Reactions

Endothermic or Exothermic?

In each reaction below energy is either absorbed or released. Decide what is happening to the energy in each reaction. Write the term *endothermic* on the space next to reactions in which energy is absorbed. Write the term *exothermic* next to reactions in which energy is released. Then list the reaction numbers from the most exothermic to the most endothermic.

______ 1. $2\ H_2 + O_2 \longrightarrow 2\ H_2O + 484{,}000\ J$

______ 2. $N_2 + O_2 + 90{,}000\ J \longrightarrow 2\ NO$

______ 3. $2\ NO_2 \longrightarrow N_2 + 2\ O_2 + 33{,}000\ J$

______ 4. $Si + 2\ H_2 + 34{,}000\ J \longrightarrow SiH_4$

______ 5. $2\ Fe + O_2 \longrightarrow 2\ FeO + 544{,}000\ J$

______ 6. $H_2 + Br_2 \longrightarrow 2\ HBr + 72{,}000\ J$

______ 7. $N_2 + 2\ O_2 + 9{,}000\ J \longrightarrow N_2O_4$

______ 8. Most exothermic to most endothermic.

Energy Changes

Each of the following problems describes a chemical reaction in which an energy change occurs. On a separate sheet of paper, solve each problem. Then write your answer on the space at the left of each problem. Hint: Use Table 9-2 in your textbook.

______ 1. How much heat energy is released when 2 g of hydrogen chloride are formed?

______ 2. How much heat energy is needed to decompose 3 g of carbon dioxide?

______ 3. How much heat energy is needed to decompose 3.5 g of hydrogen chloride?

______ 4. How much heat energy is released when 5.2 g of sulfur dioxide are formed?

______ 5. How much heat energy is needed to decompose 0.8 g of water?

______ 6. If 47,700 J of energy are released when a sample of water is formed, how many grams of water are involved?

______ 7. If 18,000 J of energy are needed to decompose a sample of hydrogen chloride, how many grams of hydrogen chloride are involved?

______ 8. If 2,760 J of energy are released when a sample of sulfur dioxide is formed, how many grams of sulfur dioxide are involved?

Using Try This

Students should be able to state that if the temperature rises, heat is released and the reaction is exothermic. Or, that if the temperature falls, heat is absorbed and the reaction is endothermic. This reaction is exothermic.

Checkpoint Answers

1. Chemical changes involve breaking or formation of chemical bonds between atoms, whereas physical changes involve physical forces of attraction between particles that are not chemically bonded. Breaking or forming chemical bonds involves more energy than that needed to overcome physical forces of attraction between molecules.
2. (b). When hydrogen and oxygen combine to form water, energy is released. When water is converted to hydrogen and oxygen, energy is absorbed.
3. An exothermic reaction is a chemical reaction in which energy is released. An endothermic reaction is a chemical reaction in which energy is absorbed.

Cross Reference

For more information about heat, refer to Chapter 18.

Teaching Suggestions 9-3

Set up the Try This activity the day you assign the section for reading. By the next day, it should be evident that the nail has begun to corrode. You might extend this experiment by having students place nails in distilled water, sodium hydroxide, potassium chromate, hydrochloric acid, and other dilute solutions. In each case, be sure students understand that corrosion reactions are examples of oxidation-reduction. The demonstration suggested for this section should help convey the concepts being taught. (See Demonstration.)

Background

Oxidation-reduction reactions are very common. They include combustion, corrosion, and electrochemical reactions in batteries.

Oxidation was originally regarded as a reaction between oxygen and another substance. The modern, broader notion includes many other reactions as well. The reaction of sodium and chlorine, for example, involves no oxygen. It is an oxidation-reduction reaction because sodium atoms lose electrons to chlorine atoms, which are reduced. Na^+ and Cl^- ions are the result:

$$2Na + Cl_2 \longrightarrow 2Na^+ + 2Cl^-$$

In this text we have not discussed the notion of oxidation involving covalent compounds whose constituent atoms do not actually lose or gain electrons. Instead, there is said to be a change in what are called oxidation numbers. For example, when carbon burns in oxygen, chemists assign an oxidation number of 2− to the oxygen atoms in the compound formed, CO_2. The carbon atom in CO_2 is assigned an oxidation number of 4+, so that the total oxidation number of the molecule is 0 ($= (2 \times -2) + 4$). All free elements are assigned an oxidation number of 0. Oxidation numbers are represented by superscripts.

$$C^0 + O_2^0 \longrightarrow C^{4+}O_2^{2-}$$

Hydrogen is assigned an oxidation number of 1+. Thus, in water the sum of the oxidation numbers is 0, as follows: $H_2^+O^{2-}$.

Demonstration

Show the reaction between zinc and copper sulfate in solution. Place a strip of clean, polished zinc in a solution of copper sulfate. Evidence of a reaction will be almost immediate, as a fine black deposit collects on the zinc. The deposit consists of fine particles of copper metal. As the reaction proceeds, the deposit will take on a reddish, cop-

9-3

Oxidation and Reduction

Goals

1. To define oxidation and reduction.
2. To give examples of oxidation-reduction reactions.

oxidation (ahk suh DAY shuhn)

reduction (rih DUHK shuhn)

The burning of fossil fuels is an example of an exothermic reaction. Because fuels combine with oxygen when they burn, burning is sometimes called *oxidation*. To a chemist, oxidation is a much broader term. In chemistry, **oxidation** is defined as the loss of electrons by an atom or ion. When sodium burns in chlorine, sodium atoms lose electrons to chlorine atoms, as you learned in Chapter 8. The sodium is oxidized. The chlorine atoms are said to be reduced. **Reduction** is the gain of electrons by an atom or ion. If one substance is oxidized, another must be reduced. The electrons lost by one substance must be taken up by another. Reactions involving oxidation and reduction are called *oxidation-reduction reactions* or redox reactions.

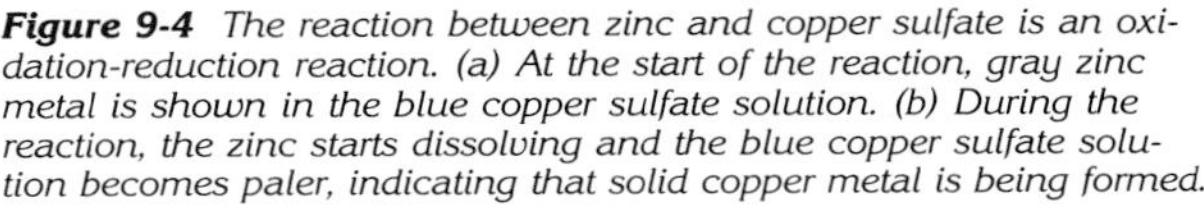

Figure 9-4 *The reaction between zinc and copper sulfate is an oxidation-reduction reaction. (a) At the start of the reaction, gray zinc metal is shown in the blue copper sulfate solution. (b) During the reaction, the zinc starts dissolving and the blue copper sulfate solution becomes paler, indicating that solid copper metal is being formed.*

An oxidation-reduction reaction takes place when zinc (Zn) is placed in a solution of copper sulfate ($CuSO_4$). A copper sulfate solution contains copper ions (Cu^{2+}), which have a double positive charge, and sulfate ions (SO_4^{2-}), which have a double negative charge. Zinc dissolves in the solution as zinc atoms lose negatively charged electrons to copper ions. When copper ions gain electrons they become copper atoms. Each copper ion accepts two electrons and becomes a neutral atom. As a result, solid copper precipitates, or separates out, from the solution. (See Figure 9-4.) The sulfate ions do not participate in the reaction. We may write the following set of chemical equations to symbolize the reaction.

Oxidation	$Zn \longrightarrow Zn^{2+} + 2e^-$
Reduction	$Cu^{2+} + 2e^- \longrightarrow Cu$
Overall reaction	$Zn + Cu^{2+} \longrightarrow Zn^{2+} + Cu$

The rusting of iron is a familiar example of a process called *corrosion*. Rusting and many other forms of corrosion are oxidation-reduction reactions. Such reactions are sometimes described as "slow combustion" reactions. Iron rusts by reacting with oxygen to form iron oxides. (See Figure 9-5.) In the process, iron gives up electrons to the oxygen, which results in formation of positive iron ions and negative oxide ions. Metals other than iron can also be oxidized. When aluminum is oxidized, the white aluminum oxide that is produced forms a coating that protects the metal from further attack.

Figure 9-5 *The rusting of iron is an example of an oxidation-reduction reaction. The iron loses electrons to oxygen.*

Checkpoint

1. Define (a) oxidation, (b) reduction.
2. Give an example of an oxidation-reduction reaction.

TRY THIS

Clean a nail with steel wool or fine sandpaper. Place it in a test tube of vinegar, so that the lower half of the nail is under the liquid. Examine the nail a day or two later. Is there any evidence of corrosion?

per color, and the blue color of the solution will fade. Allow the reaction to proceed overnight. The zinc metal is oxidized to zinc ions, while copper ions are reduced to copper metal:

$$Zn^0 + Cu^{2+} \longrightarrow Zn^{2+} + Cu^0$$

A piece of copper placed in a dilute solution of zinc sulfate, on the other hand, will not react, indicating that copper is not oxidized by zinc ions.

Teaching Tips

- Be sure students understand that oxidation is a broader concept than simply the combination of oxygen with another substance.
- Not all reactions involve oxidation-reduction. This should be pointed out to students. For example, there are no electrons lost or gained when lead iodide precipitates upon the mixing of lead nitrate and sodium iodide:

$$Pb^{2+} + 2NO_3^- + 2Na^+ + 2I^- \longrightarrow Pb^{2+}I_2^- \downarrow + 2Na^+ + 2NO_3^-$$

Acid-base reactions are another type that do not involve oxidation.

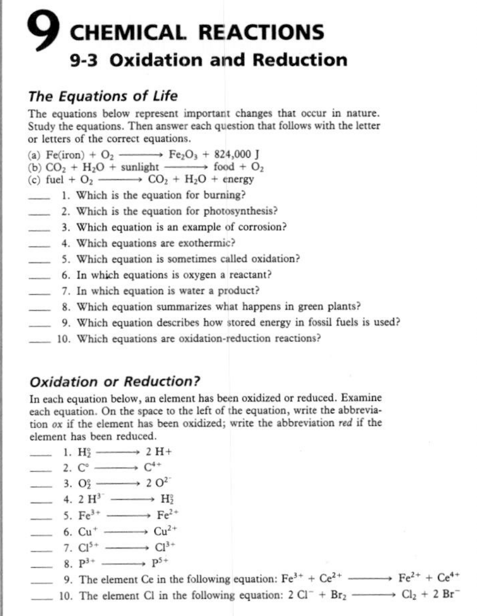

9 CHEMICAL REACTIONS
9-3 Oxidation and Reduction

The Equations of Life

The equations below represent important changes that occur in nature. Study the equations. Then answer each question that follows with the letter or letters of the correct equations.

(a) Fe(iron) + $O_2 \longrightarrow Fe_2O_3$ + 824,000 J
(b) $CO_2 + H_2O$ + sunlight $\longrightarrow$ food + O_2
(c) fuel + $O_2 \longrightarrow CO_2 + H_2O$ + energy

___ 1. Which is the equation for burning?
___ 2. Which is the equation for photosynthesis?
___ 3. Which equation is an example of corrosion?
___ 4. Which equations are exothermic?
___ 5. Which equation is sometimes called oxidation?
___ 6. In which equations is oxygen a reactant?
___ 7. In which equation is water a product?
___ 8. Which equation summarizes what happens in green plants?
___ 9. Which equation describes how stored energy in fossil fuels is used?
___ 10. Which equations are oxidation-reduction reactions?

Oxidation or Reduction?

In each equation below, an element has been oxidized or reduced. Examine each equation. On the space to the left of the equation, write the abbreviation *ox* if the element has been oxidized; write the abbreviation *red* if the element has been reduced.

___ 1. $H_2^0 \longrightarrow 2\ H+$
___ 2. $C^0 \longrightarrow C^{4+}$
___ 3. $O_2^0 \longrightarrow 2\ O^{2-}$
___ 4. $2\ H^{3-} \longrightarrow H_2^0$
___ 5. $Fe^{3+} \longrightarrow Fe^{2+}$
___ 6. $Cu^+ \longrightarrow Cu^{2+}$
___ 7. $Cl^{5+} \longrightarrow Cl^{3+}$
___ 8. $P^{3+} \longrightarrow P^{5+}$
___ 9. The element Ce in the following equation: $Fe^{3+} + Ce^{2+} \longrightarrow Fe^{2+} + Ce^{4+}$
___ 10. The element Cl in the following equation: $2\ Cl^- + Br_2 \longrightarrow Cl_2 + 2\ Br^-$

Using Try This

Students should observe corrosion where the metal touches the vinegar. Nails placed in an acidic solution, such as vinegar, generally corrode. Active metals (M), such as iron, lose electrons to hydrogen ions in solution, according to the following general formula:

$$M^0 + 2H^+ \longrightarrow M^{2+} + H_2^0$$

Checkpoint Answers

1. (a) Oxidation is the loss of electrons by an atom or ion. (b) Reduction is the gain of electrons by an atom or ion.
2. Answers will vary. One possibility is:

$$Cu^{2+} + Zn^0 \longrightarrow Zn^{2+} + Cu^0$$

Teaching Suggestions 9-4

Have students do laboratory Activity 14, Unsuspected Electrochemical Cells, student text pages 548–549. Use the method and the results of the activity to introduce the concepts covered in this section. Students will see that electricity can be generated using any two different metals and a solution containing ions that will conduct electric charge, that is, an electrolyte. Note that you will need sensitive galvanometers or microammeters to measure the currents generated, because they are quite small. Some students also may wish to do chapter-end projects 1 and 5, which both apply to this section.

Background

The electromotive series in Table 9-3 on student text page 226 can be established experimentally by testing metals in solutions of various other metallic ions. A metal will be oxidized by the ions of any metal lower in the table. Potassium, for example, will be oxidized by all other metallic ions because potassium atoms have a greater tendency to give up electrons than do atoms of any other metal. For similar reasons, zinc will give electrons to copper ions and thus be oxidized. Zinc, however, will not be oxidized in a sodium chloride solution, because sodium atoms have a much greater tendency to lose electrons than do zinc atoms. Thus, any sodium atoms formed by sodium ions accepting electrons from zinc atoms would immediately lose electrons to the less active zinc ions, re-forming zinc atoms and sodium ions.

Figure 9-6 *(a) A "D"-cell is an example of an electrochemical dry cell. (b) A lead storage battery contains several electrochemical cells.*

9-4

Electrochemical Cells

Goals

1. To describe how electrochemical cells work.
2. To explain how unwanted electrochemical reactions can cause corrosion.
3. To describe some ways of preventing corrosion.

Electrochemical (ih lek troh KEM ih kul) *cells* produce electricity that results from oxidation-reduction reactions within the cells. The dry cells shown in Figure 9-6 are electrochemical cells. The battery that is shown is made up of several electrochemical cells connected together. An electrochemical cell called the Daniell cell is shown in Figure 9-7. It produces electricity by means of the oxidation-reduction reaction between zinc and copper sulfate, which was described earlier.

Every electrochemical cell includes *electrodes* (ih LEK trohdz), typically plates through which electrons can flow into and out of the cell and at which oxidation or reduction occurs. In the Daniell cell, one electrode is a piece of copper. It is placed in a porous cup full of copper sulfate solution. The other electrode is a piece of zinc. It is placed in a zinc sulfate solution that surrounds the porous cup. The porous cup prevents the zinc metal and the copper sulfate solution from coming into direct contact. Ions can move slowly through the cup. If a wire connects the electrodes, electrons flow through the wire from the zinc electrode to the copper electrode.

At the copper electrode, copper ions pick up electrons and are reduced to copper atoms, which collect on the electrode. Atoms in the zinc electrode give up electrons and go into solution as zinc ions. Because zinc is the source of electrons, it is the negative electrode. The overall reaction is exactly the same as the reaction of zinc and copper sulfate.

Oxidation	Zn	$\longrightarrow$	$Zn^{2+} + 2e^-$
Reduction	$Cu^{2+} + 2e^-$	$\longrightarrow$	Cu
Overall reaction	$Zn + Cu^{2+}$	$\longrightarrow$	$Zn^{2+} + Cu$

Demonstrations

1. Make a simple electrochemical cell. Soak a paper towel in a concentrated ammonium chloride solution. Place the towel on a strip of shiny zinc about 10 cm by 5 cm. Sprinkle some manganese dioxide on the paper. Then put a strip of shiny copper, of the same size as the zinc, on top of the manganese dioxide. Use wires to connect each piece of metal to a sensitive ammeter or light bulb. Pressing the materials together provides a stronger current. The reactions are the same as for the Daniell cell.
2. If you have porous cups and the necessary electrodes and solutions, assemble a Daniell cell and demonstrate its capacity to produce electricity. You might want to measure

The difference between this setup and that described in the last section is that the zinc is not now in direct contact with copper ions. The only path by which the electrons released by the zinc can reach the copper ions is the wire connecting the electrodes. Like any electrochemical cell, the Daniell cell provides a means of using the energy of an oxidation-reduction reaction to obtain electricity for use outside the cell.

Different metals have different tendencies to release electrons to ions present in solution. Figure 9-8 shows that lead is oxidized by copper ions and that zinc is oxidized by lead ions. Furthermore, it shows that neither lead nor copper is oxidized by zinc ions.

By means of experiments like those illustrated in Figure 9-8, chemists have ranked metals in order of their tendency to release electrons. This ranking is commonly shown as a list called the *electromotive series*. Table 9-3 on page 226 shows part of this series. Hydrogen is included in the list as a reference point. Metals that have a greater tendency than hydrogen to release electrons are called *active metals* and occur above hydrogen in the table. Any metal in the table is able to be oxidized in the presence of ions of elements below it

Figure 9-7 *A Daniell cell produces electricity through use of the oxidation-reduction reaction of zinc and copper sulfate.*

Figure 9-8 *Metals differ in their tendencies to be oxidized, or to give up electrons. (a) Lead metal will oxidize in blue copper sulfate solution. (b) Lead metal is not oxidized in zinc nitrate solution. No reaction occurs. (c) Zinc metal is oxidized in lead nitrate solution. Lead metal precipitates from the solution. (d) Copper metal is not oxidized in lead nitrate solution. No reaction occurs.*

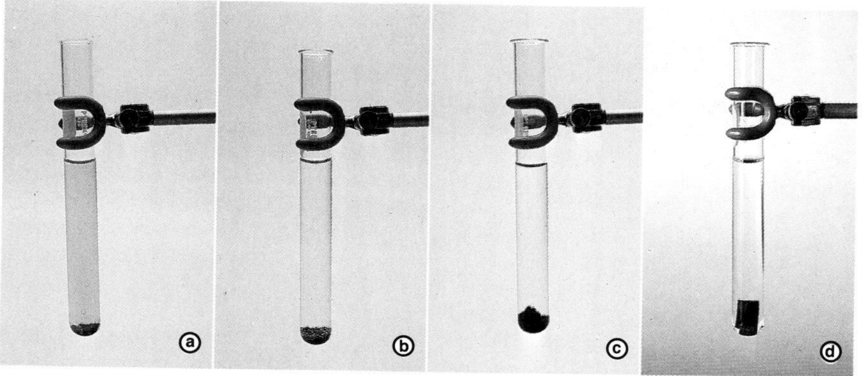

the mass of the electrodes before and after the cell has operated for 15 to 20 minutes. This will demonstrate that zinc really does dissolve and copper really does plate out of solution. For best results, use copper mesh for one electrode. After the cell has operated for some time, remove the electrodes, rinse gently in a beaker of water, and dry under a heat lamp. The zinc electrode will show a mass decrease and the mass of the copper electrode will have clearly increased.

You can show also that the cell has a voltage of about 1.1 volts before the electrodes are connected to a circuit element or to one another. Once the electrodes have been connected through a small resistance, such as an ammeter, the voltmeter will show a drop in voltage.

3. Cut a D-cell longitudinally with a hacksaw or bandsaw to show the inner structure of the cell. Students will be surprised to see that the cell continues to work as long as the black electrolyte remains moist. Touch the carbon electrode and the zinc case with wires connected to a bulb or ammeter to show the flow of charge.
4. Show the effect of a sacrificial anode on the corrosion of steel. Place a piece of steel in a beaker of synthetic sea water (add 30 g of salt to a liter of water). In a second beaker place a similar piece of steel and a piece of clean zinc. Connect the zinc and steel in the second container with a piece of wire. A sensitive ammeter may be connected between the two metals to measure the current produced.

Leave the experimental setup for several weeks, adding water occasionally to replace any lost by evaporation. Have students determine which piece of steel seems to show more corrosion—the one connected to the zinc or the one that was alone in the salt water.

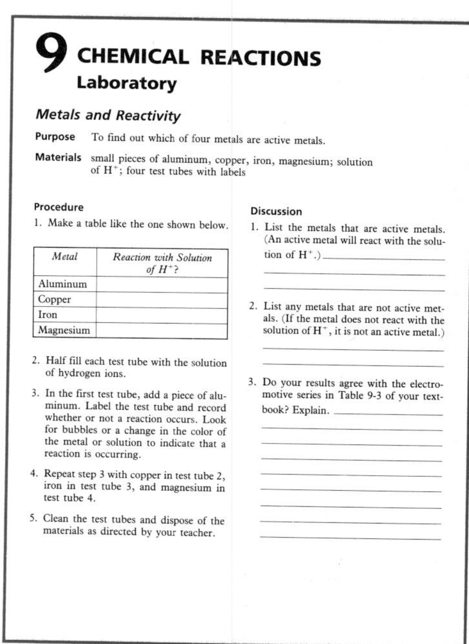

9 CHEMICAL REACTIONS
Laboratory

Metals and Reactivity

Purpose To find out which of four metals are active metals.

Materials small pieces of aluminum, copper, iron, magnesium; solution of H^+; four test tubes with labels

Procedure

1. Make a table like the one shown below.

Metal	*Reaction with Solution of H^+?*
Aluminum	
Copper	
Iron	
Magnesium	

2. Half fill each test tube with the solution of hydrogen ions.
3. In the first test tube, add a piece of aluminum. Label the test tube and record whether or not a reaction occurs. Look for bubbles or a change in the color of the metal or solution to indicate that a reaction is occurring.
4. Repeat step 3 with copper in test tube 2, iron in test tube 3, and magnesium in test tube 4.
5. Clean the test tubes and dispose of the materials as directed by your teacher.

Discussion

1. List the metals that are active metals. (An active metal will react with the solution of H^+.) ____
2. List any metals that are not active metals. (If the metal does not react with the solution of H^+, it is not an active metal.) ____
3. Do your results agree with the electromotive series in Table 9-3 of your textbook? Explain. ____

Teaching Tips

- Emphasize that electrochemical cells are simply oxidation-reduction reactions in which the reactants are kept apart, though the ions and electrons are allowed to flow.
- Be sure students understand that a battery is a series of electrochemical cells. You can build a battery by connecting several D-cells in series. A voltmeter will show that the voltage increases by 1.5-volt increments for each D-cell added.
- Write a number of oxidation-reduction reactants on the chalkboard. Ask students to complete each equation or write "NR" if there is no reaction. Some possibilities are:

$$(2)Na + Zn^{2+} \longrightarrow (2Na^+ + Zn)$$
$$Zn + Na^+ \longrightarrow (NR)$$
$$Zn + (2)Ag^+ \longrightarrow (Zn^{2+} + 2Ag)$$
$$Ag + Zn^{2+} \longrightarrow (NR)$$
$$Pb + (2)H^+ \longrightarrow (Pb^{2+} + H_2)$$

  Stress how such equations summarize oxidation-reduction reactions.
- Be sure students understand that two different metals are needed to produce electricity in electrochemical cells. The more active metal serves as the source of electrons. It is the negative electrode and undergoes oxidation. The less active metal serves as the positive electrode. Its ions are reduced by accepting electrons.

Table 9-3: Electromotive Series

Element	Reduced form		Oxidized form
potassium	K	$\longrightarrow$	$K^+ + e^-$
sodium	Na	$\longrightarrow$	$Na^+ + e^-$
magnesium	Mg	$\longrightarrow$	$Mg^{2+} + 2e^-$
aluminum	Al	$\longrightarrow$	$Al^{3+} + 3e^-$
zinc	Zn	$\longrightarrow$	$Zn^{2+} + 2e^-$
iron	Fe	$\longrightarrow$	$Fe^{2+} + 2e^-$
lead	Pb	$\longrightarrow$	$Pb^{2+} + 2e^-$
hydrogen	H	$\longrightarrow$	$H^+ + e^-$
copper	Cu	$\longrightarrow$	$Cu^{2+} + 2e^-$
silver	Ag	$\longrightarrow$	$Ag^+ + e^-$
mercury	Hg	$\longrightarrow$	$Hg^{2+} + 2e^-$
gold	Au	$\longrightarrow$	$Au^+ + e^-$

in the table. For example, zinc metal is oxidized to zinc ions in a solution such as lead nitrate, which contains lead ions. The lead ions would accept the electrons lost by the zinc, and would become neutral lead atoms.

With the use of Table 9-3, a number of electrochemical cells can be set up. The negative electrode must be a more active metal than the one used for the positive electrode. The positive electrode should be surrounded by a solution containing ions that will accept electrons from the negative electrode. Electrochemical cells are basically quite simple to make, but careful design is needed to make a really useful device. A dry cell or a battery, for example, must be designed well enough to be able to produce a large amount of electricity for a long time.

It is difficult to prevent unwanted electrochemical reactions from taking place. Most iron and steel contain impurities. When the surface of iron or steel is damp, there are ions present in the droplets of water clinging to the metal. Small electrochemical cells are thus set up. The impurities usually serve as positive electrodes in the unwanted cells. The iron usually provides the negative electrode, and so it slowly dissolves.

Some steels are made resistant to corrosion by the addition of chromium and nickel (in stainless steel) or silicon (in duriron) to the iron. To reduce corrosion, metal surfaces

Fact At Your Fingertips

- The powdered corrosion products on automobile-battery terminals are acidic. Such acids should be washed away by adding a baking-soda solution to neutralize the acids. If a battery is to be charged, the caps to the cells should be removed. Otherwise, a build-up of gas pressure within the battery could cause it to explode.

can also be covered with paint, with a ceramic enamel (such as those usually found on sinks, stoves, and refrigerators), or with an oxide coating. (See Figure 9-9.) A thick protective coating is often sprayed on the undersides of cars to prevent corrosion. Without such treatment, the undersides of cars may be corroded by unwanted electrochemical reactions, especially if the cars are driven on wet or salt-covered roads.

Another way to reduce corrosion involves the use of "sacrificial electrodes." For example, an underground steel tank that needs protection against corrosion can be made the positive electrode of a cell. Zinc, magnesium, or another active metal is used as the negative electrode. Electrons flow from the negative electrode to the positive electrode. As a result, the active metal (the sacrificial electrode) dissolves—not the tank. Hulls of steel seagoing ships are often protected in the same way.

Iron or steel pipes and steel sheet metal that need to be protected against corrosion are sometimes coated with a thin layer of zinc. The zinc serves as more than simply a protective cover. If scratches expose the iron, in the zinc-iron cells that develop, the zinc—not the iron—will act as the negative electrode and will corrode. Iron or steel coated with zinc is sometimes said to be *galvanized* (GAL vuh nyzd). The term comes from the name of an eighteenth-century Italian scientist, Luigi Galvani. Galvani was a pioneer in the development of electrochemical cells.

Figure 9-9 *Refrigerators are usually coated with ceramic enamel in order to prevent unwanted electrochemical reactions that would cause corrosion.*

Checkpoint

1. Explain how a Daniell cell works.
2. Arrange the following metals in order of decreasing tendency to be oxidized (place the one most easily oxidized at the top, then the next-most-easily-oxidized, and so on): zinc, copper, silver, lead, sodium.
3. Describe two methods for protecting metals from corrosion, and explain how the methods work.
4. An electrochemical cell has zinc and silver electrodes immersed in zinc and silver ions, respectively. (a) Make a diagram of the cell. (b) What is oxidized? (c) What is reduced? (d) Which metal is the negative electrode? (e) Which is the positive electrode?

Cross Reference

Current electricity is discussed in Chapter 19, section 19-4.

9 CHEMICAL REACTIONS

9-4 Electrochemical Cells

Energy at Your Fingertips

Match the term in Column A with its definition in Column B. Then write the letter of the definition on the space next to its term.

A	B
____ 1. electrochemical cell	a. The conductor through which electrons flow in and out of a cell.
____ 2. electrode	b. Description of the material made of iron and steel coated with zinc.
____ 3. electromotive series	c. A cell that produces electricity from an oxidation-reduction reaction.
____ 4. active metals	d. Metals that have a greater tendency than hydrogen to release electrons.
____ 5. galvanized	e. A ranking of metals in order of their tendency to release electrons.

The Daniell Cell

Use your textbook and the list next to the Daniell cell diagram to label its parts. Then write the answers to the questions that follow.

zinc electrode
copper sulfate solution
porous cup
wire connecting electrodes
copper electrode
zinc sulfate solution

1. At which electrode, copper or zinc, is oxidation occurring? ____

   What is the partial equation for the oxidation reaction? ____
2. At which electrode, copper or zinc, is reduction occurring? ____

   What is the partial equation for the reduction reaction? ____
3. What is the overall reaction for the Daniell cell? ____
4. How is the porous cup used? ____
5. How is the wire used? ____
6. For what is the Daniell cell used? ____

Checkpoint Answers

1. Answers will vary. See student text, pages 224–225.
2. Sodium, zinc, lead, copper, silver.
3. Answers will vary. See student text, pages 226–227. Two possibilities are: Painting prevents contact between metals and corroding chemicals. Galvanizing steel with zinc causes the more active zinc to serve as a source of electrons and prevents the iron from being oxidized.
4. (a) See Figure TE 9-1. (b) Zinc is oxidized: $Zn \longrightarrow Zn^{2+} + 2e^-$. (c) Silver ions are reduced: $Ag^+ + e^- \longrightarrow Ag^0$. (d) Zinc is the negative electrode. (e) Silver is the positive electrode.

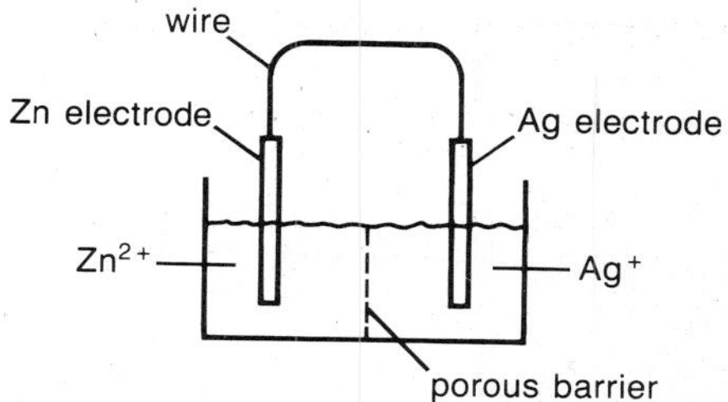

Figure TE 9-1 Diagram of an electrochemical cell with zinc and silver electrodes immersed in zinc and silver ions, respectively.

Teaching Suggestions 9-5

Laboratory Activity 15, Acids and Bases, student text pages 550–551, is a good way to begin study of this topic. Have students read the section before carrying out the laboratory activity. Use the results of the activity to help students understand the concepts covered in this section. Some students also may wish to do chapter-end projects 2, 3, and 4, which all relate to this section.

Background

Properties of acids and bases, such as taste, effect on litmus, conductivity, and so on, serve as operational definitions. These properties enable us to identify an acid or a base by observing (sensing) an effect or a characteristic. The ideas that acids form H^+ ions and bases form OH^- ions in solution are conceptual definitions. We cannot see the ions, but the theory that they exist enables us to explain the behavior of acids and bases very well.

A strong acid, such as hydrochloric acid, is one that is virtually 100% ionized. Sulfuric, nitric, and perchloric acids are all strong acids. Acids that ionize to only a small extent, such as acetic, boric, and benzoic acids, are said to be weak acids. The fact that boric acid is often used as an eyewash indicates that it is a weak acid. Only about 1 in 100,000 boric-acid molecules ionizes to form H^+ ions:

$$100{,}000H_3BO_3 \longrightarrow 1H^+ + 99{,}999H_3BO_3 + 1H_2BO_3^-$$

Note that most chemists believe that H^+ ions do not actually exist alone in solution. Each H^+ ion is always combined with a molecule of water to form H_3O^+, a positively charged hydronium ion.

Ammonia and soluble carbonates are considered to be bases because they form OH^- ions in solution, as shown on student text page 229. For similar reasons, chemicals such as ammonium chloride are often said to be acidic. They form H^+ ions in solution due to the reaction of their positive ions with water molecules:

$$NH_4^+ + Cl^- + H_2O \longrightarrow NH_4OH + H^+ + Cl^-$$

The idea of acids and bases enables us to group a large number of reactions into a single category: neutralization reactions, involving only the H^+ and OH^- ions:

$$H^+ + OH^- \longrightarrow H_2O$$

In this section we speak of acids as substances that form H^+ ions in solution, but there are other definitions. Acids are sometimes defined as substances that can donate protons; bases are then proton-acceptors. An even broader definition argues that acids are electron-pair acceptors while bases are electron-pair donors.

9-5

Acids and Bases

Goals

1. To state the properties of acids and bases.
2. To describe what happens when an acid and a base react.
3. To explain what is meant by the *p*H of a solution.

acids

If you have ever tasted lemon juice, you know that it is sour. A sour taste is one characteristic property of an acid. Generally, **acids** in solution (a) taste sour, (b) conduct electricity, (c) turn blue litmus pink, (d) contain hydrogen, and (e) react with active metals, producing hydrogen gas. (CAUTION: Never taste a substance unless you know it is safe to do so.) Some common acids are listed in Table 9-4. Note that they all contain hydrogen. In solutions, acids produce hydrogen ions (H^+). The hydrogen ions are responsible for the characteristic acid properties.

You have already learned that the gas hydrogen chloride (HCl) consists of polar molecules. When this gas dissolves in water, the molecules break up into hydrogen ions and chloride ions. The polar water molecules interact with the HCl molecules to bring about the ion formation. A solution of HCl in water is called hydrochloric acid. The H^+ ion in hydrochloric acid reacts vigorously with active metals, such as zinc, producing hydrogen gas:

$$Zn + 2H^+ \longrightarrow Zn^{2+} + H_2$$

Table 9-4: Some Common Acids

Acid	Formula
hydrochloric acid	HCl
sulfuric acid	H_2SO_4
nitric acid	HNO_3
acetic acid	$H_4C_2O_2$
citric acid	$H_8C_6O_7$
phosphoric acid	H_3PO_4
carbonic acid	H_2CO_3
boric acid	H_3BO_3

Table 9-5: Some Common Bases

Base	Formula
sodium hydroxide	NaOH
potassium hydroxide	KOH
calcium hydroxide	$Ca(OH)_2$
aluminum hydroxide	$Al(OH)_3$
ammonia	NH_3
sodium carbonate	Na_2CO_3
magnesium hydroxide (milk of magnesia)	$Mg(OH)_2$

Acids can be classified as either strong or weak. A property of strong acids is that they are good conductors of electricity. In general, a *strong acid* is one that produces a large percentage of hydrogen ions when dissolved in water. Nearly all of the HCl molecules in a water solution break up into ions, and so HCl is a strong acid.

Vinegar is a dilute solution of acetic acid in water. Acetic acid is an example of a weak acid. It reacts with zinc, producing hydrogen, but the reaction takes place very slowly. Weak acids are poor conductors of electricity. In general, a *weak acid* is one that produces a small percentage of hydrogen ions when dissolved in water. Acetic acid produces only a few hydrogen ions per thousand molecules.

indicator

Hydrogen ions in solution can be detected with an indicator such as litmus paper. An **indicator** is a substance that changes color in the presence of certain other substances. Blue litmus, for example, turns pink in an acid, as shown in Figure 9-10.

bases

Red litmus placed in a solution of sodium hydroxide turns blue. This color change indicates the presence of a base. **Bases** are compounds that (a) taste bitter, (b) conduct electricity, (c) turn red litmus blue, (d) feel slippery. (CAUTION: Never taste or touch a substance unless you know it is safe to do so.) Some common bases are listed in Table 9-5.

All of the bases listed in Table 9-5 contain hydroxide ions (OH^-) except ammonia and sodium carbonate. However, all of the bases produce hydroxide ions in water.

$$\underset{\text{ammonia}}{NH_3} + H_2O \longrightarrow \underset{\text{ammonium ion}}{NH^+} + \underset{\text{hydroxide ion}}{OH^-}$$

$$\underset{\text{sodium carbonate}}{2Na^+ + CO_3^{2-}} + H_2O \longrightarrow \underset{\text{bicarbonate ion}}{HCO_3^-} + \underset{\text{hydroxide ion}}{OH^-} + 2Na^+$$

When an acid and a base are mixed together, the hydrogen and hydroxide ions combine, forming water:

$$H^+ + OH^- \longrightarrow H_2O$$

Figure 9-10 *Litmus is an acid-base indicator.*

(a) Blue litmus paper turns pink in an acid solution.

(b) Red litmus paper turns blue in a basic solution.

Demonstrations

1. Show that grape juice is an acid-base indicator. It turns red in acids and green in bases. You can use this property of grape juice to confound your students and demonstrate indicator action. Pour some unsweetened grape juice into a glass and dilute it about 1 to 9 by adding 9 parts of water to 1 part grape juice. Now pour the diluted juice into a second glass that contains several drops of household ammonia. When the grape juice mixes with the basic ammonia solution, it turns green. Next, pour the green solution into a glass that contains 1–2 mL of vinegar, or enough vinegar to more than neutralize the ammonia, so that the solution will again be acidic. Then the solution will turn red.
2. Show that a molecular acid undergoes ionization in water. First, test some glacial acetic acid and distilled water with a conductivity apparatus (see Figure TE 8-2) to demonstrate that both distilled water and glacial acetic acid are nonconductors and therefore nonionic. Then, pour some of the glacial acetic acid into the distilled water and test the solution for conductivity. The bulb will light because the weak acid has reacted with water to form ions, as shown in the following equation:

$$CH_3COOH + H_2O \longrightarrow H_3O^+ + CH_3COO^-$$

3. Use the conductivity apparatus (Figure TE 8-2) to show that only molecular species remain at the neutralization point when an acid and a base react to form an insoluble salt. First, show that a saturated solution of barium hydroxide ($Ba(OH)_2$) is a conductor. Then, using an eyedropper, add dilute sulfuric acid to the base solution. A white precipitate of barium sulfate ($BaSO_4$) forms. Stir the solution constantly as you add acid. Just before the point of neutralization, the light will grow dim. At the point of neutralization the light will go out, because only molecular water and insoluble barium sulfate are present:

$$H_2SO_4 + Ba(OH)_2 \longrightarrow BaSO_4\downarrow + 2H_2O$$

If more acid is added, the bulb will light again, because there will then be an excess of H^+ and SO_4^- ions from the sulfuric acid.

Teaching Tips

- If students have occasion to prepare dilute acid solutions, they should be helped to remember to *always add acid to cold water*. Tell students to remember the letters "a-a-a:" "always add acid."
- Have a team of interested students set up clean collection jars at strategic points and measure the acidity of rainfall with *p*H paper. If the rainfall is more acidic than normal (about 5.7), have students investigate the source of the acidity.

The acid and base are said to *neutralize* (NOO truh lyz) each other. When hydrochloric acid and sodium hydroxide are mixed together, they react as follows:

$$H^+Cl^- + Na^+OH^- \longrightarrow Na^+Cl^- + H_2O$$

The resulting solution contains common salt, sodium chloride. Actually the sodium and chloride ions are not really involved in the reaction and can be left out of the equation. In that case the reaction is identical to one given before.

$$H^+ + OH^- \longrightarrow H_2O$$

salt

The ionic compound formed when *any* acid and base react is called a **salt**. In general:

$$\text{acid} + \text{base} \longrightarrow \text{salt} + \text{water}$$

neutralization (noo truh luh ZAY shuhn)

A reaction in which an acid and a base neutralize each other, forming a salt and water, is called a **neutralization** reaction.

Acids and bases are very commonly occurring substances. (See Figure 9-11.) Hydrochloric acid is secreted in the stomachs of human beings where it helps to break down food. Sulfuric acid is widely used in making plastics, fertilizers, and dyes. It is used in the lead storage batteries found in cars, also. The production level of sulfuric acid is considered to be a measure of a nation's industrial development.

Figure 9-11 *(a) Some commonly used acids. (b) Some commonly used bases.*

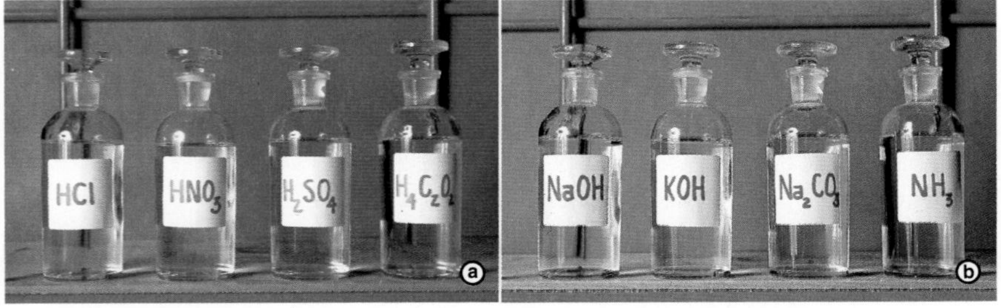

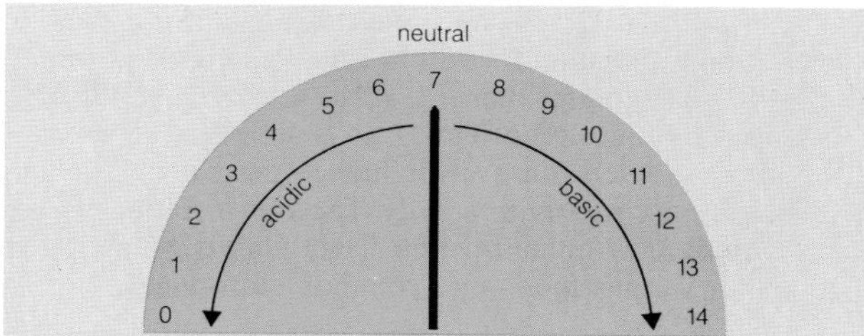

Figure 9-12 *The* pH *scale ranges from 0 to 14. Neutral solutions have a* pH *of 7.*

Sodium hydroxide (lye) is used in making soaps and detergents, cellophane, and paper. Potassium hydroxide is vital in producing fertilizers and dyes. Ammonia is used in the manufacture of explosives, synthetic fibers, medicines, and cleaning agents.

Acids and bases are commonly used to neutralize each other. Farmers use lime (CaO) to reduce the acidity of soil. (When mixed with water, the lime forms calcium hydroxide, $Ca(OH)_2$.) People take antacids to neutralize excess acid in the stomach.

pH scale

It is often important to know how acidic or basic a solution is. For example, the chlorine that is added to a swimming pool makes the water acidic. Testing the acidity of the water can reveal whether the correct amount of chlorine has been added. Chemists have devised a ***p*H scale** to measure how acidic or basic a solution is. The symbol "*p*H" is an abbreviation for "power of hydrogen ion." For dilute solutions, the range of the *p*H scale is 0–14. (See Figure 9-12.) A *p*H of 7 indicates a *neutral* solution—one with equal numbers of H^+ and OH^- ions. A *p*H below 7 indicates an acidic solution. A *p*H above 7 indicates a basic solution. Between each two successive numbers on the scale there is a tenfold difference. A solution with a *p*H of 6 has ten times as many H^+ ions as the same volume of a neutral solution. A solution with a *p*H of 8 has one tenth as many H^+ ions, or ten times as many OH^- ions, as the same volume of a neutral solution. A mixture of dyes has been devised that changes color gradually over the entire range of the *p*H scale. This universal indicator is available in the form of *p*H paper. (See Figure 9-13.)

Figure 9-13 *The mixture of dyes in* pH *paper can indicate* pH *over the entire range. The rainbowlike colors range from orange (highly acidic, low* pH*) through yellow-green (neutral) to dark blue (highly basic, high* pH*).*

Fact At Your Fingertips

- Blueberries, cranberries, azaleas, and rhododendrons are examples of plants that thrive only in distinctly acidic soils. Field crops such as corn, soybeans, tomatoes, and potatoes require slightly acidic soil. Alfalfa, clover, and asparagus grow best in neutral soils. Cliff ferns, beech trees, and salt bushes require a soil that is basic. Many plants are tolerant of a wide range of soil *p*H values.

9 CHEMICAL REACTIONS
9-5 Acids and Bases

Sour or Bitter?

Fill in the chart by using Section 9-5 and Tables 9-4 and 9-5 in your text.

Formula	*Name*	*Acid or Base?*	*Property or Use*
NH_3			
HCl			
NaOH			
H_2SO_4			

The Art of Decoding

The puzzle below must be decoded to make sense. Each set of letters in Column A will form a word when it is decoded. The clue to the code is that the alphabet has been turned around: A = Z, Z = A, B = Y, Y = B, and so forth. The ¢ sign replaces a space between words of a term. First, decode each word in Column A and write it on the space beneath the code. Then match the phrases in Column B with the decoded word in Column A by writing the letter of the phrase on the space at the left of the code.

A	B
___ 1. zxrw ________	a. A scale that indicates the "hydrogen ion power" of a solution.
___ 2. rmwrxzgli ________	b. The process in which an acid and a base combine to form water and a salt.
___ 3. yzhv ________	c. The kind of a solution that has equal numbers of hydrogen and hydroxide ions.
___ 4. mvfgizo ________	d. A substance that turns red litmus blue.
___ 5. mvfgizorazgrlm¢ivzxgrlm ________	e. A compound that changes color in the presence of substances such as acids and bases.
___ 6. kS¢hxzov ________	f. A substance that contains hydrogen ions.

9 CHEMICAL REACTIONS
9-5 Acids and Bases

A Matter of pH

For each of the four pairs of substances listed on the left, decide which substance is the more acidic and circle it. For each of the four pairs of substances listed on the right, decide which substance is the more basic and underline it. Hint: You may want to use reference books and Table 9-6 in your textbook.

1. stomach acid or cola drink
2. human blood or sea water
3. vinegar or milk of magnesia
4. oranges or apples

1. grapefruit juice or tomato juice
2. saliva or boric acid solution
3. lemon juice or bread
4. pure water or rainwater

pH and Shampoo

The following paragraphs describe what is meant by pH balanced shampoo and how the pH affects the detergent of a shampoo. Read the paragraphs. Then write the answers to the questions that follow.

The shampoo commercial promises that your hair will be cleaner,

Using Try This

Grape juice, orange juice, tea, coffee, and aspirin are acidic, while most soaps and baking soda are basic. Depending on the sensitivity of the *p*H paper, water will appear as either neutral or slightly acidic, and salt water will appear as either neutral or slightly basic. A sugar solution will be neutral.

Figure 9-14 *The discoloration of the leaves on this plant is due to excess soil acidity.*

The *p*H of soil is an important factor in the growing of crops. Some plants grow well in acidic soil. Others require soil with a *p*H greater than 7. Excess acidity in soil is often indicated by discolored leaves, as is shown in Figure 9-14. The *p*H of the soil can even influence the color of flowers.

Foods vary widely in acidity. The tart or bitter taste of some foods may enhance their flavor. Table 9-6 shows *p*H values for various foods and common substances.

Table 9-6: *p*H of Various Solutions

Substance	*p*H
stomach acid	1.0
vinegar	2.4
cola drink	2.5–3.0
grapefruit juice	3.2
tomato juice	4.3
orange juice	2.6–4.4
boric acid solution	5.0
saliva	6.6
milk	6.6–6.9
human blood	7.4
seawater	8.0
egg white	8.0
household ammonia	11–12

Checkpoint

1. What are the characteristic properties of (a) acids, (b) bases?
2. What ions are always present in solutions of (a) acids, (b) bases?
3. What happens when an acid and a base are mixed?
4. What is the *p*H scale? What does it measure?

TRY THIS

Using litmus or *p*H paper, find out whether each of the following is acidic, basic, or neutral: tap water, grape juice, and orange juice. Repeat the test on solutions made by dissolving each of the following in water: tea, coffee, aspirin, salt, sugar, soap, and baking soda (sodium bicarbonate).

Checkpoint Answers

1. (a) Acids taste sour, conduct electricity, turn blue litmus pink, contain hydrogen, and react with active metals to form hydrogen gas.
   (b) Bases taste bitter, conduct electricity, turn red litmus blue, and feel slippery.
2. (a) H^+, (b) OH^-.
3. They neutralize each other, forming a salt and water. The water results from the combining of the hydrogen and hydroxide ions.
4. The *p*H scale is a scale used to measure how acidic or basic a solution is. It measures the relative numbers of hydrogen and hydroxide ions present in the solution.

9-6

Rates of Reaction

ENRICHMENT

Goals

1. To state the factors that affect the rate of a chemical reaction.
2. To explain why these factors have their effects.

The rate of a chemical reaction is the speed at which it occurs. This can be measured in terms of how much product is formed per unit of time. The rates of chemical reactions vary greatly. A mixture of hydrogen and oxygen reacts with explosive speed when ignited with a spark. A piece of iron may take years to react completely with oxygen, forming rust. By learning what factors affect reaction rates, people can control chemical reactions. They can make reactions more safe, efficient, and economical.

The rate of a reaction depends on (a) the temperature, (b) the amounts of reactants and products in a given volume (their *concentrations*), (c) the size and surface area of the reactants, and (d) the presence of a catalyst. A **catalyst** is a substance that changes the rate of a reaction without being used up in the reaction.

catalyst
(KAT ul ist)

As a rule of thumb, a 10°C increase in temperature tends to double the rate of a reaction. As the temperature rises, molecules move faster, collide more often, and hit each other harder. This causes more of them per second to break apart and form new chemical bonds. The cooking of food involves chemical reactions. By regulating the temperature, people can control the rate at which foods cook.

If the concentration of the reactant molecules increases, the number of molecular collisions per second will also, in general, increase. This increases the rate of the reaction. If a small amount of copper sulfate is dissolved in water, the solution will react slowly with a piece of zinc. If a large amount of copper sulfate is dissolved in the same amount of water, the reaction will take place more quickly. The concentration of a solution can be increased by dissolving more of the solute in the same amount of water. The concentration of a gas can be increased by compressing it. This forces the molecules of the gas into a smaller volume, causing them to collide more often. In an automobile engine, the

Teaching Suggestions 9-6

Have students first read the section and then do the Try This activity. Ask students to predict the outcome of each part of the experiment and write their predictions on the chalkboard for reference. See Using Try This for expected outcomes.

Background

For any forward reaction there is always a reverse reaction. In some cases the reverse reaction is so slow that the forward reaction appears to go to completion. Often, though, as the concentration of products increases, the reverse reaction rate increases to the point where the two reactions are taking place at the same rate. An equilibrium is thus established and no further change in concentrations of reactants and products will occur on the macroscopic scale. At the particle level, on the other hand, both reactions are still going on, but at equal rates.

Figure 9-15 *Reaction rate depends on the amount of exposed surface area, as well as on other factors. (a) A large piece of wood tends to burn slowly in air. (b) The same amount of wood in the form of wood chips has a greater surface area and combines more rapidly with oxygen in the burning process.*

fuel-air mixture that is ignited is highly compressed. It reacts very rapidly when a spark is applied, providing power to move the car.

The surface area of a substance can be increased by breaking or grinding it into smaller particles. When this is done, the substance reacts more rapidly. This happens because more molecules of one reactant are exposed to the other reactant. A log of wood burns at a slow, steady rate. The same amount of wood in the form of chips can burn rapidly when ignited. (See Figure 9-15.)

A catalyst may provide a more efficient "pathway" for a reaction, increasing the reaction rate. Substances called *enzymes* act as catalysts within the human body. They enable reactions to take place more rapidly than they normally would at body temperature.

Demonstrations

1. Show the effects of concentration of reactants and of temperature on rate of reaction. First, prepare solution I by dissolving 15 g of potassium iodate in 1 liter of water. Label the solution. Then prepare solution II by dissolving 4 g of soluble starch in 500 mL of boiling water and adding 15 g of sodium bisulfite and another 500 mL of water. Label this solution. Before doing the experiment, dilute solution II, or both solutions, with water until there is a time delay of about 10 seconds after mixing before the color appears.

   When these two solutions react, they form, as a final product, elemental iodine. The iodine then reacts with starch and yields a dark, blue-black color.

   Have students record the time required for equal volumes of the two solutions to react and form the final product. Then show the effect of concentration on the rate of reaction by diluting Solution I with an equal volume of water and repeating the reaction. Record the time needed for reaction to occur.

   To show the effect of temperature, cool equal volumes of the two solutions to about 5–10°C in a freezer or refrigerator. After mixing, the rate will clearly be reduced.
2. Show the effect of a build-up of products on reaction rate. Let a seltzer tablet react with about 50 mL of water. Record the time needed for the reaction to be completed. Then drop a second tablet into the same container, which now also contains the products of the first reaction. The second reaction will proceed more slowly, since the accumulated products have a tendency to re-form the reactants. This cuts down the rate of the forward reaction.

Teaching Tip

- Be sure students understand how reaction rates are determined at the particle level. A combustion reaction, for example, has a very low rate until a temperature is reached at which the reacting particles collide with sufficient energy to cause old bonds to be broken and new bonds to be formed. At this point, called the ignition point, the rate suddenly increases dramatically.

Platinum is often used as a catalyst in reactions between gases. Many gases attach themselves to a platinum surface, enabling the gas molecules to react more rapidly. Platinum is used as a catalyst for the reaction between oxygen and sulfur dioxide. These gases react to form another gas, sulfur trioxide. The sulfur trioxide is then made to react with water, forming sulfuric acid. Catalysts can be used to slow down, as well as speed up, reactions. Such catalysts are sometimes called "negative catalysts."

When the products of a reaction do not escape from the reaction area, their concentrations increase. The products may re-form into reactants. The rate of this *reverse reaction*—products re-forming into reactants—tends to increase as the concentration of products increases. There is a limit to the "completeness" of the *forward reaction* under such conditions. That is, there may be reactants as well as products present at the end of the reaction. When products escape or are removed from the reaction area, a reaction can go more nearly to completion. When zinc reacts with hydrochloric acid, producing hydrogen as a product, the gas escapes. This escaped hydrogen is no longer present to re-form into the reactant hydrochloric acid. Therefore, there is no chance of a significant reverse reaction. The forward reaction tends to go to completion.

Checkpoint

1. What factors affect the rate of a chemical reaction?
2. Explain how, at the molecular level, each factor affects the reaction rate.
3. Why does compressing a gas increase its concentration?

TRY THIS

Add a seltzer tablet to half a glass of hot water. At the same time add an identical tablet to half a glass of cold water. In which case is the reaction faster?

To half a glass of cold water add a whole seltzer tablet. At the same time add a crushed seltzer tablet to an identical amount of water at the same temperature. In which glass is the reaction faster? Why?

Fact At Your Fingertips

- Pressurized gases react much faster, because squeezing the gases into a smaller volume increases the concentration of the reactant particles and the number of collisions per unit of time.

9 CHEMICAL REACTIONS

9-6 Rates of Reaction

How Fast Will It Go?

A diagram or graph appears above each set of questions below. Study each diagram or graph. Then write the answers to the questions that follow it.

A B C

1. Each circle in the diagrams represents a molecule. Each line shows the direction in which the molecules are moving. What is one difference in the molecules in the three diagrams? ______
2. In which diagram is a reaction more likely to occur? ______ Why? ______

A B

1. What is one difference between container A and container B? ______
2. In which container, A or B, is the number of molecular collisions per second greater? ______
3. In which container is the reaction proceeding more rapidly? ______

A
Particles of Cu^{+2}
Piece of zinc
B
Particles of Cu^{+2}
Piece of zinc

1. Describe at least one difference between container A and container B. ______
2. In which container, A or B, is the number of molecular collisions per second greater? ______
3. In which container, A or B, is the reaction proceeding more rapidly? ______

Using Try This

Students should predict and observe that the tablet will react faster in hot water, because increased molecular motion at the higher temperature increases the collision rates of the reactant particles. The crushed tablet will react faster because the increase in surface area will allow more water and tablet molecules to make contact.

Checkpoint Answers

1. Temperature, concentrations of reactants and products, size and surface area of reactants, and presence of a catalyst.
2. Temperature affects how quickly particles move, how often they collide, and how hard they hit. Concentrations of reactants and products affect the number of molecular collisions. Size and surface area of the reactants affect the number of molecules of one reactant that are exposed to the other reactant. A catalyst provides a more efficient "pathway" for the reaction.
3. Compressing a gas forces the molecules into a smaller volume.

Using The Feature Article

Although the gasoline engine is so common, some students will not have even a basic understanding of its operation. Thus, you should lead the class step-by-step through the description of the mechanical operation of an engine. In particular, stress that at the power stroke, the mechanical energy of the expanding hot gases is converted to the rotary motion of the drive wheel. The expanding hot gases, of course, are the products of an exothermic chemical reaction—the combustion of the fuel within the engine's cylinders. The overall process is thus the conversion of chemical energy (bound up in fossil-fuel molecules) to mechanical energy of motion, which is a type of kinetic energy.

Internal Combustion Engines

Most automobile engines are internal combustion engines. The fuel that provides the energy to move the car burns inside the engine. The energy causes pistons to move, which cause a crankshaft to turn, which then causes the wheels to turn. (See Figure 1.) The fuel is usually gasoline, a mixture of hydrocarbon compounds. When the fuel burns, the carbon and hydrogen combine with oxygen. Energy is given off when the fuel burns.

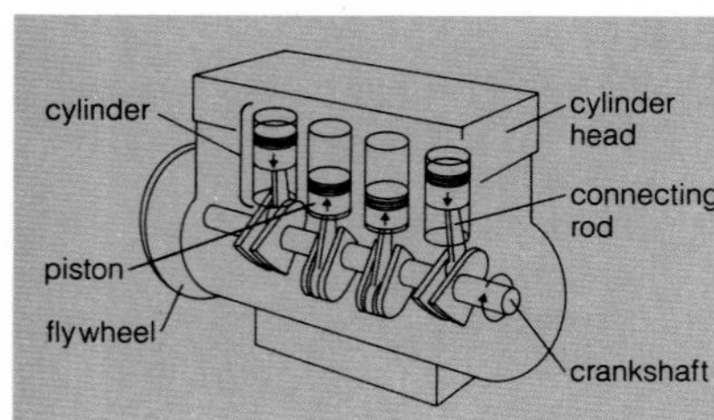

Figure 1

In an automobile, the engine is normally started by turning the ignition key. This connects the starter motor to the battery. Powered by the battery, the motor turns the crankshaft, which causes at least one piston to move downward. At the same time a mechanism opens the inlet valve of the cylinder in which the piston is located and closes the exhaust valve. As the air in the cylinder expands, its pressure decreases. As a result, a mixture of air and gasoline vapor is drawn in from the carburetor (the chamber in which the air and fuel are normally mixed). When the pressure inside the cylinder equals the air pressure outside, the inlet valve closes.

As the crankshaft completes its first full turn, the ascending piston compresses the fuel-air mixture, heating the gases in the process. At the top of the stroke, the spark plug gives off a spark. The gases ignite and burn. The high temperature produced by the combustion causes the gaseous products (mostly carbon monoxide, carbon dioxide, and water vapor) to expand. This drives the piston downward, providing power to the crankshaft. As the crankshaft completes its second turn, the ascending piston this time forces the gaseous waste products through the open exhaust valve. (See Figure 2.)

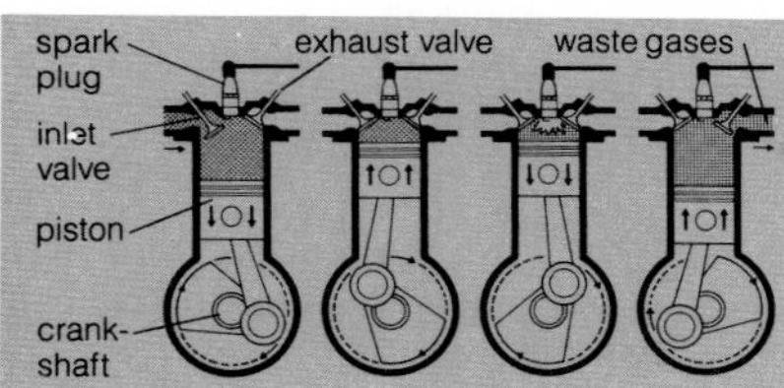

Figure 2

In most cars, there is a minimum of four cylinders. They fire so that the crankshaft gets a push (power stroke) every half turn. Each piston goes through four complete cycles between explosions within its cylinder. Once started, the cycles continue as long as fuel, air, and electricity are provided.

Understanding The Chapter

9

Main Ideas

1. Energy is needed to separate the particles of a liquid when it changes to a gas. The energy is given up when a gas changes into a liquid.
2. Chemical changes generally involve more energy than do physical changes. In chemical changes, chemical bonds are broken.
3. Exothermic reactions release energy. Endothermic reactions absorb energy.
4. Oxidation is the loss of electrons by an atom or ion in a chemical reaction. Reduction is the gain of electrons.
5. Electrochemical cells are based on oxidation-reduction reactions.
6. Corrosion often results from unwanted electrochemical reactions.
7. Active metals are those that give up electrons readily.
8. In solutions, acids produce hydrogen ions and bases produce hydroxide ions.
9. Acids have a *p*H less than 7, and bases have a *p*H greater than 7.
10. A neutralization reaction is one in which an acid and a base react with each other, forming a salt and water.
11. The rate of a chemical reaction depends on the temperature, the concentrations of reactants and products, the surface area of the reactants, and the presence of a catalyst.

Vocabulary Review

From the following list, choose the term that best completes each of the statements. Write your answers on a separate piece of paper.

acid	heat of condensation
active metal	heat of vaporization
base	indicator
catalyst	neutralization
electrochemical cell	oxidation
electrode	*p*H scale
endothermic	reduction
exothermic	salt

1. Litmus paper is a(n) __?__.
2. A(n) __?__ has a sour taste.
3. __?__ is the loss of electrons by an atom or ion.
4. A(n) __?__ reaction is one in which heat is absorbed.
5. A reading of 9 on the __?__ indicates that a substance is basic.
6. __?__ occurs when an atom or ion gains electrons.
7. A(n) __?__ produces OH^- ions in solution.
8. The burning of hydrogen is an example of a(n) __?__ reaction because heat is released.
9. The energy needed to transform a liquid into a gas at its boiling point is its __?__.
10. A reaction in which an acid and a base react to form a salt and water is called a __?__ reaction.

Vocabulary Review Answers

1. indicator
2. acid
3. oxidation
4. endothermic
5. *p*H scale
6. reduction
7. base
8. exothermic
9. heat of vaporization
10. neutralization
11. electrochemical cell
12. catalyst

Reading Suggestions

For The Teacher

Crosland, M. P. HISTORICAL STUDIES IN THE LANGUAGE OF CHEMISTRY. New York: Dover, 1979. Describes the people and the discoveries that formed the base for modern-day chemistry.

Long, Gilbert G., and Forrest C. Hentz. PROBLEM EXERCISES FOR GENERAL CHEMISTRY. New York: Wiley, 1982. Provides basic background information, tables, problems, and solutions for a general chemistry curriculum. The exercises are categorized by topic.

Preuss, Paul. "The Shape of Things to Come." *Science 83*, December 1983, pp. 81–87. An in-depth look at biochemists' ability to redesign life's molecules.

Rogers, Michael. "The Sculpture Transparent." *Science 83*, December 1983, pp. 39–45. A fascinating account of a California sculptor's projects, using the remarkable plastic polymethyl methacrylate.

Silberner, J. "Layers of Complexity in Ozone Hole." *Science News*, March 14, 1987, p. 164. Difficulties associated with models to explain the hole in the ozone layer.

For The Student

Gardner, Robert. KITCHEN CHEMISTRY: SCIENCE EXPERIMENTS TO DO AT HOME. New York: Messner, 1982. Chemical experiments related to the basic properties of matter that include some simple chemical reactions.

Sherman, Alan and Sharon J. Sherman. CHEMISTRY AND OUR CHANGING WORLD. Englewood Cliffs, NJ: Prentice Hall, 1983. An introductory textbook that illustrates the relationships among people, technology, and the environment.

Chapter Review Answers

Know The Facts

1. condensation
2. stays at about 100°C
3. large
4. greater than
5. greater than
6. copper ions
7. copper
8. oxidized
9. weak
10. hydrogen
11. bases
12. increasing the concentration of reactants

Understand The Concepts

13. 400 J
14. The attractive forces between water molecules are very large. Thus, a large amount of energy must be expended to overcome these forces and vaporize liquid water.
15. (a) Mg is oxidized; H^+ is reduced.
    (b) Fe is oxidized; Pb^{2+} is reduced.
    (c) Al is oxidized; O_2 is reduced.
16. For the drawing, see Figure TE 9-1, but substitute lead for zinc and Pb^{2+} for Zn^{2+}. Lead is the negative electrode.
17. Answers will vary. See student text, pages 226–227. See also answer to Checkpoint question 3 on page 227.

11. A(n) __?__ can be a convenient source of electricity.
12. A(n) __?__ is a substance that can change the rate of a chemical reaction without itself being used up.

Chapter Review

Write your answers on a separate piece of paper.

Know The Facts

1. The energy released when a gram of gas changes to liquid at the boiling point is the heat of __?__. (reaction, condensation, vaporization)
2. When water that is being heated starts to boil, the temperature __?__. (slowly rises, slowly falls, stays at about 100°C)
3. The forces of attraction between water molecules are __?__ compared to those acting between molecules of most other substances. (large, small, average)
4. Energy changes in chemical reactions are generally __?__ those in physical changes. (smaller than, equal to, greater than)
5. The forces between atoms *within* molecules are generally __?__ the forces *between* molecules. (less than, equal to, greater than)
6. In a Daniell cell __?__ are reduced. (zinc atoms, zinc ions, copper ions)
7. Lead is a more active metal than is __?__. (zinc, iron, copper)
8. The formation of zinc ions when zinc reacts with an acid indicates that zinc has been __?__. (reduced, oxidized, neutralized)
9. Acetic acid, which is found in vinegar, is a __?__ acid. (strong, weak, active)
10. __?__ is found in all acids. (oxygen, nitrogen, hydrogen)
11. Ammonia, sodium carbonate, and lye are all __?__. (acids, bases, salts)
12. The rate of a chemical reaction can be increased by __?__. (lowering the temperature, increasing the concentration of reactants, increasing the concentration of the products)

Understand The Concepts

13. It takes 400 J of energy to convert one gram of liquid sulfur dioxide (SO_2) to vapor at the boiling point (−10°C). How much energy is released when one gram of SO_2 gas condenses at −10°C?
14. How do you explain the fact that the heat of vaporization of water is so much higher than that of most other liquids?
15. In each of the following reactions, what is oxidized and what is reduced?

    (a) $2H^+ + Mg \longrightarrow Mg^{2+} + H_2$
    (b) $Fe + Pb^{2+} \longrightarrow Fe^{2+} + Pb$
    (c) $4Al + 3O_2 \longrightarrow 4Al^{3+} + 6O^{2-}$

16. Explain how you could build an electrochemical cell using lead and silver electrodes, lead nitrate and silver nitrate solutions, a beaker, a porous cup, and wires. Draw a sketch of the cell. Which metal is the negative electrode?
17. Describe at least two ways in which metals can be protected from corrosion.

18. How would you tell whether a liquid is (a) an acid, (b) a base?
19. What is the difference between strong and weak acids?
20. Explain why ammonia and sodium carbonate are bases even though they contain no OH^- ions.
21. Some seltzers come in tablet form and others come as powders. Which form would you expect to react faster in water? Why?
22. What increase in the rate of a reaction might you expect if the temperature were to rise by 20°C?

Challenge Your Understanding

23. How will the number of hydrogen ions per unit volume of an acid with a *p*H of 2 compare with one that has a *p*H of 6?
24. Boric acid is sometimes used as an eye rinse. Why is it safe to use boric acid this way, when other acids, such as sulfuric acid, could cause permanent blindness?
25. Explain how the connection of a piece of magnesium to a ship's iron hull can protect the hull from corrosion.
26. What would you expect to be the difference between strong and weak bases?
27. If you were to heat a solution of sodium chloride in water, you would find that the temperature would continue to rise after the solution has started to boil. How might you explain this?
28. Which of the following ions would you expect to be easiest to reduce: Cu^{2+}, Zn^{2+}, Mg^{2+}? Explain your answer.

Projects

1. Study the effects of corrosion on steel, with and without a "sacrificial electrode." Place a piece of steel in seawater (or a solution made by adding 30 g of salt to a liter of water). In a second container of salt water, place a similar piece of steel that you have connected to a piece of clean zinc by means of a wire. (A current-measuring device called an ammeter can be used to measure the current flowing between the two metals.) Allow the setup to remain for several weeks, adding water to replace any lost by evaporation. Report on the results.
2. Visit a school or community swimming pool. Find out how the *p*H of the water is tested and adjusted.
3. Talk to a farmer or your local farm bureau agent about the effect of lime on soil. Find out which crops grow best in acidic soils and which grow best in basic soils. Find out how the *p*H of soils is changed.
4. At a library, do research on acid rain. What is acid rain? What causes it? What are its effects? What can be done about it?
5. At a library, do research on the history of electric-powered vehicles. What vehicles now in use are powered by electrochemical cells? Do you think an all-purpose electric-powered automobile is likely to be developed in the near future? Why or why not?
6. Research and report on the contributions to chemistry and physics of any two of the following scientists: Luis Walter Alvarez; Mary L. Good; Mildred Cohn; Kenichi Fukui.

18. Answers will vary. One possibility is to test the liquid with litmus or *p*H paper.
19. A strong acid produces a large percentage of hydrogen ions in solution. A weak acid forms a small percentage of hydrogen ions in solution.
20. The carbonate ions and ammonia molecules react with water to form hydroxide ions.
21. The powdered form would react faster because the surface area is much greater for a powder than a tablet. Greater surface area means that more of the substance will be in contact with water. Thus, the number of molecular collisions per unit of time will be greater.
22. The reaction would be about four times as fast. Each increase of 10°C roughly doubles the reaction rate.

Challenge Your Understanding

23. The number of hydrogen ions at a *p*H of 2 will be 10,000 times greater than the number at a *p*H of 6.
24. Boric acid is too weak an acid to damage eye tissues.
25. An electrochemical cell is established between the iron and magnesium. The magnesium is the negative electrode; thus the magnesium, not the iron, is oxidized.
26. A strong base forms a large percentage of hydroxide ions in solution; a weak base forms a small percentage of hydroxide ions in solution.
27. Boiling point is elevated as NaCl concentration increases, and NaCl concentration increases as water boils off.
28. Cu^{2+}. Of the three, copper gives up electrons least readily. Thus, it can be expected to gain electrons, or be reduced, most readily.

9 CHEMICAL REACTIONS
Projects

1. At the library, find out about the life and work of each of the following scientists.
   a. Sir Humphrey Davy ______
   b. Luigi Galvani ______
   c. Michael Faraday ______
   d. Walter Nernst ______
2. Investigate the occupations associated with baking.
   a. How is bread baked commercially? ______
   b. How does a person become a baker? ______
   c. How are baking improvers such as chlorine dioxide or potassium bromate used? ______
   d. How are shelf-life improvers such as monoglycerides or calcium propionate used? ______
3. Investigate canning, a method of preserving foods.
   a. Why is the food heated? ______
   b. Why is the food sealed? ______
   c. What is the main difference between the two main methods of canning? ______

CHAPTER 10 Overview

The purpose of this chapter is to introduce basic principles of radioactivity, nuclear reactions, and nuclear energy to students. First, the decay of radioactive nuclei, yielding alpha, beta, and gamma radiation, is described, and transmutation is explained. Isotopes are then described and half-life is defined. The numerous uses of radioisotopes in science, medicine, and industry are presented in the enrichment section. Nuclear equations are then introduced, and binding energy and nuclear fission are defined and explained. This is followed by a discussion of chain reactions and their role in the generation of energy by nuclear reactors. The problem of disposal of long-lived nuclear wastes is then introduced. Finally, the process of nuclear fusion is described and its potential as an energy source is discussed.

Goals

At the end of this chapter, students should be able to:

1. describe the discovery of radioactive elements.
2. list the types of radiation given off by radioactive nuclei.
3. define transmutation.
4. define isotopes.
5. explain why isotopes of a given element have the same chemical properties.
6. compare nuclear forces with those that act between charged objects.
7. describe radioactive decay.
8. define the half-life of a radioactive isotope.
9. interpret nuclear equations.
10. explain how stable isotopes can be converted into radioactive isotopes.
11. describe some practical uses of radioactive isotopes.
12. define binding energy.
13. identify the source of binding energy.
14. explain, in words, what $E = mc^2$ means.
15. define nuclear fission.
16. explain what is meant by the critical mass of a fissionable material.
17. describe what happens in a chain reaction.
18. list the parts of a nuclear reactor and explain their functions.
19. explain what nuclear wastes are and describe how they might be safely stored.
20. define nuclear fusion.
21. state the source of the energy given off by nuclear fusion.
22. describe the major obstacle to producing a controlled fusion reaction.

Chapter 10

Nuclear Reactions

Aerial view of the Stanford University linear accelerator in California.

10-1

Radioactivity

Goals

1. To describe the discovery of radioactive elements.
2. To list the types of radiation given off by radioactive nuclei.
3. To define transmutation.

In 1896, Henri Becquerel made an amazing discovery. Becquerel had left a piece of rock that contained uranium on top of a sheet of unexposed photographic film that was covered with paper. Some time later, when Becquerel developed the film, he was surprised by the image he saw. The image had the same size and shape as the rock that had been on top of the film, as is shown in Figure 10-1. Something in the rock had given off invisible rays, or *radiation* (ray dee AY shuhn), that could pass through the film cover. It was found that the uranium in the rock was the source of the radiation. Further, the radiation was given off by the *nuclei* of uranium atoms. Nuclei that give off radiation are said to be *radioactive nuclei.*

Two years later, Marie Curie, one of Becquerel's associates, began working with pitchblende. It is a mineral even more radioactive than uranium-containing minerals. After long, painstaking work, she found that pitchblende contained two previously unknown radioactive elements. She

Figure 10-1 *In the photograph at the left, a rock that contains radioactive nuclei is resting on a piece of photographic film. Later, when the film has been developed, the rock's image is seen, as shown at the right.*

Vocabulary Preview

alpha radiation, beta radiation, gamma radiation, transmutation, isotopes, radioactive nuclei, radioactive decay, half-life, nuclear equation, cosmic rays, tracers, binding energy, nuclear fission, chain reaction, critical mass, nuclear reactors, nuclear fusion

Teaching Suggestions 10-1

An interesting way to open this lesson is to show students the operation of a cloud chamber. Simple, inexpensive cloud chambers are available from most science supply houses. Complete instructions for use are usually included. Most such cloud chambers include a source of radioactivity. The tracks left along the paths of the radioactive particles appear as tiny droplets of supersaturated vapor.

End-of-chapter project 2 applies to this section.

Background

Becquerel's discovery of radioactivity, like so many discoveries in science, was accidental. Later studies revealed the radiation from radioactive substances (elements) to be of three kinds—alpha, beta, and gamma. An alpha particle consists of a fast-moving helium nucleus that is made up of two protons and two neutrons. When an alpha particle leaves a nucleus, the mass of that nucleus decreases by four mass units, and the charge, or atomic number, decreases by two units because each proton has a charge of $+1$. Thus, after emitting an alpha particle, a nucleus with mass m and charge n will have a mass of $m - 4$ and a charge of $n - 2$.

A beta particle is a fast-moving electron. When a beta particle is emitted from a nucleus, the positive charge of the nucleus increases by 1 and the atomic number of the resulting element increases by 1. Because the mass of an electron is negligible, there is no significant change in the mass of the nucleus.

Gamma rays are high-energy electromagnetic radiation. Gamma rays have neither mass nor charge. Consequently, the emission of gamma radiation reduces the energy of a nucleus but has no effect on its mass or charge.

Teaching Tips

- Let students see as many demonstrations of radiation as you can manage.
- Be sure students understand that alpha, beta, and gamma radiation are really helium nuclei, electrons, and electromagnetic radiation, respectively. Students should also acquire an understanding of the effect each type of radiation has on the mass and charge of the radioactive nucleus from which it originates.

Fact At Your Fingertips

- From tons of pitchblende, Curie was able to extract only a tenth of a gram of pure radium.

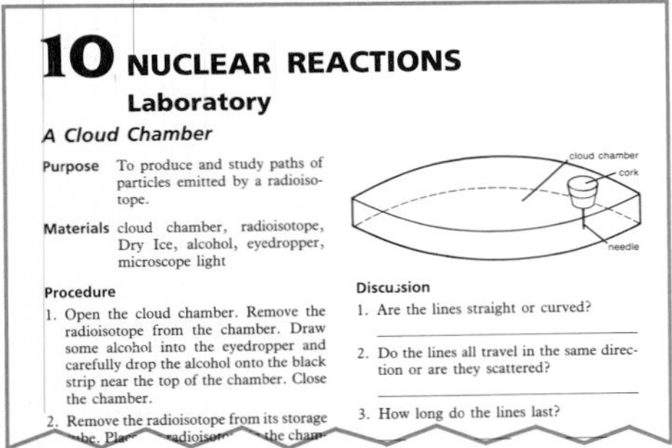

10 NUCLEAR REACTIONS

Laboratory

A Cloud Chamber

Purpose To produce and study paths of particles emitted by a radioisotope.

Materials cloud chamber, radioisotope, Dry Ice, alcohol, eyedropper, microscope light

Procedure

1. Open the cloud chamber. Remove the radioisotope from the chamber. Draw some alcohol into the eyedropper and carefully drop the alcohol onto the black strip near the top of the chamber. Close the chamber.
2. Remove the radioisotope from its storage

Discussion

1. Are the lines straight or curved? ____________
2. Do the lines all travel in the same direction or are they scattered? ____________
3. How long do the lines last?

Mainstreaming

Language cards for the hearing impaired should be prepared at the start of this chapter. Make cards for each new concept introduced and some additional cards containing concept words in phrases and questions. The abstract nature of the chapter content makes additional attention in this area especially important. In addition, provide students with as many concrete experiences as you can.

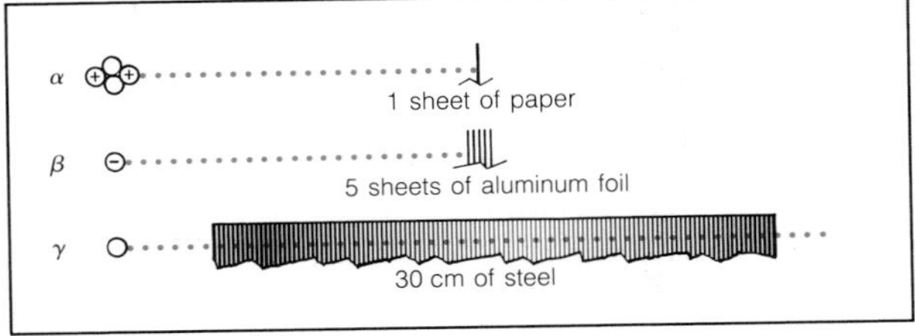

Figure 10-2 *Alpha (α), beta (β), and gamma (γ) radiation differ in ability to pass through various thicknesses of substances.*

named one of the elements polonium (Po) in honor of Poland, her native country. She named the other element radium (Ra) because of the intense radiation it gave off. From tons of pitchblende, Marie Curie was able to extract only a tenth of a gram of pure radium.

alpha radiation

beta radiation

gamma radiation

Three types of radiation are given off by radioactive nuclei. They are named after the first three letters of the Greek alphabet: *alpha* (α), *beta* (β), and *gamma* (γ). **Alpha radiation** consists of helium nuclei, which are made up of two protons and two neutrons. Because alpha particles move at relatively low speeds, they can be stopped by a sheet of ordinary writing paper. (See Figure 10-2.) **Beta radiation** consists of electrons moving at extremely high speeds. Beta particles, depending on their speeds, can be stopped by aluminum foil, 2 to 10 sheets thick. **Gamma radiation** consists of light rays that have more energy and greater penetrating power than do x rays. Some gamma rays can penetrate steel that is 30 cm thick.

transmutation

When a nucleus gives off an alpha or a beta particle, the number of protons in the nucleus changes. A nucleus that has given off an alpha particle has lost two protons and two neutrons. A nucleus that has given off a beta particle has lost a neutron and gained a proton. (Actually, a neutron turns into a proton and an electron, which escapes from the nucleus.) When the atoms of an element gain or lose protons, a new element is formed. This is called **transmutation**.

Checkpoint

1. How was radioactivity discovered?
2. What types of radiation are given off by radioactive nuclei?
3. What is transmutation?

10 NUCLEAR REACTIONS

10-1 Radioactivity

Transmutation, or Cracking the Atom

A term is missing in each of the following statements. Think of the term that best completes each statement. Then write the term on the space at the left.

____________ 1. Uranium gives off invisible rays called __1.__ that can penetrate film covers to expose film.

____________ 2. Henri Becquerel discovered that radiation was given off by __2.__ of uranium atoms.

____________ 3. Nuclei that give off radiation are called __3.__ nuclei.

____________ 4. Marie Curie discovered an element that she named __4.__ because of the intense radiation that it gave off.

____________ 5. __5.__ radiation consists of helium nuclei, a stable unit made up of two protons and two neutrons.

____________ 6. __6.__ radiation consists of electrons moving at speeds near the speed of light.

Checkpoint Answers

1. Radioactivity was discovered by Becquerel, who inadvertently left a uranium-containing rock on photographic film and later discovered the image of the rock on the film.
2. Alpha, beta, and gamma radiation.
3. Transmutation is the change of one element to another when its atoms gain or lose protons.

10-2

Isotopes

Goals

1. To define isotopes.
2. To explain why isotopes of a given element have the same chemical properties.
3. To compare nuclear forces with those that act between charged objects.

Each atom of a given element has the same number of protons. However, the atoms may have different numbers of neutrons and, therefore, different masses. Atoms of the same element that have different masses are called **isotopes**. For example, most naturally occurring hydrogen atoms have a single proton in their nucleus. Their atomic mass is 1 amu. However, some hydrogen atoms have a nucleus that consists of a proton and a neutron. (See Figure 10-3.) These atoms have a mass of 2 amu and are sometimes referred to as atoms of *heavy hydrogen* or *deuterium* (doo TEER ee um). Only two out of every 1000 naturally occurring hydrogen atoms are deuterium. (See Table 10-1.) The mass of hydrogen is given in tables as 1.008. This represents the

isotopes
(EYE suh tohps)

Table 10-1: Naturally Occurring Isotopes of Several Elements

Element	Atomic number	Masses of isotopes (amu)	Percent in nature
hydrogen	1	1	99.8
	1	2	0.2
carbon	6	12	98.89
	6	13	1.11
neon	10	20	90.92
	10	21	0.26
	10	22	8.82
chlorine	17	35	75.5
	17	37	24.5
uranium	92	234	0.01
	92	235	0.72
	92	238	99.27

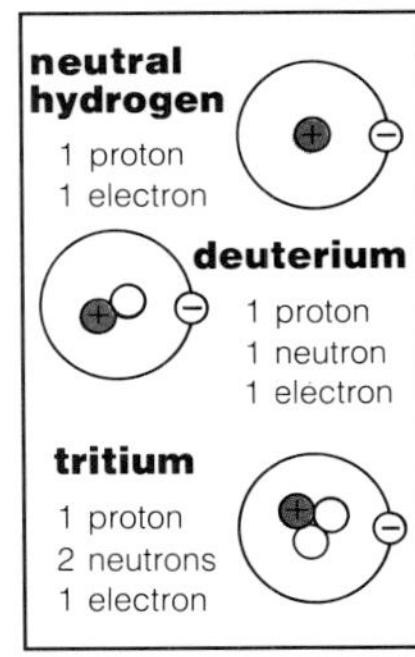

Figure 10-3 *The three isotopes of hydrogen differ in the number of neutrons present in the nucleus.*

Background

In the nineteenth century, chemists believed that atoms do not change and that all atoms of the same element are identical. The emission of alpha and beta particles from radioactive elements led scientists to distinguish radioactive elements from ordinary elements.

Later, mass spectrographs were used to sort charged atoms of the same element into different groups on the basis of mass. In a mass spectrograph, ions with the same kinetic energy enter a magnetic field that is perpendicular to the path of the ions. The paths of the more massive ions are bent less than those of the less massive ions.

It was through the use of mass spectrographs that the isotopes listed in Table 10-1 were shown to exist. Again, scientists had to amend the definition of element. Clearly, all atoms of the same element are not identical—some are more massive than others. However, the similarity in chemical properties of isotopes of the same element suggests that they share the same number of protons. The mass differences among isotopes of an element must be due to the different numbers of neutrons in their nuclei.

Teaching Suggestions 10-2

Call attention to Table 10-1 on student text page 243 and the periodic table on student text page 205. Ask students why the atomic mass of chlorine is actually about 35.5, and also how elements can have fractional atomic masses. Responses will vary. Those who have prepared well will see that chlorine is made up of two isotopes, Cl-35 and Cl-37, in a ratio of about 3 : 1, and that this produces a fractional atomic mass. The weighted average is calculated as follows:

$$\begin{aligned} 3 \times 35 &= 105 \\ 1 \times 37 &= \underline{\;37} \\ \text{Total} &= 142 \\ \text{Average} &= 142/4 = 35.5 \end{aligned}$$

If some students find this difficult, try some similar computations.

Demonstrations

1. Calculate the weighted average atomic mass for a hypothetical element X. Use nickels and pennies to represent the isotopes of this element. In a sample of 100 atoms of X, let 80 be pennies and 20 be nickels. Assuming nickels have a mass of 5 g and pennies a mass of 3 g (which is a fairly accurate estimate), you have 240 g of pennies and 100 g of nickels, as follows: $80 \times 3\ g = 240\ g$ and $20 \times 5\ g = 100\ g$, for a total of 340 g. The total number of atoms equals 100. Thus, average atomic mass is $340/100 = 3.4$ amu.
2. Show how isotopes can be separated by a mass spectrograph. Use the following simulation. Place a very strong magnet under a large sheet of glass or smooth plywood, as shown in Figure TE 10-1. Adjust the height of the level surface and the launching position of steel balls of various masses until the paths of the balls bend by the right amounts—that is, until the balls falling into the bins sort themselves by mass. Students should observe that the paths of the more massive balls are bent less than those of the less massive balls.

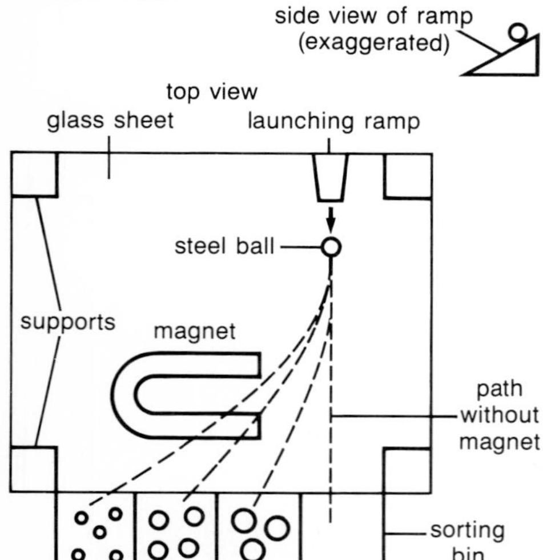

Figure TE 10-1 A simulation of how isotopes can be separated by a mass spectrograph.

mass of the naturally occurring *mixture* of hydrogen isotopes, 99.8% of which have a mass of 1 amu and 0.2% of which have a mass of 2 amu.

There is a third isotope of hydrogen, called *tritium*, which can be produced in nuclear experiments. Its nuclei contain one proton and two neutrons. Unlike the naturally occurring isotopes of hydrogen, tritium is radioactive.

Each element has several known isotopes, but each of the isotopes shows the same chemical behavior. This is because the chemical properties of an element depend on the number of electrons in its atoms. The number of electrons in a neutral atom depends, in turn, on the number of protons in the nucleus of the atom. The number of neutrons in the nucleus does not significantly affect the chemical behavior.

Each of the hydrogen isotopes has one electron, so the isotopes undergo the same chemical reactions. For example, all three isotopes react with oxygen, forming water. The water formed by causing naturally occurring hydrogen to react with oxygen includes minute amounts of *heavy water*, D_2O. ("D" stands for deuterium.) Heavy water can be separated from ordinary water, since it has a greater mass. Heavy water has properties that are useful in nuclear reactors.

In a nucleus, the particles are held together by attractive forces. The force of attraction within a stable nucleus is much stronger than the force of repulsion between protons, which have the same charge. If it were not stronger, protons in a nucleus would push away from each other, destroying the nucleus. **Radioactive nuclei** are unstable nuclei that break down, releasing particles and also large amounts of energy.

radioactive nuclei

Some nuclei are more stable than others because the number of protons and neutrons they contain fit together in an especially stable arrangement. For example, the helium nucleus that consists of two protons and two neutrons is a very stable unit. Other combinations of protons and neutrons form less-stable arrangements. For instance, the tritium nucleus (1 proton and 2 neutrons) is unstable. Tritium nuclei give off beta particles, becoming more stable as a result.

The effect of the arrangement of protons and neutrons on the stability of a nucleus is similar to the effect of the arrangement of electrons on the stability of an atom. There

Facts At Your Fingertips

- At least one radioactive isotope of every element has been found to occur naturally or has been prepared artificially.
- Elements may have more than one stable isotope. There are ten stable isotopes of tin.

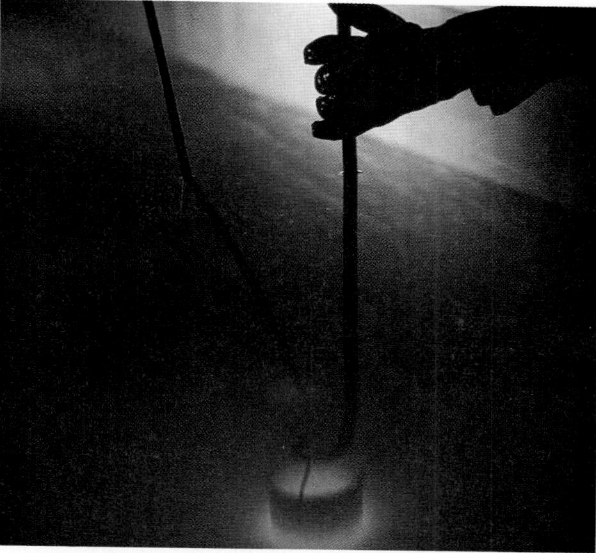

Figure 10-4 *This submerged sample of an isotope of cobalt is so radioactive it makes the water glow.*

are important differences, however. For one thing, nuclear forces are much stronger than the forces involved in electron arrangements. That is why nuclear reactions involve far more energy than do chemical reactions. For another thing, the arrangements of protons and neutrons in nuclei are more complicated than are electron arrangements in atoms. The effect of the arrangement of nuclear particles on the properties of nuclei is less well understood than is the effect of electron arrangements on chemical properties.

Checkpoint

1. What are isotopes?
2. Why do the isotopes of a given element show the same chemical behavior?
3. Use Table 10-1 to find the number of neutrons in the naturally occurring isotopes of: (a) hydrogen, (b) neon, (c) carbon, (d) uranium.

10 NUCLEAR REACTIONS

10-2 Isotopes

The Changing Neutron Number

Unscramble each term in Column A and write it on the line under the scrambled term. Then match each term with its definition in Column B and write the letter of the definition on the line next to the scrambled term. Hint: For some terms, you may need to use reference books.

A	B
1. tesiposo ____	a. Hydrogen atom with one proton and two neutrons in the nucleus.
2. erudemitu ____	b. Process by which one or more substances are changed into one or more new substances without a change in total mass.
3. imurtti ____	c. Atoms of the same element that have different masses.
	d. Hydrogen atom with one proton and one neutron in the nucleus.
4. vayhe tarew ____	e. Process in which significant amounts of matter are changed into energy.
5. macelhic tarenico ____	f. Substance that is important in some types of devices for the controlled release of nuclear energy.
6. luranec tarenico ____	

Figure It Out!

Use Table 10-1 in your textbook to answer the following questions.

____ 1. How many protons are in the heavier isotope of carbon?
____ 2. How many protons are in the lighter isotope of chlorine?
____ 3. What is the mass of the heaviest isotope of neon?
____ 4. What is the mass of the lighter isotope of chlorine?
____ 5. How many neutrons are in the lightest isotope of uranium?
____ 6. How many electrons are in the lighter isotope of chlorine?
____ 7. How many electrons are in the heaviest isotope of uranium?
____ 8. In a naturally occurring mixture of carbon, would the mass be closer to 12 or to 13? (Hint: Look at the percent abundance.)
____ 9. In a naturally occurring mixture of chlorine, would the mass be closer to 35 or to 37?
____ 10. In a naturally occurring mixture of uranium, would the mass be closer to 234 or to 238?

Checkpoint Answers

1. Isotopes are atoms of the same element that have different atomic masses because they have different numbers of neutrons.
2. They have the same number of protons and, therefore, electrons. Because chemical behavior depends on the number of electrons in an atom, isotopes of the same element behave similarly in chemical reactions.
3. (a) Hydrogen isotopes have 0 or 1 neutron; (b) neon isotopes have 10, 11, or 12 neutrons; (c) carbon isotopes have 6 or 7 neutrons; (d) uranium isotopes have 142, 143, or 146 neutrons.

Mainstreaming

Abstract ideas without direct sensory input will present a problem for some handicapped students. You may want to have students make models of isotopes out of simple materials.

10-3

Radioactive Decay

Goals

1. To describe radioactive decay.
2. To define the half-life of a radioactive isotope.
3. To interpret nuclear equations.
4. To explain how stable isotopes can be converted into radioactive isotopes.

radioactive decay

half-life

You have seen that when a radioactive isotope gives off an alpha or a beta particle, it becomes a different element. This change of a radioactive isotope into a different element is called **radioactive decay**. Tritium, the isotope of hydrogen with a mass of 3 amu, gives off beta radiation. As a result, it undergoes transmutation, becoming the isotope of helium with a mass of 3 amu. (See Figure 10-5.) Using symbols, scientists say that H-3 decays into He-3. The number 3 stands for the total number of protons and neutrons. This number, which is nearly equal to the mass in amu, is called the *mass number*. A sample of tritium that has a mass of one gram will decay to half a gram in 12.3 years. Tritium is said to have a *half-life* of 12.3 years. The **half-life** of a radioactive isotope is the time it takes for half of its atoms to decay. Half-lives range from less than a second to billions of years, as is shown in Table 10-2.

Table 10-2: Half-lives of Some Radioactive Isotopes

Isotope	Half-life
H-3	12.3 years
He-6	0.8 seconds
C-14	5570 years
P-32	14.3 days
Co-60	5.27 years
Sr-90	28 years
I-131	8.05 days
Po-210	138 days
Po-218	3.05 minutes
Ra-226	1662 years
Th-234	24.1 days
U-238	4.5 billion years

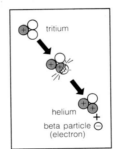

Figure 10-5 *A radioactive tritium nucleus gives off a beta particle and becomes a helium nucleus.*

Teaching Suggestions 10-3

Open this section with a discussion of half-life and radioactive decay. Be sure students understand the different types of decay reactions. Then have the class do the Try This activity (see Using Try This for expected outcomes) or Activity 16, Half-Life, student text page 552.

End-of-chapter project 4 applies to this section.

Background

Scientists have historically used the terms radioactive decay and transmutation synonymously, based on their experience with naturally occurring radioisotopes. This is because all naturally occurring radioisotopes emit either alpha or beta particles. This emission may or may not be accompanied by gamma radiation. But no naturally occurring radioisotope emits only gamma radiation. Later, it was found that some synthetic radioisotopes do emit only gamma radiation. This emission is also referred to as radioactive decay. Therefore, strictly speaking, radioactive decay and transmutation are synonymous terms only when referring to a change in the atomic number of a nucleus.

The process of radioactive decay is random. There is no way to predict when any given atom in a sample will decay. Chemists can deal with only very large samples (on the order of 10^{18} atoms) that have measurable mass or that produce measurable radiation. Thus, half-life is based on the average rate of decay of radioactive nuclei in samples containing very large numbers of atoms.

Many isotopes undergo natural transmutation to other elements by releasing alpha or beta particles. For example, uranium-238 decays naturally into thorium-234 by alpha emission, as shown in the student text on page 247. The process is accompanied by the release of energy in the form of gamma radiation, in addition to the kinetic energy of the alpha particle and the thorium atom. Thorium-234 then emits a beta particle to become protoactinium-234. But these are only the first steps in a long series of radioactive decays that terminates in the formation of a stable isotope of lead, lead-206. Each element in the series has a specific half-life, but all are short in comparison with U-238's half-life of 4.5 billion years.

The first artificial transmutation was carried out by Rutherford in 1919. He bombarded nitrogen with alpha particles and succeeded in producing oxygen-17 and a proton:

$$^{14}_{7}N + ^{4}_{2}He \longrightarrow ^{17}_{8}O + ^{1}_{1}H$$

Since that time a large number of artificial transmutations have been carried out. New elements beyond the natural ones, which end with uranium,

The uranium isotope with a mass of 238 amu (U-238) has a half-life of 4.5 billion years. That is, it takes 4.5 billion years for half of the atoms in a sample of U-238 to decay. A U-238 atom gives off an alpha particle (He-4), becoming an atom of thorium (Th-234). (See Figure 10-6.)

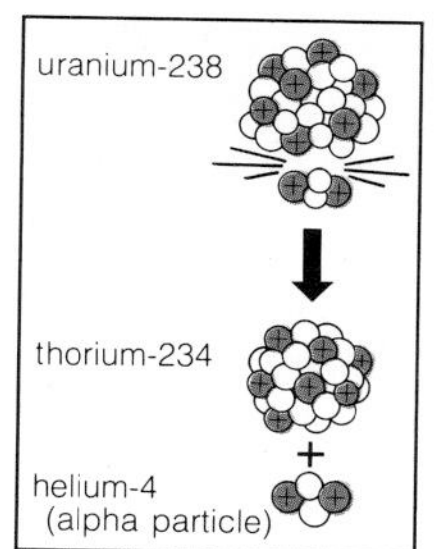

Figure 10-6 *A nucleus of uranium-238 can decay, thereby producing a nucleus of thorium-234 and an alpha particle.*

Nuclear reactions, such as the decay of a radioactive element, can be shown by means of a **nuclear equation**. The decay of U-238 into Th-234 is shown as follows:

nuclear equation

$$^{238}_{92}\text{U} \longrightarrow \, ^{234}_{90}\text{Th} + \, ^{4}_{2}\text{He}$$

The number at the upper left of each symbol in the equation indicates the mass number, or total number of protons and neutrons in the nucleus. The lower number is the charge of the nucleus. A charge of 1 unit is the amount of charge on a proton or an electron. The charge is positive unless there is a minus sign in front of the charge number. In the case of an isotope, the charge number is also the number of protons, or the *atomic number*.

In balancing nuclear equations, as in chemical equations, mass and charge are assumed to be conserved. In the equation above, the mass number to the left of the arrow is equal to the sum of the mass numbers to the right: $238 = 234 + 4$. The charge on the left is equal to the sum of the charges on the right: $92 = 90 + 2$.

The decay of U-238 to Th-234 is just the first in a long series of radioactive decays. The Th-234 nucleus is unstable, also. When Th-234 decays, it produces yet another unstable nucleus. The series of decays continues until a stable nucleus results. The U-238 decay series ends with a stable isotope of lead (Pb-206).

A stable isotope can become unstable when it is hit by a high-energy particle. High-energy particles from outer space called **cosmic rays** are constantly showering Earth. These particles include neutrons. Some cosmic-ray neutrons collide with nitrogen atoms, which are the most numerous in the atmosphere. The reaction between nitrogen-14 and a high-energy neutron produces carbon-14 and a proton, or hydrogen nucleus.

cosmic rays

$$^{14}_{7}\text{N} + \, ^{1}_{0}\text{n} \longrightarrow \, ^{14}_{6}\text{C} + \, ^{1}_{1}\text{H}$$

have been created. These are the transuranium elements—neptunium, plutonium, americium, curium, berkelium, and so on.

Demonstrations

1. Show some basic radiation phenomena using a Geiger counter. For example, demonstrate that even with no radioactive materials near the counter, it still clicks occasionally. This indicates background radiation caused by cosmic rays and radioactive materials that occur naturally. Show that when a radioactive sample is placed near the counter's probe, the count rate rises dramatically. Show how the count rate can be reduced by moving the sample away from the probe.

   Show the shielding effect of lead and other materials. Place lead sheets between the radioactive sample and the counter. Note the reduction in count rate. Try sheets of other materials, such as paper, plastic, and plywood. These materials are less effective as shields.
2. Do the Try This activity as a demonstration if students do not do it individually or in groups.
3. Show half-life using a role-playing simulation. Place all students on one side of the room to represent a small sample of radioactive isotope. Note the time and record it on the chalkboard. Then, in a random fashion, ask students to leave the sample, one at a time, at regular intervals. Suppose there are 28 students. When half of the students (14) remain, note and record the time again. The elapsed time represents the half-life of the isotope. Continue the process. Double the time-interval used. When only half of the half, or one quarter, of the students (7) remain, the "half-life" should be the same as that when 14 students remained. Some students may wonder about the outcome of the next half-life. Would 3 or 4 students remain? The answer is that we can't tell.

Teaching Tips

- Students will require practice writing nuclear equations. Show them how alpha or beta emission changes the atomic number of an atom and, in the case of alpha emission, the atomic mass as well.
- Be sure students understand that the half-life of an element is the same for all samples of the element, regardless of size.
- Show students the transuranium elements in the periodic table and point out that they were all produced by artificial transmutation.

Figure 10-7 *Particle accelerators are used to produce nuclear reactions. New elements can be made in this way.*

The carbon-14 that is produced is unstable. It decays, with a half-life of 5570 years, producing nitrogen-14 and a beta particle, or electron.

$$^{14}_{6}C \longrightarrow ^{14}_{7}N + ^{0}_{-1}e$$

The mass of an electron is shown as zero in a nuclear equation because it is very small compared to the mass of a proton or neutron.

Reactions similar to those produced by cosmic rays can occur when stable isotopes are bombarded with radiation given off by unstable isotopes. Neutrons were first detected as the result of such a reaction. The reaction was produced by the bombardment of beryllium-9.

$$^{9}_{4}Be + ^{4}_{2}He \longrightarrow ^{12}_{6}C + ^{1}_{0}n$$

Particles can be accelerated to high speeds in devices called *accelerators*. (See Figure 10-7.) Particles traveling at high speeds have the high energies needed to produce nuclear reactions. Such reactions have provided a great deal of information about the nucleus. They have also produced many very useful radioactive isotopes that do not occur naturally.

Particle accelerators have been used to produce elements that do not occur naturally on Earth. Before 1940, uranium was the element with the largest atomic number (92). Since then, more than a dozen new elements have been produced by bombarding other elements with high-speed particles. Elements with atomic numbers greater than 92 are sometimes called *transuranium elements*, since they are "beyond" uranium in the periodic table.

TRY THIS

Build a 1-m-long pendulum. Place a ruler under the bob, and note its rest position. Pull the bob 20 cm to one side. Release the bob and record the time. Calculate the pendulum's "half-life," which is the amount of time it takes for the bob to swing 10 cm from its rest position. Find out how long it will take the bob to swing 5 cm from its rest position.

Checkpoint

1. Define radioactive decay.
2. What is meant by the half-life of a radioactive isotope?
3. $^{214}_{84}Po \longrightarrow ^{210}_{82}Pb + \underline{\ ?\ }$. Complete the equation.
4. Describe two ways in which a stable isotope may be converted into a radioactive one.

Mainstreaming

All handicapped students should do the Try This activity. Other students can help the visually impaired calculate the pendulum's "half-life."

Cross References

Carbon-14 dating is discussed in Chapter 10. See Feature Article (page 260), and Chapter 11, section 11-3, for information about fossils.

10 NUCLEAR REACTIONS

10-3 Radioactive Decay

How Are Particles and Numbers Related?

Each symbol below stands for the isotope of a particle atom. In each symbol circle the mass number and underline the charge. Then answer the questions that follow.

$^{7}_{3}Li$ $^{13}_{7}N$ $^{64}_{29}Cu$ $^{120}_{50}Sn$ $^{226}_{88}Ra$ $^{239}_{94}Pu$

____ 1. What is the mass number of the lithium atom?
____ 2. What is the number of protons in the nitrogen atom?
____ 3. What is the atomic number of the copper atom?
____ 4. What is the number of neutrons in the tin atom?
____ 5. What is the number of electrons in the radium atom?
____ 6. What is the atomic number of the plutonium atom?
____ 7. What is the symbol for the isotope of carbon that has 6 protons and a mass number of 13?
____ 8. What is the symbol for the isotope of neon that has 10 protons and a mass number of 22?
____ 9. What is the symbol for the isotope of chlorine that has an atomic number of 17 and a mass number of 37?
____ 10. What is the symbol for the isotope of oxygen that has 8 protons and 8 neutrons.

Half-Lives

On a separate sheet of paper solve the following problems. Hint: Use Table 10-2 in your textbook to help you find the solutions.

____ 1. If you have a 1-g sample of Po-218, how many grams would you have after 6 minutes? After 18 minutes?
____ 2. If you have a 10-g sample of Sr-90, how many grams would you have after 28 years? After 112 years?
____ 3. After 24 days, a 10-g sample of radioactive element has decayed so that only 5 grams are left. Of the list in Table 10-2, which element could this be?
____ 4. If radioactive wastes must be stored for seven half-lives before disposal, how long must P-32 be stored?
____ 5. If radioactive wastes must be stored for eight half-lives before disposal, how long must Ra-226 be stored?

Using Try This

Based on their calculations, students should realize that it takes as long for the pendulum to swing 5 cm in one case as it does to swing 10 cm in the other.

The period of a pendulum is the time required for one complete swing. This time depends only on the length of the pendulum. That is, for a given length, the period is the same no matter how large or small the width of the swing. This situation is analogous to the relationship between the half-life of an element and the sample size. In other words, the half-life is always the same for any sample of a given radioactive element. (The "half-lives" found by different groups may differ because of variations in the frictional forces.)

Checkpoint Answers

1. Radioactive decay is the change of a radioactive isotope into a different element by the giving off of an alpha or a beta particle.
2. Half-life is the time required for one-half the atoms in a radioactive isotope to decay.
3. $^{4}_{2}He$.
4. By being hit by cosmic rays, radiation from unstable isotopes, or particles traveling at high speeds in accelerators.

10-4

Uses of Radioisotopes

ENRICHMENT

Goal

To describe some practical uses of radioactive isotopes.

Radioactive isotopes, or *radioisotopes*, can be put to practical use. There are important uses in medical treatment. Radiation can destroy or harm body cells. It can cause cancer, also. However, the ability to destroy body cells can be used to *treat* cancer. When directed at a cancer, radiation can destroy the cancer cells, while doing little damage to normal cells nearby. (See Figure 10-8.) Cobalt-60 is a radioisotope used for this purpose.

Different kinds of radioisotopes may be used to treat different cancers. For example, if cancerous tissue absorbs phosphorus, the radioisotope phosphorus-32 can be used to destroy the cancer. Similarly, since iodine is absorbed in the thyroid gland, the radioisotope iodine-131 can be used to destroy thyroid cancer. Both P-32 and I-131 have half-lives of less than 15 days. As a result, these radioisotopes do not remain in the body for long.

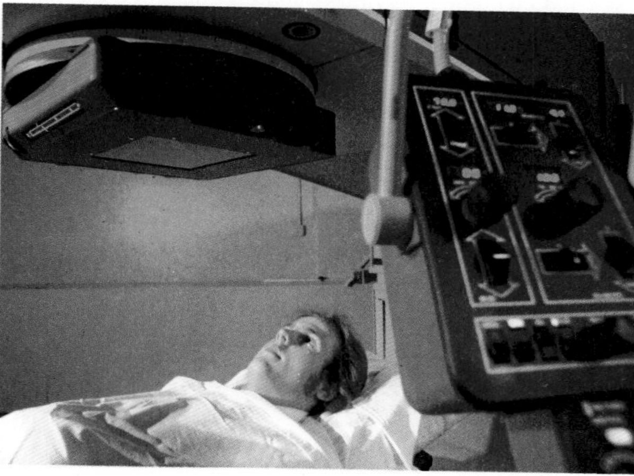

Figure 10-8 *Radioisotopes can be used to treat certain cancers. One way patients are treated is by external radiation.*

Teaching Suggestions 10-4

This enrichment section may be treated as a reading assignment. Another possibility is to use the applications given as a jumping-off point for investigating the many others that exist. Interested students may research applications and report to the class.

Be sure students realize that radioactivity is used in two ways in the treatment of cancer. A tumor may be exposed to external radiation. Also, a radioisotope may be administered internally. This is done in the treatment of thyroid cancer, in which the radioisotope I-131 is used because the thyroid gland absorbs iodine.

Chapter-end project 3 applies to this section.

Background

Most elements have more radioactive than stable isotopes. Such isotopes are widely used in research. For example, radioactive isotopes can be used in fertilizers that are absorbed by growing plants. By checking the radiation emitted from a plant, it's possible to determine how much of the fertilizer entered the plant. Such studies enable farmers to apply the proper amount and type of fertilizer for maximum growth.

Radioactive "tracers" are used to study the paths of chemical substances through the human body. Tracers also have been used to study the movement of water in underground aquifers, the wear of engines, and the amount of rubber worn from tires.

Facts At Your Fingertips

- There are 21 radioactive isotopes of iodine and 10 of cobalt.
- Rosalyn Yalow won a Nobel prize in 1977 for her pioneering work in the medical use of radioisotopes.
- Alpha radiation is not very dangerous unless ingested, because it will not penetrate skin. Beta and gamma radiation, however, do penetrate the skin.
- Lead shields are used around x-ray machines to protect people from stray radiation. The x-ray technician goes into another room to turn on the machine in order to avoid exposure to x rays.

Mainstreaming

The Try This activity should be modified for handicapped students. Use food dyes in actual examples to simulate the use of radioisotopes. For the visually impaired, heat conducted along long rods and through metal plates may be substituted.

Cross Reference

Photosynthesis is discussed in Chapter 1, section 1-5.

10 NUCLEAR REACTIONS

10-4 Uses of Radioisotopes

What Is Missing?

In each sentence below one or more terms are missing. Think of the terms that correctly complete the sentences. Write the terms on the numbered spaces at the left.

______ 1. A shorter name for radioactive isotopes is __1.__ .

______ 2. The ability to destroy __2.__ by radiation can be used to treat cancer.

______ 3. __3.__ is a radioisotope that is used to destroy cancer cells.

______ 4. Since the __4.__ gland absorbs iodine, the radioisotope __5.__ can

______ 5. be used to destroy cancer cells in that region.

______ 6. Radioisotopes can be used to find out how __6.__ reactions work.

______ 7. Radioisotopes used by scientists to trace the path followed by various reactions and products are called __7.__ .

______ 8. __8.__ can be used to measure the thickness of metal sheets.

Spot the Flaw

One statement below is true; the others are false. Read each statement. If the statement is true, circle the letter *T*. If the statement is false, circle the letter *F* and replace the italicized word with a term that makes the statement true.

______ T F 1. Radioisotopes behave the same way in *nuclear* reactions as do nonradioactive isotopes of the same element.

______ T F 2. Since *phosphorus* is absorbed by the thyroid gland, its radioisotope can be used to destroy thyroid cancer.

______ T F 3. Both P-32 and I-131 have half-lives of more than seven *years*.

______ T F 4. When scientists tagged the carbon dioxide absorbed in photosynthesis with a radioisotope of oxygen, the radioisotope showed that the oxygen released in the reaction comes from the *carbon dioxide*.

______ T F 5. *Tracers* can be used to follow the path of oil in a pipeline or the flow of water beneath the ground.

Using Try This

Suggestions will vary. (a) Add radioactive water to the soil around the plant. Then periodically measure the radiation level in various parts of the plant. (b) Place a radioactive source below the metal. Scan the metal from above with a fine Geiger-counter probe. A high count rate should be detected above any cracks.

Checkpoint Answers

1. The radioisotopes produce radiation that destroys cells. Directing the radiation at cancer cells will destroy them.
2. A tracer is a radioactive isotope that scientists use to trace a course of events. By tracking a tracer through a series of reactions, the path of a single element can be followed. For example, it is possible to trace the path of radioactive carbon during photosynthesis and thereby determine the reactions involved in photosynthesis.
3. A radioisotope can be placed beneath a metal sheet, and a radiation detector can be placed above the metal to measure the amount of radiation that passes through it. The amount of radiation that passes through the sheet depends on the sheet's thickness.

Figure 10-9 *The thickness of metal sheets can be determined through the use of radioisotopes.*

Radioisotopes behave basically the same way in chemical reactions as do nonradioactive isotopes of the same element. Therefore, radioisotopes can be used to find out how chemical reactions work. A good example involves the photosynthesis reactions. In photosynthesis the carbon from CO_2 is used as the source of carbon to make food molecules. Scientists have long wanted to know what chemical reactions were involved in this process. They needed to trace how the carbon of CO_2 ultimately becomes part of glucose. This required a radioactive form of carbon.

Around 1950 a radioisotope of carbon became available. Plants were exposed in the light to CO_2 made from the radioisotope of carbon. This radioactively "tagged" CO_2 was incorporated in the same way as nonradioactive CO_2. In fact scientists were able to show that the glucose made by these plants was radioactive. Also, they were able to detect radioactivity in several other kinds of molecules. In this way the sequence of reactions by which CO_2 becomes incorporated into food molecules was discovered. Radioisotopes used in this way are often called **tracers**, because scientists can trace a course of events by detecting the path of a radioisotope. Tracers can be used in many ways. For example, a test developed by Rosalyn Yalow, who received the Nobel Prize, uses radioisotopes to detect minute quantities of substances, like drugs or hormones, in the body.

tracers

Radiation can be stopped by a metal sheet, depending on the thickness of the metal and the energy of the radiation. A radioisotope can be placed beneath a sheet of metal that is being produced. A radiation detector above the metal can be used to measure how much radiation gets through the metal. This setup can be used to measure the thickness of the sheet. If the metal is too thick or thin, automatic controls linked to the detector can adjust the thickness.

TRY THIS

Think of a way to use a radioisotope (a) to find out how water circulates in a plant, (b) to detect cracks in a piece of metal.

Checkpoint

1. Describe how radioisotopes can be used in treating cancer.
2. What is a radioactive tracer? Give an example of the use of a tracer.
3. Describe how radioisotopes can be used to measure the thickness of metal sheets.

10-5

Nuclear Energy

Goals

1. To define binding energy.
2. To identify the source of binding energy.
3. To explain, in words, what $E = mc^2$ means.

The mass of a proton or a neutron is approximately 1 amu. However, to account for nuclear reactions, more precise values are needed. The mass of a proton is 1.00728 amu. That of a neutron is 1.00867 amu. A helium nucleus formed in a nuclear reaction contains two protons and two neutrons. You might expect that its mass would be:

$$(2 \times 1.00728) + (2 \times 1.00867) = 4.0319 \text{ amu}$$

However, its measured mass is only 4.0015 amu. It seems that 0.0304 amu is missing from the helium.

The explanation for the missing mass is that it was converted into energy when the nucleus was formed. The energy given off when a nucleus is formed from protons and neutrons is called the **binding energy**. If the nucleus breaks apart into individual protons and neutrons, energy equal to the binding energy is required. Each nucleus has its own binding energy.

binding energy

The amount of energy involved in forming or in breaking apart a nucleus is extremely great. Suppose 400 g of neutrons and protons were to combine, forming helium. The energy given off in the process could cause a lake averaging 1.6 km in diameter and 20 m in depth to boil!

Scientists no longer find it surprising that mass can be converted into energy, and vice versa. Nor do they think it strange that a small amount of mass can be transformed into an extremely large amount of energy. However, these notions seemed astounding to most people when Albert Einstein first proposed them. In 1905, Einstein published his well-known equation that shows the relationship between energy and mass: $E = mc^2$. E stands for energy, m is mass, and c is the speed of light. The equation shows that a mass (m) can be transformed into an amount of energy (E) equal to the product of the mass and the square of the speed of

Teaching Suggestions 10-5

This section may be treated as a reading assignment, but provide an opportunity for students to ask questions. Students may have difficulty grasping the idea that mass is a form of energy. It may help to point out that we once thought of electricity and magnetism as separate phenomena. But then Oersted found a magnetic field around an electric current, and Faraday generated electricity from an oscillating magnetic field.

Show students how to calculate the energy associated with a mass of 0.0304 g, the quantity of mass lost in the formation of 4.0015 g of helium from protons and neutrons.

$$0.0304 \text{ g} = 0.0000304 \text{ kg}$$
$$c = \text{about } 300{,}000{,}000 \text{ m/s}$$
$$E = mc^2$$
$$= 0.0000304 \text{ kg} \times (300{,}000{,}000 \text{ m/s})^2$$
$$= 2{,}736{,}000{,}000{,}000 \text{ J}$$

Point out to the class that this is enough energy to supply about 300 homes with electricity for a year. About 90,000 kg of coal must be burned to supply the same amount of energy that comes from 0.0304 g of converted mass.

Background

The binding energy of a nucleus is analogous to the binding energy that holds objects on Earth. To "unbind" an object resting on Earth's surface, we would have to give it enough kinetic energy to escape Earth's gravity.

Similarly, the binding energy of a nucleus is the energy we must provide to separate the nuclear particles. It is equal to the energy associated with the mass lost when protons and neutrons came together to form the nucleus.

Fact At Your Fingertips

- The water in a lake 1.6 km in diameter and 20 m deep has a volume of about 40 million cubic meters and a mass of about 40 billion kilograms.

10 NUCLEAR REACTIONS

10-5 Nuclear Energy

Einstein, the Slow Learner

Read the following paragraphs about Einstein. Then answer the questions.

Albert Einstein was born in Ulm, Germany, in 1879. When Albert was only five, his father gave him a compass. Albert was fascinated by the behavior of the compass needle. He noticed that the needle always pointed in the same direction no matter how the case around the needle was turned.

Although Einstein was Jewish, he attended Catholic parochial schools. The schools were strict and primarily concerned with memorization. Einstein's teachers and parents thought that he was a slow learner. However, Dr. Max Talmy recognized Einstein's talents. Dr. Talmy gave Einstein books on mathematics, physics, and philosophy. By the time Einstein was 15 years old, he had decided that he would specialize in mathematics and physics.

After graduating from the Swiss Polytechnic Institute in Zurich, Switzerland, Einstein worked as a patent officer. The position was ideal for Einstein, since it left him time to pursue his interests in physics. In 1905, at the age of 26, Einstein published three papers. One of the papers was on relativity, a theoretical idea that transformed the field of physics.

Now Einstein was recognized as a great theoretical physicist and was offered many teaching positions. His decisions about which positions to accept were based on the responsibilities involved rather than on pay or recognition. Einstein worked first in Switzerland, then in Germany. After the Nazis stripped him of his German citizenship and positions, Einstein went to the United States. In the United States Einstein joined the Institute for Advanced Study in Princeton, New Jersey. He became an American citizen in 1940. Einstein remained at Princeton until he died in 1955.

1. When and where was Albert Einstein born? ______
2. How would you describe Einstein's early education? ______
3. What was Einstein's first job after graduating? ______
4. Why was 1905 a very important year for Einstein? ______
5. Why did Einstein leave Germany? ______
6. What did Einstein do in the United States? ______

Checkpoint Answers

1. Binding energy is the energy given off when a nucleus is formed from protons and neutrons. It is also the energy required to break apart the nucleus into separate protons and neutrons.
2. The binding energy of a nucleus comes from the mass lost when the nucleus formed.
3. $E = mc^2$ tells us that a mass (m) can be transformed into an amount of energy (E) that is equal to the product of the mass (m) and the square of the speed of light (c^2).

Figure 10-10 *Scientist Albert Einstein first proposed that mass could be changed into energy.*

light (c^2). The speed of light is a very large number. When this number is squared, a much larger number is obtained. Thus a tiny amount of mass can be converted to a very large amount of energy.

Checkpoint

1. Define binding energy.
2. What is the source of the binding energy of a nucleus?
3. Write the equation $E = mc^2$ in words.

10-6

Nuclear Fission

Goals

1. To define nuclear fission.
2. To explain what is meant by the critical mass of a fissionable material.
3. To describe what happens in a chain reaction.

When the radioactive isotope uranium-235 is struck by a neutron, it may split into two smaller nuclei. The splitting of a nucleus into two smaller nuclei is called **nuclear fission**. A reaction for the fission of U-235 is:

nuclear fission (FISH un)

$$^{1}_{0}n + ^{235}_{92}U \longrightarrow ^{138}_{56}Ba + ^{95}_{36}Kr + 3^{1}_{0}n + \text{energy}$$

Notice that three neutrons are produced in this reaction ($3\,^{1}_{0}n$). Because neutrons are produced by the fission of U-235, a single U-235 fission reaction can give rise to as many as three others. (See Figure 10-11.) Each of these three reactions can then give rise to three others, and so on. When the products of a reaction begin new reactions, the reaction rate can increase, and a **chain reaction** can occur. A nuclear chain reaction can produce a vast amount of energy in a very short period of time.

chain reaction

Uranium compounds are present at many locations in Earth's crust. The uranium in these compounds consists of several radioactive isotopes. U-235 makes up only a small percentage of the isotopes. Since the other isotopes do not undergo fission as readily as does U-235, nuclear chain reac-

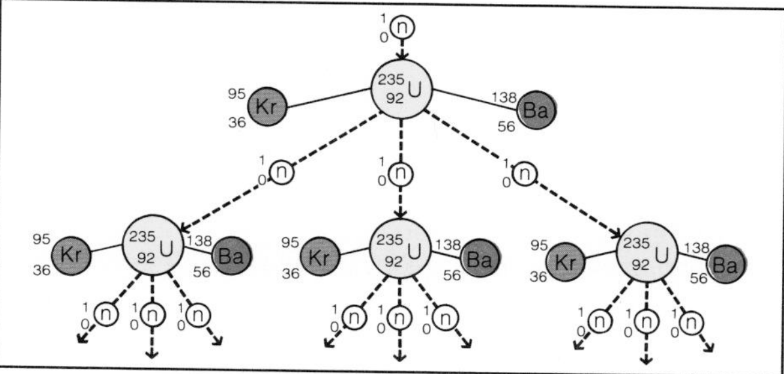

Figure 10-11 *A single fission reaction involving uranium-235 can produce many other fission reactions, in a so-called chain reaction.*

Background

Nuclear fission results in the formation of nuclei smaller than the one that was split. The binding energy per nucleon (nuclear particle) in uranium is less than that in elements with about half uranium's mass number. But binding energy is the energy that resulted from the decrease in mass when protons and neutrons fused to form the nucleus. When a uranium nucleus undergoes fission, the nuclei of the products have more binding energy (hence less mass per particle) than the uranium nucleus did. The energy equivalent of the lost mass is given by $E = mc^2$.

Demonstrations

1. Show students a mechanical simulation of a chain reaction. Cover the bottom of a cardboard box with mousetraps set to release. Place a ping-pong ball on the spring of each trap. Cut a hole in the top of the box directly above one of the traps. Then drop another ping-pong ball onto the trap under the hole. The springing trap will project its ping-pong ball, which will spring another trap, and so on, creating a chain reaction within the box that can be heard outside.
2. If students don't have dominoes at home, the Try This activity may be done as a demonstration.

Teaching Tip

- Have students do the Try This activity for additional insight into the nature of a chain reaction.

Teaching Suggestions 10-6

Begin this lesson with a dramatic simulation of a chain reaction (see Demonstrations). Then, ask students if they know how a nuclear chain reaction starts. From their reading, some will be able to answer that in a nuclear explosion two noncritical masses are pushed together to create a critical mass, which leads to a chain reaction.

Some students may realize that the huge amounts of energy released during fission reactions must be associated with a loss in mass. The products of a fission reaction, smaller atoms such as barium, krypton, and others, have slightly less combined mass than the uranium-235 that has undergone fission. The loss in mass, often called the packing effect, is converted to energy according to the formula $E = mc^2$.

10 NUCLEAR REACTIONS

10-6 Nuclear Fission

Let's Decode

The puzzle below must be decoded to make sense. The numbers in Column A represent letters. The number *1* stands for the letter *a*, the number *2* stands for *b*, and so on. First decode each word in Column A. Write the decoded word on the line below the numbers. Then match the decoded word in Column A with its phrase in Column B. Write the letter of the phrase on the space at the left of each code.

	A	B
___	14 21 3 12 5 1 18 # 6 9 19 19 9 15 14 ______	a. The smallest mass of fissionable material that will support a chain reaction:
___	3 8 1 9 14 # 18 5 1 3 20 9 15 14 ______	b. A device that uses controlled fission to produce power.
___	3 18 9 20 9 3 1 12 # 13 1 19 19 ______	c. The splitting of a nucleus into two smaller nuclei.
___	6 9 19 19 9 15 14 # 2 15 13 2 ______	d. A reaction whose products start new reactions, thereby causing an increase in the reaction rate.
___	14 21 3 12 5 1 18 # 18 5 1 3 20 15 18 ______	e. Atom bomb in which two or more masses that are smaller than the critical mass are pushed together to produce a critical mass.

Complete the Equation

In each of the following reactions a question mark shows where a symbol, number, or symbol and number are missing. Figure out what is needed to replace the question mark and complete each equation. Then write the correct symbol, number, or symbol and number on the space in front of the equation.

___ 1. $^{216}_{84}Po \longrightarrow {}^{?}_{82}Pb + {}^{4}_{2}He$

___ 2. $^{239}_{92}U \longrightarrow {}^{239}_{?}Np + {}^{0}_{-1}e$

___ 3. $^{212}_{84}Po \longrightarrow {}^{208}_{82}? + {}^{4}_{2}He$

___ 4. $^{14}_{7}H + {}^{4}_{2}He \longrightarrow {}^{?}_{8}O + {}^{1}_{1}H$

___ 5. $^{232}_{90}Th \longrightarrow {}^{228}_{88}Ra + {}^{4}_{2}?$

___ 6. $^{224}_{88}Ra \longrightarrow {}^{220}_{86}Rn + ?$

___ 7. $^{239}_{93}Np \longrightarrow {}^{239}_{94}Pu + ?$

___ 8. $^{212}_{83}Bi \longrightarrow {}^{212}_{84}Po + ?$

___ 9. $^{9}_{4}Be + {}^{4}_{2}He \longrightarrow ? + {}^{1}_{0}n$

___ 10. $^{10}_{5}B + {}^{1}_{0}n \longrightarrow ? + {}^{4}_{2}He$

Using Try This

The domino reaction is similar to a nuclear chain reaction in that one domino's fall causes others to fall. It is different in that the effect is direct; in fission, it is the release of neutrons that causes the next fission. In addition, of course, the energy released by the falling dominoes is far less than in fission, and there is no loss of mass.

Checkpoint Answers

1. Nuclear fission is the splitting of a nucleus into two smaller nuclei.
2. The critical mass is the smallest mass of fissionable material that will support a chain reaction.
3. A chain reaction is a series of reactions that occur when the products of a reaction start new reactions. A nuclear chain reaction can be controlled by controlling the number of neutrons available to produce fission.

Figure 10-12 *In the fission reactions involved in an atomic-bomb explosion, very large amounts of energy are released.*

tions do not occur naturally. In order to produce a chain reaction, uranium must contain a greater fraction of U-235 than is present in naturally occurring uranium. Even in a uranium sample that has been "enriched" with U-235, a chain reaction will not necessarily occur, however. If the mass of uranium is too small, many neutrons can escape without hitting U-235 nuclei. If the mass is made larger, fewer neutrons can escape without causing other U-235 nuclei to fission. If the mass is made large enough, a chain reaction takes place. The smallest mass of fissionable material that will support a chain reaction is called the **critical mass**.

critical mass

In a fission (or "atom") bomb, two or more masses that are smaller than the critical mass are pushed together to produce a critical mass. The chain reaction that results produces an awesome explosion, as shown in Figure 10-12. By controlling the number of neutrons available to produce fission, a fission reaction can be made to proceed at a steady rate. The energy given off by the controlled reaction can be harnessed for power production. Whether fission is controlled, as in nuclear reactors, or uncontrolled, as in a nuclear bomb, radioactive isotopes are always produced. These fission products can cause severe injury or death to living organisms.

TRY THIS

Arrange a set of dominoes so that each domino will knock over two other dominoes when it falls. Then push the first domino and watch the "chain reaction." How is this chain reaction like a nuclear chain reaction? How is it different?

Checkpoint

1. Define nuclear fission.
2. What is meant by the critical mass of fissionable material?
3. What is a chain reaction? How can a nuclear chain reaction be controlled?

10-7

Nuclear Reactors

Goals

1. To list the parts of a nuclear reactor and explain their functions.
2. To explain what nuclear wastes are and to describe how they might be safely stored.

Controlled nuclear reactions are carried out in **nuclear reactors**. Figure 10-13 shows the parts of a nuclear reactor. Refer to the diagram as each part is described.

nuclear reactors

The *fuel elements* in a reactor (usually uranium-235 or plutonium-239) are the materials that undergo fission. The fuel is the source of neutrons, also. The fuel elements are surrounded by a substance called a *moderator*, which slows down neutrons. The moderator is often graphite (a form of carbon) or heavy water. Fast-moving neutrons are slowed down by bouncing off the small nuclei in the moderator.

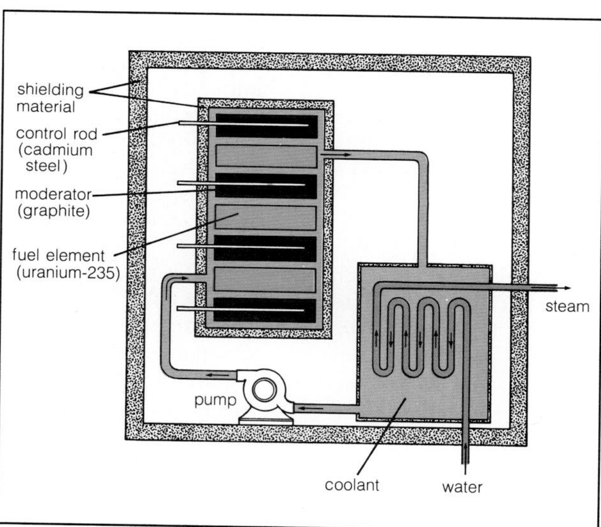

Figure 10-13 *This diagram shows the principal parts of a nuclear-fission reactor.*

Teaching Suggestions 10-7

Carefully go over Figure 10-13 on student text page 255 to be sure students understand the operation of a nuclear reactor. If possible, follow this with a field trip to a nuclear power plant.

You might also wish to stage a debate on the use of nuclear power as a source of energy. Have the debaters thoroughly explore the positive and negative economic factors of nuclear power, the potential hazards inherent in operating a nuclear power plant, and both sides of the nuclear-waste disposal issue. Some students may wish to do chapter-end projects 1, 3, and 6 in conjunction with this section.

Background

In the explosion of an atomic bomb, a very large number of uranium nuclei undergo fission, releasing vast amounts of energy in an instant. Such explosions are extremely destructive and hardly a usable energy source. It is possible, however, to control the rate of fission by surrounding the small rods of uranium-235 with graphite and using boron steel rods to absorb neutrons. In this way, energy can be released slowly to heat water to steam.

A number of such atomic power plants have already been built in various places throughout the world. They are usually sited near large bodies of water. Water is used to carry away the large quantities of heat produced by the fission reaction.

France has made a commitment to nuclear power for the generation of electricity. The French are building breeder reactors (reactors that generate nuclear fuel in excess of the fuel consumed) in an effort to reduce their overall consumption of naturally occurring fissionable fuel. Beds of U-238 around the nuclear core in a breeder reactor absorb neutrons to form Np-239. The Np-239 then decays by beta emission to form Pu-239, a fissionable fuel.

Wariness of nuclear power may have increased since the 1979 accident at the nuclear power plant at Three Mile Island in Pennsylvania. Certainly, nuclear waste and contaminated power plants are dangers, but there are also insidious hazards—pollution, mining disasters, oil slicks, and so on—associated with the use of fossil fuels.

Slow neutrons are more easily "captured" by the nuclei in the fuel than are fast neutrons. Therefore, fission occurs more readily with slow neutrons.

In order to control the rate of a nuclear reaction, the number of neutrons must be controlled. *Control rods* are used to absorb neutrons and thus control the reaction rate. These rods are made of substances such as the elements boron or cadmium that absorb neutrons strongly. When control rods are pushed into the fuel area, they absorb neutrons, thereby slowing the reaction. If the rods are pushed in far enough, they stop the reaction. When control rods are pulled out, more neutrons are available to react, and the reaction speeds up.

A *coolant* is pumped through the reactor vessel to absorb heat released by the fission reaction. The coolant is usually water or liquid sodium. After absorbing heat from the reaction, the coolant flows into the *heat exchanger*. There it gives up heat to the surrounding water, which causes the water to boil. The steam produced by the boiling water is used to drive turbines. The turning turbines run electric generators, which produce electricity.

Both the fuel and the products of a nuclear-fission reaction are radioactive. Since this radiation is harmful to living organisms, the entire reactor is surrounded by thick layers of concrete *shielding* to absorb the radiation.

Nuclear reactors can be used to produce radioactive isotopes (radioisotopes) that have practical uses in science, medicine, and industry. These radioisotopes are produced when stable isotopes are struck by the neutrons in the reactor.

A typical nuclear-energy power plant is shown in Figure 10-14. Such plants provide about ten percent of the electricity in North America. Nuclear power plants do not pollute the atmosphere with smoke or acid rain as fossil-fuel power plants do. Nuclear plants do not require the mining of coal or drilling for oil, both of which can damage the environment. However, nuclear power plants are very expensive to build. They present the danger of an accident that could release a dangerous amount of radiation into the surroundings. In addition, they produce highly radioactive *nuclear wastes* that must be disposed of in a safe manner.

Most of the radioactive isotopes in nuclear wastes decay

Figure 10-14 *This nuclear power plant provides energy by controlling fission reactions.*

rapidly. However, a small percentage of the wastes takes much longer to decay to a safe level. Some of these wastes take thousands of years to decay. Nuclear wastes can be buried in locations in which they will not be disturbed for many years. A major danger is that buried nuclear wastes might leak into underground water systems. This contaminated water might then mix with a supply of drinking water.

A number of methods have been proposed for safe disposal of nuclear wastes. A popular proposal involves the "four-barrier approach." The first step of this approach is to embed the waste in a nonradioactive ceramic material. (Ceramic materials include glass, porcelain, earthenware, and other nonmetallic substances.) The ceramic material can hold the wastes securely within it, so that they cannot easily leak out. The waste-containing ceramic material is then sealed into a ceramic or metal container. Next, an absorbent mineral filling is packed around the container. Finally, everything is securely buried underground in an area that is free from the danger of earthquakes or some other natural

Cross Reference

Solar and other sources of energy are discussed in Chapter 18, sections 18-7 and 18-8.

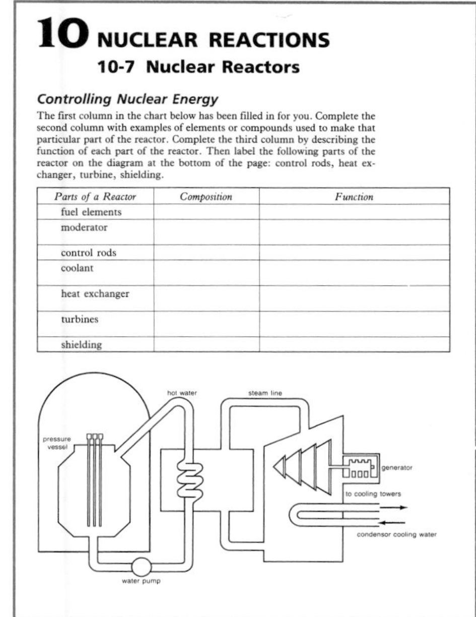

10 NUCLEAR REACTIONS

10-7 Nuclear Reactors

Controlling Nuclear Energy

The first column in the chart below has been filled in for you. Complete the second column with examples of elements or compounds used to make that particular part of the reactor. Complete the third column by describing the function of each part of the reactor. Then label the following parts of the reactor on the diagram at the bottom of the page: control rods, heat exchanger, turbine, shielding.

Parts of a Reactor	*Composition*	*Function*
fuel elements		
moderator		
control rods		
coolant		
heat exchanger		
turbines		
shielding		

Checkpoint Answers

1. The fuel elements, the moderator, the control rods, the coolant, the heat exchanger, and the shielding.
2. The fuel elements (usually uranium-235 or plutonium-239) are the materials that undergo fission and that are the source of neutrons. The moderator slows down neutrons, thereby allowing fission to occur more readily. The control rods absorb neutrons and thereby control the rate of the fission reaction. The coolant absorbs heat released by the fission reaction. The heat exchanger is the part in which the coolant gives up heat to the surrounding water, causing it to boil. The shielding absorbs radiation.
3. Nuclear wastes are short- and long-lived radioactive isotopes that arise from the fission of fuel elements in a nuclear power plant. They produce harmful radiation that can kill plant and animal cells.
4. According to some people, nuclear wastes can be stored safely by using the "four-barrier" approach. This entails embedding the waste in non-radioactive ceramic material; sealing the waste-containing ceramic material in a ceramic or metal container; packing absorbent, mineral filling around the container; and burying the container in a region not subject to natural disasters, such as earthquakes.

Figure 10-15 *According to some scientists, a salt mine, such as this, might provide a suitable burial site for nuclear wastes.*

disaster. (See Figure 10-15.) However, many people still feel that there is simply no way in which to dispose of nuclear wastes safely.

Once a disposal site has been selected, wastes must be transported to it from reactors. There are several hazards in transporting such wastes. One is that people might be exposed to dangerous amounts of radioactivity. Another is that the wastes might form a critical mass. These two hazards can be controlled by using adequate shielding. Other hazards include the possibility of fires and other accidents. These hazards can be reduced by using a combination of specially designed shipping casks. Such casks can withstand exposure to fire, immersion in water, and other rigorous conditions without leaking. It is possible that some people might try to steal radioactive waste materials in order to make a weapon of some kind. Since such an attempt is most likely to occur during transport, special care would have to be taken to guard against this possibility. Once again, the dangers involved have led many people to the conclusion that the use of fission reactions as an energy source is unwise.

Checkpoint

1. List the major parts of a nuclear reactor.
2. Explain the function of each of the major parts.
3. What are nuclear wastes? Why are they hazardous?
4. Describe a way in which nuclear wastes might, according to some people, be safely stored.

10-8

Nuclear Fusion

Goals

1. To define nuclear fusion.
2. To state the source of the energy given off by nuclear fusion.
3. To describe the major obstacle to producing a controlled fusion reaction.

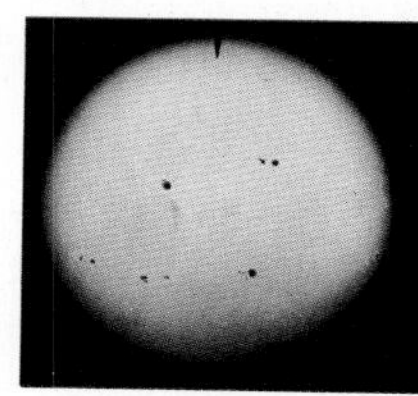

Figure 10-16 *Nuclear-fusion reactions are the source of the sun's energy.*

Combining hydrogen isotopes to form helium produces energy, just as the splitting of uranium atoms does. Indeed, the difference between the binding energies of reactants and products is much greater for such a reaction than for a fission reaction. Therefore the energy given off is much greater.

A nuclear reaction in which lighter nuclei combine to form heavier nuclei is called a **nuclear fusion** reaction. Nuclear fusion is the source of the sun's energy. In the sun, 650 million tons of hydrogen are converted to helium each second. In the process, several tons of mass are converted into energy. Hydrogen fusion will occur only at very high temperatures (about 50 million °C).

nuclear fusion
(FYOO zhun)

Scientists are trying to produce a controlled fusion reaction. This involves reaching very high temperatures without melting the containing vessel. One way of meeting this requirement is to carry out the reaction in a properly designed magnetic field that would contain the high-energy material. A device built to do this, called the *tokamak*, has produced some promising results. (See Figure 10-17.) However, a practical fusion reactor may still be decades away.

If scientists succeed in producing a practical fusion reactor, there will be plenty of fuel available for it. Earth's ocean can provide a nearly endless supply of hydrogen.

Figure 10-17 *The tokamak device can be used to produce a controlled nuclear-fusion reaction.*

Checkpoint

1. What is nuclear fusion?
2. What is the source of the energy given off by a nuclear-fusion reaction?
3. What is the major problem in producing controlled nuclear fusion?

Teaching Suggestions 10-8

Be sure students understand that a successful fusion reactor will mean that we can obtain usable energy from the abundant deuterium (heavy hydrogen) in the world's oceans. Whether the process is feasible is still not certain. Point out that a fusion reactor also will produce very few harmful radioactive wastes.

Background

The mass loss associated with the fusion of light nuclei to form heavier ones is much greater than that for fission. Thus, fusion provides much more energy per particle than does fission. This is why hydrogen bombs are so much more energetic than fission, or atomic, bombs. To detonate a hydrogen bomb, a very high temperature is required. In fact, the detonator for a hydrogen bomb is a fission bomb.

During fusion, nuclei of hydrogen fuse to form nuclei of helium. The helium nuclei formed have slightly less mass than the total mass of the hydrogen nuclei that fused. The missing mass has been transformed into energy.

If scientists and engineers can figure out a way to produce the high temperatures needed to set off a controllable fusion reaction (using lasers or another technique), fusion might well solve our energy crisis for ages. In the oceans, there is certainly an abundance of the hydrogen isotopes used in fusion. The problems involved in starting and controlling fusion, however, are difficult and numerous. The earliest date currently forecast for controlled fusion power is sometime during the first quarter of the twenty-first century.

Fact At Your Fingertips

- The fusion of a single kilogram of hydrogen can provide as much energy as the burning of 20 million kg of coal.

10 NUCLEAR REACTIONS

10-8 Nuclear Fusion

The Tokamak

Read the following paragraphs, which describe how scientists are trying to produce energy through nuclear fusion. Then answer the questions.

To imitate the nuclear fusion of the sun, conditions similar to those inside the sun must be created. These conditions include high temperature and pressure, which result in the presence of electrically charged particles called plasma. Yet this superhot plasma is very hard to keep in containers. Any normal container will turn to vapor when in contact with the plasma.

The most promising solution to the problem of containing the plasma is the tokamak. The tokamak uses strong currents to produce magnetic fields that keep the plasma in one place. The plasma is forced to remain in a donut-shaped area.

The tokamak will not be ready for commercial use for a long time. The cost of producing the magnetic fields is quite high. The magnetic fields must be at least 100,000 times the strength of the Earth's field. The plasma must be heated to a temperature of at least 100 million degrees Celsius. The plasma must be very dense and must be contained long enough to produce

Checkpoint Answers

1. Nuclear fusion is the combining of lighter nuclei to form heavier nuclei.
2. The mass lost when light nuclei fuse.
3. Producing the required high temperatures without melting the containing vessel.

Using The Feature Article

When using the C-14 dating method, scientists assume that the amount of atmospheric C-14 was the same thousands of years ago as it is today. Be sure students understand this. Confirmation of such dating results by means of tree rings is important in verifying that the assumption is valid.

Archeologists make extensive use of the C-14 method for dating artifacts. Geologists, however, who deal with much longer time periods, usually use the ratio of U-238 to Pb-206 in samples. They are thus able to measure the ages of rocks in billions, rather than thousands, of years.

10 NUCLEAR REACTIONS
Computer Program

Radiocarbon Dating

Purpose To calculate the age of unknown items.

Background Radiocarbon dating is accomplished by measuring the ratio of carbon 12 to radioactive carbon 14 and performing several calculations. Carbon 14 is produced in the atmosphere and accumulates in living organisms. It begins a steady deterioration after the organism dies. Archeologists use this method to discover the age of unknown items.

ENTER the following program:

```
10  LET K = .00012
20  INPUT "ENTER LOG OF C12 / C14 RATIO."; CR
30  LET T = (2.30 * CR) / K
40  PRINT "THE OBJECT IS"; T; "YEARS OLD."
50  INPUT "ANOTHER CALCULATION (Y OR N)?"; C$
60  IF C$ = "Y" THEN GOTO 20
70  END
```

Discussion

1. The following items were found at an archeological site and carbon 12/carbon 14 ratios were calculated. Determine the age of each. Use the answer space at the right to record your answer.

   Animal bone #1: .0645 log of ratio ______

   Animal bone #2: .3324 log of ratio ______

   Woven grass mat: .3389 log of ratio ______

   Carved wooden figure: .3394 log or ratio ______

2. Which of the above-listed artifacts seem to be from the same era?

Upgrade the Program

This program uses the variable K to represent the decay constant for carbon–14. The ratio between the number of remaining carbon–12 and carbon–14 atoms is expressed as a logarithm and provided below. (The logarithm of a number is the power of ten equal to that number.) The calculation of the age of the object occurs in line 30 of the program.

1. There is a 19-year inaccuracy in the calculation. This means that if an object tests as 1,000 years old, it may be anywhere from 981 years old to 1,019 years old. Add a calculation that prints the range for each date determination.
2. Add a calculation that will determine the log ratio, given the log value of carbon–12 and the log value of carbon–14 atoms.

Carbon-14 Dating

The carbon-14 produced by cosmic rays can be used to find the age of carbon-containing materials, such as wood and cloth. Despite the decay of C-14 to N-14, there is always a small amount of C-14 in the atmosphere due to the cosmic rays. The rate at which C-14 is produced by cosmic rays and the rate at which C-14 decays balance, keeping the amount of atmospheric C-14 constant.

Living plants use carbon dioxide (CO_2) from the atmosphere to make food by photosynthesis. A small fraction of the carbon atoms in this CO_2 is C-14. Most of the other carbon atoms in the CO_2 are C-12, and a few are C-13. Neither C-12 nor C-13 is radioactive. When a plant dies, it stops absorbing more carbon dioxide from the atmosphere. The C-14 atoms already in the plant continue to decay. The longer a plant has been dead, the less C-14 it contains. The half-life of C-14 is 5570 years. Therefore, only half of the C-14 present when a plant dies is still present after 5570 years.

A wooden object or a piece of cloth or paper made from plant fibers loses C-14 as it decays into N-14. The ages of such objects can be estimated by finding out how much C-14 is present, compared to the amount of C-12 and C-13. Dates obtained from C-14 analysis have been confirmed by other methods. For example, the age of a tree can be estimated by counting the number of annual rings. Estimates of the ages of old trees obtained by C-14 analysis are close to those obtained by counting the tree rings.

Understanding The Chapter 10

Main Ideas

1. Radioactive nuclei give off radiation in the form of alpha and beta particles and gamma rays.
2. In transmutation, an element gives off alpha or beta radiation and becomes a new element.
3. Isotopes are atoms of the same element that have different masses.
4. The half-life of a radioactive isotope is the time it takes for half the atoms in a sample to decay into another isotope.
5. Nuclear equations can be used to represent nuclear reactions.
6. Radioactive isotopes have uses in science, medicine, and industry.
7. Binding energy is the energy given off when a nucleus is formed. It results from the transformation of mass into energy.
8. Nuclear fission is the splitting of a nucleus into two smaller nuclei.
9. A chain reaction can occur during nuclear fission because the fission reaction produces neutrons. The smallest mass of fissionable material that will produce a chain reaction is the critical mass.
10. In nuclear reactors, the number of neutrons that can produce fission is controlled in order to produce a steady reaction rate.
11. Nuclear reactors produce dangerous radioactive wastes.
12. In nuclear fusion, light nuclei combine, forming heavier nuclei.

Vocabulary Review

From the following list, choose the term that best completes each of the statements. Write your answers on a separate piece of paper.

alpha radiation	isotopes
beta radiation	nuclear equation
binding energy	nuclear fission
chain reaction	nuclear fusion
cosmic rays	nuclear reactor
critical mass	radioactive decay
gamma radiation	tracers
half-life	transmutation

1. The __?__ of a radioactive isotope is the time required for half its atoms to decay to another isotope.
2. For a fission chain reaction to take place, there must be a(n) __?__.
3. The decrease in mass that occurs when protons and neutrons combine, forming a nucleus, appears as __?__.
4. The helium nuclei released when some radioactive isotopes decay are called __?__.
5. To increase its atomic number by one, a radioactive isotope must release __?__.
6. The change of one element to another when the element emits alpha particles is an example of __?__.
7. __?__ occurs when hydrogen nuclei combine to form helium nuclei.
8. Chemists use __?__ to follow the path of an element in a reaction.

Vocabulary Review Answers

1. half-life
2. critical mass
3. binding energy
4. alpha radiation
5. beta radiation
6. transmutation or radioactive decay
7. nuclear fusion
8. tracers
9. cosmic rays
10. nuclear reactor
11. isotopes
12. nuclear fission

Reading Suggestions

For The Teacher

Forward, Robert L. "Einstein's Legacy." *Omni*, March 1979, pp. 54–60. The history of nuclear physics and quantum theory.

Heppenheimer, T. A. THE MAN-MADE SUN: THE QUEST FOR FUSION POWER. Boston: Little, Brown, 1984. A very readable, informal, nontechnical discussion of the development of fusion power that conveys a sense of the frustrations and excitement of the scientists involved.

U.S. Congress. Office of Technology Assessment. THE EFFECTS OF NUCLEAR WAR. Montclair, NJ: Allenheld Osmun, 1980. A thought-provoking book that includes four case studies estimating immediate, intermediate, and long-term effects of nuclear war. The uncertainties inherent in such estimates are stressed.

For The Student

Asimov, Isaac. HOW DID WE FIND OUT ABOUT NUCLEAR POWER? New York: Walker, 1976. An easy-to-read introduction to the topic.

Hawkes, Nigel. NUCLEAR POWER. New York: Gloucester Press, 1984. A bright, colorful, readable account of the inside of the atom.

McGowen, Tom. RADIOACTIVITY: FROM THE CURIES TO THE ATOMIC AGE. New York: Watts, 1986. Provides a clear understanding of radioactive properties of matter and the uses thereof.

Moche, Dinah. RADIATION: BENEFITS/DANGERS. New York: Watts, 1979. Discusses the uses of radiation in technology, archeology, medicine, astronomy, dentistry, security, and law enforcement.

Pringle, Laurence. NUCLEAR POWER: FROM PHYSICS TO POLITICS. New York: Macmillan, 1979. A pro-nuclear account of nuclear power that stresses caution.

Weiss, Ann E. THE NUCLEAR QUESTION. New York: Harcourt Brace Jovanovich, 1981. An anti-nuclear account of nuclear power that includes an explanation of radiation and fission.

Chapter Review Answers

Know The Facts

1. moderator
2. fusion
3. control rods
4. H-1
5. gamma
6. chemical properties
7. greater than
8. 144
9. a stable isotope
10. less than
11. chemical reaction

Understand The Concepts

12. The atomic mass decreases by 4 amu and the atomic number decreases by 2 per alpha particle.
13. (a) paper, (b) steel that is greater than 30-cm thick.
14. (a) $_{-1}^{0}e$ (b) $_{2}^{4}He$ (c) $_{83}^{210}Bi$
15. The atomic number increases by 1 and the atomic mass is unchanged.
16. Answers will vary. Advantages: the plants produce little air pollution and they do not require the use of fossil fuels. Disadvantages: the plants are expensive to build, there is always a risk of accidents, and there are problems in safely disposing of the nuclear wastes that they produce.
17. The great amount of heat produced would remain in the reactor and might cause it to melt down.
18. See student text, pages 257–258. See also answer to Checkpoint question 4 on page 258.
19. Fission is the splitting of a nucleus to form two smaller nuclei. Fusion is the combining of lighter nuclei to form heavier nuclei.

9. High-energy particles that shower Earth from outer space are called __?__.
10. The rate of a fission reaction is controlled in a __?__.
11. Atoms of the same element that have different masses are called __?__.
12. The splitting of a nucleus into two smaller nuclei is called __?__.

Chapter Review

Write your answers on a separate piece of paper.

Know The Facts

1. The purpose of the __?__ in a nuclear reactor is to slow down neutrons. (moderator, coolant, shielding)
2. The __?__ of hydrogen nuclei provides the energy released by the sun. (fission, fusion, radioactive decay)
3. Adjusting the __?__ changes the rate of the fission reaction in a nuclear reactor. (shielding, turbines, control rods)
4. The most common isotope of hydrogen is __?__. (H-1, H-2, H-3)
5. __?__ radiation is like x rays, but has more energy. (alpha, beta, gamma)
6. All isotopes of an element have basically the same __?__. (half-life, chemical properties, tendency to undergo fission)
7. The amount of energy involved in a nuclear reaction is __?__ that involved in a chemical reaction. (greater than, less than, about the same as)
8. Thorium-234 nuclei contain 90 protons and __?__ neutrons. (90, 100, 144)
9. A radioactive decay series always ends with __?__. (a stable isotope, an unstable isotope, lead)
10. The fraction of U-235 in naturally occurring uranium is __?__ the fraction of U-235 in the uranium used in a nuclear reactor. (greater than, less than, about the same as)
11. A __?__ cannot be used to produce radioactive isotopes. (nuclear reactor, particle accelerator, chemical reaction)

Understand The Concepts

12. Explain what happens to the atomic mass and atomic number of an isotope that gives off alpha particles.
13. What could you use to stop (a) alpha particles, (b) gamma rays?
14. Complete the following nuclear equations:

    (a) $_{91}^{234}Pa \longrightarrow _{92}^{234}U + \underline{?}$

    (b) $_{84}^{218}Po \longrightarrow _{82}^{214}Pb + \underline{?}$

    (c) $_{82}^{210}Pb \longrightarrow \underline{?} + _{-1}^{0}e$

15. What happens to the atomic mass and atomic number of an isotope that gives off a beta particle?
16. What are some of the advantages and disadvantages of using nuclear reactors to produce electricity?
17. What might happen to a nuclear reactor if it had no coolant?
18. Describe the "four-barrier" approach to the disposal of nuclear waste.

19. What is the difference between nuclear fusion and nuclear fission?
20. What major problem is involved in producing a controlled fusion reaction?
21. Why did scientists prior to 1940 think that there could be only 92 elements?
22. Explain why a critical mass is necessary in order for a nuclear chain reaction to occur.

Challenge Your Understanding

23. An element consists of two isotopes. Isotope A has a mass of 24 amu. Isotope B has a mass of 25 amu. If 70% of the atoms of the element are A and 30% are B, what is the atomic mass of the element?
24. Strontium-90 has a half-life of 28 years. How long will it take for 100 g of strontium-90 to decay to 25 g?
25. Explain why the sum of the masses of the protons and neutrons in a nucleus is less than the sum of their masses when they exist as separate particles.
26. Determine the ratio of neutrons to protons in the stable isotopes of a wide range of elements. Then determine the same ratio for some unstable isotopes of the same elements. What can you conclude about the effect of the neutron-proton ratio on the stability of a nucleus?
27. Why might scientists who wanted to trace the flow of water under the ground choose radioisotopes with short half-lives?

Projects

1. Get information from power companies that own nuclear reactors and from organizations that oppose nuclear power. Additional information may be obtained from your school or local library. Arrange a debate on nuclear power. Let your classmates choose which side (pro or con) they wish to be on.
2. Obtain some radioactive sources from your school's physics or chemistry teacher. Under your teacher's supervision, place several sources at different points on a pack of black and white film. After a few days, remove the radioactive materials and have the film developed. What do you find?
3. Use your library to write a report on one of the following topics: (a) the threat of nuclear war, (b) useful isotopes, (c) cancer and isotopes, (d) radioactive tracers, (e) the development of nuclear reactors, (f) the breeder reactor.
4. Make a chart showing the radioactive decay series that begins with U-238 and ends with Pb-206.
5. Find out how carbon-14 has been used as a tool by archeologists.
6. Make a scale model of a nuclear reactor.
7. Research and report on the contributions to chemistry and physics of any two of the following scientists: Lise Meitner; Hedeki Yukawa; Enrico Fermi; Luisa Fernandez Hansen; Shirley Ann Jackson; Chien-Shiung Wu; Samuel Chao Chung Ting.

---

20. Producing the enormously high temperatures needed without melting the containing vessel.
21. There were only 92 elements at that time. Moreover, no one believed that new elements could be created by artificial means.
22. If the mass of fissionable material is too small, many neutrons can escape from the sample without causing fission of other atoms. In a sample of larger mass, more neutrons can strike fissionable nuclei. When a sample has critical mass, enough neutrons strike fissionable nuclei to support a chain reaction.

Challenge Your Understanding

23. $70 \times 24 = 1680$
    $30 \times 25 = 750$
    $1680 + 750 = 2430$
    $$\frac{2430 \text{ units of mass}}{100 \text{ units}} = 24.3 \text{ amu}$$
24. 56 years, or two half-lives.
25. When particles combine in a nucleus, the binding energy is released. This energy comes from mass lost when the nucleus formed.
26. As the neutron-proton ratio increases, the isotopes are more likely to be unstable. (Students will have to consult the periodic table and other sources to obtain the information needed to answer this question.)
27. So that after the tracings have been made the water will not remain contaminated by radioactive material for a long period of time.

10 NUCLEAR REACTIONS
Projects

1. At the library, research the use of x rays.
   a. How are x rays used in medical and dental diagnosis?
   b. How are x rays used in radiotherapy for cancer patients?
   c. How are x rays used in a fluoroscope? ______
   d. How are x rays used in metallography? ______

10 NUCLEAR REACTIONS
Challenge

Think About it!

1. Plutonium–238 ($^{238}_{94}Pu$) decays to give an alpha particle and a new atom. What is the atomic number of the new atom? ____ What is its mass number? ____ Identify the atom.
2. Thorium–232 ($^{232}_{90}Th$) decays to give an alpha particle and a new atom. What is the atomic number of the new atom? ____ What is its mass number? ____ Identify the atom.
3. When potassium–43 ($^{43}_{19}K$) decays, a beta particle is given off. Write the nuclear equation to show this nuclear change. ______
4. When polonium–216 ($^{216}_{84}Po$) decays, an alpha particle is given off. Write the nuclear equation to show this nuclear change. ______
5. When light-speed alpha particles bombard nitrogen–14 atoms, oxygen–17 atoms and hydrogen atoms are given. Write the nuclear equation to show this nuclear change.

Unit 2 Careers

Teaching Suggestions

Plan a field trip to a local industry or research facility that employs a number of chemists. Arrange to have students observe the chemists on-the-job. If possible, also arrange for students to talk with the chemists about their careers.

Ask students to do library research to find out about any of the following chemistry-related occupations: pharmacist, petroleum engineer, chemical oceanographer, chemotherapist, biochemist, environmental scientist, color engineer, chemical mixer, crime-lab technician, quality control technician, food tester, medical-lab technician, or chemical analysis technician. Have students prepare "Career Pamphlets" summarizing the information that they obtain. Display these pamphlets, and then keep them for future reference.

Recommend to those students who have an interest in pursuing chemistry as a career that they obtain and read the following book: *Opportunities In Chemistry* by John Woodburn, National Textbook Co., Skokie, IL, 1979. Have the students prepare a book report summarizing their findings.

Chemistry-oriented journals and magazines carry a wide variety of job openings for chemists and chemistry-related occupations. Obtain recent issues of *Chemical and Engineering News* and make copies of the employment-opportunities page. The requirements and qualifications for each position may be discussed in class. Use the advertisements in a "Chemistry Careers" bulletin-board display.

Forensic chemists carefully examine the scene of a crime for chemical clues. Most state police and F.B.I. laboratories employ chemists and assistants to test blood, paint, and fabrics to help in solving crimes. Have students contact your state crime lab or the F.B.I. to find out more about job opportunities in this fascinating field.

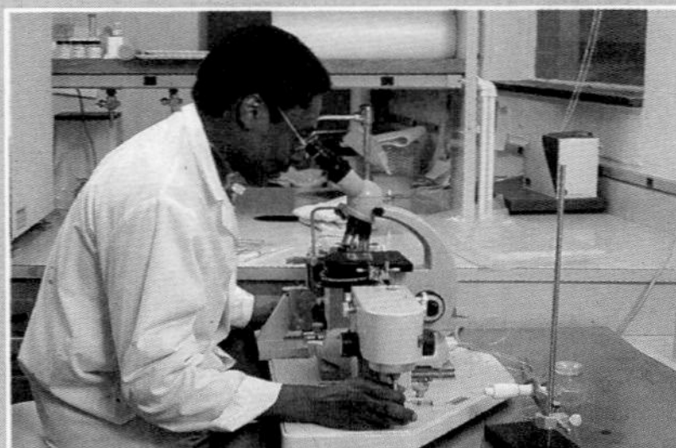

Medical Technologist

Hospital seeks medical technologists to analyze tissue and fluid samples. Technologists perform tests and gather data to assist physicians in making diagnoses. Tasks include analyzing the levels of substances in the blood and identifying bacteria in tissue samples. Technologists also may be asked to develop new tests, teach, or manage some laboratory activities.

Applicants should have a degree from a four-year college with courses in chemistry, biology, and mathematics and special training in sterile laboratory technique. Ability to perform tests precisely and accurately is essential. Technologists should expect to work some evening and weekend shifts.

Wastewater Treatment Plant Operator

Wastewater treatment plant seeks persons to train as equipment operators. Operators control the machinery that removes biological wastes and chemical pollutants from wastewater.

Applicants should have a background in chemistry, biology, and mathematics. High school graduates with some training in a technical school or a community college are preferred. Those hired will receive on-the-job training. Those who wish to become supervisors must obtain a degree in science or engineering from a four-year college.

Some outdoor work is required. Operators also should expect some night, holiday, and weekend shifts. Hazards include noise from the machinery and unpleasant odors.

Help Wanted

Additional Careers

A brief description and job outlook for three occupations related to chemistry follow:

1. *Agricultural chemical salespeople* sell chemical supplies, such as fertilizers and pesticides, to farmers. These salespeople must be able to explain the hazards, effectiveness, and correct usage of the products that they sell. In the future, as more efficient use of the land becomes necessary, off-farm occupations, such as chemical sales, will probably grow more than on-farm occupations. Thus, the job outlook for chemical salespeople is good.
2. *Health inspectors* are employed by federal, state, and local governments to enforce the laws that protect the public from health hazards. They

Lithographer

Off-set printing company seeks high school graduates to be trained as lithographers. Lithographers help prepare the metal plates used in offset-lithography, the process of printing from a flat surface. Trainees will specialize in one of several jobs: camera operator, lithographic artist, assembler, or platemaker. *Camera operators* take photographs of the material to be printed. *Lithographic artists* sharpen and reshape the images on film. *Assemblers* cut out the film and mount it on special sheets. *Platemakers* use chemicals to transfer the images from film to the metal plate.

Applicants should have a background in chemistry, optics, and photography. Artistic skill and knowledge of electronics would also be helpful. On-the-job training will be provided.

Darkroom Technician

Portrait studio has two positions open for darkroom technicians. One is for an *all-around darkroom technician* who processes both black-and-white and color film. The other position is for a *color laboratory technician*, someone who processes only color film. Applicants with prior studio experience may take photographs as well as process film.

Applicants must be high school graduates with courses in chemistry and mathematics. Good color perception and vision are important. Applicants also should be good at working with their hands. Film-processing experience as part of a hobby would be helpful. The technician's job requires doing one task over and over at a rapid pace. On-the-job training will be offered.

work with scientists to test food, water, agricultural feeds, pesticides, drugs, cosmetics, and air quality. They write reports to suggest ways of improving products and prepare evidence for legal action, as necessary. Because of reduced government spending, however, future job openings for health inspectors probably will be limited.

3. *Technical writers* with chemistry backgrounds are employed by chemical manufacturers, pharmaceutical firms, government agencies, and publishers of scientific books and journals. Technical writers must be able to research, write, and edit scientific material so that it can be readily understood by readers who do not necessarily have a scientific background. Future demand for technical writers will depend on the expected increase in research and development in the areas of energy, medicine, and environmental health.

For More Information

You or your students can write to the following organizations for more career information related to chemistry:

1. Photo Marketing Association International, 300 Picture Place, Jackson, MI 49201.
2. National Association of Trade and Technical Schools, 2021 K St., N.W., Washington, DC 20006.
3. American Chemical Society, 1155 16th St., N.W., Washington, DC 20036.
4. Chemical Manufacturers Association, 2501 M St., N.W., Washington, DC 20037.

Using Issues In Science

One cost of technological progress is increased toxic waste. After the students have read this article and answered the *Think Critically* questions, discuss the risks of hazardous waste. Then point out that there are some benefits resulting from industries that produce toxic waste—for example, employment, economic development, and products that make life more comfortable. Ask the students to think about how their lives would change if toxic waste production were prohibited.

Explain to the students that hazardous waste doesn't come just from industrial sites. Have them list the hazardous materials in their homes, including oven and drain cleaners, automotive fluids and oils, etc. Many of these materials, and even their containers, constitute toxic waste. Some communities hold annual household hazardous waste collection days. Have them research the regulations governing the hazardous waste dump that is closest to their homes.

Background

In 1970, Congress established the Environmental Protection Agency (EPA) to enforce regulations protecting land, water, and air from pollution by toxic substances. Since then, many legislative battles have been fought. An excellent set of articles on the issue of controlling toxic waste can be found in the April, 1986 volume of *Environment*, 28, No. 3, pp. 2–45.

Teaching Study Skills

Using Reference Materials

Students can research and write a report on toxic waste production, disposal, or legislation. But first they must be able to find relevant information. Knowing how and where to start researching is a valuable skill.

Demonstrate how to look up books in the card catalogue at the library. Point out that there are three kinds of cards, referenced by subject, author, or title. The listings at the bottom of these cards are related subjects to look up in the catalogue. Bibliographies in the books they find will also give additional sources to look up.

Many magazines and journals have an index at the end of each year. Some indices are found in the reference section of the library, such as the *Readers' Guide to Periodical Literature* and the *New York Times Index*. Many libraries now have periodicals indexed on microfiche or in computer data bases.

Give the students a narrow topic, such as "Legislation in the 1980s on Toxic Waste Removal," and a starting list of references. Ask them to find as many additional references as they can

ISSUES IN SCIENCE

TOXIC WASTE: What Can We Do With It?

If you drive down Interstate-44 in Missouri, you will pass an exit sign that once read Times Beach. Now it reads HAZARDOUS WASTE. In 1982, extremely high levels of a chemical called dioxin were discovered in Times Beach. The dioxin had been sprayed on unpaved roads in the 1970s, to prevent dust from blowing around. People did not notice any ill effects until several years later. Then it was too late.

The residents of Times Beach noticed a sharp increase in cancer, stomach problems, and seizures. When people found out that dioxin was responsible, they abandoned their homes, leaving behind a ghost town.

Dioxin is a toxic chemical. Toxic chemicals are poisonous; they can cause disease. The problem of toxic waste pollution can no longer be ignored. Many chemicals from sources such as sewage and industrial waste have been found in our drinking water and in the soil. Toxic chemicals, once absorbed by the environment, tend to spread out. Cleanup is difficult because often miles of shoreline or acres of land have been polluted.

What can be done about this deadly garbage? The Environmental Protection Agency (EPA) has identified more than 18,000 toxic waste sites in the U.S. Most of these are places where chemicals have contaminated the environment; others are landfills for the dumping of toxic waste. Unfortunately, the landfills themselves can leak. Often all the EPA can do is move containers that are leaking to a new site, where the same problem may occur again.

In 1980, the EPA began administering a program called Superfund to determine damage and clean up some of the worst sites. In 1986 Congress voted 9 billion dollars for this fund. But some

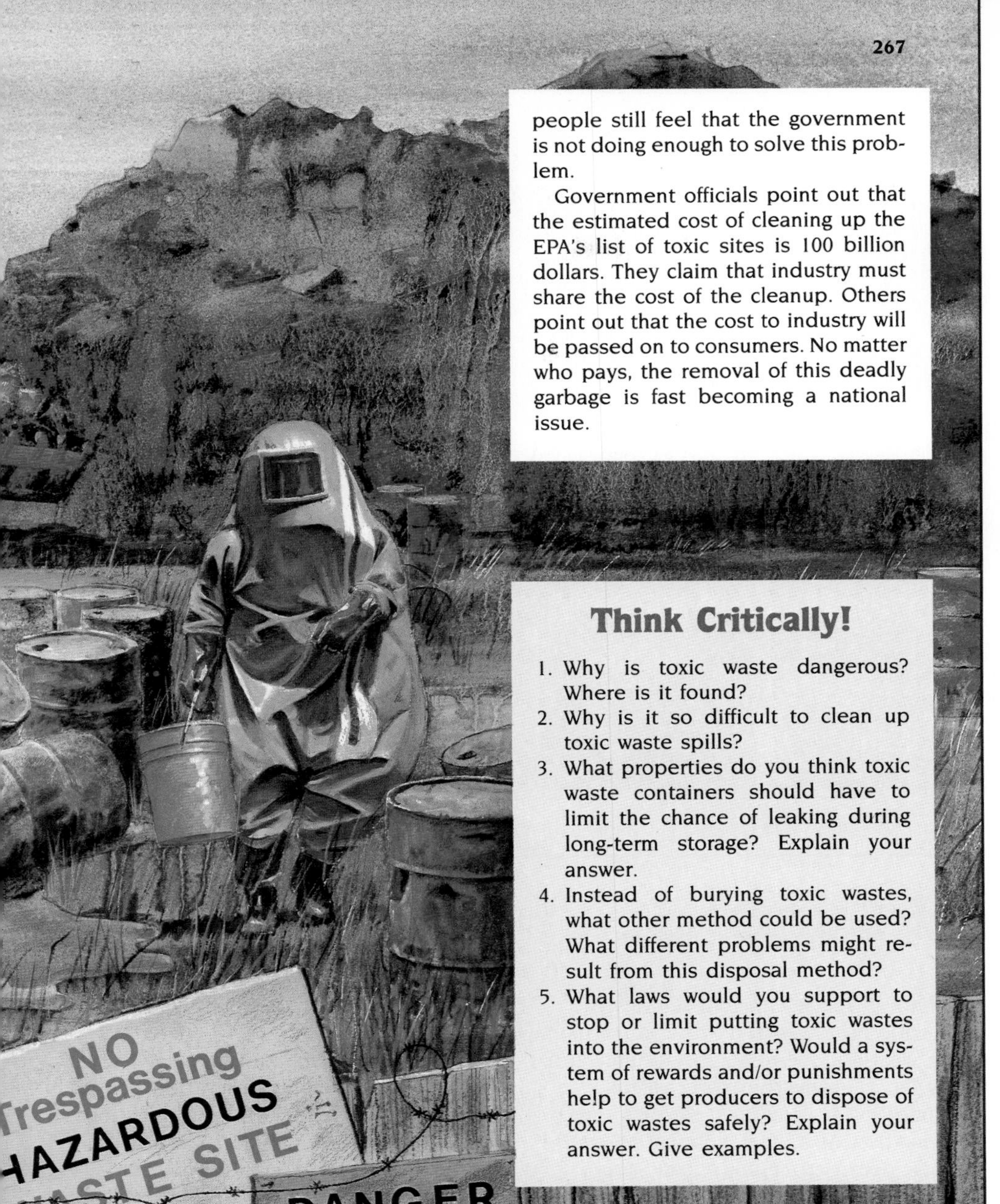

267
people still feel that the government is not doing enough to solve this problem.
Government officials point out that the estimated cost of cleaning up the EPA's list of toxic sites is 100 billion dollars. They claim that industry must share the cost of the cleanup. Others point out that the cost to industry will be passed on to consumers. No matter who pays, the removal of this deadly garbage is fast becoming a national issue.
Think Critically!
1. Why is toxic waste dangerous? Where is it found?
2. Why is it so difficult to clean up toxic waste spills?
3. What properties do you think toxic waste containers should have to limit the chance of leaking during long-term storage? Explain your answer.
4. Instead of burying toxic wastes, what other method could be used? What different problems might result from this disposal method?
5. What laws would you support to stop or limit putting toxic wastes into the environment? Would a system of rewards and/or punishments help to get producers to dispose of toxic wastes safely? Explain your answer. Give examples.
NO
Trespassing
HAZARDOUS
WASTE SITE
DANGER

Think Critically Answers

1. Toxic waste may cause cancer or other diseases. It is found in the soil, water, and air.
2. Toxic waste tends to spread out in the environment. Cleaning up large areas of pollution is expensive and time consuming.
3. Toxic waste containers should be very strong, in the event of natural disasters such as earthquakes. They should be durable, in case they have to be transported to a new site many years later. Containers should also have a very low reactivity; steel, for example, would corrode in a relatively short time.
4. Some people have suggested ocean dumping of toxic waste containers. Dumping in the ocean means that corrosion of the containers may be more likely. And any chemicals that leak will spread out over very large areas. In time the chemical concentrations could build up to a dangerous level. Organisms living in the ocean would ingest these chemicals directly from the water and put them into the food chain.
5. Laws could be passed limiting the use of certain chemicals, especially if alternatives were available. For example, the government banned DDT, in part because other pesticides were available. The government could fund research for alternative disposal methods. Also, industries could be given tax breaks for purchasing waste treatment equipment.

in the library. They might, for example, search the *Science News* index for articles on toxic waste.

Ask the students to prepare a list of the additional references they have found. They should write down all the information they would need to find each journal or book. Then have them follow through by finding three of their sources on the library shelves.

Unit 3 · Earth and Space Science

The water that covers three fourths of planet Earth can strike the shore with tremendous force.

269

Unit 3 · Earth and Space Science

The Unit In Perspective

This unit is designed to introduce basic concepts of geology, weather, oceanography, and astronomy to students and to provide an understanding of Earth's place in the solar system and universe.

Chapter 11, *Earth Materials,* discusses minerals, types of rocks, fossils and their significance, the water cycle, problems related to drinking water supplies, and the use of maps in the earth sciences.

Chapter 12, *Earth's Changing Crust,* explores the composition of soils, weathering, erosion, deposition, earthquakes, and volcanoes, and relates the theory of plate tectonics to forces affecting Earth's surface features.

Chapter 13, *Weather,* discusses the atmosphere, winds and wind patterns, basic weather phenomena, climate, and the science of weather forecasting.

Chapter 14, *Oceans,* describes the composition of seawater, wave and tidal effects, ocean currents, ocean sediments, and the ocean floor, and relates the resources of the seas to human needs and activities.

Chapter 15, *Astronomy,* establishes planet Earth's place in the solar system and universe, describes the solar system and the sun, and explores the beneficial outcomes of astronomical studies.

Students will be interested in the earth and space science careers featured on pages 392–393. The careers selected illustrate both the multidisciplinary aspects of earth and space science and the wide range of occupational possibilities.

Issues in Science, on pages 394–395, introduces the issue of sending people into space. Critical thinking questions at the end of the article focus on the process of decision making. A study skills exercise is included in the teacher pages.

Preparing To Teach The Unit

The advance preparation needed to teach this unit includes the acquisition and organization of equipment and materials for laboratory activities, planning field trips, ordering audio-visual aids, and collecting and/or placing on reserve in the library outside readings for students. A list of suggested readings will be found adjacent to the Main

Ideas section at the close of each chapter in the unit.

Equipment and Materials

The listing below identifies the equipment and materials you will need for the student Activities of this unit. The Activities are located at the back of the student book. Quantities required will depend on the nature of the Activity, the availability of materials, numbers of students, budgetary considerations, and so on. Some Activities proceed better if students work in pairs or threes. Check each Activity to make this decision.

ACTIVITY 17 (Ch. 11): *Testing Mineral Hardness* (page 553). MATERIALS: mineral samples, copper penny, iron nail, steel scissors, glass slide.

ACTIVITY 18 (Ch. 12): *Moving Land Masses* (page 554). MATERIALS: tracing paper, scissors, transparent tape, world map.

ACTIVITY 19 (Ch. 13): *Making a Simple Hygrometer* (page 556). MATERIALS: strand of human hair about 15 cm long; empty shoe box, without lid; small piece of lightweight cardboard; scissors; transparent tape; centimeter ruler.

ACTIVITY 20 (Ch. 14): *Investigating Density Currents* (page 557). MATERIALS: food coloring, test tube, artificial seawater (three different densities), meter-long plastic column with cap, ring stand and clamp, watch or clock with second hand.

ACTIVITY 21 (Ch. 15): *Seasons of the Year* (page 558). MATERIALS: styrofoam ball (8–10 cm in diameter), probe or teasing needle, flexible protractor, flexible ruler or tape measure, lamp with 25-watt bulb, several pieces of yarn long enough to encircle the ball.

Field Trips

A well planned field trip is a powerful teaching tool. Plan as many as possible, and try to have expert guest speakers address your class as well.

Make use of nearby sites. Plan a trip, if possible, to a water treatment plant, a state geological survey office, a gem and mineral exhibit at a science museum, or to view rock formations.

If you live near an area that has experienced earthquakes or volcanoes, plan a trip to observe their effects. Have students bring in photographs of weathering and erosion in your area.

A trip to a nearby weather station will give students a chance to see instruments used in weather forecasting.

If possible, have students make direct observations of the ocean. If your school is inland, make use of available equipment, such as saltwater aquaria.

Schedule some night observation sessions. Use telescopes, if available. Schedule a trip to a nearby observatory or planetarium.

Audio-Visual Aids

The list of A-V materials below provides a wide selection of topics on sound filmstrips (SFS), videocassettes (VC), and 16 mm film. Choose titles in terms of availability, student interest, coverage of content vis-à-vis field trips and other learning activities, availability of projectors, and time limitations. A list of A-V suppliers is given on page T-39. Consult each supplier catalog for ordering information and delivery times.

Chapter 11: Earth Materials

What Rocks Tell Us About the Earth, SFS, Random House.
Rocks and Minerals, SFS, National Geographic.
Geology: Our Dynamic Earth, SFS, National Geographic.
Fossils: From Site to Museum, 11 min, 16 mm, VC, Coronet.
The Water Crisis, 57 min, 16 mm, VC, Time.
Rocks That Originate Underground, 23 min, 16 mm, Britannica.
Rocks That Form On Earth's Surface, 17 min, 16 mm, VC, Britannica.

Chapter 12: Earth's Changing Crust

Weathering and Erosion, SFS, National Geographic.
The Restless Earth: Understanding the Theory of Plate Tectonics, SFS, Guidance Associates.
Face of the Earth, 17 min, 16 mm, National Film Board of Canada.
The Earth: Its Structure, 9 min, 16 mm, Coronet.
The City That Waits To Die, 60 min, 16 mm, Films, Inc.
Predictable Disaster, 32 min, 16 mm, Time.

Chapter 13: Weather

The Falling Barometer, 22 min, 16 mm, VC, Sterling.
The Restless Ocean of Air, 21 min, 16 mm, VC, Sterling.
The Weather Watchers, 30 min, 16 mm, CRM/McGraw-Hill.
Forecasting The Weather, SFS, National Geographic.
Winds and Their Causes, 11 min, 16 mm, Coronet.
Weather Forecasting, 22 min, 16 mm, Britannica.

Chapter 14: Oceans

The Science of Oceanography, SFS, National Geographic
The Earth: Its Oceans, 14 min, 16 mm, Coronet.
The Earth: Coastlines, 10 min, 16 mm, Coronet.
Weather: Understanding Storms, 10 min, 16 mm, VC, Coronet.
The Beach: A River of Sand, 21 min, 16 mm, Britannica.
Waves on Water, 16 min, 16 mm, Britannica.
How Level is the Sea? 12 min, 16 mm, Britannica.

Chapter 15: Astronomy

The Universe, SFS, National Geographic.
Mystery of the Sun, 26 min, 16 mm, VC, Carousel.
Close Up on Planets, 19 min, 16 mm, VC, Coronet.
The Stars, 12 min, 16 mm, VC, CRM/McGraw-Hill.
The Galaxies, 12 min, 16 mm, VC, CRM/McGraw-Hill.
Space Science: Comets, Meteors, and Planetoids, 10 min, 16 mm, VC, Coronet.
Space Science: Exploring The Moon, 16 min, 16 mm, VC, Coronet.

CHAPTER 11 Overview

Thanks to the persistence of certain innovative scientists and to technological advances, we know that Earth has different layers made of many different materials. These earth materials, their distribution, and the graphic way they can be represented are the topics covered in this chapter. The differences between rocks and minerals are discussed. Students can identify minerals, using simple tests. Rock characteristics are determined by the minerals they contain and by the processes that form them. Students have an opportunity to learn the differences between metamorphic, igneous, and sedimentary rocks. The rock cycle is described. Next, students are introduced to the subject of fossils: what they are, how they are preserved, where to find them, and their value to people. A later section is devoted to the water cycle. Another section discusses drinking water, its sources, and the problems associated with finding enough good, drinkable water. The enrichment section introduces students to geologic and topographic maps. Students are encouraged to use maps of their own area in order to better understand map "language."

Goals

At the end of this chapter, students should be able to:

1. explain what minerals are and how minerals can be identified.
2. identify the three classes of rocks.
3. describe the rock cycle.
4. define fossils.
5. list conditions for the preservation of fossils and describe likely fossil locations.
6. list some uses for fossils.
7. describe the water cycle and discuss some of the variations of the cycle.
8. discuss drinking water and its sources.
9. describe features and uses of geologic and topographic maps.

Chapter 11

Earth Materials

This geologist is studying mineral samples at the Smithsonian Institution in Washington, D.C.

Vocabulary Preview

crust	magma	rock cycle	water table
minerals	lava	fossils	aquifers
crystals	sedimentary rocks	water cycle	geologic map
igneous rocks	metamorphic rocks	saturated zone	topographic maps

11-1

Minerals

Goal

To explain what minerals are and how minerals can be identified.

When you hear the word *rock* you probably think of the pebbles in a driveway or rock garden, or the flat stones you have tried to make skip across the water. Rocks, stones, and pebbles are part of Earth's outermost solid layer, its **crust**. The crust includes the continents and the ocean bottoms.

crust

Everything on Earth is made from less than 100 elements. *Elements* are substances that cannot be broken down into anything simpler by chemical means. A few solid elements are found in the crust in a pure state. For example, gold can be found as individual lumps of pure gold. Most of the elements of the crust, however, are combined in *compounds*. Naturally-occurring solid elements and compounds are called **minerals**. Most rocks are mixtures of minerals.

minerals

The most abundant elements in Earth's crust are oxygen and silicon (SIL ih kun). (See Table 11-1.) They are the main ingredients of minerals called *silicates* (SIL ih kaytz). Silicates, sometimes called "rock-forming minerals," make up 87 percent of all minerals. Granite (GRAN it), for instance, is a rock made up of several different silicate minerals. Granite is chiefly biotite, feldspar, quartz, and hornblende.

Table 11-1: Most Abundant Elements in Earth's Crust

Element	*Symbol*	*Percent of crust (by mass)*
oxygen	O	46.60
silicon	Si	27.72
aluminum	Al	8.13
iron	Fe	5.00
calcium	Ca	3.63
sodium	Na	2.83
potassium	K	2.59
magnesium	Mg	2.09

Teaching Suggestions 11-1

Before starting this section, try to obtain samples of various minerals, rocks, and ores for students to examine. Before students read the text, you may wish to ask:

Where is Earth's crust?
How thick is it?
What is it made of?
Is it the same all over Earth?

Accept all answers and discuss them. Overhead transparencies or chalkboard drawings can be used to illustrate the location of the crust, and its variations in thickness (thick under continents; thin under ocean basins).

In class discussion, be sure to clarify the meanings of new words and words, such as crust, water table, and rock, that take on new meanings in an earth-science context. You may wish to review the terms "element" and "compound" by referring students back to Chapter 6, section 6-5. Describe metals by listing their characteristics (see Table 8-1). Ask students:

How do metals differ from nonmetals? (Metals are good conductors, are shiny, can be shaped, and are relatively dense.)

Discuss ways in which the metals listed in this section are used. (Iron is used to make steel; copper is used to make pipes for plumbing; aluminum is used to make pots and pans; silver is used in jewelry.) Show that some of these elements, such as calcium, do not fit the popular notion of a metal. Display ore samples, such as hematite, cinnabar, and malachite. Give students time to handle and compare the mineral samples. Discuss and demonstrate the various tests made for identifying minerals before asking students to do laboratory Activity 17, Testing Mineral Hardness, student text page 553. Use the Feature Article, Minerals and Modern Technology, in conjunction with this section. (See student text, page 287.) Also, some students can do chapter-end projects 2, 5, and 6 at this point.

11 EARTH MATERIALS

Computer Program

Mineral Classification

Purpose To describe and use some properties used to classify minerals.

Background Minerals are classified according to a variety of physical and chemical features, including hardness, color, and luster. These features are used in this program. Others used in mineral classification are crystal shape, cleavage, and specific gravity.

ENTER the following program:

```
10  INPUT "ENTER HARDNESS NUMBER."; H
20  IF H > 7 THEN GOTO 90
30  INPUT "IS MINERAL COLORLESS (Y OR N)?"; C$
40  IF C$ = "N" THEN GOTO 160
50  INPUT "IS MINERAL LUSTER GLASSY (Y OR N)?"; L$
60  IF L$ = "N" THEN GOTO 160
70  PRINT "THE MINERAL IS QUARTZ."
80  GOTO 170
90  INPUT "IS MINERAL COLORLESS (Y OR N)?"; C$
100  IF C$ = "N" THEN GOTO 160
110  PRINT "IS MINERAL LUSTER HARD AND"
120  INPUT "BRILLIANT (Y OR N)?"; L$
130  IF L$ = "N" THEN GOTO 160
140  PRINT "THE MINERAL IS A DIAMOND."
150  GOTO 170
160  PRINT "NEITHER QUARTZ NOR DIAMOND."
170  END
```

Discussion

A man was caught leaving a diamond show after an alarm sounded. He was searched and two colorless minerals were found. The man insisted they were both quartz, for use in costume jewelry. Mineral number 1 has a hardness of 7 and a glassy luster. Mineral number 2 has a hardness of 10 and a hard and brilliant luster. Is the man telling the truth about the costume jewelry? If not, which mineral is not costume jewelry?

Upgrade the Program

You will see that line 20 branches the program so that one line of questions leads to quartz and the second, starting at line 90, leads to diamond. Lines 40, 60, 100, and 130 compare the entries to make sure that the proper description for each mineral is entered. An N response would indicate a different mineral and send the program to line 170.

Rewrite the program to identify two common minerals in your area.

Background

Strictly speaking, a mineral may not necessarily be an element or a compound. A compound has a fixed chemical composition; a mineral may have a range of compositions, with various elements substituting for one another. Such variations in mineral composition stem from differences in the environment in which minerals form. Heat, pressure, and the elements that are present determine the nature of the mineral that forms. A good example of the influence of environment is found in the forms of carbon called graphite and diamond. Both are pure carbon, but have very different characteristics: diamond is much harder than graphite, and is formed under much greater heat and pressure. In addition, minerals are inorganic in origin. Thus, coal is not a mineral. It is, however, a rock.

The plane in which a mineral cleaves is determined by the bonds between its various atoms. If the bonds are weaker in only one plane, cleavage will occur in that plane. Some minerals, such as micas, cleave in only one plane; others, such as calcite, cleave in two; and others, such as galena and halite, cleave in three. Of course, some minerals do not cleave at all. Minerals, such as quartz, which fracture instead of cleave, are those in which the bonds between atoms are strong in all directions, so that a cleavage plane does not form.

You may wish to help students distinguish between specific gravity and density. Specific gravity is related to density, but is defined relative to Earth's gravitational field. Density is mass per unit volume on Earth or anywhere else in the universe. (See discussion of density in Chapter 6 and laboratory Activity 9, Measuring Density, student text pages 541–542.)

Most of the elements present on Earth are *metals*. Familiar metals include iron, copper, aluminum, and silver. In Table 11-1, all the elements listed are metals, except for oxygen. Metals, and other elements, are often obtained from rocks called *ores*. The mineral hematite is an important source of iron. Cinnabar is a source of mercury, and malachite (MAL uh kyt) is a source of copper. (See Figure 11-1.)

crystals

In order to identify minerals, you need to look for certain physical properties. For example, many minerals form as regularly-shaped solids, or **crystals**. A crystal's shape results from the regular pattern in which its basic units line up. The *shape* of its crystals is a clue that can help to identify a mineral. For example, halite (table salt) crystals are cubic in shape. (See Figure 11-2.) All crystals of halite, no matter how large or small, have the same shape.

Many minerals have identifying *colors*. Halite, for instance, is white (or colorless), malachite is green, and cinnabar is red. Some minerals come in more than one color. An example is orthoclase feldspar, which is a common mineral in some kinds of rock. Orthoclase can be white, gray, pinkish, or yellowish.

The color of a mineral sample can't always be used for identification. Impurities in a mineral can change its color. The *streak test* is a better indicator of a mineral's true color. The streak is a sample of the powdered mineral. To observe

Figure 11-1 *(a) Malachite is a compound of copper and oxygen. (b) The red mineral is cinnabar, containing mercury. (c) Hematite is iron oxide.*

Demonstrations

1. Using an unpolished porcelain tile, show how a streak test is done. Pass the tile and the material around the class, so that students can see the streak. Pyrite is a counterintuitive example that will demonstrate how the streak color can differ from simple observation of a mineral.
2. Show the difference between cleavage and fracture. Cut a cube-shaped piece of modeling clay in three different planes, showing three-way cleavage. Then cut another sample in two different planes; then, one.
3. Demonstrate the difference between metallic and nonmetallic luster with highly polished brass or copper, and a piece of glass.

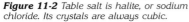

Figure 11-2 *Table salt is halite, or sodium chloride. Its crystals are always cubic.*

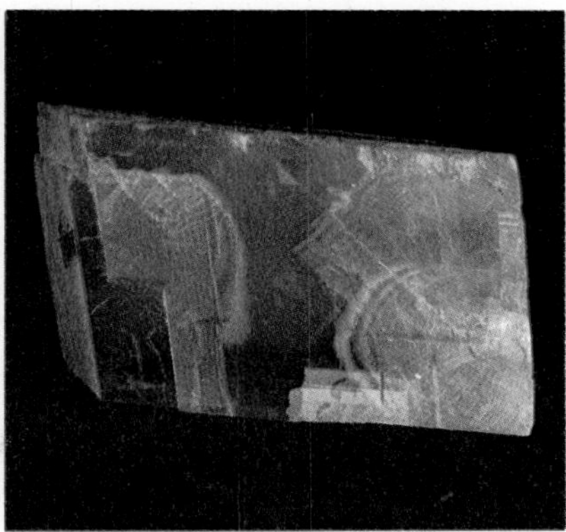

Figure 11-3 *Calcite cleaves in three directions that are not at right angles to each other.*

the color of the streak, scratch the mineral on an unpolished porcelain tile. The streak may not be the same color as the mineral. A mineral's streak color is just about the same for all samples of that mineral.

If you test a large variety of mineral samples, you will find that some have a colorless streak. Minerals that are very hard (see the next paragraph) leave no streak at all. A diamond is one of those.

Hardness is another reliable feature of a mineral. To test hardness, you can scratch one mineral with another. A harder mineral always scratches a softer mineral. On a scale of hardness, the higher the number, the harder the mineral.

Because of their crystal structure, most minerals break in a particular pattern. If a mineral breaks along flat surfaces, it is said to show *cleavage*. (See Figure 11-3.) This property can be used to help identify a mineral. The sample in Figure 11-3 has cleaved in three directions. The three directions are *not* at right angles to each other. A mineral that cleaves in three directions that *are* at right angles to each other will form cubic shapes. Galena (guh LEE nuh) is an example. The mineral mica (MY kuh) cleaves in only one direction.

The way a mineral reflects light is called its *luster*. Shiny new coins and polished silverware have a high luster. Luster is divided into two main types: metallic (like metal) and nonmetallic (like glass). The dark gray mineral galena, for

- A local university, college, or rock-and-mineral society may have gem and mineral samples or pictures that they would make available for display and/or demonstration.

Facts At Your Fingertips

- Salt is a mineral, called halite, that we can eat.
- Most minerals can be represented by a chemical formula; rocks cannot.

Mainstreaming

Students who are confined to wheelchairs can use a lapboard to facilitate their study of minerals. Be sure that such students have the opportunity to work with a nonhandicapped partner who can help, if necessary, with the handling of equipment.

Teaching Tips

- Pictures of gemstones and semiprecious stones, such as birthstones, can be used. Advertisements for precious gems are plentiful in many magazines; many mineralogy texts also have illustrations of gems.
- Ball-and-stick models of silicate crystals can be built with small styrofoam balls, beads, and toothpicks or pipe cleaners. Oxygen is a large atom, and can be represented by a styrofoam ball; the beads can represent silicon, a small atom. The beads should be small enough to rest between the "atoms" of oxygen.
- It is more important that students learn why some minerals have particular characteristics than it is for students to memorize the names of the minerals.

11 EARTH MATERIALS

11-1 Minerals

Search for Minerals

Write the name of the mineral or property that matches each of the following definitions. Then find each name in the word search below and circle it. The words may appear across, up, down, or diagonally.

_______ 1. Characteristic of a crystal that results from the regular pattern in which its basic units line up.
_______ 2. Characteristic of a mineral that can be changed by impurities in the mineral.
_______ 3. Property of a mineral that breaks along flat surfaces.
_______ 4. Mineral used in jewelry.
_______ 5. The main ingredient of a mineral.
_______ 6. Mineral that is an important source of iron.
_______ 7. Property of minerals found by scratching one mineral with another.
_______ 8. A sample of the powdered mineral.
_______ 9. A source of mercury.
_______ 10. The way a mineral reflects light.
_______ 11. A source of copper.
_______ 12. Dark gray mineral that has a metallic luster.

```
N M A E A T F Y P O E R I S S F
I E A S H C O L O R T C A N I B
U E S T R E A K H M W G T S L A
F O O N U L A B E A A A G N I C
K S A L U M E T Z T R L A H C I
E M M M U T P S B A G E M U A N
B R K W Y S C C D H J N R F T R
I Z U L L D T R U F V A A H E I
C O L Y K C L E A V A G E D V E
S R F C E I D L R Y A M I L Y E
C Y T L S N L U T E A A V R S T
H E M A T N T E F T M O U I H T
A D N T O A H E I V E T L M E U
R L R G E B P T N F C A T D M O
D N I F I A E D B S R E H L T P
N O R O H R L I G R K H I D S C
E T I S M A L A C H I T E R V D
S N O C A N S H T S W O N E R T
S C R S T V L S O I E Y U V E F
```

Figure 11-4 *(a) Galena has a beautiful metallic luster. Note the cleavage in three directions at right angles. (b) Quartz shines like glass—it has a glassy luster.*

example, shines like a highly polished metal. It is said to have a metallic luster. Quartz has a nonmetallic, glassy luster. (See Figure 11-4.)

Another test that is used to help identify minerals is their *specific gravity*, or their "heft." If you have two mineral samples of the same size and one of them feels a lot heavier than the other, you are noticing a difference in their specific gravity. Specific gravity is measured by comparing the mass of a sample to the mass of an equal volume of water.

Some rare minerals have special combinations of beauty and hardness. They are the minerals used in jewelry, and are called *gems*. Colorless diamonds can be cleaved, cut, and polished to beautiful sparkling jewels. Emeralds are dark green samples of the mineral beryl, which can be cut and polished. Sapphires are deep blue gems that are pieces of the mineral corundum.

TRY THIS

Look at a sample of granite through a magnifier. How many different kinds of mineral crystals can you see? Try to identify them.

Checkpoint

1. What is a mineral?
2. Name at least five properties of minerals that can be used to help identify them.
3. What is the name for rocks from which metals and other elements are obtained?
4. Why may the color of a mineral and its streak be different?

Using Try This

You may wish to have students do the Try This activity after discussing silicates (see student text page 271). Depending on the sample, students should be able to see three or four different minerals in the granite: quartz, feldspar, mica (biotite or muscovite), and possibly some bits of hornblende. Quartz is transparent, white to smoky gray, shiny or glassy, and breaks irregularly. Feldspar is rose, pink, light blue, milky, or colorless; chunky; has shiny facets; and breaks or cleaves angularly. Biotite mica is black or dark brown; shiny; and appears as thin flakes or plates. Muscovite mica is transparent or slightly yellowish, shiny, and flaky. Larger samples of either kind of mica may look like "books" of thin sheets. Hornblende is dull, greenish-black, and elongated.

The same material may look very different when viewed from different angles. For example, the edge view of a piece of biotite mica (book of sheets) is different from a top view (shiny, flat). Also, cleavage and fracture surfaces of the same material may look different. Feldspar fracture surfaces may appear similar to those of milky quartz, for example, but the pearly-looking cleavage surfaces of the same feldspar grain will appear shiny when light strikes them at the proper angle.

Some of the minerals may be discolored with a coating of oxide. Such a coating makes recognition difficult; therefore, students should study fresh surfaces whenever possible.

Checkpoint Answers

1. A mineral is a nonorganic, naturally-occurring solid element or compound.
2. Crystal shape, color, streak, hardness, cleavage, luster, and specific gravity.
3. Ores.
4. Impurities in a mineral can change its color. The color of the streak is more consistent.

11-2

Rocks

Goals

1. To identify the three classes of rocks.
2. To describe the rock cycle.

There are three classes of rocks, determined by the way the rocks are formed. Most of the rock that makes up the crust is igneous rock. **Igneous rocks** form from extremely hot, melted rock material, called **magma**. Magma comes from deep within Earth. The word *igneous* means "fire-formed." The differences among igneous rocks are caused by the chemical make-up of the magma and its rate of cooling. Magma may cool and harden very slowly, very rapidly, or at various rates in between.

igneous rocks (IG nee us)

magma

If magma is forced between layers of rock in the crust without ever reaching the surface, it cools very slowly. The more slowly the magma cools, the larger the crystals are. One igneous rock formed by slow cooling is *granite*. Granite, with its crystals of various colors, is frequently used for buildings. It is very hard and does not crumble easily when exposed to wind and rain, heat and cold. (See Figure 11-5a.)

Figure 11-5 *Igneous rocks: (a) granite, (b) basalt, (c) obsidian, and (d) pumice.*

Teaching Suggestions 11-2

Before starting this section, assemble samples of sedimentary rocks (sandstone, limestone, shale, conglomerate); metamorphic rocks (quartzite, marble, slate, schist, gneiss); and igneous rocks (pumice, obsidian, granite, gabbro, basalt). Obtain reference books and field guides for student use in the classroom. Discuss the characteristics of each class of rock, showing samples of each. Students can do the second Try This activity after discussing the types of igneous rock.

List the three rock types on the chalkboard. Have students fill in the characteristics of each class (at the chalkboard, student by student, or at their desks on paper) and provide an example of each type.

Give students plenty of time to handle and study rock samples. Explain that the friability (grittiness) of sandstone is due to the fact that it was formed by particles cemented together. The glassy texture of obsidian is due to its rapid cooling. The crystalline minerals in granite are present because of the relatively long time it took the granite to cool, during which the minerals had time to develop into crystals.

When discussing the rock cycle, have students make a classification chart with the following headings:

ROCK TYPE	CAN BE CHANGED TO	UNDER THESE CONDITIONS

Encourage the students to do project 2 at the end of the chapter, if they did not do it for section 11-1.

Background

Magmas may vary considerably in their chemical composition; in fact, they do so in a well-understood sequence (Bowen's Reaction Series), which depends on the atomic constituents, temperature, and pressure. Rocks formed from magma that cooled slowly below the surface are coarse-grained intrusives. Rocks formed from magma that was ejected to the surface and cooled rapidly are fine-grained or glassy extrusives. The holes in pumice are caused by the release of gases that were trapped in the thick, or viscous, lava. Hence, pumice can be considered to be a sort of hardened foam.

The sedimentary rocks described in the text are all clastic sedimentary rocks formed from particles cemented together. Nonclastic sedimentary rocks, such as limestone and gypsum, are formed from chemicals that precipitated from solution in water.

Demonstrations

1. You can demonstrate how sandstone is formed by either of two methods. One method is to punch several small holes in the bottom of a waxed paper cup and fill it about half full of gravel or pebbles. Mix

two parts of white glue to one part of water and pour the mixture into the paper cup, catching the mixture that escapes through the holes into another cup. Pour the runoff back through the paper cup several times. Let it dry overnight and then peel away the paper cup. By the second method, you heat 30 ml of silica gel in a small metal can until bubbles form. Remove it with appropriate safety measures to prevent burns from the heat, and mix in sand until a thick granular substance is formed. Pack the mixture into a paper cup. After a day, tear the cup away. With either method, there should be definite spaces between the particles. This demonstration will highlight the concept of interstitial space, which will be of help to students later in the chapter when they learn about ground water and aquifers. Students should recognize that the formation of sedimentary rocks requires two types of materials: particles and cement. Minerals dissolved in water provide the cement for sedimentary rocks; calcite, as a dissolved carbonate, is a common example.

2. You can demonstrate how pressure, or stress, can be exerted on rock layers, causing them to deform, or strain. This suggests the effect of pressure on the deformation of rocks, and the formation of metamorphic rock. Stack three or four strips of foam rubber 40 cm long, 2 cm thick, and 7–8 cm wide, one atop the other on a smooth work surface. Place one hand on each end of these strips and press gently towards the center of the strips. The strips will arch upward from this pressure, analogous to the way rocks react to compressive stress. Emphasize that rocks may take millions of years to deform in this way. Ask the students if they have seen rock layers in this kind of arched pattern (anticline). Ask students where they would expect to find metamorphic rock in such a deformation pattern.

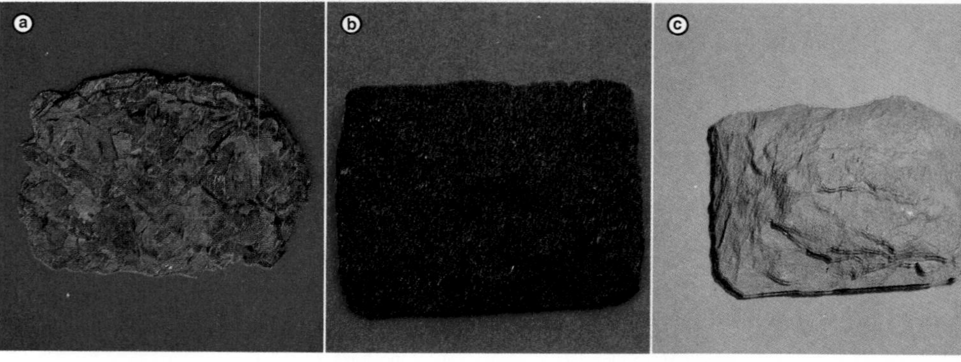

Figure 11-6 *Sedimentary rocks: (a) shale, (b) sandstone, and (c) limestone.*

lava
(LAH vuh)

sedimentary rocks

TRY THIS

Make a small batch of cooked fudge. When it is at the proper temperature, pour a small amount of boiling fudge into a cup of cold water. Pour the same amount of fudge into a small pan to allow slow cooling. When both samples are cool, observe their textures. Describe the difference. Relate it to differences in igneous rock.

Magma that comes to Earth's surface—in a volcanic eruption, for instance—is called **lava**. When lava flows onto Earth's surface, it cools fairly rapidly. Crystals form, but they don't have time to grow very large. In fact, you may need a magnifier in order to see them. *Basalt* (buh SAWLT) is an igneous rock that forms from cooling lava. It is a dark-colored, hard rock with very small crystals. (See Figure 11-5b.)

Sometimes lava cools so fast that no crystals can grow. When this happens the result is a *glass*. One rock that forms in this way is *obsidian*, a glassy rock. (See Figure 11-5c.) Some obsidian jewelry is called "Apache tears" because of its shape. Obsidian was also used for arrow points because of its hardness and sharp edges.

Another common form of igneous rock is *pumice* (PUHM is). Like obsidian, pumice cools too rapidly for crystals to form. Pumice can be considered a glass, also. Pumice is a frothy volcanic glass. It is full of holes, like foam that has become solid. Pumice is so full of holes it floats. (See Figure 11-5d.) Ground-up pumice is used in some hand soaps to help grind out dirt from soiled hands. It can be used as an abrasive also, to remove calluses from hands and feet, and to sharpen knives.

There are many types of igneous rocks besides granite, basalt, obsidian, and pumice. These four examples show how igneous rocks can differ, depending on how rapidly the molten rock cooled.

The second class of rock is the one that may be most familiar to you. **Sedimentary rocks** are rocks made of min-

(Near the axis of the anticlinal arch, where stress is the greatest.)

Using Try This

Fudge that is cooled quickly in cold water should be smooth with no crystals of sugar evident. Slowly-cooled fudge should be grainy with visible sugar crystals.

Figure 11-7 Metamorphic rocks: (a) slate, (b) quartzite, (c) marble.

eral grains that become cemented together, under pressure, forming solid rock. They usually form in water, where grains settle out in layers on the bottom of rivers, lakes, or oceans. Sometimes grains are deposited by wind, and form layers of sand in the desert.

Sedimentary rocks are known by their relative softness, their lack of crystals, and their layered appearance. Some common sedimentary rocks are *sandstone*, *shale*, and *limestone*. (See Figure 11-6.) Sandstone is made up of sand grains. Shale is made up of smaller fragments of clay and mud. Limestone is formed from the powdery remains of shells, teeth, and bones. Sedimentary rocks can be used for building materials. They do not, however, hold up under extreme weather conditions as well as igneous rocks do.

metamorphic rocks

The third class of rock making up Earth's crust is called metamorphic rock. This name means "changed" rock. Some rocks become buried deep within Earth. There large amounts of heat and pressure can change them. **Metamorphic rocks** are rocks that have been formed by changes in sedimentary or igneous rocks.

Some examples of metamorphic rock are *slate* (changed from shale), *marble* (changed from limestone), and *quartzite* (changed from sandstone). (See Figure 11-7.) These rocks are very hard and most have crystals. Many show some form of organization, such as the *banding* shown in Figure 11-8. Banding shows that similar minerals have moved and collected with others of the same kind. This happens to a rock when it gets hot enough to flow under extreme pressure.

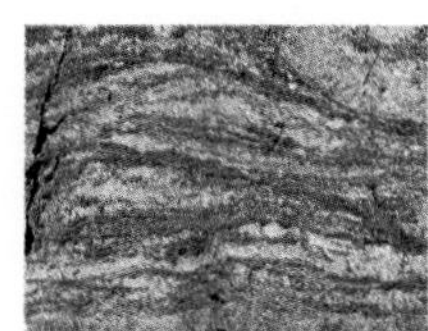

Figure 11-8 Banding in gneiss, a metamorphic rock.

Teaching Tips

- Ask the students to relate the words "ignite" and "ignition" with the word "igneous."
- Display samples of the three classes of rocks for students to study.

Facts At Your Fingertips

- Minerals are to rocks as letters of the alphabet are to words.
- Plutonic rocks are dark-colored igneous rocks formed deep underground in masses called plutons. They are named after Pluto, the Roman god of the underworld.

Mainstreaming

Visually-impaired students will benefit from feeling the textures of the different types of rocks. They can compare the glassy texture of obsidian with the frothy texture and light weight of pumice. They also may be able to distinguish coarse-grained granite from fine-grained basalt, and among conglomerate, sandstone, and limestone.

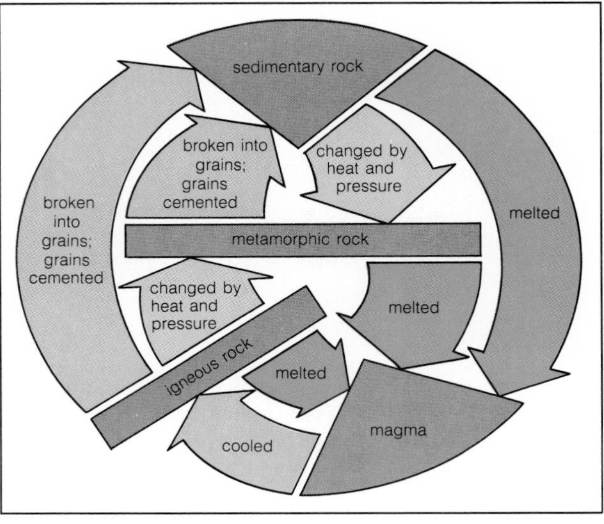

Figure 11-9 *The rock cycle. A grain of quartz from magma could become part of igneous, then sedimentary, then metamorphic rock.*

After rocks have formed, they do not remain unchanged. You have learned that metamorphic rocks can be formed from either igneous or sedimentary rocks. You may have wondered where the grains, or particles, come from that make up sedimentary rocks. They can be from igneous or metamorphic rocks, or even other sedimentary rocks. Igneous rocks can be formed from sedimentary or metamorphic rocks that become buried deep enough in the crust to form magma. The changes from one rock type to another are called the **rock cycle**. It is a never-ending process of change. The next time you pick up a rock, remember that it was not always in the form you see now. (See Figure 11-9.)

rock cycle

TRY THIS

Visit a nearby cemetery. Look for monuments and markers made of various kinds of rock. In what ways can you distinguish between granite and sandstone? limestone and marble?

Checkpoint

1. What is the difference between lava and magma?
2. How can you tell an igneous rock from a sedimentary rock?
3. If an igneous rock is glassy, what was its rate of cooling?
4. What is a metamorphic rock?
5. Describe the rock cycle.

11 EARTH MATERIALS

11-2 Rocks

Be a Rock Detective!

The table below contains clues describing rocks that make up Earth's crust. Use these clues to complete the table.

Clue	*Rock Name*	*Class*
1. The rock used for jewelry, called "Apache tears."		
2. A rock made from shells and teeth of animals.		
3. The rock that shale changes to with added heat and pressure.		
4. A rock that can float in water.		
5. A rock from which beautiful statues are made.		

Rocks Have Cycles Too!

In the sentences below, a numbered space shows that a term is missing. Find each missing term and write it on the corresponding numbered space at the left.

1. ______________
2. ______________
3. ______________
4. ______________
5. ______________
6. ______________
7. ______________
8. ______________
9. ______________
10. ______________

Rocks can change from one __1.__ to another. This change is called the rock __2.__ One example of how rocks can change is when melted rock called __3.__ is squeezed in between rock layers deep in the Earth. The heat from the melted rock changes the surrounding rock to __4.__ rock.

An example of how rocks change over long periods of time is granite, a/an __5.__ rock, changing to sandstone, a/an __6.__ rock. Weathering can wear away the minerals in granite, leaving behind particles of sand. This sand will collect on the bottom and shores of streams as it is washed down-stream. Given enough time, these layers will form __7.__

Sedimentary rock can be changed to __8.__ rock by added heat and pressure. If rocks are completely melted by the heat, the new rock will be __9.__ The characteristics of the rock will be determined by how rapidly it __10.__

Using Try This

Check to see if permission is needed for the cemetery visitation. If a cemetery is not accessible, you should be able to have students study statues or buildings that are made of rock materials, and contrast their weathering characteristics. A field guide to rocks and minerals will help students identify the different kinds of rocks.

Limestone and sandstone will show the most severe weathering. Some shales will also weather badly, although some holds up quite well. A major culprit in the severe weathering of rocks containing carbonates is acid rain. If a gravestone is very old, the markings on it may be difficult to read. Marble is harder and denser than limestone, so it won't be so badly weathered. Marble may have a crystalline, "sugary" appearance and show color banding. Granite is likely to be the most durable. Some newer grave markers are made of bronze.

Checkpoint Answers

1. Lava is molten rock that has flowed out onto the planet's surface. Magma is molten rock beneath the planet's surface.
2. An igneous rock usually shows mineral crystals (obsidian is an exception), is very hard and dense, and has no layers. A sedimentary rock is softer, less dense, has no crystals (grains are not the same as crystals), and may be layered.
3. It cooled very rapidly.
4. A metamorphic rock is formed from either igneous or sedimentary rock that has been changed by extreme heat and pressure.
5. The rock cycle is the never-ending change in rock type from one to another. A typical sequence is metamorphic or igneous to sedimentary, and back to metamorphic. Other sequences are possible.

11-3

Fossils

Goals

1. To define fossils.
2. To list conditions for the preservation of fossils and describe likely fossil locations.
3. To list some uses for fossils.

If you are interested in rock and mineral hunting, knowledge of fossils will add a new excitement to your hobby. **Fossils** are evidence of formerly-living things. Fossils can be the actual bones, teeth, or shells of animals. Or they can be indirect evidence of former life, such as footprints, trails, or imprints of shells in rocks.

fossils

Fossils are almost always found in sedimentary rocks. (See Figure 11-10.) Igneous or metamorphic rocks do not contain fossils because the heat that forms these rocks would destroy the fossils. In some rare cases fossils are found in layers of volcanic dust that have hardened to rock.

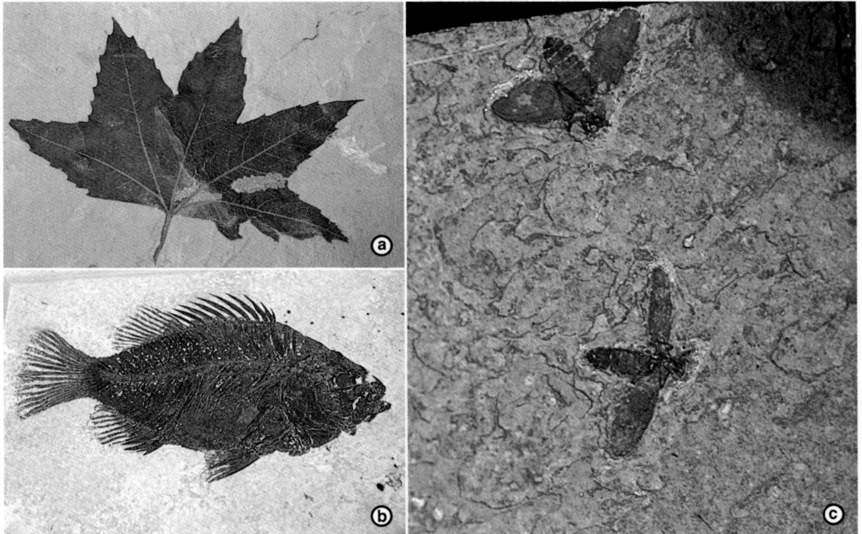

Figure 11-10 *Fossils: (a) a hardwood tree leaf; (b) a fish; (c) insects.*

Teaching Suggestions 11-3

Before starting this section, assemble as many different fossils as possible. Let students observe these fossils and the fossils shown in Figures 11-10 and 11-11. Then, ask students:

What is a fossil?

What fossils are found in this area?

How old do you think fossils are?

Accept all answers to questions and discuss. Students can do the Try This before reading the section. Students must understand the difference between casts and molds to have an idea of how various types of fossils are formed. Emphasize the age and the rarity of fossils. After discussing petrified wood, do the suggested demonstration. (See Demonstration.)

Encourage students to do project 3 at the end of the chapter. Before they do the project, however, students should obtain the proper permissions and identify prudent safety precautions.

Background

The term "fossil" applies to traces (including indirect evidence) of or preserved remains of living things. Fossils usually are of organisms that lived 10,000 or more years ago.

There are several reasons why most fossils are found in marine sedimentary rocks. Evolutionary theory suggests that life existed in the sea long before it did on land. Life always has been more abundant in the sea than on the land. Burial in marine sediments, such as mud and sand, tends to promote preservation by preventing oxidation and protecting the remains from scavengers and decomposers. In fresh water, sedimentation generally does not occur as rapidly as in the seas; consequently, freshwater fossils are less abundant than marine fossils.

On land, tar pits, often covered with water, serve as dangerous traps for unwitting, thirsty animals. In addition to the fossils mentioned in the text, fossil animals can be detected by their fossil excrement, stones from their digestive tract, or fossilized eggs.

When discussing the difference between the remains of living and nonliving things, it is usually brought up that all living things are organic (and hence contain carbon) and nonliving are not. However, plastics are made from organic compounds, while organic compounds, such as proteins and carbohydrates, can be synthesized from inorganic substances!

Demonstration

To demonstrate how a tree becomes petrified, use a dry cellulose sponge to represent the wood; the holes in the sponge represent cells. Melt enough wax to fill the holes in the sponge, and pour it over the sponge. Allow the wax to harden. The wax represents the minerals in water that fill and replace the cells of the wood.

Teaching Tips

- The paragraph on coral fossils provides a point at which to discuss the Principle of Uniformitarianism. Have pictures of representative species of each phylum on display, so that the students can compare the fossils with the morphology of living species.
- Students may enjoy researching the interesting and controversial theory that maintains that some dinosaurs were warm-blooded and agile.
- If you have enough fossil specimens, have the students classify them by phylum (see Chapter 5). This exercise is a test of their ability to identify fossils by accepted criteria.

Fact At Your Fingertips

- Fossil fuels are a source of stored solar energy that was trapped by ancient plants and locked up in the planet's crust millions of years ago.

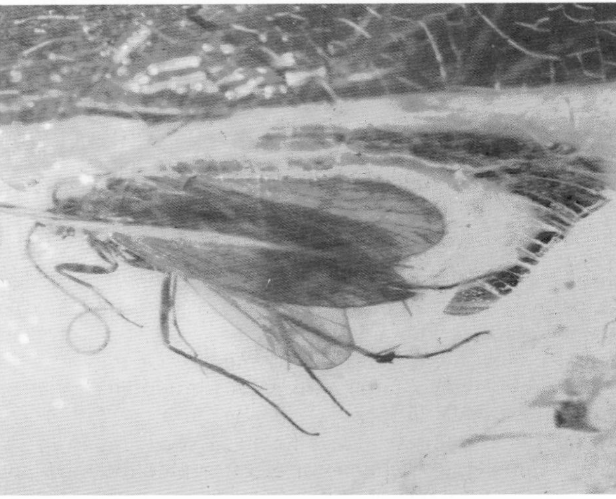

Figure 11-11 *A fossil insect in a piece of amber (hardened pine sap). It was found on the shore of the Baltic Sea and is about 30 million years old.*

There are two general conditions necessary for the formation of fossils. First, the organism should be buried rapidly, to prevent destruction by other organisms or by natural events, such as storms. Second, the organism should have hard parts, such as bones, shells, teeth, or stems, to ensure their holding up through time.

If you are going fossil hunting, look in areas where sedimentary rocks are plentiful: along the banks and cliffs near rivers, streams, or lakes. If there is a quarry nearby, you may be able to get permission from the owners to explore the area. If construction companies have been clearing land and have uncovered rocks or made cuts through sedimentary rock layers, you may find fossils there.

Fossils have been preserved in some unusual places. They have been found in frozen soil or ice, tar pits (Rancho La Brea Tar Pits, Los Angeles), and sap from trees. (See Figure 11-11.) Some organisms have become natural "mummies" from extreme drying conditions.

An unusual fossil you may have seen is *petrified wood*, or wood that has changed to stone. For wood to become like stone, a tree must be under water for a long time. The minerals dissolved in the water fill in the cells of the tree, turning the tree's structure to stone.

You are probably wondering what fossils look like, especially the ones you might find on your first expedition. Depending on your location, you are most likely to find various species of small, shelled animals that lived in water. Plant fossils are more difficult to find because they have such fragile parts. They are less likely to survive the process of being buried.

Fossils can provide valuable resources. All organisms contain carbon and hydrogen. Under certain conditions, the carbon and hydrogen in the remains of organisms combine, forming *hydrocarbons*. Hydrocarbons are the fuels in oil and natural gas. Oil and natural gas are called *fossil fuels* because they are formed from the remains of organisms. Coal is another fossil fuel. It is formed from organisms that were buried in ancient swamps.

Scientists can use fossils to learn about past environments. For instance, if coral fossils are found in a quarry, scientists know that the area must have been covered by a warm, salty sea at the time the corals were living. That is the environment needed for modern coral to survive.

Now that you know what fossils are, keep your eyes open for them. You may find them in a slab of limestone used for a sidewalk or in the foundation of a building. You can most certainly find some fossils in any nearby limestone quarry.

Checkpoint

1. What is a fossil? What conditions are needed for good fossil formation?
2. In what kind of rocks are fossils most often found?
3. How can wood become a fossil?
4. Name a fossil fuel.

TRY THIS

You can produce a fossil-like image by making an imprint of your hand or foot in modeling clay. Mix plaster of Paris and pour it into the imprint in the clay. After the plaster hardens, remove it from the clay. The plaster forms a *cast* of the "fossil" foot or hand. The imprint of the hand or foot in the clay is called a *mold* of the fossil. The cast is a three-dimensional model of the fossil.

Using Try This

The terms "mold" and "cast" are used in describing real fossils. An analogy may be useful, such as a gelatin form (the mold) and the gelatin dessert (the cast) inside the mold. Some students may confuse the term "cast" with the plaster cast used to support broken limbs. In this case, the plaster is cast around the arm, but is really an inside-out mold of the shape of the arm. This confusing example should be avoided unless raised by students.

Clean the mixing containers before the plaster sets. Warn students not to pour plaster down the drain. Have trash cans available for waste plaster.

Checkpoint Answers

1. A fossil is the actual preserved remains, or indirect evidence, of a once-living organism. The two conditions necessary for good fossil formation are rapid burial and the possession of hard parts. While soft parts may also be preserved as fossils, such a process is rare.
2. Sedimentary rock.
3. Wood can become a fossil if it becomes buried and dissolved minerals penetrate the wood and fill the cells in the wood. The minerals precipitate, and gradually replace the wood, producing a stonelike "petrified" wood.
4. Oil, natural gas, coal, or peat.

11-4

The Water Cycle

Goal

To describe the water cycle and discuss some of the variations of the cycle.

Water is a compound so familiar to all of us that we take it for granted. It covers about three quarters of Earth's surface and is abundant in the atmosphere and under the ground. You, as well as other living things, could not live long without it. In fact, water is what makes our planet unique. No other planet in the solar system has the proper range of temperature to allow water to exist in all three states: solid (ice), liquid (water), and gas (water vapor).

water cycle

The movement of water from the land to the rivers and oceans, to the atmosphere, and back to land again is called the **water cycle**. (See Figure 11-12.) This description of the water cycle is highly simplified. A number of variations of the cycle can occur. Consider, for example, a drop of water in an inland lake. This water drop could *evaporate*, or change from liquid to gas, and become part of the atmosphere as water vapor. It later could *condense*, or change from gas to liquid, as a drop in a cloud. From here the water drop could fall back to the land, and then return to the lake, completing a cycle that doesn't include the ocean.

Next, consider a water drop that evaporates off the leaf of a tree, condenses in a cloud, and falls on the ocean. In this cycle the drop doesn't enter a river before entering the ocean. Over large areas of ocean, a water drop could evaporate into the atmosphere, condense in a cloud, and fall back to the ocean. In this cycle the water drop would never reach land.

Water can remain in one place for a long period of time, thus delaying its movement through the water cycle. Imagine how long a drop of water can be trapped in the ice caps in the polar regions! On the average, a drop of water can be expected to spend about 98 out of every 100 years in the ocean. Of the remaining two years, the drop spends about 20 months as ice. It is present in lakes, ponds, and rivers for only about half a month. It spends less than a week in the atmosphere. Although only a small proportion of Earth's

Teaching Suggestions 11-4

The purpose of section 11-4 is to make students aware of the water cycle, its "paths" or patterns, and where water may be stored in nature (retained at a point in the water cycle) for extended periods of time.

Encourage discussion among the students to determine their prior knowledge of the water cycle. Use a water distilling apparatus as a model of the water cycle (see Demonstrations). Ask students:

What does distillation do to the water? (Removes impurities.)

How is this similar to the natural water cycle? (The processes of evaporation and condensation also serve to remove impurities naturally from water.)

Discuss students' responses to the questions. Have students provide examples of evaporation and condensation (of which distillation is the analogue) in the natural water cycle. Ask students why rainwater, snow, and ice are considered to be "soft" water. (They have a low mineral content.)

Encourage students to do project 4 at the end of the chapter.

Background

While studying the water cycle on Earth, it is helpful to know that the planet as a whole contains approximately 1.3 billion cubic kilometers of water. Of this amount, 97% is in the oceans, which leaves only 3% as fresh water. A little over 2% of Earth's total water is frozen in glaciers and ice caps, and is not available for uptake by living things. Of the remainder—which comprises less than 1% of water on Earth—some flows in rivers and streams; some is stored in lakes, marshes, swamps, and ponds; and some is ground water.

In the water cycle, water changes back and forth from salty to fresh. As water changes state, it loses most chemical impurities that had been in solution. This process is the basis for distillation. Conversely, one freezing process used to desalinate seawater is based on the fact that the bulk of the ice formed from a volume of salt water is free of salt.

Demonstrations

1. You or one of your students can make a drawing of the water cycle on an acetate sheet. Display the transparency with an overhead projector, and use when discussing the water cycle.
2. Set up a water distilling apparatus and demonstrate how it works.

Teaching Tips

- You may have to reinforce the concept that a given amount of water can be stored for varying periods of

Figure 11-12 *The water cycle. There are many pathways that a drop of water can follow between air, ground, and water.*

water is in circulation at any given time, there is a very large amount moving through the water cycle. Water on Earth is in constant circulation.

Checkpoint

1. Why is Earth called the "water planet"?
2. What is (a) evaporation, (b) condensation?
3. Describe the simplified water cycle and one of its variations.

TRY THIS

Study the water-cycle diagram (Figure 11-12). List as many other paths as you can for a water drop moving through the water cycle.

time at one point of the water cycle in such places as in clouds and the atmosphere, in glaciers, in the depths of the ocean, in living organisms, and in aquifers.

- Keep discussion of this section as open-ended as possible.
- If a distilling apparatus is not available, draw one on the chalkboard, explaining each component.

Facts At Your Fingertips

- Only about 6% of the water in the water cycle takes part in the direct ocean-atmosphere-land-ocean cycle.
- In the past, Mars may have had water in all three states; however, at present only Earth of all the planets in our solar system has the necessary range of temperatures to have water in all three states.
- A large oak tree may give off, or transpire, about 380 liters of water per day in warm weather.

Cross References

Information on clouds and precipitation is found in Chapter 13, section 13-4. Information on ground water is found in the next section of this chapter.

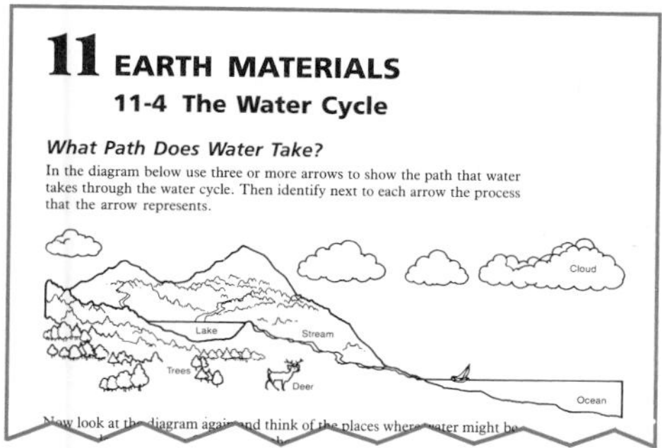
11 EARTH MATERIALS

11-4 The Water Cycle

What Path Does Water Take?

In the diagram below use three or more arrows to show the path that water takes through the water cycle. Then identify next to each arrow the process that the arrow represents.

Using Try This

There is virtually no limit to the number of paths in the water cycle. Accept all that seem possible.

Checkpoint Answers

1. Almost three fourths of Earth is covered with water. No other known planet has water in all three states.
2. (a) Evaporation is the change from liquid to gas state; (b) Condensation is the change from gas to liquid state.
3. The water cycle is the movement of water from the land to rivers and oceans, to the atmosphere, and back to land again. Variations could include from the land directly to atmosphere (cloud) and back to land, land to cloud to lake, ocean to atmosphere and back, and others.

Teaching Suggestions 11-5

To stimulate interest, question students about their drinking water:

Where does your drinking water come from? (Rain, reservoirs, and springs are some possible answers.)

How does the water get from its source into your homes? (through water mains and pipes)

Is it treated before it is distributed? (Answers will vary.)

How is the supply of usable water limited? (Some is contaminated.)

Discuss students' responses to the questions. Talk about the kinds of contaminants that can get into the water table. For example, contaminants may include chemicals dumped by manufacturing companies; chemicals spread on lawns or farm fields; waste materials from farm animals or human septic systems; chemicals leached from landfills or dump sites.

Do the porosity demonstration. (See Demonstration.) Ask the students what type of soil would have the best drainage (soil containing coarse sand and gravel); the worst (soil rich in clay).

Discuss methods that could be used to remove salt from seawater. Students can do the two Try This activities at the end of this section.

Drinking water is a topic with many potential areas of interest for students. You might want to discuss some of the following topics or suggest them to interested students for research:

1. Fights over water rights in the American West have served as a frequent movie theme. Why did fights arise?
2. Aquifer depletion and pollution in North America are topics of current concern. Find out why and what is being done about these problems.
3. Research is being done to develop food crops that are well adapted to brackish, salty water. Find out about this research.

Encourage students to do projects 1, 7, and 9 while working on this section.

11-5

Drinking Water

Goal

To discuss drinking water and its sources.

You might think that with all the ocean water on Earth there would be plenty of water for everyone. However, ocean water is salty, and people cannot safely drink salt water. There are two main sources of drinking water on Earth. One is the large bodies of fresh water on the surface, such as lakes and rivers. The other is water stored in the ground.

In inland areas that are short of water, dams may be built on rivers to form reservoirs (REZ ur vwarz). *Reservoirs* are artificial lakes that supply water for large and, in some cases,

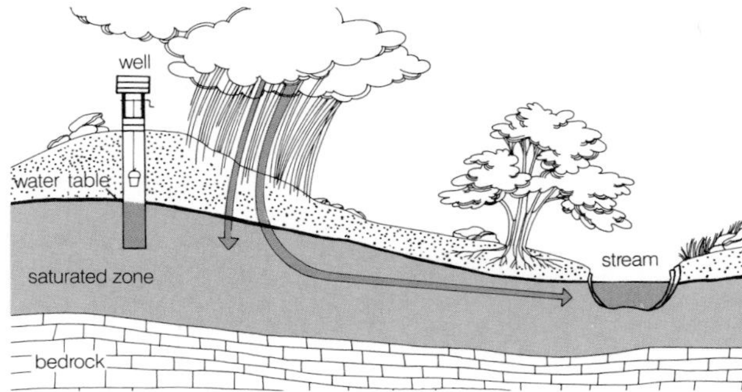

Figure 11-13 *Ground water is present in the saturated zone. The upper boundary of the saturated zone is known as the water table.*

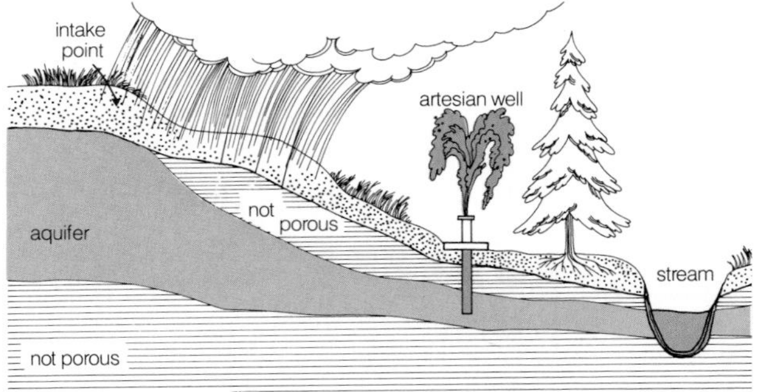

Figure 11-14 *An artesian well works because its top is below the level of the water table farther up the hill.*

Background

When people drink salt water, the fresh water in their body cells diffuses through the cell membranes toward the salt water. Too much salt water can lead to severe dehydration and could cause death.

Clay particles are smaller than sand grains, which are smaller than pebbles of gravel. A volume of gravel would have the most porosity, and would allow more water to pass through the volume than would similar volumes of finer-grained materials. It is undesirable to have fine-grained clay around the foundation of a house because clay retains water.

Aquifers can filter suspended materials from the water, but usually cannot filter dissolved materials. Polluted aquifers are hard to clean.

distant areas. In some places huge pipes called *aqueducts* (AK wuh duhktz) carry water from reservoirs across mountains to supply water for desert areas or for large cities.

Communities that don't have large bodies of surface water nearby obtain their water from the ground. When it rains, some of the water soaks into the ground, as if the ground were a sponge. This water, called *ground water*, fills the spaces between soil particles and between grains of sedimentary rock. The part of the ground that contains ground water is called the **saturated zone**. The irregular upper boundary of the saturated zone is called the **water table**. The depth of the water table determines how deep a well must be drilled to reach water. (See Figure 11-13.)

saturated zone

water table

Large volumes of ground water are stored in layers of rock called **aquifers**. To be a good aquifer, the rock must have openings, or *pore spaces*, within it. The pore spaces can be filled with air or water. Sandstone is a porous rock that makes a good aquifer.

aquifers
(AK wuh furz)

In some places an aquifer may be sandwiched between two layers of less porous rock. These keep the water trapped between them. If these layers are on a slope, the water in the aquifer builds up pressure near the lower end. When a well is sunk there, the water spurts up naturally, forming an *artesian well* (ar TEE zhun). (See Figure 11-14.)

You may have heard of a stream that is particularly clear and cold because it is "spring-fed." *Springs* (see Figure 11-15) are places where ground water flows naturally out to the

Figure 11-15 *Seven Veils Falls in Yoho National Park, British Columbia. When water flowing through porous material reaches a rock layer that is not porous, the water is forced to the surface, forming a spring.*

Acid rain, a matter of recent concern, is caused by industrial and automobile pollution from sulfur dioxide and nitrogen gases. Acid rain can pass through the soil, get into wells, and corrode pipes that supply water to homes. The acid can leach out metals from pipes, rocks, and soil. Some of these metals are toxic, such as lead, mercury, and copper.

Demonstration

Put the finest sand into one cylinder, medium-grain size in a second, and coarsest in a third, filling each cylinder to the same level. Carefully add measured amounts of water to each cylinder until the water just covers the sand. Record the amounts of water. The sand taking the most water contains the largest pore spaces.

11 EARTH MATERIALS

Laboratory

Porosity and Permeability

Purpose To observe the effects of grain size on the flow of water through the soil.

Materials beads of 3 different sizes, 2 beakers, cap or rubber stopper with drain hose, 100-mL graduated cylinder, hoseclamp or hosecock, plastic column, ringstand and clamp, screen to hold back beads

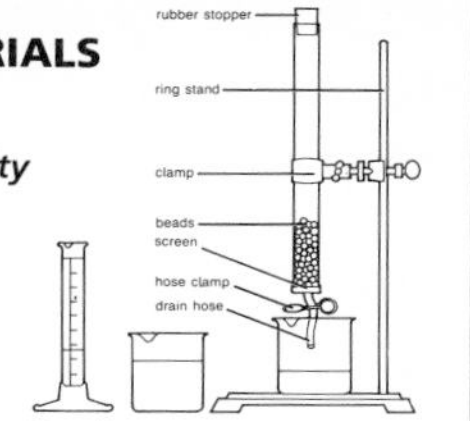

Procedure

1. Set up a column as shown in the diagram.
2. Fill the 100-mL graduated cylinder to the 100-mL mark with one size of beads.
3. Place the beads in the plastic column. Make sure the wire screen is in place to prevent the beads from running out.
4. Using a measured amount of water in the graduated cylinder, pour enough water into the plastic column to just cover the beads. Record this amount. What percent is it of the total volume, or 100 mL, of beads? This represents the percentage of space called porosity.
5. Open the clamp and allow the water to run out. Record the amount of water retained by the beads.
6. Add 300 mL of water to the plastic cylinder that holds the beads. Record the time required for all the water to drain through. This measures the permeability, or the rate at which water can pass through a porous material.
7. Repeat each of the previous steps two more times using the other two sizes of beads.

Discussion

1. How did the size of the beads affect the porosity? ______
2. How did the size of the beads affect the retention of water in the column after draining? ______
3. How did the size of the beads affect the permeability? ______

Teaching Tips

- Point out to the students that drinking water has been a primary need of all civilizations. The Romans built remarkable, extensive systems of aqueducts, many of which still exist.
- Plan a field trip to your local water pumping and treatment plant.
- Invite a speaker from a municipal or county agency that monitors water quality for your community.

Facts At Your Fingertips

- Icebergs are a possible source of fresh water for coastal desert areas. Between 1890 and 1900 several small icebergs were towed by sailing ships from southern Chile to Peru, a distance of about 3840 km. Recently, a similar plan was considered by drought-stricken Australia, but was rejected because of cost.
- Plans to tow huge icebergs have been discussed by leaders of oil-rich, but water-poor, Middle Eastern nations. Their engineers have calculated that it could be cheaper to tow icebergs than to remove the salt from similar quantities of seawater.

Cross References

For more information on salt water, see Chapter 14, section 14-1. See Chapter 9, section 9-5, to review acids.

11 EARTH MATERIALS

11-5 Drinking Water

Water, Water, Everywhere!

On the space beside each term in Column A, write the letter of its description in Column B.

A	B
____ 1. reservoirs	a. Artificial lakes used as a source of water.
____ 2. aqueducts	b. The upper boundary of the zone that contains ground water.
____ 3. ground water	c. Tiny holes in underground rocks that can contain air or water.
____ 4. water table	d. Process of removing salt from water.
____ 5. pore spaces	e. Water that fills the spaces between soil particles.
____ 6. desalinization	f. Pipes that carry water long distances.

Water Treatment

Read the paragraphs below. Using complete sentences, answer the questions that follow.

Have you ever heard someone say, "The water in our town is so hard that we must use more soap than usual"? Hard water contains dissolved minerals that have been picked up by the water as it moves through the ground. Since some areas have more minerals in the soil than others, the water in those areas contains more dissolved minerals. In an area near an iron mine, for example, the water would also contain iron. The iron would then leave yellow or brown stains on plumbing fixtures and even clog up pipes.

When water containing dissolved carbon dioxide runs through limestone, calcium and magnesium ions are set free. To soften this water, the calcium and magnesium ions must be removed. One method is to pass hard water through a mineral called zeolite. Zeolite absorbs the calcium and magnesium ions. This process "softens" the water so that soap will dissolve more easily.

1. Why does soap not dissolve easily in hard water? ____________
2. What evidence shows that the water in an area is hard? ____________
3. How can water in the soil contain dissolved carbon dioxide? ____________
4. Where do the calcium and magnesium ions come from in the example given? ____________

Using Try This

These two experiments show that when water evaporates, it leaves behind materials that it had in solution.

1. The salt should have settled out on the inside of the carton. Have students use the taste test for salt. The central mass of the ice will be free of salt. Have students wash the ice with tap water and taste the ice as it melts and dwindles in size.
2. After the water has evaporated, a crust of salt and other minerals from the water will be left in the jar.

TRY THIS

Stir up a concentrated salt solution in an empty paper milk carton. Put it in a freezer. When it has frozen solid, cut away the carton. Where do you find the salt?

surface. They are usually found on the sides of hills, where the saturated zone meets a nonporous rock layer. The water has nowhere to go but out.

Two special kinds of springs are mineral springs and warm springs. *Mineral springs* occur when water dissolves minerals from the ground as it flows through. *Warm springs* occur when ground water passes near hot rock. The hot rock can be far below the ground, or closer to the surface. A good example of the latter case is found in Yellowstone Park. There the hot layer of rock is close to the surface, and there are many warm springs.

Besides surface water and ground water, there is another source of drinking water. Some coastal cities are making use of ocean water. They must remove the salt from the water, however. This process is called desalination or, sometimes, desalinization.

Any water used for drinking purposes must not only be free of salt, but should also be free of foreign matter. Bacteria from sewage, harmful chemicals, such as fertilizers and highway salt, and waste products from industry are examples of foreign matter. Suppose a water supply comes from a sandstone aquifer. If the water has been in the aquifer long enough, it is cleansed by the filtering action of the fine-grained sandstone.

Various government agencies set the standards for drinking water. They conduct frequent tests of water supplies to ensure good quality. Chlorine can be added to water to kill bacteria. In many communities, fluorides are added to protect people from tooth decay. These additions are made by local water departments.

TRY THIS

Stir up a salt solution in a glass jar. Let it stand uncovered for several days until the water has evaporated. Has the salt evaporated?

Checkpoint

1. What are the two main sources of drinking water on Earth?
2. What is the difference between a spring and an artesian well?
3. Define (a) saturation zone, (b) water table.
4. What is the difference between an aquifer and an aqueduct?

Checkpoint Answers

1. Large bodies of fresh surface water and ground water.
2. A spring is a place where water flows out of the ground because it cannot penetrate an underlying nonporous rock or sedimentary layer. An artesian well may be found originating from a sloping aquifer that has nonporous rock both above and below it.
3. (a) The saturation zone is the part of the ground beneath the water table, which contains ground water. (b) The water table is the upper boundary of the saturated zone.
4. An aquifer is a porous rock layer that carries ground water; an aqueduct is a large pipe or other human-made conduit that carries water.

Minerals and Modern Technology

Much of modern technology depends on metals extracted from ores. At one time iron and copper were used extensively. But as technology has changed, so has the need for different metals. To increase gasoline mileage, automobile manufacturers need to use structural materials that are lighter in weight than those formerly used. Jet engines and rocket motors need metals that can withstand great heat and stress.

Unfortunately, many of the metals needed for the new technologies are not readily available in North America. Because of their importance they must be imported. These metals—niobium, manganese, tantalum, cobalt, chromium, platinum, and titanium—are known as *strategic metals.* Niobium, manganese, and chromium are used in steel production. Cobalt and titanium are used in jet engines. Strategic metals are needed in the aerospace, automobile, electronic, and chemical industries, also. The prices of these metals are rising at an ever-increasing rate.

A number of solutions to the strategic-metal problem have been offered. For one thing, recycling can be increased, in order to make more use of the strategic metals already purchased. New areas can be mined—the ocean floor, for instance, is a source of minerals. More easily available metals can be substituted for some strategic metals (for example, nickel instead of cobalt).

One of the best ways to reduce dependence on imported metals is to find substitutes. Ceramics (materials made from clay or similar minerals), plastics, and glass are being studied to find new ways of using them. Technology may have to develop in entirely new directions to make use of these materials. Who knows what jet engines of the future will be made of?

Using The Feature Article

During the study of section 11-1, have the students read the feature article. For enrichment, the following topics could be researched and used for class discussion.

1. List the technological benefits to society that have come from materials development in the space program. Are there disadvantages to society from such programs?
2. What materials are used for building space vehicles? What advantages and disadvantages do these new materials have?
3. What problems are encountered in space because of radiation and thermal energy?
4. Where do strategic materials come from? Why are they located where they are?
5. What are the advantages and disadvantages of strategic metals? What metals are being recycled?
6. What strategic materials have been found on the ocean floor? What international law, if any, regulates mining of such resources?
7. Should mining be allowed in national parks or forest preserves? Find out what mineral resources exist in the parks or forests in your state or province and how these resources are utilized, conserved, or protected.
8. What materials are being used in automobiles and airplanes to make them lighter and more fuel efficient? How else may such materials make cars and airplanes more energy efficient?
9. Find other examples of new materials replacing traditional ones. Are the new materials always better? Support your answers with specific examples.
10. Why are the strategic metals niobium, molybdenum, titanium, manganese, and chromium used in making alloys of steel? Could they be replaced by anything else?
11. Make a list of modern alloys. What are their constituents and uses?
12. Materials derived from coal and petroleum are being developed as substitutes for strategic materials. Find out what these new materials are. Describe how a shortage of coal or petroleum could affect the supply and cost of materials derived from them.

For information, the students could consult library resources, local industrialists, the state or province natural history office, United States or Canadian geologic survey regional office, or earth science instructional staff at a local university. Presentations could be made in the format of panels, individual reports, mock trials, debates, or skits.

11-6

ENRICHMENT

Geologic and Topographic Maps

Goal

To describe features and uses of geologic and topographic maps.

There are many kinds of maps. You have probably seen maps on classroom walls, for example. They generally are *political* maps that show such things as countries, states, provinces, and major cities. They may also show physical features such as rivers, lakes, mountains, and deserts. You have probably seen road maps, too. They show highways, cities, and other information useful to motorists.

geologic map

Earth scientists use special kinds of maps to help them study Earth. A **geologic map** shows the various kinds of rock found at the surface or under the soil in a particular area. (See Figure 11-16.) Geologists can learn a lot about an area from a geologic map. For instance, if there are many fine-grained igneous rocks, there may have been volcanic activ-

Figure 11-16 *The colors on this geologic map represent rocks of different kinds and ages found in northwestern Wyoming.*

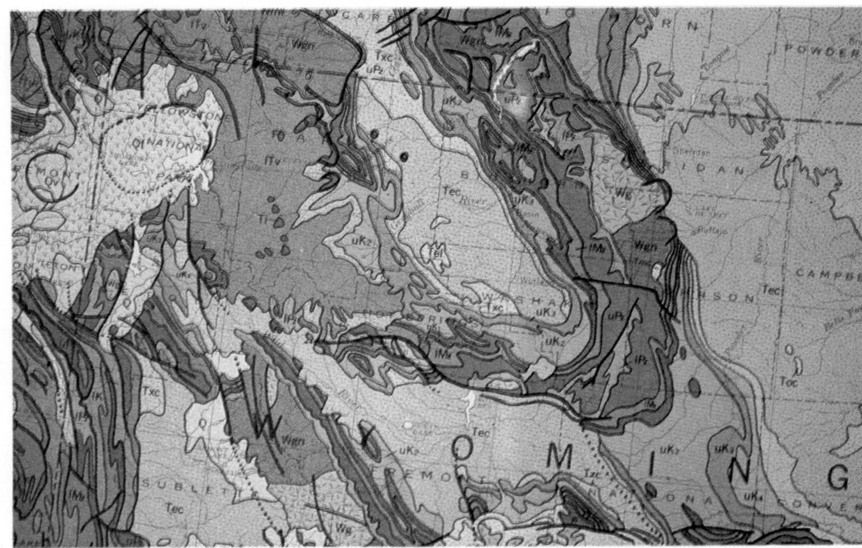

Teaching Suggestions 11-6

Acquire the maps needed for display well in advance of teaching this section. When you assign the reading of this section, emphasize the importance of studying the maps. Be sure all students understand latitude, longitude, declination, and scale notations. Have students find out the latitude and longitude of your city or town, as well as local declination. Discuss the rationale for contour lines. Ask students about the kind of terrain where contour lines would be close together (steeply sloping), or far apart (gently sloping). Have students do the Try This activity on student text page 289. Also, some students may wish to do end-of-chapter projects 8 and 10.

Background

The shorter the horizontal distance between contours, the steeper the slope represented. Except when an overhanging cliff is represented, contour lines never cross. Contour lines generally close on themselves except when a plateau or seamount is represented. Typically, contour lines enclose an area that is higher than its own value. An exception is a contour line that rings a depression. V-shaped contour lines that cross streams point upstream.

Demonstration

The concept of contour lines can be demonstrated by using a plastic shoe box, a plastic or plasticene model of a mountain that fits inside the box, water, and a centimeter scale. Place the model mountain in the shoe box. Attach the scale vertically on the side of the box. The markings on the scale represent contour intervals. Add water to the 2-cm mark. With a crayon, trace the water level on the model mountain. Add water to the 3-cm mark and trace that water level on the mountain also. Repeat this procedure until the mountain is submerged. Pour out the water. Cover the box with a stiff sheet of clear plastic. Looking straight down on the mountain model, trace onto the box cover with a crayon or marker the lines that show the various water levels. The result of this will be a topographic map of the mountain.

Teaching Tips

- Advanced scouts in the class may prove to be a useful teaching resource for this section. Map reading is a highly emphasized skill in scouting, often utilized in "orienteering" exercises.
- Check your film list for a good film on maps and map reading.

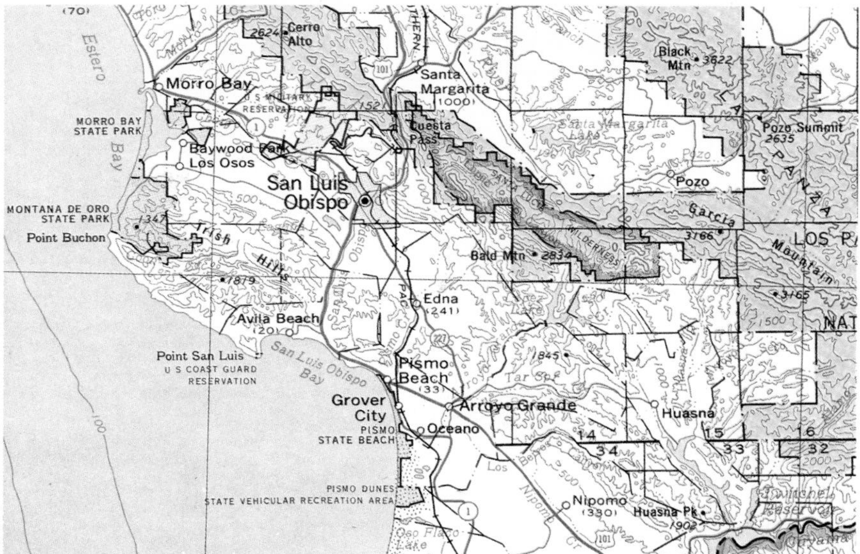

ity close by at some time in the past. If there are layers of limestone, the area must have been covered by an ocean at least once in the past. Also, knowing what kinds of rocks are found in an area is the first step in locating natural resources, such as mineral ores.

Topographic maps have a special pattern of lines to show the shape of the land surface. Points that have the same elevation above sea level are connected by lines called *contour lines*. (See Figure 11-17.) If the contour lines are close together, the land surface rises steeply. If they are far apart, the land is fairly flat. Landmarks such as schools, quarries, and radio towers are identified on topographic maps.

Figure 11-17 *A topographic map of a part of southern California. It shows land contours, altitudes, mountains, lakes, streams, highways, and towns.*

topographic maps

Checkpoint

1. What is the name of the map that shows kinds and ages of rock layers?
2. What kind of map uses contour lines?
3. What information may a geologic map provide?

TRY THIS

Using a topographic map for your area, find the elevation of several familiar places. What is the elevation of the highest point shown on the map? of the lowest?

Cross References

Addresses for obtaining topographic and geologic maps are found on page T39 of the Teacher's Edition. Also, for more information about maps, see Appendix E, page 578 of the student text.

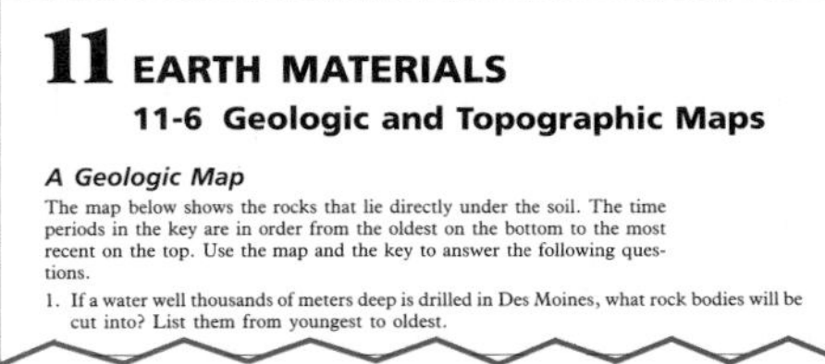

11 EARTH MATERIALS

11-6 Geologic and Topographic Maps

A Geologic Map

The map below shows the rocks that lie directly under the soil. The time periods in the key are in order from the oldest on the bottom to the most recent on the top. Use the map and the key to answer the following questions.

1. If a water well thousands of meters deep is drilled in Des Moines, what rock bodies will be cut into? List them from youngest to oldest.

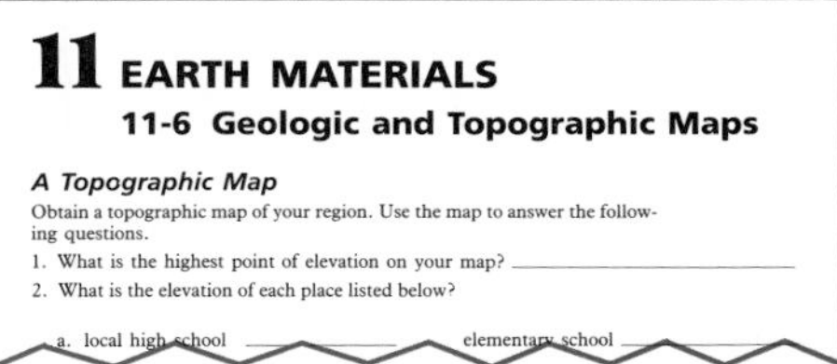

11 EARTH MATERIALS

11-6 Geologic and Topographic Maps

A Topographic Map

Obtain a topographic map of your region. Use the map to answer the following questions.

1. What is the highest point of elevation on your map? ____________
2. What is the elevation of each place listed below?

a. local high school ____________ elementary school ____________

Using Try This

Assign students the task of finding the elevations of familiar places: home, school, supermarket, favorite hangout, and so on. This exercise provides practice in reading contour lines, and actual field experience in applying the concepts of this section.

Checkpoint Answers

1. A geologic map.
2. A topographic map.
3. A geologic map will provide information about the various kind of rock layers found at or beneath the surface of the ground. It may also show such features as roadways, railroads, quarries, mines, swamps and marshes, location of benchmarks, declination, scale, streams and bodies of water, and elevations. This kind of information helps geologists to learn what an area was like in the past, and may also help in locating natural resources.

Facts At Your Fingertips

- Most maps are two-dimensional models of the world.
- Some maps, such as holographic maps, are three-dimensional.
- The Greenwich meridian was established in 1884, so that the International Date Line would fall on a sea, and not affect people on land.

Mainstreaming

Raised relief maps and topographic models are readily available and will help handicapped students better grasp the concepts of this section. Relief globes of the world would be particularly useful.

Reading Suggestions

For The Teacher

Chesterman, Charles W. AUDUBON SOCIETY FIELD GUIDE TO NORTH AMERICAN ROCKS AND MINERALS. New York: Knopf, 1979. The physical properties of minerals are described and the origin, classification, and texture of rocks are discussed.

Skinner, Brian J., ed. EARTH'S HISTORY, STRUCTURE, AND MATERIALS. Los Altos, CA: William Kaufmann, 1980. A collection of readings from *American Scientist,* dealing with Earth as a planet, geologic time, plate tectonics, and igneous activity. Well-illustrated.

Mann, Charles. "El Chichón." *Science Digest.* 91 (May 1983): 76–83. The incredible story of the eruption of the Mexican volcano in 1982 and its effect on global climate.

For The Student

Branley, Franklyn M. WATER FOR THE WORLD. New York: Crowell, 1982. Full of information about water, the water cycle, sources of water, and its pollution by humankind.

Jeffrey, David. "Annals of Life Written in Rock: Fossils." *National Geographic,* August 1985, pp. 182–191. Excellent pictures of trilobites, crinoids, fish, and dinosaur eggs.

McGowen, Tom. ALBUM OF ROCKS AND MINERALS. Chicago: Rand McNally, 1981. Illustrates more than 75 specimens in full-color and gives detailed information on where they may be found, their properties, and how they are used.

Nixon, Hershell H. and Joan Lowry Nixon. GLACIERS: NATURE'S FROZEN RIVERS. New York: Dodd, Mead, 1980. This book describes the physical features and locations of glaciers. Specialized terms are clearly defined. It is filled with beautiful illustrations.

Understanding The Chapter

11

Main Ideas

1. The outermost solid layer of Earth is made of rocks. This layer is known as Earth's crust.
2. Solid elements and compounds that occur naturally in Earth's crust are known as minerals.
3. Most of the rocks found in Earth's crust are mixtures of minerals.
4. Minerals can be identified from properties such as crystal shape, color, hardness, specific gravity, streak, and luster.
5. Rocks are classified into three groups: igneous, sedimentary, and metamorphic.
6. Fossils are evidence of formerly-living organisms.
7. Fossils can provide valuable clues to ancient climates. They are a source of oil, natural gas, and coal, also.
8. Water is one of the most important substances on Earth.
9. The movement of water from the land to the rivers and oceans, to the atmosphere, and back to land again is called the water cycle.
10. Even though water is abundant on Earth, it is a problem to find enough good drinking water.
11. Drinking water comes from sources on the surface and from under the ground.
12. Geologic and topographic maps are specialized maps that are used by scientists.

Vocabulary Review

From the following list, choose the term that best completes each of the statements. Write your answers on a separate piece of paper.

aquifers	metamorphic rock
crust	minerals
crystals	rock cycle
fossils	saturated zone
geologic map	sedimentary rock
igneous rock	topographic map
lava	water cycle
magma	water table

1. Most rocks are mixtures of _?_.
2. Earth's outermost solid layer is the _?_.
3. The shape of its _?_ is a clue that can help to identify a mineral.
4. Melted rock that is under the surface is called _?_.
5. _?_ has been made by the changing of other kinds of rock under great heat and pressure.
6. Basalt, granite, and obsidian are examples of _?_.
7. The changes in rock from one form to another are called the _?_.
8. _?_ are evidence of formerly-living things.
9. The _?_ describes the way in which a drop of water may move about on Earth.
10. Porous rock layers that store ground water are called _?_.
11. The depth of the _?_ determines how deep a water well must be drilled.

Vocabulary Review Answers

1. minerals
2. crust
3. crystals
4. magma
5. metamorphic rock
6. igneous rock
7. rock cycle
8. fossils
9. water cycle
10. aquifers
11. water table
12. geologic map
13. saturated zone
14. sedimentary rock

12. A __?__ shows the kinds of rock in an area.
13. The part of the ground that contains ground water is known as the __?__.
14. __?__ is usually formed in bodies of water, where mineral grains settle out in layers on the bottom.

Chapter Review

Write your answers on a separate piece of paper.

Know The Facts

1. Most of the rock that makes up Earth's crust is __?__. (sedimentary, igneous, metamorphic)
2. The size of the crystals in igneous rock depends chiefly on the __?__. (minerals, rocks, cooling rate)
3. Rocks are classified by __?__. (their color, their size, the way they were formed)
4. Luster is a test for a mineral. It describes __?__. (its texture, its hardness, how it reflects light)
5. Fossils are usually found in __?__ rock. (sedimentary, igneous, metamorphic)
6. Probably the most important compound on Earth is __?__. (salt, silica, water)
7. A large pipe that carries water is called a(n) __?__. (aquifer, aqueduct, water table)
8. An artificial lake that supplies water to large areas is a(n) __?__ . (reservoir, artesian well, aqueduct)
9. You could find the elevation of your home on a __?__ map. (political, geologic, topographic)
10. When water changes from gas to liquid, the process is called __?__. (condensation, saturation, evaporation)
11. Earth is an unusual planet because it has so much __?__. (atmosphere, water, igneous rock)
12. A chemical used to kill bacteria in water supplies is __?__. (chlorine, alcohol, salt)
13. Yellowstone Park has many bubbling hot springs because __?__. (it has a warm climate, there is sulfur in the ground, hot rock is close to the surface)
14. About __?__ of Earth is covered by water. (¼, ¾, ½)

Understand The Concepts

15. How can igneous rock be changed into (a) sedimentary rock, (b) metamorphic rock?
16. If an igneous rock cools slowly, how will this affect its characteristics?
17. Name two typical characteristics of sedimentary rocks that are different from those of metamorphic rocks.
18. Which would have a higher specific gravity: marble or pumice?
19. Why would you not bother to look for fossils in a granite quarry?
20. How might you use a topographic map to look for a spring?
21. Since the total amount of water on Earth doesn't change, why do you hear of water shortages?
22. Why could icebergs be used as a source of fresh water?
23. Explain how it might be possible for some igneous rocks to contain fossils.
24. Explain why the streak test for a

Chapter Review Answers

Know The Facts

1. igneous
2. cooling rate
3. the way they were formed
4. how it reflects light
5. sedimentary
6. water
7. aqueduct
8. reservoir
9. topographic
10. condensation
11. water
12. chlorine
13. hot rock is close to the surface
14. ¾

Understand The Concepts

15. (a) Particles from igneous rocks can be cemented together to form sedimentary rock. (b) Igneous rock can be changed into metamorphic rock by prolonged contact with extreme heat and pressure within Earth's crust.
16. Slow cooling allows larger crystals to form.
17. Sedimentary rocks typically are softer and have layers.
18. Marble would have a higher specific gravity than pumice.
19. Granite is an igneous rock formed from magma. The heat of the magma would have destroyed any traces or remains of living things. Therefore, no fossils would be found in granite specimens.
20. Springs usually occur on the sides of hills, so look for close-together contour lines to find a hill, and upward-pointing V-shaped contour lines on the hillside to find a stream. The origin of the stream is probably a spring.
21. The density of the populations in given regions changes over time; similarly, the size of local water resources changes, often due to the effects of weather, even though the global water resource remains reasonably constant. Streams and rivers may migrate, affecting the recharging of aquifers. Some industrial activities may remove significant quantities of water from the environment for extended periods of time.
22. Icebergs are largely fresh water. Towing icebergs may be less expensive than desalinization.
23. Some igneous rocks contain chunks of original "country rock," which may contain fossils.

24. Impurities can change a mineral's color; the color of the streak is more consistent.
25. Granite is harder, more durable.
26. The water table is the upper boundary of the saturated zone. A well must be drilled deep enough to penetrate the water table.
27. The powder formed by some minerals is colorless; some minerals may be harder than the streak plate.
28. When cut and polished, gemstones refract and reflect light in ways that humans find attractive.

Challenge Your Understanding

29. The crystal shape results from the atomic structure of the crystal. Crystals of the same minerals have the same atomic structure.
30. Animal waste could pollute the well.
31. Rivers are replenished by ground water.
32. Oxygen combines readily with other elements to form stable compounds in the crust.
33. The water cycle provides widespread distribution of water throughout the world, so that it is available for living things to use.
34. Most crystals grow very slowly because their atoms take time to group with other atoms. Usually, the slower the cooling, the larger the crystals.
35. Layers of metamorphic and igneous rocks can be formed during successive episodes of metamorphic and igneous activity. Metamorphic rock layers formed from original sedimentary rock will sometimes preserve the relative layering of the original rock.

mineral is a better indicator of the mineral's true color than is its appearance.

25. Explain why granite may be preferable to sandstone as a building material.

26. What is the water table? Why is the water table in a given area important to a person who wants to drill a water well?

27. Why do some minerals fail to produce a streak when they are subjected to a streak test?

28. What special characteristics do gems have that other minerals do not have?

Challenge Your Understanding

29. Why do all crystals of the same mineral have the same shape?

30. Why would you not want a well located on the downhill side of a barnyard where animals are kept?

31. Why don't rivers stop flowing when it stops raining?

32. Oxygen is one of the gases present in the air. It is one of the most abundant elements in Earth's solid crust, as well. What property of oxygen do you think accounts for its abundance in the crust?

33. How does the water cycle benefit all living things?

34. Why do you think a material that forms crystals forms larger ones when it cools slowly?

35. In addition to sedimentary rocks, layers are sometimes found in metamorphic and igneous rocks. In what ways might such layers be formed?

Projects

1. Study the water supply for your community. Where does the water come from? How is it protected from contamination? What is done to it before it is sent to the public?

2. Collect rocks and minerals. Put them in a display with proper identification.

3. Collect fossils from your area. Identify them using a key obtained from a library.

4. List all the ways you use or depend on water for recreation.

5. Grow crystals using a sugar solution. Get the instructions from your teacher or the library.

6. Make a model of a geyser, which is a spring that throws hot water into the air from time to time.

7. Obtain a geologic map for your area. Write a history of its geologic past. Consult experts, if possible, to check your accuracy.

8. People have increased their use of bottled water. What are people assuming when they buy bottled water? Find out if there are regulations that control the quality of bottled water.

9. Obtain a topographic map for your area. Use it to construct a model landscape out of plaster of Paris, clay, or a similar substance. Reduce the scale of the map. For example, you could let one millimeter in your model represent one meter on the map.

10. Research and report on the contributions to geology of any one of the following scientists: Florence Bascom; Alice Wilson; Inge Lehmann.

11 EARTH MATERIALS
Projects

1. Visit, call, or write your local water purification plant to find the following information.
   a. What is the source of your water supply? ______
   b. What tests are made on the water? ______
   c. What are the standards for purity? ______
   d. What additives are put into the water? Why? ______
2. Contact two or three local industries, such as textile mills, manufacturing plants, and power plants. Find out how water is used in manufacturing their products.
   a. How much water is required? ______
   b. Does the industry use city water supplies? ______ If not, what is the source?

11 EARTH MATERIALS
Challenge

1. In the table below, either the name of the gemstone or a description of its mineral composition is missing. The hardness is also missing. Use reference books to find the missing information. Then complete the table.

Gemstone	*Hardness*	*Mineral Composition*
1. Diamond		
2.		beryl
3.		corundum
4. Topaz		
5.		conchiolin
6. Garnet		
7.		quartz
8. Moonstone		
9. Turquoise		

Chapter 12

Earth's Changing Crust

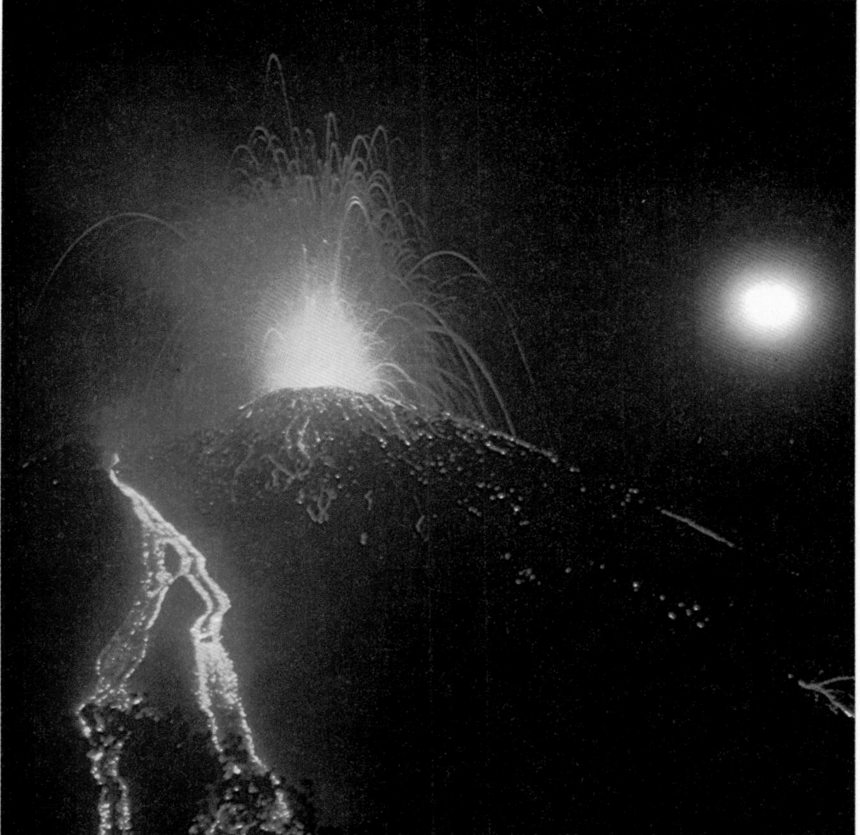

The eruption of Mt. Fuego, Guatemala, is an example of a rapid, dramatic change in Earth's crust.

Vocabulary Preview

weathering
soil
erosion
glacier
deposits
meanders
deltas
moraines
mid-ocean ridges
inner core
outer core
mantle
plate tectonics
earthquake
fault
focus
magnitude
intensity
seismograph
P waves
S waves
epicenter
volcano
crater
cone

CHAPTER 12 Overview

This chapter demonstrates that Earth is constantly changing. Minor changes to its crust due to weathering and erosion occur over tens or hundreds of years; major changes can take place over millions of years. But changes caused by earthquake and volcanic activity can occur in minutes or hours. The weathering of rock materials, a slow process that helps form soil, is discussed early in the chapter. Students learn how soil forms and why soil types differ from one region to another. The agents of erosion—water, wind, temperature, and glaciers—are discussed. Students also learn how scientists have been able to differentiate the layers of Earth, which are defined in the context of the theory of plate tectonics. This mechanism is used to explain continental drift, earthquakes, and volcanoes.

Goals

At the end of this chapter, students should be able to:

1. define weathering.
2. list some of the factors that affect weathering.
3. describe how soil is formed.
4. list the factors that affect soil development.
5. define erosion.
6. list and describe the agents of erosion.
7. describe various types of depositional features on Earth's surface.
8. recognize events that caused these features.
9. define plate tectonics.
10. discuss some evidence in support of the plate theory.
11. describe the layers of Earth.
12. describe earthquakes.
13. discuss the causes of most earthquakes.
14. describe earthquake waves.
15. describe the causes of volcanoes.
16. describe the kinds of volcanoes.
17. identify places where volcanoes can be found.

Teaching Suggestions 12-1

Before starting this section, collect samples of weathered and unweathered rocks to display in class. On the bulletin board, hang up pictures showing the effects of weathering on buildings, sidewalks, hillsides, and natural outcrops of bedrock.

Students have probably heard the word "weathering," but may not be fully aware of its geologic implications. Ask the class to define weathering, and guide the discussion toward a comprehensive understanding.

Show students samples of weathered and unweathered rock. You might bring a geologist's pick to class and ask students why it is such an important tool. (It is used to expose a fresh, unweathered surface on a rock.) Encourage students to photograph or draw examples of weathering and erosion and bring the photographs or drawings into class for a bulletin-board display.

Background

In many rocks, physical weathering begins long before their exposure at the surface. While still deeply buried, rocks are under great pressure. As overlying rock layers are eroded away, the pressure on the rocks below is lessened, and they expand upward. Cracks form, providing pathways for water to move through the rocks and initiate weathering.

In deep mines, such as the gold mines in South Africa, rocks release stored stress with such explosive violence that miners may be killed by flying fragments. Quarry workers occasionally witness sudden pressure releases that cause an upheaval of the quarry floor.

12-1

Weathering

Goals

1. To define weathering.
2. To list some of the factors that affect weathering.

All around you are examples of materials breaking down. Sidewalks and driveways crack, paint on buildings peels, and metal on cars and bicycles rusts. The rocks in Earth's crust go through a similar process. When rocks break down, the process is called **weathering**.

weathering

There are two types of weathering, physical and chemical. Through *physical weathering*, rocks are broken into smaller chunks. One way this can happen is by alternate freezing and thawing. There is usually some water trapped in the cracks in rocks. Since water expands when it freezes, freezing widens the cracks. This puts pressure on the rock on either side. Eventually the rock breaks into smaller pieces. If this happens high on the side of a cliff, pieces of rock that break off can fall on other rocks, causing them to break, too. You may have traveled through an area where signs are posted saying "Danger—Falling Rocks." (See Figure 12-1.)

Chemical weathering takes place when chemicals in air or water react with rocks to change their composition. For example, carbon dioxide in the air combines with water, forming a weak acid called *carbonic acid*. This acid is strong enough

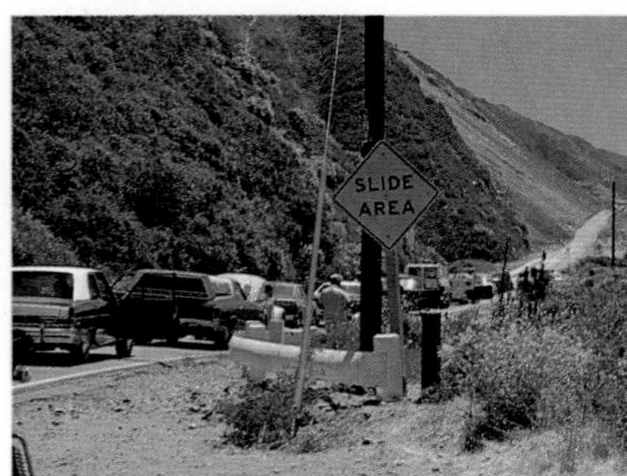

Figure 12-1 *Physical weathering caused this rock slide in California.*

Figure 12-2 *Physical and chemical weathering combined can ruin a stone statue.*

Demonstration

Using a dropper, drop dilute hydrochloric acid on a piece of limestone, as shown in Figure TE 12-1. The "fizzing" that is observed is the release of carbon dioxide gas from the reaction of the acid with the carbonate rock.

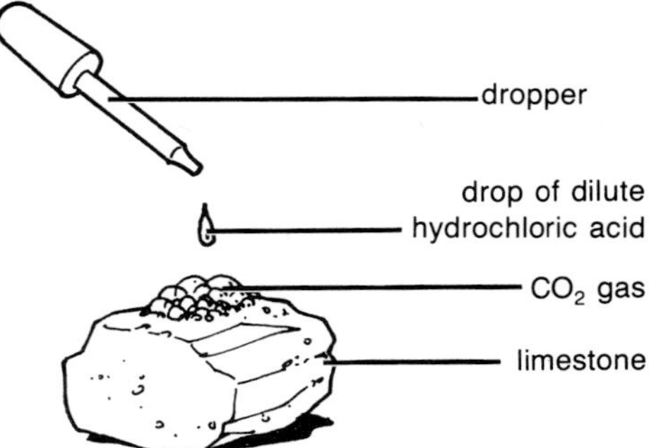

Figure TE 12-1 Chemical weathering.

to dissolve parts of some rocks. (See Figure 12-2.) Chemical weathering also occurs when oxygen in the air reacts with certain minerals in the rocks to form new minerals called *oxides*. Iron oxide, or rust, is a common example. The reddish color of some soils is usually caused by the presence of iron oxide.

Most of the time physical and chemical weathering work together. Physical weathering breaks rocks into smaller pieces. This process exposes more surface area of the rock, allowing chemical weathering to take place more quickly. An increase in surface area can cause physical changes to take place more quickly, too. For example, crushed ice in water melts faster than ice cubes would. With crushed ice, the water comes into contact with more ice surface area, causing it to melt faster.

Several factors determine how rapidly weathering takes place. A large amount of rainfall or extreme temperature variations can increase the rate of weathering. The type of rock in an area has an effect on weathering, also. Some kinds of sedimentary rock, such as soft sandstone, weather faster than igneous rock, such as granite. The type and quantity of vegetation growing in an area are another factor. They determine how moist the ground will be and how much protection leaves may provide for surface rocks. Whatever factors are involved, it usually takes a long time for a large amount of weathering to occur.

Checkpoint

1. Define weathering.
2. Give an example of physical weathering.
3. What is chemical weathering?
4. Name three factors that affect the rate at which weathering occurs.

TRY THIS

Fill two cups or bowls with equal amounts of water. The water in each container should be at the same temperature. Place a cube of sugar in one container and a spoonful of sugar in the other. Stir both at the same time. Which dissolves faster? Why? Relate this to the weathering of rocks.

Cross Reference

Chapter 6 reviews chemical and physical changes in matter.

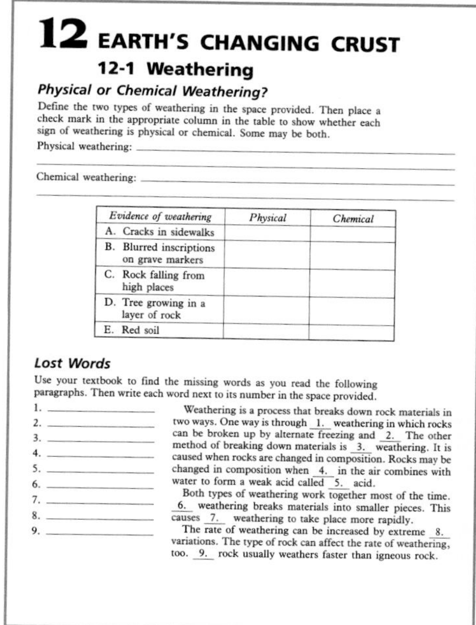

12 EARTH'S CHANGING CRUST

12-1 Weathering

Physical or Chemical Weathering?

Define the two types of weathering in the space provided. Then place a check mark in the appropriate column in the table to show whether each sign of weathering is physical or chemical. Some may be both.

Physical weathering: ______

Chemical weathering: ______

Evidence of weathering	*Physical*	*Chemical*
A. Cracks in sidewalks		
B. Blurred inscriptions on grave markers		
C. Rock falling from high places		
D. Tree growing in a layer of rock		
E. Red soil		

Lost Words

Use your textbook to find the missing words as you read the following paragraphs. Then write each word next to its number in the space provided.

1. ______
2. ______
3. ______
4. ______
5. ______
6. ______
7. ______
8. ______
9. ______

Weathering is a process that breaks down rock materials in two ways. One way is through __1.__ weathering in which rocks can be broken up by alternate freezing and __2.__ The other method of breaking down materials is __3.__ weathering. It is caused when rocks are changed in composition. Rocks may be changed in composition when __4.__ in the air combines with water to form a weak acid called __5.__ acid.

Both types of weathering work together most of the time. __6.__ weathering breaks materials into smaller pieces. This causes __7.__ weathering to take place more rapidly.

The rate of weathering can be increased by extreme __8.__ variations. The type of rock can affect the rate of weathering, too. __9.__ rock usually weathers faster than igneous rock.

Using Try This

The spoonful of sugar dissolves faster because the grains of sugar expose more surface area to the water than does the cube of sugar. This demonstrates how physical weathering accelerates chemical weathering.

Checkpoint Answers

1. Weathering is the breaking down of rock materials by natural processes.
2. Alternate freezing and thawing of water trapped in rock crevices causes the rock to break up.
3. Chemical weathering occurs when substances in air or water react with rocks to change their composition, and to weaken them.
4. Amount of precipitation, relative acidity of precipitation, temperature, type of rock, type and quantity of vegetation, humidity.

Teaching Tips

- Emphasize that physical and chemical weathering occur at the same time.
- Students may be interested in discussing the cost of weathering. Buildings, bridges, roads, ships, and other costly constructs of our society require maintenance, and eventually, replacement.

Facts At Your Fingertips

- Sulfur dioxide and oxides of nitrogen in the air combine with water to form sulfuric and nitric acids, the major acid components in acid rain.
- Cleopatra's Needle, a 3000-year-old granite monument moved to New York City's Central Park from Egypt in 1880, has weathered noticeably in New York City's environment.

Teaching Suggestions 12-2

Before starting this section, collect and display illustrations and/or photographs showing soil profiles that illustrate stages of soil development, from immature to mature (see Background). Collect materials needed for the demonstration (see Demonstration) and do the demonstration early in the teaching of this section, so that the completed model can be displayed during later discussions.

Background

Immature soils occur when soil development has been retarded because of the hardness of rock, rapidity of erosion, or lack of time. They are common in mountainous regions, where all of these conditions exist simultaneously. Soils in areas of frequent flooding also are commonly immature, because new parent material is deposited with each flood, giving the soil horizons little time to develop. Conversely, mature soils are a product of a weathering pattern in which climate and vegetation have been able to work their full effect.

In colonial days, settlers found that many leached forest soils could not support their crops. Squanto, an Indian, showed the earliest settlers in what is now Massachusetts how to bury a dead fish near each hill of corn to supply nutrients to the plants as the fish decomposed.

During the westward migration of the nineteenth century, land was rapidly cleared and crops were planted with no thought of their effect on the fertility of the soil. When the fertility became exhausted, settlers moved farther west to seek more fertile land.

Moving to find more fertile land was a common practice throughout the world until people began to understand the needs and limits of the soil. Lime and other fertilizers are now commonly added to soils to replace nutrients that have leached out.

ENRICHMENT

12-2 Soils

Goals

1. To describe how soil is formed.
2. To list the factors that affect soil development.

soil

One of the most valuable products of the weathering process is soil. **Soil** is a mixture of weathered rock material and decayed *organic* (once-living) material. Soil provides the proper nutrients for plant growth.

The development of soil begins with solid, unweathered rock, called *bedrock*, or *parent material*. The bedrock weathers and breaks up into smaller and smaller pieces. In time, small plants and animals start to live in the layer of weathered rock. Plants help the weathering process and also help prevent the soil from being washed away. As the organisms die, their decayed remains, called *humus* (HYOO muhs), mix with the weathered rock. This forms the uppermost layer of soil, the *topsoil*. Developing soil consists of topsoil above a layer of weathered rock fragments. It takes centuries to develop several centimeters of topsoil.

As weathering of the bedrock continues, the soil develops further. Dissolved minerals are carried downward from the topsoil by ground water. So are particles of *clay*, which is made up of very small, closely packed grains of certain minerals. Eventually a middle layer of soil develops. This middle layer, the *subsoil*, is harder to plow than topsoil because of its clay content. (See Figure 12-3.)

A number of factors affect the development of soil. The

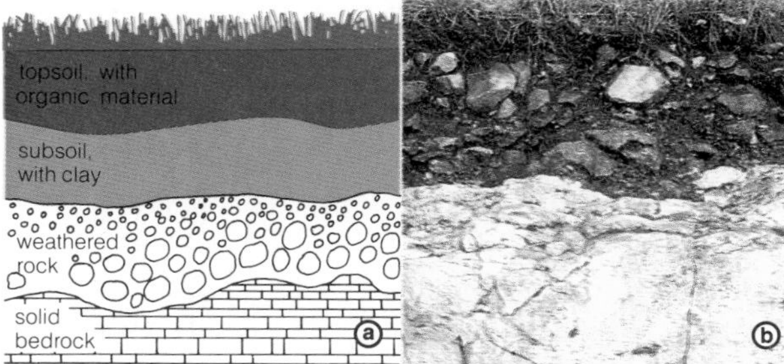

Figure 12-3 *(a) A diagram of soil. (b) The photograph shows topsoil, subsoil, and the less weathered material the soil is forming on.*

Demonstration

In a tall, transparent container such as a battery jar, build up layers of rocks and soil to create a model of a typical mature soil profile. Place the coarsest rock on the bottom, and put less coarse rock on top of that. Next, put sand, then clay, then brown soil, and then dark soil. Finally, place humus on top. This model can be used to illustrate how soil forms from weathered rock and humus.

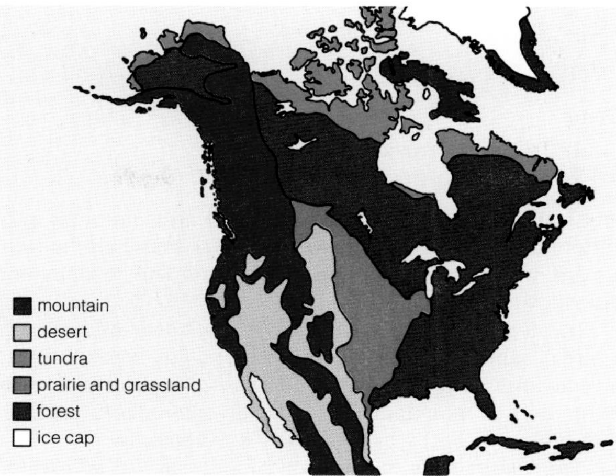

Figure 12-4 *North American soils. Mountain soils are thin and rocky. Forest soils are thin with little humus. Grassland and prairie soils have a thick layer of rich, black topsoil with more humus than forest soils. Desert soils are high in minerals because there is little rain to wash the minerals downward. The lower layers of tundra soil remain frozen, reducing drainage in the top layers.*

type of bedrock in an area affects the soil that develops from it. The kinds of organisms that live in an area affect soil development, also. Since soil needs time to develop, time is another factor. The most important factor in the development of soil, however, is climate. In warm, moist climates rocks weather rapidly, and soil develops quickly. Desert areas, on the other hand, have little rainfall. Rock weathers much more slowly in deserts.

The North American continent has many different types of soil. Soils differ because of where they occur, and what the local climates are. (See Figure 12-4.)

TRY THIS

Find a spot in the schoolyard or at home where you can dig a hole deep enough to expose the depth of the topsoil. How does topsoil differ from the layer beneath it? How thick is the topsoil? Replace the soil when you are finished.

Checkpoint

1. What is soil?
2. How is topsoil formed?
3. How is subsoil formed?
4. What is the major influence on the type of soil that develops in an area?

12 EARTH'S CHANGING CRUST

12-2 Soils

Another Way to Classify Soils

Use reference materials in the library to find the information you need to fill in the table below and to answer the questions that follow it.

Soil Type	*Composition*	*Environment Needed*	*Location*
Pedalfer			
Pedocal			
Laterite			

1. What is a residual soil? ______
   Give an example. ______
2. What is a transported soil? ______
   Give an example. ______
3. List the transporting agents for soil. ______
4. What is the most important factor in determining the soil of an area? ______
5. Describe the A horizon of the soil. ______
6. Describe the B horizon of the soil. ______
7. Describe the C horizon of the soil. ______
8. What is a mature soil? ______
9. How long does it take to develop a mature soil? ______
10. What is a soil profile? ______

Using Try This

This activity is best done as a class project in the schoolyard, if appropriately sited, so that all can share in the observation. If students choose to do the activity at home, have them photograph or draw a diagram of their findings. Students in the inner city may find soil displays in museums, or can research and photocopy illustrations from texts in the library.

The type of soil in your area can be determined roughly from soil maps. If your soil seems different from what you expected, discuss factors that might account for the difference.

Checkpoint Answers

1. Soil is a mixture of weathered rock and decayed organic material.
2. Topsoil is formed when the decayed remains of organisms, called humus, mix with weathered rock material.
3. Subsoil is formed when clay and dissolved minerals are carried with water downward from the topsoil, forming a middle layer of soil.
4. Climate.

Teaching Tips

- Have a committee of students locate soil maps in the library and describe them to the class.
- Interested students could research soil conservation methods, such as contour plowing, minimum tillage, and windbreaks.

Facts At Your Fingertips

- It takes approximately 100 years to develop two centimeters of topsoil.
- Glacial deposits have contributed to the rich farmland in the northern part of the U.S. and Canada.

Teaching Suggestions 12-3

Before starting this section, collect pictures showing erosion along shorelines or on barren hillsides. Display them while discussing this section.

Students can do the Try This activity after discussing the paragraph on solutions and suspensions. Other examples of suspensions are salad dressings, puddings, paint, liquid shoe polish. Have students add to this list. You may wish to refer back to Chapter 6, where the concepts of suspension and solution were first introduced.

Encourage students to search the community for examples of natural or human-caused erosion. Have students list ways to prevent erosion of topsoil. Discuss ways to avoid some of the hazards of erosion prior to constructing a building, such as considering the soil, topography, and underlying rock structure of a building site. Discuss the hazards of building a home or city on land below a dam, or near a nuclear power plant in a known fault area.

Encourage students to do project 10 at the end of the chapter. Even if a soil conservation service is not active locally, soil maps can be obtained from a library and discussed in class.

Background

The ability of a medium to transport particles is sometimes said to be roughly proportional to its viscosity, or resistance to flow. Gases are much less viscous than liquids; hence, the differing abilities of air and water to transport particles, and the need by air of higher flow velocities to transport particles easily moved by comparatively slow-moving water.

Because gravity is the force that makes water flow downhill, gravity is behind the erosion of materials by streams and rivers. As rivers drain water toward the ocean from the continents, the water moves according to a highly involved system of physical and hydraulic relationships. Resisting this movement is the friction (shearing stress) exerted on the water by the bed and banks of the channel, as well as by internal resistance to flow within the water itself. When these forces are small, the water moves smoothly, in sheet, or laminar, flow. But when the movement accelerates, such as when water goes over a falls, or when a fast-moving stream strikes a large rock or negotiates a bend, the flow becomes turbulent. Turbulent flow erodes much more forcefully than does laminar flow.

The erosional capabilities of great masses of glacial ice are profound. Glaciers round out valleys that were once V-shaped, pluck up great boulders and move them many kilometers, and scratch and polish the bedrock over which they slowly move.

12-3

Erosion

Goals

1. To define erosion.
2. To list and describe the agents of erosion.

erosion

Physical and chemical weathering of Earth materials precede erosion. **Erosion** is the process by which Earth materials are moved from one place to another. They usually end up at lower levels of elevation.

If you have a large load to move, such as a fallen tree or a pile of sand, your task is much easier if you break the load up into smaller pieces or smaller loads. This is what weathering does to mountains. It breaks up the rocks into pieces that are more easily carried away by erosion.

The "movers" in the erosion process are called *agents of erosion*. The agents of erosion are water, ice, wind, organisms, and gravity. The one that moves the most material is water.

Earth materials can be moved by running water in three different ways. One way is by *pushing* the water along the bottom or sides of a stream. (Have you ever used running water to push dirt off a sidewalk?) Another way that water can carry particles is as a cloud of mud and sand called a *suspension*. River water is usually cloudier than lake water

Figure 12-5 *The dramatic features that occur along the Oregon coast were produced by wave erosion.*

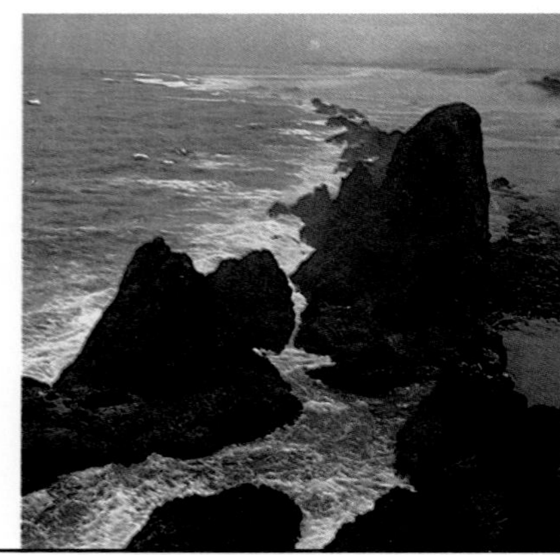

Figure 12-6 *The dark streaks on this Alaskan glacier are rock fragments weathered from valley walls.*

because there are more particles suspended in the moving water. The third way that water can carry materials is in *solution*. When in solution, particles are dissolved in water, just as salt may be dissolved in water. Dissolved particles are not visible.

Waves pounding against a shoreline over long periods of time are powerful agents of erosion. If you live by the ocean or a large lake, you have seen the erosion done by waves. Depending on the shape of the shoreline (smooth or jagged) and on the kind of rock found there (hard or soft), wave erosion can make some interesting patterns. (See Figure 12-5.)

In areas with cold climates, glaciers may act as agents of erosion. A **glacier** is a large, slowly moving mass of ice, many meters to thousands of meters thick. Glaciers form high in mountain valleys. As they grow they move downward, scraping and scouring the sides and bottom of the valley. Glaciers push and carry large quantities of rock debris with them. They also can act as a "conveyor belt." Rocks that fall onto the top of the glacier, are carried along. (See Figure 12-6.) Unlike running water, glaciers move very slowly. Their movement is measured in meters per year.

glacier
(GLAY shur)

Wind is another agent of erosion. Although wind cannot move pieces as large as those carried by water or ice, it can carry sand and dust long distances. In dry regions, where vegetation is scarce, wind does a considerable amount of erosion. (See Figure 12-7.) Farmers are encouraged to leave their fields unplowed until spring because of wind erosion. Plowing loosens the soil. If farmers plow in the fall, strong winter winds may blow away much of the rich topsoil.

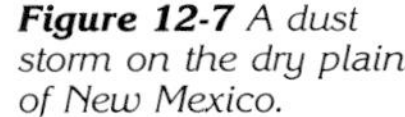

Figure 12-7 *A dust storm on the dry plains of New Mexico.*

Teaching Tips

- Discuss why parts of Earth are still higher than sea level even though erosion has been occurring since the earliest storms.
- You may wish to expand the discussion of glaciers, especially if your school is in an area that was affected by glaciation.

12 EARTH'S CHANGING CRUST

Computer Program

Annual Erosion Loss

Purpose To calculate erosion loss along shallow hillsides.

Background Erosion leads to the loss of thousands of tons of soil each year from farms, construction sites, river banks, and beaches. Since it takes years to produce soil, there are long-term effects of severe erosion.

ENTER the following program:

```
10  PRINT "ENTER ERODIBILITY FACTOR"
20  INPUT "FOR THE SOIL."; EF
30  INPUT "ENTER SIZE OF AREA IN SQUARE METERS."; A1
40  LET A2 = A1 / 4047
50  LET L1 = 228 * EF * A2
60  LET L2 = L1 * .7646
70  PRINT "EROSION LOSS = "; L2; " CUBIC METERS."
80  INPUT "ANOTHER CALCULATION (Y OR N)?"; C$
90  IF C$ = "Y" THEN GOTO 10
100  END
```

Discussion

1. Two farmers and a land developer are concerned about erosion loss from their lands. Farmer A has a 101,175 square meter pasture with soil that has a heavy gravel constant. The erodibility factor for the soil is 0.20. How much soil will be lost in a year?

____

2. Farmer B has 59,500 square meters under cultivation. The soil in this area has an erodibility factor of 0.40. How much soil will be lost in three months? Hint: Remember the program displays loss for one year.

____

3. The developer has an 18,320 square meter oceanside parking lot. The erodibility factor for the sandy soil is 0.30. How much soil will be lost in one year?

____

Upgrade the Program

The program uses lines 20 and 30 for entry of essential data. Line 40 converts the area, in square meters, to acres. The calculation of actual erosion loss occurs in line 50, with the answer being expressed in cubic yards of soil. In line 60, the loss is converted to cubic meters of soil.

1. Ask the user how erosion loss should be displayed, by month or year.
2. Adapt the program to print a graph of erosion loss on the screen.

Facts At Your Fingertips

- It is estimated that more than 1.5 billion kilograms of sediment, including topsoil, flow from the mouth of the Mississippi River daily.
- The Great Lakes, Niagara Falls, and the St. Lawrence River are products of glacial erosion.
- In the Plains states, telephone poles are often partly covered with sheet metal to protect them from abrasion by wind-borne particles.
- Erosion is the response of weathered materials to the force of gravity. Indirectly, even wind erosion is a response to gravity, since gravity brings about the movement of air masses.

12 EARTH'S CHANGING CRUST
12-3 Erosion

Moving Earth Materials

Complete the following statements by unscrambling the letters underlined to form the correct words. Then write the words on the spaces at the left.

______ 1. osireno is the process that moves Earth materials from one place to another.
______ 2. Water, wind, and ice are called ganets of erosion.
______ 3. heewatnigr aids erosion by breaking rocks into smaller pieces.
______ 4. The most important mover of Earch materials is tewra.
______ 5. River water is cloudy because of all the material in pinesnsous.
______ 6. Ice can carry heavier materials than niwd can.
______ 7. When particles are in listonou, they are invisible.
______ 8. clariges can act like a conveyor belt as they carry rock material down a mountain.
______ 9. sawev on a shore can erode the rocks into some interesting shapes.
______ 10. In dry areas, where tevotigena is scarce, wind can cause serious erosion problems.
______ 11. If farmers plow their fields in the fall, they could lose valuable pilosto from wind erosion.
______ 12. linama paths can cause erosion on hillsides.
______ 13. yritvag aids erosion by causing rivers to flow and rocks to fall.
______ 14. A delisudm is an example of erosion that occurs rapidly enough to make the headlines.

Weathering or Erosion?

For each statement below, decide whether the process of weathering or the process of erosion is being described. Then write the appropriate term, *weathering* or *erosion*, on the answer space at the left.

______ 1. Oxygen in the air reacts with minerals in the rocks to form oxides.
______ 2. Interesting rock formations appear, as waves pound the shore.
______ 3. Topsoil is lost when land is overgrazed by livestock.
______ 4. Where vegetation is scarce, wind carries soil away.
______ 5. Alternate freezing and thawing cause rocks to break into smaller pieces.

Using Try This

The salt gradually disappears during stirring. The dirt will be mixed throughout the water, making the water dark and cloudy. After settling for a while, the salt water will remain clear, with no salt visible. In the other glass, dirt will settle on the bottom, with relatively clear water on the top. The dirt and water combination is a suspension; the salt and water combination is a solution.

Checkpoint Answers

1. Weathering is the breaking down of Earth materials; erosion is the process by which they are carried away.
2. Water, ice, wind, organisms, and gravity.
3. Water.
4. People can accelerate erosion through poor soil management.

Figure 12-8 *Flooding of the St. John's River, Maine.*

You might think it strange that organisms are listed as agents of erosion. By stripping the land of vegetation when mining minerals, or by allowing the overgrazing of land by livestock, people have caused a considerable loss of topsoil in some areas. Animals such as cows or sheep can cause erosion on hillsides as they carve out paths where they walk.

Gravity is an agent of erosion because it is the force that causes materials to fall to lower levels of elevation. Gravity helps the other agents of erosion do their work. It causes rivers to flow and rocks to fall and strike other rocks.

When erosion occurs rapidly it makes the headlines. (See Figure 12-8.) Landslides, mudslides, and raging floods are examples of rapid erosion. People can protect themselves from most of these hazards, however. They can avoid building houses on hillsides where mudslides frequently occur or in areas subject to flooding.

TRY THIS

Fill two glasses about ⅔ full of water. Stir a spoonful of salt into one glass and a spoonful of dirt into the other. Describe what you observe. Let the glasses sit undisturbed until all the material has stopped moving. Which one is a suspension?

Checkpoint

1. What is the major difference between erosion and weathering?
2. List several agents of erosion.
3. Which agent of erosion has the greatest effect?
4. Why can people be considered agents of erosion?

12-4

Deposition

Goals

1. To describe various types of depositional features on Earth's surface.
2. To recognize events that caused these features.

Materials moved by erosion and left in another place are called **deposits**. Different agents cause different forms of deposits, or *depositional features* (dep uh ZISH uhn ul).

deposits

Running water leaves behind several recognizable features. One feature is a buildup of sandbars along the inside curves of moving streams. Sandbars form there because the stream slows down on the inside curve as it goes around a bend. (The faster-moving water is on the outside of the curve.) As the water slows down, the heavier material being carried is dropped. The stream continues to move in this way, cutting out the far side of the curve and depositing debris on the inside curves. Eventually, it develops a series of **meanders**, or looping curves. A meander may loop so much that it closes back on itself, forming an *oxbow lake*. (See Figure 12-9.)

meanders (mee AN durz)

Another feature formed by deposits from running water is a delta. **Deltas** are deposits formed at the mouths of rivers as they flow out into lakes, oceans, or bays. The river slows down as it joins a still body of water, and drops the materials it carries. Deltas typically have a triangular shape. (See Figure 12-10.)

deltas

Figure 12-9 Erosion can cut through the land between meanders, forming an oxbow lake.

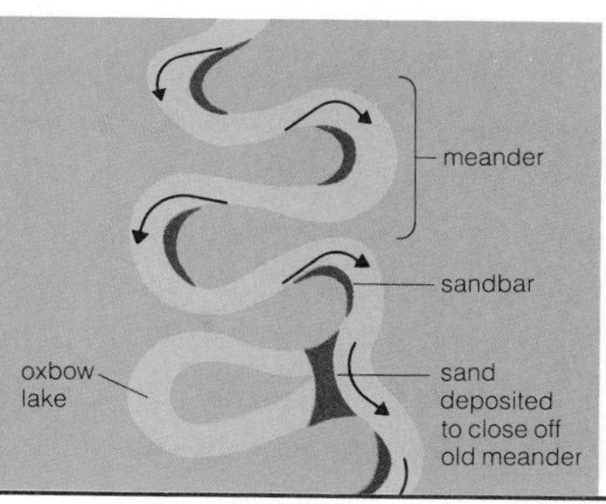

Figure 12-10 The red triangle is the delta of the Nile River, photographed from space.

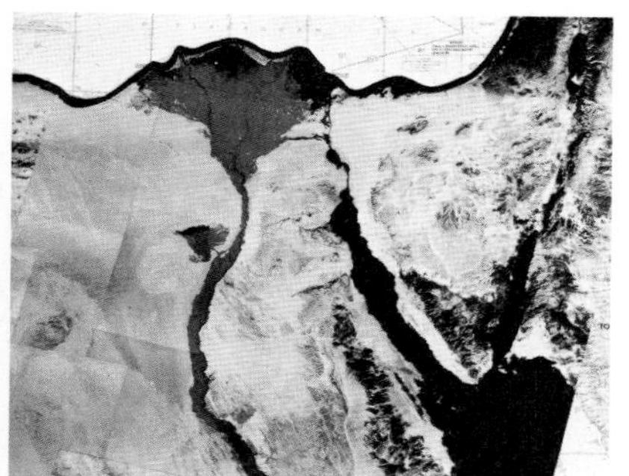

Background

Understanding sedimentation is of great importance to geologists. Much of geologic history is recorded in sedimentary rocks, and virtually all of the fossil record is found in sedimentary rocks. Sedimentary rocks are formed from sediments deposited by the slow-moving water in streams, rivers, and shallow seas. Because the makeup of the sediments carried by moving water varies as the water erodes different kinds of bedrock, the sedimentary rock formed from such materials also varies.

A rock formation is created from materials deposited during the same time interval, but changes in rock makeup may occur within a rock formation. Such variations can include changes in color, texture, grain size, or even rock type. Significant areas of the formation exhibiting a certain variation are called facies.

Certain fossils are indicative of a period of deposition. These fossils, called index fossils, are used to identify particular formations. Where index fossils are associated with economically valuable rock strata, such as oil-containing strata, they can be key elements of intensive research.

The term *drift* in the context of deposition generally includes everything carried by a glacier: boulders, till, gravel, sand, and clay. Till is the unsorted, nonstratified sediment carried or deposited by a glacier. Other depositional features left by glaciation are drumlins (large streamlined hills of unsorted drift), kettles (depressions filled with drift), kames (steep-sided hills of sorted drift), and eskers (ridges, often in groups, of stratified drift).

Teaching Suggestions 12-4

Display pictures of depositional features in the classroom before starting this section. These can be the focus of many discussions. As with section 12-3, have students look for depositional features in their community. Such features can be large or small. Pictures of talus slopes and deltas can help guide urban students to similar features on a small scale on park grounds, for example, or along streets bounded with earthen "cuts"; photos of glacial deposits can serve as field guides for students in glaciated terrain.

When discussing glacial deposits be sure students understand the difference between sorted and unsorted materials and what causes each to form.

Demonstration

Build a delta in a stream table by having a stream of water flow into a "lake." By changing the path of the stream, show how the shape of the delta can be changed. After the demonstration, carefully drain the water from the lake and allow the sand to dry out. Then slice through the delta to show the cross-bedding of the layers.

The stream table can also be used to demonstrate a meandering stream and the formation of ox-bow lakes. Use a minimal slope—just enough to keep the water flowing.

Teaching Tips

- The first law of Newton's laws of motion (see Chapter 16) can be invoked to explain the movement of water in streams. Water tends to flow in a straight line until acted upon by a shoreline, a rock, or barrier. This tendency causes erosion on the outside of a curve, and a buildup of sediment on the inside of the curve, as the water differentially speeds up on the outside and slows down on the inside while rounding the curve.
- The U.S. or Canadian Geological Survey can assist in identifying depositional features in your community. Local highway and water supply departments can also help.
- In urban areas that have been glaciated, erratic boulders may sometimes be found in parks or in the lawn at zoos, as well as in major excavations.
- Discuss with students what has happened since the completion of the Aswan Dam. The Lower Nile no longer deposits mud, and the delta is eroding, not building.

There is a feature common to all materials deposited by running water. As the water slows down and drops its load, the largest particles are dropped first, then the next largest, and so forth. This process is called *sorting*, because the deposits are grouped according to size.

A depositional feature found at the base of mountains and cliffs is a collection of loose rocks and boulders called a *talus slope* (TAY luhs). Talus slopes are caused by gravity pulling down loose material from above.

Glaciers leave a number of recognizable depositional features. Since ice can carry heavier material than running water can, the deposits left by ice are quite different. When a glacier begins to melt, the materials it is carrying—inside the ice, underneath, or on top—are dumped. There is no sorting of this material as there is with running water. Large boulders and small particles are mixed together.

All the unsorted, loose rock material deposited by a glacier is called *glacial drift*. Glacial drift can form many different-shaped features. Deposits of drift dropped in front of a glacier or along its sides are called **moraines**. A *terminal moraine* is formed at the end of a glacier. (See Figure 12-11.) A *lateral moraine* is formed along the sides of a glacier. Moraines form large irregular hills. These hills may display the shape of the glacier that produced them.

moraines
(muh RAYNZ)

Glaciers can carry huge boulders and leave them hundreds of kilometers from their sources. Such boulders are not like the rocks that are most common in the area, and are called *erratics*. (Have you seen any of these rocks in your area?)

Figure 12-11 *The loose material these people are standing on is part of a terminal moraine.*

Facts At Your Fingertips

- Wind can carry materials over long distances. Volcanic dust can be carried by upper level winds hundreds of kilometers from the volcano.
- Streams leave cone-shaped deposits, called alluvial fans, where they run onto a plain or meet a slower-running stream or a pond.

Wind also leaves characteristic depositional features. Sand dunes are probably the best example. Desert sand dunes have a typical "boomerang" shape. The tips of the "boomerang" always point away from the wind. (See Figure 12-12.) Snow drifts have a similar shape.

Figure 12-12 *Sand dunes in the Monahans Sandhills of Texas. The tips show that the wind blows from "right" to "left."*

Checkpoint

1. Which agents of erosion sort materials as they deposit them?
2. On which side of a curve do sandbars develop?
3. What is a delta?
4. How do glacial deposits differ from deposits left by running water?
5. What is an erratic?

TRY THIS

Partially fill a long, plastic tube with water. Hold it in an upright position, preferably using a ring stand. Drop spoonfuls of unsorted sand into the water. Allow each spoonful to settle before adding the next one. What do you observe happening to the sediment in the bottom? How does this compare to what happens in a river delta?

12 EARTH'S CHANGING CRUST

12-4 Deposition

Sort This Out!

As you read each definition below, think of the term being described. Write the letters of the term on the spaces at the right. Then choose from among the terms as you label the illustrations at the bottom of the page.

1. Materials that have been moved by erosion and left in another place. ________
2. The looping curves in a stream. ________
3. A collection of loose rock and boulders at the base of mountains. ________
4. A small lake formed when a stream loops back on itself. ________
5. Huge boulders carried by glaciers. ________
6. Loose, unsorted material deposited by a glacier. ________
7. A deposit found at the mouth of a river. ________
8. A deposit found where a glacier stopped moving forward. ________
9. Materials being deposited by size. ________
10. A typical deposit left by wind. ________

What word do the double underlined letters spell? ________

A ________ B ________ C ________

Using Try This

As the sand settles, the larger particles fall first and the finest, last, creating layers. This is similar to what happens in a delta, except that the materials would be sloped outward into a lake, an ocean, or a bay.

Checkpoint Answers

1. Running water and wind.
2. On the inside of the curve.
3. A delta is a deposit that forms at the mouth of a river or stream as it flows into a lake, an ocean, or a bay.
4. Glacial deposits left by a rapidly melting, or receding, glacier are unsorted; however, the deposits left by streams flowing from a slowly-melting glacier will deposit sorted materials just as would any stream. Glaciers can carry much larger and heavier objects than can running water.
5. An erratic is a boulder left by a glacier; erratics are usually unlike native "country" rock types.

Teaching Suggestions 12-5

Before starting this section, obtain maps of the ocean floor for display. Help familiarize students with the trenches and mid-ocean ridges; the maps should provide depths of these features. Collect materials for Activity 18, Moving Land Masses, student text page 554.

The students can form a model of the ocean floor in a stream table or large sandbox, showing its various features. Commercial three-dimensional maps and globes are also available.

Some students may wish to do chapter-end project 5 at this point.

Background

The theory of plate tectonics has produced the most unifying concept in earth science since uniformitarianism. While a mainstay of earth science in the Western world, however, the theory has not been accepted everywhere. Consequently, your students are living in the midst of a scientific revolution, with global controversial overtones!

In addition to evidence discussed in this section, other evidence in support of the theory of plate tectonics includes: paleomagnetic polarity in ocean-floor rock, fossil correlation among continental land masses, correlation among distinct geologic features on continents, relationships or distinctions among living life forms on modern continents, and the intuitive "fit" of continental shelf outlines. The vehement and emotional attacks on Wegener have not been mentioned in the text. These came mainly from American geologists. One reason geologists opposed Wegener's ideas was that he was a meteorologist, and they could not accept such a sweeping, innovational perspective from a nongeologist. Students should consider that meteorology is today considered a part of earth science, but it was not during Wegener's time.

Scientists discovered the layers of Earth by studying earthquake waves.

12-5

Plate Tectonics

Goals

1. To define plate tectonics.
2. To discuss some evidence in support of the plate theory.
3. To describe the layers of Earth.

In 1912 a scientist named Alfred Wegener (VAY guh nur) came up with a startling idea. He proposed that all the continents had at one time been together, but have been slowly drifting apart for many years. (See Figure 12-13.) He based his hypothesis on the obvious way that the continents could fit together like a jigsaw puzzle, especially South America and Africa. He also discovered matching rock layers on continents separated by the Atlantic Ocean.

Wegener's idea was given support when scientists learned more about the ocean floor. They discovered a system of underwater mountain chains, or **mid-ocean ridges**, that rise hundreds to thousands of meters above the ocean floor. The youngest ocean floor rocks are near the mid-ocean ridges, and the oldest are near the edges of the ocean basins.

mid-ocean ridges

The youngest rocks on the sea floor are near the mid-ocean ridge because the sea floor is spreading apart at the mid-ocean ridges. The ridges are made up of many volcanoes that allow hot, molten rock to flow out onto the ocean floor. Horizontal motion of the sea floor carries the new rock away from the ridge, just as if the floor were on a huge con-

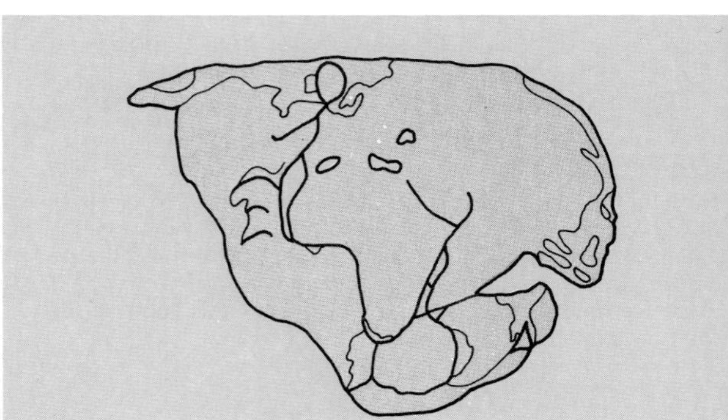

Figure 12-13 *This is Alfred Wegener's map of the land mass he thought broke up into the present continents.*

The mantle is essentially solid, being only about 1% molten. Within the mantle is a layer called the asthenosphere, or zone of weakness (see section 12-7), which is about 10% molten. The asthenosphere is at the base of the plates. Students often have difficulty understanding how flow and movement can occur within an essentially solid medium. This occurs under great pressure and over long periods of time and at rates that are on the order of only a few centimeters per year.

Heat within Earth's mantle probably is generated by radioactivity and by the compressive force of overburden. The resultant convection is considered by many scientists to be the mechanism responsible for plate movement.

veyor belt. When the sea floor reaches a continental boundary, it is forced downward beneath the continent. Sea floor *trenches,* long grooves in the ocean floor thousands of meters deep, are formed where the ocean floor is pulled down under the continents.

In order to understand how parts of the crust can move about, it helps to step back and look at Earth as a whole. Earth's interior consists of several layers, each having different properties. (See Figure 12-14.) The **inner core** is believed to be solid, composed mainly of nickel and iron. The **outer core** is a hot liquid. The **mantle** is a thick layer, denser than the crust, that is mostly solid. The outer layer is the crust.

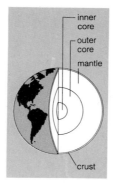

Figure 12-14 *The layers in Earth's interior.*

Since the rocks in the crust are less dense than the material in the mantle, the crust "floats" on the mantle. Where continents are thicker, as in mountain ranges, the crust sinks deeper into the mantle. Where the crust is thinner, as under the ocean, it doesn't sink into the mantle as much. Thinner parts of the crust are similar to an empty ship floating high on the water. The thicker parts, like a ship full of cargo, "float" at a lower level.

inner core

outer core

mantle

plate tectonics
(playt tek TAHN iks)

The crust and the uppermost part of the mantle, with a thickness of about 100 km, are divided into segments called *plates.* (See Figure 12-15.) There are six major plates, and several small ones. According to the theory called **plate tectonics**, these plates move about on Earth, carrying continents and ocean floor with them. Shifting plates account for continental movement and sea floor spreading. When plates move apart, sea floor spreading takes place between them. When plates slip alongside each other, as they do

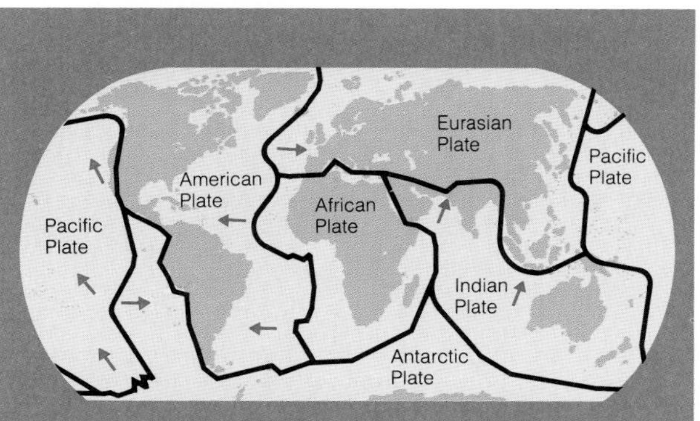

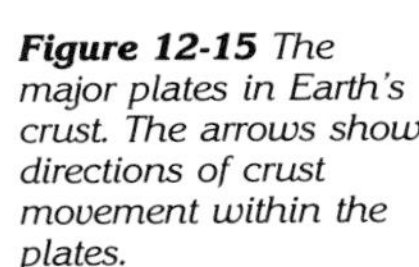

Figure 12-15 *The major plates in Earth's crust. The arrows show directions of crust movement within the plates.*

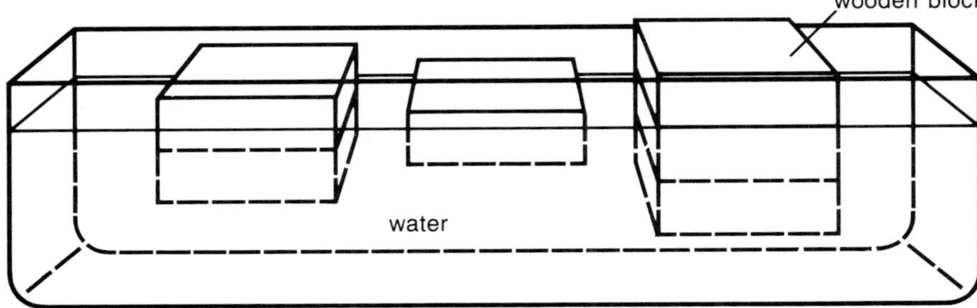

Figure TE 12-2 A model of how Earth's crust floats on the mantle.

Demonstrations

1. Cut a hard-boiled egg in half to provide a model for the layers of Earth. The shell is analogous to the crust; the white is the mantle; the yolk is the core. (The inner core will not be represented, unless the inner part of the yolk is colored.)
2. To show how the crust floats on the mantle, float a layer of wood blocks on water in an aquarium. Pile on more blocks at one end of the aquarium, as shown in Figure TE 12-2. This demonstrates that the crust is thicker under the continents, and that relative to mean sea level, the thickest parts of continents extend both above and below that mean.

Teaching Tips

- You may want to review or introduce the concept of density in conjunction with the discussion of Earth's layers. Refer to Chapter 6.
- A good topic for discussion is the proposal that all wastes be dumped into the deep-sea trenches along continental shelves. From here the waste materials—including nuclear wastes, toxic chemical wastes, and obsolete military biological and chemical ordnance and wastes—would be carried deep into the planet and "naturally" recycled. What are the advantages and disadvantages of such a plan? What are the uncertainties and risks?

Facts At Your Fingertips

- The word tectonics has its origin in a Greek word that means "carpenter" or "builder," and thus refers to the building or creating of features on Earth's surface.
- The 80-year-old Simplon rail tunnel, the world's longest (12 kilometers long), which runs beneath the Alps between Switzerland and Italy, is being rebored. It has been getting

smaller because of the movement of the plates below and the weight of the mountains above.

- The Atlantic Ocean is growing in width; the Pacific is narrowing.

Cross References

Chapter 14 provides additional background on the ocean floor in section 14-6. Chapter 12, sections 12-6 and 12-7, are closely related to this section.

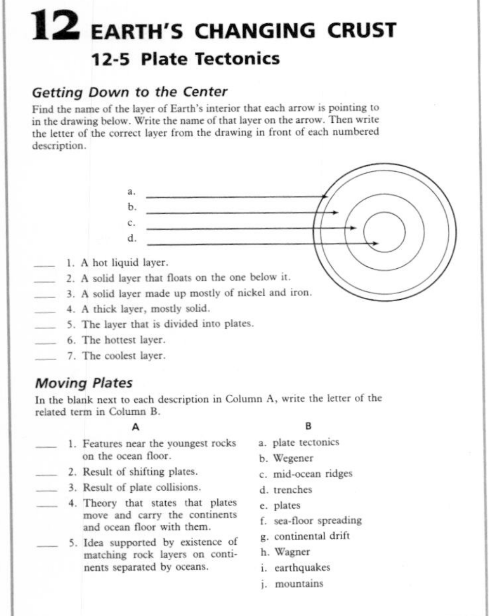

12 EARTH'S CHANGING CRUST

12-5 Plate Tectonics

Getting Down to the Center

Find the name of the layer of Earth's interior that each arrow is pointing to in the drawing below. Write the name of that layer on the arrow. Then write the letter of the correct layer from the drawing in front of each numbered description.

a.
b.
c.
d.

____ 1. A hot liquid layer.
____ 2. A solid layer that floats on the one below it.
____ 3. A solid layer made up mostly of nickel and iron.
____ 4. A thick layer, mostly solid.
____ 5. The layer that is divided into plates.
____ 6. The hottest layer.
____ 7. The coolest layer.

Moving Plates

In the blank next to each description in Column A, write the letter of the related term in Column B.

A	B
____ 1. Features near the youngest rocks on the ocean floor.	a. plate tectonics
	b. Wegener
____ 2. Result of shifting plates.	c. mid-ocean ridges
____ 3. Result of plate collisions.	d. trenches
____ 4. Theory that states that plates move and carry the continents and ocean floor with them.	e. plates
	f. sea-floor spreading
____ 5. Idea supported by existence of matching rock layers on continents separated by oceans.	g. continental drift
	h. Wagner
	i. earthquakes
	j. mountains

Using Try This

This is a good activity to show how new material is extruded from the mid-ocean ridge and is moved away from the ridge on a conveyor-belt-like mechanism. Students should understand that the youngest material is that which is closest to the ridge.

along the San Andreas Fault in California, earthquakes occur. When one plate slides under another, as they do at continental margins, earthquakes and volcanoes occur. Where plates collide, mountains are pushed up.

Continental motion has been measured at from 1–5 cm per year. As the plates continue to move about, this will produce a slow change in Earth's geography. Each year, for instance, the Atlantic Ocean becomes slightly wider.

Checkpoint

1. Define plate tectonics.
2. Name the layers of Earth, from the inside to the outside.
3. Where are the youngest rocks found on the sea floor? How does this fact provide support for the plate theory?

TRY THIS

Tape two sheets of paper together, making one long sheet. Move two desks or tables together. Fold the paper in half where you have taped it. Place it so that it hangs down between the desks, as in Figure 12-16. The paper represents molten rock before it comes out of the mid-ocean ridge.

Pull the paper up on both sides about 3 or 4 cm and draw a line on the paper at the edge of each desk. Put a "1" on the paper on each side of the gap between the desks. Pull the paper up again, about the same distance as before. Draw another pair of lines and number the "new rock" as "2."

Repeat this until all the paper is on top. Study the pattern that has developed. The lower the number, the older the rock layer. Where do you find the youngest rock?

Figure 12-16 *This model shows why younger rock layers are closer to the mid-ocean ridge than older rock layers.*

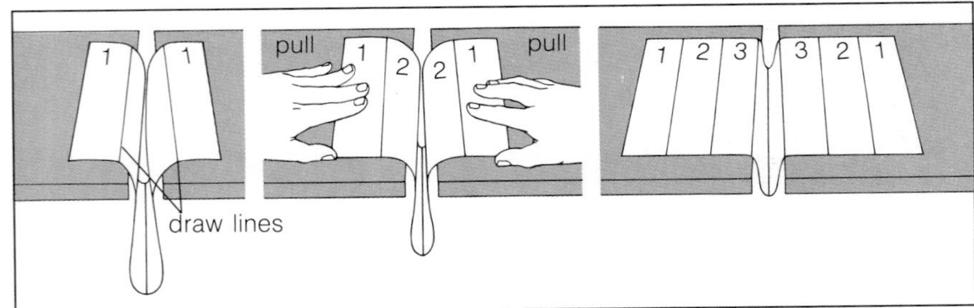

Checkpoint Answers

1. Plate tectonics is the theory used to explain continental drift and sea-floor spreading. The crust and the upper mantle are divided into six major plates that move, and which carry continents and ocean floors with them.
2. Inner core, outer core, mantle, crust.
3. Near the mid-ocean ridges. If the ocean is spreading apart at the ridges, the oldest rock should be found near the edges of the ocean basins. This is what has been found.

12-6

Earthquakes

Goals

1. To describe earthquakes.
2. To discuss the causes of most earthquakes.
3. To describe earthquake waves.

The plates that make up Earth's crust move very slowly and over long periods of time. It is impossible for us to feel this slow overall motion. But sometimes there is a sudden shaking of the crust at a particular place—an **earthquake**. Some earthquakes are so mild that they can be sensed only by the most delicate instruments. Other earthquakes are strong enough to cause widespread damage and destruction. (See Figure 12-17.)

earthquake

Most earthquakes are associated with faults. A **fault** is a zone of weakness in the crust, along which some movement of rock takes place. Pressure within the crust, often caused by moving plates, causes stress to build up along the fault. When the stress becomes great enough, the rocks slip along the fault. Energy is released and an earthquake occurs.

fault

The sudden release of stress in an earthquake can be compared to the breaking of a tightly wound watch spring. A great deal of stress builds up in a watch spring when it is wound tightly. If the watch is allowed to run down normally, the stress is relieved steadily over time. But if the spring is given one turn past its maximum, the stress becomes too much and the spring suddenly breaks.

Figure 12-17 *The 1971 earthquake in California destroyed parts of the Golden State Freeway.*

Teaching Suggestions 12-6

Before beginning this section, collect pictures and newspaper clippings about earthquakes. Have these on display during the study of this section. You may wish to consider screening "The City That Waits to Die," a film about San Francisco, which assesses the likelihood of a major earthquake occurring along the San Andreas Fault. Encourage students to do project 8 at the end of the chapter.

Most of us have never experienced a major earthquake, so that most learning must come from pictures, films, and reading. Using the Guide to Periodical Literature, have students research tsunamis. Have them report to the class on recent occurrences. In addition, some students may wish to do project 7 at this point.

Background

Observations of earthquakes around the world have provided a pattern that supports the theory of plate tectonics. The pattern shows that earthquakes tend to occur most frequently around the boundaries of the plates.

The release of stored stresses during earthquakes typically causes faulting, or rupture, of the crust, along with a trembling and vibration that may be strong enough to shake an entire continent. Severe shocks may last for several minutes; lighter shocks may last only for a few seconds. Erosion is likely to be accelerated in a fault zone, and fault valleys are common features.

When an earthquake fault occurs under the ocean, the disturbance may cause a tsunami, a giant wave. A tsunami may travel thousands of kilometers at speeds on the order of 750 kilometers per hour—about the speed of a jet airliner! They may be 18 meters higher than mean sea level. They have nothing to do with tides, and the term "tidal wave" is a misnomer.

A disastrous tsunami occurred when Krakatoa erupted in 1883. Some 36,000 people living on islands in the Pacific Ocean near what is now Indonesia drowned in the huge wave. A tsunami caused by the 1964 Anchorage, Alaska earthquake propagated virtually to all parts of the North Pacific.

As terrifying as tsunamis are, the sheer destructive force of an earthquake is more likely to have effect when its focus is beneath a continent. During the twentieth century, great earthquakes have taken many lives. Two occurred in Italy, in 1908 and 1915, taking 105,000 lives. Kansu Province, China was hit by an earthquake in 1920 that took 120,000 lives. In 1960, 12,000 people died in a Moroccan earthquake.

Much of the danger from earthquakes lies not in the shaking of Earth itself, but from the fires caused by the shaking. Some fires occur from overturned stoves and lamps in less-developed regions; electrical fires may result in more developed places. Water pipes rupture, making fire-fighting difficult.

The major source of information about Earth's interior has been records of waves that propagated through the planet from natural earthquakes or human-caused explosions. The energy released in a natural earthquake may be as great as that released by a million atomic (fission) bombs.

By studying the arrival times of P waves (primary compression) and S waves (shear) caused by the vibrations of sudden, violent disturbances as they pass through the interior of the planet, scientists have been able to reconstruct the paths of these waves. It has been found that P waves can penetrate the entire planet and are received at seismograph stations on the opposite side of Earth from the focus of the disturbance. However, S waves cannot penetrate the outer core of Earth, and are not received by instruments located on the opposite side of Earth from the focus of the disturbance. From these observations, scientists have concluded that the outer core of Earth is fluid, because S waves cannot propagate through fluid media.

L waves, which propagate through the crust and travel about three kilometers per second, arrive after the P waves and S waves, as shown in Figure TE 12-3. It is the L waves that do the tremendous damage.

Demonstration

Use a spring (such as "Slinky" brand toy) or a long, flexible rope to demonstrate the motion of S and P waves. To create a model of S-wave motion, have one student hold one end of the spring or rope, while another student vibrates the spring or rope side-to-side or up-and-down at right angles to the line of motion of the wave. The resulting sinusoidal pattern can serve as a model of the movement of an S wave. To create a model of a P-wave motion, have one student hold one end of the spring or rope, while another student

Figure 12-18 *A map of earthquake locations. Compare this map with the one in Figure 12-15.*

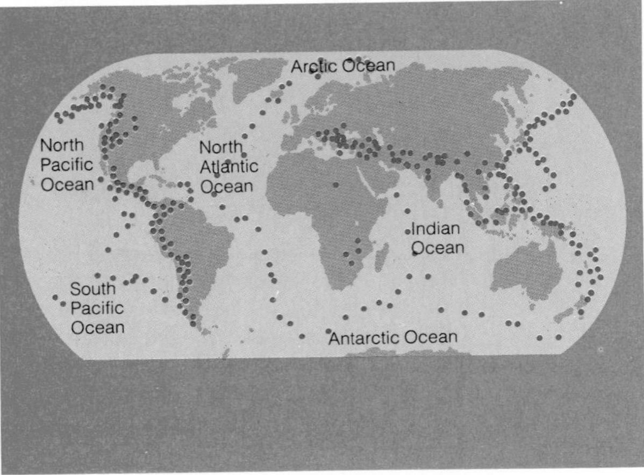

More than half of all earthquakes take place along the edges of plates. (See Figure 12-18.) Many earthquakes occur in areas that are not densely populated, so they get little publicity. For instance, there is much activity along the mid-ocean ridges.

focus

The area in the crust where energy is released during an earthquake is called the **focus**. It is the point of origin of the quake. The focus can be shallow (0–70 km below the surface), intermediate (70–300 km), or deep (over 300 km). The deepest earthquakes occur about 700 km below the surface. Most destruction is caused by earthquakes of shallow or intermediate depth.

magnitude

The relative size of an earthquake—the amount of energy released—is called its **magnitude**. Magnitude is expressed in terms of the *Richter scale* (RIK tur). This scale uses numbers from 1 up, with each number indicating a magnitude ten times stronger than the number below. For example, if an earthquake has a rating of 8 on the Richter scale, its magnitude is ten times as great as an earthquake with a magnitude of 7. A magnitude of 7 or higher indicates a major earthquake. If there were an earthquake with a magnitude of 10, it would be felt all over Earth. An earthquake in Chile, in 1960, had the highest magnitude ever recorded, 9.5.

intensity

There is another scale used to describe earthquakes. The *Modified Mercalli Scale* is used to rate **intensity**, the

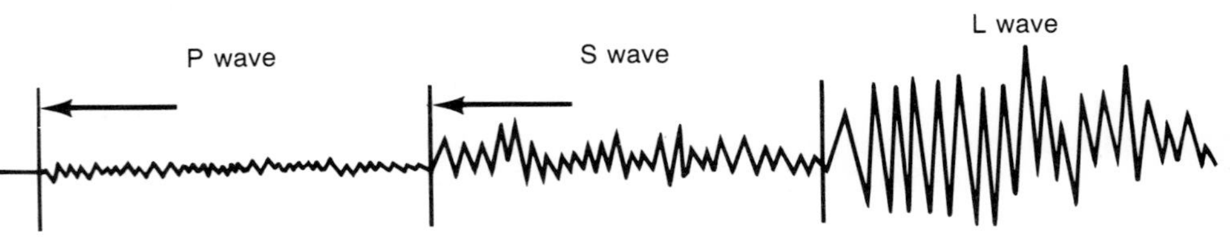

Figure TE 12-3 An illustration of a seismogram.

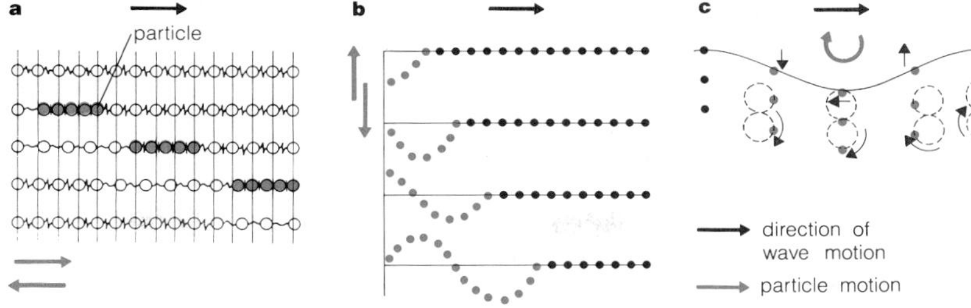

strength of an earthquake at a particular place. This scale uses Roman numerals from I to XII to describe the damage done by the quake. A rating of XII would indicate total destruction.

An earthquake can have only one magnitude. But the amount of damage it does, its intensity, can vary depending on the location. An earthquake of magnitude 6.3 might cause a lot of damage and have a high intensity at one place, but cause little or no damage 200 km away.

Figure 12-19 *Three kinds of earthquake waves. (a) A compressional wave. (b) A shear wave. (c) A Rayleigh wave. The black arrows show the direction of wave motion; the blue ones show particle motion.*

The energy released by an earthquake travels away from the focus in waves. The waves, which travel through and around Earth, are detected by a recording device called a **seismograph**. There are four kinds of waves sent out from the focus of an earthquake. The first waves to arrive at a seismograph are **P waves**, also called primary or *compressional* waves. Compressional waves cause the individual particles in a substance to move back and forth in the direction of the motion of the wave. (See Figure 12-19a.) They can travel through solids, liquids, or gases.

seismograph
(SYZ muh graf)

P waves

The second waves to arrive at a seismograph are **S waves**, also called secondary or *shear* waves. Shear waves cause rock particles to move from side to side at right angles to the direction of wave motion. (See Figure 12-19b.) They cannot pass through liquids.

S waves

The third and fourth kinds of earthquake waves, *Love waves* and *Rayleigh waves*, travel along the surface. Love waves involve no vertical movement of the surface. Rayleigh waves involve both horizontal and vertical movement, as indicated in Figure 12-19c. Surface waves cause most of the damage.

The arrival times of P and S waves, recorded on seismographs at three or more different locations, can be used to

compresses several coils at the other end, and then releases them. The resulting to-and-fro compressive pulse can serve as a model of the motion of a P wave.

Teaching Tips

- Students may confuse meanings of the terms epicenter and focus. An epicenter is a point on the surface of the planet projected on a radius of the planet directly above the focus. The focus is, ideally, the point of origin of the seismic waves within Earth.
- You may wish to point out that small earthquakes relieve built-up stresses gradually. Several small earthquakes would be far less hazardous than one major one.
- Earthquakes that have shallow foci—up to 64 kilometers deep—predominate. They account for about 85% of all earthquakes. In addition, they are more destructive than deep-focus earthquakes, which have foci as deep as 644 kilometers or more.
- Earthquake shocks that affect unconsolidated materials are likely to cause more damage than those that affect bedrock.
- Students may have difficulty understanding the exponential Richter scale. Have them calculate how an earthquake with a magnitude of nine would compare in destructive force to one with a magnitude of one.

Facts At Your Fingertips

- Earthquakes can be caused by people. For example, the pressure of a dam and reservoir on underlying rock has been known to cause small earthquakes. An underground nuclear explosion may cause a major local earthquake; in fact, the seismic waves caused by such explosions are the primary means of the detection of secret nuclear testing.
- A shadow zone, which receives neither P nor S waves, exists on the side of Earth opposite an earthquake. The shadow zone is a doughnut-shaped zone that occurs between 102 and 143 degrees along Earth's circumference from the epicenter.
- The Richter scale is actually open at both extremes. Thus, earthquakes may have magnitudes of less than 1 or greater than 10.

12 EARTH'S CHANGING CRUST
Laboratory

Locating the Epicenter of an Earthquake

Purpose To locate the epicenter of an earthquake using a travel-time graph.

Materials paper, pencil

Background

Different types of earthquake waves travel at different speeds. P waves travel faster than S waves. This difference in travel speed can be used to find out how far a given place is from an earthquake's epicenter.

Procedure 1

1. The graph shows P- and S-wave travel times. Suppose an epicenter is 3,000 km away from the place where P and S waves are detected. What is the travel time of the P waves? ______ The S waves? ______
2. How much sooner do the P waves arrive? ______

Discussion 1

If you know the arrival times of P and S waves at a detector, how could you use the graph to find the distance to the epicenter?

______

Procedure 2

1. A seismograph at station A on the map recorded P and S waves that arrived 6 min., 40 s apart. To find the distance to the earthquake's epicenter, first line up a piece of paper along the left edge of the graph. Make marks on the paper at 0 min. and at 6 min., 40 s. Then slide the paper to the right until the distance between the marks on the paper is exactly equal to the distance between the curves for the P and S waves. Read the corresponding epicenter distance on the bottom scale of the graph. Record the distance.
2. A seismograph at station B received P and S waves 9 min., 20 s apart. At station C, the P and S waves arrived 7 min., 40 s apart. Use the graph to find and record the distance of the epicenter from B and from C.

______

3. On the map, what is the radius of the partially shown circle around station A? ______ Around station B? ______ Around station C? ______ How does the radius of each circle compare to the graph reading of the distance from each station to the epicenter?

______

Cross Reference

Refer to the description of wave motion in Chapter 20, section 20-4.

12 EARTH'S CHANGING CRUST
12-6 Earthquakes

All Shook Up!

Complete the crossword puzzle below with terms from Section 12-6 of your textbook.

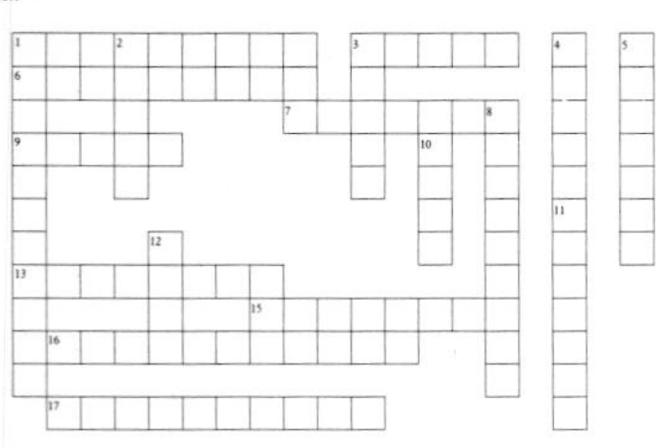

Across

1. Another name for shear waves.
3. The point of origin of an earthquake.
6. A point on the surface above the earthquake's origin.
7. An ocean wave caused by an earthquake.
9. Secondary wave.
13. A surface wave with both vertical and horizontal movements.
15. The scale that measures the strength of an earthquake.
16. Waves that cause particles to move back and forth in the direction of the wave.
17. The best known fault in North America.

Down

1. A device that records earthquakes.
2. A site of many earthquakes is near mid-____ ridges.
3. A zone of weakness in the crust.
4. Divisions in Earth's crust.
5. Measures magnitude of an earthquake.
8. The damage done by an earthquake.
10. Surface waves with no vertical movement.
11. An earthquake that is 0–70 km below the surface.
12. An earthquake that is more than 300 km under the surface.

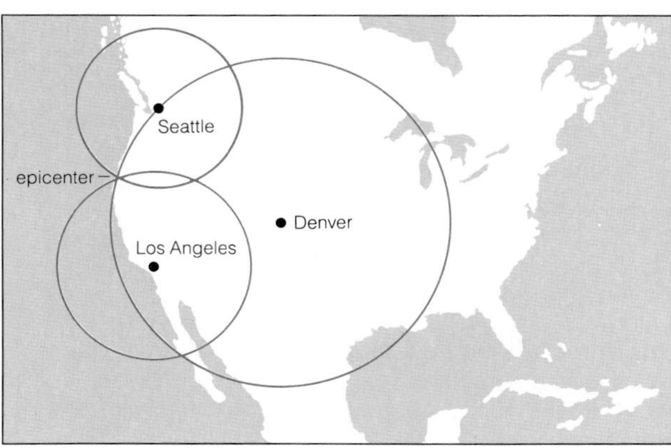

Figure 12-20 *How to find the epicenter of an earthquake. Each circle represents the distance of the epicenter from a seismograph station.*

epicenter
(EP uh sen tur)

find the epicenter of an earthquake. (See Figure 12-20.) The **epicenter** is the point on the surface directly above the focus of an earthquake. People at each seismograph station can figure out how far away the epicenter is, but they don't know which direction the waves came from. They draw a circle representing that distance on a map. Then the circles of two other stations can be drawn on the map. The epicenter is where all three circles cross.

When earthquakes occur on the ocean floor, energy is passed through the water, resulting in huge sea waves, or *tsunamis* (tsoo NAH meez). These waves move extremely fast, up to 750 km/h. This is similar to the speed of a jet plane. When a tsunami reaches shallow water near land, a wave as high as 30 m may build up. These waves do great damage to the shorelines where they release their energy.

Checkpoint

1. What is the cause of most earthquakes?
2. How do P waves differ from S waves?
3. What scale is used to measure the magnitude of an earthquake?
4. What scale is used to measure the intensity of an earthquake?
5. What is a tsunami?

Checkpoint Answers

1. Natural earthquakes are usually caused by the slippage of rocks along a fault line, rupturing and releasing built-up stress. Other earthquakes may be caused by sudden, violent events, such as a dam collapse, or a human-caused explosion.
2. P waves move faster, can move through solids, liquids, and gases, and are compressional waves. S waves move more slowly, and cause rock particles to move at right angles to the direction of wave motion. S waves cannot pass through liquids or gases.
3. Richter scale.
4. Modified Mercalli Scale.
5. A tsunami is a huge sea wave caused by an earthquake on the ocean floor.

Earthquake Prediction

Scientists would like to be able to predict where and when earthquakes will occur, and at what magnitudes. Earthquake-prone areas in which no quakes have occurred for a long time are likely to have major earthquakes fairly soon. Scientists are concentrating their activities in such areas. The San Andreas Fault in California—shown in the photograph—is one example. The last great earthquake along this fault was the San Francisco earthquake of 1906. Since there are many large population centers near this fault, it is being monitored to detect the slightest movement.

Chinese scientists have had some success with earthquake prediction. They have noted changes in water levels in wells prior to a quake. They have observed changes in the amount of a certain gas in the well water, also. They have even noted the behavior of animals, since animals often act in highly unusual ways just before earthquakes. For example, animals that normally live underground, like snakes and mice, leave their burrows up to several days before a quake. Fish that normally stay on the ocean bottom begin to appear near the surface. Dogs, horses, and pheasants become very noisy. Chinese scientists gave advance warning of a major quake in Haicheng, in 1975. As a result, many lives were saved.

In earthquake-prone areas, preparations for a major quake are being made. In California, Civil Defense teams hold regular exercises to prepare them for such a disaster. Many localities have strengthened local building codes to make buildings more quake-proof. New methods of reducing damage to buildings are being tested. One method is to place springs between the walls and the footings of buildings. This method is intended to prevent buildings from breaking off at their foundations when earthquakes occur.

Many of the observations concerning earthquakes seem to be fitting together like pieces of a jigsaw puzzle. They suggest models to explain why earthquakes occur. Only time will tell how successfully such models can predict earthquakes.

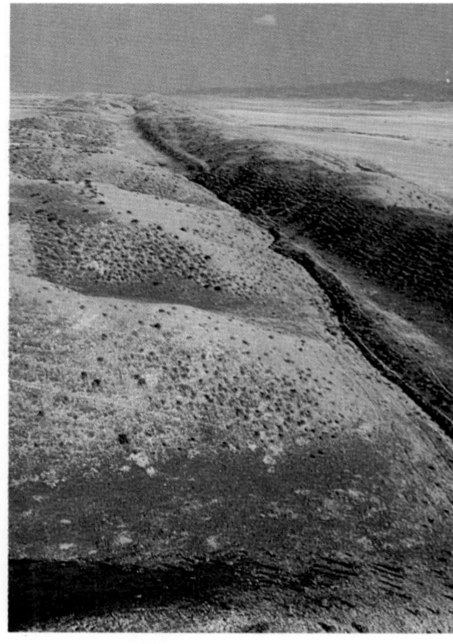

Using The Feature Article

After students have read the feature article, discuss the efforts being made to predict earthquakes. Have students research answers to the following questions and discuss their findings in class: Why might water levels in wells change before a quake? Why might unusual gases be found before a quake? Why might animal behavior change? What kinds of instruments would be needed to detect and measure slight rises in the crust that could presage an earthquake? What kinds of engineering arrangements could minimize a building's sensitivity to earthquake tremors?

Research can be done in the library on more recent attempts to predict earthquakes, which involve satellites, tiltmeters, extensometers, and lasers. The U.S. or Canadian Geological Survey could provide information, but because weeks might be required for the organization to respond, inquiries should be sent as soon as possible.

Students should be given time to brainstorm in small groups. In addition to responding to the suggested questions, ask them to list ways that people who live in earthquake zones could prevent damage to buildings and injury to people. Motivated students can make models of earthquake-proof structures, using simple materials. Some students can research the tsunami and earthquake incidents cited in the Background for section 12-6. Have some students write to the Office of Civil Defense in an earthquake zone and ask for details on procedures used to minimize damage and injury during an earthquake. Share this information with the class. Any students who have experienced an earthquake may wish to share their recollections with the class.

At the conclusion, have students list and combine the reasons they have determined that make the prediction of earthquakes so difficult. They should agree that earthquakes are phenomena in which stored stresses in the crust are suddenly released. It is virtually impossible to determine with great accuracy both the magnitude of the stored stresses and the breaking point, or rupture, of the crustal masses involved. Earthquake prediction is necessarily based on the best available empirical evidence, which may range from the nonquantifiable strange behavior of animals to the precisely laser-measured micro-movements of the crust. Any system of prediction is no more reliable than its least reliable basic data.

Teaching Suggestions 12-7

Before starting this section, collect pictures and news stories about volcanic eruptions, particularly recent ones. Have these on display while the students are studying this section. You also may wish to display a model volcano (see Demonstration).

If you live in an area where there is or has been earthquake or volcanic activity, plan a field trip with the class to observe the effects. The remains of extinct volcanoes can be found as remnants of their pipes, sills, and cross-cutting relationships with other rocks throughout different parts of North America. Ask highway departments about likely roadcuts through ancient igneous bedrock.

Several good films are available that document volcanic eruptions at Surtsey, Mount St. Helens, and Paricutín, as well as less violent eruptions in Hawaii. Encourage students, individually or in research groups, to choose among projects 1, 2, 3, 4, 6, and 9 at the end of this chapter.

Background

The temperature of molten lava can be between 600 and 1200 degrees Celsius. As magma rises through the crust, some of its dissolved gases bubble out, and may escape from the ground well in advance of a lava flow or eruption. For example, carbon dioxide gas is given off from volcanoes in large quantities. Although nontoxic, carbon dioxide can cause suffocation. This danger is enhanced by the fact that carbon dioxide is heavier than air, and sinks to the ground. Volcanoes also give off water vapor, as has been discussed, as well as carbon monoxide, hydrogen sulfide, hydrogen chloride, and hydrogen fluoride. A strong smell of sulfur is characteristic of many volcanoes.

The effect of volcanic activity on acid rain is thought to be significant by many scientists. It is estimated that 25% more sulfur dioxide has been

12-7

Volcanoes

Goals

1. To describe the causes of volcanoes.
2. To describe the kinds of volcanoes.
3. To identify places where volcanoes can be found.

volcano

A **volcano** is an opening in the crust through which magma from within Earth flows out as lava. Most of the magma that comes to the surface originates in a partly molten "zone of weakness" in the mantle. This zone of weakness is at the base of the plates. The plates put great pressure on the zone of weakness, and this pressure forces magma upward through cracks and other openings in the crust.

The locations of major volcanoes around the world are shown in Figure 12-21. The pattern should look familiar to you. Most volcanoes are near plate boundaries, in the same regions where earthquakes occur. The area around the Pacific Ocean is sometimes called the "Ring of Fire" because so many volcanoes and earthquakes are found there. Plate boundaries are areas of movement in the crust, where most of the cracks and weaknesses in the crust are located. This makes plate boundaries areas where magma is likely to find its way to the surface. There are exceptions to this rule. The Hawaiian Islands, for example, are major volcanoes that are not on a plate boundary.

When lava comes out of a volcano, the pressure on it is released. Gases that were dissolved in the lava are given off. Sometimes a huge cloud of steam, the most common of

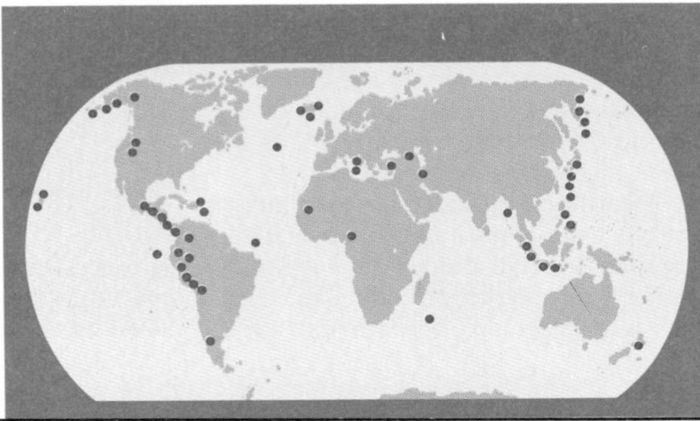

Figure 12-21 *Active and recently extinct volcanoes are located on this map. Compare it with Figure 12-18.*

emitted by the 1982 eruptions than by North American industry in a year's time. How this increased potential for acid rain will affect North American rivers and lakes is uncertain. The effects could be spread over several years, depending on how long it takes the sulfur dioxide to be brought back to Earth with precipitation.

Explosive eruptions are thought to be the result of new activity from old plugged vents. Very little rock material is ejected during such violent, explosive eruptions. In contrast, the eruption of shield volcanoes is mild, despite dramatic displays of lava flows, fountains, and lakes.

The eruptions of Mount Pelée on Martinique in 1902 and of Krakatoa in 1883 are commonly discussed explosive events. Mauna Loa and Kilauea, both on the big island of Hawaii in the state

these gases, accompanies a volcanic eruption. Harmful gases, such as carbon monoxide and sulfur dioxide, may be given off also. Besides lava and gases, a volcano can throw out chunks of rock of various sizes. These can range from small particles of volcanic dust to large chunks called *bombs* or *blocks*. Many of these form as the lava hits the cool air, solidifying the molten rock immediately.

The opening at the top of a volcano is called a **crater**. After an eruption is over, the lava settles back down into the crust and a solid cap, or *plug*, forms in the crater. The plug is similar to the hard crust formed on top of warm pudding. A volcano that is quiet or inactive is said to be *dormant*. Before its eruption in 1980, Mount St. Helens was dormant for more than 100 years. A volcano that has not erupted within historic times is called *extinct*.

There are three main types of volcanoes, classified by the shape of their cones. The **cone** of a volcano is the pile of material that collects around the opening. The shape of the cone depends on how the lava flows out of the volcano. A *shield cone*, typical of the Hawaiian Island volcanoes, forms when the lava runs out quietly from the opening and forms gently sloping, almost flat layers. A *cinder cone* volcano is typical of an explosive type of eruption and is built up by layers of ash and cinders. It has steeper sides than a shield cone. *Composite cone* volcanoes are built by both quiet and explosive eruptions and have alternating layers of lava and cinders. (See Figure 12-22.)

In the case of Mount St. Helens in 1980 (Figure 12-23), one side of the mountain swelled up before the explosion

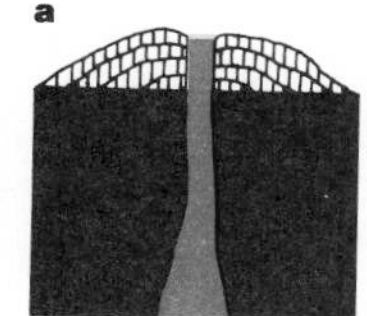

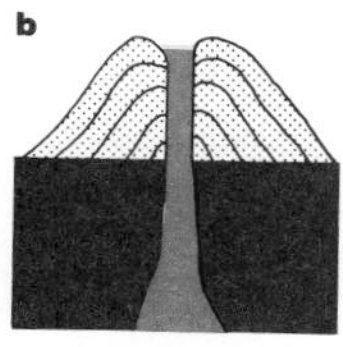

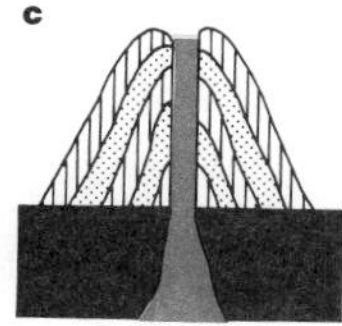

Figure 12-22 *Types of volcanoes: (a) shield cone, (b) cinder cone, (c) composite cone.*

crater

cone

Figure 12-23 *Mount St. Helens, before and after erupting.*

of Hawaii, are nonexplosive, active, and well-studied.

Demonstration

Follow these instructions to create a simple model of a volcano. Melt a crayon and coat a piece of string with the melted crayon. With the crayon-covered string, tie several pieces of crayon in a bundle, leaving at least 5 centimeters of free string hanging from the knot. Make sure the bundle of crayons is small enough to fit into a paper cup without touching the sides. Fill the cup with a thick mixture of plaster of Paris and water. Push the bundle of crayons into the cup full of plaster of Paris and water. The crayon bundle should not touch the sides of the cup, and both ends of the string should hang out over the side of the cup. Tap the cup on the table to remove air bubbles. Allow the plaster to harden overnight. Be sure it is completely set, and then cut the string close to the surface of the plaster and tear away the cup. Put the plaster in a pan of boiling water, string-end up, and leave it there until the crayon inside melts, as shown in Figure TE 12-4. The crayons, which melt inside the plaster, exert pressure as they expand. The string in the crayon-and-plaster model represents a zone of weakness in Earth's crust. The melted crayon comes to the surface like "magma" along the zone of weakness and is extruded from the plaster model.

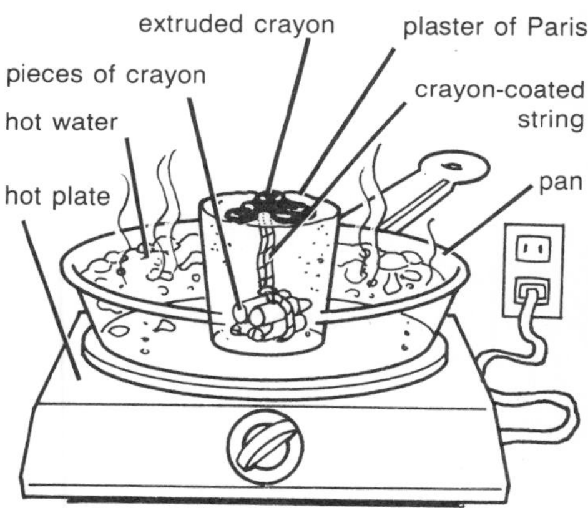

Figure TE 12-4 A model of a volcano.

Teaching Tip

- Remind students that the steam rising from lava is condensed from the water vapor in the lava.

Facts At Your Fingertips

- Volcanic ash and lava bring up new minerals that help enrich the soil.
- The scattering of light by volcanic dust particles may produce colorful sunsets.
- Volcanoes thought to be "extinct" can become active if pressure within a magma chamber builds up again. Mt. Lassen in California was thought to be extinct until it erupted in 1914.

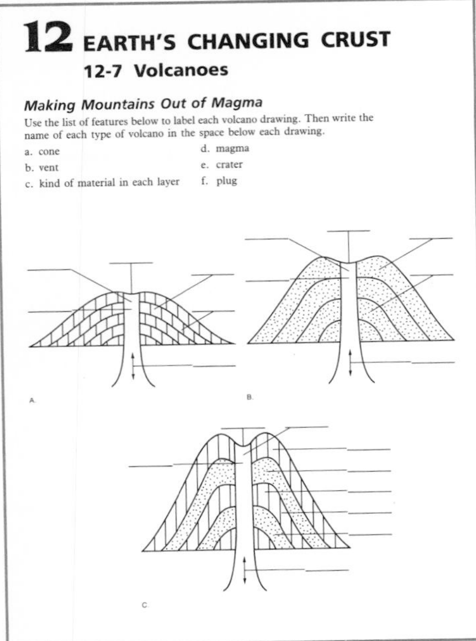

12 EARTH'S CHANGING CRUST

12-7 Volcanoes

Making Mountains Out of Magma

Use the list of features below to label each volcano drawing. Then write the name of each type of volcano in the space below each drawing.

a. cone
b. vent
c. kind of material in each layer
d. magma
e. crater
f. plug

A

B

C

Using Try This

If your supply of modeling clay is limited, have a team of students do the activity as a class demonstration. You may want to cut the clay volcanoes in half to provide a cross-sectional view.

Checkpoint Answers

1. A volcano is an opening in Earth's crust through which magma from within Earth flows out as lava. The occurrence of volcanoes is caused by plates putting great pressure on zones of weakness within Earth's crust. The pressure forces magma upward through openings in the crust.
2. At plate boundaries.
3. Shield, cinder, and composite.
4. Steam.

that blew off the top of the mountain. Scientists have been watching active volcanic areas for that and other clues to help them predict when a volcano is going to erupt. Many times there is mild earthquake activity in the area around a volcano before an eruption. The plug inside the crater may steam and swell. Predictions of volcanic eruptions are very difficult because there are so many factors that must be considered.

Although volcanoes can cause great destruction to land and life, some of their effects can benefit people. Volcanic ash may spread for hundreds of kilometers around a volcano, creating rich soil for crops. (See Figure 12-24.) In Iceland, which is located on a mid-ocean ridge, hot springs are an important heat source. Twenty-eight towns, including the capital, Reykjavik (RAY kyuh veek), heat their homes directly with water from hot springs. In one town, a bakery uses hot springs to heat its ovens.

Figure 12-24 *Gardening in the rich volcanic soil of Iceland.*

Throughout history there have been some spectacular volcanic eruptions. In A.D. 79 the eruption of Mt. Vesuvius (vuh SOO vee us) buried the city of Pompeii (pahm PAY), Italy. In 1883, the island of Krakatoa near Java was almost completely blown up by a volcanic eruption. The dust from this eruption was carried around the world by winds in the upper atmosphere. One result was extremely colorful sunsets for several years.

By reducing the amount of sunlight reaching Earth's surface, dust and ash from volcanic eruptions can affect weather conditions. The explosion of the Tambora volcano on Java in April, 1815 was one of the largest known. The debris from this explosion is thought to have caused some unusually cold weather. During the summer of 1816, snow fell and temperatures were often below freezing in eastern North America and Northern Europe.

TRY THIS

Using modeling clay of different colors, make a model of each type of volcanic cone. Use powder between the layers of different-colored clay so they can be separated at the end of the activity.

Checkpoint

1. What is a volcano? What causes the occurrence of volcanoes?
2. Where do the majority of volcanoes occur?
3. What are the three main types of volcanoes?
4. What is the most common of the gases given off by volcanoes?

Understanding The Chapter

12

Main Ideas

1. Weathering is a process in which rocks are broken down. Weathering can be physical or chemical.
2. Soil is a mixture of weathered rock material and decayed organic material. The North American continent has many types of soil.
3. Through erosion, weathered materials are carried to lower levels of elevation.
4. Agents of erosion are water, ice, wind, organisms, and gravity.
5. Materials that have been moved through erosion and left in another place are called deposits.
6. Running water, glaciers, and wind are responsible for most of the depositional features on Earth.
7. Earth consists of an inner core, an outer core, a mantle, and a crust.
8. The plate tectonic theory states that Earth's outer layers are divided into six major plates that are moving.
9. Continental drift and sea floor spreading are evidence of the plate theory.
10. Earthquakes and volcanoes are the result of stresses that build up inside Earth.
11. Most earthquakes and volcanoes occur in the same locations on Earth, at the edges of plates.
12. The energy that is released by an earthquake travels in the form of waves.

Vocabulary Review

From the following list, choose the term that best completes each of the statements. Write your answers on a separate sheet of paper.

cone	mantle
crater	meanders
deltas	mid-ocean ridges
deposits	moraines
earthquake	outer core
epicenter	P waves
erosion	plate tectonics
fault	S waves
focus	seismograph
glacier	soil
inner core	talus slope
intensity	volcano
magnitude	weathering

1. __?__ is the process in which rocks break down.
2. Triangular-shaped deposits at the mouths of rivers are called __?__.
3. A(n) __?__ is an opening in Earth's crust from which lava flows.
4. Large hills of material left by a glacier are called __?__.
5. A(n) __?__ is used to detect earthquakes.
6. The point on Earth's surface directly above the origin of an earthquake is called the __?__.
7. The most important agent of __?__ is water.
8. __?__ explains how continents can move.
9. __?__ are mountain ranges on the ocean floor.

Vocabulary Review Answers

1. weathering
2. deltas
3. volcano
4. moraines
5. seismograph
6. epicenter
7. erosion
8. plate tectonics
9. mid-ocean ridges
10. crater
11. magnitude
12. soil

Reading Suggestions

For The Teacher

Boraiko, Allan A. "Earthquake in Mexico." *National Geographic*, May 1986, pp. 655–675. Dramatic pictures and story of the earthquake in Mexico City in 1985. Discusses how tall buildings withstand earthquakes.

Eiby, G. A. EARTHQUAKES. New York: Van Nostrand Reinhold, 1980. A clear, accurate introduction to earthquakes and plate tectonics.

Fodor, Ronald V. EARTH AFIRE! VOLCANOES AND THEIR ACTIVITY. New York: Morrow, 1981. A vigorous and fresh treatment of volcanoes, with excellent illustrations.

Fodor, Ronald V. EARTH IN MOTION: THE CONCEPT OF PLATE TECTONICS. New York: Morrow, 1978. A clear, accurate, easy-to-read introduction to this difficult topic.

For The Student

Asimov, Isaac. HOW DID WE FIND OUT ABOUT VOLCANOES? New York: Walker and Co., 1981. Part of a series designed to instruct readers in methods of science and increase their scientific understanding.

Berger, Melvin. DISASTROUS VOLCANOES. New York: Watts, 1981. An excellent description of the causes of volcanoes, how they work, and famous eruptions such as of Vesuvius, Kraharva, Paracutín, and Mount St. Helens. Well-illustrated.

Gore, Rick. "Our Restless Planet: Earth." *National Geographic*, August 1985, pp. 142–181. A survey of Earth's changes through geologic time.

McDowell, Bart. "Eruptions in Colombia." *National Geographic*, May 1986, pp. 640–653. A pictorial essay on damage done from volcanic mudflows.

Navarra, John. EARTHQUAKE. Garden City, NY: Doubleday, 1980. An up-to-date and well-illustrated book that covers plate tectonics, faults, seismographs, warning systems, and earthquake prediction.

Chapter Review Answers

Know The Facts

1. physical
2. running water
3. erosion
4. delta
5. P wave
6. magma
7. carbonic acid
8. composite cone
9. boundary of the Pacific Ocean
10. inner core
11. solution
12. plates are moving apart
13. iron oxide
14. reddish color in rocks
15. magnitude
16. desert
17. subsoil

Understand The Concepts

18. Water in the cracks of bedrock would tend to freeze. When water freezes, it expands. The expansion of the freezing water would wedge rocks apart, causing more rock falls than in a dry climate.
19. The snow would prevent strong winter winds from blowing away the topsoil.
20. S waves cannot pass through fluids, so part of Earth's interior behaves like a fluid in this respect.
21. The sea floor is on a moving plate that carries older material to the outer edge of the ocean basin as new material comes up near the mid-ocean ridges.
22. Volcanoes and earthquakes both occur at the edges of plates, where the crust is weak. Magma finds a path to the surface along these zones of weakness.
23. A talus slope is a deposit of loose material at the base of a mountain. It is not deposited by water, as a delta is. The two are similar in that they are both depositional features that occur as a result of weathering and erosion.
24. Wind causes more erosion in areas where large amounts of loose material are on the ground, for example, in deserts or plowed fields.
25. The pointed ends of a sand dune point away from the origin of the wind.

10. A(n) __?__ is the opening at the top of a volcano.
11. The relative size of an earthquake is called its __?__.
12. __?__ is a mixture of weathered rock and decayed organic material.

Chapter Review

Write your answers on a separate piece of paper.

Know The Facts

1. Rocks are broken into smaller chunks by means of __?__ weathering. (chemical, physical, organic)
2. The agent of erosion that moves the most material is __?__. (wind, ice, running water)
3. Landslides are an example of __?__. (weathering, erosion, glaciation)
4. An example of a depositional feature is a __?__. (delta, glacier, fault)
5. The kind of earthquake wave that can travel through all kinds of Earth materials is a(n) __?__. (L wave, S wave, P wave)
6. Melted rock still under Earth's surface is called __?__. (lava, magma, dormant)
7. One of the main direct causes of chemical weathering is __?__. (carbonic acid, carbon dioxide, water)
8. A volcanic cone made up of alternate layers of lava and cinders is called a __?__. (shield cone, cinder cone, composite cone)
9. The "Ring of Fire" is another name for __?__. (the boundary of the Mediterranean Sea, the boundary of the Pacific Ocean, the middle of the Atlantic Ocean)
10. Earth's __?__ is composed of solid nickel and iron. (inner core, outer core, mantle)
11. When particles in water are not visible, they are in __?__. (suspension, a mixture, solution)
12. Sea floor spreading seems to be happening because __?__. (continents are drifting, plates are moving apart, plates are colliding)
13. The reddish color in some soils is caused by __?__. (iron oxide, carbon dioxide, nitrogen)
14. An example of chemical weathering is __?__. (ice breaking up rocks, reddish color in rocks, rocks being moved by running water)
15. The Richter scale is used to indicate an earthquake's __?__. (intensity, magnitude, focus)
16. __?__ soil forms where there is little rainfall. (tundra, desert, forest)
17. The layer of soil where water collects and further changes the materials is the __?__. (subsoil, topsoil, parent material)

Understand The Concepts

18. Why would a mountainous area with abundant rainfall tend to have more rockfalls than a mountainous area in a dry climate?
19. How does a thick snow cover in the winter aid in soil conservation?
20. S waves from an earthquake do not pass all the way through Earth. What does this tell you about the interior of Earth?
21. What accounts for the oldest rocks on the sea floor being near the edge of the ocean basin?

26. Physical weathering breaks up rocks into smaller pieces, exposing more surface area to chemical weathering.
27. Water in the cracks of pavement expands as it freezes. Condensed moisture beneath pavement also expands as it freezes. Together, these freezing actions heave the road material upward and crack the material itself.
28. The terminal moraine has the shape of the forward edge of a glacier, having been deposited as the glacier melted back. It is, therefore, a natural record of the most advanced position of the leading edge of the glacier.
29. Gases produced within the lava escaped as the lava was solidifying.

22. Why are volcanoes often found in an earthquake zone?
23. How is a talus slope different from a delta? How are the two similar?
24. In what kind of area would you expect wind to be the most important agent of erosion?
25. How does the shape of a sand dune indicate wind direction?
26. How does physical weathering aid chemical weathering?
27. Why do streets and highways show more damage in winter than in summer?
28. Explain how a terminal moraine shows where a glacier stopped moving forward.
29. Why do some kinds of solidified lava have bubblelike holes in them?

Challenge Your Understanding

30. As a glacier moves slowly down a mountain valley, what part of the glacier moves at the fastest rate? Why?
31. Why would a river erode more of the riverbed during floods?
32. How can gravity be said to be the cause of all erosion?
33. Why did scientists conclude that Earth's inner core is solid?
34. Compare a map of the world showing major mountain systems with the map showing the six major plates in Earth's crust. Try to explain how some mountains are formed.
35. Other than at the mouths of rivers, where would sediments be piling up and forcing lower layers down?

Projects

1. Report on Pompeii and the volcano Vesuvius.
2. Research the effect volcanoes have had on Earth's weather.
3. Report on how Iceland has made use of the heat from volcanoes. Make a small model showing how the people use this heat.
4. Report on the geysers and bubbling springs in Yellowstone Park. How do they relate to geothermal energy?
5. Make a model of Earth, showing the layers of the interior. Describe each layer. Explain how scientists have been able to learn about Earth's interior.
6. Make a model of a volcano. Show the structures underneath as well as outside.
7. Use a library to find out where major earthquakes have taken place in the last 30 years. Draw or make a copy of a world map. Mark the earthquake locations on the map. Note whether the locations fall into any noticeable patterns. If they do, try to explain why such patterns may exist.
8. Find out if you live in an earthquake-prone zone. Has your community made any provisions for earthquakes in its building codes? What plans, if any, does it have for coping with an earthquake, if one should occur?
9. Report on Mount St. Helens. What caused it to erupt?
10. Find out whether there are any soil conservation programs in your area. If so, learn how they operate.

Challenge Your Understanding

30. The middle of the glacier moves fastest, because this part of the glacier does not contact the ground. This condition is similar to the flow in a liquid stream.
31. Erosion is a result of the kinetic energy of water. The faster the water and the greater the volume, the more rapidly erosion occurs.
32. Gravity is the force that causes all matter to fall to lower levels. It is the force behind running water, moving glaciers, falling rocks, winds, and precipitation.
33. P waves slow down when they penetrate a liquid, and speed up when they go from a liquid to a solid. Studies of many P-wave records show that P waves slow down at the outer core and speed up when they meet the inner core, slowing down again as they leave the inner core and reenter the outer core.
34. Existing mountain systems are at plate boundaries, indicating that mountains are formed as plates collide. Some ancient mountain systems, which exist today only as low-relief features, are evidence of ancient plate boundaries.
35. Sediments pile up in shallow seas. Such sediments from ancient, vanished seas, and which have changed to rock, contain fossilized evidence of life that once existed in these seas.

12 EARTH'S CHANGING CRUST
Projects

1. Put a layer of sand about 15 cm thick in a plastic dishpan. Bury several ice cubes in the sand. After several hours, when the ice cubes have melted, observe what has happened to the area where the ice cubes had been.
   a. What do you observe? ______
   b. What kind of a glacial feature would this be similar to? ______
2. Fill a small jar with soil from your yard. Dump the soil on a large sheet of paper. With a magnifier, carefully look through the soil. Put any living organisms in one container and any pieces of wood, leaves, twigs, and roots in another. Make a list of the living organisms you found. Make another list of the remains of former living things you found. Carefully examine the rest of the soil.
   a. Is the soil light or dark in color? ______

12 EARTH'S CHANGING CRUST
Challenge

Acid Rain

Use library resources to research the problem of acid rain.

1. What is acid rain? ______
2. What is its major cause? ______
3. Is your area one where acid rain is found? ______
4. How does acid rain affect the soil? ______

CHAPTER 13 Overview

In this chapter students will come to understand what causes various weather phenomena. Topics covered include the general composition and structure of the atmosphere, atmospheric pressure, relative temperature, the relationship of pressure systems to air currents, the effect of Earth's rotation on wind patterns, and the effect of jet streams on weather. Condensation, the process that causes clouds to form, also is explained. Types of clouds are described and the relationship of the kind of precipitation to cloud type and temperature variations is discussed. Storms of various kinds are related to the wind patterns and cloud types that accompany them. Lightning and thunder are explained. In addition, students are introduced to the data and methods used to predict weather. In the final section of the chapter, the factors that produce the major climates of the world are described.

Goals

At the end of this chapter, students should be able to:

1. describe the different layers of the atmosphere.
2. describe how the atmosphere affects people.
3. define wind and explain what causes it.
4. describe wind patterns of Earth and their causes.
5. describe different kinds of clouds and how they form.
6. describe different types of precipitation.
7. describe different kinds of storms.
8. discuss some factors that can be used to predict weather.
9. describe how moving air masses affect the weather.
10. define climate, climate zones, and the factors that determine the climate of an area.

Chapter 13

Weather

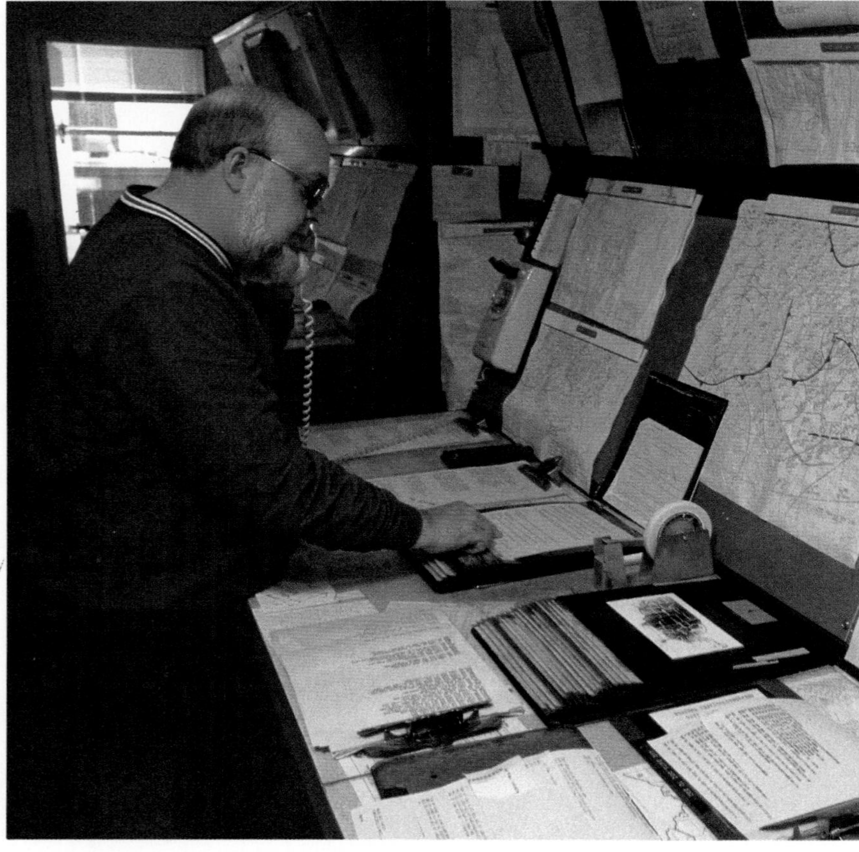

This meteorologist at Dulles International Airport provides information about atmospheric conditions.

Vocabulary Preview

atmosphere
atmospheric pressure
barometer
troposphere
stratosphere
ozone layer
ionosphere
greenhouse effect
Coriolis effect
jet streams
dew point
condensation nuclei
stratus clouds
cumulus clouds
cirrus clouds
cyclones
air mass
isotherms
cold front
warm front
occluded front
relative humidity
isobars
climate
windward
leeward

13-1

The Atmosphere

Goals

1. To describe the different layers of the atmosphere.
2. To describe how the atmosphere affects people.

Earth is surrounded by a blanket of gases about 900 km thick. This is Earth's **atmosphere**. The atmosphere helps to hold in Earth's heat. In addition, it protects Earth's surface from harmful rays of the sun. The atmosphere is difficult to detect on a clear day. You don't sense its presence unless the wind is blowing.

atmosphere

atmospheric pressure

barometer (buh RAHM uh tur)

The gases of the atmosphere push down on Earth's surface with great force. Force applied on a given area of surface is called *pressure*. The force exerted by the atmosphere on a given area of Earth's surface is called **atmospheric pressure**, or air pressure. Atmospheric pressure at Earth's surface is sometimes referred to as *one atmosphere* of pressure.

You notice differences in air pressure when it changes rapidly. Think of what happens when you are on a fast-moving elevator. The plugged-up feeling in your ears is caused by a change in air pressure. Air pressure decreases as height above Earth's surface increases. The low pressure at high altitudes makes it necessary for airplanes to have pressurized cabins.

Atmospheric pressure is measured with an instrument called a **barometer**. In the 1600's, Otto von Guericke (GAY rih kuh) made a giant barometer to demonstrate air pressure. (See Figure 13-1.) He attached a 10-m pipe almost full of water to the side of his house. The bottom of the pipe was in a tub of water. When air pressure increased on the water in the tub, the water rose in the pipe. When the pressure decreased, the level in the pipe dropped. Von Guericke tried to predict weather as the water-level rose and fell. Figure 13-2 on the next page shows a modern barometer.

Figure 13-1 *Otto von Guericke's famous barometer.*

Scientists have found that the atmosphere consists of several layers. The layers differ in such physical properties as pressure and the types of gases. The most significant characteristic of each layer, however, is its pattern of temperatures.

Teaching Suggestions 13-1

Before studying this section, obtain a barometer or barograph (a recording barometer) to have on display. Students will be able to observe how weather changes relate to changes in the barometric readings. It can be used for their weather observations while they study the rest of Chapter 13. You also may wish to do the first suggested demonstration (see Demonstrations) while discussing air pressure.

While studying this section, it would be helpful to have groups of students make large charts showing the layers of the atmosphere. The best charts can be displayed on the bulletin board. The major characteristics of each layer should be included on the chart. Some students may wish to do projects 1 and 2 at the end of the chapter.

Background

The average air pressure at sea level is 1013 millibars, which is equal to about 1 kilogram per square centimeter. (The word root "-bars" in millibars relates to barometer.) At 10 km above sea level, the air is only about one fourth as dense as it is at sea level. This makes it easier and thus more efficient for jet planes to fly at this elevation. However, humans cannot live at this level because of the low oxygen supply and reduced pressure. The atmosphere extends many hundreds of kilometers above the surface. But beyond 80 km of altitude, the atmosphere is almost a perfect vacuum.

The rapid rotation of Earth at the equator accounts for the atmosphere being thicker at this point than at the poles. The rotation opposes the force of gravity, so the air and water are "flung" away from Earth near the equator.

Temperature changes inversely with height in the atmosphere. The average drop in temperature is about 6.5° C/km. This is called the average lapse rate. The air is not moving upward; only the thermal energy is being transferred upward. When air masses do rise, a different lapse rate occurs. This is called the adiabatic lapse rate, and it is due to the cooling of the expanding, rising air. If air is falling to earth, it becomes compressed and warms up.

Temperatures begin to rise again in the ionosphere because of the interaction between atoms of gases and the ultraviolet rays of the sun. Ionization of gases occurs, causing rapid movement and increasing temperatures.

Ozone in the stratosphere is beneficial to living things on Earth because it absorbs ultraviolet energy from the sun. This reduces mutations and skin cancer. However, when ozone is found in large quantities (more than 1 part per 10 million parts of air) in the troposphere, it is considered a pollutant. Ozone in smog harms growing plants and is an irritant to the eyes.

The ozone layer in the stratosphere reacts with hydrofluorocarbons emitted from some spray cans, destroying the beneficial ozone. Restrictions have been placed on the use of these kinds of spray cans. However, exhaust from aircraft and nuclear explosions can contribute to the destruction of the ozone layer also.

Demonstrations

1. Connect a large can or plastic bottle to a vacuum pump. Allow the pump to remove the air from inside the container. Ask students to observe what happens and then have them explain why. (The container collapses because the pressure outside the container is greater than the pressure inside.)
2. Set up the equipment shown in Figure TE 13-1. The screw clamp will be open to begin with. Ask the students to predict how high the mercury will go when the vacuum pump is turned on. (The mercury will rise as high as air pressure will allow when the pump is on.) What happens when the pump is turned off? (When the pump is turned off, some mercury remains in the tube.)

troposphere (TROH puh sfeer)

The layer of the atmosphere closest to Earth is called the **troposphere**. (See Figure 13-3.) It contains about 90% of the total mass of the atmosphere. The troposphere is about 16 km thick at the equator. It thins toward the poles. The temperature of the troposphere is highest at Earth's surface. It decreases gradually with height. The lowest temperature in the troposphere is found at its upper boundary—the *tropopause*.

Almost all of Earth's weather occurs in the troposphere. For this reason, it is the part of the atmosphere that has the most noticeable effect on our daily lives.

stratosphere (STRAT uh sfeer)

The layer above the troposphere is called the **stratosphere**. The stratosphere begins about 16 km above Earth's surface. It is about 32 km thick. In this layer there is a steady increase in temperature due to the absorption of high-energy ultraviolet rays from the sun. (Ultraviolet rays are invisible rays with higher energy than visible light.) You may have heard of the **ozone layer**. This is the part of the stratosphere where absorption of ultraviolet rays occurs. The ozone layer is a broad band of gas that extends through most of the stratosphere. A molecule of oxygen is O_2. When O_2 molecules are hit by ultraviolet rays, they break apart into separate oxygen atoms. Some of these atoms then combine with other O_2 molecules, forming ozone, O_3.

ozone layer (OH zohn)

Figure 13-2 *(a) A typical modern barometer contains a box (b) with a very thin top and bottom. Inside is a spring. High pressure squeezes the box; low pressure allows the box to expand.*

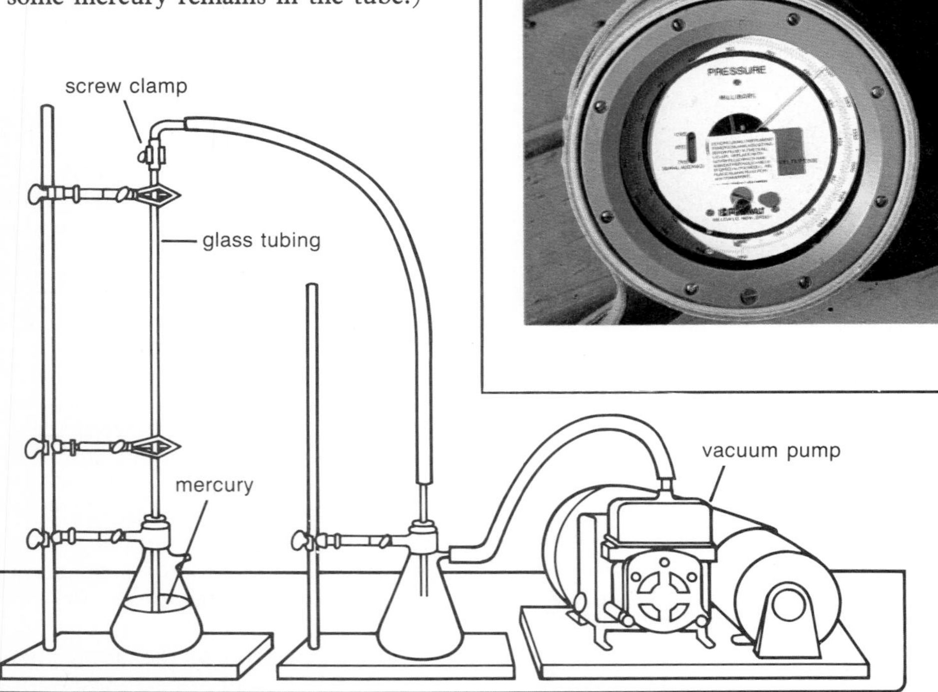

Figure TE 13-1 A setup to demonstrate air pressure.

3. To set up a siphon, you will need two large containers (beakers, flasks, or jars), a siphon tube (flexible plastic tubing about one centimeter in diameter and one-half meter long), and water. Set one container about 25 centimeters higher than the other. Fill the higher container with water. Now fill the siphon tube with water, and plug both ends to keep air out of the tube. Submerge one

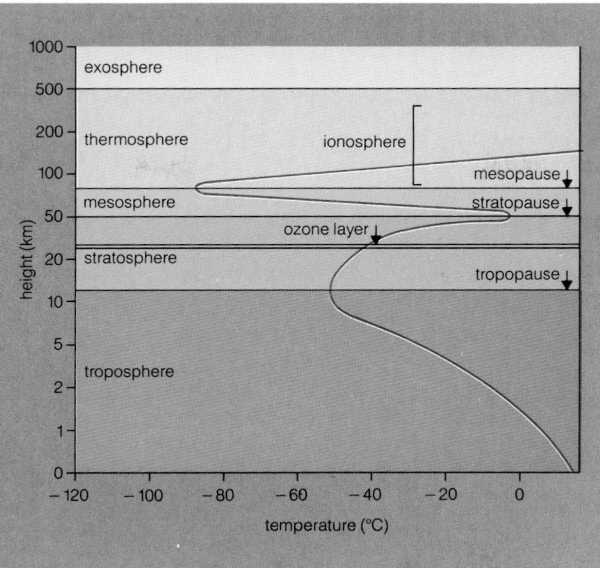

Figure 13-3 *The layers of the atmosphere. The curving line shows how much the temperature varies in the different layers.*

Scientists worry that the ozone layer can be destroyed by some of the gases used in spray cans. Some of these gases react with ozone. This reduces the amount of ozone in the ozone layer, allowing more ultraviolet rays to reach Earth's surface. Ultraviolet rays can be harmful to living things. For example, people who are exposed to large amounts of ultraviolet rays are more likely to get skin cancer. This is one reason people are cautioned to limit exposure to the sun, or to use special lotions that block out ultraviolet rays.

The upper boundary of the stratosphere is called the *stratopause*. Beyond this the temperature decreases again. The layer above the stratosphere is called the *mesosphere*. In the mesosphere the temperature decreases to − 100°C. The *thermosphere* lies above the mesosphere. Here the temperature increases to several thousand degrees. Ultraviolet rays from the sun bombard the gases in the thermosphere. This ionizes (charges) them. These ionized gases form a layer within the thermosphere called the **ionosphere**. The ionosphere has electrical properties, and because of these, the

ionosphere
(eye AHN uh sfeer)

Teaching Tips

- Let students hold objects with a mass of one kilogram. Examples include an average textbook, two large grapefruits, or a can of coffee. This will give students an idea of how much pressure air exerts on one centimeter.
- Have students give examples from their own experience as to when the air has a different color or odor. For example, after a thunderstorm the air smells cleaner (because of ozone). Also, certain pollutants, such as from industrial smokestacks or volcanoes, add color and odor.

Facts At Your Fingertips

- As a person goes to a higher elevation, the outside air pressure decreases. This will cause the pressure inside the head to push the eardrum outward. By swallowing, with the mouth open, the Eustachian tube leading to the ear from inside the throat is opened and equalizes the pressure. This makes the ear feel normal again.
- As forests are cut down, termite populations go up. They give off more carbon dioxide than fossil fuel burning. An increase in carbon dioxide in the atmosphere could cause warmer climates on Earth, and a rising sea level from melting icecaps.

end of the tube in the water inside the higher container, and put the other end in the empty, lower container. Carefully remove the plug in the submerged end, being careful to keep the tube end submerged. Now unplug the lower end of the tube. The water will flow from the higher container to the lower one through the siphon, as shown in Figure TE 13-2.

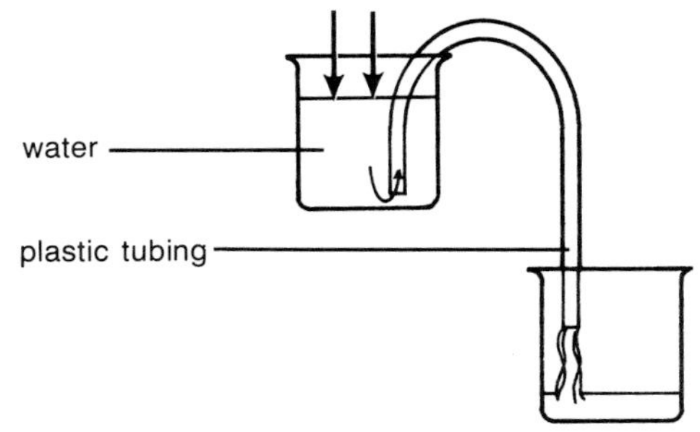

Figure TE 13-2 A simple siphon.

Mainstreaming

More time may have to be spent explaining this section's vocabulary to handicapped students. Make use of charts and models to help explain the layers of the atmosphere.

Cross References

Air pressure and breathing are discussed in section 3-3 of Chapter 3. For information about light and energy, refer to Chapter 20, section 20-4.

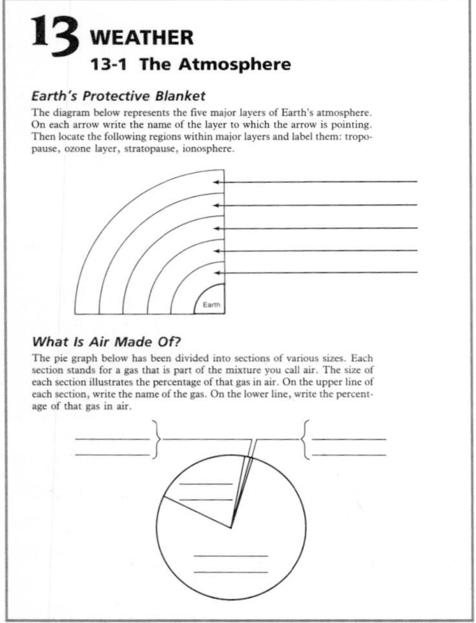

13 WEATHER

13-1 The Atmosphere

Earth's Protective Blanket

The diagram below represents the five major layers of Earth's atmosphere. On each arrow write the name of the layer to which the arrow is pointing. Then locate the following regions within major layers and label them: tropopause, ozone layer, stratopause, ionosphere.

Earth

What Is Air Made Of?

The pie graph below has been divided into sections of various sizes. Each section stands for a gas that is part of the mixture you call air. The size of each section illustrates the percentage of that gas in air. On the upper line of each section, write the name of the gas. On the lower line, write the percentage of that gas in air.

Checkpoint Answers

1. The atmosphere holds in Earth's heat and shields Earth from harmful ultraviolet rays from the sun.
2. Troposphere.
3. If the ozone layer is decreased, more harmful ultraviolet energy from the sun would reach Earth's surface causing more skin cancer. If the ozone layer is increased, it would give more protection to life on Earth.
4. Nitrogen.

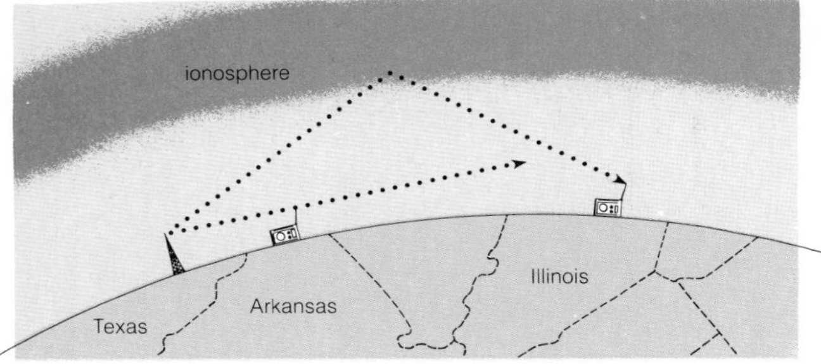

Figure 13-4 *People in Arkansas can listen to the Texas radio station at any time. Because Earth is curved, people in Illinois can't. They can get the radio waves only if the waves bounce off the ionosphere.*

ionosphere is used to relay radio communications around the world. Radio waves can be reflected off parts of the ionosphere. This allows them to travel farther around Earth. (See Figure 13-4.)

The outermost part of Earth's atmosphere extends from about 480 km out to 960 km. This part is called the *exosphere*. There is no real boundary between the exosphere and outer space. The gas particles just get farther and farther apart. Some may be kilometers apart.

The composition of the atmosphere varies. The two lower levels (the troposphere and stratosphere) contain the mixture of gases that we call *air*. The most abundant gases in air (by volume) are nitrogen (78%), oxygen (21%), argon (0.9%), and carbon dioxide (.03%). There are a number of other gases, including water vapor, in air. These gases, however, make up less than 0.1% of the air.

Air contains many solid particles also. Dust-sized bits of matter from industry and volcanic eruptions are examples. In addition, the ocean adds salt particles to the air as water evaporates from ocean spray. Substances that enter the atmosphere may travel far from their sources. They may eventually affect the weather.

Checkpoint

1. Name two ways in which the atmosphere affects people.
2. In what layer of the atmosphere does weather occur?
3. How could a change in the ozone layer affect life on Earth?
4. What is the most abundant gas in air?

13-2

Wind

Goal

To define wind and explain what causes it.

Wind is air in motion. You have probably observed how changeable wind can be. Sometimes there is hardly any wind. At other times, it is difficult to move against it. Wind is something you can't see, but you can see its work.

The major cause of wind is the unequal heating of Earth. Energy from the sun is not received equally all over Earth's surface. The greatest amount of energy is received where the sun's rays striking the surface are *vertical*, that is, at right angles to the surface. A vertical ray delivers more energy per square meter of surface than does a slanting ray. (See Figure 13-5a.) As a result, areas receiving vertical rays are warmer. Because Earth's surface is curved, some places receive vertical rays while other places receive slanting rays. (See Figure 13-5b.)

Other factors too affect how much of the sun's energy is received by Earth. As the sun's energy nears Earth, it is affected by the atmosphere. Some energy is reflected back by clouds. Some is absorbed in the stratosphere or by water vapor in the troposphere. And some is scattered by dust particles in the air. A little less than half of the sun's energy that reaches the atmosphere arrives at Earth's surface.

Figure 13-5 *(a) The same amount of light is spread over a larger area in the oval than in the circle. Thus a given area within the oval gets less light than the same area within the circle. (b) Vertical rays provide more energy to a given area than slanting rays do.*

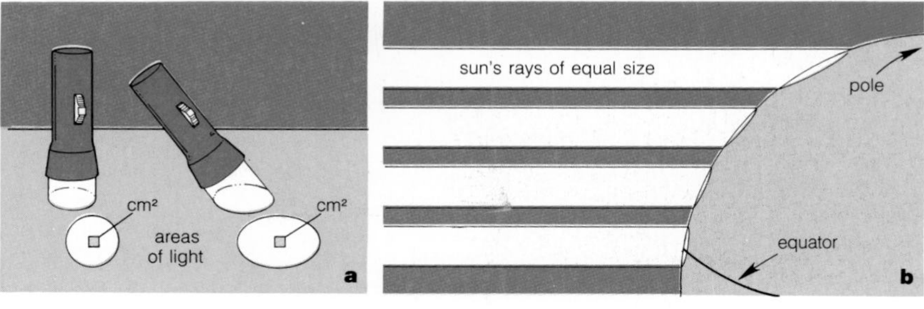

as a practical energy source. Alternately, obtain information from government agencies as well as copies of alternate energy magazines for the class to see. Some students may wish to do project 8 at the end of this chapter.

Background

Temperature changes within a parcel of air cause instabilities that lead to movement of the parcel, both internally and relative to nearby parcels. The dry adiabatic lapse rate determines whether a parcel of air is unstable or stable. If the environmental lapse rate is greater than the dry adiabatic lapse rate, the air will become unstable and rise. If the opposite is true, the air will become stable and will tend to settle back to the level from which it was displaced. Such air is colder and denser than the surrounding air.

A temperature inversion occurs when the air increases in temperature as height above the surface increases. This leads to a stable condition with little convective air motion. If pollutants are in the air at the surface, a temperature inversion can cause serious difficulties for those who have respiratory problems. The air at the surface remains cool and will not move upward. Unstable lapse rates lead to turbulent air with strong vertical currents.

The speed of the wind is measured with an anemometer. The direction from which wind blows is measured with a wind vane. Wind speed indicators should be set out in the open away from buildings, as buildings cause eddies, which will give false readings.

Teaching Suggestions 13-2

Collect pictures that show the damage done by strong winds and display them while discussing this section. Also, show pictures of hang-gliders using breezes along the seashore.

Have students research and draw the standard "station symbols" used by the National Weather Service. These symbols are used to denote weather conditions on weather maps.

When students study the section on the vertical rays of the sun, do the suggested demonstration (see Demonstration.) If possible, consider contacting a person who is using wind power to generate electricity. Arrange for this person to visit the class to tell students the pros and cons of using wind power

Demonstration

Project a beam from a strong light source (representing a bundle of rays of the sun) vertically on the equator of a large globe. Point out that areas north and south of the equator receive light that slants at angles. The light becomes increasingly spread out, or less intense, away from the equator. Show how the spot of light shining on a flat piece of cardboard, held at the same distance from the light source as the globe and at right angles to the beam, will be about the same intensity all over the cardboard. The roundness of Earth results in the unequal heating of its surface by the sun.

Teaching Tip

- Wind speed can be measured by the effect of wind on smoke, leaves, tree branches, whole trees, shingles on houses, or other materials. Wind speed ranges from calm to hurricane and tornado velocities.

Fact At Your Fingertips

- Convection cells are the basis for heating in passive solar-heated homes.

Some heat is added to the atmosphere by the ground. Energy from the sun comes to the surface of Earth in short, high-energy waves. These are absorbed by the ground. Earth, in turn, radiates long, low-energy waves back to the atmosphere. In this way, the ground adds a small amount of energy to the atmosphere. The long-wave energy radiated by Earth is absorbed by carbon dioxide in the atmosphere. This "trapped" energy causes the atmosphere to become warmer. This is called the **greenhouse effect**, because greenhouses work on the same principle. (See Figure 13-6.) The greenhouse effect insulates Earth, keeping it warmer than it would be without an atmosphere.

greenhouse effect

As a result of unequal heating, air is warmer in some places than in others. Warm air expands, becomes less dense, and rises. Cooler air then flows in to replace the warm air. The result is wind. The warmer air cools as it rises, becomes denser, and sinks again. This motion of air is called a *convection current* or *cell*. The principle of convection is used in some heating systems in homes.

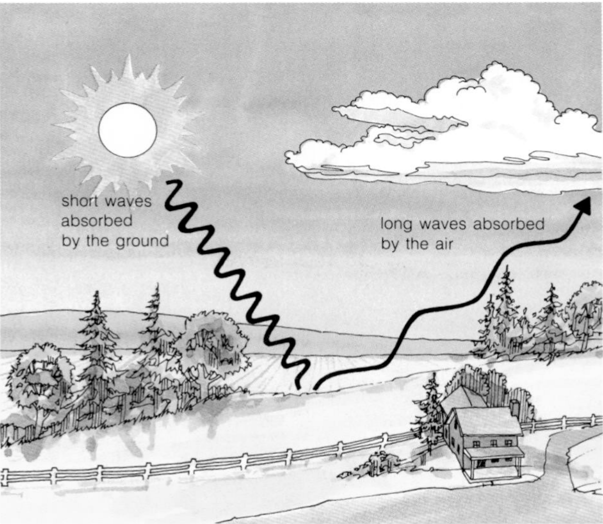

Figure 13-6 *The greenhouse effect. The sun warms the ground, and the ground radiates long waves that warm the atmosphere.*

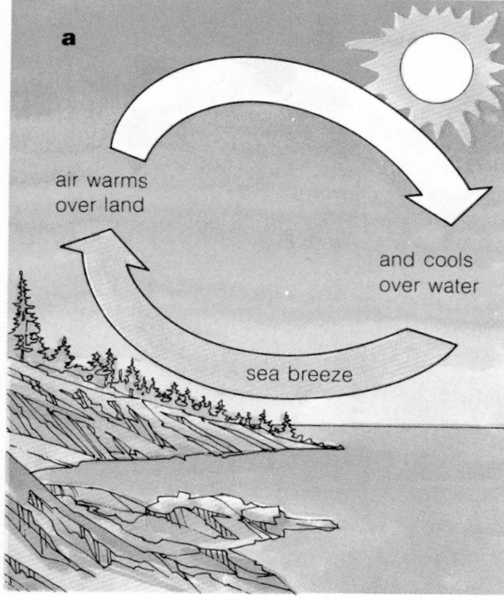

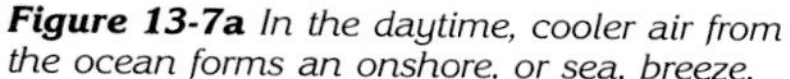

Figure 13-7a *In the daytime, cooler air from the ocean forms an onshore, or sea, breeze.*

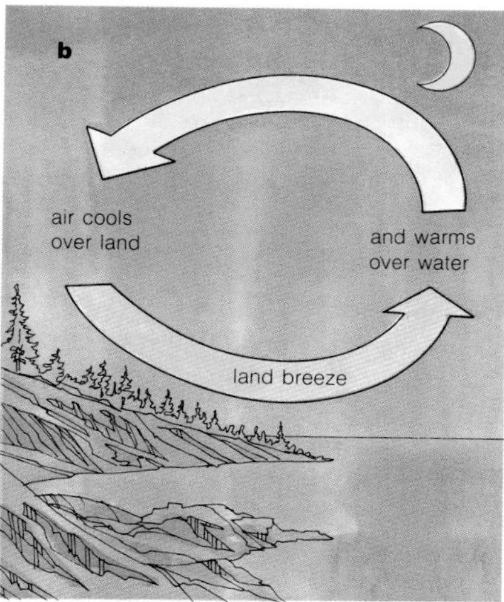

Figure 13-7b *At night, cooler air from the land forms an offshore, or land, breeze.*

A common example of a local air convection cell occurs along the seashore. Land heats up faster in the daytime than water does. During the day, warmer air over the land rises and flows toward the water. Cooler air from over the water flows in to replace the warm air. The result is a cool onshore breeze, or *sea breeze*. (See Figure 13-7a.) At night, the reverse takes place. Land cools more rapidly than water. The warmer air over the water rises and moves toward the land. Cooler air from the land flows in to replace the warm air. The result is an offshore breeze, or *land breeze*. (See Figure 13-7b.)

Checkpoint

1. What is wind? What is the major cause of wind?
2. About how much of the sun's energy that reaches Earth's atmosphere actually arrives at Earth's surface?
3. Describe the greenhouse effect.
4. In which direction does the wind on a lakeshore usually blow during a sunny day? Why?

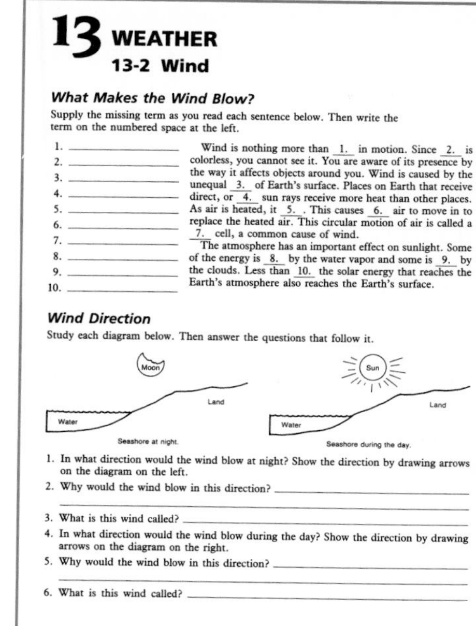

13 WEATHER

13-2 Wind

What Makes the Wind Blow?

Supply the missing term as you read each sentence below. Then write the term on the numbered space at the left.

1. ________
2. ________
3. ________
4. ________
5. ________
6. ________
7. ________
8. ________
9. ________
10. ________

Wind is nothing more than __1.__ in motion. Since __2.__ is colorless, you cannot see it. You are aware of its presence by the way it affects objects around you. Wind is caused by the unequal __3.__ of Earth's surface. Places on Earth that receive direct, or __4.__ sun rays receive more heat than other places. As air is heated, it __5.__ . This causes __6.__ air to move in to replace the heated air. This circular motion of air is called a __7.__ cell, a common cause of wind.

The atmosphere has an important effect on sunlight. Some of the energy is __8.__ by the water vapor and some is __9.__ by the clouds. Less than __10.__ the solar energy that reaches the Earth's atmosphere also reaches the Earth's surface.

Wind Direction

Study each diagram below. Then answer the questions that follow it.

1. In what direction would the wind blow at night? Show the direction by drawing arrows on the diagram on the left.
2. Why would the wind blow in this direction? ________
3. What is this wind called? ________
4. In what direction would the wind blow during the day? Show the direction by drawing arrows on the diagram on the right.
5. Why would the wind blow in this direction? ________
6. What is this wind called? ________

Checkpoint Answers

1. Wind is air in motion. The major cause of wind is the unequal heating of Earth.
2. A little less than half of the sun's energy reaches Earth's surface.
3. Short wavelengths pass into our atmosphere from the sun, are absorbed at the surface, and reradiated as long wavelengths, which cannot get back through the atmosphere.
4. The wind blows from water to land. The air over land heats up faster than the air over the water during the daytime, rises, and is replaced by air from over the water.

Cross Reference

To review the definition of density, see Chapter 6, section 6-1.

13-3

Wind Patterns

Goal

To describe wind patterns of Earth and their causes.

If Earth were not rotating, air heated at the equator would rise, spread toward each pole, and sink as it cooled. Cool air would then move along the surface toward the equator. The winds at the surface would always be from the north or south. (See Figure 13-8.) However, Earth's rapid rotation from west to east affects air currents. Earth's rotation causes air to be deflected in a clockwise direction in the Northern Hemisphere and in a counterclockwise direction in the Southern Hemisphere. This change in the path of movement caused by Earth's rotation is called the **Coriolis effect**.

Coriolis effect (kor ee OH lis)

In the region between 30°N and 30°S the surface winds are mostly from the east. These are the *trade winds*, or tropical easterlies. Between 30°N and 60°N and 30°S and 60°S, the surface winds are from the west, the so-called *prevailing westerlies*. Finally, between 60°N and 90°N and 60°S and 90°S, the winds are from the east—*polar easterlies*. (See Figure 13-9.) This is the global pattern of winds on Earth. Notice that winds are named according to the direction *from which* they blow.

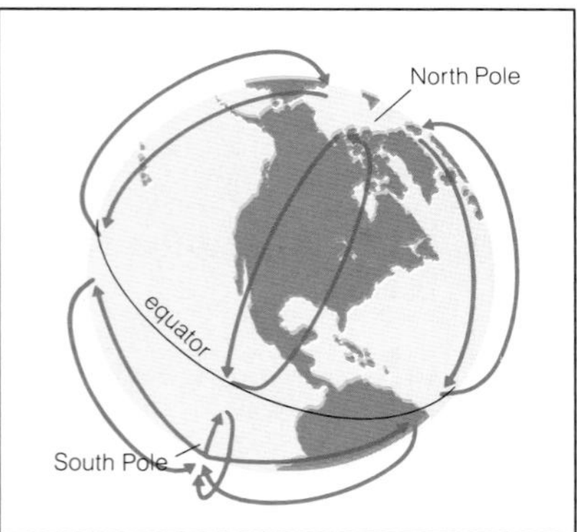

Figure 13-8 *If Earth did not rotate, the air would flow in this simple pattern. Air would rise near the equator and sink at the poles.*

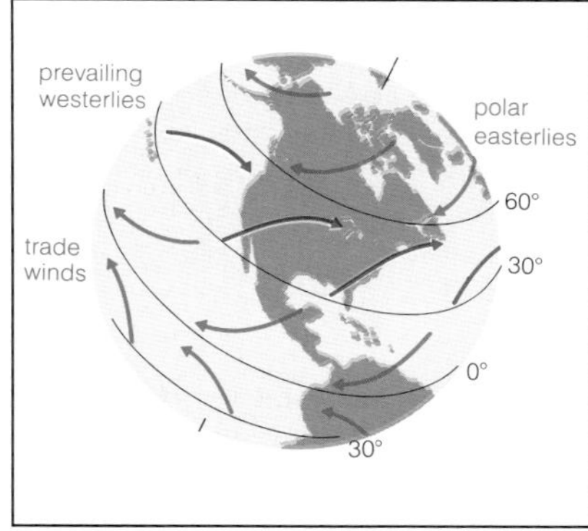

Figure 13-9 *Because Earth rotates, the temperature differences and rotation cause this pattern of easterly and westerly winds.*

Teaching Suggestions 13-3

The subject matter in this section will probably be new to most students. It is difficult to demonstrate the concepts because air is invisible, and its motion cannot be directly observed. Films are useful when explaining the effects of Earth's rotation on wind patterns.

Water can be used to demonstrate air motion. But because the viscosity of water is greater than that of air, it tends to simplify the patterns. As with any model, emphasize that patterns set up in water are similar to those in the air, but not identical. A large globe can be used to show global wind patterns.

Some students may wish to do chapter-end project 7 at this point.

Background

Pilot balloons are still used to monitor high-level winds. They carry metal targets that can be detected easily by radar. The balloons are gas filled and are easily carried along by the strong winds at high altitudes. The usual height for balloon-lofted instruments is 8 to 9 km; they can be used at altitudes of up to 20 km.

Radiosondes are miniature electronic devices carried by balloons. They measure temperature, pressure, and humidity at upper levels and radio the data to a ground-based receiver. When the balloon bursts, the package of instruments parachutes safely back to Earth. Radiosondes are accurate up to about 32 km.

To gather data from higher levels, rockets are used to carry instrument packages. These will go about 100 km up into the atmosphere. Remote sensing from satellites is an even more effective technique that is commonly used in weather forecasting. Students are likely to be familiar with televised weather forecasts based on satellite data that have been processed by computer.

Teaching Tips

- If possible, get a hot-air balloonist to describe ballooning and the principle of buoyancy to your class, and/or show pictures of ballooning when you are studying the density of air at different temperatures.
- When discussing prevailing winds, ask students to follow weather forecasts to determine from which direction most of the weather approaches your area. Relate this direction to the prevailing winds at your latitude.

Facts At Your Fingertips

- The jet streams are faster in the winter than in the summer.
- Jet streams were discovered by pilots during World War II.
- The Coriolis effect is named after a French physicist, G. G. Coriolis, who first described the force mathematically in the nineteenth century.

Warm air rises because it is less dense than cool air. Cooler air, being more dense, exerts a higher pressure than warm air. Thus, when cool air flows in to replace rising warm air, the air flow is from a region of higher pressure to a region of lower pressure. Regions of high and low pressure, caused by unequal heating of the atmosphere, determine which way the wind will blow.

Wind blowing outward from a high pressure area in the Northern Hemisphere is deflected by the Earth's rotation. The wind moves clockwise. (See Figure 13-10.) Wind blowing inward around a low pressure area in the Northern Hemisphere is deflected in a counterclockwise direction. (In the Southern Hemisphere, these directions are reversed.) The greater the difference in pressure and the closer these systems are together, the faster the winds will blow.

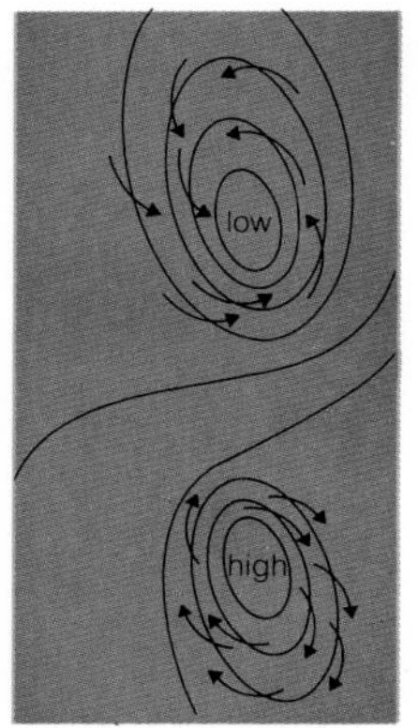

Figure 13-10 *Winds move outward and clockwise around a Northern Hemisphere high, and inward and counterclockwise around a low.*

High above Earth's surface are upper-level winds called **jet streams**. These are about 10 km above Earth's surface in the middle latitudes. The jet streams are caused by extreme differences in pressure at this level of the atmosphere. The major jet streams in the Northern Hemisphere blow from west to east.

High-level winds in the atmosphere are not slowed down by contact with Earth's surface. The air is less dense at this altitude, so winds can reach very high speeds—as great as 400 to 500 km/h.

jet streams

By observing the path of a jet stream, *meteorologists* (scientists who study the weather) can predict the paths of highs and lows on the surface. High pressure areas, or *highs*, on the ground are usually under highs at upper levels. Low pressure areas, or *lows*, on the ground are under lows in a jet stream. Pressure systems on Earth's surface behave like shadows of the jet streams above them.

Checkpoint

1. What is the name for the change in the path of movement of air caused by Earth's rotation?
2. How does air pressure affect wind?
3. What are jet streams? From which direction do they blow?
4. How does knowing the path of a jet stream help weather prediction?

TRY THIS

Using a child's toy pinwheel, how many different things can you do to make it turn? How does each method relate to winds on Earth?

13 WEATHER
13-3 Wind Patterns

Find the Error!
In each sentence below, one term makes the sentence false. Read each sentence; then underline the term that makes the sentence false. On the numbered space beside the sentence, write the term that makes the sentence true.

______ 1. Warm air rises because it is more dense than cool air.
______ 2. Air flows from a high pressure area to an area where pressure is variable.
______ 3. Jet streams are found in the upper latitudes.
______ 4. Biochemists are trained to predict the paths of highs and lows on Earth's surface.
______ 5. Dense air moves more rapidly than less dense air.
______ 6. High-pressure areas on the ground are usually under lows at the upper level.

Using Try This

Students can obtain pinwheels from a toy department, or make them from construction paper. By holding the pinwheel over a heat source, such as a radiator, students can observe the effect of a convection current on air motion. (CAUTION: If an open flame is used, keep the pinwheels well away from the flame.) To make the pinwheels turn, students may blow on them, move them through the air, or hold them in a wind.

Checkpoint Answers

1. Coriolis effect.
2. In the Northern Hemisphere, wind blows outward in a clockwise pattern from a high-pressure area. Wind moves in a counterclockwise pattern inward around a low-pressure area in the Northern Hemisphere. (These directions are reversed in the Southern Hemisphere.) Wind always blows from high- to low-pressure areas.
3. Jet streams are strong upper-level winds, which occur about 10 km above Earth's surface. In the Northern Hemisphere, they blow eastward as fast as 500 km/hr.
4. The movement of highs and lows on the surface of Earth is determined by the path of the jet stream. Studying the jet stream will help predict these movements, and therefore weather on the surface.

- All objects that are not attached to Earth's surface are affected by the Coriolis effect, which is most noticeable when moving at airplane-like speeds in a N–S direction. Navigators on planes must correct for the rotation of Earth underneath the plane.

Cross References

See Chapter 14, section 14-4, for further discussion of the Coriolis effect. See Chapter 15, section 15-1, for a discussion of Earth's rotation. Refer to Chapter 16, section 16-3, for a reference to the relationship between acceleration and mass. Refer to Appendix E, student text page 578, for a definition of latitude.

Teaching Suggestions 13-4

Clouds are actually made of tiny droplets of liquid water; a common misconception is that they are made of water vapor. Clouds are familiar to students and make a good topic for daily discussion. Take the class outside and have them make observations and records of cloud types while studying this section. An illustrated cloud chart would make good bulletin board material. Have students bring to class newspaper clippings about precipitation. Post the most dramatic of these on the bulletin board. Show the class a movie film that includes time-lapse pictures of cloud growth and change. (See audio-visual materials for Chapter 13 on page 269.)

Since dew point is described at the beginning of the section, students should do the Try This activity early. Chapter-end projects 4 and 6 apply to this section.

Background

Questions often arise as to what causes rainbows, sun-dogs, and a "ring" around the moon or sun. Rainbows are caused by sunlight passing through water droplets in the air. The water droplets act like a prism and break, or refract, the light into its component wavelengths of color. Rainbows are always seen in the sky opposite the sun. (West in the morning and east in the late afternoon or evening.) Sun-dogs are rainbowlike spots seen on either side of the sun during cold weather. These are caused by the sunlight passing through ice crystals in high cirrus clouds. The ice crystals act like prisms, causing a rainbow effect. Temperatures are usually near 0°C when sun-dogs appear. These also occur only in the morning or late afternoon. Rings around the moon or sun are caused by the refraction of their light by the water droplets in high cirrus clouds. Since cirrus clouds often precede rain, superstition has it that a ring around the moon means that there will be precipitation within 24 hours.

13-4

Clouds and Precipitation

Goals

1. To describe different kinds of clouds and how they form.
2. To describe different types of precipitation.

dew point

condensation nuclei

stratus clouds

cumulus clouds (KYOO myuh lus)

cirrus clouds (SIHR us)

Clouds form when warm, moist air rises. As it rises, the warm air expands and cools. As air cools it loses its ability to hold water vapor. As warm, moist air rises and cools, it reaches a temperature called the **dew point**. At this point the water vapor it is holding starts to *condense*, or change to liquid water. Water condenses on a solid surface. (You may have seen this happen on the outside of a glass of cold water.) Water in the atmosphere condenses on dust or salt particles. These are called **condensation nuclei**. Clouds are made of billions of water droplets that have condensed out of the air.

There are many different kinds of clouds, but they are classified into three main types. (See Figure 13-11.) **Stratus clouds** are layered clouds that usually cover the whole sky. **Cumulus clouds** are billowy clouds that look like huge heaps of cotton. **Cirrus clouds** are high, thin, wispy clouds.

Figure 13-11 *Cloud types: (a) stratus, (b) cirrus, (c) cumulus.*

Why do some places almost never have rain? In order to have rain, clouds must be present. For clouds to form, air must be lifted until it cools to its dew point—when its water vapor condenses into water droplets. If air is not being lifted, or if it doesn't carry enough water vapor, no clouds will form. This cloudless condition prevails in the major high-pressure areas on Earth, particularly near the 30-degree North and South latitudes, where the major desert areas of the world are located. Other factors, such as prevailing winds, will have an influence on rainfall, particularly along coastlines.

Once you know the three main cloud types, a few other terms are all you need to help you name clouds. For example, *nimbus*, which is Latin for "rain," can be added to cloud names. *Cumulonimbus clouds* (kyoo myuh loh NIM bus), also called *thunderheads*, are huge, towering clouds that form when a cumulus cloud expands into a vertical rain cloud. (See Figure 13-12.) Cumulonimbus clouds bring winds, lightning, showers, and sometimes hail. *Nimbostratus clouds* are sheetlike clouds that usually bring rain. The prefix *alto* is often used to identify clouds of intermediate height. *Altostratus* and *altocumulus* are "middle-level" clouds. A special term is used to describe stratus clouds at ground level—*fog*.

Cloud droplets are very tiny. It is estimated that it takes a million cloud droplets to make one raindrop! When enough cloud droplets accumulate, a drop becomes heavy enough to fall as *precipitation*.

The kind of precipitation that falls depends on many factors. The four major factors are the type of cloud from which it falls, the temperature of the cloud, the temperature of the air through which it falls, and the temperature of the ground.

Rain is probably the most common form of precipitation. It can fall from any cumulonimbus or nimbostratus cloud. *Sleet* is rain that has fallen through a cold layer of air and is frozen by the time it hits the ground. Sleet usually originates in high stratus clouds. *Drizzle* is the slow fall of very

Figure 13-12 *Cumulonimbus clouds have strong vertical currents. They cause rainstorms, thunder, and lightning.*

13 WEATHER
Computer Program

Cloud Seeding Program

Purpose To compute the amount of seeding particles needed to make rain.

Background Cloud seeding is a process for making rain, a first step in weather control. A moisture-laden cloud is sprayed with particles of either solid carbon dioxide, often called Dry Ice, or silver iodide. The particles permit droplets of water to collect in the cloud and fall as either rain or snow. The seeding particles are usually released from airplanes flying through the clouds.

ENTER the following program:

```
10  PRINT "ENTER THE VOLUME OF THE CLOUD"
20  INPUT "IN CUBIC METERS."; V
30  PRINT "WHICH SEED WILL YOU USE?"
40  PRINT "1) CARBON DIOXIDE"
50  PRINT "2) SILVER IODIDE"
60  INPUT "ENTER THE NUMBER OF YOUR CHOICE."; C
70  IF C = 1 THEN GOTO 100
80  IF C = 2 THEN GOTO 140
90  GOTO 30
100  LET S1 = V / 100
110  PRINT "RELEASE"; S1; "GRAMS OF"
120  PRINT "CARBON DIOXIDE."
130  GOTO 170
140  LET S2 = V / 100000
150  PRINT "RELEASE"; S2; "GRAMS OF"
160  PRINT "SILVER IODIDE."
170  END
```

Discussion

1. What is the weight of carbon dioxide required to seed cloud A of 132,000 cubic meters?

____

2. What weight of silver iodide is required to seed cloud B of 947,000 cubic meters?

____

Upgrade the Program

This program checks the user's entry for the variable C in lines 70 and 80. If the user has entered a number other than 1 or 2, the GOTO statement in line 90 sends the computer back to line 30 to ask the question again.

Adapt the program to permit the user to enter the three dimensions of the cloud in meters and a calculation for the volume of the cloud.

Facts At Your Fingertips

- About 540 calories of energy are released for every gram of water vapor changed to liquid.
- The energy released within clouds as water vapor condenses produces strong air currents that typically make cumulus clouds build upwards. In cumulonimbus clouds, winds are strong enough to tear the wings off small aircraft. Even large planes often avoid well-developed cumulus clouds because of their "bumpy" air.
- Cumulonimbus clouds can build up in altitude until they reach the top of the troposphere. The tops will then level off, forming the anvil shape often seen in thunderstorms.

Teaching Tips

- Make sure students know the difference between water vapor and the water droplets in clouds. Water vapor is an invisible gas. Students should realize that fog is a form of cloud; they should be familiar with the droplets of water in fog, which diffuse and reflect automobile headlights at night, making driving in a nighttime fog dangerous.
- Relate the method by which water vapor condenses inside a cloud to the way water vapor condenses on the outside of a glass of cold water.

Mainstreaming

Handicapped students may need help with the vocabulary in this section. For the Try This activity, visually-impaired students can work with a sighted student, but they should be able to feel the condensation with their fingers.

Cross Reference

Referring to Chapter 6, section 6-2, review how gases cool as they expand and heat up when compressed and review the energy exchange that occurs during evaporation and condensation.

13 WEATHER

13-4 Clouds and Precipitation

A Cloudy Problem

A term is missing in each sentence below. Fill in each missing term. Then solve the puzzle by answering the question in number 8.

1. _ _ _ _ _ _ _ clouds are layered clouds that usually cover the whole sky.
2. The temperature at which water vapor changes to liquid is called the _ _ _ _ _ _ _ _.
3. A form of frozen rain that can cause much damage is _ _ _ _.
4. It takes a million _ _ _ _ _ droplets to make one raindrop.
5. Rain that freezes as it falls through a cold layer of air is called _ _ _ _ _.
6. The word _ _ _ _ _ _ means "rain" in Latin.
7. High, thin, wispy _ _ _ _ _ _ clouds usually precede precipitation.
8. What word do the double underlined letters spell? Hint: You must unscramble the letters first. ______________________

Find the Words

In the word search below there are 14 terms related to precipitation. Find the terms and circle them. Then write each term on one of the numbered spaces at the left.

1. ____________
2. ____________
3. ____________
4. ____________
5. ____________
6. ____________
7. ____________
8. ____________
9. ____________
10. ____________
11. ____________
12. ____________
13. ____________
14. ____________

```
R A I N C C N C A P R S D T
L A B T U I I I L O A L N R
U P W H R M R R U T G I I C
S T R A T U S R O F O G P U
U H G I N S C U J P F G I M
T C R L O C K S W Y D V A U
J H O H N T H E M O S B A L
S C U O T R D R I Z Z L E U
R A L N O H S H E E L A I S
C A H P D U N I H N E L R T
M Y A O B E H A S F L T E A
E V E U S T R V E C L O U D
N I M B U S C H S O C T A U
L E T U S S E A E M Y S H O
C I S C O B L C K A M A L I
E V E S L E E T L Y D N C O
```

Using Try This

The outside of the can will take on a foggy appearance as water vapor condenses on it. When students rub a finger across the surface, the "fog" disappears for a moment, and then reappears, showing that the surface was covered with small drops of water.

Checkpoint Answers

1. Clouds form when warm, moist air cools and condenses, forming water drops on particles in the chilled air.
2. Stratus, cumulus, and cirrus.
3. Thunderhead.
4. Sleet is rain that freezes after it leaves the cloud. Snow is water vapor that changes directly to ice inside the cloud before falling.
5. Hailstones get larger the longer they circulate within a cumulonimbus cloud before falling.
6. Low-lying stratus clouds.

Figure 13-13 *If you cut one of these hailstones open, you would be able to see its layers.*

tiny drops of water. They sometimes stay suspended in the air as a mist. Drizzle forms in low-lying stratus clouds. If the temperature of the ground is below freezing, the drizzle freezes as it hits the ground.

If the clouds are high and cold enough, the water freezes in the cloud and starts out as *snow*. If it melts on the way down, it lands as rain. If the temperatures are low enough all the way down, the precipitation remains snow.

Hail is another form of frozen precipitation. It usually accompanies thunderstorms and comes from cumulonimbus clouds. If the clouds are piled up extremely high, the winds inside the cloud form strong convection cells. In these cells, raindrops are lifted to heights where it is so cold they freeze into hailstones. Hailstones can be tossed up and down within the cloud many times. They pick up layers of ice as they move. When they get heavy enough, they fall. The largest recorded hailstone was 19 cm in diameter, about the size of a volleyball. A hailstorm can do a tremendous amount of damage in a few minutes. Hailstorms often flatten growing crops such as grain.

The amount of precipitation varies around the world. Some places, such as the Atacama Desert in Chile, have never recorded any rainfall. In contrast, Mt. Waialeale in Hawaii receives about 1200 cm each year!

Checkpoint

1. What causes a cloud to form?
2. What are the three main types of clouds?
3. What is another name for a cumulonimbus cloud?
4. What is the difference between sleet and snow?
5. What causes hailstones to vary in size?
6. What kind of cloud produces drizzle?

TRY THIS

Find the dew point of the air in the classroom by slowly adding small pieces of ice to a container of water. Use a shiny can for the container, if possible. Gently stir the mixture with a thermometer. Record the temperature when drops of water begin to condense on the outside of the can. Repeat this several times to make sure you get the correct temperature. This temperature is the dew point.

13-5

Storms

Goal

To describe different kinds of storms.

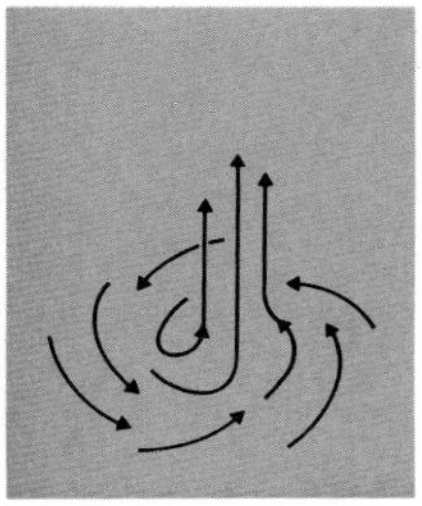

Storms can be exciting to observe. Victims of powerful storms, however, might not call them exciting. Tremendous amounts of energy, in the form of wind, are released in storms. On seacoasts this wind can cause huge waves that cause damage along the shore. By understanding how storms arise and being able to predict their behavior, people can save many lives.

Figure 13-14 Winds in a cyclone curve inward and upward.

Most violent storms are caused by circulating winds blowing inward around a low pressure center. These regions of low pressure with their circulating winds are called **cyclones**. In a cyclone, the winds move upward as they circle in toward the center of the low. (See Figure 13-14.) The severity of these storms depends on the speed of the winds.

cyclones (SY klohnz)

A *tornado* is a small cyclone with a diameter less than a few hundred meters. Tornadoes travel along the ground for only about 25 km. They generally contact the ground for only a few minutes, but often do much damage in that time. They are the most powerful storms on Earth. The winds inside a tornado can reach speeds up to 400 km/h. (See Figure 13-15.)

Figure 13-15 A tornado touching the central plains of the United States. Tornadoes usually touch ground for only a few minutes.

Teaching Suggestions 13-5

Depending on your location, students may never have experienced some of the storms described in this section. Films can be used to illustrate the various storms. In addition, students may do library research to gather more information on specific hurricanes, tornadoes, or other storms. Display pictures of the damage done by tornadoes and hurricanes. Pictures of tornado clouds make good subjects for discussion. Invite a meteorologist to talk to your class about storms and the precautions necessary to avoid injury during violent storms. Some students may wish to do chapter-end projects 3, 5, 10, and 11, which pertain to this section.

Background

Sometimes a squall line can be seen approaching ahead of thunderstorms. From the ground, a squall line looks like a wall of dark clouds. It is composed of dozens of thunderstorms. Squall lines form 50 to 250 km in front of a cold front that is moving into warm, muggy air. When the squall line hits an area, it is followed by violent winds and rain, which may last up to about a half hour. After the storm, the air is noticeably cooler than before the storm.

Great efforts have been made to know more about hurricanes so that their paths can be predicted and people and property can be spared. Students can research the work of "hurricane hunters," daring scientists who fly in aircraft through hurricanes to better understand them by close observation. With satellite photos and other detection equipment, scientists have been able to add to the knowledge gained by the "hurricane hunters." It has been shown that hurricanes are a necessary part of the huge heat "engine" that distributes energy from the equator to northern and southern latitudes. Some scientists fear that more drastic results might occur if they try to control these storms. In some cases the storms act as "safety valves" that release energy that has built up in one area. Besides the damage done by storms, they provide the benefits of cleansing the air of pollutants and replenishing the water on the surface of Earth.

Demonstration

Fill a large clear container about two thirds full with water. Using a long-handled spoon or other stirrer, stir the water vigorously. The movement of the whirling, coneshaped vortex formed in the water is similar to the motion of air in a tornado or hurricane.

Teaching Tips

- Ask students to share experiences they may have had with storms. Students in warmer regions might find accounts of blizzards quite dramatic! And a student who has experienced a tornado is likely to find a ready audience.
- Ask students if they have ever observed leaves or dust being blown around in little circles. These are miniature tornadoes.

Facts At Your Fingertips

- Tornadoes usually will not occur unless the temperature has reached at least 16°C before the storm.
- A thunderstorm can release about 475 million liters of water and give off enough energy to supply the whole U.S. or Canada with electricity for 20 minutes. A hurricane can have 12,000 times more energy than a tornado.
- Most thunderstorms move at a speed of 35 to 40 km/hr. Based on this fact, it may start to rain about 15 to 20 minutes after you hear the first thunderclap.

Figure 13-16 *Lightning often creates spectacular displays.*

If you are ever caught in tornado weather, keep tuned to radio and television for watches and warnings. If you're inside, stay away from windows and go to the lowest level, preferably to a closet or bathroom. If you're in a car, get out and lie flat in a ditch or ravine. Cars, mobile homes, and other vehicles can be picked up by a tornado. Following these simple rules could save your life. Tornadoes can be predicted, but not as accurately as hurricanes.

Hurricanes are the same kind of storm as tornadoes, but they cover more territory. They occur only over or near water, and they last much longer. A hurricane is shaped like a doughnut, with a "hole" in the middle of the circulating winds. During a hurricane, strong winds will blow for several hours. Then it is quiet as the "eye" or "hole" passes. As the storm moves, winds blow strongly again from the opposite direction. Hurricanes are 480–960 km in diameter, and their winds can approach 320 km/h. (Compare the size of a hurricane with the distance from Chicago to Nashville—about 750 km.) The eye of a hurricane can be 24–40 km across (less than the distance run in a marathon).

Hurricanes begin in tropical oceans. There temperatures are high and large amounts of water are evaporated. During evaporation, energy is stored in the water vapor. When this water vapor condenses in clouds, the stored energy is released, producing wind. When a hurricane reaches land it starts to lose its power because there is much less water vapor to condense.

A more common type of storm is the *thunderstorm*. Thunderstorms originate in cumulonimbus clouds. There the winds are strong, but not as great as the winds in hurricanes or tornadoes. Thunderstorm winds are accompanied by thunder and lightning. (See Figure 13-16.) Lightning can be within parts of the cloud, or between cloud and land, or between two clouds. Thunder is sound produced by the violent expansion of air being heated by the lightning. Heavy rain often falls during a thunderstorm.

To protect yourself from lightning, try to stay indoors during a thunderstorm. Stay away from open windows and doors. Do not touch metal pipes, including faucets. Try not to use electric appliances, such as a television or an iron. Outdoors, stay away from tall trees, poles, or power lines. Never swim during a storm. If you are in a boat, head for land. If you are in a car, stay there, but do not touch any metal parts. The metal of the car will conduct any electricity to the ground.

Checkpoint

1. Describe a cyclone.
2. What do cyclones, tornadoes, and hurricanes have in common?
3. Why does a hurricane start to die out when it reaches land?

TRY THIS

Blow a small paper bag up and twist the top so the air will stay inside. Holding the top of the bag in one hand, quickly break the bag between your hands. What did you observe? How is this similar to thunder?

13 WEATHER
13-5 Storms

Can You Tell One Storm from Another?

Each numbered item below describes a particular kind of storm. Decide what kind of storm is being described. Then write the name of the kind of storm on the space beside the description.

________ 1. The most powerful kind of storm on Earth.
________ 2. Storm that occurs only over water.
________ 3. Description of a family of storms.
________ 4. Storm that has a quiet period as the "eye" passes.
________ 5. Storm in which expansion of air from lightning causes a loud noise.
________ 6. Storm that cannot be predicted as well as other kinds of storms.
________ 7. Storm whose name means "winds around a low".
________ 8. Storm that can contact the ground, doing much damage.

El Niño

Read the paragraphs below. Then answer the questions that follow.

The winter of 1982–83 brought several destructive storms to the coast of California. An unusual warming of the eastern Pacific Ocean, called El Niño, caused these storms to hit California. The El Niño, or Christ child, usually occurs around Christmas every two to seven years.

In early 1983 the ocean warmed up more quickly than usual. This warming caused the jet stream to speed up and move farther south. As the jet stream moved south, the surface storm systems were lured toward California rather than toward Canada and Alaska.

Scientists believe that the unusual warming of the ocean may have been caused by volcanic dust in the atmosphere. The dust prevented cooling of the ocean by increasing the greenhouse effect.

1. What is the El Niño? ________
2. How was the El Niño unusual in the winter of 1982–83? ________
3. What do scientists believe caused the El Niño to be different in 1982–83? ________
4. What change does volcanic dust produce on the ocean? ________

Using Try This

A lunch bag is good for this activity. Students should hear a loud noise as the bag breaks. The noise is caused by the sudden rush of air out of the bag as it breaks. Similarly, thunder is caused by air expanding rapidly after being heated by lightning.

If you have access to a Van de Graaff generator, it can be used to demonstrate the sound generated by a high-voltage mini-lightning bolt. The example is more closely related to the natural process of thunder than the paper-bag analogy.

Checkpoint Answers

1. A cyclone is any low-pressure storm with counterclockwise winds in the Northern Hemisphere, or clockwise winds in the Southern Hemisphere.
2. Cyclones, tornadoes, and hurricanes are all low-pressure storms.
3. The rate of evaporation is much less over land than over the sea, so when over land, the hurricane loses most of its source of energy.

Teaching Suggestions 13-6

During this section, you may wish to give students the opportunity to predict the next day's weather. If the class hasn't made meteorological observations themselves, they can use daily weather maps from local newspapers, together with television and radio weather forecasts to predict the next day's weather. Students should be able to explain clearly why they agree or disagree with a published or aired weather forecast. Their predictions will have to wait only a day to be corroborated or disproved. If wrong, they should try to explain why their forecasts were incorrect.

If you haven't yet had a meteorologist talk to your class, now would be a good time to invite one. Ask the speaker to include an explanation of the equipment used in making observations and predictions about weather. If possible, follow this visitation with a field trip to a weather station at an airport, a nearby National Weather Service station, or a university.

Drawings on the chalkboard or transparency overlays on the overhead projector are helpful when explaining abstractions like air masses, weather fronts, and the resulting weather conditions.

Collect as many weather instruments as possible (if you haven't been using them all along) to show students how various weather data are collected. Activity 19, Making a Simple Hygrometer, student text page 556, should be done in conjunction with this section. Also chapter-end projects 9, 10, and 11 apply to this section.

13-6

Weather Prediction

Goals

1. To discuss some factors that can be used to predict weather.
2. To describe how moving air masses affect the weather.

Farming, aviation, and construction are industries that rely on accurate weather predictions. Weather forecasting is a big business. The movement of weather patterns on Earth's surface is determined by winds high in the atmosphere. Thus, data from upper layers of the atmosphere are collected by weather balloons and satellites. These data, collected in all areas of the country, are sent to the National Weather Service. Forecasts are then sent to newspapers and radio and television stations for local use. In small communities, some individuals collect weather information and send it to area forecasters. Computers have speeded up and improved the collection and interpretation of weather data.

Some data are needed to successfully predict weather. These are temperature, atmospheric pressure, humidity, wind speed and direction, precipitation, cloud type, sky condition, and dew point. Because there are so many factors, weather prediction is a complicated process.

air mass

The first step in weather prediction is identifying the size, location, and movement of an air mass. An **air mass** is a huge volume of air that has the characteristics of the area over which it started out. For instance, an air mass that originates over the ocean contains a lot of water vapor. One that starts out over land is dry. An air mass that originates in the northern latitudes (Northern Hemisphere) is cold. One that starts out in the south is warm. These characteristics can be mixed. Thus, cold-dry, cold-wet, warm-dry, or warm-wet air masses occur.

isotherms (EYE suh thurmz)

Air masses do not stay in one place very long. Earth's rotation sets them in motion. By studying **isotherms**, lines on a weather map that connect areas of equal temperature, the boundaries of air masses can be located.

When one air mass overtakes another, the leading edge

Background

Relative humidity can be measured with a wet-dry bulb thermometer. The wet bulb gives the temperature at which water is evaporating. The dry bulb gives the air temperature. These two temperature readings can then be related to a standard table that gives the relative humidity. A sling psychrometer is another instrument that can be used to find relative humidity.

After considering all the weather data, forecasters (often using computers) will calculate the chance of precipitation for an area. A 30% chance of rain also means that there is a 70% chance that it won't rain!

The movement of weather fronts is determined by the movement of high-of the approaching air mass is called a *front*. Each type of front has a typical weather pattern associated with it. If a cold air mass moves into a warm air mass, a **cold front** is formed. (See Figure 13-17a.) Because cold air is denser than warm air, cold air in motion has more momentum. The cold air pushes the warm air up. This upward motion causes cumulonimbus clouds to form, and often a short period of precipitation follows. After the cold front passes, the weather becomes clear, dry, and cooler.

cold front

If a warm air mass overtakes a cold air mass moving in the same direction, a **warm front** is formed. (See Figure 13-17b.) The warm air slowly moves up and over the denser cold air. High cirrus clouds announce the approach of a warm front. As time goes by, the clouds get thicker. A long period of precipitation usually occurs. After the front passes, the weather takes the characteristics of the warm air mass.

warm front

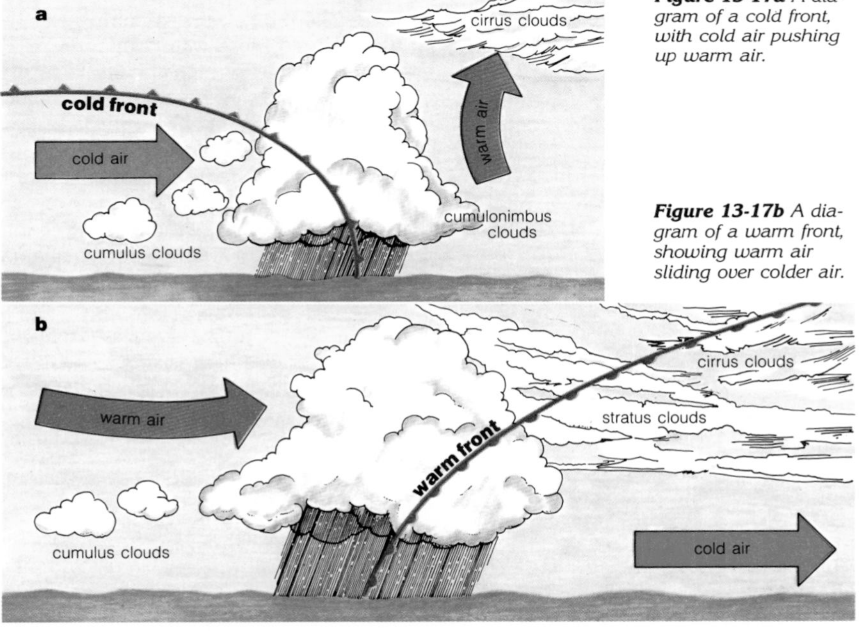

Figure 13-17a *A diagram of a cold front, with cold air pushing up warm air.*

Figure 13-17b *A diagram of a warm front, showing warm air sliding over colder air.*

Teaching Tips

- When defining relative humidity, you can use the analogy of a bucket of water to represent water-laden air. When the air is saturated, the bucket is full to the brim. If the temperature of such air is lowered, the bucket "contracts," so that the same amount of water becomes relatively large, and some water may "spill" out (analogy for precipitation). If the air temperature increases, the bucket expands, and has an increased capacity to hold water.
- Besides the industries listed on page 334, what other jobs depend directly or indirectly on the weather? Have students brainstorm to compile a list of 100 occupations or more.

13 WEATHER Laboratory

Weather Observation

Purpose To observe and record daily weather observations; to relate certain weather characteristics to other weather observations; to predict the weather.

Materials large sheet of paper on which to record data, outside thermometer, barometer, weather vane, windspeed indicator, cloud chart, rain gauge

Date / Time									
Air temperature									
Atmospheric pressure									
Wind direction, speed, and sky condition / Cloud type / Weather									
Precipitation									

Procedure

1. Make a large chart like the one shown on this page. Put the chart on the bulletin board.
2. Each day, gather the needed information from your observations and the instruments. On the chart, record temperature, barometric pressure, wind speed, wind direction, sky condition, cloud type, precipitation, and relative humidity.
3. At the end of each observation period, predict the weather for the next day. Base your prediction on your past observations. Check your success the next day.

Discussion

1. Did you find certain weather conditions that are often observed together? If so, list them. ______
2. What conditions preceded a rainy day? ______
3. What conditions preceded a clear day? ______
4. How accurate was your weather prediction? ______

pressure and low-pressure centers on the surface. These highs and lows are guided by the highs and lows in the jet streams far above Earth's surface.

The prefix "iso-" in isotherm is derived from the Greek word that means "equal." Familiar words with the prefix "iso-" are isosceles, an equal-sided triangle; isotonic exercise, equal tension; and isometric exercise, equal (force) measurement. Isotherm refers to the lines on a weather map connecting points of equal temperature. Isobar refers to the lines showing equal barometric pressure; and isohyet refers to the lines connecting points of equal rainfall.

Facts At Your Fingertips

- When humidity is high, we are uncomfortable because the wet air cannot take on more moisture, and consequently our bodies cannot cool off efficiently by the evaporation of perspiration.
- Humidity will tend to be higher with low temperatures because it takes less water vapor to saturate cold air.
- An air mass can be large enough to cover half a continent.

Mainstreaming

Ask handicapped students to listen to weather forecasts before and during the study of this section. They may need extra help with the vocabulary.

occluded front

An **occluded front** results when a warm air mass is lifted upward as two cold air masses meet. The warm air is "cut off" from the ground by cold air. (*Occluded* means "cut off.") Occluded fronts produce long periods of precipitation. Can you remember a period when it rained for several days? Still another kind of front is a *stationary front*. This is caused by a front that stops moving. It can cause precipitation that lasts as long as the front remains stationary. (Figure 13-18 shows the symbols used on weather maps for these four kinds of fronts.)

relative humidity

Knowing how much water vapor is in the air is useful in predicting weather. An important measure of water vapor is **relative humidity**. The amount of water vapor in the air compared to the maximum amount the air *can* hold at a given temperature is called relative humidity. It is usually expressed in terms of percent. If the relative humidity is 100%, air is said to be *saturated*. It cannot hold any more water vapor at that temperature. If the temperature of the air goes down, water vapor will condense out of the air. If the temperature of the air goes up, the relative humidity will be lower. The warmer air can hold more water vapor.

isobars (EYE suh barz)

In order to predict wind direction, high and low pressure areas must be located. Lines on a weather map that connect areas of equal pressure are called **isobars**. The pattern of the isobars shows where the centers of the highs and lows are located. (See Figure 13-19.) Remember, in the Northern Hemisphere winds move counterclockwise around a low and clockwise around a high. By knowing the wind direction, you can better predict the type of weather for your area. Wind direction is measured with a *wind vane* and wind speed with an *anemometer*. (See Figure 13-20.)

High pressure areas are characterized by clear skies and drier air. The air at the center of a high is denser than the surrounding air. In addition, it is sinking toward the surface.

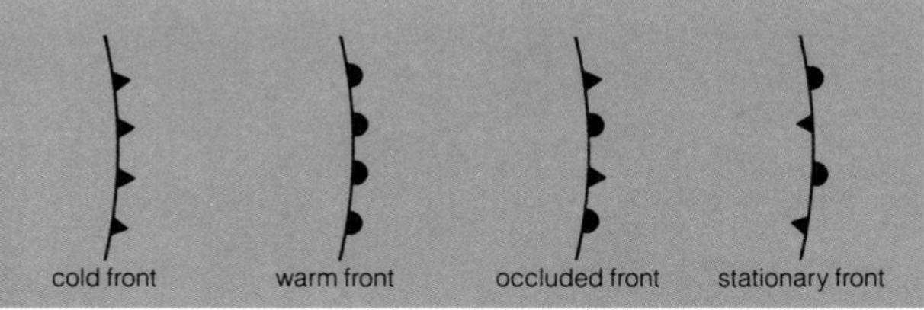

Figure 13-18 *The symbols for weather fronts.*

Figure 13-19 *On this weather map, the black isobars outline highs and lows. The blue arrows show upper air currents, which are above the surface weather pattern. Highs and lows tend to move along with the upper air currents. Fronts are shown in red.*

A low pressure region has warmer air and is usually cloudy. When the barometer is steady, you can expect weather to stay the same. When the barometer is either rising or falling, you can expect a change in the weather.

Checkpoint

1. A hot, dry air mass would have its origin over what part of Earth?
2. What causes an air mass to move?
3. What is the name of the lines on a weather map that connect points of (a) equal temperature, (b) equal pressure?
4. What weather pattern is typically associated with (a) a cold front, (b) a warm front?

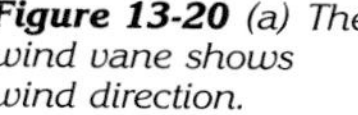

Figure 13-20 *(a) The wind vane shows wind direction.*

(b) Wind turns the cups of the anemometer to register wind speed.

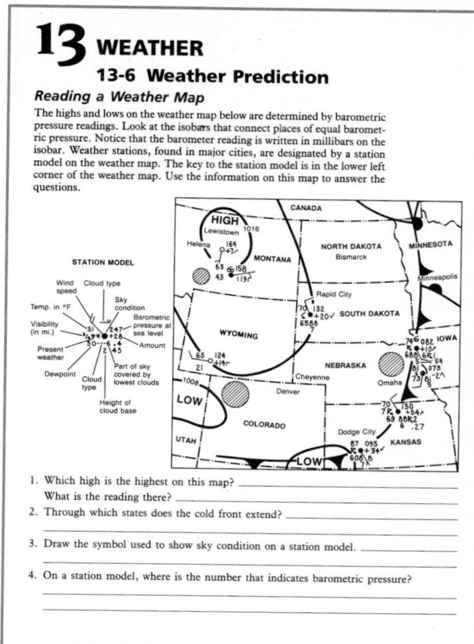

13 WEATHER

13-6 Weather Prediction

Reading a Weather Map

The highs and lows on the weather map below are determined by barometric pressure readings. Look at the isobars that connect places of equal barometric pressure. Notice that the barometer reading is written in millibars on the isobar. Weather stations, found in major cities, are designated by a station model on the weather map. The key to the station model is in the lower left corner of the weather map. Use the information on this map to answer the questions.

1. Which high is the highest on this map? ______
   What is the reading there? ______
2. Through which states does the cold front extend? ______
3. Draw the symbol used to show sky condition on a station model. ______
4. On a station model, where is the number that indicates barometric pressure? ______

Checkpoint Answers

1. Probably a hot desert region.
2. Rotation of Earth; convection within the atmosphere.
3. (a) Isotherms, (b) isobars.
4. (a) A short period of precipitation followed by clear, dry, cooler weather; (b) a long period of precipitation followed by warmer weather.

Cross Reference

Momentum, which tends to keep moving air masses in motion, is equal to the mass of a body times its velocity. Hence, the more mass there is, the greater is the momentum for a given velocity. See Chapter 16, section 16-2.

Teaching Suggestions 13-7

Large maps of the world showing global climates, vegetation, topography, ocean currents, and precipitation can all be helpful aids to use while studying this section. Films depicting sociological as well as meteorological differences among climatic areas can help students visualize life under varying climatic conditions.

Before starting this enrichment section, you could divide the class into small "climate research" groups or committees. Each group would be responsible for reporting on the way people live in a particular kind of climate and geographical setting. The groups can use the library or consult other resources (including social studies teachers). Encourage the students to use audio-visual materials. Refer students back to Chapter 5 for information about the communities that characterize different climate zones.

Background

Chinook winds are warm, dry winds on the leeward side of the Rocky Mountains in the western U.S. The cold air composing these winds falls rapidly from the tops of the mountains. The resulting winds, which sometimes reach speeds of 90 km/h, warm up rapidly and dry out as they descend. As they are drawn into the low-pressure region on the leeward side of the mountains, chinook winds bring to the Plains states welcome relief from the cold winter.

In general, annual precipitation is highest in the tropics and lowest in the polar regions. The warmth of the tropics better promotes evaporation and the formation of clouds than does the cold climate of the polar regions.

In addition to temperature, topography, and precipitation, the presence of large bodies of water, such as oceans or lakes, affects climate. Water is slow to change its temperature, and tends to moderate nearby air temperature. Chapter 14 tells how ocean currents also affect climate (see page 354).

13-7

ENRICHMENT

Climate

Goal

To define climate, climate zones, and the factors that determine the climate of an area.

climate

Climate, the average weather in an area over a long period of time, affects living things. Climate determines the types of plants and animals that can survive. It also helps to determine the jobs people have, their forms of recreation, the houses they live in, and the clothing they wear.

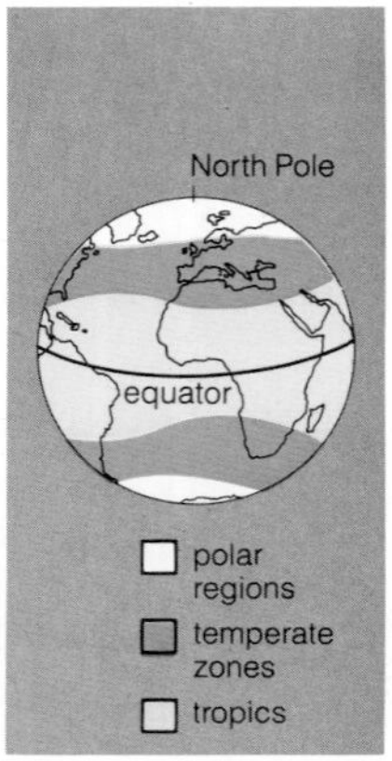

Many factors affect climate. Temperature, however, is the major factor that determines the climate of an area. Based on temperature alone, climates of the world are divided into three major zones: the *tropics*, the *middle latitudes* or *temperate zones*, and the *polar regions*. (See Figure 13-21.)

North Pole
equator
polar regions
temperate zones
tropics

Figure 13-21 *Earth's major climate zones.*

The tropics are bounded on the north and south by the line of 18°C average daily temperature. This is the lowest temperature at which tropical plants can grow. The tropics do not have much contrast between summer and winter. They receive large amounts of energy from the sun year around.

The middle latitudes, or temperate zones, are regions where hot air masses from the equator meet cold air masses from the poles. The result is a wide variety of weather. Most of the world's population lives in the temperate zones.

The polar regions have long, cold winters. They do not support much life. The average temperature in the warmest month is only 10°C, compared to 22°C for the temperate zones. Little energy is received from the sun, due to the low angle of the sun's rays.

A second factor affecting the climate of an area is *topography* (tuh PAHG ruh fee), the shape of the landscape. For example, the higher the *altitude* (height above sea level) of a place, the cooler the climate. Air at high altitudes is less dense. Therefore, it has less of an insulating effect.

windward

Topography combines with a third factor, precipitation, in determining the climate of an area. If there are mountains, the **windward** side—the side where the wind hits first—usually has more rainfall. The air is forced up on the windward side. Then it flows over the top of the mountains.

The temperature of ocean water is largely determined by where the water originates. Circulation is clockwise in the Northern Hemisphere (counterclockwise in the Southern Hemisphere), so that water on the east coast of North America (the Gulf Stream) flows from the Equator, and is warm. On the west coast of the U.S., the water in the California Current flows from the north, and is cold.

Teaching Tip

- Have students research evidence for major climatic changes that have occurred around the world during the last millenium. According to one of the more well-documented theories, numerous episodes of unusually cold winters during the past few centuries suggest that we are approaching another ice age. Students can consult the Readers' Guide to

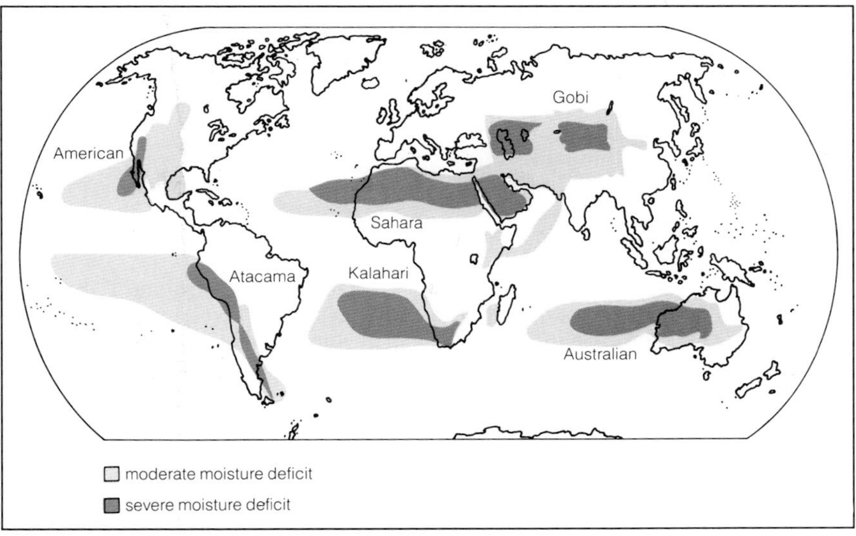

Figure 13-22 *The regions of dry climate around the world. Each of the dry regions contains a desert that is named on the map.*

As warm air containing moisture moves upward, it cools. Cool air can hold less water vapor, so the water condenses and falls as precipitation. On the **leeward** side—the side of the mountain away from the wind—the air dries out as it descends. Deserts are usually found on the leeward sides of mountains. (See Figure 13-22.)

leeward

The interiors of large continents are drier than the coasts. Because of this, there are great temperature differences between summer and winter. Within the interior of the Europe-Asia continent, for example, there is an average annual temperature range of 60°C!

Checkpoint

1. Define climate.
2. Name the three major climate zones of the world.
3. List three factors that influence the climate of an area.
4. In a mountain area, where would the most rain fall?

Periodical Literature for pertinent articles.

Facts At Your Fingertips

- By "average," a climatologist means an expected, repeated, general annual pattern of weather that tends to predominate throughout a region.
- Beijing in the People's Republic of China has cold winters and warm summers, while Valdivia, Chile has about the same temperature all year long. While their climates are obviously different, their average annual temperatures are the same.

Mainstreaming

Visually-handicapped students can gain information about climatic areas from the soundtracks of films. Before showing a film, screen it for informative and complete dialogue that doesn't depend on charts, photographs, or other visual cues.

Cross References

Review how length of day is affected by Earth's tilt and revolution around the sun. See Chapter 15, section 15-1.

Dry air warms up faster and cools off faster than does humid air, because with less water, the effect of water's high specific heat is diminished. See Chapter 18, section 18-8.

13 WEATHER
13-7 Climate

Climate of a Continent on Planet Y

Here is a map of a continent on an imaginary planet. The continent is surrounded by water. The imaginary planet is similar to Earth in size, composition, and distance from its sun. Apply what you have learned in Section 13-7 of this chapter to answer the questions below the map.

90°
60°
30°
0°
30°
60°
90°

1. Where are the three major zones of climate? (Hint: These zones are based on temperature.) To answer this question, label the zones in the left-hand margin of the map.
2. What is the direction of the major wind patterns over the oceans in each hemisphere? To answer this question, use arrows to indicate the direction.
3. Which side of the mountains in each hemisphere would have the most rainfall? ______ Why? ______
4. Which hemisphere would probably have more desert area? ______ Why? ______
5. Which hemisphere could support a greater population? ______ Why? ______
6. Where would the greatest difference between average high and low temperatures occur? ______ Why? ______

Checkpoint Answers

1. Climate is an average of daily weather conditions, compiled over a long period of time.
2. Tropics, temperate zone, and polar regions.
3. Altitude, latitude, and topography; also, predominant wind direction and nearness to large bodies of water.
4. The windward side of mountains would have more rain.

Using The Feature Article

The feature article can be used as a culminating activity at the end of Chapter 13. It refers to concepts, such as air mass, that aren't introduced until later in the chapter.

After assigning "Air Quality" as a reading assignment, briefly discuss the whole article with the class. Then divide the class into work groups and give each one a topic, which might include any of the following:

1. How much do natural sources of air pollution, such as volcanoes and forest fires, contribute to total air pollution?
2. How much do human-caused kinds of air pollution from motor vehicles, industry, and burning fuels contribute to total air pollution? What is being done to control these kinds of air pollution?
3. What are the good effects of weather on air pollution?
4. What are the bad effects of weather on air pollution?
5. Why can ozone be considered both a pollutant and a desirable substance in the atmosphere?

Have the members of each work group discuss their topic among themselves and record as many facts as they can. When they have exhausted their own resources, they can go to the library for more information. When they have had enough research time (one to two class periods), give them time in the classroom to organize their reports. Have each group present the results of their research to the class.

If your city or county has a health or environmental department that monitors air quality, invite someone from that department to visit your class. Have the speaker describe local problems, relate them to other regions of the nation, and describe the measures being pursued to help. As a finale, have the class make a list of things that they and their families can do personally to improve air quality.

Air Quality

In recent years people have become increasingly aware of health problems caused by polluted air. Air pollution can cause watery, burning eyes, raw throats, and painful breathing. Extremely high levels of pollution can endanger lives. People with asthma, or other problems involving the lungs or heart, are especially at risk.

According to the Pollution Standards Index (P.S.I.), air quality is classified as good, moderate, or unhealthful. The last category is further subdivided into alert, warning, and emergency conditions. At the emergency level, lives can be in danger.

There are two basic sources of air pollution—natural events and human activities. Erupting volcanoes are natural events. They add particles and poisonous gases, such as carbon monoxide and sulfur dioxide, to the air.

Vehicles and industrial plants are major sources of the air pollution produced by human activities. Pollutants include carbon monoxide, ozone, sulfur dioxide, and carbon dioxide.

Air quality is affected by weather conditions. Pollutants build up to higher levels in air masses that are stationary, or stagnant. A highly polluted, stagnant air mass is most likely to occur on hot, humid days when there is little wind.

People cannot do much about the weather or the natural sources of pollution. But choices can be made about the pollution caused by human activities. In many places people are trying to decrease the amount of pollution produced by their activities.

Pollen-producing plants also can affect air quality, although pollen is not usually considered to be a "pollutant." A lot of pollen in the air (a high pollen count) can cause severe problems for people who have allergies. Pollen counts are usually highest in mid-central North America. They are lowest on the Pacific coast and in southern Florida. Pollen counts are highest during the day—particularly in dry, windy weather in the summer. They are lowest after a good rain.

Understanding The Chapter

13

Main Ideas

1. The atmosphere is a mixture of gases that exerts pressure on Earth's surface.
2. Weather occurs in the troposphere. The ozone layer is in the stratosphere.
3. The major cause of wind is the unequal heating of Earth.
4. Unequal heating of the atmosphere causes high and low pressure air masses to form.
5. Earth's rotation affects air currents and is responsible for the pattern of winds on Earth.
6. The jet streams affect the path of surface winds.
7. Clouds form when water vapor condenses in the atmosphere. The three main types of clouds are stratus, cumulus, and cirrus.
8. The kind of precipitation depends on the type of cloud from which it falls. Other factors are the temperature of (1) the cloud, (2) the air through which the precipitation falls, and (3) the ground.
9. Most severe storms are caused by the winds around a low pressure center.
10. Warm and cold fronts have typical weather characteristics that accompany them.
11. Climate is the average of weather in a region over a long period of time.
12. The major climate zones are the tropics, the temperate zones, and the polar regions.

Vocabulary Review

From the following list, choose the term that best completes each of the statements. Write your answers on a separate piece of paper.

air mass	dew point
atmosphere	greenhouse effect
atmospheric pressure	ionosphere
barometer	isobar
cirrus clouds	isotherm
climate	jet streams
cold front	leeward
condensation nuclei	occluded front
Coriolus effect	ozone layer
cumulonimbus clouds	relative humidity
cumulus clouds	stratosphere
cyclones	stratus clouds
	troposphere
	warm front
	windward

1. The __?__ is the portion of the atmosphere that contains the ozone layer.
2. The change in the path of wind movement caused by Earth's rotation is called the __?__.
3. The instrument used to measure air pressure is called a(n) __?__.
4. Upper-level winds high above Earth's surface are the __?__.
5. The temperature at which water vapor condenses is the __?__.
6. The atmosphere traps heat being radiated from Earth. This is called the __?__.
7. A thunderstorm is most likely to occur if __?__ are overhead.

Vocabulary Review Answers

1. stratosphere
2. Coriolis effect
3. barometer
4. jet streams
5. dew point
6. greenhouse effect
7. cumulonimbus clouds
8. leeward
9. troposphere
10. cyclones
11. cirrus clouds
12. climate

Reading Suggestions

For The Teacher

Breuer, Georg. WEATHER MODIFICATION: PROSPECTS AND PROBLEMS. New York: Cambridge University Press, 1980. An analysis of the consequences of weather modification and the impact on society.

Field, Frank. DR. FRANK FIELD'S WEATHER BOOK. New York: Putnam's, 1981. A meteorologist, who is also a television weathercaster, explains the art of forecasting, techniques of weather modification, and theories of climate change. Full of anecdote.

Grove, Noel. "Air: An Atmosphere of Uncertainty." *National Geographic*, April 1987, pp. 502–537. An excellent, well-illustrated article on pollutants and what is being done to reduce the hazards to living things.

Miller, Peter. "Tornado." *National Geographic*, June 1987, pp. 690–715. An update on tornado "chasers." Good pictures and diagrams of tornadoes.

Schaefer, Vincent J. and John A. Day. A FIELD GUIDE TO THE ATMOSPHERE. Boston: Houghton Mifflin Co., 1981. Covers global atmosphere, clouds, color in the atmosphere, severe storms, and weather modification. There are many photographs and an appendix with experiments on atmospheric phenomena.

For The Student

Compton, Grant. WHAT DOES A METEOROLOGIST DO? New York: Dodd, Mead, 1981. A well-illustrated guide to atmospheric science and the activities and tools of a meteorologist.

Fisher, Ron and others. NATURE ON THE RAMPAGE: OUR VIOLENT EARTH. Washington, D.C.: National Geographic Society, 1986. A collection of articles about winds, thunder, lightning, floods, earthquakes, and volcanoes. Good pictures and descriptions.

Smith, Norman E. WIND POWER. New York: Coward, McCann and Geoghegan, 1981. A clear presentation of the causes and uses of wind.

Reading Suggestions

Know The Facts

1. it holds in the heat
2. it is less dense than cold air
3. trade winds
4. from a high- to a low-pressure area
5. particles in the atmosphere
6. cumulus clouds
7. condensation of water vapor slows down
8. rapidly expanding air
9. nitrogen
10. as it falls to the ground
11. warm and wet
12. isotherm
13. water is slow to change in temperature

Understand The Concepts

14. Pilots use the jet streams to give their aircraft added speed.
15. Warm air has expanded, so that a volume of warm air contains fewer molecules of atmospheric gases and thus is less dense than is cold air.
16. It would be cold and humid.
17. The angle of the sun's rays with the surface of Earth at the poles is very low, so that the sun does not give much heat to the surface.
18. This is where the warm air masses from the south meet the cold air masses from the north.
19. Central Asia is in the middle of a large continent, so air masses are generally dry. Dry air will cool off faster and warm up more quickly than will humid air, because it takes more energy to change the temperature of water.
20. Because of the planet's curved surface, the rays of the sun are vertical, or direct, in only one region at one time.
21. The air over deserts is dry; air must contain water vapor for clouds to form. In many deserts the air is falling back to Earth in a high-pressure area.

8. The side of a mountain where deserts are found is the ? side.
9. Weather occurs in the layer of the atmosphere called the ? .
10. Winds circulating around low pressure centers form ? .
11. ? are high, thin, wispy clouds.
12. ? is the average of weather in an area over a long period of time.

Chapter Review

Write your answers on a separate piece of paper.

Know The Facts

1. The atmosphere acts as a blanket for Earth because ? . (it is not very thick, it holds in the heat, it covers the whole Earth)
2. Warm air rises because ? . (it is less dense than cold air, it is more dense than cold air, it can hold less water vapor than cold air)
3. The prevailing winds near the equator are called ? . (prevailing westerlies, trade winds, polar easterlies)
4. Winds always blow ? . (from a low to a high pressure area, back and forth between highs and lows, from a high to a low pressure area)
5. Condensation nuclei are ? . (in the middle of a convection cell, particles in the atmosphere, necessary for a dust storm)
6. The billowy clouds that look like cotton are ? . (cumulus clouds, cirrus clouds, stratus clouds)
7. The primary reason a hurricane loses its strength when it reaches land is that ? . (the winds die down, condensation of water vapor slows down, it cools down)
8. Thunder is caused by ? . (two clouds hitting each other, hail being tossed around in a cloud, rapidly expanding air)
9. The most abundant gas in air by volume is ? . (nitrogen, oxygen, carbon dioxide)
10. Sleet is precipitation that freezes ? . (in a cold cloud, as it falls to the ground, after it hits the ground)
11. An air mass originating over the Gulf of Mexico would be ? . (cold and dry, warm and dry, warm and wet)
12. A line on a weather map that connects areas of equal temperature is called an ? . (isobar, isotherm, isocline)
13. Oceans have a modifying effect on climate because ? . (water is slow to change in temperature, there is a lot of wind, the humidity is low)

Understand The Concepts

14. How do jet pilots make use of the jet streams?
15. What makes warm air less dense than cold air?
16. What would be the characteristics of an air mass that originated over the Arctic Ocean?
17. Why is it so cold at the poles, even though the sun never sets at the poles for part of the year?
18. Why is there such a variety of weather in the middle latitudes?
19. What causes the great range of tem-

Challenge Your Understanding

22. The coldest month in the Northern Hemisphere is not the same as the coldest month in the Southern Hemisphere because when the Northern Hemisphere is having winter, the Southern Hemisphere is having summer. This is due to the tilt of the axis of Earth.
23. Relative humidity can be high in low temperatures because it takes less water vapor to saturate cold air. Cold air can't hold as much water as can warm air.
24. Students' drawings must include the ear trumpet, ear canal, ear drum, eustachian tube, and oral and nasal openings. Swallowing equalizes the air pressure inside passages in the head with outside

perature between summer and winter in central Asia?

20. How does the curved surface of Earth contribute to the unequal heating of the surface?
21. Why would you not expect to find clouds over desert areas?

Challenge Your Understanding

22. Is the coldest month in the Northern Hemisphere the same as the coldest month in the Southern Hemisphere? Explain.
23. How can the relative humidity of the outside air remain high when the temperature is low?
24. Make a rough drawing to show why your ears "pop" when you change your elevation in the mountains, in an elevator, or in an airplane. Why does swallowing make your ears feel normal again?
25. Why is an apparent ring around the moon an indicator of precipitation?
26. Why do pilots avoid flying through large cumulus clouds?
27. How can you account for the fact that some areas on the equator have snow?
28. With what kind of front would drizzle and fog be associated?

Projects

1. Research what is being done in your community to monitor and control air quality.
2. Research the importance of the ionosphere in shortwave radio communications.
3. Learn all you can about tornadoes.
4. Report on the process of cloud seeding. Does it work? What are some arguments for and against seeding?
5. Is it a good idea to be able to control hurricanes? Debate this question or have a panel discussion.
6. Research the effect of jet planes on cloud formation in the upper atmosphere.
7. Make a model showing how the Coriolis effect works.
8. Build a model to demonstrate the greenhouse effect.
9. Report on how satellites are used in meteorology.
10. Research the effect that weather has on human behavior.
11. Ask various people for their weather superstitions. Example: "If the leaves on the trees are turned upside down, it is going to rain." After collecting several, analyze each for its accuracy.

13 WEATHER
Projects

1. Keep a daily record of barometric pressure. The data may be collected from a barometer in the classroom, from a home barometer, or from a daily weather report on radio, television, or newspaper. Each day for two to three weeks, take the reading at about the same time of day. Graph the results with barometric pressure on the vertical axis and the date of your observation on the horizontal axis. For the same two to three-week period, write a daily description of the general weather conditions.
   a. What type of weather were you having when the barometric pressure reading was high? ____
   b. What type of weather were you having when the barometric pressure reading was low? ____
   c. Can you predict temperature changes by changes in the barometric pressure reading? ____ Explain how. ____
   d. Was the sky more cloudy when the barometric pressure reading was high or when it was low? ____
2. Visit a weather station at an airport or an observation station in your area.
   a. What types of instruments are used to observe the weather? ____
   b. What part do weather satellites play in predicting weather? ____
   c. Is this weather station government operated? ____
   d. How wide an area does the weather station serve? Identify the area where you live. ____
   e. In case of a bad storm, does the station have a method for alerting the public? ____ If so, what is it? ____

air. When the altitude increases, outside air pressure decreases; the opposite is true when the altitude decreases.

25. The ring around the moon is caused by the refraction of "moonlight" in the water droplets in a layer of cirrus clouds, which are forerunners of a cold front. Such a front may produce cumulonimbus clouds and precipitation.
26. Inside large cumulus clouds are strong vertical winds that are caused by the release of energy as water vapor is condensing. These winds are strong enough to tear the wings off an airplane.
27. Even in Equatorial regions, mountainous regions such as in Equador can have snow due to their high altitudes and cold climates.
28. A warm front.

CHAPTER 14 Overview

This chapter focuses on Earth's oceans. The first section explains the difference between oceans and seas. Salinity is defined and the processes that cause salinity to vary are discussed. In addition, ocean temperature variations are described. Examples of oceanographic methods are given. In the second section, waves and their effects upon shorelines are explained. The third section covers the causes and effects of tides. Next, the major ocean currents are treated.

The fifth section of the chapter discusses the sediments of the ocean floor. The sixth section describes the major features of the ocean floor, including the continental shelf, continental slope, continental rise, trench, mid-ocean ridge, and abyssal plain. The last section covers the many natural resources of the ocean.

Goals

At the end of this chapter, students should be able to:

1. discuss the composition and temperature of seawater.
2. describe a method used to study seawater.
3. identify the parts of waves.
4. describe how a wave moves.
5. explain what causes waves.
6. explain tides and their causes.
7. describe the factors that influence the height of tides.
8. describe the different kinds of currents.
9. describe the methods by which ocean currents are detected.
10. describe the types of sediment found on the ocean floor.
11. describe the methods used for studying the layers of the ocean floor.
12. describe the general landscape and some specific features of the ocean floor.
13. describe a method used to map the ocean floor.
14. discuss the natural resources of the ocean.

14 Oceans

This is one of the Society Islands in the Pacific Ocean.

Vocabulary Preview

oceans
seas
salinity
inland seas
Nansen bottle
crest
trough
spring tides
neap tides
ocean currents
organic sediments
inorganic sediments
nodules
continental shelf
continental slope
continental rise
abyssal plain
mid-ocean ridge
sonar
mariculture
desalination

14-1

Composition of Seawater

Goals

1. To discuss the composition and temperature of seawater.
2. To describe a method used to study seawater.

Earth is unusual because it is the only planet in our solar system with so much liquid water on its surface. Almost 75% of the planet is covered by water. These large areas of water are called **oceans**. Since water circulates freely throughout the oceans, they are often called simply "the ocean." Smaller areas of the ocean, sometimes partly surrounded by land, are called **seas**.

oceans

seas

Have you ever noticed how easily you float in salt water? Ocean water has a higher salt content than the water in most rivers and lakes. The salty water is denser than fresh water, so it holds you up better. (See Figure 14-1).

Rivers carry dissolved salts picked up by the water as it moves across and through the land. These salts eventually end up in the ocean. The most abundant salt in the ocean is

Figure 14-1 *A vacationer floating in the Dead Sea, Israel.*

Teaching Suggestions 14-1

A globe model of Earth will be helpful during the study of this section. Emphasize the vastness of water areas compared to land areas on Earth. If you live in an area where students have not had the opportunity to see or be near an ocean, films on physical oceanography, as well as on marine life, can help provide needed background. If your school has both fresh- and salt-water aquaria, you could use them to demonstrate the differences between salt-water and fresh-water organisms. Students should do the Try This activity early in this section in order to demonstrate the density differences between salt and fresh water. Plan a field trip to a nearby body of water to demonstrate the use of some simple oceanographic devices, such as the Secchi disc, water sampler, and tide gauge. (See Demonstration.) If you have access to a college or university where oceanography is taught, invite an expert to visit the class and tell more about the study of the oceans. Chapter-end projects 1, 5, and 9 apply to this section.

Background

It is surmised that the oceans started out as pure water from the condensed water vapor from volcanic action early in Earth's formation. Later, continents formed, and as water drained off the continents, it carried chemicals into the oceans. As life developed in the ocean, the organisms took some chemicals out of the water (calcium, silicon, and carbon dioxide) and added others (oxygen and calcium carbonate). These chemical cycles continue today.

More chemical changes take place at the site of mid-ocean ridges and plate boundaries where volcanic activity is high. In spite of all these changes, oceanic composition remains relatively constant, and probably has remained so for the past 700 million years.

It is interesting to note that the salt content of human blood is remarkably close to the salt content of the ocean. Some scientists speculate that this suggests that our distant forebears lived in the ocean. Another observation is that the elements hydrogen, oxygen, carbon, nitrogen, and phosphorus, all found in ocean water, are necessary ingredients of DNA, the chemical compound that is the basis of life. Could this support the idea that life on Earth originated in the ocean?

The Mediterranean Sea is one of the saltiest seas in the world. Because of the warm Mediterranean climate, evaporation removes 70,000 tons of water per second from that body of water. In the eastern part of the Mediterranean, the salinity reaches 39 parts per thousand. The Dead Sea and Red Sea are other inland seas with high salinity.

Great Salt Lake in Utah is an example of a small inland body of water with no outlet. As sediments collect from surrounding mountains and as the water evaporates, the salinity of the lake increases.

Demonstration

Show students how to use some simple oceanographic measuring devices. You or your students can use Figure TE 14-1 to construct a Secchi disc; Figure TE 14-2 to make a water sampler; and Figure TE 14-3 to build a tide gauge.

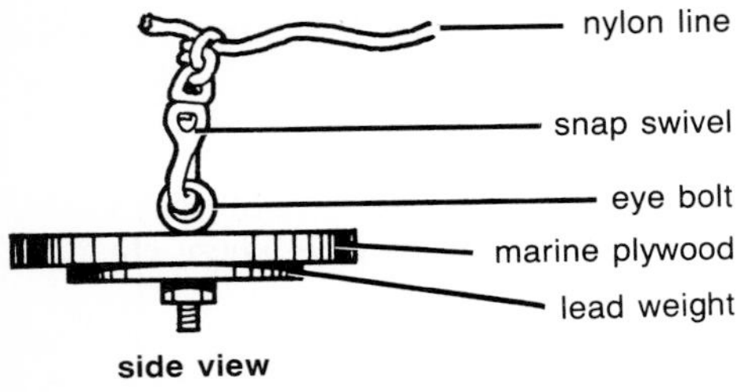

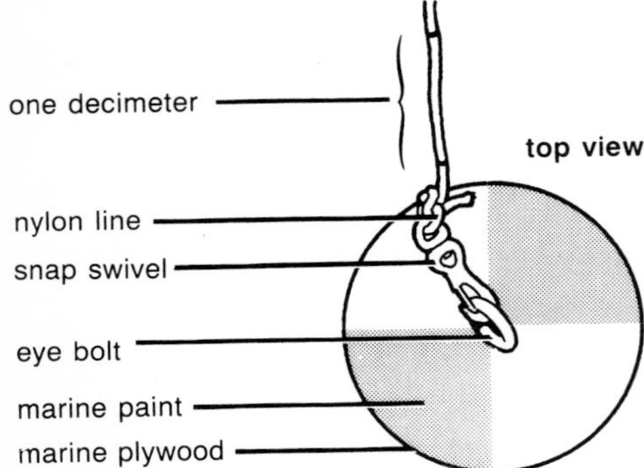

Figure TE 14-1 A Secchi disc.

salinity (suh LIN uh tee)

inland seas

sodium chloride, or table salt. The amounts of other salts in ocean water, or seawater, are given in Table 14-1. The amount of salt in seawater is called its **salinity**. The average salinity of seawater by mass is about 35 grams per kilogram of water, or 3.5%. A large pinch of salt in a drinking glass full of water produces a similar salinity.

The salinity of seawater depends on several factors. If fresh water is added, by precipitation such as rain or snow, for example, the salinity decreases. If water evaporates, the salt is left behind and the salinity increases. Where evaporation is greater than precipitation, the salinity is high. For example, **inland seas**, seas that are almost completely surrounded by land, have higher salinity because evaporation exceeds precipitation. (The Red Sea and the Mediterranean Sea are considered inland seas.) Temperature too can affect salinity. Warm water can hold more salt than cold water. So warmer ocean water has a higher salinity than cold water.

Seawater also contains gases, such as nitrogen, oxygen, and carbon dioxide, dissolved from the atmosphere. The gases are mixed with the water by waves at the surface. They are used by plants and animals in the ocean.

The temperature of ocean water varies from place to place. It ranges from 0°C to 30°C (from freezing to quite warm). In general, it is warmest at the surface, where the effect of heat from the sun is greatest, and coldest at the ocean floor. Temperature varies with latitude, also. Ocean water is warmer near the equator than near the poles.

Collecting ocean water samples at various depths and measuring the temperature at those depths was once a diffi-

Table 14-1: Dissolved Salts in Seawater

Salt	*g/kg of Water*
sodium chloride	27.2
magnesium chloride	3.8
magnesium sulfate	1.7
calcium sulfate	1.3
potassium sulfate	0.9
calcium carbonate	0.1
magnesium bromide	0.1
Total	35.1

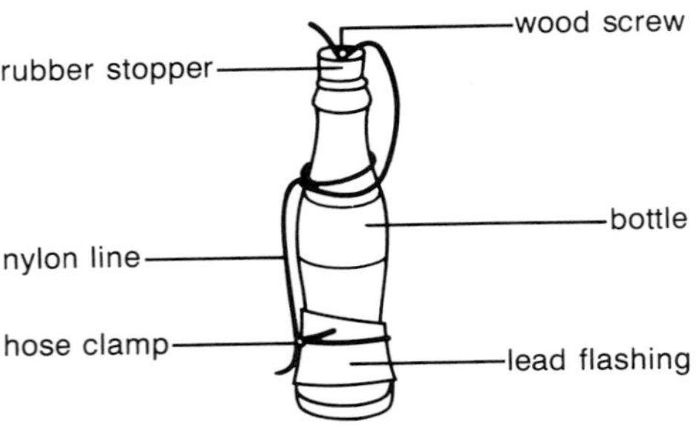

Figure TE 14-2 A water sampler.

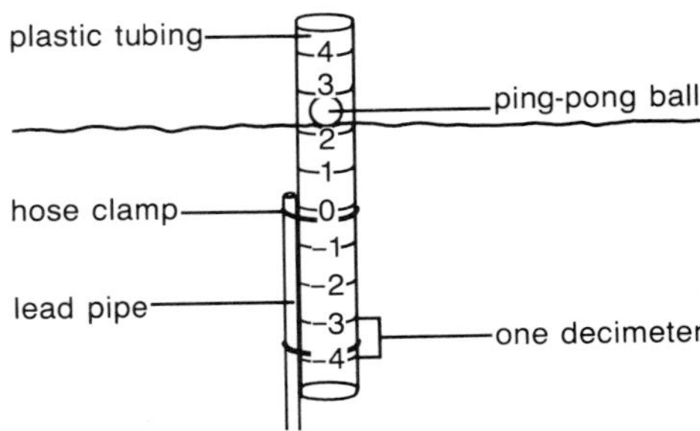

Figure TE 14-3 A tide gauge.

Figure 14-2 *This scientist is checking a Nansen bottle that has returned to the surface.*

cult task. Early *oceanographers*, scientists who study oceans, had to lower bottles and thermometers into the water. This task was made much simpler by the invention of the **Nansen bottle**. A thermometer is attached to the outside of each Nansen bottle. Several bottles are hung along a wire and lowered into the water. Each bottle can sample a different depth. When the bottles are in place, a small weight, or *messenger*, is sent down the wire. It trips the first bottle, causing it to turn over. In the process, the valves on the bottle close to collect a water sample. At the same time, the temperature is permanently recorded by the thermometer. As each Nansen bottle is tripped, it releases another messenger that slides down the wire to trip the next bottle. (See Figure 14-2.) Samples from many different places have shown that the composition of ocean water varies around the world.

Nansen bottle

Checkpoint

1. How much of Earth's surface is covered by water?
2. What causes seawater to be salty?
3. Why does the Red Sea have a high salinity?
4. What are two main reasons for variation in ocean temperatures?
5. What is a Nansen bottle?

TRY THIS

Dissolve 35 g of table salt in 1 L of water. Fill one small plastic bag with the salt solution and another with the same amount of tap water. Seal the bags with twist ties. Place the bags *gently* on the surface of a bucket or sink full of tap water. Explain what happens.

Teaching Tips

- Ask students to speculate on the origin of the salt in the ocean. Ask them if the salt in the sea is the same as the salt we use on our food.
- Have students look at the side of a globe with the Pacific Ocean on it: they will see very little land. The same is true if students look directly at the South Pole.
- Remind students that when water evaporates or freezes, it leaves behind (precipitates) anything it held in solution.

Facts At Your Fingertips

- One liter of water has a mass of one kilogram.
- The salinity of the Red Sea is a little more than 4%.

- It has been suggested that the oceans serve as a "vacuum cleaner" of sorts for the atmosphere, to clear it of excess carbon dioxide.

Mainstreaming

Visually-handicapped students will have problems with ordinary charts and tables. Such materials could be described orally by a sighted student. Alternatively, a relief map and globe would help visually-handicapped students detect the shape and relative locations of the continents and other land masses. Pair handicapped students with non-handicapped students when your class does the Try This activity.

14 OCEANS

14-1 Composition of Seawater

What Makes Ocean Water Salty?

In each example below there are two factors that may affect the salinity in ocean water. Circle the factor that would tend to result in a higher salinity.

1. High precipitation, low precipitation
2. Deep water, shallow water
3. Oceans at 80 degrees N. latitude, oceans at 10 degreees N. latitude

What in the Ocean Is This?

Each series of numbers in Column I stands for a term. Decode the numbers in Column I by letting *1* stand for *Z*, *2* stand for *Y*, *3* stand for *X*, and so forth. Write the term on the space under the code. Then find the description of that term in Column II and write its letter next to the term. Hint: Start by writing the letters of the alphabet on another sheet of paper. Below each letter write its number. Use your alphabet to decode the numbers.

I II

Using Try This

The bag of salt water will sink, while the bag of tap water will stay on top, because the salt water is more dense than the fresh water.

Checkpoint Answers

1. Almost 75%.
2. Rivers with dissolved salts empty into the oceans.
3. Evaporation is greater than precipitation in the Red Sea.
4. Two reasons for variation in ocean temperatures are depth and latitude.
5. A Nansen bottle is a device used to gather water samples and measure the temperature of water at various depths.

14-2

Waves

Goals

1. To identify the parts of waves.
2. To describe how a wave moves.
3. To explain what causes waves.

If you drop a hollow rubber ball into a river or stream, the current carries it along. Suppose you drop the ball into a lake or pond and watch it for a while. You will see that the ball moves up and down as waves pass.

crest

trough (trawf)

Waves can be described by defining the parts of a wave. (See Figure 14-3.) The highest point in a wave is the **crest**. The lowest point is the **trough**. The distance between two crests (and between two troughs) is called the *wavelength*. *Wave height* is the distance between the top of a crest and the bottom of a trough.

The water in a wave does not move forward. The energy of the wave travels forward, but the water moves up and down, "in place." People have studied the motion of water in a wave by hanging objects at various levels in the water. Objects floating on the surface travel in circles. The diameters of the circles equal the wave height. As the depth increases, the circular motion of the objects gets smaller and smaller. At a depth equal to half the wavelength, no motion can be seen. (See Figure 14-4a.)

Figure 14-3 *Locate the crest, the trough, the wavelength, and the wave height in this illustration.*

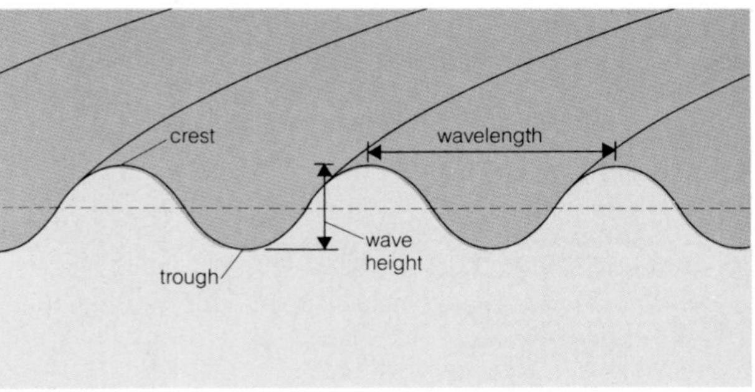

Teaching Suggestions 14-2

If you live near a coastline, plan a field trip to the shore to observe the effects of waves upon the shoreline. Use one or all of the demonstrations to show wave action. (See Demonstrations.) Collect pictures of the damage done by tsunamis or other waves. Display these while studying this section. Students can do the Try This activity after discussing the work done by waves. Chapter-end projects 10 and 11 apply to this section.

Background

The forecasting of waves was begun during World War II to aid the landing of amphibious craft on the beaches of Normandy, West Africa, and the Pacific islands. Now such forecasts are used for oil-drilling rigs, ship routing on the ocean, and recreational surfing.

The approximate height of waves can be forecast if good weather maps are available. If the speed and direction of the winds over the ocean are known, the arrival time of local waves and the intensity of high surf conditions on the beach can also be forecast. Bad wave conditions in the open ocean some distance from the storms can be predicted, too.

Wavelike motion can occur in closed or partially closed basins in a lake or harbor. Such motion is similar to the movement of water in a pan or bathtub. The period of the wave depends on the depth of the water and the length of the basin. This wave motion, called a seiche, is a form of a standing wave. One of the most common ways a seiche can start is by the piling up of water at one end of a lake by the wind of a storm. When the wind dies down, the piled-up water begins to move back. This begins a back-and-forth motion within the lake.

A breeze of 0.5 km/h can set up a chain of ripples on the water. Real waves can be generated by a 6-km/h breeze. Waves can only grow to a height of about one-seventh the distance between crests without falling over in whitecaps. This means that when waves are close together, they can't get very big. No matter how hard the wind blows across a narrow bay, it cannot build up very high waves. On the open ocean, a rule of thumb is that the height of a wave will usually be no more than half the wind's speed.

Demonstrations

1. If your school has a swimming pool, arrange to use it for one class period in order to demonstrate how waves move through water. You can also illustrate how a floating object will not move forward with the wave. The students can dress in their swim suits and can make waves in several ways. Sitting on the side

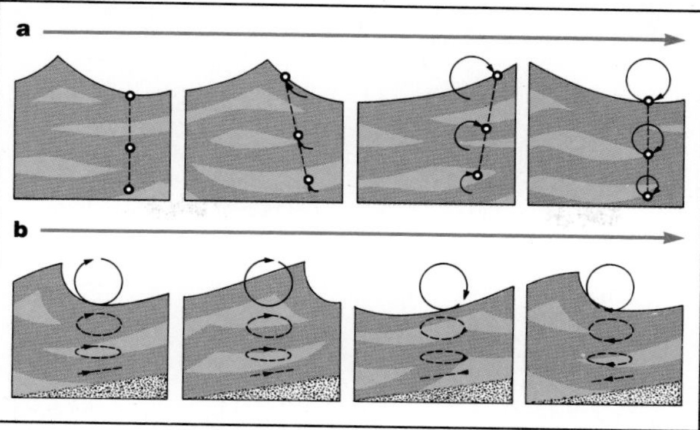

Figure 14-4 *Particle movement in waves. (a) In deep water, the circular paths of particles get smaller with greater depth. (b) In shallow water, the bottom of a wave drags on the ocean floor, causing its circular path to flatten out.*

As a wave approaches the shore, the bottom of the wave "drags" on the sloping beach and causes the circular motion to flatten out. This, in turn, forces the top of the wave to fall forward. This process produces *breakers*. (See Figure 14-4b.)

Waves do much work carving out features on the shoreline. (See Figure 14-5.) You may wonder where waves get all their energy. Most of this energy comes from the wind. Wave size depends on three factors: the speed of the wind, how long the wind has been blowing, and the distance over

Figure 14-5 *Waves can carve out many forms in a rocky shoreline. How many different features do you see?*

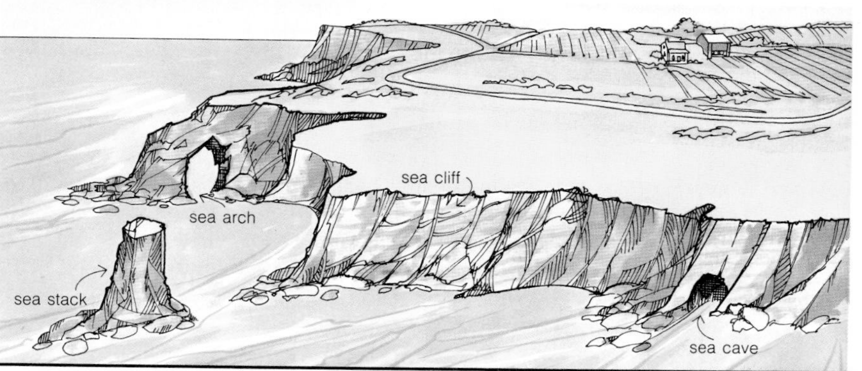

Teaching Tips

- Remind students that the terms used to describe the parts of a water wave are the same as those used for sound waves and light waves. Wave height is the vertical distance between the crest and trough.
- Ask students to bring to class pictures that show the work done by waves.
- Using the three factors that cause waves to be the highest, have students look at a globe or map and predict the areas of the ocean that would have the highest waves. Recall the prevailing wind patterns of the world, and point out that parts of the Pacific Ocean have the largest areas of open water.

Facts At Your Fingertips

- Waves can be caused by volcanic eruptions or landslides on the ocean floor or at the shoreline.
- Tsunamis are sometimes erroneously called "tidal waves." However, these waves are caused by submarine earthquakes and have nothing to do with tides or the moon.

Cross References

Refer to Chapter 12, section 12-6, for an explanation of earthquakes and tsunamis. The relationship between the oceans and weather is discussed in Chapter 13. The similarities between water waves and light waves are considered in section 20-4, Chapter 20.

with their feet in the water, students can move their feet back and forth to make small waves. To make larger waves, have them stand in shallow water facing the end of the pool. While hanging onto the gutter or end with their hands, have a row of students pull themselves toward the edge of the pool and then push away strongly in unison. Repeating this action several times in a row will increase the size of the waves.

2. A ripple tank is useful for demonstrating how waves break around barriers of different shapes.
3. A stream table can be used to show wave action. Fill it with water and have an electric fan blow across the surface from one end. This is good for showing fetch.

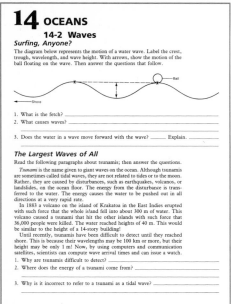

14 OCEANS

14-2 Waves

Surfing, Anyone?

The diagram below represents the motion of a water wave. Label the crest, trough, wavelength, and wave height. With arrows, show the motion of the ball floating on the wave. Then answer the questions that follow.

Ball

Shore

1. What is the fetch? ____________
2. What causes waves? ____________
3. Does the water in a wave move forward with the wave? ______ Explain. ____________

The Largest Waves of All

Read the following paragraphs about tsunamis; then answer the questions.

Tsunami is the name given to giant waves on the ocean. Although tsunamis are sometimes called tidal waves, they are not related to tides or to the moon. Rather, they are caused by disturbances, such as earthquakes, volcanos, or landslides, on the ocean floor. The energy from the disturbance is transferred to the water. The energy causes the water to be pushed out in all directions at a very rapid rate.

In 1883 a volcano on the island of Krakatoa in the East Indies erupted with such force that the whole island fell into about 300 m of water. This volcano caused a tsunami that hit the other islands with such force that 36,000 people were killed. The water reached heights of 40 m. This would be similar to the height of a 14-story building!

Until recently, tsunamis have been difficult to detect until they reached shore. This is because their wavelengths may be 100 km or more, but their height may be only 1 m! Now, by using computers and communication satellites, scientists can compute wave arrival times and can issue a watch.

1. Why are tsunamis difficult to detect? ____________
2. Where does the energy of a tsunami come from? ____________
3. Why is it incorrect to refer to a tsunami as a tidal wave? ____________

Using Try This

Keep this activity open-ended. Students should find that larger waves do more work on the coastlines. If a stream table is not available, students can use a paint tray or dish pan.

Checkpoint Answers

1. Crest is the highest point of a wave; trough is the lowest point of a wave; wavelength is the distance between two successive crests or troughs; wave height is the distance between a successive crest and trough.
2. The water moves up and down in a circular motion. Wave motion does not move the water itself forward with the wave.
3. As the water at the bottom of the wave drags against the bottom of the beach, the top of the wave falls forward, causing the breaker.
4. Wind.

Figure 14-6 *Waves become parallel to the shore as they approach.*

which the wind has blown, the *fetch*. Where the wind blows over a large area for long periods of time, waves can get as high as 12 m!

Spectacular and damaging waves called *tsunamis* (tsoo NAH meez) are caused by earthquakes on the ocean floor. A tsunami can travel as fast as 750 km/h (as fast as a jet plane) and have a wavelength as great as 60 km! On a boat in the middle of the ocean, you would not notice a tsunami. But when it hits shore, a tsunami can be over 30 m high.

Have you ever had the opportunity to watch long waves approach a shore? You may have observed that sometimes the waves that are farther out come in at an angle. Yet when they reach the shore, they hit in parallel rows. (See Figure 14-6.) This happens because the part of the wave in shallow water tends to slow down as it drags on the bottom. The part of the wave in deeper water maintains its speed and catches up with the other end. This is like what happens in a game of "crack the whip."

TRY THIS

Build a model shoreline in a stream table and then add water. Make waves in the water by tapping on the surface with your finger. Observe the work done to the shoreline. Try changing the shape of the shoreline. Also try making larger waves.

Checkpoint

1. Define the following terms: crest, trough, wavelength, wave height.
2. Describe the motion of water in a wave.
3. What causes breakers on a shoreline?
4. What is the major cause of waves?

14-3

Tides

Goals

1. To explain tides and their causes.
2. To describe the factors that influence the height of tides.

Tides may have little meaning for people who do not live on a seacoast. If you live there, you know that the level of the sea rises and falls at regular intervals throughout each day. It is important to know the exact times of these changes when you are planning a beach picnic or a hike to find seashells. Knowledge of tidal activity is important when planning buildings and other structures near the shore. The design of docks and harbors also depends on knowledge of tidal behavior.

The tides are caused by the gravitational pull of the moon and sun on the water. The moon has more influence on the tides than the sun because the moon is closer to Earth.

As Earth rotates, the amount of pull by the moon at any one place on Earth changes. The moon's pull is strongest on the side of Earth that faces it. (See Figure 14-7.) The water there bulges outward, and a high tide occurs. At the same time, there is a high tide on the side of Earth farthest away from the moon. The solid part of Earth is between the moon and the water on the side away from the moon. The moon tends to pull Earth away from the water. That causes the water on the side away from the moon to bulge outward, too. High tide occurs about once every 12 hours.

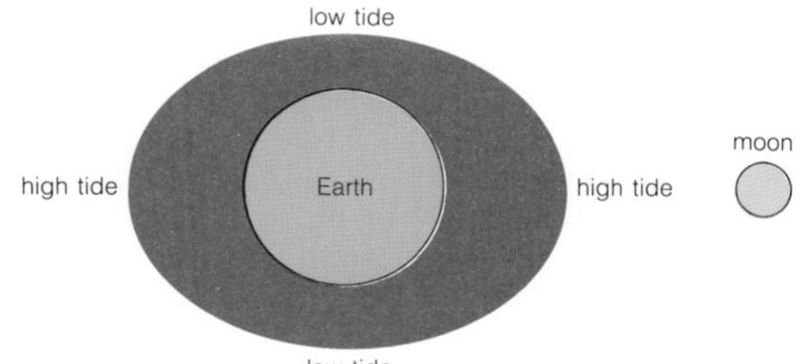

Figure 14-7 *The high tides on Earth are in line with the moon. Low tides occur at a 90° angle to that line.*

Teaching Suggestions 14-3

Before starting this section, obtain a set of tide tables for the current time period. You can usually get tide tables from the National Oceanic and Atmospheric Administration in Washington, D.C., or from marine supply stores. Show students how to read the tide tables. Then, have students graph the height of tides in one general area versus days in a month. Display the graphs on the bulletin board. They can be used for reference while studying the ocean. Students can observe the phases of the moon and compare the occurrence of the phases with the height of tides.

Students can do the Try This activity after they understand the major factors that affect the height of tides. Some students may also wish to do project 10 at the end of the chapter. Other students may wish to prepare a report on how tides affect organisms in the intertidal zone.

Contact Navy, Coast Guard, or Marine personnel in your community who might visit your class and talk about tides and their importance to the military.

Background

The general location of the highest tides changes slightly during the year because the moon—the main cause of tides—does not follow the same path around Earth every day. During a year, the moon's orbit shifts its maximum from 28.5° north to 28.5° south. The bulge of the tide changes with the position of the vertical position of the moon in relation to Earth. Tides occur about 50 minutes later each day because it takes the moon longer to revolve around Earth than it does for Earth to rotate once on its axis.

A tidal bore is a wall of foaming, turbulent water that rushes up a long, shallow, sloping river mouth or funnel-shaped bay in response to the pull of the moon. A well-known tidal bore occurs in the Petitcodiac River, which empties into the Bay of Fundy, toward the city of Moncton. The tidal bore in the Petitcodiac River can be several meters high. Tidal bores also occur in the Amazon River and in the Tsientang Kiang River in the People's Republic of China.

The unusually high tide in the Bay of Fundy not only creates a tidal bore, but also causes a waterfall in the St. John River to change direction and flow upstream. This phenomenon is known as the "Reversing Falls" of the St. John River.

Demonstration

When describing neap and spring tides, have one student represent Earth, a second represent the moon, and a third, the sun. Put them in the proper positions for neap and spring tides.

Teaching Tips

- Have students research the part tides have played in military strategies.
- Have students research how tides are being used and considered as alternate energy sources.
- Have students think of other factors that might affect tides (storms, tsunamis, distance from the sun).

Facts At Your Fingertips

- Tides occur in the atmosphere and the lithosphere (solid part of Earth) as well as in the oceans.
- Some places on Earth have only one high tide and one low tide daily because of the shape of the tidal basin and the position of the moon in relation to Earth.
- The two days when the tides are the highest each month are when the moon is full and when it is new. The moon will be in first- or third-quarter phase when the moon and sun are at right angles to Earth.

Mainstreaming

Use models to help visually-impaired students understand the concepts developed in this section. Visually-impaired students doing the Try This should work with sighted students.

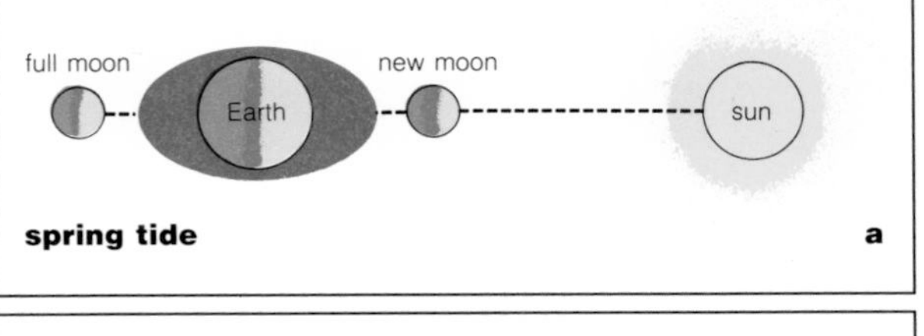

Figure 14-8a *A spring tide occurs when the sun and moon are in line with Earth.*

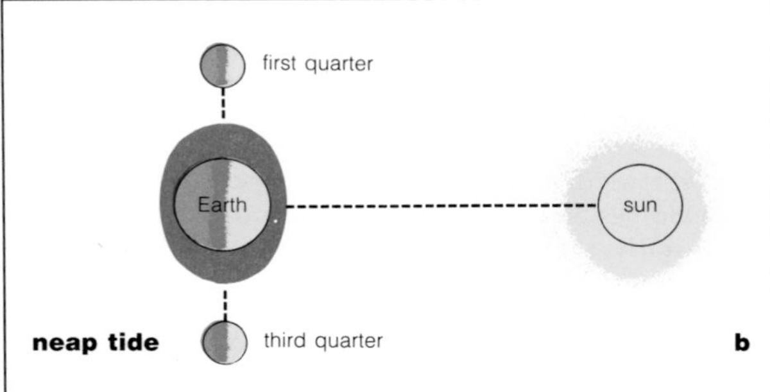

Figure 14-8b *A neap tide occurs when the moon and sun are at 90° angles to Earth.*

The gravitational pull of the moon is weakest on those parts of Earth that are 90° out of line with the moon. These are the places where low tides occur. In a given location, low tides occur about halfway between high tides.

Twice each month the tides are the highest they can be, because the sun and moon are in a direct line with Earth. The sun's pull is added to the moon's pull at these times. (See Figure 14-8a.) These highest high tides are called **spring tides**. On days when the high tides are highest, the low tides are lowest.

spring tides

When the moon and sun are at right angles to Earth, their pulls partly cancel each other out. This causes the **neap tides**, the lowest of the high tides and the highest of the low tides. (See Figure 14-8b.)

neap tides (neep)

Other factors affect the height of the tides. If ocean depth changes very gradually over a broad expanse, there may be little difference between high and low tides. However, if the coastline is jagged and the ocean bottom has deep valleys, the difference between high and low tides may be dramatic. The Bay of Fundy, between Nova Scotia and New Brunswick, is a good example. (See Figure 14-9.)

Figure 14-9 *High and low tides in the Bay of Fundy.*

Another factor that affects the height of the tides is the distance between the moon and Earth. The moon does not move around Earth in a perfect circle. It is closer to Earth at some times than at others. When it is closer, it pulls more strongly on Earth's ocean water.

Checkpoint

1. What factors cause the tides on Earth?
2. About how much time is there between two high tides in one place?
3. What causes the spring tides? neap tides?
4. How does the shape of the coastline affect tides?

TRY THIS

In a stream table or a plastic tray with sides, use clay to build a model shoreline. Shape it to look like the flat coastline of Florida. Add water for an ocean to one side. Gently tip the table or tray so that the water resembles high tide. (Plan how to tip the table or tray to the same point each time.) Then lower it, making a low tide. Now change the shoreline so that it has deep valleys to resemble the Bay of Fundy. Notice the difference in the heights of high and low tides in each model.

14 OCEANS

14-3 Tides

A Tug of War Between Earth and Moon

A set of questions follows each of the diagrams below. Study each diagram; then answer the questions that follow it.

Key
z = Point about which Earth and Moon rotate.
Y_1, Y_2 = Location of high tides

1. Which high tide is due to the moon's attraction? __________
2. Which high tide is due to centrifugal force? __________
3. On the diagram above show the locations of low tides with the labels X_1 and X_2.

1. What kind of tide is represented in 14-5a? __________
   In 14-5b? __________
2. How many times during a month would the tide in Figure 14-5a occur?
   ____
3. Which diagram shows the lowest of the low tides? __________
4. Which diagram shows the lowest of the high tides? __________
5. Which diagram shows the highest of the low tides? __________
6. Which diagram shows the highest of the high tides? __________

Using Try This

See that students control all variables—amount of water, amount of clay, degree of tipping in each direction—in both parts of this activity. Students may be able to create a model of the extra high tides in the Bay of Fundy by carefully timing the sloshing of water in the tray.

Checkpoint Answers

1. The gravitational pull of the sun and moon.
2. Usually, twelve hours.
3. Spring tides occur when the moon, sun, and Earth are lined up. Neap tides occur when the moon and sun are at right angles to Earth.
4. A jagged coastline with a steep, narrow dropoff may create dramatic differences between low and high tides. A flat, shallow coastline decreases the differences between low and high tides.

Teaching Suggestions 14-4

Use a large world map to show the location of the various ocean currents and the pattern of their movement. Obtain smaller world maps for each student; have all students chart the major ocean currents on their maps. Arrange for a meteorologist to visit the class and explain in detail how ocean currents affect the climate.

Students can do the Try This activity during the discussion of density currents. Students should also do laboratory Activity 20, Investigating Density Currents, in conjunction with this section. (See page 557 of the student text.) Chapter-end projects 1, 6, 9, and 10 also apply to this section.

Background

The Coriolis effect (Chapter 13, section 13-3) was first defined in 1835 by Gaspard Gustave de Coriolis, an assistant professor of analysis and mechanics at the Ecole Polytechnique in Paris. Coriolis made many of his observations while playing billiards. He found that any object moving over the surface of a rotating round body would appear to be deflected from its original path due to the changing velocity of the curved surface rotating under the object. The Coriolis effect applies to ocean currents as well as bullets, rockets, airplanes, and winds. In the Northern Hemisphere, deflection is toward the right; in the Southern Hemisphere, deflection is toward the left of the original path.

The ocean currents carry vast amounts of heat from the equator to the poles. This process moderates climate and prevents the formation of continent-sized ice chunks at the poles. The currents also keep the equatorial temperatures cooler.

Ocean currents that are caused by surface winds move in a direction to the right of the wind. The velocity of this current decreases with depth, but continues to turn to the right. The spiral pattern caused by such currents is called the Ekman spiral after the Swedish physicist Vagn Walfrid Ekman, who described it mathematically.

The first ocean-going vessels were powered by the wind and ocean currents. The era of great sailing ships ended around 1910. Today, because of the rising cost of diesel fuel, updated, computerized sailing vessels are being evaluated to supplement and possibly replace much of the need for diesel power. A computer controls the sails and masts, and also aids in navigation, with the aid of satellite data.

14-4

Currents

Goals

1. To describe the different kinds of currents.
2. To describe the methods by which ocean currents are detected.

ocean currents

Ocean currents are movements of water similar to huge "rivers" running through the oceans. The motion of these currents is caused by several factors. Surface currents are controlled mainly by the prevailing winds. Thus surface current patterns are similar to prevailing wind patterns. The surface currents in the Northern Hemisphere have a clockwise motion. Those in the Southern Hemisphere have a counterclockwise motion. (See Figure 14-10.)

The Gulf Stream on the east coast of North America is a large current flowing north. This warm current is responsible for keeping climate moderate on the east coast. It also causes icebergs from the cold Labrador Current to melt in the north Atlantic. Foggy weather in London, England is due to the warm water in the Gulf Stream meeting cold air at this latitude.

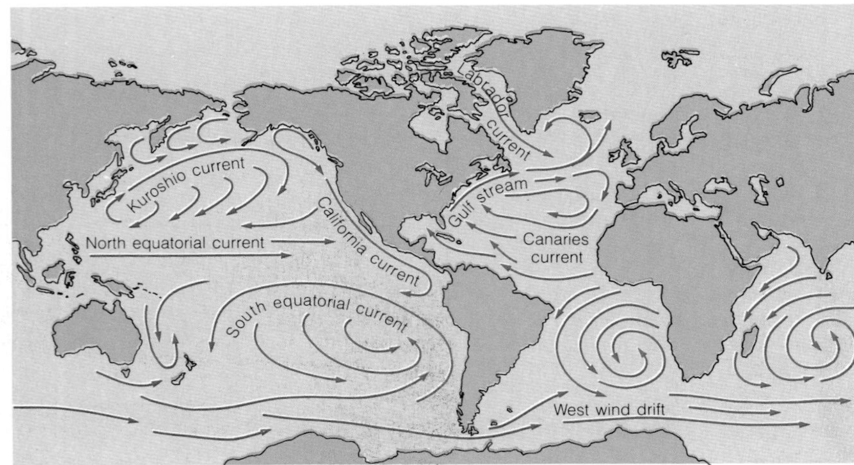

Figure 14-10 *The major surface ocean currents of the world. Note that there is an equatorial counter (opposite) current in the Pacific.*

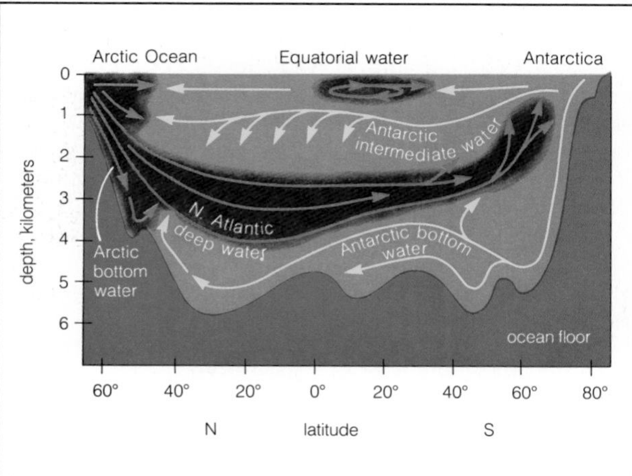

Figure 14-11 *A north-south cross section of the Atlantic Ocean, showing the movement of currents at different depths.*

Around 1750, Benjamin Franklin's cousin, Timothy Folger, was captain of a whaling ship. Sailors on these ships noticed a strong eastward-flowing current in the North Atlantic. The whalers had learned to sail with the current on trips from North America to England. On return trips, they chose a more northerly route to avoid fighting the current. Benjamin Franklin published the first map of the Gulf Stream, based on a map drawn by Captain Folger.

The California Current is a large cold current flowing from the north toward the equator on the west coast of North America. It keeps the west coast cooler than the east coast at the same latitude. The Gulf Stream and the California Current are part of the clockwise pattern in each ocean.

The strong easterly trade winds cause the equatorial currents at the equator. Early navigators in their sailing ships used these powerful currents to get them to their destinations in less time. Modern ocean vessels can use them to help conserve fuel.

Deeper currents in the ocean are caused by density differences. Denser water sinks beneath less dense water, causing a *density current*. It has been found that there are three levels of density currents in the Atlantic Ocean. Two of these come from the Antarctic and one from the Arctic. (See Figure 14-11.)

Facts At Your Fingertips

- The coldest water in the oceans is at the bottom under polar ice, and is 4°C. Water between 4°C and 0°C begins to expand, so will be closer to the surface.
- Captain Timothy Folger (mentioned on student text page 355) and his crew were the first people to discover the Bounty mutineers on Pitcairn Island.

Teaching Tips

- Water drains out of bathtubs and wash basins in a clockwise pattern in the Northern Hemisphere, and counterclockwise in the Southern Hemisphere. Have students check the validity of the part of the statement appropriate for your location.
- Have students locate and describe the Sargasso Sea. What did sailors used to believe it was?

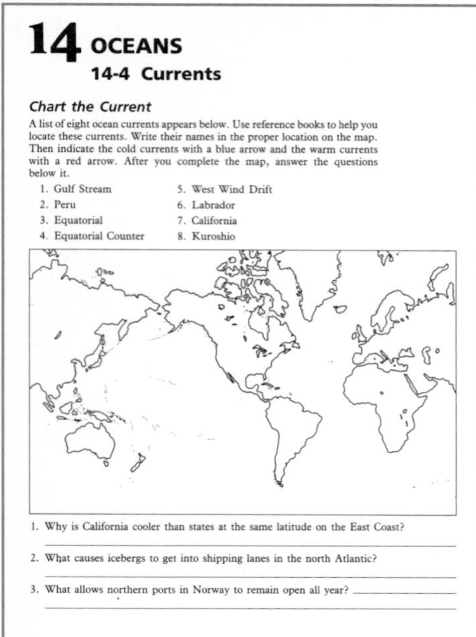

14 OCEANS
14-4 Currents

Chart the Current

A list of eight ocean currents appears below. Use reference books to help you locate these currents. Write their names in the proper location on the map. Then indicate the cold currents with a blue arrow and the warm currents with a red arrow. After you complete the map, answer the questions below it.

1. Gulf Stream
2. Peru
3. Equatorial
4. Equatorial Counter
5. West Wind Drift
6. Labrador
7. California
8. Kuroshio

1. Why is California cooler than states at the same latitude on the East Coast? ____________
2. What causes icebergs to get into shipping lanes in the north Atlantic? ____________
3. What allows northern ports in Norway to remain open all year? ____________

Using Try This

The students should observe the colored cold water streaming from the bottom of the cup and moving along the bottom of the shoe box toward the far end. It will rise at the opposite end, replacing the surface water that has moved over to the cup side. The descent of the denser cold water starts the flow. The next day the color will be evenly distributed throughout the container, showing that the temperature throughout the water has equalized. The model is an analog of a polar ocean, in which the cold water from melting ice sinks to the bottom.

Checkpoint Answers

1. Prevailing winds, or Coriolis effect.
2. Density differences due to variations in temperature or salinity.
3. Once the information card is returned, scientists can tell how long it took the drift bottle to travel a certain distance. That speed is the speed of the current in which the drift bottle traveled.
4. A current meter, swallow float, or drift bottle.

A good example of density currents occurs at the Straits of Gibraltar. The dense water of the Mediterranean Sea flows out to the Atlantic at a low level. Less dense water from the Atlantic flows into the Mediterranean at a higher level. During World War II, German submarine captains used these density currents. They often shut off the engines of the submarines and floated in and out of the Straits of Gibraltar, undetected.

The motion of ocean currents can be detected with special instruments called *current meters.* A current meter can be anchored in the water at any depth. It measures the speed and direction of flow of the water.

Drift bottles are used to measure surface currents. They float just below the surface, where they aren't affected by winds. Each bottle contains a card with information about where and when the bottle was released, plus instructions to the finder. After the finder returns the card, the direction and speed of the current can be determined.

A *swallow float* anchored in the water is used to measure the flow of deep currents. A swallow float is an electronic device that sends out "pings" that are picked up by a ship's receiver. The "pings" show the speed and direction of the currents.

Checkpoint

1. What is the major cause of surface currents?
2. What causes currents deep in the ocean?
3. How are drift bottles used to measure the speed and direction of surface currents?
4. Name an instrument that is used to measure speed and direction of deep currents.

TRY THIS

Fill a clear plastic shoe or sweater box half full of water at room temperature. Punch some tiny holes in the bottom of a small styrofoam cup and tape the cup into one corner of the box. Only the bottom of the cup should be under the water surface. Fill the cup with ice cubes. Put several drops of food coloring over the ice. Observe what happens. Can you detect a current in the water? What direction is it flowing? Let the water stand overnight. What do you observe the next day?

14-5

Sediments

Goals

1. To describe the types of sediment found on the ocean floor.
2. To describe the methods used for studying the layers of the ocean floor.

If you have ever waded in a river or lake, you have felt the *sediment* on the bottom squeeze up between your toes. This sediment is mostly sand and clay. Sediments are formed from weathered rock material that has settled down to the bottom of bodies of water. The floor of the ocean is covered by sediments, too.

It is much more difficult to study the sediments on the ocean floor because they are so far below the surface. However, oceanographers have divided ocean-floor sediments into two main categories: organic and inorganic. **Organic sediments** come from the remains of organisms that lived in the sea. Bits of shells and other tiny pieces from animals and plants form some organic sediments. Other kinds, called *oozes*, are made up mainly of hard parts of one-celled organisms.

organic sediments

inorganic sediments

nodules (NAHJ oolz)

Inorganic (nonliving) **sediments** come from several different sources. Mineral grains and rock fragments that form clay on the ocean floor are inorganic sediments. So are meteorite dust, volcanic ash, and nodules. **Nodules** are formed when chemicals from the water collect in lumps. Nodules are an excellent source of pure manganese, copper, nickel, and cobalt. In Figure 14-12, you can see manganese nodules on the floor of the Pacific Ocean.

Oceanographers have found sediments to be 600 m thick in the Atlantic and 300 m thick in the Pacific Ocean. The reason for the difference is that there are fewer rivers draining into the Pacific Ocean. Rivers carry sediments from the land to the sea. Some sediments brought to the Pacific are deposited in deep *trenches* close to the shorelines. Trenches are deep, narrow valleys on the sea floor. They are the deepest parts of the ocean. The Atlantic Ocean does not have trenches, so more sediments collect on the ocean floor of the Atlantic.

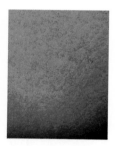

Figure 14-12 *Manganese nodules on the floor of the Blake Plateau, off the southeastern coast of North America.*

Teaching Suggestions 14-5

Models can be very helpful in demonstrating how materials settle out of water. (See Demonstration 2.) Maps of the ocean floor show such features as trenches and mouths of rivers. Try to obtain some maps of the ocean floor and post them on the bulletin board for reference during the teaching of this section. Other teaching aids can include films (see page 269) showing how sediments accumulate, how samples are collected, and how they are used to tell the history of a part of the ocean. You also may wish to invite a member of a college or university Geology or Oceanography Department to talk to your class about the information provided by sediments.

Students can do the Try This activity after discussing core samplers.

Chapter-end projects 1 and 5 pertain to this section.

Background

The earliest core samples taken of marine sediments were only about 30 cm long, but since the development of the piston corer, cores up to 70 times that size have been recovered. Sediments are classified by their origin. Thus, terrigenous sediments come from the continents, and result from the weathering of rocks or from vulcanism, and are widely distributed by ocean currents. They can be found in the red deep-sea clay. Gravel and stone sediments probably came from icebergs that melted into the ocean. Wind can also carry terrigenous sediments into the ocean. Dust from the Sahara Desert has been found in the sediments west of the Cape Verde Islands. Biogenous sediments are organic in origin. Halmyrogenous sediments are new formations of minerals that were chemically precipitated out of ocean water. Deposits of iron and manganese oxides are examples. Cosmogenous sediments come from outer space and consist of small ball-shaped objects, generally about 0.2 mm in diameter. They are believed to be remnants of meteors, or parts of the moon. Sand also has been found over large parts of the ocean floor. It is now believed that the sand is distributed by density currents.

Demonstrations

1. Soak a few egg shells in a dilute vinegar solution. They will become very soft, showing how the organic shells of dead animals can change to an ooze-like sediment on the ocean floor, especially if the water is slightly acidic. Carbon dioxide dissolved in ocean water makes a weak carbonic acid solution.
2. Stream tables can be used to show how sediments settle to form a delta and how finer particles are carried farther away from land.

Teaching Tip

- Ask students to explain why it is easier to detect meteoric dust on the ocean floor than on the continents.

Facts At Your Fingertips

- Organic materials cannot descend to the great depths of the ocean because of the movement of the ocean and the solvent action of seawater.
- Columbia University's Lamont Geological Observatory has one of the world's largest deep-sea core libraries. The advances and retreats of glaciers for the past one and one-half million years have been documented by means of these cores.

Mainstreaming

New vocabulary is introduced quickly in this section, and handicapped students may need extra help with new words and concepts. Visually-impaired students should work with sighted classmates on the Try This activity.

Cross References

Chapter 11 tells about the formation of sedimentary rocks in section 11-2 and tells about fossils in section 11-3. Chapter 12 discusses weathering (12-1), erosion (12-3), deposition (12-4), and plate tectonics (12-5), including the origin of ocean trenches.

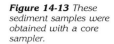

14 OCEANS
14-5 Sediments

Valuable Mineral Nodules

The paragraphs that follow describe mineral nodules on the ocean floor. Read the paragraphs about the nodules; then answer the questions.

Scattered over the ocean floor is a fortune in valuable minerals called nodules. They are just waiting for someone to invent the technology that would allow their inexpensive removal from the ocean depths. The minerals contained in nodules are manganese, cobalt, nickel, and copper. These minerals come from leaching of the continents or from volcanic eruptions on the ocean floor. The minerals may also precipitate out of the water itself. As the nodules settle out of the water, they attach to irregular solids on the ocean floor, such as shark's teeth, the bones of a whale's ear, or a lump of clay. Nodules grow at a very slow rate, several millimeters in a million years! Nodules get to be the size of potatoes and are dark in color.

Manganese is important in the production of steel. Cobalt, nickel, and copper are used in electronics, aeronautics, and the aerospace industries. These minerals are not available in large quantities in our country. They must be imported from other places and this makes them more expensive.

Using Try This

The equipment mentioned is only a suggestion. Other possibilities will also work. Keep methods flexible. The layers of clay in the core are in the same order as they were put down. Students should discuss how oceanographers can tell when a sediment was deposited, what kind of materials formed the sediment, and what the environment was like at the time of deposition.

Checkpoint Answers

1. Organic and inorganic. Examples of organic sediments include animal shells and plant remains; examples of inorganic sediments include rock fragments and meteoritic dust.
2. Atlantic Ocean.
3. Core sampler.

Figure 14-13 *These sediment samples were obtained with a core sampler.*

Samples of sediment layers can be obtained with a *core sampler.* (See Figure 14-13.) A core sampler is a long tube that is forced down into the sediment. When it is brought up, the tube is filled with layers of sediment in the order in which they were formed. (See Figure 14-14.) Much can be learned about the ocean by studying these sediments. The kinds of organisms that once lived in a part of the ocean are clues to the temperature of the water and the available food supply in that place. By observing the layers in sediments, scientists can learn about previous periods of volcanic activity. Certain fossils are clues to the location of oil and other minerals.

Other tools used by oceanographers to gather samples of sediments from the ocean floor are dredges and *box corers.*

Checkpoint

1. What are the two categories of ocean sediment? Give two examples of each.
2. Which ocean seems to have the thickest sediment layers?
3. Which tool can an oceanographer use to get samples of undisturbed layers of sediment?

Figure 14-14 *Coring tubes are used to collect samples of ocean bottom sediments.*

TRY THIS

Place layers of different colored modeling clay in a milk carton with the top cut off. (Use talcum powder between the layers, so they can be separated later.) For a coring tube, use 1.2-cm diameter electrical conduit the same length as the height of the milk carton. Force the coring tube down through the layers of clay and pull the tube out. With a round wood stick to fit, push the "core sample" out on the counter. Observe the layers. They are in the same order as they were put down. How can oceanographers use this information?

14-6

Features of the Ocean Floor

Goals

1. To describe the general landscape and some specific features of the ocean floor.
2. To describe a method used to map the ocean floor.

By looking just at the surface of the ocean, you might assume that the ocean floor is uniformly flat. But this is not the case. The ocean floor has as varied a landscape as the continents do.

As shown in Figure 14-15, the **continental shelf** is an extension of the continent that is covered by water. The shelf can be very narrow or it can extend out about 1000 km from the shoreline. It slopes very gently down to a depth of about 200 m. Next comes the **continental slope**, which drops off more sharply, about as steeply as the aisle in a movie theater. It could be called the wall of the ocean basin. At the bottom of the continental slope is a gently sloping collection of sediments from land called the **continental rise**. Beyond the continental rise is true ocean bottom—the **abyssal plain**. This flat floor of the ocean covers 60% of the ocean floor.

In the middle of the Atlantic Ocean there is a chain of rugged mountains, the **mid-ocean ridge**. This ridge is part of a system of ridges extending through the Indian Ocean into part of the Pacific. These ridges form the longest feature

continental shelf

continental slope

continental rise

abyssal plain (uh BIS ul)

mid-ocean ridge

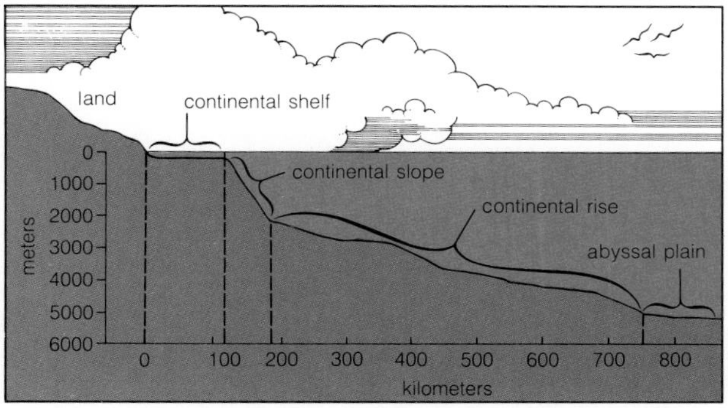

Figure 14-15 *A cross section of a typical ocean floor.*

Teaching Suggestions 14-6

Display maps and pictures of the ocean floor while studying this section. Encourage students to build a three-dimensional model of the ocean floor (see chapter-end project 3). This will give students valuable experience in understanding the size and shape of the features described in this section. Invite a geologist or an oceanographer to visit your class to discuss the topography of the ocean floor, and also the latest research methods and results of benthic (deep-sea) research.

Students can do the Try This activity—a classical physics demonstration—after completing this section. Chapter-end projects 2, 3, and 11 apply to this section.

Background

Between 1769 and 1779, Captain James Cook led three scientific oceanographic expeditions. During these expeditions, ocean depths and temperature variations were measured. In 1840, Sir James Clark Ross obtained soundings of more than 2600 fathoms (about 5 km) in the Antarctic. In those days, soundings were made by manually lowering and raising a hemp line. Each sounding took hours.

In December, 1872, H.M.S. *Challenger* sailed from England on a three and one-half year oceanographic voyage. By 1895 a team of scientists had prepared a 50-volume report of the work that was done by the *Challenger* team, and their interpretations of the data gathered.

In 1882, the U.S. Commission of Fish and Fisheries launched the *Albatross*, the first ship ever built in this nation specifically for oceanographic research. Alexander Agassiz (the son of Louis Agassiz, the father of the study of glaciation) led the research effort aboard this ship.

In 1923, the first echo sounder was used to study the profiles of the ocean floor. In 1930, the first bathysphere was constructed, enabling scientists to descend safely to great depths. And in 1938, camera systems were developed to photograph the ocean floor.

The most recent innovation in ocean exploration is charting the topography of the ocean surface by satellite. This measurement is accomplished by determining the round-trip travel time of pencil-thin beams of microwaves or laser beams between the satellite and the ocean surface. Such a system aboard a satellite can determine the distance between the satellite and the ocean surface with an accuracy of about 5.5 cm—the same degree of accuracy as measuring the length of a city block to the width of a hair! One of the surprising finds of satellite measurement was that the ocean surface rises over submarine mountains and falls over submarine trenches.

Teaching Tips

- When discussing continental drift, remind students that the edges of continental shelves make a better "fit" with each other than do the modern sea-level boundaries. Sea level changes with time.
- Refer often to the maps of the ocean floor that show depths of various features.
- Ask students to write a short report on the history and use of sonar. Navy films on sonar would be likely to stimulate class interest.

Facts At Your Fingertips

- Sound travels faster in water than in air.
- The average height of the continents above sea level is only 0.8 km.
- Charles Darwin, while on his oceanic expedition aboard the H.M.S *Beagle,* studied coral reefs to find out why they were found in deep water, where living coral can't survive. He concluded that the sea floor sank, carrying coral reefs down with it.
- Coral atolls and fringing reefs in the Pacific Ocean were good defensive positions during World War II.
- Pressure in the ocean increases by one atmosphere for every ten meters of depth.
- Volcanic oceanic islands in the Pacific Ocean are often found in linear groups, such as the Hawaiian, Society, and Marquesan Islands. These island groups and their ridges are built of huge lava platforms that rise from submarine ridges.

on Earth—75,000 km long and as much as 1500 km wide in some places. The mountains stand 2 to 4 km high off the ocean floor. This is similar to the height of the Rocky Mountains in western North America. (See Figure 14-16.)

Figure 14-16 *The deep basins, mountain ranges, and zones of movement on the Atlantic Ocean floor.*

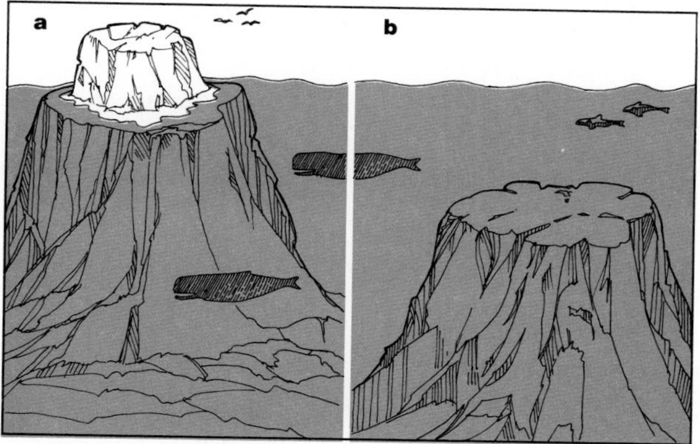

Figure 14-17 *The formation of a guyot. (a) An island is eroded by ocean waves. (b) After the island has been eroded, it becomes submerged.*

All oceans do not have the same pattern of features. Deep sea trenches, for example, are found in the Pacific Ocean but not in the Atlantic. The deepest point in the oceans is about 11.5 km deep, in the Mindanao trench near the Philippine Islands. Compare this depth with the height of the tallest mountain on Earth, Mt. Everest, which is 8.85 km high.

Distributed around the Pacific floor are hundreds of underwater volcanic mountains called *seamounts*. Some seamounts have unusual flat tops. These flat-topped seamounts, called *guyots*, were formed when waves cut across the tops of the mountains. They are evidence that sea level has changed. (See Figure 14-17.) Some underwater volcanoes form mountains high enough to protrude above the surface, forming islands. The Hawaiian Islands are examples of such volcanoes. Other volcanoes occur in groups and form *island arcs*, or chains of islands. The Aleutian Islands and the Philippine Islands are examples of island arcs.

Another feature of the ocean floor is coral reefs. Corals are small animals that live in shallow warm water near the shore. They live in large colonies. *Reefs* are made of coral skeletons. If the shore happens to be on a sinking volcanic island, corals will grow on the skeletons of former corals. So the living corals are always near the water's surface.

There are three kinds of reefs. *Fringing reefs* are on shorelines of land. *Barrier reefs* are found away from land, separated from the land by water. An *atoll* is a complete circle of

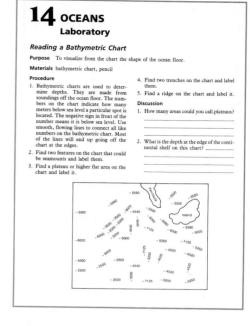

14 OCEANS
Laboratory

Reading a Bathymetric Chart

Purpose To visualize from the chart the shape of the ocean floor.

Materials bathymetric chart, pencil

Procedure

1. Bathymetric charts are used to determine depths. They are made from soundings off the ocean floor. The numbers on the chart indicate how many meters below sea level a particular spot is located. The negative sign in front of the number means it is below sea level. Use smooth, flowing lines to connect all like numbers on the bathymetric chart. Most of the lines will end up going off the chart at the edges.
2. Find two features on the chart that could be seamounts and label them.
3. Find a plateau or higher flat area on the chart and label it.
4. Find two trenches on the chart and label them.
5. Find a ridge on the chart and label it.

Discussion

1. How many areas could you call plateaus? ______
2. What is the depth at the edge of the continental shelf on this chart? ______

Mainstreaming

Special-needs students are likely to need extra help with new vocabulary words. Provide more illustrations for them to reinforce the concepts. The Try This activity should be useful for these students.

Cross References

Refer to Chapter 5, section 5-7, for a description of the deep-water zone. Also, you may wish to reread the Feature Article in Chapter 5: Life in the Deep. For more information on continental drift, see Chapter 12, section 12-7. See Chapter 20, section 20-7, for a description of sonar.

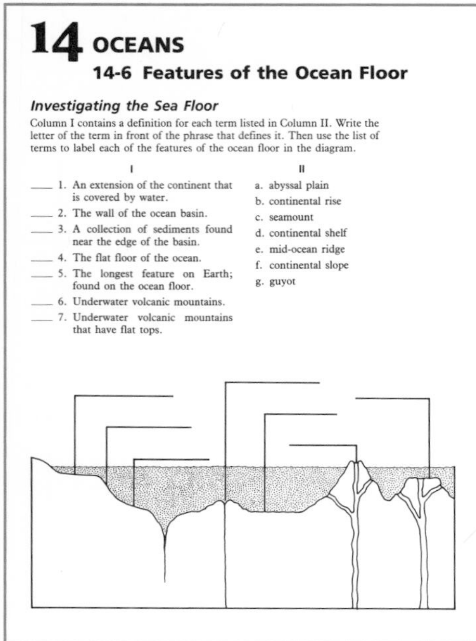

14 OCEANS
14-6 Features of the Ocean Floor

Investigating the Sea Floor

Column I contains a definition for each term listed in Column II. Write the letter of the term in front of the phrase that defines it. Then use the list of terms to label each of the features of the ocean floor in the diagram.

I	II
___ 1. An extension of the continent that is covered by water.	a. abyssal plain
___ 2. The wall of the ocean basin.	b. continental rise
___ 3. A collection of sediments found near the edge of the basin.	c. seamount
___ 4. The flat floor of the ocean.	d. continental shelf
___ 5. The longest feature on Earth; found on the ocean floor.	e. mid-ocean ridge
___ 6. Underwater volcanic mountains.	f. continental slope
___ 7. Underwater volcanic mountains that have flat tops.	g. guyot

Using Try This

Use a small nail so that the holes will be the same size, and not too large. Students should observe that water from the lower holes will spurt out farther because of the greater force produced by the higher "head" of water than for holes higher on the carton. A "head" of water is the relative height of the water above an equilibrium level: it is a measure of the effect of gravity upon the column of water.

Checkpoint Answers

1. Continental slope.
2. Abyssal plain.
3. Deep-sea trenches.
4. A seamount is an underwater volcanic mountain; a guyot is a flat-topped seamount.
5. Scanning with sonar, or sound waves. (The old method of sounding with a rope or chain is also an acceptable answer).
6. An atoll is an above sea-level circle of coral that forms as the level of the sea changes relative to the heights of an old volcano and the coral around it.

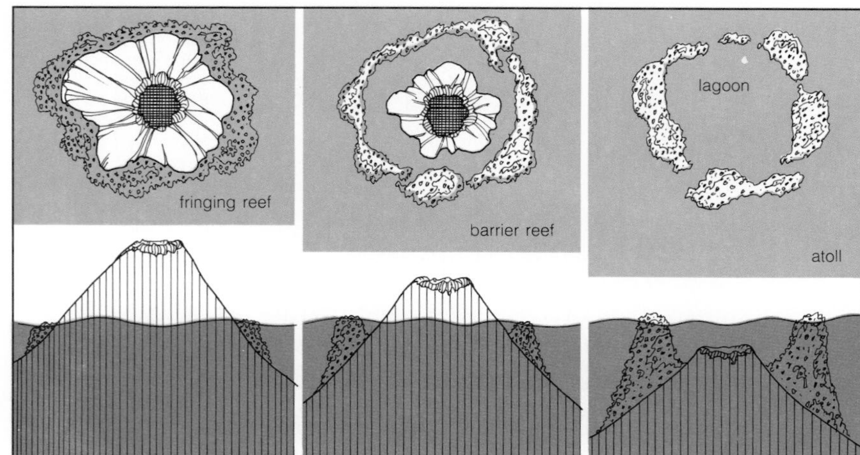

Figure 14-18 *The three stages in the development of an atoll, as an old volcano sinks and the coral reefs grow upward.*

coral with no land visible. Atolls form as the sea level changes relative to the heights of land and coral. Figure 14-18 shows how an atoll develops.

It is difficult to explore the ocean depths because of the extreme pressure. Water pressure increases with depth. At 20 m deep the pressure is twice the air pressure we experience at sea level. Imagine what it is at 100 m! Most of the ocean floor has been mapped using **sonar** (**so**und **n**avigation **a**nd **r**anging). Sound waves are sent from a ship to the ocean floor. The time it takes for the sound to go to the bottom and return depends on the depth of the water. The average depth of all the oceans is 3.8 km.

sonar
(SOH nar)

TRY THIS

Fill an empty milk carton with water. Place it on the edge of a sink. With a nail, punch holes in the carton along the sink side, from top to bottom. Does each stream of water shoot out at the same angle? Explain.

Checkpoint

1. What ocean feature serves as the wall of the ocean basin?
2. What is the name for the flat floor of the ocean?
3. Name an ocean feature found in the Pacific Ocean but not in the Atlantic.
4. What is the difference between a seamount and a guyot?
5. Name a method used to map the ocean floor.
6. How is an atoll formed?

14-7

Resources of the Ocean

ENRICHMENT

Goal

To discuss the natural resources of the ocean.

The ocean, or "sea," has been a source of fish and other seafoods for centuries. Recently, there has been progress made in more efficient farming of the sea. Seafood is an excellent source of protein. By concentrating food for fish in small areas, large numbers of fish can be raised. The same thing has been done for oyster beds. (See Figure 14-19.) This farming of the sea is called **mariculture**. The prefix "mari" comes from *mare*, the Latin word for sea.

mariculture
(MA ruh kuhl chur)

Salt, used in preparing and preserving food, has long been taken from the sea. By evaporating water from seawater, large amounts of salt are obtained. A cubic kilometer of seawater contains enough salt to supply the world's needs for nine years!

Kelp, another food being cultivated, is a form of seaweed. It can be used for food because it is rich in minerals and other nutrients. Kelp can be ground up into flour to make bread or it can be eaten as a vegetable. Kelp also is used in the production of certain drugs and cosmetics.

Supplies of fresh water are getting smaller because of increasing demands for water by industry. So scientists are trying to improve the process of **desalination**, removal of salt from seawater. If this process can be done efficiently, the sea can become an excellent source of fresh water.

desalination
(dee sal uh NAY shuhn)

Figure 14-19 *Oysters are grown in a number of "farms" such as this one in Japan.*

Teaching Suggestions 14-7

Before starting this enrichment section, collect pictures of as many ocean resources as you can. You might enlist the help of your students in this effort, perhaps for extra credit. Make a bulletin board display. Depending upon your location, you could invite a specialist in mariculture, oil-well operations, or marine biochemistry to speak to your class about how their field of specialty applies to ocean resources.

To start students thinking about their dependence upon the ocean, they can do chapter-end project 4. They should do the Try This activity early in this section, so that results can be discussed before you finish. In addition to project 4, chapter-end projects 1, 7, and 11 apply to this section.

14 OCEANS
Computer Program

Oyster Cultivation

Purpose To calculate the oyster harvest number.

Background Oyster harvest depends on the number of shells in the bed for growth of young, the quantity of food in the water, the absence of predators and parasites, and the length of time for development.

Procedure

ENTER the following program:

```
10  PRINT "ENTER THE NUMBER OF YOUNG OYSTERS"
20  INPUT "SEEDED INTO THE BED."; YO
30  INPUT "ENTER NUMBER OF SHELLS IN BED."; S
40  INPUT "ENTER FOOD LEVEL NUMBER."; F
50  INPUT "ENTER PREDATION RATE."; P
60  INPUT "ENTER YEARS TO HARVEST (2, 3, 4)."; Y
70  IF Y < 2 THEN GOTO 60
80  IF S < YO THEN LET YO = S
90  IF F > 1 THEN LET F = 1
```

Background

Commercial fishermen don't want oil rigs placed near good fishing grounds because fish and fish breeding grounds could be destroyed if oil were to leak from the well into the ocean. Oil leaks also endanger other valuable wildlife, such as shellfish and birds. And, oil leaks damage beaches, thus harming tourism.

Since scientists have developed the technology to explore the deep-ocean floor with submersible vehicles, many new, unforeseen discoveries have been made. On the seabed near rifts, vents have been found from which hot water is rising. This water is rich in minerals such as iron, sulfur, zinc, and others. When the hot water mixes with cold seawater, the minerals precipitate onto the ocean floor. It is believed that cold seawater flows down into the faults in the rift zone, becomes heated by high temperatures as it approaches the mantle, and then is forced upward to form the vents or "smokers," as they are called. The water dissolves and picks up minerals as it moves through the fault zones.

There is as yet no legal way for a country to stake an exclusive claim in the open sea and proceed to gather seafloor minerals for commercial purposes. In 1982, the United Nations adopted the Law of the Sea Treaty, but the U.S. voted against it, because the federal government objected to the restrictions placed on the mining of manganese nodules. For the present, mining is concentrated in territorial waters within 320 km of the coastline.

Teaching Tips

- Bring in samples of edible seaweed products. These can be purchased at an oriental food store.
- Challenge students to add more items to the list of resources of the ocean named in the text. Offer a prize (for example, a seafloor map) to the student who adds the most items.

Facts At Your Fingertips

- Much of the salt we use is mined from land sources. These sources were deposited by ancient seas that once covered the land.
- Some people have to limit their intake of salt (sodium) because it can cause high blood pressure and increased water retention.
- Seafood is high in iodine. Iodized salt is recommended for people who live inland and don't eat much seafood. Iodine is necessary for thyroid health.
- Algin, a gelatin-like substance made from kelp, is used in making ice cream.

14 OCEANS

14-7 Resources of the Ocean

Hidden Treasures of the Ocean

Some terms are missing in the statements below. Think of the term that best completes each statement and write that term on the corresponding numbered space at the left.

1. ______
2. ______
3. ______
4. ______
5. ______
6. ______
7. ______
8. ______
9. ______
10. ______
11. ______
12. ______
13. ______
14. ______
15. ______

The sea has long been a source of __1.__, an excellent source of protein. The food additive __2.__, used to make food more tasteful, is another resource from the sea. One cubic __3.__ of seawater contains enough of this substance to supply the whole world for __4.__ years. __5.__, a form of seaweed, can be used for food as well as in the production of drugs and cosmetics. When people cultivate or farm the sea, their work is called __6.__

The continental shelves of many continents have proven to be good sources of __7.__ and __8.__ Off-shore drilling __9.__ are a common sight in the Gulf of Mexico.

__10.__ earth is another product of the sea. It is formed from the silica shells of microscopic __11.__ It makes good filter material for swimming pools.

In order to make seawater suitable for drinking, the __12.__ must be removed, a process called __13.__ If this process can be done efficiently, the ocean could become an excellent source of __14.__

Most of the resources of the sea are replenished by deposition of materials from the __15.__ When the land runs low on resources, they cannot be replaced so easily.

Farming the Ocean

The topic Modern Methods of Mariculture is related to the field of oceanography. To research the topic, use reference books from the library or contact one of the institutions that your teacher will suggest. Then answer the questions.

1. What different kinds of food can be produced by the method of mariculture? ______
2. What is the basic principle used in mariculture? ______
3. Where is mariculture being used the most? ______

Figure 14-20 *Off-shore drilling rigs, like this one off the coast of Yucatan, Mexico, are a common sight in many places.*

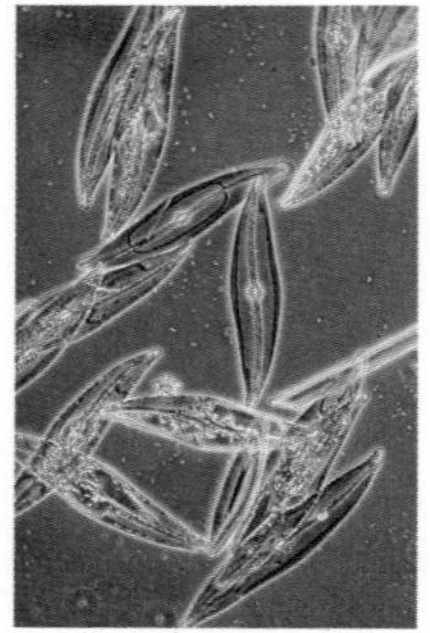

Figure 14-21 *This photograph of diatoms was taken through a microscope.*

The crust beneath the ocean has proven to be an excellent source of oil and natural gas. These fuels are in great demand. Supplies of oil and gas have been found on the continental shelves of many continents. Off-shore drilling rigs are a common sight in the Gulf of Mexico, off the coast of California, at Cook Inlet (Alaska), and in many other places. (See Figure 14-20.)

Diatoms (DY uh tahmz), microscopic algae in the ocean, have transparent shells of silica. When diatoms die, these shells form thick deposits called *diatomaceous* (dy uh tuh MAY shus) *earth*. (See Figure 14-21.) Because of its strength and porous quality, diatomaceous earth makes excellent filter material for swimming pools. Drug firms also use it to strain out bacteria in the production of medicines. It is used to strengthen concrete, to produce a flat or semigloss finish in paint, and also as an abrasive in toothpaste. The thickest deposits of diatomaceous earth are in California.

A unique feature of many ocean resources is that they are being replenished by deposition of materials from the land and by processes within the ocean. When resources are taken from the land, they cannot be replaced so quickly.

TRY THIS

Dissolve 3.5 g of salt in 100 ml of water in an open flat container. Let this stand at room temperature until all the water has evaporated. What do you find in the dish? Relate this process to the formation of salt flats in Utah, or to salt mines in Russia.

Checkpoint

1. List some of the food sources in the sea.
2. On what part of the ocean floor are supplies of oil and gas found?
3. Name some uses of diatomaceous earth.
4. How can the ocean be used as a source of salt and fresh water?

Using Try This

The solution described in the text has a salinity similar to the average salinity of seawater. Use any shallow container that will allow the water to evaporate readily (a large surface area is desirable). A layer of salt will be found on the bottom of the container after evaporation. Students can appreciate how much water would have to be evaporated to form a layer of salt 50 m thick or more.

Checkpoint Answers

1. Fish, oysters, kelp.
2. On the continental shelf.
3. Diatomaceous earth is used to filter materials, to strengthen concrete, and to produce paints.
4. Through desalination.

FEATURE ARTICLE

Ocean Exploration

Modern technology has allowed people to explore the oceans with more ease and to collect more data in the past 30 years than was ever possible before. In order to study the oceans, several problems had to be overcome. For example, the water at extreme depths exerts great pressure. So submarine vessels had to have strong walls. Divers had to wear heavy suits to protect themselves against the cold water if they wanted to leave the vessel. The suits, like the JIM suit used by U.S. Navy deep-sea divers, also had to be designed to maintain normal pressure at extreme ocean depths. Special breathing apparatus was needed. In addition, there is very little light in deep water.

Figure 1

Figure 2

Small, submarinelike vessels called *submersibles* were specially designed to help overcome the difficulties of ocean research. Examples of these early laboratories are the U.S. *Tektite*, the *Trieste*, and the small submersible *Alvin*. But now a new generation of ocean explorers is sending robots to the bottom of the ocean. Electronic devices do the sensing and recording. Messages are sent back to surface ships by fine fiber optic cables. One such system is called *Argo-Jason*. (See Figure 1.) From *Argo* a smaller, self-propelled vehicle, *Jason*, can be sent to get a closer look and gather specimens. Television images can be transmitted by satellite all over the world.

Another new vehicle is a helicopter-like, one-person submersible. *Deep Rover* has mechanical arms, a pressurized cabin, enough oxygen for 100 hours, and safety devices designed so that if anything goes wrong, it will pop back up to the surface like a cork. (See Figure 2.)

Using The Feature Article

After students have read the feature article, hold a class discussion to clear up any confusion about terms or ideas that they may not understand. During the discussion, make a list of topics related to ocean exploration that students would like to know more about. Typical topics could be:

1. What different kinds of data do modern oceanographers collect? What tools do they use to collect the data?
2. What were the first scientific diving vessels like?
3. What effect does water pressure have on deep-sea divers?
4. What kind of photography equipment and observation lights have been developed for deep-sea exploration?
5. Describe and obtain pictures of the U.S. *Tektite*, *Trieste*, and *Alvin*.
6. Describe the new robots that are being used for work on the ocean floor.
7. How do submersibles communicate with the surface? Of what value to submarine communications are lasers that produce a powerful blue-green beam?
8. What is the latest use of satellites for ocean navigation and exploration? What is NAVSTAR?
9. Is the *Deep Hawk* still operating? Or is there a newer model?

Each student, or small group of students, can be responsible for researching one of these topics. You may wish to allow students to spend one or two class periods doing research in the school library. Each student or group should deliver their final report to the class. Encourage the use of supplemental aids, such as pictures, diagrams, or other media, to illustrate the presentations. Encourage questions and answers after each of the presentations.

Reading Suggestions

For The Teacher

Aksyonov, Andrei and Alexander Chernov. EXPLORING THE DEEP. New York: Watts, 1980. A profusely-illustrated account of human efforts to explore the depths of the ocean, with a final chapter on conservation of the ocean's resources.

Goldin, Augusta. OCEANS OF ENERGY: RESERVOIR OF POWER FOR THE FUTURE. New York: Harcourt Brace Jovanovich, 1980. Well-written and well-illustrated survey of possible energy resources.

Marx, Wesley. THE OCEANS: OUR LAST RESOURCE. San Francisco: Sierra Club, 1981. A discussion of the myth of the ocean's inexhaustibility.

Ross, David A. INTRODUCTION TO OCEANOGRAPHY. 3d ed. Englewood Cliffs, NJ: Prentice Hall, 1982. Up-to-date, readable coverage of the job of an oceanographer.

For The Student

Ballard, Robert D. "A Long Last Look at the Titantic." *National Geographic,* December 1986, pp. 698–727. Pictures taken from the submersible Alvin. Diagrammatic sketches and story of what happened to the Titanic.

Ballard, Robert D. "NR-1: The Navy's Inner-Space Shuttle." *National Geographic,* April 1985, pp. 450–458. What it is like to live and work on a nuclear submarine.

Cook, Jan Leslie. THE MYSTERIOUS UNDERSEA WORLD. Washington, DC: National Geographic Society, 1980. A pictorial album, with sections on the edge of the sea, coral reefs, ocean depths, sunken treasure, and sea-bed habitats. Includes suggested classroom activities.

Lambert, David. THE OCEANS. New York: Warwick, 1980. A vividly-illustrated, well-written, up-to-date survey of the marine world.

Understanding The Chapter

14

Main Ideas

1. Rivers carry dissolved salts from the land to the ocean.
2. The most abundant salt in seawater is sodium chloride.
3. The energy of a wave moves forward, but the water does not.
4. Most waves get their energy from wind.
5. Tides are caused by the gravitational pull of the sun and moon.
6. There are usually two low tides and two high tides per day.
7. Ocean currents are caused by winds, Earth's rotation, and density differences.
8. Ocean sediments fall into two categories: organic and inorganic.
9. Sediments are thicker in the Atlantic Ocean than in the Pacific.
10. By studying core samples, scientists can learn about the history of the ocean.
11. About 60% of the ocean floor is a plain.
12. All oceans do not have the same pattern of features. Deep sea trenches are found in the Pacific Ocean, but not in the Atlantic, which has a mid-ocean ridge.
13. Many islands are chains of volcanic mountains that formed on the ocean floor.
14. Coral reefs are a common feature of warm, shallow water.
15. The sea has a multitude of natural resources.

Vocabulary Review

From the following list, choose the term that best completes each of the statements. Write your answers on a separate piece of paper.

abyssal plain	Nansen bottle
continental rise	neap tides
continental shelf	nodules
continental slope	ocean currents
crest	organic sediments
desalination	salinity
inland seas	seas
inorganic sediments	sonar
mariculture	spring tides
mid-ocean ridge	trough
	tsunamis

1. The highest tides are __?__.
2. __?__ is used to find the depth of the ocean.
3. Part of the continent that is under water is the __?__.
4. __?__ found on the ocean floor are a good source of manganese and other minerals.
5. The percentage of salt in seawater is called __?__.
6. The process of removing salt from seawater is called __?__.
7. Some __?__ are also called oozes.
8. The __?__ covers 60% of the ocean floor.
9. The __?__ is the longest feature on Earth.
10. The farming of the sea by people is called __?__.
11. The highest point in a wave is the __?__.

Vocabulary Review Answers

1. spring tides
2. sonar
3. continental shelf
4. nodules
5. salinity
6. desalination
7. organic sediments
8. abyssal plain
9. mid-ocean ridge
10. mariculture
11. crest

Chapter Review

Write your answers on a separate piece of paper.

Know The Facts

1. Most of the dissolved salts in seawater come from the __?__. (continents, atmosphere, sea-dwelling organisms)
2. A swimmer floats higher in seawater than in fresh water because __?__. (the density of seawater is less, the density of seawater is greater, seawater is a bigger body of water)
3. Which of the following affects the salinity of surface seawater at a given location? (depth, density, evaporation)
4. Most of the gases in seawater come from the __?__. (continents, atmosphere, life in the sea)
5. What happens to an object floating at the surface when a wave passes? (It moves horizontally, but not up and down; it moves rapidly in the direction of wave travel; it moves in a circle, but ends up in its original position)
6. What is the principal cause of waves in the ocean? (wind, temperature, density differences)
7. What is the primary cause of vertical currents in the ocean? (wind, density differences, shape of the ocean floor)
8. Low tides occur on Earth in places that are __?__ the moon. (opposite, 90° out of line with, 180° out of line with)
9. Oceanographers use __?__ to map the ocean floor. (Nansen bottles, core samplers, sonar)
10. A device used to find the temperature of ocean water at various levels is a __?__. (current meter, core sampler, Nansen bottle)
11. Samples of sediment layers at the bottom of the ocean are brought up to the surface using a __?__. (Nansen bottle, core sampler, barrel)
12. Tsunamis are caused by __?__. (wind, earthquakes, gravity)
13. The most abundant salt in seawater is __?__. (magnesium chloride, calcium chloride, sodium chloride)
14. The moon affects the tides more than the sun because __?__. (the sun is too big, the moon is closer, the moon is bigger)
15. Trade winds north of the equator cause the __?__. (Kuroshio current, south equatorial current, north equatorial current)
16. There are more sediments in the Atlantic Ocean than in the Pacific because __?__. (the Atlantic is bigger, the Atlantic is deeper, the Atlantic has more rivers emptying into it)

Understand The Concepts

17. You know what time a high tide occurs in one place. Explain how you would know approximately when the next low tide will be.
18. Why are there no atolls in the North Atlantic?
19. What happens to the salinity of ocean water when evaporation takes place?
20. Why is the water along the California coast colder than the water along the east coast at the same latitude?

Chapter Review Answers

Know The Facts

1. continents
2. the density of seawater is greater
3. evaporation
4. atmosphere
5. it moves in a circle, but ends up in its original position
6. wind
7. density differences
8. 90° out of line with
9. sonar
10. Nansen bottle
11. core sampler
12. earthquakes
13. sodium chloride
14. the moon is closer
15. north equatorial current
16. the Atlantic has more rivers emptying into it

Understand The Concepts

17. The low tide will occur about six hours after high tide. High tides are about twelve hours apart; lows occur between highs.
18. The water is too cold in the North Atlantic for coral to live, and thus for atolls to form.
19. Salinity increases.
20. California currents come from the north; the warm Gulf Stream current on the East Coast comes from the equator.

21. Densest water would be at the bottom of the ocean near the polar regions.
22. The pattern of currents causes the change in temperature between western and eastern sides of a basin.
23. The west-to-east rotation of Earth on its axis makes currents flow in a clockwise pattern in the Northern Hemisphere, and a counterclockwise pattern in the Southern Hemisphere.
24. Hawaiian Islands, Aleutian Islands, most islands in the South Pacific.
25. The level of the sea has changed in relation to the coral reefs.
26. Tides are caused by the moon and sun and Earth's rotation. Generally, two high and two low tides occur daily. Currents are continuous movements of water in the oceans that flow in a definite pattern throughout the ocean. They are caused by winds and density differences.
27. The warmest ocean water would be at the surface near the equator. The coldest water would be just under the ice in the polar regions.
28. The ships going to Europe sailed in the Gulf Stream, which added speed to the wind speed. On the return trip, they had to resist the Gulf Stream, as well as the winds.

Challenge Your Understanding

29. The kind of sediment in a layer will indicate the type of climate in which it was formed. Limestone that is made from shells of warm-water shellfish would indicate a formerly warm sea and climate.
30. The Gulf Stream brings warm water to the Norwegian ports.
31. The height of a tsunami in mid-ocean is negligible, while its length is very great. The wave builds as it nears shore and drags on the bottom.
32. Organic sediments would be rare on the abyssal plain because of the solvent action of seawater; inorganic sediments would be more common.

21. Where would you expect to find the densest water in the ocean?
22. What causes the difference in temperature between currents on the eastern and western sides of an ocean basin?
23. What factor has the greatest influence on ocean surface current patterns?
24. Give an example of a chain of volcanic islands.
25. Explain how coral reefs are found in deep water where temperatures are too cold for coral to live.
26. How do currents differ from tides?
27. Where would you expect to find the warmest ocean water? Where would you expect to find the coldest unfrozen ocean water?
28. Benjamin Franklin, when a postmaster, noticed that sailing ships going from North America to Europe took less time than those coming from Europe. What is one good reason for this?

Challenge Your Understanding

29. How can past climates be determined from core samples of ocean sediments? Give an example.
30. Why are ports in Norway near the Arctic Circle open to ships all year long?
31. If you were on the ocean in a ship, why would you not be aware of a tsunami, even if one passed beneath you?
32. Which kind of sediment—organic or inorganic—would be more common on the abyssal plain? Explain your reasoning.

Projects

1. Research the new technology used for getting data about the oceans and the seas of the world.
2. Report on the different kinds of life that have been found on the ocean floor near the mid-ocean ridge.
3. Build a model of the ocean floor showing features such as a continental shelf, a continental slope, a continental rise, an abyssal plain, a mid-ocean ridge, and a trench.
4. Walk through a supermarket and list all of the items that had their source in the ocean.
5. Trace the path of a nearby river to where it enters an ocean. What kind of materials would be carried from your region to the ocean?
6. Read *Kon-Tiki* by Thor Heyerdahl. Report to your class.
7. Interview a newspaper reporter, an environmental lobbyist, or government official about laws governing the use of the oceans for fishing, mining, and oil exploration.
8. Make a model water wheel. List and describe the different machines that could use the energy supplied by the water wheel.
9. Make a drift bottle out of an empty bottle or fruit jar. Place a card in it with your name and address, and instructions for the finder. Put the bottle in a nearby river or stream and wait for an answer.
10. Research the latest ideas for harnessing the energy of tides, currents, or waves.
11. Make a scrapbook of articles about the ocean from magazines and newspapers.

14 OCEANS
Projects

1. Research the latest technology for collecting data about the ocean.
   a. What are the names of the latest deep-sea submersible vehicles?
   b. How are satellites used in ocean exploration?
   c. What role do computers play in oceanography?
   d. Identify any international programs for studying the oceans.

14 OCEANS
Challenge

The questions below are designed to test your ability to apply what you have learned in this chapter. Read each question; then write the answer to the question on the lines that follow it.

1. How does the height of Mt. Everest compare with the depth of the Marianas Trench?
2. If 90% of atmospheric oxygen is produced by tiny ocean plants, of what importance are these algae to other living things?
3. How did so much salt get into the oceans?

Chapter 15

Astronomy

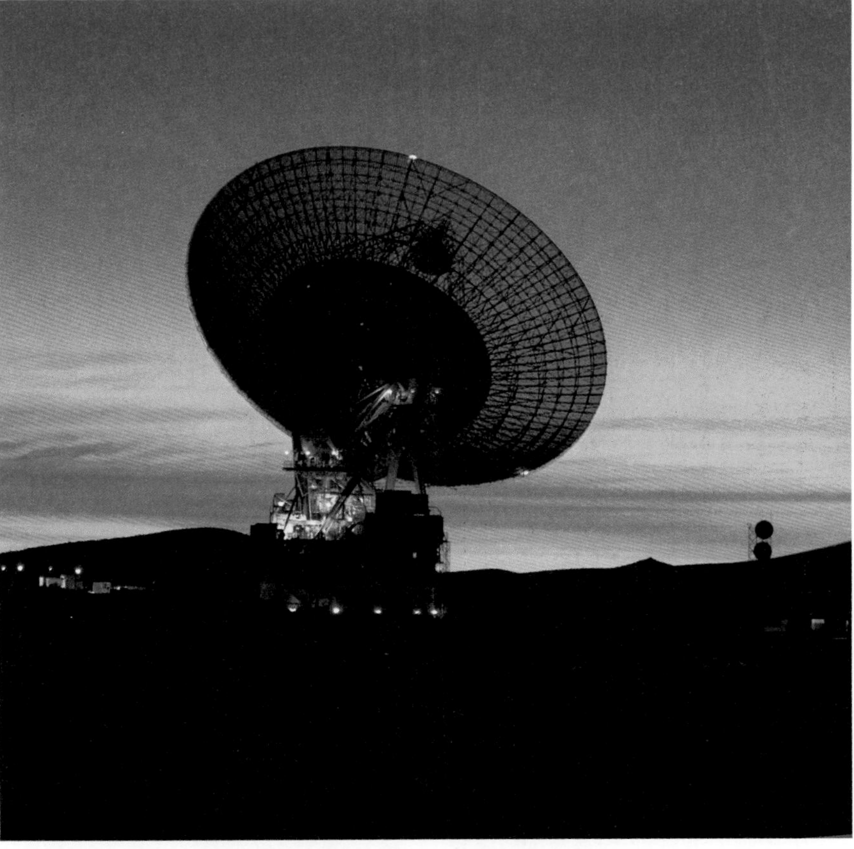

This radio telescope in the Mojave Desert gathers radio waves from space.

Vocabulary Preview

universe
astronomy
rotation
revolution
constellations
satellite
phases
solar system
asteroids
comets
meteoroid
meteors
meteorite
galaxies
nuclear fusion
sunspots
solar eclipse
lunar eclipse
magnitude
nebulas
red giant
white dwarf
supernova
neutron star
black hole
light-year
visual telescopes
radio telescopes
spectroscope

CHAPTER 15 Overview

The chapter first describes planet Earth's shape and motions. Earth's seasons are related to the tilt of Earth's axis. Then, the chapter focuses on lunar surface features, motions, phases, and tidal effects upon Earth. The third section focuses on the components of our solar system: the nine planets as well as the many moons, asteroids, comets, meteoroids, and meteors. In the fourth section, students learn that the sun is typical of an average star, fueled by nuclear fusion. Its motions, its relative size, its life expectancy, and its role in solar and lunar eclipses are explained. Some general characteristics of stars, and the means of studying their chemistry are briefly described in the fifth section. The apparent stages of development among stars are presented and interstellar measurement in light-years is discussed. The last section provides an overview of telescopes. Visual and radio telescopes are described, as well as the use of bright-line spectra, absorption spectra, and spectroscopes in astronomy.

Goals

At the end of this chapter, students should be able to:

1. describe Earth's shape.
2. describe the rotation and revolution of Earth.
3. identify major surface features of the moon.
4. describe some effects of the moon's motions.
5. identify the general features of the planets and other members of the solar system.
6. describe important properties of the sun.
7. compare a solar eclipse to a lunar eclipse.
8. describe different properties and life cycles of stars.
9. describe various kinds of telescopes.
10. explain how a spectrograph is used to analyze starlight.

Teaching Suggestions 15-1

Encourage the students to give their impressions of the universe, the solar system, and astronomy in general, before starting this section. If the students bring up astrology, point out that there is no scientific basis for astrology. Ask them if they can provide evidence that Earth is moving. This discussion is likely to stimulate their curiosity about how scientists learned these things.

Before discussing constellations, obtain or make transparencies of the seasonal constellations to help students get an idea of the pattern to look for. Be sure to include the polar constellations because they are visible all year.

If the night sky is clear enough to see stars, encourage students to observe the stars and try to identify the seasonal constellations. Have on hand star maps and a globe showing the stars in relation to Earth. Alternatively, you could visit a planetarium or use a constellation projection apparatus.

Activity 21, Seasons of the Year, on student text pages 558–559, should be done while the class discusses the seasons. Chapter-end projects 2, 4, and 7 also apply to this section.

Background

There are several ways to determine that the Earth is round. One way is to observe the curved shadow of Earth on the moon during an eclipse. However, the opportunity for using this method occurs only occasionally, and requires a knowledge of eclipses. Another method was developed by Eratosthenes, an ancient Greek mathematician, in 230 B.C. He had read that on one day during each year, the sun shone down to the bottom of a deep well that was located 900 km south of Alexandria, Egypt. He observed that in Alexandria vertical objects, such as fence posts and buildings, cast shadows on the same day. This indicated to him that while the sun was directly over the distant well on this day, it was not directly over Alexandria. Eratosthenes concluded that the Earth must be curved between these two places. He proceeded to calculate the circumference of the Earth using the following equation: A (angular distance around Earth, 360 degrees) divided by a (angle of shadow cast in Alexandria) equals D (unknown distance around Earth) divided by d (distance between well and Alexandria, 900 km). Eratosthenes knew that the angle cast by the shadow would be equal to the angle between the well and Alexandria at the center of Earth. He applied the geometric theorem stating that when parallel lines are cut by a transversal, the alternate interior angles are equal. He assumed that the sun's rays strike Earth in parallel lines, as shown in Figure TE 15-1. He then calculated that Earth's circumference was 39,250 km, which is close to the modern-day calculation of 40,075 km.

15-1

Planet Earth

Goals

1. To describe Earth's shape.
2. To describe the rotation and revolution of Earth.

universe

The **universe** is everything that exists. It includes all of the stars that you can see on a clear night, for example. It has been said that there are as many stars as there are grains of sand on all the beaches in the world!

The sun is one of the billions of stars in the universe. Moving around the sun are nine large objects called *planets*, one of which is Earth. The study of everything in the universe that can be observed from Earth or space is called **astronomy**.

astronomy

Pictures taken from space show that Earth is "round." It has a spherical shape, like a baseball or basketball. (See Figure 15-1.) However, Earth is not a perfect sphere. The distance around Earth, its *circumference*, is somewhat greater at the equator than around the poles. The average circumference of Earth's "flattened" sphere is 40,008 km. (If you drove that distance nonstop at a speed of 80 km/h, it would take you about 21 days.)

Figure 15-1 *Try to locate Africa, the Arabian Peninsula, and the South Pole.*

Figure 15-2 *Earth's rotation, as recorded by a camera pointed at the North Star.*

Earth is not a stationary object. It is constantly in motion. But its motion is difficult to detect on Earth. When you are riding in a car, you know you are in motion because you see trees or other fixed objects "whizzing" by. But there are no obvious fixed *reference objects* by which to detect Earth's motion. As a result, people thought for centuries that Earth was stationary.

rotation

We now know that Earth moves in two basic ways. One is **rotation**, the spinning or turning about a straight line, or *axis*. (A moving wheel also rotates about an axis.) Earth's axis is an imaginary straight line that runs through Earth from the North Pole to the South Pole. Earth completes one rotation every 24 hours.

Earth's rotation produces changes that you can see. The sun appears to rise each day in the east and set in the west. The sun's *apparent* motion results from Earth's rotation. The sun does not move relative to Earth. But a different part of Earth's surface faces the sun at any given time, due to Earth's rotation.

Similarly, Earth's rotation accounts for changes that you can see in the night sky. Early in the evening, find a group of stars that you can easily recognize in the eastern sky. An hour or two later, look for the same stars. You will find that they appear to have moved toward the west. Pictures have been taken focusing the camera on the North Star and using a time exposure of several minutes. The circular streaks of light are the star trails made as Earth rotates underneath the stars. (See Figure 15-2.)

revolution

The other basic motion of Earth is its **revolution**, or its movement around the sun. Earth completes one *orbit*, or revolution, in one year. Other planets take a longer or a shorter time to revolve around the sun.

You can tell that Earth revolves around the sun by looking at the sky at night. We see stars at night, when we are on the side of Earth that faces away from the sun. The stars can serve as "fixed" reference points to help us observe Earth's motion.

constellations

People have known for many years that certain patterns of stars, or **constellations**, are visible only at certain times of the year. Imagine sitting at a window table in a slowly-revolving rooftop restaurant. From the restaurant, the view of the city lights below is constantly changing. Similarly, from

The ecliptic is the term used to define the path the sun makes across the sky in a year's time due to the Earth's 23½° tilt from the place of its orbit. The vertical ray of the sun travels from 23½° N (Tropic of Cancer) about June 21 to the equator around September 21, then to 23½° S (Tropic of Capricorn) on December 21 and to the equator about March 21. By June 21st the vertical ray will return to 23½° N. An analemma, the diagram shaped like a figure eight that is often seen on globes or maps, gives the latitude of the vertical ray of the sun on any day of the year.

Other evidences of Earth's rotation are the changing pattern of the Foucault pendulum found in many museums; the existence of the Coriolis effect; and the variations in the force of gravity. The force of gravity is stronger at the poles and weaker at the equator due to the planet's rotation.

The North Star, or Polaris, is located above the North Pole, or the north end of Earth's axis of rotation. The North Star does not appear to move in relation to the stars around it. All stars in the northern sky appear to rotate around Polaris. There is a similar point in the Southern Hemisphere.

Demonstration

You can demonstrate Eratosthenes' reasoning using a large piece of oak tag. Glue two golf tees, large end down, about 30 cm apart, onto the oak tag. These will represent two tall buildings or fence posts at two locations in Egypt, one 30.5 cm "north" of the other. Turn out the room lights. Using a projector lamp, demonstrate that if the oak tag is held flat and perpendicular to the light, no shadow will be cast by either stick. As you bend the oak tag, keeping one stick perpendicular to the light, you can show that only one of the "buildings" casts a shadow.

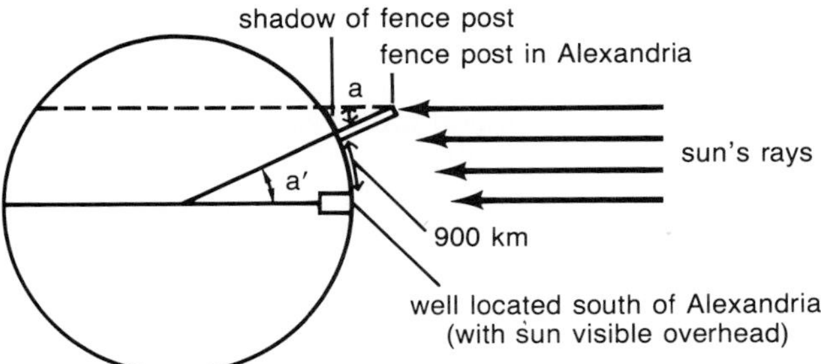

Figure TE 15-1 Method used by Eratosthenes to calculate Earth's circumference.

Teaching Tips

- Make sure students know the difference between astronomy and astrology. Astrology is the nonscientific association of the heavenly bodies with one's personality and future.
- Students might be interested in contacting the tongue-in-cheek International Flat Earth Research Society, Box 2533, Lancaster, CA 93534.
- Point out to students that prevailing scientific opinion since the time of Eratosthenes has been that Earth is spherical. Explorers did not "discover" that Earth is round; they were operating under that assumption from the outset.

Facts At Your Fingertips

- The use of the sun's apparent motion to tell time is believed to date from around 3500 B.C.
- A sundial has been built into the sidewalk in front of the Alberta Government Telephone building in Calgary. On a sunny day, the shadow of a pedestrian standing on the indicated spot falls on an elliptical arrangement of numbers, which correspond to daylight hours.
- Earth does have other motions: it "wobbles" on its axis; it moves along with the sun in the galaxy; and it moves with the galaxy in the universe.
- When a ship sails beyond the horizon, it "disappears" beneath the horizon relative to its point of embarkation. This is a demonstration of the roundness of the sea's surface.

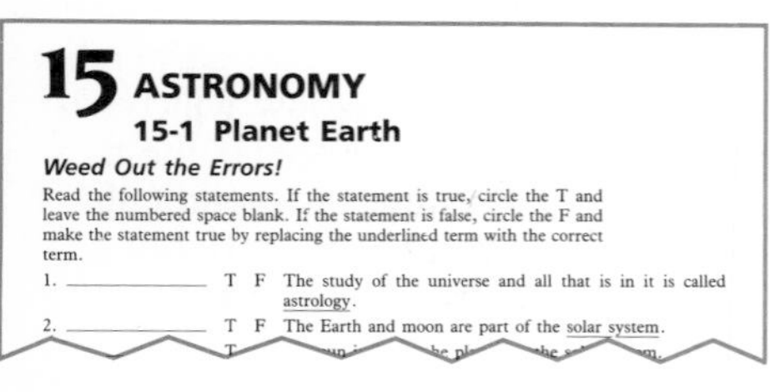
15 ASTRONOMY

15-1 Planet Earth

Weed Out the Errors!

Read the following statements. If the statement is true, circle the T and leave the numbered space blank. If the statement is false, circle the F and make the statement true by replacing the underlined term with the correct term.

1. ________ T F The study of the universe and all that is in it is called astrology.
2. ________ T F The Earth and moon are part of the solar system.

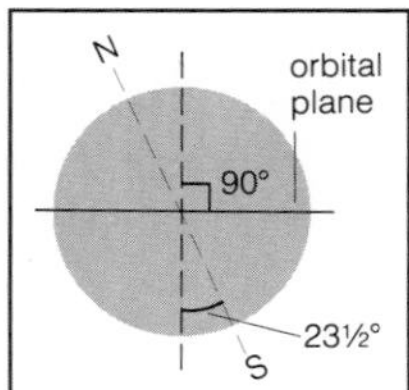

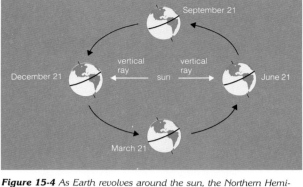

Figure 15-3 *Earth's axis is not at right angles to its orbital plane.*

Figure 15-4 *As Earth revolves around the sun, the Northern Hemisphere receives more vertical rays during summer than winter.*

the revolving Earth, the view of the sky changes with time. Each season we see a different view. As a result, we speak of winter, spring, summer, and fall constellations.

The imaginary flat surface, or plane, in which Earth revolves around the sun is called Earth's *orbital plane.* Earth's axis is not at right angles to its orbital plane. Instead, it is tilted 23½° from the right-angle (90°) position. (See Figure 15-3.)

If Earth were not tilted, vertical rays of sunlight would be received only at its equator. Areas receiving vertical rays are warmer. Because of Earth's tilt, vertical rays are shifted north of the equator part of the year and south of it at a different part of the year. (See Figure 15-4.) The 23½° tilt determines how far north and south of the equator the vertical rays of the sun reach. As the location of the vertical rays changes, the amount of energy received at each latitude changes. Thus, the seasons change. Summer begins in June in the Northern Hemisphere (Earth's northern half) and in December in the Southern Hemisphere.

TRY THIS

In a darkened room, shine a bright light onto a globe. The light source should be 1–3 m from the globe. Rotate the globe slowly. Compare the amount of light that falls on each hemisphere as it is (a) tilted toward, (b) tilted away from the source.

Checkpoint

1. Describe Earth's shape.
2. What motion of Earth causes day and night?
3. Explain why we have seasons.

Using Try This

Turn the globe slowly, so students can see that during the summer in the Northern Hemisphere there are more hours of daylight than night. Just the opposite is true during the winter.

Checkpoint Answers

1. Flattened sphere.
2. Rotation on its axis.
3. The tilt of Earth's axis allows a vertical ray of the sun to reach farther north and south of the equator as Earth moves around the sun.

15-2

Earth's Moon

Goals

1. To identify major surface features of the moon.
2. To describe some effects of the moon's motions.

A **satellite** is an object that revolves around a planet. Earth's natural satellite is the moon. It revolves around Earth, as Earth revolves around the sun. The planet Mercury has no moon. Saturn has at least 17! As moons go, Earth's is large. The circumference of our moon is more than one fourth that of Earth.

satellite

When you look at the full moon you can see contrasting light and dark areas. These areas make a pattern, often called the face of the "man in the moon." The light areas are mountains, and the dark areas are flat parts of the moon. (See Figure 15-5.) The moon was first viewed through a telescope by Galileo, in 1609. He thought the dark areas were seas. So he called them *maria* (MAH ree uh), which is the plural of *mare* (MAH ray). *Mare* means "sea" in Latin.

The moon is dotted with many steep-walled circular depressions called *craters*. Most of the craters are located in the mountain regions. Some are more than 100 km across. Some craters have *central peaks*, which are small mountains on the crater floors. (See Figure 15-6.) Most of the craters on

Figure 15-5 *The light areas on the moon are mountains and craters.*

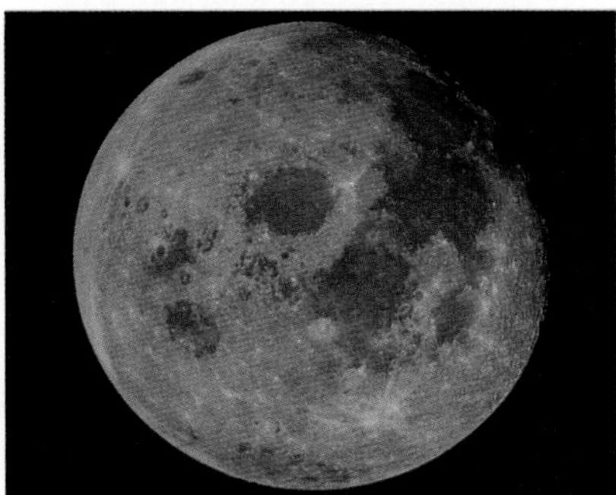

Figure 15-6 *Craters on the moon. Note the central peaks, and smaller craters.*

Teaching Suggestions 15-2

Large maps or a globe model of the moon should be available for students while studying this section. Give students time to become familiar with the names of maria and major craters, and the heights and diameters of craters. NASA has good films from the Apollo flights to the moon. These show the surface features of the moon, as well as the effect of zero gravity in space and the low gravity of the moon upon the astronauts. The use of animated films and models is an effective way of showing the moon's motions and phases. Encourage students to try project 6 and to bring the model to class. Students should do the Try This activity while discussing the moon's motions. For an extended assignment, have the students observe the moon nightly as it goes through one full set of phases. Weather permitting, have them record its shape, height above the horizon, azimuth (horizontal degree from North), and the time of their observations. Cloudy skies may interfere some nights, and some waning phases will not be visible before midnight. Have students use binoculars or a telescope to see the moon's surface more clearly. How many features can they identify?

Background

The lunar month (29½ days) is longer than the moon's revolution period (27⅓ days) because of Earth's movement around the sun. Earth moves around the sun about one degree a day. The moon has to revolve two extra days around Earth before it is positioned in "new moon" position. Because the moon's rotation period is about the same as its revolution period, daylight at any one place on the moon is about two weeks long. The same is true for night. Since the moon is not massive enough to hold an atmosphere capable of moderating temperatures, the long periods of daylight and night create the extremes in temperature.

From the Apollo flights to the moon, we know that the moon was formed about the same time as Earth and is made of rock material similar to basalt rocks found on Earth. However, the moon is less dense than Earth. The Apollo astronauts left behind many instruments on the moon that measure seismic activity, radiation of various kinds, magnetic variations, and solar wind, to name a few. We have been able to learn more about the origin of the solar system from the study of the moon.

From a study of mass concentrations on the moon, it has been found that there are higher concentrations of mass associated with the maria. There is no firm understanding about the cause or significance of these mass anomalies. Since there are no maria on the back side of the moon, it has been suggested that the more massive side of the moon is more strongly attracted to Earth than is the less massive side of the moon. Thus, we see only one lunar side.

Demonstration

Use two students. Have one be "Earth" and the other be a nonrotating "moon." As "moon" moves around "Earth," it will become apparent that the moon has to rotate for people on Earth to see only one side of the moon. The moon's period of rotation must be the same as its period of revolution for this condition to exist.

Teaching Tip

- The phases of the moon can be described as waxing and waning. Waxing phases occur when the moon is getting larger, between new and full moon. Waning phases occur when the moon is getting smaller, between full and new moon.

Facts At Your Fingertips

- During a waxing crescent phase, the whole moon will be dimly lit by reflected light from the Earth. This light is called earthshine.
- When the moon is near the horizon, it looks extra large because its light is being refracted—as if by a magnifying glass—in Earth's thicker lower atmosphere.
- For comparison to lunar conditions, the lowest temperature recorded in Antarctica was at the Soviet Antarctic station in August, 1960: −88.3°C.
- Our word month is derived from one "moonth," the period of time for one full set of phases of the moon.

Figure 15-7 *The phases of the moon. Compare these photographs with Figure 15-8.*

the moon were caused by objects hitting the moon. Other craters are of volcanic origin.

A moon day, the time it takes for the moon to rotate once about its axis, is 27⅓ Earth days. This is the same time it takes the moon to *revolve* around Earth. As a result, the same side of the moon always faces Earth. The far side of the moon was a complete mystery until pictures of it were taken from space in 1959.

The moon's long days and nights help to explain the great temperature extremes on its surface. Temperatures range from about −173°C at night to more than 100°C in the daytime. Since the moon has no atmosphere, there is nothing to filter the powerful rays of the sun. Nor is there an insulating blanket of atmosphere to slow the escape of heat at night.

The impact of objects hitting the moon, strong radiation from the sun, and extreme temperature changes combine to break up rock material over time. As a result, the moon's surface is covered with a layer of fine dust and rock fragments.

The moon does not produce its own light. The moon is visible because it reflects sunlight. We see the moon in different parts of the sky during a month because of its motion

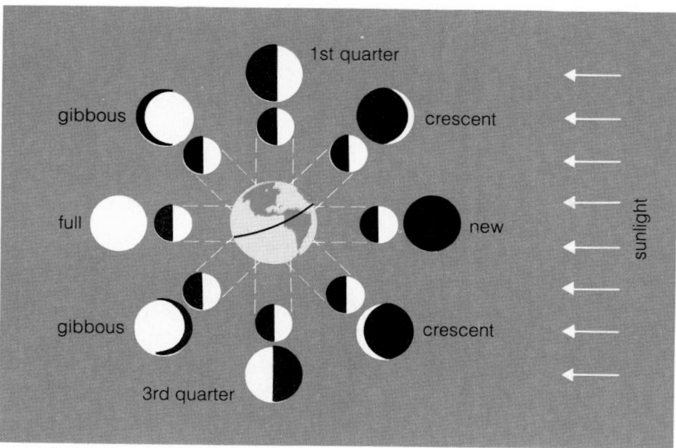

Figure 15-8 *As the moon revolves around Earth, the sun shines on the same side of the moon. But we see the lighted part from different angles.*

around Earth. The apparent shapes of the moon that we see as it completes one revolution are called **phases**. It takes about 29½ days, a lunar month, for the moon to go through a full set of phases. (See Figure 15-7.) The relative positions of sun, moon, and Earth determine what phase we see at any given time. (See Figure 15-8.)

phases

Like all objects in space, Earth and the moon attract each other. So do Earth and the sun. The most noticeable effect of this attraction is seen in the tides. The moon has a greater effect on the tides because it is closer to Earth.

Checkpoint

1. What is a satellite?
2. Describe the main features of the moon's surface.
3. How long does it take the moon to go through a full set of phases?
4. Why does the moon have a greater effect on the tides than does the sun?

TRY THIS

Use a globe or large ball to represent Earth and a smaller ball to represent the moon. Make some markings on one side of the "moon" and move it around "Earth" *without* rotating it. Do you see only one side of the "moon" from "Earth"? Now rotate the "moon" so that the same side is always toward "Earth." What can you say about the speed of the "moon's" rotation?

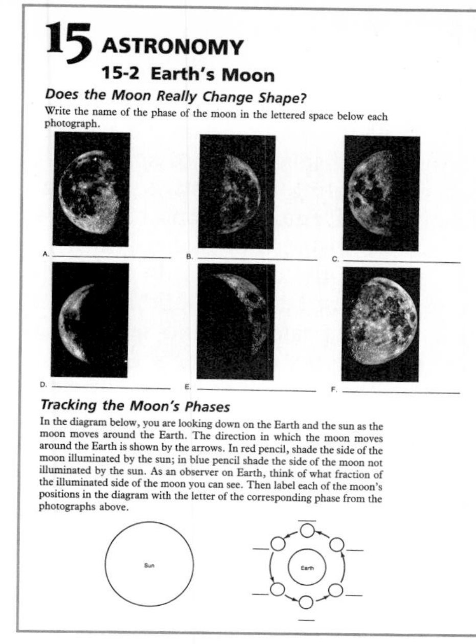

15 ASTRONOMY

15-2 Earth's Moon

Does the Moon Really Change Shape?

Write the name of the phase of the moon in the lettered space below each photograph.

A. ____ B. ____ C. ____

D. ____ E. ____ F. ____

Tracking the Moon's Phases

In the diagram below, you are looking down on the Earth and the sun as the moon moves around the Earth. The direction in which the moon moves around the Earth is shown by the arrows. In red pencil, shade the side of the moon illuminated by the sun; in blue pencil shade the side of the moon not illuminated by the sun. As an observer on Earth, think of what fraction of the illuminated side of the moon you can see. Then label each of the moon's positions in the diagram with the letter of the corresponding phase from the photographs above.

Using Try This

Before doing this activity, ask students: Does the moon rotate on its axis? Let them discover the answer by doing the activity. If the equipment is not available for students to do this, students can do the activity without the ball models. Students should discern that the moon must rotate for us to see only one of its sides all the time.

Checkpoint Answers

1. A satellite is a body that orbits a larger body.
2. Mountains, maria, craters, dust.
3. About one month (29.5 days).
4. The moon is much closer, so that its mass, although smaller than that of the sun, has more effect upon Earth.

Cross References

Refer back to Chapter 12, section 12-7, for a discussion of volcanoes. Tides are discussed in Chapter 14, section 14-3. Meteoroids are discussed in the next section of this chapter.

Teaching Suggestions 15-3

Display pictures of the planets and a chart of the solar system. To help students visualize the comparative size of the planets, display a set of spheres, each representing the relative size of a planet. (See Demonstrations.) To show the relative distance of planets in the solar system, the class can do a modeling experiment on an athletic field. Have students calculate and set up a scale distance for the planets that is based on the placement of the sun at one end of the field and Pluto (or Neptune, if there is not enough space) at the other. The other planets should be located at appropriate distances from the "sun." Two or three students should stand at each planet location.

In order to obtain the most recent data on planets, assign a team of students to each planet and object in the solar system. They might spend two class periods in the library collecting information from periodicals and books about their topics. After giving them time to organize their reports, have each team report their findings to the whole class. Encourage the use of films, slides, and pictures from magazines to augment the reports.

Encourage students to look for planets in the night sky. Have them do the Try This at the end of the section. Chapter-end projects 1, 3, and 5 apply to this section.

Background

Meteor showers that occur at regular times during the year are caused when Earth passes through the path of a previous comet. The names, dates, and locations for these meteor showers are the Quadrantids, January 3 near Ursa Major; the Lyrids, April 21–22 near Lyra; the Eta Aquarids, May 4 near Aquarius; the Delta Aquarids, July 27–29 near Aquarius; the Perseids, August 11–12 near Perseus; the Orionids, October 20–21 near Orion; and the Geminids, December 12–14 near Gemini. The best viewing is after midnight on these dates. It is possible to see as many as 40 meteors per hour.

Some meteors are huge fireballs; others are only tiny specks. It is said that meteors hit the Earth's atmosphere at a rate of one million an hour. However, only about 150 of these statistically survive atmospheric heating and impact on the Earth. Dust from burned-up meteors is found in layers on the ocean floor. Occasionally, but rarely, meteors will hit houses, cars, or people. Twice in eleven years a meteorite hit in the town of Wethersfield, Connecticut.

Scientists study meteorites that fall on the ice in Antarctica, because the frigid and unpolluted conditions leave the object relatively unchanged. Such meteorites can provide clues to what lies in outer space.

15-3

The Solar System

Goal

To identify the general features of the planets and other members of the solar system.

solar system

The sun and all the objects in orbit around it make up the **solar system**. Besides Earth and its moon, the solar system includes eight other planets, at least 47 other moons, and numerous smaller objects.

The planets are divided into two groups, according to their characteristics. One group is the rocky planets: Mercury, Venus, Earth, Mars, and Pluto. The second is the gaseous group: Jupiter, Saturn, Uranus, and Neptune. The rocky planets are more dense and smaller than the giant gaseous planets. (See Figure 15-9.) The gaseous planets have thicker atmospheres and more moons than the rocky planets.

Figure 15-9 *Facts about the planets. The minutes, hours, days, and years are Earth time units. The relative size of the sun is shown in yellow.*

Planet	Size	Radius	Time to rotate	Length of year
Mercury		0.38	58.7 d	88 d
Venus		0.95	243 d	225 d
Earth		1.00	23 h 56 m	365.25 d
Mars		0.53	24.7 h	687 d
Jupiter		11.18	9 h 50.5 min	11.9 y
Saturn		9.42	10 h 14 min	29.5 y
Uranus		3.84	17 h 14 min	84 y
Neptune		4.10	17.8 h	165 y
Pluto		0.24	6.39 d	248 y

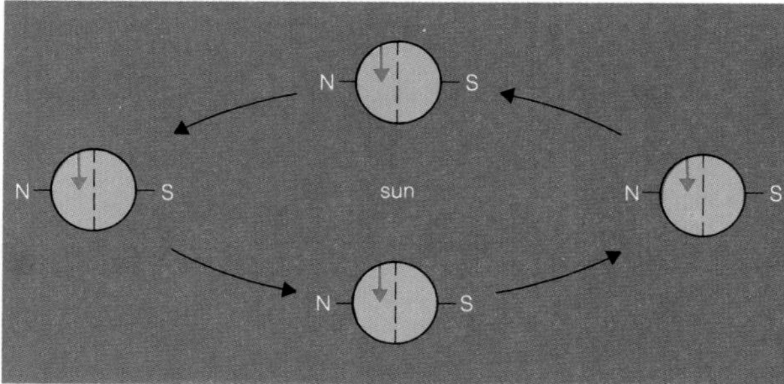

Figure 15-10 *Uranus seems to be lying on its side as it moves around the sun.*

All planets revolve around the sun in about the same plane. The ones closer to the sun move faster than those that are farther away. The length of time it takes a planet to revolve is called its *year*. If you want to look for a planet in the sky, always look on the same path that the sun took during the day.

Each planet rotates around a tilted axis. The tilt of most planets is about the same as Earth's, except for Uranus. The tilt of Uranus' axis is so great that the axis lies nearly in the planet's orbital plane. It looks like a spinning top that has fallen over on its side. (See Figure 15-10.) Uranus and Venus rotate in a direction opposite to Earth's rotation.

It is difficult to have up-to-date data about the planets in any book. Sometimes it seems as though new finds are reported almost daily. For years, astronomers thought Saturn was the only planet with rings. Recently they have discovered rings around Uranus and Jupiter, too. *Rings* are frozen bits of matter that revolve around some planets. Matter in the rings is frozen because the planets the rings encircle are so far from the sun.

The **asteroids**, or minor planets, are found mainly between the orbits of Mars and Jupiter. Asteroids are rocky and range in size from less than 1 km to 1000 km across. There are about 100,000 of them. They may be material that never combined to form a planet.

Far out near the edge of the solar system is a nearly spherical cloud of objects. These objects, called **comets**, average only a few kilometers across. They consist chiefly of fine dust and frozen water. Occasionally, one of the billions of comets travels toward the sun.

TRY THIS

With binoculars or a small telescope, watch the planet Jupiter for several nights. Use a sky chart to help you locate the planet. Draw the planet and the positions of the four largest moons as you see them each night.

asteroids

comets

Using Try This

Remind students to brace their arms well when using binoculars. A tripod to support the binoculars would be ideal. They will be seeing the moons of Jupiter as Galileo did. These four moons are called the Galilean satellites of Jupiter. Local newspapers or astronomy magazines list planet positions.

Teaching Tips

- Make sure students understand the scale difference between the universe and solar system.
- Point out to students that the terms "terrestrial," meaning Earth-like, and "Jovian," meaning Jupiter-like, are sometimes used, respectively, instead of the terms "rocky" and "gaseous" to describe the planets.

Facts At Your Fingertips

- Seasons on Uranus are very long, as are days and nights. One pole is pointed at the sun approximately half of the orbital period, which is as long as 42 Earth years. Each season is about 20 Earth years long. Days and nights are longer in the winter and summer than they are in spring and fall.
- *Comet* in Latin means "long-haired."
- Spacecraft must be protected from the intense heat on re-entering Earth's atmosphere. The tiles on the Space Shuttle serve as heat shields.
- Meteors are described as stony, iron, or stony-iron. ("Iron" meteors also contain nickel.)
- The largest known meteor lies in Namibia in southwestern Africa; it weighs more than 60 tons—as much as 35 full-size cars.

15 ASTRONOMY
Computer Program

Trip to Mars

Purpose To determine time needed to travel from Earth to Mars.

Background Have you ever dreamed of traveling from Earth to Mars? The shortest distance between the two planets is 56,300,000 km. It takes light, at 299,792 km per second, more than three minutes to traverse the distance.

ENTER the following program:

```
10 LET D = 56300000
20 PRINT "ENTER THE VELOCITY OF YOUR ROCKET"
30 INPUT "SHIP IN KM PER HOUR."; V
40 LET T = D / V
50 PRINT "THE VOYAGE WOULD TAKE"
60 PRINT T; "HOURS."
70 INPUT "ANOTHER ENTRY (Y OR N)?"; C$
80 IF C$ = "Y" THEN GOTO 20
90 END
```

Demonstrations

1. Display a set of spheres, each representing a planet in our solar system. Sizes will vary from small (1½ cm) styrofoam balls to large globes.
2. Set up a model of the sun with a planet moving around the sun with its axis tilted to the correct angle. This is especially helpful to understand Uranus.

■ The largest meteorite on exhibit is at the American Museum of Natural History in New York City. It weighs 34 tons and was found in Cape York, Greenland in 1897.

15 ASTRONOMY

15-3 The Solar System

The Sun's Family

In library reference books find the information missing from the chart below. Then fill in the chart and answer the questions that follow it.

Planet	*Diameter Compared to Earth*	*Length of day*	*Length of year*	*Number of Moons*	*Distinguishing Characteristics*
Mercury					
Venus					
Earth					
Mars					
Jupiter					
Saturn					
Uranus					
Neptune					
Pluto					

1. How are the planets arranged in this chart? ______
2. Where would the asteroids be on this chart? ______
3. Name two objects other than asteroids, planets, and moons that are considered part of the solar system. ______
4. Identify the following terms that relate to planet characteristics: a. oblateness, b. retrograde motion, c. astronomical unit, d. albedo, e. perturbation, f. eccentric orbit. ______

Checkpoint Answers

1. Rocky planets are more dense and smaller than gaseous planets.
2. Uranus.
3. Comets, asteroids, meteoroids.
4. A meteor is a burning piece of matter passing through Earth's atmosphere; a meteorite is the remainder of a meteor that impacted upon Earth's land or sea.

Figure 15-11 *Halley's comet, in 1910.*

Figure 15-12 *This meteorite fell in Arizona.*

When a comet gets as near to the sun as Jupiter's orbit, it starts to warm up. As it gets closer to the sun, the original comet, or *nucleus*, begins to boil off gases and dust. These stream out behind the nucleus in a *tail* that may be millions of kilometers long. (See Figure 15-11.) The nucleus becomes surrounded by a glowing cloud called the *coma*.

Since comets lose matter each time they near the sun, they may not last through many orbits. Some, however, have returned as bright objects in the sky many times. Halley's Comet is famous because it was the first comet known to appear at regular intervals. It returns every 76 years.

meteoroid

Any small piece of rock or mineral matter traveling through space is known as a **meteoroid**. Meteoroids may be small, stray asteroids or debris from a comet. If meteoroids enter Earth's atmosphere, they usually burn up due to friction. The fiery streaks that are produced in this way are called **meteors**. You may see meteors in the sky on a clear night. They are often called "falling stars." A meteoroid that does not burn up in the atmosphere and hits Earth is called a **meteorite**. (See Figure 15-12.) If large enough, meteorites may make craters in Earth's surface. The last large meteorite fell in Siberia in 1908.

meteors

meteorite

Checkpoint

1. How do rocky planets differ from gaseous planets?
2. Which planet's axis tilts the most?
3. Name some members of our solar system other than planets and moons.
4. How do a meteor and a meteorite differ?

15-4

The Sun

Goals

1. To describe important properties of the sun.
2. To compare a solar eclipse to a lunar eclipse.

Most stars in the universe are found in groups of billions of stars called **galaxies**. Our sun is an average star, one of the many stars in the Milky Way Galaxy.

galaxies

The sun is a huge, gaseous ball. It gives off tremendous amounts of energy. The source of the sun's energy is **nuclear fusion**, the *fusing* of hydrogen nuclei to form helium nuclei. Nuclear fusion occurs only at very high temperatures (millions of degrees Celsius). Such temperatures are produced in the dense interiors of stars.

nuclear fusion (NOO klee ur FYOO zhun)

The sun accounts for nearly 99.9% of the total mass of the solar system. Most of the sun's mass is hydrogen (about 75%) and helium (about 24%). Other elements make up no more than 1–2% of the sun's mass. The sun's volume is about one million times that of Earth. It is estimated that there is enough hydrogen in the sun to continue the fusion process for another 4–5 billion years.

In Figure 15-13, hotter areas of the sun show up as lighter colors. The darker areas are cooler. The leaping *prominence*, or flamelike body, shows that the sun's surface is very active.

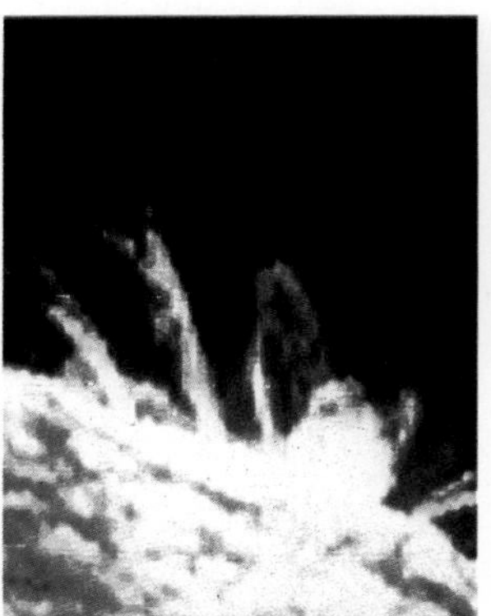

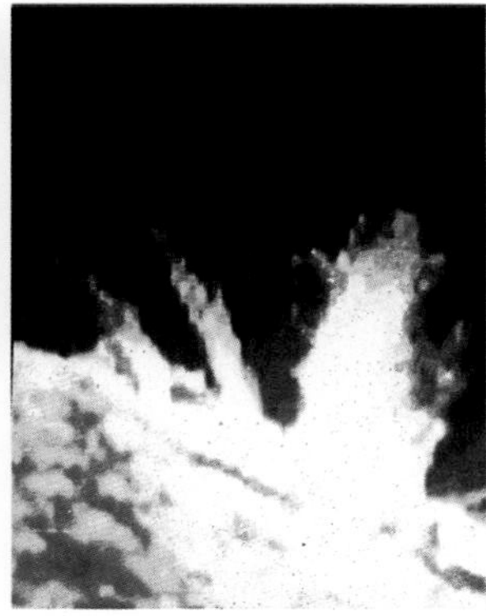

Figure 15-13 *Great explosions of hot gases make the sun's surface "boil." These two photographs, taken a short time apart, show an erupting prominence.*

Teaching Suggestions 15-4

Show a film (see page 269) about the sun and sunspots. If your school has a telescope, use it to project the image of the sun onto a sheet of paper. The class can then safely view and measure the size of sunspots in comparison to the diameter of the sun's image. While some years are better than others for sunspot study, it is rare to find no sunspots.

Display articles about how solar energy, including wind energy, is being used for the production of heat and electricity. Invite an architect to visit and explain how solar energy is used to heat homes. If possible, take a field trip to see how solar energy is being used in your community.

Students can do the Try This when discussing eclipses. There also are good films available so that students can study a real eclipse by means of the film. Some students may wish to do chapter-end project 8.

Background

There is no actual surface to the sun. It is a huge ball of hot gas; the gas becomes thinner away from the center of the sun. The outer part of the sun is divided into three layers. The photosphere is the surface of the sun that gives off the light (thus, the prefix "photo," or light). The chromosphere is just above the photosphere, and it gives off red light (thus, the prefix "chromo," meaning color). The chromosphere is visible just before totality as the moon moves across the sun during a total solar eclipse. The third layer of the sun's atmosphere is the corona. The corona is the outermost atmosphere, and can only be seen during a total solar eclipse.

Temperatures are quite different among the three layers of the sun's atmosphere. The photosphere varies between 4500°C at the top to 6800°C at the bottom. The chromosphere increases from 4500°C at the bottom to 100,000°C at the top. Temperatures in the corona are measured in millions of degrees. The corona is very low in density. It is believed that the high temperatures here are caused by shock waves produced by matter that moves up from the photosphere. Because of the low density in the corona, the heat content is low, despite the corona's high temperatures.

Sunspots appear dark because they are cooler areas on the sun's surface. Energy escapes the sun from these spots, leaving cooler areas behind. Prominences, which are great tonguelike masses of gases leaving the sun's surface, are often seen near sunspots. Sunspots and prominences are related to magnetic activity on the sun. Sunspots travel in pairs, one spot having a positive magnetic pole, the other negative. After each 11-year sunspot cycle, the polarity of the sunspots changes.

Scientists have learned a great deal about the sun from the Sky Lab experiments and the Space Shuttle. Much remains to be learned. As we learn about the sun, we will know more about other stars.

Teaching Tips

- Interested students might like to research mythological explanations for the northern lights.
- Tell students that whenever they view a solar eclipse they should use the indirect method, in which the image of the sun is projected onto a sheet of paper from the eyepiece of a telescope. Another method is to allow the sunlight to come through a pinhole in one sheet of thin cardboard and be projected onto another piece of paper held parallel to the cardboard, a few meters away.
- Have students check an advanced astronomy book or encyclopedia for a timetable of future eclipses.
- Have students research the effects of sunspot activity on weather, climate, and radio and television signals.

Facts At Your Fingertips

- The northern lights are also known as the aurora borealis (in the north) and aurora australis (in the south). One has to be in high northern or southern latitudes to view the auroras. South of 40°N (or north of 40°S), the auroras probably have never been seen.
- The sun's motion was determined by observing the motion of sunspots.

Cross Reference

See Chapter 10, section 10-5, for more information on nuclear energy.

sunspots

The sun has dark spots on its surface called **sunspots**. Sunspots have a lower temperature than the rest of the sun's surface. Astronomers have found that sunspot activity (the number of sunspots) increases and decreases over an average period of 11 years. Recent periods of high sunspot activity were 1968–69 and 1979–80.

During periods of high sunspot activity, the solar wind increases. The *solar wind* is a continuous flow from the sun of ionized (charged) gases, chiefly hydrogen and helium. The solar wind extends far out beyond Earth. When the solar wind increases, it causes noticeable effects on Earth. Bright-colored areas in the sky, the *northern and southern lights*, are seen more often. (See Figure 15-14.) Interference with radio and television is more frequent. You may have heard the announcement, "Audio (or video) difficulties are due to atmospheric conditions." This interference could be due to the charged particles from the sun entering Earth's upper atmosphere.

solar eclipse

Scientists have found that the sun rotates on its axis once every 27 Earth days. Its equator rotates faster than its polar regions, showing that it is not a solid. The sun also revolves around the center of the Milky Way at a speed of about 250 km/s. The fact that it takes 200 million years for one revolution at that rate shows just how big our galaxy is.

A **solar eclipse** occurs when the moon passes directly between Earth and the sun. The moon can either partially or totally block out the sun's light. (See area *a* in Figure 15-15.)

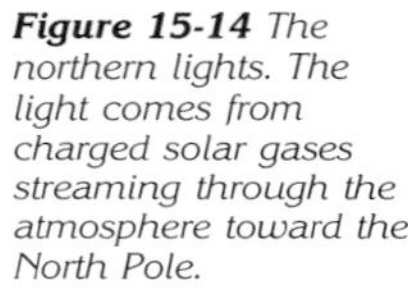

Figure 15-14 *The northern lights. The light comes from charged solar gases streaming through the atmosphere toward the North Pole.*

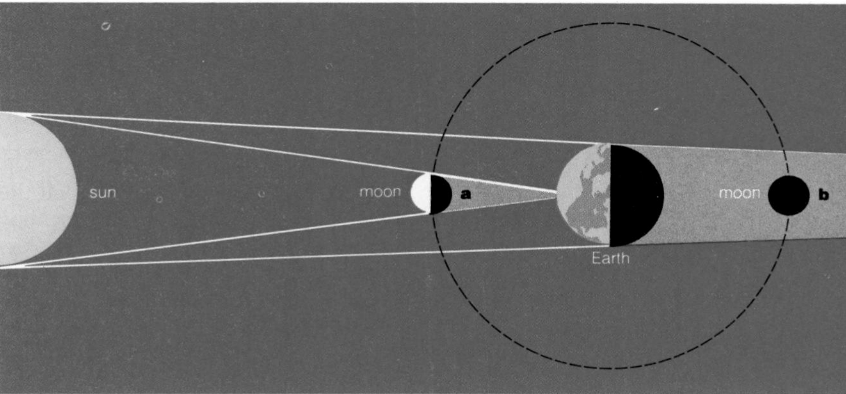

Figure 15-15 *When the moon passes between Earth and the sun, there is a solar eclipse (a). When the moon is in Earth's shadow, the moon is eclipsed (b).*

If you ever have the opportunity to see a solar eclipse, never look at the sun directly. In fact, *at no time* should you view the sun directly. Its powerful rays can permanently damage your eyes.

During a total solar eclipse, daylight gradually disappears, and stars become visible. A solar eclipse must have been very frightening before people understood what was happening. A solar eclipse lasts only a few minutes, because the moon and Earth are both moving very rapidly. A total solar eclipse can be seen from only a small portion of Earth's surface, because the moon's shadow on Earth is quite small.

A **lunar eclipse** occurs when Earth passes directly between the sun and the moon. Earth's shadow is then cast on the moon, as shown in area *b* in Figure 15-15. Lunar eclipses are easier to see than solar eclipses. Everyone on the night side of Earth can see them. Lunar eclipses last longer, too.

lunar eclipse (LOO nur)

Checkpoint

1. What process is the source of the sun's energy?
2. How does high sunspot activity affect Earth?
3. What is the difference between a solar eclipse and a lunar eclipse?

TRY THIS

Set up a model to show how eclipses occur. Hold a coin so that it blocks all the light from a light bulb. What represents the sun in this model? the moon? Earth? Is this a solar or lunar eclipse? Show how the other kind of eclipse occurs.

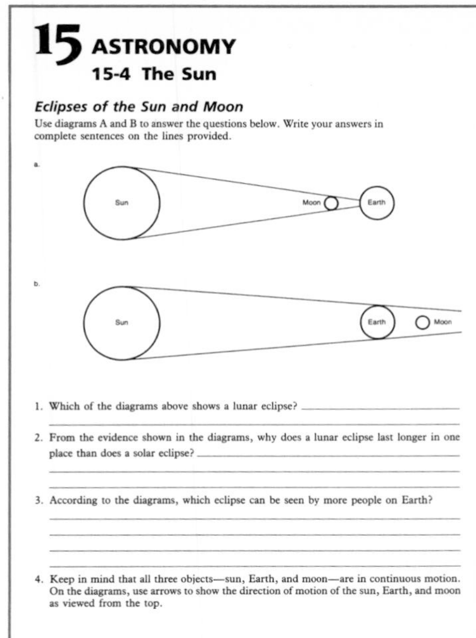

15 ASTRONOMY
15-4 The Sun

Eclipses of the Sun and Moon

Use diagrams A and B to answer the questions below. Write your answers in complete sentences on the lines provided.

a. Sun — Moon — Earth

b. Sun — Earth — Moon

1. Which of the diagrams above shows a lunar eclipse? ______
2. From the evidence shown in the diagrams, why does a lunar eclipse last longer in one place than does a solar eclipse? ______
3. According to the diagrams, which eclipse can be seen by more people on Earth? ______
4. Keep in mind that all three objects—sun, Earth, and moon—are in continuous motion. On the diagrams, use arrows to show the direction of motion of the sun, Earth, and moon as viewed from the top.

Using Try This

The sun is represented by the light. The coin, which is held between a student's eye and the light, is the "moon." A student's head is "Earth." This activity models a solar eclipse. For a lunar eclipse, a student's head would be between the coin and the light.

Checkpoint Answers

1. Nuclear fusion.
2. We may see northern lights or experience radio and television interference.
3. A solar eclipse occurs when the sun's light is blocked by the moon. A lunar eclipse occurs when the moon's light is blocked by Earth.

Teaching Suggestions 15-5

This section is very abstract. To stimulate interest, mention some of the special effects in recent science-fiction films. Relate these special effects to the stellar life cycle and other star characteristics. Ask students if the "warp speed" (greater than the speed of light) depicted in various ways in science-fiction films is possible.

Encourage students to do the Try This activity. In addition, students may be able to make some observations of stars that have different magnitudes and colors, if atmospheric conditions permit. But, most of the subject matter in this section is beyond their direct observation. Check the audio-visual materials that are suggested for this chapter (see page 269) to find films that relate to this section.

Small groups of students can report on each of the stages in a star's life cycle. Assign groups and give them time to research their topic in the library, collect pictures, organize the presentation, and then deliver their reports to the class.

Invite a local astronomer to tell the class some of the latest discoveries about galaxies and the universe. Students will probably have lots of questions to ask.

Background

Science fiction writers as well as scientists are speculating about the possibility of other habitable planets existing in the universe. According to typical calculations, in our own galaxy, the Milky Way (and there are billions of other galaxies), there are about 400 billion stars. Most of these are too young to have planets, or are binary systems. Assuming that 100 billion stars in the galaxy are similar to the sun, and that each such star has one planet, there would be 100 billion planets in our galaxy. To find the planets that could support life, we must find ones that are similar to Earth in their distance from their star, their mass, and their composition. A reasonable number would lie between one and ten billion planets that could support life, but how far away are they? Given the distribution of stars in our galaxy, it turns out that the nearest possibly habitable planet would be 250 light-years away! With today's science and technology, we have no chance of exploring that planet.

Another unit of distance used in astronomy is the parsec, which is equal to 3.26 light-years. This equals the distance a star would have to be from Earth to have a parallax angle of one second (1/3600 of a degree) as measured from extreme positions of Earth's orbit. Proxima Centauri, the closest star to Earth in the Southern Hemisphere, is 4.6 light-years away; and Sirius, the closest star in the Northern Hemisphere, is 8 light-years away. Parallax can be used to measure distances of up to 20 parsecs.

15-5

The Universe

Goal

To describe different properties and life cycles of stars.

As you look at a constellation, you may notice that all the stars in it are not equally bright. This might be because they are not all at the same distance from Earth. Or, the stars might not all be the same size. It might be a combination of both factors. For example, a large star may look no brighter than a small, dim star because it is much farther away. The observed brightness of a star is called its **magnitude**.

magnitude

Stars differ in color as well as in brightness. The colors indicate the temperature of the stars. The hottest stars are blue, and the coolest stars are red. In between are yellow, orange, and white stars. The sun is a yellow star.

nebulas (NEB yuh luhz)

Stars change over long periods of time. They form in huge clouds of gas and dust called **nebulas**. (See Figure 15-16.) A nebula may be the remains of a former star that exploded. Some of the gas and dust may be drawn together into a clump by gravitational attraction, just as objects on Earth are drawn toward its center. As the cloud of gas and dust contracts, it heats up. When the temperature gets high enough nuclear fusion begins, and a new star is born.

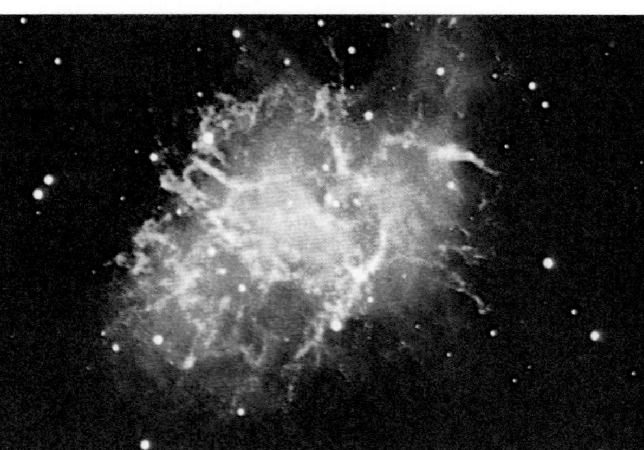

Figure 15-16 *The Crab Nebula is a cloud of dust and gas among the stars.*

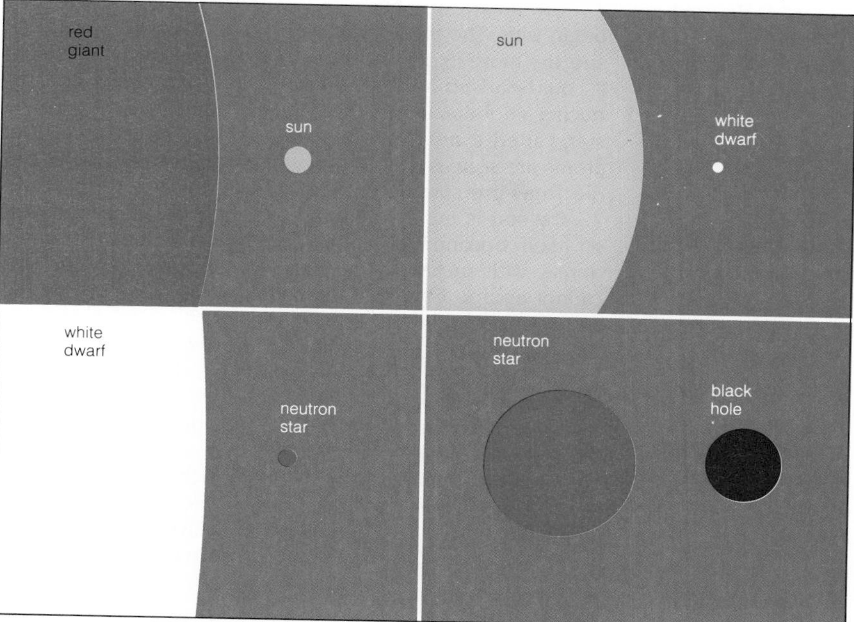

Figure 15-17 *Relative sizes of different kinds of stars, from the largest to the smallest. Notice how they compare with the sun. Black holes are extremely small and dense.*

As long as nuclear fusion takes place at a steady rate, a star is said to be stable. But the hydrogen fuel cannot last forever. When the rate of fusion slows down, the star changes. The sun, for instance, will begin to expand in a few billion years. It will probably become a **red giant**, a star hundreds of times bigger and cooler than the sun is now. (See Figure 15-17.) A red giant eventually becomes smaller as the force of gravity pulls its parts closer together.

red giant

After the red-giant stage, which lasts for millions of years, the sun may go through a variable-star stage. In this stage it would flicker like a light bulb that is burning out. In any case, the sun will probably become a **white dwarf**—a very small, dense, hot star. Even this stage cannot last forever. Eventually, a white dwarf becomes a cold, dark mass in space. This stage is sometimes called a *black dwarf*.

white dwarf

Teaching Tip

- Encourage students to study the stars at night. Have them look for stars that are brighter than others. How many different-colored stars can they observe? Using star charts, identify these stars and confirm their color by using books from the library. As an alternative, you may be able to arrange a trip to a planetarium. Small-star (planetarium) projectors are available from a number of science equipment suppliers. They are very useful for classroom or auditorium use in star study.

Facts At Your Fingertips

- Variable stars do not give off energy at a constant rate. They are in a dying stage, similar in a sense to a flickering candle that is burning out.
- No observable black dwarf has yet been found.
- The more massive the star, the faster nuclear reactions would take place within it, and thus the shorter the time it would be likely to exist.
- Supernovas can be seen from Earth. The Crab Nebula is the remains of a supernova that was observed in A.D. 1054.
- In 1983, the General Conference on Weights and Measures adopted a fixed value for the speed of light: 299,762,458 m/s.

Demonstration

Use four or five small lamps to represent stars. Put a bulb of different wattage in each. Line them up on the counter in front of the students. Turn them on. This arrangement represents how astronomers can determine the absolute magnitude, or true brightness, of stars. When stars are all at the same distance, it is easy to see their true brightness. By placing the brighter lamps at greater distances from students, show how it is impossible to know the true brightness of a star without also knowing the distance to the star. Emphasize that stars' brightness, uncorrected for distance, is called apparent brightness. With real stars, apparent brightness is a measure of how bright the star appears from Earth.

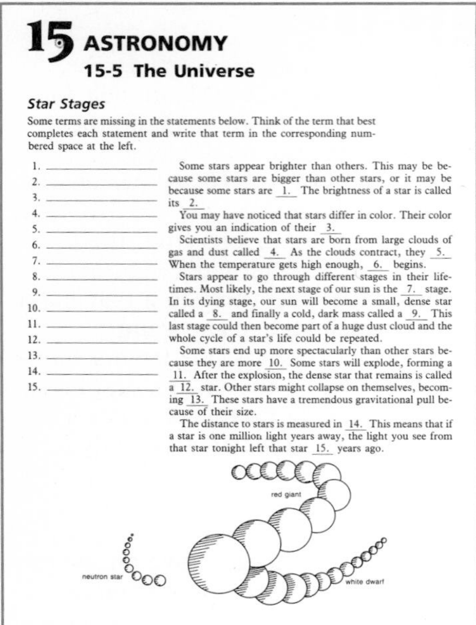

15 ASTRONOMY

15-5 The Universe

Star Stages

Some terms are missing in the statements below. Think of the term that best completes each statement and write that term in the corresponding numbered space at the left.

1. ________________
2. ________________
3. ________________
4. ________________
5. ________________
6. ________________
7. ________________
8. ________________
9. ________________
10. ________________
11. ________________
12. ________________
13. ________________
14. ________________
15. ________________

Some stars appear brighter than others. This may be because some stars are bigger than other stars, or it may be because some stars are _1._ The brightness of a star is called its _2._

You may have noticed that stars differ in color. Their color gives you an indication of their _3._

Scientists believe that stars are born from large clouds of gas and dust called _4._ As the clouds contract, they _5._ When the temperature gets high enough, _6._ begins.

Stars appear to go through different stages in their lifetimes. Most likely, the next stage of our sun is the _7._ stage. In its dying stage, our sun will become a small, dense star called a _8._ and finally a cold, dark mass called a _9._ This last stage could then become part of a huge dust cloud and the whole cycle of a star's life could be repeated.

Some stars end up more spectacularly than other stars because they are more _10._ Some stars will explode, forming a _11._ After the explosion, the dense star that remains is called a _12._ star. Other stars might collapse on themselves, becoming _13._ These stars have a tremendous gravitational pull because of their size.

The distance to stars is measured in _14._ This means that if a star is one million light years away, the light you see from that star tonight left that star _15._ years ago.

Using Try This

The results may vary from 100 to 3000, depending on sky conditions and what areas were chosen for the trials. Have students compare their results. The figures for city-based observations will differ from those made in the country. City lights, haze, and high thin clouds will all interfere with the counts. The factor of 700 is used because it has been calculated that the area of the sky visible through the tube is about 1/700 of the area of the sky above the horizon.

Checkpoint Answers

1. Size and distance from Earth.
2. A supernova is a huge star that explodes, leaving only a neutron star. A black hole is the remnant of a huge star that has collapsed upon itself.
3. Our sun will probably next become a red giant, then a variable star, and finally will end up as a white dwarf.
4. The distance light travels in ten years. (The speed of light equals about 300,000 km/s.)

The length of a star's life depends on how big it is to begin with. The more massive a star, the sooner it will die, and the more likely it will die in a very spectacular fashion. It could end up as a **supernova**, a star that explodes, producing a nebula. Within this nebula a very small, very dense star, called a **neutron star**, remains. In a neutron star the atoms are squeezed together so tightly that all protons and electrons are combined, forming neutrons.

supernova

neutron star

If a star is massive enough, it may completely collapse on itself, becoming a **black hole**. A black hole is extremely dense, with such a strong gravitational pull that even light cannot escape from it. Black holes can be detected by the way they affect matter around them.

black hole

The distance between stars is so great that it is measured in special units. A **light-year** is the distance light travels in one year. Light travels at about 300,000 km/s. (At this rate you could make nearly eight trips around Earth in one second!) The nearest star that we can see from the Northern Hemisphere is Sirius (SIHR ee us), which is 8.7 light-years away. This means that the light we see when we look at Sirius left that star 8.7 years ago. Distances within the universe can range up to millions of light-years.

light-year

Checkpoint

1. What two factors determine the brightness of a star as seen from Earth?
2. What is a supernova? a black hole?
3. What stages will our sun probably go through before it dies out?
4. What is meant by a distance of 10 light-years?

TRY THIS

Choose a clear night. Using a cardboard tube (such as one from inside a roll of paper towels) for a viewing tube, look at one part of the night sky. Count the number of stars you see through the tube. Call this "Trial 1" and write down the number of stars you see. Do this nine more times, looking at a different part of the sky each time. Record the number of stars seen each time. Calculate the average number you saw from the ten trials. Multiply this number by 700. This is the average number of stars in our galaxy visible to the unaided eye, from one position on Earth.

15-6

Tools of Astronomy

ENRICHMENT

Goals

1. To describe various kinds of telescopes.
2. To explain how a spectrograph is used to analyze starlight.

Everything we know about the stars has been learned from the energy that they give off. Most instruments astronomers use to study stars are light-gathering devices. The greater the amount of light captured, the better an object in space can be seen. Binoculars and telescopes are examples of such instruments.

visual telescopes

Visual telescopes gather visible light, so that objects can be directly seen or photographed. There are two major kinds of visual telescopes. *Refracting* telescopes contain lenses that gather light. *Reflecting* telescopes use curved mirrors for this purpose. (See Figure 15-18.) Reflecting telescopes are easier to build. They can be made larger than refracting telescopes, also.

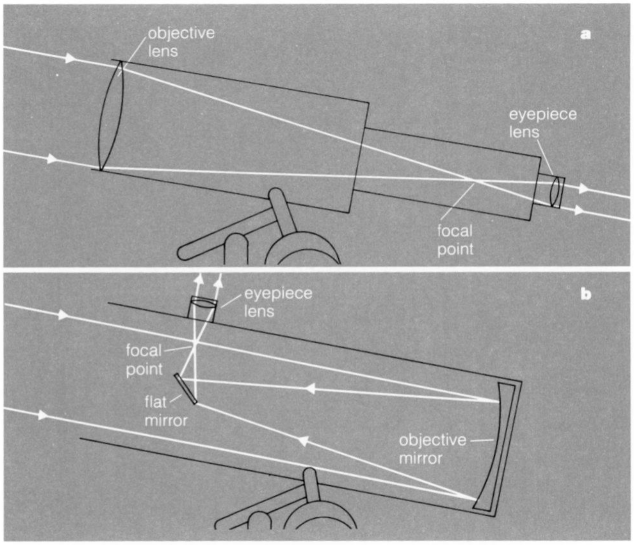

Figure 15-18 *The two kinds of visual telescopes. (a) A refracting type, that uses only lenses. (b) A reflecting telescope, with a concave mirror.*

Encourage students to do the Try This activity early in this section while discussing binoculars as a tool of astronomy. Chapter-end projects 1 and 7 apply to this section.

Background

A new telescope is being developed at Kitt Peak National Observatory in Tucson, Arizona. It consists of several large parabolic mirrors with the light-gathering power of a single mirror 20 m across. It will be six times more powerful than the Soviet 600-cm telescope and nine times more powerful than the 508-cm telescope at Palomar Observatory in California. It will have a light-collecting area double that of the world's 20 largest telescopes combined. This new telescope will allow astronomers to study distant galaxies in detail as they are forming. The astronomers will also be able to see stars and their planetary systems as they are being created out of dust and gas.

The unusual solar telescope at Kitt Peak reflects images of the sun to a point that is more than 100 m underground. There the image is projected on a large screen where astronomers can study the surface features of the sun.

The largest refracting telescope has a 102-cm-diameter lens. It is at Yerkes Observatory, Williams Bay, Wisconsin.

Some of the most famous large radio telescopes are the 76-m dish at Jodrell Bank, England; the Parkes dish in Australia (64 m); and the 91-m dish at the Observatory in Greenbank, West Virginia. Australia, France, Canada, Italy, the Netherlands, the U.S.S.R., and the U.S. all have arrays of dishes.

Teaching Suggestions 15-6

While studying this enrichment section, give students an opportunity to use or see a reflecting telescope, refracting telescope, binoculars (see Try This), and a spectroscope, if possible. Refer students to Figures 15-18, 15-19, and 15-20 as you discuss these tools of astronomy. If you can, visit a nearby observatory. Alternatively, show films about the tools of astonomy. (See audio-visual materials for Chapter 15, page 269.) Inexpensive spectroscopes are available from science supply houses. If you can get some spectroscopes, have the students look at incandescent bulbs, fluorescent bulbs, the sky, and electrified tubes of gases. Students can use spectrum charts and can compare what they see with the spectra shown in Figure 15-20.

15 ASTRONOMY Laboratory

Make an Astrolabe

Purpose To construct a simple astrolabe for use in measuring the horizon of classroom objects and objects in the sky.

Materials plastic soda straw, 15-cm protractor, 20-cm piece of string, a washer or nut, some tape

Procedure

1. Use tape to attach the straw to the back side of the protractor. (See diagram.) The straw should be lined up with the 90-degree line and the hole in the protractor. The protractor should be placed close to one end of the straw.
2. Put one end of the string through the hole in the protractor and tape the string to the back side. Attach the washer or weight to the other end of the string.

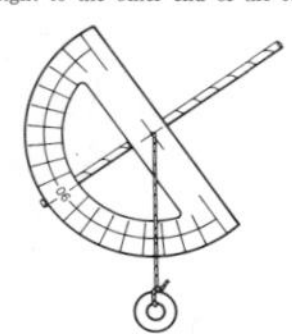

3. Use the astrolabe to measure the altitude above the horizon of some objects in the classroom. While looking through the protractor end of the straw, sight an object, such as the top of the wall clock, through the straw. Allow the string to hang free until you have the object sighted. Then put a finger on the string and hold it against the protractor until you read the elevation in degrees. Record this in the chart below.

Object Measured	*Altitude*
1.	
2.	
3.	
4.	
5.	
6.	
7.	
8.	

4. Now use the astrolabe on the night sky. Measure the altitude of some familiar stars: Polaris, Spica, Betelgeuse, Antares, Sirius. Take these measurements at different times during one evening.

Discussion

1. Compare the star altitudes that you recorded in your chart. What is different about Polaris? ____________
2. Measure the altitude of the moon at different hours in the night. What do you observe? ____________

Teaching Tips

- Students may have telescopes of their own that they would be willing to share with the class.
- Contact a local astronomical society, if one exists at a reasonable distance. Try to involve your students in their viewing sessions, and invite members to lead class discussions.

Facts At Your Fingertips

- One use of radio telescopes is to listen for radio waves from deep space, to learn whether other scientifically advanced beings exist.
- Some spectroscopes use a diffraction grating, a piece of glass or plastic with many fine grating lines inscribed in it, instead of a prism.
- Chemical compounds in stars can also be identified, but their spectra are more complicated.

Cross Reference

Refer to Chapter 20, sections 20-2 and 20-3, for more information on the use of mirrors and lenses in telescopes.

radio telescopes

Radio telescopes gather radio waves from space. Radio waves can pass more readily than light waves through dust clouds in space and through clouds in Earth's atmosphere. Radio waves can be received 24 hours a day, while visual telescopes can be used only when it is dark. Radio astronomy has come into use only in the last 50 years.

Since radio waves are longer than light waves, the antennas, or *dishes*, of radio telescopes must be very large. One of the more spectacular radio telescopes is the 305-m dish at Arecibo (ah ray SEE boh), Puerto Rico. The dish is in the crater of an extinct volcano. As Earth rotates, the dish picks up radio waves from space. (See Figure 15-19.) Another way of making a large radio telescope is to have groups of dishes spread over the ground. There are many such groups around the world.

spectroscope

The chemical elements present in a star can be identified by analyzing starlight. If a narrow beam of light from a glowing gas is passed through a prism, a colorful *spectrum* is produced. (See Figure 15-20.) A **spectroscope** is used to produce a spectrum and measure the wavelengths of light in it. If the gas is an element, the spectrum is a series of narrow lines. Every element has its own *bright-line spectrum.*

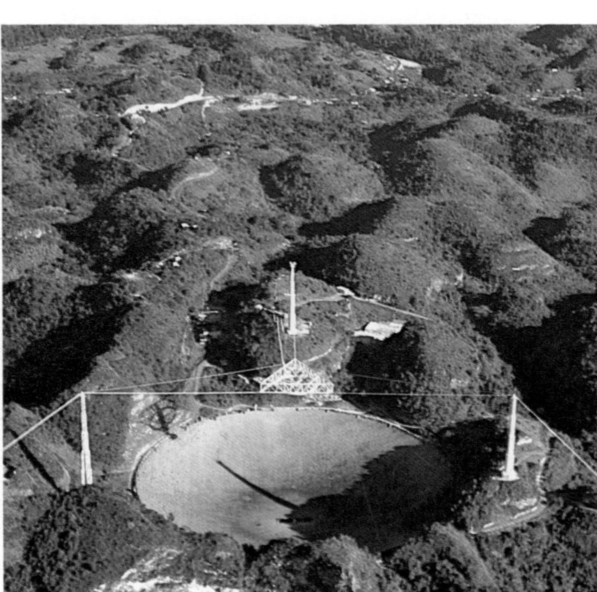

Figure 15-19 *The radio telescope at Arecibo, Puerto Rico.*

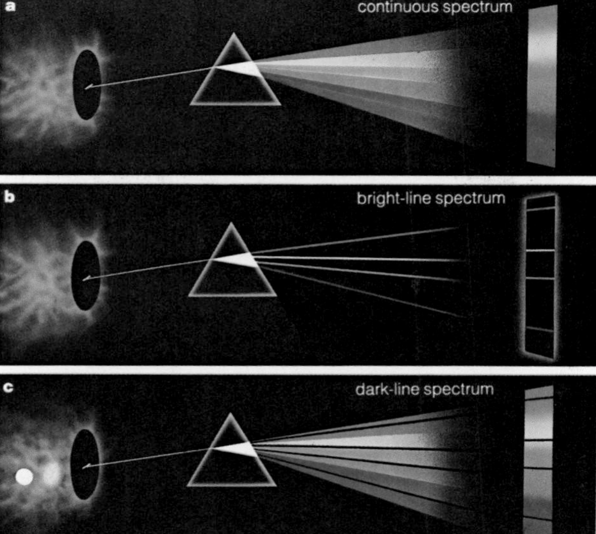

Figure 15-20 *Three kinds of spectra. (a) Light from glowing gas gives a continuous spectrum. (b) A gas of a single element gives a bright-line spectrum. (c) Light passing through cooler gas gives an absorption spectrum of dark lines.*

Its line spectrum can serve as a "fingerprint" to identify the presence of the element.

Most stars don't produce bright-line spectra. The hot gases in the interior of the star radiate a continuous spectrum. As the radiations pass through the star's atmosphere, each gaseous element absorbs the wavelengths of light that appear in its bright-line spectrum. So, a certain number of black lines appear in the spectrum. The *absorption spectrum* of an element can also serve as its "fingerprint."

TRY THIS

On a clear night, go outside and find the Big Dipper in the northern sky. Look closely at the second star from the end of the "handle." What do you see? Now take a pair of binoculars and look again at the same area. What do you see now?

Checkpoint

1. What two types of visual telescopes are in use?
2. How does a visual telescope differ from a radio telescope?
3. What information can a spectroscope provide?
4. What can astronomers learn about stars by using a spectroscope?

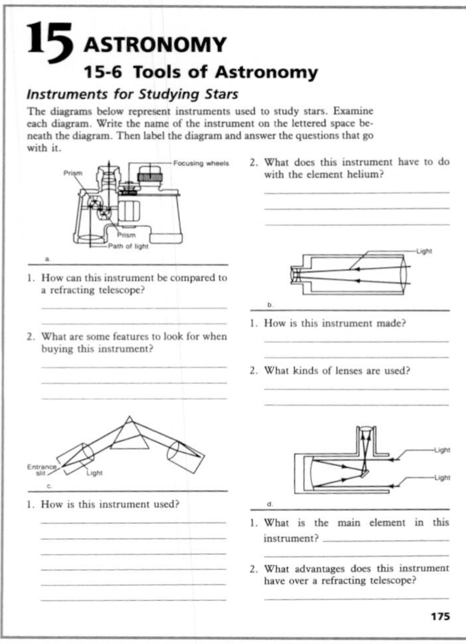

15 ASTRONOMY
15-6 Tools of Astronomy
Instruments for Studying Stars

The diagrams below represent instruments used to study stars. Examine each diagram. Write the name of the instrument on the lettered space beneath the diagram. Then label the diagram and answer the questions that go with it.

a. ______

1. How can this instrument be compared to a refracting telescope?
2. What are some features to look for when buying this instrument?

c. ______

1. How is this instrument used?

2. What does this instrument have to do with the element helium?

b. ______

1. How is this instrument made?
2. What kinds of lenses are used?

d. ______

1. What is the main element in this instrument? ______
2. What advantages does this instrument have over a refracting telescope?

175

Using Try This

With the unaided eye, students should see two stars in this location. Their names are Mizar and Alcor. When binoculars (or a telescope) are used, they should be able to see more stars in this area. Mizar and Alcor are each double stars, so if the binoculars are strong enough, they may be able to see four stars.

Checkpoint Answers

1. Refracting and reflecting.
2. Visual telescopes use lenses or mirrors to capture light waves; radio telescopes use a dishlike antenna to capture radio waves from space.
3. A spectroscope breaks light into a spectrum and enables us to measure the wavelengths produced.
4. They can find out what elements are present in stars.

Using The Feature Article

It would be best to assign the feature article at the conclusion of the chapter, since tools of astronomy are discussed in the final section of the chapter. After students have read the feature article have them make a list of questions based on or suggested by the material. Have them list at least a dozen topics. If students need help with ideas for topics, you could suggest the following:

1. When will the Space Telescope be launched? How long will it be in orbit? Will the Space Shuttle be able to rejoin the Space Telescope at a later date?
2. What are some of the special tasks the Space Telescope will perform?
3. Will other nations have access to the Space Telescope? If so, what is the agreement with them? Are they partially responsible for maintenance of the telescope?
4. How long has it taken to build the Space Telescope? Who built it and where was it built? What caused delays in the original schedule.
5. How will the Space Telescope compare with other ventures such as the studies made of the sun from Sky Lab, and the Orbiting Astronomical Observatories in the late 1960s up to 1981?
6. Where can we find pictures of the Space Telescope showing its various instruments and how it will be launched from the Space Shuttle?
7. What is LODE—the Large Optics Demonstration Experiment—designed to do? What government agency is developing LODE and why?

Assign these topics to small groups of students. Allow one or two class periods for students to research the topics. Current information can be found in recent scientific magazines and journals. One good comprehensive article is in *Scientific American*, July 1982: "The Space Telescope," by John N. Bahcall and Lyman Spitzer, Jr.

When the research is completed, have students return to class and organize their reports. They then can present their findings to the class.

Telescopes in Space

The study of astronomy will take a giant step forward when a telescope is launched into space. It is possible that a space telescope will help us to learn more about the universe in a few years than has been learned in all of past history! The greatest advantage of a space telescope is that it will be outside Earth's atmosphere. Earth's thick atmosphere limits the distance telescopes on Earth can "see" into space. The atmosphere also absorbs and refracts much of the light from the stars.

The space telescope that will be put into orbit by the Space Shuttle will be 5 m wide and 15 m long, with a mass greater than 10,000 kg. Its mirror will be 30 cm thick and 240 cm in diameter. Along with the telescope there will be two cameras, two spectrographs, and two instruments to measure distances between stars and changes in their brightness. All of this equipment will be in orbit 500 km above Earth. This is far above the thickest part of Earth's atmosphere. The largest telescopes on Earth can detect objects two billion light-years away. The space telescope will be able to detect light from galaxies 14 billion light-years away. This will extend our view of the universe by 350 times.

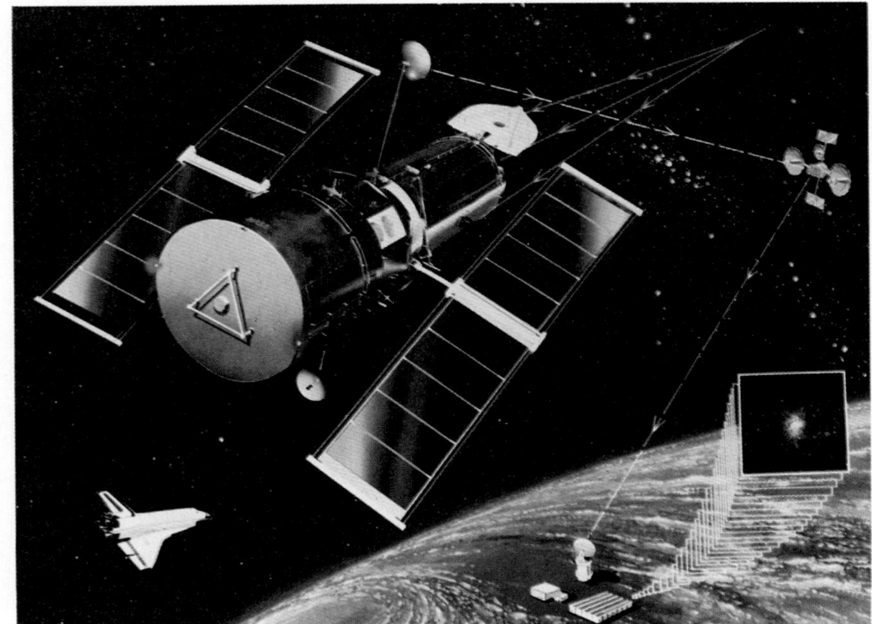

Understanding The Chapter 15

Main Ideas

1. Earth is spherical, but it is not a perfect sphere.
2. Earth rotates on its axis and revolves around the sun.
3. Earth's seasons depend on the tilt of its axis.
4. The moon's major surface features are maria, craters, and mountains.
5. The moon's phases are caused by the positions of the moon, the sun, and Earth relative to each other.
6. Planets, moons, comets, meteoroids, and asteroids are members of the solar system.
7. Earth is a part of the solar system. The solar system is in the Milky Way Galaxy. The Milky Way Galaxy is one of many galaxies in the universe. The universe includes all the matter that exists.
8. The sun is an average star. It produces its energy by nuclear fusion, as do all stars.
9. Solar and lunar eclipses may be observed by people on Earth when the sun, Earth, and moon are lined up with each other.
10. Each star has a characteristic size, temperature, brightness, and color.
11. Stars go through a series of changes in their lifetimes. The stages they go through depend on their original mass.
12. Starlight can be studied using visual telescopes, radio telescopes, and spectroscopes.

Vocabulary Review

From the following list, choose the term that best completes each of the statements. Write your answers on a separate piece of paper.

asteroid	orbit
astronomy	phase
black hole	radio telescope
comet	red giant
constellation	revolution
galaxy	rotation
light-year	satellite
lunar eclipse	solar eclipse
magnitude	solar system
meteor	spectroscope
meteorite	sunspot
meteoroid	supernova
nebula	universe
neutron star	visual telescope
nuclear fusion	white dwarf

1. A(n) __?__ is a huge cloud of gas and dust.
2. When the moon moves between the sun and Earth, a(n) __?__ occurs
3. A(n) __?__ is sometimes called a "falling star."
4. The apparent shape of the moon as we see it from Earth is its __?__.
5. High __?__ activity causes the northern lights to increase.
6. A group of billions of stars is called a(n) __?__.
7. The moon is a natural __?__ of Earth.
8. The observed brightness of a star is called its __?__.
9. A(n) __?__ is made of dust and ice and has a long tail.

Vocabulary Review Answers

1. nebula
2. solar eclipse
3. meteor
4. phase
5. sunspot
6. galaxy
7. satellite
8. magnitude
9. comet
10. asteroid
11. light-year
12. visual telescope
13. radio telescope
14. meteoroid

Reading Suggestions

For The Teacher

Dixon, Don. THE UNIVERSE. Boston: Houghton Mifflin Co., 1981. Investigates problems of the origins of the cosmos and of life itself.

Kaufmann, William J., III. UNIVERSE. New York: W. H. Freeman, 1985. A college astronomy textbook. A good reference for teachers. Easy to read and well-illustrated.

Seeds, Michael A., ed. ASTRONOMY: SELECTED READINGS. Menlo Park, CA: Benjamin Cummings, 1980. Readings from *Astronomy* which focus on the solar system, stellar and galactic evolution, cosmology, and the possibility of intelligent extraterrestrial life.

Trefil, James F. SPACE, TIME, INFINITY. Washington, D.C., Smithsonian Books, 1985. Well-illustrated summary of the past, present, and future of astronomy.

For The Student

Frazier, Kendrick. OUR TURBULENT SUN. Englewood Cliffs, NJ: Prentice Hall, 1982. Recent developments on sunspots, geomagnetic storms, and other solar topics.

Frazier, Kendrick. SOLAR SYSTEM. Alexandria, VA: Time-Life Books, Planet Earth series, 1985. Illustrated stories about ancient and modern astronomy, with essays on various astronomical subjects.

Gore, Rick. "Uranus: Voyager Visits a Dark Planet." *National Geographic*, August 1986, pp. 178–195. Photos and diagrams are excellent. Gives reader a look into the geological past of the moons of Uranus.

Heidman, Jean. EXTRAGALACTIC ADVENTURE: OUR STRANGE UNIVERSE. New York: Cambridge University Press, 1982. Explains astronomical measurements, cosmic evolution, galaxies, quasars, black holes, and extraterrestrial objects.

Chapter Review Answers

Know The Facts

1. basketball
2. west
3. the tilt of its axis
4. the moon rotates as it moves around Earth
5. smaller
6. supernova
7. move around the sun
8. is producing energy by nuclear fusion
9. telescope
10. yellow star
11. northern lights
12. distance
13. original mass
14. constellation
15. radio telescope

Understand The Concepts

16. The next high will be 1990–91. In 1985, sunspots will be near a minimum; from 1985–1990 they will be on the increase. After 1991, the number will decrease again.
17. Daylight on the moon is half the rotational period, or about two weeks.
18. We only see one side of the moon.
19. See Figure 15-18.
20. The rays of the sun are so indirect at the poles that little energy is received.
21. The moon has no water and no atmosphere.
22. The elements present in the star.
23. About one week.
24. One sequence might be yellow star, red giant, variable star, white dwarf.
25. They are less dense, larger, have thicker atmospheres, and more moons than other planets.

Challenge Your Understanding

26. The planets are in the same plane as Earth's orbit around the sun.
27. Five degrees.
28. Cool temperatures and a large size.
29. 88 Earth days per Mercury year, divided by 59 Earth days per Mercury day equals 1.5 Mercury days per Mercury year.

10. A(n) ? is an object that orbits the sun between Mars and Jupiter.
11. The unit used to measure distances between stars is called a(n) ?.
12. A(n) ? uses a lens or a mirror to gather light from a star.
13. A "dish" is the antenna for a(n) ?.
14. A small piece of rock or mineral matter traveling through space is a(n) ?.

Chapter Review

Write your answers on a separate piece of paper.

Know The Facts

1. Earth's shape is most like a(n) ?. (football, basketball, egg)
2. Because of Earth's rotation, the stars appear to move to the ? during the evening. (east, west, north)
3. The fact that Earth has seasons is due primarily to ?. (its revolution around the sun, the tilt of its axis, its distance from the sun)
4. We always see the same side of the moon because ?. (Earth is moving around the moon, the moon rotates as it moves around Earth, the moon does not rotate as it moves around Earth)
5. The rocky planets are ? than the gaseous planets. (larger, smaller, less dense)
6. A star that explodes, producing a nebula, is called a ?. (nova, supernova, black hole)
7. Comets, planets, and asteroids are similar because they all ?. (contain ice, move around the sun, have atmospheres)
8. The sun differs from other members of the solar system because it ?. (is made up of gases, is rotating slowly, is producing energy by nuclear fusion)
9. An instrument that helps us see stars because it gathers light is a ?. (microscope, spectroscope, telescope)
10. The sun is a ?. (red giant, white dwarf, yellow star)
11. During periods of high solar activity, Earth experiences ?. (strong winds, northern lights, meteor showers)
12. A light-year is a unit of ?. (time, distance, mass)
13. The length of time a star will "live" depends on its ?. (original mass, temperature, motion)
14. A group of stars that forms a recognizable pattern in the sky and is visible to the unaided eye is called a ?. (galaxy, solar system, constellation)
15. A ? can receive radiation from space on a cloudy day. (radio telescope, refracting telescope, reflecting telescope)

Understand The Concepts

16. The most recent period of high sunspot activity was in 1979–80. What part of the sunspot cycle is the sun in now?
17. How long are periods of daylight on the moon?
18. What evidence is there that the moon rotates on its axis?
19. Draw a diagram to show how images are produced in a refracting telescope.

30. A lunar eclipse lasts longer than a solar eclipse because Earth's shadow is much larger than the moon's, and more time is needed for the moon to pass through it than is needed for Earth to move through the moon's shadow during a solar eclipse. Students' drawings should resemble those shown in Figure TE 15-2.

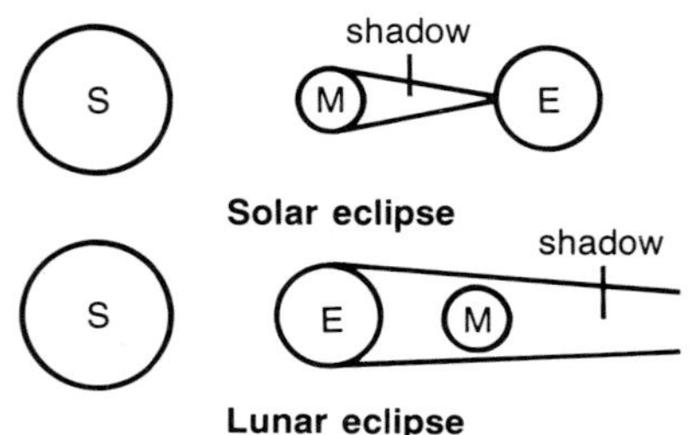

Figure TE 15-2 Diagrams of a solar eclipse and a lunar eclipse.

20. During part of every year the sun never sets at the poles. Is it hot in those areas at that time? Why or why not?
21. Why doesn't it rain on the moon?
22. What information can be learned about a star by studying its absorption (black-line) spectrum?
23. How long is it between a full moon and a third-quarter moon?
24. Describe the life cycle of a star such as our sun.
25. How do Jupiter, Saturn, Uranus, and Neptune differ from other planets in our solar system?

Challenge Your Understanding

26. In order to find the planets in the night sky, why do we look along the same path followed by the sun during the day?
27. If you took a 20-minute time exposure picture of the sky, focusing on the North Star, through how many degrees would the stars move in that time?
28. Name two characteristics of a planet that would allow it to have an atmosphere.
29. The planet Mercury rotates about its axis once every 59 Earth days. It revolves in the same direction around the sun in 88 Earth days. How many Mercury days are there in a Mercury year?
30. Explain why a lunar eclipse lasts longer than a solar eclipse. Use drawings to show what is happening with each.
31. Explain, using a diagram, why the North Star is always visible in the Northern Hemisphere.

Projects

1. Find out what astronomers mean by the term "light pollution" and what they are doing to cope with this problem.
2. Observe the motion of Venus in relation to the sun for a period of three months. Make a large drawing showing what motion Venus is making in relation to Earth and the sun during this time.
3. Build a sundial that keeps good time. Books in the library will give helpful hints.
4. With a constellation chart for the current season, plot the paths of meteors you observe during a meteor shower. Your teacher can give you the dates for yearly meteor showers. The best viewing time is after midnight. You might see up to 25 per hour.
5. In the library, find a photograph of the moon that has major landscape features labeled. Using materials such as clay or plaster of Paris, make a model of a portion of the lunar landscape.
6. Research the topic of celestial navigation used by ships, planes, submarines, and space vehicles.
7. Read *A Connecticut Yankee in King Arthur's Court* by Mark Twain. Plan a class skit based on the "eclipse" scene.
8. Research and report on the contributions to astronomy of any two of the following scientists: Annie Jump Cannon; Sarah F. Whiting; Benjamin Banneker; Maria Mitchell; Eleanor Burbidge; Cecilia Payne-Gaposchkin; Henrietta Leavitt.

15 ASTRONOMY
Projects

1. In a reference book look up the term *light pollution*.
   a. What is light pollution? ______
   b. List the locations of some observational telescopes around the world.
   c. What do the locations of many of these observatories have in common?
   d. Why do you suppose these locations were chosen? ______
2. Research the seasonal migration of birds.
   a. Complete the chart below by listing several kinds of birds and the places where they migrate to and from.

Bird	*Summer Home*	*Winter Home*

   b. What conditions give birds the idea that it is time to migrate? ______
3. Because of Earth's slightly flattened spherical shape, how will the weight of an object at the equator differ from its weight at the poles? ______
   Why? ______
4. Research, through interviews and reference books, the topic of U.F.O.'s.
   a. What is a U.F.O.? ______

31. The North Star is always above the horizon in the Northern Hemisphere. The degree above the horizon will depend on the latitude of your location. It will vary from 0° at the equator (on the horizon) to 90° at the North Pole. Students' diagrams should resemble the one shown in Figure TE 15-3.

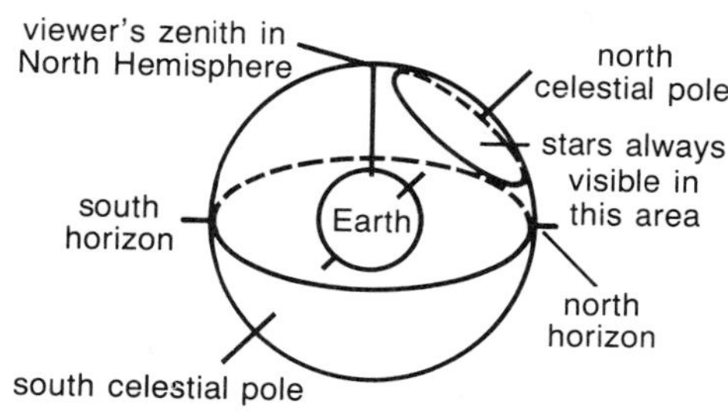

Figure TE 15-3 The North Star is always visible in the Northern Hemisphere.

Unit 3 Careers

Teaching Suggestions

Show students some audio-visual materials that tell about careers related to earth science. Each of the following filmstrips comes with cassettes and a teacher's guide: *What Is Science? Earth Science Career Challenges*, McGraw-Hill Films, Del Mar, CA; *Careers in Energy*, Pathescope Educational Media, Mount Kisco, NY; and *Careers in Marine Science and Industries*, Pathescope Educational Media, Mount Kisco, NY.

Have students do library research to find out about the following occupations that are related to earth science: hydrologist, oil treater, geochemist, meteorologist, marine biologist, volcanologist, coal miner, seismic engineer, heavy machinery operator, oil rotary driller, mining technician, astronomer, marine geologist, and metal miner. Using the information that they obtain, have students prepare "Earth-Science Career Pamphlets." Use the pamphlets for a discussion of the prerequisite training and the rewards and drawbacks for each occupation.

Plan a field trip to a nearby airport, weather service station, or local television or radio station to observe meteorologists or weather reporters at work. Ask the meteorologists or weather reporters to explain to students how they do their job.

Have groups of students research career opportunities in different earth-science-related fields. The most recent edition of the U.S. Government's *Occupational Outlook Handbook* is a good starting place for students' research. Other relevant books include *Career Opportunities in Geology and the Earth Sciences* by Lisa Rossbacher, ARCO, New York, 1983; *Getting Started in Marine Occupations* by Suzan Prince, Personnel Publications, Huntington, NY, 1982; and *Opportunities in Landscape Architecture* by Ralph Griswald, VGM Career Horizons, Skokie, IL, 1978. Follow up students' research by visiting local businesses, universities, government agencies, or research facilities that employ people in occupations that require an earth-science background. For example, you might plan a trip to the offices of a local landscape architect, real estate agency, oil company, mining company, engineering firm, seismic research station, or oceanographic institute.

Mining Engineer

Coal mining company is in search of engineers to plan and improve mining operations. Of special concern are mine safety, pollution control, and environmental restoration. Some time will be spent searching for new coal deposits. Experience with systems to bring air and water into underground mines is required. Most of the work involves creative problem-solving.

The position requires an engineering degree with courses in earth science, physics, chemistry, and mathematics. Those hired will begin by learning about coal production and doing routine work under close supervision. Most of the work will be done as part of a team working outside or in mines. Work can be hazardous. Opportunities exist for advancement.

Geologist

Federal government has several positions open for geologists to assist in resource exploration, construction site analysis, and geologic research. Projects include recording plate movements, improving methods of earthquake prediction, monitoring volcanic activity, analyzing ground stability, and searching for oil and minerals. Applicants must be able to travel worldwide and draw maps of their findings. The positions involve working indoors in a laboratory and working outdoors in the field.

A college degree in geology is required with course work in earth science, physics, chemistry, and mathematics. Physical stamina is needed for some of the field work. Applicants must be able to work well with others as part of a team.

Help Wanted

Additional Careers

A brief description and job outlook for three occupations related to earth science follow:

1. *Cartographers* use satellite data and aerial photos to develop maps and charts of Earth's surface. Although most cartographers are employed by the federal government in the Departments of Defense and Interior,

Agricultural Technician

Farm agency seeks technicians to help research scientists who are studying ways to improve farming methods. Research involves how to increase the resistance of farm organisms to disease, how to best breed and feed farm animals, and how to conserve and improve soil. Technicians will talk with farmers and help them apply research findings. The job involves a lot of travel and work outdoors. It also requires the ability to work well with many different people.

Candidates with college or technical school training are preferred. Courses should include earth science, botany, biology, chemistry, and agricultural sciences. On-the-job training will be provided. As experience is gained, there will be opportunities for advancement to a supervisory level.

Aeronautical Technician

Aerospace company has openings for aeronautical technicians to assist engineers and scientists. Technicians perform tests, gather data, write reports, and make sketches that relate to the design of missiles, rockets, and airplanes. Technicians must be able to use highly technical instruments and take readings accurately.

Candidates must have a high school diploma, and preferably some college or technical school training as well. Courses in earth and space science, physics, mathematics, computers, and drafting would be helpful.

Some on-the-job training will be provided. Technicians should be able to work well with others in small teams. When deadlines are near, overtime can be expected.

many work for research organizations, private industry, or colleges and universities. The accurate maps they prepare are used by textbook publishers, real estate developers, and manufacturing firms. The demand for cartographers with good quantitative and computer skills should remain strong throughout the 80's.

2. *Meteorologic technicians* apply the more theoretical knowledge of meteorologists to actual situations in which weather predictions are made. Technicians help gather data and keep weather instruments working properly. They also do research and aid in the development of accurate instruments. As more technically advanced instruments are used to improve forecasting, a larger number of meteorologic technicians probably will be called upon to work with scientists and engineers.
3. *Land surveyors* work for federal and state governments, architectural and engineering consulting firms, construction companies, and oil industries. A lot of their work is done outside. They work as part of a team that measures construction and mineral sites, sets boundaries, and collects information for maps. They often must walk long distances, carrying heavy equipment. The rest of their work is done indoors at a desk, drawing maps and writing reports. As construction of highways, roads, buildings, housing developments, and recreational areas continues, demand for additional surveyors should rise.

For More Information

You or your students can write to the following organizations for more information about careers related to earth science:

1. American Geological Institute, 5202 Leesburg Pike, Falls Church, VA 22041.
2. Association of American Geographers, 1710 16th St., N.W., Washington, DC 20009.
3. National Weather Service, Manpower Utilization Staff, Gramax Bldg., 8060 13th St., Silver Spring, MD 20910.
4. American Society of Limnology and Oceanography, I.S.T. Bldg., Great Lakes Research Division, University of Michigan, Ann Arbor, MI 48109.

Using Issues In Science

After the students read this article, hold a classroom debate about one of these topics:

- The space program should be speeded up to prevent other countries from gaining a technological advantage over the U.S.
- The space station program should be delayed several decades to make it fully automated rather than risk human lives.

This *Issue* can also be a basis for a discussion of how to set priorities for spending money. Ask students how they make decisions about spending their own money. How do they think the government makes decisions about which programs to finance? Since any budget has limits, students should be able to relate their personal finances to the government's spending concerns.

Background

There have been several temporary laboratories in orbit. The Soviet Union successfully occupied the *Salyut* space stations for periods of up to seven months. *Skylab* was operated by the U.S. in 1973 and 1974; more recently, *Spacelab*, designed by the *European Space Agency*, was flown aboard the Shuttle in 1983. One of the primary objects of these missions was to determine the effects of prolonged space flight on humans.

Some scientists claim that the United States should not have focused on the one-time goal of sending people to the moon. They think that we would already have a permanent space station like the Soviets if we had made that our goal.

ISSUES IN SCIENCE

PEOPLE IN SPACE:

IS IT WORTH THE RISK?

Imagine living in a space station. You can't go out in the sunlight without protection because the strong radiation would damage your skin and eyes. You would suffocate in the vacuum of space if you left your life support systems. And long periods of weightlessness would cause your muscles to weaken.

Does it sound dangerous? It is. The 1986 accident of the Space Shuttle *Challenger* made it very clear that first getting people into space is dangerous. Yet, despite the dangers, the United States is continuing towards its goal of a permanent Space Station orbiting the earth, where people will live and work for months at a time.

The Space Station will have cost at least $14 billion by the time it is completed in the 1990s. In addition, more money must be spent to perfect the Space Shuttle necessary for carrying supplies to build and operate the Space Station.

Many scientists question the costs and risks involved in sending people into space. Some think that most of the operations in space could be performed more safely and at a lower cost by space stations remotely controlled from earth. They point out the success of the Voyager and Pioneer missions which explored the outer planets. But others argue that putting people into space is far more effective. They are concerned that we are falling behind the Soviet Union, which already has people working in a space station called *Mir*.

Supporters of the Space Station point out that the costs will be more than repaid by large profits. The Center for Space Studies predicts that by the year 2000 there will be 40 billion dollars in profits from renting the Space Station to industry.

Scientists and industries are interested in developing new technologies that may only be possible with people

Teaching Study Skills

Writing An Outline

Issues in Science can be used to practice the skill of writing an outline. Outlining can help students understand the organization of the text, pick out main ideas, and determine which ideas are more important.

- Have the students find the topic sentence from each paragraph in the Feature.
- Tell them to take each topic sentence and summarize it in a phrase; then list these phrases in the order they were found in the text.
- Next have them inspect the phrases to see which are most important; these will be the headings, designated by capital Roman numerals.
- The subheads indicate secondary thoughts. They are indented and are designated by capital letters A, B, C, etc. Any further subheads are indented again and labelled with Arabic numbers. There are always a minimum of two subheads.

working in space. For example, biologists could separate substances more easily in zero gravity, and make new hormones and drugs under these conditions. Chemists could make purer materials for computer chips by using the low temperatures and vacuum of space. Better computer chips mean faster and more powerful computers.

Even though the public is divided on this issue, there are few who deny that space is the new frontier. With or without people, we will have to explore the possibilities for new technologies in space.

Think Critically!

1. What are the major disadvantages to the space station program? What are the advantages?
2. What conditions in space can not be easily created on earth?
3. What kinds of products and services might the space station provide? Would the space station be more or less effective with people on board? Explain.
4. You are a scientist who wants to do experiments on the space station. Would you choose to go yourself, even though it is risky? Or would you rather conduct the experiments remotely from earth? Give your reasons.
5. Do you think non-scientists should go into space? If so, for what reasons and under what conditions? If not, why not? Explain.

Now make a complete outline of this *Issue* on the blackboard, soliciting the students' suggestions. The outline might look like this:

I. People in Space: Cons
   A. Dangers of Space
      1. Strong radiation
      2. Vacuum
      3. Muscles weaken
      4. Accidents like *Challenger*
   B. Space Station is costly
II. Space Station planned despite dangers.
III. People disagree about the need for people in space
IV. People in Space: Pros
   A. Large potential profits
   B. New technologies possible
      1. New medicines
      2. Better computer chips

The students can practice this skill by outlining newspaper or magazine articles that deal with science issues. They can also develop outlines to write their own articles.

Think Critically Answers

1. Major disadvantages to the space station program include: risks to human lives; high costs; the space station's dependency on the shuttle. Advantages include new technologies and potential profits.
2. Special conditions in space include: zero gravity; low temperatures; the near vacuum of space.
3. The space station could provide weather prediction and satellite repair services, crystals for semiconductors, and new medicines. It would be more effective to have people on board the space station, assuming that serious accidents are infrequent, because people can perform tasks that are impossible for machines.
4. Answers will vary. Some students may want to be present to monitor the experiment. Other students may think the risks are unnecessary, citing the success of remotely controlled experiments in space.
5. Some students may feel that space travel is too risky for civilians, citing the Shuttle *Challenger* tragedy. Others might feel that people such as artists, poets, and historians are entitled to the experience of space travel as much as scientists. The students' answers may change if they think about the future: whether people will live in space, or just visit for extended periods.

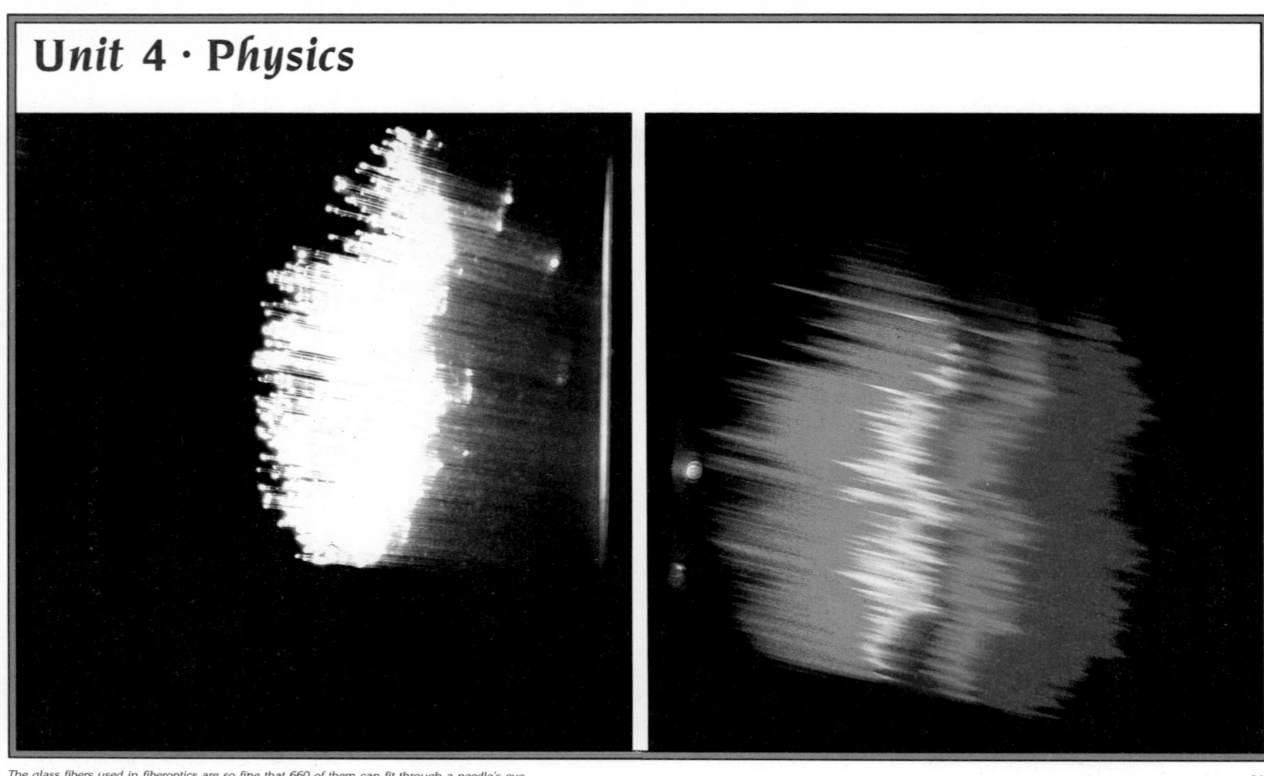

The glass fibers used in fiberoptics are so fine that 660 of them can fit through a needle's eye.

397

Unit 4 · Physics

The Unit In Perspective

The purpose of this unit is to introduce basic concepts of mechanics, heat, electricity and magnetism, and light and sound to students, and to provide an understanding of the place of physics in modern science and technology.

Chapter 16, *Motion and Force,* introduces Newton's laws of motion, gravitational force, weight and mass, and circular motion.

Chapter 17, *Work and Energy,* explains the concept of work and relates it to simple and other machines. In addition, efficiency is introduced and work, energy, and power are discussed.

Chapter 18, *Heat,* describes the relationship between heat and temperature and temperature and pressure, and discusses the transfer of heat, solar and other energy sources, space heating and cooling, and insulation.

Chapter 19, *Electricity and Magnetism,* defines electric charge, discusses the electrical nature of matter, and describes static and current electricity, magnetism, electricity from magnetism, using electricity at home, and conserving electricity.

Chapter 20, *Light and Sound,* discusses the nature of light, describes image formation by mirrors and lenses, discusses the electromagnetic spectrum and color, and compares the physical basis of sound with that of light.

Students will be interested in the physics-oriented careers featured on pages 526–527. The careers chosen illustrate the wide range of occupational opportunities in the physical sciences.

Issues in Science, on pages 528–529, presents the issue of increasing usage of robots in factories. Questions at the end of the article focus on critical thinking skills.

Preparing To Teach The Unit

The advance preparation required for this unit includes the acquisition and organization of equipment and materials for laboratory activities, planning field trips, ordering audio-visual aids, and collecting and/or placing on reserve in the library outside readings for students. A list of suggested readings will be found adjacent to the Main Ideas section at the close of each chapter.

Equipment and Materials

The listing below identifies the equipment and materials you will need for the student Activities of this unit. The Activities are located at the back of the student book. Quantities required will depend on the nature of the Activity, the availability of materials, numbers of students, budgetary considerations, and so on. Some Activities proceed better if students work in pairs or threes. Check each Activity to make this decision.

ACTIVITY 22 (Ch. 16): *Measuring Walking Speed* (page 560). MATERIALS: meterstick or metric measuring tape, metronome or stopwatches, distance markers.

ACTIVITY 23 (Ch. 16): *Measuring Forces of Friction* (page 561). MATERIALS: smooth, heavy wooden blocks; smooth, level board; thumbtacks; paper clips; string; sandpaper; aluminum foil; waxed paper; newspaper; wide rubber bands; spring scale.

ACTIVITY 24 (Ch. 17): *Measuring Power* (page 563). MATERIALS: Meterstick, stopwatch, stairs.

ACTIVITY 25 (Ch. 18): *Testing Insulation* (page 564). MATERIALS: thermometer (0–100°C); one small and one large container made of plastic, glass, or paper; clock or watch; small and large pieces of paper, to cover containers, insulating materials, such as mineral wool, vermiculite, and Styrofoam.

ACTIVITY 26 (Ch. 19): *Conductors of Electricity* (page 565). MATERIALS: four D cells or 6 volt battery, insulated electrical wires, two or more alligator clips (optional), light bulb, bulb holder, solid objects (wood, metal, plastic, paper, chalk, glass, or some other material).

ACTIVITY 27 (Ch. 20): *A Look into a Plane Mirror* (page 566). MATERIALS: small plane mirror, small straight pins, white paper, cardboard sheet, large straight pin, ruler, clay.

Field Trips

A well planned field trip is an excellent teaching tool. Plan as many as your schedule permits, and try to have expert guest speakers address your class as well.

Arrange a visit to a local playground and have students look for the simple machines. Or go to a construction site, and do the same for any large construction machinery in use. Visit a science museum that has a Foucault pendulum.

If one is nearby, visit a firm engaged in solar energy research. If possible, look into both solar heat collection and the production of electricity by solar energy.

Many science museums have exhibits and programs on electricity. Check with the nearest one and set up a trip. Visit an electric power plant to learn (a) the source of the energy being converted to electricity, (b) what processes are used to convert this energy to electricity, and (c) the hazards, if any, to the environment.

Visit a nearby observatory to learn how lenses and mirrors are used in large telescopes.

Audio-Visual Aids

The list of A-V materials below provides a wide selection of topics on sound filmstrips (SFS), videocassettes (VC), and 16 mm film. Choose titles in terms of availability, student interest, coverage of content vis-à-vis field trips and other learning activities, availability of projection equipment, and time limitations. A list of A-V suppliers is given on page T-39. Consult each supplier for ordering information and delivery times.

Chapter 16: Motion and Force

Mass and Weight, 9 min, 16 mm, VC, Coronet.
Galileo's Laws of Falling Bodies, 6 min, 16 mm, Britannica.
Friction: A First Film, 7 min, 16 mm, VC, Phoenix.
Gravity, Weight, and Weightlessness, 11 min, 16 mm, VC, Phoenix.

Chapter 17: Work and Energy

Simple Machines: Work and Mechanical Advantage, 14 min, 16 mm, VC, Coronet.
Energy and Work, 11 min, 16 mm, Britannica.
Simple Machines: Wheels and Axles, 5 min, 16 mm, VC, Coronet.
Simple Machines: Levers, 6 min, 16 mm, VC, Coronet.
Simple Machines: Inclined Plane, 6 min, 16 mm, VC, Coronet.
Perpetual Motion, 11 min, 16 mm, VC, Coronet.

Chapter 18: Heat

Energy Alternatives: What is Safe and Affordable?, SFS, Guidance Associates.
The World of Energy, Parts I, II, III, SFS, National Geographic.
Energy: Harnessing the Sun, 20 min, 16 mm, VC, Sterling.
Learning About Heat, 15 min, 16 mm, Britannica.
Wind and Water Energy, 25 min, 16 mm, VC, Time.
The Invisible Flame, 57 min, 16 mm, VC, Time.

Chapter 19: Electricity and Magnetism

Magnets, Magnetism, and Electricity, 9 min, 16 mm, VC, CRM/McGraw-Hill.
Magnetic, Electric, and Gravitational Fields, 11 min, 16 mm, Britannica.
Learning About Electric Current, 15 min, 16 mm, Britannica.
Electricity and Magnetism, SFS, National Geographic.
Energy: The Facts, the Fears, the Future, 55 min, 16 mm, VC, Phoenix.
Electronics: An Introduction, 9 min, 16 mm, VC, Phoenix.

Chapter 20: Light and Sound

The Nature of Sound, 14 min, 16 mm, Coronet.
The Nature of Light, 16 min, 16 mm, VC, Coronet.
The Nature of Color, 10 min, 16 mm, VC, Coronet.
Light of the 21st Century, 57 min, 16 mm, VC, Time.
The Simple Lens: An Introduction, 12 min, 16 mm, VC, Phoenix.

Chapter 16 Overview

After speed, velocity, and acceleration have been defined, Newton's three laws of motion are explained. The resistance to motion that results from inertia and from friction are discussed. The ability of a force to produce an acceleration is described. Mass is defined.

Gravitational force is then introduced and its relationship to weight is explained. The effect of air resistance on freely falling objects is then discussed. Finally, circular motion and the manner in which centripetal force can be used to create an artificial gravity are discussed.

Goals

At the end of this chapter, students should be able to:

1. find average speed from the distance traveled in a given period of time.
2. distinguish between speed and velocity.
3. define acceleration and give some examples.
4. explain what forces are.
5. recognize the role of friction in slowing motion.
6. state Newton's first law of motion and give examples that illustrate the law.
7. state Newton's second law and give examples to illustrate it.
8. explain how force, mass, and acceleration are related in the motion of objects.
9. state Newton's third law of motion and give examples that illustrate it.
10. give evidence that Earth attracts objects at or near its surface.
11. recognize that all objects attract each other with a force that depends on the mass of each object and the distance between them.
12. explain the difference between weight and mass.
13. explain how mass can be measured on a balance.
14. explain why an object's weight is less on the moon than on Earth.
15. explain why all objects fall with the same acceleration.
16. describe the effect of air resistance on falling objects.
17. recognize that the horizontal and vertical motions of falling objects are independent of each other.
18. recognize that an inward force is required to make objects move in a circle.
19. explain weightlessness.

Vocabulary Preview

speed
velocity
acceleration
force
first law of motion
inertia
friction
mass
second law of motion
third law of motion
force of gravity
weight
centripetal force

Chapter

16

Motion and Force

Capturing the motion of a speeding car in a photograph is a challenge.

16-1

Motion

Goals

1. To find average speed from the distance traveled in a given period of time.
2. To distinguish between speed and velocity.
3. To define acceleration and give some examples.

It seems that people have always been fascinated by motion. For centuries scientists have followed the paths of the planets across the sky. Every year many people run in marathons, races first held in ancient Greece. Every day people gather somewhere to watch cars or horses race around tracks.

Galloping horses and sleek cars show the beauty and grace of motion. However, in races, beauty and grace are not as important as speed. **Speed** is the distance traveled in a unit of time. You can find average speed by dividing the distance traveled by the time taken.

speed

velocity
(vuh LAHS uh tee)

$$\text{average speed} = \frac{\text{distance}}{\text{time}}, \text{ or } s = \frac{d}{t}$$

EXAMPLE

Find the average speed of an automobile that travels 150 kilometers in two hours.

$d = 150$ km $\qquad s = \frac{d}{t}$

$t = 2$ h $\qquad s = \frac{150 \text{ km}}{2 \text{ h}}$

$s = 75$ km/h

Velocity is the speed an object has in a particular direction. An airplane flying due north at a speed of 400 km/h has a velocity of 400 km/h north. Velocities can be represented by arrows. (See Figure 16-1.) The arrowhead shows the direction and the length of the arrow represents speed. Because you must travel in the right direction to reach a chosen destination, it is more useful to know your velocity than your speed.

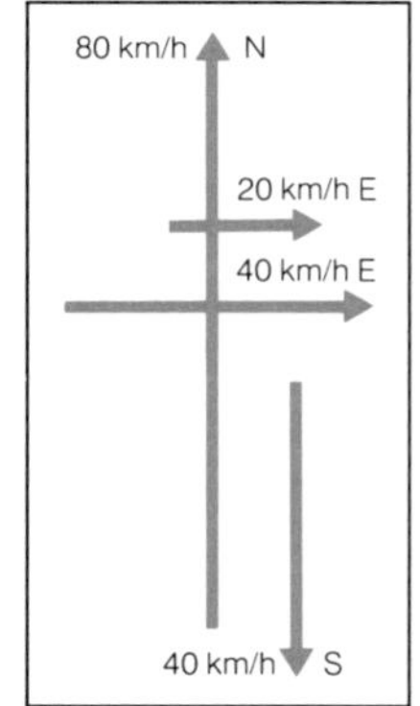

Figure 16-1 *Velocity arrows show direction and speed. The longer the arrow, the greater the speed.*

Teaching Suggestions 16-1

Students will intuitively have an understanding of, as well as practical experience with, velocity and speed. However, the difference between these two rates will likely be new to them. Once the distinction has been grasped, introduce acceleration.

Have students do Activity 22, Measuring Walking Speed, student text page 560. Chapter-end projects 2, 5, and 8 also apply to this section.

Background

Students should quickly understand the concept of speed: the rate at which distance is covered. Thus, if a body travels 100 meters in 10 seconds, its speed is 10 m/s. However, unlike speed, velocity has direction as well as a numerical component (speed). To show the direction and speed of a moving object, physicists use an arrow, which is called a vector. The direction in which the arrow points is the direction of the velocity; the length of the arrow represents the speed at which the body is moving. For example, with a scale of 1 cm = 1 m/s, a vector 10 cm long pointing straight downward could represent a velocity of 10 m/s south; an arrow 20 cm long pointing straight up would then represent a velocity of 20 m/s north.

Any quantity that has both a numerical component and direction can be represented by a vector. Thus, velocity, force, and acceleration are all vector quantities. Speed, time, and mass, which have no direction, are not vector quantities; they are called scalars.

The sum, or resultant, of two vectors can be found graphically. See Figure TE 16-1. Draw the two vectors to the same scale and with a common endpoint. Starting at the arrowhead of the one vector, draw a vector parallel and equal in length to the other vector. Repeat for the second vector. The result will be a parallelogram. The diagonal of the parallelogram that originates at the common endpoint of the two original vectors is the resultant, or vector sum.

The change in velocity per unit of time is called acceleration. For example, if you are traveling at a velocity of 10 m/s north and your velocity increases to 20 m/s north in 2 seconds, your average acceleration is the difference between the initial and final velocities, divided by the time required for the change in velocity: (20 − 10) m/s north divided by 2 s, or 5 m/s/s, which is more commonly written as 5 m/s^2.

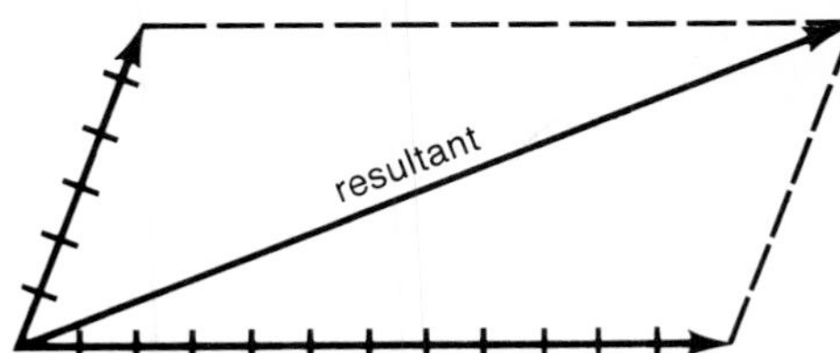

Figure TE 16-1 Parallelogram method: finding the resultant of two vectors.

Teaching Suggestions 16-2

An effective way to begin this lesson is to walk briskly across the classroom, dropping a ball as you walk. The ball has the same horizontal velocity as you do. Therefore, it will continue to move along with you as it falls. Consequently, you should be able to catch the ball without stopping. If you then stop walking immediately after dropping the ball, students will see the ball bounce on ahead of you.

Ask students why, without seatbelts, they will continue to move forward in a car that stops suddenly. (Since they are not fastened to the car, they will continue to move at the same velocity, because no restraining force has acted on them.) Films showing mannequins during crashes may often be obtained from the driver's education department.

Challenge students to think of examples of the first law of motion. (For example, a ball set in motion along the floor continues to roll long after you stop pushing it. The same is true of a ball that is thrown or hit, of a hockey puck that is shot or passed, or of any coasting motion where frictional forces are small.)

Have students do laboratory Activity 23, Measuring Forces of Friction, student text pages 561–562. Demonstration 2 on page 403 could be substituted for the activity.

Chapter-end project 1 applies to this section.

16-2

Newton's First Law of Motion

Goals

1. To explain what forces are.
2. To recognize the role of friction in slowing motion.
3. To state Newton's first law of motion and give examples that illustrate the law.

Isaac Newton was an English scientist who did outstanding work in many areas of science. In 1664, at the age of 22, Newton gave an explanation of motion that was true for all objects. By the time of his death, in 1727, his theories about motion were accepted as laws.

force

Newton's *first law of motion* is about objects moving along a straight line. It involves the idea of **force**, a push or a pull on an object. You apply a force when you hit a tennis ball with a racquet, or a baseball with a bat, or move a boat by pulling on an oar.

Like velocities, forces can be represented by arrows. The length of the arrow indicates the strength of the force. In Figure 16-3 the arrows show that the tractor pulls on the rock with a larger force than the horse.

Often, high-flying aircraft are able to take advantage of the forces available in the jet streams — winds far above

Figure 16-3 *A horse and a tractor can pull the object, but the tractor can exert a much greater force.*

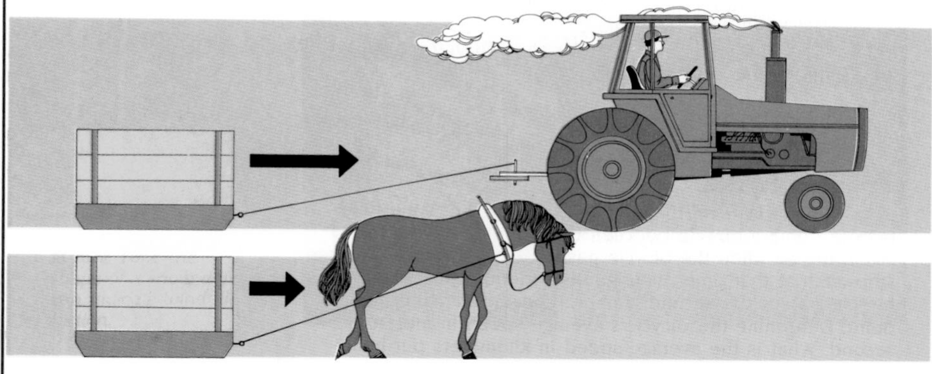

Background

It was Galileo who first conceived of motion along a frictionless plane. The ability to imagine such a condition was a giant step forward in the conceptualization of motion. Newton's laws of motion were developed in the framework of an idealized situation in which friction does not exist.

The notion that a body at rest will remain at rest unless acted upon by an outside force is common to our experience. But the idea that a body in motion will remain in motion unless acted upon by an outside force is less easily accepted. This may be so because friction is not an intuitively obvious force. Generally, it is only when friction is greatly reduced that we become aware of its significance.

Figure 16-4 *If this rider pedals into a strong wind, he will have to work harder to maintain his speed.*

Earth's surface that move at very high speeds. These winds push on the aircraft, adding to the force applied by the plane's engines. Of course, a plane that tries to fly into a jet stream *against* the wind will be slowed down. It would be much like a bike being pedaled into a very strong wind. (See Figure 16-4.)

first law of motion

Newton's **first law of motion** states that *an object in motion at constant velocity along a straight-line path will maintain that motion unless acted on by an outside force. Also, an object at rest will remain at rest unless acted on by an outside force.*

Safety engineers, knowing about Newton's first law of motion, design automobile seat belts to protect you. If the automobile you are riding in suddenly stops, you continue forward. You could bump into or even go through the windshield. Seat belts, fastened around you and attached to the automobile's frame, hold you firmly in place and keep you from continuing forward. An object tends to keep moving forward because of its **inertia**, its resistance to change in its motion. When an automobile stops, you keep moving because you are not part of the automobile.

inertia
(in UR shuh)

Because of its inertia, an object at rest will remain at rest unless acted on by a force. This is the secret of the magician's trick of pulling a tablecloth from under dishes. With a very smooth tablecloth, there is little force on the dishes

Demonstrations

1. To demonstrate the first law of motion, place a small piece of cardboard on the cap of a sealed, narrow-mouthed bottle, such as a soda or pop bottle. Then place a marble on the cardboard. Use a finger to snap the card out from under the marble. The marble should remain at rest on the bottle's mouth.
2. Place a wooden block on a wide board. Raise the board at one end until the block slides. Now modify the surface of the block in contact with the board. Use rubber bands, felt strips, newspapers, or some other material. Compare the amount of frictional force generated in each case by recording the height to which the board must be raised for the block to slide.
3. The motion of carts along an air track will illustrate frictionless motion very well.

Teaching Tips

- The newton (N) is introduced here as a unit of force. Show your class a spring scale that has been calibrated in newtons.
- Have students consider what life would be like in a frictionless world.
- Students may have noticed that a bird flying into the wind may appear to be standing still. That is because the net force on the bird is zero. Pedaling a bicycle uphill or into a stiff headwind is an experience familiar to many students. In this situation their pedaling force is nearly equalled by the forces of the wind, gravity, and friction. The result is that velocity drops greatly.

Mainstreaming

Visually-impaired students can manually apply a pushing force to a wood block on various surfaces, including rubber, wood, and paper. Vegetable oil on waxed paper can be used to demonstrate "frictionless" motion.

16 MOTION AND FORCE

16-2 Newton's First Law of Motion

The Matching Game

Match each term in Column A with its definition in Column B by writing the letter of the definition on the space in front of the correct term.

	A	B
____	1. Newton's first law	a. A push or a pull.
____	2. arrow	b. Unit in which force is measured.
____	3. friction	c. Kind of diagram that represents force.
____	4. inertia	d. Resistance to change in motion.
____	5. force	e. Statement about objects moving along a straight line.
____	6. newton	f. Force that opposes the motion of one surface over another.

Force or Friction?

The following phrases describe the reduction or increase of friction or the reduction or increase of force. Read each phrase and decide what is happening to friction or force. On the space at the left of each phrase write the term *reduces* or the term *increases* and the term *friction* or the term *force* to describe what is happening. The first example is done for you.

increases force	1. Running with the wind.
____________	2. Drying your hands to open a jar lid.
____________	3. Putting up the sail to catch the wind.
____________	4. Using crutches with rubber tips.
____________	5. Using a skate board with ball-bearing wheels.
____________	6. Wearing boots with lug soles.
____________	7. Waxing skis.
____________	8. Using a car jack, compared to lifting without a jack.

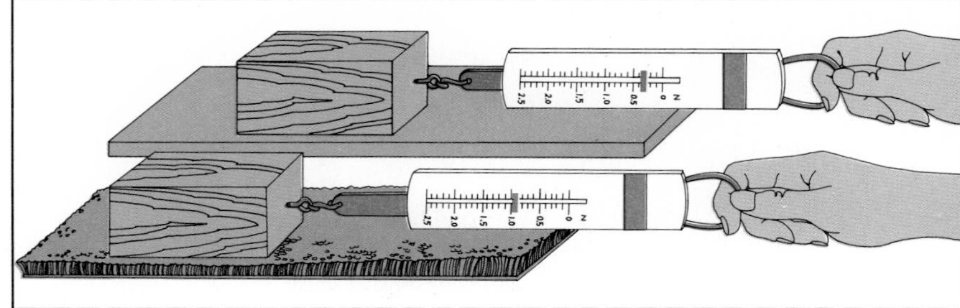

Figure 16-5 *With a spring scale, you can measure how many newtons of force it takes to move an object. Less force is needed to move it over the smoother surface.*

when the cloth moves. Because of their inertia, the dishes remain at rest.

Whenever one object moves over another the motion is opposed by a force. This force, called **friction**, acts against the motion of one surface over another. When you move a box resting on the floor by giving it a push, it soon slows down and stops. Frictional force between the bottom of the box and the floor is what stops the box.

friction

Without friction, an object in motion along a horizontal surface would move at constant velocity forever. There would be no force to oppose its motion. Friction between your shoes and concrete enables you to walk along a sidewalk. An ice-covered path provides very little friction, so it is difficult to walk on.

To reduce friction in machinery and promote freer motion, grease or oil is placed between moving parts. Without these *lubricants* the heat produced by friction could cause the moving parts to wear or even to melt.

The force of friction can be measured with a spring scale. (See Figure 16-5.) A spring scale includes a spring that stretches evenly. The greater the force pulling on the spring, the more it stretches. The scale shows how much force is pulling on the spring. Force is measured in units known as *newtons* (N).

Checkpoint

1. What is a force? Use an arrow to represent (a) a large force to the left, (b) a small force upward.
2. What is friction? How does it affect motion?
3. Give an example to illustrate Newton's first law.

Checkpoint Answers

1. A force is a push or pull on an object.
   (a) ←――――――――――
   (b) ↑
2. Friction is a force that acts against the motion of one object sliding over another.
3. Answers will vary. For examples, see student text page 403.

16-3

Newton's Second Law of Motion

Goals

1. To state Newton's second law and give examples to illustrate it.
2. To explain how force, mass, and acceleration are related in the motion of objects.

According to Newton's first law, an object in motion keeps moving at the same velocity unless it is acted on by an outside force. Frictional force, however, is always present. It slows down and stops any moving object, unless a force equal to or greater than friction opposes it.

You can probably think of situations in which frictional force is quite small. (See, for example, Figure 16-6.) By using extremely smooth objects and surfaces, we can produce such a situation. When this is done, a simple relationship between the force acting on an object and its resulting acceleration can be demonstrated. A force applied to an object at rest causes it to accelerate in the direction of the applied force. If the force remains constant, the object will move with a constant acceleration. That is, its velocity will

Figure 16-6 *A hockey puck will slide a long distance over the ice because there is little friction between the ice and the puck.*

Teaching Suggestions 16-3

It is likely that students will intuitively understand force to be proportional to acceleration, and therefore acceleration to be inversely proportional to mass. Students should have the opportunity to observe the effects of varying force on a given mass, and a given force on different masses. Demonstration 3 on page 406 is useful for this purpose.

The effects of Newton's second law of motion in space, which is a frictionless environment, can be viewed in various films available from the National Aeronautics and Space Agency (NASA) in Washington, D.C.

Newton's second law of motion is behind the "downsizing" of automobiles that has occurred since the mid-1970's. Ask students to explain the benefits of downsizing. (Less mass is easier to move, and therefore less force is needed. Smaller, more economical engines can do the job.) Ask students why professional bicycle racers choose the lightest possible frames and components; why runners choose the lightest shoes appropriate for particular races; and so on.

Background

Newton's second law of motion is described by the equation $F = ma$, where F is the net force, m is the mass, and a is the acceleration. Solving the equation for m, mass is seen to be equal to the ratio of net force to acceleration: $m = F/a$. Because for a given object m is unvarying, the value of F/a must be constant. Therefore, for a given mass, the greater the force, the greater the acceleration.

Newton defined inertia, or resistance to acceleration, for a free body in just this way. As mass increases, the acceleration resulting from the same force must decrease, because the product $m \times a$ must equal F. Mass and acceleration are inversely proportional; thus, if mass is doubled, acceleration is halved.

Force may be measured in terms of a unit called the newton, which is defined as the force required to give a one-kilogram mass an acceleration of one meter per second per second. (1 m/s/s).

Demonstrations

1. With a spring balance, pull a student in a wagon. Increase the force and have the class observe the increase in acceleration. Repeat, using lighter and heavier students, to show the effect of mass on acceleration.
2. If you can obtain an air track, have students observe the effect on acceleration of changing force and mass in a frictionless environment.
3. Obtain small physics demonstration carts, which have low-friction wheels. Load a cart with as many bricks or weights as will fit. Pull the cart with a rubber band that is allowed to stretch a constant amount. Students can see that the cart accelerates under the influence of a constant force. If two rubber bands are used in a similar way, the acceleration is roughly twice as great.

   Have students predict what will happen to the acceleration of the cart if some mass is removed. Take out half the bricks or weights, and repeat the experiment using a single rubber band. The acceleration will be about twice as great. Have students record the data and graph the results.

Teaching Tips

- Plot a graph of velocity versus time for a constant force and a given mass. See Figure TE 16-3. On the same graph, show the effect of doubling the force, and of doubling the mass. Note that the slope of the function (its vertical "rise" divided by its horizontal "run") is a measure of the acceleration of the body, the change in its velocity per unit of time.
- Emphasize that mass, which is often considered to be a measure of the amount of matter in a body, has been scientifically defined as the ratio of force to acceleration.
- Net force is the force applied in a given direction, less friction.

increase at a constant rate. For example, if the velocity at the end of one second were 1 cm/s, it would be 2 cm/s at the end of two seconds, 3 cm/s at the end of three seconds, and so on.

Suppose the force applied to the object is doubled. Then the acceleration will double, too. The velocity will be 2 cm/s at the end of one second, 4 cm/s at the end of two seconds, and so on.

Experiments of this kind show that the acceleration of an object increases when the force acting on it increases. There is another factor to consider, however. Suppose the inertia of the object were twice as great. You would have to double the applied force to produce the same acceleration as before.

mass

Newton used the relationship between force and acceleration to measure inertia. He called the measure of inertia *mass*. He defined an object's **mass** as the ratio of an applied force to the resulting acceleration of the object.

$$\text{mass} = \frac{\text{force}}{\text{acceleration}}, \text{ or } m = \frac{F}{a}$$

Figure 16-7 *When the force on the car is doubled (b), the acceleration is doubled. If the mass is then doubled (c), the acceleration is halved.*

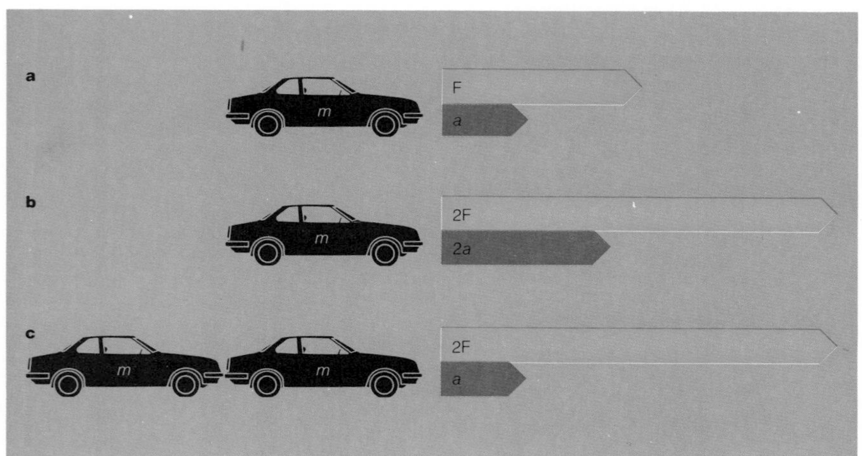

- When the velocity of a body is constant, the net force must be zero; that is, the applied force must be exactly equal to friction; otherwise, the body would speed up or slow down.

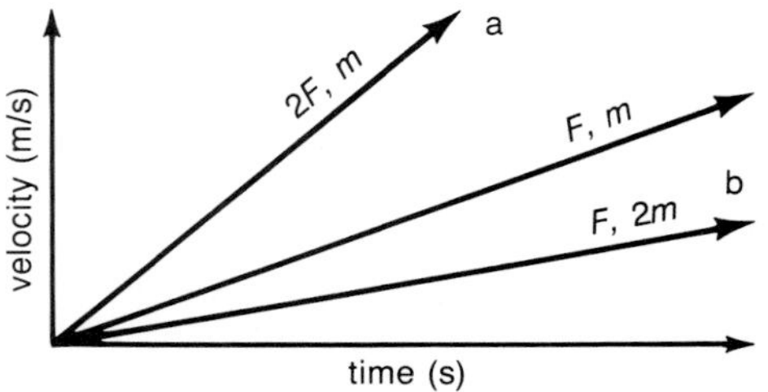

Figure TE 16-3 Effect on acceleration of (a) doubling the force and (b) doubling the mass.

Two objects are of equal mass if the same force applied to each object produces the same acceleration.

The effects of mass and force on acceleration may be summarized as follows:

1. For a given mass, acceleration is greater when the force acting on it is greater.
2. For a given force, acceleration is less when the mass on which the force acts is greater.

Figure 16-7 illustrates the effects of force and mass on acceleration.

Newton summarized his results in the **second law of motion**. *When a force is applied to an object, the object accelerates in the direction of the applied force. The acceleration is greater when the force is greater. The acceleration is less when the mass of the object is greater.*

second law of motion

Newton's first two laws can help you understand the interaction of force and motion that takes place when you ride in a car. A car at rest will not move until a force at least equal to the force of friction is applied. If the applied force is just equal to friction, the car moves at a constant velocity. For a car to accelerate, the force applied must exceed friction. It is the *net* force, the force in excess of friction, that causes acceleration. Once a car is in motion, a force greater than friction must continue to be applied to maintain its acceleration. As the car's velocity increases, the retarding force of air resistance increases, too. When the forces tending to move the car forward and those tending to resist its motion are equal, the car moves at a constant velocity.

Checkpoint

1. What is Newton's second law of motion? Give an example to illustrate this law.
2. To make an object accelerate, what must be true of the force applied to the mass?
3. A force that is 5 N greater than the force of friction is applied to an object. The object accelerates. A force 10 N greater than friction is applied to an object of equal mass. How will the accelerations of the objects compare?

Mainstreaming

Visually-impaired students can gain experience with the effects of Newton's second law of motion by pushing and pulling carts holding various masses. A rough sense of the effect of doubling, tripling, and quadrupling the mass can be gained from one run to the next.

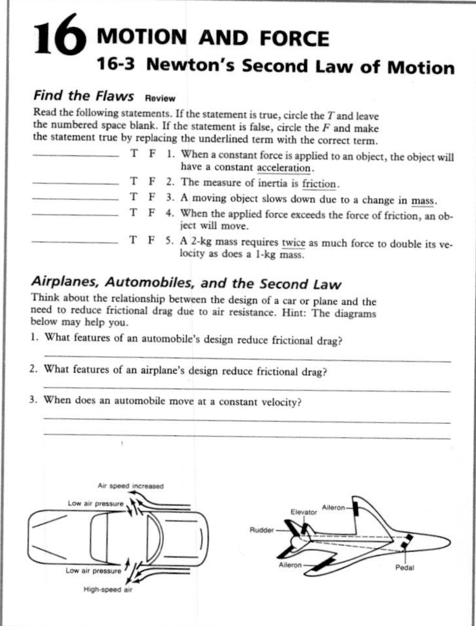

16 MOTION AND FORCE

16-3 Newton's Second Law of Motion

Find the Flaws **Review**

Read the following statements. If the statement is true, circle the *T* and leave the numbered space blank. If the statement is false, circle the *F* and make the statement true by replacing the underlined term with the correct term.

________ T F 1. When a constant force is applied to an object, the object will have a constant acceleration.

________ T F 2. The measure of inertia is friction.

________ T F 3. A moving object slows down due to a change in mass.

________ T F 4. When the applied force exceeds the force of friction, an object will move.

________ T F 5. A 2-kg mass requires twice as much force to double its velocity as does a 1-kg mass.

Airplanes, Automobiles, and the Second Law

Think about the relationship between the design of a car or plane and the need to reduce frictional drag due to air resistance. Hint: The diagrams below may help you.

1. What features of an automobile's design reduce frictional drag?
2. What features of an airplane's design reduce frictional drag?
3. When does an automobile move at a constant velocity?

Checkpoint Answers

1. Newton's second law is: force = mass × acceleration. Examples will vary. One possibility is that, having overcome friction, the harder you push an object, the faster it moves.
2. The applied force must be greater than the force of friction.
3. The acceleration due to the 10-N force will be twice as great as that due to the 5-N force.

Facts At Your Fingertips

- Inertial mass is ordinarily defined on Earth as the ratio of force to acceleration along a horizontal plane. However, it can be determined in space as well, because gravity is not involved in its determination. *Gravitational mass* is determined on a balance, which works only in the presence of a gravitational field.
- Force in newtons is equal to mass in kilograms × acceleration in meters per second per second. Force in dynes = mass in grams × acceleration in centimeters per second per second. Force in pounds = mass in slugs × acceleration in feet per second per second.

Teaching Suggestions 16-4

Ask students to suggest some situations illustrating the third law of motion. While some examples may be familiar, such as jet or rocket propulsion, as in the release of an inflated balloon, others may not be so obvious; for example, the explosive push from the ignition of a fuel/air mixture in a cylinder produces, ultimately, the movement of an automobile.

Ask students to consider how a horse and wagon can move, since the horse pulls forward on the wagon with a force equal to the force with which the wagon pulls back on the horse. (The horse also pushes against the ground, which pushes back on the horse with an equal but oppositely directed force. Thus, the net forward force on the horse exceeds the backward force exerted by the wagon, and the horse and wagon move forward.)

Chapter-end project 3 applies to this section.

Background

Newton's third law of motion is one that typically is not perceived as part of everyday experience. When someone pushes on a building, for example, neither the person nor the building moves, although equal and opposite forces are present. However, if the person were on roller skates, the resulting acceleration due to a reaction force would be obvious in the person's movement away from the building. Still, the building would not appear to move, because it is rigid, extremely massive, and firmly attached to Earth. The total mass of the building/Earth system is so huge that its acceleration is practically zero with the relatively weak force applied by any single individual.

To observe accelerations due to the third law, we must act on masses comparable to our own, and in relatively friction-free situations. For example, two people pushing and pulling each other while ice skating, on skateboards, or on roller skates would experience accelerations due to equal and opposite forces as described by the third law.

16-4

Newton's Third Law of Motion

Goal

To state Newton's third law of motion and give examples that illustrate it.

When you push on a friend while both of you are ice-skating, your friend slides away from you. But you also move. (See Figure 16-8.) When you push on your friend, you get a push in the opposite direction. A similar thing happens when burning fuel is directed out of the engines of a rocket. The burning fuel pushes on the rocket and the rocket pushes back with equal force. If you've ever watched a rocket blast off, you know that the burning fuel and rocket move in opposite directions.

Figure 16-8 *When one skater pushes on another, the first skater moves, too—backward.*

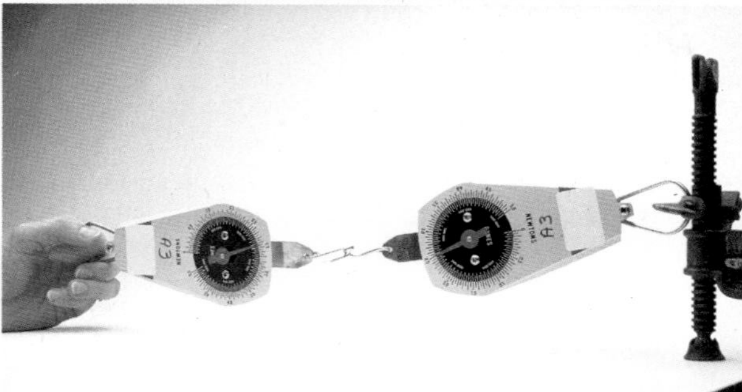

Figure 16-9 *If a person pulls on one scale, and the other is attached to a stationary object, the scales show that the person and the object exert the same force.*

Demonstrations

1. Have two students sit in separate wagons. Ask them to try to push and pull the other so that they move in the same direction across the classroom or schoolyard. After some practice, they should be able to make slow progress. It is difficult to move because the third law of motion is acting.

Figure 16-10 *This swimmer is moving fast because the water pushes on her as she pushes on it.*

Newton's third law of motion explains what happens when one object applies a force to another object. The **third law of motion** states that *for every action force there is an equal but opposite reaction force*. This means that if you push or pull on an object, the object will push or pull back with an equal force. (See Figure 16-9.) When an automobile leaves the road and hits a tree, the tree pushes back with the same amount of force. Every time you take a step you have to push against Earth. Earth pushes back with an equal force in the opposite direction, and you move forward. Earth moves also, but its mass is so huge that its acceleration cannot be measured.

third law of motion

Newton's third law explains why it is unwise to step from a small boat that is not tied to the dock. As you step from the boat, it will move away from the dock and your first step could be into the water. Then you might have to swim to shore. Swimming, too, involves Newton's third law. When you take a stroke, pushing against the water, the water pushes back with equal force. With repeated strokes you move forward. (See Figure 16-10.)

TRY THIS

Build a model of a rocket by attaching a balloon to a soda straw on a string as shown in Figure 16-11. As the balloon forces air to the right, the air pushes back on the balloon, forcing it to the left.

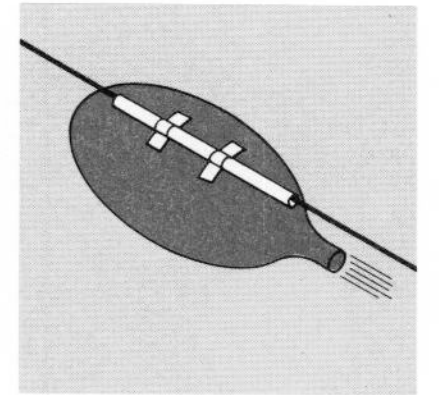

Figure 16-11 *Your balloon "rocket" should look like this.*

Checkpoint

1. State Newton's third law.
2. Explain the act of walking in terms of the third law.
3. Explain how the launching of objects into space illustrates the third law.

2. Hold two physics demonstration carts together with a compressed spring between them. Release the spring and ask the class to explain why the carts move apart. (The third law of motion is acting.)

Teaching Tips

- Emphasize the fact that third-law forces are equal but oppositely directed, and that each force acts on a different object. The accelerations produced are proportional to the mass of the object on which each force acts.
- To understand it fully, students should experience the third law of motion directly, and through demonstrations.
- Be certain students understand that the third law relates to forces, not activities. There may be no movement as the result of an action or reaction force.

Mainstreaming

Students in wheelchairs can be effective third-law demonstrators.

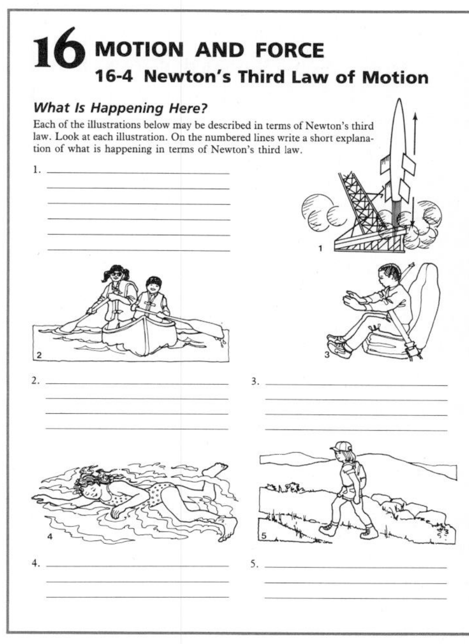

16 MOTION AND FORCE
16-4 Newton's Third Law of Motion

What Is Happening Here?

Each of the illustrations below may be described in terms of Newton's third law. Look at each illustration. On the numbered lines write a short explanation of what is happening in terms of Newton's third law.

1. ____________________
2. ____________________
3. ____________________
4. ____________________
5. ____________________

Using Try This

Relate the behavior of the balloon to the discussion of rocket propulsion on student text page 408.

Checkpoint Answers

1. Newton's third law states that for every action force there is an equal and opposite reaction force.
2. You push downward and backward against Earth; Earth pushes upward and forward, causing you to move forward.
3. The products of the oxidizing fuel escape from the ship with terrific force. The force of these gases can drive the ship forward with an equal, but opposite, reaction force.

Teaching Suggestions 16-5

One way to help students begin to understand gravity is to have them lift a heavy object with their eyes closed. Now tell them to imagine that the object is attached to the Earth by an invisible spring: as they lift the object, the spring stretches, pulling the object back toward the center of the Earth.

Use Demonstration 1 to show that all objects, no matter how light or heavy, fall with the same constant acceleration. (See Demonstrations section.) With a spring balance and a set of kilogram masses, you can also show that doubling the mass doubles the weight.

Explain that the force of gravitational attraction is proportional to the masses of the bodies involved. Since the mass of Earth is so huge, Earth's pull dominates all other gravitational forces on Earth's surface.

Chapter-end project 4 applies to this section.

Background

Galileo discovered that objects fall toward Earth, not with constant speed, as Aristotle had supposed, but with constantly increasing speed (constant acceleration).

Newton showed that the laws of motion apply not only to all earthly motion but also to all motion anywhere in the universe. The same force—gravity—that makes objects fall toward Earth also holds the moon in orbit around the Earth and the planets in orbit about the sun, and makes the galaxies spin.

Despite his ability to quantify the effects of the force of gravity, Newton was unable to explain why or how two masses separated in space should attract each other. In his view, gravity was a force that acts at a distance. The difficulty with this idea was obvious to Newton but he could not resolve it. In the modern view, a gravitational field curves space, causing a deflection of the path of moving objects.

16-5

Gravity

Goals

1. To give evidence that Earth attracts objects at or near its surface.
2. To recognize that all objects attract each other with a force that depends on the mass of each object and the distance between them.

When you drop something, it falls to the ground. This is true of all objects near the surface of Earth. A ball, a skydiver, raindrops, and snowflakes all fall toward Earth. Perhaps Newton's most important contribution was explaining how objects fall toward Earth.

Newton was born in the same year that Galileo Galilei, an Italian scientist, died. This was not the only way in which their lives crossed, however. Galileo found that all objects

Figure 16-12 *If the mass of one object is doubled (b), the force of gravity between the objects doubles. If the distance is doubled instead (c), the force becomes one quarter as great.*

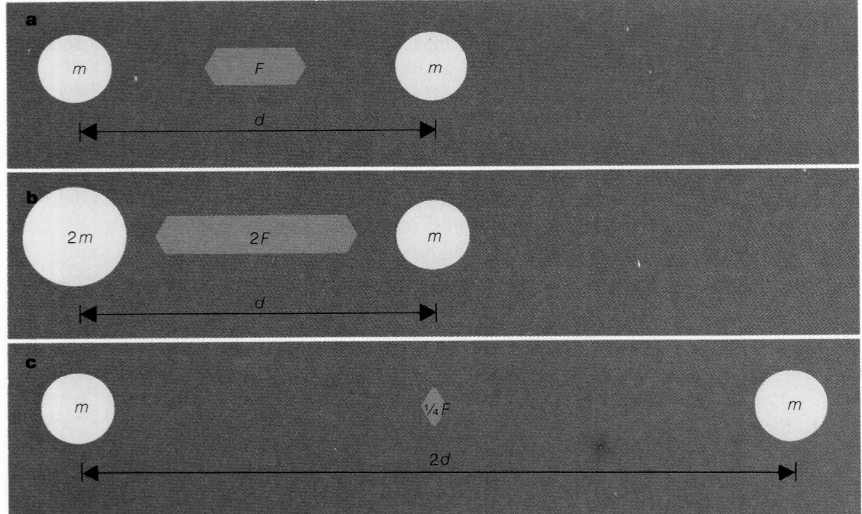

Gravitational force obeys an inverse square law: two masses attract each other with a force proportional to the product of their masses and inversely proportional to the square of the distance between them. Thus, while two people who are 2 m apart attract each other with a negligible gravitational force, larger bodies, such as planets, exert considerable gravitational force on spacecraft. Such forces must be taken into account when calculating a course for spacecraft whose missions will take them close to planets or large moons. The calculation of satellite orbits around Earth requires taking into account Earth's gravitational force.

accelerate at the same rate when they fall toward Earth. Galileo tested falling objects when he was a mathematics professor at the University of Pisa, in about 1590. He dropped objects made of heavy metals (gold, lead, and copper) along with light objects, all from the same height. All of them hit the ground at the same time. (You may have heard or read that Galileo dropped objects from the Leaning Tower of Pisa. Historians have not found any proof that he did.)

Using Galileo's experiments with falling objects, Newton suggested that the acceleration must be due to a force. The force, he believed, was the attraction Earth has for other objects.

According to Newton's third law, the force pulling an object toward Earth is equal to the force pulling Earth toward the object. Newton suggested that all objects attract each other in the same way. He was able to explain the motions of Earth (and other planets) around the sun. The explanation was based on the mutual attraction between each planet and the sun. This force of attraction between objects is called the **force of gravity**. The greater the mass of either object, the greater the force of attraction. The greater the distance between the objects, the smaller the attraction. (See Figure 16-12.)

force of gravity

Compared to objects near its surface, Earth is very large. The distances from its center to all objects near its surface are very nearly the same. Since the size of Earth and its mass remain the same, the force of gravity acting on any object on or near Earth depends only on the mass of the object.

Checkpoint

1. Why do objects fall toward Earth?
2. How did Newton know there was a force pulling falling objects?
3. Bill's father's mass is twice Bill's. How will the gravitational force acting on Bill compare with that acting on his father?
4. If Earth's orbit were closer to the sun, would the force of gravity between the sun and Earth be greater or less than it is now?

Demonstrations

1. Drop a baseball and a tennis ball simultaneously from the same height; repeat with a large steel ball bearing and a basketball. (They will hit the floor at the same time. Even a piece of paper formed into a ball will fall at the same rate as a baseball over short distances.)
2. Attach wooden blocks to a 3-m length of rope or string. The blocks should be set at intervals of 1, 4, 9, and 16 units of distance from the floor. When dropped, they will strike so as to make sounds that are evenly spaced in time. This indicates that the mass that fell 4 units took only twice as long to fall as the one that fell 1 unit of distance. The final velocity for falls from increasing heights must therefore be increasing. Some students may even see that the distance through which an object falls is proportional to the square of the time.

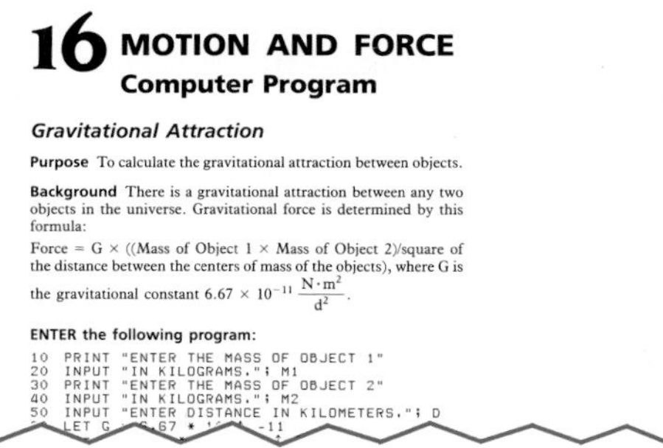

16 MOTION AND FORCE
Computer Program

Gravitational Attraction

Purpose To calculate the gravitational attraction between objects.

Background There is a gravitational attraction between any two objects in the universe. Gravitational force is determined by this formula:

Force = G × ((Mass of Object 1 × Mass of Object 2)/square of the distance between the centers of mass of the objects), where G is the gravitational constant $6.67 \times 10^{-11} \frac{N \cdot m^2}{d^2}$.

ENTER the following program:

```
10  PRINT "ENTER THE MASS OF OBJECT 1"
20  INPUT "IN KILOGRAMS."; M1
30  PRINT "ENTER THE MASS OF OBJECT 2"
40  INPUT "IN KILOGRAMS."; M2
50  INPUT "ENTER DISTANCE IN KILOMETERS."; D
```

Facts At Your Fingertips

- The mass of Earth is 6 septillion kg (6 followed by 24 zeros).
- A body's center of mass is the point at which its entire mass may be considered to be concentrated. For a planet, the difference between center of mass and geometric center could be a critical factor in the calculation of a spacecraft's orbit around the planet.

16 MOTION AND FORCE
16-5 Gravity

Earth's Attraction

In each of the following sentences, a term is missing. Read each sentence and think of the missing term. Then write it on the space at the left.

______ 1. The mutual attraction between Earth and the sun is called the __1.__

______ 2. Newton suggested that there is a mutual force of attraction between all __2.__

______ 3. Newton suggested that the acceleration of falling bodies must be due to a __3.__

______ 4. Galileo found that every object __4.__ at the same rate when falling toward Earth.

______ 5. The force of gravity acting on an object on or near Earth depends only on the __5.__ of the object.

Galileo

Checkpoint Answers

1. The force of gravity pulls objects toward Earth.
2. Since falling objects accelerate at a constant rate, a constant force must be acting on them, according to Newton's second law of motion.
3. The gravitational force acting on Bill will be half as much as that acting on his father.
4. Greater.

Teaching Suggestions 16-6

Ask students to imagine that they can move quickly from Earth to the moon. Then do the demonstrations. (See Demonstrations section.)

Emphasize that the mass of an object does not change, but its weight depends on the force of gravity. Since the mass of the moon is much less than the mass of Earth, gravity on the moon is much weaker and objects weigh much less.

Background

People are often confused by the terms "mass" and "weight." Mass is the inertia of a body; that is, its resistance to motion. A large truck has a lot more mass, or inertia, than a small car. The mass of an object is the same anywhere in the universe—on the moon, on Earth, Venus, Mars, Jupiter, or in a spaceship far from any other masses.

Weight is the force with which a body is pulled, or attracted, because of gravity. Thus, for a body with a mass of 100 kg, its inertia, or resistance to motion, will be the same wherever you place it. But its weight will depend on the gravitational force acting on it. On Earth, it will weigh 980 newtons, because Earth's gravity pulls on it with a force of 9.8 N for every kilogram of mass. On the moon, the force of gravity is only about one sixth as great as on Earth, because of the moon's smaller mass. Thus, on the moon, the 100-kg mass will weigh only about 164 N, because the moon pulls on each kilogram with a force of only about 1.64 N.

16-6

Weight and Mass

Goals

1. To explain the difference between weight and mass.
2. To explain how mass can be measured on a balance.
3. To explain why an object's weight is less on the moon than on Earth.

weight

The gravitational force acting on an object is called the object's **weight**. The greater the object's mass, the greater the gravitational force. If the mass of object A is twice that of object B, its weight will be twice B's also. Earth will pull on A with twice as much force as on B. If the weights of two objects, measured at the same location, are the same, they must have the same mass. If the objects are given the same horizontal push, they will have the same acceleration.

Like other forces, weights are measured in newtons (N). An apple weighs about 1 N. You probably weigh between 400 and 1000 N. A spring scale can be used to measure weight. (See Figure 16-13.)

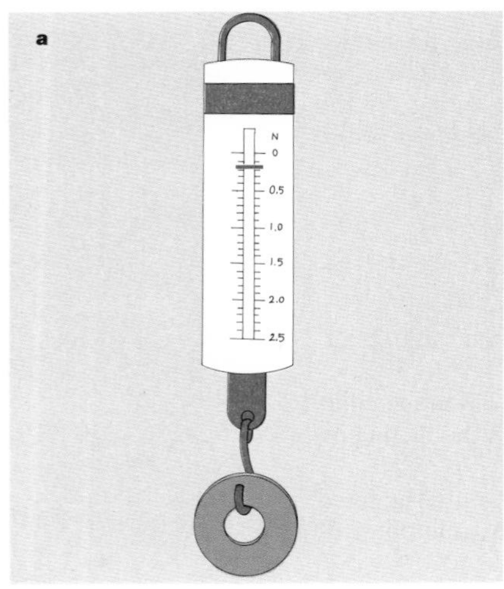

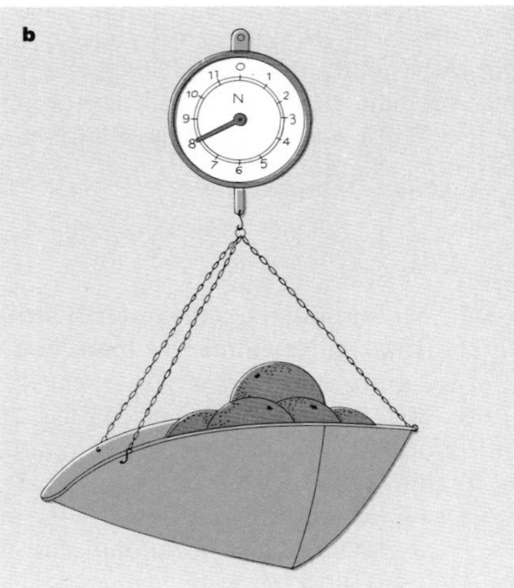

Figure 16-13 *(a) This spring scale measures the force of gravity on a washer. (b) Groceries can be weighed on a similar type of spring scale.*

Demonstrations

1. Show students what a 1-kg mass weighs on Earth and on the moon. You can simulate the moon weight by using a metal can you have rigged to weigh 1.64 N.
2. Have students measure the mass of objects on an equal-arm balance. Then let them measure the weights of these objects on a spring scale calibrated in newtons. (If the scales are graduated in grams or kilograms, cover the scale markings with tape and replace the 100-g markings with 1-N marks; or replace 1000-g or 1-kg marks with 10-N marks.) After students have measured their masses and weights, ask them how results would differ on the moon.

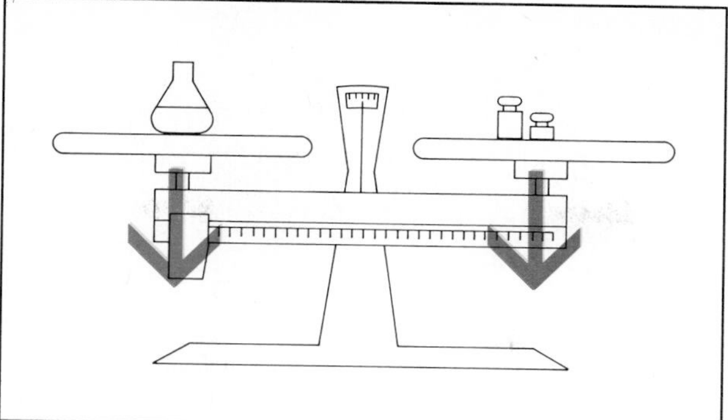

Figure 16-14 *An equal-arm balance. When the mass of the object on the left pan equals the sum of the standard masses on the right pan, the pointer is at zero.*

Mass is measured in *kilograms* (kg). A mass of one kilogram weighs 9.8 N near Earth's surface. Since 9.8 N is almost 10 N, an object that weighs 5 N has a mass of about 0.5 kg, while an object that weighs 20 N has a mass of about 2 kg.

Normally you would find the mass of an object with the use of an equal-arm balance like the one in Figure 16-14. The object of unknown mass is placed on the left pan of the balance. Then standard, or known, masses, such as kilograms or grams, are added to the right pan. When the gravitational force acting on each pan is the same, the balance is level, and the pointer is at the center of the scale. (See Figure 16-14.) This shows that the masses are equal.

Suppose you went to the moon. Suppose, further, that you brought along a balance, standard masses, and a spring scale. With this equipment, you could compare the masses and weights of objects on the moon with the masses and weights of the same objects on Earth.

At your moon station, you would find the mass of any object to be the same as on Earth. But the *weight* of an object on the moon is not the same as on Earth. The spring scale would show that the weight of an object is only about one sixth of its weight on Earth. This is because the moon pulls on an object at or near its surface with only about one sixth as much gravitational force as Earth does. If your mass is 70 kg, you weigh nearly 700 N on Earth. On the moon, you would weigh only about 120 N. Since your muscle strength is the same on the moon as it is on Earth, you would appear to be much stronger on the moon. You could easily lift your

Facts At Your Fingertips

- The force acting upon astronauts during liftoff is calculated in g's, units of force equal to the force of gravity.
- There is no atmosphere on the moon, because the moon's gravity is not strong enough to hold gases. Jupiter's gravity, on the other hand, is strong enough to hold even light gases such as hydrogen.

16 MOTION AND FORCE
Laboratory

The Inclined Plane and Force

Purpose To discover how the angle of an inclined plane is related to the force needed to move an object up the plane.

Materials piece of 0.25-in. (0.62-cm) thick plywood that is about 6 in. (15 cm) by 18 in. (25 cm), spring scale calibrated in newtons, 4 books of the same size and thickness, protractor, piece of string, small metal toy car or other object on wheels

Procedure

1. With string, attach the object on wheels to the spring scale. Then weigh the object by pulling it straight up with the spring scale. In the chart, record the weight of the object. To make sure your reading is correct, weigh the object and record its weight two more times.
2. Use the plywood board and one book to make an inclined plane.
3. Measure the angle that the inclined plane forms with the surface on which you placed the board. As you measure, be sure that the center of the protractor is at the point where the surface meets the bottom edge of the board.
4. Place the object on the bottom of the inclined plane. Pull the object up the inclined plane by pulling on the scale. (Some friction with the top of the board will result, but it will be the same for each trial and will not affect the results.) Record the force in newtons required to pull the object up the inclined plane. Follow the procedure at least three more times and record the force each time.
5. Increase the angle of the inclined plane by adding two more books. With this larger angle, repeat steps 3 and 4.

Discussion

1. How would you compare the size of the force needed to pull the object straight up with the force needed to move the object up the one-book inclined plane?
____
____
2. In this experiment, could the force ever be greater than the weight? __________
____
Why or why not? __________
____
____

Angle	*Trial Number*	*Force (or Weight) in Newtons*
90°	1	
	2	
	3	
angle 1 = (1 book)	1	
	2	
	3	
angle 2 = (3 books)	1	
	2	
	3	

Teaching Tip

- Sports-minded students may wish to calculate what various world records in track-and-field events might be if the events were held on the moon. For example, a high-jump of 2 meters on Earth would be equivalent to one of 12 meters on the moon; a hammer-throw of 20 meters on Earth would be equivalent to one of 120 meters on the moon.

16 MOTION AND FORCE
16-6 Weight and Mass

Choose a Term

Each sentence below is incomplete. Choose the term in parentheses that best completes the sentence. Then write that term on the space at the left.

______ 1. The gravitational force that Earth exerts on an object is called the object's (mass, weight, density).

______ 2. Weights are measured in (kilograms, kilometers, newtons).

______ 3. The gravitational force exerted on an object depends on the object's (mass, weight, density).

______ 4. Mass is measured in (liters, kilograms, newtons).

______ 5. The greater the object's mass, the greater the (gravitational, nuclear, magnetic) force.

______ 6. Two objects with the same weight must have the same (density, volume, mass).

______ 7. A mass of one kilogram weighs almost (ten, twenty, thirty) newtons near Earth's surface.

______ 8. The moon pulls on an object on or near its surface with about (one-sixth, one-third, six times) as much gravitational force as Earth does.

______ 9. You can find the mass of an object by using a(n) (spring scale, equal-arm balance, graduated cylinder).

______ 10. You can find the weight of an object by using a(n) (buret, spring scale, equal-arm balance).

Some Weighty Problems

Read each problem below and think what has to be done to solve it. Then solve the problem in the space provided.

1. What is the mass of an object that weighs 3N? ______
2. What is the weight of a 16.5-kg mass? ______
3. If your mass is 50 kg, how much do you weigh? ______ How much would you weigh on the moon? ______

Using Try This

The ratio of an object's weight in newtons to its mass in kilograms is very close to 9.8 for all masses. This ratio is about 6 times larger than the ratio that would be found if the same experiment were done on the moon.

Checkpoint Answers

1. The mass of A is twice as great as the mass of B, assuming both have been weighed under the same conditions.
2. Since a 1-kg mass weighs about 9.8 N on Earth, a 10-kg mass would weigh about 10×9.8 N, or 98 N.
3. The gravitational forces on each side of the balance are equal when the masses held by the two pans are equal.
4. Your mass would be the same on Earth and the moon, but your weight would be about one sixth as great on the moon as on Earth.

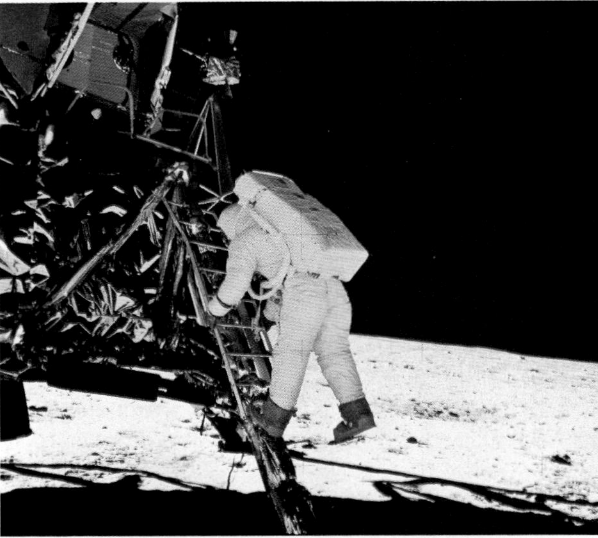

Figure 16-15 *The backpack carried by an astronaut on the moon has a large mass. But, because its weight on the moon is so small, it can be carried easily.*

own mass, for example. The moon's smaller force of gravity allowed astronauts on the moon to move about easily even though they carried backpacks almost equal to their own mass. (See Figure 16-15.)

Checkpoint

1. If object A weighs twice as much as object B, how do their masses compare?
2. What is the weight of a 10-kg mass on Earth?
3. Explain how an equal-arm balance is used to measure mass.
4. How would your mass on the moon compare with your mass on Earth? How about your weight?

TRY THIS

Using a spring scale, find the weights, in newtons, of various known masses. What is the ratio of each object's weight to its mass (divide weight by mass)? Is it the same for all the masses you tried? How would the ratio on Earth compare with the ratio on the moon?

16-7

Falling Objects

Goals

1. To explain why all objects fall with the same acceleration.
2. To describe the effect of air resistance on falling objects.
3. To recognize that the horizontal and vertical motions of falling objects are independent of each other.

Suppose you dropped a marble and a baseball at the same time, from the same height. You would find that they hit the ground at the same time. (See Figure 16-16.) Falling rocks and pebbles, bricks and tennis balls, ice cubes and refrigerators, all cover about the same distance in the same amount of time.

At first, this behavior may seem strange. You might expect the larger of two masses to fall faster, since the gravitational force acting on it is greater. But the larger mass *requires* a larger force to accelerate it. (Remember Newton's second law.)

The same situation applies when different cars are accelerated. Imagine two cars, one having a mass of 500 kg and the other a mass of 1000 kg. Now imagine that a force of 100 N is applied to each of the cars. The more massive car accelerates half as fast as the less massive one. However, if the force acting on the 1000-kg car is doubled, this car accelerates at the same rate as the 500-kg car. Similarly, a 2-kg object that is falling requires twice as much force as a 1-kg object to give it the same acceleration. But the gravitational force (weight) acting on a 2-kg object *is exactly* twice as great as that on a 1-kg object. Thus, the acceleration is the same for both objects.

As an object falls faster, the air resistance to its fall becomes greater. However, air resistance is not the same for all falling objects. A leaf has a larger surface compared to its mass than a stone has. As a result, a falling leaf is slowed down more by air resistance than a falling stone is. For this reason, a leaf and a stone will not fall with the same acceleration. As the stone falls faster and faster the air resistance to its fall becomes greater. Soon air resistance equals the force

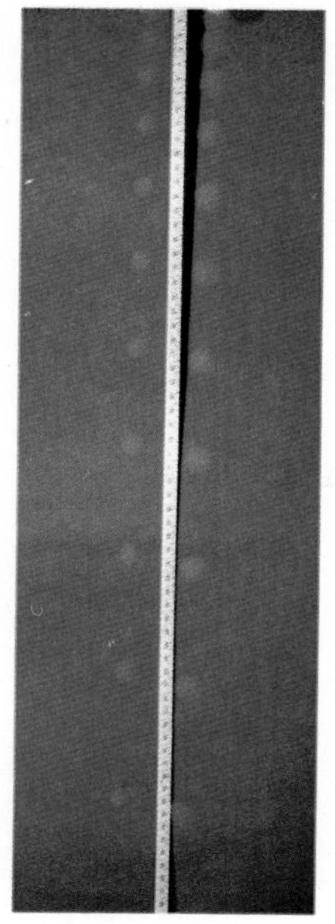

Figure 16-16 *The larger ball has the greater mass. But it does not fall faster than the smaller one.*

Teaching Suggestions 16-7

Review the fact that objects fall with the same constant acceleration regardless of their mass. A demonstration similar to Demonstration 1 in section 16-5, in which you drop a heavy ball and a light ball at the same time from the same height, will reinforce this idea.

It is more difficult to convey why this is true, so be prepared to go through the logic. With a spring scale, you can show that the gravitational force on a 2-kg mass is about $(2 \times 9.8)\ \text{N} = 19.6\ \text{N}$, while the gravitational force on a 1-kg mass is 9.8 N. Recall that acceleration is the ratio of the force to the mass ($a = F/\text{m}$). In both cases the ratio is 9.8; hence, in free fall, both masses accelerate at 9.8 m/s/s.

The same is true of projectiles launched with a horizontal velocity. The only force acting after a projectile has been launched is gravity; hence, the only acceleration is downward. This idea is easy to show with Demonstration 1. (See Demonstrations section.) Figure 16-18 on student text page 416 can also be very useful. Also, have students do the Try This activity on student text page 417.

To show the effect of air resistance on objects in free fall, use Demonstration 3. (See Demonstrations section.)

Chapter-end project 6 applies to this section.

Background

The force of gravity acting on a body is always proportional to the mass, whether on Earth, the moon, or any other planet. Thus, ignoring air resistance, all bodies of equal mass will fall side-by-side in a given gravitational field, because their accelerations due to gravity will be identical. The acceleration of a body is always the ratio of the force acting on it to the mass of the body.

On Earth, the gravitational force on a 1-kg mass is about 9.8 N; hence, in free fall, its acceleration toward Earth is about 9.8 m/s/s. Regardless of its mass, any body will accelerate at 9.8 m/s/s near Earth's surface. On the moon, the gravitational force on a 1-kg mass is only about 1.64 N; thus, on the moon, all objects in free fall accelerate toward the moon's surface with an acceleration of 1.64 m/s/s. Because the moon has no atmosphere, air resistance is not a factor.

While Galileo had shown that horizontal and vertical motions of projectiles near Earth's surface are independent of one another, it was Newton who explained why this was true. After a projectile has been fired, the only force acting on it (except for air resistance) is gravity. Since all masses have the same acceleration in free fall, an object fired horizontally will fall vertically at the same rate as a body that falls with no horizontal velocity. By Newton's first law of motion, the horizontal movement will persist because there are no forces (except air resistance) acting in a horizontal direction.

Demonstrations

1. If possible, obtain a commercial physics demonstration device designed to launch one projectile horizontally at the same time another is dropped. Use the device to illustrate that the vertical accelerations of the two objects are the same. Listen carefully to determine that both objects strike the floor at the same time.
2. While walking across the classroom, drop a ball that will bounce well, such as a handball or lacrosse ball. The fact that you can catch the ball as it rebounds demonstrates clearly that the ball retains its horizontal velocity. Indeed, if you abruptly stop walking after dropping the ball, the ball's horizontal velocity will cause it to bounce on ahead of you.
3. To show the effect of air resistance on a falling body, drop a book and a piece of paper simultaneously. The book will, of course, hit the ground first. Now place the paper on the book, so that the book shields the paper from air resistance. The book and paper will now tend to fall together.

of gravity. Then the stone falls at constant velocity. A leaf reaches constant velocity much earlier in its fall.

Newton predicted that if a feather and a hammer were to fall in a place where there was no air, both would fall with the same acceleration. Three hundred years later astronauts on the moon tested Newton's prediction and found it to be true. Of course, by then no one doubted that he was correct.

A feather or leaf very quickly reaches constant velocity as it falls through the air. No matter what its mass, an object will reach constant, or *terminal*, velocity if it falls long enough. Skydivers take advantage of this fact to perform stunts as they fall. As a skydiver falls faster and faster, the air resistance increases. After 12 s, the frictional force between skydiver and air becomes equal to the force of gravity. To increase air resistance, the diver may "spread-eagle" to increase the surface exposed to air. (See Figure 16-17.) When the air resistance and gravitational force are equal, the skydiver falls at the terminal velocity. For skydivers, terminal velocity is about 50 m/s. When the skydiver's parachute has unfurled, the large area of the open chute increases the air resistance. The increased upward force on the parachute decelerates the skydiver to a new, smaller, terminal velocity.

Figure 16-18 shows two balls that were released at the same time. One was allowed to fall straight down. The other was pushed horizontally. The horizontal lines in the illustration represent the distance fallen in a given period of time.

Figure 16-17 *Skydivers can increase air resistance and slow their fall by spreading out their arms and legs.*

Figure 16-18 *One ball is pushed horizontally while the other drops straight. Yet they both fall toward Earth at the same rate.*

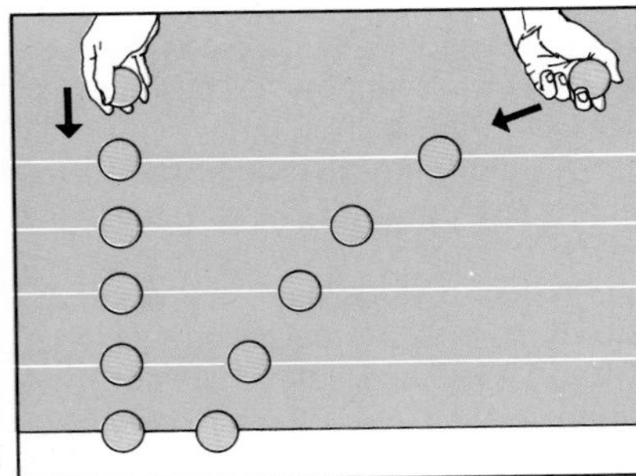

Teaching Tips

- To be certain students grasp the ideas in this section, do as many demonstrations as possible.
- Give many examples to show that g (acceleration due to gravity) is the same for all bodies.
- When using Figure 16-18 on student text page 416, point out that the horizontal velocity doesn't change. There are no significant horizontal forces acting on the object after it has been launched. By using small rulers, students can confirm the fact that the horizontal displacements for the ball are equal for equal time periods. The lines make it quite clear that the vertical displacements are equal for all time periods.

The lines show that both balls fell the same distance during the same time. Horizontal motion had no effect on the vertical motion. The two motions are, as Galileo discovered, independent of one another. Once both balls started to fall, the force of gravity was the only force acting on them. Since gravity gives all falling objects the same acceleration, the balls reached the floor at the same time.

Checkpoint

1. Explain why a marble and a baseball will fall side-by-side.
2. Why do all objects that fall through air eventually reach a terminal velocity?
3. A bullet is fired horizontally over level ground. At the moment the bullet leaves the gun, another bullet is dropped from the same height as the gun barrel. Will both land at the same time? Explain.

TRY THIS

Place two coins 5 cm apart near the edge of a table. Place a flexible ruler as shown in Figure 16-19. Release the ruler, knocking both coins off the table. Listen for the coins to strike the floor. Do both coins reach the floor at the same time? What does this tell you about the *downward* acceleration of the two coins?

Figure 16-19 *For the activity to work, the ruler and coins must be in these positions.*

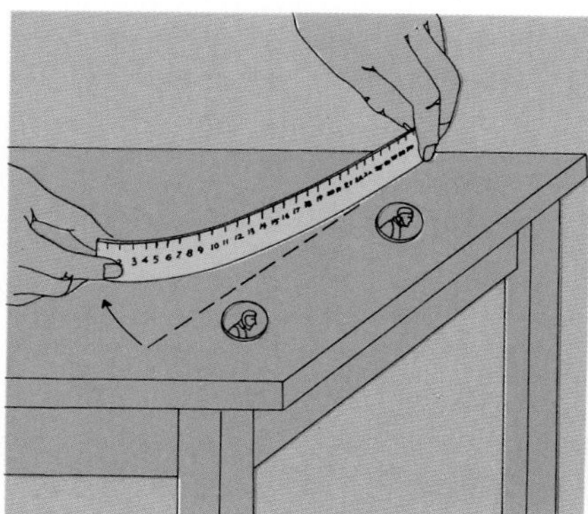

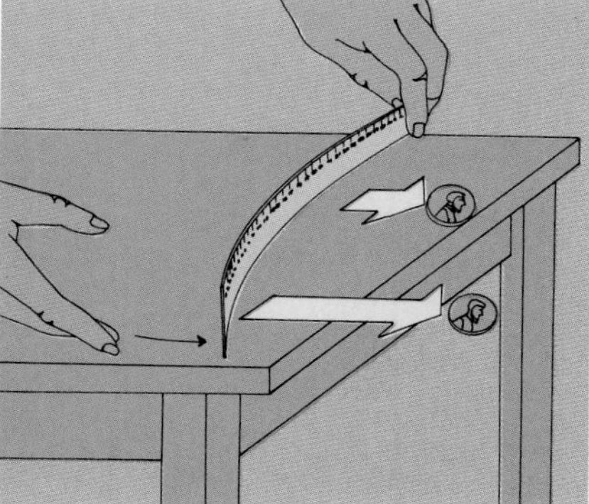

16 MOTION AND FORCE

16-7 Falling Objects

Search for Errors

Some of the statements below are true; others are false. Read each statement and decide whether it is true or false. If it is true, circle the T and leave the numbered space blank. If it is false, circle the F and replace the underlined word or phrase with a word or phrase that makes the sentence true.

________ T F 1. A tennis ball and a basketball are dropped at the same time from the same height above a floor. They will reach the floor at the same time.

________ T F 2. The gravitational force acting on a 10-kg object is twice as great as that acting on a 5-kg object.

________ T F 3. As an object falls in air, the air resistance to its fall remains the same.

________ T F 4. If an object falls far enough through the air, it will reach a constant velocity.

________ T F 5. Two identical objects fall to the ground from the same height at the same time, but one of the objects is given a push in a horizontal direction. The objects will not reach the ground at the same time.

Graphing Motion

The distance travelled by an object falling to Earth can be predicted by the following equation: $d = 9.8\ t^2$, where d = distance and t = time. The graph below represents several seconds of any object's fall, with air friction ignored. Examine the graph carefully. Then answer the questions.

Distance in meters: 0, 10, 20, 30, 40, 50, 60, 70, 80, 90, 100
Time in seconds: 0, 1, 2, 3, 4, 5, 6

1. What is represented on the horizontal scale? ________
   What does each mark on the horizontal scale represent? ________
   What does the vertical scale represent and what does each mark represent?
   ____
2. What is represented by the point at which time and distance are zero?
   ____
3. Estimate how far the object will fall in 2.5 seconds.
   ____
4. What is your estimate of how long it would take the object to fall 100 meters?
   ____

Using Try This

The coins should strike the floor simultaneously. This indicates that the downward accelerations of both coins are the same.

You might do this Try This as a demonstration, particularly if materials are difficult for students to obtain.

Checkpoint Answers

1. Since the ratio F/m is the same for both objects, they will have the same acceleration.
2. After a time, the air resistance creates a braking force that is as great as the weight of the falling body. When this happens, the net force is 0, and the acceleration therefore becomes 0.
3. Yes. Both bullets have the same vertical acceleration due to gravity, which is independent of any horizontal acceleration.

Facts At Your Fingertips

- The time (t) required for an object in free fall to reach the ground is related to the height (d) through which it falls by the equation

$$d = \tfrac{1}{2}gt^2.$$

- Air resistance increases as the cube of the velocity of an object.

16-8

ENRICHMENT

Circular Motion

Goals

1. To recognize that an inward force is required to make objects move in a circle.
2. To explain weightlessness.

The hammer-thrower in Figure 16-20 has to pull inward on the revolving mass to keep it moving in a circle. If he didn't, the hammer would fly off along a straight-line path, as it does when he lets go (Newton's first law). The inward-directed force that keeps an object moving in a circle is the **centripetal force**.

centripetal force (sen TRIP uh tuhl)

When a car goes around a curve, the tires push outward against the road, and the road pushes inward against the tires (Newton's third law). The inward force keeps the car moving along its curved path. Should the car move onto ice along the curve, the friction between tires and road that provides the centripetal force might become very small. The car would then follow a straight-line path off the road.

Earth *satellites*, including the moon, are in orbit because of Earth's gravity. Newton explained how Earth's gravity provides the centripetal force that keeps the moon in orbit. If gravity could be "shut off," satellites would fly off into space along straight-line paths just as a hammer does when it is released by a thrower. The moon and other Earth satellites are falling toward Earth. But as they fall, they also move sideways. Their sideways speed is so great that when combined with their fall, the satellites follow a curved path around Earth.

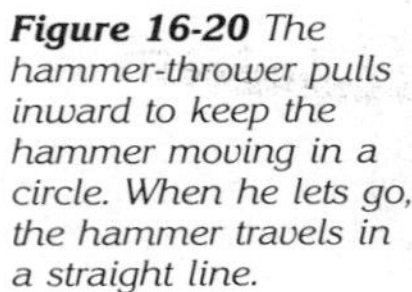

Figure 16-20 *The hammer-thrower pulls inward to keep the hammer moving in a circle. When he lets go, the hammer travels in a straight line.*

Teaching Suggestions 16-8

You may treat this lesson as a reading assignment. Or you may choose to demonstrate centripetal force and circular motion. (See Demonstrations section.) Ask students which way the cork will move if you swing an accelerometer in a circle. You may have to review with them the fact that the cork always moves in the direction of the acceleration or force. Since the force that you exert on the jar is inward, the acceleration must be inward as well.

Ask the class to name the force that keeps satellites, including the moon, in orbit around Earth. The answer, of course, is gravity, a centripetal force that acts toward the center of Earth. If a satellite moves fast enough or accelerates strongly, as a spacecraft would, its horizontal velocity will carry it forward as it falls from the straight-line path it would otherwise follow. When the rate of falling coupled with the horizontal velocity matches the curvature of Earth, then the object will orbit Earth, neither gaining nor losing altitude from orbit to orbit.

Background

Newton explained that gravity is the source of the centripetal force that keeps planets in orbit around the sun. Without gravity, satellites would move in straight-line paths away from their parent planet. This inward force creates an inward acceleration. The resultant of this inward acceleration and the perpendicular component of the horizontal velocity keeps the satellite in orbit. Suppose the inward force were removed. Then, by the first law of motion, the orbiting body would leave its circular path and fly off on a straight-line path tangent to the circle.

Demonstrations

1. Swing a pail of water in a vertical circle. When you swing the pail fast enough, the water will not spill out because you are supplying enough centripetal force to overcome the effect of gravitational force on the pail and its contents.
2. Use an accelerometer (see Figure TE 16-2, page 400) to show that there is an inward acceleration when the device is swung in a circle in a horizontal plane. In clear view of the students, allowing space around you, hold an accelerometer in your hand. As you swing the jar, it will be clear that the cork moves inward, toward the center of the circle.

Teaching Tips

- Explain that the circular motions of Earth's satellites, including the moon, and the motions of the planets about the sun are all examples of the effect of centripetal forces that arise from gravity.

Figure 16-21 *This is a NASA design for a space station. It will rotate to create artificial gravity.*

Astronauts on an orbiting space shuttle are said to be "weightless." They do not feel the effects of gravity that are felt on Earth. They float freely in the ship's cabin because gravity causes the astronauts and the ship to fall toward Earth at the same rate.

Long periods of weightlessness might weaken the muscles of astronauts. This could occur because they do not use their muscles to support their weight. For this reason, some plans for large space stations where people will live, work, and play for long periods of time call for the space station to rotate. (See Figure 16-21.) The outside of the rotating station will be accelerated inward, creating an artificial gravity. People walking along the outer rim will feel the ship pushing inward against their feet. Thus, "down" in this artificial gravity will be "out" toward the outer edge of the spinning space station. "Up" will be "in" toward the ship's center.

Checkpoint

1. What is a centripetal force?
2. Why do astronauts in orbit feel "weightless"?
3. How might artificial gravity be created in a space station?

Facts At Your Fingertips

- Due to its inertia, a mass that is moving in a circle exerts a reaction force on the object that is supplying the centripetal force. This reaction force, which is equal in magnitude and opposite in direction to the centripetal force, is sometimes called centrifugal force.
- Lack of gravitation has produced serious "spacesickness"—nausea and dizziness—in some astronauts. A cure is being sought.
- People living in a weightless environment, such as a space station, will lose their muscle power unless they exercise vigorously on a regular basis. The resistance to muscle movement usually supplied by gravity must be artificially simulated by devices utilizing springs and hydraulics. Isometric exercises are also effective under such conditions.

Cross References

Acceleration is discussed in section 16-1 and the first law of motion in section 16-2.

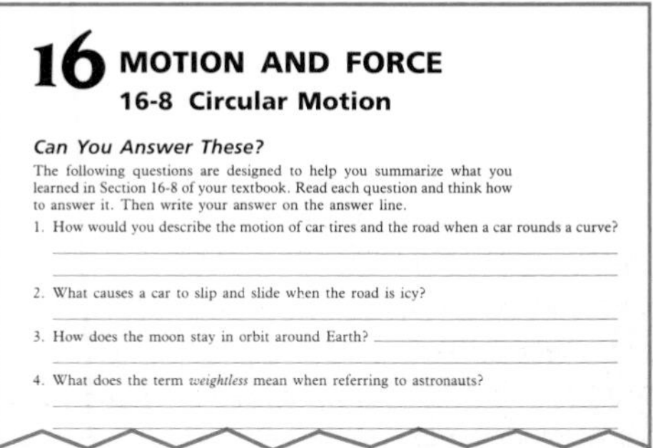

16 MOTION AND FORCE
16-8 Circular Motion

Can You Answer These?

The following questions are designed to help you summarize what you learned in Section 16-8 of your textbook. Read each question and think how to answer it. Then write your answer on the answer line.

1. How would you describe the motion of car tires and the road when a car rounds a curve?
2. What causes a car to slip and slide when the road is icy?
3. How does the moon stay in orbit around Earth?
4. What does the term *weightless* mean when referring to astronauts?

Checkpoint Answers

1. An inward-directed force that keeps an object moving in a circular path is centripetal force.
2. They are accelerating toward Earth at the same rate as their spaceship.
3. Rotation of the station creates an artificial gravity.

- Ask students to draw the path that a satellite of Earth would follow if the force of gravity suddenly disappeared.
- Point out that weightlessness in space is similar to conditions in a free-falling elevator. Ask what would be perceived by someone in a free-falling elevator if that person were to drop a ball? (Since the person and the ball are falling with the elevator, the ball would appear to float.)
- Relate the force exerted by the hammer-thrower in Figure 16-20 on student text page 418 to the gravitational force exerted by Earth on a satellite.

Using The Feature Article

Relate Newton's three laws of motion to the flight of the Space Shuttle. The first law of motion explains why the Shuttle needs a force to lift off from Earth. (An object at rest tends to stay at rest unless acted on by an outside force.) The second law of motion explains why the required propulsive force is so huge. (The force is equal to the product of the large mass of the Shuttle and the large acceleration needed to achieve orbital velocity.) The third law of motion explains why the propulsive action must be a downward exhaust stream. (The reaction force, or thrust, on the Shuttle will be opposite in direction, or upward.)

Students may enjoy discussing this article in relation to the latest Shuttle flights. Newspaper and magazine articles may be clipped and displayed. Models of the Shuttle may be built by interested students. A press kit describing Shuttle systems and mission objectives may be obtained from NASA or the U.S. Air Force.

Space Shuttle

Its powerful engines blazing, the Space Shuttle lifts off on another flight to orbit. The Space Shuttle Orbiter is the first spaceship to make round trips from Earth to space to Earth and back to space. Its large cargo bay and robot arm allow the Shuttle to carry payloads as large as 29,500 kg and place them into orbit. Satellites can now be carried into space or recovered and returned to Earth to be repaired or modified. Satellites and spaceships can now make more than one voyage into space.

At liftoff, the two rocket boosters and the Orbiter's main engines fire together. The force accelerates the ship to an altitude of 43 km and a speed of 5100 km/h in slightly more than two minutes. The rocket boosters are then released. They parachute into the ocean, where they are recovered.

The Orbiter's engines, fueled by the giant external fuel tank (ET), continue to push the ship higher and faster. At a speed of 28,000 km/h and an altitude of 106 km, the empty fuel tank is released. Fuel needs beyond this point are relatively small. The 44 rockets in the ship's orbital maneuvering system (OMS) and two small rear engines provide the forces needed to bring the ship to a stable orbit.

To return the Orbiter to Earth, one or more of the OMS rockets are fired in the direction of the ship's motion. This produces an equal and oppositely directed force against the Orbiter (in accordance with Newton's third law), causing it to slow down. It re-enters the air that blankets Earth. Here, friction further slows down the ship. Heat produced by the friction would turn the Orbiter into a flaming meteor were it not for the insulating tiles that cover its surface. After its hot flight through the atmosphere, the spacecraft glides to a landing under the control of the astronauts. The Shuttle is then prepared for another voyage into space.

Understanding The Chapter

16

Main Ideas

1. Speed is the distance traveled in a unit of time. Velocity is speed in a particular direction. Acceleration is the change in velocity in a unit of time.
2. An object at rest will remain at rest, and an object moving in a straight line will maintain its motion unless acted on by an outside force.
3. Friction acts against the motion of one surface over another.
4. If enough force is applied to an object, the object accelerates in the direction of the applied force.
5. For every action force there is an equal but opposite reaction force.
6. The force of gravity depends on the mass of each attracting object and the distance between them.
7. Weight is the gravitational force acting on an object.
8. An object has the same mass on the moon as it does on Earth, but it weighs less on the moon.
9. In the absence of air resistance, all objects fall to Earth with the same acceleration.
10. Objects that are falling in air reach a terminal velocity if they fall far enough.
11. The horizontal and vertical motions of falling objects are independent of one another.
12. Inward-directed forces, called centripetal forces, keep objects moving in circular paths.

Vocabulary Review

From the following list, choose the term that best completes each of the statements. Write your answers on a separate piece of paper.

acceleration	speed
centripetal force	velocity
force	weight
force of gravity	first law of motion
friction	second law of motion
inertia	
mass	third law of motion

1. A(n) __?__ is a push or a pull.
2. The force required to move an object in a circle is called __?__.
3. When one surface slides over another there is a force called __?__ which slows the movement.
4. According to the __?__ there is an equal and opposite reaction force for every action force.
5. An object's resistance to a change in its motion is its __?__.
6. A 1-kg ball will fall with the same __?__ as a 2-kg ball.
7. The __?__ between two objects depends on each object's mass and the distance between them.
8. The distance an object travels in a unit of time is the object's __?__.
9. A 1-kg __?__ weighs 9.8 N near Earth's surface.
10. An object traveling south at 70 km/h has a(n) __?__ of 70 km/h south.
11. __?__ is the distance traveled in a unit of time.

Vocabulary Review Answers

1. force
2. centripetal force
3. friction
4. third law of motion
5. inertia
6. acceleration
7. force of gravity
8. speed
9. mass
10. velocity
11. third law of motion.

Reading Suggestions

For The Teacher

Goodwin, Peter. PHYSICS WITH COMPUTERS. New York: Arco, 1985. Sixteen physics experiments that can be simulated with a computer by programming in BASIC.

McKay, David W. and Bruce G. Smith. SPACE SCIENCE PROJECTS FOR YOUNG SCIENTISTS. New York: Watts, 1986. Useful as a source of demonstrations and experiments for space-related activities.

Silberstein, Evan P. "Graphically Speaking." *The Science Teacher*, May 1986, pp. 41–45. Graphing as a way to improve student understanding.

Walker, Jearl. "The Essence of Ballet Maneuvers is Physics." *Scientific American*, June 1982, pp. 146-148. An analysis of the laws of the physics of motion applied to certain movements in ballet.

For The Student

Aylesworth, Thomas G. SCIENCE AT THE BALL GAME. New York: Walker, 1977. Principles of physics explained through baseball. Pitching, catching, and stealing bases are some of the examples used.

Fisher, David E. THE IDEAS OF EINSTEIN. New York: Holt, Rinehart and Winston, 1980. Presents the basic ideas of relativity, including time dilation, space-time curvature, and black holes in terms comprehensible to junior-high-school students.

Smith, Howard E., Jr. BALANCE IT! New York: Four Winds, 1982. Explains the principles behind familiar happenings, with emphasis on balance, symmetry, and center of gravity. Easy reading.

White, Jack R. SATELLITES OF TODAY AND TOMORROW. New York: Dodd, Mead, 1985. A "you are there" approach to understanding space science.

Chapter Review Answers

Know The Facts

1. weight
2. first
3. force
4. friction
5. second
6. 30 N
7. 7 N
8. equal to
9. less than
10. 9.8

Understand The Concepts

11. 60 km/h.
12. 10 kg.
13. 50 N.
14. Yes. The force (thrust) exerted on the plane by the exhaust flow pushes the plane forward (reaction).
15. 800 N.
16. The gravitational force of Earth exerts an inward (centripetal) force on the moon.
17. Air resistance is less for the paper ball than for the sheet of paper, since the surface area of the ball is smaller than that of the sheet.
18. Since muscle power is the same on the moon as on Earth, an astronaut pushing against the moon acquires an upward force as great as would be needed to jump on Earth. But because the downward force due to lunar gravity is only one sixth that on Earth, the net upward force is much greater and the astronaut's leap is much higher than it would be on Earth.
19. When bicycling uphill, the rider must overcome the forces of friction and wind as well as the gravitational force acting to pull the bicycle down the hill. When bicycling downhill, the rider must still overcome the forces of friction and wind, but now gravitational force is working to help pull the bicycle down the hill.

Chapter Review

Write your answers on a separate piece of paper.

Know The Facts

1. An object's _?_ is greater on Earth than it is on the moon. (weight, mass, velocity)
2. Newton's _?_ law of motion states that an object in motion will maintain that motion unless acted on by an outside force. (first, second, third)
3. Mass is the ratio of the applied _?_ to the resulting acceleration. (weight, speed, force)
4. You can walk because of the force of _?_ between your shoes and the ground. (gravity, inertia, friction)
5. The acceleration that results when you and your friends push a stalled car with a given force can be predicted by Newton's _?_ law of motion. (first, second, third)
6. A pushes on B with a force of 30 N. B pushes back on A with a force of _?_. (20 N, 30 N, 10 N)
7. There is a 10-N force acting on an object. The frictional force on the object is 3 N. The net force on the object is _?_. (10 N, 5 N, 7 N)
8. You drop a baseball at the same time a baseball leaves the pitcher's hand. The force of gravity acting on your baseball is _?_ the force of gravity acting on the pitched ball. (less than, more than, equal to)
9. The terminal velocity of a falling oak leaf is _?_ the terminal velocity of a falling acorn. (less than, equal to, more than)
10. A 1-kg rock falls with an acceleration of 9.8 m/s/s. A 2-kg rock will fall with an acceleration of _?_ m/s/s. (4.9, 19.6, 9.8)

Understand The Concepts

11. If a car travels 15 km in 15 min, what is its average speed in km/h?
12. An object weighs 98 N. What is its mass?
13. The force of friction acting on a car is 100 N. You apply a horizontal force of 150 N. What is the net horizontal force acting on the car?
14. Does the third law of motion apply to the motion of jet planes? Explain.
15. A person who weighs 800 N stands on a floor. What force does the floor exert on the person?
16. Explain why the moon moves in a nearly circular path around Earth.
17. Why does the terminal velocity of a piece of paper increase when it is squeezed into a ball?
18. Why could astronauts jump higher on the moon than on Earth?
19. Why is it more difficult to bicycle uphill than downhill?
20. Explain why it takes longer to bring an automobile to a stop on a wet or icy road.
21. Why are astronauts orbiting in a space vehicle "weightless"?
22. What force do skiers utilize in skiing downhill? What force can skiers decrease by waxing their skis?
23. Explain how the third law of motion applies to the rowing of a boat or the paddling of a canoe.

20. The force of friction caused by the application of an automobile's brakes must be transmitted by the tires to the road surface in order to slow down the car. When the road is wet or icy, there is less friction between the tires and the road surface. Therefore, the automobile requires more space and time to transmit its friction.
21. Astronauts orbiting in a space vehicle are weightless because gravity causes them to fall toward Earth at the same rate as their ship.
22. Skiers utilize gravity when skiing downhill. Waxing reduces friction.
23. The action of the oar or paddle in pushing back on the water creates an equal and opposite reaction that pushes the boat or canoe forward.

Challenge Your Understanding

24. An object is accelerating at a rate of 10 km/h/s. What would its acceleration be (a) if its mass were doubled, (b) if the force acting on it were doubled, (c) if both mass and force were doubled at the same time?

25. Explain how an equal-arm balance works.

26. A car traveling at 50 km/h is given an acceleration of 2 km/h/s. How fast will the car be going 10 s after it starts to accelerate?

27. An object weighs 49 N on Earth. What is the object's mass? About how much will it weigh on the moon if the moon's force of gravity is one sixth that of Earth?

28. Two boxes are at rest on the same floor. One box weighs twice as much as the other one. If each of the boxes is pushed across the floor, for which one will the force of friction be greater? How do you explain this?

29. Describe several swimming strokes, such as the backstroke and breaststroke. Explain how each stroke causes the swimmer to move.

30. The sun and the moon both pull on Earth's oceans, causing tides. Explain why the moon's pull has the greater effect.

Projects

1. Try to pull a sheet of plastic or smooth cloth out from under some unbreakable objects, such as plastic dishes or silverware, without disturbing the objects.

2. Find what the posted speed limits are in your neighborhood or along a highway near your home. What do you think are the reasons for the different speed limits?

3. While skating with a friend, stand facing each other. Ask your friend to give you a push. In which direction does your friend move? In which direction do you move? Who moves faster? Try this experiment with someone much bigger or smaller than you. What effect does your partner's mass have on the result?

4. Prepare a library report on the life and work of Galileo or Newton.

5. Interview some people at your local highway department. Find out how they decide on the location of highway markers such as those that advise motorists to slow down or that tell them where to get on or off a highway. Report your findings to the class.

6. Punch a hole in the side of a styrofoam cup near the bottom of the cup. Fill the cup with water and hold it above a sink. Notice how a stream of water emerges from the side of the cup. What happens to the stream when you let go of the cup and it falls to the sink? Try to explain what you observe.

7. Use a library to find the minimum speed a satellite must have to orbit Earth. How do such speeds affect astronauts?

8. Check government regulations on the crash resistance of automobile bumpers. Why do you think these regulations exist? Do you think they help save lives?

Challenge Your Understanding

24. (a) 5 km/h/s, (b) 20 km/h/s, (c) 10 km/h/s.

25. When the forces due to gravity are equal on each pan, the net force causing the pans to move is zero. Since force is proportional to mass, at the same location the masses must be equal if the gravitational forces on the pans are equal.

26. 70 km/h.

27. The object's mass is 5 kg. On the moon, the 5-kg mass will weigh about 8.2 N.

28. The heavier one will experience more friction. According to the second law, force is proportional to mass.

29. Each swimming stroke functions because of the concept described by Newton's third law of motion. The swimmer's hands, arms, and legs push against the water with great force; the water pushes back against the swimmer's hands, arms, and legs with an equal but opposite force. That force causes the water to push the swimmer along (since the hands, arms, and legs are all attached to the swimmer).

30. The moon's pull has a much greater effect on Earth's ocean (and land) tides than does the sun, despite the sun's much greater mass. This is because the moon is closer to one side of Earth than the other. The force it exerts on the near side is greater than the force exerted on the far side.

16 MOTION AND FORCE
Projects

1. If you have a machine shop in your school, have your teacher arrange an interview with the machine shop teacher. Talk with the shop teacher about how friction is reduced by using different kinds of bearings and lubricants.
   a. Make a list of several kinds of bearings and how each is used.
   b. List two lubricants. How are they used?
2. Isaac Newton was one of the greatest scientists. He formulated many important ideas about motion. Several books and encyclopedia articles have been written about him. Go to the library and find some of this information.
   a. Write a short biographical sketch of Isaac Newton. Include information about the place and time in which he grew up and the influences on his scientific development.
   b. Briefly describe two of Newton's scientific discoveries.

16 MOTION AND FORCE
Challenge

Vectors

Sometimes a wind blows at right angles to the direction in which an airplane is flying. This causes the plane to move sideways. Whenever this happens, two forces are at work—the force provided by the engines of the plane and the force of the wind itself. Each force has a direction. Any force with a direction is called a vector. In the example, one vector is the force and direction of the plane; the other vector is the force and direction of the wind. Vectors are represented with arrows. The length of an arrow represents the amount of force.

Look at the diagram below. Two forces are acting on a given object. One force (2 N) is vertical and the other force (7 N) is at a right angle to it. Here the arrows representing the forces are used as two sides of a rectangle and the other two sides of the rectangle are drawn in. A diagonal from the bottom left-hand corner to the upper right-hand corner of the rectangle is the result of combining two forces. The diagonal represents the net force (7.3 N) acting on the object. Since the two forces (2 N and 7 N) are pulling against each other at right angles, the net force (7.3 N) is less than the sum

CHAPTER 17 Overview

This chapter establishes the relationship among work, power, and energy. Work is defined and the concept of a machine is introduced. The six simple machines are explained and described. Complex machines are defined and their advantages are discussed. Mechanical advantage is defined and computed. The concept of efficiency of machines is discussed. The impossibility of building a perpetual motion machine is explained. Energy and power are defined in terms of work. Potential and kinetic energy are defined and the conversion of one to the other is described.

Goals

At the end of this chapter, students should be able to:

1. define and give examples of work.
2. define a machine and give some reasons why people use machines.
3. explain how a lever works.
4. explain the difference between simple and complex machines.
5. give examples of simple machines.
6. define the mechanical advantage of a machine.
7. define the efficiency of a machine.
8. explain why complex machines are less efficient than simple machines.
9. recognize that efficiency is not the only measure of a machine's usefulness.
10. define energy and distinguish between potential and kinetic energy.
11. give examples of objects that possess kinetic or potential energy.
12. describe one type of perpetual motion machine and explain why such machines cannot work.
13. define power and explain its importance.

Chapter 17

Work and Energy

Machines are used to do work.

Vocabulary Preview

work
joule
machine
fulcrum
simple machines
complex machine
mechanical advantage
efficiency
energy
kinetic energy
potential energy
power
watt

17-1

Work

Goal

To define and give examples of work.

Every day you do things that take time and effort. You sweep the floor, mow the lawn, complete your school lesson, and plan tomorrow's activities. But not all of these things fit the scientific definition of work. To do work in the scientific sense, you must apply a force that causes an object to move through a distance. You do work when you push a mower across a lawn or lift your school books. You do not do much work in the scientific sense when you solve your math problems or read your history lesson. (Of course, you do push your pencil and turn a few pages.)

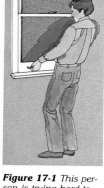

Figure 17-1 *This person is trying hard to lift the window. Because it does not move, no work is done.*

If the force acting on an object is less than the friction opposing its motion, no work is done. This is because the object does not move. The distance it travels is zero. If you push or pull on an object, and it does not move, no work is done. (See Figure 17-1.)

If you know the force acting on an object and the distance the object moves, you can find out how much work is done. **Work** is the force acting on an object multiplied by the distance the object moves.

work

work = force × distance, or $W = F \times d$

EXAMPLE

You apply a horizontal force of 40 N to push a mower a distance of 30 m across a lawn. Find the work done.

$$W = F \times d$$
$$W = 40\text{ N} \times 30\text{ m}$$
$$W = 1200\text{ N} \cdot \text{m}$$

Notice that the unit for work is the newton-meter (N • m). This unit has a special name. The work done when a force of 1 N acts over a distance of 1 m is 1 N • m, or one **joule** (J). The unit was named in honor of James Joule, a 19th-century English scientist.

joule (jool)

A person who walks up a flight of stairs must do work to

Teaching Suggestions 17-1

Review the concept of force, which was discussed in Chapter 16. Explain that both force and distance are needed to establish the concept of work.

Lift a heavy object. As you hold it, ask the class which behavior—lifting or holding—involves doing work. (No work is done in simply holding the object, because the force does not move through a distance. Work was done in lifting the object, because the force acted through the vertical distance between the floor and the raised position.)

Expect students to resist the counterintuitive idea that simply holding even a very heavy object is not work. Explain that work has a special meaning in physics. Thus, while considerable force is needed to hold a heavy object, no work is done as long as the object is motionless.

Background

Work is defined as the product of the force acting on an object and the distance through which that force acts. Only that component of the force directed along the path of the moving object is involved in calculating the work done. If a force is at a right angle to the direction of motion, no work is done by that force.

Work is used as a measure of the energy transferred from or to an object. For example, when an object falls, as in a pile driver, the work done by gravity (weight of object × distance it falls) is transferred to the falling object. If the object strikes Earth, the work done in bringing it to rest is converted into thermal energy, or heat. Such a conversion occurs when a meteorite strikes the surface of the moon. The result of the impact can be highly explosive, forming a crater and spreading debris for kilometers around.

Demonstrations

1. Use rubber bands, strings, or rope to pull a lab cart or wagon. Vary the angle at which you pull in order to show that the cart or wagon will not move (no work will be done) unless some part (component) of the force is directed along the direction of motion. See Figure TE 17-1.

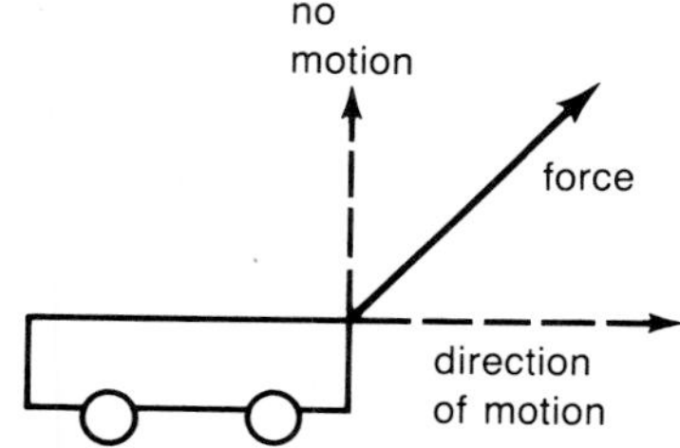

Figure TE 17-1 Demonstration 1.

2. Do some work by lifting an object of known mass through a known distance. Have students calculate the work done.

3. Allow an unbreakable object to fall. Ask students the origin of the force that does the work. (gravity) Ask them how much work has been done. (weight of object $\times$ distance through which it falls)

Teaching Tips

- Make sure students understand that in ascending a flight of stairs, no work is done along the horizontal portion of the path, because the force (weight) acting is vertical.
- Review carefully the computation of work. Spend time working through sample problems in the text.
- Point out that the supportive walls of a building do no work, because there is no distance through which the force acts.
- Use Demonstration 1 to show that no work is done unless the applied force has a component in the direction of the motion.

Fact At Your Fingertips

- Because friction is always associated with motion, some work must always be done simply to overcome friction. For this reason, the work input must exceed the work output for any machine or organism.

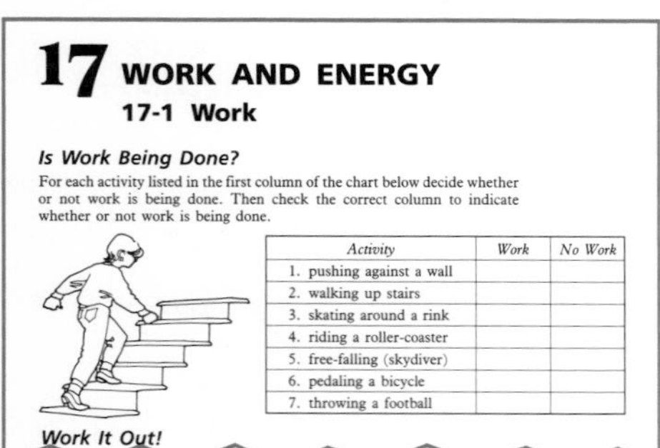

17 WORK AND ENERGY

17-1 Work

Is Work Being Done?

For each activity listed in the first column of the chart below decide whether or not work is being done. Then check the correct column to indicate whether or not work is being done.

Activity	Work	No Work
1. pushing against a wall		
2. walking up stairs		
3. skating around a rink		
4. riding a roller-coaster		
5. free-falling (skydiver)		
6. pedaling a bicycle		
7. throwing a football		

Work It Out!

overcome gravitational force. Remember that the amount of this gravitational force equals the person's weight. Since gravitational force is directed vertically downward, the force exerted on the person must be directed vertically upward. Thus, the work that the person does is the product of the force (the person's weight) and the vertical distance through which the person "is lifted" (the vertical height of the stairs).

The amount of work done in climbing straight up a ladder is the same as that done in walking up a very gentle incline. This is so because the vertical distance covered is the same in each case. Of course, with a gentle incline, the work is "spread out" over a longer distance. So walking up the incline is easier than climbing the ladder.

EXAMPLE

You use a vertical force of 10 N to lift an object to a height (distance) of 1 m. Find the work done, in joules.

$$W = F \times d$$
$$W = 10 \text{ N} \times 1 \text{ m}$$
$$W = 10 \text{ N} \bullet \text{m} = 10 \text{ J}$$

EXAMPLE

A person who weighs 420 N walks up a flight of stairs that has a vertical height of 5 m. What work, in joules, does the person do?

$$W = F \times d$$
$$W = 420 \text{ N} \times 5 \text{ m}$$
$$W = 2100 \text{ N} \bullet \text{m} = 2100 \text{ J}$$

TRY THIS

Use your mass in kilograms to find your weight in newtons. (Remember, a 1-kg mass weighs 9.8 N.) Figure out how much work you do when you walk up a certain flight of stairs.

Checkpoint

1. What is the scientific definition of work?
2. While rearranging furniture, you apply a force of 70 N to move a couch 6 m. How much work is done in moving the couch? (Remember, $W = F \times d$.)
3. A person who weighs 500 N walks up a flight of stairs that has a vertical height of 2 m. How much work is done in climbing the stairs?

Using Try This

Weight in newtons = mass in kilograms $\times$ 9.8. Work in joules = weight in newtons $\times$ vertical height of stairs in meters.

Checkpoint Answers

1. Work is the product of the net force acting on an object and the distance the object moves.
2. 420 J.
3. 1000 J.

17-2

Machines

Goals

1. To define a machine and give some reasons why people use machines.
2. To explain how a lever works.

machine

A **machine** changes the direction or strength of a force. Machines can save time and effort. Using machines, such as a bicycle or roller skates, you can get where you are going more quickly than if you walk. Lifting heavy loads can be made easier with machines like a jack or a crane. (See Figure 17-2.) With a small jack, you can lift an automobile.

A seesaw is a familiar machine. It can change both the direction and the strength of a force. By making one side of the seesaw longer than the other, a smaller person's weight on one end of the plank can balance or lift a larger person on the other end. (See Figure 17-3.) The *downward* force acting on the woman can produce an upward force on someone at the opposite end of the seesaw. The distance the woman moves downward is more than the distance the man is lifted. The work (force × distance) done *by* the woman is very nearly the same as the work done *on* the man.

The seesaw is an example of a basic type of machine called the *lever*. You can use another form of the lever to move a heavy rock that you cannot budge without using

Figure 17-2 *This crane is a machine that can easily lift heavy freight.*

Figure 17-3 *This seesaw is balanced because the smaller person is farther from the middle.*

Teaching Suggestions 17-2

It is essential that students understand the physicist's concept of work before studying machines. Machines are designed to do work, which is defined as the movement of a force through a distance. If necessary, review section 17-1.

You might begin with a demonstration involving a large lever. (See Demonstration 1.)

Students tend to think of machines only as cars, planes, heavy industrial equipment, and the like. Point out to them that such machines are really combinations of simple machines like the lever. Simple and complex machines are discussed in section 17.3.

Background

A machine is a device to which energy is applied at one point and from which a more useful form of energy is delivered at another point. Note that a machine need not amplify effective force as a lever does. Instead, some machines simply increase the speed with which a task is done.

For any lever, the product of the force acting on one side of the fulcrum and the distance between that force and the fulcrum is equal to the product of the force on the other side of the fulcrum and the distance of that force to the fulcrum:

$$F_1 d_1 = F_2 d_2.$$

The product of force and distance is called the moment of a lever.

Demonstrations

1. Place a strong board or plank on a fulcrum. (A split piece of firewood could serve as the fulcrum.) Have students stand on the board to demonstrate that a small force (a light student) can lift a larger resistance (a heavier student). The fulcrum should be closer to the heavier student. Move the fulcrum until the students balance. The ratio between the effort and resistance arms will be the reciprocal of the ratio of their weights; that is,

$$\frac{d_1}{d_2} = \frac{w_2}{w_1}$$

   The results will not be ideal because there is some friction on the fulcrum, and because the weight of the board is equally distributed only when the fulcrum is centered.
2. Show students that a lever actually rotates about its fulcrum.

Teaching Tips

- Explain to students that balances used for determining mass are finely made levers. For an equal-arm balance, the resistance and effort arms are of equal length.
- Show students that for a lever where the friction about the fulcrum is small, the product of the effort force and the effort arm equals the product of the resistance arm and the resistance, or load.
- Have students study the insertion of the biceps muscle in the human arm, as shown in Figure 17-6 on student text page 429.

fulcrum

some kind of machine. (See Figure 17-4.) Levers turn, or pivot, about a point called the **fulcrum**. The fulcrum divides the lever into two parts or *arms*. The *effort arm* is the distance from the fulcrum to the *effort force*, the force applied to the machine. The *resistance arm* is the distance from the fulcrum to the *resistance force*, or *load*, the force resisting the machine's effort.

A small effort force applied to a lever can match a large resistance force. This happens when the fulcrum is placed so that the effort arm is much longer than the resistance arm. Levers can be used to increase the effect of an effort force or to make things more convenient. Many persons with handicaps depend on machines based on the lever to make their lives easier. (See Figure 17-5.)

The human body contains a number of levers. When you lift an object, your lower arm acts as a lever. The biceps muscle, which is attached to the shoulder at one end and to

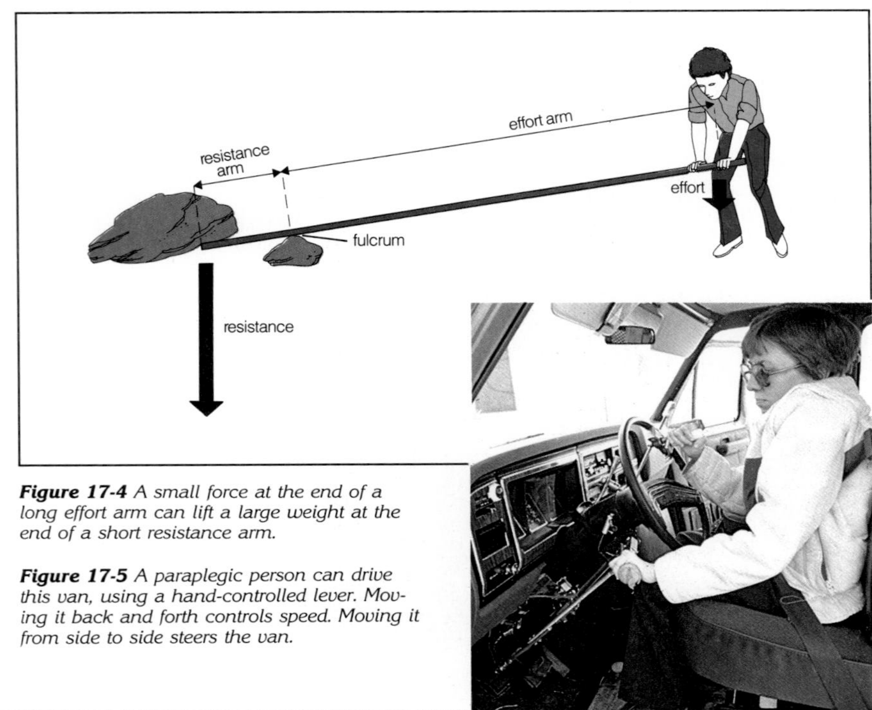

Figure 17-4 *A small force at the end of a long effort arm can lift a large weight at the end of a short resistance arm.*

Figure 17-5 *A paraplegic person can drive this van, using a hand-controlled lever. Moving it back and forth controls speed. Moving it from side to side steers the van.*

Fact At Your Fingertips

- Archimedes, a third-century B.C. Greek mathematician and inventor, claimed that with a long enough lever, he could move the planet. Ask students to explain what he meant and think about the difficulties involved in the task.

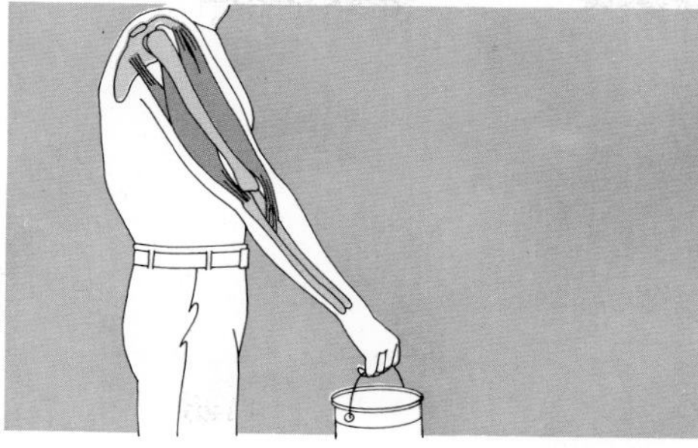

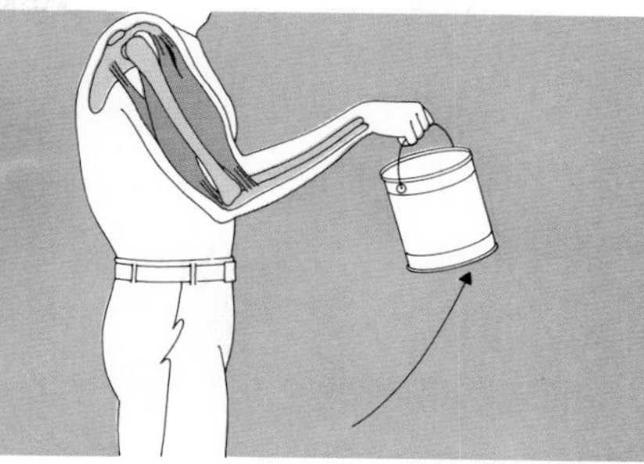

one of the bones of the lower arm at the other end, contracts. This causes a force to be applied to the lower arm. (See Figure 17-6.) The work done in picking up the object is the product of the object's weight and the height through which it is raised. The lower-arm lever does not increase the effect of the force applied by the biceps muscle. Instead it increases distance. The muscle contracts a relatively short distance compared to the distance the load moves.

Figure 17-6 *A human forearm acts as a lever when it lifts an object.*

Checkpoint

1. What is a machine? Why do people use machines?
2. Use a diagram to show where you would place the fulcrum of a lever to move a heavy rock. Identify the effort arm and the resistance arm.
3. How do the effort arm and resistance arm compare when a child is balancing an adult on a seesaw? Explain.

TRY THIS

Place a large book at the edge of a table. Slide a wooden ruler or pencil several centimeters under the book. Let the rest of the pencil or ruler extend beyond the table surface. Push down on the ruler at a point several centimeters out from the table's edge. (The edge of the table will serve as a fulcrum.) Note how hard you have to push to lift the book with the ruler. Now push on the ruler at a point farther out from the fulcrum. What happened to the force needed to lift the book as you moved the force farther from the fulcrum?

Mainstreaming

Students with artificial limbs may be able to explain how various prostheses extend their physical capacity.

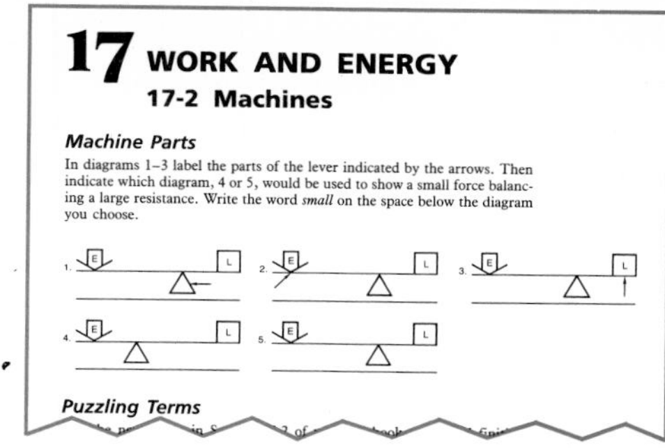

17 WORK AND ENERGY

17-2 Machines

Machine Parts

In diagrams 1–3 label the parts of the lever indicated by the arrows. Then indicate which diagram, 4 or 5, would be used to show a small force balancing a large resistance. Write the word *small* on the space below the diagram you choose.

Puzzling Terms

Using Try This

The strength of the force needed to lift the book should decrease as the point of its application is moved farther from the fulcrum.

You may also have students lift the end of the ruler, creating a second-class lever. The amount of effort required to move the book this way is the same as that needed when pushing down on the ruler.

Checkpoint Answers

1. A machine changes the direction or strength of a force. People use machines to save time and effort.
2. See Figure TE 17-2.
3. The length of the effort arm on the child's side of the balance is longer than the resistance arm on the adult's side. The work (force × distance) done by the child is equal to the work done on the adult.

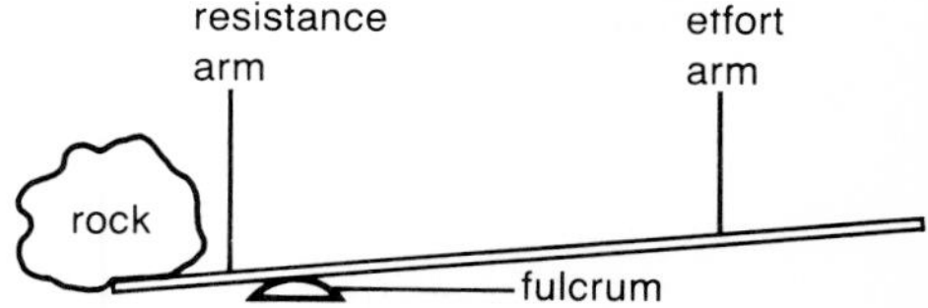

Figure TE 17-2 Using a lever to lift a heavy rock.

Teaching Suggestions 17-3

Show students that an inclined plane reduces the force needed to raise a load. Measure the force needed to pull a cart up an incline. Compare this force with the total weight of the loaded cart. (See Demonstration 4.)

Students should be able to calculate the work done in moving the cart up the incline. (Measure the length of the board in meters and multiply this length by the force in newtons needed to pull the cart up the incline.) Then, they should calculate the work needed to lift the weight the same height. (Multiply the maximum height of the incline by the total weight of the load.) This demonstration should make it clear that although less force is needed to raise a load on an inclined plane, the work done has not been reduced. In fact, if friction is appreciable, the actual work done with the inclined plane is greater than the work required to lift the object straight up.

Chapter-end projects 3, 4, and 7 apply to this section.

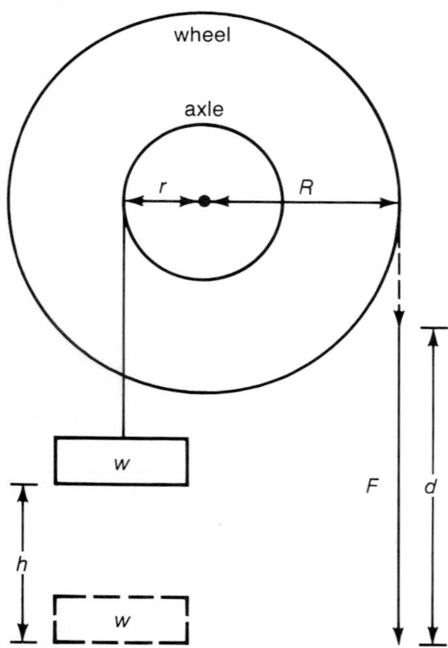

Figure TE 17-3 Using a wheel and axle to lift a load.

17-3

Simple Machines

Goals

1. To explain the difference between simple and complex machines
2. To give examples of simple machines.

simple machines

complex machine

The **simple machines** are devices from which all of the other machines are made. The lever is one of six simple machines. The others are the pulley, the wheel and axle, the inclined plane, the wedge, and the screw. A **complex machine** is made up of two or more simple machines.

You can change the direction of a force with a *fixed pulley*. You can lift an object by pulling downward as shown in Figure 17-7a. This is easier than bending to pick up the object in your arms. With a *movable pulley* the effort can be increased. A force less than the weight of the load can be used to lift it. (See Figure 17-7b.) Fixed and movable pulleys can be used together in a *block and tackle*, as shown in Figure 17-7c. Block and tackle machines are very important on a sailboat. Using them the crew can quickly change sails to take advantage of a shift in the wind. (See Figure 17-8.)

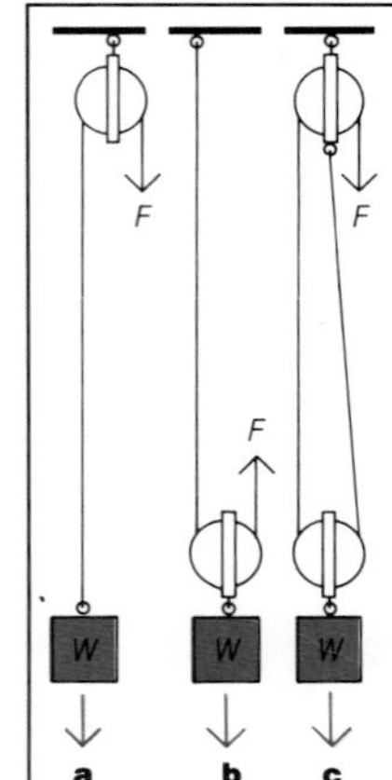

Figure 17-7 A pulley is a simple machine. (a) Fixed pulley. (b) Movable pulley. (c) Block and tackle.

Figure 17-8 Sails can be lifted by a block and tackle. The fixed pulley is on the left and the moving one on the right.

Background

Similarities exist among the principles behind the simple machines. For example, a wheel and axle may be considered to involve the principle of the lever. In Figure TE 17-3, a load w is lifted by a rope wrapped around an axle of radius r. A force F is applied to an attached wheel of larger radius R. The load is lifted a distance h. Neglecting friction, the work input, $F \times d$, is equal to the work output, $w \times h$. The work input and work output are equal, but by making the smaller input force F move through a long distance, we can lift a load that is much larger than F; however, the load moves only a short distance.

The effectiveness of a fixed pulley is similar to that of a lever with equal arms. Assuming no friction, the input

A *wheel and axle* machine consists of a wheel or crank attached to an axle, or shaft. A doorknob is a familiar wheel and axle. When you rotate the knob (wheel), the shaft (axle) to which it is attached turns, also. The wheel and axle shown in Figure 17-9a can be used to lift a weight. The effort force applied to the crank is increased by the machine to match the resistance force (the weight). A pencil sharpener is a wheel and axle machine. (See Figure 17-9b.) The gears in a car or bicycle are more complicated forms of the wheel and axle. The pedals on a bicycle connect to a gear that turns smaller gears through a connecting chain. (See Figure 17-9c.) This produces the rotation of the rear wheel about its axle.

An *inclined plane* makes it easier to raise an object through a height. A ramp is an inclined plane. (See Figure 17-10.) Less force is needed to lift an object using an inclined plane than that needed to lift the object straight up. However, the distance through which the force must be applied is greater. Either way, the same amount of work is done on the object.

If you put two inclined planes together, as shown in Figure 17-11, you will have a *wedge*. Chisels, axes, knives, and nails are wedges. An inclined plane wound around a central

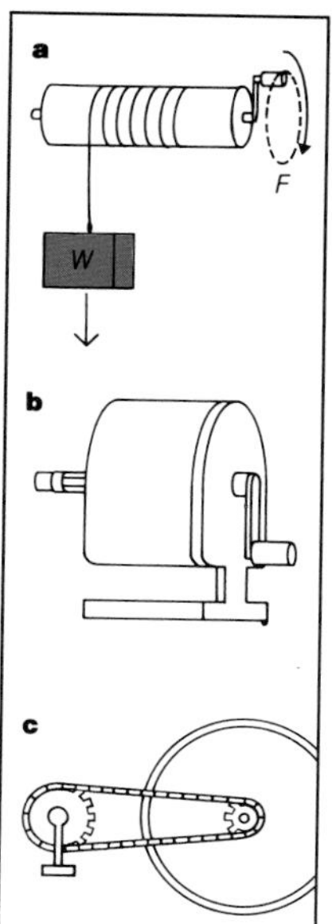

Figure 17-9 *(a) Simple wheel and axle. (b) Pencil sharpener. (c) Bicycle drive-system.*

Figure 17-10 *A ramp helps this man get his load into the van. He will use less force than he would need to lift the load straight up.*

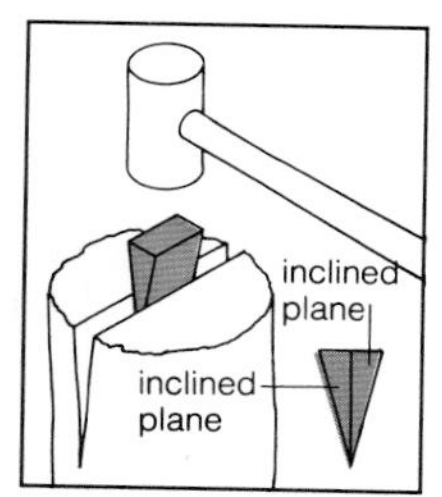

Figure 17-11 *A wedge is like two inclined planes put together.*

force equals the output force. Despite this, a fixed pulley is often used, because a downward force is frequently more convenient to apply than an upward force.

With a movable pulley (assuming no friction), an input force F can be used to move a load w that weighs $2F$. To raise w a height h, both supporting ropes must be shortened by the same amount. Thus, the force F moves up a distance $d = 2h$, and since the weight of the load lifted is twice as great as the input force, the work input and output are equal.

For the block and tackle (neglecting friction), tension throughout the rope must be the same. The tension in each of the parallel sections of rope holding up the load w, which includes the lower block, is equal to the applied force F. The total upward force, which must equal w, is equal to the product of F and the number of parallel rope sections supporting w. Therefore, for n rope sections, a load equal in weight to n times the applied force may be lifted.

An inclined plane makes it possible to raise an object by applying a force that is far less than the weight of the body. Only the vertical component of the force used to move a body up an inclined plane actually lifts it. Ignoring friction, very little additional force need be applied. Of course, this force F must move through a much greater distance than the height through which the body is raised. Thus, the work input equals or exceeds the work output.

Demonstrations

1. Use a spring scale to measure the force required to lift a load with a simple fixed pulley. Set up a movable pulley. Show that only half as much force is required when two ropes support the load.
2. Try to turn a screw by holding the metal shaft of the screwdriver. Then try it while grasping the handle. Ask the class to explain why it is easier to turn the screw by holding the handle. (The handle provides a longer effort arm, and thereby reduces the effort force needed.)
3. Examine one or more complex machines in order to find the simple machines within.
4. Make an inclined plane with the same large board used to make a lever in section 17-2. If the board is wide enough, you could place a student on a cart or in a wagon at the base of the incline. If the board is too narrow for this, use a small physics-lab cart, or something similar, loaded with weights. With a spring scale, measure the force needed to pull the cart up the incline. Compare this force with the total weight of the loaded cart.

17 WORK AND ENERGY

17-3 Simple Machines

Simple or Complex?

Decide whether each of the machines listed below is a simple machine or a complex machine. Then write the word *simple* on the space next to the names of machines that are simple machines; write the word *complex* on the space next to the names of machines that are complex machines.

________ 1. wind-up clock
________ 2. hammer
________ 3. tape recorder
________ 4. wheelchair ramp
________ 5. flagpole pulley
________ 6. pliers
________ 7. electric drill
________ 8. scissors

From Simple to Complex

For each complex machine listed or illustrated below, decide which of the six simple machines it is made of. Then write the names of the simple machines on the numbered space at the left.

________________ 1.
________________ 2. scissors
________________ 3.
________________ 4.
________________ 5. pencil sharpener

Using Try This

As more turns are added, it becomes easier to resist the pull of the two students holding the sticks, because the force is shared by each turn of the rope.

Checkpoint Answers

1. A simple machine is one of the devices from which all other machines are made. A complex machine is a machine that consists of two or more simple machines.
2. The six simple machines are the lever, pulley, wheel and axle, inclined plane, wedge, and screw. Examples will vary. A crowbar is a lever; a block and tackle consists of fixed and movable pulleys; a doorknob is a wheel and axle; a ramp is an inclined plane; an axe is a wedge; and a dowel with threads is a screw.
3. The force needed to move an object up an inclined plane is less than the force needed to lift the object straight up.

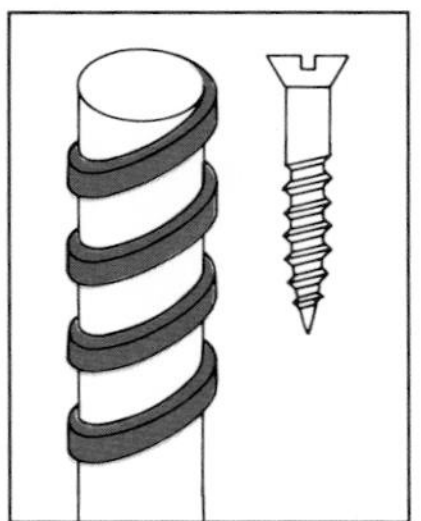

Figure 17-12 *A wood screw is made by wrapping an inclined plane around a cylinder.*

cylinder forms a *screw*. (See Figure 17-12.) The smaller the distance between the threads of a screw, the easier it is to turn the screw. Just as with an inclined plane, you need less force if you apply the force through a greater distance.

Most machines are complex machines. Automobiles, airplanes, and bicycles are complex machines. If you look at them closely, you will find that they are made up of many simple machines.

Checkpoint

1. What is a simple machine? a complex machine?
2. Name the six simple machines. Give an example of each one.
3. Explain why it is easier to raise an object using an inclined plane than to lift it to the same height.

TRY THIS

Ask two friends to hold onto two sticks, as in Figure 17-13. With your "block and tackle" you will be able to move the sticks together no matter how hard they pull. What happens as you add more turns (pulleys) to the sticks?

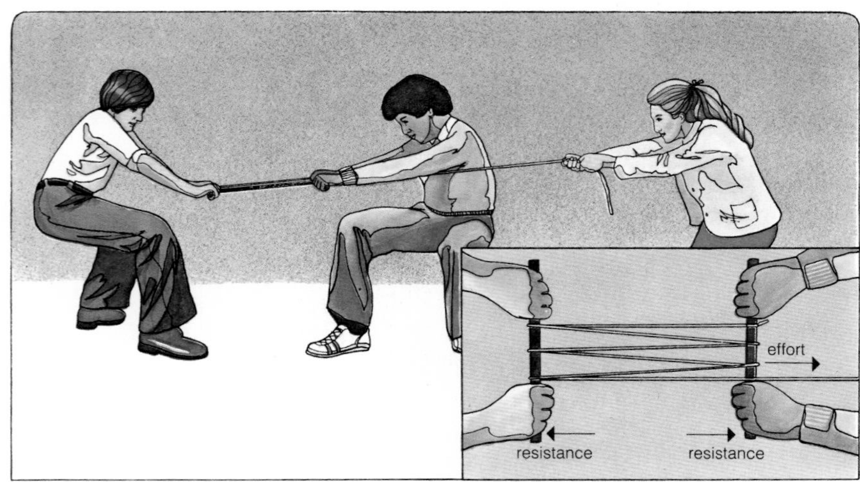

Figure 17-13 *This is the way to test your "block and tackle."*

17-4

Efficiency of Machines

Goals

1. To define the mechanical advantage of a machine.
2. To define the efficiency of a machine.
3. To explain why complex machines are less efficient than simple machines.
4. To recognize that efficiency is not the only measure of a machine's usefulness.

Machines are designed for different purposes. Some are designed so that a force can move a larger load. Other machines are designed to decrease the time needed to do the work. For example, a fishing rod is designed to give you great speed of movement at its end. (See Figure 17-14.)

A fishing rod is a third-class lever. With a third-class lever, a small movement of the effort arm can result in a large movement of the resistance arm. There are also first- and second-class levers. The class of a lever depends on the relative positions of the effort, resistance, and fulcrum.

Figure 17-14 *When a fishing rod is used to cast a lure, its far end moves much faster than its near end. The great speed flings the lure far out into the water.*

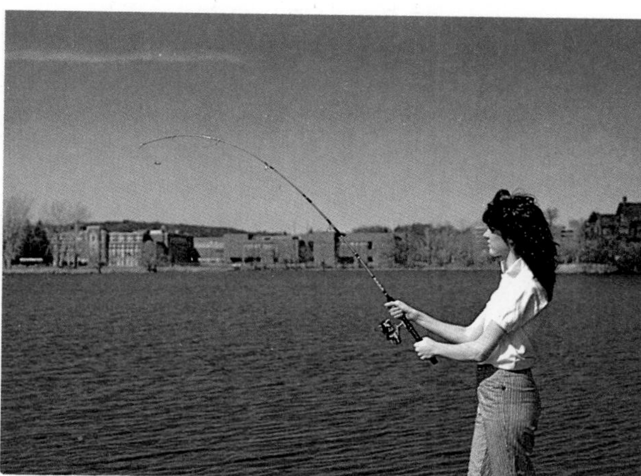

Background

The concept of efficiency goes one step beyond the concept of mechanical advantage. While mechanical advantage is simply the ratio of resistance force to effort force, efficiency is the ratio of work output to work input. Thus, efficiency addresses the distance through which the forces act, as well as the strength of the forces. While mechanical advantage ranges from less than 1 to greater than 1, the efficiency of a machine is always less than 1 (100%).

The work put into a machine always exceeds the work the machine puts out. Some of the work input is converted into heat by friction.

For any machine, the effort force must overcome both the resistance force and the force of friction. Since efficiency is the ratio of work output to work input, and work input is always greater than work output, efficiency must be less than 1. Most complex machines have typical efficiencies *much less* than 1, because overall efficiency is equal to the product of the efficiencies of all the component machines.

Teaching Suggestions 17-4

Have students determine the mechanical advantage and efficiency of a simple machine. (See Demonstrations.)

Chapter-end projects 2, 5, and 6 apply to this section.

Demonstrations

1. Measure the mechanical advantage for first-, second-, and third-class levers. Use a meter stick, fulcrum, spring scale, and weights.
2. Show that the efficiency of any machine is less than 1. Measure the work input and work output for a machine. For example, the efficiency of a fishing rod/reel assembly can be measured by determining the height (h) that the line lifts a weight (w) as a force (F) on the reel handle is turned through a distance (d) equal to the circumference of the handle multiplied by the number of turns; that is, for a fishing rod/reel, efficiency is equal to

$$\frac{w \times h}{F \times d}.$$

Figure 17-15 *A crowbar is a first-class lever, a wheelbarrow is a second-class lever, and a fishing rod is a third-class lever.*

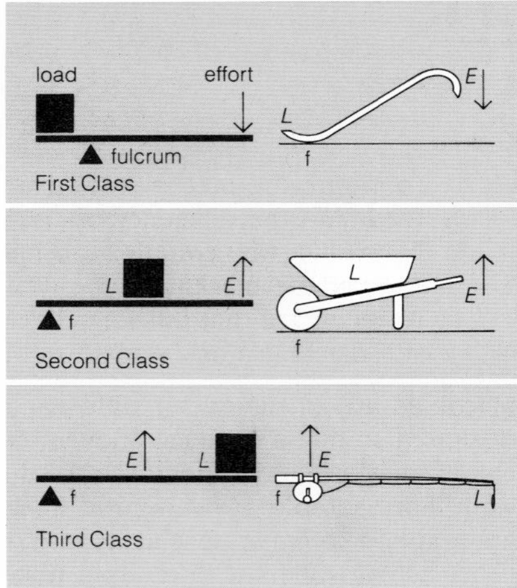

Figure 17-16 *With the lever drawn here, the effort force is one-fifth the resistance force. The lever has a mechanical advantage of 5.*

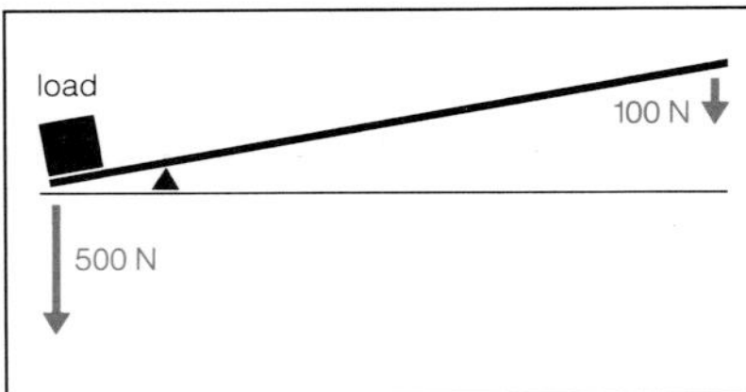

Figure 17-15 shows first-, second-, and third-class levers, and some common applications of them.

It is often important to know how much force a machine will produce when a certain amount of force is applied. The **mechanical advantage** (MA) of a machine is the ratio of the resistance force to the effort force. The *ideal* mechanical advantage (IMA) is the ratio that would be obtained in the absence of friction. Under ideal conditions, the lever shown in Figure 17-16 exerts a resistance force of 500 N when an effort force of 100 N is applied.

mechanical advantage

$$\text{IMA} = \frac{\text{resistance force}}{\text{effort force}} = \frac{500 \text{ N}}{100 \text{ N}} = 5$$

A simple first-class lever can come very close to its ideal mechanical advantage. The friction involved in its use is very small. But for most machines, the amount of effort needed to overcome friction must be taken into account. Suppose the force of friction for the lever shown in Figure 17-16 were 25 N. The effort force would then become 100 N + 25 N, or 125 N. The mechanical advantage (MA) of the lever would be 4.

$$\text{MA} = \frac{\text{resistance force}}{\text{effort force}} = \frac{500 \text{ N}}{125 \text{ N}} = 4$$

Teaching Tips

- It is important that students realize that mechanical advantage is very different from efficiency. A machine has a large mechanical advantage if the ratio of resistance force to effort force is much greater than 1; but the efficiency of a machine depends also on the distances over which these forces move, and it is always less than 1 (<100%).
- Help students understand that mechanical advantage is of significance for many machines, such as first- and second-class levers. However,

You might think that the greater the mechanical advantage, the more desirable the machine. But this is not always the case. A fishing rod has a mechanical advantage of less than 1. Its real advantage is the speed of motion at its end. All third-class levers have a mechanical advantage of less than one.

It is often more helpful to compare the input and output of a machine in terms of work rather than force. The work input for a lever is the effort force multiplied by the distance through which it moves ($W = Fd$). The work output is the product of the resistance force and the distance through which the load moves.

efficiency

The **efficiency** of a machine is the ratio of work output to work input. In the ideal case, the work output and input are equal, and the machine's efficiency is 100%. But because there is always some friction to overcome, the efficiency of any machine is less than 100%.

In general, complex machines are far less efficient than simple machines. They are made up of one or more simple machines, each of which contributes some friction. Even with the use of lubricants to reduce friction, complex machines are less efficient than most simple machines. But the efficiency of a machine may be less important to people than other advantages it may provide, such as convenience or safety.

Hand-operated can openers come in two types, one simple and the other more complex. One type is a second-class lever with a sharp edge on the resistance arm. (See Figure 17-17a.) The second type contains gears, a wheel and axle, a lever, and a sharp wheel. (See Figure 17-17b.)

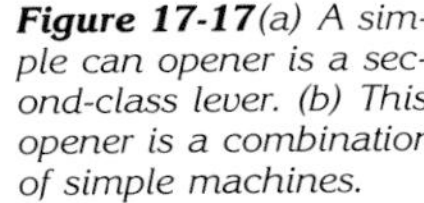

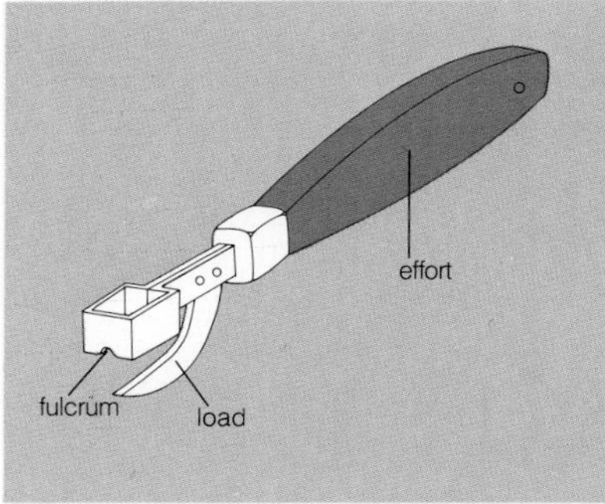

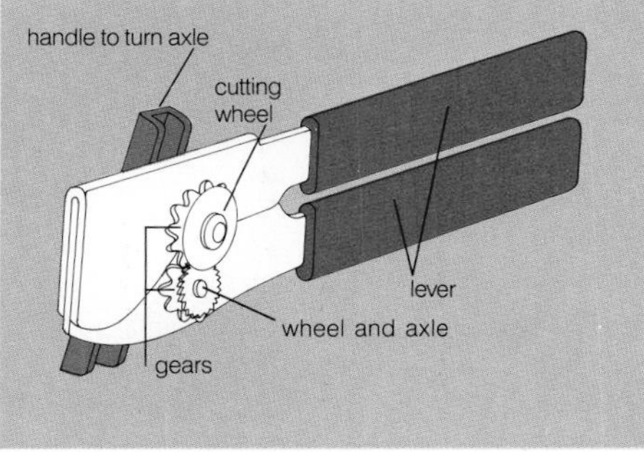

Figure 17-17*(a) A simple can opener is a second-class lever. (b) This opener is a combination of simple machines.*

Facts At Your Fingertips

- Not all machines are designed with a mechanical advantage greater than 1. Machines with a mechanical advantage less than 1 would increase the distance over which they move a load.
- The efficiencies of some typical complex machines are listed below in Table TE 17-1.

Table TE 17-1: Efficiencies of Complex Machines

Machine	*Efficiency*
diesel engine	35%
steam turbine	30%
automobile engine	20%
human body	18%
steam locomotive	10%
airplane	10%

for other machines, such as third-class levers, the mechanical advantage may be less than 1. The usefulness of such machines derives from their ability to increase the speed of a movement.

- Have students measure the mechanical advantage of a third-class lever, such as their own forearm/biceps muscle system, or a crane hoist.
- By examining actual machines and calculating their mechanical advantages and efficiencies, students will gain familiarity with these difficult but important concepts.

17 WORK AND ENERGY

17-4 Efficiency of Machines

Complete It!

Statements 1–5 below are incomplete. To complete the statements, choose from among the possible endings in a–g. Then write the appropriate ending on the line that follows each incomplete statement.

a. force is increased.
b. distance is increased.
c. speed is increased.
d. efficiency decreases.
e. friction decreases.
f. force and distance are not increased.
g. force is decreased.

1. When the effort arm of a lever is longer than the resistance arm, ____________
2. When simple machines are combined to form complex machines, ____________
3. When the resistance arm is long and the effort arm is short, ____________
4. When the fulcrum is far from the effort and close to the load, ____________
5. When the fulcrum is the same distance from the load and the effort, ____________

First, Second, or Third Class?

First-, second-, and third-class levers are represented in the diagrams below. Study each diagram and decide whether it is a first-, second-, or third-class lever. Then think whether the lever is designed to increase force or decrease the time needed to do the work. On the line under each diagram write the number of the class of lever and either the word *force* (increase force) or the word *time* (decrease time needed), depending on the design of the lever illustrated in the diagram.

Using Try This

Each student will need a pulley, string, spring scale, meterstick, and weights. To calculate the work output, multiply w by h. To calculate the work input, multiply F by d. Efficiency is then calculated by dividing work output by work input:

$$\frac{w \times h}{F \times d}.$$

Checkpoint Answers

1. The mechanical advantage of a machine is the ratio of resistance force to effort force. Ideal mechanical advantage is mechanical advantage calculated by ignoring friction.
2. Examples will vary: (a) crowbar, (b) wheelbarrow, and (c) fishing rod.
3. The efficiency of a machine is the ratio of work output to work input. Efficiency is always less than 100%, because there is always some friction to overcome.

The first type of opener is crude, but quite efficient. Although there is some friction between the resistance arm and the can being opened, most of the effort goes into opening the can. But this type of opener is difficult and, at times, dangerous to use. The second type of opener is more convenient and safer to use. But when this machine is used, some of the effort goes into overcoming friction in the gears and at the axle. Less of the effort goes into the useful work of opening the can.

Checkpoint

1. What is mechanical advantage? ideal mechanical advantage?
2. Give an example of a machine that illustrates (a) a first-class lever, (b) a second-class lever, (c) a third-class lever.
3. What is the efficiency of a machine? Why do machines have efficiencies of less than 100%?
4. Why are complex machines less efficient than simple ones? Why do people use complex machines?

TRY THIS

Build a simple pulley like the one shown in Figure 17-18. First find the weight of the object being lifted. Then measure the force (F) that you apply to the string to lift the object. Also measure the height (h) to which the object is lifted and the distance (d) through which you apply force F. Find the efficiency of this simple machine.

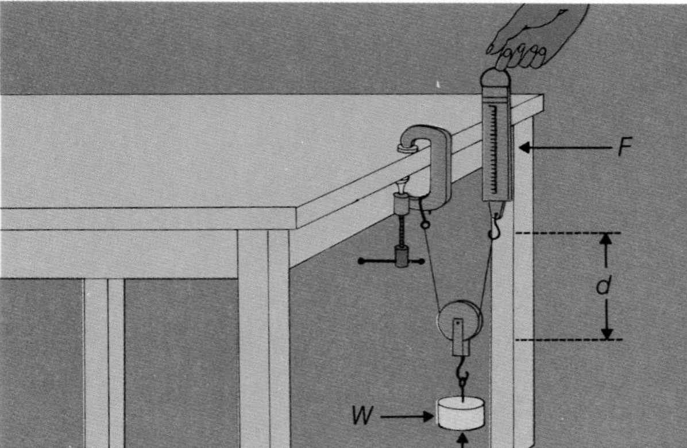

Figure 17-18 *This is the setup for finding the efficiency of a simple pulley.*

4. The simple machines that compose a complex machine all contribute some friction. Complex machines are used for reasons of convenience or safety.

17-5

Work and Energy

Goals

1. To define energy and distinguish between potential and kinetic energy.
2. To give examples of objects that possess kinetic or potential energy.

Energy can be defined as the ability, or capacity, to do work. Anything that can do work has energy. Wind can turn the blades of a windmill. Falling water can turn a water wheel or the turbines in a hydroelectric power plant. Windmills, water wheels, and turbines are all wheel and axle machines. (See Figure 17-19.)

energy

You can give energy to an object by "doing work on it." For example, you can push a billiard ball and cause it to move. A moving ball has energy. It can cause another ball, which is not moving, to move when the moving ball hits it. The energy that a moving object has due to its motion is called **kinetic energy**.

kinetic energy

If you push a billiard ball harder, it will move faster. A faster-moving ball has more energy than one that moves more slowly. Any ball that it hits will go farther, as a result. If a marble and a billiard ball were traveling at the same speed, however, any ball hit by the billiard ball would move

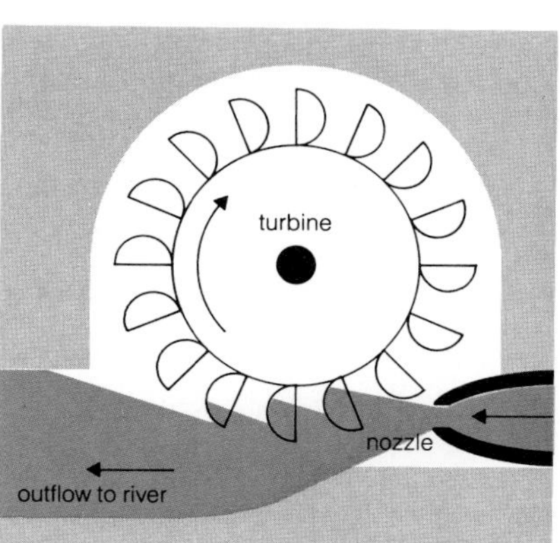

Figure 17-19 *The drawing shows a turbine in the base of a dam, as if you were looking down on it. The photograph shows turbines in a hydroelectric power station.*

After establishing that work can be converted into kinetic energy, you may do Demonstration 1, in which carts of various weights collide with a brick. (See Demonstrations.) Students should see that the kinetic energy possessed by the cart enables it to do work. In this case, it does work on the brick.

Background

There is no simple definition of energy. Energy has been defined as that which is involved in getting jobs done—jobs that require the use of fuel. Energy exists in many forms and is frequently converted from one form to another. A simple pendulum, for example, converts the mechanical work done in raising the bob into the potential energy stored in the raised bob. When the bob is released, this stored, or potential, energy is converted into kinetic energy as the bob swings downward. After it reaches the lowest point in its arc, the bob's kinetic energy is converted to potential energy again as it swings upward. Of course, some of the energy is converted to thermal energy, or heat, as the bob swings. Eventually it stops swinging, because all the kinetic and potential energy has been lost. In its place is the random, kinetic energy of the air molecules with which the bob collided.

Teaching Suggestions 17-5

Load a lab cart with bricks and place it on a ramp. (To avoid spreading brick dust, wrap the bricks in plastic wrap.) Hold the cart firmly to prevent it from moving. The class should see that the cart has potential energy: it can roll down the ramp because of gravitational force. Release the cart, and as it begins rolling, explain that its potential energy is being converted into kinetic energy.

Now place the cart on a flat table in clear view of the students. Ask them what must be done to put the cart in motion. They should see that a force, such as you supply with your hand, must act on the cart through a distance. In other words, some work must be done on the cart to make it move. The work done on the cart is converted into kinetic energy as the cart moves.

Demonstrations

1. Use two lab carts, one loaded with wrapped bricks or other objects, and one empty. The heavier cart can represent a truck, and the lighter one, a car. Roll both carts down the same ramp so they move at the same speed, and allow each to collide with different, but identical, bricks. It should be clear that the truck, which has more kinetic energy because of its greater mass, can do more work.
2. Let a long pendulum swing. Have students describe the energy transfers involved. Be sure students see that the pendulum bob obtains its initial potential energy from your raising it.

17 WORK AND ENERGY
Laboratory

The Pendulum

Purpose To find out how the length of the pendulum and the mass on the pendulum affect the number of swings.

Materials piece of string, about 1.5 m in length; several masses of about 10 g, 20 g, and 30 g each; meterstick; clock with second hand or stop watch; tape

Procedure

1. Attach a 10-g mass to the end of a 10-cm string.
2. Tape one end of the string to a table top so that the string and the 10-g mass hang over the edge.
3. Pull the mass to the left, all the way up to the table edge. Then let it swing. Count the number of complete swings in 5 seconds. Do this three times and record each observation in a chart like the one on this page. Estimate fractions of a swing. From your three observations, find the average number of swings in 5 s. Record.
4. Lengthen the string to 40 cm and repeat step 3.
5. Lengthen the string to 100 cm and repeat step 3.
6. Using the three lengths of string, change the mass to 20 g and repeat the experiment.
7. Change to the 30-g mass and repeat step 6.

Discussion

1. When the string was lengthened and the mass stayed the same, what happened to the number of swings per time period?
____
____
2. What happened to the number of swings per time period when the mass was increased, for each length of string?
____
____
3. Complete the following statements.
   a. If the length of a pendulum increases,
   ____
   ____
   ____
   b. If the mass on a specific length of a pendulum increases, ___________
   ____
   ____

Observation Number	*Mass*	*String Length*	*Number Swings Per 5 Seconds*	*Average Number Swings Per 5 s*
1				
2				
3				

faster than one hit by the marble. A great variety of experiments has shown that the energy of a moving object depends on both its mass and its speed. Furthermore, its speed, or velocity, has the greater effect on a moving object's energy.

$$\text{kinetic energy} = \tfrac{1}{2} \times \text{mass} \times \text{velocity} \times \text{velocity}$$
$$\text{k.e.} = \tfrac{1}{2}mv^2$$

A truck traveling at the same velocity as a car has more kinetic energy than the car because it has much more mass. But a fast-moving car may have as much kinetic energy as a slow-moving truck. When the velocity of a car doubles, its kinetic energy becomes four times as great. A car going 100 km/h can do four times as much "work," or damage, as the same car going 50 km/h.

Falling objects, like falling water, can do work. The amount of work that they can do depends on their height above the ground. Suppose an acrobat climbs to a high platform. The work done to reach the platform is the product of the platform's height and the acrobat's weight. In jumping from the platform, the acrobat converts this work into kinetic energy. The kinetic energy can be converted into work if the acrobat lands on one end of a seesaw. (See Figure 17-20.) The acrobat standing on a platform has the *potential* to do work. The acrobat "stored" energy by climbing to the platform. Stored energy, which an object may possess due to its position, is called **potential energy**.

potential energy

There are a number of ways of storing potential energy. It can be stored in a diving board, for example, or in the string of a bow. (See Figure 17-21.) Until recently, all watches

Figure 17-20 *The acrobat stores potential energy by climbing, converts it to kinetic energy in jumping, and converts that into work on the seesaw.*

Teaching Tips

- Discuss the various forms of energy described in the section. Ask students to name other forms of energy. (heat, nuclear energy, solar, chemical, etc.) Ask how these are similar to the energy of motion. (They can all be thought of as forms of potential energy until harnessed and used, thus becoming kinetic energy.)
- It is important that students understand that the more work done on a body, the more energy it acquires. Thus, the higher an object is lifted, the more potential energy it gains. The harder or farther an object is pushed, the more kinetic energy it acquires. This can easily be demonstrated. (See Demonstrations.)
- Emphasize that in any transfer of energy, some of the energy will be lost as thermal energy, or heat, due to friction.

Figure 17-21 *Energy stored in the bowstring can be used to shoot the arrow.*

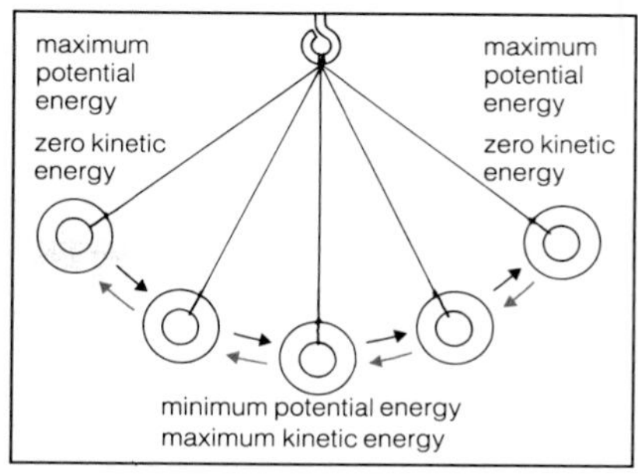

Figure 17-22 *A pendulum bob stores energy as it swings up.*

were run by springs. A wound watch spring contains stored energy. As the watch spring unwinds, a portion of its energy is transformed into the kinetic energy of the moving hands.

A pendulum (PEN ju lum) stores energy in its bob every time it swings to the top of its path. (See Figure 17-22.) When it swings back it converts potential energy to kinetic energy. Then as it moves again in its upward path it gains potential energy. In an ideal pendulum (no friction), the kinetic energy the bob acquires on its downward swing equals the potential energy it loses. The potential energy the bob has at its highest position equals the work done moving the bob up to that point. In most pendulums, frictional forces are very small, so the pendulum can keep swinging for a long time.

TRY THIS

To build a simple pendulum, screw a cup hook into an old golf ball. Tie one end of a thread to the hook. Tie the other end to a paper clip and attach the paper clip to the top of a door frame. Does the ball return to its original position after making a swing?

Checkpoint

1. What two factors determine the kinetic energy of an object?
2. Give two examples of things that have (a) kinetic energy and (b) potential energy.
3. A force of 10 N is exerted in lifting a certain object to a height of 2 m above the ground. How much work was done on the object? (Remember, $W = F \times d$.) How much potential energy is stored in the object at that height?

Facts At Your Fingertips

- There are many other kinds of potential energy besides the potential energy stored in a gravitational field. Energy can be stored in springs, in the separation of molecules or atoms, in the separation of electrical charge in capacitors, and in the chemical bonds of fuels.
- James Joule, for whom the joule, or newton-meter, is named, was instrumental in establishing the law of conservation of energy. This law states that energy can be neither created nor destroyed, merely converted from one form to another.

17 WORK AND ENERGY
17-5 Work and Energy

Potential or Kinetic?

In each example below, decide whether potential energy or kinetic energy is being described. On the numbered space write the term *potential energy* if potential energy is being described; write the term *kinetic energy* if kinetic energy is being described.

_______ 1. Water in a town water tower.
_______ 2. Lava flowing down a mountainside.
_______ 3. A car that is braking to a stop.
_______ 4. A swinging baseball bat.
_______ 5. A skydiver standing in the door of an airplane.
_______ 6. Water behind a dam.
_______ 7. Pounding nails with a hammer.
_______ 8. Pushing against a wall.

Energy from Start to Finish

Think about each event described below in terms of the energy involved. Then answer the questions that follow the description of the event. Hint: Keep in mind that the event begins when motion starts and ends when motion stops.

1. A baseball is thrown straight up in the air. After the ball is thrown, where is the point of maximum potential energy? _______ Maximum kinetic energy? _______
2. An archer draws her arrow back in her bow. The arrow is shot at the target and hits the bull's eye. Where did the energy to propel the arrow come from? _______ When did the arrow have kinetic energy? _______
3. A person is chopping wood. In each stroke of chopping, where is the point of maximum potential energy? _______ Where is the point of maximum kinetic energy? _______

Using Try This

The ball will return to very nearly the point from which it was released. Some of the potential energy was used to overcome air resistance, and some to overcome the internal friction in the string. Alternatively, some tape may be used to attach the paper clip to the top of the door frame.

Checkpoint Answers

1. The mass and the square of the velocity of a moving body determine its kinetic energy.
2. Answers will vary. Some examples are: (a) any moving body, and (b) a wound watch spring, food, and fossil fuel.
3. 20 J of work was done. The object has 20 J of potential energy.

Using The Feature Article

The article serves as a good review of various energy forms studied in this chapter. It also introduces the idea, discussed in greater detail in the next chapter, that all forms of energy can be transformed into thermal energy.

Have students analyze the transformation of energy in other sports. Have one student write a list of sports on the chalkboard. Then have each student choose one to analyze. The energy analyses may be given extemporaneously and left open to class comments. Then ask the students to suggest at least five ways that an athlete could improve performance in any of the sports discussed. Ask also if any physical traits might give one talented, well-trained athlete an advantage over another. (For example, height would be an advantage in the javelin throw, but not in marathon running. Muscle bulk would be helpful in weight lifting, but not necessarily in swimming or bicycle racing.)

Pole Vaulting

Pole vaulting involves a number of energy transformations. A pole vaulter begins by sprinting down a track to develop kinetic energy. When he places the pole firmly against the ground, he heads skyward. Some of the pole vaulter's kinetic energy is stored in his body as potential energy due to his position above the ground. Some is stored in the bent, springy pole, as shown in Figure 1. As the pole straightens, potential energy is converted into kinetic energy that helps to propel the vaulter over the cross bar. At this point, (see Figure 2), the vaulter pushes against the pole to get the maximum possible leverage.

When the pole vaulter reaches the highest point in his path, his potential energy is at its maximum. At this point, the force of gravity causes the vaulter to begin to accelerate downward. As he falls, his kinetic energy increases and his potential energy decreases. When the vaulter hits the cushioned pads on the ground, all of the kinetic energy that he gained in his fall is lost. Most of it is transformed into heat.

Figure 1

Figure 2

17-6

Perpetual Motion Machines

ENRICHMENT

Goal

To describe one type of perpetual motion machine and explain why such machines cannot work.

Many people think a perpetual motion machine is something that will keep moving forever. But that is not the kind of perpetual motion machine that countless people have spent years trying to build. The machine these dreamers dream about is one that could provide a work output greater than the work input.

One machine that has been proposed is shown in Figure 17-23. The machine is a series of levers on a wheel. All of the levers are the same length and have the same "hammerhead" mass on one end. The red levers on the right apply *effort forces* that are larger than the *resistance forces* exerted by the blue levers on the left. Because of the greater force of the right-hand levers, the wheel, once started, would continue to turn, according to its designer. It would even gain speed as it rotated.

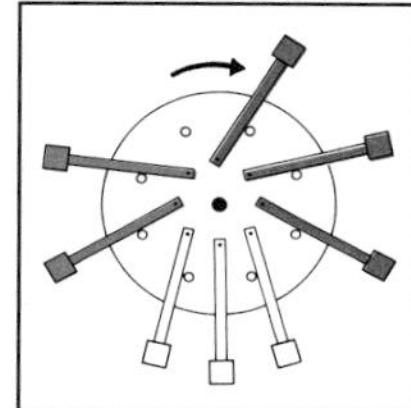

Figure 17-23 *This perpetual motion machine is based on a design from the 13th century.*

Before you rush off to build this machine and beat the original designer to the patent office, take another look. All work done by the falling hammers on the right is balanced by the work required to lift the hammers on the left. Inventors have devised a number of similar machines, none of which has ever produced "perpetual motion."

No machine has ever been designed that will put out more work than that put in. Indeed, a moving object cannot even retain the kinetic energy that it has. A rolling marble will eventually stop because of friction. Even satellites close to Earth eventually fall back to Earth. (The moon is far enough away to be in nearly empty space, where friction is extremely small.)

Checkpoint

1. What was the advantage of the perpetual motion machine described in this section *supposed* to be?
2. Why didn't the machine work as planned?
3. Why is any machine less than 100% efficient?

Teaching Suggestions 17-6

You may begin this lesson by setting in motion a pendulum and a bicycle wheel equipped with ball bearings. Ask students to guess how long these moving bodies will retain their kinetic energy.

Encourage students to design their own perpetual motion machines. Then help them to find the flaws in their designs.

Background

Despite the belief of some people, perpetual motion machines are an impossibility. They violate the law of conservation of energy, because they propose to create energy—a phenomenon that has been shown by experiments to be impossible.

The concept of the conservation of energy is embodied in the first law of thermodynamics (sometimes referred to as the "you can't win" law). The second law of thermodynamics is often called the "you can't even break even" law, because it says that in the transfer of energy, some useful energy is always lost as thermal energy.

Teaching Tips

- Challenge your students to measure the deceleration of a swinging pendulum or spinning bicycle wheel and calculate when motion is likely to cease.
- Exhibit a radiometer. Ask if it is a perpetual motion machine. (It isn't, because it derives its kinetic energy from the thermal energy in sunlight.)

17 WORK AND ENERGY

17-6 Perpetual Motion Machines

Passing Energy Along

The perpetual motion machine and some reasons it never worked are described in the following paragraphs. Read the paragraphs. Then write answers to the questions.

The ancients observed that if energy is used, there must be a source of energy. For example, a windmill only generates electricity if the wind supplies the force needed to move the blades of the wind generator. The wind gets its energy from the sun, due to unequal heating of the atmosphere.

It may seem strange, but many people have tried to develop a machine that once put into action would never stop. They believed that the machine would supply its own energy and continue operation endlessly. This so-called perpetual motion machine has been around, in one form or another, for hundreds of years.

Some of the machines that people invented were very clever, but perpetual motion never worked. There are two reasons for this: 1) energy cannot be created, and 2) energy cannot be destroyed. When energy seems to be destroyed, it is actually changed to another form. In 1775, therefore, the French Academy of Science decided to give no further consideration to work on perpetual motion machines.

Pivoting-Ball Machine

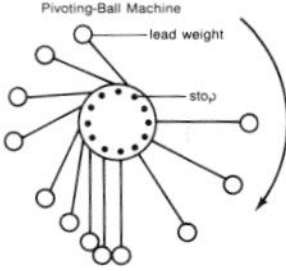

1. What would be the efficiency of a perpetual motion machine? Why do you think such an efficiency is impossible? ______
2. Explain why satellites close to Earth eventually fall back to Earth rather than continue endless operation. ______
3. Does the U.S. Patent Office accept applications for patents on perpetual motion machines? Explain why you think it should or should not. ______

Checkpoint Answers

1. Its kinetic energy was supposed to increase as it turned.
2. The work input must equal the work output and overcome frictional losses, as well.
3. In the operation of any machine, some of the work input must overcome friction.

Teaching Suggestions 17-7

Begin with a brief discussion of power. Be sure students understand that power is the ratio of work to time. Students may enjoy carrying out Activity 24, Measuring Power, student text page 563. Students will find that the heavier people are generally the more powerful. For example, a 1000-N person who can climb a flight of stairs as quickly as a 500-N student is twice as powerful as the lighter student.

Students will be able to relate the abstract concept of power to its concrete application in race cars. When mounted in lightweight racing cars, the powerful engines used in heavier passenger cars are able to achieve much greater speeds and acceleration. Ask students to explain why this is true. (By the second law of motion, $a = F/m$. For a constant value of F and a decreased value of m, a will have an increased value.)

Be sure students understand the difference between work and power. Just as speed is the rate of covering distance, so power is the rate of doing work.

Chapter-end project 1 applies to this section.

17 WORK AND ENERGY
Computer Program

Measuring Power

Purpose To calculate work in joules, and power in watts and horsepower.

Background This program performs the series of calculations for activity number 24. The program requires the entry of the height of the stairs, the weight of the student in newtons, and the time.

Background

A machine's capacity to do work rapidly is often the reason it is used. Such a machine can provide an amount of power that is not available from human labor alone. Machines can also perform repetitious tasks with greater consistency and accuracy.

Teaching Tips

- Lift a kilogram mass a height of one meter, and ask the class how much work you have done. (9.8 newton-meters, or 9.8 joules) Then ask what additional information they need to calculate your power. (elapsed time) Do the work slowly. Have a student time your effort. If you do the work in 10 seconds, the power can be calculated as follows:

$$P = 9.8\ \text{N} \times 1\ \text{m}/10\ \text{s} = 9.8\ \text{N·m}/10\ \text{s} = 0.98\ \text{J/s} = 0.98\ \text{W}$$

- Ask students to determine the horsepower equivalent of a 100-W bulb. (100/746 = 0.134 hp)

Facts At Your Fingertips

- The idea of measuring power in units of horsepower was developed by James Watt, who invented an efficient steam engine. Watt found that a horse could lift a 150-pound

17-7

Power

Goal

To define power and explain its importance.

It is often useful to know not only how much work is done, but how quickly it is done. For example, a farmer can move a bale of hay from the ground to a hayloft more quickly with a block and tackle than by carrying it up. **Power** is the amount of work done in a certain length of time.

power

$$\text{power} = \frac{\text{work}}{\text{time}}, \text{ or } P = \frac{W}{t}$$

EXAMPLE

A motor does 10,000 J of work in 10 s. What is its power?

$$P = \frac{W}{t}$$

$$P = \frac{10{,}000\ \text{J}}{10\ \text{s}} = 1000\ \text{J/s}$$

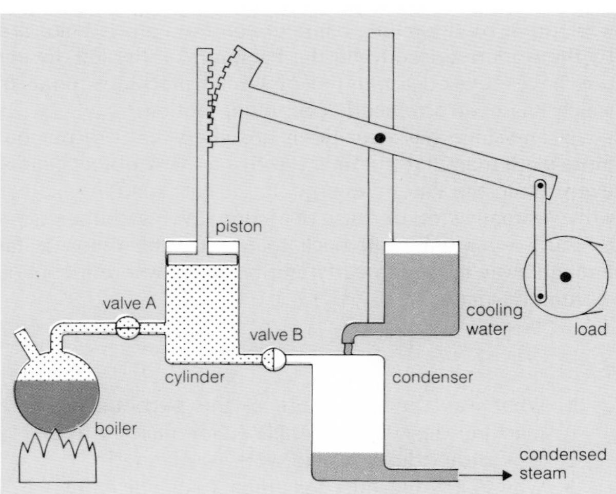

Figure 17-24 A simplified diagram of James Watt's steam engine.

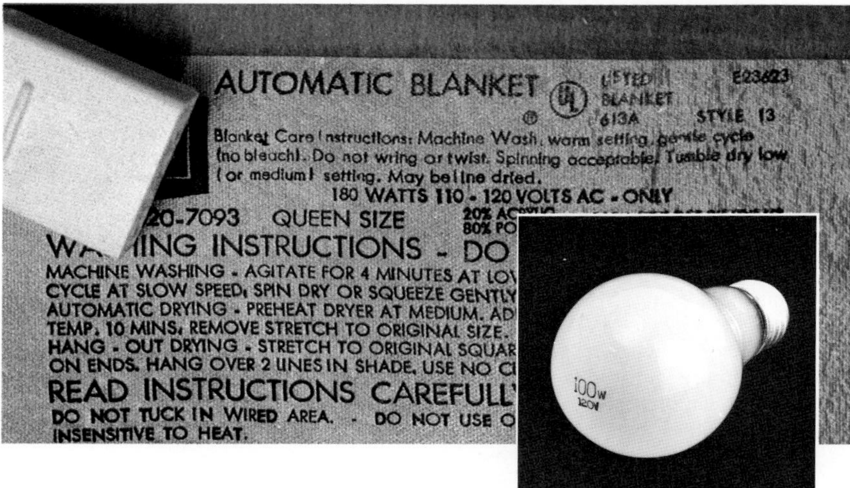

Figure 17-25 Light bulbs and appliances, such as this electric blanket, are marked to show their power ratings.

The joule per second has a special name. It is called a **watt** (W). The watt is named after James Watt, an 18th-century Scottish engineer who invented a practical steam engine. (See Figure 17-24.) You have probably noted that light bulbs are rated in watts. A 60-W light bulb converts 60 J of electric energy to light and heat every second. The power of most appliances is printed somewhere on them. (See Figure 17-25.) The power produced in large electrical machines is often expressed in *kilowatts* (kW). One kilowatt equals 1000 watts.

watt

The power of trucks, cars, airplanes, and motors is often given in *horsepower* (hp). This unit was defined in earlier days when horses, rather than machines, did much of the work. One horsepower is equal to 746 W.

Checkpoint

1. What is power?
2. What important information does *power* give that *work* does not give?
3. You do the same amount of work against gravitational force whether you walk or run up a flight of stairs. In which case do you develop more power?

TRY THIS

Examine a number of light bulbs and appliances around your home. Find the power required to operate each one. You can usually find the power rating printed somewhere on the machine or appliance.

weight about 3.7 feet in one second while working at a steady rate. The product of 150 pounds and 3.7 feet/second is about 550 foot-pounds per second. Watt, therefore, defined 1 horsepower to be 550 ft·lb/s, which is equivalent to about 746 watts.

- Table TE 17-2 shows the power developed by an average person doing various kinds of work for different periods of time.

Table TE 17-2: Power Generated by People

Type of work	*Time period*	*Power*
Run up stairs.	10 s	1.00 hp
Climb treadmill.	30 s	0.65 hp
Climb steep mountain.	1 h	0.20 hp
Climb hill.	1 day	0.10 hp

- Machines can develop large amounts of power, as indicated in Table TE 17-3.

Table TE 17-3: Power of Machines

Machine	*Power (hp)*
modern private plane	65
compact car	70
touring motorcycle	105
Lindbergh's plane	223
fuel dragster	700
cruise ship	60,000
Boeing-747 jetplane	656,000
Space Shuttle (at launch)	14,000,000

Cross Reference

Electric power is discussed in Chapter 19.

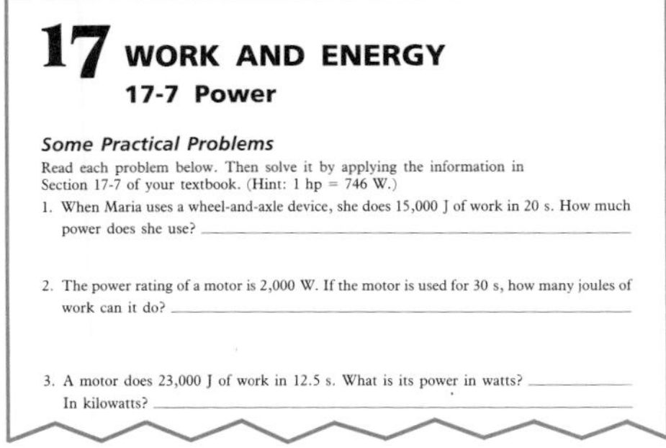

17 WORK AND ENERGY

17-7 Power

Some Practical Problems

Read each problem below. Then solve it by applying the information in Section 17-7 of your textbook. (Hint: 1 hp = 746 W.)

1. When Maria uses a wheel-and-axle device, she does 15,000 J of work in 20 s. How much power does she use? ______
2. The power rating of a motor is 2,000 W. If the motor is used for 30 s, how many joules of work can it do? ______
3. A motor does 23,000 J of work in 12.5 s. What is its power in watts? ______ In kilowatts? ______

Using Try This

After students have completed this activity, ask them to report the most and least powerful device or bulb. Ask them to suggest ways to decrease their use of electricity.

Checkpoint Answers

1. Power is the amount of work done in a period of time.
2. It gives the rate at which work is done.
3. You develop more power if you run up the stairs, because you do the same work in a shorter time.

Reading Suggestions

For The Teacher

Graf, Rudolf F., George J. Whalen, and the editors of *Popular Science*. HOW IT WORKS, ILLUSTRATED: EVERYDAY DEVICES AND MECHANISMS. New York: Van Nostrand Reinhold, 1979. Detailed, easily understood explanations, with cutaway drawings of 86 common appliances.

Patton, William J. KINEMATICS. Reston, VA: Reston, 1979. An interesting treatment of how gears, cams, ball bearings, and linkage mechanisms work. Requires knowledge of geometry and trigonometry.

Van Cleave, Janice Pratt. TEACHING THE FUN OF PHYSICS. Englewood Cliffs, NJ: Prentice Hall, 1985. A good reference for teachers that provides a wealth of experiments and demonstrations to help make physics interesting.

Van Hise, Yvette A. "They Want Me to Teach Physics!" *The Science Teacher,* May 1986, pp. 52–53. Where the novice physics teacher can find help.

For The Student

Eldridge, Frank R. WIND MACHINES. Second edition. New York: Van Nostrand Reinhold, 1982. Surveys the history, as well as present uses, of wind machines. Emphasizes the potential of wind machines as an energy source, while admitting the economic, environmental, and social constraints on large-scale applications.

McDonald, Lucile. WINDMILLS: AN OLD-NEW ENERGY SOURCE. New York: Lodestar, 1981. A history of the development of wind-machines, their replacement by other sources of energy, and the current revival of interest in wind power.

Stwertka, Albert and Eve Stwertka. PHYSICS FROM NEWTON TO THE BIG BANG. New York: Watts, 1986. A clear, factual, engaging account of the development of physics as a science.

Understanding The Chapter

17

Main Ideas

1. Work is the force acting on an object multiplied by the distance the object moves.
2. A machine changes the direction or strength of a force.
3. Complex machines are made up of two or more simple machines.
4. The six simple machines are the pulley, wheel and axle, lever, inclined plane, wedge, and screw.
5. There are three classes of levers. The class of a lever depends on the positions of the effort, resistance, and fulcrum.
6. The mechanical advantage of a machine is the ratio of the resistance force to the effort force.
7. The efficiency of a machine is the ratio of its work output to its work input.
8. Despite their inefficiency, complex machines save time and effort.
9. Anything that has the capacity to do work has energy.
10. An object's stored energy is called its potential energy.
11. An object's energy due to its motion is called kinetic energy.
12. Potential energy can be transformed to kinetic energy and vice versa.
13. Machines cannot produce output work greater than input work.
14. Power is the rate at which work is done.

Vocabulary Review

From the following list, choose the term that best completes each of the statements. Write your answers on a separate piece of paper.

complex machine	mechanical advantage
efficiency	potential energy
energy	power
fulcrum	simple machines
joule	watt
kinetic energy	work
machine	

1. _?_ is the net force acting on an object multiplied by the distance the object moves in the direction of the force.
2. _?_ is the work done divided by the time taken.
3. Anything that has the capicity to do work has _?_.
4. An object gains _?_ as it falls.
5. A(n) _?_ is a unit of power that is equal to 1 J/s.
6. _?_ are devices from which all other machines are made.
7. A lever turns about its _?_.
8. A machine's _?_ is the ratio of its work output to its work input.
9. A wound watch spring has _?_.
10. A(n) _?_ changes the direction or strength of a force.
11. The _?_ of a machine is the ratio of the resistance force to the effort force.
12. _?_ are made up of two or more simple machines.

Vocabulary Review Answers

1. work
2. power
3. energy
4. kinetic energy
5. watt
6. simple machines
7. fulcrum
8. efficiency
9. potential energy
10. machine
11. mechanical average
12. complex machine

Chapter Review

Write your answers on a separate piece of paper.

Know The Facts

1. The _?_ is a simple machine. (pulley, automobile, bicycle)
2. The effort force applied to a lever is used to overcome the _?_. (resistance, power, fulcrum)
3. The kilowatt is a unit used to measure _?_. (energy, friction, power)
4. Most machines have an efficiency _?_ 100%. (equal to, less than, more than)
5. A 10-N weight at rest 1.0 m above the floor has 10 J of _?_ with respect to the floor. (kinetic energy, potential energy, power)
6. An object's _?_ can be found using the mathematical expression $\frac{1}{2}mv^2$. (potential energy, power, kinetic energy)
7. Perpetual motion machines _?_. (create energy, have more work output than work input, are impossible)
8. If an input force of 100 N lifts a load of 400 N, the machine has a mechanical advantage of _?_. (1, 4, 8, 16)
9. One watt is _?_ (equal to, less than, greater than) one joule per second.
10. A machine's work input of 500 J produces a work output of 400 J. The machine's efficiency is _?_. (40%, 100%, 80%, 125%)
11. An example of a wheel and axle machine is a _?_. (block and tackle, windmill, chisel)
12. At the top of its swing, a pendulum has its maximum _?_. (kinetic energy, potential energy, power)
13. The distance from the fulcrum of a lever to the load that the lever is moving is the lever's _?_. (effort arm, resistance arm, efficiency)

Understand The Concepts

14. You push on a wall with a force of 100 N. The wall does not move. How much work did you do on the wall?
15. An effort force of 100 N moves through a distance of 0.5 m. Find the work input.
16. Describe some objects having (a) potential energy, (b) kinetic energy.
17. Why does a ball rolling on a horizontal surface come to rest?
18. Explain the energy transformations that occur when a rubber ball falls from your hand to the floor and bounces once.
19. Explain how a screw resembles an inclined plane.
20. What advantages might a complex machine have that would compensate for its low efficiency?
21. Explain how one machine could have both a lower efficiency and a higher power rating than another.
22. A power rating of 120 W is how many J/s? kW? hp?
23. Explain why the perpetual motion machine described in this chapter does not work.
24. Describe how you would adjust the fulcrum of a seesaw so that a small person can balance a larger one. Where would each person sit?

Chapter Review Answers

Know The Facts

1. pulley
2. resistance
3. power
4. less than
5. potential energy
6. kinetic energy
7. are impossible
8. 4
9. equal to
10. 50 J
11. windmill
12. potential energy
13. resistance arm

Understand The Concepts

14. None.
15. $\frac{400 \text{ J}}{500 \text{ J}} = \frac{4}{5} = 80\%.$
16. Answers will vary. Some examples are:
 (a) a ball on a roof, a wound watch spring, and a stretched spring.
 (b) a falling ball, a running watch (mechanical or electronic), and a running athlete.
17. Friction exerts a retarding force.
18. As it falls, the ball loses potential energy and gains kinetic energy; the kinetic energy is converted to potential energy as the ball hits the floor and compresses, storing most of the energy and losing some to internal friction and friction against the floor; the elastic nature of the ball converts potential energy back to kinetic energy, as the ball rebounds and bounces back upwards; at the top of the bounce, the kinetic energy is completely converted to potential energy.
19. A screw is really an inclined plane wound around a central cylinder, in a manner similar to the stairs on a spiral staircase.
20. It might do dangerous tasks, and perform with great speed, consistency, accuracy, and power.
21. Suppose a machine can do 800 J of work in one second while another can do only 80 J in the same time. The first machine's power rating is 800 watts; the second machine's is 80 watts. But if the first machine requires an input of 2000 J to do its 800 J of work, its efficiency is only 40%. If the second machine requires only 100 J of input to do 80 J of work, its efficiency is 80%.
22. 120 W = 120 J/s = 0.12 kW = 0.16 hp
23. The work input on one side of the machine equals the work output on the other side; hence, the friction in the machine will make the efficiency less than 1.0.
24. The fulcrum should be nearer the large person.

25. A fixed pulley changes the direction of the force, enabling you to pull down rather than up, or up rather than down. A movable pulley provides mechanical advantage.
26. A complex machine is made up of many simple machines, each of which involves friction. The frictional force for the complex machine comes from the constituent simple machines.
27. A 100-W bulb will require

$$100 \text{ J/s} \times 3600 \text{ s} = 360{,}000 \text{ J}.$$

A 60-W bulb will require
$60 \text{ J} \times 3600 \text{ s} = 216{,}000 \text{ J}.$

Challenge Your Understanding

28. 4 m.
29. The force needed to roll the barrel is less than the force needed to lift it because of the mechanical advantage provided by the incline. A force that is only a fraction of the barrel's weight is needed to move the barrel along the incline.
30. (a) Kinetic energy doubles; (b) kinetic energy is only ¼ as great; (c) kinetic energy is 8 times as great.
31. $500 \text{ J}/ 5 \text{ s} = 100 \text{ J/s} = 100$ watts.
32. IMA $= 150 \text{ N}/ 75 \text{ N} = 2$.
33. You do $10 \text{ N} \times 5 \text{ m} = 50 \text{ J}$ of work. It has 50 J of potential energy. It can acquire 50 J of kinetic energy in falling 5 m.
34. As the skier ascends the slope, his/her potential energy increases. During the downhill run, the skier loses potential energy and gains kinetic energy. Some of the potential energy is converted to thermal energy as the skis slide over the snow; the rest is lost to wind resistance.

25. What is the advantage of using a fixed pulley to raise an object rather than lifting it yourself? What is the advantage of using a movable pulley rather than a fixed pulley?
26. Explain why complex machines are usually much less efficient than simple machines.
27. Which light bulb requires more energy to burn for one hour, a 60-W bulb or a 100-W bulb? Explain your answer.

Challenge Your Understanding

28. An effort force of 100 N is applied through a distance of 20 m. What is the maximum distance that the effort force can move a load of 500 N?
29. Explain why it is easier to roll a heavy barrel up an inclined plane than to lift it straight up.
30. Explain how the kinetic energy of a moving object changes if (a) the mass is doubled but the velocity remains the same, (b) the velocity is halved but the mass remains the same, (c) both the mass and the velocity are doubled.
31. In 5s, a machine does 500 J of work. Find the power developed by this machine.
32. The effort force on a lever is 75 N. If the effort force moves a resistance of 150 N, what is the ideal mechanical advantage of the lever?
33. You lift a 10-N weight to a height of 5 m. How much work do you do on the weight? How much potential energy does the weight have after you raise it? How much kinetic energy will the weight have just before it hits the floor?
34. Describe the changes in potential and kinetic energy experienced by a skier. Start with the skier about to go up a slope. End when the skier has arrived at the bottom of the slope again, after skiing down.

Projects

1. Use the library to investigate the origin of "horsepower" as a unit of power.
2. Build a lever, a pulley, and an inclined plane. Test their mechanical advantages and efficiencies.
3. Find as many simple machines as you can in your home and school.
4. Go to the library and find out what kinds of machines were used to build the pyramids in Egypt.
5. Ask your school's driving instructor what lubrication is recommended for automobiles. Why do you think the engine oil should be changed periodically?
6. Find the mechanical advantage for a seesaw in your local playground.
7. Take apart a complex machine, such as an old clock. How many simple machines can you find?

17 WORK AND ENERGY
Projects

1. Turbine engines power jet airplanes. This is a complex machine that is based on a simple machine. The design is complex, but the principle is not.
   a. What is a turbojet? ______________________
   b. Make a diagram of a turbojet engine.
2. Each day people use many machines that rely on one type of motion to get another type of motion.

17 WORK AND ENERGY
Challenge

Problems to Ponder

The problems below relate to the ideas presented in this chapter. Read each problem. Then apply what you learned in the chapter to solve it.

1. Find the work done in lifting a 10-kg mass to a height of 10 m. Hint: A 1-kg mass weighs 9.8 N. ______________________
2. If 1,000 N · m of work have been done in raising an object 50 m, how much does the object weigh? ______________________
3. If the work input of a pulley is 240 N · m and the work output is 180 N · m, what is the efficiency of the pulley? ______________________

Chapter 18

Heat

This solar furnace in France produces a great deal of heat.

Vocabulary Preview

heat
temperature
absolute zero
pressure
specific heat
conduction
conductor
insulator
convection
radiation
law of conservation of energy
heat engine
solar energy
renewable energy source
hydropower
geothermal energy

CHAPTER 18 Overview

This chapter first develops the concepts of heat and temperature and then explains how temperature is measured. The effect of temperature on the pressure and volume of a confined gas is described. The concept of specific heat is introduced and the specific heats of some common substances are compared. The moderating influence on climate of large bodies of water is shown to be a consequence of water's high specific heat. The mechanisms by which heat is transferred are then described. Energy transformations are discussed and the nature of heat engines is explained. The law of conservation of energy is stated and explained. Practical applications of solar energy and other renewable energy sources are examined. A practical overview of space heating and cooling systems and the role of insulation in building design completes the chapter.

Goals

At the end of this chapter, students should be able to:

1. define heat and temperature.
2. describe a common thermometer.
3. compare the Celsius and Kelvin temperature scales.
4. define pressure and explain the effect of temperature on gas pressure.
5. define specific heat.
6. describe how the specific heat of certain substances can affect local climates.
7. describe three ways in which heat is transferred.
8. describe some energy transformations.
9. state the law of conservation of energy.
10. discuss the nature of heat engines.
11. define solar energy and give examples of how it can be used to decrease the use of fossil fuels.
12. describe alternatives to fossil fuels other than solar energy.
13. describe various heating and cooling systems.

Teaching Suggestions 18-1

Show the students a thermometer that has no markings on it. Ask if such a thermometer is of any use. Students who have read the lesson are likely to suggest calibrating the thermometer by using ice water and boiling water to establish two points of known temperature. Thus, 0°C and 100°C can be indicated on the thermometer and equal intervals between them marked to establish a scale.

Distinguish carefully between the concepts of heat and temperature. Point out that heat is the total kinetic energy of all the constituent particles (molecules) of a body or substance; temperature is the measure of the average kinetic energy of its particles. The terms *warm, cool, hot* and *cold* are relative terms, indicating only a temperature differential and containing no information about heat content or temperature. For example, 10 g of water at 15°C contains twice as much heat as 5 g of water at the same temperature. To someone exposed to 0°C outdoor temperatures, the water will feel warm; to some one stepping out of a hot shower, it will feel cool.

Chapter-end project 1 applies to this section.

Background

Heat is the thermal energy transferred from one object or substance to another. According to the kinetic theory of heat, the thermal energy of a body is the sum of the kinetic energies of all its constituent molecules. Temperature is a measure of the average kinetic energy of its molecules. Whether a body gains or loses heat depends on its temperature. If a body is warmer than its surroundings, that is, at a higher temperature, it will give up thermal energy, or heat, to its surroundings. If a body is colder than its surroundings, then its surroundings will transfer heat to the body.

The Kelvin scale is a more logical scale than either the Fahrenheit or the Celsius scale. The minimum Kelvin temperature, or 0 K, is absolute zero. Note that the word *degree* is not used with Kelvin temperatures. For example, 25 K is read, 25 Kelvin.

Demonstrations

1. If you do not have enough noncalibrated thermometers for every student or student team, calibrate a thermometer as a demonstration. If you have only calibrated thermometers, cover the markings with masking tape.

   First place the unmarked thermometer in a mixture of crushed ice and water. Mark the mercury level after the liquid in the tube has stopped contracting. Then place the thermometer in boiling water and

18-1

Heat and Temperature

Goals

1. To define heat and temperature.
2. To describe a common thermometer.
3. To compare the Celsius and Kelvin temperature scales.

Have you ever rubbed your hands together briskly to warm them up? If so, you were using friction to produce heat. The work done to overcome the friction between the surfaces of your hands appeared as heat. In this case, you wanted to produce heat. Often, however, the heat produced by friction can be a nuisance. When a car traveling at high speed stops suddenly, the brakes must provide a tremendous amount of stopping force. The friction involved produces a great deal of heat. For this reason, the brakes on a car must include heat-resistant materials. Disc brakes transfer heat to the air more rapidly than do other types of brakes. (See Figure 18-1.)

heat

Heat is the energy transferred *to* an object when it warms or *from* an object when it cools. An object can gain heat when work is done on it or when heat is transferred from a hotter object. Rubbing your hands together warms them. Putting your hands in hot water warms them, also.

Figure 18-1 *The metal disc in a disc brake is attached to a wheel and rotates with it. Braking occurs as two friction pads are squeezed against the spinning disc.*

All matter is made of extremely small particles. These particles are in constant motion. When heat is transferred to an object, the motion of these particles increases. The faster a particle moves, the greater its kinetic energy. To measure the average kinetic energy of the particles in a substance is to measure how hot, or cold, the substance is. The **temperature** of a substance is a measure of the average kinetic energy of the particles in it.

temperature

If water in a small glass is at the same temperature as water in a large pitcher, the average kinetic energy per water particle, or molecule, is the same in each container. But there is more water in the pitcher. Thus there are more molecules in that water than in the glass of water. It follows that the total energy of the water in the pitcher is greater than that of the water in the glass. Temperature is not a measure of total energy. It is the measure of the average kinetic energy *per particle*.

Temperature is usually measured with a thermometer. Many thermometers contain a liquid. The liquid expands when it gains heat and contracts, or shrinks, when it loses heat. The liquid is usually mercury or alcohol. Mercury has a silver color. Alcohol is colorless, so a red dye is normally added to it, to make it show up. The liquid is sealed into a narrow glass tube. The tube is narrow so that a small change in the space taken up by the liquid in the bulb causes a noticeable change in the height of the liquid in the tube.

A common thermometer is shown in Figure 18-2. Its scale is marked off in *Celsius* (SEL see us) degrees (°C). The Celsius scale is named after Anders Celsius, a Swedish scientist who invented the scale in the early 1700's. On the Celsius scale the temperature at which pure water freezes is 0°C. The temperature at which water boils is 100°C. Pure water is readily obtainable and it is relatively easy to freeze or boil. Thus it is a good substance on which to base a thermometer scale.

To make a Celsius thermometer, a liquid-containing tube is put in a mixture of ice and water. The 0°C mark is made on the tube, at the top of the liquid column. Next, the tube is placed in boiling water. The liquid in the tube rises and a mark is made for 100°C. Then the distance between the 0°C mark and the 100°C mark is divided into equal parts. The Celsius degree is one one-hundredth of the distance between the 0°C and 100°C marks.

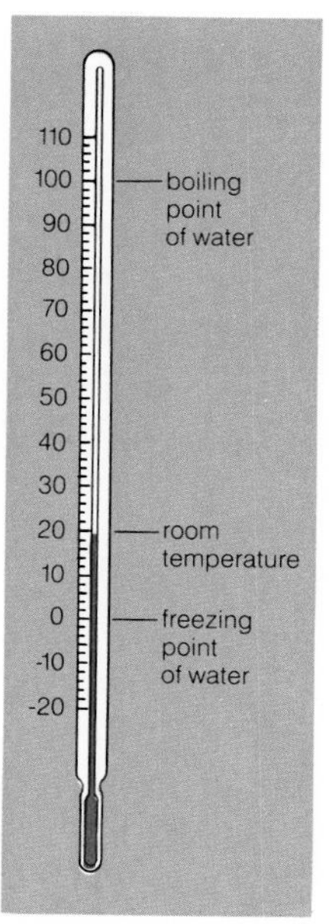

***Figure 18-2** A typical laboratory thermometer with a Celsius scale.*

mark the mercury level once expansion has ceased. Have students devise a method to divide the scale into 10-degree intervals.

2. Make a thermometer similar to Galileo's. See Figure TE 18-1. Invert the flask so that the attached glass tubing is inserted in a container filled with colored water. Warm the flask to drive out some air. When the flask cools, water will rise in the glass tubing. Then you can show, by heating or cooling the flask with your hands, a burner, or wet cloths, that a change in temperature will cause a change in the liquid level. The column will rise or fall as a result of the expansion or contraction of the air in the flask.

   Leave the apparatus set up for several days to show that this unsealed thermometer is affected by air pressure as well as by temperature. (Decreasing air pressure will make the liquid fall, and increasing air pressure will make it rise.)
3. If your school has a mechanical model designed to show the motion of molecules, use it to show the greater average velocity of molecules at high temperatures.

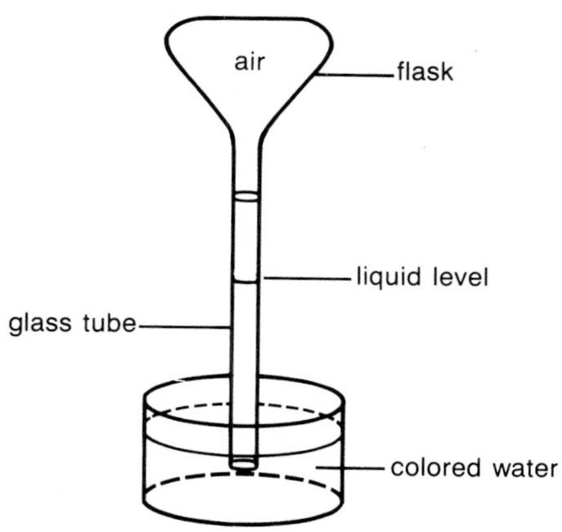

Figure TE 18-1 Galileo's thermometer.

Facts At Your Fingertips

- The first thermometer is believed to have been made by Galileo in 1592.
- Mechanical thermometers contain a metal strip that contracts or expands in response to changes in temperature. Liquid-crystal thermometers exhibit changes in polarity as a result of temperature changes. Thermistors are solid-state devices that change their resistance to the flow of electric current as the temperature changes.
- At one time, heat was thought to be a fluid that flowed from warmer to cooler bodies. While this is still a useful model for the flow of thermal energy, we now understand that thermal energy is the kinetic energy of the molecules of a substance.
- The very careful experimental work done by James Joule revealed the equivalence between heat and mechanical energy. He demonstrated that any form of energy—kinetic, potential, chemical, light, etc.—could be transformed into thermal energy.

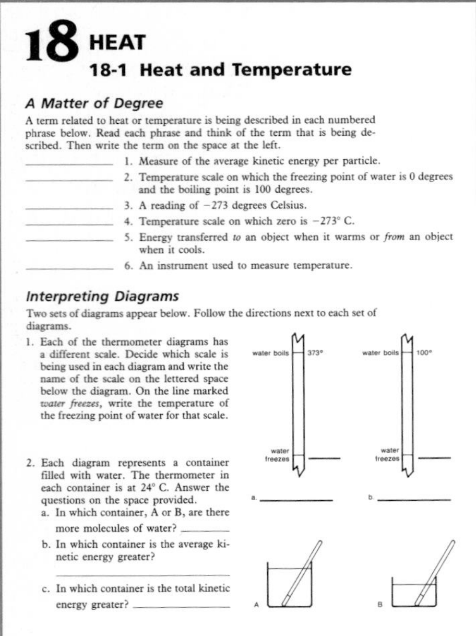

18 HEAT
18-1 Heat and Temperature

A Matter of Degree

A term related to heat or temperature is being described in each numbered phrase below. Read each phrase and think of the term that is being described. Then write the term on the space at the left.

__________ 1. Measure of the average kinetic energy per particle.
__________ 2. Temperature scale on which the freezing point of water is 0 degrees and the boiling point is 100 degrees.
__________ 3. A reading of −273 degrees Celsius.
__________ 4. Temperature scale on which zero is −273° C.
__________ 5. Energy transferred *to* an object when it warms or *from* an object when it cools.
__________ 6. An instrument used to measure temperature.

Interpreting Diagrams

Two sets of diagrams appear below. Follow the directions next to each set of diagrams.

1. Each of the thermometer diagrams has a different scale. Decide which scale is being used in each diagram and write the name of the scale on the lettered space below the diagram. On the line marked *water freezes*, write the temperature of the freezing point of water for that scale.

2. Each diagram represents a container filled with water. The thermometer in each container is at 24° C. Answer the questions on the space provided.
   a. In which container, A or B, are there more molecules of water? ______
   b. In which container is the average kinetic energy greater? ______
   c. In which container is the total kinetic energy greater? ______

Using Try This

1. Temperatures will be cooler near the floor and warmer near the ceiling, because warm air rises. It will probably be cooler on the north side of the house, which receives no direct sunlight.
2. The hand that was in cold water will feel warm; the hand that was in hot water will feel cool.

Checkpoint Answers

1. Heat is the energy transferred to an object when it warms or from an object when it cools.
2. Temperature is a measure of the average kinetic energy of the particles that make up a substance.
3. A common thermometer contains a liquid, such as mercury or alcohol, in a glass tube. The liquid in the thermometer expands when heated, rising up the thin bore in the tube; it contracts when cooled, shrinking down the bore.
4. Kelvin temperature is calculated by subtracting 273 from the temperature in degrees Celsius.
5. Absolute zero is the lowest temperature possible.

With the size of the Celsius degree established, the scale can be extended above 100°C and below 0°C, as shown in Figure 18-2. There is apparently no limit to how hot a substance can become. But there is a limit to how cold anything can be. At −273°C, the particles in a substance have their lowest possible energy. No experiments have produced a temperature as low as −273°C, although scientists have come within 0.001°C of this temperature. Since nothing can be colder than −273°C, this temperature is called **absolute zero**.

absolute zero

For scientists doing calculations, a temperature scale without negative numbers is useful. The Kelvin scale has absolute zero as its lowest point. The divisions of the Kelvin scale, called *kelvins* (K), are the same size as those of the Celsius scale. That is, one kelvin is equal to one Celsius degree. (See Figure 18-3.) The Kelvin scale is named after Lord Kelvin (William Thomson), a pioneer in the science of heat.

Kelvin Celsius
373 100° boiling point of water
273 0° freezing point of water
0 -273° absolute zero

Figure 18-3 *The Kelvin scale compared to the Celsius scale.*

Checkpoint

1. What is heat?
2. What is temperature?
3. Describe a common thermometer and explain how it works.
4. Compare the Celsius and Kelvin temperature scales.
5. What is absolute zero?

TRY THIS

1. Using a thermometer, look for warm and cold places in and around your home or school. How do the temperatures near a ceiling and floor compare? How about the north and south sides of a building on a sunny day?

2. Put one hand in a pan of ice water. Put the other hand in very warm water. Leave both hands in for about one minute. Then plunge both hands into a pan of lukewarm water. How warm does each hand feel?

18-2

Temperature and Pressure

Goal

To define pressure and explain the effect of temperature on gas pressure.

A person standing on snow sinks into it less when wearing skis (or snowshoes). (See Figure 18-4.) The force that the person exerts on the snow is the same with or without skis. But, with skis, the force is spread over a greater area, and the force on any given area is less. **Pressure** is the ratio of a force to the area over which it acts.

pressure

$$\text{pressure} = \frac{\text{force}}{\text{area}}, \text{ or } P = \frac{F}{A}$$

Most substances expand when they are heated. In a gas, such as air, there is more space between particles than in liquids or solids. So gas particles move more independently, and gases expand more noticeably than liquids or solids when heated.

Figure 18-4a *The weight of a person puts great pressure on the snow directly below each foot.*

Figure 18-4b *When a person wears snowshoes, the force is spread out. The pressure on any one spot is much less.*

Teaching Suggestions 18-2

A dramatic introduction to this section can be provided by demonstrating the ability of air pressure to bend metal. (See Demonstration 1.)

Challenge the class to explain why the can buckles. (Atmospheric pressure crushes the can when the water vapor condenses, greatly reducing the gas pressure inside the can.)

Background

Pressure is defined as force per unit area. The total force exerted by two objects may be the same, but if one of the forces is distributed over a greater area, the object will exert less pressure.

The pressure exerted by a gas depends on the speed at which the gas molecules are moving. The pressure of a gas in a closed container will increase if the temperature of the gas increases. The increased thermal energy is transferred to the molecules of the gas, endowing them with increased kinetic energy. They strike the walls of the container with greater force per unit area, and therefore greater pressure.

Demonstrations

1. You will need an empty, thoroughly clean, 4-liter metal can with a tight screw cap. (Cans that paint thinner is sold in work well.) Pour about a cup of water into the can and place the *open* can on a hot plate. (CAUTION: Be sure the can is uncovered during this step. Do *not* heat over an open flame.) Allow the water to boil vigorously. Using gloves or a hot pad, carefully remove the can from the hot plate. Quickly screw the cap tightly onto the can. Turn off the hot plate and put it safely out of the way. To speed up the cooling of the can, you may pour cold water on it.
2. Carry a barometer up a hill or several flights of stairs. Students will see that even over such small changes in altitude, there is a measurable change in air pressure.

Teaching Tip

- Analogies can be used to explain the phenomenon of pressure. For example, the more people there are in the hallways of a school, the more likely it is that they will collide with one another and with the walls and lockers. The more cars there are on the highway, the more likely it is that accidental collisions will occur. In these examples, the collisions are analogous to the collisions of particles in hot confined gases.

Facts At Your Fingertips

- The equation $pV = kT$ states the relationship among pressure (p), volume (V), and temperature (T) for all gases; k is a constant.
- When compressed air is allowed to escape rapidly from a pneumatic tire valve, the temperature of the remaining compressed air decreases. This expansion/cooling effect is the principle behind air conditioning, as well.

18 HEAT

Computer Program

Temperature and Pressure

Purpose To calculate temperature changes in gas as pressure is altered.

Background The temperature of a specific volume of gas is closely related to atmospheric pressure. The temperature change can be predicted according to the following equation: Pressure 1/Pressure 2 = Temperature 1/Temperature 2

ENTER the following program:

```
10  PRINT "ENTER THE STARTING PRESSURE"
20  INPUT "IN ATMOSPHERES."; P1
30  PRINT "ENTER THE FINAL PRESSURE"
40  INPUT "IN ATMOSPHERES."; P2
50  PRINT "ENTER THE STARTING TEMPERATURE"
60  INPUT "IN DEGREES CELSIUS."; T1
70  LET T1 = T1 + 273
80  LET T2 = (T1 * P2) / P1
90  PRINT "FINAL TEMPERATURE = "; T2; " KELVINS."
```

Cross Reference

See Appendix B, student text page 570, for information on measurement.

18 HEAT

18-2 Temperature and Pressure

Pressure Increase or Pressure Decrease?

In each example described below, decide whether pressure will be increased or decreased. Then complete each statement by writing the word *increased* if pressure will be increased or *decreased* if pressure will be decreased.

1. An air-tight can containing room temperature air is heated to 100° C. The pressure of the air in the can will be ______________
2. A firefighter is trying to rescue a person who has fallen through thin ice. The firefighter lies on a thin sheet of plywood, 1 m by 2 m, and slides toward the person in the hole. Compared to the firefighter lying directly on the ice, the plywood causes the pressure on the ice to be ______________
3. An air-tight can containing room temperature air is placed in a freezer overnight. When the can is observed the next day, it is found to be crushed inward. The pressure in the can had ______________
4. Automobile tires use air to pressurize them. It is recommended that tire pressures be checked when the tires are cold. After a car has been driven, the tires get warm. If you

Figure 18-5 *When air inside a balloon warms, the air particles move faster and with more force. The pressure on the inside surface increases and the balloon expands.*

A balloon may burst if the air inside heats up enough. The tiny particles in the air inside the balloon are constantly colliding with the balloon's inside surface. (See Figure 18-5.) As the air in the balloon gains heat, the particles move faster. They hit the inside surface of the balloon more often and with more force. As a result, the pressure on the inside surface of the balloon increases. There is a limit to how much the balloon itself can expand, however. Eventually, the balloon bursts.

Similarly, the pressure exerted by the air inside an automobile tire will rise when the air gets warmer. The air in the tire warms up as the car is driven. Thus its pressure increases. That is why tire air pressure should be measured "cold."

Checkpoint

1. What is pressure?
2. What happens to its volume when gas is heated?
3. How are gas pressure and temperature related?

TRY THIS

Find the air pressure in the tires of a car that has not been driven for an hour or more. Check the pressure again, immediately after the car has been driven for an hour or so. What is the change in the air pressure?

Using Try This

The pressure of the air in the tire will increase after the car has been driven, because friction will have heated the air within the tire.

Checkpoint Answers

1. Pressure is the force per unit of area.
2. The volume of an unconfined gas will increase if the gas is heated. If the gas is confined, the pressure exerted by the gas on its container will increase.
3. As the temperature of a confined gas increases, the pressure it exerts on its container will increase.

18-3

Specific Heat

ENRICHMENT

Goals

1. To define specific heat.
2. To describe how the specific heat of certain substances can affect local climates.

People who live near the ocean in temperate climates enjoy winter and summer temperatures that are relatively moderate. Temperatures along a coastline generally do not vary as much day to day or season to season as temperatures inland. The coast is warmer in the winter and cooler in the summer than inland locations at the same latitude. The differences between day and night temperatures are usually greater inland at any season. People who live near large lakes experience a similar effect. This is because a lot of heat is needed to warm a large body of water.

The amount of heat needed to increase the temperature of a substance depends on the *amount* of the substance. The more particles the substance contains, the greater the amount of heat needed to increase their average kinetic energy. It takes more heat to increase the temperature of a pot of water than to increase the temperature of a cupful by the same amount.

The amount of heat needed to increase the temperature of a substance depends on the *nature* of the substance, also. The type of particles the substance consists of and their arrangement are important factors. It takes more heat to raise the temperature of 1 g of water 1°C than to raise the temperature of 1 g of iron 1°C. The particles in water are different from the particles that make up iron. It takes more heat to raise the temperature of 1 g of liquid water 1°C than to raise the temperature of 1 g of ice 1°C. The arrangement of the particles is different in water and ice.

The amount of heat needed to raise the temperature of 1 g of any substance 1°C is called the **specific heat** of the substance. It takes 4.2 J of heat to raise the temperature of 1 g of water 1°C. Thus the specific heat of water is 4.2 J/g/°C. (Read this as 4.2 joules per gram per degree Celsius.) Table 18-1 shows the specific heats of various substances.

specific heat

Teaching Suggestions 18-3

If necessary, review the distinction between heat and temperature, as explained in section 18-1. Students should understand what it means for a substance to have a high specific heat. Changing the temperature of a given mass of the substance will require more heat than would be needed for a substance with a lower specific heat. For example, a given mass of cooking oil will be warmed up more quickly by a given amount of heat than will an equal mass of water, because the specific heat of cooking oil is lower than that of water.

Background

Thermal energy is the sum of the kinetic energies of all the molecules of a substance. Temperature is a measure of the average kinetic energy of the molecules of a substance. Specific heat may be thought of as a kind of "thermal inertia," which, like mass, is a characteristic property of a substance. Specific heat is a measure of how much heat must be added to a given mass of a substance to raise its temperature a given amount.

Adding a certain amount of heat to a fixed amount of water will raise the temperature a given amount. Doubling the amount of heat added will double the increase in temperature. Doubling the amount of water will halve the temperature increase.

The product of the mass of an amount of water and its temperature change will increase proportionately to an increase in added heat. This relationship has been used to define the *calorie* as the amount of heat required to increase the temperature of 1 gram of water 1 degree Celsius. (One joule of heat will raise the temperature of 1 gram of water 0.24 degree Celsius.)

Demonstrations

1. Using an immersion heater, heat 100 g of water in an insulated cup for 30 s. Show that with twice as much water and twice as much heat (keep heater plugged in for 1 min), the temperature change is the same; and that with twice the heat and the same amount of water, the temperature increase is doubled.
2. Establish the change in temperature that results from heating 100 g of water with an immersion heater for 30 s. Now use the same amount of heat to warm 100 g of ethylene glycol (antifreeze). The antifreeze should exhibit a temperature change about 1.5 times greater than the water; it takes only about ⅔ as much heat to raise its temperature 1°C.

Teaching Tip

- Before presenting Table 18-1 on student text page 454, encourage students to speculate about the relative specific heats of the substances listed in the table. After they have studied the table, ask the class which substance would require the most heat to increase its temperature 1°C. (Water) Which substance would require the least heat? (Lead or gold)

Fact At Your Fingertips

- The British thermal unit, or Btu, is the amount of heat needed to raise the temperature of 1 pound of water by 1 degree Fahrenheit.

18 HEAT

18-3 Specific Heat

Energy and Specific Heat

In each example below two processes are described. Decide which process requires more energy at room temperature; then explain your answer. Hint: It may be helpful to refer to Table 18-1 in your textbook.

1. Raising the temperature of 100 mL of water 10° C or raising the temperature of 200 mL of water by the same amount. ______
2. Increasing the temperature of 1 g of aluminum 10° C or increasing the temperature of 1 g of lead by the same amount. ______
3. Raising the temperature of an iron pot 10° C or raising the temperature of a copper pot of the same mass by the same amount. ______

Some Heated Problems!

The problems below are related to specific heat. Use Table 18-1 in your textbook to help you solve the problems.

1. A cup that contains 250 g of water is heated from 15° C to 50° C. How many joules of heat are required? ______
2. Some solar hot water heaters use 10,000 g of glycol as a heat transfer fluid. How many joules are required to raise this amount of glycol 10° C? ______
3. Which requires more energy: raising the temperature of 10 g of water 1° C or raising the temperature of 20 g of olive oil by the same amount, if both are at room temperature at the start? ______

Using Try This

The metal with the greater specific heat will melt the farthest through the ice. Thus, since the washers will probably be made of steel, they will cause more melting than the copper pennies. Aluminum washers would cause even more melting than the steel ones.

Table 18-1: Specific Heats of Various Substances

Substance	*Specific heat* (J/g/°C)	*Substance*	*Specific heat* (J/g/°C)
aluminum	0.92	iron	0.46
brass	0.38	lead	0.13
copper	0.39	olive oil	2.0
glass	0.84	silver	0.24
glycol	2.6	water	4.2
gold	0.13	zinc	0.38
ice (0°C to −20°C)	2.1		

Of all the substances in Table 18-1, water has the largest specific heat, by far. That is why large bodies of water have moderating effects on temperature changes. A large lake or ocean is slow to warm up in spring and slow to cool off in fall. Air moving over the lake or ocean is warmed or cooled by the water. This air in turn keeps the land near the water warmer or cooler than land away from the water.

The specific heat of sandy soil is small. As a result, desert temperatures may quickly rise to very high levels (40°C or more) during the day. But at night temperatures quickly fall to 0°C or below.

Checkpoint

1. It takes 0.24 J to warm 1 g of a metal 1°C. What is the specific heat of the metal? What metal could it be?
2. The temperatures of equal masses of aluminum, copper, gold, and zinc were all raised from 10°C to 30°C. To which metal was the most heat transferred?
3. Why do the temperatures of coastal cities generally not vary as much as those of inland cities?

TRY THIS

Place a stack of pennies on a flat piece of ice. Place an equal mass of metal washers on a similar piece of ice. When the washers stop sinking into the ice, remove them. Which metal melted more ice? Which metal has the larger specific heat?

Checkpoint Answers

1. The specific heat of the metal is 0.24 J/g/°C. The metal could be silver.
2. The most heat was transferred to the aluminum, because aluminum has the greatest specific heat.
3. Large bodies of water, such as oceans or large lakes, are slow to warm up in the spring and slow to cool down in the fall, due to the high specific heat of water. The temperature of the nearby air is influenced by the water temperature, and so is cooler in the spring and warmer in the fall than the air inland.

18-4

Transferring Heat

Goal

To describe three ways in which heat is transferred.

Heat always "flows" from a warmer to a cooler body. If someone placed one end of a metal spoon in a flame, that person would soon let go of the other end. It would get too hot to hold. Heat from the flame caused the kinetic energy of the particles to increase in the part of the spoon in the flame. This energy was then transferred to nearby particles along the length of the spoon. The "flow" of energy quickly reached the person's hand, raising its temperature.

The transfer of heat from particle to particle is called **conduction** of heat. A material through which heat passes readily is called a **conductor**. Metals are good conductors of heat. We use metal pots and pans for cooking, because heat passes quickly from the heat source to the food being cooked.

conduction

conductor

insulator

A material through which heat passes very slowly is called an **insulator**. We pour hot drinks into cups made of ceramic or plastic materials. Because these materials are good insulators, the cups are easier to hold than metal cups. In addition, such cups keep their contents warm longer. Air is a good insulator. In fact, most good insulators are filled with tiny air spaces. But air can transfer heat in a way that is different from conduction.

When a mass of air is heated it expands. As a result it occupies a larger space, or a larger *volume*. The expanded gas, however, is less *dense* because there are fewer particles in a given volume of the gas. A less dense substance will rise above a denser substance. For example, if you pour salad oil into water, the oil will soon float to the top because it is less dense than water. Heated air will rise above cooler air. A hot air balloon rises because the hot air in the balloon is less dense than the air through which the balloon is rising. (See Figure 18-6.)

Figure 18-6 *When the burner is going, the air inside a hot air balloon is warmed. The air expands, becomes less dense, and the balloon rises.*

Your hand does not get hot when you hold it near the *side* of a flame. But if you hold it quite far *above* a flame, you will feel warm air. The air that is warmed directly by the flame expands, becomes less dense, and rises. This creates

Teaching Suggestions 18-4

The section can be initiated by referring to some familiar examples of heat transfer. Conduction is transfer by direct contact: if you touch a cold object, your hands feel cold. Thermal transfer by convection is the basis of the familiar idea that warm air rises, making upper floors warmer than lower floors in a centrally-heated building. Thermal transfer by radiation will be familiar to students who have experienced the warmth from a campfire, from a radiant electric heater, or from the sun.

You may wish to explain how the three types of heat transfer occur when a building is heated. Fast-moving molecules of heated air impact the walls, ceilings, and floors. If they have sufficient kinetic energy, they may transfer some of that energy to the building by conduction. Through leaks around doors and windows, colder outside air can mix with warmer inside air by way of convection currents. To chase the chill, electric space heaters, which radiate heat from a hot filament, may be used.

Background

Conduction is the transfer of heat by the collision of molecules. In the process, kinetic energy is transmitted by direct impact. Most metals are good conductors of heat. On the other hand, most liquids and gases are good insulators, and so are usually heated by convection currents instead. This process involves the gradual heating of a portion of a fluid, which then expands and rises away from the heat source, allowing another portion to be heated. The third method of heat transmission, called radiation, is the propagation of infrared waves. Radiation can occur even in a vacuum, since it does not rely on a material medium. Rough texture and dark color are characteristics of good radiators of heat.

Demonstrations

1. If students are not assigned the Try This activity, it may serve as a demonstration. For added interest, slowly add a few drops of cold, colored water and watch convection currents form.
2. Fill two flasks with cool water. Wrap aluminum foil around one of the flasks, and a rough, black cloth around the other. Place both flasks in bright sunlight or under a heat lamp. After a period of time, measure the temperature of the water in both flasks. (The water in the aluminum-foil wrapped flask will be cooler, because the shiny foil reflects heat well, while the rough, black cloth absorbs heat well.)

Teaching Tip

- Students should understand that the radiation of heat is a process similar to that by which light is transferred from the sun to Earth. Radiant energy can travel through empty space, while both conduction and convection require a material medium.

Facts At Your Fingertips

- The so-called Hadley cells that define the global wind zones are an example of convection on a planetary scale.
- The U.S. Navy has tried to melt huge icebergs with radiant heat. Navy aircraft have dumped carbon black on icebergs in an effort to maximize the absorption of solar thermal energy. However, the method has not been effective in eliminating icebergs as a hazard to shipping.
- Heat and light are transmitted though space as electromagnetic waves.
- Thermal radiation peaks in the infrared. Thus, devices sensitive to the infrared portion of the electromagnetic spectrum are used to detect heat. Rooftop survey instruments sense the leakage of heat from a building. On-board guidance systems enable missiles to seek out "hot" infrared sources, such as the exhaust from enemy aircraft or missiles.
- Infrared photographic film can be used to record the relative infrared radiation from everyday objects, sometimes producing interesting visual effects.

a movement of warm air upward from the flame. The movement of gases or liquids due to differences in density is called **convection**. Convection currents in a liquid are shown in Figure 18-7. Heat is transferred within a room by convection currents. Figure 18-8 shows how warm air near a room heater rises, cools, settles downward, and returns to the heater.

convection (kahn VEK shuhn)

The sun's heat cannot reach Earth by conduction because there is nothing in the vast emptiness of space to conduct heat. The heat cannot be transferred by convection because there are no liquids or gases that link the sun to Earth. Heat moves from the sun to Earth in the same way that light does, but the heat is invisible. This type of heat transfer is known as radiation. In general, **radiation** is a transfer of energy that does not require matter to transmit it. It can occur in empty space.

radiation

The heat given off by hot objects, such as the sun, is often called *thermal* radiation. A fire in a fireplace gives off thermal radiation. We cannot see thermal radiation, but we can detect it. It affects some types of film in the same way that light affects ordinary film. Film sensitive to thermal radiation can be used to detect heat leaking away from a building. (See Figure 18-9.)

Figure 18-7 *The dye in this container of hot water shows that convection currents form in liquids as well as in air.*

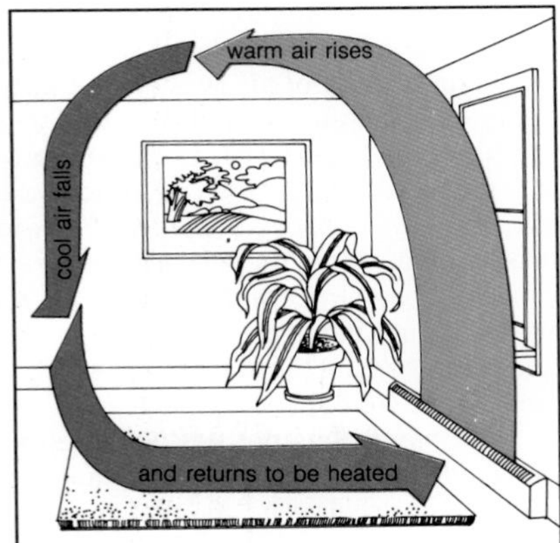

Figure 18-8 *Warm air next to a radiator or from a warm air duct will start a convection current in a room.*

Mainstreaming

The sensing of heat transfer is a capability that hearing- or vision-impaired students can master. (CAUTION: Provide safeguards when nonsighted students work with a heat source.)

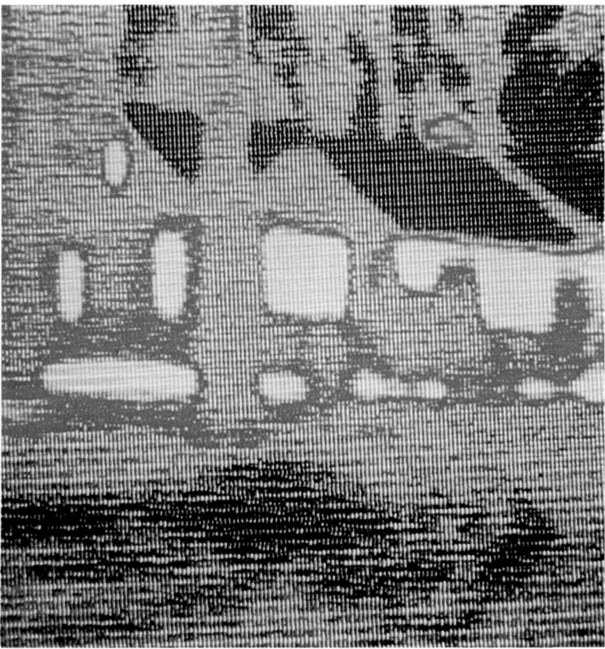

Figure 18-9 *This photograph was taken with special film to detect heat leaking from the house. Different colors on the film indicate different degrees of heat loss.*

Checkpoint

1. Describe three ways in which heat is transferred.
2. What are conductors?
3. What are insulators? Give two examples.
4. Explain why heated air rises.
5. Describe the circulation of air in a typical heated room.
6. How can thermal radiation be detected?

TRY THIS

Add a few drops of food coloring to a small amount of hot water. Using a medicine dropper, slowly add a few drops of the hot, colored water to the bottom of a transparent container of cold water. Watch the convection currents form as the warm water rises.

Cross References

Chapter 6, section 6-1, provides a detailed discussion of the effects of density differences.

Electromagnetic waves are treated in depth in Chapter 20, section 20-5.

18 HEAT

18-4 Transferring Heat

Find That Term!

A term is missing in each of the following statements. Think of the term that best completes each statement. Then write the term on the space at the left.

______ 1. When heat is transferred from particle to particle, the method of heat transfer is called __1.__

______ 2. When air is heated, it occupies a larger volume because it has __2.__

______ 3. The heat given off by the sun is called __3.__

______ 4. The heat you feel above a burning candle is partially due to thermal radiation, but is mostly due to __4.__

______ 5. A gas that has expanded occupies a larger volume and is less __5.__ than before.

Ways to Transfer Heat

In each example below, a method of heat transfer is described. Read each example and decide which method of heat transfer is represented. Then write the name of that method on the space at the left. Hint: Since there may be more than one method of transfer represented in some examples, write the name of the main method.

______ 1. You are holding a marshmallow on a metal fork over a fire. The fork gets hot.

______ 2. You put your hand on a 4-L carton of ice cream that was just removed from the freezer. Your hand feels cold.

______ 3. Heat given off by the sun is unable to pass through windows to which a film of "sun-screening" material has been applied.

______ 4. You place a glass front on a fireplace, but you can still feel heat.

______ 5. On a bright summer day you step onto the edge of a concrete pool and it feels hot on your feet.

______ 6. Heat from the sun can give you a sunburn if you are exposed to sunlight too long.

______ 7. You hold your hands above a fire in order to warm them.

______ 8. You heat a pan and its contents on a hot stove.

Using Try This

The small volume of warm water will rise in the cooler, denser water.

Checkpoint Answers

1. Heat can be transferred by conduction, convection, and radiation.
2. Conductors are materials through which heat passes readily.
3. Insulators are materials through which heat moves very slowly. Examples will vary. Some possibilities are air, ceramics, and plastic.
4. Heated air rises, because it expands, and therefore is less dense than cooler air.
5. Heated air rises, cools as it strikes walls and ceiling, and then sinks to the floor, where it is warmed again, repeating the cycle.
6. Thermal radiation can be felt as heat by the skin or detected by special types of film sensitive to infrared radiation.

Teaching Suggestions 18-5

Point out to students that energy transformations occur constantly. For example, water is heated, motors spin, light is generated, refrigerators cool food, and automobile engines are started because electrical energy can be converted to other energy forms. You may wish to have students list common energy conversions. You could ask such questions as: How does the energy stored in food become kinetic energy involved in body movement? How could thermal energy from sunlight be made to produce electricity?

Stress that when energy is converted to another form, none is destroyed and none is created. This fact is the essence of the law of conservation of energy. Be sure students understand, though, that energy conversions always involve "losses" of useful energy as nonproductive heat.

Chapter-end project 5 applies to this section.

Background

It is interesting to note that the law of conservation of energy is based on experimental data. For example, consider that to make a pendulum swing, some work must be done. Raising the bob gives it potential energy. Releasing the bob results in the conversion of that potential energy to kinetic energy and thermal energy (frictional losses). Of course, the pendulum eventually does stop. The cause of the gradual cessation of movement is the friction resulting from the pendulum's movement through the air.

In a similar way, the chemical energy in food is transformed into kinetic energy when you run on a level surface and into gravitational potential energy when you run up a hill. Some of the chemical energy is, of course, used in the bodily processes that keep you alive, as well as in overcoming the friction within the body's joints and muscles, between the feet and the ground, and between the body's surface and the air. Some is even transformed into the electrical energy associated with the transmission of nerve impulses.

Students will recall from Chapter 17 that it is because of the existence of the law of conservation of energy that a perpetual motion machine is an impossibility. The energy input of a machine must overcome nonproductive losses, as well as the productive, desired process for which the machine was designed.

18-5

Transformation of Energy

Goals

1. To describe some energy transformations.
2. To state the law of conservation of energy.
3. To discuss the nature of heat engines.

Count Rumford (Benjamin Thompson), an American-born scientist, was one of the first to recognize that work can be converted into heat. In the 1780's he was supervising the boring of cannons in Munich (in what is now West Germany). He noticed that when the cannons were bored great amounts of heat were released. By boring a cannon in water, Rumford showed that heat was produced as long as work was done. In fact, with enough work he could even make the water boil.

Working about 50 years after Rumford, James Joule made careful measurements of the heat produced when work was done. One of Joule's experiments showed that the potential energy of a slowly falling mass could be transformed into heat. (See Figure 18-10.) The falling mass caused a stirrer to turn in a well-insulated container of water. The thermometer

Figure 18-10 In this setup, the falling mass makes the stirrer turn, which heats the water. Potential energy is converted into heat.

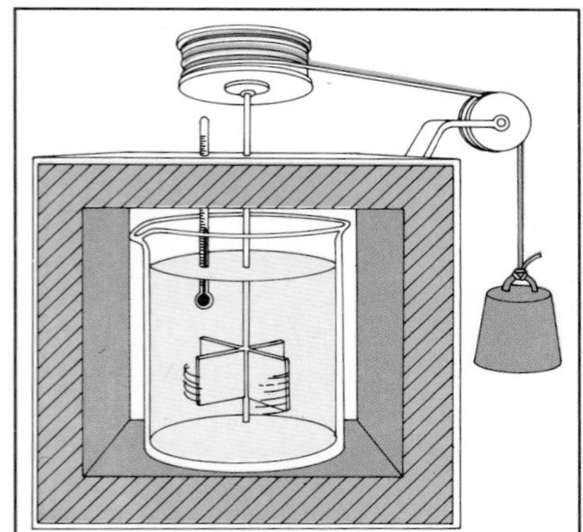

Figure 18-11 This cutaway view of a typical modern water heater clearly shows its well-insulated wall.

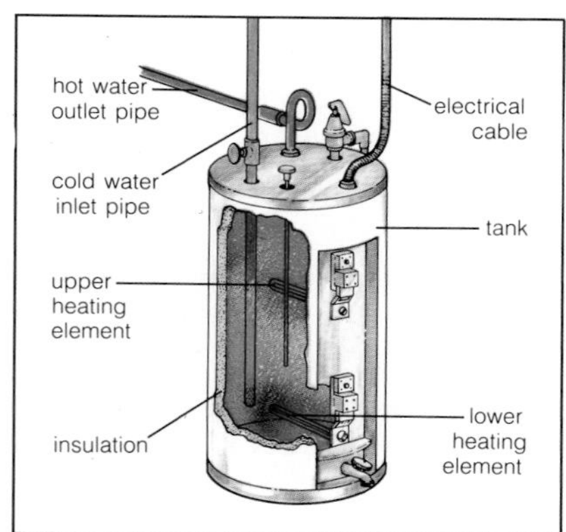

Figure 18-12 An electric motor can be used to transform electric energy into potential energy stored in a raised object.

measured the temperature of the water. The amount of energy gained by the water could be calculated from the temperature increase. The energy lost by the falling mass could be calculated, too. Joule showed that the energy lost by the mass was *proportional* to the energy gained by the water. In other words, the more work done, the more heat produced, and the greater the temperature increase.

Joule also used electricity to heat water. He found that the water's temperature increase depended on the amount of electric energy used. In many homes, water is heated by the use of electricity. Figure 18-11 shows a typical electric water heater. Note the insulation around the tank to help keep heat from escaping.

Electric energy can be used to produce work. An electric motor can be used to lift a heavy object, as shown in Figure 18-12. Some of the electric energy is transformed into potential energy as the object is raised. Some is transformed into heat, which raises the temperature of the motor and its surroundings.

Many experiments have shown that whenever one form of energy disappears, an equal amount of another form, or forms, appears. Energy can be transformed, but the total amount of energy does not change. This is known as the **law of conservation of energy**.

law of conservation of energy

Demonstrations

1. By rubbing their hands together, students can provide a familiar example of the transformation of kinetic energy into heat energy.
2. Transform electricity to potential mechanical energy by using an electric motor to lift an object. If possible, power a small electric motor with electricity produced by a photovoltaic (solar) cell.
3. Show the class how a small bicycle light generator can be used to charge a battery, which can then be used to power a light. Thus, kinetic energy is converted into chemical energy and then into radiant energy (light). However, significant losses are incurred in this process.

Teaching Tips

- Students familiar with the heat produced by a high-speed drill penetrating metal stock will be able to corroborate the story about Count Rumford's drilling of cannon bores. When the heat produced by machining becomes very great, cutting oils are used to lessen the nonproductive friction at the cutting zone and to disperse the heat, as well.
- Stress that while all forms of energy can be transformed into heat, not all the heat from an energy source can be transformed into a useful form of energy.

Fact at Your Fingertips

- The generation of electricity within an electrical generator is a very efficient process, yielding in excess of 90% efficiency.

Mainstreaming

Students who utilize powered wheelchairs may enjoy explaining how the chemical energy in the chair's battery is converted into kinetic energy.

Cross References

Chapter 17, section 17-4, contains a discussion of efficiency. Chapter 19, section 19-4, contains a discussion of the production of electricity from batteries.

18 HEAT

18-5 Transformation of Energy

Energy into Heat

In each statement below a term or phrase is missing. Think of the term or phrase that best completes the statements. Then write it on the space at the left.

______ 1. James Joule showed that mechanical energy can be transformed into __1.__

______ 2. A heat engine is a machine that transforms heat energy to __2.__

______ 3. Energy can be transformed, but the total amount of energy __3.__

______ 4. When one form of energy seems to disappear, an equal amount of energy appears in __4.__

______ 5. An electric motor can be used to transform electric energy into __5.__

Heat Engines

When water is boiled in a covered pan, the expanding steam causes the cover to rattle about noisily. In other words, some of the energy of the steam is converted into mechanical energy. Heat engines harness the force produced by expanding gases to do useful work. Read about one of the heat engines listed in Table 18-2 of your textbook and explain how heat is converted into useful work in that engine.

Using Try This

Some of the work done in bending the paper clip was converted into thermal energy. At its point of bending, the paper clip will become noticeably hot.

Checkpoint Answers

1. Joule studied the transformation of potential energy into heat, and the transformation of electrical energy into heat.
2. Energy can be neither created nor destroyed, but only transformed from one form to another.
3. A heat engine is a machine that changes heat into mechanical energy.

Figure 18-13 *A steam engine provides energy to move the locomotive. A steam engine is a heat engine.*

heat engine

Work can be transformed completely into heat, as Joule's experiments showed. But there are limits to the amount of heat that can be transformed into work. Despite these limits, heat engines are widely used. A **heat engine** is a machine that changes heat into mechanical energy. Gasoline engines and steam engines are heat engines. Table 18-2 shows the efficiencies of some heat engines.

Table 18-2: Efficiencies of Some Heat Engines

Heat engine	*Efficiency*
airplane engine	10%
steam locomotive	10%
automobile engine (gasoline)	22%
steam turbine	30%
diesel engine	35%

Note that even the diesel engine is only 35% efficient at converting the energy in fuel to useful work. The efficiencies of heat engines are necessarily low, no matter how well they are designed and maintained. The human body uses fuel much more efficiently. But when you want to get somewhere quickly, speed can be more important than efficiency.

TRY THIS

Hold a paper clip to your lips. How warm does it feel? Now bend the paper clip back and forth rapidly 10 times. Again hold it to your lips. How warm does it feel now? What was the effect of the work you did on the paper clip?

Checkpoint

1. Name two of the energy transformations that Joule studied.
2. State the law of conservation of energy.
3. What is a heat engine?

18-6

Solar Energy

Goal

To define solar energy and give examples of how it can be used to decrease the use of fossil fuels.

Have you ever had to shovel the walk after a heavy snowstorm? If so, you might envy a ski-lodge owner in Colorado. He owns a solar "snow plow." Sunlight reflecting off a series of curved mirrors melts any snow that collects on the sidewalk or steps. (See Figure 18-14.) This is just one way energy from the sun, or **solar energy**, can be used.

solar energy

Buildings today often have many south-facing windows to capture winter sunlight. Thick concrete floors or walls, or large containers of water, are used to absorb and store the sun's heat. This stored energy is then available when there is no sunlight.

Solar energy can be used to generate electricity. *Photovoltaic cells* produce electricity when they absorb solar energy. They are now used in remote places, including outer space. This means of producing electricity is too expensive for more general use today. But the cost of photovoltaic cells has been decreasing dramatically.

Figure 18-14 *The solar "snow plow" in Colorado. (a) The house, with the curved mirror over the front door. (b) A close-up view of the mirror.*

Teaching Suggestions 18-6

Stress that the sun provides heat, electricity, and even the potential energy in fossil fuels. Ask the class if wind could be considered a form of solar energy, as well. (Yes.) They should recall that wind is the result of convection within the atmosphere, produced by the differential heating of Earth by the sun. Warmed air rises, while cooler, denser air rushes underneath it, forcing the warm air out of the way and creating winds.

Chapter-end project 2 applies to this section.

Background

Because of the increasing cost of electricity, due to the growing costs of generation and distribution, more and more homeowners and businesses are utilizing one or another form of solar energy. Some are installing solar collectors, devices that absorb solar thermal energy and transfer it to an air or water medium to provide free heat for water- and space-heating. Others are utilizing passive architectural features such as larger expanses of double- or triple-paned glass on sun-facing surfaces. Design features, such as longer roof overhangs to provide shade during hot summer days, when the sun is highest in the sky, are also being used.

Demonstrations

1. Devices utilizing photovoltaic cells to drive small electric motors can be used to demonstrate that sunlight can do useful work.
2. Connect a voltmeter or milliammeter to a photovoltaic cell. Measure the voltage or current the cell produces when exposed to various intensities of light.

Teaching Tips

- Obtain from the library several solar-energy trade magazines, which address the practical aspects of solar collector design and the general utilization of solar energy. If possible, display some in class.
- Students may know of homes or businesses in the community that are using solar energy. Ask them to find out how the systems operate.
- Point out that "energy conservation" is not identical to the law of conservation of energy. Ask students to explain what energy conservation really means. (Curbing our use of nonrenewable energy sources.)

Facts At Your Fingertips

- 50% of the solar energy reaching Earth is reflected back into space by the atmosphere and 8% is reflected by the world's oceans. About 5% is absorbed by soil, 2% by marine vegetation, and 0.2% by land plants.
- In Israel, solar hot-water heaters have been widely used for decades.

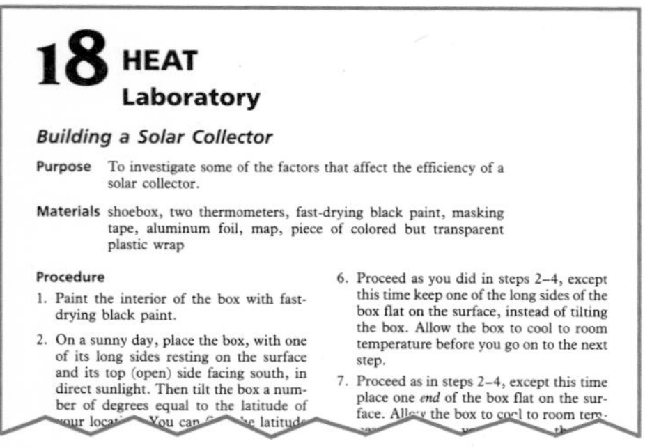

18 HEAT
Laboratory

Building a Solar Collector

Purpose To investigate some of the factors that affect the efficiency of a solar collector.

Materials shoebox, two thermometers, fast-drying black paint, masking tape, aluminum foil, map, piece of colored but transparent plastic wrap

Procedure

1. Paint the interior of the box with fast-drying black paint.
2. On a sunny day, place the box, with one of its long sides resting on the surface and its top (open) side facing south, in direct sunlight. Then tilt the box a number of degrees equal to the latitude of

6. Proceed as you did in steps 2–4, except this time keep one of the long sides of the box flat on the surface, instead of tilting the box. Allow the box to cool to room temperature before you go on to the next step.
7. Proceed as in steps 2–4, except this time place one *end* of the box flat on the surface.

Cross References

In order to compare the advantages and disadvantages of solar energy with those of nuclear power, see Chapter 10, section 10-7, for a discussion of nuclear power. The electromagnetic spectrum in relation to sunlight is discussed in Chapter 20, section 20-5.

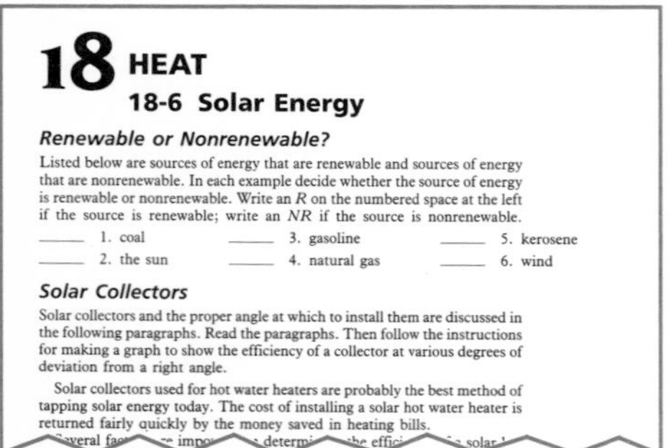

18 HEAT
18-6 Solar Energy

Renewable or Nonrenewable?

Listed below are sources of energy that are renewable and sources of energy that are nonrenewable. In each example decide whether the source of energy is renewable or nonrenewable. Write an *R* on the numbered space at the left if the source is renewable; write an *NR* if the source is nonrenewable.

____ 1. coal	____ 3. gasoline	____ 5. kerosene
____ 2. the sun	____ 4. natural gas	____ 6. wind

Solar Collectors

Solar collectors and the proper angle at which to install them are discussed in the following paragraphs. Read the paragraphs. Then follow the instructions for making a graph to show the efficiency of a collector at various degrees of deviation from a right angle.

Solar collectors used for hot water heaters are probably the best method of tapping solar energy today. The cost of installing a solar hot water heater is returned fairly quickly by the money saved in heating bills.

Using Try This

The dark water will become warmer than the clear water. Dark material absorbs more of the infrared part of the solar spectrum—the part that contains the most heat energy.

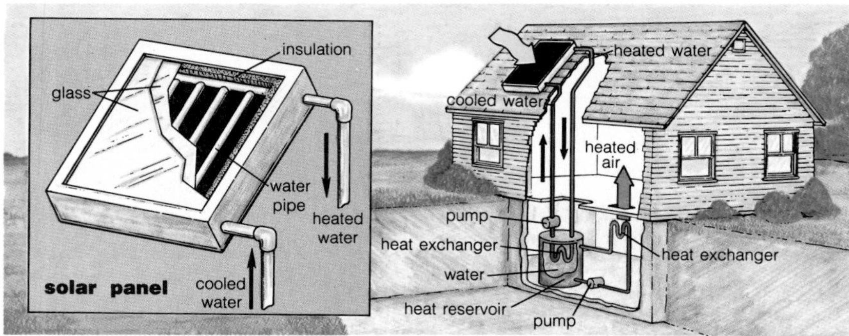

Figure 18-15 *The parts of a solar heating system. This one uses water to transfer the heat.*

Most of the energy now used to heat homes and water, provide transportation, generate electricity, and turn the wheels of industry comes from *fossil fuels*. They are mainly coal, oil, and natural gas. Fossil fuels are the remains of organisms that lived millions of years ago. While living, plants transformed energy that they received from the sun into stored energy called *chemical energy*. When fossil fuels burn they release energy from the sun that was captured by living organisms millions of years ago.

renewable energy source

To reduce the use of expensive fossil fuels, people are turning directly to solar energy to heat their homes and water. A solar heating system uses black collector plates in solar panels mounted on a south-facing roof to absorb heat from the sun. The heat is transferred to cool air or water passing over the plates. The warmed air or water is then used to heat the building. (See Figure 18-15.)

Once fossil fuels have been burned, they are gone forever. But solar energy keeps coming to Earth in the same amounts no matter how much is used by humans. Solar energy will last as long as the sun shines, probably another 10 billion years. An energy source that is produced about as fast as it is used is called a **renewable energy source**. Solar energy is renewable. Fossil fuels are not.

TRY THIS

Fill a clear glass or plastic bottle with water and cap it. Fill an identical bottle with water that has been darkened with black ink. Place both bottles in the same sunny place. After an hour or so find the temperature of the water in each bottle. Try to explain your results.

Checkpoint

1. What are fossil fuels?
2. Describe two ways in which solar energy is used.
3. Explain what a renewable energy source is.

Checkpoint Answers

1. Fossil fuels, such as coal, oil, and natural gas, are the remains of organisms that lived and died in the far distant past.
2. Solar energy can be captured to provide heat or used to generate electricity by way of photovoltaic cells.
3. A renewable energy source is one that can be replaced in a relatively brief period of time.

18-7

Other Energy Sources

Goal

To describe alternatives to fossil fuels other than solar energy.

Figure 18-16 *A modern power-generating windmill at Clayton, New Mexico.*

Approximately two percent of the solar energy that reaches Earth is transformed into the kinetic energy of winds. (Winds are created by the unequal heating of Earth.) Winds can turn windmills, which can generate electricity and do other work. (See Figure 18-16.) Sweden has plans to build 3300 wind generators. They will produce a substantial portion of that country's electricity.

hydropower

Solar energy contributes to **hydropower**, the use of falling water to do work. Water on Earth absorbs solar energy and evaporates. When it condenses some water falls back to Earth at higher levels as rain or snow. Eventually the water flows downhill in streams or rivers. Water flowing through tunnels in a dam across a river can turn turbines and generate electricity. (See Figure 18-17.)

In some regions sunlight warms the upper layer of the ocean to temperatures well above those of the cold depths below. These temperature differences could be used to produce *ocean thermal energy*. Heat from the warm layer could

Figure 18-17 *Roosevelt Dam near Phoenix, Arizona. Such dams help produce hydropower.*

Teaching Suggestions 18-7

This section may be treated primarily as an extension of section 18-6, because these alternative energy sources all result from the work done by sunlight.

Because the indirect energy products of solar energy are so diverse, they serve as excellent areas of research for students. The class may be divided into research teams that report on current progress in one of the alternative forms of energy.

Teaching Tips

- Consider field trips to sites on which any of the alternative energy sources described in this unit are being used.
- Alternative energy demonstrations can serve as topics in science fair competitions or exhibitions. Some of your students might be sufficiently motivated to pursue such a project.
- Ask the class why wind, hydropower, ocean thermal power, wood and crops used to produce fuels such as alcohol are really examples of solar power. (These alternative energy sources exist as the result of the effect of solar energy on Earth's environment.)

Fact At Your Fingertips

- Experimental ocean-thermal plants were built and operated over 50 years ago by U.S. scientists. The experiments were abandoned because of lack of both funding and interest.

evaporate a liquid such as ammonia. The gas produced could be used to turn a turbine. Electricity produced by the turbine could be sent ashore via cables or used to decompose water into oxygen and hydrogen. The hydrogen can be used as a fuel, also.

Plants are an energy resource with great potential. Wood stoves are becoming a familiar sight again in many areas where wood is plentiful. Wood can be a renewable energy source if forests are carefully managed and harvested. Other plants, such as sugar cane and sugar beets, are being grown for fuel. Sugar can be fermented to produce alcohol, which can be burned in engines instead of fossil fuels. In Brazil, alcohol has already begun to replace gasoline as a fuel for motor vehicles.

In some places solid waste is a serious problem. Many cities have used up their burial sites for trash and garbage. Foods and paper products make up much of this waste. Munich, West Germany obtains one-eighth of its electricity by burning garbage and trash. Baltimore, Maryland has a similar recovery program.

The tides that rise and fall twice each day can be a source of energy. The energy that causes the ocean to bulge

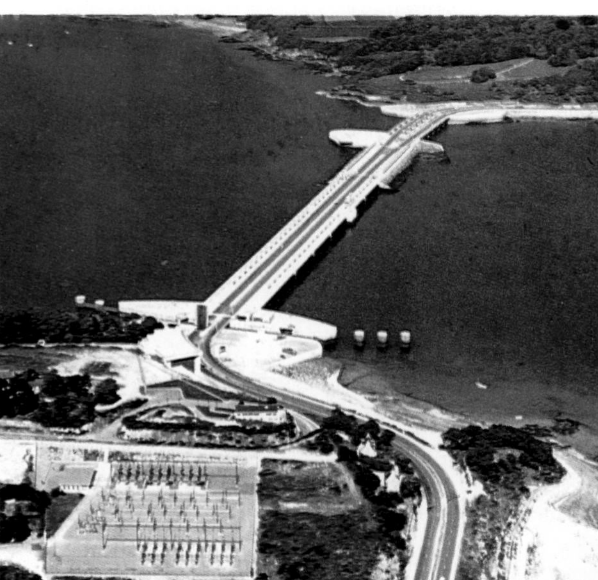

Figure 18-18 *The tidal power plant at Saint-Malo, France. As the tide rises and falls, water moves first into the river then out into the ocean through openings in the dam. The flow of water turns the turbines.*

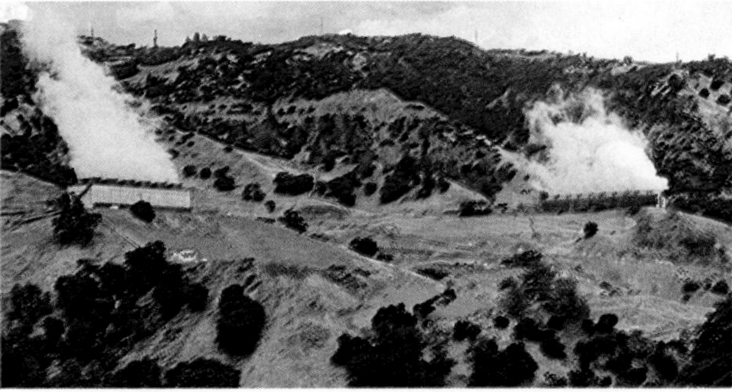

Figure 18-19 *Geothermal energy is readily available in some regions of Earth. This plant in northern California uses geothermal energy to generate electricity.*

outward comes from the gravitational pull of the moon (and, to a lesser extent, the sun). By harnessing the ebb and flow of tidal water to turbines, it is possible to generate electricity. The French have built a plant that uses tidal energy on the coast of Brittany. (See Figure 18-18.) For fifty years there have been proposals to build a similar energy plant in North America in the Bay of Fundy.

Hot springs, geysers (GY zurz), and the molten rocks from volcanoes reveal the huge storehouse of energy within Earth. This is called **geothermal energy**. It can be used to heat buildings, generate electricity, and power industry. Iceland uses geothermal energy to heat two-thirds of its homes. Nearly all the greenhouses, which produce most of that country's fresh vegetables, are heated this way, too.

geothermal energy (jee oh THURM ul)

The potential for tapping geothermal energy is limited to those regions where Earth's crust is thin. Nevertheless, many such sites exist. Canada is one of a number of nations planning to build industrial plants or electric power plants on these sites. Canadians are also considering using geothermal heat as a source of space heating for remote villages in the Northwest Territories.

Checkpoint

1. Name the three renewable sources of energy that are "powered" by solar energy.
2. What is the source of the energy that produces Earth's tides?
3. Give two practical uses of geothermal energy.

18 HEAT

18-7 Other Energy Sources

What Type of Energy?

Match each term in Column A with the correct type of energy listed in Column B. Then write the letter of the type of energy on the numbered space at the left of each term.

A	B
____ 1. geysers	a. plant energy
____ 2. evaporation of ammonia	b. ocean thermal energy
____ 3. windmill	c. wind energy
____ 4. dam	d. geothermal energy
____ 5. wood	e. energy of falling water

Investigating Energy Alternatives

Wind generators are an effective method of getting electricity from wind energy. You know that when wind energy is transformed to electrical energy, less than 100% of the wind energy will be transformed to electrical energy. For wind generators, a mathematical rule called Betz's Law will tell you the available power in the wind. Study the following table for a propellor-type wind generator. Then answer the questions.

Available Power in the Wind in Kilowatts According to Betz's Law

	Rotor Diameter in Feet									
Wind Speed, Mph	*2*	*4*	*6*	*8*	*10*	*12*	*14*	*16*	*18*	*20*
6	.001	.008	.018	.033	.051	.074	.100	.131	.166	.205
8	.005	.019	.044	.078	.121	.175	.238	.311	.393	.485
10	.009	.038	.085	.152	.237	.341	.464	.607	.768	.948
12	.016	.066	.148	.262	.409	.590	.802	1.048	1.327	1.638
14	.026	.104	.234	.416	.650	.936	1.274	1.665	2.107	2.601
16	.039	.156	.350	.622	.970	1.398	1.902	2.485	3.145	3.883
18	.055	.222	.498	.885	1.381	1.990	2.708	3.538	4.478	5.528

1. When the size of the rotor increases from 8 feet to 16 feet, what happens to the number of kilowatts of available power? ____________________
2. If your average wind speed is 12 mph and you want to have about 600 watts of power available, what rotor size would you use? Hint: You must first change watts to kilowatts.
____

Checkpoint Answers

1. Renewable sources of energy that are powered by solar energy are wind, hydropower, and biomass (wood and other organic combustibles.)
2. The energy that causes the tides is the moon's gravitational force, and, to a lesser extent, the sun's gravitational force.
3. Geothermal energy can be harnessed to heat buildings and to produce steam for the generation of electricity by steam-turbine-powered generators.

18-8

Heating and Cooling Systems

Goal

To describe various heating and cooling systems.

Colonial homes are noted for large fireplaces that were used to heat them. Few homes today use the fireplace as a primary heat source. In fact, in many homes the warm air drawn up the chimney when a fireplace is used actually increases the rate at which heat is lost from the house.

The wood stove, which was North America's primary source of heat for homes through most of the 19th century, is again finding wide use. A centrally located modern stove is about 50% efficient in transferring the energy stored in the wood to heat in the home. The stove radiates heat, conducts heat to air in contact with it, and sets up convection currents as the heated air rises and cooled air falls.

Most homes and businesses are heated by circulating hot air or water. The air or water gains heat from the burning of fuel oil or natural gas by conduction. Then convection currents carry heat to the rooms. During this process some heat is transferred by radiation, also. (See Figure 18-20.) A good oil or gas furnace transfers about 80% of the energy in the fuel to the building. The rest of the energy goes up the chimney.

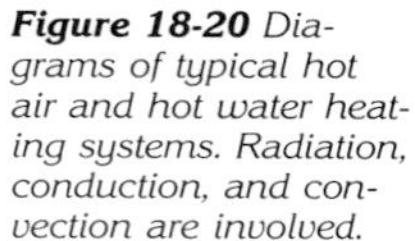

Figure 18-20 *Diagrams of typical hot air and hot water heating systems. Radiation, conduction, and convection are involved.*

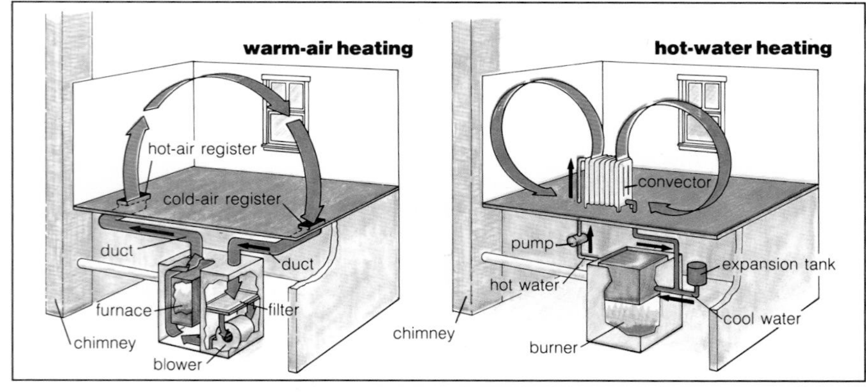

Teaching Suggestions 18-8

Take your students on a tour of the school's heating (and cooling) plant. While most schools have oil-fired furnaces for heating hot water and interior space, some may have supplementary solar heating systems, as well.

If you act as tour guide, be sure you are familiar not only with the location of the furnace, fuel tanks, hot and cold pipes and ducts, etc., but also with potential hazards to students' safety.

Chapter-end project 4 applies to this section.

Background

The condition of a furnace greatly influences the efficiency with which it transforms the potential energy in a fuel into useful heat that can warm living space or domestic water. Adjustment of input and output gas flow, of fuel/air mixture, and of ignition-system contacts; cleaning of the combustion chamber; replacement of the fuel nozzle; and other maintenance operations and upgrades can significantly improve furnace efficiency. Table TE 18-1 indicates the approximate savings per $100 of fuel oil that can be expected from oil-fired heating systems at various furnace efficiencies.

Table TE 18-1: Savings at Various Heating Efficiencies

Efficiency before adjustment	*Efficiency after adjustment*			
	70%	75%	80%	85%
50%	$29	$33	$38	$41
60%	14	20	25	29
70%	—	7	12	18
80%	—	—	—	6

Wood stoves are becoming increasingly popular in the northern states, especially where wood is plentiful, because of the increasing price of fuel oil and other fossil fuels. However, most people are unaware of the serious pollution problems that led to the conversion of heating systems from wood to fossil fuel. As wood stoves proliferate, new technologies are helping to cut down on the smoke pollution problem. For example, catalysts are used in stoves to promote much higher combustion efficiencies, thereby cutting down on undesirable, incompletely burned products that cause air pollution and also potential fire hazards when they condense on the chimney.

Demonstration

Have students measure the temperatures at various heights and in various parts of the classroom. They will find higher temperatures near the ceiling and lower ones near the floor, indicating that there are convection currents in the classroom. When sufficient data have been gathered, have students make a thermal map of the classroom.

Some homes are heated by electric coils in baseboards, floors, or ceilings. Since all the electric energy is transformed into heat, the efficiency is 100%. However, the generation of electricity at the power plant is probably only about 30% efficient. This is why electric heat is generally more expensive than other forms.

In some parts of North America, cooling is a more important concern than heating. To cool buildings, energy is extracted from the air inside and transferred to the outside. This requires work, since heat does not flow naturally from a cool area to a warm one. Electricity is commonly used as the energy source for cooling buildings. To reduce the costs of air conditioning, many homeowners install fans to pull warm air out of the house and set up cooling convection currents.

Several office buildings are being cooled by circulating ice water. During winter months a snowmaking machine produces a huge mound of snow in a lined pit that is then covered with insulation. In the summer the melted snow is circulated through nearby buildings to absorb excess heat.

In moderate climates, ***heat pumps*** are increasingly being used to both warm and cool buildings. A refrigerator is a type of heat pump. If you put your hand behind a refrigerator that is running, you can feel warm air. A refrigerator extracts energy from the air inside it and transfers the energy to the air outside it.

In the summer, a heat pump works like a refrigerator. It transfers heat from a building to the outside. In the winter, a heat pump transfers heat from the outside air, water, or ground to the building. Even "cold" matter has energy that can be extracted as heat. But work must be done to extract it. (See Figure 18-21.)

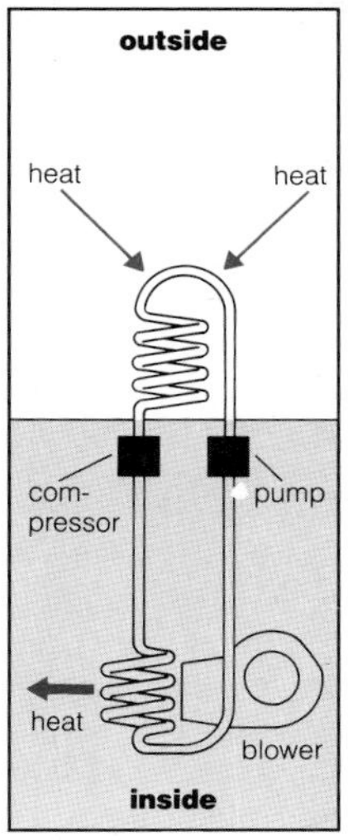

Figure 18-21 *A diagram of a heat pump. A liquid is pumped outside, where it expands and absorbs heat. Then it is moved inside, where it is compressed and gives off heat.*

Checkpoint

1. Name some energy sources that can be used to heat buildings.
2. Describe some common heating systems.
3. Why is electric heating generally more expensive than heating with oil or natural gas?
4. Which energy source is commonly used to cool buildings?

Teaching Tips

- Students may be interested in understanding why the use of a fireplace may result in increased fuel consumption by the central-heating system. Close to the fire, the warmth resulting from radiant heating will be experienced. However, the convection currents set up by the hot air rising through the chimney pull warm room air up the chimney.
- Conduct a class survey to determine the types of fuel used to heat and cool homes in your community.

Facts At Your Fingertips

- The stockpiling and use of snow for cooling buildings is a technique developed by scientists at Princeton University.
- The cost of home heating oil has climbed over 500% during the past 10 years in some parts of the nation.

18 HEAT

18-8 Heating and Cooling Systems

Fill in the Gaps!

A term is missing from most sentences in the paragraphs below. Read the paragraphs and fill in the missing terms. Then write each term on the corresponding numbered spaces at the left.

______ 1.
______ 2.
______ 3.
______ 4.
______ 5.
______ 6.
______ 7.
______ 8.

In North America in the nineteenth century __1.__ heated houses. Today, energy costs are making the stoves practical again. Central heating and cooling systems usually __2.__ hot air or water. Some buildings are heated by __3.__ coils in the baseboards, floors, or ceilings. In areas of North America where winters are mild heat pumps are used to both __4.__ and __5.__ buildings.

Each method of heating and cooling has a different efficiency. A gas or oil furnace is about __6.__ percent efficient. Although electric energy can be completely transformed to heat in buildings, the efficiency of producing electricity at a power plant is only about __7.__ percent. A typical wood stove is about __8.__ percent efficient.

Can You Explain It?

Use Section 18-8 in your textbook and your knowledge of how heat is transferred to explain each of the situations described below.

1. Some fireplaces can actually increase heat loss from a house. Explain how.
______
2. An attic fan is able to cool an entire house. Explain how.
______
3. Snow can be used to cool buildings. Explain how.
______
4. The heat given off by the burning of fuel in a furnace is circulated throughout a building. Explain how. ______

Checkpoint Answers

1. Answers will vary. Buildings are commonly heated with fuel oil, natural gas, coal, electricity, and wood. Some homes use solar energy to supplement other energy sources.
2. Common heating systems are based on circulating (forced or convection) air and water, electric resistance heating, wood stoves, steam, and heat pumps.
3. The process of generating electricity in an electric utility is only about 30% efficient. In comparison, the production of heat from combustion furnaces in buildings can be 80% efficient.
4. Electricity is typically used to cool buildings, by powering air conditioners and heat pumps.

Using The Feature Article

Typically, insulation is used around hot-water pipes, over ceilings or under the outer layers of roofing, and in the walls of newer buildings. Insulation can also take the form of air space between the panes of thermal glass. (It is the air spaces among the fibers that make goose down an extremely effective insulator when used in cold-weather clothing and bedding.)

CAUTION: Students should not handle or in any way disturb glass-fiber or asbestos-containing insulation materials. Inhaling the air-borne fibers of such insulation is hazardous to health. Asbestos has been proved to be a carcinogen.

A vacuum is a superb insulator, as only radiant energy can cross a true vacuum. Thus, the common vacuum bottle found in lunch kits utilizes a permanent vacuum between two walls of a glass or plastic container. See Figure TE 18-2.

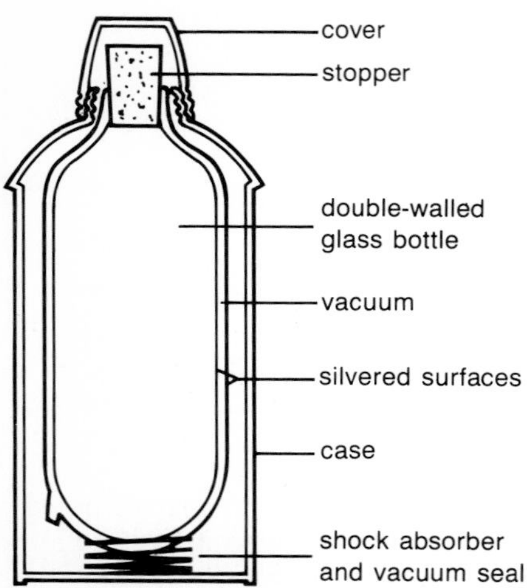

Figure TE 18-2 A vacuum bottle.

You may wish to invite a representative of a construction company to speak to the class on the various types of insulation used in residential and commerical construction. A builder specializing in solar-heated homes will be able to explain the critical role of insulation to prevent heat loss from such buildings.

Insulation

Soaring energy costs make it necessary to reduce the amount of energy used to heat or cool buildings. Insulation is one of the best ways to reduce the transfer of heat between the inside and outside of a building. *Insulation* slows the transfer of heat into or out of a building.

In most wooden buildings the space between the inside and outside walls is filled with mineral wool batts or cellulose (tiny pieces of paper). In brick buildings large sheets of rigid insulation are usually placed between the bricks and the inner wall. All such insulating materials are filled with tiny air spaces that slow the conduction and convection of heat. Because warm air rises, the most important area to insulate is the ceiling. Most new homes have at least 22 cm of insulation in the attic over the ceilings. The illustration shows insulation being installed in an attic.

The insulating quality, or resistance to heat flow, of various materials is expressed as their *R-values*. The larger the R-value, the better the insulating quality.

During the winter the warm air inside a house holds much more moisture than the cold air outside. If moisture in the air flows from the living area of a home into the cooler walls, it may condense. When wet, insulation conducts heat more quickly. This reduces its effectiveness. To avoid this, the inner side of the insulation is covered with a thin plastic sheet or thick paper. This cover, called a *vapor barrier*, prevents water vapor from reaching the insulation.

To provide more insulation, many modern homes have double outer walls. The windows often are made of two or more sheets of glass with air space between them. To further reduce heat transfer through glass windows, many homes have insulated shades or heavy drapes that can be drawn over the glass.

Many new homes are being built at least partially underground. Because the ground itself is a good insulator, underground temperatures remain constant throughout the year. Underground houses are therefore less expensive to heat or cool.

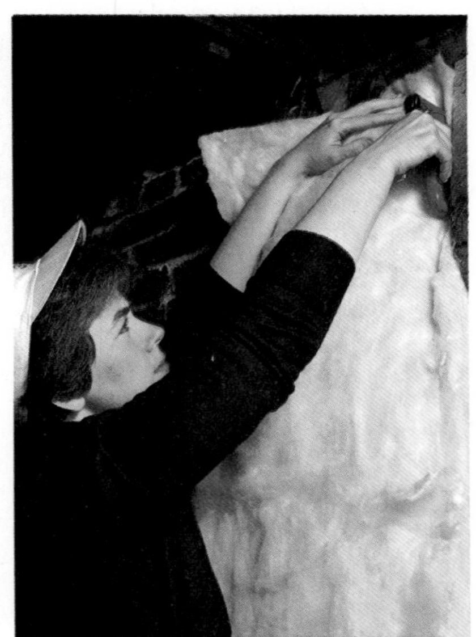

Laboratory Activity 25, Testing Insulation, student text page 564, may be done in conjunction with this article.

Chapter-end projects 3, 6, and 7 apply to this feature article.

Understanding The Chapter

18

Main Ideas

1. Heat is the energy transferred to an object when it warms or from an object when it cools.
2. Matter is made of tiny particles that are in constant motion. When an object gains heat the kinetic energy of these tiny particles increases.
3. Temperature is a measure of the average kinetic energy of the particles in a substance.
4. The scales on thermometers are established by fixed points, such as the melting and boiling points of water.
5. Pressure is the ratio of a force to the area over which it acts.
6. The heat needed to change the temperature of a fixed mass of a substance depends on the nature of the substance.
7. Heat can be transferred by conduction, convection, and radiation.
8. Energy can be transformed from one form into another, but some of the energy transferred always appears as heat.
9. Energy can be transformed, but the total amount does not change.
10. A heat engine changes heat into mechanical energy.
11. An energy source that is produced about as fast as it is used is a renewable energy source.
12. Heating and cooling systems for buildings still depend primarily on energy from fossil fuels.

Vocabulary Review

From the following list, choose the term that best completes each of the statements. Write your answers on a separate piece of paper.

absolute zero
conduction
conductor
convection
geothermal energy
heat
heat engine
hydropower
insulator
law of conservation of energy
pressure
radiation
renewable energy source
solar energy
specific heat
temperature

1. Energy that is transferred from a warm body to a cooler one is called __?__.
2. The lowest possible temperature is __?__.
3. __?__ is the ratio of a force to the area over which it acts.
4. Heat may be transferred from place to place by __?__, __?__, and __?__.
5. A thermometer is used to measure __?__.
6. The amount of heat required to raise the temperature of one gram of a substance one degree Celsius is the __?__ of that substance.
7. __?__ is energy from within Earth.
8. The tides can be thought of as a __?__.
9. A material through which heat passes very slowly is a(n) __?__.
10. A __?__ is a machine that changes heat into mechanical energy.
11. A metal is a good __?__ of heat.

Vocabulary Review Answers

1. heat
2. absolute zero
3. pressure
4. conduction, convection, radiation
5. temperature
6. specific heat
7. geothermal energy
8. renewable energy source
9. insulator
10. heat engine
11. conductor

Reading Suggestions

For The Teacher

Gardner, Robert. IDEAS FOR SCIENCE PROJECTS. New York: Watts, 1986. More than one hundred science projects in a variety of subjects ranging from astronomy to zoology.

Goldin, Augusta. GEOTHERMAL ENERGY: A HOT PROSPECT. New York: Harcourt Brace Jovanovich, 1981. A concise, interesting account of geothermal energy and its use from ancient times to the present. (Could also be read by students.)

Hedley, Don. WORLD ENERGY: THE FACTS AND THE FUTURE. New York: Facts on File, 1981. Summarizes energy production and consumption patterns around the world and attempts to predict what the future will hold.

Maycock, Paul D. and Edward N. Stirewalt. PHOTOVOLTAICS: SUNLIGHT TO ELECTRICITY IN ONE STEP. Andover, MA: Brick House Publishing Co., 1981. A useful text on solar energy for general audiences.

For The Student

Asimov, Isaac. HOW DID WE FIND OUT ABOUT SOLAR POWER? New York: Walker, 1981. Describes the discoveries involved in the development of solar power, and clearly shows the solar origin of almost all our energy.

Gardner, Robert. ENERGY PROJECTS FOR YOUNG SCIENTISTS. New York: Watts, 1987. A variety of projects, all of which involve various forms of energy.

Sachwell, John. ENERGY AT WORK. New York: Lothrop, Lee & Shepherd, 1981. Defines energy and discusses various forms, as well as energy measurement and conservation.

Sachwell, John. FUTURE SOURCES. New York: Watts, 1981. Summarizes energy history and suggests alternative energy sources.

Chapter Review Answers

Know The Facts

1. temperature
2. hydropower
3. absolute zero
4. insulator
5. increases
6. convection
7. heat
8. 4.2
9. radiation
10. solar energy
11. oil furnace

Understand The Concepts

12. The work done by the egg beater will be transferred to the water as heat.
13. Gasoline comes from petroleum. Petroleum contains the energy stored by ancient plants. These plants received their energy from sunlight, by the process of photosynthesis.
14. 273 K.
15. It can be used to both heat and cool.
16. Copper.
17. Answers will vary. Some examples are a toaster, which converts electrical energy to heat, and a hair dryer, which also converts electrical energy to heat.
18. There are no significant concentrations of gases between the sun and Earth.
19. Heat.
20. Answers will vary. Renewable energy sources include solar, geothermal, wind, and hydropower. Nonrenewable energy sources include coal, oil, and natural gas.
21. Descriptions will vary. Some examples are south-facing windows and roof-mounted collector plates that transfer heat to air or water.
22. Wind is the result of the differential heating of the atmosphere by the sun. Hydropower exists because the sun evaporates some water, which then condenses and falls to Earth as rain or snow and flows downhill.

Chapter Review

Write your answers on a separate piece of paper.

Know The Facts

1. _?_ is a measure of the average kinetic energy of the particles that make up an object. (heat, temperature, pressure)
2. _?_ is a renewable energy source. (natural gas, coal, hydropower)
3. The lowest point on the Kelvin scale is _?_. (−100°C, the freezing point of water, absolute zero)
4. Air is a good _?_. (conductor, insulator, fuel)
5. The pressure exerted by an enclosed gas _?_ when the temperature of the gas increases. (increases, decreases, stays the same)
6. _?_ is a form of heat transfer involving the movement of gases or liquids due to density differences. (conduction, convection, radiation)
7. _?_ is energy transferred from one object to another. (heat, pressure, temperature)
8. The specific heat of water is _?_ J/g/°C. (0, 1, 4.2)
9. Energy is transferred from the sun to Earth by _?_. (conduction, convection, radiation)
10. Ocean thermal energy arises from _?_. (solar energy, fossil fuels, tidal power)
11. A(n) _?_ is the most efficient heating system of the three listed. (oil furnace, wood stove, fireplace)

Understand The Concepts

12. How could you use an eggbeater to warm water?
13. Explain why solar energy is the ultimate source of the energy stored in gasoline.
14. The temperature of an object is 0°C. What is its temperature on the Kelvin scale?
15. What is the main advantage of a heat pump compared to other heating systems?
16. You are given equal masses of silver, gold, copper, and lead. Which metal will require the most heat to raise its temperature 10°C? (See Table 18-1.)
17. Think of two appliances or machines that you use every day. Name them. What energy transformations occur in each one?
18. Why can't the sun's heat be transferred from the sun to Earth by convection?
19. You pedal a bicycle up a hill. The work you do increases the potential energy of both you and the bicycle. What other form of energy can you be sure has been produced, too?
20. List three renewable sources of energy and three nonrenewable sources.
21. Describe a home heating system that is based on solar energy.
22. Explain how wind and hydropower depend on solar energy.
23. How can plants such as sugar cane reduce the use of fossil fuels?

Challenge Your Understanding

24. The particles in the air inside a tire strike the walls of the tire, exerting pressure. Since there are particles in the air outside of the tire, also, why does the tire stay inflated?

23. Sugar can be fermented to alcohol, which can be used as a fuel.

Challenge Your Understanding

24. The pressure inside the tire is greater than the pressure outside, because more molecules of air per unit of volume are squeezed into the tire by a pump. The tire valve closes tightly because of the pressure inside the tire, thus sealing off this pressurized air.

25. How do hot air balloonists make their balloons ascend and descend?

26. Why is an increase in the price of fossil fuels often accompanied by increases in electric bills?

27. If 420 J of work are transformed completely into heat that is absorbed by 100 g of water, what temperature change will occur in the water?

28. Equal masses of water at 20°C and 30°C are mixed. What will be the final temperature of the mixture?

29. Why are there gaps between the ends of joining sections of a metal bridge?

30. A one-gram sample of anthracite coal, when burned, produces 30,000 J of heat. If the coal is used to generate electricity, each gram burned produces only 10,000 J of electricity. What is the efficiency of this energy transformation?

Projects

1. Find an unmarked thermometer or cover the scale of a thermometer with masking tape. Then invent your own temperature scale. How can you convert readings on your scale to temperatures in degrees Celsius?

2. Try to obtain permission to visit some solar homes in your vicinity. What fraction of the heating needs of each of the homes you visit is obtained from solar energy?

3. Visit a store that sells insulation and examine the various kinds of insulation. How is the thickness of a particular type of insulation related to its R-value?

4. See if you can find out what factors are responsible for heat losses from a building. Then see if you can estimate the amount of heat that is lost from your home or school in one year. Try to find ways to reduce these heat losses.

5. Use your library to write a brief biography of Count Rumford or Sir James Joule.

6. Have a contest with your classmates. See who can design the best ice-cube keeper. You will want to develop some rules first. Obviously, the keeper cannot be placed in a cold place, and you will want to establish a standard-size cube. Then go to work and see who can build a device that will keep an ice cube from melting for the longest time.

7. Try to find out how the R-values of insulating materials are determined experimentally.

18 HEAT
Projects

1. Plan a visit to a heating and cooling contractor. Since 1973 people have been more concerned about the amount of energy used in both heating and cooling. Discuss with the contractor how equipment has changed.
   a. What is the EER on air conditioners? ____________
   b. Are there new furnaces that are more efficient than those of several years ago? If so, how efficient are they? ____________
   c. Are the new furnaces made differently? If so, how? ____________
2. Contact someone who is selling and installing solar equipment in your area. Have your teacher help you arrange an interview to discuss a career in the solar energy industry.
   a. What type of education is necessary to enter the business of selling and installing solar equipment? ____________
   b. Is it possible for someone to start a business in solar equipment, or must the solar business be combined with some other business to be successful? ____________

18 HEAT
Challenge

Pressure Problems

The following problems involve the equation for pressure given in Section 18-2 of your textbook. Read each problem. Then apply the pressure equation to solve it.

1. A box that weighs 100 N is sitting on the ground on its 10-cm by 50-cm side. What pressure is being exerted by the box on the ground? ____________
2. Assume that a runner who weighs 800 N lands on an area of 40 cm^2 as each foot hits the ground. How much pressure is exerted on the ground with each footfall? Give your answer in newtons per square centimeter. ____________
3. A woman who weighs 480 N is standing on the end of a diving board. She exerts a pressure of 8 N/cm^2 on the diving board. What total surface area of her feet is in contact with the diving board? ____________

25. Hot-air balloonists cause their balloon to climb by burning gas from a gas burner directly under the open bottom of their balloon. The hot air rises into the balloon, displacing denser, cooler air, and thus making the balloon more buoyant. The balloon will lose altitude as its gas cools. Alternatively, a flap on the top of the balloon can be opened, allowing the warm air to escape quickly and the balloon to descend.
26. These fuels are often used in the generation of electricity.
27. 420 J/100 g = 4.2 J/g. This is enough heat to raise the temperature of the water 1°C.
28. 25°C.
29. The space allows the metal to expand in warm weather.
30. 10/30 = 33⅓% efficiency.

CHAPTER 19 Overview

The purpose of this chapter is to introduce basic concepts of electricity and magnetism. The chapter opens with a discussion of electric charge and a description of the electrical nature of matter, which is then related to static electricity and electrical conductivity. Charging by induction is described, followed by a definition of grounding and a discussion of related safety considerations. The fundamentals of current electricity are then introduced, and series and parallel circuits are described. A discussion of magnetism, magnetic fields, and the relationship between a magnetic field and an electric current follows. The differences between a generator and an electric motor are explained. Finally, the many uses of electricity in the home are described, followed by a discussion of conservation measures.

Goals

At the end of this chapter, students should be able to:

1. explain how electric charges are produced.
2. describe the behavior of charged objects.
3. explain how the atomic model of matter is related to electric charges.
4. use the atomic model to explain static electricity.
5. define and illustrate induction and grounding.
6. describe some practical uses of static electricity.
7. describe the parts of a simple electric circuit.
8. explain Ohm's law.
9. compare series and parallel circuits.
10. explain how fuses work.
11. describe magnetic fields, including Earth's magnetic field.
12. show that there is a magnetic field around a current-carrying wire.
13. explain how electricity can be produced by a changing magnetic field.
14. explain how an electric motor works.
15. explain how the amount of electrical energy used is measured.
16. list ways people can reduce their use of electricity.

Chapter 19

Electricity and Magnetism

This photo of the effects of electric charge was taken at the Ontario Science Center in Canada.

Vocabulary Preview

negative charge
positive charge
conductors
insulators
static electricity
induction
grounding
electric current
electric circuit
amperes
volts
resistance
ohms
series circuit
battery
parallel circuit
magnets
magnetic field
electromagnet

19-1

Electric Charge

Goals

1. To explain how electric charges are produced.
2. To describe the behavior of charged objects.

You reach for a door knob after walking across a rug and a spark "jumps" between your hand and the knob. When you take off your sweater in a dark room, you see small flashes of light and hear crackling sounds. During the winter, you notice that your hair tends to "fly away." If you brush it, you may see sparks snapping between your hair and the brush. All of these events result from the buildup of *electric charge.* For example, your shoes moving across a rug cause electric charge to build up on your body. If the charge is large enough, a spark will jump between your hand and a metal object, such as a door knob.

People were aware of the effects of electric charge long before they understood it. The ancient Greeks knew that amber (shown in Figure 19-1) could be rubbed vigorously to produce a charge. In fact, our word *electricity* is derived from the Greek word *elektron,* which means amber.

Suppose two objects are charged in the same way, say by rubbing them with a woolen cloth. They will then move away from, or *repel,* each other. (See Figure 19-2a.) Figure 19-2b shows two objects that were charged in different ways. These two objects move toward, or *attract,* each other.

Charged objects always either attract or repel each other. This is because there are only two kinds of charge.

Figure 19-1 *A piece of amber. Rubbed amber was known to attract light objects (because it had a charge) as long ago as about 300 B.C. Amber is the hardened sap of extinct pine trees.*

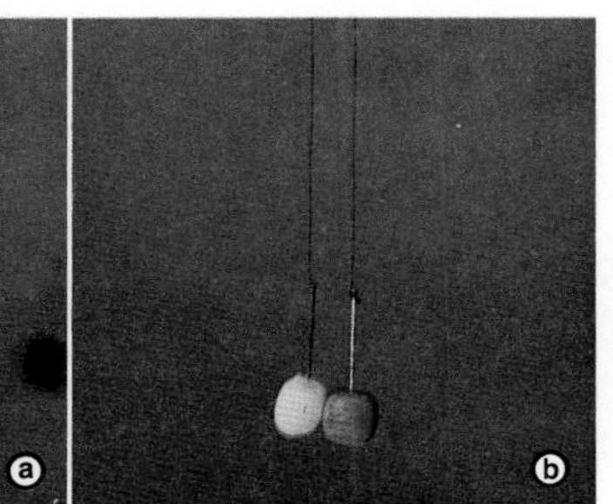

Figure 19-2 *(a) When two pith balls have the same charge, they repel each other. (b) Oppositely charged pith balls attract each other.*

Teaching Suggestions 19-1

Ask students to relate their own experiences of "electric shocks." Explain that such "shocks" are caused by the build-up of electric charge. Then explain and describe positive and negative electric charge.

Be sure students understand that opposite charges attract and like charges repel. Stress that the force of attraction or repulsion depends on two factors: the amount of charge and the distance between the charges. A demonstration of the behavior of charged objects can be very helpful in teaching this section. (See Demonstration.)

Background

Benjamin Franklin assigned a "negative" charge to a rubber rod rubbed with fur, since two such rods repelled each other. Because a glass rod rubbed with silk attracted a rubber rod rubbed with fur, he knew the charges on the two rods were different. He called the charge on the glass rod "positive." These definitions of charge are still in use today.

Static electricity is discussed in greater detail in section 19-3.

Demonstration

Show how charged objects behave. Charge a rubber rod by rubbing it with a piece of fur. Hang the rod in a stirrup made of plastic-covered wire, or some other insulating material. Then bring an identically charged rod near one end of the suspended rod. The two rods will repel each other. Then approach the suspended rod with a glass rod that has been rubbed with silk; the two rods will attract each other. Test other charged objects, such as plastic rulers or pens rubbed with paper. If you cannot obtain rubber or glass rods, substitute plastic strips, etc. Note that this demonstration will work only if the air is dry; don't try it on a humid day.

Teaching Tips

- Discuss static-electric effects your students have experienced firsthand, such as sparks, shocks from static build-up, and static on machine-dried clothing.
- Point out the similarities between electrical and gravitational forces. Both decrease markedly with physical separation and both are proportional to the product of two factors—the two masses in the case of gravitational force, the two charges in the case of electrical force.

Facts At Your Fingertips

- Static-electric effects are most often seen during the winter months, when the air is very dry. Moist air carries away charge.
- Coulomb's law for electric charges, like Newton's law of universal gravitation, is an inverse square law: two point charges attract or repel each other with a force that is proportional to the product of the charges and inversely proportional to the square of the distance between them.

19 ELECTRICITY AND MAGNETISM

19-1 Electric Charge

Attraction or Repulsion?

Decide whether each pair of charges shown below will attract or repel each other. If the pair of charges will attract each other, write an *A* on the space between the charges. If the pair of charges will repel each other, write an *R* on the space.

1. (+) ____ (−)
2. (+) ____ (+)
3. (−) ____ (−)
4. (−) ____ (+)

Otto von Guericke

The life and work of Otto von Guericke are described in the paragraphs below. Read the paragraphs. Then write the answers to the questions.

Otto von Guericke was a German scientist. He was born in Magdeburg in 1602. While investigating motion, Guericke found that he was building up an electric charge on certain objects. This led him to investigate some experiments performed by William Gilbert, an English scientist who lived from 1540–1603. Gilbert had found that an electric charge could be built up on objects when they were rubbed with certain materials. That is, an electric charge could be deliberately produced by means of friction.

Guericke built the first machine designed to build up an electric charge by means of friction. The machine was made up of a sphere of the element sulfur. When Guericke rotated the sphere while rubbing his hand across its surface, he built up a large electric charge on the sphere. The charge could be removed in the form of sizable electric sparks. Guericke's work led to further experiments with electric charges, such as the experiments of Benjamin Franklin a century later.

1. What did Guericke learn from Gilbert? ________
2. How did Guericke change the way in which electric charge was produced? ________
3. About a century after Guericke's experimentation, what notable American was doing important experiments with electric charge? ________

Using Try This

Students should be able to state that the balloon sticks to the wall because it has been charged by rubbing. If the air is dry, these charges will leak away slowly, and the balloon will adhere to the wall for a few minutes, or longer.

Figure 19-3 *Benjamin Franklin proved lightning is electricity. With a kite in a storm cloud, he detected electricity at a key hanging from the end of the kite string. He was very lucky—he might have been killed.*

Benjamin Franklin (Figure 19-3), well known for his early studies of electricity, called the two types of charge *positive* and *negative*. Two positively charged objects repel each other. Two objects that are negatively charged also repel each other. But suppose one object is positively charged and the other is negatively charged. Then the objects attract each other. The important thing to remember is that like charges repel each other and unlike charges attract.

Two factors determine how much two objects attract or repel each other. The first factor is the amount of charge. If the charge is large on one or both objects, the attraction or repulsion too is large. If the charge is small, the attraction or repulsion is small. The second factor is the distance between the objects. If they are close, the attraction or repulsion is large. If they are far apart, the attraction or repulsion is small.

TRY THIS

Rub a balloon on your clothing. "Stick" it to a wall or ceiling. What holds it in place? How long does it stay attached?

Checkpoint

1. Give at least one way that charge can be built up on an object.
2. How many kinds of charge are there? What are they called?
3. If two objects have the same type of charge, will they repel or attract each other? If the charges are different, how will the two objects react?

Checkpoint Answers

1. Answers will vary. Rubbing is one method.
2. There are two types of charge—positive and negative.
3. Like charges repel; opposite charges attract.

19-2

The Electrical Nature of Matter

Goal

To explain how the atomic model of matter is related to electric charges.

All matter is made up of tiny particles called *atoms*. Atoms are too small to be seen, even with the best microscopes available. Atoms themselves are made up of even smaller particles. These are called *protons, neutrons,* and *electrons*.

Protons and neutrons are found in the central core, or *nucleus*, of an atom. Neutrons carry no charge. Thus, they are *neutral*. Protons carry a positive charge. The particles called electrons are in constant motion around the nucleus. Each electron carries a negative charge equal to the positive charge of a proton. The force of attraction between positive protons and negative electrons gives the atom its structure. A diagram of a model atom is shown in Figure 19-4. In a real atom, electrons are much farther from the nucleus than in the model.

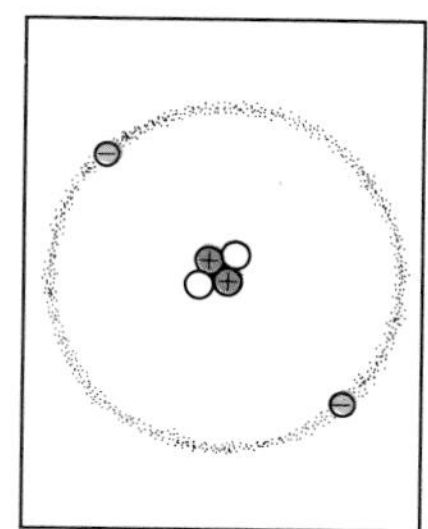

Figure 19-4 *A diagram of a model helium atom. It has two protons and two neutrons in the nucleus and two electrons outside. The + and − show the charges of the particles.*

Different kinds of atoms have different numbers of protons, neutrons, and electrons. Normally, whatever the number of particles is, an atom contains equal numbers of protons and electrons. The total positive charge balances the total negative charge, so the atom itself is neutral.

Atoms can gain or lose electrons. For example, if you rub a glass rod with a piece of silk, the silk will gain electrons and become negatively charged. The glass rod loses electrons. Because its atoms now have more protons than electrons, the glass is positively charged. A **negative charge** results from an excess of electrons. A **positive charge** results from too few electrons to balance the protons in the atoms of an object.

negative charge

positive charge

Checkpoint

1. Name the three main particles that make up an atom. What is the charge on each particle?
2. Why is an atom normally neutral?
3. How can an object become negatively charged? positively charged?

Teaching Suggestions 19-2

If students have already studied the chapters on chemistry, particularly Chapter 7, Atoms and Molecules, this section may be treated as a reading assignment. If not, you will want to discuss the arrangement of electrons, protons, and neutrons in atoms, and describe the relationship between this structure and electrical phemonena. Be sure students can answer the Checkpoint questions.

Background

This section is designed to show students that the charges that collect on objects are due to the gain or loss of electrons, which are more easily removed from atoms than are protons. Because protons are at the core of the atom and very tightly bound within the nucleus, they are very difficult to remove. Electrons, however, reside at the periphery of the atom and are not as tightly bound.

Some students may wonder why atomic nuclei don't fly apart, for the positively charged protons in the nucleus repel each other. The answer is that there are nuclear forces of attraction that are far stronger than the electrical forces of repulsion.

Fact At Your Fingertips

- The nuclear, or strong, force is 100 times greater than electrical force. The strong force is a force of attraction only.

Cross Reference

For a more detailed description of the structure of the atom, see Chapter 7, Atoms and Molecules.

19 ELECTRICITY AND MAGNETISM

19-2 The Electrical Nature of Matter

It's a Small World!

Some statements below are ture; others are false. Read each statement. If the statement is true, circle the letter *T*. If the statement is false, circle the letter *F* and replace the underlined word with a term that makes the statement true.

1. ________ T F All matter is made up of tiny particles called atoms.
2. ________ T F A neutron is a positively charged particle that is present in the nucleus of an atom.
3. ________ T F The electron is a negatively charged particle that is present in an atom.
4. ________ T F Electrons are in constant motion about the nucleus.
5. ________ T F A neutral atom must contain equal numbers of protons and neutrons.
6. ________ T F If a neutral atom loses an electron, the atom becomes negatively charged.

Checkpoint Answers

1. The three major atomic particles are positively charged protons, negatively charged electrons, and neutral neutrons.
2. An atom is normally neutral because it has the same number of negative and positive charges: that is, the number of protons equals the number of electrons.
3. An object becomes negatively charged when its atoms gain electrons. An object becomes positively charged when its atoms lose electrons.

Teaching Suggestions 19-3

Begin this section by reviewing atomic structure. Point out that the creation of static charge is possible because electrons in atoms are loosely held. They can often be stripped away by friction between two objects. The result is an excess of electrons on one object and a deficiency on the other.

Using Demonstration 1 is an excellent way to convey the concepts of this section. (See Demonstrations. If you do not have an electroscope, your high school physics department may be able to provide one.) Explain how an electroscope works. Describe the steps involved in charging the device by induction.

Laboratory Activity 26, Conductors of Electricity, student text page 565, may be carried out in conjunction with this section.

Background

The electrons in a metal may be thought of as a "sea" of charge, because they can flow easily from one atom to another. Only a relatively small force is needed to push electrons along in a metal. In material such as glass or plastic, on the other hand, the electrons are more tightly held. Thus large forces are required to push them off their atoms.

A battery or generator separates charge onto electrodes. If circuit elements are connected to these electrodes, the repelling forces on the electrons at the negative electrode cause them to flow along the circuit to the positive electrode, which attracts them.

If two metal bars are insulated, as shown in Figure 19-5 on student text page 476, charge can flow along the metal, but not to Earth. Under such conditions, separation of charge into positive and negative parts occurs. The presence of an outside charge is needed to maintain such an uneven distribution of charge.

Lightning, which is a very rapid discharge, occurs when atmospheric charge is not neutralized quickly enough. Lightning rods drain away charge built up in the atmosphere during a storm. If you examine a lightning rod, you will see that its tip is pointed. Charges tend to collect on points and charge flows readily to or from points. If the atmospheric charge is negative, it flows through the rod to the ground, which can accommodate an enormous amount of charge. If the atmospheric charge is positive, electrons flow from Earth to the tip of the rod, and from there to the clouds carrying the positive charge.

Grounding involves connecting electric circuits to Earth. Because Earth has a huge capacity for holding charge, practically all charge on an object will flow to Earth if connected to it. The ground wires in electrical devices provide a path, should the normal path be broken, that allows charge to flow to

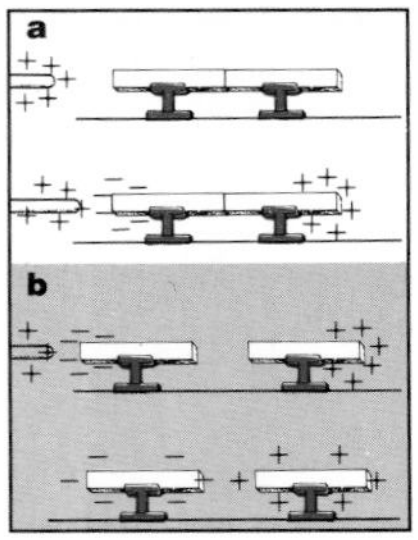

19-3

Static Electricity

Goals

1. To use the atomic model to explain static electricity.
2. To define and illustrate induction and grounding.
3. To describe some practical uses of static electricity.

Figure 19-5 *(a) A positively charged rod close to a pair of neutral metal bars that are in contact attracts negatively charged particles. (b) When the bars are separated, they are oppositely charged.*

conductors

insulators (IN suh layt uhrz)

static electricity

Electrons are more easily removed from some atoms than from others. If you rub a plastic comb with cloth, the comb will become charged. If you rub a metal rod with a cloth, however, it does not become charged. This is because electrons move easily along a metal rod. Any charge that builds up flows away immediately. If the metal rod is in your hand, the extra electrons flow into your body. Electrons move easily along some materials. Such materials are called **conductors** of electricity. Metals are good conductors.

Materials that keep electrons from moving are electrical **insulators**. Examples of electrical insulators are rubber, glass, wool, and many plastics. Because insulators do not conduct electricity, any charge they gain stays in one place. Anything that stays in one place is said to be *static*. Thus, charges that stay on an object are called **static electricity**.

When electricity "flows," it is because electrons move. They tend to move from areas that have an excess of electrons to areas that have too few. Electrons can also move to areas that are neutral, that is, to areas neither positively nor negatively charged.

Figure 19-5 shows a positively charged rod being held near two neutral metal bars on insulating holders. Note that the bars are in contact. The electrons in the metal bars are attracted to the positive rod. Because dry air is a good insulator, however, the electrons cannot leave the bars and move to the rod. Instead, the electrons collect on the metal bar nearest the positive rod. This bar becomes negatively charged. Because electrons have moved away, the other bar becomes positively charged.

induction (in DUHK shuhn)

If the two bars are now separated, one carries a positive charge, the other a negative charge. When a neutral object is charged by being placed near a charged object, the process is called charging by **induction**. These metal bars were charged by induction.

Figure 19-6 *A bolt of lightning striking the Empire State Building in New York City. It will not harm the building because it is striking a lightning rod at the top of the roof-top tower.*

If a negatively charged conductor touches an uncharged conductor, charge will move from the charged body to the neutral one. Movement stops when the charges are spread over both conductors. This leaves the original charged object with a fraction of its original charge. Suppose the neutral object is Earth. Earth, of course, is very large compared to ordinary objects. Thus, once the charge is distributed, the small object is left with practically no charge at all. This process of a charge flowing into Earth is called **grounding**.

grounding

The most spectacular example of grounding is the action of a lightning rod during a storm, as shown in Figure 19-6. A lightning rod is a pointed metal conductor placed at the top of a building. Lightning rods are connected by heavy wires to metal pipes buried in the ground. When lightning strikes, the charges are grounded when they travel along the heavy wires to the metal pipe in the ground. This grounding protects the building from damage.

Earth through a wire rather than through the user's body.

Demonstrations

1. Show students how an electroscope works. Try to charge the electroscope with a rubbed metal rod held in your hand. Alert students will realize that the metal is conducting any charge through your body to the ground, and thus cannot charge the electroscope. Repeat, this time using a glass rod rubbed with silk. The leaves of the electroscope will separate, indicating the presence of charge. See Figure TE 19-1.

   The electroscope will have acquired a negative charge. Proceed to determine the charge on a variety of charged objects. If an object brought near the plate attracts charge onto the plate, the leaves will fall. In that case, the object must carry a charge opposite to the charge on the electroscope. If, on the other hand, the leaves move farther apart, the object must be driving charge from the plate onto the leaves. In that case, the object carries the same charge as the electroscope.

   Show students that a charged electroscope can be grounded by connecting it to Earth through your body.
2. Show that water molecules are polar (see Chapter 7, section 7-7). Hold a comb or plastic ruler that has been rubbed with cloth near a very narrow stream of falling water. The polar water molecules will be attracted by the charged object, causing the stream to bend.

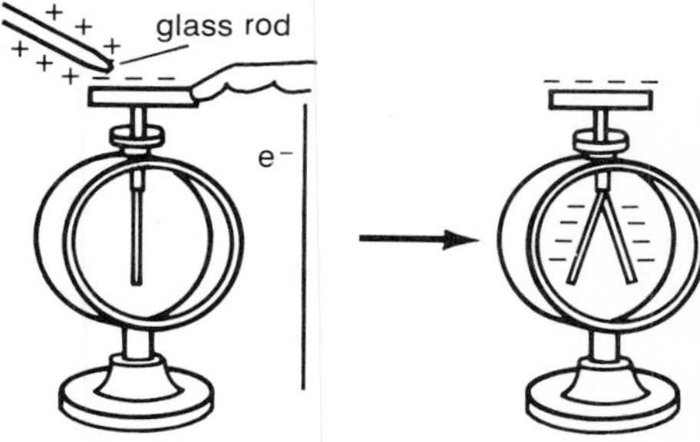

Figure TE 19-1 Charging an electroscope negatively by induction.

Teaching Tips

- Be sure students understand that an electroscope is charged by induction. When the positively charged glass rod is near the plate of the electroscope, and your finger is in contact with the plate, negative charges are induced to flow through your body to the electroscope because they are attracted to the positively charged glass rod. When you then remove your finger, the excess negative charges on the electroscope are trapped, with no path of escape.

- Be sure students understand that grounding discharges an object. Connecting the object to Earth enables its charge to spread over the entire Earth.
- Because charge tends to leak away in moist air, the experiments and demonstrations of this section will work only in dry air. Plan, if possible, to do this chapter during the winter, when the inside air is drier than usual, at least in colder regions.

Facts At Your Fingertips

- In most buildings the ground wire is connected to Earth through the pipe that carries water into the building.
- In Britain and some parts of Canada, grounding is called "earthing."

Mainstreaming

Handicapped and special-needs students will require assistance and will benefit from hands-on experience.

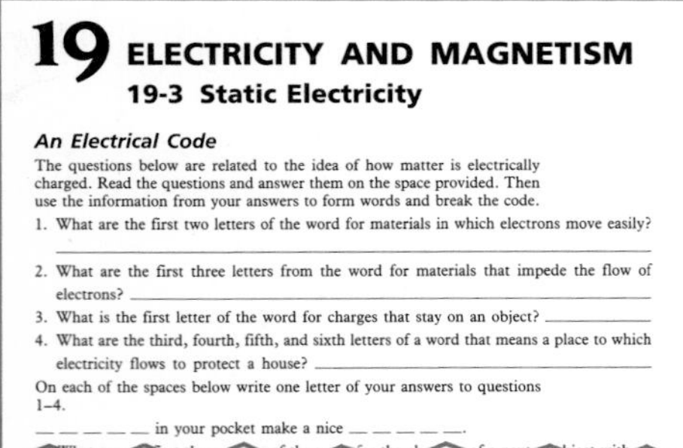

19 ELECTRICITY AND MAGNETISM

19-3 Static Electricity

An Electrical Code

The questions below are related to the idea of how matter is electrically charged. Read the questions and answer them on the space provided. Then use the information from your answers to form words and break the code.

1. What are the first two letters of the word for materials in which electrons move easily? ____________
2. What are the first three letters from the word for materials that impede the flow of electrons? ____________
3. What is the first letter of the word for charges that stay on an object? ____________
4. What are the third, fourth, fifth, and sixth letters of a word that means a place to which electricity flows to protect a house? ____________

On each of the spaces below write one letter of your answers to questions 1–4.

_ _ _ _ _ in your pocket make a nice _ _ _ _ _.

Using Try This

Students should observe that the uncharged comb has no effect on the paper. When the comb is charged, it will attract the tiny pieces of paper by inducing charge on them.

Checkpoint Answers

1. Electric charge that collects on an object but does not flow is called static electricity.
2. Answers will vary. Any metal is a good conductor.
3. An electrical insulator does not conduct charge. Answers will vary. For example, glass, plastic, dry air, and rubber are insulators.
4. Lightning rods attract electric charges, which then travel along heavy wires to a metal pipe in the ground.
5. Answers will vary. Two examples are its use in Cottrell precipitators and in painting automobiles.

Figure 19-7 *These smokestacks contain Cottrell precipitators to catch smoke particles. At the left, the precipitators are turned off. At the right, they are turned on.*

Charge tends to collect on the outside surface of materials, especially metals. During a thunderstorm, you are safest if you are inside a building or car. Stay away from high ground, tall trees, power lines, and open boats. Never swim during a thunderstorm.

Static electricity can be bothersome and even dangerous, but it can be useful too. For example, static electricity is used in painting automobiles. The surface of the car is given a charge. The paint is then given the opposite charge. When the auto body is sprayed or dipped, the paint droplets are attracted to the body's surface. The result is a strong bond between paint and metal.

Cottrell precipitators use static electricity to control industrial air pollution. (See Figure 19-7.) They are placed in the chimneys of plants that burn coal. Each precipitator is made up of pairs of screens bearing static charges. The charged screens attract oppositely charged soot and smoke particles that would otherwise escape.

TRY THIS

Bring a plastic comb or ruler near a pile of tiny pieces of paper. Be careful not to touch the paper. What happens? Now rub the comb or ruler through your hair or on some cloth to put a charge on the plastic. Again, bring the comb or ruler near the paper. What happens this time?

Checkpoint

1. What is static electricity?
2. Name one conductor of electricity.
3. What is an electrical insulator? Give two examples.
4. Explain how lightning rods protect buildings.
5. Describe one way in which static electricity can be useful.

19-4

Current Electricity

Goals

1. To describe the parts of a simple electric circuit.
2. To explain Ohm's law.

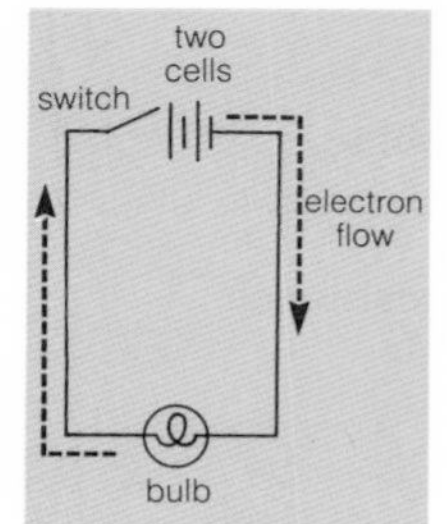

Figure 19-8 *A diagram of a simple electric circuit.*

Electrons flowing in wires can be used to make bulbs light, motors run, heaters heat, and television sets form images. A stream of flowing electrons is called an **electric current**. It takes more than wire to make a current flow. Electrons must move along a path from a region that has an excess of electrons to an area where there are too few. This path is called an **electric circuit**.

electric current

electric circuit

Figure 19-8 is a diagram of a simple electric circuit. The *switch* controls the circuit. When the switch is closed, the circuit is complete and electrons can flow freely. They flow from the negative pole of one *cell* (an area of excess electrons) to the positive pole of the other. (What most people call a flashlight "battery," scientists call a *cell*.) Along the way, the electrons flow through the bulb. In the bulb, some of the energy of the electrons is converted to light and heat.

Figure 19-9 shows symbols used in circuit diagrams. An *ammeter* is a device that measures current. In effect, it counts electrons flowing by one point in a circuit. The ammeter measures current in units called **amperes**.

amperes
(AM pihrz)

Figure 19-9 *There is a symbol for each part of an electric circuit.*

symbol	meaning	symbol	meaning
+ −	One cell. The long line is the positive (+) side of the cell. It can also be drawn with the short line on the left and the long one on the right.		Resistor. This is a wire that gets hot when electrons travel in it. There is no special number of zigzags.
		V	Voltmeter.
+ −	Three cells connected together. The positive (+) side of each cell is connected to the negative (−) side of the next cell.	A	Ammeter.
		M	Electric motor.
	Switch, always shown open in diagrams.		Alternating current electric energy source. This is what you plug into when you use electricity in your house.
	Connecting wire.		
	Lamp.		Fuse.

Teaching Suggestions 19-4

Introduce the concept of an electric circuit by pointing out that you have progressed from static to moving electric charges. Then diagram on the chalkboard the simple circuit shown in Figure 19-8 on student text page 479. Discuss closed and open circuits in terms of electron flow. Draw additional simple circuits on the chalkboard. Be sure students thoroughly understand the symbols for circuit elements and the fact that electrons flow only in a closed circuit. The Try This activity on student text page 480 works well in conjunction with the introduction to simple circuits. See Using Try This for expected results.

Be sure students thoroughly understand the concepts of current, voltage, and resistance before introducing Ohm's law. Analyzing and discussing a variety of circuit arrangements will help students grasp the basic concepts.

Background

The force that drives electrons along a circuit arises from the accumulation of charge at the electrodes of a battery or generator. One pole of a battery has an excess of electrons, or negative charge, while the other pole has an excess of positive charge. Chemical reactions within the cells of the battery are responsible for the separation of charge. In moving from the negative to the positive electrode, electrons are forced over a distance. The work done on the electrons constitutes the energy that they can provide in the circuit to operate bulbs, appliances, and so on.

A voltmeter measures the energy per coulomb (C) of charge. One coulomb is equivalent to 6.25×10^{18} electrons. When the voltmeter reads 1 volt (V), each coulomb of electrons that flows through the circuit releases 1 joule (J) of energy. A voltmeter reading of 10 V indicates that each coulomb of charge releases 10 J of energy to the circuit.

Electric current, measured in amperes (A), is the rate of flow of electric charge. One ampere is a flow of charge equal to one coulomb per second.

The ratio of voltage to current for any circuit element is called the resistance of that circuit element. A resistance of 10 ohms (Ω) means that a voltage of 10 V is required for a current of 1 A to flow through that element. In other words, the electrons at one end of the element must have 10 J more energy than they will have at the other if 6.25×10^{18} electrons are to flow through that part of the circuit in 1 second.

Demonstration

Set up some simple circuits. Demonstrate that a closed path is required for current to flow.

Teaching Tip

- Allow students to examine and use simple circuits containing batteries, wires, bulbs, ammeters, and voltmeters. Such hands-on experience is very valuable.

Facts At Your Fingertips

- Technically, a flashlight "battery" is an electric cell. A battery consists of two or more electric cells. Thus, a flashlight that contains several D- or C-cells does contain a battery. Automobile batteries consist of several cells arranged in series. Students may have heard an auto mechanic say, "One of the cells is dead," when referring to a battery.
- In an electrolytic solution, current consists not of moving electrons but rather of the motion of both positive and negative ions. These ions move toward the positive and negative electrodes, where electrons are either lost or gained.

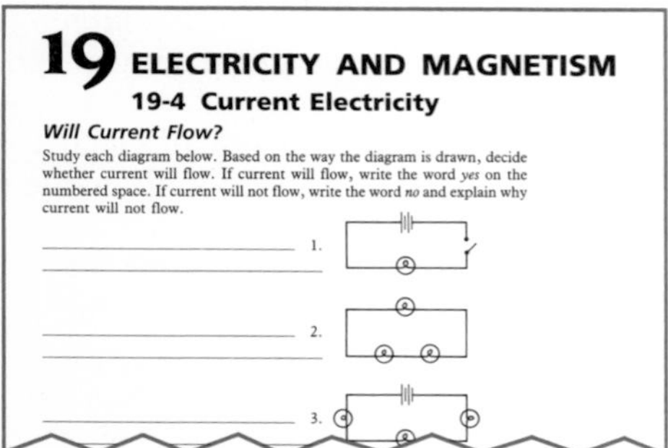

19 ELECTRICITY AND MAGNETISM

19-4 Current Electricity

Will Current Flow?

Study each diagram below. Based on the way the diagram is drawn, decide whether current will flow. If current will flow, write the word *yes* on the numbered space. If current will not flow, write the word *no* and explain why current will not flow.

1. ______

2. ______

3. ______

Using Try This

Students should record all the arrangements that make the bulb light. They will find that one end of the D-cell must be connected to the tiny knob at the bottom of the bulb and that the other pole of the D-cell must be in contact with the metal side of the base.

Checkpoint Answers

1. An electric current is a stream of moving electrons.
2. An ammeter measures current by counting the number of electrons flowing by one point in a circuit.
3. A voltmeter measures the energy of electrons flowing in a circuit.
4. Answers will vary. A toaster and an iron are two examples.
5. Ohm's law states that the current in a circuit equals the voltage divided by the resistance.

Another kind of meter, the *voltmeter*, is used to measure the energy of the electrons that flow in the circuit. This energy is measured in units called **volts**. The number of volts measured by the meter is called the *voltage*.

volts

Some types of wire allow a large current to flow. Others offer more **resistance** to electron flow. Wires that allow a large current to flow are said to have a low resistance. Wires that allow only a small current to flow have a high resistance. In most cases, a wire's resistance increases if the wire is made thinner or longer.

resistance

The filaments in most light bulbs are thin tungsten wires with a high resistance. As a result of this resistance, electric energy is converted to light and heat inside a bulb. A toaster works the same way. Inside the toaster, electric energy is converted to heat, which browns the bread. Electric heating systems in buildings also depend on resistance.

Georg Ohm, a German physicist, studied the relationship between voltage, current, and resistance. He found that the current in any circuit is equal to the voltage divided by the resistance. This relationship is called Ohm's law. It means that for a constant voltage, current is less when resistance is greater. Or, for a constant voltage, current is greater when resistance is less. Units of resistance are called **ohms**, after Georg Ohm.

ohms

$$\text{current (amperes)} = \frac{\text{voltage (volts)}}{\text{resistance (ohms)}}$$

Checkpoint

1. What is an electric current?
2. How is electric current measured?
3. What does a voltmeter measure?
4. Give an example of a household appliance that must have a large electric resistance.
5. State Ohm's law.

TRY THIS

Using one D-cell, a flashlight bulb, and a 15-cm length of copper wire, see if you can light the bulb. How many different ways can you find to make the bulb light? Make a circuit diagram for each way that works.

19-5

Series and Parallel Circuits

Goals

1. To compare series and parallel circuits.
2. To explain how fuses work.

A basic circuit is made up of three parts: a power supply, an energy-consuming device, and a control, such as a switch.

series circuit

In a **series circuit** there is only one path the electrons can follow from the power supply and back. The power supply and the energy consumers are placed one after the other, in a series. (See Figure 19-10a.) When two or more cells are connected this way, they form a **battery**. A battery has more voltage than a single cell. When cells are connected in series, the negative pole of one is connected to the positive pole of the next.

battery

parallel circuit

In a **parallel circuit**, electrons have more than one path to follow (Figure 19-10b). When two cells are connected in parallel, their negative poles and their positive poles are connected.

Look again at the circuit in Figure 19-10a. If you open either switch A or B, both lights go out. In this circuit the bulbs are in series. Any current that flows through A must flow through B.

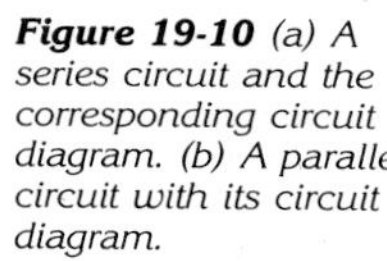

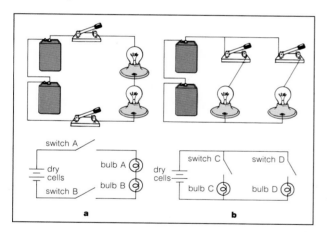

Figure 19-10 (a) A series circuit and the corresponding circuit diagram. (b) A parallel circuit with its circuit diagram.

Teaching Suggestions 19-5

Begin by reviewing the elements needed to make up a complete circuit. Also review the symbols used to represent circuits and their constituent parts. Then diagram and, if the equipment is available, use the first demonstration (see Demonstrations.) to display the differences between series and parallel circuits. This is a good way to both review simple circuits and introduce series and parallel wiring. Demonstrating the circuits shown in Figure 19-10 on student text page 481 will work well, also.

Be sure students understand why parallel wiring is used in household circuits, and why fuses, or circuit breakers, and adequate grounding of appliances are an absolute necessity.

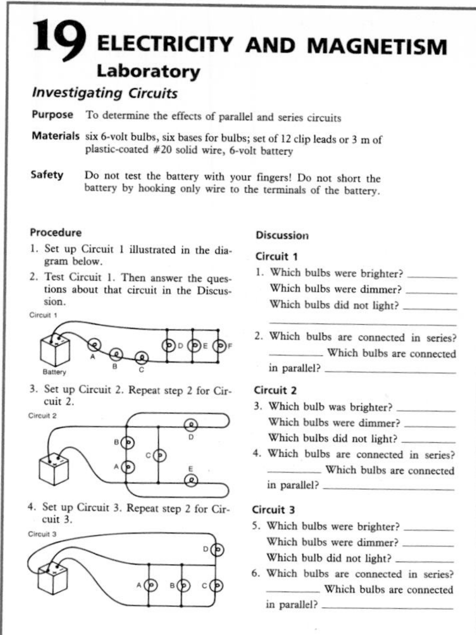

19 ELECTRICITY AND MAGNETISM
Laboratory

Investigating Circuits

Purpose To determine the effects of parallel and series circuits

Materials six 6-volt bulbs, six bases for bulbs; set of 12 clip leads or 3 m of plastic-coated #20 solid wire, 6-volt battery

Safety Do not test the battery with your fingers! Do not short the battery by hooking only wire to the terminals of the battery.

Procedure

1. Set up Circuit 1 illustrated in the diagram below.
2. Test Circuit 1. Then answer the questions about that circuit in the Discussion.

Circuit 1 (Battery; A, B, C, D, E, F)

3. Set up Circuit 2. Repeat step 2 for Circuit 2.

Circuit 2 (A, B, C, D, E)

4. Set up Circuit 3. Repeat step 2 for Circuit 3.

Circuit 3 (A, B, C, D)

Discussion

Circuit 1

1. Which bulbs were brighter? ______
   Which bulbs were dimmer? ______
   Which bulbs did not light? ______
   ______
2. Which bulbs are connected in series? ______ Which bulbs are connected in parallel? ______

Circuit 2

3. Which bulb was brighter? ______
   Which bulbs were dimmer? ______
   Which bulbs did not light? ______
4. Which bulbs are connected in series? ______ Which bulbs are connected in parallel? ______

Circuit 3

5. Which bulbs were brighter? ______
   Which bulbs were dimmer? ______
   Which bulb did not light? ______
6. Which bulbs are connected in series? ______ Which bulbs are connected in parallel? ______

Background

In a series circuit, there is only one path for the electrons to follow. The current is the same in all parts of the circuit. If several resistors are connected in series, an ammeter will read the same when connected in series with any of the individual resistors. However, for a given voltage, the total current in a series circuit diminishes as resistors are added to the circuit. The total resistance of the circuit is equal to the sum of the individual resistances connected in series. The circuit voltage is divided up among the individual resistors according to the amount of resistance each presents. That is, the voltage drop per resistor, as measured by a voltmeter connected across each resistor in a series circuit, is proportional to the value of the resistance. To sum up, for a series circuit with n resistors:

$$I = I_1 = I_2 = \ldots = I_n$$
$$R = R_1 + R_2 + \ldots + R_n$$
$$V_1 = IR_1, V_2 = IR_2, \ldots, V_n = IR_n$$

Parallel circuits offer electrons a choice of paths. The voltage across any two points in the circuit, as measured by a voltmeter connected in parallel, is always the same. The current through each of the elements connected in parallel is inversely proportional to the resistance of the element, that is, it is greater for a smaller resistance, and vice versa. The total current in the circuit is equal to the sum of these individual currents. The total circuit resistance is less than the resistance of any one of the individual resistors. To sum up, for a parallel circuit with n resistors:

$$V = V_1 = V_2 = \ldots = V_n$$
$$I = I_1 + I_2 + \ldots + I_n$$
$$\frac{1}{R} = \frac{1}{R_1} + \frac{1}{R_2} + \ldots + \frac{1}{R_n}$$

The electricity produced by practically all power plants is 60-cycle alternating current (AC). That is, the current oscillates back and forth 60 times per second. The simple circuits described in this section are direct current (DC) circuits. In these, the current flows in only one direction at all times.

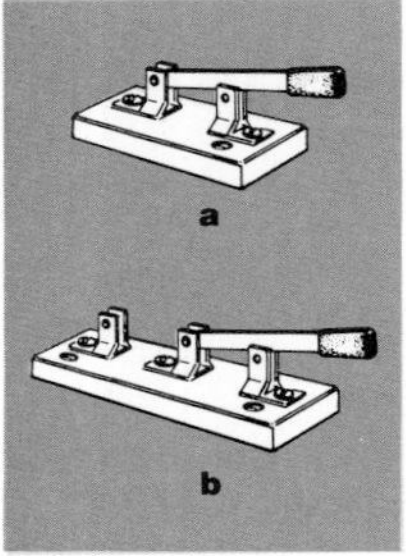

Figure 19-11 *(a) A toggle switch with one handle, or pole. The pole can be "thrown" only one way. (b) A double-throw switch. The single pole can move two ways. This is a three-way switch in that it can be closed one way, opened, and closed the other way.*

Now look at Figure 19-10b. If you open switch C with D closed, only bulb C goes out. If you open switch D with C closed, only bulb D goes out. In this circuit the bulbs are wired in parallel. Current from the power supply divides. Some electrons flow through bulb C at the same time that other electrons flow through bulb D. In a home, circuits are wired in parallel. In series, all the lights and other appliances would have to be either on or off at the same time.

Several kinds of switches are used in home-wiring circuits. (See Figure 19-11.) The simplest is the *toggle switch*, or single-pole, single-throw switch. The single-pole, double-throw switch is more complicated. It is a three-way switch. It might be used in a circuit with a light at the top of a stairway. With two single-pole, double-throw switches, the light can be controlled at the top and bottom of the stairs.

Suppose you are upstairs and the light is on, as shown in Figure 19-12a. Current is flowing through the right-hand wire between the switches. Then you walk downstairs and flip the lower switch. (See Figure 19-12b.) The circuit is broken, and the light goes out. Later on, you want to go back upstairs. So you flip the lower switch, which closes the circuit again. The light goes on. (See Figure 19-12c.) Then you walk up the stairs and flip the upper switch. The circuit is broken, and the light goes off. (See Figure 19-12d.)

Electricity from a power plant goes to homes in parallel circuits. The electricity travels from the power plant at very high voltage. This voltage is much higher than the voltage in your house. In fact, if this voltage were used with your appliances, it would damage them. Thus, before the current enters your house, it passes through a *transformer*. The transformer lowers the voltage to 110 V or 220 V. Transformers can be used to increase voltage, also.

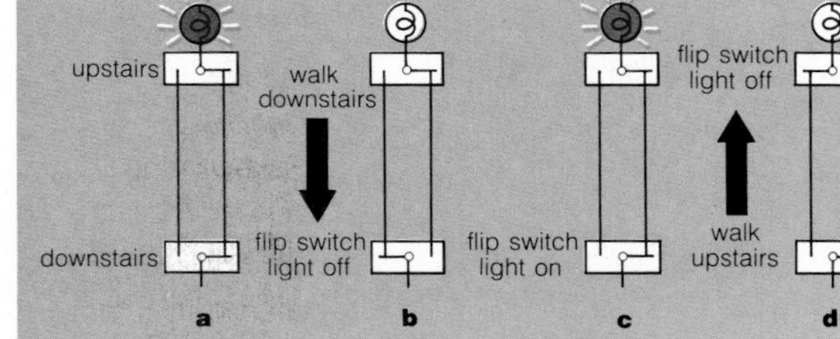

Figure 19-12 *Double-throw switches make it possible to control the light above a stairway from the top or bottom of the stairs.*

Demonstrations

1. Show the nature of a simple circuit and the effects of series and parallel wiring. Set up a circuit consisting of several D-cells in series, an ammeter in series with a flashlight bulb and a switch, and a voltmeter in parallel with the bulb. (**NOTE:** Always connect ammeters in series and voltmeters in parallel.) Show students that the current and voltage vary with the size of the battery (the number of cells in series). Show what happens when more bulbs are placed in series with the first bulb and the voltmeter is still connected in parallel with the first bulb. (The ammeter and voltmeter readings both drop.) Show what happens when two bulbs are wired first in series and then in parallel, keeping the voltmeter connected in parallel with one bulb and the ammeter connected in series with the voltage source. (The ammeter and voltmeter readings are both higher in the parallel circuit.) Have students suggest other circuit variations; then try the variations.
2. Show examples of series and parallel circuits. The circuits shown in Figure 19-10 on student text page 481 will serve this purpose quite well.
3. Show students a variety of household outlets and switches. Have them determine where connections are made and how the switches work.

Household electric circuits have fuses or circuit breakers in them in order to protect the circuits from carrying too much current. In a *fuse*, a metal connector will melt if too large a current flows through it. Once the connector has melted, the circuit is broken. A *circuit breaker* is like a switch. It opens when too much current flows through it.

In a home, too many electric appliances are sometimes plugged into one circuit and used at the same time. Then a large current flows in the circuit. The wires overheat, and a fuse burns out or a circuit breaker trips. The fuse or circuit breaker opens the circuit. This prevents the possibility of a fire being produced by overheated wires. Fuses are designed for your safety. Never replace a fuse with a penny. A penny can carry a current large enough to cause a fire.

Your body is a reasonably good conductor, especially when it is wet. Suppose you touch an uninsulated wire or switch when you are connected to the water system. You then become part of a circuit. You might be standing in a bathtub or touching a faucet. A current large enough to kill you can go through your body to Earth. Current can pass from the wire through you to the faucet or tub and then to the ground.

Many modern appliances have plugs with three prongs, as shown in Figure 19-13. The round prong carries the ground wire. If the appliance develops a defect, grounding provides an easy path for the current to pass rapidly to Earth.

Never use an appliance if the cord appears to be worn. Wires in the cord might make contact with grounded objects, injuring you or starting a fire. If you receive a shock from an electrical appliance, shut it off. Do not use it again until it has been checked by an electrician and repaired.

Figure 19-13 *A three-pronged plug. The round middle prong is for the ground connection.*

Checkpoint

1. Make sketches showing a light bulb and a motor wired in series and in parallel.
2. How do you know that the electric lamps in your home are wired in parallel?
3. Why are fuses used in circuits?
4. List some safety procedures you should follow to avoid electrical shocks or fires.

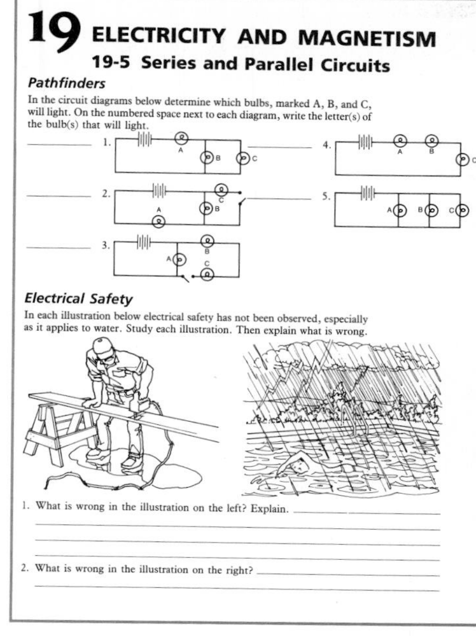

19 ELECTRICITY AND MAGNETISM

19-5 Series and Parallel Circuits

Pathfinders

In the circuit diagrams below determine which bulbs, marked A, B, and C, will light. On the numbered space next to each diagram, write the letter(s) of the bulb(s) that will light.

Electrical Safety

In each illustration below electrical safety has not been observed, especially as it applies to water. Study each illustration. Then explain what is wrong.

1. What is wrong in the illustration on the left? Explain. ______
2. What is wrong in the illustration on the right? ______

Checkpoint Answers

1. See Figure TE 19-2.
2. You know they are wired in parallel because you can turn off one electrical appliance without affecting the others.
3. Fuses or circuit breakers are used to prevent circuit overloads, which can lead to overheating of wires and may start a fire.
4. Answers will vary. Some examples are: do not overload circuits; do not touch electrical devices when wet; use only electrical appliances with three-pronged plugs; do not use appliances with worn cords; do not replace fuses with pennies.

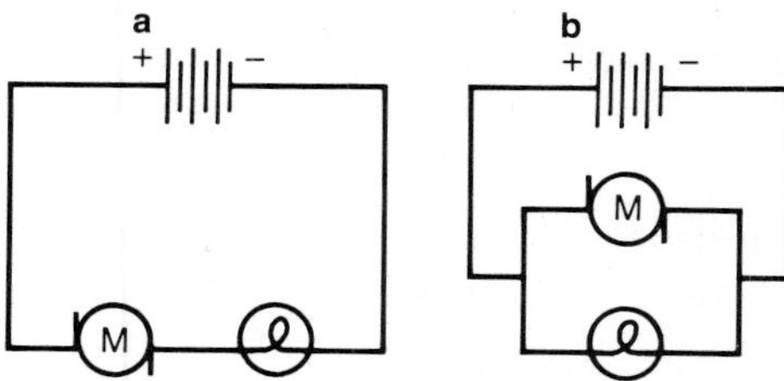

Figure TE 19-2 Motor and light bulb connected in (a) series and (b) parallel.

Teaching Tips

- Stress the information on the safe use of household electricity. Provide adequate time to discuss this material thoroughly and to answer any questions students may have.
- Be certain students understand that a transformer simply either "steps up" or "steps down" voltage.

Mainstreaming

Hands-on work is essential for special-needs and handicapped students.

19-6

Magnetism

Goal

To describe magnetic fields, including Earth's magnetic field.

You have probably seen **magnets**, objects that attract anything containing iron. If you put a bar-shaped magnet into a box of paper clips, some clips stick to it. Most of the clips stick to one end or the other of the magnet. These regions where the magnetic attraction is strongest are called *poles*.

Magnets can attract or repel objects without touching them. In other words, the magnet can exert a force (a push or pull) on an object. If you place a paper clip on a piece of paper and then move a magnet under the paper, the paper clip will move. A clearer way to show this is to pour iron filings on a paper covering a magnet. The iron filings line up, making a picture of the invisible force around the magnet. The region in which a magnet can affect an object is called its **magnetic field**. Figure 19-14 shows the magnetic *lines of force* that make up a magnetic field.

Earth itself has a magnetic field. It has a north and a south magnetic pole, also. Any magnet, freely suspended and far from other magnetic materials, lines up along a north-south direction. Knowing this, you can use a magnetic compass to determine direction. The needlelike indicator is a small magnet. (See Figure 19-15.) The end of the magnet that points north is called a *north-seeking pole* (N). The oppo-

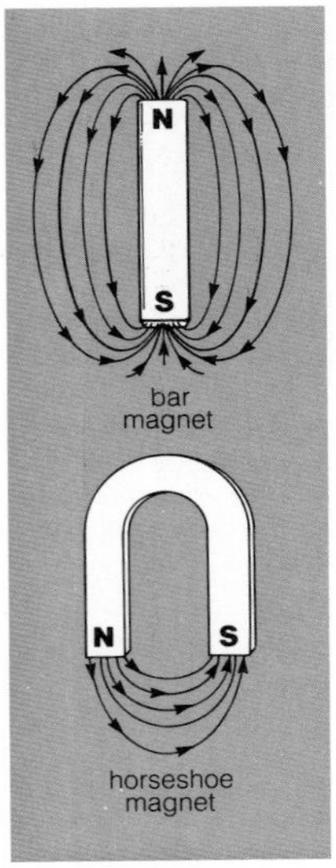

Figure 19-14 *Magnetic lines of force around a bar magnet (above) and a horseshoe magnet (below).*

magnets

magnetic field

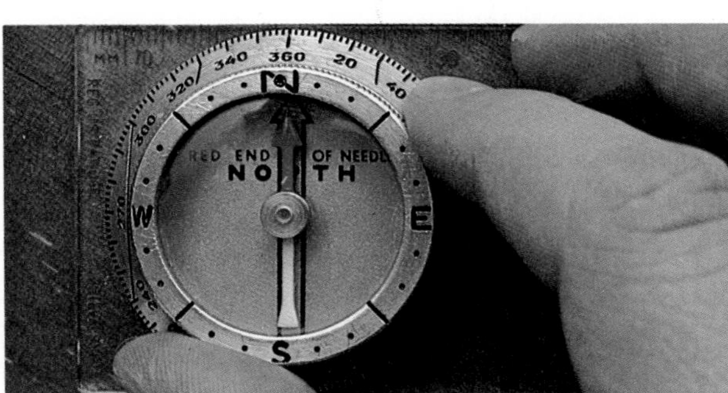

Figure 19-15 *A person using a magnetic compass.*

Teaching Suggestions 19-6

Most students are somewhat knowledgeable about magnets and the fact that they will attract magnetic materials such as iron. The concept of a magnetic field, however, is subtle. It is best to treat this idea lightly. A good way to begin is to ask students if they are aware of Earth's magnetic field. Responses will vary. Some students may mention the functioning of a magnetic compass. To demonstrate Earth's magnetic field, you may wish to do Demonstration 1. (See Demonstrations.)

When students accept the fact that there is a magnetic field around Earth near its surface, extend the concept to magnets. Referring to Figure 19-14 on student text page 484, discuss the lines of magnetic force that surround a bar magnet or a horseshoe magnet. This phenomenon, too, is worth demonstrating. Use the simple demonstration described in the student text on page 484 or do Demonstration 2.

Explain to students that the iron filings around a magnet indicate that its effect extends beyond the magnet itself. Physical contact with the magnet is not required.

Background

Magnetism always arises from electric currents. Without moving charges there is no magnetism. In ordinary bar magnets, it is the motion of electrons in atoms that constitutes the charges in motion. The fact that an electric current is surrounded by a magnetic field is discussed in section 19-7. In section 19-6, magnets are introduced in a descriptive way, without explanation of the underlying causes of magnetism.

Demonstrations

1. Show the presence of Earth's magnetic field near its surface. Hang several bar magnets from the ceiling in various parts of the room. Be sure the magnets are well removed from any other magnetic materials. Ask students why all the magnets are aligned parallel to each other. (Presence of Earth's magnetic field accounts for alignment.)
2. Show the magnetic field around a bar magnet. Place a bar magnet on an overhead projector. Put a glass plate over the magnet. Sprinkle iron filings on the glass and tap the plate with your finger. The filings will line up along the field lines because the pieces of metal are tiny magnets.
3. Show that opposite magnetic poles attract and that like poles repel. Suspend a bar magnet on a string. Bring both poles of another bar magnet near one end of the suspended bar magnet. Have students note the resulting attraction or repulsion.
4. Show how a compass responds to magnetic field lines by using a compass with glass on both top and bot-

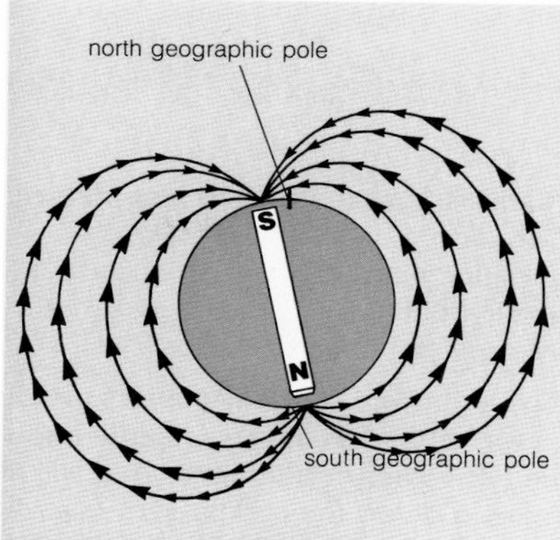

Figure 19-16 *Magnetic lines of force around Earth. They are drawn as they would be if there were no streams of particles from the sun to bend them.*

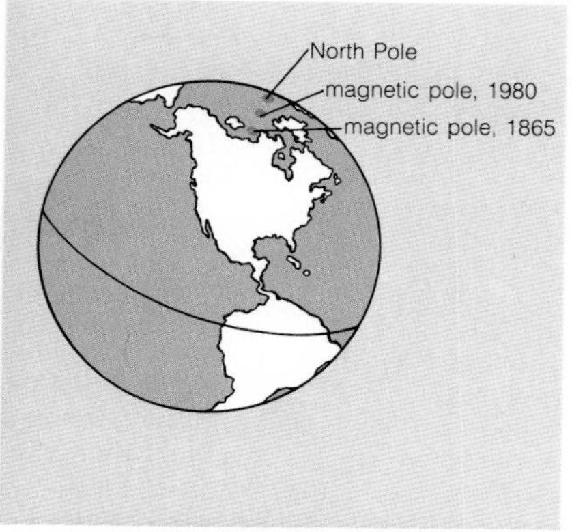

Figure 19-17 *The positions of Earth's magnetic poles have changed. This shows how far the north (south-seeking) magnetic pole has moved since 1865.*

site end of the magnet is called a *south-seeking pole* (S). A compass needle lines up in a north-south direction. The north-seeking pole of one magnet attracts the south-seeking pole of another magnet. Like magnetic poles repel each other.

Maps of Earth's magnetic field lines have the general shape shown in Figure 19-16. Earth can be thought of as a sphere with a giant bar-shaped magnet inside. The south-seeking pole of this very large magnet is in northern Canada. The poles of Earth's magnetic field do not coincide with its geographic poles. Therefore, compass needles in most locations do not point to true (geographic) north. Earth's magnetic field changes slightly every year. (See Figure 19-17.)

Checkpoint

1. What can be learned from a picture of a magnetic field?
2. What happens when the north-seeking poles of two bar magnets are brought close to one another?
3. How do scientists know that Earth's magnetic pole in northern Canada is a south-seeking pole?

TRY THIS

Stroke a paper clip with a bar magnet. Do the same to several other paper clips. How do these paper clips now behave?

tom. Move the compass around a bar magnet. Have students observe its reaction to the field of the magnet.

Teaching Tips

- Be sure students see a demonstration of the magnetic field around a bar magnet.
- Students are often confused by the fact that Earth's magnetic pole in northern Canada is a south-seeking pole. Be sure they understand the reason for this.
- Ask students if they can suggest a way to determine if a substance is magnetic. (Is the substance attracted, or repelled, or unaffected by the pole of a magnet?)

Facts At Your Fingertips

- Earth's magnetic field is not uniform. It is constantly changing, although the change occurs very slowly. There is geological evidence that the polarity of Earth's magnetic field has reversed several times during its history. The causes of these reversals are not known.
- Earth's magnetic field is thought to arise from the movement of charged matter within its molten metal core.

19 ELECTRICITY AND MAGNETISM

19-6 Magnetism

A Hidden Substance

Each phrase below describes a term related to magnetism. Write the letters of each term on the spaces to the left of each description. The letters in the boxes will spell out the name of a magnetic substance.

1. Portions of a magnet where magnetic force is greatest.
2. The end of a bar magnet that points north.
3. The region in which a magnet can affect an object.
4. What like magnetic poles do.
5. Earth's north magnetic pole.
6. Lines that represent a magnet's field of force.

The name of the hidden magnetic substance is **lodestone.**

Magnetism Through History

Read the paragraphs below. Then write the answers to the questions.

The natural forces of magnetism were known to the ancient Greeks. Thales of ancient Greece knew that iron ores containing lodestone attract iron.

In medieval times Europeans learned that when the Chinese hung a piece of lodestone from a string, the lodestone pointed north and south. This primitive compass gave explorers an improved method of navigating.

In the early 1600's William Gilbert, the English scientist, published a paper in which he demonstrated that Earth is a giant magnet. Gilbert later showed that there are two kinds of poles: south-seeking and north-seeking. He also showed that like poles repel each other and unlike poles attract each other.

1. How did the Chinese demonstrate a property of lodestone?
2. What did Gilbert find out about the poles of a magnet?

Using Try This

After steel paper clips have been stroked with a bar magnet, they behave like small magnets.

Checkpoint Answers

1. Answers will vary. For example, a picture of a magnetic field shows the region in which a magnet can affect objects.
2. They repel each other.
3. Because the north-seeking poles of all magnets point toward the pole in northern Canada, and because opposite poles attract, the pole in Canada must be a south-seeking pole.

Teaching Suggestions 19-7

Describe how a current flowing through a wire deflects a nearby compass needle. Sketch Figure TE 19-3 on the chalkboard, but leave out the compass needles. Ask students what will happen to the needles when current is flowing in the wire. Answers will vary. Some students may know that the needles will be deflected, and that they will point in the direction of the lines of magnetic force around the wire. You might want to set up a very simple demonstration. Attach a large loop of wire to a dry cell through a switch and have small groups of students hold a compass near the wire while current flows.

Follow with a description of how a changing magnetic field can induce a current. Refer to Figure 19-19 on student text page 487. Stress that the magnetic field must be changing for a current to be induced. The induction of a current by a moving magnet is described in Demonstration 3. (See Demonstrations.)

The ideas in this section may be difficult for some students. However, they must be mastered if students are to understand how electricity is produced by generators. Proceed slowly, do as many demonstrations as possible, provide as much hands-on experience as equipment and time allow, and encourage questions from students.

Chapter-end projects 1, 3, 4, 5, and 6 apply to this section.

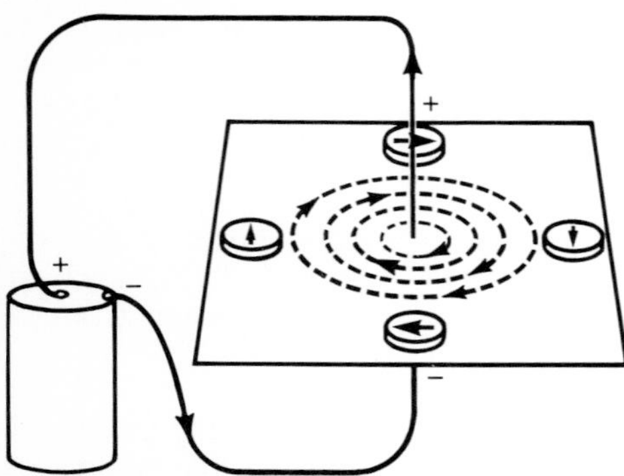

Figure TE 19-3 Effect of current flow on compass needle.

19-7

Electricity from Magnetism

Goals

1. To show that there is a magnetic field around a current-carrying wire.
2. To explain how electricity can be produced by a changing magnetic field.
3. To explain how an electric motor works.

Current flowing through a wire will cause a compass needle to turn. If you place a wire above a compass needle and in line with it, the needle turns when current flows. (See Figure 19-18.) If the wire is placed below the compass, the needle turns in the opposite direction. If you reverse the connections to the battery poles, the direction that the needle turns is also reversed.

These experiments show that a current-carrying wire is surrounded by a magnetic field. By winding the wire into a coil, you can produce a magnetic field similar to that around a bar magnet. Placing a soft iron core within the coil makes the field even stronger. A wire coil wound around an iron core is an **electromagnet**. It behaves like a magnet only when electricity flows through the coil.

electromagnet (ih lek troh MAG nit)

After people learned that electricity can produce a magnetic field, they tried to turn the process around. They tried to produce electricity from magnets. At first, attempts to

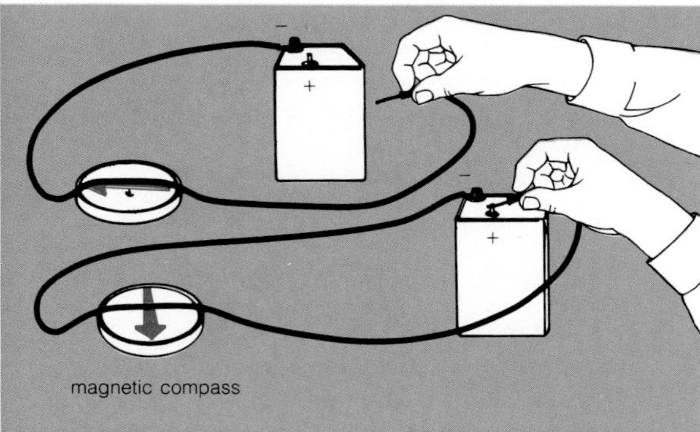

Figure 19-18 *When current flows, a compass needle turns as shown.*

Background

One of the greatest scientific breakthroughs of all time was Michael Faraday's discovery that electricity can be produced by passing a magnetic field through a coil of wire. Out of this discovery came the electric power plants of today. The energy per charge produced in this way depends only on the rate at which the magnetic field changes. If the field changes rapidly, a large voltage is induced. If the field doesn't change at all, no voltage is induced—no matter how strong the field.

produce electricity from strong magnets were not successful. But in 1831, Michael Faraday, a self-taught English scientist, uncovered the secret. For electricity to be produced in a coil of wire, the magnetic field through the coil must be changing.

Look at Figure 19-14 (on page 484) again. Notice that the number of lines in the field is much greater at the poles than at the center of the magnet. The number of lines in an area indicates how strong the magnetic force is—more lines, stronger force.

If a bar magnet is pushed into and out of a coil of wire, the number of lines of force affecting the coil is constantly changing. When this happens, a current is induced in the coil. The current can be detected with a sensitive ammeter. (See Figure 19-19.) A current is also induced when the magnet is stationary and the coil is moved.

Most of the electricity used in homes and industry is produced at power plants. Power plants make use of electric *generators*. These consist of large magnets turning inside giant wire coils. Current flows as the magnetic field through the wire coils changes.

Figure 19-19 *A coil of wire can be attached to an ammeter. If a magnet is moved back and forth through it, the ammeter will show that a current has been induced. A current also may be induced by moving the coil back and forth around a stationary magnet.*

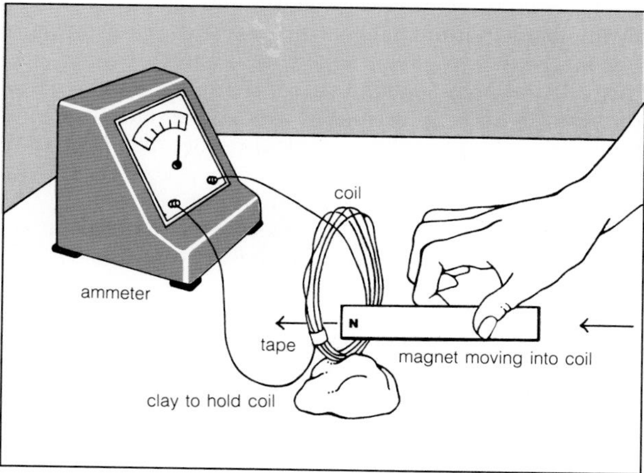

Demonstrations

1. Show how an electric current deflects a compass needle. Use a compass with glass on both sides. Place the compass just above a long wire resting on an overhead projector. When the wire is connected to a battery, the compass needle will be deflected. Reversing the leads to the battery, or placing the wire above the compass, will deflect the needle in the opposite direction.
2. Demonstrate that a wire coil behaves like a bar magnet. Connect a wire coil or solenoid (check with your physics department) to a battery or small power source. Then move a compass around the coil. Have a student team sketch the lines of force around the coil, as indicated by the changes in the direction of the compass needle.
3. Show that current is induced by a magnetic field only when the field is changing. Connect a solenoid by means of long wires to a sensitive ammeter or galvanometer. Move a magnet into and out of the coil. Current will flow only when the magnet is moving; that is, when the magnetic field relative to the coil is changing.
4. Have an interested student or students make a simple DC motor, using the design shown in Figure TE 19-4, and demonstrate the motor to the class.

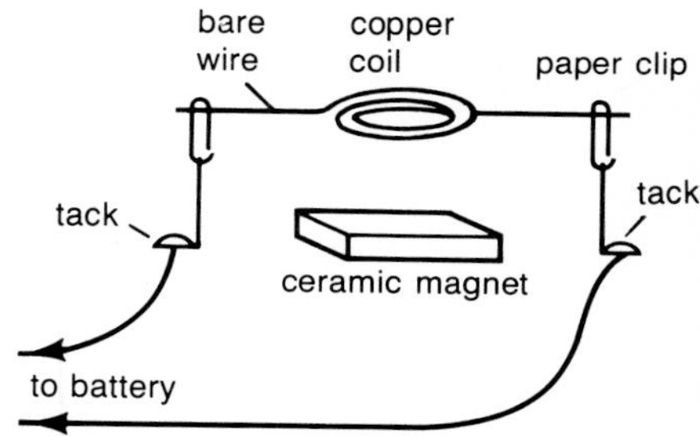

Figure TE 19-4 Simple DC motor.

Teaching Tips

- Follow the logic in this section carefully with students. Allow ample time for discussion and questions.
- Do as many demonstrations and provide as much hands-on experience as possible.

Facts At Your Fingertips

- The left-hand rule can be used to indicate the direction of the magnetic field around a current-carrying wire. Point the thumb of the left hand in the direction of electron flow in the wire. The fingers will then curl around the wire in the direction of the magnetic field lines.
- In the early days of electric power, Thomas Edison was an advocate of direct current. George Westinghouse, who realized that transporting alternating current over long distances would be cheaper, proposed alternating-current generators.
- Whether or not an individual atom is magnetic depends on how the magnetic fields of all the electrons line up. Any piece of matter in which the magnetic fields of the electrons do not cancel each other will be magnetic. Iron, nickel, and cobalt are three elements that exhibit magnetic properties.
- If the atoms in a substance are arranged so that their magnetic fields reinforce each other, strong magnetic properties will result.

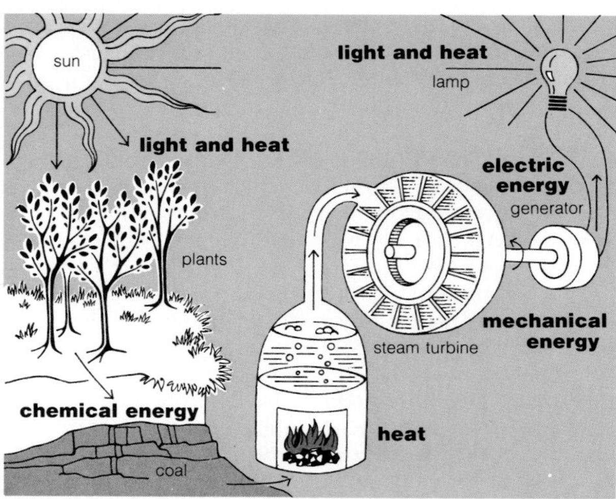

Figure 19-20 *Modern light and heat come from ancient light and heat. Energy from the sun was stored by ancient plants, which then became coal. Today, burning coal gives off heat that is transformed into the energy used by a lamp.*

The generator converts kinetic energy to electrical energy. The kinetic energy comes from the spinning magnet, which is powered by a turbine. (See Figure 19-20.) The turbine can be driven by water flowing through a dam, as in hydroelectric power stations. The potential energy of water behind the dam is converted to kinetic energy, then to electricity. More commonly, generators are driven by steam produced from burning fossil fuels (coal, oil, or natural gas) or from the heat released by a nuclear reactor. So, in a typical power plant, chemical or nuclear energy is converted to kinetic energy and then to electricity.

The current produced in these generators changes direction as the magnetic field changes its direction through the coils. The back-and-forth movement of the current in the coils and wires leading from the coils is called *alternating current* (AC). This current differs from the current produced by flashlight cells or car batteries. Such current is called *direct current* (DC) because it flows in only one direction.

Kinetic energy from steam or flowing water can be used to produce electric energy. Electric energy, in turn, can be used to produce kinetic energy. *Electric motors* transform electric energy to kinetic energy. The operation of an electric motor is just the reverse of the way a generator works. In fact, electric motors can be used as generators and generators can be made to run as motors. To make an electric motor, a magnet is suspended inside a coil. Then an alternating current is sent through the coil. This produces a magnetic field around the coil. This magnetic field is constantly changing in strength and its poles are constantly switching.

If the switching is timed correctly, the magnet is alternately attracted and repelled. This makes it spin. The spinning movement is kinetic energy that can be used to do work, such as turning a fan.

The motor just described is a very primitive one. Most of the motors you use do not contain permanent magnets. In most cases both the coil and the spinning part are electromagnets.

Checkpoint

1. What happens if you place a magnetic compass over a current-carrying wire?
2. Describe an electromagnet.
3. What did Faraday discover about producing electricity from magnets?
4. What does a generator do? What does an electric motor do?
5. What is the difference between alternating and direct current?

TRY THIS

Wrap 2–3 m of enameled copper wire around an iron (or steel) nail. (See Figure 19-21.) Be sure to wrap the wire in one direction only. Sandpaper the ends of the wire to remove the enamel insulation. Connect the ends of the wire to the poles of a battery and you will have an electromagnet. (**CAUTION**: Don't leave the wires connected for very long or you will wear out the battery quickly.)

How many paper clips can your electromagnet lift? How can you make the electromagnet stronger? weaker?

How strong is your magnet when you wind half the wire turns one way and the other half in the opposite direction? How can you explain this?

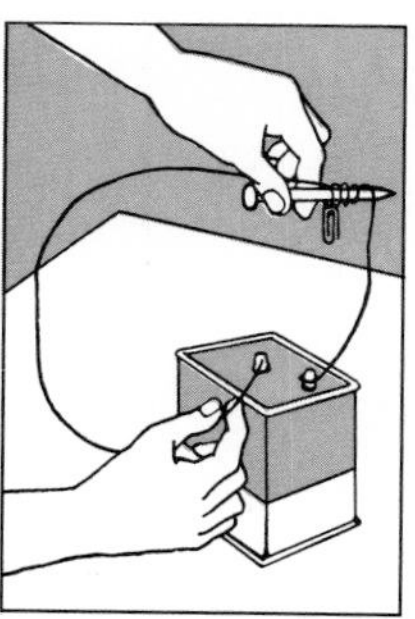

Figure 19-21 *This illustration shows how to make an electromagnet.*

19 ELECTRICITY AND MAGNETISM
19-7 Electricity From Magnetism

May the EMF Be with You!

Match the term in Column A with its definition in Column B. Then write the letter of the definition on the space next to its term.

A	B
___ 1. electric motor	a. The back-and-forth movement of current in coils or wires.
___ 2. electric generator	b. The current that flows from a battery.
___ 3. alternating current	c. Type of energy that a generator converts to electrical energy.
___ 4. magnetic field	d. A device whose operation is the reverse of a generator.
___ 5. direct current	e. Area that surrounds a wire through which current flows.
___ 6. turbine	f. A device that consists of large magnets turning inside giant wire coils.
___ 7. electromagnet	
___ 8. kinetic energy	

Using Try This

Students should discover that the electromagnet can be made stronger by using more turns of wire or a battery with more cells. Conversely, fewer turns of wire or a battery with fewer cells will reduce the strength of the electromagnet. Winding half of the wire one way and half the other way causes the fields to cancel each other, and so there will be no magnetic field.

Checkpoint Answers

1. The compass needle turns, indicating the presence of a magnetic field surrounding the wire.
2. An electromagnet consists of a wire coil that surrounds a soft iron core. When current flows in the coil, the coil and core behave like a magnet.
3. To produce electricity from a magnet, the magnetic field must be in motion, or changing.
4. A generator converts kinetic energy into electricity. An electric motor converts electricity into kinetic energy.
5. Alternating current moves back and forth. Direct current flows in one direction only.

Teaching Suggestions 19-8

Your class might benefit from observing the electric meters in the school building. Once students have seen what meters look like, they can observe the ones at their houses or apartments. Some may enjoy keeping records of the energy used in their homes and checking power-company charges.

Be sure students understand how to convert watts to kilowatts, and how to calculate energy use for a given period of time in kilowatt-hours.

Find out your power company's average charge per kilowatt-hour. With this figure, students can determine (a) the approximate cost of electricity in their homes, and (b) the approximate cost per year of operating some of the appliances listed in Table 19-1 on student text page 491. It will probably be necessary to work out several sample problems for students. They should then try a few on their own.

Chapter-end project 7 fits in well with this section.

19 ELECTRICITY AND MAGNETISM
Computer Program

Home Electricity Usage

Purpose To calculate electricity usage, in kilowatts.

Background Electric power arrives in most homes at both 120 and 240 volts. It passes through an electric meter and into a service panel with fuses or circuit breakers. Lights and other common household items require 120 volts of current. An electric water heater, furnace, dryer, stove, or large air conditioner usually requires 240 volts of current. You can calculate the electricity requirements of a home by measuring fuses or circuit breakers in the service panel.

ENTER the following program:

```
10  PRINT "ENTER AMPS OF EACH FUSE OR"
20  PRINT "CIRCUIT BREAKER AT 120 VOLTS."
30  INPUT "TYPE 999 WHEN FINISHED."; A1
40  IF A1 = 999 THEN GOTO 70
50  LET T1 = T1 + A1
60  GOTO 30
70  PRINT "ENTER AMPS OF EACH FUSE OR"
80  PRINT "CIRCUIT BREAKER AT 240 VOLTS."
90  INPUT "TYPE 999 WHEN FINISHED."; A2
100  IF A2 = 999 THEN GOTO 130
110  LET T2 = T2 + A2
120  GOTO 90
130  LET W1 = 120 * T1
140  LET W2 = 240 * T2
150  LET TW = (W1 + W2) / 1000
160  PRINT "MAXIMUM USAGE = "; TW; " KILOWATTS."
170  END
```

Discussion

A home has two 15-amp fuses and one 20-amp fuse for 120-volt current and two 50 amp fuses for 240-volt current. What is this home's maximum kilowatt usage?

Upgrade the Program

Lines 30 and 90 permit entry of amp rating of a circuit breaker or fuse. Lines 40 and 100 check to see if 999 has been entered, signifying the last entry. Lines 50 and 110 keep a running total of the amp ratings. Calculation of watts occurs in lines 130 and 140, multiplying the voltage by total number of amps. Line 150 totals watt values and divides by 1000, so that the answer can be expressed in kilowatts. Add questions about percent of kilowatts used and price per kilowatt-hour in order to figure monthly electric bill.

Figure 19-22 *Electric meters like this register the kilowatt-hours of electricity used in a home.*

19-8

Using Electricity at Home

Goal

To explain how the amount of electrical energy used is measured.

The *electric meter* in Figure 19-22 is a familiar sight. Every customer serviced by an electrical utility company has a meter. As the meter's disc turns, it indicates the rate at which electricity is being used. Thus, an electric meter is similar to a speedometer in an automobile.

The dials on an electric meter resemble an automobile's odometer. Suppose you want to find how far a car has traveled in a period of time. You simply subtract an earlier odometer reading from the present reading. Similarly, your utility company can determine how much electrical energy you have used by reading the dials on your meter. Utility companies measure and charge for electrical energy in *kilowatt-hours* (kW · h). A kilowatt is a unit of power equal to 1000 watts.

Power is energy used per unit of time. If power is multiplied by time, the product will be energy.

$$\text{power} \times \text{time} = \text{energy}$$

or

$$\text{kW} \times \text{h} = \text{kW} \cdot \text{h}$$

For example, to find the electrical energy transformed into heat by a toaster, multiply the toaster's power by the time it operates. A 1.2 kW toaster uses 1.2 kW · h in 1 hour, 2.4 kW · h in 2 hours, 3.6 kW · h in 3 hours, and so on. If the toaster is used for 15 minutes each day, the energy required for the toaster each day is:

$$1.2 \text{ kW} \times 0.25 \text{ h} = 0.3 \text{ kW} \cdot \text{h}$$

The total electric bill each month is calculated by multiplying the total kilowatt-hours of energy used by the cost of electricity per kilowatt-hour.

Background

Electric energy flows from the power source to one's home or the place of its use through an electric meter. As the electricity flows through the meter, it turns a disc inside that is visible from the outside. The turning disc is connected by means of a gear train to the dials that record the amount of electrical energy passing through the meter.

The more energy used, the faster the disc turns. The right-hand dial indicates kilowatt-hours, the next one shows tens of kilowatt-hours, the next one hundreds of kilowatt-hours, and so on. Some meters have a digital display rather than dials. On all meters, the consumption of electrical energy is simply the difference between the readings at the beginning and the end of the billing period.

Table 19-1: Energy Use of Common Home Appliances

Electric appliance	*Average wattage*	*Average hours used per year*	*Kilowatt-hours per year*
Kitchen			
broiler	1400	70	98
coffee maker	900	120	108
dishwasher	1200	300	360
range	12,200	100	1220
microwave oven	1450	130	189
toaster	1200	35	42
refrigerator (340 L)	240	3000	720
frostless refrigerator (340 L)	320	3800	1216
frostless freezer (425 L)	440	4000	1760
Laundry			
clothes dryer	4800	200	960
iron	1000	140	140
washing machine	500	200	100
water heater	2500	1600	4000
quick-recovery water heater	4500	1000	4500
Lighting			
light bulbs	660	1500	990

Electric appliance	*Average wattage*	*Average hours used per year*	*Kilowatt-hours per year*
Comfort			
air conditioner	900	1000	900
electric blanket	180	830	149
dehumidifier	250	1500	375
fan (window)	200	850	170
humidifier	180	900	162
Health/Beauty			
hair dryer	750	50	38
shaver	14	80	1
sun lamp	280	60	17
Entertainment			
radio	70	1200	84
B/W television			
tube	160	2200	352
solid-state	55	2200	121
color television			
tube	300	2200	660
solid-state	200	2200	440
Housewares			
clock	2	8760	18
vacuum cleaner	630	75	47
sewing machine	75	140	11

Table 19-1 lists the average power, hours of use, and energy used for a number of common appliances throughout a one-year period. To find the average annual operating cost of any appliance, multiply the annual energy required in kilowatt-hours by the cost per kilowatt-hour.

Checkpoint

1. What does an electric meter measure?
2. How could you calculate the energy used by an electric appliance?

Teaching Tips

- Encourage students to read the electric meters at their homes.
- Be sure students understand that power is measured in kilowatts and that 1000 watts equal one kilowatt. Because power is energy, or work, per unit of time, we must multiply power by time to obtain energy. To find the cost of electrical energy, multiply the number of kilowatt-hours by the cost per kilowatt-hour.
- Confusion about units of electrical energy and power is widespread. Have students look for errors in newspapers and other print media.
- Show students an electric bill. You might duplicate copies of your own bill. Point out that the prices charged by the utility company depend on how much electric energy is used. It might be \$0.10 per kW·h for the first 100 kW·h, \$0.08 for the next 200 kW·h, \$0.075 for the next 300 kW·h, and so on. Check the calculation and discuss any additional charges, such as the fuel-adjustment charge.
- Have a team of students monitor the school's use of electric energy. Ask them to suggest reasons for fluctuations.

Fact At Your Fingertips

- From 1960 to 1973 world energy use almost doubled. During the same period, the use of energy in the United States grew by about 57%. Through conservation, U.S. energy use has diminished somewhat in recent years.

19 ELECTRICITY AND MAGNETISM

19-8 Using Electricity at Home

Measuring Electrical Energy

The questions below relate to the use of electrical energy by household appliances. Read each question. Then use Table 19-1 in your textbook to help you answer the question. Hint: Electrical energy is measured in kilowatt hours.

1. Which appliance uses the most energy in one year? ________
2. Which appliance uses the least energy in one year? ________
3. If you are going to buy a refrigerator and you want one that uses a minimum of energy, what type should you buy? ________
4. If you want to reduce the energy used for a television set, but do not want to decrease your viewing time, what type of television set would save you the most energy? ________

The Cost of Power

The problems below relate to the cost of electrical power. Read each problem. Then apply what you have learned in section 19-8 to solve it. Hints: Assume that the cost per kilowatt hour is \$0.08. Use Table 19-1 in your textbook for the number of kilowatt-hours an appliance is used each year.

1. What does it cost to operate a quick-recovery water heater for one year? ________
2. If you could lower the wattage of the bulbs in your house by 33%, what would you expect to pay for the use of energy for light bulbs? ________
3. The television set at a particular house uses 276 kilowatt-hours per year. If the input power is 200 watts, what is the average number of hours of use per day? ________

Checkpoint Answers

1. It measures the kilowatt-hours of electric energy used.
2. Multiply its power rating in kilowatts by the time in hours that the appliance is in operation.

Teaching Suggestions 19-9

Ask students to list the different ways they use electricity. Choose several of the most common usages and analyze them. Encourage students to voluntarily identify essential and nonessential uses. The ability to analyze and evaluate electric energy usage is a prerequisite for reducing costs. The Try This activity involves students in an energy-saving study and so does chapter-end project 8.

Background

Any effort to find ways to reduce the use of electric energy is worthwhile. The cost of electricity is on a virtually indefinite upward path. Nuclear power increases in cost every year, and the costs of fossil fuels, such as coal, oil, and natural gas, are increasing as well. The only break in this trend, according to most experts, will come when fusion power is practical. This achievement, however, lies in the future.

In addition to careful analysis of electricity usage and appropriate cost-cutting, off-peak usage is a possibility worth considering. There is far less demand for electricity at certain times—during the night, for example—so some utility companies offer much lower rates at such times.

ENRICHMENT

19-9 Saving Electricity

Goal

To list ways people can reduce their use of electricity.

The rising cost of fossil fuels is causing the price of electricity to soar. To keep costs down, people can use certain *energy conservation* measures. These reduce the use of electricity, and thus its cost.

If you use an electric stove, be sure to cover pans when you cook food in boiling water. To make the best use of the available heat, select pans that cover the entire heating element. Those are two ways to get as much heat as possible to the food. Turn off heating elements, or the oven, several minutes before food is thoroughly cooked. The food will continue to cook while the elements are cooling. Do not open the oven to check on food while it is cooking. Use heating elements on the top of the stove, rather than using the oven, whenever possible. The heat is transferred more directly to the food.

If you have an automatic dishwasher, do not run it unless you have a full load. After the rinse cycle, let the dishes air dry instead of using the heating element.

To keep a refrigerator and freezer efficient, the inside of the refrigerator should be no colder than 4°C. The doors of the refrigerator should close tightly. If the rubber strips that seal the door are in good condition, they should grip a strip of paper firmly. A freezer that is full of food is more efficient than one that is half empty. The freezer or refrigerator should be well insulated so that heat from the surroundings does not easily enter the cold space. Because cold air tends to sink, a chest type freezer is more efficient than an upright model. A frostless model of either of these appliances needs more energy to operate.

Figure 19-23 *Drying clothes on a clothesline saves the energy needed to operate a dryer.*

To wash clothes, use cold or warm rather than hot water whenever possible. Use the washer and dryer only when you have enough clothes for a full load. Dry large volumes of clothing in consecutive loads. That way the dryer stays hot from one load to the next. Dry clothes on a clothesline whenever possible. You can do this either outside or inside. Wear wash-and-wear clothing that does not require ironing.

Teaching Tip

- Interested students might like to design experiments to test such statements as "covering pans while heating water reduces energy use" and "taking quick showers instead of taking baths saves energy."

Figure 19-24 *The old TV set with tubes uses much more energy than a modern solid-state set.*

There are other important ways in which people can conserve electrical energy:

1. Insulate hot-water heaters and pipes.
2. Turn off the electric water heater when going away for a day or more.
3. Lower the thermostat on the water heater to the temperature at which the water is normally used.
4. Take quick showers instead of baths.
5. Turn off lights and appliances that are not being used.
6. Use a single high-wattage light bulb in multi-bulb fixtures, rather than many low-wattage bulbs.
7. Buy a solid-state television set (Figure 19-24) with as small a screen as can be reasonably used in the selected location.

TRY THIS

List the appliances used in your home throughout one day. After studying the ways in which electricity is used in your home, think of ways to reduce your family's use of electricity. Try your ideas for one month. Is there a difference in the electric bill?

Checkpoint

1. Why is it more energy efficient to use the elements on the top of the stove instead of the oven?
2. What is the best operating temperature for a refrigerator?
3. List three ways to make a hot-water heater more efficient.

19 ELECTRICITY AND MAGNETISM

19-9 Saving Electricity

How Can You Conserve Energy?

For each of the practices described below, decide whether electricity is conserved or not conserved. For each practice that conserves electricity, write the letter *C* on the numbered space at the left. For each practice that does not conserve electricity, write the letters *NC* on the space.

____ 1. Keeping the refrigerator temperature at 4° C or higher.
____ 2. Leaving the refrigerator door open longer than necessary.
____ 3. Using a dishwasher to wash a handful of dishes.
____ 4. Turning down the temperature setting on an electric hot water heater.
____ 5. Leaving a television set on when no one is watching it.
____ 6. Opening the door of an electric oven while food is cooking in it.
____ 7. Turning off lights in a room that no one is using.
____ 8. Insulating an electric hot water heater.
____ 9. Using warm or cold water, rather than hot water, when washing clothes.
____ 10. Drying clothes on a clothesline when possible.
____ 11. Using a 100–W light bulb where a 60–W bulb would do.
____ 12. Wearing wash-and-wear clothing that does not require ironing.

Using Try This

Depending on the ideas produced, students should find a reduction in electricity usage. Another approach is to compare the test month with the same month the previous year. Because prices may have changed, students should compare amounts of energy used, not costs, for this approach.

Checkpoint Answers

1. Less energy is required to cook the food because the heat is transferred more directly to the food.
2. About 4°C.
3. Insulate it, lower the thermostat, and insulate the pipes that lead from the tank.

Facts At Your Fingertips

- A black-and-white TV set is more energy efficient than a color set.
- Solid-state TV sets are more energy efficient than sets with vacuum tubes.
- When the cost of heating water is taken into account, it is more energy efficient to use an electric shaver than a blade razor.

Using The Feature Article

Students should enjoy reading and discussing this article. Some may want to do research on magnetic materials in organisms and report details of their findings to the class.

Magnetic Organisms

Scientists have assumed for many years that pigeons can sense Earth's magnetic field, and use that sense to help them fly in the right direction. In 1978, a scientist discovered that pigeons have small grains of magnetite in their heads. Magnetite is the kind of iron mineral that is a magnet itself. The grains are in a pocket at the base of the beak.

Scientists believe magnetite helps pigeons tell the difference between north and south. Perhaps the grains are affected by Earth's magnetic field, and nerve cells sense the changes.

In the last few years, magnetite has been found in other animals, including dolphins, whales, tuna, mice, and monkeys. It is easy to see why birds, fish, and dolphins would find a magnetic sense of direction useful in their travels. But can you imagine magnetic bacteria?

Magnetic bacteria were first found in 1975, in marshes along the east coast of North America. (See Figure 1.) They seem always to move toward the North Pole. Scientists tested some bacteria by putting them in a reversed magnetic field. Sure enough, they moved in a reverse direction.

Other magnetic bacteria have been found in the Southern Hemisphere, in New Zealand and Tasmania. They always move toward the South Pole.

Each magnetic bacterium contains a string of magnetic particles. The particles are very big and heavy compared to the rest of the organism. When the particles react to a magnetic field, the rest of the bacterium has to go along.

Earth's lines of force are parallel to Earth's surface in the middle latitudes. But at higher latitudes, the lines of force angle downward. Bacteria that follow the lines swim downward into the bottom mud from the ocean above. The bottom mud is where they find the best living conditions.

One of the latest discoveries is that Monarch butterflies have magnetite in their bodies. (See Figure 2.) Perhaps it helps them on their yearly migration from Canada to Mexico.

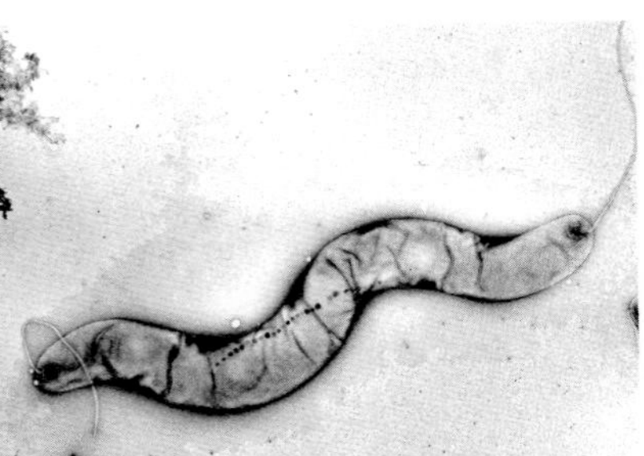

Figure 1

Figure 2

Understanding The Chapter

19

Main Ideas

1. Friction between objects can create static electricity.
2. Like charges repel one another. Unlike charges attract.
3. Atoms can gain or lose electrons.
4. Electrons flow along materials that are good conductors of electricity.
5. A neutral object held near a charged object can be charged by induction.
6. Grounding can protect buildings and people.
7. An electric circuit provides a complete path for electrons. The flow of electrons along a circuit is an electric current.
8. Circuits may be wired in series or parallel. Household circuits are always wired in parallel.
9. A magnet is surrounded by a magnetic field that can be represented by lines of force.
10. There is a magnetic field around a current flowing in a wire.
11. When the magnetic field through a coil of wire changes, an electric current is produced.
12. A generator turns kinetic energy into electricity; a motor reverses the process.
13. Use of electric energy is measured in kilowatt-hours.
14. Wise use of home appliances can reduce the amount of electric energy used.

Vocabulary Review

From the following list, choose the term that best completes each of the statements. Write your answers on a separate piece of paper.

amperes	magnetic field
battery	negative charge
conductor	ohms
electric circuit	parallel circuit
electric current	positive charge
electromagnet	resistance
grounding	series circuit
induction	static electricity
insulator	voltmeter
magnets	volts

1. A(n) _?_ is a stream of flowing electrons.
2. Resistance in an electric circuit is measured in units called _?_.
3. A poor conductor of electric charge is a(n) _?_.
4. A(n) _?_ is a circuit with only one path for electrons.
5. Like poles of _?_ repel.
6. Charges that don't move are _?_.
7. _?_ is one way of generating a static electric charge.
8. A(n) _?_ is the kind of circuit used in homes.
9. The rate of electron flow in a circuit is measured in units called _?_.
10. Wires that allow a large current to flow have a low _?_.
11. Your body is a reasonably good _?_ of electricity.

Vocabulary Review Answers

1. electric current
2. ohms
3. insulator
4. series circuit
5. magnets
6. static electricity
7. induction
8. parallel circuit
9. amperes
10. resistance
11. conductor
12. magnetic field
13. negative charge
14. electromagnet
15. grounding

Reading Suggestions

For The Teacher

Branley, Franklyn M. THE ELECTROMAGNETIC SPECTRUM: KEY TO THE UNIVERSE. New York: Harper & Row, 1979. A well-written introduction to electromagnetic waves from an astronomer's point of view.

Goldberg, Joel. FUNDAMENTALS OF ELECTRICITY. Englewood Cliffs, NJ: Prentice Hall, 1981. An easy-to-read yet sound introduction, from batteries to the electronic control of power.

Math, Irwin. WIRES AND WATTS: UNDERSTANDING AND USING ELECTRICITY. New York: Scribner, 1981. An excellent introduction to electricity, covering basics, sources, measurements, tools, power and light, magnetism, motors, and games. Construction projects of increasing difficulty are provided.

Taffel, Alexander. PHYSICS: ITS METHODS AND MEANING. Newton, MA: Allyn and Bacon, 1986. An excellent secondary textbook that provides a strong background in the basic concepts of electricity.

Taubes, Gary. "An Electrifying Possibility." *Discover*, April 1986, pp. 22–37. Swedish radiologist Björn Nordenström theorizes that the human body houses the equivalent of electric circuits that promote healing.

For The Student

Gardner, Robert. PROJECT ELECTRICITY. New York: The Boston Museum of Science Series, Messner, 1987. A history of electricity that includes activities the readers may do to gain a better understanding of electrical concepts. Stresses that electricity *is* matter.

Gutnik, Martin J. SIMPLE ELECTRICAL DEVICES. New York: Watts, 1986. Descriptions of electrical devices followed by related projects.

Math, Irwin. WIRES AND WATTS: UNDERSTANDING AND USING ELECTRICITY. New York: Scribner, 1981. Clear descriptions of working models and simple experiments.

Chapter Review Answers

Know The Facts

1. glass
2. induction
3. protons and neutrons
4. ammeter
5. wire
6. an excess of
7. magnetic compass
8. magnetic north
9. voltage
10. changing
11. the magnet's magnetic field
12. 1.2
13. ohms

Understand The Concepts

14. 500 kW·h of energy would be saved, for a cost saving of $50.
15. The socks may acquire opposite static electric charges and attract each other.
16. The electricity in lightning must have a good conducting path to ground or it will follow other conductors into the building.
17. Charge is conducted away or to ground along a conductor.
18. Charge tends to collect on the outside surface of metal structures such as cars.
19. A fuse will break the circuit if too much current flows, but a penny can conduct enough current to cause a fire.
20. A compass needle held nearby will be deflected when the current flows.
21. Turning magnets within wire coils will cause a current to be induced in the coils.

12. A compass needle lines up with a(n) _?_.
13. A positive charge attracts a(n) _?_.
14. A wire coil wound around an iron core is a(n) _?_.
15. A lightning rod used for conducting and _?_ electric charge.

Chapter Review

Write your answers on a separate sheet of paper.

Know The Facts

1. To obtain electric charge, rub a(n) _?_ rod with silk. (glass, copper, aluminum)
2. A charged object will attract opposite charges in a neutral object by means of _?_. (induction, insulation, magnetism)
3. The particles in an atom's nucleus are _?_. (electrons and protons, protons and neutrons, neutrons and electrons)
4. Electric current can be measured with a(n) _?_. (voltmeter, ammeter, generator)
5. A(n) _?_ conducts charge in a circuit. (open switch, wire, insulator)
6. Electrons move from areas that have _?_ electrons. (an excess of, too few, an equal amount of)
7. A(n) _?_ can be used to determine direction. (ammeter, voltmeter, magnetic compass)
8. A magnetic compass points to the _?_. (geographic north, magnetic north, south)
9. A transformer decreases or increases _?_. (electric charge, resistance, voltage)
10. To produce electricity in a coil of wire, the magnetic field through the coil must be _?_. (constant, changing, weakened)
11. A small compass is placed near a bar magnet. The direction of the north-seeking pole of the compass needle gives the direction of _?_. (the magnet's magnetic field, Earth's magnetic pole, north)
12. The energy, in kilowatt-hours, required to operate a 120 watt light bulb for 10 hours is _?_. (1.2, 12, 1200)
13. Resistance in a wire is measured in _?_. (volts, ohms, kilowatt-hours)

Understand The Concepts

14. Use Table 19-1 to find how much energy the average family would save in one year if it bought a regular refrigerator instead of a frostless model. At 10¢ per kilowatt-hour, how much money would be saved?
15. Explain why socks may stick together when you take them out of a clothes dryer.
16. Why is it important that lightning rods be well grounded?
17. Explain why insulators develop static electricity but conductors don't.
18. Why are you safe in an automobile during a thunderstorm, even though you are surrounded by metal?
19. You should never replace a fuse with a penny. What is the difference between a fuse and a penny that makes the fuse safe?
20. How do scientists know there is a magnetic field around an electric current?
21. How can magnets be used to produce electricity?

22. A meter reader from a utility company reads an electric meter once and reads it again one month later. The reading is 35678 the first month and 36678 for the second. If the company charges 10¢ per kW·h, what would the bill be?
23. Name five ways someone who heats water with electricity could save energy.
24. If two objects that are rubbed together become charged, will the charges be the same or different? Will the objects repel or attract?

Challenge Your Understanding

25. Draw a diagram showing an electric circuit with a lamp and a motor wired in parallel. On your diagram show where you would place an ammeter to measure the current through (a) the lamp and (b) the motor.
26. A fuse is needed to protect several appliances wired into the same circuit. Should the fuse be wired in series or in parallel with the rest of the circuit? Why?
27. Explain what is happening to electrons when you rub a balloon on your clothing and then stick the balloon to a wall.
28. Think of a way to clean the particle collectors of a Cottrell precipitator. Describe your cleaner.
29. Explain how using an appliance with a worn cord could cause a fire.
30. Two small charged spheres each carry a positive charge. They repel each other with a force *F*. How will *F* change if: (a) the charge on one sphere is doubled, (b) the distance between the charges is doubled?

Projects

1. Report on Faraday's work in electricity and magnetism.
2. Do some library research to answer these questions:
   (a) How does a transformer work?
   (b) What is a thermocouple? How can it be used to generate electricity?
   (c) What are ferromagnetic and paramagnetic materials?
   (d) What is lodestone? How is it related to magnetism? Where was it first used?
3. Visit an electric power plant and see electricity being produced on a large scale.
4. Find a magnetic compass, a battery, and a piece of wire. What happens when you place the compass needle over and parallel to the wire when the wire is connected to the battery? What happens if you put the compass under the wire?
5. Using the materials shown in Figure 19-19, demonstrate Faraday's discovery.
6. Consult reference books to learn how to build a simple electric motor. Then build the motor.
7. Find the power ratings of several appliances in your home. Keep a record of the time they are in use. Then estimate the energy required to run them for one year. Estimate the cost to operate them for one year.
8. Contact your local electric company. Ask them for information about ways to save electricity in your home. Most electric companies have such information.

22. 36678 − 35678 = 1000 kW·h. $0.10 per kW·h × 1000 kW·h = $100.
23. Answers will vary. Some possibilities are: take showers, not baths; wash clothes only when there is a full load; insulate hot-water pipes and tanks; lower the water-heater's thermostat; wash clothes in cold or warm water.
24. The objects will attract each other because the charges will be different.

Challenge Your Understanding

25. See Figure TE 19-5.
26. The fuse should be wired in series with the several appliances. If it were in parallel, a large current could flow through one appliance and very little current through the fuse.

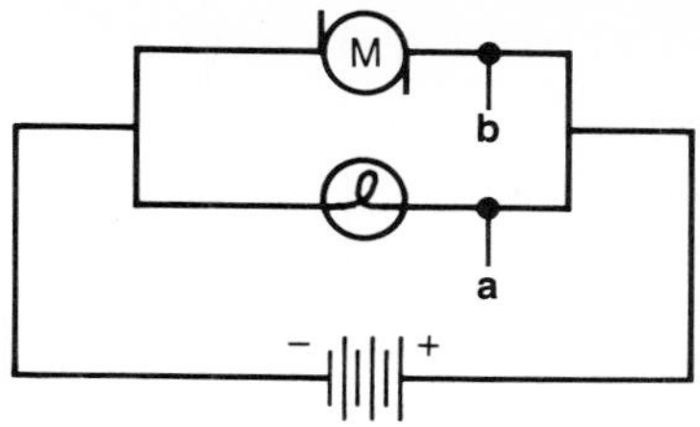

Figure TE 19-5 Locating an ammeter to measure current through (a) a lamp and (b) a motor.

27. Electrons stripped off the surface of one object collect on the other. The charged balloon, when placed on the wall, induces an opposite charge on the wall. These opposite charges attract, holding the balloon to the wall.
28. Answers will vary. One way is to move charged bodies of first one charge and then the other near the collectors.
29. The two wires in the cord, carrying current to and from the appliance, might touch. Without the resistance of the appliance, a large current would flow and might generate enough heat to start a fire.
30. (a) Force *F* will double (b) Force *F* will be quartered.

19 ELECTRICITY AND MAGNETISM
Projects

1. Find out about job opportunities for people in the electrical field.
   a. Your school or town library may have the *U.S. Government Occupational Outlook Handbook.* Use this reference to make a list of jobs in the electrical field that are expected to become increasingly plentiful in the future.
   b. Pick one type of electrical job and write a brief summary explaining salary, the part of the country where the job is available, and what is involved in the typical day of a worker.

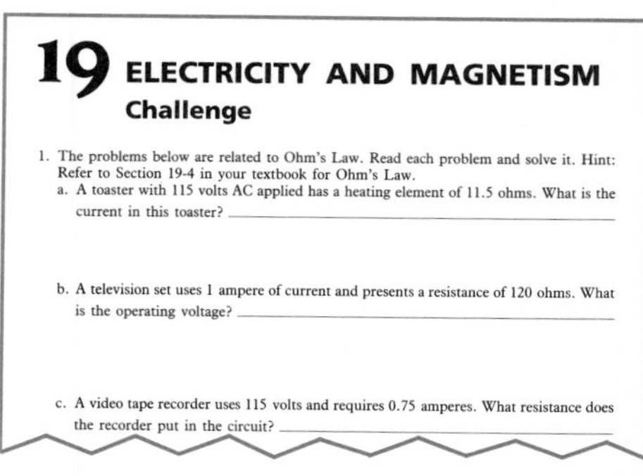

19 ELECTRICITY AND MAGNETISM
Challenge

1. The problems below are related to Ohm's Law. Read each problem and solve it. Hint: Refer to Section 19-4 in your textbook for Ohm's Law.
   a. A toaster with 115 volts AC applied has a heating element of 11.5 ohms. What is the current in this toaster?
   b. A television set uses 1 ampere of current and presents a resistance of 120 ohms. What is the operating voltage?
   c. A video tape recorder uses 115 volts and requires 0.75 amperes. What resistance does the recorder put in the circuit?

CHAPTER 20 Overview

The purpose of this chapter is to introduce students to basic phenomena of light and sound. The chapter opens with a description of light sources and how light travels. Reflection, refraction, and the formation of images by mirrors and lenses are then discussed. Next, the wave and particle models of light are considered in terms of light's properties, and the electromagnetic spectrum and color phenomena are described. The nature of sound, how sound is produced, and properties of sound are then introduced. Finally, properties of light and sound are compared.

Goals

At the end of this chapter, students should be able to:

1. describe some sources of light and the way light travels from a source.
2. describe what may happen to light when it strikes an object.
3. explain the formation of images by plane and concave mirrors.
4. state some practical uses for plane, convex, and concave mirrors.
5. define refraction of light.
6. explain how concave lenses and convex lenses form images.
7. state some practical uses of lenses.
8. describe wave motion in terms of wavelength and frequency.
9. compare wave and particle models of light.
10. describe the visible spectrum and how it can be produced.
11. identify various components of the electromagnetic spectrum.
12. explain how people see color.
13. describe how lights or pigments can be combined to produce colors.
14. describe how sound is produced.
15. discuss some properties of sound that depend on its frequency or loudness.
16. state similarities and differences between sound and light.
17. describe how sound waves move through air.

Chapter 20

Light and Sound

Still water reflects light the way a mirror does, forming images.

Vocabulary Preview

reflection
transparent
translucent
opaque
plane mirror
concave mirror
principal focus
focal length
real image
virtual image
convex mirror
refraction
lenses
concave lens
convex lens
wavelength
frequency
visible spectrum
electromagnetic spectrum
infrared radiation
ultraviolet radiation
longitudinal wave

20-1

Introduction to Light

Goals

1. To describe some sources of light and the way light travels from a source.
2. To describe what may happen to light when it strikes an object.

Light is a form of energy. A source of light must produce or transform energy. Powerful nuclear reactions in the sun provide the energy that causes it to give off light. Energy released when substances burn causes light to be given off by candles, kerosene lamps, and wood fires. Electric energy from power plants, batteries, and solar cells is transformed into light by light bulbs.

Light from a source, such as the sun or a bare light bulb, generally spreads out in all directions, in straight lines. Light given off by a bulb can be concentrated into a *beam* by a flashlight or spotlight. (See Figure 20-1.)

Figure 20-1 *Performing musicians are often highlighted by beams from spotlights.*

Teaching Suggestions 20-1

Open discussion of light by asking students to volunteer their own ideas of what light is. A wide variety of answers will be given, some possibly close to either the wave or the particle model. Allow students to hypothesize freely, making certain only that their hypotheses are consistent with what we know about light.

Be sure students understand that light travels in straight lines. Discuss what happens when light strikes an object, and be sure students grasp the meanings of reflection, transparent, translucent, and opaque. Stress that objects are visible because light is reflected from them to an observer's eyes. See the Demonstration section for some ways of concretizing the concepts being taught.

Background

Light is a form of energy. It is converted to heat when absorbed by dark-colored objects, such as solar collectors. Light also can be converted directly into electricity by what is called the photovoltaic effect, as in solar cells. Light energy is converted to the kinetic energy of moving electrons in a sensitive photoelectric cell, such as a light meter.

Light is produced by the sun and other stars when nuclear energy is transformed into light energy, other kinds of electromagnetic waves, and heat. Light is also produced when the electrons in atoms return to lower energy levels after excitation by electricity, atomic collisions, or energetic electrons or photons. The potential energy lost in dropping down from a higher to a lower energy level appears in the form of light.

Light intensity falls off with distance according to an inverse square law. At twice the distance, the intensity of the light becomes $\frac{1}{4}$ as great, at three times the distance the intensity is $\frac{1}{9}$, at four times the distance the intensity is $\frac{1}{16}$, and so on.

This effect can be explained by considering light to be small particles (photons) that move at high speeds in straight lines. Imagine a small point source emitting photons at an equal rate in all directions, and a sphere of radius r with the light source at its center. Suppose N photons are emitted each second. The intensity of light at the surface of the sphere will be N particles divided by the area of the sphere ($4\pi r^2$). Consider a second, concentric imaginary sphere with twice the radius of the first one. The same N particles pass through this sphere each second, but its area is four times as great ($4\pi(2r)^2 = 16\pi r^2$). Thus, the intensity of the light is $\frac{1}{4}$ as great at twice the distance.

Demonstration

Show students some samples of transparent materials (glass and cellophane), translucent materials (frosted glass and waxed paper), and opaque matter (black paper and cardboard). Stress what happens to light that strikes all three types of matter.

Teaching Tip

- With a flashlight in a darkened classroom, you can easily show students that the only objects they can see clearly are those that reflect the light from the flashlight back to their eyes.

Facts At Your Fingertips

- The moon and the planets do not produce light. They are visible only because they reflect sunlight to Earth.
- Early Greek philosophers held the belief that the light rays reflected by visible objects back to the eye actually originated in the eye.

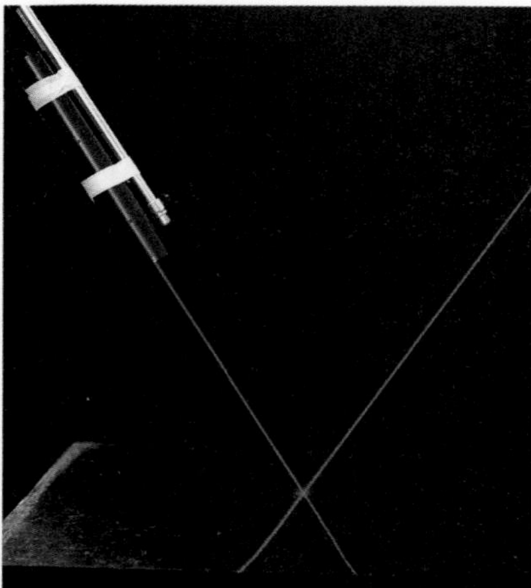

Figure 20-2 *A beam of light that strikes a mirror bounces off the reflective surface.*

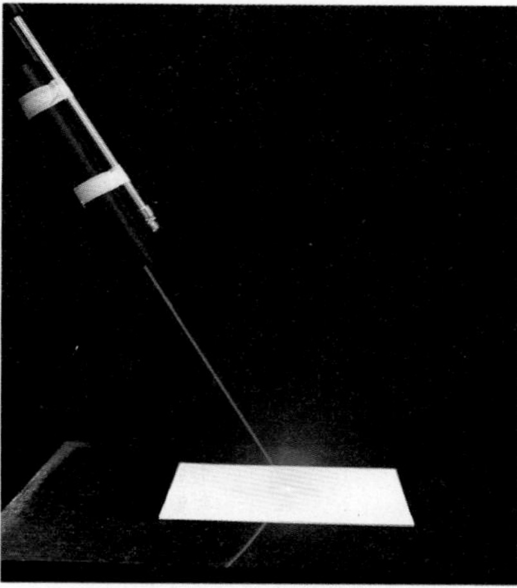

Figure 20-3 *Light is scattered by a translucent object*

reflection

When a beam of light strikes a mirror or some other smooth surface, most of it bounces off. (See Figure 20-2.) The bouncing of light off an object is called **reflection**. Most objects reflect at least some light.

Reflection changes the direction in which light is moving. You see objects that reflect light toward you. You see the moon because it reflects sunlight. When you shine a flashlight in the dark, you see only those objects that reflect its light toward you.

transparent

When light strikes a **transparent** object, most of the light passes through. Clear glass and plastic are typical transparent materials. Windows provide light inside a building in the daytime, since window glass is transparent.

translucent (tranz LOO sunt)

Most of the light that strikes a **translucent** object is *scattered* (reflected in many different directions). (See Figure 20-3.) You cannot see things clearly through a translucent object. Frosted glass, waxed paper, and many lampshades are translucent.

opaque (oh PAYK)

Most of the light that strikes an **opaque** object is absorbed. You cannot see through opaque materials. They are usually dark in color and often rough in texture. Wood, brick, and soil are opaque materials. The light absorbed by

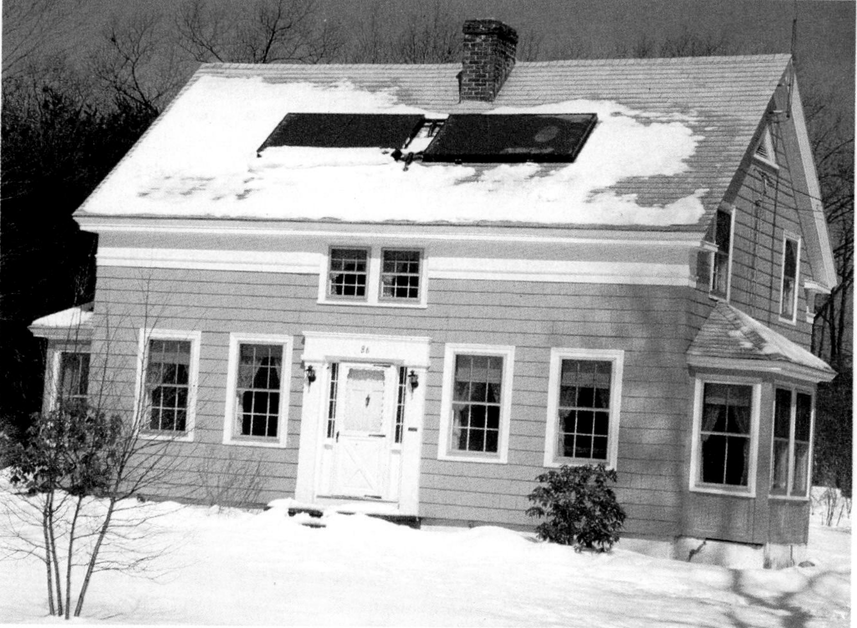

Figure 20-4 *Notice how the black solar collector on this house contrasts with the lighter roof.*

opaque materials is transformed into heat. The parts of solar collectors that absorb sunlight are usually black. (See Figure 20-4.)

Most objects are not totally reflective, transparent, or opaque. For example, transparent materials such as glass and water reflect some light. You have probably seen your reflection in still water or window glass.

Checkpoint

1. Give some examples of light sources. Identify the form of energy that enables them to give off light.
2. Why can people see some objects but not others?
3. Give an example of a material that is (a) transparent, (b) translucent, (c) opaque.

TRY THIS

Make a list of transparent, translucent, opaque, and reflecting objects that you see during a day.

Mainstreaming

Handicapped and special-needs students will require assistance and much hands-on experience to master the concepts presented in this chapter.

20 LIGHT AND SOUND

20-1 Introduction to Light

The Way Light Behaves

In each of the following sentences a term is missing. Read each sentence and think of the missing term. Then write it on the numbered space at the left.

______ 1. Since light can pass through window glass, we know that the glass is __1.__

______ 2. You cannot see clearly through a frosted glass window; therefore, the window must be __2.__

______ 3. Light cannot pass through a brick wall because the wall is __3.__

______ 4. When light is absorbed by a dark color, much of the light energy is changed into __4.__ energy.

______ 5. It is difficult to see through a translucent object such as a stained glass window because the light is __5.__

______ 6. When you watch a movie in a theatre, you are seeing the light that the screen __6.__

______ 7. __7.__ is the bouncing of light off an object.

______ 8. A solar collector usually has a black surface to transform light energy to __8.__

Classify!

Each of the materials listed below is either transparent, translucent, or opaque. As you read the list, decide how each material should be classified. On the space at the left write the term *transparent, translucent, or opaque* to correctly classify the material.

______	1. waxed paper	______	6. window pane
______	2. plywood	______	7. diamond
______	3. sheet of newspaper	______	8. white t-shirt
______	4. cardboard	______	9. leather coat
______	5. ice on a window	______	10. cellophane

Using Try This

Students should be able to come up with a rather lengthy list for each of the terms given. All the objects listed reflect light, even the transparent ones, for otherwise they would not have been visible.

Checkpoint Answers

1. Answers will vary. Some examples are: Light from the sun comes from nuclear energy. Light from light bulbs comes from electrical energy.
2. Objects can be seen only if they either emit light or reflect light to our eyes.
3. Answers will vary. Some examples are: (a) glass or clear plastic; (b) waxed paper or frosted glass; (c) black paper, coal, concrete, and cardboard.

Teaching Suggestions 20-2

Ask students where they think the image seen in a plane, or flat, mirror is formed. When you have collected one or two reasonable hypotheses, refer students to Figure 20-6 on student text page 503. Explain how such an image can be constructed based on the facts that (a) light travels in straight lines and (b) the angle at which light strikes a reflective surface, called the angle of incidence, is equal to the angle at which it is reflected, called the angle of reflection. (See Figure 20-5 on student text page 502.) Be sure students understand that the image formed by a plane mirror is a virtual image; that is, it cannot be displayed on a screen. This is a good point at which to have students do laboratory Activity 27, A Look into a Plane Mirror, student text page 566. The demonstration, too, shows image formation in a plane mirror. (See Demonstration.)

Chapter-end project 1 applies to this section.

Move on to a discussion of image formation by curved mirrors. Be sure students understand the conditions that control image formation by concave mirrors. Use the Try This activity to help convey this. Stress the differences between the images formed by plane, convex, and concave mirrors.

Background

Plane and convex mirrors form only virtual images, but concave mirrors can also produce real images. The virtual images produced are always right side up, but the size of the image is variable: enlarged for concave mirrors, reduced for convex mirrors, and any size, depending on distance, for plane mirrors. The real images formed by concave mirrors are always upside down and may be enlarged, reduced, or same size, depending on distance.

A virtual image appears to be behind a mirror or lens. The divergence of rays of light after reflection or refraction accounts for its formation. The divergent rays seem to be emanating from beyond the mirror. A real image results when rays of light that diverged after leaving an object are made to converge (brought back together again), producing an image that can be projected on a screen. Any image is dimmer than the object itself, because part of the light coming from the object is absorbed by the mirror.

20-2

Mirrors

Goals

1. To explain the formation of images by plane and concave mirrors.
2. To state some practical uses for plane, convex, and concave mirrors.

Mirrors are everywhere: in cars, purses, bathrooms, bedrooms, and lobbies. You find them at amusement parks and in telescopes. You find them in unexpected places too—on a calm lake, the hood of a car, or the surface of a spoon.

Ancient Egyptians learned to make mirrors by polishing metals. Coated glass mirrors were first made in Venice in the 1500's. Mercury was placed on tinfoil and covered with paper, making a "sandwich." A sheet of glass was placed on the paper. When the paper was carefully withdrawn, the mercury and tin formed a reflecting layer that stuck to the glass. In 1835, von Liebig (vahn LEE bik) discovered that a solution containing silver nitrate, when heated, would deposit a thin layer of silver on glass. This process is similar to the method used today to make ordinary mirrors.

When a beam of light strikes a mirror, the angle at which it is reflected is equal to the angle at which it strikes the surface. (See Figure 20-5.) This behavior makes it possible to explain how mirrors form images.

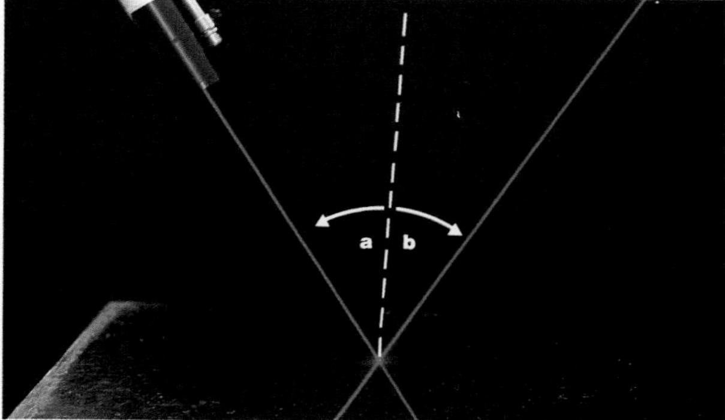

Figure 20-5 *The angle at which light strikes the surface of a mirror (a) is the same as the angle at which it is reflected (b).*

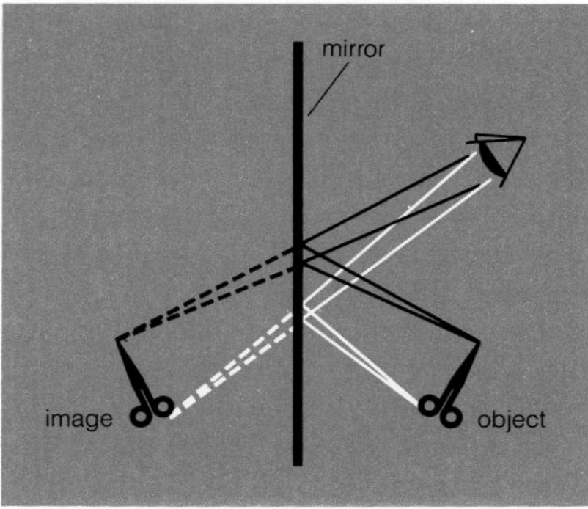

Figure 20-6 *The observer sees the light reflected by a plane mirror as if it had traveled in straight lines from behind the mirror.*

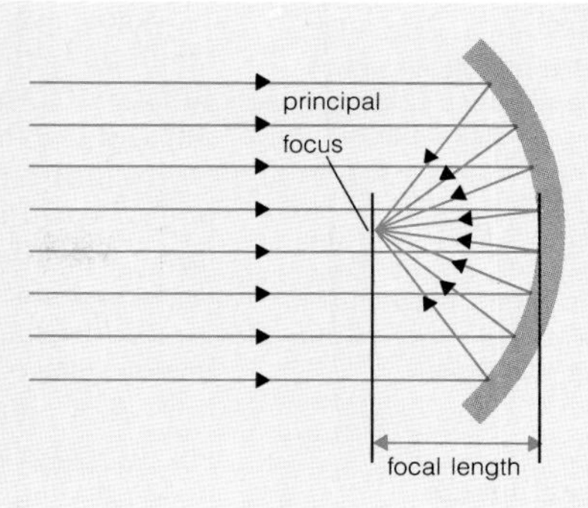

Figure 20-7 *A concave mirror focuses parallel light rays at a point called the principal focus.*

plane mirror

An ordinary flat mirror is called a **plane mirror**. Figure 20-6 shows how images are seen in plane mirrors. The light reflected by the object and the mirror is represented by imaginary lines called *rays*. The rays represent only a few of the countless "paths" that the reflected light takes. The light reflected by the object strikes the mirror at an angle and is reflected by the mirror at an equal angle. The viewer sees the reflected light as if it had traveled in straight lines from behind the mirror.

Your image, as seen in a plane mirror, looks just like you (except that left and right are reversed). But if you look at your image in an amusement park mirror, you may find it stretched, inverted, thin, fat, even unrecognizable. This happens because these mirrors are curved instead of flat.

concave mirror (kahn KAYV)

Both sides of a shiny metal soup spoon can act as a mirror. A **concave mirror** has a curved reflecting surface similar to the soup-holding side of a spoon. Figure 20-7 shows light rays striking a concave mirror. The rays shown striking the mirror are *parallel*. (They remain the same distance apart and never touch or cross.) A concave mirror will *focus* (bring together) parallel rays. The point where these reflected rays meet is called the **principal focus**. The distance between a mirror and its principal focus is called the **focal length** of the mirror. The more curved the mirror, the shorter the focal length.

principal focus

focal length

Teaching Tips

- Students should find the uses to which plane and curved mirrors are put a help in understanding how such mirrors form images.
- Have students experiment with concave and convex mirrors, if such mirrors are available.
- Stress the differences between real and virtual images.
- Caution students that a focused image of the sun can cause serious burns.
- Caution students that they should never look directly at the sun. The sun's direct rays can seriously injure the eye, or even cause blindness.

Demonstration

Show the positions of images in a large plane mirror. Use the same technique introduced in Laboratory Activity 27, but with string instead of pins to establish lines of sight. The strings may pass over the mirror and intersect behind it.

Facts At Your Fingertips

- Concave mirrors are used in telescopes, searchlights, and vehicle headlamps.
- The principal focus of a curved mirror is sometimes called the focal point.

Mainstreaming

Team visually-impaired students with other students for the activities of this section.

TRY THIS

Use a concave mirror to produce real images. Stand back away from a window. Use the mirror to reflect light from the outside onto a "screen," such as a white piece of paper or index card.

A concave mirror can do one of three things. It can form a beam of light. It can focus an image onto a screen or other surface. It can produce a magnified image. What it does depends on the position of the object whose light it reflects.

To form a beam, a small light is placed *at the principal focus* of a concave mirror. The mirror directs the light as shown in Figure 20-8. That is why concave mirrors are used in flashlights and automobile headlamps.

real image

To focus an object's image onto a screen with a concave mirror, place the object *beyond* the principal focus. Light from each point on the object is focused by the mirror, forming an image in front of the mirror. (See Figure 20-9.) The image is called a **real image** because it can be displayed on a screen. The size of the image depends on how far the object is from the mirror.

Light from any distant object is very nearly parallel. When light from the sun strikes a concave mirror, a small, hot, real image of the sun forms at the principal focus of the mirror. The temperature of an object can be increased by placing it at the mirror's principal focus. A solar cooker works this way. (See Figure 20-10.)

virtual image

A concave mirror produces a magnified image of an object placed *between* the mirror and its principal focus. Makeup and shaving mirrors are concave mirrors. The image seen in the mirror is called a **virtual image** because it cannot be displayed on a screen. The images formed by plane mirrors are virtual images, also.

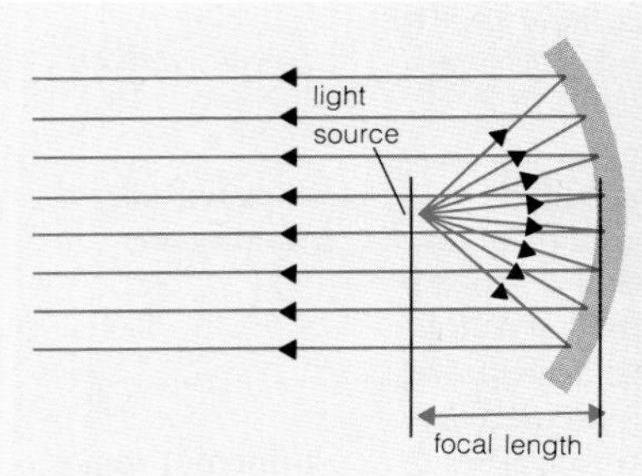

Figure 20-8 *If you place a light source at the principal focus of a concave mirror, the light is reflected as a beam.*

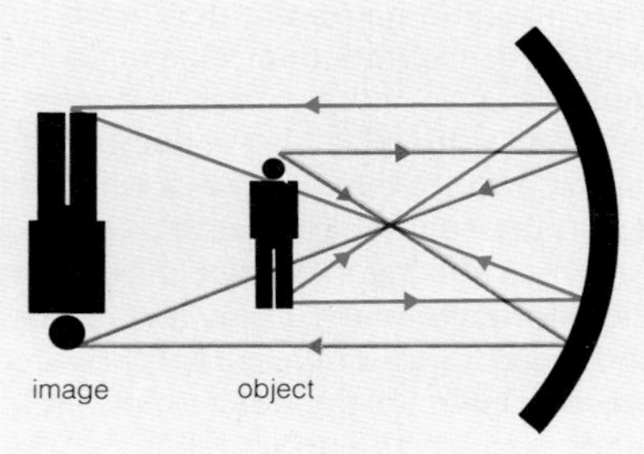

Figure 20-9 *A real, upside down image is formed by a concave mirror if you place an object beyond the principal focus.*

Figure 20-10 *You can use sunlight to cook if you place the food at the principal focus of a concave mirror.*

Figure 20-11 *(a) A convex mirror reflects a larger view with distorted images. (b) In a plane mirror, images are not distorted.*

A curved reflecting surface shaped like the side of a spoon that does not hold soup is called a **convex mirror**. Images formed by a convex mirror are always virtual, right side up, and smaller than the objects they reflect. Convex mirrors are used as rearview mirrors on trucks or cars. They are also used in stores to help detect shoplifters. Convex mirrors reflect a larger view than comparable plane mirrors, but the images they form are distorted. (See Figure 20-11.)

convex mirror
(KAHN veks)

Checkpoint

1. Use a light-ray diagram to show (a) how a plane mirror forms a virtual image, (b) how a concave mirror forms a real image.
2. What is the difference between real images and virtual images?
3. State one use each for plane, concave, and convex mirrors.
4. Describe how a solar cooker works.

20 LIGHT AND SOUND

20-2 Mirrors

Focus In!

Each of the statements in Column B describes a type of mirror in Column A. Read each statement and decide which mirror it describes. On the space next to the name of each mirror in Column A write the letter of the statement that describes it from Column B. Hint: One type of mirror is described by more than one statement.

A	B
______ 1. concave mirror	a. Can form real images.
______ 2. convex mirror	b. Can form an image that is much like you, except that left and right are reversed.
______ 3. plane mirror	c. Can form images that can be focused onto a screen.
	d. Produces reflected images that are virtual and smaller than the object.
	e. Magnifies images of an object placed between the mirror and its principal focus

Using Try This

Students will enjoy this activity. Note that to see the images clearly the room lights must be dimmed. Draw all but one shade and have students stand back from the exposed window. They should discover that it is essential that the screen be outside the focal point of the mirror in order to display the image. Make certain students realize that any images so formed are real and upside down.

Checkpoint Answers

1. (a) See Figure 20-6 on student text page 503. (b) See Figure 20-9 on student text page 504.
2. Real images can be displayed on a screen. They are formed by light rays that come together to form an image. Virtual images cannot be displayed on a screen. They arise from the extension backward of reflected rays to a point behind a mirror.
3. Answers will vary. Some examples are: plane—in bathrooms and dressing rooms in stores; concave—in flashlights and vehicle headlamps; convex—auto rearview mirrors, mirrors in stores to watch for shoplifters.
4. A solar cooker is a concave reflecting surface that focuses light from the sun onto a substance, raising its temperature enough for cooking to occur.

Teaching Suggestions 20-3

Open discussion of the section with the concept of refraction. Develop the notion that the path of light bends as it passes from one transparent medium to another. Then ask students to relate personal experiences with refraction phenomena. You should get a variety of responses. To help establish the concept, sketch Figure TE 20-1 on the chalkboard. First, explain why in (a) the eye cannot see the coin. (It is below the path of the reflected rays from the coin that emerge from the cup.) Then explain what happens (b) after water has been poured into the cup. (Reflected rays from the coin are bent downward due to refraction as they pass from water to air, enabling them to reach the eye.) You may want to demonstrate this effect or have students try it for themselves. (See Demonstrations.)

Take the time to be sure students understand image formation by concave and convex lenses. Refer to Figures 20-14a and 20-14b on student text page 507. Students will be interested in the applications of lenses given in the text.

End-of-chapter project 3 applies to this section.

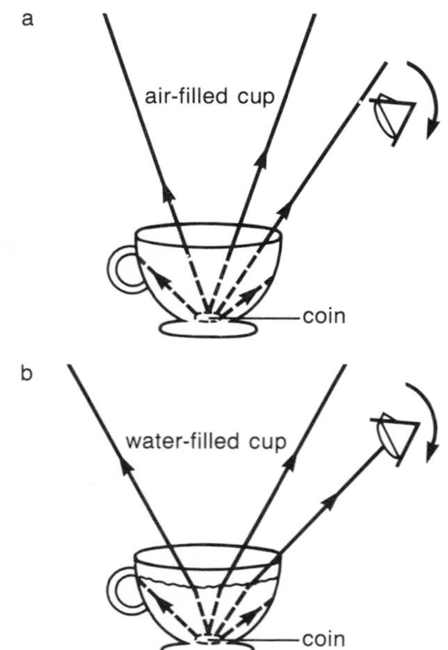

Figure TE 20-1 Refraction effect.

20-3

Lenses

Goals

1. To define refraction of light.
2. To explain how concave lenses and convex lenses form images.
3. To state some practical uses of lenses.

Figure 20-12 *A beam refracted by glass. The beams that go up are reflections.*

When light passes from one transparent material into another, it changes direction. Figure 20-12 shows that its path bends as light passes from air into glass and from glass into air. The bending of its path as light passes from one transparent material into another is called **refraction**.

refraction

Refraction produces some rather strange effects. For example, if you look at a ruler that is partly under water, it looks bent. (See Figure 20-13.) This happens because the light reflected from the part of the ruler that is under water is refracted when it passes into the air. Since the light reflected from the part of the ruler that is above the water line is not refracted, the ruler appears to bend at the water line.

Refraction of light is put to practical use in a number of devices, including cameras, telescopes, and movie projec-

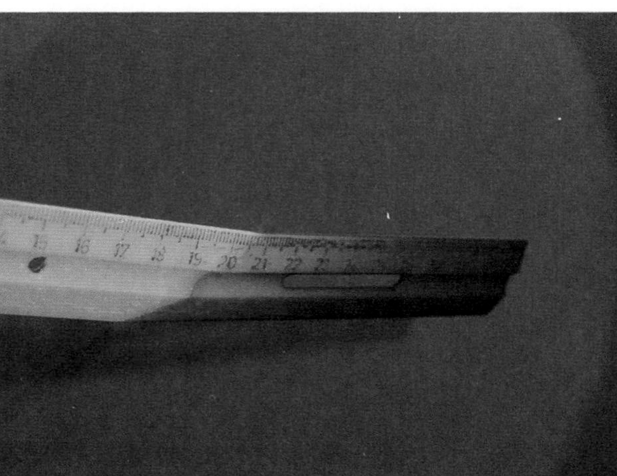

Figure 20-13 *When you put something partly into water, it appears to bend. The lower end of this ruler is not where it seems to be.*

Background

The law governing refraction is more complicated than that governing reflection. The ratio of the sines of the angles of incidence and refraction is equal to a constant called the index of refraction, which is a characteristic property of transparent materials.

Refraction occurs because the speed of light is different in different media.

Light travels at a constant speed in air but is slowed down upon entering an optically denser medium. This speed difference is measured by the index of refraction, which is equivalent to the ratio of the speed of light in air to the speed of light in the other medium.

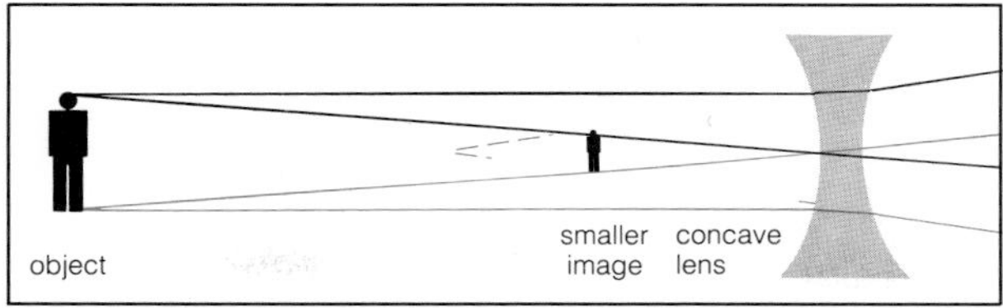

Figure 20-14a *The light rays spreading from a concave lens make the image look as if it were where the dashed lines go on the diagram.*

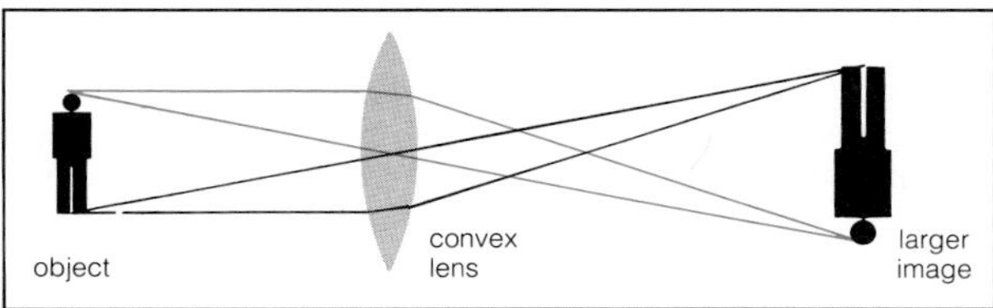

Figure 20-14b *A convex lens forms an image that is upside down and larger than the object.*

tors. These devices use **lenses**, transparent objects with at least one curved surface, to control light. Lenses are usually made of clear glass or plastic. Light is refracted both when it enters a lens and when it leaves the lens. The path of light as it leaves a lens depends on the shape of the lens.

A **concave lens** is thicker at the ends than in the middle. It spreads light, producing images that are right side up and smaller than the objects. (See Figure 20-14a.) A **convex lens** is thicker in the middle than at the ends and brings light rays together. (See Figure 20-14b.)

A single convex lens is often used as a magnifier. The earliest microscopes were nothing but powerful magnifying glasses. Two convex lenses can be used together to make a more powerful microscope. Figure 20-15 shows how a two-lens microscope works.

The two-lens (or *compound*) microscope can magnify objects much more than a single-lens microscope can. But using two lenses together has a drawback. If you have used a microscope, you know that the image is reversed and upside down. Also, when you move the object to the right and toward you, the image moves left and away.

lenses

concave lens

convex lens

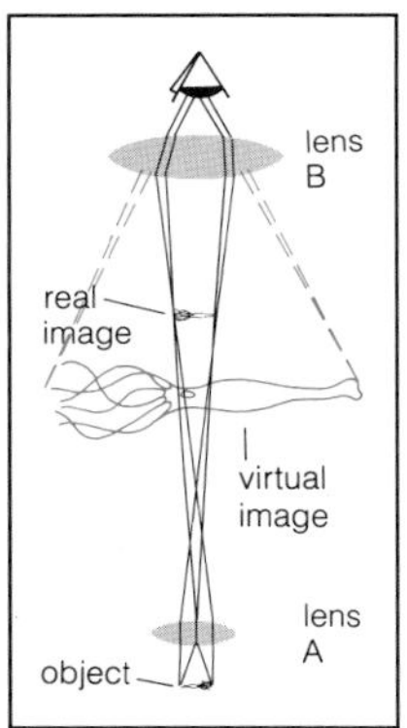

Figure 20-15 *The eyepiece lens, B, magnifies the image from the objective lens, A.*

Demonstrations

1. Show the refraction of light as it passes from water to air. (See Figure TE 20-1.) Place a coin in a teacup. Ask a student to look at the coin and then lower his or her head until the coin just disappears from view. Have the student maintain that position while you slowly add water to the cup. The coin will reappear in the student's line of sight.
2. Show how water can be used as a lens. Fill a round-bottomed flask with water. Use it as a magnifier.
3. Fill several vials with colored water of various hues. Place the vials, one at a time, over each of the words BOX, CHOICE, OXIDE, RED, CAT, DOG, FAT, BOY, and GIRL. (Be sure to write or type the words using only capital letters.) The water-filled vials will seem to have no effect on the first three words, but the remaining words will appear to be inverted. Ask students what is responsible for this. Elicit that all the images are inverted, but the appearance of the first three words is the same right side up as it is upside down.

Teaching Tips

- Discuss the illustrations in this section very thoroughly to be sure students understand the concepts presented.
- Try to obtain convex and concave mirrors. Have students use them to form images and assemble simple microscopes and telescopes.

Facts At Your Fingertips

- Most microscopes contain several convex lenses.
- The inverted images formed on the retina of the human eye are interpreted by the brain as being right side up. In one experiment, a subject wore glasses containing prisms so that retinal images would be right side up, rather than inverted. At first, the world appeared upside down to the subject. After a while, however, the brain adapted to the situation and the world looked right side up again. After removing the prisms, the world again appeared inverted, and the adaptation process was gone through all over again.
- Light is not refracted if it strikes the boundary between two transparent media with an angle of incidence of 0°, that is, if it strikes perpendicular to the surface as it passes into the second medium.
- Diffraction is an optical phenomenon distinct from and unrelated to refraction. Diffraction is the spreading of light after it passes through a narrow opening.

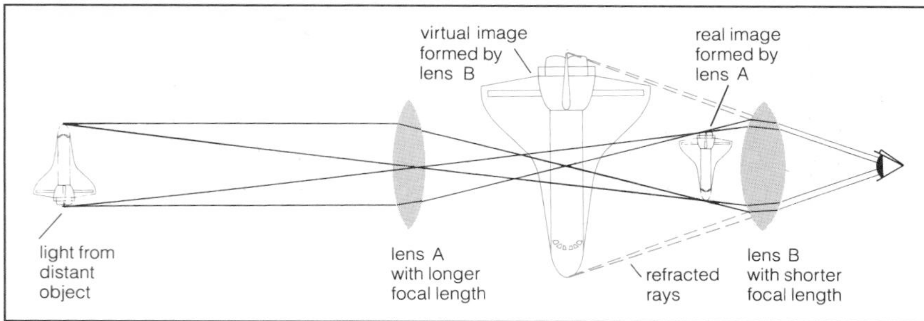

Figure 20-16 *A telescope's objective lens has a long focal length to make a large image.*

Two convex lenses can be put together to make a telescope, also. (See Figure 20-16.) Because the object to be magnified is far away, the objective lens (lens A) has a long focal length. This is to make the image it forms as large as possible. Lens A should be wide, to collect enough light to make a bright image.

The human eye contains a convex lens. This lens forms real images on the *retina* (RET n uh), the inner lining of the eye. (See Figure 20-17.) These images stimulate nerve endings in the retina. Nerves carry the resulting impulses to the brain, where the images are perceived. The image on the retina is upside down. In the brain, the image is turned right side up.

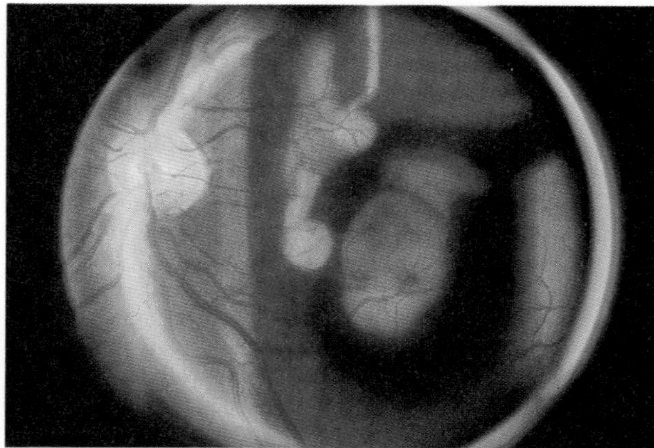

Figure 20-17 *This photograph shows the inside of a human eye. Notice the upside down image on the retina.*

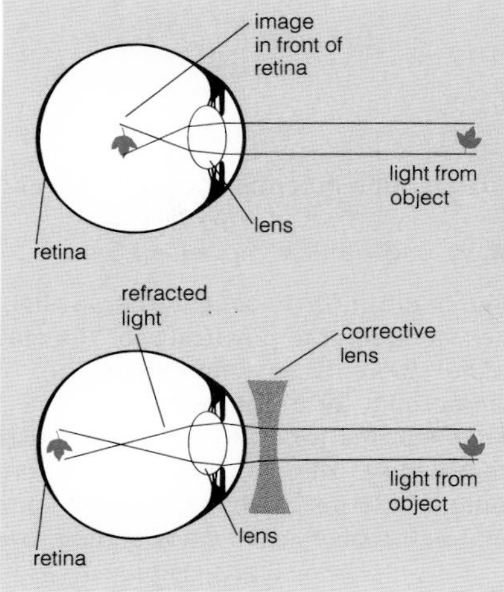

Figure 20-18 *Above: The eye of a near-sighted person. The image forms in front of the retina, and the person's vision is not clear. Below: A corrective lens causes the image to form on the retina.*

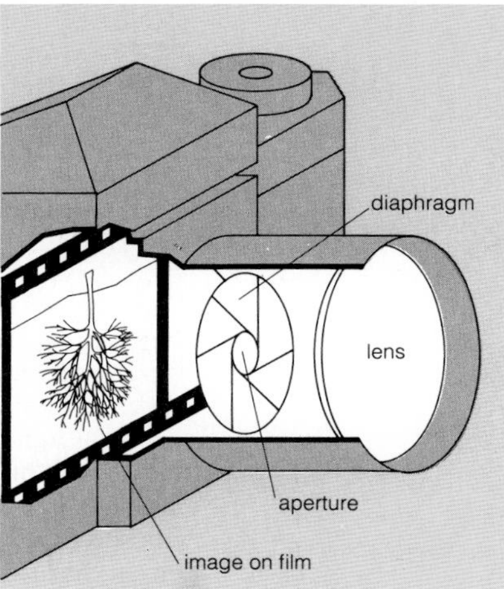

Figure 20-19 *A camera is similar to a person's eye. Light goes through the lens, which focuses the image on the film. The diaphragm is like the pupil of your eye, controlling the amount of light.*

If the lens does not focus light into clear images on the retina, corrective lenses can be used. These may be contact lenses or in eyeglasses. Concave lenses are used to correct the vision of nearsighted people. (See Figure 20-18.) Convex lenses can correct farsightedness.

A camera works very much like the human eye. In place of a retina there is sensitive film to make a permanent record of images. (See Figure 20-19.)

Checkpoint

1. Explain what is meant by refraction. Use an example to illustrate your explanation.
2. How can you tell convex from concave lenses?
3. Using diagrams, explain how images are formed by convex and concave lenses.
4. List some uses for convex lenses.

TRY THIS

Place a tiny drop of water on a letter of a word found on a shiny page of a magazine. Does the drop act as a lens? How can you tell? Try to explain what you observe.

20 LIGHT AND SOUND

20-3 Lenses

Make It True!

Read the following statements. If the statement is true, circle the *T* and leave the numbered space blank. If the statement is false, circle the *F* and write on the space the term to replace the underlined term and make the statement true.

1. ________ T F As light travels from one transparent material into another, we can expect the direction of the light to be changed.
2. ________ T F A concave lens is thicker in the middle than at the ends.
3. ________ T F A convex lens makes light rays come together.
4. ________ T F Two concave lenses can be used as a simple microscope.
5. ________ T F The human eye contains a convex lens that focuses light on the pupil of the eye.
6. ________ T F The lens system of a camera focuses light on film to record images.

Sorting Things Out

Unscramble each term in Column A and write it on the line under the scrambled term. Then match each term with its definition in Column B and write the letter of the definition on the line next to the scrambled term.

A	B
1. ainotcfrre ___	a. Transparent object used to control light.
2. anocecv seln ___	b. Uses two convex lenses to magnify distant objects.
3. veconx seln ___	c. Uses two convex lenses to magnify small objects.
4. nodomcup romisepocc ___	d. Brings light rays together.
5. poceleset ___	e. Spreads light, producing images that are smaller than the objects.
6. seln ___	f. The bending of light's path as it passes from one transparent material into another.

Using Try This

Students should observe a magnifying effect, since the shape of a drop of water is similar to that of a convex lens. Some students may want to investigate to determine which makes a better magnifier—a drop of water or a drop of alcohol.

Checkpoint Answers

1. Refraction of light is the bending of light as it passes from one transparent medium to another. An example is the apparent bending of a partially submerged ruler.
2. Convex lenses are thicker in the middle than at the edges. Concave lenses are thicker at the edges than at the middle.
3. See Figures 20-14a and 20-14b on student text page 507.
4. Answers will vary. Some examples are: eyeglasses, magnifiers, telescopes.

Teaching Suggestions 20-4

Ask students to describe wave motions they have observed. Responses will vary. Some possibilities are water waves, standing waves in a rope, and the waves in a spring. Follow with a description of transverse waves, the class to which water waves belong, and be sure students understand the terms wavelength and frequency. Refer to Figures 20-23 and 20-24 on student text page 511 to show students that (transverse) waves exhibit reflection and refraction. A ripple tank is very useful for displaying these effects. (See Demonstrations.)

Be sure students understand that both the wave model and the particle model are needed to account fully for the behavior of light. Either one alone produces an incomplete picture. The second demonstration can be used to show how reflection and refraction can be accounted for by the particle model. (See Demonstrations.)

Some students may be interested in doing chapter-end project 5 in conjunction with this section.

Background

When the scientific study of light began, both a particle model and a wave model were proposed. In the seventeenth century, Isaac Newton developed a particle theory, and Christian Huygens worked out a wave theory. Newton's particle theory satisfactorily explained the three characteristics of light then known: straight-line propagation, reflection, and refraction. The wave theory ran into difficulty when it attempted to account for straight-line propagation. However, in the nineteenth century Newton's particle theory was abandoned. The discovery of interference of light—a wave phenomenon—and its use to explain diffraction (the spreading of light after passing through a narrow slit) tipped the scales toward the wave theory. The wave theory received another boost later in the nineteenth century. It was demonstrated that the speed of light is less in an optically denser medium. Newton had assumed that the opposite was true.

A modified particle theory made a comeback at the turn of the twentieth century with the discovery of the photoelectric effect. Einstein successfully explained this effect, that is, the ability of light to cause the release of electrons, by assuming that light consists of discrete bundles of energy called photons. The wave model cannot explain the photoelectric effect. The modern theory is that light has a dual nature, wave and particle.

20-4

The Nature of Light

Goals

1. To describe wave motion in terms of wavelength and frequency.
2. To compare wave and particle models of light.

You have studied some of the *properties* of light (the way light behaves) and how they can be put to practical use. You may have wondered whether there is any mental picture, or *model*, of light that can account for its properties.

After light had been studied for many years, scientists came to think of it as wavelike. The results of their experiments could be explained in terms of "light waves." No one can see light waves. But water waves have many of the properties that light has, and water waves are visible.

If you drop an object into still water, waves spread out, or radiate, in all directions from the point of impact. (See Figure 20-20.) Some of the object's energy is transformed into wave energy. A cork floating on water bobs up and down as waves pass by. The waves provide the force needed to move the cork upward. Similarly, light energy radiates in all directions from a source, such as a light bulb or the sun.

Figure 20-21 shows a side view of water waves. The highest points on the waves are called *crests*, and the lowest

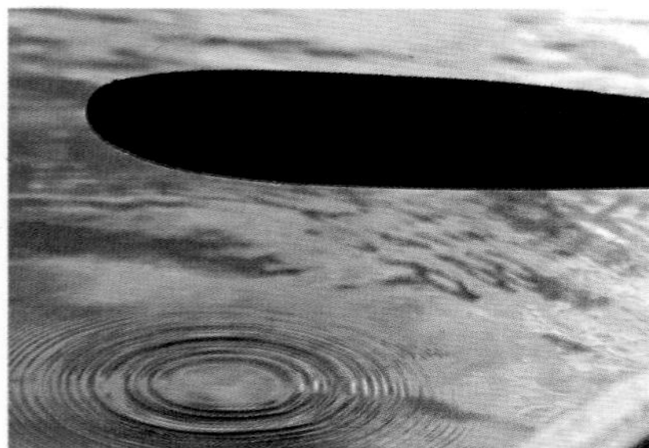

Figure 20-20 *Waves radiate out from the point at which a drop of water from the paddle hits the surface of still water.*

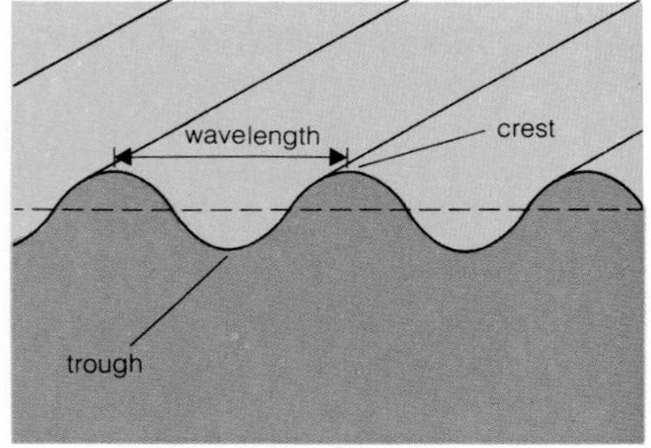

Figure 20-21 *This cutaway view of water waves indicates how the wavelength is determined.*

points are called *troughs* (trawfs). The distance from the crest of one wave to the next is the **wavelength**. The number of wavelengths, or waves, that pass a given point in one second is the **frequency**. Frequency is measured in units called *hertz* (Hz). A frequency of 1 Hz is 1 wave per second, 2 Hz is 2 waves per second, and so on.

wavelength

frequency

A wave can be identified by either its wavelength or its frequency. As shown in Figure 20-22, the frequency of a wave decreases as its wavelength increases and increases as its wavelength decreases.

Water waves can be produced and photographed. Figure 20-23 shows water waves striking a smooth surface. The angle at which the waves are reflected is the same as the angle at which they strike the surface. This is what happens when light is reflected from a smooth surface.

As waves pass from deep to shallow water, their path bends, or refracts. (See Figure 20-24.) Their path is similar to that of light moving from air to glass.

Light and water waves have a number of similar properties. In general, a wave model can explain many of the properties of light. But light has some properties that do not fit the wave model.

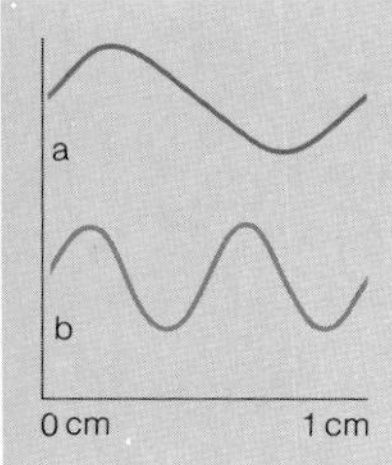

Figure 20-22 *The wavelength of wave b is half as long as that of wave a. Thus its frequency is twice that of wave a.*

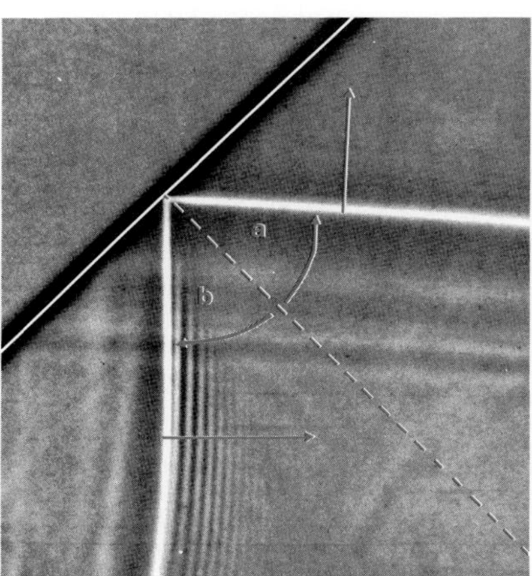

Figure 20-23 *In this photograph, the angle of reflection (a) is the same as the angle at which the waves strike the surface (b).*

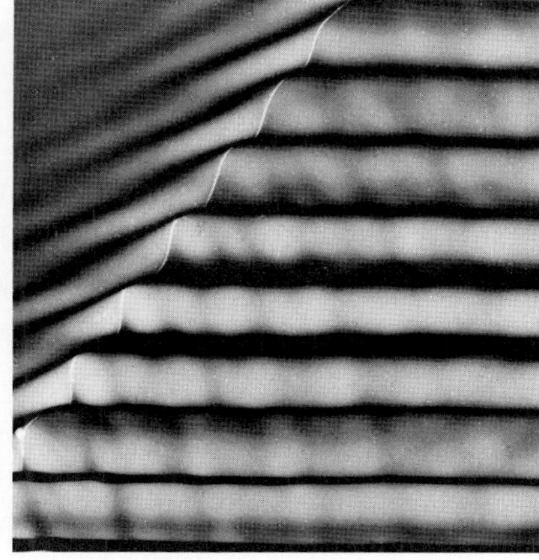

Figure 20-24 *Water waves are refracted when they strike shallow water. The waves bend where the depth changes.*

Teaching Tips

- Point out to students that no one has ever actually seen light waves or photons. They are simply models, useful in explaining the behavior of light. Review the idea of a scientific model by showing students a sealed box with a marble and a metal rod inside. Ask students to devise explanations, or models, to explain the phenomena they perceive when the box is turned, shaken, or tipped.
- Use a ripple tank to make it clear to students how the wave model explains reflection and refraction.
- Students may have difficulty accepting the idea that both a wave and a particle model are needed to explain the behavior of light. Do not belabor this issue.

Facts At Your Fingertips

- The amplitude of a wave is half the distance between a crest and a trough.
- The velocity (v) of any wave is equal to the product of the wavelength (λ) and the frequency (f): $v = \lambda f$.
- The energy of a quantity of light is proportional to both the number of photons and the frequency. A photon of blue light has more energy than a photon of red light, since blue light has a higher frequency than red light. The energy (E) associated with any photon is given by the formula $E = hf$. The h in the equation is a constant known as Planck's constant and has the value 6.62×10^{-34} J·s.

Demonstrations

1. Display the behavior of (transverse) waves by means of a ripple tank. (If you do not have a ripple tank, check with your physics teacher.) Demonstrate reflection and refraction, as shown in Figures 20-23 and 20-24 on student text page 511. Instructions for using a ripple tank are usually provided with the device.
2. Show how the particle theory can be used to explain reflection and refraction. Rolling steel balls against a smooth, flat wall illustrates the law of reflection, that is, that the angle of incidence equals the angle of reflection. To simulate the refraction of particles, give a rolling ball a push as it passes a line on the floor. The force will produce a change in direction.

20 LIGHT AND SOUND

20-4 The Nature of Light

Wave Motion

Use Diagram X to complete sentences 1–4. Use Diagrams X and Y to complete sentences 5–6. Write each term on the space at the left.

________ 1. Point A to point B is one 1.

________ 2. If two waves, or wavelengths, pass point A in two seconds, the frequency is 2.

________ 3. Points A , B, and C are called 3.

________ 4. Point D is called a 4.

________ 5. Compare the two diagrams, X and Y. Diagram 5. shows a longer wavelength.

________ 6. Diagram 6. shows the higher frequency wave.

A B C D

Diagram X

Diagram Y

Investigating Waves

The questions below are based on the wave diagrams A, B, C, and D on this page. Examine the diagrams. Then answer the questions.

________ 1. If Diagram A shows a frequency of 100 Hz, what frequency is shown in Diagram B?

________ 2. Which diagram shows the highest frequency waves?

________ 3. If Diagram A shows a frequency of 200 Hz, then what is the frequency of Diagram C?

________ 4. If Diagram D shows a frequency of 1,000 Hz, then what is the frequency of Diagram A?

________ 5. Which diagram shows the lowest frequency?

________ 6. If Diagram C shows a frequency of 250 Hz, then what is the frequency of Diagram B?

A B C D

Using Try This

Students should discover that the center, or origin, of the reflected waves appears to be behind the "mirror." Thus, the image of the point source—the drop of water—is found behind the mirror. It is a virtual image.

Checkpoint Answers

1. See Figure 20-21 on student text page 510.
2. Frequency is the number of waves, or wavelengths, that pass a given point in one second. A frequency of 2000 Hz means that 2000 waves pass a given point in one second.
3. Both the wave model and the particle model are used in describing light.

Figure 20-25 *The array in the photograph contains photovoltaic cells, which convert sunlight into electricity.*

In some cases, light energy absorbed by a material can cause electrons in the material to move as if they had been struck by tiny bullets. Sometimes the electrons are knocked right out of the material. In order to explain this property of light, it is necessary to think of light as a stream of tiny bullet-like particles. These particles are called ***photons***. The transfer of light energy to electrons is the basis for devices that convert sunlight into electricity. (See Figure 20-25.)

Both a wave model ***and*** a particle model are needed to explain the way light behaves. Neither model can account for all its properties.

Checkpoint

1. Draw a side view of a series of water waves. Label crests, troughs, and wavelength.
2. What is the frequency of a wave? Explain what is meant by a frequency of 2000 Hz.
3. Name two different models used in describing light.

TRY THIS Put some water in a shallow pan or a sink. One end of the pan can represent a "mirror." Use a medicine dropper to let a drop of water fall a few centimeters in front of the "mirror." Watch the waves spread out from this point source of "light." Watch the circular waves that bounce off the mirror. Find the center or origin of these reflected waves. Where is the "image" of the "object" in front of the "mirror"?

20-5

The Electromagnetic Spectrum

Goals

1. To describe the visible spectrum and how it can be produced.
2. To identify various components of the electromagnetic spectrum.

When sunlight passes through droplets of rain in the air, a rainbow may result. Similarly, "white" light given off by a bulb can be spread out into a rainbowlike spectrum of colors. One way to do this is to pass the light through a wedge-shaped, transparent object called a *prism* (PRIZ um). (See Figure 20-26.) If the wavelengths, or frequencies, of the separate colors are measured, they are found to be different. For example, red light has a lower frequency than violet light. The colors in white light can be spread out by refraction because each frequency, or wavelength, is refracted to a different extent.

The band of colors produced when white light passes through a prism is called the **visible spectrum**. It is a spectrum of colors, and also of wavelengths or frequencies. Since each frequency corresponds to a different energy, it is an energy spectrum, as well.

visible spectrum

Visible light given off by hot, glowing gases in the sun spreads out, or radiates, in all directions. It consists of constantly varying electromagnetic waves, which can travel through empty space at the highest possible speed (300,000 km/s).

Figure 20-26 *When white light passes through a prism, it is separated into the colors of a rainbow.*

Teaching Suggestions 20-5

Begin this section with a discussion of the visible spectrum. Refer to Figure 20-26 on student text page 513 to explain how refraction accounts for the separation of white light into the colors of the rainbow. Students will enjoy seeing this effect demonstrated. (See Demonstration.) Follow with a description of the entire electromagnetic spectrum. See Figure 20-27 on student text page 514.

End-of-chapter project 2 applies to this section.

Background

All electromagnetic waves travel at the same speed—3×10^8 m/s, or 3×10^5 km/s—in empty space. However, the speed of electromagnetic radiation is proportional to wavelength and is lower in optically denser media.

When light is refracted, the angle of refraction varies with the wavelength. Light of shorter wavelengths is refracted more than light of longer wavelengths. It is this dispersion according to wavelength that accounts for the visible spectrum produced when white light is refracted by a prism.

Demonstration

Show the spectrum produced when white light passes through a prism. Produce a narrow beam of light by shining a light bulb through a thin slit in a piece of dark construction paper. Place a prism in the path of the beam to form the visible spectrum. A white sheet of paper or screen beyond the prism will display the spectrum clearly.

Replace the prism with a transparent box filled with water. Position a corner of the box in the path of the light beam. The water will refract light in the same way the prism does, confirming that water bends light that enters it from air.

Teaching Tips

- Be sure students grasp that visible light is only a portion of the entire electromagnetic spectrum, and that all electromagnetic waves travel at the same speed, known as the speed of light.
- Be sure students understand that the visible spectrum is formed by a prism because white light consists of various wavelengths, which are all refracted at slightly different angles.
- Explain how a rainbow is formed. The sun must shine directly on a large mass of raindrops. When sunlight enters a raindrop, it is refracted, reflected internally, and dispersed, to be similarly refracted and dispersed by others. The result is the visible spectrum in an arc of a circle, with red along the outer edge.

Facts At Your Fingertips

- Birds are able to perceive parts of the electromagnetic spectrum that humans cannot detect. Tree swallows and house wrens, for example, avoid areas of high radiation when nest-building and breeding.
- Ozone (O_3) in the upper atmosphere screens out much of the ultraviolet radiation from the sun. This is fortunate, since ultraviolet radiation is a common cause of skin cancer.

Mainstreaming

A light sensor developed by Lawrence Hall of Science, University of California, Berkeley, produces audible sounds for light of different energies. Light of equal brightness but different frequency is translated into sound of different pitch. Even with such a device, however, visually-impaired students should work with partners.

20 LIGHT AND SOUND

20-5 The Electromagnetic Spectrum

Find the Answer!

The following questions relate to information about the properties of light covered in Section 20-5 of your textbook. Use your textbook to help you in answering the questions.

1. The distance around Earth is about 40,000 km. About how many times could light travel around Earth in one second? ____
2. Why is white light spread out into a spectrum of colors when it passes through a prism? ____
3. The electromagnetic spectrum is made up of all the kinds of electromagnetic radiation. Why are the properties of different kinds of radiation so different? ____
4. Why is the visible spectrum also said to be an energy spectrum? ____
5. What three types of radiation are given off by the sun? ____
6. If you buy suntan lotion and the clerk says that it screens out infrared, why don't you want to buy it? ____

Wavelengths

Find the range of wavelengths, or frequencies, that make up each of the following types of radiation: radio waves, television, x rays, infrared, radar, microwaves, visible light, gamma rays. Then write the name of each type of radiation in the space beneath its range of frequencies on the spectrum bar.

Wavelength	0.00000000001m	0.000000001m	0.000001m	0.0001m	0.1m	1m-100,000m
Wave, ray						

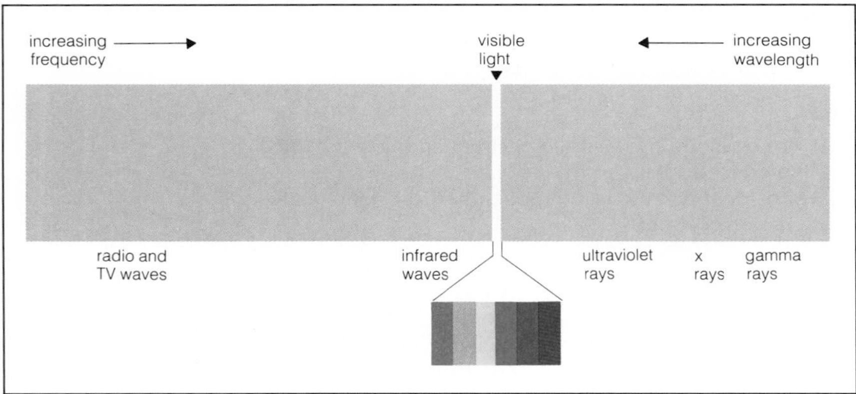

Figure 20-27 *Visible light is only a small portion of the entire electromagnetic spectrum.*

electromagnetic spectrum (il lek troh mag NET ik)

infrared radiation (in fruh RED)

ultraviolet radiation (uhl truh VY uh lit)

Visible light is only a tiny portion of a broad spectrum of electromagnetic radiation called the **electromagnetic spectrum**. (See Figure 20-27.) Different types of electromagnetic radiation have different wavelengths, or frequencies, and thus different properties. The radiation ranges from very low frequency radio waves to very high frequency gamma rays.

Sunlight contains invisible radiation as well as visible light. Radiation with somewhat lower frequencies than those of red light is called **infrared radiation**. Heat radiation consists of infrared rays. Radiation with somewhat higher frequencies than those of violet light is called **ultraviolet radiation**. The ultraviolet rays in sunlight are the rays that can give you a sunburn. Sun screen products can help prevent a serious burn.

Checkpoint

1. What is the visible spectrum? How can it be produced?
2. Compare the wavelength and frequency of visible, infrared, and ultraviolet radiation.
3. Which type of electromagnetic radiation has longer wavelengths, radio waves or x rays?

Checkpoint Answers

1. The visible spectrum is that portion of the electromagnetic spectrum that humans can see. It is the band of colors produced when white light passes through a prism.
2. The wavelength of visible light is shorter than that of infrared and longer than that of ultraviolet radiation. The frequency of visible light is greater than that of infrared and less than that of ultraviolet.
3. Radio waves have longer wavelengths than x rays.

20-6

Color

ENRICHMENT

Goals

1. To explain how people see color.
2. To describe how lights or pigments can be combined to produce colors.

You see an object because the light reflected from that object reaches your eyes. The colors that you see depend on what wavelengths of light are reflected by the object.

When white light strikes an object, the light may be totally reflected. If so, the color that you see is white. If white light is completely absorbed by an object, the object looks black. No light is reflected from the object.

White light is a mixture of different wavelengths, or colors, of light. When white light strikes an object, some of these colors may be absorbed and others reflected. For example, if all the colors except red are absorbed by an object, only red light is reflected. As a result, the object looks red.

One way to produce red light is to pass white light through a transparent red object, such as red cellophane. All colors except red are absorbed by the object. An object of this kind is called a *filter*. It "filters out" the unwanted colors and allows only the desired color to pass through.

Figure 20-28 shows what happens when equal amounts of red, blue, and green light shine on a white object. A white

Figure 20-28 *When the primary colors (red, blue, and green) are combined in equal amounts, they produce white light. Combining the primary colors in varying amounts produces other colors.*

Teaching Suggestions 20-6

Open discussion of color phenomena by asking students to relate their experiences with the alteration of color perception by environmental factors such as dim light (colors fade), standard incandescent bulbs, mercury vapor lamps, fluorescent lights, and sodium vapor lamps. Follow with an explanation of the primary colors of light. Be sure students understand how white light and other colors are produced by combining the primary colors of light. An easy way to show these effects is provided in Demonstration 1.

Next, develop the relationship between the primary colors of light and the primary colors of pigments. Refer to Figures 20-29a and 20-29b on student text page 516 to explain how combining pigments produces different colors. These effects, too, are easily demonstrated. (See Demonstrations.)

Background

The primary colors of light are the colors at both ends and in the center of the visible spectrum: Red and blue are at the ends and green is in the center.

Figure TE 20-2 shows a triangular arrangement of the primary colors for light and for pigment, with white at the center. The primary colors for light are located at the corners of the triangle. The primary colors for pigment are located in between, along the sides of the triangle.

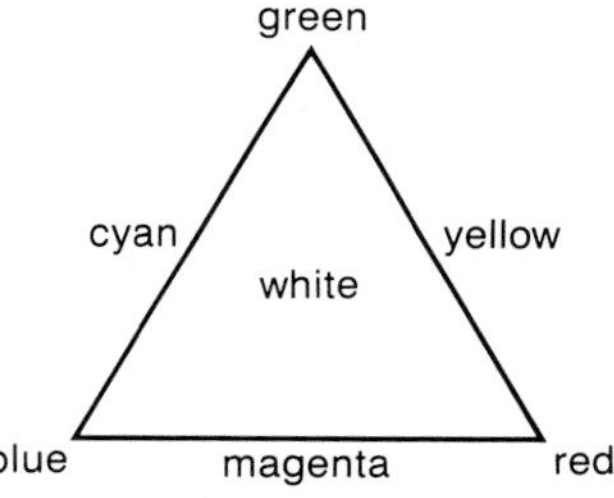

Figure TE 20-2 Color triangle.

The mixing of colored light to form other colors and white is an additive process. Combining the three primary colors of light in the proportion 1 : 1 : 1 produces pure white light. All the other colors of light can be produced by combining two or three of the primaries in varying proportions. Each primary color and the composite color directly opposite it are called complementary colors. When combined, a pair of complementary colors for light, such as red and cyan, produce white light.

The mixing of pigments to form colors is a subtractive process. The subtractive primaries—cyan, magenta, and yellow—appear along the sides of the color triangle. Mixing equal amounts of any two subtractive primaries will produce the light, or additive, primary that is located at the corner between them. For example, mixing cyan and yellow pigments produces green. Cyan pigment absorbs red light and reflects green and blue light. Yellow pigment absorbs blue light and reflects red and green light. When cyan and yellow pigments are mixed, the only color of light not absorbed is green. Green light is reflected back to the eye by the mixture, which therefore appears green. The pigments absorb, or subtract, certain colors of light from the light that falls on them. When all colors of light are absorbed, or subtracted, an object appears black. Black pigment is produced by combining equal amounts of all three subtractive primaries.

The human eye is a complex organ that sees different colors in terms of the portions of the spectrum that strike the retina. The mechanism of color vision is very complicated. See Figure 20-28 on student text page 515 for a summary of the effects of color addition (where light of different colors is combined). See Figures 20-29a and 20-29b on student text page 516 for an illustration of color subtraction (where light of different colors is absorbed by pigments).

Demonstrations

1. Show students the effects of mixing colored lights. Mount red, green, and blue plastic filters on cardboard. Use three overhead projectors or slide projectors to project colored circles on a screen or white wall. Overlap the circles from the three sources to show that red plus green produces yellow light, red plus blue produces magenta light, green plus blue produces cyan light, and a mixture of all three light primaries produces white light.

   Produce colored shadows by holding a hand near the screen so as to block out certain primary colors. Or place some wooden dowels in a large test tube holder and move them in front of the screen. A beautiful, constantly changing array of colors results.
2. Show the effects of subtractive mixing. Use cyan, magenta, and yellow filters. Cover the platform of an overhead projector with a piece of cardboard containing a small hole. Over the opening, place pairs of different colored filters to produce the three light primaries—red, green, and blue. When all three filters are used, no light will pass through, producing black. However, inexpensive filters may allow some leakage of light.

color is seen where all three colors overlap. Red and blue combine to produce a color called *magenta* (muh JEN tuh). Blue and green combine to produce a color called *cyan* (SY an). Red and green combine to produce yellow. By varying the amounts and combinations of red, green, and blue light, any color can be produced. Therefore, red, green, and blue are called the *primary colors of light.*

A *pigment* is a substance that is added to other substances to produce a desired color. Yellow, cyan, and magenta are called the *primary colors of pigments* because they can be mixed to produce any color.

When white light shines on a mixture of magenta and cyan pigments, red and green light are absorbed and blue light is reflected. (See Figure 20-29a.) Similarly, a mixture of yellow and cyan pigments absorbs red and blue light and reflects green light when white light shines on the mixture. (See Figure 20-29b.) When white light shines on a mixture of yellow and magenta pigments, red light is reflected because blue and green light are absorbed.

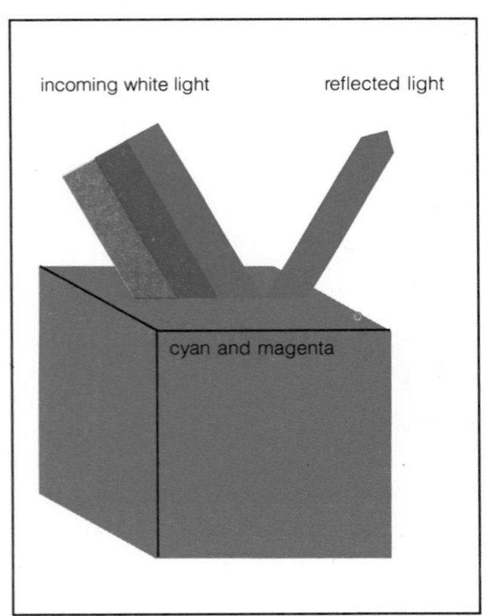

Figure 20-29a *A mixture of magenta and cyan pigments absorbs red and green light and reflects blue light.*

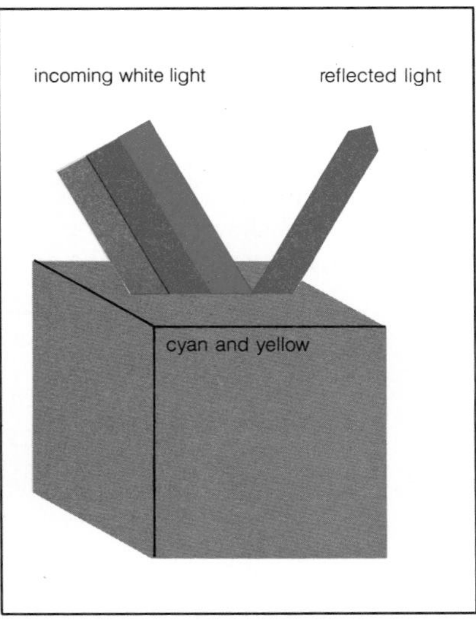

Figure 20-29b *A mixture of yellow and cyan pigments absorbs red and blue light and reflects green light.*

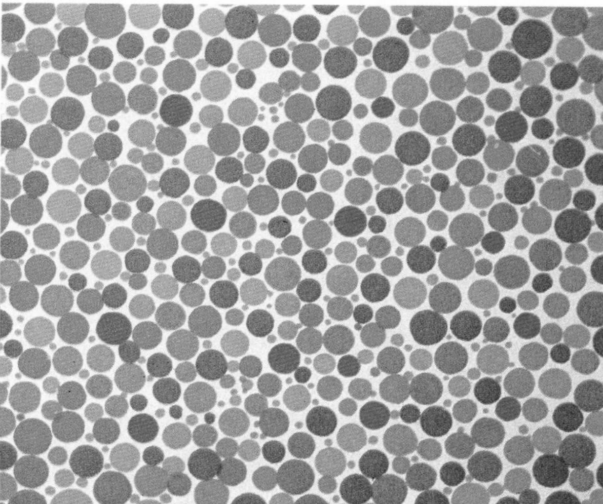

Figure 20-30 *This is a test for red-green colorblindness. Can you see the number?*

The light that reaches your eyes is not the only factor that determines what colors you see. The way your eye responds to the light is also a factor. People who do not see the same colors that most other people see are said to be *colorblind*. The ability to perceive colors accurately is important to many people. Electricians must identify color-coded wires and photographers must control the color in their pictures. You can take the test in Figure 20-30 to see if you have red-green colorblindness.

Checkpoint

1. Explain how a person with normal color vision sees an object as "red."
2. What two colors of light would you shine onto a white piece of paper to produce a yellow color?
3. What primary pigments would you mix to produce a green color?
4. What colors would you expect the pigments in a black sweater to absorb?

TRY THIS

Draw a picture using crayons, paints, or pens. Look at the drawing under different colors of light in an otherwise dark room. (Make color filters, if necessary.) Try to explain why the colors in the drawing seem to have changed.

Teaching Tip

- Students will probably find it fairly easy to understand how different colored lights can be mixed to produce other colors. The mixing of pigments to produce different colors may be a bit more difficult for them to understand.

Facts At Your Fingertips

- Color blindness is far more common in men than in women, because it is a sex-linked trait; that is, it is carried on the X chromosome.
- Black is not a true color. It is the name for the visual sensation that results from the complete absence of reflected light.

20 LIGHT AND SOUND
20-6 Color

Primary Colors

Each of the following questions is related to the primary colors. Read each question. Write the answer to the question on the space provided.

1. What are the primary colors of light? Why are they called primary? Explain how you might use the primary colors of light if you were in charge of stage lighting for a drama club. ______________________

red
yellow
green
white
magenta
blue-green
blue

Combining primary colors

2. What are the primary colors of pigments? Why are they called "primary"? Explain how you might use the primary colors if you were printing full-color advertising brochures.
____

3. Explain why an automobile driver's red-green color blindness can be dangerous.
____

Using Try This

The effect of a colored light will be identical to that of a filter of the same color. All pigments will appear either black or the color of the light or filter used. In the case of colored light, only light of that color will be reflected. In the case of a color filter, only light of that color will be transmitted. Pigments that absorb the particular color used will appear black. For example, under red light or with a red filter, green and blue appear black but red and yellow appear red.

Checkpoint Answers

1. A red object reflects back to the eye only red light and absorbs all other colors.
2. Red and green lights produce yellow.
3. Yellow and cyan pigments produce green.
4. The pigments in a black object absorb all colors of the visible spectrum.

Using The Feature Article

Students generally find the phenomenon of internal reflection of light interesting. If you can obtain some optical fibers, you may want to use them to demonstrate internal reflection. If not, you can use the setup shown in Figure TE 20-3. Fill a large jar with water. Cover the sides with black paper and punch a small hole in the cover, near the edge. Shine a slide projector lamp through the base of the jar. Place a plane mirror perpendicular to the narrow stream of water emitted.

Plant biologists at Stanford University in California have found that seedlings of oak, corn, and mung bean plants have tissues that can pipe light from the stem to light-sensitive pigments elsewhere in the plants. The stem acts like an antenna, gathering and transmitting light, even where it is below the soil surface.

Fiber Optics

Optics, the science that deals with light, could change our world in many ways. Scientists working in *fiber optics* have developed a way to send light through thin strands of glass no thicker than the width of a hair.

Each flexible glass strand, or fiber, consists of a glass *core* coated with a thin layer of glass called the *cladding*, as shown in Figure 1. The glass in the cladding is different from the glass in the core. As a result, when light strikes the core-cladding boundary at a great enough angle, it is reflected back into the core. This behavior is shown by the arrows in Figure 1. This *internal reflection* allows light to travel from one end of an optical fiber to another, as shown in Figure 2. Light will even travel around curves and corners in this manner. Doctors and dentists are using this property of optical fibers to bring light to hard-to-see places within the body. (See Figure 3.)

Fiber optics may be the answer to the growing need for communications pathways. The major advantage to using optical fibers is that they carry many more messages at one time than the metal wires now in use. Optical fibers can be used in telephone cables, to transmit television pictures, and to link computers. Someday soon, optical fibers may even link a computer in your home to such things as weather reports, your bank account, or even the latest baseball standings.

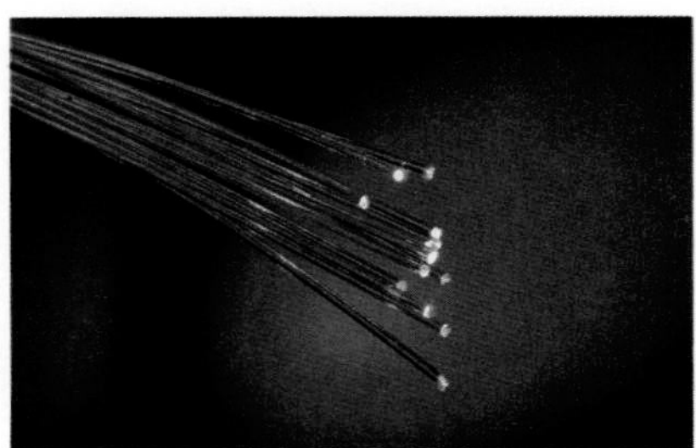

Figure 2

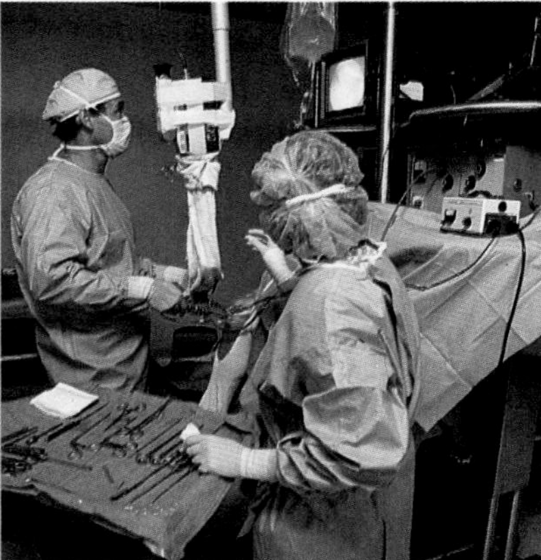

Figure 3

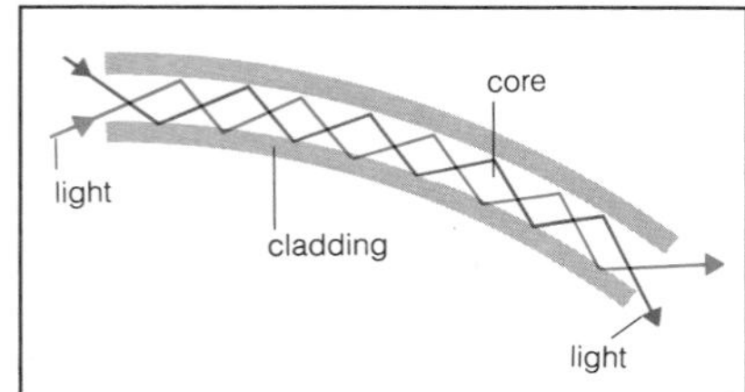

Figure 1

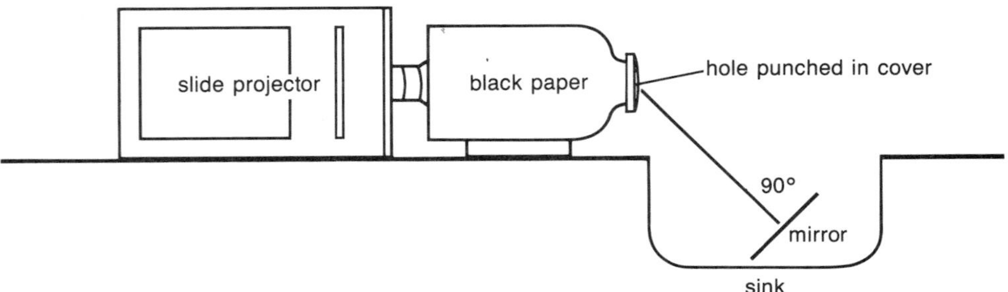

Figure TE 20-3 Light transmitted by internal reflection.

20-7

Sound

Goals

1. To describe how sound is produced.
2. To discuss some properties of sound that depend on its frequency or loudness.

Sound can help us to understand our surroundings. A doctor listens to the sound of a heartbeat through a stethoscope. A mechanic listens to the sounds of an engine. In each case, sound tells them about something they cannot see.

You commonly use sound to communicate. Within your larynx (LA ringks), or Adam's apple, are vocal cords that *vibrate* (move back and forth) when air from your lungs moves over them. If you put your fingers over your larynx as you talk you can feel the vibrations.

Vibration is the source of most sound. The number of vibrations per second, or frequency, determines the *pitch* of the sound that is produced. (The sounds that a canary makes have a higher pitch than those made by a bullfrog.) The greater the frequency, the higher the pitch. The vibration frequency of vocal cords depends on their length, thickness, and tautness, or tension. This is true for similar sources of sound, such as guitar strings.

Ideally, humans can hear sounds with frequencies of 20 to 20,000 vibrations per second (20–20,000 Hz). Some animals can hear higher frequencies. "Silent" dog whistles produce a pitch so high that it cannot be heard by humans, but it can be heard by dogs. Frequencies greater than 20,000 Hz are called *ultrahigh frequencies*.

Bats can produce and hear ultrahigh frequency sounds. By responding to the echoes of these sounds, bats can locate objects that they cannot see. Some ships use a batlike device called *sonar* (*so*und *n*avigation *a*nd *r*anging) to detect and locate underwater objects. Another use of ultrahigh frequencies is to produce "pictures" of objects inside the human body. For example, doctors can check the development of a fetus within its mother's body. The appropriate use of ultrahigh frequencies for this purpose can be extremely valuable. However, there is some concern that use on a routine basis may be unwise.

Teaching Suggestions 20-7

Generate interest by asking students for individual reactions to very loud music and very loud environmental noises, such as sirens, jackhammers, subway trains, and jet airplanes. Try to elicit individual differences in reaction in terms of both loudness and pitch.

Be sure students understand that sound is produced by some sort of vibration. Distinguish between the meanings of pitch and frequency. Pitch is a perceived quality and frequency is the rate of vibration. Stress the danger to the ears of prolonged exposure to extremely loud sounds. Point out that there is concern about possible hearing loss from amplified music.

Background

Sound, like light, is a wave phenomenon; however, sound travels much more slowly. Sound depends on a material medium for its transmission, whereas light can travel in a vacuum. In gases, such as helium, that have a greater average molecular speed than air, the speed of sound will be greater. Since speed is proportional to frequency, this accounts for the increase in the pitch of one's voice after the inhalation of helium.

The loudness of sound is measured by the decibel (db), which is a logarithmic function. A sound that is barely audible is assigned a value of 0 db. A sound 10, or 10^1, times as loud is said to have a volume of 10 db. One that is 100, or 10^2, times as loud has a volume of 20 db; 30 db is 1000, or 10^3, times as loud; and so on. Thus, a sound with a loudness of 70 db is 10 times as loud as one with a loudness of 60 db. Loudness levels of 80 db, 90 db, and 100 db are 100, 1000, and 10,000 times as loud as 60 db.

Demonstrations

1. Show how length and tension on a string affect the pitch of sound. Use the simple "guitar" shown in Figure TE 20-4. Vary the length (by moving the wooden block) and tension (by pinching the rubber band) on the "string." Have students note differences in pitch when the "string" is plucked. (The tauter the string, the higher the pitch; the longer the string, the lower the pitch.) Show also that the thickness of the "string" (use a heavier rubber band) affects pitch. (The thicker the string, the lower the pitch.)

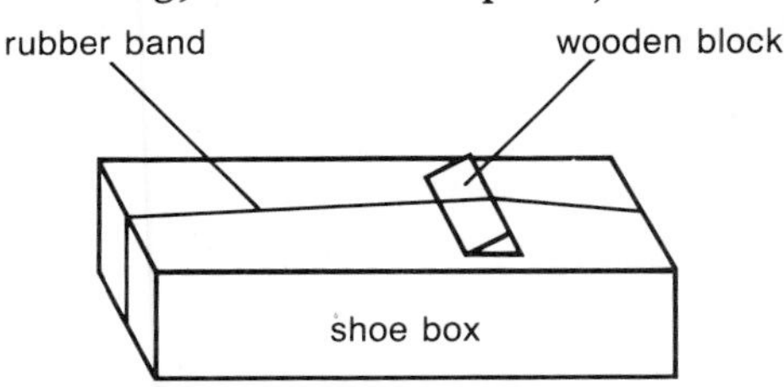

Figure TE 20-4 Simple "guitar."

2. Show the wave patterns of sounds on an oscilloscope screen. If necessary, borrow an oscilloscope from the physics department. Adjust the frequency and gain so that when sound is input by means of a microphone, the wave pattern appears on the screen. If there is someone in the class who can hold a note, a set of evenly spaced waves can be made to appear on the screen. Conversation will appear as a series of waves of changing wavelength, and louder sounds will produce a greater wave amplitude (height).

Teaching Tip

- Your music department may be able to supply tuning forks, pitch pipes, and musical instruments useful in teaching this section. If you use an oscilloscope, be sure to show the wave patterns produced by these devices.

Fact At Your Fingertips

- Prolonged or repeated exposure to 80 decibels of sound can produce hearing loss. It is not uncommon for amplified music to reach a level of 115 decibels.

20 LIGHT AND SOUND

20-7 Sound

What Do You Hear?

Each observation stated below is related to a term or concept about sound. Read each observation. Then complete the statement that follows the observation by writing the correct term on the numbered space at the left.

_______ 1. The pitch of a whistle increases. The __1.__ of the sound must be increasing, also.

_______ 2. A stereo set is producing sound at 25 db, when someone adjusts it to produce 85 db. The sound becomes much __2.__

_______ 3. A slack guitar string is tightened. As a result, there is a change in the __3.__ of the sound produced by the string when it is plucked.

_______ 4. A dog ran to the girl who had blown a whistle that she herself could not hear. The whistle produced sound whose frequency was greater than 20,000 Hz, or in the __4.__ frequency range.

20 LIGHT AND SOUND

Computer Program

Echo Sounding

Purpose To calculate depth of water.

Background Echo sounding is similar to SONAR in that high frequency sound signals are sent through water at 1,531 meters per second and an echo is returned from the ocean floor. The difference between transmission time and echo reception time is used to determine water depth.

Sound can cause problems as well as pleasure. Loud noises can be disturbing and even frightening. They can damage your ears, also. Loudness of sound is measured in units called *decibels* (db). The loudness of ordinary conversation is about 60 db. A sound of 120 db, which is not twice but a million times as intense as 60 db, can produce permanent damage to your ears. How much damage is done depends on the sound's loudness, frequency, and duration. Even at 80 db, long or repeated exposure can cause hearing loss. High-pitched sounds are more harmful than low-pitched sounds. The noise levels of some common sounds are listed below in Table 20-1.

Table 20-1: Noise Levels

Source of sound	*Noise level (db)*
faintest sound detectable	0
rustling leaves	20
whisper	25
conversation	60
street traffic	70
subway train	100
amplified music	115
jet plane 30 m away	140

Checkpoint

1. What is the cause of most sounds?
2. What range of frequencies can humans normally hear?
3. List two uses of ultrahigh frequency sound.
4. What factors can determine whether a sound may damage your ears?

TRY THIS

Press one end of a wooden ruler firmly against a desktop, with the free end extending horizontally. Set the free end of the ruler vibrating by plucking it with your other hand. Repeat the experiment several times. Each time move more of the ruler onto the desk so that less of the ruler is free to vibrate. In what way does the frequency of vibration change as the length of the ruler that is free to vibrate changes?

Using Try This

Students should observe that as the length of the ruler decreases, the frequency of its vibrations increases.

Checkpoint Answers

1. Vibrations produce sounds.
2. Humans can hear frequencies in the range of 20–20,000 Hz.
3. Answers will vary. Two examples are in sonar and dog whistles.
4. Loudness, frequency, and duration of sound are the factors that determine if it will damage the ears.

20-8

Sound and Light

Goals

1. To state similarities and differences between sound and light.
2. To describe how sound waves move through air.

Like light, sound can be reflected. That is why we can hear echoes. Sound shares other properties with light, including refraction. But sound cannot be the same as light. We see light from the sun and stars that has passed through nearly empty space. But we do not *hear* the raging fires of the sun. Sound cannot travel across empty space. The vibrations of a source of sound must be "passed along" through some medium.

The sounds that we hear most commonly travel through air. Air is made up of many tiny invisible particles (molecules) that are constantly moving about at high speeds. On the average, however, a given region of air contains a fairly uniform, or even, distribution of molecules. When a plucked guitar string vibrates, it pushes together, or *compresses*, the air nearby. (See Figure 20-31.) As the guitar string moves in the opposite direction, a region of *rarefaction* (rair uh FAK shun), in which there are fewer molecules than average, results. Some of the molecules from the region of *compression* move back into the region of rarefaction. As the pulses of alternating compression and rarefaction continue in response to the

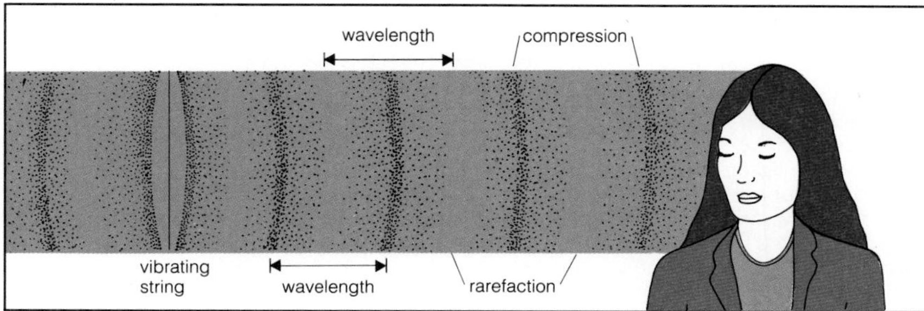

Figure 20-31 A vibrating guitar string causes waves to move through the air.

Teaching Suggestions 20-8

Use this section to review what students have learned earlier in the chapter. The similarities and differences between light and sound provide a good vehicle for such a review. Be sure students understand the difference between transverse and longitudinal waves, and that sound travels by longitudinal waves. If possible, show how transverse waves and longitudinal waves differ. (See Demonstration.)

Background

Some students may be interested in the method for determining the distance from an observer to a lightning strike. Because the speed of light is so great, it can be assumed that the occurrence and sighting of a lightning strike are simultaneous. That is, the time required for the light to reach the observer is 0 seconds. The time it takes for the sound to reach the observer is not negligible, and will be about 331 m/s × time in seconds. In 3 seconds the sound will travel 331 m/s × 3 s = 993 m, which is very close to 1000 m, or 1 km. Thus, for every 3 seconds of elapsed time, the lightning strike is about 1 km distant. If the time delay is 6 seconds, the strike is about 2 km distant, and so on.

Demonstration

Show the characteristics of transverse and longitudinal waves with a spring toy, as shown in Figure TE 20-5.

transverse waves

longitudinal waves

Figure TE 20-5 Waves in a spring toy.

Teaching Tips

- Discuss Figure 20-31 on student text page 521 to be sure students understand how sound waves are formed and transmitted through air.
- Be sure students are convinced that the speed of light is much greater than the speed of sound.

Facts At Your Fingertips

- The speed of sound varies with temperature. Sound is propagated by the movement of molecules in a medium. As the temperature rises, the molecules of a substance move faster, and so the speed of sound increases.
- Echoes occur because sound is reflected. However, human beings do not hear all reflected sounds. Having

responded to one sound, the human ear needs about 0.10 second before it can respond to another sound. An echo reaching the ear before 0.10 second has elapsed will not be heard. Because sound travels a little over 30 m in 0.10 second, the reflecting surface must be at least 15 m away for an echo to be heard. Thus echoes are not heard in small rooms. In large rooms, echoes will occur unless walls and ceilings are designed to absorb sound or are lined with sound absorbing materials.

- Until after World War II, it was feared that any plane flying at or above the speed of sound would break up. When the sound barrier was finally broken, it was discovered that flight was possible at supersonic speeds, although accompanied by a loud sonic boom.
- Until the 1970's, supersonic planes were limited to military use. Then the Concorde SST (supersonic transport) began making passenger flights across the Atlantic. The Soviet Union also built an SST for civilian use.

20 LIGHT AND SOUND
20-8 Sound and Light

Unscramble and Complete

Unscramble each word in Column A and write it on the space beneath the scrambled word. Then match the word with the sentence in Column B that it completes. Write the letter of that sentence on the space at the right of the unscrambled word.

A	B
1. iladugotlnni ___	a. Echoes are caused by the _a._ of sound.
2. peytm ___	b. Sound cannot travel through _b._ space.
3. elconiterf ___	c. The movement of a vibrating guitar string _c._ the air nearby.
4. somspeercs ___	d. When the guitar string in the previous example (c) moves in the opposite direction, a region of _d._ results.
	e. As a sound wave travels through air, _e._ in the air move back and forth

20 LIGHT AND SOUND
Laboratory

What Is the Speed of Sound?

Purpose To measure the speed of sound.

Materials measuring tape, loud whistle, stopwatch that reads to 1/100 s, flashlight that uses two "D" cells, clipboard (optional)

Procedure

1. To measure sound by this method, at least two people are needed. Therefore, get a partner. One of you will use the flashlight and the other will use the stopwatch and record observations.
2. With a measured pace or a long measuring tape, measure a distance of about 500 m.
3. The whistle person should stand at one end of the measured distance. This per-

doing the experiment? ___

2. What conditions were not controlled? Would you get different results if the temperature were 15 degrees colder or warmer? ___

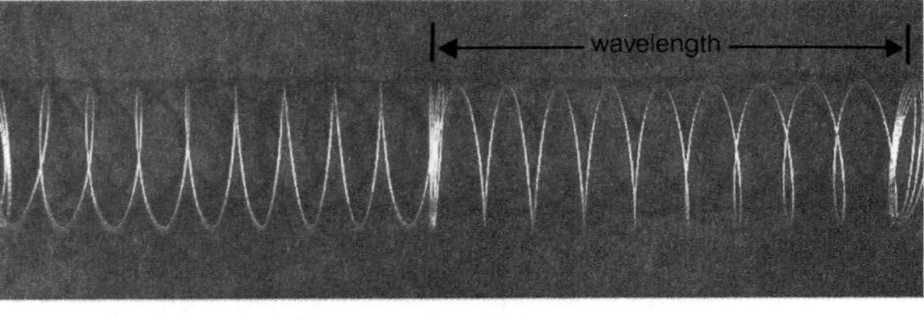

Figure 20-32 *This photograph of a spring shows two complete moving waves.*

motion of the guitar string, they cause pulses in the air next to them. In the same way, these pulses produce other pulses, creating a wave of alternating regions of compression and rarefaction. As a result, waves travel out in all directions from the vibrating string. The molecules move back and forth within a small region of space. It is the wave that travels from a guitar string to your ear, *not* the molecules.

Sound waves are similar to water waves (ripples), since it is the waves that travel, while the medium only vibrates. Water vibrates (up and down) at right angles to the direction of wave motion. Waves of this type are called *transverse waves*. Light waves are transverse waves. When sound waves travel in air, the molecules move back and forth (vibrate) in the direction of wave motion. Waves that travel in the same direction as the vibrations of the medium are called **longitudinal waves**. Longitudinal waves can be shown in a coiled spring. (See Figure 20-32.)

longitudinal waves (lahn juh TOOD un ul)

During a thunderstorm, you see lightning before you hear thunder because light travels much faster than sound. The speed of sound in air is only 331 m/s at 0°C. Light travels at about 300,000 km/s in air. Generally, a thunderstorm is 1 km away for every 3s between the time you see lightning and hear the resulting thunder. You can estimate how far away a storm is by counting (at one-second intervals) from the time you see lightning until you hear thunder.

TRY THIS

Hold the base of a tuning fork or a ticking watch a few centimeters from your ear. Then hold the sound source against a wooden door. Does wood transmit sound? To see if metals conduct sound, repeat the experiment using a metal doorknob in place of the wood.

Checkpoint

1. State two properties of sound that are (a) similar to (b) different from those of light.
2. What is a longitudinal wave? Give an example of longitudinal wave motion.
3. Explain how you can estimate the distance between yourself and a thunderstorm.

Using Try This

Students will observe that both wood and metal are better conductors of sound than is air.

Checkpoint Answers

1. Answers will vary. Some examples are (a) reflection, refraction; (b) longitudinal waves, lower speed.
2. Longitudinal waves are waves that travel in the same direction as the vibrations of the particles of the medium. Some examples are sound waves and waves in a coiled spring.
3. When you see a lightning strike, begin counting at 1-second intervals. For every 3 seconds that pass before you hear the thunderclap, the storm is 1 km distant.

Understanding The Chapter

20

Main Ideas

1. Light given off by a source generally radiates outward in all directions, in straight lines.
2. Light may be reflected, transmitted, or absorbed when it strikes an object.
3. Plane mirrors and convex mirrors reflect light, forming images, but convex mirrors can reflect light from a larger area.
4. Concave mirrors can collect and focus light to produce real images or parallel beams.
5. Light is refracted when it passes from one transparent material into another.
6. Concave and convex lenses can be used to form images in a variety of practical devices.
7. A wave model can explain many properties of light. A particle model is needed to explain other properties of light.
8. White light can be separated into a spectrum of colors by a prism.
9. The frequencies and wavelengths of the various types of electromagnetic radiation form a broad spectrum. This spectrum is known as the electromagnetic spectrum.
10. The color of an object is the color of the light it reflects.
11. Most sounds are produced by vibrating objects.
12. Important properties of sound include frequency and loudness.

Vocabulary Review

From the following list, choose the term that best completes each of the statements. Write your answers on a separate piece of paper.

concave (lens, mirror)	plane mirror
convex (lens, mirror)	principal focus
electromagnetic spectrum	real image
focal length	reflection
frequency	refraction
infrared radiation	translucent
lens	transparent
longitudinal wave	ultraviolet radiation
opaque	virtual image
	visible spectrum
	wavelength

1. The bending of light as it passes from one transparent material into another one is called __?__.
2. A __?__ can be produced by a concave mirror and captured on a screen.
3. The number of wavelengths per second is the __?__.
4. __?__ materials let light pass through them.
5. We can hear echoes due to the __?__ of sound.
6. The __?__ is the distance between a concave mirror and its principal focus.
7. Light is reflected or absorbed completely by __?__ materials.
8. When white light is passed through a prism it spreads out into a band of colors called the __?__.

Vocabulary Review Answers

1. refraction
2. real image
3. frequency
4. transparent
5. reflection
6. focal length
7. opaque
8. visible spectrum
9. electromagnetic spectrum
10. longitudinal wave
11. convex
12. virtual image
13. translucent
14. infrared radiation

Reading Suggestions

For The Teacher

Asimov, Isaac. HOW DID WE FIND OUT ABOUT THE SPEED OF LIGHT? New York: Walker, 1986. Includes times required for light to reach us from various stars.

Gardner, Robert. MAGIC THROUGH SCIENCE. New York: Doubleday, 1978. Contains a number of demonstrations appropriate for this chapter.

Rossing, Thomas D. THE SCIENCE OF SOUND: MUSICAL, ELECTRONIC, ENVIRONMENTAL. Reading, MA: Addison-Wesley, 1981. The physics of music and musical instruments, sound reproduction, noise and vibration control, and speech.

Scientific American. LIGHT AND ITS USES: MAKING AND USING LASERS, HALOGRAMS, INTERFEROMETERS, AND INSTRUMENTS OF DISPERSION. San Francisco, CA: W. H. Freeman, 1980. Twenty-six articles from "The Amateur Scientist" section of *Scientific American*, with experiments, practical illustrations, and clear explanations.

For The Student

Burkig, Valerie C. PHOTONICS: THE NEW SCIENCE OF LIGHT. Hillside, NJ: Enslow, 1986. An overview of modern optics.

Kettelkamp, Larry. LASERS: THE MIRACLE LIGHT. New York: Morrow, 1979. Recounts the discoveries that led to the development of the laser and explains its construction and operation. Describes applications in medicine, industry, fusion, and information science. Covers fiber optics, optical video discs, and holography.

Knight, David C. SILENT SOUND: THE WORLD OF ULTRASONICS. New York: Morrow, 1980. Clear, easy-to-read, and well illustrated introduction to ultrasonics, including applications in science, industry, oceanography, and medicine.

Ward, Alan. EXPERIMENTING WITH LIGHT AND ILLUSIONS. London: Batsford/Dyad (dist. by David & Charles), 1985. Activities illustrate the nature of light and illusions.

Chapter Review Answers

Know The Facts

1. reflected
2. rapid
3. concave
4. at the principal focus of the mirror
5. yellow
6. equal to
7. sound
8. convex
9. sound
10. concave
11. less than
12. loudness
13. refracted
14. decreases

Understand The Concepts

15. Real images are formed when light rays coming from an object are brought together. Real images can be displayed on a screen. Virtual images are formed when light rays only *seem* to be coming from points behind a mirror or lens. Virtual images cannot be displayed on a screen.
16. Use the lens to concentrate the sun's rays at the focal point, where the flammable matter is situated.
17. Cats can hear higher frequency sounds than humans.
18. See Figure 20-8 on student text page 504.
19. Mix cyan and magenta, since both reflect blue light. Because the cyan pigment absorbs red and the magenta absorbs green, only blue will remain to be reflected.

9. X rays, ultraviolet rays, and radio waves are part of the _?_.
10. A sound wave is a _?_.
11. A _?_ lens is thicker in the middle than at the ends.
12. When you look in the mirror you see a _?_.
13. Waxed paper is an example of a _?_ material.
14. Heat radiation consists of _?_, which has lower frequencies than those of red light.

Chapter Review

Write your answers on a separate piece of paper.

Know The Facts

1. Light that strikes a plane mirror is _?_. (refracted, reflected, absorbed)
2. High-pitched sounds arise from _?_ vibrations. (rapid, slow)
3. _?_ lenses should be used to correct nearsightedness. (concave, convex)
4. A concave mirror will produce a parallel beam of light if a small light bulb is placed _?_. (at the principal focus of the mirror, between the principal focus and the mirror, beyond the principal focus)
5. When red and green light shine on the same part of a white screen at the same time, the color you see on the screen is _?_. (yellow, magenta, cyan)
6. When a ray of light strikes a plane mirror, the angle at which it is reflected is _?_ the angle at which it strikes the mirror. (greater than, less than, equal to)
7. _?_ is *not* transmitted through empty space. (light, sound, infrared radiation)
8. _?_ mirrors are commonly used in stores to detect shoplifters. (plane, concave, convex)
9. The electromagnetic spectrum does *not* include _?_. (sound, light, x rays)
10. _?_ mirrors are used in flashlights to produce a light beam. (plane, convex, concave)
11. The speed of sound is _?_ the speed of light. (less than, greater than, equal to)
12. The _?_ of sound is measured in decibels. (pitch, loudness, speed)
13. The wavelengths that are present in white light are _?_ to different extents when the light passes through a prism. (reflected, refracted, absorbed)
14. If the frequency of a wave traveling at a certain speed increases, the wavelength _?_. (increases, decreases, stays the same)

Understand The Concepts

15. Explain the difference between real and virtual images.
16. Explain how a convex lens can be used to start a campfire.
17. Account for the fact that cats seem to hear sounds that you cannot hear.
18. Draw a diagram to show how a concave mirror can be used to make a spotlight.
19. What primary colors of pigments would you mix to produce a blue color? Explain how the blue color is produced.

20. A black brick absorbs all colors of light. Thus, it will absorb more energy than a white brick, which reflects all colors.
21.

Eye	*Camera*
lens	lens
retina	film
pupil	aperture

20. Which would you expect to absorb more heat from the sun, a black- or white-painted brick? Explain.
21. Compare the human eye with a camera.
22. At a race the starter, who is far from the timekeeper, uses both a flag and a gun to signal the start of the race. How do you explain this?
23. An echo of your shout reflected from a cliff wall returns one second after you shouted. How far away is the cliff?
24. Explain how a solar cooker works.
25. Most of the light that strikes still water passes through it. What else happens to the light that strikes the water?

Challenge Your Understanding

26. Draw a diagram to show how you would position two plane mirrors in order to see the back of your head.
27. Explain why sound travels faster through iron than through air at the same temperature.
28. A large source of light, such as the sun, often produces fuzzy shadows when it strikes small objects. Show, using a diagram, why this should be so.
29. Draw a diagram to show how an image is formed by a convex mirror.
30. Explain why you can be heard farther away if you use a megaphone, or simply cup your hands around your mouth.
31. You can see a rainbow when sunlight passes through the fine spray produced by a garden hose. Explain why.

Projects

1. Set two plane mirrors together at angles of 120°, 90°, 45°, and so on, and find the number of images formed in each case.
2. Find out how many different ways you can produce a visible spectrum. Demonstrate each one.
3. Cameras, slide projectors, and movie projectors all use lenses to produce images. Examine one or more of these devices closely. Try to figure out how they work.
4. Bees are able to communicate without the use of sound. Do some library research to find out how bees communicate.
5. Optical illusions, such as mirages, can be explained in terms of the properties of light. Find out how they are produced.
6. Research and write a report on the history of artificial lighting (the use of light sources other than the sun or moon).
7. Find out what regulations your community has concerning noise pollution.
8. Find out what means of communication are available to people in your community who have impaired sight or hearing.

22. The racers can see the flag fall before they hear the sound of the gunshot. Thus, those farther from the starter are not penalized by hearing the sound later.
23. Sound travels at 330 m/s. It took 0.50 second for the sound to reach the wall and 0.50 second for it to return. The distance to the wall must be 330 m/s × 0.50 s = 115 m.
24. A solar cooker is a concave mirror that reflects the sun's rays to a focal point, where the food is placed.
25. Some light is reflected and some is absorbed.

Challenge Your Understanding

26. See Figure TE 20-6.

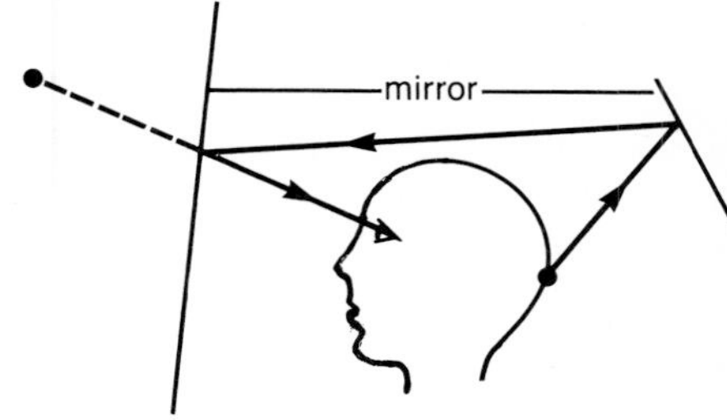

Figure TE 20-6 Viewing the back of one's head.

27. The speed of sound in iron is greater than it is in air because atoms of iron move only a very short distance before striking adjacent atoms and transmitting the sound.
28. See Figure TE 20-7.

light source

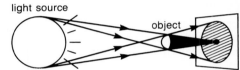

object

Figure TE 20-7 Fuzzy shadow produced by a large light source.

29. See Figure TE 20-8.

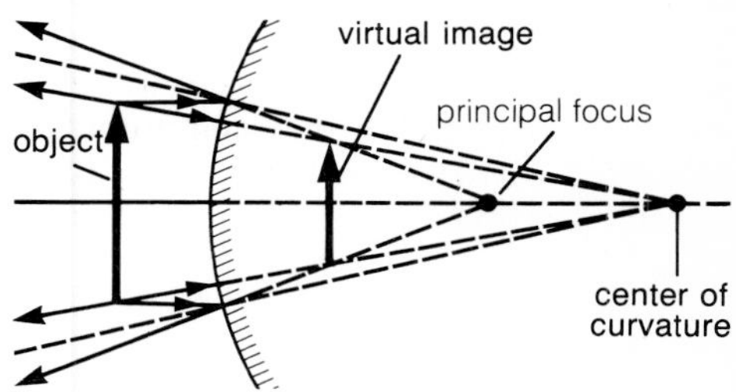

Figure TE 20-8 Formation of a virtual image by a convex mirror.

30. The megaphone reflects sound into a beam, much as a concave mirror produces a beam of light.
31. The droplets reflect and refract light.

20 LIGHT AND SOUND
Projects

1. Use library reference books to find out about the lives and work of Anton van Leeuwenhoek and Galileo Galilei.
   a. Briefly describe Leeuwenhoek's background: when and where he lived, what he did for a living.

Unit 4 Careers

Teaching Suggestions

Show students audio-visual materials that tell about careers related to physics. Each of the following filmstrips comes with cassettes and a teacher's guide: *What Is Science? Physics Career Challenges,* Prentice-Hall Media, Tarrytown, NY, and *Exploring Careers in Science,* EMC Publishing, St. Paul, MN. If you have access to a film projector, you also may wish to show *Electronics: Your Bridge to Tomorrow* (16mm), Electronics Association, Washington, DC.

Manufacturers launch new products everyday and need to know if they will perform well and safely. In-depth analysis and testing is necessary and is done at an independent testing laboratory. Locate the testing laboratory nearest you and plan to visit with your class. Arrange to speak with the scientists and technicians who work there.

Plan a field trip to a manufacturer that employs workers in physics-related areas. For example, take your class to visit an automobile assembly plant, an aircraft assembly plant, an appliance manufacturing plant, or a communication-product manufacturing plant. If possible, arrange for students to meet with some of the workers to discuss their jobs.

Several career areas have great potential for growth throughout the 80's and 90's. They are robotics, lasers, fiber optics, ceramics, microprocessors, energy systems, and hazardous waste management. Have students locate articles in science and business magazines about these exciting fields and the companies and people who are presently involved. Display the results on the bulletin board. If possible, follow up students' research by visiting local companies that are involved in these career areas, or inviting guest speakers to talk about opportunities in these areas.

Industrial Engineer

Positions are available for industrial engineers with a large consulting firm that advises banks, insurance companies, hospitals, and government agencies. Engineers analyze a client's operations and future plans and suggest ways of reducing costs and increasing efficiency, productivity, and safety. Some engineers will assist clients in making relocation plans. Applicants should demonstrate creativity and flexibility and should enjoy working as part of a team. Much time will be spent in travel.

A degree in industrial engineering is required with courses in physics, mathematics, and business. Candidates will begin by assisting experienced engineers. As experience is gained, there will be opportunities for advancement.

Computer Service Technician

Computer company with offices nationwide seeks technicians to install, maintain, and service computers. Maintenance involves adjusting, oiling, and cleaning all parts of the computer. Servicing requires quickly identifying the cause of a breakdown and repairing it. Service technicians must also explain to customers how computers work and respond tactfully to complaints.

Candidates need one to two years of training in electronics after high school. Courses in physics and mathematics also are necessary. Technicians will be trained for one year. After that, technicians should expect to work a forty-hour week with some emergency repairs done during odd hours. As experience is gained, opportunities exist for travel nationwide.

Help Wanted

Additional Careers

A brief description and job outlook for three occupations related to physics follow:

1. *Physicists* do basic research and develop theories in specific areas, such as optics, acoustics, nuclear science, solar energy, alloys, and space. This usually requires hours of independent work. Physicists also teach and solve problems for industry, government, and medicine. In the future, demand for people with a physics background should remain strong in areas such as secondary-school teaching, engineering, and computer science.
2. *Ophthalmic lab technicians* use precision instruments to grind and polish the lenses that are used in eyeglasses. They follow the

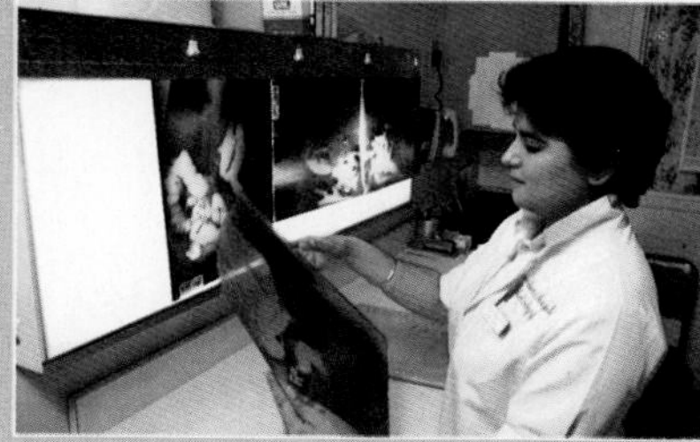

Radiologic Technician

Local hospital has openings for several kinds of radiologic technicians. *X-ray technicians* take x rays of patients. *Radiation therapists* treat cancer patients with prescribed doses of radiation. *Nuclear medicine technologists* prepare radioactive solutions and administer them to patients. The technologists then prepare radiographs, or photographs produced by the radiation.

Technicians must keep accurate records and follow safety procedures. All work is supervised by physicians. The job requires standing for long periods, and often lifting or supporting patients.

A degree from a two-year accredited program in radiology is necessary. A bachelor's or master's degree will improve chances for advancement.

Technical Writer

Technical writers are needed by a large company that makes electronic parts for computers. Writers help produce research reports, user manuals, annual reports, and new product information for customers. Clear, concise writing is essential. Writers must be able to absorb and organize large amounts of technical data. Writers also must be able to explain technical information in terms that can easily be understood by nontechnically-oriented readers.

Applicants should be college graduates with a major in electronics and a minor in writing. Experience with computers is necessary. Applicants should be able to work well as part of a team. Most writers should expect to work a regular forty-hour week, except during rush periods.

specifications of eye doctors and optometrists. The job requires a background in both science and mathematics. With technological and computer advances, lens production probably will become more automated in the future. However, an aging population and increasing concern for good eyesight will mean the need for more eyeglasses, and, therefore, more technicians.

3. *Electrical technicians* work for electrical power companies, manufacturers of electrical equipment, government inspection departments, and private companies that are responsible for installing electrical systems. Electrical technicians play an important part in the design, assembly, testing, installation, maintenance, and repair of electrical equipment. They often work as part of a team under the supervision of electrical engineers. The employment outlook for electrical technicians is favorable, especially for those with some job experience. Education beyond a two-year program will enhance opportunities for advancement to manager, instructor, or electrical engineer.

For More Information

You or your students can write to the following organizations or agencies for more information about careers related to physics:

1. National Institute of Ceramic Engineers, 65 Ceramic Dr., Columbus, OH 43210.
2. U.S. Office of Education, Div. of Vocational/Technical Education, Washington, DC 20202.
3. American Institute of Physics, 335 East 45th St., New York, NY 10017.
4. Society of Women Engineers, 345 East 45th St., New York, NY 10017.
5. National Association of Trade and Technical Schools, 2021 K St., N.W., Washington, DC 20006.

Using Issues in Science

The purpose of this *Issue* is to think about the impact, both positive and negative, of modern technological advance. Start by discussing the purpose of a simple machine (see section 17.3). Then go on to the purpose of complex machines. Students may give definitions that are more socially oriented for complex machines—e.g., a car makes it easier to transport goods. Then ask the students what makes robots different from other machines. (Robots can be programmed to perform tasks independently.)

Make a list on the blackboard using the students' answers to *Think Critically* question 3. The choice of items will reveal the students' preconceptions. Ask one student to analyze the list from an employer's perspective. Have another student analyze from an employee's perspective. This illustrates the subjective nature of the issue.

Try this class activity: ask the students to write a precise set of instructions for making a peanut butter-and-jelly sandwich, as if they were programming a robot to do the task. Then pick a few of the best results, and follow the directions *exactly*, in front of the class. This activity illustrates the difficulty in programming a robot to do even the simplest function.

Background

Robots are getting better at functioning without the need for frequent reprogramming and monitoring by people. At the National Bureau of Standards (NBS) in Gaithersburg, Md. there is a machine shop experiment called the Automated Manufacturing Research Facility. The robots being tested there can do many different tasks without being reprogrammed. The factory uses dozens of computers and video cameras to monitor production.

ISSUES IN SCIENCE

ROBOTS: Helpful or Harmful?

You are sitting at home doing your homework when your mom comes home early from her job at the factory. "Is anything wrong?" you ask her. "I just lost my job to a robot," she says sadly.

Does this sound far-fetched? Already many thousands of robots are being used in factories all across the country. More are being installed each day. Some new factories are completely automated. All of the production is done by machines.

One result of this trend towards automation is increased unemployment in industrial areas. Factory workers in these regions are concerned. They are demanding a slowdown in the pace of the "robot revolution."

Even when jobs are kept, workers complain that robots change the nature of the work. They say their jobs become much less interesting and creative. Skilled machinists, for example, complain that now they only "babysit" for the new machines.

Other people think that robots will benefit both factory owners and the work force. Company spokespersons are quick to point out that robots often perform jobs that people would rather not do, jobs that can be boring or dangerous. For example, a robot may paint thousands of automobiles without being affected by fumes that might be hazardous to humans. So the use of robots for "dirty work" can improve the working conditions for humans.

For many factory tasks, robots are simply cheaper to operate and more efficient than humans. Many factory owners think that they must either use robots or lose business to companies that do. As one business analyst put it,

Teaching Study Skills

Writing A Report

Assign a report on an issue related to robots. Give the students a choice of two topics. Possibilities include:

- Do robots give factories a competitive advantage?
- Are robots necessary for improving human working conditions?

Prepare students by using the *Teaching Study Skills* exercises on *Using Reference Materials* (TE pages 266–267) and *Writing an Outline* (TE pages 394–395). After students have completed their research and written an outline, you might present the following approach to writing a convincing report about an issue:

- Paragraph 1—State the issue clearly and the position taken.

"Are you going to reduce your work force by 25 percent by putting in robots, or by 100 percent by going out of business?"

Some industry experts feel that robots will create many new jobs. After all, people will have to build, maintain, and fix the machines. But other experts point out that although using robots may create more jobs in the long run, they certainly cause job losses in the short run. Such large-scale unemployment may offset the benefits of automation for the economy and society.

Think Critically!

1. What are the advantages of using robots instead of people for assembly line work?
2. Why are some people against the use of robots in factories?
3. Make a list of five factory tasks. For each task, state whether you feel it could be best done by robots or by humans. Explain your reasoning.
4. What do you think would be the results of moving ahead rapidly with installing robots in factories? Think about both short-term and long-term effects.
5. You are asked by a group of workers and owners to decide about using more and more robots in factories. Propose a compromise. Would it be fair to both groups? Explain your thinking.

- Paragraph 2—Present the strongest argument for the position, supported with facts.
- Paragraphs 3 and 4—Present two more arguments with supporting details.
- Paragraph 5—Summarize and make a statement to persuade the reader to adopt this position.

After completing a first draft, the students should revise their reports. This is the time for them to take a fresh look at their report, correct spelling errors, and work on the clarity of expression.

Students should hand in their outlines, and both drafts. Instead of grading at this stage, make further suggestions for improvement. The students can then make a final draft. With the author's name deleted, copy and distribute the first and last drafts of a report that showed the greatest improvement. Have the students compare the two drafts for clarity and content. In this way they can see the improvement that comes with revision.

Think Critically Answers

1. Using robots is often cheaper and more efficient; robots can perform jobs that are dangerous or boring.
2. Using robots in factories may cause higher unemployment; jobs may become less interesting.
3. Here are some examples of factory tasks: a. Fixing a machine—humans are better at this task because it requires advanced reasoning; b. Arc welding—robots can be better at this because it is a repetitive, dangerous task; c. Painting cars—robots are suited to this because it is dangerous and the process can be programmed; d. Selecting colors of different materials—better done by humans, because it is a visual and aesthetic process.
4. In the short run, moving rapidly ahead with installing robots means that people would lose jobs, but it may be necessary for the survival of industries in which robots give a competitive advantage. It takes a lot of money to install robots, and this cost must be balanced by increased profits. In the long run, there may be many jobs created in new industries for building, supplying, and repairing robots.
5. Answers will vary. A compromise might be to set up a committee of employers and employees to limit and regulate the introduction of robots. If workers had to be laid off, the owners could compensate them beyond normal severance pay and offer outplacement services. A fair compromise will meet the needs of both parties.

Activities

Contents

Working Safely

The activities in this book can be done safely if you follow directions and act responsibly. The following general guidelines can help you to perform the activities safely.

1. Maintain a serious attitude at all times. Fooling around can be dangerous for you and your classmates.
2. Never perform unauthorized procedures. Always check with your teacher before trying anything different from what you are directed to do.
3. Always read the assigned activities carefully before you do them. Try to understand the reasons for the given procedures. If you have any questions about how to proceed, check with your teacher first.
4. Wear safety goggles whenever directed to do so. Follow all other rules for safe procedure that you are given.
5. Never taste or touch unfamiliar substances unless you are directed to do so. Do not directly inhale any fumes given off by a substance.
6. Keep any materials that will burn away from open flames and other heat sources.
7. Clean up any chemical spills at once. If you spill something on yourself, rinse it off immediately with lots of water. In either case, report the spill to your teacher as soon as possible.
8. Always keep your work area clean and well-organized. Make sure that nothing, such as gas or electricity, is left on when not in use.
9. Know the location of safety equipment, such as fire extinguishers, safety showers, and first aid kits.
10. Report any accident or injury to your teacher immediately.

ACTIVITY 1 Introduction

This activity is intended for use in conjunction with Chapter 1, *Introduction to Life*. Learning to use microscopes can help students appreciate the microorganisms they will study in that chapter. The students will also use microscopes in Activities 2, 3, and 4.

Before students do this activity, it is strongly recommended that they read about the microscope in Appendix A. They should learn the names of the parts of a microscope and understand the function of each part.

A description of how a microscope produces images is given in Chapter 20 (section 20-3). You might want to refer any students who may be curious about the topic to that section.

Before students use the microscopes, be sure they understand precisely how to go about it. Emphasize, especially, the CAUTION statements given in Steps 3, 5, 6, and 8. You may want to demonstrate how to lower the low-power objective toward the stage, as illustrated in Figure 2. Emphasize that students should never use the coarse adjustment with the high-power objective.

Safety Precaution

Point out that the mirror below the stage of the microscope should not reflect direct sunlight (as noted in the CAUTION statement in step 3) because reflected sunlight can damage the retina of the eye.

Teaching Suggestions

In Step 2, you may want to demonstrate the preparation of a slide for the students. A slide prepared in this way is often referred to as a "wet mount." Make sure students include a letter, such as "e", which will clearly show that the compound microscope produces a reversed, inverted image.

Some microscopes have more than two objectives. You may want students to view the slide with more than two

ACTIVITY 1

Using a Microscope

PURPOSE To learn how to make a slide and view it under a microscope.

MATERIALS microscope, dropper, water, slide, cover slip, piece of newspaper with writing

Procedure

1. Hold the slide by the edges so that you do not get fingerprints on it.
2. Put the piece of newspaper in the middle of the slide. Place one or two drops of water on the newspaper. Pick up the cover slip by the edges and lower it carefully on the newspaper as shown in Figure 1. Try not to get air bubbles under the cover slip. Tap on the cover slip very gently with the tip of a pencil to get out any air bubbles you may have trapped. Set the slide aside.
3. Move the low-power objective into position over the diaphragm. You should hear a click when the objective is in place. Open the diaphragm to the widest possible setting. Adjust the mirror so that you can see a bright circle of light when looking through the eyepiece. **CAUTION**: Do not let direct sunlight fall on the mirror.
4. Raise the lens using the coarse adjustment. Place the slide you prepared on the stage. Position the slide so that the object you will be viewing is over the diaphragm. Move the clips so they hold the slide in place.
5. While looking at the microscope from the side—*not* through the eyepiece—use the coarse adjustment to lower the low-power objective toward the stage. (See Figure 2.) Bring the objective to within 0.5 cm of the slide. **CAUTION**: Do not let the objective touch the slide.
6. Look through the eyepiece. (Keep

Figure 1

objectives, if your microscopes are so equipped. In any case, the low-power objective, used in Step 3, is the one with the smallest magnification value written on its housing. Typically, the value is "10 × ," which means that an image produced by the objective is ten times as large as the object being viewed. The high-power objective is the one labeled with the largest magnification—typically, "43 × ." Refer students to Appendix A for a definition and discussion of *magnification*.

Students will notice in Step 7 that an image seen through the microscope is reversed and inverted compared to the object being viewed. This is a disadvantage of the compound microscope, as compared to a "single-lens" magnifier, such as a magnifying glass. However, the larger image produced by a compound microscope is more desira-

both eyes open.) Focus the lens by slowly raising the objective using the coarse adjustment. When the image is nearly in focus, use the fine adjustment to bring it clearly into focus. **CAUTION**: Never use the coarse adjustment to lower the objective while looking through the eyepiece. You could break the slide and damage the lens.

If you cannot see the object when you look through the eyepiece, check to see if it is centered above the diaphragm and under the lens. Lower the lens while watching from the side and start over again.

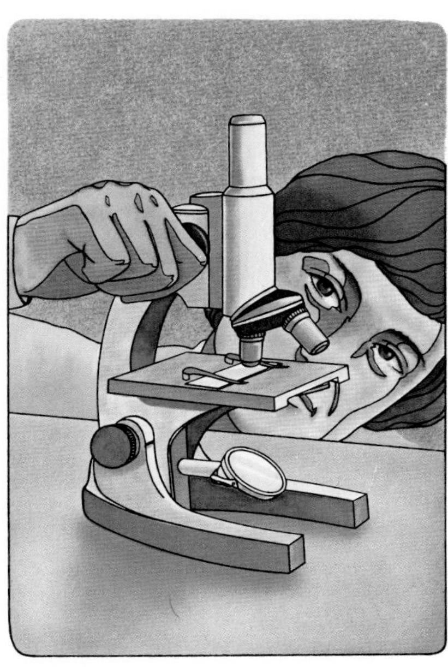

Figure 2

7. Try keeping both eyes open while looking at the slide. Draw a picture of the object as it appears under low power.
8. Without moving the coarse or fine adjustment, switch the objective to high power. **CAUTION**: Watch from the side to be sure that the lens does not touch the slide. Use the fine adjustment to focus the image. Draw a picture of the object as it appears under high power.
9. Observe what happens when you move the slide slowly right and left on the stage. What happens when you move it towards you and away from you?

Discussion

1. Compare the views of the object under low and high power. How are they alike? How are they different?
2. Which objective lens magnified more? What was the total magnification of the object viewed through that lens?
3. Did you see more of the object under low or high power? Explain.
4. When you moved the slide to your right, which way did the object appear to move when viewed through the microscope?
5. When the object on the slide was right-side up, how did it look to you when you viewed it through the microscope?

ble than a smaller image—even if the smaller image is neither inverted nor reversed.

Expected Results

Students' drawings should resemble the ones shown here. Their drawings, however, may vary somewhat, depending on the magnifications that were used.

"e" under low power

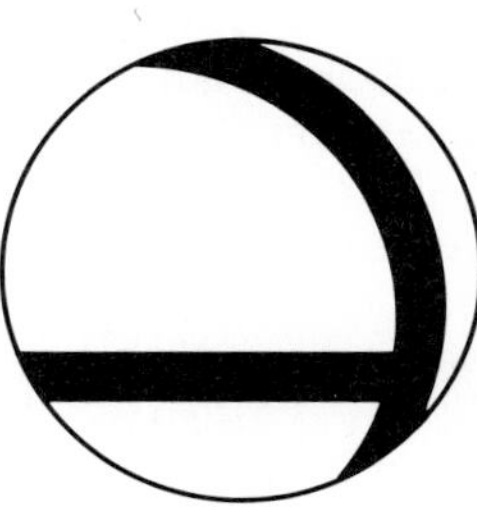

"e" under high power

Discussion

1. The object is not as large under low power as it is under high power. However, more of the object can be seen under low power than under high power. (Students may also note that the field of view appears brighter under low power.)
2. The lens with the greater (or greatest) magnification marked on it magnified more (the most). If the low-power objective is marked "10 × " and the high-power objective is marked "43 × ," the latter will magnify more. The total magnification is the product of the magnification by the eyepiece and the magnification by the objective. If the magnification of the eyepiece is 10 × , the total magnification under low power in this example is 10 × 10 = 100 × . Under high power, the total magnification is 10 × 43 = 430 × . (Appendix A includes directions on how to find total magnification.)
3. Students will see more of the object under low power (as they may have noted in response to the first Discussion question). Because the object is magnified more under high power, students see a smaller portion of the magnified image in the field of view.
4. When the slide was moved to the right, the object appeared to move to the left.
5. When the object on the slide was right side up, it was inverted, or upside down, when viewed through the microscope.

ACTIVITY 2 Introduction

This activity is intended for use in conjunction with section 1-2. Before doing this activity, students should have read Appendix A and completed Activity 1, or else they should have the equivalent preparation.

Advance Preparation

Prepare the iodine stain by dissolving 1.5 g of potassium iodide and 0.3 g of iodine in a liter of water. Store the stain in a dark bottle.

Safety Precautions

Use caution in handling iodine. It can cause burns or skin irritations. Its vapor is irritating and can be toxic if inhaled in large amounts. If students spill iodine stain on themselves, have them rinse immediately with water.

Teaching Suggestions

In Step 2, you may want to demonstrate the technique for obtaining cells from the inside of the cheek. In Step 6, students should draw what they see under high power. The illustration below shows what students' drawings should look like.

In Step 9, you may want to demonstrate the peeling away of a piece of onion skin. Under low power, students should be able to identify the nuclei and the cell walls of the onion skin cells. Under high power, they should be able to see small dots (nucleoli) within the nuclei. The illustrations below show what students' drawings should look like.

Discussion

1. Both kinds of cells have nuclei, cytoplasm, and cell membranes. Onion cells are larger, more rectangular, and have cell walls.
2. Protection is the main function of both kinds of cells.

ACTIVITY 2

Observing Cells

PURPOSE To observe human cheek cells and onion skin cells.

MATERIALS microscope, slide, cover slip, iodine solution, water, dropper, toothpick, onion section, tweezers

Procedure

1. Place a drop of water in the middle of your slide.
2. Gently scrape the inside of your cheek with the flat part of a toothpick.
3. Spread the cheek scrapings in the drop of water on the slide.
4. Stain the cells to make their structure show up better. Do this by adding a drop of iodine solution to the cheek scrapings.
5. Add a cover slip and examine the slide under low power, then high power.
6. Draw the cells as they appear under the microscope. Label the cell parts on your drawing.
7. Wash and dry the slide and cover slip.
8. Place a drop of water in the middle of the slide.
9. Pull away the thin skin of the onion section with your tweezers, as shown in the illustration. Do not let the onion skin curl or fold over.
10. Place the onion skin so that it is lying flat in the drop of water on the slide. Add a drop of iodine solution and cover the skin with a cover slip.
11. Observe the slide under low power. Draw one onion skin cell as it appears under the microscope. Label the cell parts on your drawing. Switch to high power. Draw and label one cell.
12. Wash and dry the slide and cover slip.

Discussion

1. Compare your drawings of the two kinds of cells. How are the cells alike? How are they different?
2. What is the main function of cheek cells? of onion skin cells?

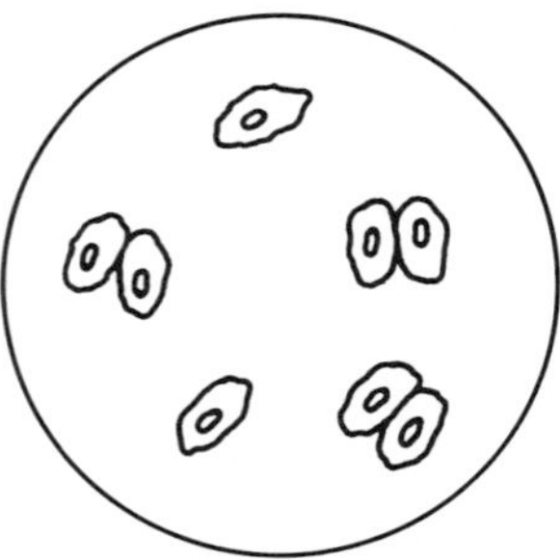

human cheek cells

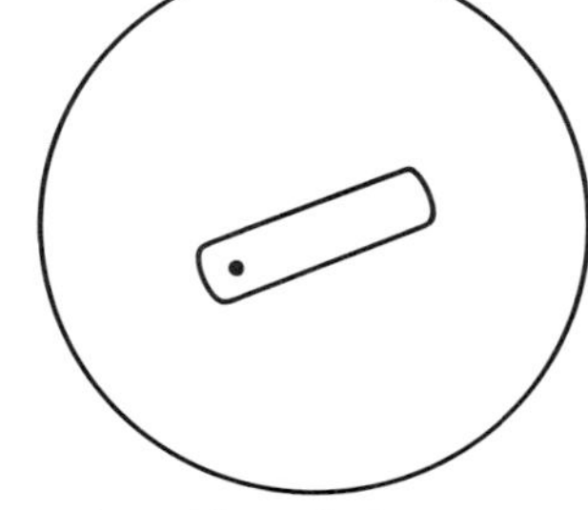

onion skin cell—low power

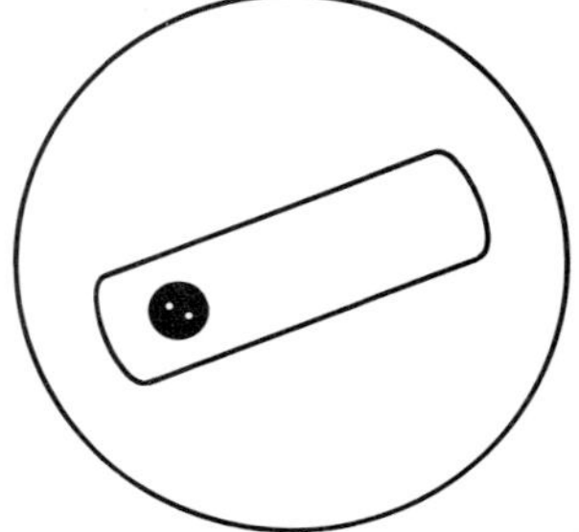

onion skin cell—high power

ACTIVITY 3

Observing One-Celled Organisms

PURPOSE To compare the structure and means of movement of different one-celled organisms.

MATERIALS microscope, slide, cover slip, dropper, paramecium culture, amoeba culture

Procedure

1. Place a drop of the paramecium culture on your slide. Cover the drop carefully with a cover slip.
2. Examine the slide through a microscope, using low power. Move the slide around slowly until you locate one or more of the paramecia. Observe the paramecia for a few minutes. Watch how they move. Paramecia move very quickly. You may have to move the slide to follow their paths.
3. Switch to high power. Observe and draw one paramecium. Label the parts on your drawing.
4. Use the dropper to return the paramecia to the culture. Be sure you put them back in the correct container.
5. Wash the dropper, slide, and cover slip.
6. Obtain a drop from the amoeba culture and repeat the procedure outlined in Steps 1–5 above. As you observe an amoeba, note how its structure and movement differ from those of the paramecium you observed earlier.

Discussion

1. Compare your drawings of a paramecium and an amoeba. How are the organisms alike? How are they different?
2. Compare the way a paramecium and an amoeba move.
3. How does the way an amoeba moves make it well-suited to life at the bottom of a pond?

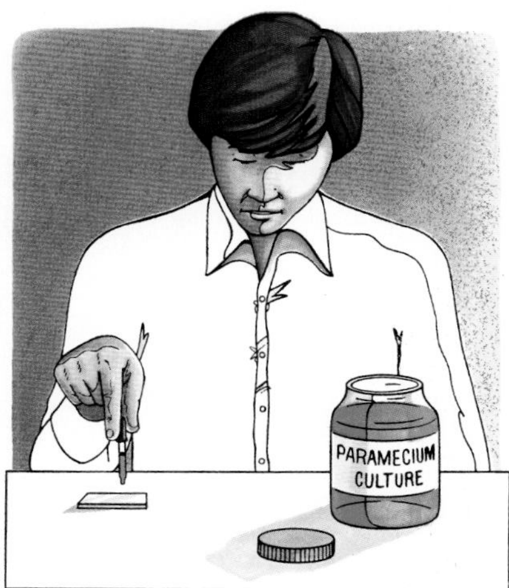

1

ACTIVITY 3 Introduction

This activity is intended for use in conjunction with section 1-3. Students will compare the movement of paramecia with the movement of amoebas. If students are not familiar with the use of the microscope and the preparation of a wet mount, refer them to Appendix A and have them complete Activity 1.

Teaching Suggestions

In Step 1, tell students to include some "gray scum" in the culture drop in order to get several paramecia. Remind students in Step 2 that when following a moving organism on the slide, they must move the slide in the direction opposite to that of the organism's movement. If the paramecia are moving too quickly to observe easily under high power, students can place a few cotton fibers or a drop of methyl cellulose solution on a slide with a culture drop, to slow the organisms. Students' paramecium drawings should resemble the one shown below, but they will not see all the labeled structures.

In Step 6, tell students to take the sample from the *bottom* of the culture dish in order to get several amoebas. The illustration below shows what students' drawings of an amoeba should look like. All the structures will not be visible, however.

Discussion

1. Both a paramecium and an amoeba have one large nucleus, cytoplasm, food vacuoles, and a cell membrane. A paramecium has a definite shape, cilia, an oral groove, a mouth, a gullet, and a second small nucleus. An amoeba has a changing shape and pseudopods.
2. A paramecium "swims" quickly through the water by means of "beating" cilia; an amoeba moves slowly, in a flowing motion, by means of pseudopods.
3. By forming pseudopods, an amoeba can move over a solid surface.

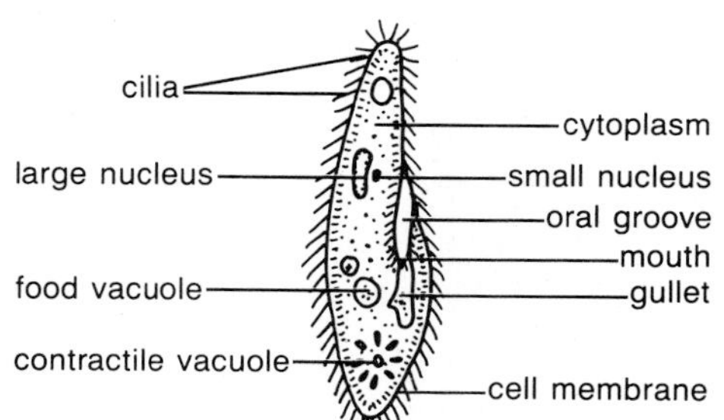

paramecium

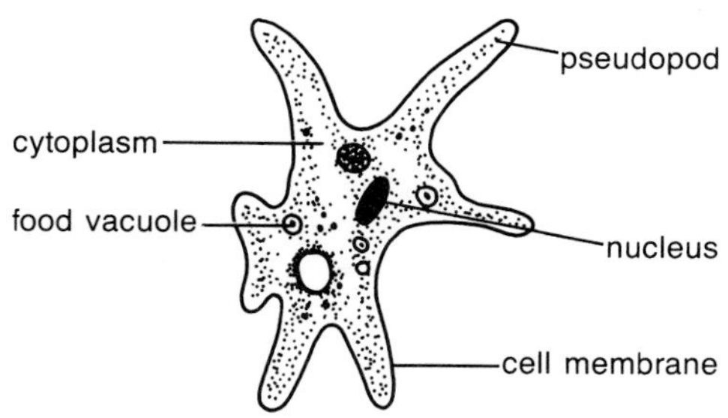

amoeba

ACTIVITY 4 Introduction

This activity is intended for use in conjunction with section 2-2. A root tip is an ideal place to view cells undergoing mitosis because it grows quickly and usually contains many cells in various stages of mitosis. Point out to students that cell division in plant cells differs from cell division in animal cells. Plant cells undergoing mitosis divide by formation of new cell membranes and new cell walls across the cell, rather than by constriction of the existing cell membrane.

Students are expected to be familiar with the use of the microscope. If they are not, refer them to Appendix A, and have them complete Activity 1.

Teaching Suggestions

In Step 1, you may want to assist students in locating the region of rapidly dividing cells (just behind the root cap). It is most easily found under low power. The dividing cells contain thick, darkly stained chromosome threads, which are not distinct in nondividing cells. In Step 3, tell students that they should try to include the four stages of mitosis in their drawings, as in Figure 2-3 c, d, e, and f on page 44. They need not include the period before mitosis, as in Figure 2-3 a and b. The illustration below shows what students' drawings should look like. Keep in mind, however, that all four stages are not always visible.

Discussion

1. Continuous mitosis indicates that the root tip is constantly growing.
2. Mitosis takes place in any growing tissue, such as a stem tip, young leaf, and flower bud.
3. Mitosis takes place in any tissue that is growing or repairing itself, such as the skin.
4. Mitosis permits growth and repair.

ACTIVITY 4

Observing Onion Root-Tip Cells

PURPOSE To observe the stages of mitosis in plant cells.

MATERIALS microscope, prepared slides of onion root-tip cells

Procedure

1. Examine each section of the slide under low power and high power. Look for cells containing stained chromosomes. These cells were undergoing mitosis when the slide was made.
2. Locate an area on the slide where several cells are in different stages of mitosis.
3. Prepare drawings of each stage of mitosis from the cells you see through the microscope. Label the chromosomes, cell wall, and spindle fibers on your drawing. Number your drawings in the order in which the stages occur during mitosis.

Discussion

1. Mitosis is going on continually in onion root-tip cells. What does this indicate about that part of the plant?
2. In what other areas of a plant is mitosis probably taking place?
3. Where is mitosis likely to occur in humans?
4. What is the function of mitosis?

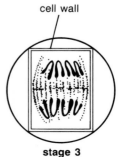

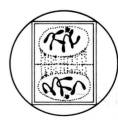

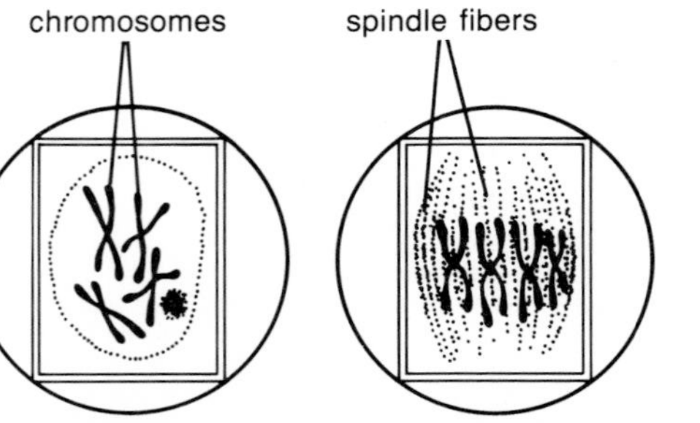

ACTIVITY 5

Measuring Lung Volume

PURPOSE To measure the volume of air your lungs can hold.

MATERIALS large jar, water, bucket, cardboard, rubber tube, graduated cylinder, wax marking pencil, ruler

Procedure

1. Make a table for recording normal and deep breath volumes.
2. Put water in the bucket to a height of about 6 cm.
3. Fill the jar with water. While holding the cardboard over the opening of the jar, turn the jar upside-down and place it in the bucket as shown.
4. Remove the cardboard. Slip one end of the rubber tube in the mouth of the jar. Take a normal breath, hold your nose, and breathe out normally into the rubber tube. **CAUTION**: Don't breathe in while the tube is in your mouth.
5. Straighten the jar and mark the top of the water with a wax pencil.

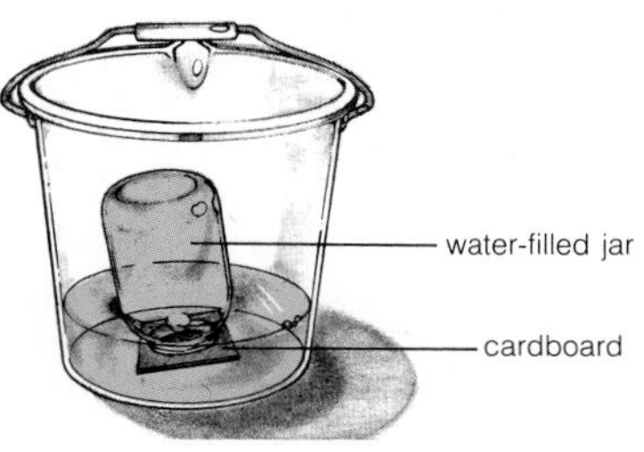

6. Turn the jar right-side up. Empty water from the jar until the water level reaches the pencil line. Measure the volume of the remaining water using a graduated cylinder. (This volume is the same as that of the air that you breathed into the tube.) Record the volume in your table.
7. Now find the volume of air you can force out after a deep breath. Repeat the procedure, only this time take a *deep* breath and exhale as much air as you can into the rubber tube. Record the results.

Discussion

1. Compare the exhalations following your normal and deep breaths. About how many normal exhalations would be equal in volume to one forced exhalation following a deep breath?
2. Compare your lung volumes with those of other members of the class. Which type of exhalation shows greater variation in volume?
3. Under what conditions do you use more lung volume than usual?

3

ACTIVITY 5 Introduction

This activity is intended for use in conjunction with section 3-3. Students should work in groups of three. Students will discover that during normal breathing, they inhale and exhale only a small portion (roughly 8-13%) of the volume of air that they can force out of their lungs following a deep breath.

The volume of air forcibly exhaled following a deep breath is *not* the maximum (total) volume the lungs can hold. A residual volume of air, which normally cannot be exhaled, remains in the lungs. The residual volume represents about 23% of the lungs' total capacity. A normal breath contains less than 10% of the total volume of air that the lungs can hold. The average lung capacity of females is about 77% that of males.

If students are not familiar with the use of a graduated cylinder, refer them to Appendix B.

Teaching Suggestions

In Step 1, you may want to draw a table like the one below on the chalkboard.

Volume of Exhaled Air (mL)

Normal Breath	Deep Breath

In Step 3, you may want to demonstrate how to invert the jar in the bucket. Emphasize that students should try not to spill water or get large air bubbles in the jar while inverting it. Warn students that they may have to support the bucket to prevent it from tipping after the jar is placed inside it. Have students in each team take turns performing Steps 3-7.

Emphasize the CAUTION statement in Step 4, to prevent students from drawing water into their lungs. Students should wipe the rubber tube with alcohol between turns in Step 4. Later, in Step 7, one team member should always support the jar to keep the mouth of the jar from rising out of the water. If students need to raise the jar to mark the water level after the deep breath, warn them not to lift the mouth of the jar out of the water.

Expected Results

Results will vary widely. The amount of air exhaled following a normal breath can range from 300 mL to 500 mL. The amount of air forced out following a deep breath can range from 3000 mL to 5000 mL.

Discussion

1. Answers can range from 7.5 to 12.
2. The volume of air exhaled following a deep breath varies more.
3. You use more of your lung volume at times of greater activity.

ACTIVITY 6 Introduction

This activity is intended for use in conjunction with section 3-4. Students should work in pairs. Students will prepare "maps" of the tongue showing the regions of the four kinds of taste receptors—sweet, salty, sour, and bitter.

Advance Preparation

Prepare the solutions to be tested as follows. To make 5% sugar solution, dissolve 10 g of sugar in 190 mL of tap water. To make 5% salt solution, dissolve 10 g of salt in 190 mL of tap water. To make 10% lemon-juice solution, combine 20 mL of lemon juice and 180 mL of tap water. To make 1% quinine solution, dissolve 2 g of quinine sulfate in 198 mL of tap water.

Because these solutions are to be placed in students' mouths, be sure all equipment is clean. Label each container and test tube with the name of the solution.

Teaching Suggestions

Tell students not to dip the swabs directly in the solutions in Step 2. For each taste, students should test the tip, sides, front center, middle center, and back center of the tongue.

Expected Results

Students' maps should resemble those shown below. X's should appear in the areas that are shaded.

Discussion

1. None of the taste areas covers the entire tongue; each tends to cover a different portion of the tongue.
2. Sweet, salty, and sour taste areas may overlap; sour and bitter taste areas also may overlap.
3. The very middle of the tongue has no taste receptors.
4. Receptors for a given taste are found in specific areas.

ACTIVITY 6

Mapping Taste Areas of the Tongue

PURPOSE To map the location of taste areas on the tongue.

MATERIALS 8 cotton-tipped swabs, dropper, 1 test tube each of sugar solution, salt solution, lemon-juice solution, and quinine solution, test-tube rack

Procedure

1. Make four outline drawings of a tongue. Label one *sweet*, another *salty*, another *sour*, and the last one *bitter*.
2. Use the dropper to place some sugar solution on one cotton-tipped swab. Touch the swab to one spot on your partner's tongue. If your partner could taste the sugar solution, make an X on the tongue drawing labeled *sweet* to show where the taste was detected. If your partner could not taste it, place an O on the tongue map. (See the illustration.)
3. Continue to put small amounts of sugar solution on different areas of your partner's tongue, recording the results after each trial, until you have a "map" of the areas that can taste *sweet*.
4. Now have your partner map the sweet-tasting areas of your tongue. Use a fresh cotton-tipped swab and a new tongue drawing.
5. Rinse the dropper and repeat the procedure, using salt solution, fresh cotton-tipped swabs, and the tongue drawing labeled *salty*.
6. Repeat again using the lemon juice, then the quinine solution. Be sure to rinse the dropper and use fresh swabs for each solution.

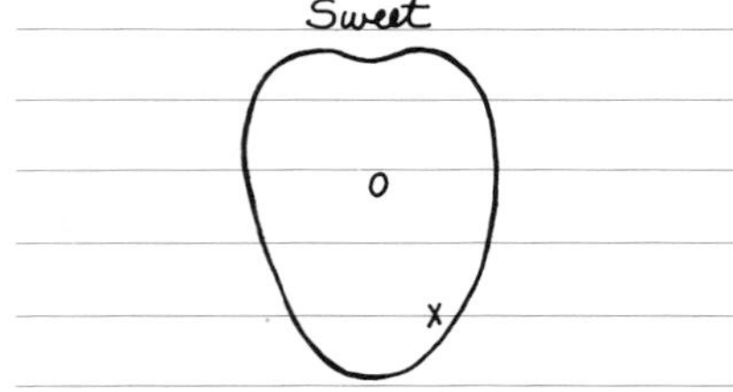

Discussion

1. Compare the four tongue maps you made. How are the maps alike? How are they different?
2. Do any taste areas overlap?
3. Are there places on the tongue that appear to have no taste receptors?
4. What conclusions can you draw as a result of looking at your tongue maps?

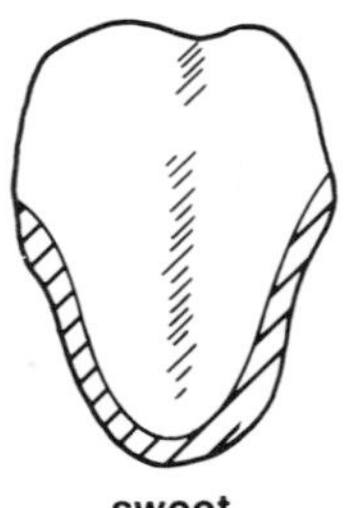
sweet

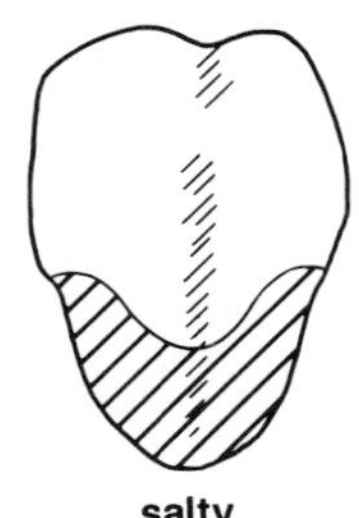
salty

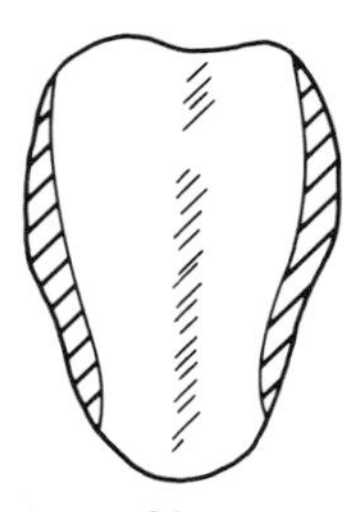
sour

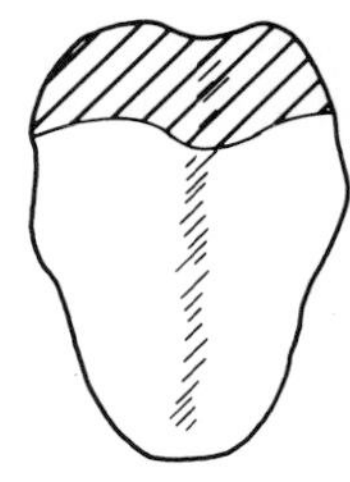
bitter

ACTIVITY 7

Life in a Square Meter of Ground

PURPOSE To look for living organisms and evidence of life in a square meter of ground.

MATERIALS string, stakes, small garden tools, meterstick, dissecting microscope (optional)

Procedure

1. Get permission to investigate a square meter of ground in your back yard, school yard, or a wooded area. Mark it off with stakes and string, as in the illustration.
2. Write descriptions and make sketches of the plants in the square meter. Do the same for any evidence you might find of animal life, such as a bird feather or snail shell.
3. If permissible, dig out a cube of soil with sides one decimeter long. Look through this soil sample for living animals, such as worms, snails, spiders, and insects. Describe and sketch any that you find.
4. If a dissecting microscope is available, use it to look at some of the soil. The soil should be moist, but not wet. With the microscope you can see smaller animals than those you saw before. Describe and sketch any animals that you may see through the microscope.

Discussion

1. Try to identify each of the living plants and animals that you found. Make a list of all that were in your square meter.
2. List any evidence that animals other than those you saw had been in your square meter.
3. Do your lists include all the living organisms present in your square meter? Explain.
4. How do the plants and animals you found obtain food?
5. What consumer orders are represented by the animals in your square meter? Give examples to support your answer.

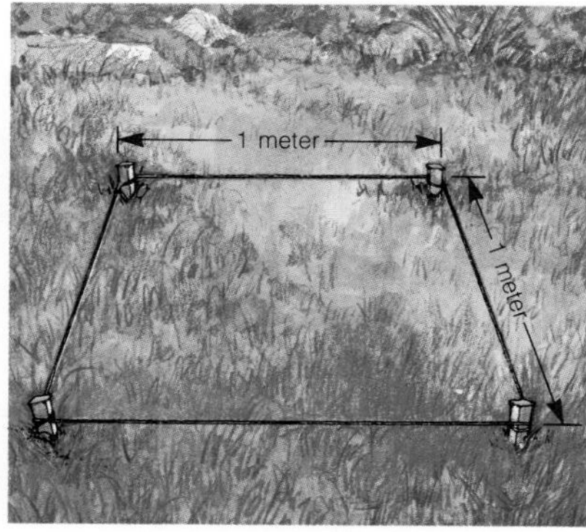

ACTIVITY 7 Introduction

This activity is intended for use in conjunction with section 4-4. You may want students to work in groups of three or four. Students will observe a small community of organisms in their natural environment and will gain an appreciation for the diversity of life in a small area.

Teaching Suggestions

In Step 1, instruct each team to choose a plot of ground that contains as varied a group of plants as possible. A wooded area is ideal for this activity. You may want the class to select plots in a variety of locations, for comparison.

In Step 2, ask students how the physical features of their plots and recent weather conditions might affect the forms of life found in the plots. Students may want to use hand lenses to look at the ground more closely for evidence of life.

In Step 3, tell students to try to remove the cube of soil in one piece, to minimize disturbance of the organisms living in it. They should place the soil sample on newspaper or a paper plate for observation. Students can use pencils to separate debris and particles of soil.

If you do not wish to have students bring soil samples back to the classroom in Step 4, students can examine their soil samples "on location," using hand lenses. If soil samples are to be brought back to the classroom, they should be labeled and placed in separate containers.

Discussion

1. Answers will vary. Plants listed might include various trees, seedlings, shrubs, flowers, ferns, and mosses. Students may also list lichens and fungi (such as mushrooms and molds). Animals listed might include earthworms, sowbugs, snails, slugs, insects, millipedes, centipedes, spiders, mites, roundworms, salamanders, toads, lizards, small snakes, moles, shrews, and mice.
2. Answers will vary. Lists may include burrows, buried reptile eggs, snail shells, bird tracks, and molted snake skins.
3. Not all organisms present are listed because microorganisms, such as bacteria, could not be seen without a microscope.
4. Answers will vary, but will probably include photosynthesis, decomposition, eating other animals, and eating dead leaves and soil debris.
5. Answers depend on kinds of organisms listed. Students will probably include first- and second-order consumers.

ACTIVITY 8 Introduction

This activity is intended for use in conjunction with section 5-2. Students will discover how various bird beaks are adaptations for different ways of getting and eating food.

The illustration in the activity shows the beaks of the following birds:

(a) black skimmer	(f) flamingo
(b) hummingbird	(g) parrot
(c) merganser	(h) toucan
(d) woodpecker	(i) whippoorwill
(e) hawk	(j) crossbill

Stress that the diversity of bird beaks is one factor contributing to the world-wide distribution of birds.

Teaching Suggestions

Encourage students to think of the beaks as tools and to imagine what "tool" pictured would be best suited to each feeding behavior listed. You may want to list the following tools on the chalkboard as a hint, and either ask students to match the tools to the beaks or tell them which tool each beak resembles.

Tool	Beak	Tool	Beak
skimmer	A	hoe	F
medicine dropper	B	nutcracker	G
pliers	C	paring knife	H
pick or drill	D	net	I
shredder	E	screwdriver	J

Expected Results

The feeding behaviors should be matched with the following beaks:

1. d	3. c	5. i	7. g	9. a
2. j	4. b	6. e	8. h	10. f

Discussion

The following are the reasons for the beak choices:

1. The beak must be able to get at insects through holes in tree trunks.
2. The beak must be able to twist apart pine-cone scales to expose the seeds.
3. The beak must be able to grip slippery fish.
4. The beak must be able to reach the nectar deep inside flowers.
5. The beak must have a means of trapping flying insects (bristles).
6. The beak must be able to tear meat.
7. The beak must be able to crack the hard shells of seeds.
8. The beak must be able to slice through the tough skin of fruit.
9. The beak must be able to skim from the water's surface.
10. The beak must be able to dig in mud.

ACTIVITY 8

Bird Beak Adaptations

PURPOSE To match bird beaks with the feeding behaviors for which the beaks are adapted.

MATERIALS pencil, paper

Procedure

Match the beaks shown on this page with the following feeding behaviors.

1. Feeds on insects that live in the trunks of trees.
2. Pries seeds out of the inside of pine cones.
3. Catches slippery fish in its beak.
4. Drinks nectar from flowers.
5. Catches flying insects.
6. Feeds on small animals.
7. Eats seeds that have hard shells.
8. Eats fruit that has tough skin.
9. Scoops insects from the surface of the water in lakes or ponds.
10. Digs up organisms that live in the mud under shallow water.

Discussion

Compare your answers with those of your classmates and discuss the reasons for the choices made.

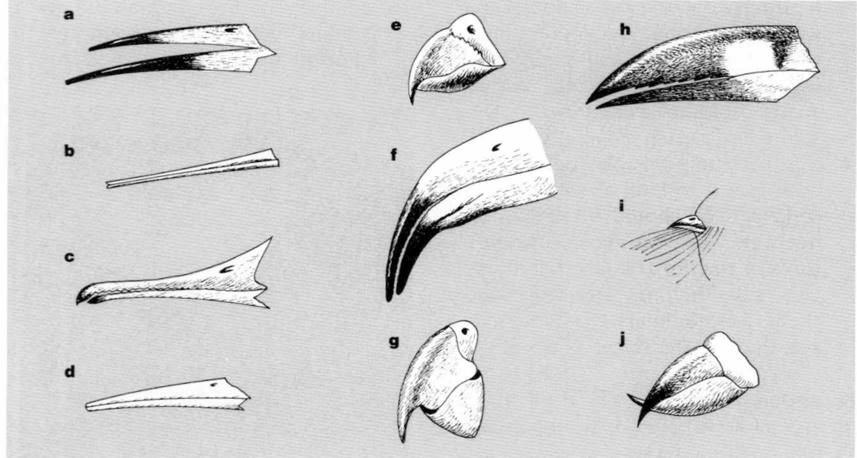

ACTIVITY 9

Measuring Density

PURPOSE To determine the density of some solids and liquids.

MATERIALS laboratory balance with standard masses, thread, centimeter ruler, graduated cylinder, solid and liquid samples

Procedure 1

1. Obtain a solid sample from your teacher. If it is cubic or rectangular, make a table like the one in Figure 1. If it is irregularly shaped, make a table like the one in Figure 2. As you obtain data, record it in the appropriate table.
2. Use a laboratory balance to determine the mass of the sample.
3. If the sample is cubic or rectangular, use a ruler to measure its length (*L*), width (*W*), and height (*H*).
4. If the sample is irregularly shaped, you can find its volume by measuring the amount of water that it displaces. First, half-fill a graduated cylinder with water. Read and

Density of a Cubic or Rectangular Solid

Sample Number	Mass (g)	Dimensions L(cm)	W(cm)	H(cm)	Volume L X W X H (cm³)	Density (g/cm³)

Figure 1

Density of an Irregularly Shaped Solid

Sample Number	Mass (g)	Volume 1 (mL)	Volume 2 (mL)	Volume of Sample (mL)	Density (g/mL)

Figure 2

ACTIVITY 9 Introduction

This activity is intended for use in conjunction with section 6-1. Before doing this activity, students should be familiar with the use of a laboratory balance and graduated cylinder. It is recommended that students read Appendix B, Measurement, before doing this activity. If they are not familiar with metric units, they should read Appendix C, Metric Units, also. Although data tables are provided for this activity, you may want students to read the first part of Appendix D, Organizing and Displaying Data, as well. You might want to have students analyze the data table given in Appendix D or those given in this activity in order to help them develop skill in recording their data in a well-organized, readily retrievable form.

Advance Preparation

Gather cubic or rectangular solids made of wood, glass, plastic, metal, or other materials. Children's building blocks of various sizes are appropriate. If rectangular solids are not otherwise available, they could be prepared in school wood- or metal-working shops.

Suitable irregularly-shaped solids include large screws and bolts, wire, and other insoluble objects, made of various metals or alloys. Marble (calcium carbonate) chips and mossy zinc or tin are other possibilities. Geology-oriented students might enjoy finding the density of rock and mineral samples.

Liquid samples might include water, milk, and orange juice (which are referred to in section 6-1) or other liquid foods. Household cleaners, automobile cooling system anti-freeze, gas-line anti-freeze, and similar liquids are suitable, also. However, be sure to check the labels on any such products for potential safety hazards. Liquids, such as ethanol, methanol, and acetone, are appropriate, also, but be sure to warn against any potential safety hazards.

In order to keep track of any "unknown" materials, it is suggested that you number them. After students have determined densities of unknown materials, you could give them the correct density or tell them what the material is and have them look it up in a handbook, such as the *Handbook of Chemistry and Physics*. If not given in handbooks or on labels, the average class result can serve as the correct value.

Procedure 1 Teaching Suggestions

In Step 4, be sure students lower the sample into the liquid on a thread (as illustrated in Figure 3). If this is done, the chances of damage to the cylinder are decreased. Objects that are not easily handled by this method (such as marble chips) can be made to slide gently down the side of a slightly inclined cylinder.

Discussion 1

1. For a rectangular block of walnut wood, typical data and computations follow:

$$L = 5.8 \text{ cm} \quad \text{Volume} = L \times W \times H$$
$$W = 5.2 \text{ cm} \quad = 150.8 \text{ cm}^3$$
$$H = 5.0 \text{ cm} \quad \text{Density} = \frac{101.0 \text{ g}}{150.8 \text{ cm}^3}$$
$$\text{Mass} = 101.0 \text{ g} \quad = 0.67 \text{ g/cm}^3$$

2. For a brass screw, typical data and computations follow:

Mass = 12.6 g
Volume 1 = 10.0 mL
Volume 2 = 11.5 mL
Volume of sample = 1.5 mL

$$\text{Density} = \frac{12.6 \text{ g}}{1.5 \text{ mL}} = 8.4 \text{ g/mL}$$

Note that in the preceding example the accuracy of the result depends on how accurately the volume is estimated. Thus, similar student results may vary significantly from accepted values. Note, further, that the answers given in the preceding examples are rounded off to two significant figures. You may want to discuss the topic of significant figures at this point.

Densities of solids are conventionally reported in grams per cubic centimeter, but it is acceptable to report them in grams per milliliter.

Procedure 2 Teaching Suggestions

In Step 1, you may want students to record room temperature in addition to the data specified in Figure 4. The volume of a liquid changes as its temperature changes. It is therefore important to specify the temperature at which a liquid density is determined. The volume of a solid also changes with temperature, but the change is smaller than that of a liquid, given the same change in temperature. Thus it is less important to specify the temperature at which a solid density is determined.

In Step 3, caution students to wipe off any liquid that may have spilled onto the outside of a graduated cylinder before they place the cylinder and its contents on the pan of a balance.

record (as Volume 1) the volume of water in the cylinder. Then carefully lower the sample into the cylinder. If possible, tie the sample to the end of some thread and lower the sample into the water as shown in Figure 3. Make sure that the sample is entirely under water. Then read and record (as Volume 2) the volume of the water plus the sample.

Discussion 1

1. Compute the volume of each cubic or rectangular sample you tested by multiplying the length times the width times the height ($L \times W \times H$). Then compute the density by dividing the mass by the volume.
2. Compute the volume of each irregularly shaped object you tested by subtracting Volume 1 from Volume 2. Then compute the density of each object by dividing the mass by the volume.

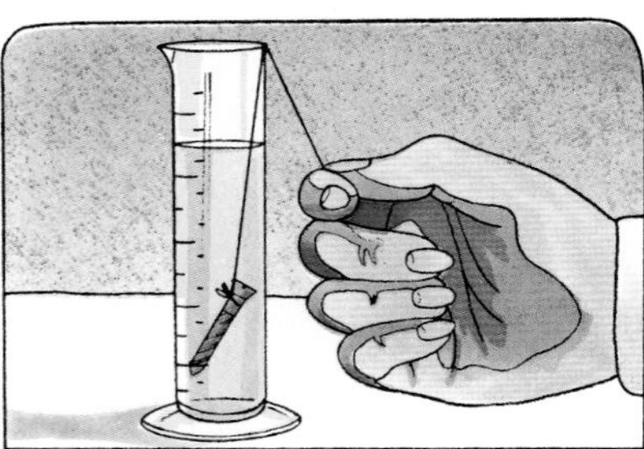

Figure 3

Procedure 2

1. Make a table like the one in Figure 4.
2. Find the mass of a graduated cylinder and record it as Mass 1.
3. Pour a liquid sample into the cylinder. Find and record (as Mass 2) the mass of cylinder plus sample.
4. Read and record the volume of the liquid in the cylinder.
5. Dispose of the liquid as your teacher directs. Wash and dry the cylinder.

Discussion 2

1. Compute the mass of each liquid sample you tested by subtracting Mass 1 from Mass 2.
2. Compute the density of each sample by dividing the mass of the sample by the volume.

Density of a Liquid

Sample Number	Mass 1 (g)	Mass 2 (g)	Mass of Sample (g)	Volume of Sample (mL)	Density g/mL

Figure 4

6

Discussion 2

1. For ethanol, or ethyl alcohol, at room temperature, typical data and computations follow:

Mass 1 = 112.4 g
Mass 2 = 120.3 g
Mass of 10-mL sample = 7.9 g

2. For ethanol:

$$\text{Density} = \frac{7.9 \text{ g}}{10 \text{ mL}} = 0.79 \text{ g/mL}$$

ACTIVITY 10

Paper Chromatography

PURPOSE To separate mixtures using paper chromatography.

MATERIALS 2–3 large test tubes, test-tube rack, filter paper, solvent, food coloring, 2–3 toothpicks, scissors, ruler, 2–3 small paper clips, aluminum foil

Procedure

1. Cut two strips of filter paper, about 2 cm × 15 cm. Cut one end to form a point, as in the illustration. Fold the strips lengthwise down the middle.
2. Put two dry test tubes in the rack. Add just enough solvent to fill the curved part of each tube.
3. Use the small end of a toothpick to put two small spots of food coloring on one piece of filter paper, as in the illustration.
4. In a paper cup, mix together one drop each of two or three different colors of food coloring, using a toothpick. With the toothpick, put two small spots of the mixture on the second piece of filter paper.
5. After the spots on both strips of filter paper have dried for at least three minutes, place the strips in separate test tubes, as in the illustration. The spots must not be below the top of the solvent. Cover the test tubes with aluminum foil.
6. Observe the paper in the test tubes every 5 min for at least 30 min.

Discussion

1. What do you see in each test tube?
2. Is the food coloring you used by itself a mixture? How can you tell?

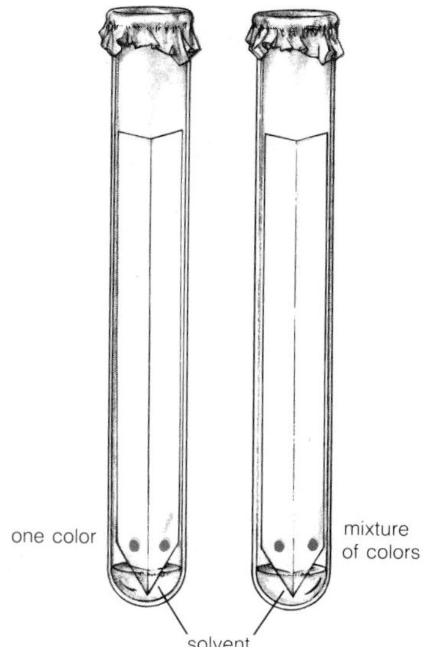

ACTIVITY 10 Introduction

This activity is intended for use in conjunction with section 6-4. Paper chromatography is useful for separating small amounts of mixtures. Components of a mixture are carried along a strip of filter paper by a solvent. Each of the components may travel at a different rate, depending on the nature of the solvent and of the paper along which the dissolved components travel. As a result, the various components are separated, appearing at different distances along the paper strip. If the components of the mixture are colored, they produce areas of different colors along the paper strip.

The word *chromatography* literally means a "colored array," but it refers to a variety of separation techniques that involve colorless as well as colored substances. Absorption may be on various surfaces besides paper and may involve gases, as well as liquids. Chromatography is today one of the most important methods used for the separation and analysis of mixtures.

Advance Preparation

Beforehand, you should prepare either of the following two solutions. To prepare Solution 1, mix together three parts, by volume, of 70% isopropyl alcohol (rubbing alcohol); one part, by volume, of ethanol (ethyl alcohol); and one part, by volume, of household ammonia. To prepare Solution 2, mix together three parts, by volume, of 70% isopropyl alcohol with two parts, by volume, of white vinegar.

Teaching Suggestions

In Step 3, each spot could be a different color. Green food coloring is usually a mixture of yellow and blue dyes. Blue food coloring usually is not a mixture. In Step 4, a good mixture to use is one consisting of equal volumes of red, blue, and green food coloring.

Discussion

1. Students will notice that a solution of solvent and food coloring moves up the paper strip (due to capillarity). If a spot of food coloring consists of a single dye, only one color will be seen on the strip. If a spot is a mixture of dyes, different colors will be concentrated at different distances along the strip.
2. If a given color of food coloring is a mixture of dyes, those dyes will appear separately on the paper strip. (The mixing together of different colored substances to produce a single color is discussed in section 20-6.)

ACTIVITY 11 Introduction

This activity can be used prior to section 7-3 as a means by which students can discover the law of conservation of mass, or it can be used to demonstrate that law after section 7-3 has been completed. The activity could be done as a demonstration, if desired.

You might want to review the definitions of physical and chemical change given in section 6-5. Students are expected to be familiar with the use of laboratory balances. If they are not, refer them to the section on mass in Appendix B.

Advance Preparation

Prepare 0.1 M lead nitrate solution by dissolving 331 g of $Pb(NO_3)_2$ in enough distilled water to make one liter of solution. Prepare 0.1 M sodium iodide solution by dissolving 150 g of NaI in enough distilled water to make a liter of solution. Put the solutions into small bottles for student use.

Teaching Suggestions

In Step 1 of Procedure 1, you may want to ask students where the water that forms on the outside of the vials comes from. (Water vapor in the air condenses on the cool vial surfaces. Review section 6-2 if necessary.) You might also ask why it is important to dry the surface of a vial before finding its mass. (The water vapor from the air will add to the mass of the vial and its contents, giving an incorrect result.)

Discussion 1

1. The two masses should be the same, within experimental error.
2. Students should conclude that, on the basis of this experiment, the mass of matter does not change when it undergoes a physical change.

ACTIVITY 11

Conservation of Mass

PURPOSE To demonstrate the law of conservation of mass for both a physical and a chemical change.

MATERIALS laboratory balance and set of standard masses, vial about one-third full of lead nitrate solution, vial about one-third full of potassium iodide solution or sodium iodide solution, vial full of crushed ice, three covers for the vials

Procedure 1

1. Wipe away any condensed liquid on the covered vial of ice as you determine its mass.
2. After the ice has melted, carefully dry the surface of the covered vial and find the mass again.

Discussion 1

1. How does the mass of the vial of melted ice compare with the mass of the vial of ice before it melted?
2. What can you conclude about whether the mass of matter changes when it undergoes a physical change?

Procedure 2

1. Determine the total mass of the two covered vials that contain different solutions.
2. Remove the covers and carefully pour the contents of one vial into the other.
3. Put the covers back on the vials and determine their total mass again.

Discussion 2

1. What evidence of a chemical reaction did you see when you mixed the two solutions?
2. How does the mass of the matter before the reaction compare with the mass after the reaction?

Discussion 2

1. When the two solutions are mixed a yellow solid precipitates out. This indicates that a chemical reaction has taken place. (You may want to review the signs of chemical change given in section 6-5.)
2. The two masses should be the same, within experimental error. Students can conclude that mass does not change during a chemical reaction.

ACTIVITY 12

Some Model Reactions

PURPOSE To demonstrate the law of definite proportions using atomic and molecular models.

MATERIALS paper fasteners and washers (at least six of each), laboratory balance and set of standard masses

Procedure 1

1. Use washers to represent atoms of element "W" and paper fasteners to represent atoms of element "F."
2. Combine as many of your "atoms" as you can to make "molecules" of the compound "WF." (You may have an excess of one of the elements.)
3. Determine the total mass of all the molecules of WF.
4. "Decompose" the molecules of WF by separating the molecules into atoms of elements W and F.
5. Determine the total mass of element W that was in compound WF.
6. Determine the total mass of element F that was in compound WF.

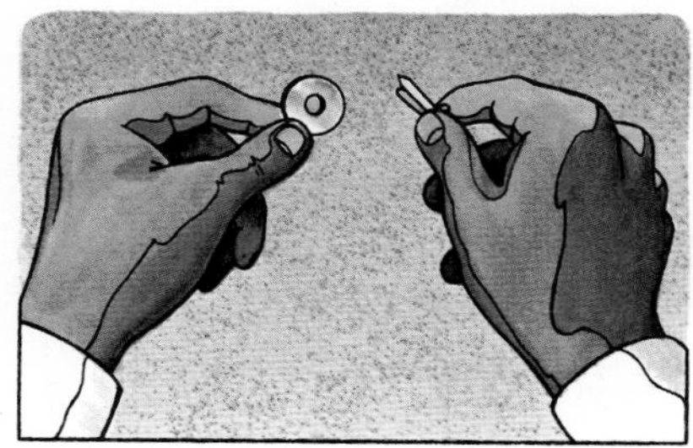

Discussion 1

1. What was the percent of element W in WF? To find out, divide the mass of W by the mass of WF and multiply this result by 100.
2. What was the percent of element F in WF? To find out, divide the mass of F by the mass of WF and multiply this result by 100.
3. Compare your results with those of your classmates. Does the percentage composition of WF depend on the number of molecules in the WF sample?

Procedure 2

1. Use washers and paper fasteners to make molecules of W_2F.
2. Repeat Procedure 1 using compound W_2F instead of WF.

Discussion 2

1. Find the percent of W and of F in W_2F.
2. Divide the percent of W by the percent of F in (a) W_2F, (b) WF. How does the ratio of W to F in W_2F compare with that in WF? How do you account for this ratio?

ACTIVITY 12 Introduction

This activity is intended for use with section 7-3 or subsequent to it. Although the activity demonstrates the law of definite proportions, given in section 7-1, students should have some understanding of chemical symbols (section 7-2) and equations (section 7-3) before performing it.

Advance Preparation

Obtain enough washers and paper fasteners for use in all your classes. If these are unavailable, obtain any two objects of different masses that can be fitted together in the specified combinations.

Teaching Suggestions

Students will need twice as many W atoms as F atoms for Procedure 2. Thus, if they are given enough washers and paper fasteners for both procedures at the start of Procedure 1, they will have an excess of W atoms in Step 2 of Procedure 1, which requires equal numbers of W and F atoms.

In Step 3 of Procedure 1 students should put all the WF molecules that they have made on the balance pan and find the total mass of the molecules.

Discussion 1

Actual results will vary. Sample data and computations follow.

1. Mass of WF molecules = 75 g
   Mass of W atoms = 50 g
   Mass of F atoms = 25 g
   $$\text{Percent of W in WF} = \frac{50}{75}(100) = 67\%$$
2. $$\text{Percent of F in WF} = \frac{25}{75}(100) = 33\%$$
3. The percentage composition does not depend on the number of molecules in the sample.

Discussion 2

Actual results will vary. Sample data and computations follow.

1. Mass of W_2F molecules = 125 g
   Mass of W atoms = 100 g
   Mass of F atoms = 25 g
   $$\text{Percent of W in } W_2F = \frac{100}{125}(100) = 80\%$$
   $$\text{Percent of F in } W_2F = \frac{25}{125}(100) = 20\%$$
2. (a) 80%/20% = 4; (b) 67%/33% = 2. All W atoms have the same mass and all F atoms have the same mass. Since the ratio of W atoms to F atoms is twice as great in W_2F as in WF, the ratio of the mass of W to F is also twice as great in W_2F.

ACTIVITY 13 Introduction

This activity is intended for use in conjunction with section 8-1. Students will investigate two physical properties of metals and nonmetals: electrical conductivity and melting temperature. Encourage students to notice other physical properties given in Table 8-1 in section 8-1. For example, students may note that copper wire is shiny, easily bent, and denser than some of the other substances that they are to test.

Students may or may not be familiar with simple electric circuits, such as the one used in this activity. Electric circuits are discussed in Chapter 19. Since students are probably familiar with flashlights, they should readily accept the fact that a small bulb like the ones used in flashlights can be made to light with the use of a battery. Have students try to light the bulb by touching the bare ends of the wire together. The bulb lights because electricity can flow from one end of the battery to the other and, thus, through the bulb. Then have them move the two bare ends of the wire a few centimeters apart. The bulb will go out, indicating that electricity cannot flow through the circuit because it cannot flow across the air space between the ends of the wire. Thus, if a substance that does not conduct electricity is placed between the ends of the wire, the bulb will not light. If a substance that conducts electricity is placed between the ends of the wire, the bulb will light. This should be most easy to accept when the substance being tested is a metal wire, since students have already seen that electricity passes through metal wires.

If students do not know how to operate an alcohol burner or a Bunsen burner, demonstrate the operation of these burners prior to the activity.

Safety Precautions

Be sure students wear safety goggles when working with flames. Ask them to roll up any long sleeves and tie back any long hair. Warn them against reaching over a flame. Remind them to turn off the gas jets when they have finished using the Bunsen burners. Also, students should work in a well-ventilated area, and they should not let the sulfur burn. It gives off an irritating gas when it burns.

Discussion 1

1. Students should find that candle wax, sulfur, and sugar do not conduct electricity. Thus they are nonmetals. Tin, copper, aluminum, iron, and platinum conduct electricity. Thus they are metals.
2. Students should conclude that the nonmetals would have lower melting points than the metals have. Table

ACTIVITY 13

Metals and Nonmetals

PURPOSE To investigate the electrical conductivity and melting temperature of some metals and nonmetals.

MATERIALS candle, Bunsen burner, alcohol burner, small pieces of candle wax, sugar cubes, roll sulfur (pea-size pieces), tin (foil or mossy), copper wire, iron wire, aluminum foil, platinum wire, pair of tongs, C or D cell, connecting wires, rubber band, small bulb, bulb holder, matches

Procedure 1

1. Make a table like the one shown in Figure 1.
2. Use a setup like the one shown in Figure 2 to test each assigned sample. Be sure that both wires are firmly in contact with the sample being tested. If the bulb lights, the sample you are testing is a conductor of electricity.

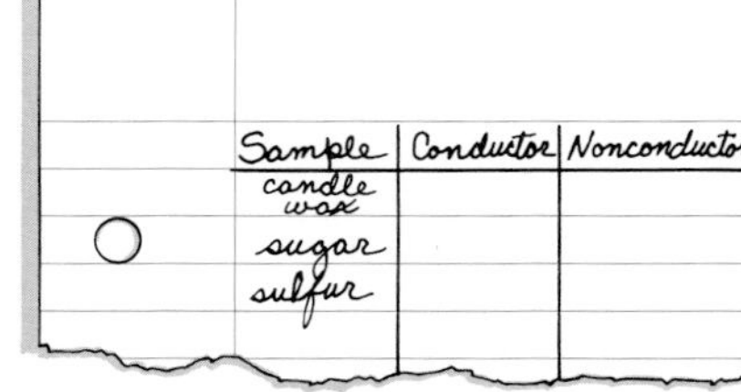

Figure 1

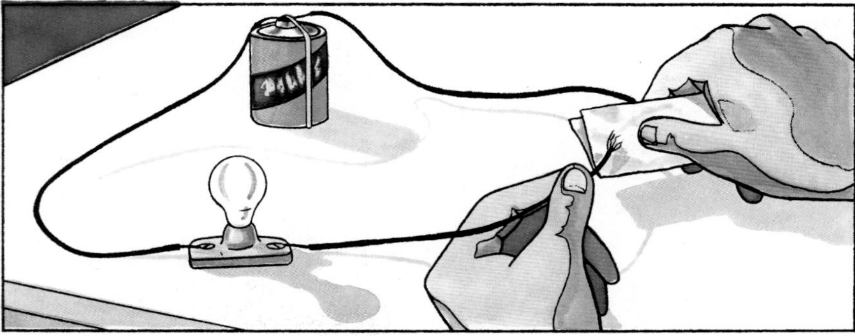

Figure 2

Discussion 1

1. On the basis of your data, which of the substances do you think are metals? nonmetals?
2. Which of the substances do you think have low melting points? high melting points?

Procedure 2

1. Make a table like the one shown in Figure 3.
2. Put on your SAFETY GOGGLES. Then use a pair of tongs to hold each of the assigned samples in a candle flame. (See Figure 4.) Note whether or not each sample melts in the flame. Record the results in your table.
3. Repeat Step 2 using (a) an alcohol burner and (b) a Bunsen burner.

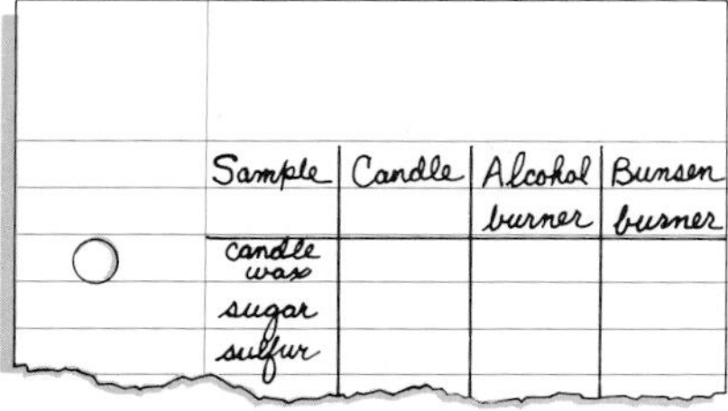

Figure 3

Discussion 2

1. Which flame has the lowest temperature? the highest temperature?
2. How do the melting temperatures of the nonmetals you tested compare with the melting temperatures of the metals you tested?
3. Ask your teacher for the melting temperatures of the substances you tested. Compare them with your results. Use this comparison to estimate the range within which the temperature of each flame you used lies.

Figure 4

8-1 indicates that metals tend to be solids at 20°C, while many nonmetals are liquids and gases at that temperature. From this information, students should infer that, in general, metals have higher melting points than do nonmetals. (If necessary, students should review the material on melting points in section 6-2.)

Discussion 2

1. The candle flame has the lowest temperature. The Bunsen burner flame has the highest temperature.
2. The melting temperatures of the nonmetals are lower than the melting temperatures of any of the metals tested.
3. The melting points of the substances tested are as follows:

candle wax	52–54°C
roll sulfur	113°C
sugar (sucrose)	185–186°C
tin	232°C
aluminum	660°C
copper	1083°C
iron	1535°C
platinum	1774°C

Student results should confirm the order of the melting points through aluminum. Depending on experimental conditions, they probably will not be able to melt copper. Neither iron nor platinum is melted by a Bunsen burner flame. Thus students probably will not be able to distinguish among copper, iron, and platinum from their results.

Sugar will melt in a candle flame, but tin probably will not. Thus students should infer that the temperature of the flame is at least 185°C, but less than 232°C.

Tin will melt in an alcohol burner flame, but aluminum probably will not. Thus students should infer that the temperature of the flame is at least 232°C, but less than 660°C.

Aluminum will melt in a Bunsen burner flame, but copper probably will not. Thus students should infer that the temperature of the flame is at least 660°C, but less than 1083°C.

Some students may want to test the melting points of various other substances. If so, caution them not to test any substance without your prior approval, since some substances could be dangerous in a flame.

ACTIVITY 14 Introduction

This activity is intended for use in conjunction with section 9-4. Students will discover that lemons and other types of fruit, together with metal electrodes, can function as electrochemical cells.

A property of lemons with which students are likely to be familiar is their sour taste. As students will learn when they study section 9-5, a sour taste is a property of acids. The sour taste of fruits is due to the presence of hydrogen ions in their juices. Students will learn more about the properties of hydrogen ions in section 9-5. At this point, it is only necessary for them to realize that lemons and other fruits contain ions that can move about within the fruit juices. These ionic solutions enable the fruits to function as electrochemical cells. You might want to remind students that the electrochemical cells described in section 9-4 require solutions of ions in order to function. Positive ions are able to accept electrons that build up on the cathode. The drift of positive and negative ions through the solution prevents an electric charge from building up around the electrodes, as the anode releases positive ions into solution and the cathode removes positive ions from solution by neutralizing them. Negative ions (anions) drift toward the anode and positive ions (cations) drift toward the cathode. As a result, electrical neutrality is maintained about the electrodes.

The amount of current produced by an electrochemical cell made from a fruit is quite small. Therefore it is necessary to use a microammeter or a sensitive galvanometer to measure the current. The advantage of using a galvanometer is that it is easier to see which way the current is flowing when a cell is set up, since the zero point of the meter is in the middle.

If a microammeter is used, students should connect one of the wires from the microammeter to one of the electrodes, then touch the other wire to the other electrode briefly, in order to determine the direction in which current is flowing. If the pointer moves in a negative direction, connect the wires to the electrodes in the opposite way to which they were connected when a negative current was indicated. When the electrodes have been connected so as to produce a positive current, read the current immediately and then disconnect an electrode from the microammeter, in order to avoid an unnecessary drain on the electrochemical cell.

Students should make sure that the electrodes are clean before they use them. The electrodes will need to be cleaned after each use.

Discussion

1. Expected results are as follows:

ACTIVITY 14

Unsuspected Electrochemical Cells

PURPOSE To show that electrochemical cells can be made using common fruits and metal electrodes.

MATERIALS lemons and other available fruits, such as apples, pears, olives, pickles, and oranges; galvanometer or microammeter; two insulated wires, preferably with alligator clips; copper, aluminum, and iron nails; zinc strips; centimeter ruler; paper towels and newspapers

Procedure

1. Make a table like the one shown in Figure 1. Measure the lengths of two electrodes, a zinc strip and a copper nail. Record the lengths (as Length 1) in the table.
2. Insert the electrodes into a lemon, as in Figure 2. Insert both to the same depth. Measure the length of each electrode above the surface of the lemon. Record the lengths (as Length 2) in the table. (The depth is the difference between Lengths 1 and 2.)
3. Measure and record the distance between the electrodes.
4. Connect the electrodes to a microammeter or galvanometer, as shown in Figure 2. In order to get a positive reading on the meter, you must connect the negative electrode to the negative side of the meter, and the positive electrode to the positive side. The connections that give you a positive meter reading, therefore, indicate the type of charge (positive or negative) on each electrode. Record the

Electrode pair	Length 1 (cm)	Length 2 (cm)	Depth (cm)	Distance apart (cm)	Electrode charge (+, -)	Cell current (microamperes)
1. zinc copper						

Figure 1

electrode charges. Read and record the current. Disconnect the electrodes from the meter.

5. Investigate how the electrode depth affects the amount of current produced. Keeping the distance between electrodes the same, change the depth of insertion. Connect the electrodes to the meter. Read and record the current. Repeat the procedure for other electrode depths.
6. Investigate how the distance between electrodes affects the amount of current produced. Keeping the depths the same, change the distance between the electrodes. Connect the electrodes to the meter. Read and record the current. Repeat the procedure for other electrode distances.
7. Replace the zinc electrode with a second copper nail. Connect the electrodes to the meter. Read and record the current.

Figure 2

8. Compare other electrode pairs to the copper-zinc pair. Pairs that you could try include copper and aluminum, copper and iron, and aluminum and zinc. All electrode pairs should be compared at the same depth of insertion and distance apart. For each electrode pair, note and record (a) the charge on each electrode and (b) the amount of current produced. After each trial, clean the electrodes.
9. Make electrochemical cells using fruits other than lemons. Find the amount of current that each produces, trying to keep electrode depth and distance apart the same for each fruit.

Discussion

1. Compare the charges that you found for each of the electrode pairs that you tested. How might this data enable you to predict the charges of the electrodes in pairs that you did not test?
2. What do you conclude about the use in an electrochemical cell of two electrodes that are made of the same metal?
3. How is the current produced by a cell affected by (a) the depth to which the electrodes are inserted, (b) the distance between the electrodes?
4. What does a fruit contain that enables it to act as an electrochemical cell when electrodes are added?

9

(a) zinc-copper:
zinc −, copper +
(b) copper-copper:
no result
(c) aluminum-copper:
aluminum −, copper +
(d) iron-copper:
iron −, copper +
(e) aluminum-zinc:
aluminum −, zinc +

You may want to have students compare their results with the electromotive series given in Table 9-3. They will find that the negative electrode in each pair is made of an element that is higher in the electromotive series than is the element of which the positive electrode is made. The negative electrode has a greater tendency to release electrons (become oxidized) than does the positive electrode.

You might also want students to compare the currents that they measured for the various electrode pairs with the relative positions in the electromotive series of the elements from which the electrodes were made. They will find that the greater the distance separating two elements in the series, the greater the current that is produced by the electrode pair.

2. Little or no current was produced when both electrodes were made from the same metal. This was due to the fact that each of the electrodes had the same tendency to release electrons. As was seen in the case of the other electrode pairs, the greater the difference between the electrodes' tendency to lose electrons, the greater the amount of current produced. When both electrodes have the same tendency to release electrons, neither electrode tends to gain or lose electrons.
3. (a) The greater the depth to which the electrodes were inserted, the greater the amount of current that was produced. (b) The shorter the distance between the electrodes, the greater the amount of current that was produced (provided the electrodes were not brought into contact).
4. A fruit contains ions in its juice. It is necessary to have something in an electrochemical cell that contains ions and allows the ions to move about. (Most electrochemical cells include solutions of ions in water. Dry cells, however, contain a moist paste through which ions can move.)

ACTIVITY 15 Introduction

This activity is intended for use in conjunction with section 9-5. Students will use indicators to determine whether various solutions are acids or bases. They will use an indicator to find the endpoint of a neutralization reaction, also. Students will prepare the solid product of the neutralization reaction (sodium chloride, or common salt). Finally, students will test some common substances for acidity or basicity.

Advance Preparation

Prepare 0.01 M sodium hydroxide solution by dissolving 0.40 g of NaOH in enough distilled water to make one liter of solution. Similarly, prepare 0.01 M potassium hydroxide solution by dissolving 0.56 g of KOH per liter of solution and 0.005 M calcium hydroxide solution by dissolving 0.37 g of $Ca(OH)_2$ per liter of solution. (Calcium hydroxide dissolves more readily in cold water than in hot water.) Prepare 0.10 M hydrochloric acid by diluting 8.6 mL of concentrated (11.7 M) HCl to one liter. The acid is ten times as strong as the sodium hydroxide solution so that it will take only about 20 drops (1 mL) of acid to neutralize the 10 mL of base. The calcium hydroxide solution is half as concentrated as the other base solutions because each mole of $Ca(OH)_2$ furnishes two moles of hydroxide ion.

Safety Precautions

Students should wear safety goggles throughout the activity. Ideally, they should also wear laboratory aprons to protect their clothing from any spills. If students spill acid or base on themselves, they should rinse it off immediately with water. They should report any spills as soon as practical to you. Caution students not to taste or touch any substance unless directed to do so.

ACTIVITY 15

Acids and Bases

PURPOSE To use indicators to test for acids and bases and to investigate an acid-base reaction.

MATERIALS stirring rod, 4 test tubes, medicine dropper, evaporating dish, graduated cylinder, phenolphthalein solution, red and blue litmus paper, *p*H paper (optional), safety glasses, dilute hydrochloric acid, dilute solutions of sodium, potassium, and calcium hydroxide, heat lamp

CAUTION: Be sure to wear SAFETY GOGGLES throughout this activity.

Procedure 1

1. Pour about 5 mL of each hydroxide solution into a separate test tube. Test each solution. Use a stirring rod to remove a drop of solution from the test tube and place it on litmus paper, as shown in Figure 1. Use both red and blue litmus paper and, if available, *p*H paper. Record your results.
2. Add a drop of phenolphthalein solution to each hydroxide solution. Record your results.
3. Pour about 5 mL of hydrochloric acid into a fourth test tube. Test the acid with litmus paper and, if available, *p*H paper. Then add a drop of phenolphthalein solution to the acid. Record your results.

Discussion 1

1. What can you say about the *p*H of each hydroxide solution? of the acid solution?
2. What is the color of phenolphthalein solution in an acid? in a base?

Figure 1

Procedure 1 Teaching Suggestions

In Step 1, you may want to demonstrate the technique used for testing an acid or base solution with litmus or *p*H paper. Simply remove a drop of the solution on the end of a stirring rod. Then touch the drop to the litmus or *p*H paper, as shown in Figure 1, and note the color produced.

Discussion 1

1. Each of the hydroxide solutions is basic, as shown by the fact that they turn red litmus blue. If *p*H paper is used, then each of the hydroxide solutions will be found to have a *p*H of approximately 12. The acid solution will be found to turn blue litmus red. When tested with *p*H paper, the *p*H of the acid solution

Procedure 2

1. Put about 10 mL of sodium hydroxide solution in an evaporating dish. Add 2 drops of phenolphthalein. Stir slowly as you carefully add hydrochloric acid, drop by drop, until the solution suddenly changes from pink to colorless. (See Figure 2.) Test the solution with *p*H paper, if available, or with red and blue litmus paper. Record your results.
2. Evaporate the solution to dryness under a heat lamp, as shown in Figure 3. Note the appearance of the substance that remains in the dish after all of the liquid has been evaporated.

Discussion 2

1. Explain why the phenolphthalein changed color when you added hydrochloric acid to the sodium hydroxide solution. Why did you get the results you noted with *p*H or litmus paper?
2. What is the white solid that was left after you evaporated the water from the solution? Write an equation for the reaction that took place between the acid and the base.

Procedure 3

Test some common substances to find out whether they are acidic, basic, or neutral. Try to determine how strongly acidic or basic each acid or base is, as well. Use red and blue litmus paper and, if available, *p*H paper. You might test substances such as tap water, rainwater, lemon juice, grapefruit juice, quinine water, and a stomach "alkalizer" tablet dissolved in water. **CAUTION**: Be sure to check with your teacher before testing any substance.

Discussion 3

1. On the basis of the class results, classify each substance that was tested as acidic, basic, or neutral.
2. Rank the (a) acidic and (b) basic substances in order of strength.

Figure 2

Figure 3

will be found to be approximately one.

2. Phenolphthalein solution is colorless in acidic solutions and pink in basic solutions.

Procedure 2 Teaching Suggestions

In Step 1, students should stir as they add the acid, because they might overshoot the endpoint otherwise. (The endpoint is the point at which *one* drop of acid changes the solution from pink to colorless.) As they near the endpoint, students should stir thoroughly after each drop is added.

The resulting solution should be neutral. It will cause neither red nor blue litmus to change color. When tested with *p*H paper, it will be found to have a *p*H of approximately seven.

In Step 2, a heat lamp is used to give slow, steady evaporation of the water, without any spattering of the solution from the evaporating dish.

Discussion 2

1. Phenolphthalein changed color because the last drop of acid that was added neutralized the base, and a tiny amount of excess acid was present. Phenolphthalein is colorless in acid solution. Since the acid just neutralized the base, the *p*H of the resultant solution was approximately seven, which is neutral.
2. The white solid left when the solution was evaporated is sodium chloride. The equation for the reaction can be written as follows:
$$H^+Cl^- + Na^+OH^- \longrightarrow Na^+Cl^- + H_2O$$
The Na^+Cl^- is sodium chloride, a white solid that is made up of sodium ions and chloride ions.

Procedure 3 Teaching Suggestions

If students want to test any common substances other than those listed in Procedure 3, caution them to check first with you. They should bring in the container so that you can check the label to see what the substance contains. It is impractical to test any substance that has an acid or base strength greater than 1*M*. The *p*H of a 1*M* acid is 0; that of a 1*M* base is 14.

Discussion 3

1. Results will depend on what substances are tested. Of the substances listed in Procedure 3, the fruit juices will be acidic, the alkalizer solution and quinine water will be basic, and the tap water and rainwater will vary.
2. Answer depends on what substances are tested.

ACTIVITY 16 Introduction

This activity is intended for use in conjunction with section 10-3. Students use dice or marked cubes as a model for radioactive isotopes. It is arbitrarily assumed that one out of six isotopes will decay in a given time period. Using the model, students are able to simulate the decay of radioactive isotopes and analyze the data gathered in the course of the decay.

Each model decay is carried out five times to obtain an average decay rate for the sample. The average of five trials is a more representative rate than would be the result of a single trial.

The model isotopes that have decayed represent a new element. Removal of the decayed isotopes from the undecayed ones is analogous to the separation of two elements.

Expected Results

Students' graphs should look like the one shown below. The half-life (50 cubes decayed) lies in the fourth time period.

Discussion

1. Students should predict that approximately three isotopes (cubes) would be expected to decay in the tenth time period. (The "average" is 3.2.) Students should note from their data that approximately one sixth of the cubes decayed in each trial.
2. There is no way to predict *which* cubes will decay in any time period, since the decay is completely random. Only the fraction of cubes that will decay can be predicted. Similarly, there is no way to predict which radioisotopes will decay; only the fraction of radioisotopes that will decay can be predicted.

ACTIVITY 16

Half-Life

PURPOSE To observe the "decay rate" of cubes and determine their "half-life."

MATERIALS 100 cubes (sugar or wooden), each with a dot on one face, or 100 dice; container to hold cubes or dice

Procedure

1. Put 100 cubes in the container. Then pour the cubes onto a table. Count all the cubes that have the side with one dot on it face up. Record this number. Each of the cubes with the dot face up represents a decay.
2. Put all the cubes back in the container. Repeat Step 1 four times.
3. Add the total number of cubes with one dot face up in all five trials and divide by five. This represents the average number of "radioactive atoms" that decayed in the first time period. Remove this number of cubes from the container. Record the number of cubes that remain. They represent atoms that have not decayed.
4. Pour the remaining cubes onto a table, as in Step 1. Count the number of cubes with one dot face up. Repeat the procedure four times.
5. Average the five results as before. Remove this number of cubes from the container and record the number that remain in the container.
6. Continue the procedure until 20 or fewer cubes are left in the container.
7. Make a bar graph to display your results. Plot time periods on the horizontal axis and the number of atoms that remain undecayed at the end of a time period on the vertical axis. In what time period does the half-life lie?

Discussion

1. How many cubes will "decay" in the tenth time period?
2. Can you predict *which* cubes will "decay" in the tenth time period?

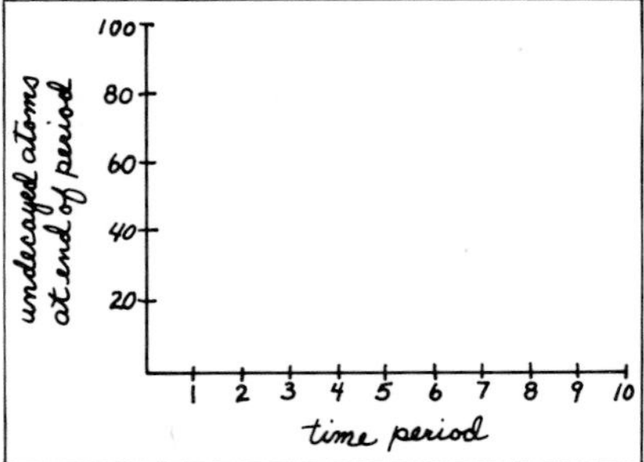

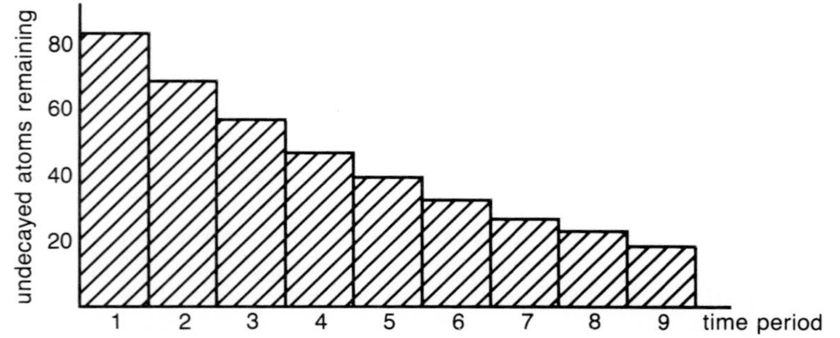

ACTIVITY 17

Testing Mineral Hardness

PURPOSE To compare the hardness of mineral samples.

MATERIALS mineral samples, copper penny, iron nail, steel scissors, glass slide

Procedure

Test each of the mineral samples supplied by your teacher for hardness. Find and record the hardness in terms of the hardness scale given on this page.

Discussion

1. Your teacher will give you the accepted hardness values for the samples you tested. How close were your estimates?
2. How can the hardness test help you in identifying a mineral sample? What are some other properties of minerals that can be used to tell them apart?

Number	*Hardness Test*
1	Easily scratched by fingernail.
2	Barely scratched by fingernail.
3	Barely scratched by copper penny.
4	Scratched by iron nail.
5	Scratched by steel scissors.
6	Easily scratches glass.
7	Easily scratches steel and glass.
8	Scratches all other common minerals.
9	Scratches topaz.
10	Hardest mineral (diamond).

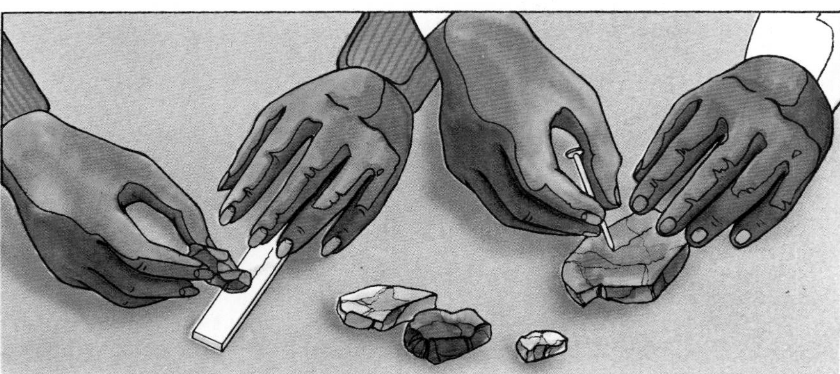

ACTIVITY 17 Introduction

This activity is intended for use in conjunction with section 11-1. The activity concentrates on testing one of the several mineral properties described in the section. If you wish to expand the activity, you could have students test other properties, such as density (see Activity 9), color, streak, or luster. Stress that identification of a mineral sample involves the results of several tests. Among the most helpful tests are those for hardness, color, and cleavage or fracture.

Advance Preparation

You can use mineral samples from mineral kits, if your school has them. If not, try to get mineral samples from the geology department of a nearby college or university, or purchase samples from a scientific supply company. Before students do the activity, look up the hardness value for each sample in a mineralogy text.

You may want to place each mineral in a separate, numbered container. Plastic containers for margarine are suitable. Direct students to make sure that they put each mineral back into the proper container. You will probably want to check from time to time during the activity to make certain that they have done so. If you make up a key that matches mineral names, numbers, and hardness values beforehand, you can more easily keep track of the minerals and readily give students the accepted hardness values required for the first question in the Discussion.

Discussion

1. Students should be able to identify the correct hardness number for each mineral having a hardness close to that represented by one of the numbers on the hardness scale. If the hardness of a mineral corresponds to a value intermediate between numbers on the hardness scale, students should be able to place the hardness of the mineral in the correct range (for example, 5–6).

   Some hardness values, according to Mohs' scale, are as follows:

1—talc	6—orthoclase
2—gypsum	7—quartz
3—calcite	8—topaz
4—fluorite	9—corundum
5—apatite	10—diamond

2. Hardness test results can be used to eliminate many minerals from consideration. Other properties, such as color, streak, luster, density, and cleavage or fracture, can then be used to choose among the remaining possibilities.

ACTIVITY 18 Introduction

This activity is intended for use in conjunction with section 12-5. Students will fit together several of the land masses that are believed to have been part of one ancient land mass, called Pangaea. In doing so, they will make use of some of the evidence that supports the Pangaea theory.

The oldest evidence is provided by maps of the land masses on either side of the Atlantic Ocean. The shapes of the masses suggest that they may have once fit together, like pieces in a jigsaw puzzle. This evidence was noted as early as 1620, by Francis Bacon.

Later, it was noted that certain mountain chains in the Western Hemisphere seemed to match up with similar chains on the other side of the Atlantic. Similarities were also noted between rocks and fossils found on opposite sides of the Atlantic. Both of these observations suggested that the land masses on opposite sides of the Atlantic Ocean had once been part of a larger land mass.

Scratches left on rocks by ancient glaciers occur in both Africa and South

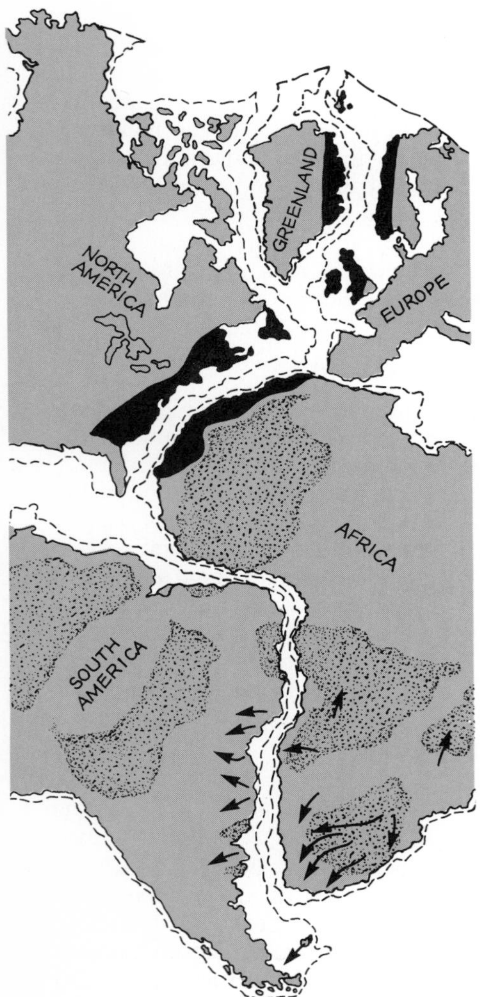

ACTIVITY 18

Moving Land Masses

PURPOSE To use diagrams of land masses to determine how these masses may have fit together in the past.

MATERIALS tracing paper, scissors, transparent tape, world map

Procedure

1. Put a piece of tracing paper over the next page. Trace the diagrams on the page, including all the details. Use the diagrams that you have traced for all of the following steps.
2. Write the name of each land mass on it. (Use a world map, if necessary, to identify the land masses.) Using scissors, cut out each land mass diagram. Cut along the broken lines, where they exist. If there is no broken line, cut along the outermost solid line.
3. Arrange the land masses on a flat surface in their present location, as shown on a map.
4. Move the land masses together, matching up similar areas. Consider the shape of the continental margins, the similarity of mountains and rocks, and directions of glacial markings in the matching process. When you have finished moving the land masses together, use small pieces of tape to hold them in position.

Discussion

1. Between what land masses do the continental margins match closely?
2. Between what continents do the glaciers appear to have moved? In what general direction would you expect the glaciers to have moved? (Hint: In what general direction do such glaciers move today?)
3. Between what land masses do the (a) ancient mountains, (b) rocks of similar ages match up?
4. Did any of the land masses rotate as they moved into their present positions? If so, which ones?
5. List as much evidence as you can to support the theory that these land masses were together at one time.

Key to the Land Mass Diagrams

The broken line around each land mass represents the continental margin. The arrows show the directions in which glaciers moved. The darkly shaded areas represent ancient mountains that were formed at about the same time. The lightly shaded areas represent rocks of about the same age.

12

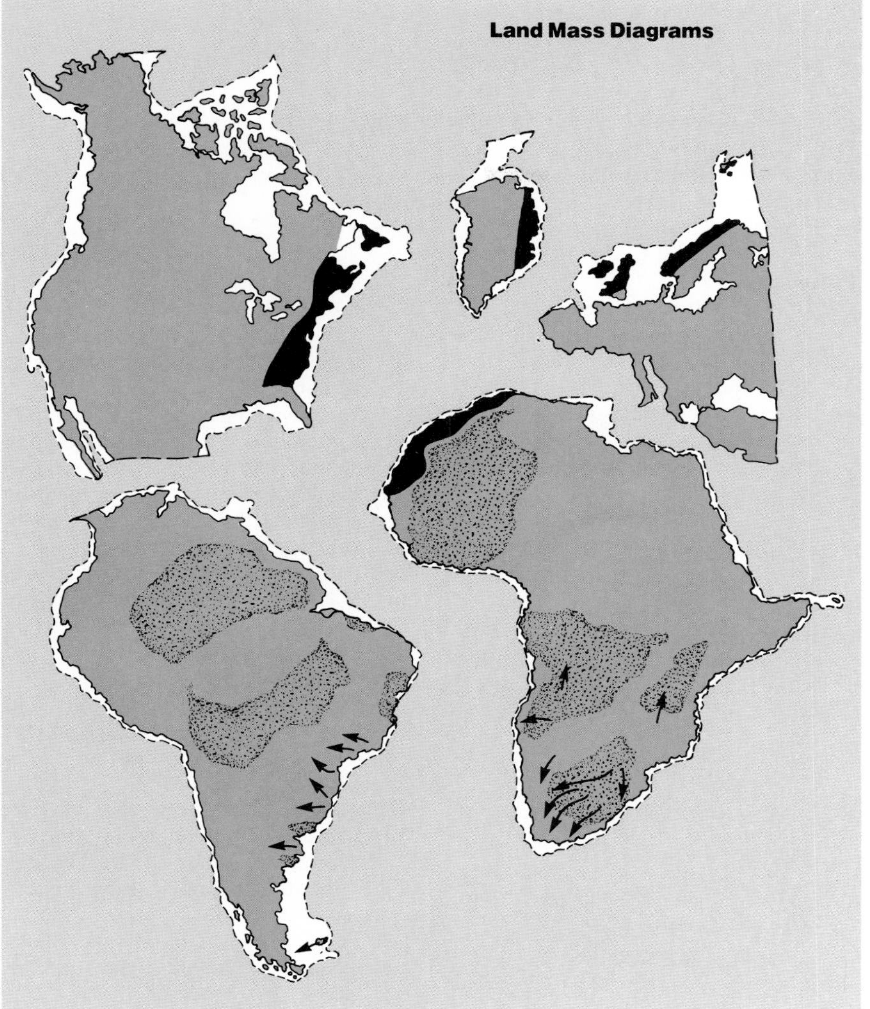

America. The pattern of glacial movement suggested by these scratches is best explained by assuming that these glaciers once covered both land masses. This suggests that the two masses were once joined.

The ancient glaciers that left traces in Africa and South America left traces on other land masses, as well, including Antarctica and Australia. It is believed that these land masses were joined to Africa, also, as part of Pangaea. This activity focuses on the land masses that now lie on either side of the Atlantic Ocean, but similar evidence exists for the entire ancient land mass called Pangaea. Emphasize that additional evidence, such as that obtained from exploration of the ocean floor, is crucial to the concept of drifting land masses and the plate theory that explains how drifting occurs.

Teaching Suggestions

In Step 2, emphasize that the land masses should be labeled before they are cut out. (The land mass that includes Europe also includes a portion of modern Asia, but *Europe* is the most appropriate label for the land mass.) If students cut around the continental margins carefully, the masses will fit together much better.

Students might wonder why Central Amercia and much of Mexico are missing from the North American land mass. Presumably, these regions were produced later, probably by volcanic action and the upward arching of Earth's mantle.

Expected Results

The illustration shows how the land masses should fit together.

Discussion

1. Continental margins match closely between South America and Africa, and between Northern Europe and Greenland.
2. Glaciers appear to have moved between Africa and South America. Glaciers tend to move out from (or back toward) a central location, such as the South Pole. Apparently there was a similar location on the African land mass.
3. Ancient mountains match up between North America and North Africa, and between Northern Europe and Greenland. Rocks of similar ages match up between Africa and South America.
4. Yes. South America, Africa, and Greenland rotated as they moved to their present locations.
5. Students should list the following evidence: the continental shelves fit together; mountain chains on different land masses match up; rock layers of similar ages match up; scratches made by glaciers match up.

ACTIVITY 19 Introduction

This activity is intended for use in conjunction with section 13-6. Make sure that students understand the definition of relative humidity given in that section.

An instrument used to measure relative humidity is called a *hygrometer*. The prefix "hygro-" is from the Greek word *hugros*, which means "wet or moist." The suffix "-meter," indicates a measuring device. Some hygrometers use strands of human hair to produce relative humidity readings in a manner that is essentially the same as that employed in the simple hygrometer that students will make. Human hair is very responsive to humidity. When it absorbs moisture, it stretches. When it dries out, it contracts. With proper calibration, then, a human hair can be used in a device that produces direct readings of relative humidity. Students will record relative humidity readings for their area each day, along with the corresponding length of the strand of hair in their hygrometer. When enough hair length/relative humidity comparisons have been made, students will have, in effect, a direct reading hygrometer.

This activity will work best when the relative humidity changes over a wide range during its course.

Advance Preparation

Several days before the activity is to be performed, ask students to bring in empty shoe boxes.

Discussion

1. When the relative humidity increases, the reading on the human hair hygrometer increases. When the relative humidity decreases, the reading on the hygrometer decreases. (Depending on weather conditions, it may take several days for the hair to change from short to long to short again.)
2. When the relative humidity increases, the hair absorbs moisture. As a result, its length increases. When the relative humidity decreases, the hair gives off moisture. As a result, its length decreases.

Students may be interested in comparing the changes in length exhibited by different colors of hair. Hair of lighter color usually stretches more when it absorbs moisture than does hair of darker color.

ACTIVITY 19

Making a Simple Hygrometer

PURPOSE To make a simple hygrometer that will indicate changes in relative humidity.

MATERIALS strand of human hair, about 15 cm long; empty shoe box, without lid; small piece of lightweight cardboard; scissors; transparent tape; centimeter ruler

Procedure

1. On the bottom of the inside of a shoe box, make a 25-cm scale like the one in the illustration. Then stand the shoe box on end.
2. From a small piece of lightweight cardboard, cut an arrow like the one in the illustration. Tape the arrow to one end of a strand of human hair that is about 15 cm long. Then tape the other end of the strand to the top of the shoe box, so that it hangs straight down, as in the illustration.
3. You have constructed a human-hair hygrometer. The length of the strand will change as the relative humidity changes. You can read the amount of change using the scale of your hygrometer.
4. Put your hygrometer in a safe place. Read and record the length of the strand of hair. Obtain the relative humidity from your teacher and record that, also.
5. Record the relative humidity and the reading on your hygrometer each day for several days.

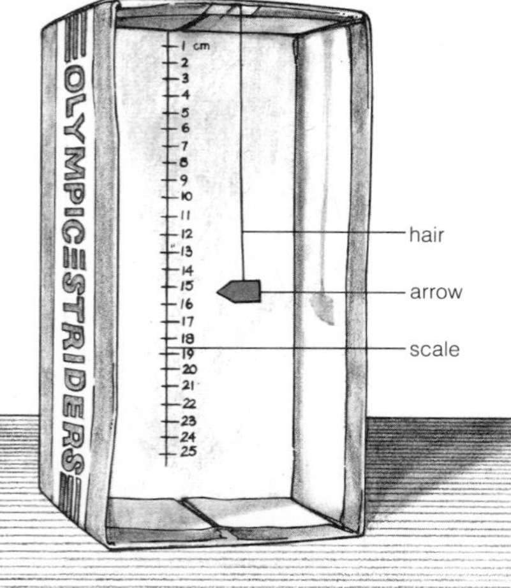

Discussion

1. Compare the changes in your hygrometer readings with the changes in relative humidity. How are the two related?
2. How do changes in relative humidity affect the length of the strand of hair?

13

ACTIVITY 20

Investigating Density Currents

PURPOSE To observe the behavior of solutions of different densities when they are put in less dense water.

MATERIALS food coloring, test tube, artificial seawater (different densities), meter-long plastic column with cap, ring stand and clamp, watch or clock with second hand

Procedure

1. Set up the equipment as shown in the drawing. Fill the plastic column two-thirds full of tap water.
2. Put some artificial seawater in a test tube. Add three or four drops of food coloring. Swirl the solution to mix uniformly.
3. Pour the seawater into the water in the plastic column. Record the time in seconds needed for the leading edge of the colored solution to reach the bottom of the column.
4. Repeat Steps 2 and 3, using fresh tap water in the column and a different salt solution.
5. Repeat Step 4, using the third salt solution.

Discussion

1. Which of the seawater solutions took the longest time to reach the bottom of the tube? Which took the least time?
2. Find out from your teacher which of the artificial seawater solutions was the most dense and which one was the least dense. Use this information to explain your results.

ACTIVITY 20 Introduction

This activity is intended for use in conjunction with section 14-4. Students will observe the formation of density currents when artificial seawater is poured into tap water. Density currents form when seawater flows into fresh water, when denser (more saline) seawater flows into less dense seawater, and when water that is carrying a lot of sediment flows into water that holds less sediment (for example, when a stream flows into a lake). In general, density currents may form whenever a more dense liquid flows into a less dense liquid.

Density currents may also be produced when cold water flows into warmer water, since the cold water is more dense. Such density currents carry cold water from Earth's polar regions to the equatorial region. Some students may want to try to produce density currents by pouring cold water into warm water.

Advance Preparation

You will need to prepare three salt (sodium chloride) solutions, each of a different density. These solutions will serve as the samples of artificial seawater to be used in the activity. Students will determine the relative densities of the three solutions by observing how rapidly they sink to the bottom of a tube of water, when poured into it. Students should not, therefore, know the relative densities of the solutions before they perform the activity. The solutions can be designated Solutions 1, 2, and 3, or the equivalent.

The volume of salt solutions needed depends on the class size and the amount of salt solution used in each trial. For 15 student teams using 20 mL of each salt solution per trial, you will need a minimum of 300 mL of each solution. The following amounts of sodium chloride (NaCl) are suggested per 100 mL of tap water:

10 g NaCl for Solution 1
5 g NaCl for Solution 2
2 g NaCl for Solution 3

Discussion

1. "Solution 3" will take the longest time to reach the bottom of the tube. "Solution 1" will take the least amount of time to reach the bottom.
2. Students should conclude that the most dense solution sinks at the fastest rate and the least dense solution sinks at the slowest rate. This illustrates how readily a density current may form when there is a difference in density between two masses of water.

ACTIVITY 21 Introduction

This activity is intended for use in conjunction with section 15-1. Students will find the activity very valuable in helping them to understand how Earth's tilt and the revolution of Earth about the sun relate to the phenomena of the seasons experienced on Earth. Students will need some time for experimentation before they manage to produce the proper alignments between the sun and Earth. Schedule enough time to allow each student to perform the activity individually, even if the number of lamps is limited.

Advance Preparation

Styrofoam balls may be available locally, or they can be ordered from a scientific supply house. A flexible protractor may be one made of plastic, heavy paper, or metal. However, it must be flexible enough to lie flush against the surface of a sphere.

Teaching Suggestions

After applying the equator to "Earth," students can use a tape measure in Step 1 to locate the North and South Poles. Each pole is a point equidistant from all points on the equator.

In Step 2, students should find that the Tropics of Cancer and Capricorn are circles parallel to the equator located at 23.5°N and 23.5°S, respectively. The Arctic and Antarctic Circles are located at 66.5°N and 66.5°S, respectively. Latitude and longitude are discussed in Appendix E, Maps.

To be precise, degrees of latitude should be measured from the center of Earth. However, the method given in Step 3 gives results that are close enough for the purposes of this activity. (If students want to measure from the center of "Earth," they can cut the styrofoam ball in half, measure the desired latitudes, and then stick the two halves of the ball together again.) If a suitable protractor is not available, students can locate the various latitude lines with the use of a tape measure, given that the Arctic and Antarctic Circles are approximately 25.7% of the distance from a pole to the equator, and the Tropics of Cancer and Capricorn are approximately 71.4% of the distance from a pole to the equator.

In Step 5, you might want to direct the students to keep the North Pole of Earth pointed toward an imaginary North Star as they cause Earth to revolve around the sun. If students do not keep Earth pointed in the same direction throughout its revolution, they will not obtain the desired results.

Expected Results

Students' drawings should look like the ones in the illustration. The views are from above, but not directly over, the North Pole, reflecting the tilt of

ACTIVITY 21

Seasons of the Year

PURPOSE To investigate the relative positions of the sun and Earth at various times of the year.

MATERIALS Styrofoam ball (8-10 cm diameter), probe or teasing needle, flexible protractor, flexible ruler or tape measure, lamp with 25-watt bulb, several pieces of yarn large enough to encircle the ball, world map or globe

Procedure

1. Use an unmarked Styrofoam ball to represent Earth. Wrap a piece of yarn around the ball to represent the equator. (The yarn should stick to the ball without the aid of tape or glue.) Locate and mark the North and South Poles.
2. Use a map or globe to find the latitudes (in degrees) of the Tropic of Cancer, Tropic of Capricorn, Arctic Circle, and Antarctic Circle.
3. Use a protractor lined up along the equator of your model Earth to locate these latitudes. Use a piece of yarn to represent each of them. Your model Earth should look like the one in Figure 1.
4. Stick a probe into your model Earth to serve as a handle. If you stick the probe in somewhere along the Antarctic Circle, the model will tilt at the proper angle (23½°) when you hold the probe straight up, as shown in Figure 2.
5. Use a lamp without a shade to represent the sun. Move your model Earth around the lighted lamp to represent Earth's revolution around the sun. Be sure to keep the amount and direction of the tilt constant. (Do not try to rotate Earth about its axis.)
6. As you move Earth around the sun, note the direction of Earth's tilt relative to the sun. On the first day of summer the Northern Hemisphere tilts directly toward the sun. On the first day of winter it tilts directly away from the sun. On the first day of spring and fall it tilts neither toward nor away from the sun.
7. Make a drawing of Earth, as seen from above, for each of the days given in Step 6. In each drawing, indicate the North Pole with a dot. Draw in any portions of the yarn circles that can be seen from above. Show which portions of Earth are lighted by the sun and which ones are in shadow.

Discussion

1. Where is the top of the shadow on

the first day of summer? How much of Earth above the Arctic Circle is lighted? Where is the bottom of the shadow? How much of Earth below the Antarctic Circle is lighted?

2. On the first day of winter, where is the shadow line in each hemisphere? Which polar area is fully lighted?
3. On the first day of spring and fall, where does the shadow line appear in relation to the poles? How does this affect the number of hours of daylight and darkness over the globe?
4. Predict what would happen to the shadow lines if the tilt of the axis were 45° or 0°. If time permits, repeat the activity to test your predictions.

Earth's axis. The direction of the tilt is away from the viewer.

Discussion

1. In the Northern Hemisphere, it is on the 66.5°N latitude line, on the side away from the sun. All of Earth above the Arctic Circle is lighted. The bottom of the shadow is on the 66.5°S latitude line, on the side facing the sun. None of Earth below the Antarctic Circle is lighted.
2. In the Northern Hemisphere, it is on the 66.5°N latitude line, on the side facing the sun. In the Southern Hemisphere, it is on the 66.5°S latitude line, on the side away from the sun. The area below the Antarctic Circle is lighted.
3. The shadow line goes through both poles, and divides Earth in half. The hours of daylight and darkness are equal all over the globe.
4. If the tilt of Earth's axis were 45°, the shadow lines on the first day of summer and winter would be on the 45°N and 45°S latitude lines. The shadow lines on the first day of spring and of fall would be the same as now.

   If the tilt of Earth's axis were 0°, the shadow lines would always be as they are now on the first day of spring and fall. No changes of seasons would occur.

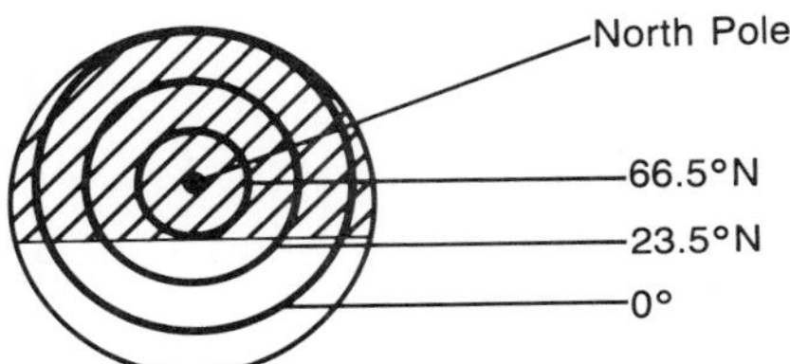

first day of winter

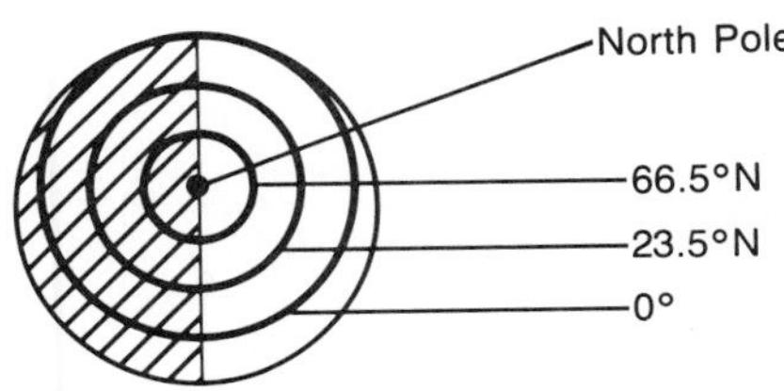

first day of spring

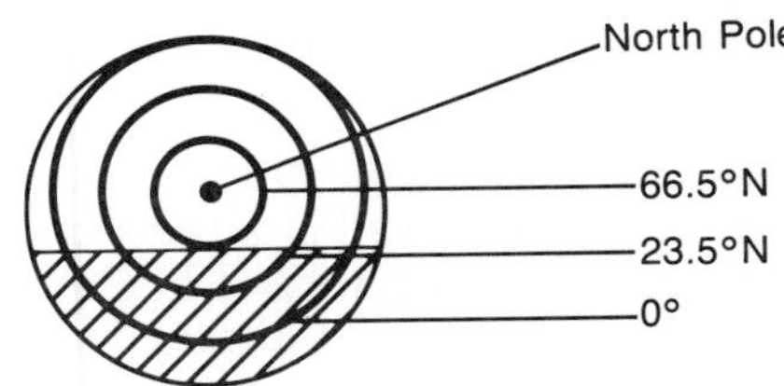

first day of summer

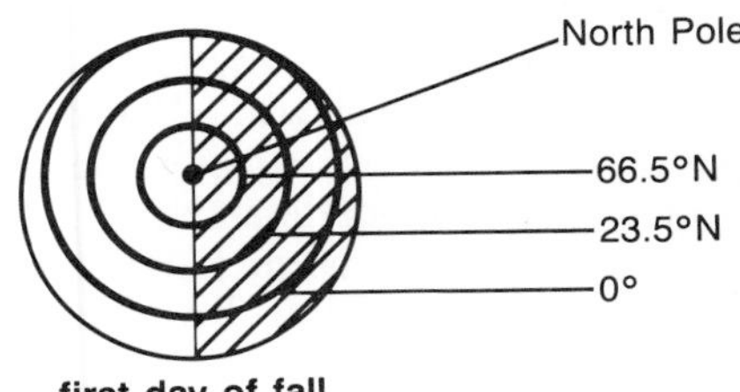

first day of fall

ACTIVITY 22 Introduction

This activity is intended for use in conjunction with section 16-1. By determining the average speed of a person walking over a given distance and the person's speed over shorter intervals along the way, students will develop a better understanding of average and instantaneous speed.

Students will probably want to compare their walking rates, so you may want to have several students walk the same course. Students could answer the Discussion questions on the basis of their own or another's walking speeds.

Expected Results

See Table TE 1.

Table TE 1: Typical Student Results

Distance from starting line(m)	Time to walk distance(s)	Speed (m/s)
10	6.2	1.6
20	12.4	1.61
30	18.8	1.60
40	25.1	1.59
50	31.3	1.60

Discussion

1. Students will probably find that each walker walked at a fairly constant speed. (See Table TE 1.) You might want to ask students to give examples of courses over which walking speed would not be expected to be nearly constant. Possible responses include: up and down hill, a course with many turns, and a crowded city sidewalk. You might want to use the examples to develop the concept of acceleration.
2. Average speed = 50 m/31.3 s = 1.60 m/s, using Table TE 1 results.

ACTIVITY 22

Measuring Walking Speed

PURPOSE To determine the speed at which someone walks.

MATERIALS meterstick or metric measuring tape, metronome or stopwatches, distance markers

Procedure

1. Choose a level surface, such as a hallway or sidewalk, on which a volunteer will walk. Use a meterstick or tape measure to mark off 10-m intervals along the walkway. Station a volunteer at the end of each interval to record the time at which the walker just passes that checkpoint. (See Figure 1.)
2. As the walker crosses the starting line, the timers start their stopwatches. When the walker passes the checkpoints, the timers stop their watches. (Alternatively, you could start a metronome, set to sound once each second, when the walker crosses the starting line.)
3. When the walker has completed the course, collect the times from the timers. Record them in a table like the one in Figure 2.

Discussion

1. Compute the walker's speed over each interval. Did the walker walk at a fairly constant speed?
2. Compute the walker's average speed over the entire course.
3. (Optional) Plot a graph of distance vs. time from the data. Let the vertical axis represent distance and the horizontal axis represent time.

Figure 1

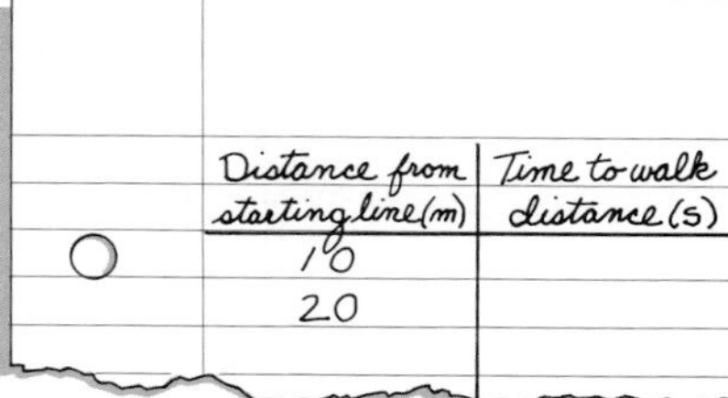

Figure 2

16

3. See illustration. (Before asking students to draw the graph, you may want to refer them to Appendix D, Organizing and Displaying Data.)

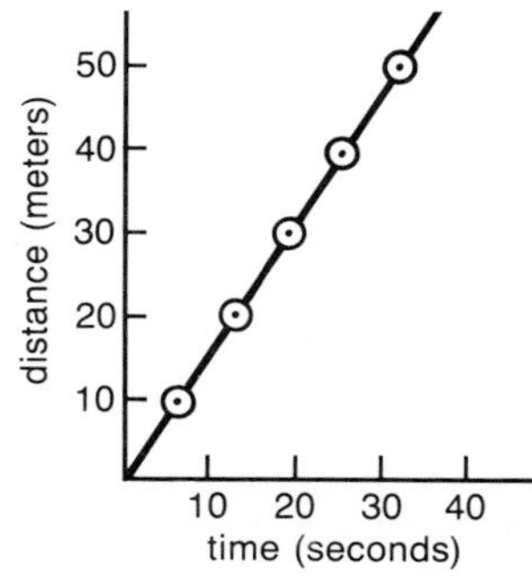

ACTIVITY 23

Measuring Forces of Friction

PURPOSE To measure the force of friction between different surfaces.

MATERIALS smooth, heavy wooden blocks; smooth, level board or other surface; thumbtacks; paper clips; string; sandpaper; aluminum foil; waxed paper; newspaper; wide rubber bands; spring scale

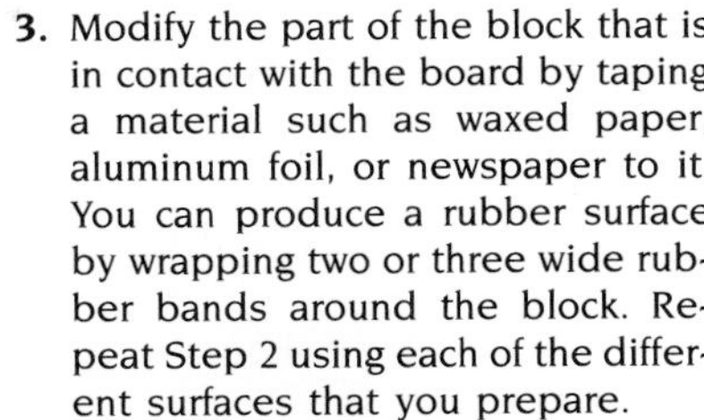

Procedure 1

1. Open a paper clip, so that the two portions are at right angles to each other. Hold one portion flat against one face of a wooden block. Push a thumbtack into the block to hold the paper clip, as in Figure 1.
2. Place the block at one end of a smooth, level board. Attach the hook of a spring scale to the paper clip, as in Figure 2 (on the next page). Pull on the other end of the scale with just enough force to make the block move along the board at a steady rate. Read and record the force shown on the scale.

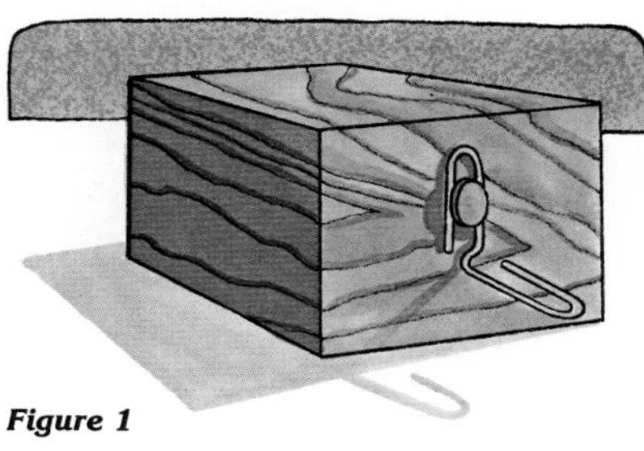

Figure 1

3. Modify the part of the block that is in contact with the board by taping a material such as waxed paper, aluminum foil, or newspaper to it. You can produce a rubber surface by wrapping two or three wide rubber bands around the block. Repeat Step 2 using each of the different surfaces that you prepare.

Discussion 1

1. What does the force that you have measured represent?
2. Rank the forces of friction for the different block surfaces that you tested, from greatest to least.

Procedure 2

1. Find out what happens when you increase the force with which the plain wooden block presses down on the board. You can do this by placing a heavy object on top of the block. You could use a small container half-full of washers, for example. Find and record the force of friction between the block and the

ACTIVITY 23 Introduction

This activity is intended for use in conjunction with section 16-2, or later in the chapter. Students will determine the frictional force that acts when a wooden block is pulled across a smooth, level wooden surface. Then they will alter the part of the block that is in contact with the other surface, in order to see how the nature of the block surface affects the force of friction between the block and the board. Other variables that students will investigate are the amount of the block's surface area that is in contact with the board, and the amount of force with which the block pushes down on the board (that is, the weight of the block). The term *weight* is not used in this activity because weight is not introduced until section 16-6. The more general term *force* is used, since students are introduced to the concept of force in section 16-2.

Even the smoothest-looking surface, when sufficiently magnified, will be seen to be rough and uneven. As a result, there is a retarding interaction between any two surfaces when they slide across each other. This interaction is called *sliding friction*. Sliding friction is what students will be investigating in this activity. The amount of friction that is produced when an object is starting from rest is somewhat greater than that produced when it is sliding. In this activity, the difference between starting and sliding friction should not be a factor, since students will determine the amount of frictional force acting on a block while it is sliding across a surface at a steady rate. When an object is sliding across a surface at a steady rate, the forces acting on the object are balanced. Thus, the force needed to just balance the retarding force of friction must be equal to that frictional force.

The retarding force acting on an object is less if the object has wheels. The rolling friction that acts on an object with wheels is significantly less than the sliding friction that would be acting on the same object if it were moving over the same surface without wheels.

Another retarding force that is often encountered by a moving object is air resistance. This force becomes greater as the speed of the moving object increases. An automobile designer must take into account the retarding force of the air through which the automobile will move.

Advance Preparation

Smooth, heavy wooden blocks and a smooth, level wooden board on which to slide them may be available from the school wood shop or from a lumber supply store, if you do not have such materials available. It is preferable to use rectangular, rather than cubic, blocks for this activity, so that the surface area of the blocks can be readily changed in Step 2 of Procedure 2.

Spring scales may be available from the school physics department. If not, they can be ordered from a scientific supply house. Try to get spring scales that are calibrated in newtons, since the newton is the preferred unit of force in the SI metric system. Some commercially available spring scales are calibrated in grams. Since grams are units of mass, rather than force, grams should not be used in measuring force. If your spring scale is calibrated in grams, you can cover over the gram scale with the equivalent scale calibrated in newtons. A mass of 1 g weighs 0.0098 N near Earth's surface. Thus, to convert a mass in grams to the corresponding force in newtons, multiply by 0.0098.

Teaching Suggestions

In Step 2 of Procedure 1, make sure that students pull the block at a steady rate and keep the spring scale parallel to the surface of the board as they do so.

Discussion 1

1. The force measured by the students is the force necessary to just balance the force of friction acting between the two surfaces. Thus, the measured force is equivalent to the force of friction between the two surfaces.
2. The range of frictional forces, from greatest to least, for the different block surfaces will probably be as follows: rubber, newspaper, aluminum, waxed paper, and wood. Students' results may vary depending on experimental conditions.

   You may want to ask students how they might reduce friction between the surfaces even more. (Possible answers include: wax or varnish the surfaces; coat the surfaces with a thin film of soap or oil; put wheels on the block.) You might want to ask the students whether they can imagine a situation in which there would be no friction between two surfaces. Point out that this is an idealized case. By imagining frictionless surfaces, Newton was better able to understand and explain the nature of forces. He was not able to experiment with frictionless surfaces, of course.

Discussion 2

1. The force of friction increases when the force with which the block presses down on the board is increased.
2. Changes in the amount of surface area that is in contact between two surfaces have little or no effect on the amount of frictional force acting between the surfaces.

board. Repeat the procedure, this time with the container full of washers. Again read and record the force of friction.

2. Find out what happens when you change the amount of surface area of the block that is in contact with the board. If you have a rectangular block, you can simply stand it on a different face to change the surface area. Find and record the force of friction for at least one surface area that is different from the one you used in the preceding steps.

Discussion 2

1. What happened to the force of friction when you increased the force with which the block pressed down on the board?
2. What happened to the force of friction when you changed the area of block surface that was in contact with the board?
3. What factor(s) must you keep the same when you are comparing the forces of friction that are produced by objects moving over the same surface?

Figure 2

16

ACTIVITY 24

Measuring Power

PURPOSE To measure the power produced by a person in running up a flight of stairs.

MATERIALS meterstick, stopwatch, stairs

Procedure

1. Use a stopwatch to measure the amount of time it takes for one of your classmates to run up a flight of stairs.
2. Measure the vertical height of the flight of stairs, as indicated in the illustration.
3. Repeat Step 1 for several other classmates.

Discussion

1. Determine the weight, in newtons, of each person who ran up the stairs. Use the weight to compute the work, in joules, done in running up the stairs.
2. Compute the power, in watts, that was produced by each person in running up the stairs. You may want to use this result to compute the power in terms of horsepower, also.
3. Compare the amount of power produced by your classmates who ran up the stairs. What factors were most important in determining how much power was produced? What physical characteristics produce the highest horsepower rating in this activity?

ACTIVITY 24 Introduction

This activity is intended for use in conjunction with section 17-7. Students should enjoy the activity, especially the competition to find out who can generate the most power in running up a flight of stairs.

Safety Precaution

Check to see if any students have medical histories that would preclude participation in the exercise portion of the activity. If so, they can serve as timekeepers and participate fully in the activity in that capacity.

Discussion

1. If students do not know their weight in newtons, they can calculate it from their mass. The gravitational force acting on a 1-kg mass at or near Earth's surface is given by:

$$F = ma = (1\text{ kg})(9.81\text{ m/s}^2) = 9.81\,\frac{\text{kg}\cdot\text{m}}{\text{s}^2} = 9.81\text{ N}$$

Thus, a student whose mass is 50 kg has a weight equal to:

$$(50\text{ kg})(9.81\text{ N/kg}) = 490\text{ N}$$

If students know their weight in pounds, they can convert that weight to the equivalent weight in newtons. A weight of 1 lb is equivalent to a weight of 4.45 N. A 100-lb student, then, weighs:

$$(100\text{ lbs})(4.45\text{ N/lb}) = 445\text{ N}$$

If this student runs up a flight of stairs that has a vertical height of 2.0 m, the work done is:

$$W = Fd = (445\text{ N})(2.0\text{ m}) = 890\text{ J}$$

2. The power, in watts, produced by a 445-N person who runs up a flight of stairs 2.0 m high in 2.2 s is:

$$P = \frac{W}{t} = \frac{890\text{ J}}{2.2\text{ s}} = 404\text{ W}$$

Expressed as horsepower, this is:

$$(404\text{ W})\left(\frac{1\text{ hp}}{746\text{ W}}\right) = 0.54\text{ hp}$$

3. The most important factors in determining the amount of power produced are the vertical height of the stairs, the weight of the person who is running, and the time taken to run up the stairs. The highest horsepower ratings are produced by the persons of the greatest weight who can run the fastest.

ACTIVITY 25 Introduction

This activity is intended for use in conjunction with section 18-4 or the Feature Article, "Insulation." Students should read the Feature Article before they do the activity. Also, if students are unfamiliar with thermometer scales, have them read about them in Appendix B before doing the activity.

Advance Preparation

Some insulating materials can be obtained at hardware stores. Some, such as vermiculite, are often used as packing materials and may be retrieved from packages.

Safety Precautions

Caution students to always remember that they are working with hot materials when they perform this activity. They should take the proper precautions to avoid burning themselves through carelessness.

Teaching Suggestions

The object of Step 1 is to provide a "control." Students can compare the cooling rates of water in the insulated container with the rate in the uninsulated container.

You may want to have all students agree on a suitable starting temperature before they begin the activity, so that all class members will start cooling processes at the same temperature.

Caution students to be sure the bulb of the thermometer is entirely in the liquid when they make a temperature reading, instead of being in contact with the bottom or the side of the container.

In Step 3, a piece of paper might not be strong enough to hold the insulating material above the container of water. If so, cardboard or a similar material can be used instead of paper.

ACTIVITY 25

Testing Insulation

PURPOSE To compare the effect of different insulating materials on the rate of cooling of warm water.

MATERIALS thermometer (0–100°C); one small and one large container made of plastic, glass, or paper; clock or watch; small and large pieces of paper, to cover containers; insulating materials, such as mineral wool, vermiculite, and Styrofoam

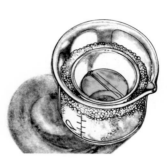

Procedure

1. Pour 100 mL of hot water into the small container. Let the water cool to a starting temperature between 35°C and 40°C. Record the starting temperature. Let the water cool for 15 min more. Then again measure and record the temperature of the water.
2. Put an insulating material into the large container to a depth of 3 to 4 cm. Place the small container on the material. Surround it with more material, as in the illustration.
3. Pour 100 mL of hot water into the small container. When the water has cooled to the starting temperature used for Step 1, cover the small container with a piece of paper. Then fill the large container to the top with more insulating material. Place a large piece of paper on top.
4. Let the water cool for 15 min. Then measure and record its temperature.
5. Repeat Steps 2–4 for each of the other insulating materials that you are testing.

Discussion

1. How many degrees cooler did the water in the uninsulated container become in 15 min?
2. How many degrees cooler did the water become when insulated by each of the materials?
3. How effective was each type of insulation in slowing the rate at which the water cooled?

18

Discussion

1. Students report here the temperature change for the "control."
2. Student results will vary, depending on the type of insulation used. In each case, however, the amount of cooling should be less than that of the "control."
3. Results will depend on the types of insulation that were used. You might want to ask the students what factors besides insulating ability might be important in selecting an insulating material. (Answers might include cost, availability, ease of handling, and potential health hazards.)

ACTIVITY 26

Conductors of Electricity

PURPOSE To test various solid materials in order to find out how well they conduct electricity.

MATERIALS 4 D cells or 6-V battery, insulated electrical wires, 2 or more alligator clips (optional), light bulb, bulb holder, solid objects (made of wood, metal, plastic, paper, chalk, glass, or some other material)

Procedure

1. Use wires to connect a battery and other objects, as shown in the illustration. This setup is called an *electric circuit*. The wires and the bulb are conductors of electricity. If the object being tested is a conductor, the circuit is said to be *complete*. That is, electrons can flow from the negative terminal of the battery, through the wires and other objects, back to the positive terminal of the battery. They flow in this way due to chemical reactions that take place in the battery. If the circuit is complete, the bulb lights. If not, the bulb does not light.
2. Test various objects. Be sure the bare wire ends or clips are firmly in contact with the object being tested. Note whether or not the bulb lights. If it lights, note how brightly it lights. Record your results.

Discussion

1. What materials that you tested conduct electricity? Do some of these materials conduct electricity better than others? Give evidence for your answers.
2. What materials that you tested do not conduct electricity? Give evidence for your answer.
3. What name is used to describe materials that do not conduct electricity?

ACTIVITY 26 Introduction

This activity is intended for use in conjunction with section 19-3, but it could be used just prior to section 19-3, in order to allow students to discover that different materials conduct electricity to different extents. It could be used after students have studied electric circuits. However, this activity does not require familiarity with electric circuits.

Advance Preparation

Beforehand, collect all necessary materials. Either a 6-V battery or four D-cells connected in series can be used as a source of electricity. In either case the voltage produced is 6 V. A Number 48 GE bulb works well in the activity, but any bulb with low power requirements, such as a flashlight bulb, could be used.

Teaching Suggestions

In Step 1, you may want students to demonstrate that the bulb will light when the circuit is complete by touching the wire ends or clips together. The bulb will light, demonstrating that electricity is flowing through the circuit. Then have students hold the wire ends or clips a few centimeters apart. The light will go out, demonstrating that current is not flowing through the circuit. You might ask at this point whether air is a good conductor of electricity. (Students should answer that it is not, since electricity does not seem to flow across the air-filled gap between the wire ends or clips.)

Not all conductors conduct equally well in Step 2. Some conductors will cause a bulb to glow brightly; others will cause it to glow less brightly. Even among metals, there are fairly large differences in the ability to conduct current. For example, copper conducts about four times as well as does zinc and about 60 times as well as does bismuth.

Discussion

1. Answers will vary, depending on the materials tested. Generally, metallic materials are good conductors, although all are not equally good. Good conductors cause the bulb to glow brightly. Poorer conductors cause a weaker glow.
2. Answers will vary, depending on the materials tested. Generally, nonmetallic materials, such as paper, plastic, wood, glass, and chalk, conduct poorly or not at all. Nonconductors produce no glow in the bulb.
3. Materials that do not conduct electricity are called insulators.

ACTIVITY 27 Introduction

This activity is intended for use in conjunction with section 20-2. Students will view the shape and size of an image in a plane mirror and will determine the apparent location of the image behind the mirror, with the use of sight lines. This activity enables students to gain a better understanding of what is meant by a virtual image. It will also give them an opportunity to gain a better understanding of ray diagrams. Students will find that the reflected rays of light appear to be coming from a point behind the mirror—from the virtual image. However, they will realize that there is no image behind the mirror. Their eyes locate the image there only because the rays reflected from the mirror seem to be coming from behind the mirror. Actually, what the students see is light that was reflected by the object and then reflected toward the students by the mirror.

Students learned in section 20-1 that they see objects when the objects reflect light toward them, and they also learned that light travels in straight lines. Thus, they see the pin's image because the light reflected by the pin toward the mirror is reflected toward them by the mirror. When students see their own image in a mirror, they see light reflected by their body toward the mirror that is reflected back toward them by the mirror. In this case, the image that they see is reversed from left to right. Students should be able to explain why this is so when they have completed this activity.

ACTIVITY 27

A Look into a Plane Mirror

PURPOSE To find out how to locate an image in a plane mirror.

MATERIALS small plane mirror, small straight pins, white paper, cardboard sheet, large straight pin, ruler, clay

Procedure 1

1. Place a sheet of white paper on top of a piece of cardboard. Then stick a large pin into the center of the paper.
2. Place two small pins 7 to 10 cm apart in a straight line with the large pin. Do this by sighting along the pins as shown in the illustration. Remove the small pins and use a ruler to draw a *sight line* through the pin holes to the large pin. Repeat the procedure to make two more sight lines.

Discussion 1

1. Where do the sight lines that you drew cross?
2. How can you locate an object's position using sight lines?

Procedure 2

1. Draw a straight line about one-quarter of the distance from one end of a sheet of white paper. Place the paper on the cardboard. Use clay to set a small mirror upright on the line, facing the large portion of the paper.
2. Stick the large pin into the paper a few centimeters in front of the mirror. Use two small pins to make a sight line to the large pin's IMAGE. Repeat the process to obtain two more sight lines.
3. Remove the mirror. Extend the sight lines until they meet. Then measure the distance from this point to the mirror line. Measure the distance from the large pin to the mirror line, also.

Discussion 2

How is the distance an image *appears* to be behind a plane mirror related to the distance from the object to the mirror?

20

Discussion 1

1. The sight lines cross at the location of the large pin.
2. An object's position is the point at which two or more sight lines drawn to the object cross.

Discussion 2

The distance an image appears to be behind a plane mirror is equal to the distance of the object from the mirror. You may want to have students draw a diagram to show why this should be so. Have them draw two or more incident rays from a point on the object to the mirror and the corresponding reflected rays, remembering that the angle of incidence and the angle of reflection are equal. Then have them extend the reflected rays behind the mirror, in straight lines. These extensions of the reflected rays should cross at a point behind the mirror. This point is the point on the virtual image that corresponds to the point on the object from which the incident rays were drawn.

APPENDICES

Contents

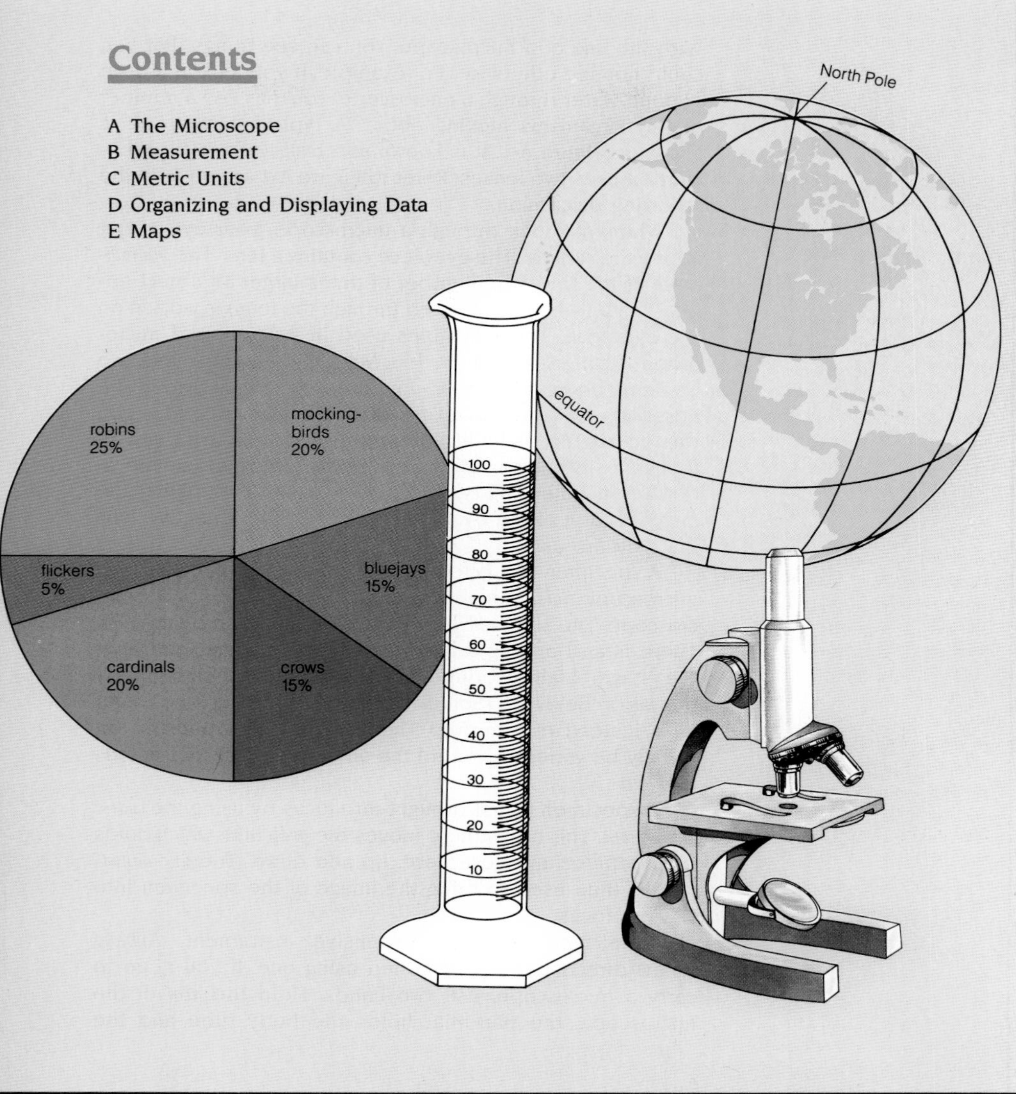

Teaching Suggestions A

Appendix A (or the equivalent) is a prerequisite for student use of a microscope. Students use a microscope in Activities 1–4; the use of a dissecting microscope is optional in Activity 7. Before students do their first microscope activity, allow enough time for them to complete Appendix A. Be sure students are familiar with the parts of a microscope, and their functions, before permitting them to handle microscopes.

If you plan to use the dissecting microscope in this course, you may at this time want to explain how it functions and how it differs from the monocular compound microscope.

Appendix A can be used in conjunction with the Chapter 1 Feature Article, "Zeroing In On Life's Secrets," which discusses the differences between simple, compound, and electron microscopes. Further information about different microscopes is given here, under Background.

Explain that a microscope can be used to photograph, as well as to view, tiny objects. Point out that many photographs in this book were taken through a microscope. You can have students look for examples of photomicrographs and electron micrographs in Unit 1. (See Cross References.)

Background

The typical school microscope is a type of *light* microscope, which is a microscope that uses light to convey the image of the specimen to the observer's eye. Light microscopes have been in use for over 300 years.

The *compound* light microscope uses two lenses—one, the *objective,* that magnifies the specimen, and a second, the *eyepiece,* that further magnifies the magnified image produced by the objective. Actually, both the eyepiece and the objective consist of more than one lens. For simplicity, each is often discussed as if it were a single lens. Typical compound microscopes used in schools magnify up to about 430 times.

A The Microscope

With the use of a microscope, you can see things that you could not see otherwise. For example, if you look at a drop of pond water through a microscope, you may see a number of tiny organisms moving about. A typical microscope is shown in Figure A-1. It is known as a *compound microscope* because it uses two lenses. Refer to Figure A-1 as you read the following discussion.

When you look through a microscope, your eye is just above the *eyepiece*. The eyepiece contains a lens. The *magnification* of the lens (the number of times larger an object appears to be when it is viewed through the lens) is written on the eyepiece. The other lens used in a compound microscope is called the *objective lens*. Most microscopes have two or more objective lenses, mounted on the *nosepiece*. The nosepiece can be moved to select the objective lens you want to use. The magnifications of these lenses are given on their housings, also. You can compute the total magnification of an object that you are viewing by multiplying the magnification of the eyepiece by that of the objective lens that you are using.

A specimen that is to be viewed is usually mounted on a microscope *slide*. The slide is placed on the *stage* of the microscope. The slide is held in place on the stage by the *clips*. There is an opening called the *diaphragm* in the middle of the stage. Light is directed through the diaphragm by the *mirror* below it. (Some microscopes have a small light bulb instead of a mirror.) The size of the diaphragm can usually be varied to control the amount of light that passes through.

A specimen is first brought into focus by using the *coarse adjustment*. This adjustment moves the *body tube*, which holds the eyepiece and nosepiece, up and down. The *fine adjustment* is then used to bring the image of the specimen into sharper focus.

A microscope is an expensive instrument. Always follow directions carefully when using one. If you need to carry a microscope, use two hands. Hold the *arm* of the microscope, the part that holds the body tube and the

Many types of cells and microorganisms can be seen clearly at this magnification.

Another light microscope often used in schools is the *dissecting* microscope, or binocular compound microscope. It provides a three-dimensional image at lower magnifications than those available through the monocular compound microscope. The dissecting microscope is especially useful for observing specimens too large for the monocular compound microscope and for help in the manipulation of minute specimens.

A compound light microscope can be used to magnify objects only up to about 1000 times. Greater magnifications would make objects look larger but more blurred, without revealing more detail. Because of its limited readjustment knobs, in one hand. Use your other hand to support the *base*. Always carry a microscope in an upright position.

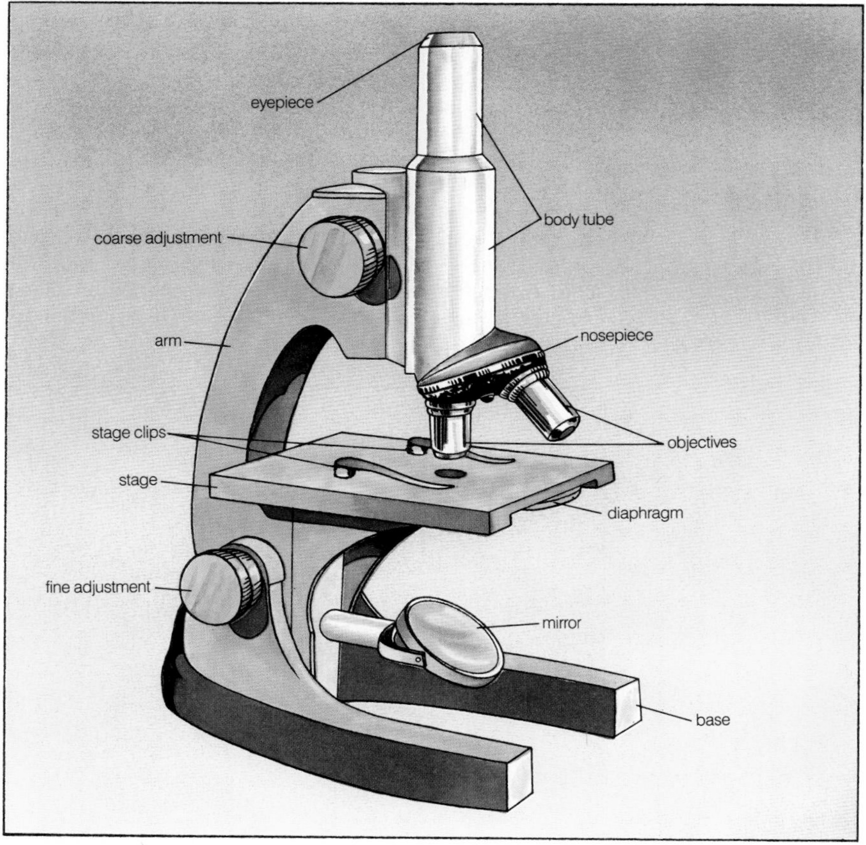

Figure A-1

solving power, a compound light microscope is useless for observing many subcellular objects.

The *transmission electron microscope* (TEM), invented in the 1930's, eliminates the light microscope's problem of limited resolving power. The resolving power of the TEM is 500 times greater than that of the light microscope. This means that the TEM can resolve objects at magnifications hundreds of times greater than those at which objects can be resolved by a light microscope. Because specimens observed with a TEM must be dead, specially stained, and sliced extremely thin, information about three-dimensional structure is not easily obtained with the TEM.

A more recently perfected device called the *scanning electron microscope* (SEM) is used today to view or to photograph surfaces of microscopic objects. The resulting images look strikingly three-dimensional, with great depth of field. While the SEM can produce high magnifications (up to 50,000 times), it has less resolving power than the TEM. The SEM is useful for viewing the surface structure of relatively large specimens.

Cross References

A general discussion of lenses, and their use in a compound microscope, appears in Chapter 20 (section 20-3). The Chapter 1 Feature Article contains a discussion of different types of microscopes in use today. Students use a compound microscope in Activities 1–4 at the back of the textbook.

Photomicrographs, as well as electron micrographs, can be found in Chapter 1 (sections 1-2, 1-3, 1-4, 1-7, and 1-8, and the Feature Article), Chapter 2 (sections 2-1 and 2-3 and the Feature Article), and Chapter 3 (sections 3-5, 3-8, 3-9, and 3-10).

Teaching Suggestions B

Appendix B (or the equivalent) is a prerequisite for student use of a graduated cylinder, thermometer, and laboratory balance. Students use a graduated cylinder in Activities 5, 9, and 15; a laboratory balance in Activities 9, 11, and 12; and a thermometer in Activity 25. Allow enough time for students to complete Appendix B before they first use any of these pieces of equipment. Before permitting students to handle laboratory balances, be sure they are familiar with the parts of a balance and their functions.

See Appendix C for a discussion of metric units of measurement, or SI (the International System of Units).

Length

Give students practice in using both metric rulers and metersticks to measure length. The use of a meterstick will help students to understand metric units as multiples of 10 (1 meter = 100 cm = 1000 mm). However, in a laboratory situation in which relatively small objects are being measured at a desk, metric rulers have the advantage of being less cumbersome.

Metersticks can be used to determine the area of the classroom in m^2 ($1 \times w$). You may also wish to have students use a meterstick to find their height in centimeters. List the heights on the board and find a class mean (average), mode (the height occurring most often), and range (tallest–shortest).

While one student is seated with eyes closed, another can tap a glass with a spoon nearby. Have the seated student first guess the distance to the sound in meters and then measure it with a meterstick to determine accuracy. Distribute copies of optical illusions and have students use metric rulers to measure and compare the true lengths of lines within the illustrations.

B Measurement

Length

Length is commonly measured with a ruler. Suppose you want to measure the length of a cereal box to find out whether it can be stored upright in a cupboard. First, line up the zero mark of the ruler with one end of the box. Then, while holding the ruler along the edge of the box, as in Figure B-1, note where the opposite end of the box lines up with the scale on the ruler. In Figure B-1, the ruler shows that the length of the box is between 30 and 35 cm. Since the end of the box is about two-fifths of the way between 30 and 35, the estimated length is 32 cm.

If you needed to make a more precise measurement of length, you could use a ruler with more marks on it. For example, if a ruler has a mark for each centimeter, you can read the number of centimeters directly from the ruler. You could estimate the length between the centimeter marks and get a reading to the nearest tenth of a centimeter (0.1 cm).

Figure B-1

Volume

You can find the volume of a box if you know the length of each of its sides. Suppose you found the lengths of the sides of a cereal box to be 32 cm, 24 cm, and 8 cm. The volume of the box is the product of the lengths of the sides: $32 \text{ cm} \times 24 \text{ cm} \times 8 \text{ cm} = 6144 \text{ cm}^3$, or 6144 cubic centimeters. The volume of any regularly shaped container, such as a box, a cylinder, or a sphere, can be found in a similar way.

What is the volume of a milk jug? It would be very difficult to find out by making measurements of its dimensions. The best way to find its volume is to fill it full of liquid, then pour the liquid into a special container like the one shown in Figure B-2. This container is called a *graduated cylinder*. The scale marked on the graduated cylinder enables you to read the volume of the liquid in the cylinder.

The scale on a graduated cylinder is read in essentially the same way as is the scale on a ruler, but there is an important difference. A liquid in such a container does not form a flat surface. Instead, it is curved, as shown in Figure B-3. The scale is read at the bottom of the curved portion, called the *meniscus*. When you read a volume, your eye should be at the same level as the bottom of the meniscus. Otherwise, the reading will be inaccurate.

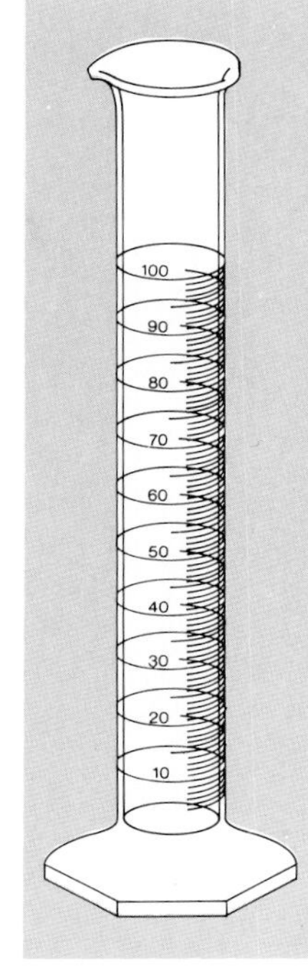

Figure B-2

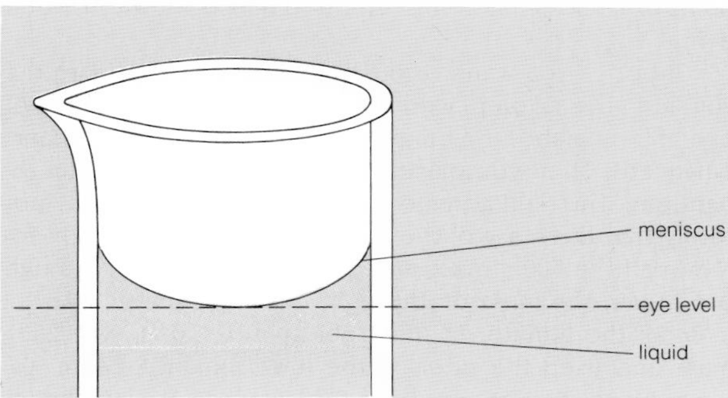

Figure B-3

Volume

Some students may have difficulty conceptualizing a cubic centimeter and the number of cubic centimeters there are in, for example, a box measuring $4 \text{ cm} \times 2 \text{ cm} \times 2 \text{ cm}$. Inform students that a small sugar cube is slightly more than 1 cm^3 in volume. You can provide them with enough sugar cubes so they can construct a box with the previously stated dimensions. Having students change the lengths of the sides while using the same total number of sugar cubes will help to reinforce the concept of cubic centimeters and measuring volume in metric units.

Give students practice in using a graduated cylinder to measure various volumes of water. You may want to explain that the curved surface of water visible while reading the level in a graduated cylinder is the result of capillary action. Adhesion between the molecules of water and the glass of the cylinder tends to pull the liquid up the sides of the container. In contrast, mercury would form a convex surface since it does not wet the glass and possesses a strong cohesive force.

Temperature

Stress that whenever students use thermometers, they must take special care to avoid breakage. When the thermometer is not actually being used to measure temperature during an activity, students could place it between the pages of an open book (if it is clean and dry), to prevent it from rolling off the desk. Instruct students to hold onto the thermometer as they take readings from a small container to avoid tipping the container.

Give students an opportunity to measure and record room temperature (20°C), the temperature of boiling water (100°C), and body temperature (37°C) with a Celsius thermometer to improve their familiarity with the Celsius scale. (Body temperature should be measured by placing the thermometer under the arm, *not* in the mouth.)

Since most students can guess outdoor temperature in Fahrenheit with some accuracy, you may want them to develop a similar ability using the Celsius scale. Have them measure and record both Fahrenheit and Celsius outdoor temperatures on a daily basis for two weeks. Then check their ability to guess the outdoor air temperature in Celsius before checking it with a thermometer.

Mass

Emphasize that the mass of an object equals the sum of the masses that exactly balance the object. Discuss the difference between mass and weight and the reason for using a balance instead of a spring scale. (Chapter 16, section 16-6, discusses this topic.)

A spring scale measures *weight;* it is sensitive to changes in the pull of gravity. It would indicate a slightly greater weight for an object weighed at the poles than for the same object weighed at the Equator (where it is farther from Earth's center). Weight also varies with the altitude. Of course, great variations in weight would occur on other bodies in space.

A balance measures *mass*. The mass of an object remains constant regardless of gravitational pull. When mass is equal on both pans of a balance, the pans balance one another at any location on Earth.

Give students practice in using a laboratory balance to find the masses of various solids and liquids. Show students how to account for the mass of the container when finding the mass of a liquid or powdery or grainy solid.

Point out that one cubic centimeter (cm^3) of distilled water equals one milliliter (mL) and has a mass of precisely one gram (g) at 4°C and approximately one gram over the range 0–100°C. Have the students use a balance to obtain the mass of a graduated cylinder. They should then place exactly 10 mL of distilled water in the cylinder and find its mass again. The mass of the

Temperature

Temperature is usually measured with a thermometer. Common thermometers contain a liquid in a very narrow tube. The liquid expands or contracts as the temperature of its surroundings increases or decreases.

Thermometer scales are usually marked off in units called degrees, such as degrees Celsius (°C). The scales are read in essentially the same way as are those on rulers. Your eye should be level with the top of the liquid column when you read the scale. The temperature shown by the thermometer in Figure B-4 is 28°C.

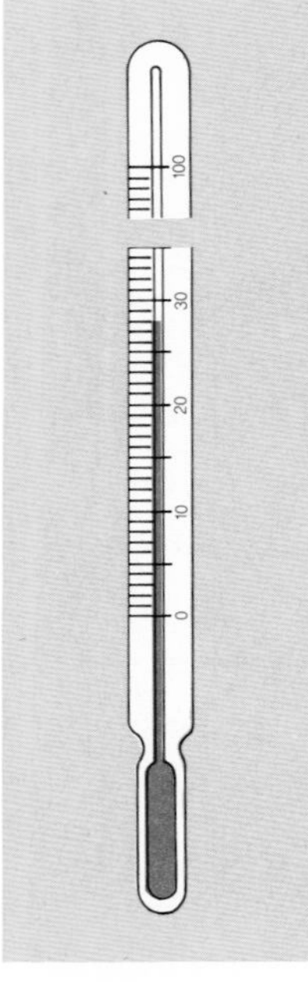

Figure B-4

Mass

The mass of an object is commonly measured on a *balance*, such as the one shown in Figure B-5. The object of unknown mass is placed on the left pan of the balance—the pan on your left as you face the balance. Known masses are added to the right side of the balance until the known and unknown masses just balance. Known mass may be added by placing *standard masses*, objects of known mass, on the right pan or by moving to the right the *riders* on the beams just below the balance pans. When the known and unknown masses balance, the pointer will come to rest at the midpoint of its scale, as in the empty balance shown in Figure B-5.

Suppose you wanted to find the mass of a tomato that you had grown. You placed the clean, dry tomato on the left pan of the balance (first making sure that the empty balance balanced). Then you placed a 500-g standard mass on the right pan. The right pan moved lower than the left pan, indicating that the mass of the tomato was less than 500 g. You removed the 500-g mass and put a 200-g mass on the right pan. Again the mass on the right pan was greater than the mass of the tomato. At this point, you removed the 200-g mass and used the rider on the lower balance beam. You moved it to the right until the right pan sank below the left one, then moved the rider back a notch. Then you moved the rider on the upper beam to the right until the two pans were in balance. The reading on the lower beam was 110 g and the reading on the upper beam was 3.7 g, as shown in Figure B-6. The mass of your tomato was 113.7 g.

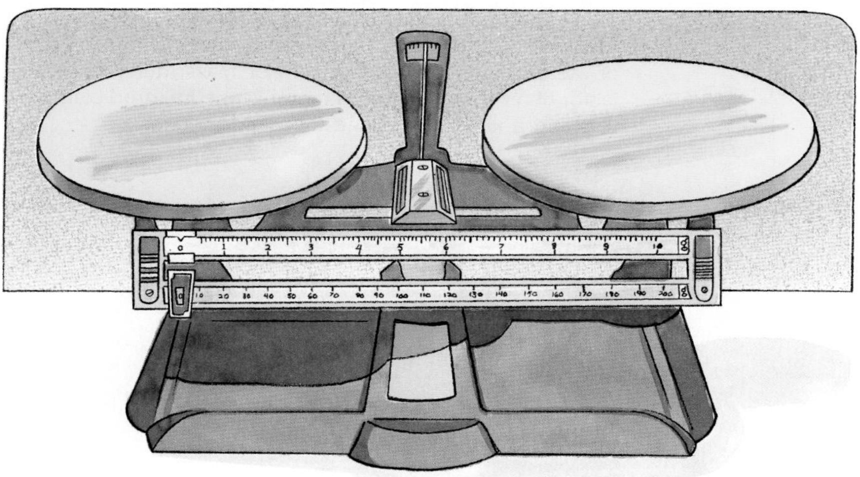

Figure B-5

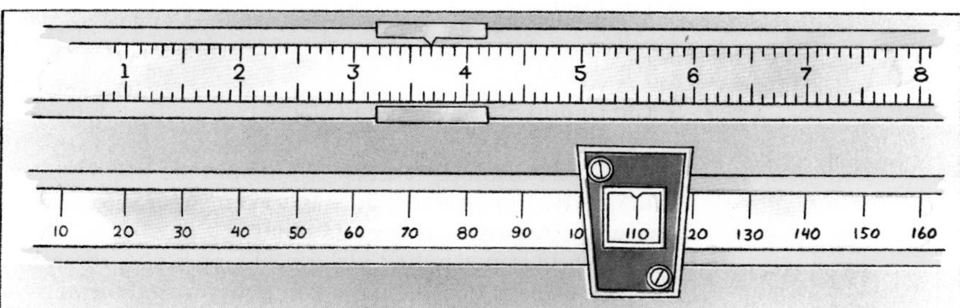

Figure B-6

10 mL of water can be obtained by subtracting the mass of the cylinder from the second mass. If the resultant class data are listed on the board, the students should infer that 10 mL (or cm^3) of water has a mass of 10 g. Allow students to measure other volumes of water in the graduated cylinder and determine their mass.

Classroom balances should provide years of accurate service if students are reminded to observe the following rules:

1. Hold and carry the balance with both hands on the base. Do not allow the pans to move up and down wildly.
2. Use the balance on a level surface.
3. Be sure the riders are on zero before using the balance. Move them to zero before returning the balance to storage.
4. Notify the teacher to make the fine adjustments necessary to zero the balance.
5. Always use a piece of paper under chemicals placed on the pan. Be sure to account for the mass of the paper when calculating the mass of a chemical.
6. Remember that standard masses will become inaccurate if they are dropped. Masses smaller than a gram should be handled only with tweezers.

Cross References

Students use a graduated cylinder in Activities 5, 9, and 15, at the back of the textbook. They use a laboratory balance in Activities 9, 11, and 12. A thermometer is used in Activity 25.

For a discussion of mass and weight, see Chapter 16, section 16-6. The International System of Units (SI) is discussed in Appendix C.

Teaching Suggestions C

An understanding of metric units (SI) is necessary before students undertake activities in this textbook that require the taking of measurements. Because the textbook gives all measurements in metric units, students will benefit most from reading Appendix C at the start of this course.

Some students may question the necessity of learning a system of measurement they may not use outside the science classroom and one whose implementation in the United States has been slow and vague. This should provoke an interesting and perhaps controversial discussion concerning advantages and disadvantages of SI. Explain that SI is the numerical language of science in all countries and that more and more industries in the U.S. are converting to SI every year. Point out that the use of multiples of ten in SI enables quicker and more accurate calculations.

You may wish to use the following approximate comparisons to help students learn common units in the metric system:

millimeter	thickness of a dime
centimeter	width of a fingernail
meter	distance from doorknob to floor
gram	mass of a paper clip
kilogram	mass of a hardcover textbook
cubic centimeter	volume of a sugar cube
liter	one quart

The best way for students to become comfortable with measuring in SI is to actually use SI standards in a variety of activities. If necessary, spend a few days familiarizing students with measuring length, volume, temperature, and mass in metric units. (See Appendix B Teaching Suggestions.)

Background

The metric system of measurement was first used in France in 1795. At that time, the standard meter was inaccurately defined as one ten-millionth of the distance between the North Pole and the Equator along an imaginary line passing through Paris. Technological advancements have enabled scientists to become more precise. Today a meter is defined as 1,650,763.73 × the wavelength of orange-red light emitted by krypton-86 in a vacuum.

SI has been used for years in the U.S. in science, photography, the pharmaceutical and tobacco industries, and international sporting competition. Although the system has helped companies involved in international trade, the general public in the U.S. has resisted changing to the metric system.

C Metric Units

The units used in scientific measurements are normally those of the metric system. Scientists in all parts of the world have long used metric units. In recent years, the metric system has been updated. The updated metric system is called the International System of Units—SI, for short.

The table on this page includes some of the commonly used metric units. Note that the units used for a given type of measurement, such as length or mass, differ by some factor of ten. For example, a meter is one hundred centimeters and a kilometer is one thousand meters.

Some Commonly Used Metric Units

	Unit	*Symbol*	*Equivalent*
Length	centimeter	cm	0.01 m
	meter	m	
	kilometer	km	1000 m
Volume	cubic centimeter	cm^3	1 mL
	milliliter	mL	0.001 L
	liter	L	
Mass	gram	g	0.001 kg
	kilogram	kg	
Temperature	degree Celsius	°C	

The United States is the only large country that does not use metric units in everyday life. In terms of units used in the United States, the meter is 3.3 feet and the liter is 1.056 liquid quarts. The weight of a 1-kilogram mass at sea level is 2.2 pounds. A Celsius degree is nearly twice as large as a Fahrenheit degree. On the Celsius scale, pure water freezes at 0°C and boils at 100°C. On the Fahrenheit scale, the corresponding temperatures are 32°F and 212°F. Room temperature is about 21°C, or 70°F.

D Organizing and Displaying Data

People use graphs when they want to compare two sets of numbers. The numbers can be observations or measurements made during an experiment. The graph helps anyone looking at it to understand exactly what was happening during the experiment.

Suppose you wanted to find out how the mass of water changes when its volume is increased. In order to do so, you poured water into a container, 10 mL at a time, finding the mass of the container and its contents after each addition. Before you did this, you made a table in which to record your data. The table with the data that you obtained was as follows:

Volume of water in container (mL)	*Mass of container plus contents (g)*	*Mass of water in container (g)*
0	150	0
10	160	10
20	170	20
30	180	30
40	190	40
50	200	50
60	210	60
70	220	70
80	230	80
90	240	90
100	250	100

The results show that the amount of mass added is equal to the volume of water added. For example, 10 mL of water has a mass of 10 g.

Suppose you decided to display your results in the form of a graph. In order to do this, you first drew the x- and y-axes for the first quadrant of a coordinate system. (The other quadrants involve negative values, which you did not obtain.) You made each axis just long enough to be able to include all your data.

You chose to plot volume along the vertical (y) axis and mass along the horizontal (x) axis. You labeled each axis, so that a person looking at the graph can tell what was being

Teaching Suggestions D

Emphasize to students that when they conduct a scientific experiment and collect data, it is often helpful to construct a graph of the data. A graph presents a clear picture of the quantitative results obtained in an experiment. Careful examination of experimental data presented in a line, bar, or circle graph makes it easier to arrive at a meaningful conclusion.

Stress that students should first construct tables in which to record experimental data in an organized manner. Students should then use a straight edge and graph paper to graph the data.

Some students may need help determining which groups of data to assign to the x- and y-axes. Help them also to determine the range of numbers and the intervals to use on the axes. Emphasize that a graph is not complete until labels are assigned to both the x-axis and y-axis and the graph has a title.

Explain that line graphs are often used to represent changes in a substance over time. Time is displayed in equal intervals along the horizontal axis. If more than one substance is measured over an equal amount of time, all the substances may be displayed on a single graph for purposes of comparison. In order to differentiate them, different symbols should be used to plot the data, or the lines should be drawn with different colors. Each line must have an identifying key alongside the graph.

Cross References

Various activities involving measuring in SI are described in the Teaching Suggestions for Appendix B.

Cross References

Graphs and graphing activities can be found in Chapter 4 in Figures 4-9, 4-10, and 4-11. In Chapter 6, a graph of solubilities appears in Figure 6-8. A graph of ionization energies can be found in Figure 8-16 in Chapter 8.

Students construct a graph of "half-life" data in Activity 16 at the back of the textbook. Activity 22 includes construction of a graph of data on the walking speed of students.

The section of the *Teacher's Resource Book* called *Science Graphing Skills*, on pages GS-1 to GS-47, includes many activities designed to develop fundamental graphing skills.

measured and the unit in which it was measured. Then you marked each axis with numbers to show the scale.

When you plotted your data, you obtained a graph like the one shown in Figure D-1. The best line through the plotted points, which are circled to make them show up better, is a straight line that passes through the origin.

The graph in Figure D-1 is a line graph. Other kinds of data might be more suitably displayed using a different type of graph. Here is an example: On a certain day, a birdwatcher recorded the following sightings: 3 bluejays, 4 cardinals, 3 crows, 1 flicker, 4 mockingbirds, and 5 robins. These data are displayed in the form of a bar graph in Figure D-2. Another common way of displaying data is to use a circle, or pie, graph like the one shown in Figure D-3. Each "slice" of the pie shows a percentage or fraction of a total.

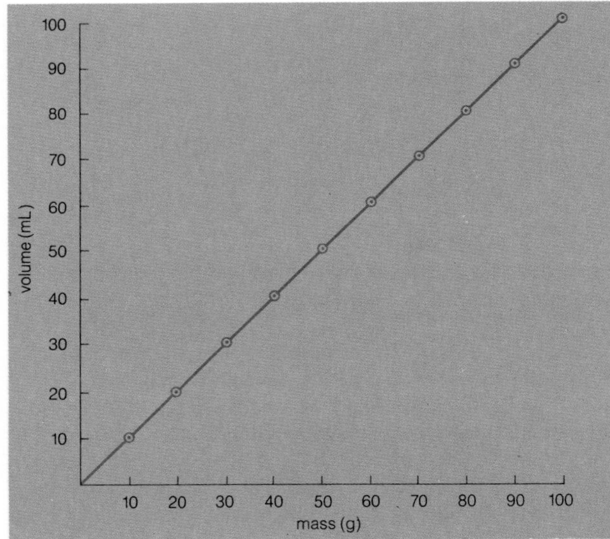

Figure D-1

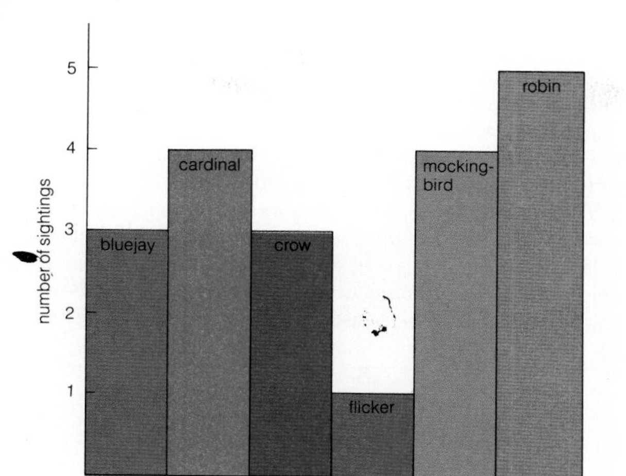

Figure D-2

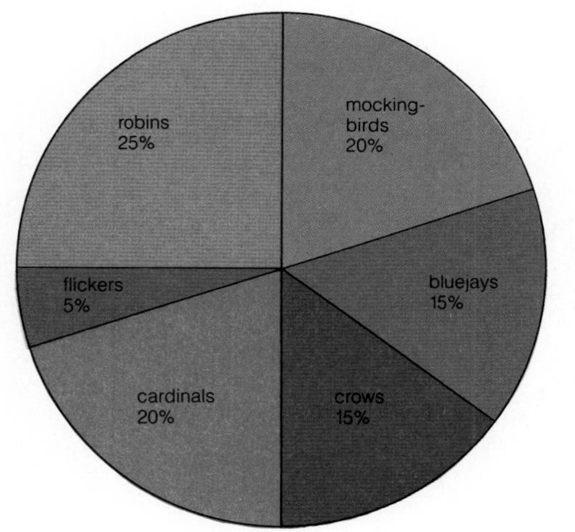

Figure D-3

Teaching Suggestions E

Appendix E (or the equivalent) is a prerequisite for student use of a world map or globe to find a latitude or longitude. Students find latitudes on a world map or globe in Activity 21. Before students do this activity, allow enough time for them to complete Appendix E.

Appendix E can be used in conjunction with section 11-6 in Chapter 11, which discusses the features and uses of geologic and topographic maps. Further information about maps is given on pages 288–289 of the Teacher's Edition.

Be sure students understand the difference between latitude and longitude, and how they aid in locating a point on Earth. Give students practice in finding various cities on a world map or globe based on their latitude and longitude. Have students use a world map or globe to determine the latitude and longitude of your school's city or town and of other cities around the world. You can have students also practice using scales on maps to estimate distances between various points.

You may want to display a variety of flat-map projections and discuss the uses and advantages and disadvantages of each type. (See Background for information on flat-map projections.)

Background

The objective of a *cartographer*, or mapmaker, is to communicate clearly and accurately information about Earth's surface in a graphic form. A map reduces and generalizes the spatial characteristics of a large area, allowing observation of geographic relationships that are outside the normal range of vision.

The most accurate graphic representation of Earth is a spherical globe. On a globe map, Earth's features are greatly reduced in size, but the true geometric properties of area, angle, distance, and direction are retained. A globe map has disadvantages, however.

E Maps

You can locate any point on Earth if you know its *latitude* and *longitude*. Latitude and longitude lines are shown in Figure E-1. The latitude lines run east-west around the globe. The longitude lines run north-south, from pole to pole.

The equator is halfway between the poles. The other latitude lines represent distances north or south of the equator. These distances are measured in degrees. Figure E-2 shows the relationship between the equator and the 60° north latitude line.

Longitude lines are measured with respect to the Prime Meridian. This meridian passes through Greenwich, Eng-

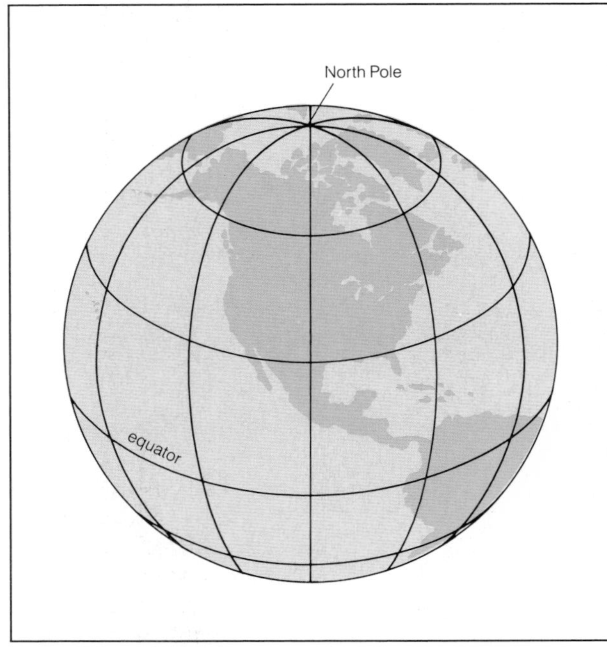

Figure E-1

Often, the scale of a globe is too small to show details of an area. Also, a globe's round shape makes it difficult to handle, store, and reproduce, and allows less than half of Earth's surface to be viewed at a time.

The disadvantages of a globe map can be eliminated by systematically transforming, or *projecting*, the lines and points of a curved globe onto a flat surface. This cannot be done, however, without producing some error, or *distortion*, in at least one of the geometric properties of area, angle, distance, and direction.

Because every transformation method results in some form of distortion, cartographers must be careful to choose a projection that comes closest to accurately portraying the properties

land—an arbitrary choice. Other meridians, or longitude lines, are drawn to the east and west of the Prime Meridian. Figure E-3 shows the relationship between the Prime Meridian and 60° east longitude. As before, the angle is measured from the center of Earth. The meridian on the side of Earth opposite the Prime Meridian is 180° east *and* west.

Maps of Earth, including lines of latitude and longitude, can be transferred, or projected, onto two-dimensional surfaces, such as a flat piece of paper. Such maps can accurately represent locations and distances on Earth's surface.

Locations may be indicated by the intersection of latitude and longitude lines. For example, New Orleans, Louisiana is located where the 30°N latitude line and the 90°W longitude line cross. Its location can be stated as 30°N,90°W. Similarly, Columbus, Ohio is at 40°N,83°W. Can you find Perth, Australia on a map? Its location is 32°S,116°E.

Most maps include a scale for estimating distances between points. For example, a scale might indicate that a distance of 1 cm on the map represents an actual distance of 100 km.

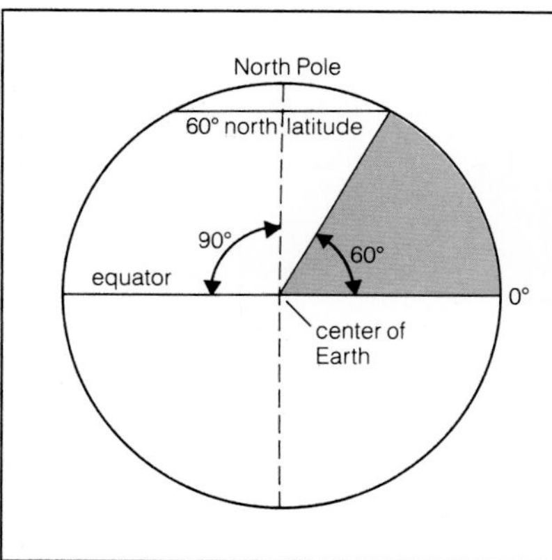

Figure E-2

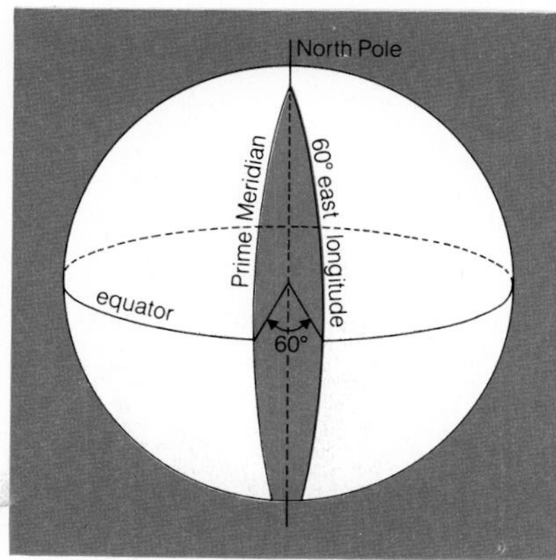

Figure E-3

required for a particular map subject. A geographer or historian, for example, is likely to be concerned with maintaining the relative sizes of areas, while a navigator would require accurate representation of angles and distances. Many of the most commonly used projections display several useful characteristics by limiting the distortion of one or more properties at the expense of the others.

An understanding of projections and their characteristics is necessary for accurate interpretation and use of maps. The nature of the distortion inherent in a particular projection can be determined by comparing the latitude and longitude grid of the flat map with the five fundamental characteristics of a globe grid:

1. All longitude lines are equal in length and meet at the poles.
2. All latitude lines are parallel.
3. Latitude lines decrease in length around the globe from the Equator to the poles.
4. Distances along all longitude lines between any two latitude lines are equal.
5. All lines of latitude and longitude meet at right angles.

Wherever the flat-map grid disagrees with one or more of the five fundamental characteristics of a globe grid, there is distortion. On a Mercator projection, for example, the longitude lines do not converge at the poles and the latitude lines do not decrease in circumference toward the poles, resulting in considerable exaggeration of area in the higher latitudes.

Cross References

Chapter 5 includes a map of Earth's major climate zones (Figure 5-7) and a map of the world's biomes (Figure 5-8). For a discussion of the features and uses of geologic and topographic maps, see Chapter 11, section 11-6. Additional special-subject, or *thematic*, maps appear in Chapters 12, 13, and 14. Students find latitudes on a world map or globe in Activity 21 at the back of the textbook.

Pronunciation Guide

Some words in this book may be unfamiliar to you or hard to pronounce. These words are followed by a pronunciation guide in parentheses. The pronunciation guides appear either in the margins or in the text. The respelled words are divided into syllables. The guide syllable printed in small capital letters should be given the greatest emphasis when you say the word. The following list contains some of the symbols used in the pronunciation guides in this book, along with examples of words that contain the sounds represented by those symbols.

You can find an example of a pronunciation guide for the word "mitosis" in the margin of page 44: (my TOH sis). When you say the word, the sound "toe" (TOH) should be given the greatest emphasis. You may wish to sound out the various pronunciation symbols listed below to make sure you understand them.

Symbol	Sound
a	p*a*t
ah	f*a*ther, p*o*t
air	c*are*
ar	c*ar*
aw	p*aw*
ay	p*ay*
e	p*e*t
ee	b*ee*
eer	*ear*
eh	*e*ssay
ehr	*err*and
eye	*i*ron
i	p*i*t
ih	d*i*vide
ihr	*irr*itate
oh	t*oe*
oo	b*oo*t
or	f*or*
ow	*ou*t
oy	n*oi*se
th	*th*in
<u>*th*</u>	*th*is
u	p*u*t
uh	c*u*t, *a*bout, secti*o*n
ur	*ur*ge, butt*er*
y	p*ie*
zh	vi*s*ion

Glossary

When an important science word is introduced in a unit, the word is printed in **boldface** type and defined. This glossary lists, in alphabetical order, all these important science words. Words used to define them that are also in the glossary appear in SMALL CAPITAL LETTERS. The chapter and section number where the word is found in the book appear in parentheses.

absolute zero: temperature (−273°C) at which the particles in a substance have their lowest possible energy. (18-1)

abyssal plain: flat floor of the OCEAN covering 60% of the ocean bottom. (14-6)

acceleration: change in VELOCITY divided by the period of time it takes for the change to happen. (16-1)

acid: COMPOUND that produces hydrogen IONS in SOLUTIONS. (9-5)

active metals: METALS that have a greater tendency than hydrogen to release ELECTRONS. (9-4)

adaptation: any TRAIT that enables an ORGANISM to live in a particular ENVIRONMENT. (5-2)

air mass: huge parcel of air that has the characteristics of the area over which it originated. (13-6)

alkali metals: group of ELEMENTS with ATOMIC NUMBERS always one more than those of the NOBLE GASES. (8-4)

alpha radiation: helium NUCLEI given off by RADIOACTIVE NUCLEI. (10-1)

alveoli: thin-walled air sacs in the lungs surrounded by tiny CAPILLARIES. (3-3)

amperes: units of ELECTRIC CURRENT. (19-4)

Animal Kingdom: multi-celled ORGANISMS that get food by eating other organisms. (1-7)

aquifers: layers of rock in which large VOLUMES of ground water are stored. (11-5)

arteries: blood vessels that carry oxygen-rich BLOOD away from the heart toward the CELLS. (3-2)

asexual reproduction: type of REPRODUCTION in which new ORGANISMS are produced by a single parent. (2-3)

asteroids: minor planets. (15-3)

astronomy: study of everything in the UNIVERSE that can be observed from Earth or space. (15-1)

atmosphere: blanket of gases, about 900 km thick, that surrounds Earth. (13-1)

atmospheric pressure: FORCE caused by the ATMOSPHERE pressing down on Earth. (13-1)

atom: smallest particle of an individual ELEMENT that has all the PROPERTIES of that element. (7-1)

atomic mass: number of PROTONS plus the number of NEUTRONS in an ATOM. (7-4)

atomic number: number of PROTONS in the NUCLEUS of an ATOM. (7-4)

balanced equation: CHEMICAL EQUATION in which the REACTANTS and PRODUCTS contain the same number of each type of ATOM. (7-3)

barometer: device that measures ATMOSPHERIC

PRESSURE. (13-1)

base: COMPOUND that produces hydroxide IONS in water. (9-5)

battery: two or more ELECTROCHEMICAL CELLS connected in SERIES, with the negative pole of one connected to the positive pole of the next. (19-5)

beta radiation: ELECTRONS moving near the speed of light, given off by RADIOACTIVE NUCLEI. (10-1)

binding energy: ENERGY given off when a NUCLEUS is formed from PROTONS and NEUTRONS. (10-5)

biomes: large areas of land, scattered over the globe, that contain similar COMMUNITIES. (5-3)

black hole: star completely collapsed into itself with a gravitational pull so great that light cannot escape from it. (15-5)

blood: LIQUID that carries substances to and from body CELLS. (3-2)

boiling point: TEMPERATURE at which a LIQUID changes to a GAS. (6-2)

capillaries: narrow, thin-walled blood vessels that connect the ARTERIES and VEINS. (3-2)

catalyst: substance that changes the rate of a CHEMICAL REACTION without being used up itself. (9-6)

cell: smallest unit able to carry out life activities of an ORGANISM. (1-2)

cell membrane: thin layer that surrounds the parts of a CELL. (1-2)

cell wall: structure, providing protection and support, found only in plant CELLS. (1-2)

centripetal force: inward-directed FORCE that keeps an object moving in a circle. (16-8)

cerebellum: part of the brain just below the CEREBRUM. (3-5)

cerebrum: largest part of the brain. (3-5)

chain reaction: series of reactions that occur when the PRODUCTS of one reaction start new reactions. (10-6)

chemical changes: changes in MATTER in which new substances, with different PROPERTIES, are formed. (6-5)

chemical equation: set of symbols representing the changes that take place in a CHEMICAL REACTION. (7-3)

chemical formula: series of CHEMICAL SYMBOLS and numbers representing the number and type of ATOMS in a MOLECULE. (7-2)

chemical reaction: process in which CHEMICAL CHANGES take place. (7-3)

chemical symbol: letter or pair of letters standing for the name of an ELEMENT. (7-2)

chlorophyll: green substance that enables plants to produce food. (1-2)

chloroplasts: CHLOROPHYLL-containing structures in plants. (1-2)

chromosomes: long, thin fibers containing instructions that control the activities of a CELL. (1-2)

cilia: tiny beating hairs that propel paramecia. (1-3)

cirrus clouds: high, thin, wispy clouds. (13-4)

classification: system for separating objects into groups by similarities. (1-6)

climate: average weather of an area over a long period. (5-3, 13-7)

climax community: COMMUNITY that does not change except through natural disasters, sudden large CLIMATE changes, or human intervention. (4-5)

cold front: front formed when a cold AIR MASS moves into a warm air mass. (13-6)

comets: small objects near the edge of the SOLAR SYSTEM made up of fine dust and frozen water. (15-3)

community: interacting ORGANISMS in an ENVIRONMENT. (4-1)

complex machine: device made up of two or more SIMPLE MACHINES. (17-3)

compounds: pure substances that are combinations of two or more ELEMENTS and have different PROPERTIES from these elements. (6-5)

concave lens: LENS that is thicker at the edges than in the middle. (20-3)

concave mirror: mirror with a bowl-shaped surface that focuses or brings together parallel light rays. (20-2)

condensation: change in state from GAS to LIQUID. (6-2)

condensation nuclei: dust or salt particles on which water in the ATMOSPHERE condenses. (13-4)

conduction: transfer of HEAT from particle to particle. (18-4)

conductor: material through which HEAT passes quickly; material along which ELECTRONS move easily. (18-4, 19-3)

cone: pile of material that collects around the opening of a VOLCANO. (12-7)

constellations: patterns of stars visible only at certain times of the year. (15-1)

consumers: ORGANISMS that do not produce their own food. (4-2)

continental rise: gently sloping collection of sediments from land at the bottom of the CONTINENTAL SLOPE. (14-6)

continental shelf: extension of a continent that is covered by water. (14-6)

continental slope: steeply sloped area of the ocean floor beyond the CONTINENTAL SHELF. (14-6)

control group: group not subjected to the EXPERIMENTAL FACTOR in a CONTROLLED EXPERIMENT. (P-2)

controlled experiment: EXPERIMENT, consisting of an EXPERIMENTAL GROUP and a CONTROL GROUP, that measures the effect of only the factor being tested. (P-2)

convection: transfer of HEAT through movement of GASES or LIQUIDS due to differences of DENSITY. (18-4)

convex lens: LENS that is thicker in the middle than at the edges. (20-3)

convex mirror: dome-shaped mirror curving outward like part of the surface of a ball. (20-2)

Coriolis effect: change in the path of air currents caused by the ROTATION of Earth. (13-3)

cosmic rays: extremely high-ENERGY particles that shower Earth from outer space. (10-3)

covalent bonds: chemical bonds formed by sharing valence ELECTRONS. (8-6)

crater: opening at the top of a VOLCANO. (12-7)

crest: highest point in a wave. (14-2)

critical mass: smallest mass of fissionable material that will support a CHAIN REACTION. (10-6)

crust: Earth's outermost SOLID layer. (11-1)

crystal: regular repeated arrangement of the particles of which it is made; regularly simple solid. (8-1, 11-1)

cumulus clouds: billowy clouds that resemble heaps of cotton. (13-4)

cyclones: winds circulating around a low pressure center. (13-5)

cytoplasm: matter that makes up most of the body of a CELL. (1-2)

data: measurements, changes observed, or other bits of information acquired during an EXPERIMENT. (P-2)

deciduous: shedding leaves annually in response to shortening periods of daylight. (5-4)

decomposers: ORGANISMS that break down dead organisms and animal waste and use them as food. (4-2)

deltas: DEPOSITS formed at the mouths of rivers. (12-4)

density: ratio of the MASS of a sample of MATTER to its VOLUME. (6-1)

deposits: materials that have been moved through EROSION and left in another place. (12-4)

desalination: removal of SALT from seawater. (14-7)

dew point: TEMPERATURE at which water vapor in air CONDENSES. (13-4)

diaphragm: strong, flat muscle, located at the base of the ribs below the lungs. (3-3)

diffusion: movement of a substance from an area where there is more of it to an area where there is less. (1-2)

digestion: breaking down of food into smaller and simpler substances. (3-4)

DNA: deoxyribonucleic acid, substance in CHROMOSOMES that contains information that controls all CELL activities. (2-7)

dominant gene: GENE that is always expressed when present in an ORGANISM. (2-6)

earthquake: sudden shaking of Earth's CRUST at a particular place. (12-6)

ecosystem: COMMUNITY and the nonliving parts of its ENVIRONMENT. (4-1)

efficiency: ratio of WORK output to work input. (17-4)

electric circuit: path from a region that has an excess of ELECTRONS to one where there are too few. (19-4)

electric current: stream of flowing ELECTRONS. (19-4)

electrochemical cell: device that produces electricity through OXIDATION-REDUCTION reactions. (9-4)

electrode: device by which ELECTRONS can flow into or out of an ELECTROCHEMICAL CELL. (9-4)

electromagnet: wire coil wound around an iron core which behaves like a MAGNET when electricity flows through the coil. (19-7)

electromagnetic spectrum: broad band of electromagnetic RADIATION. (20-5)

electron: NEGATIVELY CHARGED particle present in all MATTER. (7-4)

element: substance that cannot be separated into simpler substances during PHYSICAL and CHEMICAL CHANGES. (6-5)

endangered species: species that are close to being EXTINCT. (4-7)

endocrine glands: structures that produce HORMONES. (3-6)

endothermic: gaining ENERGY, describing a CHEMICAL REACTION that requires energy in order to take place. (9-2)

energy: ability or capacity to do WORK. (17-5)

energy levels: ENERGIES that an ELECTRON can have in an ATOM. (7-5)

environment: an ORGANISM'S surroundings, consisting of living and nonliving things. (4-1)

epicenter: point on the surface directly above the FOCUS of an EARTHQUAKE. (12-6)

erosion: process by which Earth materials are moved from one place to another, usually ending at lower levels of elevation. (12-3)

esophagus: long tube that connects the throat and the stomach. (3-4)

evaporation: change in state from LIQUID to GAS taking place below the BOILING POINT. (6-2)

evergreen forest: forest made up mostly of trees that stay green all year long. (5-4)

exothermic: releasing ENERGY, describing a CHEMICAL REACTION in which energy is lost. (9-2)

experiment: way of testing that makes use of measurements or observations in the natural world. (P-2)

experimental factor: factor being tested in a CONTROLLED EXPERIMENT. (P-2)

experimental group: group subjected to the EXPERIMENTAL FACTOR in a CONTROLLED EXPERIMENT. (P-2)

extinct: no longer existing. (4-7)

fault: zone of weakness in Earth's CRUST along which movement of rock takes place. (12-6)

fertilization: process by which a sperm CELL and an egg cell fuse together. (2-4)

focal length: distance between a mirror or LENS and its PRINCIPAL FOCUS. (20-2)

focus: area in Earth's crust where ENERGY is released in an EARTHQUAKE. (21-6)

food chain: set of eating interactions within an ECOSYSTEM. (4-2)

food pyramid: illustration of the way energy is passed along in an ECOSYSTEM. (4-3)

food web: FOOD CHAINS that overlap. (4-2)

force: push or pull. (16-2)

force of gravity: FORCE of attraction between objects. (16-5)

fossils: evidence of formerly living things. (11-3)

frequency: number of WAVELENGTHS or waves that pass a given point in one second. (20-4)

friction: FORCE that acts against the motion of one surface over another. (16-2)

fulcrum: point about which a lever turns or pivots. (17-2)

Fungi Kingdom: plant-like ORGANISMS, such as mushrooms and molds, that do not have CHLOROPHYLL and so must obtain food from other organisms. (1-7)

galaxies: groups of billions of stars. (15-4)

gametes: sex CELLS. (2-4)

gamma radiation: rays similar to x rays but with greater ENERGY and penetrating power, given off by RADIOACTIVE NUCLEI. (10-1)

gas: one of the STATES OF MATTER; a substance without definite shape and VOLUME. (6-2)

gene: segment of a CHROMOSOME. (2-6)

geologic map: map showing the various kinds of rock found at the surface or under the soil in a particular area. (11-6)

geothermal energy: ENERGY stored inside Earth. (18-7)

glacier: large, slowly moving mass of ice. (12-3)

greenhouse effect: warming of the ATMOSPHERE due to absorption by atmospheric carbon dioxide of long-wave ENERGY radiated from Earth. (13-2)

grounding: process by which a charge flows into Earth. (19-3)

group: vertical column in the PERIODIC TABLE. (8-5)

habitat: place where an ORGANISM lives in an ECOSYSTEM. (4-1)

half-life: time needed for half the ATOMS of a radioactive ISOTOPE to decay. (10-3)

halide: compound of a HALOGEN and another ELEMENT. (8-3)

halogens: group of ELEMENTS whose ATOMS have one less ELECTRON in their outermost shells than those of the NOBLE GASES. (8-3)

heat: ENERGY transferred to an object when it warms or from an object when it cools. (18-1)

heat engine: MACHINE that changes HEAT into mechanical ENERGY. (18-5)

heat of condensation: ENERGY given up when a GAS condenses to a LIQUID at its BOILING POINT. (9-1)

heat of vaporization: ENERGY required to change one gram of a LIQUID to a GAS at its BOILING POINT. (9-1)

hibernation: sleeplike state in which an animal's body TEMPERATURE, heartbeat and breathing rate decrease. (5-4)

hormones: chemicals released directly into the BLOOD by the ENDOCRINE GLANDS and responsible for changes in specific areas of the body. (3-6)

hybrid trait: TRAIT expressed when one of the GENES controlling it is DOMINANT and the other is RECESSIVE. (2-6)

hydrocarbon: COMPOUND that contains only hydrogen and carbon. Many fuels are hydrocarbons. (8-7)

hydropower: use of falling water to do WORK. (18-7)

hypothesis: tentative explanation for a scientific problem. (P-2)

igneous rocks: rocks formed from extremely hot melted rock deep within Earth. (11-2)

impulse: "message" that travels through a NEURON or nerve CELL. (3-5)

indicator: COMPOUND that changes color in the presence of certain substances. (9-5)

induction: charging of an object by its placement near a charged object. (19-3)

inertia: resistance to change in motion. (16-2)

infrared radiation: invisible RADIATION contained in sunlight, with somewhat lower FREQUENCIES than those of red light. (20-5)

inland seas: seas almost completely surrounded by land. (14-1)

inner core: Earth's innermost layer, believed to be SOLID and composed mostly of nickel and iron. (12-5)

inorganic sediments: sediments on the ocean floor that come from nonliving sources. (14-5)

instincts: behaviors an ORGANISM is born with. (5-2)

insulator: material through which HEAT passes very slowly; material that keeps ELECTRONS from moving. (18-4, 19-3)

intensity: strength of an EARTHQUAKE at a particular place. (12-6)

involuntary muscle: muscles that are not under conscious control. (3-1)

ion: ATOM or group of atoms that has gained or lost one or more ELECTRONS. (8-3)

ionic bond: bond between oppositely charged IONS. (8-6)

ionization energy: ENERGY required to remove an ELECTRON from an ATOM. (8-5)

ionosphere: layer of ionized (charged) GASES in the upper ATMOSPHERE. (13-1)

isobars: lines on a weather map connecting areas of equal ATMOSPHERIC PRESSURE. (13-6)

isotherms: lines on a weather map connecting areas of equal TEMPERATURE. (13-6)

isotopes: ATOMS of the same ELEMENT that have different ATOMIC MASSES. (10-2)

jet streams: upper level winds found in middle latitudes about 10 km above Earth's surface. (13-3)

joint: point at which two bones meet. (3-1)

joule: WORK done when a FORCE of one newton acts over a distance of one meter. (17-1)

kidneys: bean-shaped ORGANS that filter waste

and water from the BLOOD. (3-4)

kilogram: standard unit of MASS equal to 1000 grams. (6-1)

kinetic energy: ENERGY of a moving object due to its motion. (17-5)

lava: MAGMA that comes to Earth's surface when a VOLCANO erupts. (11-2)

law of conservation of energy: law stating that ENERGY can be transformed but the total amount of energy remains the same. (18-5)

law of conservation of mass: law stating that MASS is neither created nor destroyed in a CHEMICAL REACTION. (7-3)

law of definite proportions: law stating that the ELEMENTS in a COMPOUND are combined in definite percentages by MASS. (6-5)

leeward: facing away from the wind. (13-7)

lens: TRANSPARENT object with at least one curved surface. (20-3)

ligaments: bands of tough, stretchy TISSUE that hold bones together. (3-1)

light-year: distance that light travels in one year. (15-5)

limiting factor: any factor that keeps a POPULATION from growing indefinitely. (4-4)

liquid: one of the STATES OF MATTER; something whose shape changes but whose VOLUME does not. (6-2)

liter: unit of VOLUME equal to 100 cm^3. (6-1)

longitudinal waves: waves that travel in the same direction as the vibrations of the medium through which the waves move. (20-8)

lunar eclipse: blocking out of the moon when Earth passes directly between it and the sun, casting Earth's shadow on the moon. (15-4)

machine: device that changes the direction or strength of a FORCE. (17-2)

magma: extremely hot melted rock material deep within Earth. (11-2)

magnet: object that has attraction for anything containing iron. (19-6)

magnetic field: region in which a MAGNET can affect an object. (19-6)

magnitude: amount of ENERGY released in an EARTHQUAKE; observed brightness of a star. (12-6, 15-5)

mantle: thick, mostly SOLID layer between Earth's OUTER CORE and CRUST. (12-5)

mariculture: farming of the sea. (14-7)

mass: measure of the amount of material in an object; ratio of an applied FORCE to the resulting ACCELERATION of an object. (6-1, 16-3)

matter: anything that takes up space and has MASS. (6-1)

meanders: looping curves in a river. (12-4)

mechanical advantage: ratio of resistance FORCE to effort force in a MACHINE. (17-4)

medulla: part of the brain, located at its base, which controls involuntary activities. (3-5)

meiosis: type of CELL division resulting in cells with only half the usual number of CHROMOSOMES. (2-4)

melting point: TEMPERATURE at which a SOLID changes to a LIQUID. (6-2)

metals: one of two basic groups of ELEMENTS having PROPERTIES in common that are different from those of NONMETALS. (8-1)

metamorphic rocks: rocks formed by changes in SEDIMENTARY or IGNEOUS ROCKS caused by large amounts of HEAT and pressure. (11-2)

meteor: fiery streak produced when a METEOROID burns up on entering Earth's ATMOSPHERE. (15-3)

meteorite: METEOROID that does not burn up in the ATMOSPHERE as it hits Earth. (15-3)

meteoroid: small piece of rock or mineral matter traveling through space. (15-3)

mid-ocean ridges: underwater mountain chains rising above the OCEAN floor. (12-5, 14-6)

minerals: naturally occurring solid ELEMENTS and COMPOUNDS. (11-1)

mitosis: process by which CHROMOSOMES duplicate themselves, then separate into two identical sets. (2-2)

mixture: physical combination of two or more substances. (6-3)

molecule: particle of MATTER made up of two or more chemically joined ATOMS. (7-1)

Moneran Kingdom: smallest one-celled ORGANISMS. (1-7)

moraines: deposits of glacial drift dropped in front of a GLACIER or along its sides. (12-4)

mutation: change in a GENE or CHROMOSOME. (2-8)

Nansen bottle: device used to collect a water sample at a specific depth and record its temperature. (14-1)

natural selection: the process by which the fittest survive. (5-1)

neap tides: lowest high tides each month. (14-3)

nebulas: huge clouds of GAS and dust in which stars form. (15-5)

negative charge: electric charge on an object that results from an excess of ELECTRONS. (19-2)

neurons: nerve CELLS. (3-5)

neutralization: CHEMICAL REACTION in which an ACID and BASE are mixed together forming a SALT and water. (9-5)

neutron: uncharged particle in the NUCLEUS of an ATOM. (7-4)

neutron star: very small, very dense star within the NEBULA produced by the explosion of a SUPERNOVA. (15-5)

Newton's first law of motion: objects remain in motion or at rest unless acted on by outside FORCES. (16-2)

Newton's second law of motion: an object ACCELERATES in the direction of an applied FORCE; the acceleration is greater when the force is greater and less when the MASS of the object is greater. (16-3)

Newton's third law of motion: for every action there is an equal but opposite reaction. (16-4)

niche: way of living of a species within a HABITAT. (4-1)

noble gases: helium, neon, argon, krypton, xenon, and radon, "inert" GASES with low MELTING and BOILING POINTS. (8-2)

nodules: solid particles formed when chemicals in seawater collect in lumps. (14-5)

nonmetals: one of two basic groups of ELEMENTS having PROPERTIES in common that are different from those of METALS. (8-1)

nuclear equation: set of symbols representing the process of RADIOACTIVE DECAY. (10-3)

nuclear fission: splitting of a NUCLEUS into two smaller nuclei. (10-6)

nuclear fusion: nuclear reaction in which lighter NUCLEI combine to form heavier nuclei; source of the sun's ENERGY. (10-8, 15-4)

nuclear reactor: device in which a controlled nuclear reaction is carried out. (10-7)

nucleus: "control center" of a CELL; center of an ATOM containing all its POSITIVE CHARGE and most of its MASS. (1-2, 7-4)

observations: measurements, changes observed, or other bits of information acquired during an EXPERIMENT. (P-2)

ocean currents: river-like movements of water running through the OCEANS. (14-4)

oceans: large areas of water covering almost 75% of Earth's surface. (14-1)

occluded front: front formed when a warm AIR MASS is lifted upward as two cold air masses meet. (13-6)

ohms: units of electrical RESISTANCE. (19-4)

opaque: the capacity to absorb light so that it cannot pass through a substance. (20-1)

organ: two or more different types of TISSUE functioning together. (1-4)

organic sediments: sediments on the ocean floor that come from the remains of ORGANISMS that lived in the sea. (14-5)

organism: any living thing. (1-1)

outer core: hot LIQUID surrounding Earth's INNER CORE. (12-5)

oxidation: loss of ELECTRONS by an ATOM or ION. (9-3)

ozone layer: that part of the STRATOSPHERE where absorption of ULTRAVIOLET rays occurs. (13-1)

P waves: primary waves, waves of ENERGY released by an EARTHQUAKE that cause particles in a substance to move back and forth in the direction of wave motion. (12-6)

parallel circuit: ELECTRIC CIRCUIT in which ELECTRONS have more than one path to follow. (19-5)

period: horizontal row in the PERIODIC TABLE. (8-5)

periodic table: arrangement of ELEMENTS in which those with similar PROPERTIES are grouped together. (8-5)

permafrost: permanently frozen SOIL below the surface in the TUNDRA BIOME. (5-3)

***p*H scale:** scale used to measure ACID or BASE strength of a SOLUTION. (9-5)

phases: apparent shapes of the moon as it completes one REVOLUTION around Earth. (15-2)

photosynthesis: process by which plants and certain other ORGANISMS make their own food. (1-5)

physical changes: changes in MATTER in which no new substances are formed. (6-5)

pioneer community: first COMMUNITY to develop in a bare area. (4-5)

plane mirror: ordinary flat mirror. (20-2)

Plant Kingdom: multi-celled ORGANISMS that make their own food by PHOTOSYNTHESIS and have CELL WALLS. (1-7)

plasma: clear LIQUID, making up about half of the BLOOD; consists mostly of water. (3-2)

plate tectonics: theory that Earth's CRUST and upper MANTLE are divided into large, moving segments called plates. (12-5)

polar molecule: MOLECULE in which the ELECTRONS are not distributed evenly. (8-6)

pollination: process by which a pollen grain is transported from the stamen to the pistil of a flower. (2-4)

pollution: accumulation of harmful substances in an ENVIRONMENT. (4-6)

population: number of ORGANISMS of the same kind living within a particular area at the same time. (4-4)

positive charge: electric charge on an object that results from too few ELECTRONS to balance the PROTONS in an ATOM. (19-2)

potential energy: stored ENERGY of an object due to its position. (17-5)

power: amount of WORK done in a certain length of time. (17-7)

predators: ORGANISMS that catch and eat other organisms. (4-4)

pressure: ratio of a FORCE to the area over which it acts. (18-2)

prey: ORGANISMS that are caught and eaten by PREDATORS. (4-4)

principal focus: point at which rays from a CONCAVE MIRROR or LENS meet. (20-2)

producers: ORGANISMS that manufacture food. (4-2)

property: any quality that can help in identifying something. (6-1)

Protist Kingdom: one-celled ORGANISMS that have nuclear membranes. (1-7)

proton: POSITIVELY CHARGED particle in the NUCLEUS of an ATOM. (7-4)

pure trait: TRAIT expressed when both GENES of a pair are identical, i.e., both DOMINANT or both RECESSIVE. (2-6)

radiation: transfer of ENERGY that does not require MATTER to transmit it. (18-4)

radioactive decay: change of a radioactive ISOTOPE into a different ELEMENT by the giving off of an alpha or a beta particle. (10-3)

radioactive nuclei: NUCLEI that give off RADIATION. (10-1)

radio telescopes: telescopes that gather radio waves from space. (15-6)

real image: image focused on a screen by a LENS or a CONCAVE MIRROR. (20-2)

recessive gene: GENE expressed in an ORGANISM only when the DOMINANT form of the gene is absent. (2-6)

red giant: stage in the life of a star caused by the slowing of NUCLEAR FUSION in which the star grows cooler and much larger. (15-5)

reduction: gain of ELECTRONS by an ATOM or ION. (9-3)

reflection: bouncing of light off an object. (20-1)

reflex: automatic response to a STIMULUS, controlled by the spinal cord, not the brain. (3-5)

refraction: bending of the path of light as it passes from one TRANSPARENT material into another. (20-3)

relative humidity: amount of water vapor in the air compared to the maximum amount the air could hold at that TEMPERATURE. (13-6)

renewable energy source: ENERGY source produced about as fast as it is used. (18-6)

replicate: to copy, to make a duplicate, as CHROMOSOMES do of themselves just before MITOSIS. (2-2)

reproduction: process by which ORGANISMS produce new organisms. (2-3)

resistance: limitation on the amount of ELECTRIC CURRENT that can flow through a CONDUCTOR. (19-4)

respiration: the process by which ENERGY is made available from food. (1-2)
revolution: movement of Earth around the sun. (15-1)
rock cycle: never-ending process of change from one rock type to another. (11-2)
rotation: spinning or turning about a straight line or axis. (15-1)
S waves: secondary waves, waves of ENERGY released by an EARTHQUAKE that cause rock particles to move from side to side at right angles to direction of wave motion. (12-6)
salinity: amount of salt in seawater. (14-1)
salt: ionic COMPOUND formed when an ACID and BASE react. (9-5)
satellite: object that revolves around a planet. (15-2)
saturated zone: that part of the ground that contains ground water. (11-5)
savannas: grasslands that contain scattered trees, found in South America and Africa. (5-6)
scientific law: precise statement of a relationship in nature that can be expressed mathematically. (P-3)
scientific method: planned, logical way to develop and test explanations for scientific problems. (P-1)
scientific model: explanation of something unfamiliar or invisible in terms of something familiar or visible. (7-1)
seas: smaller areas of OCEAN, sometimes partly surrounded by land. (14-1)
sedimentary rocks: rocks make of MINERAL grains that become cemented together under pressure. (11-2)
seismograph: recording device that detects waves of ENERGY released by an EARTHQUAKE. (12-6)
series circuit: ELECTRIC CIRCUIT in which there is only one path that ELECTRONS can follow from the power supply and back to it. (19-5)
sex chromosomes: pair of CHROMOSOMES that contain GENES determining the sex of the ORGANISM. (2-6)
sex-linked trait: trait controlled by GENES found on the SEX CHROMOSOMES. (2-6)
sexual reproduction: creation of a new ORGANISM by joining reproductive, or sex CELLS from two parents. (2-4)
simple machine: any of six devices, for example, the lever, from which all other MACHINES are made. (17-3)
soil: mixture of WEATHERED rock material and decayed formerly-living material which provides nutrients for plants. (12-2)
solar eclipse: blocking out of the sun's light when the moon passes directly between Earth and the sun. (15-4)
solar energy: ENERGY from the sun. (18-6)
solar system: sun and all objects in orbit around it. (15-3)
solid: one of the STATES OF MATTER; something whose VOLUME and shape do not change. (6-2)
solubility: amount of a substance in a given amount of solvent at a given TEMPERATURE. (6-3)
solution: MIXTURE in which the smallest particles of two or more substances are evenly distributed. (6-3)
sonar: device that maps the ocean floor by recording the time it takes for sound waves to be reflected from it. (14-6)
specific heat: amount of HEAT needed to raise the TEMPERATURE of 1 g of a substance 1°C. (18-3)
spectroscope: instrument used to produce a spectrum and measure the WAVELENGTHS of light in it. (15-6)
speed: distance traveled in a period of time. (16-1)
spring tides: highest high tides each month. (14-3)
states of matter: the three states in which a substance can exist: GAS, LIQUID, and SOLID. (6-2)
static electricity: charges that stay on an object. (19-3)
stimulus: any "message" that produces a response by the body. (3-5)
stratosphere: layer of the ATMOSPHERE above the TROPOSPHERE. (13-1)
stratus clouds: layered clouds that usually cover the whole sky. (13-4)
structural formula: diagram showing the ar-

rangement of ATOMS in a MOLECULE. (8-7)

succession: process by which an ECOSYSTEM changes over time. (4-5)

sunspots: dark spots on the surface of the sun. (15-4)

supernova: star that explodes, producing a NEBULA. (15-5)

synapse: gap between the end of one NEURON or nerve CELL and the beginning of another. (3-5)

system: ORGANS functioning together, such as the digestive system, circulatory system, etc. (1-4)

temperature: measure of the average KINETIC ENERGY of the particles of a substance. (18-1)

tendons: bands of TISSUE that attach muscles to bones. (3-1)

theory: HYPOTHESIS supported by experimental results. (P-3)

theory of evolution: theory that new species develop over time from earlier species. (5-1)

tissue: group of CELLS that are alike in appearance and function. (1-4)

topographic map: map showing the shape of the land surface by a system of contour lines. (11-6)

tracers: radioactive ISOTOPES used to trace a course of events. (10-4)

trachea: windpipe, tube that carries air from the nose and mouth to the lungs. (3-3)

traits: specific features of an ORGANISM. (2-1)

translucent: having the capacity to scatter light so that things cannot be seen clearly through a substance. (20-1)

transmutation: process by which ATOMS of an ELEMENT gain or lose PROTONS, forming a new element. (10-1)

transparent: having the capacity to allow most light to pass through so that objects can be seen clearly through a substance. (20-1)

troposphere: layer of the ATMOSPHERE closest to Earth. (13-1)

trough: lowest point in a wave. (14-2)

tundra: BIOME in the polar CLIMATE zone with low TEMPERATURES, little rainfall, and long periods with no sunlight. (5-3)

ultraviolet radiation: invisible RADIATION contained in sunlight, with somewhat higher FREQUENCIES than those of violet light. (20-5)

universe: everything that exists. (15-1)

urine: LIQUID produced by the KIDNEYS from waste products and water brought to them by the BLOOD. (3-4)

vascular plants: plants that have a system of transport tubes. (1-7)

veins: blood vessels that carry BLOOD toward the heart. (3-2)

velocity: SPEED of an object in a particular direction. (16-1)

villi: fingerlike projections lining the walls of the small intestine through which digested food MOLECULES diffuse into the BLOOD. (3-4)

virtual image: images seen in a mirror but not able to be seen on a screen. (20-2)

visible spectrum: that part of the ELECTROMAGNETIC SPECTRUM consisting of the band of colors produced when white light passes through a prism. (20-5)

visual telescopes: telescopes that gather visual light. (15-6)

volcano: opening in Earth's CRUST through which MAGMA flows out as LAVA. (12-7)

volts: units of ENERGY of the electrons in an ELECTRIC CIRCUIT. (19-4)

volume: space taken up by an object. (6-1)

voluntary muscles: muscles that function when you want them to. (3-1)

warm front: front formed when a warm AIR MASS moves into a cold air mass. (13-6)

water cycle: movement of water from the land to the rivers and OCEANS, to the ATMOSPHERE, and back to the land again. (11-4)

water table: irregular upper boundary of the SATURATED ZONE. (11-5)

watt: unit of POWER equal to one JOULE per second. (17-7)

wavelength: distance from the CREST of one wave to the CREST of the next. (20-4)

weathering: process by which rocks break down. (12-1)

weight: gravitational FORCE that Earth exerts on an object. (16-6)

white dwarf: stage in the life of a star following the RED GIANT stage. (15-5)

windward: facing the wind. (13-7)

work: FORCE acting on an object multiplied by the distance the object moves. (17-1)

zygote: fertilized egg, formed by the fusing of two GAMETES, a sperm CELL and an egg cell. (2-4)

Index

D

E

F

G

H

I

J

K

L

M

Q

R

S

T

U

V

W

Illustration Credits

Arthur Ciccone: 1-4, 1-12, 13-1; **Marsha Dohrman:** 3-6, 3-11, 3-14, 3-15, 3-16, 3-17, 3-18, 3-19, 3-20, 3-24, 3-26; **Frank Fretz:** 1-14, 2-12, 2-13, 4-6, 4-17, 11-12, 13-6, 13-7, 14-5, 14-16, 14-17, 14-18; **John Hamberger:** 5-1, 5-3; **Bob Jackson:** 2-7, 11-13, 11-14, 16-3, 16-5, 16-13, 16-19, 17-1, 17-4, 17-6, 17-17, 18-8, 18-10; **Omnigraphics:** 1-3, 1-7, 1-13, 2-2, 2-3, 2-8, 2-14, 2-15, 2-16, 2-18, 2-19, 4-7, 4-9, 4-10, 4-11, 4-18, 5-7, 5-11, 6-2, 6-3, 6-6, 6-8, 6-12, 6-13, 7-2, 7-3, 7-6, 7-13, 7-15, 8-8, 8-9, 8-10, 8-11, 8-13, 8-15, 8-16, 8-17, 8-18, 8-20, Feature 8-1, Feature 8-2, 9-1, 9-2, 9-12, Feature 9-1, Feature 9-2, 12-3, 12-4, 12-9, 12-13, 12-14, 12-15, 12-18, 12-19, 12-20, 12-21, 12-22, 13-2, 13-3, 13-8, 13-9, 13-10, 13-14, 13-18, 13-19, 13-21, 13-22, 14-3, 15-3, 15-4, 15-8, 15-9, 15-10, 15-15, 15-17, 15-18, 16-1, 16-7, 16-11, 16-12, 16-14, 17-7, 17-9, 17-11, 17-12, 17-15, 17-16, 17-18, 17-19, 17-22, 17-23, 17-24, 18-2, 18-3, 18-21, 20-6, 20-7, 20-8, 20-9, 20-14, 20-15, 20-16, 20-18, 20-19, 20-21, 20-22, 20-27, 20-28, 20-29, 20-31, Feature 20-1, Appendix Figures B-2, B-3, B-4, D-1, D-2, D-3, E-1, E-2, E-3; **Roget Ink:** 4-15, 7-4, 7-5, 7-9, 7-10, 7-14, 10-2, 10-3, 10-5, 10-6, 10-11, 10-13, 11-9, 12-16, 13-4, 13-5, 13-17, 14-4, 14-7, 14-8, 14-10, 14-11, 14-15, 16-18, 18-5, 18-11, 18-15, 18-20, 19-4, 19-5, 19-8, 19-9, 19-10, 19-11, 19-12, 19-14, 19-16, 19-17, 19-18, Activity Section Figures 5, 6, 7, 9-1, 9-2, 9-4, 10, 13-1, 13-3, 14-1, 14-2, 16, 18, 19, 22-2, 25, Appendix Figure B-1; **James Teason:** 1-16, 1-21, 2-9, 2-10, 2-11, 4-1, 4-2, 4-3, 4-4, 4-12, 5-5, 5-6, 5-13, 5-14, 5-17, 5-18, 5-19, 5-20, 5-21, 5-22, Activity Section Figure 8; **Gary Torrisi:** 6-18, 17-13, 17-20, Activity Figures 1-1, 1-2, 2, 3, 4, 9-3, 11, 12, 13-2, 13-4, 15-1, 15-2, 15-3, 17, 20, 21, 22-1, 23-1, 23-2, 24, 26, 27, Appendix Figures A-1, B-5, B-6; **Lydia Vickers:** 1-2, 3-1, 3-2, 3-3, 3-4, 3-12, 3-13, 3-21, 3-22, 3-23, 3-25, 19-19, 19-20, 19-21; **Paul Weiner:** 15-20

Teacher's Edition
Boston Graphics, Inc./Paul S. Foti

Photo Credits

Animals, Animals: *Zig Leszczynski* 6-10, *Breck P. Kent* 5-4, *C.W. Perkins* 1-20c, *Carl Roessler* 1-19a, *Leonard Lee Rue III* 1-20e, 2-21; **Peter Arnold, Inc.:** *Fred Bavendam* 1-19f, *Dieter Blum* 14-19, *Bonnie Geller* 14-1, *Phillip A. Harrington* 1-4a, *W.H. Hodge* 1-18h, *Manfred Kage* 1-1a, 1-9, 1-10c, 1-11a, 1-18a, Feature 1-1, Feature 1-2, 3-10, 7 Opener, *Yoram Kahama* Unit 3 Careers b, *Malcom S. Kirk* 10-7, *Stephen J. Krasemann* 12-6, *John MacGregory* 1-20d, *C. Allan Morgan* 4 Feature; **D. Balkwill, D. Maratea:** Feature 19-1; **Lester V. Bergman:** 3-8; **Bettmann Archives:** 1-5, 1-10, 5-2; **Brown Brothers:** 7-1, 19-3, 19-6; **Bruce Coleman:** *Don Brewster* 8-6, *Vic Beunza* 5 Opener, *Bob and Clara Calhoun* 5-8b, *E.R. Degginger* 3-5b, 3-5c, 13-15, *Phil Degginger* 3-5a, *Nicholas de Vore III* 14 Opener, *Robert Donne* 1-11b, 1-15a, 1-15b, 1-18d, *Jessica Ehlers* 14-9a, 14-9b, *Michael Fogden* 5-8a, *Barry Parker* 13-16, *Mark Sherman* 2-22, *Norman Owen Tomalin* 13-13; **David Conklin:** 8 Opener, 16-4, 16-16, 17-5, 19-2a, 19-2b, 19-24, 20-1, 20-30; **Paul Conklin:** 1-17a, 1-18b, 1-18e, 1-20f, 2-1, 2-23b, 3-7a, 3-7b, Feature 3-2, 4-14, 4-16, Unit 1 Careers b, Feature 6, 6-7a, 6-7b, 6-14a, 6-14c, 6-15a–f, 6-16a, 6-16c, 7-7a, 7-7b, 7-7c, 7-8, Feature 7-1, Feature 7-2, Feature 7-3, 8-1, 8-2, 8-3, 8-4, 8-7a, 8-7b, 8-7c, 8-14, 8-19, 9 Opener, 9-4a, 9-4b, 9-5, 9-6a, 9-6b, 9-7, 9-8a–d, 9–10a, 9-10b, 9-11a, 9-11b, 9-13, 9-14, 10-1a, 10-1b, 10-14, Feature 10, Unit 2 Careers d, 11 Opener, 11-1a, 11-1b, 11-1c, 11-2, 11-3, 11-4a, 11-4b, 11-5a–d, 11-6a, 11-6b, 11-6c, 11-7a, 11-7b, 11-7c, Feature 11, 11-16, 12-5, 12-7, 13 Opener, 13-2a, 13-11a, 13-20a, 13-20b, Feature 13, 15-12, Unit 3 Careers d, 16-9, 16-10, 17-2, 17-8a, 17-8b, 17-21, 17-25b, 18-4a, 18-4b, 18-16, 18-17, 19-1, 19-13, 19-22, 19-23, 20 Opener, 20-4, 20-13, 20-25, 20-30, Unit 4 Careers b, c, d; **John D. Cunningham:** 5-12; **Deep Ocean Engineering, Inc.:** Feature 14-2; **E.R. Degginger:** 1-11c, 1-19d, 2-6, 2-17, 2-23a, 4-5, 5-9, 5-10b, 6-4, 13-15, 16-2, 18-7; **Phil Degginger:** 4-13a, 4-13b, 12-23a, 12-23b; **Earth Scenes:** *Adrienne T. Gibson* 1-18g; **Courtesy, Dr. Edgerton/M.I.T.:** 16-18; **William E. Ferguson:** 4 Opener, 5-15, 5-16, 11-8; **Hanson Planetarium:** 15-13; **Courtesy, D.C. Heath and Co.:** 20-23, 20-24; **Grant Heilman:** *Barry L. Runk* 1-18f, 1-22b, 3-9, 6-14d, 11-11, 14-13, *Runk/Schoenberger* 1-1b, 1-6, 1-10a, 1-10b, 2-5, 3-27a, 3-27b; **Randall Hyman:** 12-24; **Index Stone:** Unit 3 Opener; **Paul A. Johnson:** Unit 1 Careers a, c, d, Unit 2 Careers a, b, c; **Peter Kane:** Feature 5; **Lick Observatory, University of California, Santa Cruz:** 10-16, 15-17a–f, 15-11, 15-16; **Life Magazine:** 10-4, *Huntington Potter and David Dressler* Feature 2-2; **Los Alamos National Laboratory:** 10-12; **Michael Phillip Manheim:** 19 Opener; **David Muench:** 6 Opener; **Josef Muench:** 11-15; **NASA:** 12-10, 15-1, Feature 15, 16-15, 16-21, Feature 16; **Lennart Nilsson:** 20-17; **Photo Researchers:** *Robert C. Herms* 1-8, 1-19e, *R. Knauft* 14-21, Roger Tory Peterson 2 Opener; **The Picture Cube:** *Jeff Albertson* 12-2, *Barkam* 20-20, *L. Lorisso* 1-17b; **Rainbow:** *Dan McCoy* Feature 2-1, 10-17, 15-19, *Pierce* 10-8; **J.H. Robinson:** 5-7, 5-8c, 5-9a; **Jeffrey Rotman:** 1 Opener, 1-19b, 1-19c, 1-19f, 1-19g, 1-20a, 1-20b, 1-20g; **Dave Schaefer:** Unit 2 Opener, 6-1, 6-7c, 6-16b, 6-17, 7-11a, 7-11b, 7-11c, 7-12a, 7-12b, 8-5, 8-12a, 8-12b, 9-9, 9-15a, 9-15b, 16-6, 16-8, 17-3, 17-10, 17-14, 18-1, 18-12, Feature 18, 19-15, 20-2, 20-3, 20-5, 20-12, Unit 4 Careers a; **Spectra-Action:** *Lewis Portnoy* 3 Opener, Feature 17-1, Feature 17-2; **Dr. Fred M. Spiess, Scripps Institute of Oceanography, University of California, San Diego:** 6-9; **Tom Stack and Associates:** *Don Benson* 15-14, *John Gerlack* Feature 19, *Steve Martin* 12-17, *Gary Milburn* 19-7; *Spencer Swanger* 15-2; **Stock Boston:** *Michael Collier* 10 Opener, *Ira Kirschenbaum* 12-11, *Peter Menzel*

Feature 12, 15 Opener, 18 Opener, 18-19, *Stacy Pick* 17 Opener, *Cary Wolinsky* Unit 1 Opener, Feature 4; **Michael Sullivan and Associates:** *George* O. *Miller* 12-12, *Patricia Smith-Fox* 12-1; **Sygma:** *Chuck Baker* 18-14, *Tom Zimberhoff* 18-6b; **Taurus Photos Inc.:** *Charles Marden Fitch* Unit 3 Careers b, *Eric Kroll* 15-5, 15-6, Unit 3 Careers a, Feature 20-3, T.R. *Leibel* 13-12; **Alfred Owczarzak** 1-1c, 2-4, *Alfred Pasieka* Unit 4 Opener, *Martin* M. *Rotker* 1-16a, Feature 3-1, *James Somers* 1-16b, 1-22a, 1-22c; **Uniphoto:** *Leah Bellas* 18-6a, R.H. *Confer* 6-5, *Dallas Heaton* 18-13, *Jerry Irwin* 16-17, *Richard Mata* 12 Opener, *Voscar* 12-8; **United States Steel Corporation:** 10-9; **USDA Soil Conservation Service:** 12-3b; **U.S. Department of Energy:** 10-15, 18-9, 18-18; **Valan Photos:** *Albert Kunigk* 1-18c, Brian Milne 5-17; **Mark Warner:** 13-11b; **Water Thermal and Chemical Technology Center, University of California, Berkeley:** 6-11; **Woods Hole Oceanographic Institute:** 14-2, 14-12, 14-14, Feature 14-1

TEACHER NOTES

TEACHER NOTES

TEACHER NOTES